Achieve Your Potential

e authors have developed specific content in MyMathLab®
ensure you have many resources to help you achieve success
nathematics - and beyond! The MyMathLab features described
'e will help you:

Review math skills and concepts you may have forgotten

Retain new concepts as you move through your math course

Develop skills that will help with your transition to college

daptive Study Plan

ne Study Plan will help you study
ore efficiently and effectively.

our performance and activity are
sessed continually in real time,
oviding a personalized experience
sed on your individual needs.

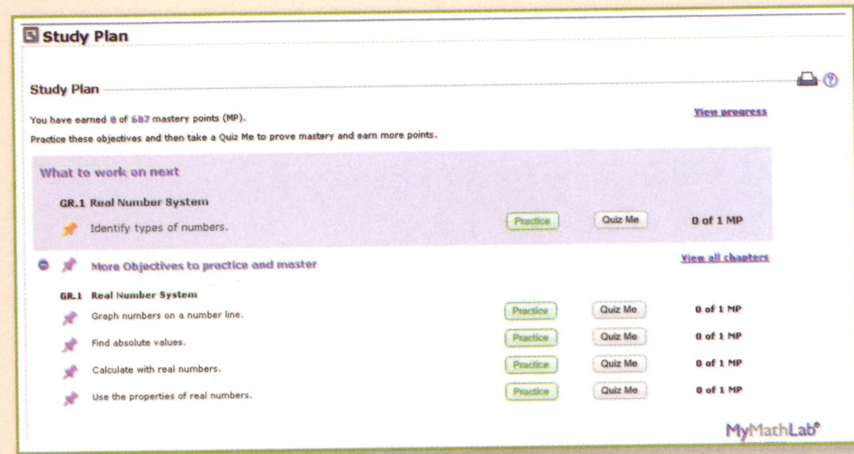

Skills for Success

The Skills for Success Modules support your continued success in college. These modules provide tutorials and guidance on a variety of topics, including transitioning to college, online learning, time management, and more.

Additional content is provided to help with the development of professional skills such as resume writing and interview preparation.

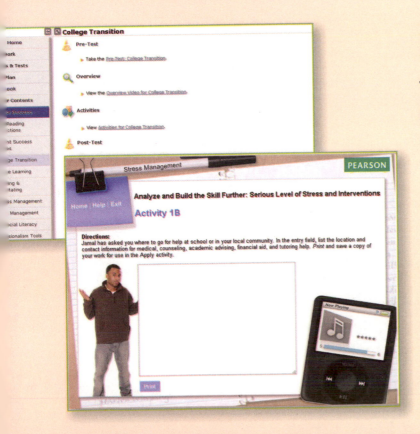

Getting Ready

Are you frustrated when you know you learned a math concept in the past, but you can't quite remember the skill when it's time to use it? Don't worry!

The authors have included Getting Ready material so you can brush up on forgotten material efficiently by taking a quick skill review quiz to pinpoint the areas where you need help.

Homework

Due	Assignment
05/08/14 11:59pm	Section P.1 Homework
05/08/14 11:59pm	Section P.2 Homework
05/08/14 11:59pm	Section P.3 Homework
05/08/14 11:59pm	Section P.4 Homework
05/08/14 11:59pm	Chapter P Mid-Chapter Check Point Homework
05/08/14 11:59pm	Section P.5 Homework
05/08/14 11:59pm	Section P.6 Homework
05/08/14 11:59pm	Chapter P Review Homework
06/20/14 11:59pm	Getting Ready for Chapter 1 Homework
06/20/14 11:59pm	Section 1.1 Homework
06/20/14 11:59pm	Section 1.2 Homework
06/20/14 11:59pm	Section 1.3 Homework
06/20/14 11:59pm	Section 1.4 Homework
06/20/14 11:59pm	Section 1.5 Homework
06/20/14 11:59pm	Chapter 1 Mid-Chapter Check Point Homework
06/20/14 11:59pm	Section 1.6 Homework
06/20/14 11:59pm	Section 1.7 Homework
06/20/14 11:59pm	Chapter 1 Review Homework
08/02/14 11:59pm	Getting Ready for Chapter 2 Homework

Then, a personalized homework assignment provides additional practice on those forgotten concepts, right when you need it.

▶ Skill Maintenance

Simplify. [3.1]

77. $(1 - 4i)(7 + 6i)$ **78.** $\dfrac{2 - i}{3 + i}$

Find the x-intercepts and the zeros of the function.

79. $f(x) = 2x^2 - 13x - 7$ [3.2]

80. $h(x) = x^3 - 3x^2 + 3x - 1$ [4.4]

81. $h(x) = x^4 - x^2$ [4.1]

82. $g(x) = x^3 + x^2 - 12x$ [4.1]

Solve.

83. $x^3 + 6x^2 - 16x = 0$ [4.1]

84. $3x^2 - 6 = 5x$ [3.2]

Skill Maintenance

As you work through your math course, these MyMathLab® assignments support ongoing review to help you maintain essential skills.

The ability to recall important math concepts as you continually acquire new mathematical skills will help you be successful in this math course and in your future math courses.

College Algebra

EDITION

5

College Algebra

Judith A. Beecher

Judith A. Penna

Marvin L. Bittinger

Indiana University Purdue University Indianapolis

PEARSON

Boston Columbus Hoboken Indianapolis New York San Francisco
Amsterdam Cape Town Dubai London Madrid Milan Munich Paris Montréal Toronto
Delhi Mexico City São Paulo Sydney Hong Kong Seoul Singapore Taipei Tokyo

Editorial Director	Chris Hoag
Editor in Chief	Anne Kelly
Sponsoring Editor	Kathryn O'Connor
Editorial Assistant	Judith Garber
Program Manager	Tatiana Anacki
Project Manager	Kathleen A. Manley
Program Management Team Lead	Marianne Stepanian
Project Management Team Lead	Christina Lepre
Media Producer	Erica Lange
TestGen Content Manager	Marty Wright
MathXL Content Manager	Kristina Evans
Marketing Manager	Peggy Lucas
Marketing Assistant	Justine Goulart
Senior Author Support/ Technology Specialist	Joe Vetere
Rights and Permissions Project Manager	Diahanne Lucas Dowridge
Procurement Specialist	Carol Melville
Associate Director of Design	Andrea Nix
Program Design Lead and Cover Design	Barbara T. Atkinson
Text Design, Art Editing, and Photo Research	The Davis Group, Inc.
Editorial and Production Coordination	Martha Morong/Quadrata, Inc.
Composition	Lumina Datamatics Ltd.
Illustrations	Network Graphics, William Melvin, and Lumina Datamatics Ltd.
Cover Image	Petals of a flower, © irin-k/ Shutterstock

Acknowledgments of third party content appear on page A-50, which constitutes an extension of this copyright page.

Library of Congress Cataloging-in-Publication Data

Beecher, Judith A.
College algebra / Judith A. Beecher, Judith A. Penna, Marvin L. Bittinger, Indiana University Purdue University Indianapolis, 5th edition.
pages cm
Includes indexes.
ISBN 0-321-96957-X (student edition)
1. Algebra—Textbooks. I. Penna, Judith A. II. Bittinger, Marvin L. III. Title.
QA152.3.B44 2016
512.9dc23

2014022118

8 18

www.pearsonhighered.com

SE ISBN-13: 978-0-321-96957-6
ISBN-10: 0-321-96957-X

Contents

Preface

This *College Algebra* textbook is known for enabling students to "see the math" through its

- focus on visualization,
- early introduction of functions,
- complete, optional technology coverage, and
- connections between math concepts and the real world.

With the new edition, we continue to innovate by positioning the review material as a more effective tool for teachers and students. Chapter R from the previous edition has been condensed into 25 Just-In-Time review topics that are placed at the back of the book. This new review feature is designed to give each student the opportunity to be successful in this course by providing a quick review of topics from intermediate algebra that will be built upon in new college algebra topics. The review can be used in an individualized instruction format since some students will require more review than others. Treating the review in this manner will allow more time to cover the college algebra topics in the syllabus.

On the other hand, some instructors might choose to review some or all of the topics with the entire class at the beginning of the course or in a just-in-time format as each is needed. We think instructors will appreciate the flexibility that the Just-In Time feature offers them.

Additional resources in the MyMathLab courses reflect the themes of just-in-time review and concept retention. For example, new Cumulative Review assignments allow students to synthesize and retain concepts learned throughout the course.

Our overarching goal is to provide students with a learning experience that will not only lead to success in this course, but also prepare them to be successful in the mathematics courses they take in the future.

Content Changes to the Fifth Edition

- **Just-In-Time Review** Review of prerequisite algebra topics is now presented when students need it most.
    - A set of 25 numbered short review topics creates an efficient review of intermediate algebra topics.
    - This feature is placed at the back of the text. Just-In-Time icons are positioned throughout the text next to the example where review of an intermediate algebra topic would be helpful.
    - Even more just-in-time review resources are available in the MyMathLab course for *College Algebra with Integrated Review* and in the Getting Ready MyMathLab exercises.
- **Informed Exercises** We have analyzed the MyMathLab usage data, which has helped us revise our exercises for this new edition. The goal is to ultimately improve the quality and quantity of exercises that matter the most to instructors and students.
- **Symmetry and Transformations** These topics are now presented in two sections rather than one.

Emphasis on Functions

Functions are the core of this course and are presented as a thread that runs throughout the course rather than as an isolated topic. We introduce functions in Chapter 1, whereas many traditional college algebra textbooks cover equation-solving in Chapter 1. Our approach of introducing students to a relatively new concept at the beginning of the course, rather than requiring them to begin with a review of material that was previously covered in intermediate algebra, immediately engages them and serves to help them avoid the temptation to not study early in the course because "I already know this."

The concept of a function can be challenging for students. By repeatedly exposing them to the language, notation, and use of functions, demonstrating visually how functions relate to equations and graphs, and also showing how functions can be used to model real data, we hope to ensure that students not only become comfortable with functions but also come to understand and appreciate them. You will see this emphasis on functions woven throughout the other themes that follow.

Classify the Function Exercises With a focus on conceptual understanding, students are asked periodically to identify a number of functions by their type (linear, quadratic, rational, and so on). As students progress through the text, the variety of functions with which they are familiar increases and these exercises become more challenging. The "classifying the function" exercises appear with the review exercises in the Skill Maintenance portion of an exercise set. (See pp. 266 and 356.)

Visual Emphasis

Our early introduction of functions allows graphs to be used to provide a visual aspect to solving equations and inequalities. For example, we are able to show students both algebraically and visually that the solutions of a quadratic equation $ax^2 + bx + c = 0$ are the zeros of the quadratic function $f(x) = ax^2 + bx + c$, as well as the first coordinates of the x-intercepts of the graph of that function. This makes it possible for students, particularly visual learners, to gain a quick understanding of these concepts. (See pp. 182, 185, 227, 285, and 344.)

Visualizing the Graph Appearing at least once in every chapter, this feature provides students with an opportunity to match an equation with its graph by focusing on the characteristics of the equation and the corresponding attributes of the graph. (See pp. 143, 198, and 280.) In addition to this full-page feature, many of the exercise sets include exercises in which the student is asked to match an equation with its graph or to find an equation of a function from its graph. (See pp. 145, 146, 236, and 330.) In MyMathLab, animated Visualizing the Graph features for each chapter allow students to interact with graphs on an entirely new level.

Side-by-Side Examples Many examples are presented in a side-by-side, two-column format in which the algebraic solution of an equation appears in the left column and a graphical solution appears in the right column. (See pp. 176, 290–291, 360, and 361.) This enables students to visualize and comprehend the connections among the solutions of an equation, the zeros of a function, and the x-intercepts of the graph of a function.

Technology Connections This feature appears throughout the text to demonstrate how a graphing calculator can be used to solve problems. The technology is set apart from the traditional exposition so that it does not intrude if no technology is desired. Although students might not be using graphing calculators, the graphing calculator windows that appear in the Technology Connection features enhance the visual element of the text, providing graphical interpretations of solutions of equations, zeros of functions, and x-intercepts of graphs of functions. (See pp. 21, 181, and 360.) A graphing calculator manual providing keystroke-level instruction, written by author Judy Penna, is available online.

Making Connections

Zeros, Solutions, and x-Intercepts We find that when students understand the connections among the real zeros of a function, the solutions of its associated equation, and the first coordinates of the x-intercepts of its graph, a door opens to a new level of mathematical comprehension that increases the probability of success in this course. We emphasize zeros, solutions, and x-intercepts throughout the text by using consistent, precise terminology and including exceptional graphics. Seeing this theme repeated in different contexts leads to a better understanding and retention of these concepts. (See pp. 176 and 185.)

Connecting the Concepts This feature highlights the importance of connecting concepts. When students are presented with concepts in visual form—using graphs, an outline, or a chart—rather than merely in paragraphs of text, comprehension is streamlined and retention is enhanced. The visual aspect of this feature invites students to stop and check their understanding of how concepts work together in one section or in several sections. This check in turn enhances student performance on homework assignments and exams. (See pp. 73, 185, and 253.)

Annotated Examples We have included over 730 annotated examples designed to fully prepare the student to work the exercises. Learning is carefully guided with the use of numerous color-coded art pieces and step-by-step annotations. Substitutions and annotations are highlighted in red for emphasis. (See pp. 179 and 352.)

Now Try Exercises Now Try Exercises are found after nearly every example. This feature encourages active learning by asking students to do an exercise in the exercise set that is similar to the example the student has just read. (See pp. 182, 272, and 328.)

Synthesis Exercises These exercises appear at the end of each exercise set and encourage critical thinking by requiring students to synthesize concepts from several sections or to take a concept a step further than in the general exercises. For the Fifth Edition, these exercises are assignable in MyMathLab. (See pp. 32, 255, 333, and 380.)

Real-Data Applications We encourage students to see and interpret the mathematics that appears every day in the world around them. Throughout the writing process, we conducted an energetic search for real-data applications, and the result is a variety of examples and exercises that connect the mathematical content with everyday life. Most of these applications feature source lines and many include charts and graphs. Many are drawn from the fields of health, business and economics, life and physical sciences, social science, and areas of general interest such as sports and travel. (See pp. 39 ("Food Stamp Program"), 66 ("Words in Languages"), 133 ("Peace Corps Volunteers"), 187 ("Funding for Afghan Security"), 236 ("Vinyl Album Sales"), 331 ("Alternative-Fuel Vehicles"), 406 ("Cosmetic Surgery"), 415 ("Top Auction Art Sales"), 494 ("The Ellipse at the White House"), and 546 ("The Economic Multiplier; Super Bowl XLVII").)

Ongoing Review

The most significant change to the Fifth Edition is the new Just-in-Time Review feature, designed to provide students with efficient and effective review of basic algebra skills.

New! Just-in-Time Review Chapter R has been condensed into 25 numbered short review topics to create an efficient review of intermediate algebra topics. This feature is placed at the back of the book.

- Just-In-Time icons are placed throughout the text next to the example where review of an intermediate algebra topic would be helpful. (See pp. 35, 99, 115, 171, 232, and 319.)
- The coverage of each topic contains worked-out examples and a short exercise set. Answers to all exercises appear at the back of the book.

- Worked-out solutions to all exercises are included in the *Student Solutions Manual*.
- Students can find additional review support in the MyMathLab course for College Algebra with Integrated Review and in the Getting Ready MyMathLab exercises.

Mid-Chapter Mixed Review This review reinforces understanding of the mathematical concepts and skills covered in the first half of the chapter before students move on to new material in the second half of the chapter. Each review begins with at least three true/false exercises that require students to consider the concepts they have studied and also contains exercises that drill the skills from all prior sections of the chapter. These exercises are assignable in MyMathLab. (See pp. 125–126 and 256–257.)

Collaborative Discussion and Writing Exercises appear in the Mid-Chapter Mixed Review as well. These exercises can be discussed in small groups or by the class as a whole to encourage students to talk about the key mathematical concepts in the chapter. They can also be assigned to individual students to give them an opportunity to write about mathematics. (See pp. 202 and 257.)

A section reference is provided for each exercise in the Mid-Chapter Mixed Review. This tells the student which section to refer to if help is needed to work the exercise. Answers to all exercises in the Mid-Chapter Mixed Review are given at the back of the book.

Study Guide This feature is found at the beginning of the **Summary and Review** near the end of each chapter. Presented in a two-column format and organized by section, this feature gives key concepts and terms in the left column and a worked-out example in the right column. It provides students with a concise and effective review of the chapter that is a solid basis for studying for a test. In MyMathLab, these Study Guides are accompanied by narrated examples to reinforce the key concepts and ideas. (See pp. 214–220 and 381–387.)

Exercise Sets There are over 5040 exercises in this text. The exercise sets are enhanced with real-data applications and source lines, detailed art pieces, tables, graphs, and photographs. In addition to the exercises that provide students with concepts presented in the section, the exercise sets feature the following elements to provide ongoing review of topics presented earlier:

- **Skill Maintenance Exercises.** These exercises provide an ongoing review of concepts previously presented in the course, enhancing students' retention of these concepts. These exercises include **Vocabulary Reinforcement**, described below, and **Classifying the Function** exercises, described earlier in the section "Emphasis on Functions." A section reference is provided for each exercise. This tells the student which section to refer to if help is needed to work the exercise. Answers to all Skill Maintenance exercises appear in the answer section at the back of the book. (See pp. 133, 210, 283, and 347.)
- **Enhanced Vocabulary Reinforcement Exercises.** This feature checks and reviews students' understanding of the vocabulary introduced throughout the text. It appears once in every chapter, in the Skill Maintenance portion of an exercise set, and is intended to provide a continuing review of the terms that students must know in order to be able to communicate effectively in the language of mathematics. (See pp. 84, 154, 214, and 283.) These are now assignable in MyMathLab and can serve as reading quizzes.
- **Enhanced Synthesis Exercises.** These exercises are described under the Making Connections heading and are also assignable in MyMathLab.

Review Exercises These exercises in the **Summary and Review** supplement the Study Guide by providing a thorough and comprehensive review of the skills taught in the chapter. A group of true/false exercises appears first, followed by a large number of exercises that drill the skills and concepts taught in the chapter. In addition, three multiple-choice exercises, one of which involves identifying the graph of a function, are included in the Review Exercises for every chapter. Each review exercise is accompanied by a section reference that, as in the Mid-Chapter Mixed Review, directs students to the section in which the material being reviewed can be found. Collaborative Discussion and Writing exercises are also included. These exercises are described under the Mid-Chapter Mixed Review heading on p. xiv. (See pp. 220–223 and 388–390.)

Chapter Test The test at the end of each chapter allows students to test themselves and target areas that need further study before taking the in-class test. Each Chapter Test includes a multiple-choice exercise involving identifying the graph of a function. Answers to all questions in the Chapter Tests appear in the answer section at the back of the book, along with corresponding section references. (See pp. 223–224 and 391–392.)

DOMAIN

REVIEW SECTION 1.2

Review Icons Placed next to the concept that a student is currently studying, a review icon references a section of the text in which the student can find and review topics on which the current concept is built. (See pp. 267 and 324.)

Acknowledgments

We wish to express our heartfelt thanks to a number of people who have contributed in special ways to the development of this textbook. Our editor, Kathryn O'Connor, encouraged and supported our vision. We are very appreciative of the marketing insight provided by Peggy Lucas, our marketing manager, and of the support that we received from the entire Pearson team, including Kathy Manley, project manager, Barbara Atkinson, cover designer, Judith Garber, editorial assistant, and Justine Goulart, marketing assistant. We also thank Erica Lange, media producer, for her creative work on the media products that accompany this text. And we are immensely grateful to Martha Morong for her editorial and production services, and to Geri Davis for her text design and art editing, and for the endless hours of hard work they have done to make this a book of which we are proud. We also thank Mike Rosenborg for his meticulous accuracy checking and proofreading of the text.

The following reviewers made invaluable contributions to the development of the Fifth Edition and we thank them for that:

Holly Ashton, *Pikes Peak Community College*
Stacie Bardran, *Embry-Riddle Aeronautical University*
Kim Berges, *Morrisville State College*
Sherry S. Biggers, *Clemson University Department of Mathematical Sciences*
Nadine Bluett, *Front Range Community College*
Gary Brice, *Lamar University*
Christine Bush, *Palm Beach Community College, Lake Worth*
Shawn Clift, Ph.D., *Eastern Kentucky University*
Walter Czarnec, *Framingham State College*
Joseph De Guzman, *Riverside College, Norco Campus*
Douglas Dunbar, *Okaloosa-Walton Community College*
Wayne Ferguson, *Northwest Mississippi Community College*
Joseph Gaskin, *State University of New York, Oswego*
Sunshine Gibbons, *Southeast Missouri State University*
Dauhrice K. Gibson, (retired), *Gulf Coast Community College*
Jim Graziose, *Palm Beach Community College*
Joseph Lloyd Harris, *Gulf Coast Community College*
Dr. Mako E. Haruta, *University of Hartford*
Susan K. Hitchcock, *Palm Beach Community College*
Sharon S. Hudson, *Gulf Coast Community College*
Patricia Ann Hussey, *Triton College*
Jennifer Jameson, *Coconino Community College, Flagstaff*

Cheryl Kane, *University of Nebraska*
Marjorie S. LaSalle, *DeKalb College*
Valerie LaVoice, *Community College System of New Hampshire*
Susan Leland, *Montana Tech of the University of Montana*
Jeremy Lyle, *The University of Southern Mississippi*
Bernard F. Mathon, *Miami-Dade College*
Debra McCandrew, *Florence-Darlington Technical College*
Barry J. Monk, *Macon State College*
Claude Moore, *Cape Fear Community College*
Darla Ottman, *Elizabethtown Community College*
Vicki Partin, *Lexington Community College*
Martha Pate, *Lamar State College—Port Arthur*
Leslie Richardson, *College of the Mainland*
Kathy V. Rodgers, *University of Southern Indiana*
Lucille Roth, *Tech of the Low Country*
Abdelrida Saleh, *Miami-Dade College*
Nicholas Sedlock, *Framingham State University*
Pavel Sikorskii, *Michigan State University*
Russell Simmons, *Brookhaven College*
Rajalakshmi Sriram, *Okaloosa-Walton Community College*
Corwin Stanford, *The University of Southern Mississippi*

J.A.B.
J.A.P.
M.L.B.

BREAKTHROUGH
To improving results

Get the most out of MyMathLab®

MyMathLab creates personalized experiences to help each student achieve success and provides powerful tools so instructors can create the perfect learning experiences for their courses.

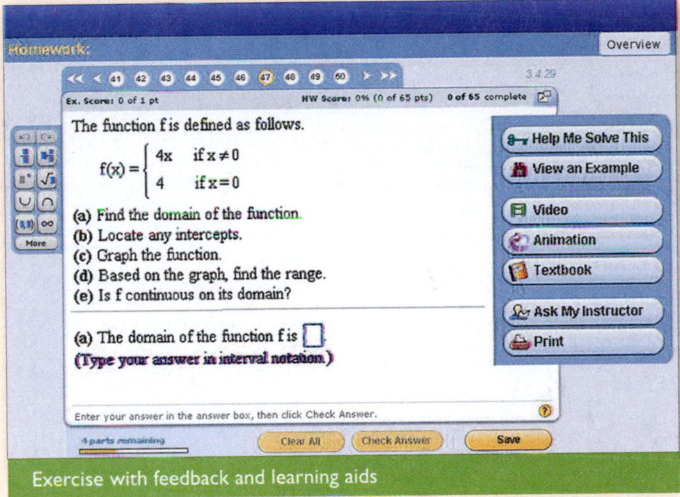

Exercise with feedback and learning aids

Personalized Support *for* Students

- MyMathLab comes with many learning resources—eText, animations, videos, and more—all designed to support you as you complete your assignments.
- Whether you're doing homework or working from the adaptive study plan, you'll receive immediate feedback, so you'll know exactly where you need help.

Data-Driven Reporting *for* Instructors

- MyMathLab's comprehensive online gradebook automatically tracks students' results on tests, quizzes, homework, and in the study plan.
- The Reporting Dashboard makes it easier than ever to identify topics where students are struggling or specific students who may need extra help.

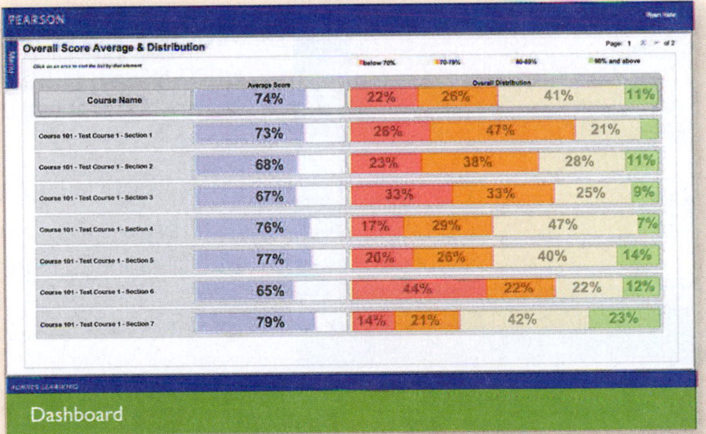

Dashboard

www.mymathlab.com

Resources for Success

MyMathLab® Online Course (access code required)

MyMathLab delivers **proven results** in helping individual students succeed. It provides **engaging experiences** that personalize, stimulate, and measure learning for each student. And, it comes from an **experienced partner** with educational expertise and an eye on the future. **MyMathLab** helps prepare students and gets them thinking more conceptually and visually through the following features:

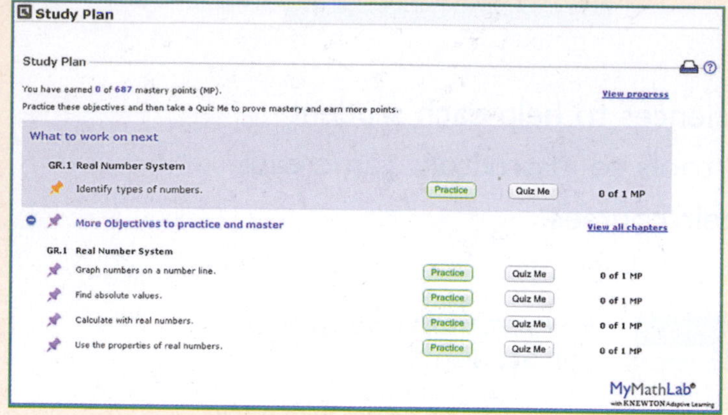

Adaptive Study Plan

The Study Plan makes studying more efficient and effective for every student. Performance and activity are assessed continually in real time. The data and analytics are used to provide personalized content–reinforcing concepts that target each student's strengths and weaknesses.

Getting Ready

Students refresh prerequisite topics through assignable skill review quizzes and personalized homework integrated within **MyMathLab**.

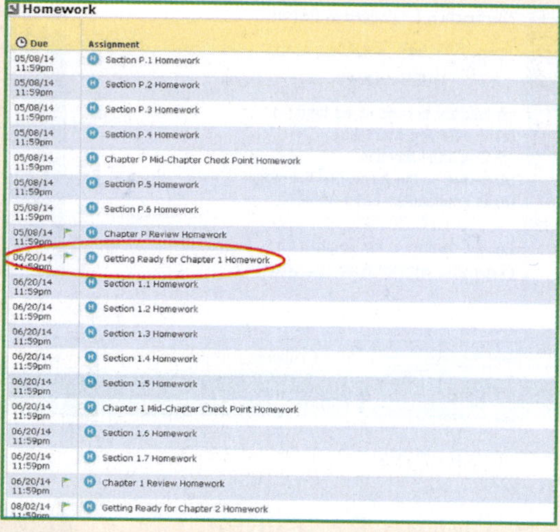

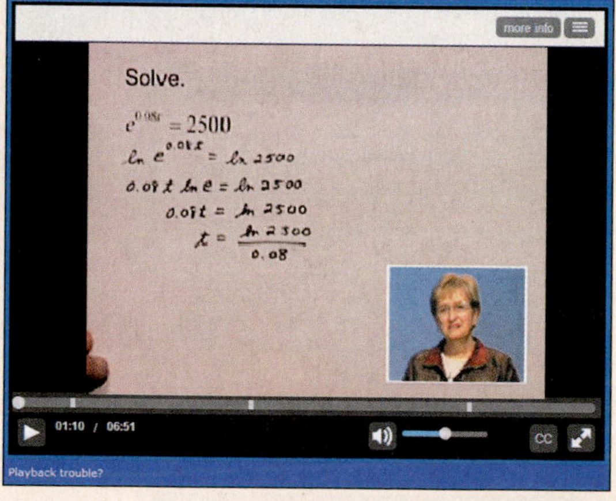

Video Assessment

Video assessment is tied to key author example videos to check students' conceptual understanding of important math concepts.

Enhanced Graphing Functionality

New functionality within the graphing utility allows graphing of 3-point quadratic functions, 4-point cubic graphs, and transformations in exercises.

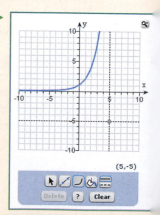

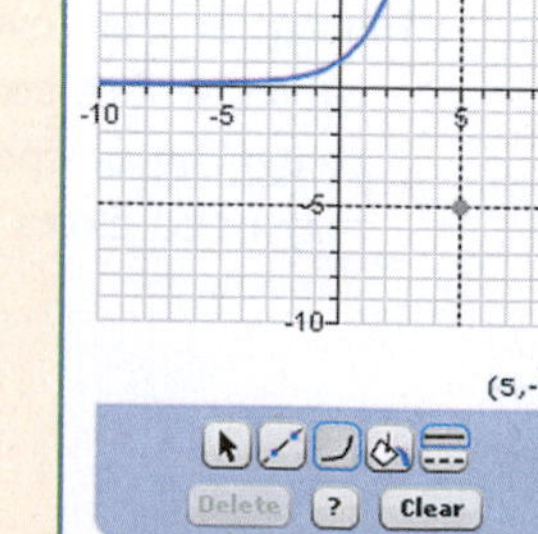

Skills for Success Modules are integrated within the **MyMathLab** course to help students succeed in collegiate courses and prepare for future professions.

Skill Maintenance These exercises support ongoing review at the course level and help students maintain essential skills.

Instructor Resources

Additional resources can be downloaded from **www.pearsonhighered.com** or hardcopy resources can be ordered from your sales representative.

Ready to Go MyMathLab® Course

Now it is even easier to get started with **MyMathLab**. The Ready to Go **MyMathLab** course option includes author-chosen preassigned homework, integrated review, and more.

TestGen®

TestGen® (www.pearsoned.com/testgen) enables instructors to build, edit, print, and administer tests using a computerized bank of questions developed to cover all the objectives of the text.

PowerPoint® Lecture Slides

Feature presentations written and designed specifically for this text. These lecture slides provide an outline for presenting definitions, figures, and key examples from the text.

Annotated Instructor's Edition

Includes all answers to the exercise sets, usually on the page on which the exercises appear. Sample homework assignments are indicated by a blue underline within each end-of-section exercise set and may be assigned in MyMathLab.

Instructor's Solutions Manual (Download Only)

Written by Judy Penna, this resource contains worked-out solutions to all exercises in the exercise sets, Mid-Chapter Mixed Reviews, Chapter Reviews, and Chapter Tests, as well as solutions for all the Just-In-Time exercises.

Online Test Bank (Download Only)

Contains four free-response text forms for each chapter following the same format and having the same level of difficulty as the test in the main text and two multiple-choice test forms for each chapter. It also provides six forms of the final examination, four with free-response questions and two with multiple-choice questions.

Student Resources

Additional resources to help student success.

Author Example Videos

Ideal for distance learning or supplemental instruction, these videos feature authors Judy Beecher and Judy Penna working through and explaining examples in the text. Assignable in MyMathLab with new Video Assessment questions.

New! Video Notebook

The new Video Notebook contains fill-in-the-blank work-sheets to accompany the video examples presented by the authors. Key definitions, theorems, and procedures are also included. After filling in the worksheet while watching the video, the student has an excellent study guide for review and test preparation. This is available in print or as a PDF or Word document in MyMathLab.

Student's Solutions Manual

Written by author Judy Penna, this resource contains completely worked-out solutions with step-by-step annotations for all the odd-numbered exercises in the exercise sets, Mid-Chapter Mixed Reviews, and Chapter Reviews, as well as solutions for all the Chapter Test exercises and the Just-In-TIme exercises.

Graphing Calculator Manual

Contains keystroke level instruction for the Texas Instruments TI-84 Plus using MathPrint OS. Mirrors the topic order in the main text to provide a just-in-time mode of instruction.

To the Student

GUIDE TO SUCCESS

Success can be planned. Combine goals and good study habits to create a plan for success that works for you. The following list contains study tips that your authors consider most helpful.

Skills for Success

- **Set goals and expect success.** Approach this class experience with a positive attitude.
- **Communicate with your instructor** when you need extra help.
- **Take your text with you to class and lab.** Each section in the text is designed with headings and boxed information that provide an outline for easy reference.
- **Ask questions in class, lab, and tutoring sessions.** Instructors encourage them, and other students probably have the same questions.
- **Begin each homework assignment as soon as possible.** If you have difficulty, you will then have the time to access supplementary resources.
- **Carefully read the instructions** before working homework exercises **and include all steps.**
- **Form a study group** with fellow students. Verbalizing questions about topics that you do not understand can clarify the material for you.
- After each quiz or test, **write out corrected step-by step solutions** to all missed questions. They will provide a valuable study guide for the midterm exam and the final exam.
- **MyMathLab has numerous tools to help you succeed.** Use MyMathLab to create a personalized study plan and practice skills with sample quizzes and tests.
- **Knowing math vocabulary is an important step toward success.** Review vocabulary with Vocabulary Reinforcement exercises in the text and in MyMathLab.
- If you miss a lecture, **watch the video in the Multimedia Library** of MyMathLab that explains the concepts you missed.

In writing this textbook, we challenged ourselves to do everything possible to help you learn the concepts and skills contained between its covers so that you will be successful in this course and in the mathematics courses you take in the future. We realize that your time is both valuable and limited, so we communicate in a highly visual way that allows you to learn quickly and efficiently. We are confident that, if you invest an adequate amount of time in the learning process, this text will be of great value to you. We wish you a positive learning experience.

Judy Beecher
Judy Penna
Marv Bittinger

CHAPTER 1

Graphs, Functions, and Models

PPLICATION This problem appears as Exercise 67 in Exercise Set 1.5.

Together, Italy, Spain, and the United States consume 58% of the world's olive oil. The percentage consumed in Italy is $3\frac{3}{4}$ times the percentage consumed in the United States. The percentage consumed in Spain is $\frac{2}{3}$ of the percentage consumed in Italy. (*Source*: www.OliveOilEmporium.com) Find the percent of the world's olive oil consumed in each country.

1.1 Introduction to Graphing

- Plot points.
- Determine whether an ordered pair is a solution of an equation.
- Find the x- and y-intercepts of an equation of the form $Ax + By = C$.
- Graph equations.
- Find the distance between two points in the plane and find the midpoint of a segment.
- Find an equation of a circle with a given center and radius, and given an equation of a circle in standard form, find the center and the radius.
- Graph equations of circles.

Graphs

Graphs provide a means of displaying, interpreting, and analyzing data in a visual format. It is not uncommon to open a newspaper or a magazine and encounter graphs. Examples of bar, line, and circle graphs are shown below.

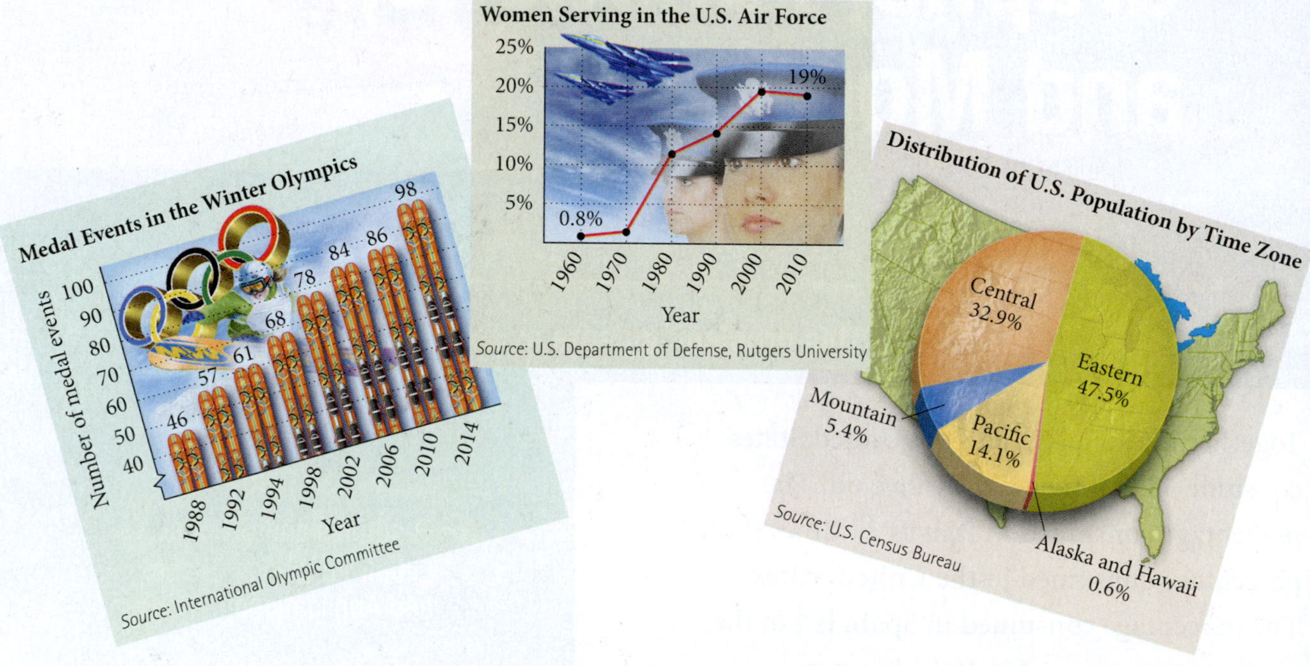

Many real-world situations can be modeled, or described mathematically, using equations in which two variables appear. We use a plane to graph a pair of numbers. To locate points on a plane, we use two perpendicular number lines, called **axes**, that intersect at (0, 0). We call this point the **origin**. The horizontal axis is called the ***x*-axis**, and the vertical axis is called the ***y*-axis**. (Other variables, such as a and b, can also be used.) The axes divide the plane into four regions,

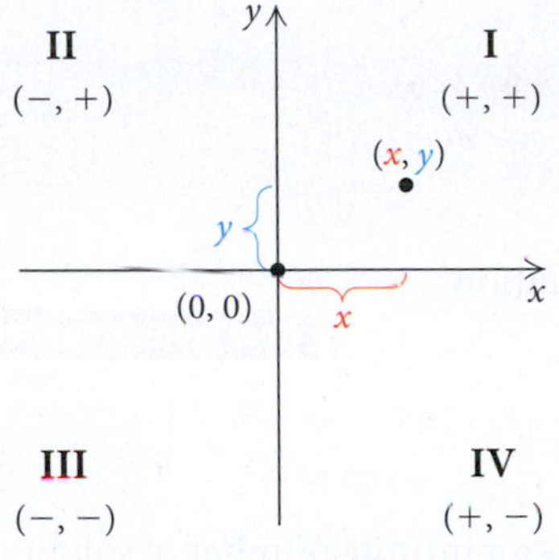

called **quadrants**, denoted by Roman numerals and numbered counterclockwise from the upper right. Arrows show the positive direction of each axis.

Each point (x, y) in the plane is described by an **ordered pair**. The first number, x, indicates the point's horizontal location with respect to the y-axis, and the second number, y, indicates the point's vertical location with respect to the x-axis. We call x the **first coordinate,** the **x-coordinate**, or the **abscissa**. We call y the **second coordinate,** the **y-coordinate**, or the **ordinate.** Such a representation is called the **Cartesian coordinate system** in honor of the French mathematician and philosopher René Descartes (1596–1650).

In the first quadrant, both coordinates of a point are positive. In the second quadrant, the first coordinate is negative and the second is positive. In the third quadrant, both coordinates are negative, and in the fourth quadrant, the first coordinate is positive and the second is negative.

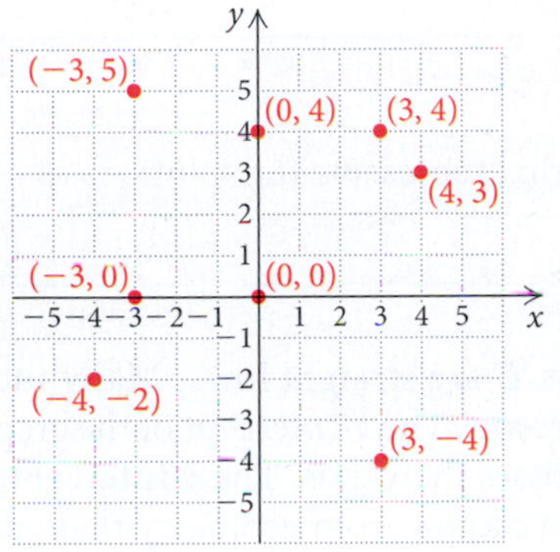

EXAMPLE 1 Graph and label the points $(-3, 5)$, $(4, 3)$, $(3, 4)$, $(-4, -2)$, $(3, -4)$, $(0, 4)$, $(-3, 0)$, and $(0, 0)$.

Solution To graph or **plot** $(-3, 5)$, we note that the x-coordinate, -3, tells us to move from the origin 3 units horizontally in the negative direction, or 3 units to the left of the y-axis. Then we move 5 units up from the x-axis.* To graph the other points, we proceed in a similar manner. (See the graph at left.) Note that the point $(4, 3)$ is different from the point $(3, 4)$.

Now Try Exercise 3.

Solutions of Equations

Equations in two variables, like $2x + 3y = 18$, have solutions (x, y) that are ordered pairs such that when the first coordinate is substituted for x and the second coordinate is substituted for y, the result is a true equation. The first coordinate in an ordered pair generally represents the variable that occurs first alphabetically.

EXAMPLE 2 Determine whether each ordered pair is a solution of the equation $2x + 3y = 18$.

a) $(-5, 7)$

b) $(3, 4)$

Solution We substitute the ordered pair into the equation and determine whether the resulting equation is true.

a)

$$\begin{array}{r|l} 2x + 3y = & 18 \\ \hline 2(-5) + 3(7) \; ? & 18 \\ -10 + 21 & \\ 11 & 18 \quad \text{FALSE} \end{array}$$

We substitute −5 for x and 7 for y (alphabetical order).

The equation $11 = 18$ is false, so $(-5, 7)$ is not a solution.

*Here the notation $(-3, 5)$ represents an ordered pair. This notation can also represent an open interval. See Just-In-Time 6 review. The context in which the notation appears usually makes the meaning clear.

b)

$$\begin{array}{r|l} 2x + 3y & = 18 \\ \hline 2(3) + 3(4) & ?\ 18 \\ 6 + 12 & \\ 18 & 18 \quad \text{TRUE} \end{array}$$

We substitute 3 for x and 4 for y.

The equation $18 = 18$ is true, so $(3, 4)$ is a solution.

Now Try Exercise 11.

Graphs of Equations

The equation considered in Example 2 actually has an infinite number of solutions. Since we cannot list all the solutions, we will make a drawing, called a **graph**, that represents them. On the following page are some suggestions for drawing graphs.

> **TO GRAPH AN EQUATION**
>
> To **graph an equation** is to make a drawing that represents the solutions of that equation.

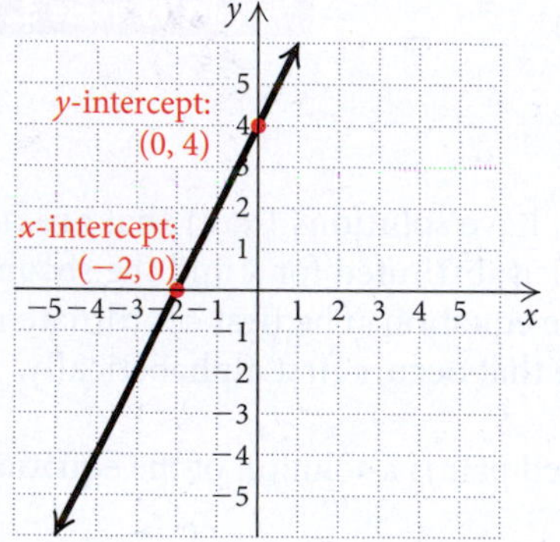

Graphs of equations of the type $Ax + By = C$ are straight lines. Many such equations can be graphed conveniently using intercepts. The ***x*-intercept** of the graph of an equation is the point at which the graph crosses the x-axis. The ***y*-intercept** is the point at which the graph crosses the y-axis. We know from geometry that only one line can be drawn through two given points. Thus, if we know the intercepts, we can graph the line. To ensure that a computational error has not been made, it is a good idea to calculate and plot a third point as a check.

> ***x*- AND *y*-INTERCEPTS**
>
> An ***x*-intercept** is a point $(a, 0)$. To find a, let $y = 0$ and solve for x.
>
> A ***y*-intercept** is a point $(0, b)$. To find b, let $x = 0$ and solve for y.

EXAMPLE 3 Graph: $2x + 3y = 18$.

Solution The graph is a line. To find ordered pairs that are solutions of this equation, we can replace either x or y with any number and then solve for the other variable. In this case, it is convenient to find the intercepts of the graph. For instance, if x is replaced with 0, then

$$\begin{aligned} 2 \cdot 0 + 3y &= 18 \\ 3y &= 18 \\ y &= 6. \end{aligned}$$

Dividing by 3 on both sides

Thus, $(0, 6)$ is a solution. It is the *y-intercept* of the graph. If y is replaced with 0, then

$$2x + 3 \cdot 0 = 18$$
$$2x = 18$$
$$x = 9. \quad \text{Dividing by 2 on both sides}$$

Thus, $(9, 0)$ is a solution. It is the *x-intercept* of the graph. We find a third solution as a check. If x is replaced with 3, then

$$2 \cdot 3 + 3y = 18$$
$$6 + 3y = 18$$
$$3y = 12 \quad \text{Subtracting 6 on both sides}$$
$$y = 4. \quad \text{Dividing by 3 on both sides}$$

Thus, $(3, 4)$ is a solution.

We list the solutions in a table and then plot the points. Note that the points appear to lie on a straight line.

x	y	(x, y)
0	6	$(0, 6)$
9	0	$(9, 0)$
3	4	$(3, 4)$

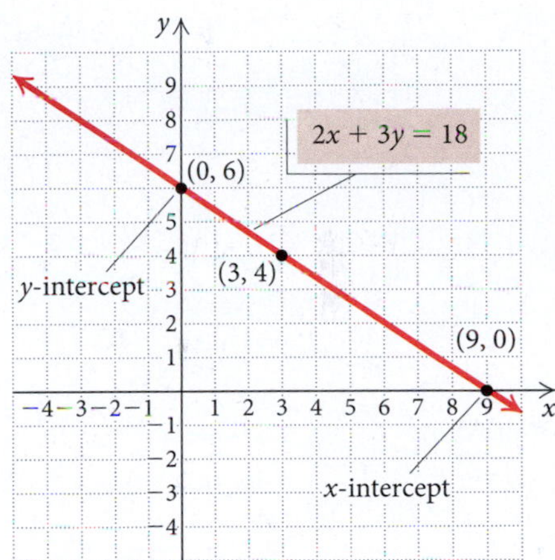

Suggestions for Drawing Graphs

1. Calculate solutions and list the ordered pairs in a table.
2. Use graph paper.
3. Draw axes and label them with the variables.
4. Use arrows on the axes to indicate positive directions.
5. Scale the axes; that is, label the tick marks on the axes. Consider the ordered pairs found in part (1) above when choosing the scale.
6. Plot the ordered pairs, look for patterns, and complete the graph. Label the graph with the equation being graphed.

Were we to graph additional solutions of $2x + 3y = 18$, they would be on the same straight line. Thus, to complete the graph, we use a straight-edge to draw a line, as shown in the figure. This line represents all solutions of the equation. Every point on the line represents a solution; every solution is represented by a point on the line.

Now Try Exercise 17.

When graphing some equations, it is convenient to first solve for y and then find ordered pairs. We can use the addition and multiplication principles to solve for y.

EXAMPLE 4 Graph: $3x - 5y = -10$.

Solution We first solve for y:

$$3x - 5y = -10$$
$$-5y = -3x - 10 \quad \text{Subtracting } 3x \text{ on both sides}$$
$$y = \tfrac{3}{5}x + 2. \quad \text{Multiplying by } -\tfrac{1}{5} \text{ on both sides}$$

By choosing multiples of 5 for x, we can avoid adding and subtracting fraction values when calculating y. For example, if we choose -5 for x, we get

$$y = \tfrac{3}{5}x + 2 = \tfrac{3}{5}(-5) + 2 = -3 + 2 = -1.$$

The following table lists a few points. We plot the points and draw the graph.

x	y	(x, y)
-5	-1	$(-5, -1)$
0	2	$(0, 2)$
5	5	$(5, 5)$

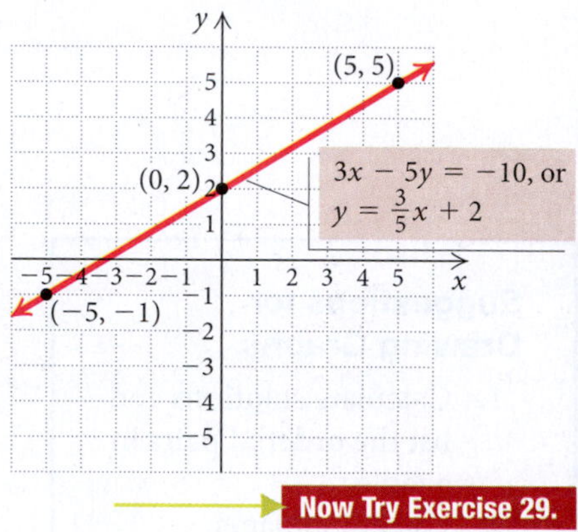

Now Try Exercise 29.

In the equation $y = \tfrac{3}{5}x + 2$ in Example 4, the value of y *depends* on the value chosen for x, so x is said to be the **independent variable** and y the **dependent variable**.

Technology Connection

Plot1 Plot2 Plot3
\Y1 = $\frac{3}{5}$X+2
\Y2=
\Y3=
\Y4=
\Y5=
\Y6=
\Y7=

We can graph an equation on a graphing calculator. Many calculators require an equation to be entered in the form "$y =$." In such a case, if the equation is not initially given in this form, it must be solved for y before it is entered in the calculator. For the equation $3x - 5y = -10$ in Example 4, we enter $y = \tfrac{3}{5}x + 2$ on the equation-editor, or $y =$, screen in the form $y = (3/5)x + 2$, as shown in the window at left.

Next, we determine the portion of the xy-plane that will appear on the calculator's screen. That portion of the plane is called the **viewing window**.

WINDOW
Xmin = −10
Xmax = 10
Xscl = 1
Ymin = −10
Ymax = 10
Yscl = 1
Xres = 1

The notation used in this text to denote a window setting consists of four numbers [L, R, B, T], which represent the **L**eft and **R**ight endpoints of the x-axis and the **B**ottom and **T**op endpoints of the y-axis, respectively. The window with the settings $[-10, 10, -10, 10]$ is the **standard viewing window**. On some graphing calculators, the standard window can be selected quickly using the ZSTANDARD feature from the ZOOM menu.

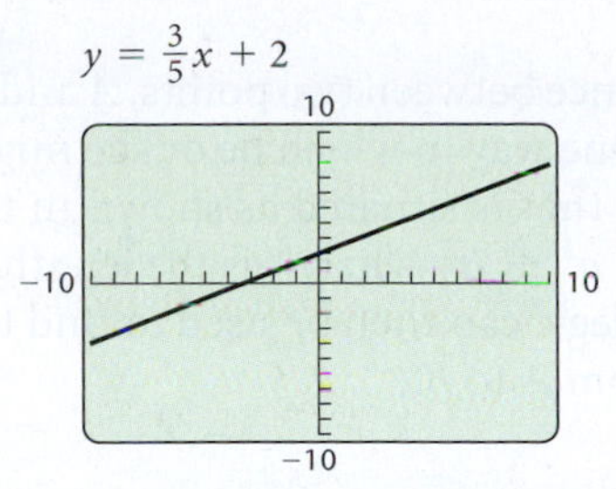

Xmin and Xmax are used to set the left and right endpoints of the x-axis, respectively; Ymin and Ymax are used to set the bottom and top endpoints of the y-axis, respectively. The settings Xscl and Yscl give the scales for the axes. For example, Xscl = 1 and Yscl = 1 means that there is 1 unit between tick marks on each of the axes. In this text, scaling factors other than 1 will be listed by the window unless they are readily apparent.

After entering the equation $y = (3/5)x + 2$ and choosing a viewing window, we can then draw the graph shown at left.

JUST IN TIME 7

EXAMPLE 5 Graph: $y = x^2 - 9x - 12$.

Solution Note that since this equation is not of the form $Ax + By = C$, its graph is not a straight line. We make a table of values, plot enough points to obtain an idea of the shape of the curve, and connect the points with a smooth curve. It is important to scale the axes to include most of the ordered pairs listed in the table. Here it is appropriate to use a larger scale on the y-axis than on the x-axis.

x	y	(x, y)
−3	24	(−3, 24)
−1	−2	(−1, −2)
0	−12	(0, −12)
2	−26	(2, −26)
4	−32	(4, −32)
5	−32	(5, −32)
10	−2	(10, −2)
12	24	(12, 24)

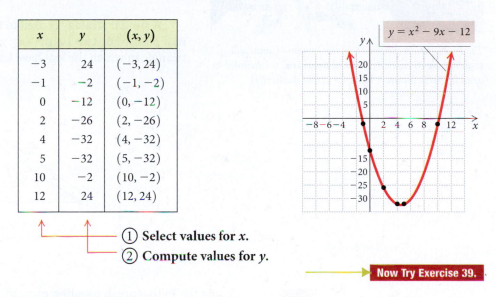

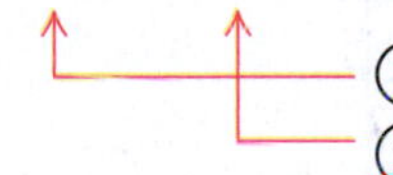

① **Select values for x.**
② **Compute values for y.**

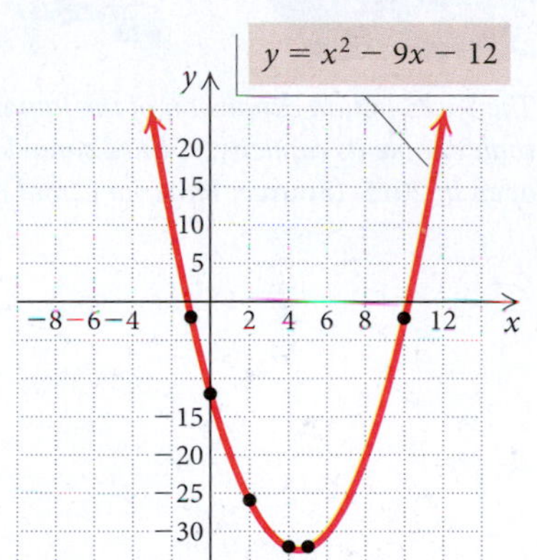

Now Try Exercise 39.

Technology Connection

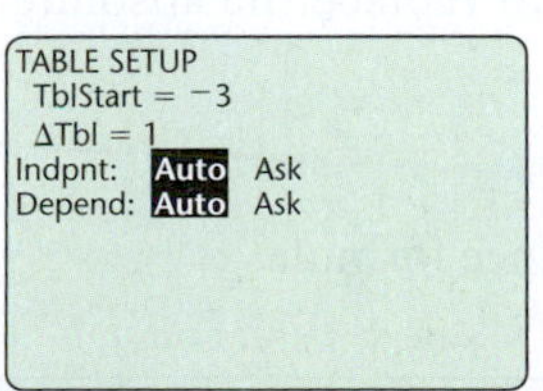

X	Y1
−3	24
−2	10
−1	−2
0	−12
1	−20
2	−26
3	−30

X = −3

A graphing calculator can be used to create a table of ordered pairs that are solutions of an equation. For the equation in Example 5, $y = x^2 - 9x - 12$, we first enter the equation on the equation-editor screen. Then we set up a table in AUTO mode by designating a value for TBLSTART and a value for ΔTBL. The calculator will produce a table starting with the value of TBLSTART and continuing by adding ΔTBL to supply succeeding x-values. For the equation $y = x^2 - 9x - 12$, we let TBLSTART = −3 and ΔTBL = 1. We can scroll up and down in the table to find values other than those shown here.

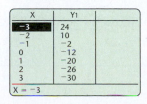

The Distance Formula

Suppose that an architect must determine the distance between two points, A and B, on opposite sides of a lane of the Panama Canal. One way in which he or she might proceed is to measure two legs of a right triangle that is situated as shown in the following figure. The Pythagorean equation, $c^2 = a^2 + b^2$, where c is the length of the hypotenuse and a and b are the lengths of the legs, can then be used to find the length of the hypotenuse, which is the distance from A to B.

The $5.25 billion expansion of the Panama Canal will soon double its capacity. A third canal lane is scheduled to open in 2015. (Source: *Panama Canal Authority*)

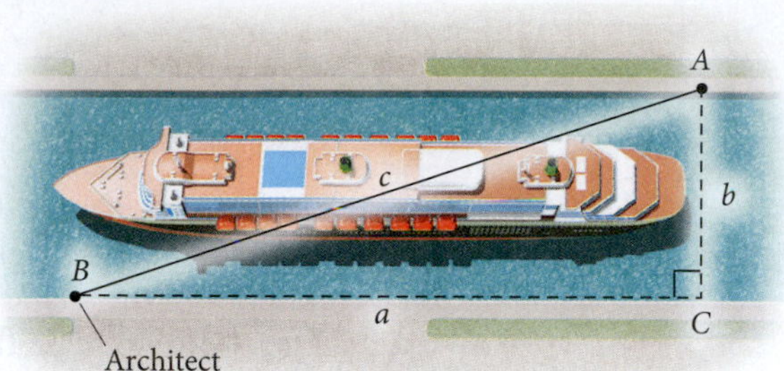

A similar strategy is used to find the distance between two points in a plane. For two points (x_1, y_1) and (x_2, y_2), we can draw a right triangle in which the legs have lengths $|x_2 - x_1|$ and $|y_2 - y_1|$.

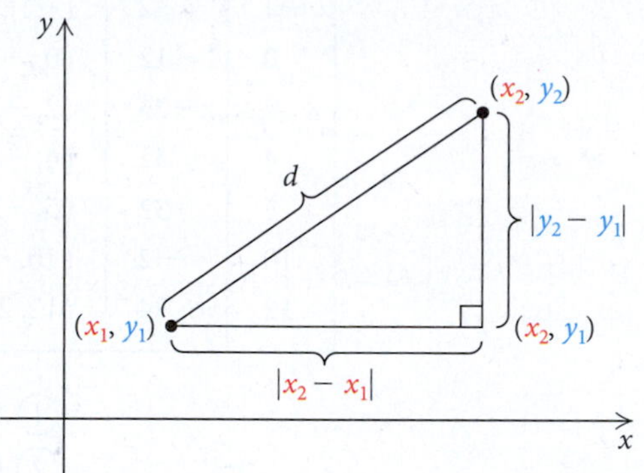

Using the Pythagorean equation $c^2 = a^2 + b^2$, we have

$$d^2 = |x_2 - x_1|^2 + |y_2 - y_1|^2.$$

Substituting d for c, $|x_2 - x_1|$ for a, and $|y_2 - y_1|$ for b in the Pythagorean equation

Because we are squaring, we can use parentheses to replace the absolute-value symbols:

$$d^2 = (x_2 - x_1)^2 + (y_2 - y_1)^2.$$

Taking the principal square root, we obtain the distance formula.

THE DISTANCE FORMULA

The **distance d** between any two points (x_1, y_1) and (x_2, y_2) is given by

$$d = \sqrt{(x_2 - x_1)^2 + (y_2 - y_1)^2}.$$

The subtraction of the x-coordinates can be done in any order, as can the subtraction of the y-coordinates. Although we derived the distance formula by considering two points not on a horizontal line or a vertical line, the distance formula holds for *any* two points.

EXAMPLE 6 Find the distance between each pair of points.

a) $(-2, 2)$ and $(3, -6)$ **b)** $(-1, -5)$ and $(-1, 2)$

Solution We substitute into the distance formula.

a)

$$\begin{aligned} d &= \sqrt{(x_2 - x_1)^2 + (y_2 - y_1)^2} \\ &= \sqrt{[3 - (-2)]^2 + (-6 - 2)^2} \\ &= \sqrt{5^2 + (-8)^2} = \sqrt{25 + 64} \\ &= \sqrt{89} \approx 9.4 \end{aligned}$$

b)

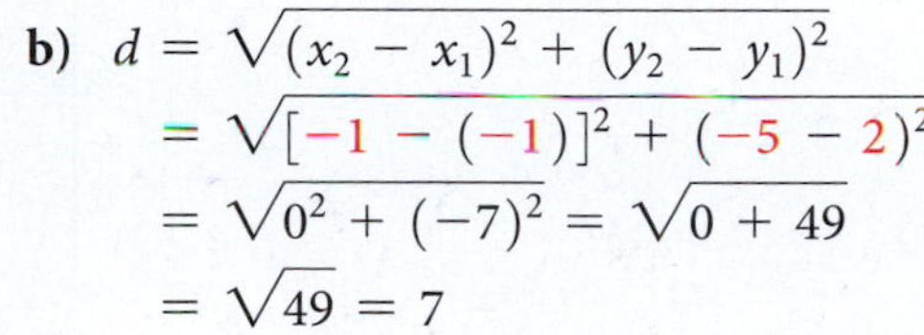

$$\begin{aligned} d &= \sqrt{(x_2 - x_1)^2 + (y_2 - y_1)^2} \\ &= \sqrt{[-1 - (-1)]^2 + (-5 - 2)^2} \\ &= \sqrt{0^2 + (-7)^2} = \sqrt{0 + 49} \\ &= \sqrt{49} = 7 \end{aligned}$$

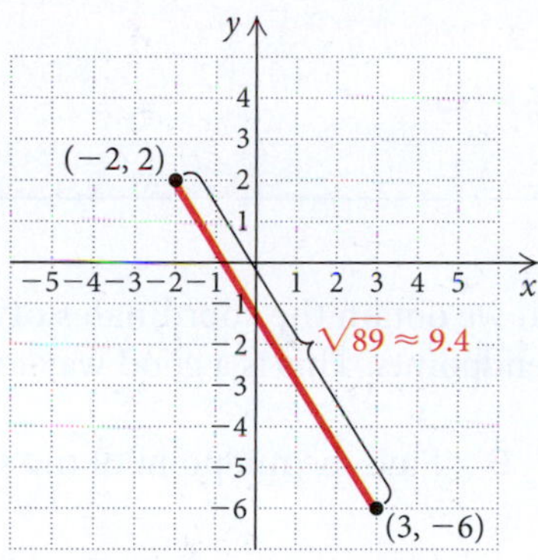

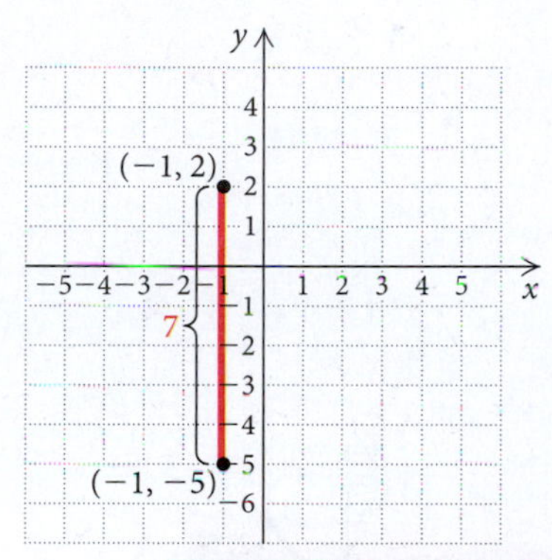

Now Try Exercises 41 and 49.

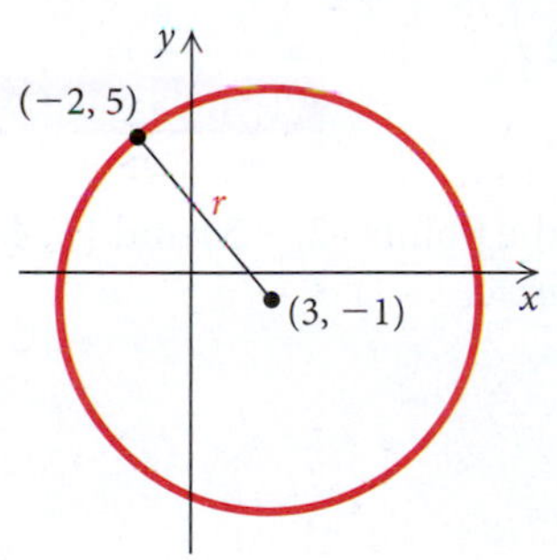

EXAMPLE 7 The point $(-2, 5)$ is on a circle that has $(3, -1)$ as its center. Find the length of the radius of the circle.

Solution Since the length of the radius is the distance from the center to a point on the circle, we substitute into the distance formula:

$$d = \sqrt{(x_2 - x_1)^2 + (y_2 - y_1)^2}$$

$$r = \sqrt{[3 - (-2)]^2 + (-1 - 5)^2}$$

Substituting r for d, $(3, -1)$ for (x_2, y_2), and $(-2, 5)$ for (x_1, y_1). Either point can serve as (x_1, y_1).

$$\begin{aligned} &= \sqrt{5^2 + (-6)^2} = \sqrt{25 + 36} \\ &= \sqrt{61} \approx 7.8. \end{aligned}$$

Rounded to the nearest tenth

The radius of the circle is approximately 7.8.

Now Try Exercise 55.

Midpoints of Segments

The distance formula can be used to develop a method of determining the *midpoint* of a segment when the endpoints are known. We state the formula and leave its proof to the exercises.

THE MIDPOINT FORMULA

If the endpoints of a segment are (x_1, y_1) and (x_2, y_2), then the coordinates of the **midpoint** of the segment are

$$\left(\frac{x_1 + x_2}{2}, \frac{y_1 + y_2}{2}\right).$$

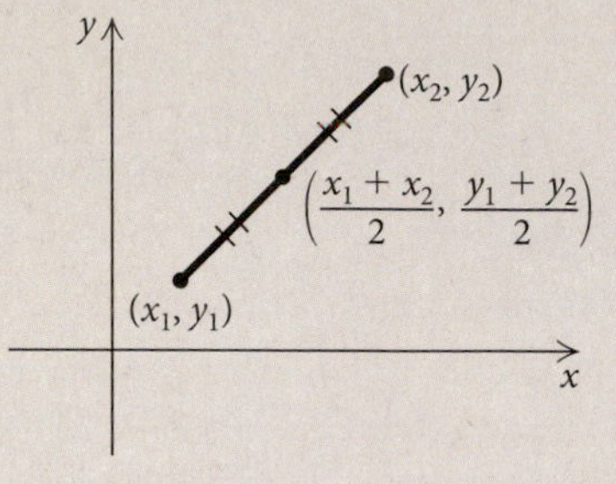

Note that we obtain the coordinates of the midpoint by averaging the coordinates of the endpoints. This is a good way to remember the midpoint formula.

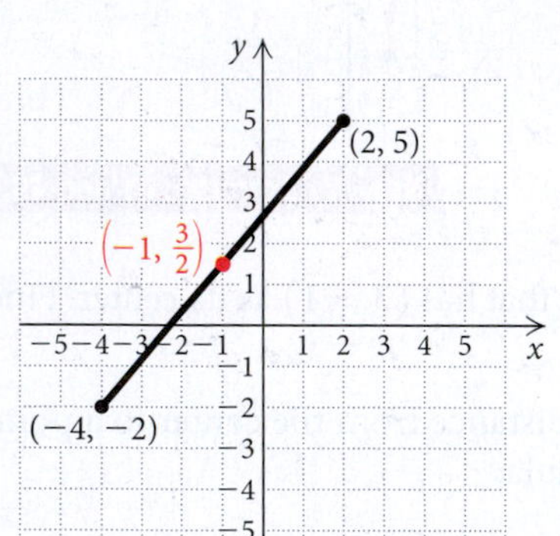

EXAMPLE 8 Find the midpoint of the segment whose endpoints are $(-4, -2)$ and $(2, 5)$.

Solution Using the midpoint formula, we obtain

$$\left(\frac{-4 + 2}{2}, \frac{-2 + 5}{2}\right) = \left(\frac{-2}{2}, \frac{3}{2}\right) = \left(-1, \frac{3}{2}\right).$$

Now Try Exercise 61.

EXAMPLE 9 The diameter of a circle connects the points $(2, -3)$ and $(6, 4)$ on the circle. Find the coordinates of the center of the circle.

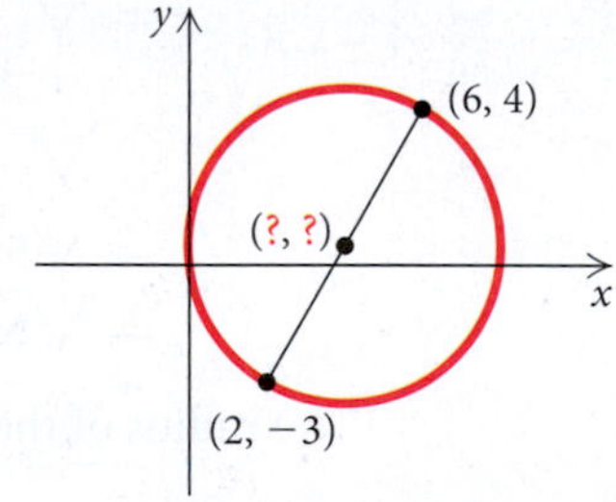

Solution Since the center of the circle is the midpoint of the diameter, we use the midpoint formula:

$$\left(\frac{2 + 6}{2}, \frac{-3 + 4}{2}\right), \text{ or } \left(\frac{8}{2}, \frac{1}{2}\right), \text{ or } \left(4, \frac{1}{2}\right).$$

The coordinates of the center are $\left(4, \frac{1}{2}\right)$.

Now Try Exercise 73.

Circles

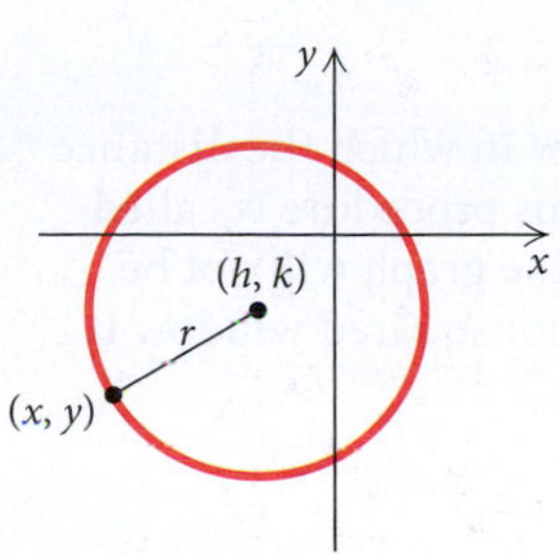

A **circle** is the set of all points in a plane that are a fixed distance r from a *center* (h, k). Thus if a point (x, y) is to be r units from the center, we must have

$$r = \sqrt{(x - h)^2 + (y - k)^2}.$$ Using the distance formula, $d = \sqrt{(x_2 - x_1)^2 + (y_2 - y_1)^2}$

Squaring both sides gives an equation of a circle. The distance r is the length of a *radius* of the circle.

> **THE EQUATION OF A CIRCLE**
>
> The standard form of the equation of a circle with center (h, k) and radius r is
>
> $$(x - h)^2 + (y - k)^2 = r^2.$$

EXAMPLE 10 Find an equation of the circle having radius 5 and center $(3, -7)$.

Solution Using the standard form, we have

$$[x - 3]^2 + [y - (-7)]^2 = 5^2$$ Substituting

$$(x - 3)^2 + (y + 7)^2 = 25.$$

Now Try Exercise 75.

EXAMPLE 11 Graph the circle $(x + 5)^2 + (y - 2)^2 = 16$.

Solution We write the equation in standard form to determine the center and the radius:

$$[x - (-5)]^2 + [y - 2]^2 = 4^2.$$

The center is $(-5, 2)$ and the radius is 4. We locate the center and draw the circle using a compass.

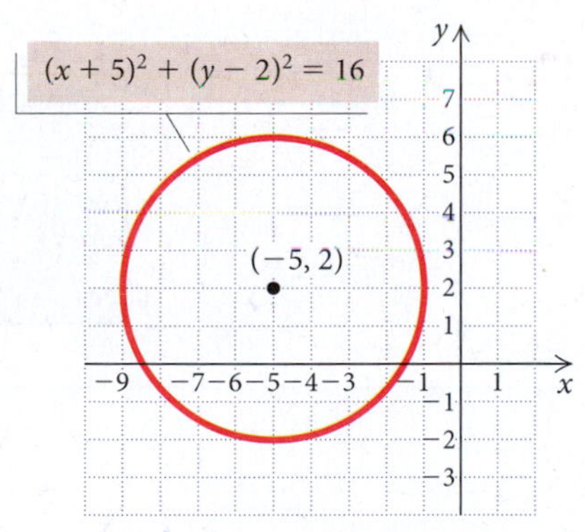

Now Try Exercise 87.

Technology Connection

When we graph a circle, we select a viewing window in which the distance between units is visually the same on both axes. This procedure is called **squaring the viewing window**. We do this so that the graph will not be distorted. A graph of the circle $x^2 + y^2 = 36$ in a nonsquared window is shown in Fig. 1.

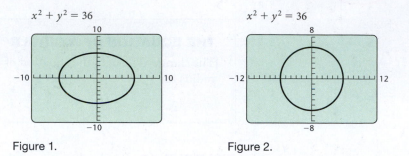

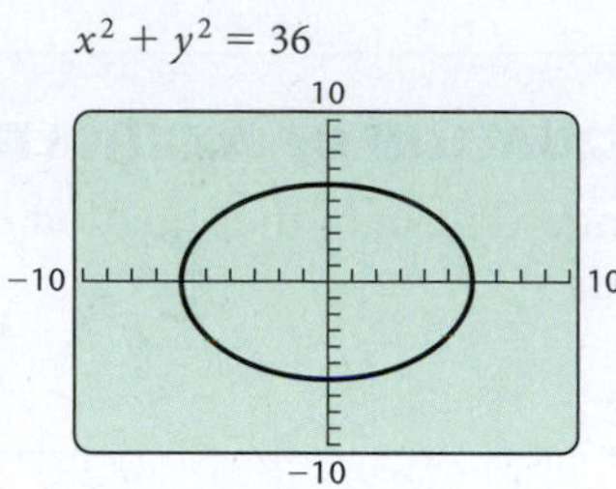

Figure 1.

$x^2 + y^2 = 36$

8
−12
12
−8

Figure 2.

On many graphing calculators, the ratio of the height to the width of the viewing screen is $\frac{2}{3}$. When we choose a window in which Xscl = Yscl and the length of the y-axis is $\frac{2}{3}$ the length of the x-axis, the window will be squared. The windows with dimensions $[-6, 6, -4, 4]$, $[-9, 9, -6, 6]$, and $[-12, 12, -8, 8]$ are examples of squared windows. A graph of the circle $x^2 + y^2 = 36$ in a squared window is shown in Fig. 2. Many graphing calculators have an option on the ZOOM menu that squares the window automatically.

To graph a circle, we select the CIRCLE feature from the DRAW menu and enter the coordinates of the center and the length of the radius. The graph of the circle $(x - 2)^2 + (y + 1)^2 = 16$ is shown here. For more on graphing circles with a graphing calculator, see Section 7.2.

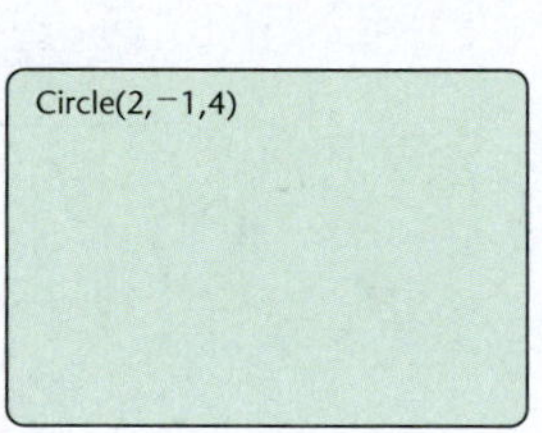

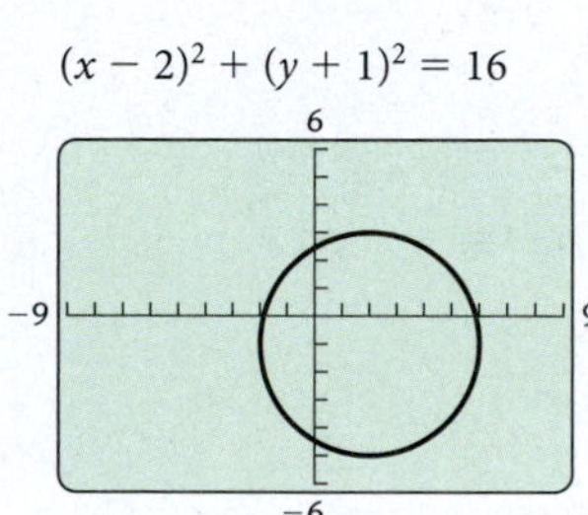

Visualizing the Graph

Match the equation with its graph.

1. $y = -x^2 + 5x - 3$

2. $3x - 5y = 15$

3. $(x - 2)^2 + (y - 4)^2 = 36$

4. $y - 5x = -3$

5. $x^2 + y^2 = \dfrac{25}{4}$

6. $15y - 6x = 90$

7. $y = -\dfrac{2}{3}x - 2$

8. $(x + 3)^2 + (y - 1)^2 = 16$

9. $3x + 5y = 15$

10. $y = x^2 - x - 4$

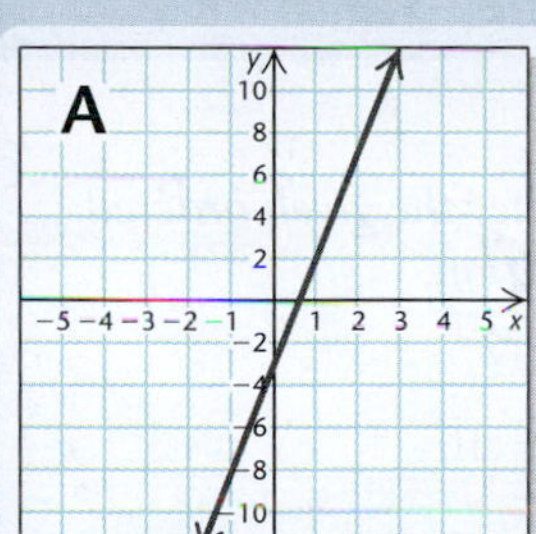

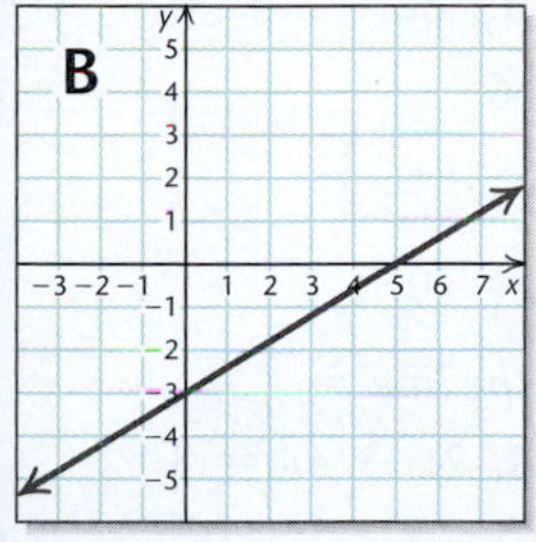

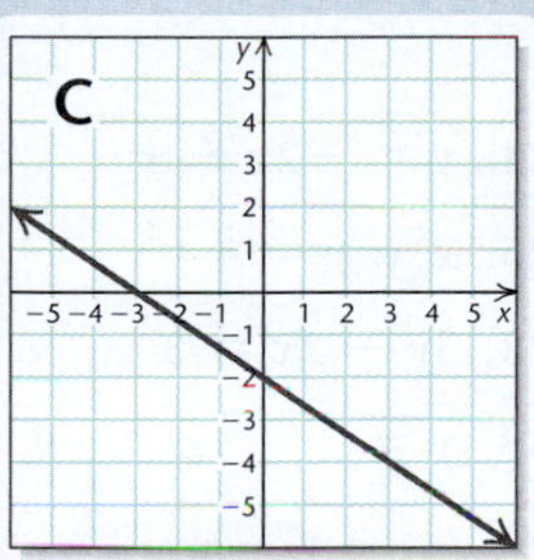

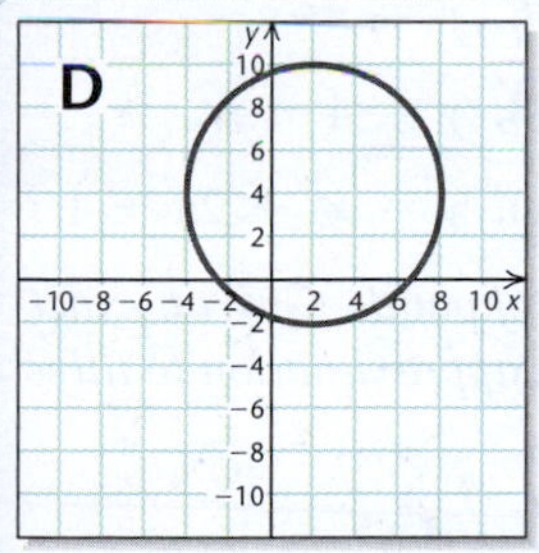

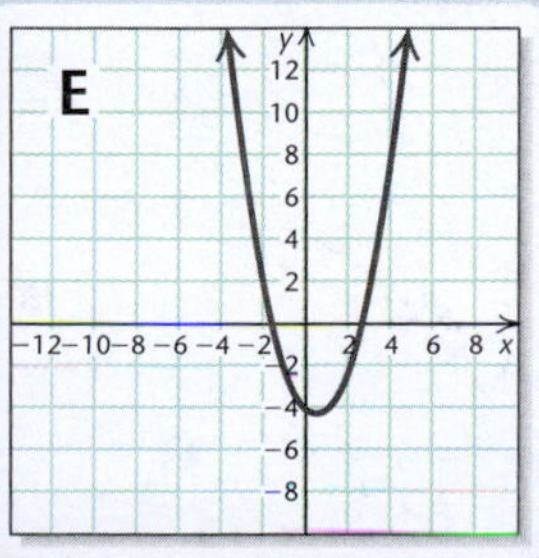

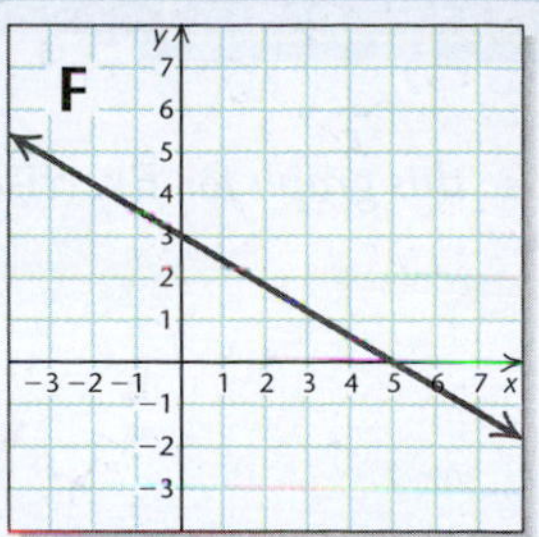

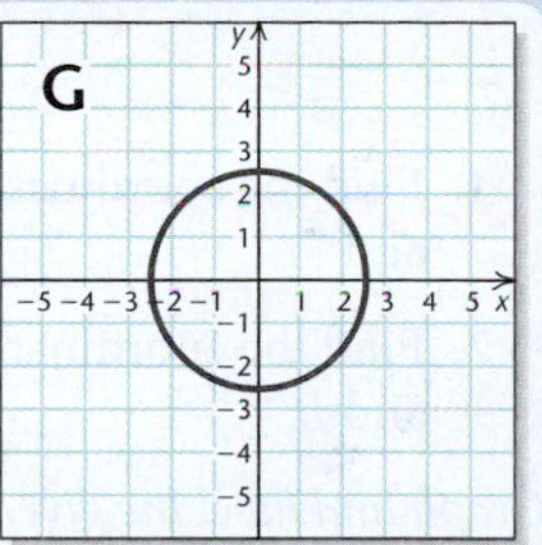

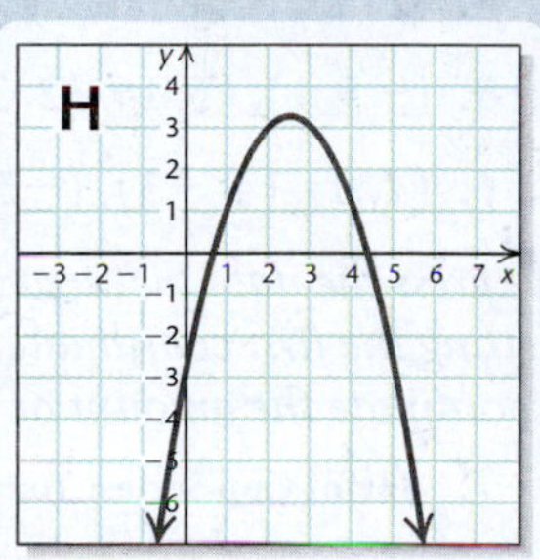

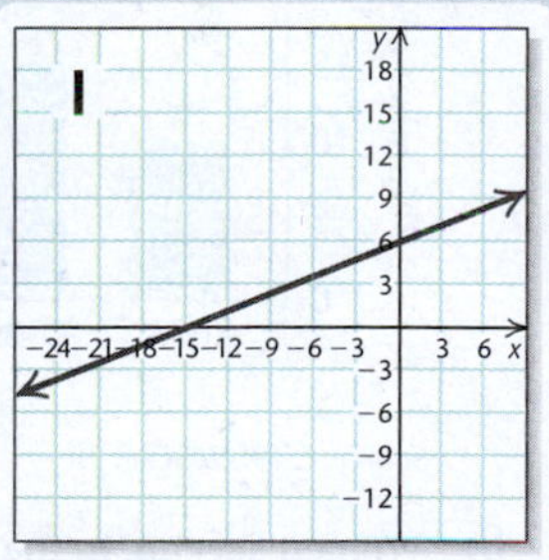

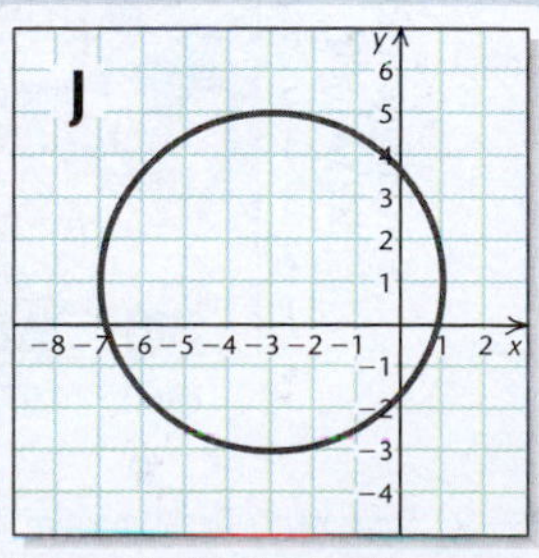

Answers on page A-1

1.1 Exercise Set

Use this graph for Exercises 1 and 2.

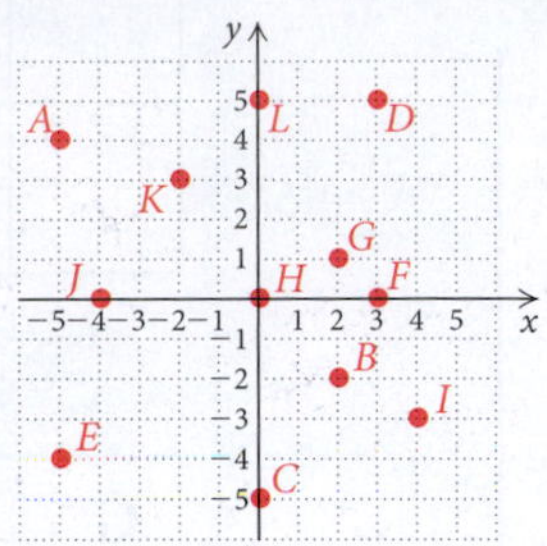

1. Find the coordinates of points A, B, C, D, E, and F.

2. Find the coordinates of points G, H, I, J, K, and L.

Graph and label the given points.

3. $(4, 0), (-3, -5), (-1, 4), (0, 2), (2, -2)$

4. $(1, 4), (-4, -2), (-5, 0), (2, -4), (4, 0)$

5. $(-5, 1), (5, 1), (2, 3), (2, -1), (0, 1)$

6. $(4, 0), (4, -3), (-5, 2), (-5, 0), (-1, -5)$

Express the data pictured in the graph as ordered pairs, letting the first coordinate represent the year and the second coordinate the amount or percent.

7. Sprint Cup Series: Tony Stewart in the Top 5

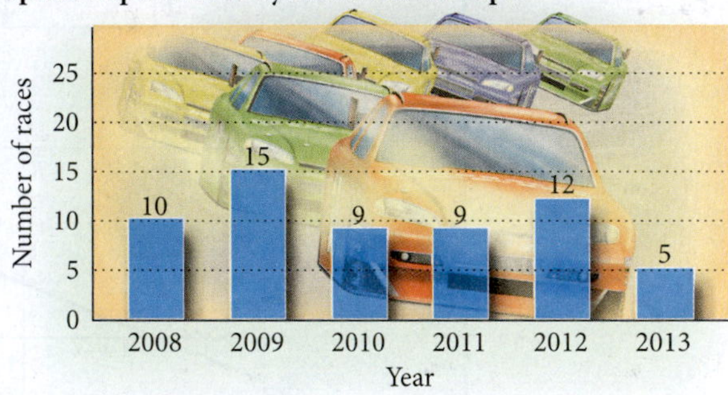

Source: ESPN NASCAR

8. Women Serving in the Marines

Source: U. S. Department of Veterans Affairs, Rutgers University

Use substitution to determine whether the given ordered pairs are solutions of the given equation.

9. $(-1, -9), (0, 2)$; $y = 7x - 2$

10. $\left(\frac{1}{2}, 8\right), (-1, 6)$; $y = -4x + 10$

11. $\left(\frac{2}{3}, \frac{3}{4}\right), \left(1, \frac{3}{2}\right)$; $6x - 4y = 1$

12. $(1.5, 2.6), (-3, 0)$; $x^2 + y^2 = 9$

13. $\left(-\frac{1}{2}, -\frac{4}{5}\right), \left(0, \frac{3}{5}\right)$; $2a + 5b = 3$

14. $\left(0, \frac{3}{2}\right), \left(\frac{2}{3}, 1\right)$; $3m + 4n = 6$

15. $(-0.75, 2.75), (2, -1)$; $x^2 - y^2 = 3$

16. $(2, -4), (4, -5)$; $5x + 2y^2 = 70$

Find the intercepts and then graph the line.

17. $5x - 3y = -15$

18. $2x - 4y = 8$

19. $2x + y = 4$

20. $3x + y = 6$

21. $4y - 3x = 12$

22. $3y + 2x = -6$

Graph the equation.

23. $y = 3x + 5$

24. $y = -2x - 1$

25. $x - y = 3$

26. $x + y = 4$

27. $y = -\frac{3}{4}x + 3$

28. $3y - 2x = 3$

29. $5x - 2y = 8$

30. $y = 2 - \frac{4}{3}x$

31. $x - 4y = 5$

32. $6x - y = 4$

33. $2x + 5y = -10$

34. $4x - 3y = 12$

35. $y = -x^2$

36. $y = x^2$

37. $y = x^2 - 3$

38. $y = 4 - x^2$

39. $y = -x^2 + 2x + 3$

40. $y = x^2 + 2x - 1$

Find the distance between the pair of points. Give an exact answer and, where appropriate, an approximation to three decimal places.

41. $(4, 6)$ and $(5, 9)$

42. $(-3, 7)$ and $(2, 11)$

43. $(-11, -8)$ and $(1, -13)$

44. $(-60, 5)$ and $(-20, 35)$

45. $(6, -1)$ and $(9, 5)$

46. $(-4, -7)$ and $(-1, 3)$

47. $\left(-8, \frac{7}{11}\right)$ and $\left(8, \frac{7}{11}\right)$

48. $\left(\frac{1}{2}, -\frac{4}{25}\right)$ and $\left(\frac{1}{2}, -\frac{13}{25}\right)$

49. $\left(-\frac{3}{5}, -4\right)$ and $\left(-\frac{3}{5}, \frac{2}{3}\right)$

50. $\left(-\frac{11}{3}, -\frac{1}{2}\right)$ and $\left(\frac{1}{3}, \frac{5}{2}\right)$

51. $(-4.2, 3)$ and $(2.1, -6.4)$

52. $(0.6, -1.5)$ and $(-8.1, -1.5)$

53. $(0, 0)$ and (a, b)

54. (r, s) and $(-r, -s)$

55. The points $(-3, -1)$ and $(9, 4)$ are the endpoints of the diameter of a circle. Find the length of the radius of the circle.

56. The point $(0, 1)$ is on a circle that has center $(-3, 5)$. Find the length of the diameter of the circle.

The converse of the Pythagorean theorem is also a true statement: If the sum of the squares of the lengths of two sides of a triangle is equal to the square of the length of the third side, then the triangle is a right triangle. Use the distance formula and the Pythagorean theorem to determine whether the set of points could be vertices of a right triangle.

57. $(-4, 5)$, $(6, 1)$, and $(-8, -5)$

58. $(-3, 1)$, $(2, -1)$, and $(6, 9)$

59. $(-4, 3)$, $(0, 5)$, and $(3, -4)$

60. The points $(-3, 4)$, $(2, -1)$, $(5, 2)$, and $(0, 7)$ are vertices of a quadrilateral. Show that the quadrilateral is a rectangle. (*Hint*: Show that the quadrilateral's opposite sides are the same length and that the two diagonals are the same length.)

Find the midpoint of the segment having the given endpoints.

61. $(4, -9)$ and $(-12, -3)$

62. $(7, -2)$ and $(9, 5)$

63. $\left(0, \frac{1}{2}\right)$ and $\left(-\frac{2}{5}, 0\right)$

64. $(0, 0)$ and $\left(-\frac{7}{13}, \frac{2}{7}\right)$

65. $(6.1, -3.8)$ and $(3.8, -6.1)$

66. $(-0.5, -2.7)$ and $(4.8, -0.3)$

67. $(-6, 5)$ and $(-6, 8)$

68. $(1, -2)$ and $(-1, 2)$

69. $\left(-\frac{1}{6}, -\frac{3}{5}\right)$ and $\left(-\frac{2}{3}, \frac{5}{4}\right)$

70. $\left(\frac{2}{9}, \frac{1}{3}\right)$ and $\left(-\frac{2}{5}, \frac{4}{5}\right)$

71. Graph the rectangle described in Exercise 60. Then determine the coordinates of the midpoint of each of the four sides. Are the midpoints vertices of a rectangle?

72. Graph the square with vertices $(-5, -1)$, $(7, -6)$, $(12, 6)$, and $(0, 11)$. Then determine the midpoint of each of the four sides. Are the midpoints vertices of a square?

73. The points $(\sqrt{7}, -4)$ and $(\sqrt{2}, 3)$ are endpoints of the diameter of a circle. Determine the center of the circle.

74. The points $(-3, \sqrt{5})$ and $(1, \sqrt{2})$ are endpoints of the diagonal of a square. Determine the center of the square.

Find an equation for a circle satisfying the given conditions.

75. Center $(2, 3)$, radius of length $\frac{5}{3}$

76. Center $(4, 5)$, diameter of length 8.2

77. Center $(-1, 4)$, passes through $(3, 7)$

78. Center $(6, -5)$, passes through $(1, 7)$

79. The points $(7, 13)$ and $(-3, -11)$ are at the ends of a diameter.

80. The points $(-9, 4)$, $(-2, 5)$, $(-8, -3)$, and $(-1, -2)$ are vertices of an inscribed square.

81. Center $(-2, 3)$, tangent (touching at one point) to the y-axis

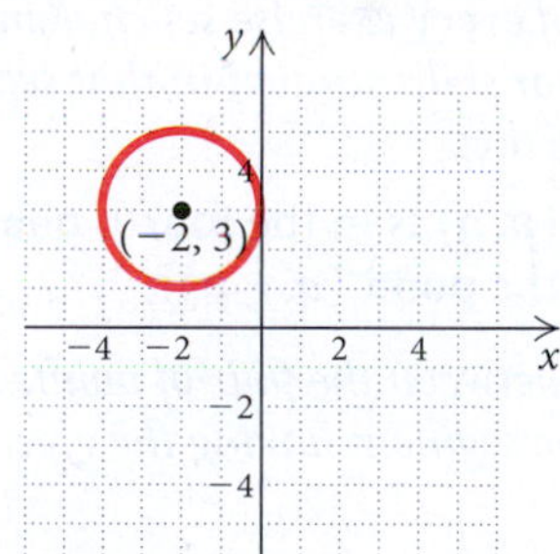

82. Center $(4, -5)$, tangent to the x-axis

Find the center and the radius of the circle. Then graph the circle.

83. $x^2 + y^2 = 4$

84. $x^2 + y^2 = 81$

85. $x^2 + (y - 3)^2 = 16$

86. $(x + 2)^2 + y^2 = 100$

87. $(x - 1)^2 + (y - 5)^2 = 36$

88. $(x - 7)^2 + (y + 2)^2 = 25$

89. $(x + 4)^2 + (y + 5)^2 = 9$

90. $(x + 1)^2 + (y - 2)^2 = 64$

Find the equation of the circle. Express the equation in standard form.

91.

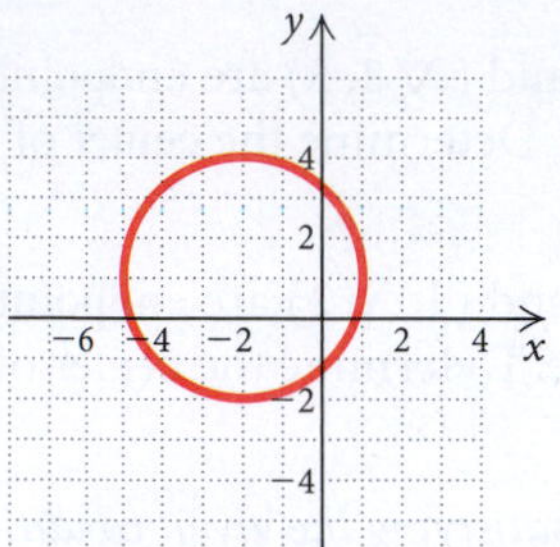

92.

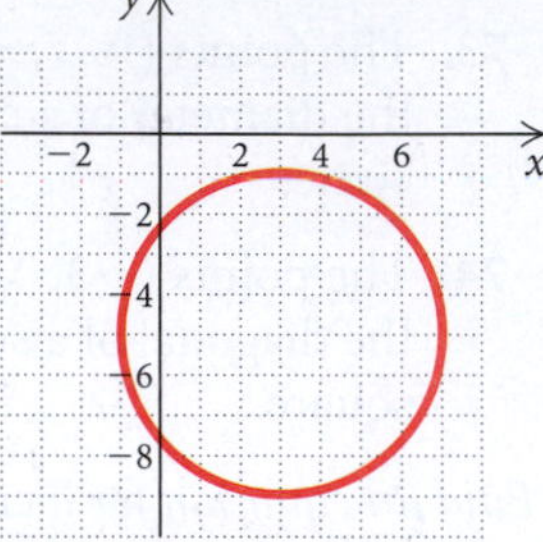

93.

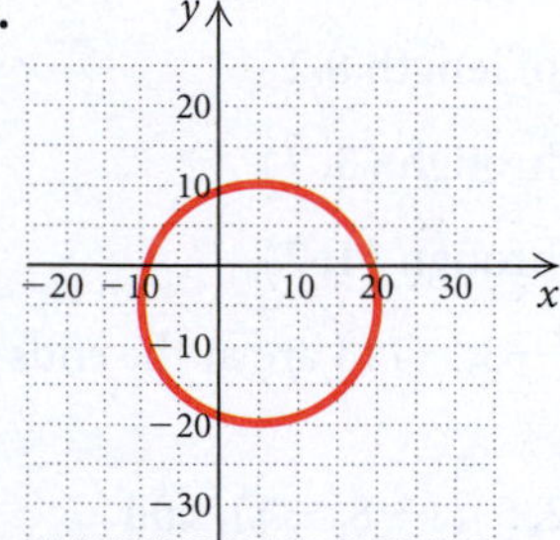

94.

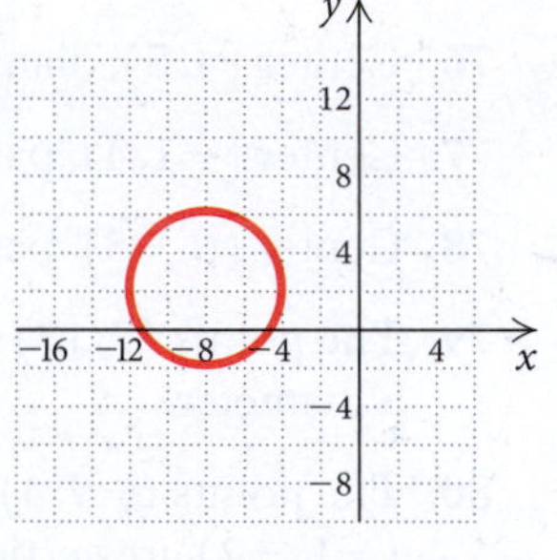

▶ Synthesis

To the student and the instructor: *The Synthesis exercises found at the end of every exercise set challenge students to combine concepts or skills studied in that section or in preceding parts of the text.*

95. If the point (p, q) is in the fourth quadrant, in which quadrant is the point $(q, -p)$?

Find the distance between the pair of points and find the midpoint of the segment having the given points as endpoints.

96. $\left(a, \frac{1}{a}\right)$ and $\left(a + h, \frac{1}{a + h}\right)$

97. $(a, \sqrt{a})$ and $(a + h, \sqrt{a + h})$

Find an equation of a circle satisfying the given conditions.

98. Center $(-5, 8)$ with a circumference of 10π units

99. Center $(2, -7)$ with an area of 36π square units

100. Find the point on the x-axis that is equidistant from the points $(-4, -3)$ and $(-1, 5)$.

101. Find the point on the y-axis that is equidistant from the points $(-2, 0)$ and $(4, 6)$.

102. Determine whether the points $(-1, -3)$, $(-4, -9)$, and $(2, 3)$ are collinear.

103. *An Arch of a Circle in Carpentry.* Matt is remodeling the front entrance to his home and needs to cut an arch for the top of an entranceway. The arch must be 8 ft wide and 2 ft high. To draw the arch, he will use a stretched string with chalk attached at an end as a compass.

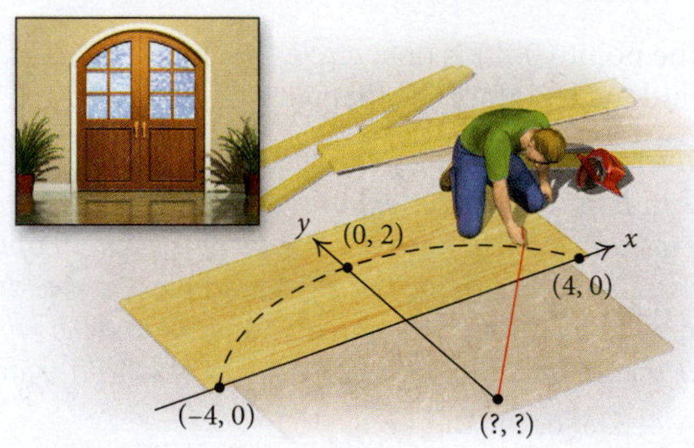

a) Using a coordinate system, locate the center of the circle.

b) What radius should Matt use to draw the arch?

104. Consider any right triangle with base b and height h, situated as shown. Show that the midpoint of the hypotenuse P is equidistant from the three vertices of the triangle.

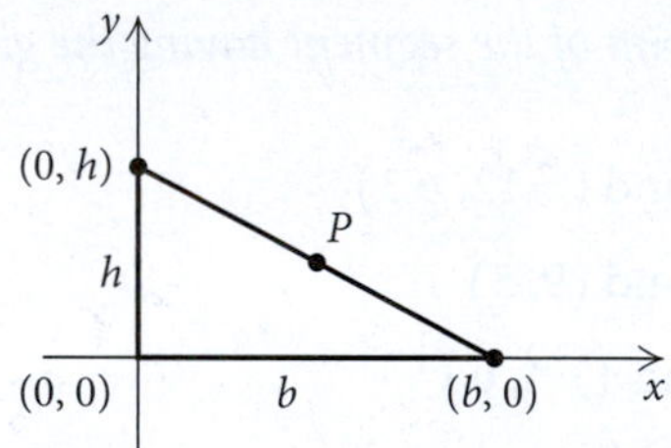

Determine whether each of the following points lies on the ***unit circle***, $x^2 + y^2 = 1$.

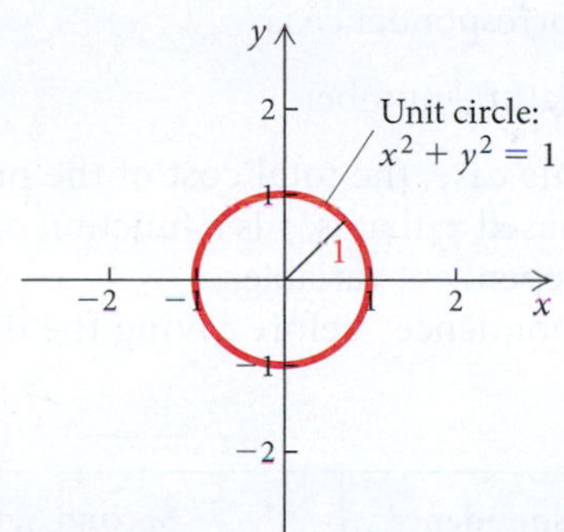

105. $\left(\frac{\sqrt{3}}{2}, -\frac{1}{2}\right)$

106. $(0, -1)$

107. $\left(-\frac{\sqrt{2}}{2}, \frac{\sqrt{2}}{2}\right)$

108. $\left(\frac{1}{2}, -\frac{\sqrt{3}}{2}\right)$

109. Prove the midpoint formula by showing that $\left(\frac{x_1 + x_2}{2}, \frac{y_1 + y_2}{2}\right)$ is equidistant from the points (x_1, y_1) and (x_2, y_2).

1.2 Functions and Graphs

- Determine whether a correspondence or a relation is a function.
- Find function values, or outputs, using a formula or a graph.
- Graph functions.
- Determine whether a graph is that of a function.
- Find the domain and the range of a function.
- Solve applied problems using functions.

We now focus our attention on a concept that is fundamental to many areas of mathematics—the idea of a *function*.

Functions

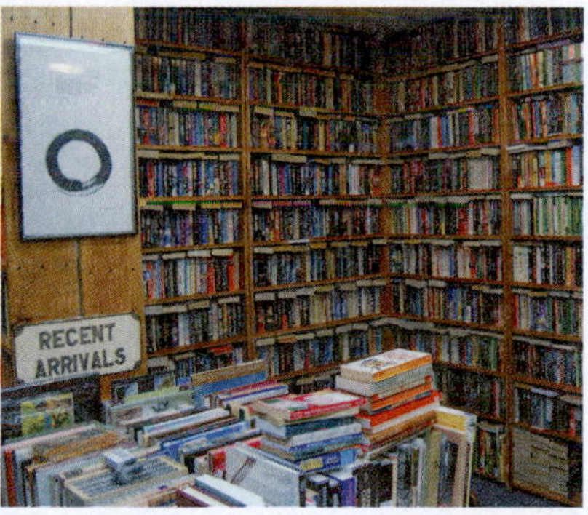

Used-Book Co-op. A community center operates a used-book co-op, and the proceeds are donated to summer youth programs. The total cost of a purchase is \$2.50 per book plus a flat-rate surcharge of \$3. If a customer selects 6 books, the total cost of the purchase is

$$\$2.50(6) + \$3, \quad \text{or} \quad \$18.$$

We can express this relationship with a set of ordered pairs, a graph, and an equation. A few ordered pairs are listed in the following table.

x	y	Ordered Pairs: (x, y)	Correspondence
1	5.50	(1, 5.50)	1 → 5.50
2	8.00	(2, 8)	2 → 8
4	13.00	(4, 13)	4 → 13
7	20.50	(7, 20.50)	7 → 20.50
10	28.00	(10, 28)	10 → 28

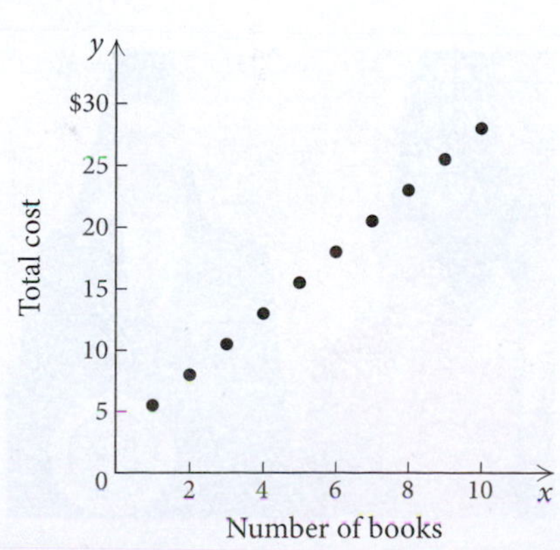

The ordered pairs express a relationship, or a correspondence, between the first coordinate and the second coordinate. We can see this relationship in the graph as well. The equation that describes the correspondence is

$$y = 2.50x + 3, \quad \text{where } x \text{ is a natural number.}$$

This is an example of a *function*. In this case, the total cost of the purchase y is a function of the number of books purchased x; that is, y is a function of x, where x is the independent variable and y is the dependent variable.

Let's consider some other correspondences before giving the definition of a function.

First Set	Correspondence	Second Set
To each person	there corresponds	that person's DNA.
To each blue spruce sold	there corresponds	its price.
To each real number	there corresponds	the square of that number.

In each correspondence, the first set is called the **domain** and the second set is called the **range**. For each member, or **element**, in the domain, there is *exactly one* member in the range to which it corresponds. Thus each person has exactly *one* DNA, each blue spruce has exactly *one* price, and each real number has exactly *one* square. Each correspondence is a *function*.

FUNCTION

A **function** is a correspondence between a first set, called the **domain**, and a second set, called the **range**, such that each member of the domain corresponds to *exactly one* member of the range.

It is important to note that not every correspondence between two sets is a function.

EXAMPLE 1 Determine whether each of the following correspondences is a function.

a) $-6 \rightarrow 36$
$6 \rightarrow 36$
$-3 \rightarrow 9$
$3 \rightarrow 9$
$0 \rightarrow 0$

b)

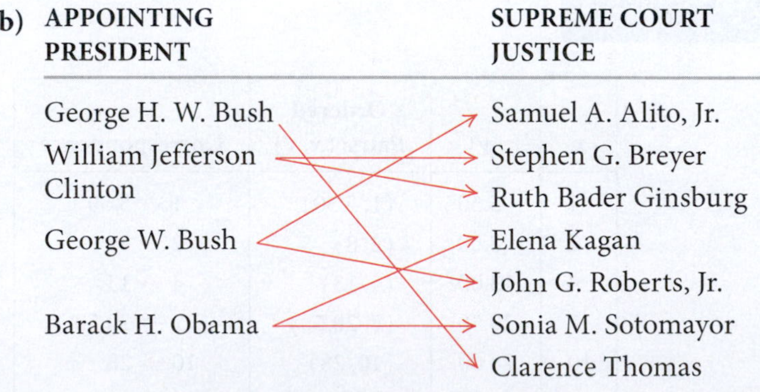

Solution

a) This correspondence *is* a function because each member of the domain corresponds to exactly one member of the range. Note that the definition of a function allows more than one member of the domain to correspond to the same member of the range.

b) This correspondence *is not* a function because there is at least one member of the domain who is paired with more than one member of the range (William Jefferson Clinton with Stephen G. Breyer and Ruth Bader Ginsburg; George W. Bush with Samuel A. Alito, Jr., and John G. Roberts, Jr.; Barack H. Obama with Elena Kagan and Sonia M. Sotomayor). **Now Try Exercises 5 and 7.**

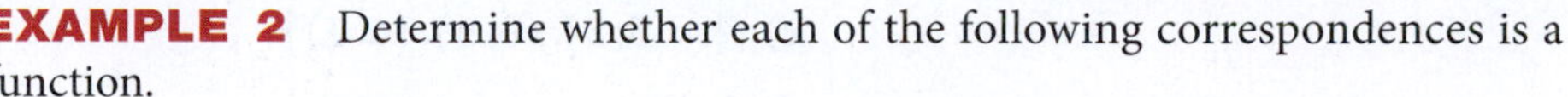

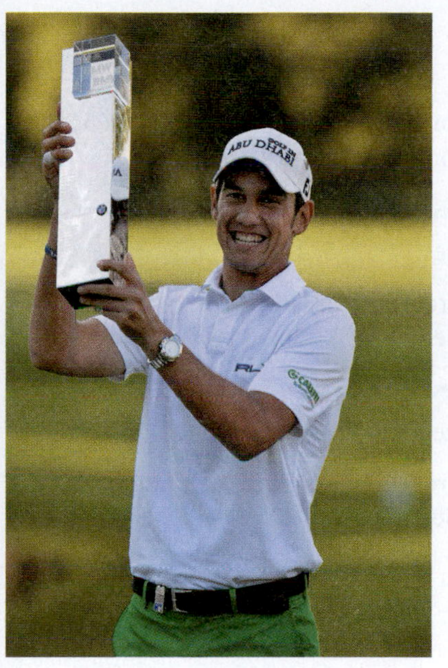

EXAMPLE 2 Determine whether each of the following correspondences is a function.

	DOMAIN	CORRESPONDENCE	RANGE
a)	Years in which a presidential election occurs	The person elected	A set of presidents
b)	All automobiles produced in 2014	Each automobile's VIN (Vehicle Identification Number)	A set of VINs
c)	The set of all professional golfers who won a PGA tournament in 2013	The tournament won	The set of all PGA tournaments in 2013
d)	The set of all PGA tournaments in 2013	The winner of the tournament	The set of all golfers who won a PGA tournament in 2013

Solution

a) This correspondence *is* a function because in each presidential election *exactly one* president is elected.

b) This correspondence *is* a function because each automobile has *exactly one* VIN.

c) This correspondence *is not* a function because a winning golfer could be paired with more than one tournament.

d) This correspondence *is* a function because each tournament has only one winning golfer. **Now Try Exercises 11 and 13.**

When a correspondence between two sets is not a function, it may still be an example of a **relation**.

> **RELATION**
>
> A **relation** is a correspondence between a first set, called the **domain**, and a second set, called the **range**, such that each member of the domain corresponds to *at least one* member of the range.

All the correspondences in Examples 1 and 2 are relations, but, as we have seen, not all are functions. Relations are sometimes written as sets of ordered pairs (as we saw earlier in the example on the total cost of a purchase of used books) in which elements of the domain are the first coordinates of the ordered pairs and elements of the range are the second coordinates. For example, instead of writing $-3 \longrightarrow 9$, as we did in Example 1(a), we could write the ordered pair $(-3, 9)$.

EXAMPLE 3 Determine whether each of the following relations is a function. Identify the domain and the range.

a) $\{(9, -5), (9, 5), (2, 4)\}$
b) $\{(-2, 5), (5, 7), (0, 1), (4, -2)\}$
c) $\{(-5, 3), (0, 3), (6, 3)\}$

Solution

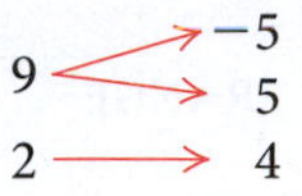

$9 \to -5$, $9 \to 5$, $2 \to 4$

Figure 1.

a) The relation *is not* a function because the ordered pairs $(9, -5)$ and $(9, 5)$ have the same first coordinate and different second coordinates. (See Fig. 1.)

The domain is the set of all first coordinates: $\{9, 2\}$.

The range is the set of all second coordinates: $\{-5, 5, 4\}$.

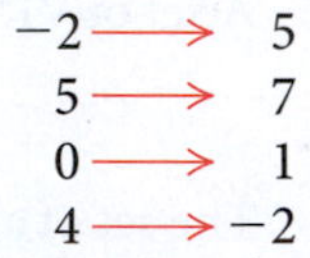

$-2 \to 5$, $5 \to 7$, $0 \to 1$, $4 \to -2$

Figure 2.

b) The relation *is* a function because *no* two ordered pairs have the same first coordinate and different second coordinates. (See Fig. 2.)

The domain is the set of all first coordinates: $\{-2, 5, 0, 4\}$.

The range is the set of all second coordinates: $\{5, 7, 1, -2\}$.

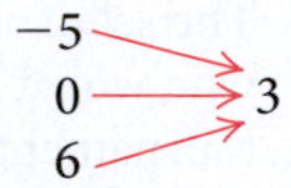

$-5 \to 3$, $0 \to 3$, $6 \to 3$

Figure 3.

c) The relation *is* a function because *no* two ordered pairs have the same first coordinate and different second coordinates. (See Fig. 3.)

The domain is $\{-5, 0, 6\}$.

The range is $\{3\}$.

Now Try Exercises 15 and 17.

Notation for Functions

Functions used in mathematics are often given by equations. They generally require that certain calculations be performed in order to determine which member of the range is paired with each member of the domain. For example, in Section 1.1 we graphed the function $y = x^2 - 9x - 12$ by doing calculations like the following:

$y = f(x) = x^2 - 9x - 12$

$$\text{for } x = -2,\ y = (-2)^2 - 9(-2) - 12 = 10,$$
$$\text{for } x = 0,\ y = 0^2 - 9 \cdot 0 - 12 = -12, \quad \text{and}$$
$$\text{for } x = 1,\ y = 1^2 - 9 \cdot 1 - 12 = -20.$$

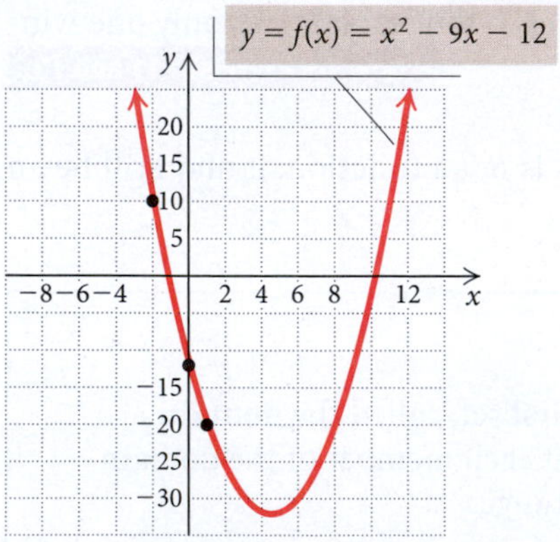

A more concise notation is often used. For $y = x^2 - 9x - 12$, the **inputs** (members of the domain) are values of x substituted into the equation. The **outputs** (members of the range) are the resulting values of y. If we call the function f, we can use x to represent an arbitrary *input* and $f(x)$—read "f of x," or "f at x," or "the value of f at x"—to represent the corresponding *output*. In this notation, the function given by $y = x^2 - 9x - 12$ is written as $f(x) = x^2 - 9x - 12$ and the above calculations would be

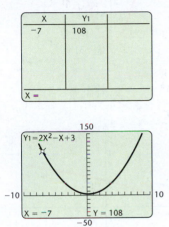

$$f(-2) = (-2)^2 - 9(-2) - 12 = 10,$$
$$f(0) = 0^2 - 9 \cdot 0 - 12 = -12, \quad \text{and}$$
$$f(1) = 1^2 - 9 \cdot 1 - 12 = -20.$$

Keep in mind that $f(x)$ *does not* mean $f \cdot x$.

Thus, instead of writing "when $x = -2$, the value of y is 10," we can simply write "$f(-2) = 10$," which can be read as "f of -2 is 10" or "for the input -2, the output of f is 10." The letters g and h are also often used to name functions.

Technology Connection

We can find function values with a graphing calculator. Below, we illustrate finding $f(-7)$ from Example 4(b), first with the TABLE feature set in ASK mode and then with the VALUE feature from the CALC menu. On both screens, we see that $f(-7) = 108$.

X	Y1
−7	108

X =

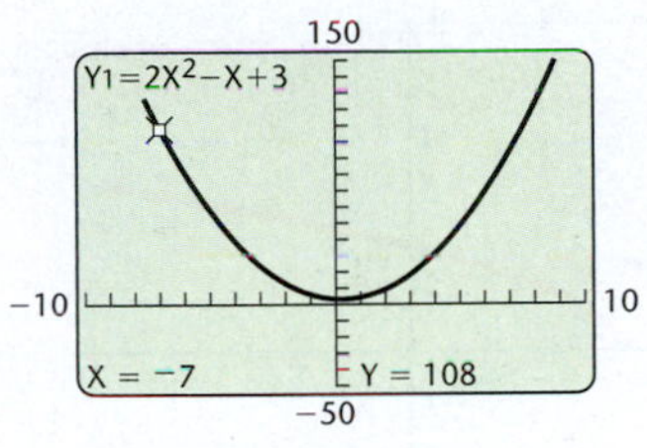

EXAMPLE 4 A function f is given by $f(x) = 2x^2 - x + 3$. Find each of the following.

a) $f(0)$ **b)** $f(-7)$

c) $f(5a)$ **d)** $f(a - 4)$

Solution We can think of this formula as follows:

$$f(\square) = 2(\square)^2 - (\square) + 3.$$

Then to find an output for a given input, we think: "Whatever goes in the blank on the left goes in the blank(s) on the right." This gives us a "recipe" for finding outputs.

a) $f(0) = 2(0)^2 - 0 + 3$
$= 0 - 0 + 3 = 3$

b) $f(-7) = 2(-7)^2 - (-7) + 3$
$= 2 \cdot 49 + 7 + 3 = 108$

c) $f(5a) = 2(5a)^2 - 5a + 3$
$= 2 \cdot 25a^2 - 5a + 3$
$= 50a^2 - 5a + 3$

JUST IN TIME 12

d) $f(a - 4) = 2(a - 4)^2 - (a - 4) + 3$
$= 2(a^2 - 8a + 16) - (a - 4) + 3$
$= 2a^2 - 16a + 32 - a + 4 + 3$
$= 2a^2 - 17a + 39$

Now Try Exercise 21.

Graphs of Functions

We graph functions in the same way that we graph equations. We find ordered pairs (x, y), or $(x, f(x))$, plot points, and complete the graph.

EXAMPLE 5 Graph each of the following functions.

a) $f(x) = x^2 - 5$

b) $f(x) = x^3 - x$

c) $f(x) = \sqrt{x + 4}$

Solution We select values for x and find the corresponding values of $f(x)$. Then we plot the points and connect them with a smooth curve.

a) $f(x) = x^2 - 5$

x	$f(x)$	$(x, f(x))$
-3	4	$(-3, 4)$
-2	-1	$(-2, -1)$
-1	-4	$(-1, -4)$
0	-5	$(0, -5)$
1	-4	$(1, -4)$
2	-1	$(2, -1)$
3	4	$(3, 4)$

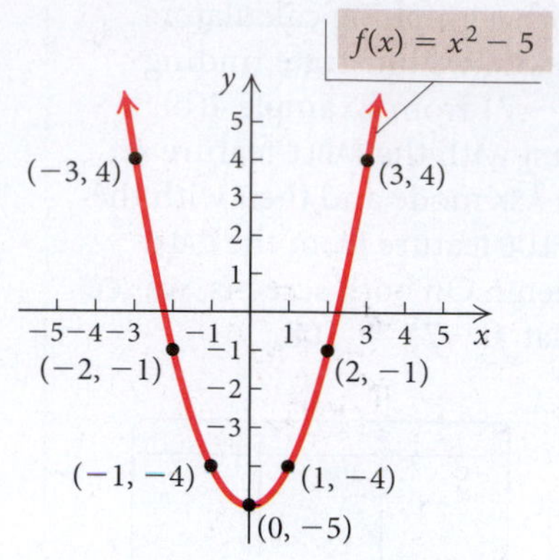

b) $f(x) = x^3 - x$

c) $f(x) = \sqrt{x + 4}$

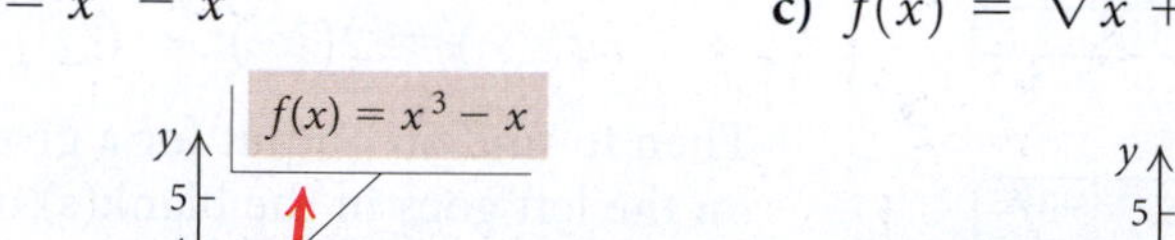

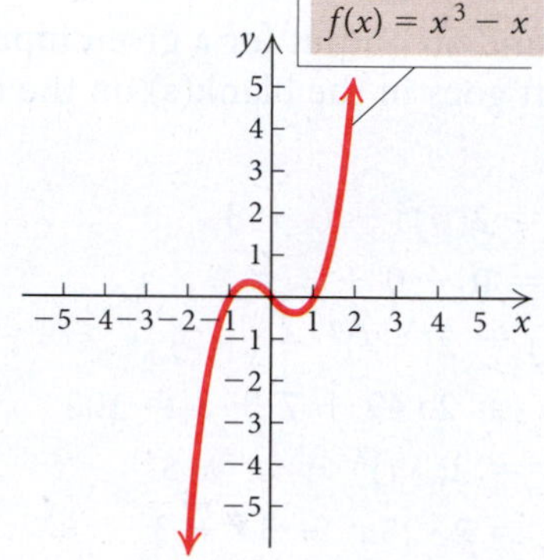

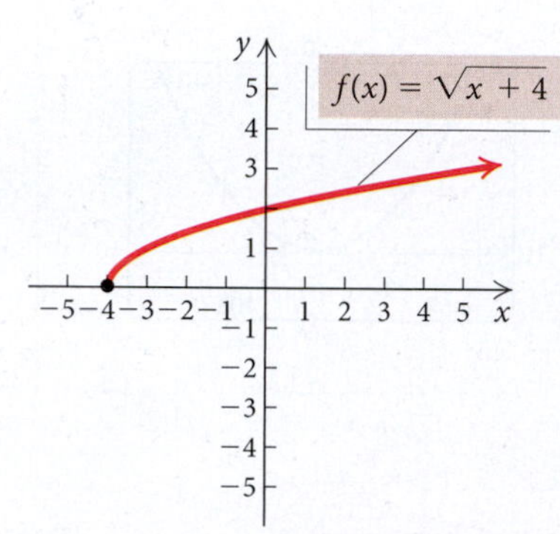

Now Try Exercise 31.

Function values can also be determined from a graph.

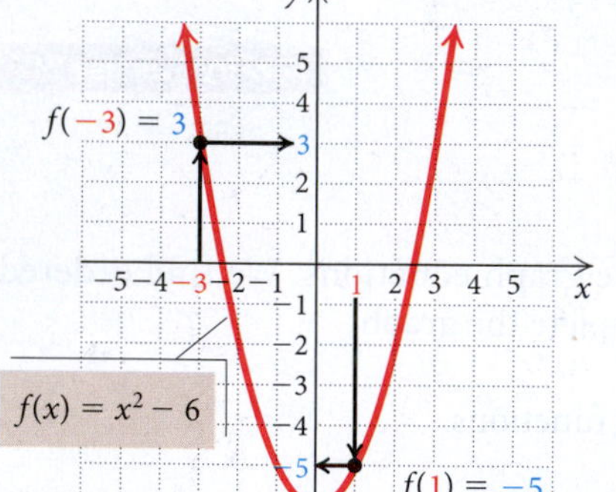

EXAMPLE 6 For the function $f(x) = x^2 - 6$, use the graph at left to find each of the following function values.

a) $f(-3)$

b) $f(1)$

Solution

a) To find the function value $f(-3)$ from the graph, we locate the input -3 on the horizontal axis, move vertically to the graph of the function, and then move horizontally to find the output on the vertical axis. We see that $f(-3) = 3$.

b) To find the function value $f(1)$, we locate the input 1 on the horizontal axis, move vertically to the graph, and then move horizontally to find the output on the vertical axis. We see that $f(1) = -5$.

Now Try Exercise 35.

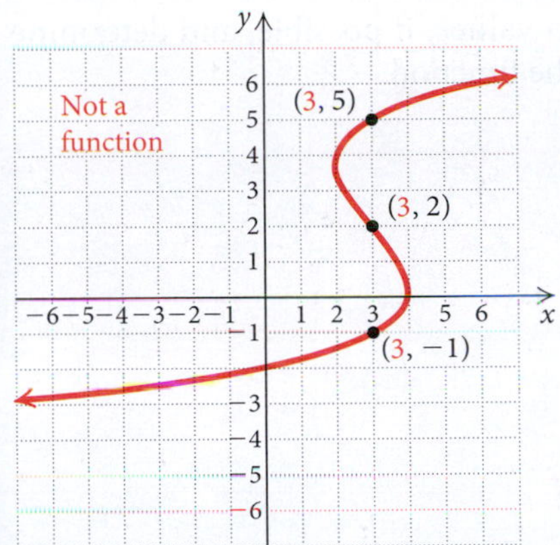

Since 3 is paired with more than one member of the range, the graph does not represent a function.

We know that when one member of the domain is paired with two or more different members of the range, the correspondence *is not* a function. Thus, when a graph contains two or more different points with the same first coordinate, the graph cannot represent a function. (See the graph at left. Note that 3 is paired with -1, 2, and 5.) Points sharing a common first coordinate are vertically above or below each other. This leads us to the *vertical-line test.*

THE VERTICAL-LINE TEST

If it is possible for a vertical line to cross a graph more than once, then the graph *is not* the graph of a function.

To apply the vertical-line test, we try to find a vertical line that crosses the graph more than once. If we succeed, then the graph is not that of a function. If we do not, then the graph is that of a function.

EXAMPLE 7 Which of graphs (a)–(f) (in red) are graphs of functions? In graph (f), the solid dot shows that $(-1, 1)$ belongs to the graph. The open circle shows that $(-1, -2)$ does *not* belong to the graph.

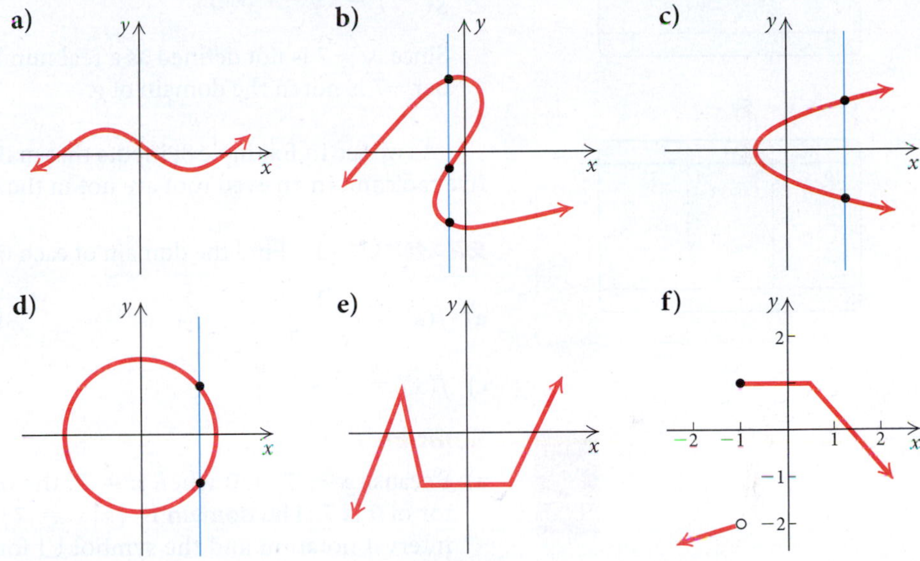

Solution Graphs (a), (e), and (f) are graphs of functions because we cannot find a vertical line that crosses any of them more than once. In (b), the vertical line drawn crosses the graph at three points, so graph (b) is not that of a function. Also, in (c) and (d), we can find a vertical line that crosses the graph more than once, so these are not graphs of functions.

Now Try Exercises 43 and 47.

Finding Domains of Functions

When a function *f* whose inputs and outputs are real numbers is given by a formula, the *domain* is understood to be the set of all inputs for which the expression is defined as a real number. When an input results in an expression that is not defined as a real number, we say that the function value *does not exist* and that the number being substituted *is not* in the domain of the function.

Technology Connection

When we use a graphing calculator to find function values and a function value does not exist, the calculator indicates this with an ERROR message.

In the following tables, we see in Example 8 that $f(3)$ for $f(x) = 1/(x - 3)$ and $g(-7)$ for $g(x) = \sqrt{x} + 5$ do not exist. Thus, 3 and -7 are *not* in the domains of the corresponding functions.

$y = \dfrac{1}{x - 3}$

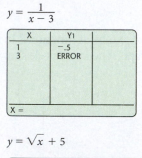

X	Y1
1	−.5
3	ERROR

X =

$y = \sqrt{x} + 5$

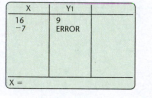

X	Y1
16	9
−7	ERROR

X =

EXAMPLE 8 Find the indicated function values, if possible, and determine whether the given values are in the domain of the function.

a) $f(1)$ and $f(3)$, for $f(x) = \dfrac{1}{x - 3}$

b) $g(16)$ and $g(-7)$, for $g(x) = \sqrt{x} + 5$

Solution

a) $f(1) = \dfrac{1}{1 - 3} = \dfrac{1}{-2} = -\dfrac{1}{2}$

Since $f(1)$ is defined, 1 is in the domain of f.

$f(3) = \dfrac{1}{3 - 3} = \dfrac{1}{0}$

Since division by 0 is not defined, $f(3)$ does not exist and the number 3 is not in the domain of f.

b) $g(16) = \sqrt{16} + 5 = 4 + 5 = 9$

Since $g(16)$ is defined, 16 is in the domain of g.

$g(-7) = \sqrt{-7} + 5$

Since $\sqrt{-7}$ is not defined as a real number, $g(-7)$ does not exist and the number -7 is not in the domain of g.

As we see in Example 8, inputs that make a denominator 0 or that yield a negative radicand in an even root are not in the domain of a function.

EXAMPLE 9 Find the domain of each of the following functions.

a) $f(x) = \dfrac{1}{x - 7}$

b) $h(x) = \dfrac{3x^2 - x + 7}{x^2 + 2x - 3}$

c) $f(x) = x^3 + |x|$

d) $g(x) = \sqrt[3]{x - 1}$

JUST IN TIME 6, 13, 16

Solution

a) Because $x - 7 = 0$ when $x = 7$, the only input that results in a denominator of 0 is 7. The domain is $\{x | x \neq 7\}$. We can also write the solution using interval notation and the symbol $\cup$ for the **union**, or inclusion, of both sets: $(-\infty, 7) \cup (7, \infty)$.

b) We can substitute any real number in the numerator, but we must avoid inputs that make the denominator 0. To find those inputs, we solve $x^2 + 2x - 3 = 0$, or $(x + 3)(x - 1) = 0$. Since $x^2 + 2x - 3$ is 0 for -3 and 1, the domain consists of the set of all real numbers except -3 and 1, or $\{x | x \neq -3 \textit{ and } x \neq 1\}$, or $(-\infty, -3) \cup (-3, 1) \cup (1, \infty)$.

c) We can substitute any real number for x. Thus the domain is the set of all real numbers, $\mathbb{R}$, or $(-\infty, \infty)$.

d) Because the index is odd, the radicand, $x - 1$, can be any real number. Thus x can be any real number. The domain is all real numbers, $\mathbb{R}$, or $(-\infty, \infty)$.

Now Try Exercises 55, 57, and 61.

Visualizing Domain and Range

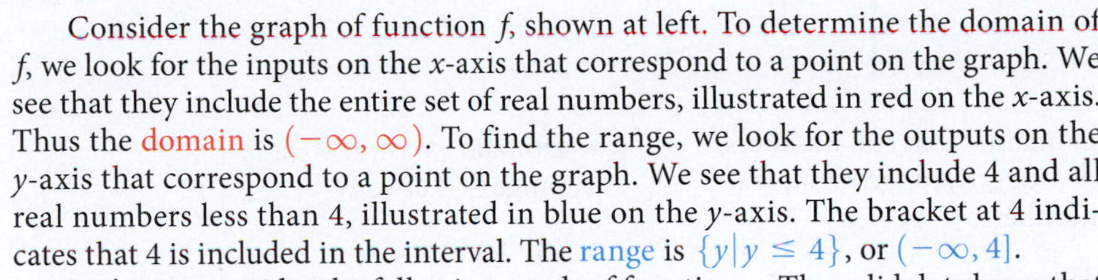

Keep the following in mind regarding the *graph* of a function:

Domain = the set of a function's inputs, found on the horizontal axis (x-axis);

Range = the set of a function's outputs, found on the vertical axis (y-axis).

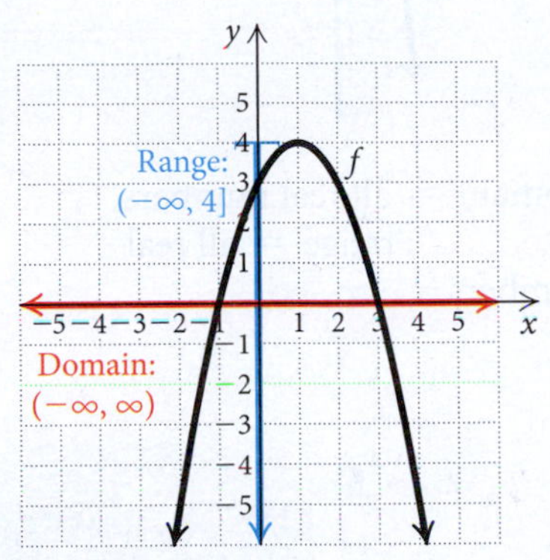

Consider the graph of function f, shown at left. To determine the domain of f, we look for the inputs on the x-axis that correspond to a point on the graph. We see that they include the entire set of real numbers, illustrated in red on the x-axis. Thus the domain is $(-\infty, \infty)$. To find the range, we look for the outputs on the y-axis that correspond to a point on the graph. We see that they include 4 and all real numbers less than 4, illustrated in blue on the y-axis. The bracket at 4 indicates that 4 is included in the interval. The range is $\{y|y \leq 4\}$, or $(-\infty, 4]$.

Let's now consider the following graph of function g. The solid dot shows that $(-4, 5)$ belongs to the graph. The open circle shows that $(3, 2)$ does *not* belong to the graph.

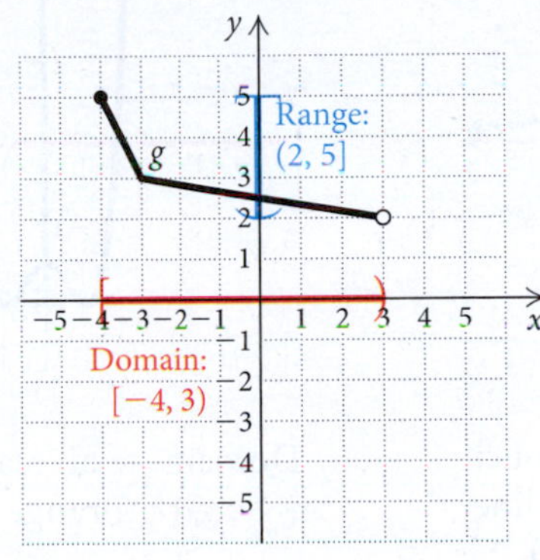

We see that the inputs of the function include -4 and all real numbers between -4 and 3, illustrated in red on the x-axis. The bracket at -4 indicates that -4 is included in the interval. The parenthesis at 3 indicates that 3 is not included in the interval. The domain is $\{x|-4 \leq x < 3\}$, or $[-4, 3)$. The outputs of the function include 5 and all real numbers between 2 and 5, illustrated in blue on the y-axis. The parenthesis at 2 indicates that 2 is not included in the interval. The bracket at 5 indicates that 5 is included in the interval. The range is $\{y|2 < y \leq 5\}$, or $(2, 5]$.

EXAMPLE 10 Using the graph of the function, find the domain and the range of the function.

a) $f(x) = \frac{1}{2}x + 1$

b) $f(x) = \sqrt{x + 4}$

c) $f(x) = x^3 - x$

d) $f(x) = \frac{1}{x - 2}$

e) $f(x) = x^4 - 2x^2 - 3$

f) $f(x) = \sqrt{4 - (x - 3)^2}$

Solution

a)

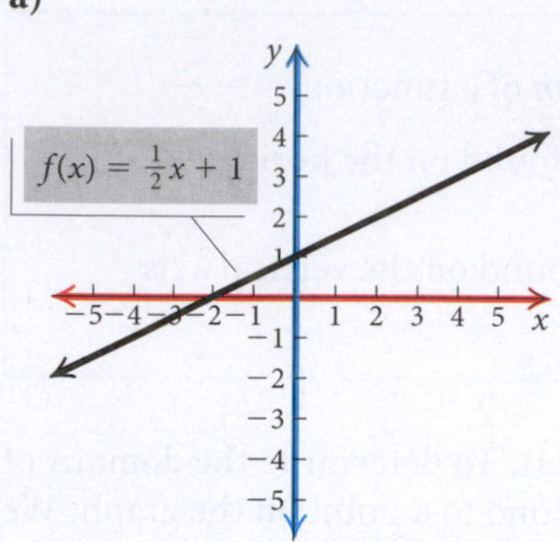

Domain = all real numbers, $(-\infty, \infty)$; range = all real numbers, $(-\infty, \infty)$

b)

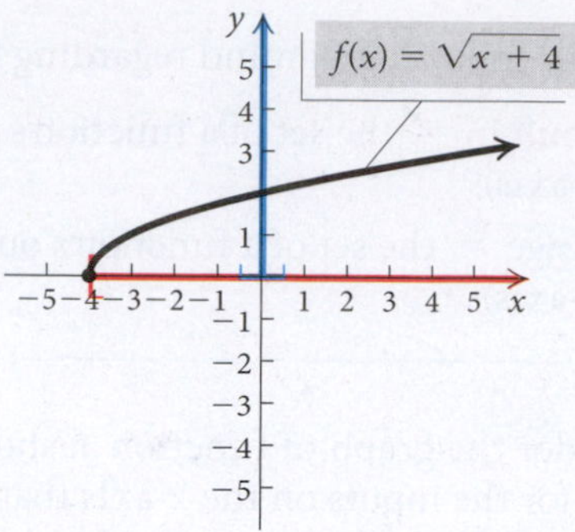

Domain = $[-4, \infty)$; range = $[0, \infty)$

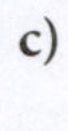

c)

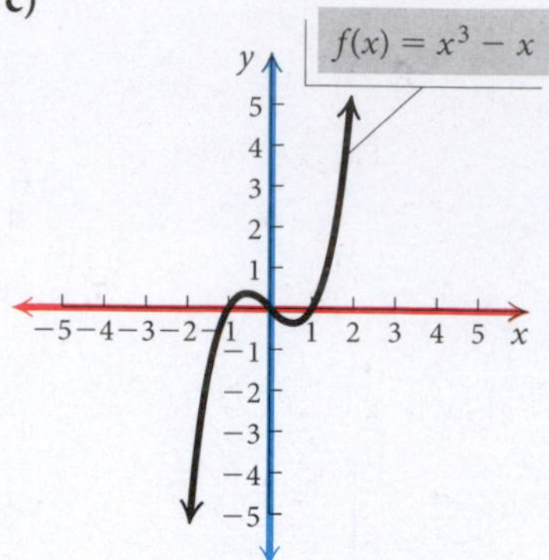

Domain = all real numbers, $(-\infty, \infty)$; range = all real numbers, $(-\infty, \infty)$

d)

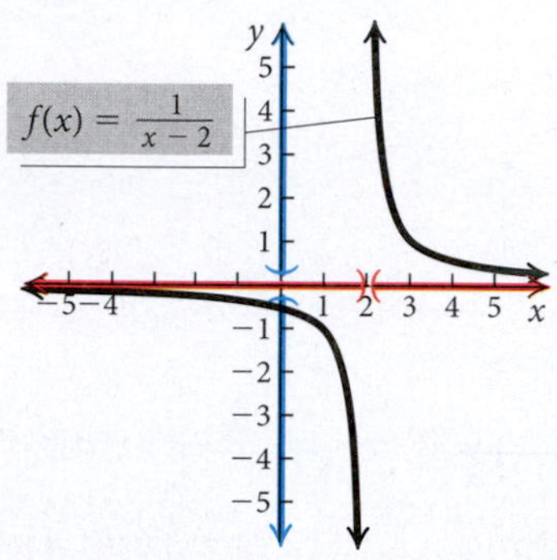

Since the graph does not touch or cross either the vertical line $x = 2$ or the x-axis, $y = 0$, 2 is excluded from the domain and 0 is excluded from the range.
Domain = $(-\infty, 2) \cup (2, \infty)$; range = $(-\infty, 0) \cup (0, \infty)$

e)

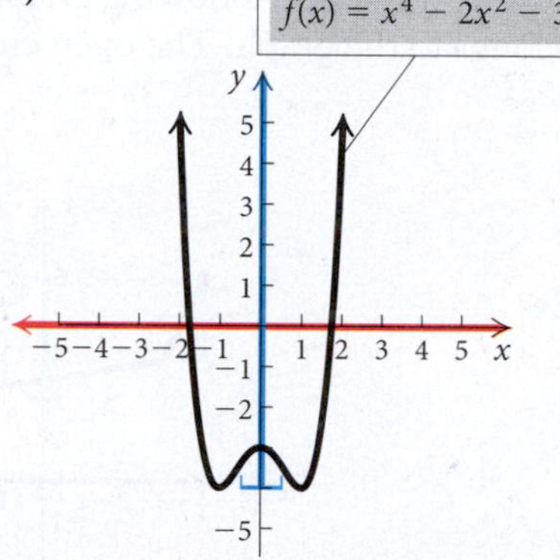

Domain = all real numbers, $(-\infty, \infty)$; range = $[-4, \infty)$

f)

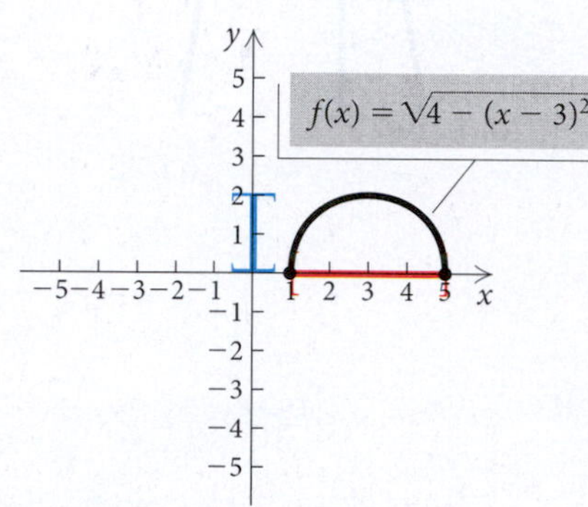

Domain = $[1, 5]$; range = $[0, 2]$

Now Try Exercises 71 and 77.

Always consider adding the reasoning of Example 9 to a graphical analysis. Think, "What can I input?" to find the domain. Think, "What do I get out?" to find the range. Thus, in Examples 10(c) and 10(e), it might not appear as though the domain is all real numbers because the graph rises steeply, but by examining the equation we see that we can indeed substitute any real number for x.

Applications of Functions

EXAMPLE 11 ***Linear Expansion of a Bridge.*** The linear expansion L of the steel center span of a suspension bridge that is 1420 m long is a function of the change in temperature t, in degrees Celsius, from winter to summer and is given by

$$L(t) = 0.000013 \cdot 1420 \cdot t,$$

where 0.000013 is the coefficient of linear expansion for steel and L is in meters. Find the linear expansion of the steel center span when the change in temperature from winter to summer is 30°, 42°, 50°, and 56° Celsius.

Solution Using a calculator, we compute function values. We find that

$$L(30) = 0.5538 \text{ m},$$
$$L(42) = 0.77532 \text{ m},$$
$$L(50) = 0.923 \text{ m}, \quad \text{and}$$
$$L(56) = 1.03376 \text{ m}.$$

Now Try Exercise 85.

CONNECTING THE CONCEPTS

FUNCTION CONCEPTS

Formula for f: $f(x) = 5 + 2x^2 - x^4$.

For every input, there is exactly one output.

$(1, 6)$ is on the graph.

For the input 1, the output is 6.

$f(1) = 6$

Domain: set of all inputs $= (-\infty, \infty)$

Range: set of all outputs $= (-\infty, 6]$

GRAPH

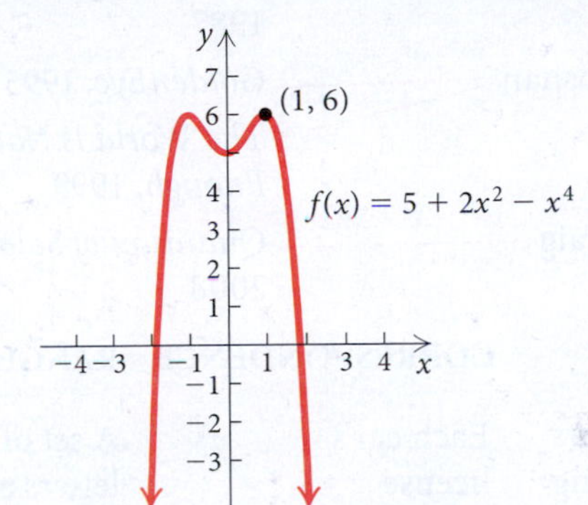

1.2 Exercise Set

In Exercises 1–14, determine whether the correspondence is a function.

1.
$a \to w$
$b \to y$
$c \to z$

2.
$m \to q$
$n \to s$
$o \to r$

3.
$-6 \to 36$
$-2 \to 4$
$2 \to 4$

4.
$-3 \to 2$
$1 \to 4$
$1 \to 6$
$5 \to 6$
$9 \to 8$

5.
$m \to A$
$m \to B$
$n \to C$
$r \to D$
$s \to C$

6.
$a \to r$
$b \to t$
$c \to r$
$d \to s$

7.

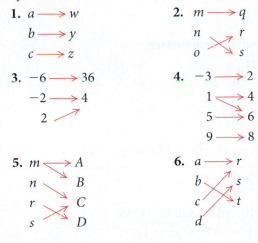

PAINTING	ARTIST
Night Watch	
Old Guitarist	Vincent van Gogh
Irises, Saint-Remy	
Starry Night	Claude Monet
The Water-Lily Pond	Pablo Picasso
Sunflowers	
Mona Lisa	Rembrandt van Rijn
Woman with a Parasol	Leonardo da Vinci
An Elephant	

8.

ACTOR PORTRAYING JAMES BOND	MOVIE TITLE
Sean Connery	*Goldfinger*, 1964
George Lazenby	*On Her Majesty's Secret Service*, 1969
Roger Moore	*Diamonds Are Forever*, 1971
	Moonraker, 1979
Timothy Dalton	*For Your Eyes Only*, 1981
	The Living Daylights, 1987
Pierce Brosnan	*GoldenEye*, 1995
	The World Is Not Enough, 1999
Daniel Craig	*Quantum of Solace*, 2008

Sean Connery → *Goldfinger*, 1964; Sean Connery → *Diamonds Are Forever*, 1971; George Lazenby → *On Her Majesty's Secret Service*, 1969; Roger Moore → *Moonraker*, 1979; Roger Moore → *For Your Eyes Only*, 1981; Timothy Dalton → *The Living Daylights*, 1987; Pierce Brosnan → *GoldenEye*, 1995; Pierce Brosnan → *The World Is Not Enough*, 1999; Daniel Craig → *Quantum of Solace*, 2008

	DOMAIN	CORRESPONDENCE	RANGE
9.	A set of cars in a parking lot	Each car's license number	A set of letters and numbers
10.	A set of people in a town	A doctor a person uses	A set of doctors
11.	The integers less than 9	Five times the integer	A subset of integers
12.	A set of members of a rock band	An instrument each person plays	A set of instruments
13.	A set of students in a class	A student sitting in a neighboring seat	A set of students
14.	A set of bags of chips on a shelf	Each bag's weight	A set of weights

Determine whether the relation is a function. Identify the domain and the range.

15. $\{(2, 10), (3, 15), (4, 20)\}$

16. $\{(3, 1), (5, 1), (7, 1)\}$

17. $\{(-7, 3), (-2, 1), (-2, 4), (0, 7)\}$

18. $\{(1, 3), (1, 5), (1, 7), (1, 9)\}$

19. $\{(-2, 1), (0, 1), (2, 1), (4, 1), (-3, 1)\}$

20. $\{(5, 0), (3, -1), (0, 0), (5, -1), (3, -2)\}$

21. Given that $g(x) = 3x^2 - 2x + 1$, find each of the following.

a) $g(0)$ **b)** $g(-1)$
c) $g(3)$ **d)** $g(-x)$
e) $g(1 - t)$

22. Given that $f(x) = 5x^2 + 4x$, find each of the following.

a) $f(0)$ **b)** $f(-1)$
c) $f(3)$ **d)** $f(t)$
e) $f(t - 1)$

23. Given that $g(x) = x^3$, find each of the following.

a) $g(2)$ **b)** $g(-2)$
c) $g(-x)$ **d)** $g(3y)$
e) $g(2 + h)$

24. Given that $f(x) = 2|x| + 3x$, find each of the following.

a) $f(1)$ **b)** $f(-2)$
c) $f(-x)$ **d)** $f(2y)$
e) $f(2 - h)$

25. Given that

$$g(x) = \frac{x - 4}{x + 3},$$

find each of the following.

a) $g(5)$ **b)** $g(4)$
c) $g(-3)$ **d)** $g(-16.25)$
e) $g(x + h)$

26. Given that

$$f(x) = \frac{x}{2 - x},$$

find each of the following.

a) $f(2)$ **b)** $f(1)$
c) $f(-16)$ **d)** $f(-x)$
e) $f\left(-\frac{2}{3}\right)$

27. Find $g(0)$, $g(-1)$, $g(5)$, and $g\left(\frac{1}{2}\right)$ for

$$g(x) = \frac{x}{\sqrt{1 - x^2}}.$$

28. Find $h(0)$, $h(2)$, and $h(-x)$ for

$$h(x) = x + \sqrt{x^2 - 1}.$$

Graph the function.

29. $f(x) = \frac{1}{2}x + 3$

30. $f(x) = \sqrt{x} - 1$

31. $f(x) = -x^2 + 4$

32. $f(x) = x^2 + 1$

33. $f(x) = \sqrt{x - 1}$

34. $f(x) = x - \frac{1}{2}x^3$

In each of Exercises 35–40, a graph of a function is shown. Using the graph, find the indicated function values; that is, given the inputs, find the outputs.

35. $h(1)$, $h(3)$, and $h(4)$

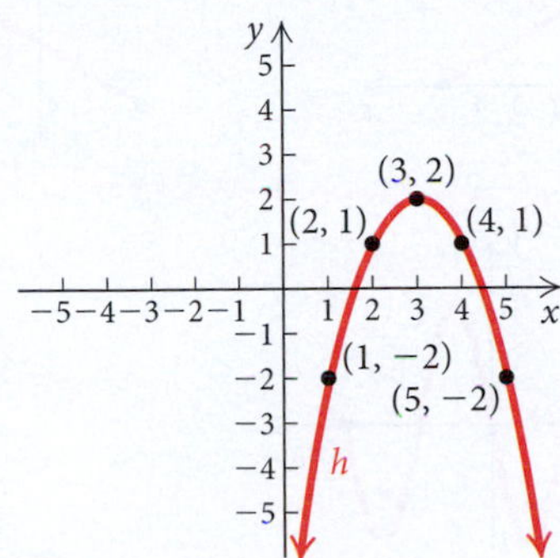

36. $t(-4)$, $t(0)$, and $t(3)$

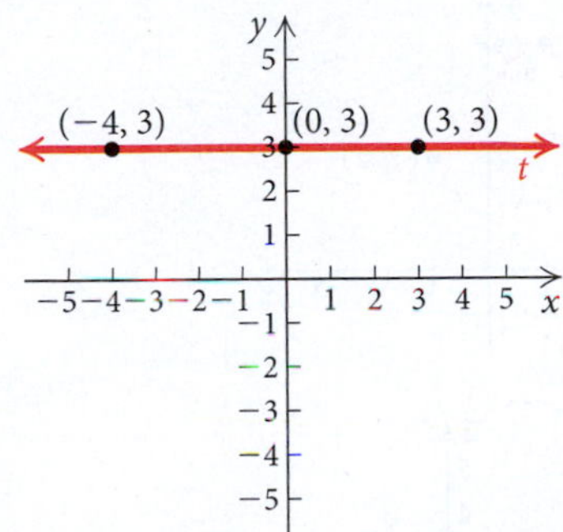

37. $s(-4)$, $s(-2)$, and $s(0)$

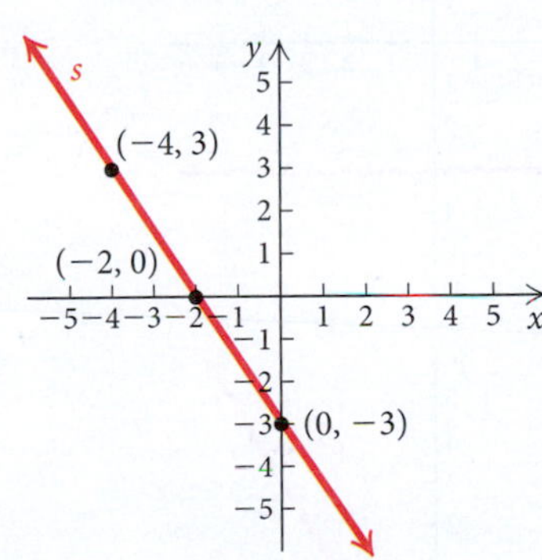

38. $g(-4)$, $g(-1)$, and $g(0)$

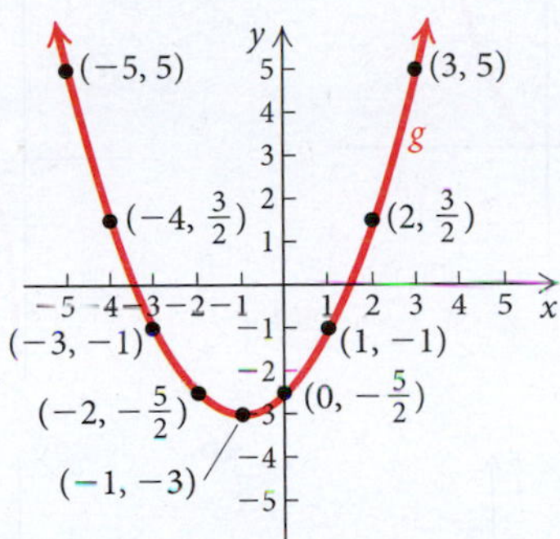

39. $f(-1)$, $f(0)$, and $f(1)$

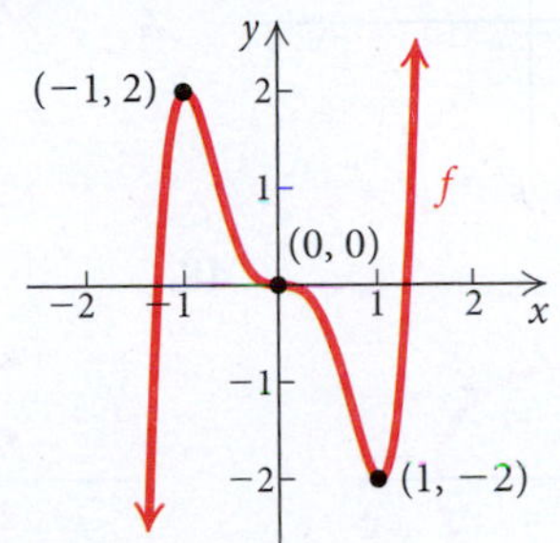

40. $g(-2)$, $g(0)$, and $g(2.4)$

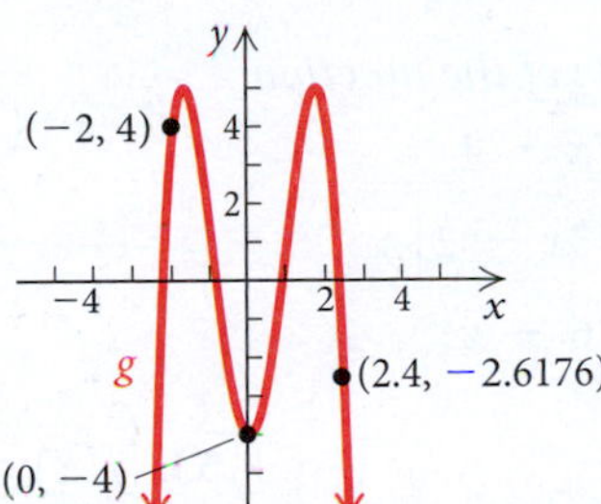

In Exercises 41–48, determine whether the graph is that of a function. An open circle indicates that the point does not belong to the graph.

41.

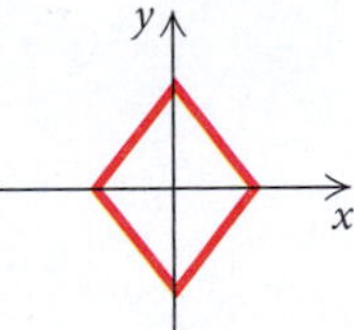

42.

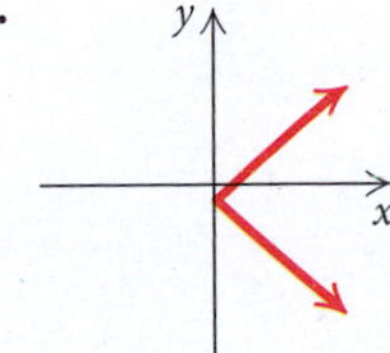

43.

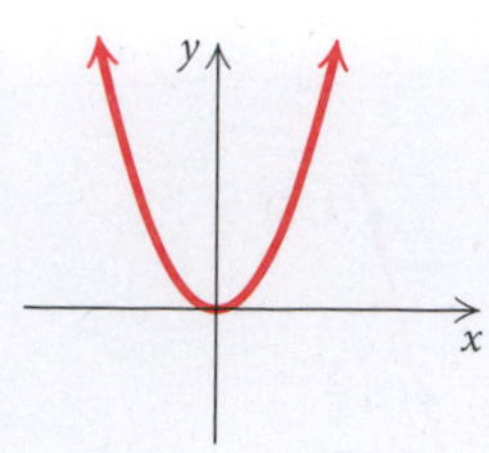

44.

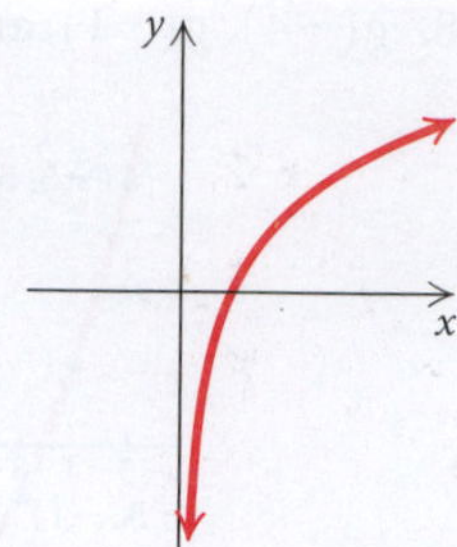

45.

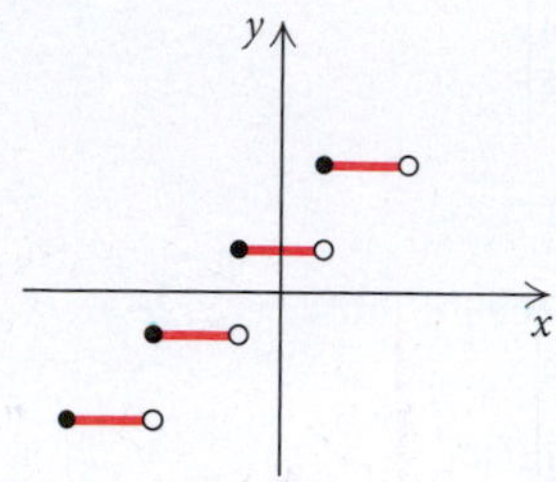

46.

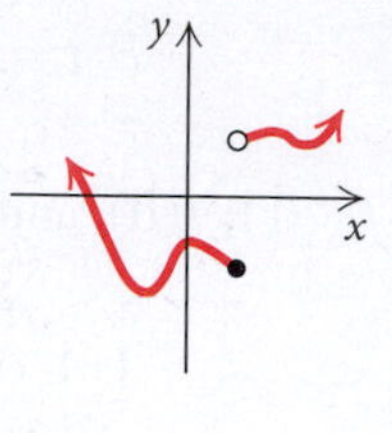

47.

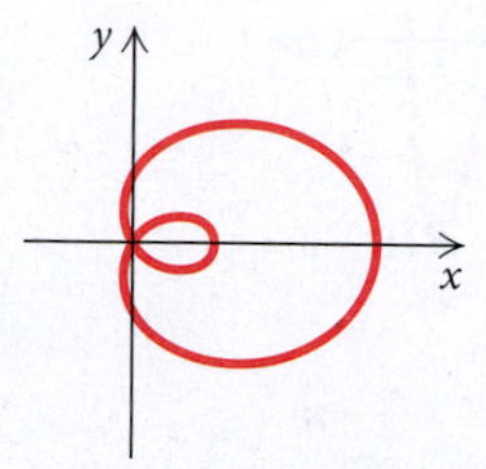

48.

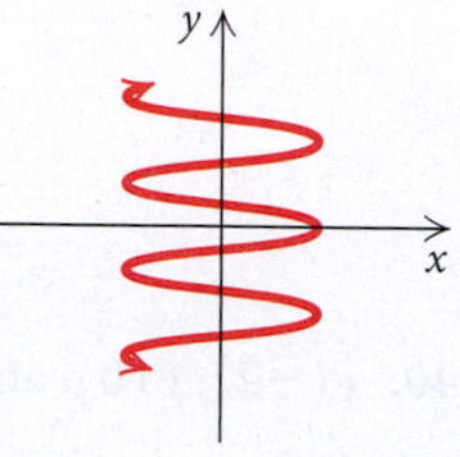

Find the domain of the function.

49. $f(x) = 7x + 4$

50. $f(x) = |3x - 2|$

51. $f(x) = |6 - x|$

52. $f(x) = \dfrac{1}{x^4}$

53. $f(x) = 4 - \dfrac{2}{x}$

54. $f(x) = \dfrac{1}{5}x^2 - 5$

55. $f(x) = \dfrac{x + 5}{2 - x}$

56. $f(x) = \dfrac{8}{x + 4}$

57. $f(x) = \dfrac{1}{x^2 - 4x - 5}$

58. $f(x) = \dfrac{(x - 2)(x + 9)}{x^3}$

59. $f(x) = \sqrt[3]{x + 10} - 1$

60. $f(x) = \sqrt[3]{4 - x}$

61. $f(x) = \dfrac{8 - x}{x^2 - 7x}$

62. $f(x) = \dfrac{x^4 - 2x^3 + 7}{3x^2 - 10x - 8}$

63. $f(x) = \dfrac{1}{10}|x|$

64. $f(x) = x^2 - 2x$

In Exercises 65–72, determine the domain and the range of the function.

65.

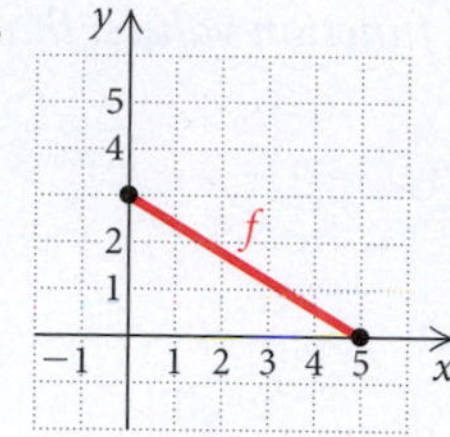

66.

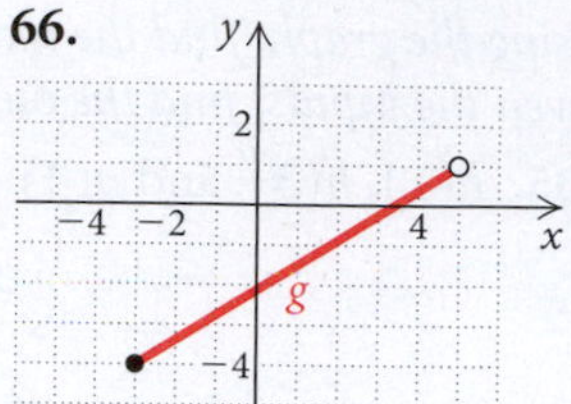

67.

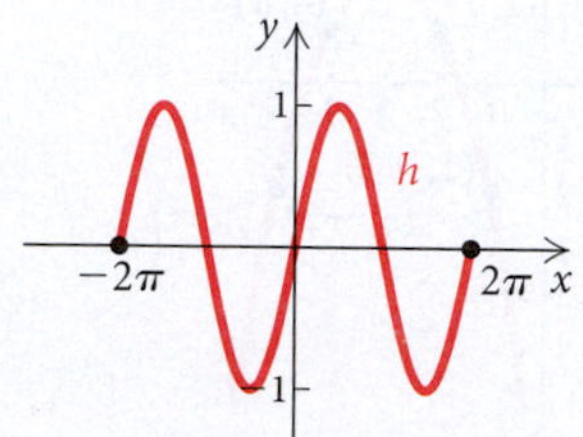

68.

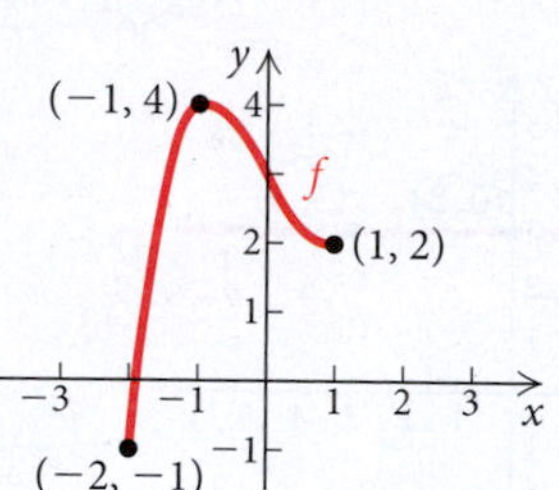

69.

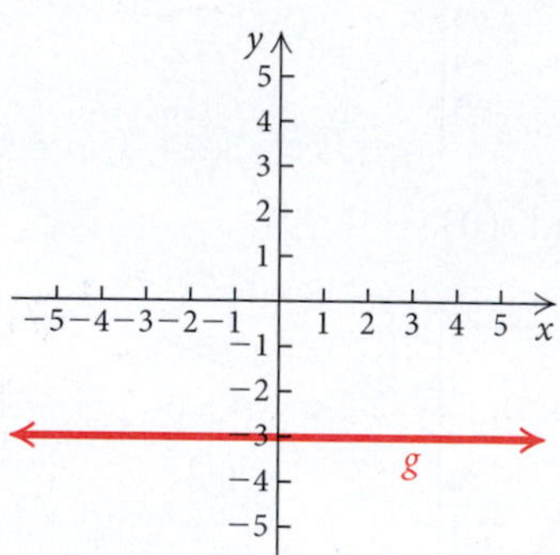

70.

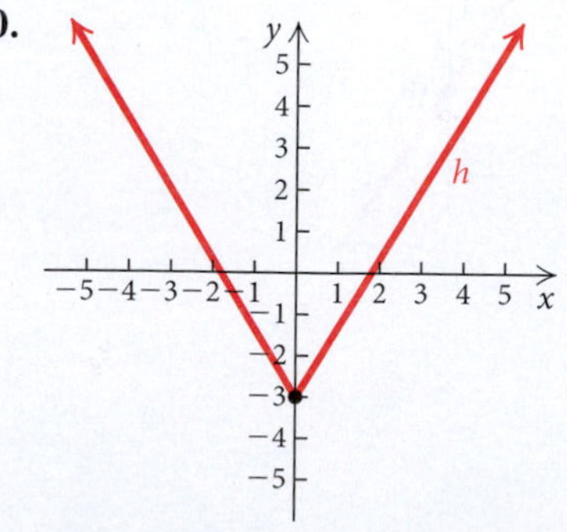

71.

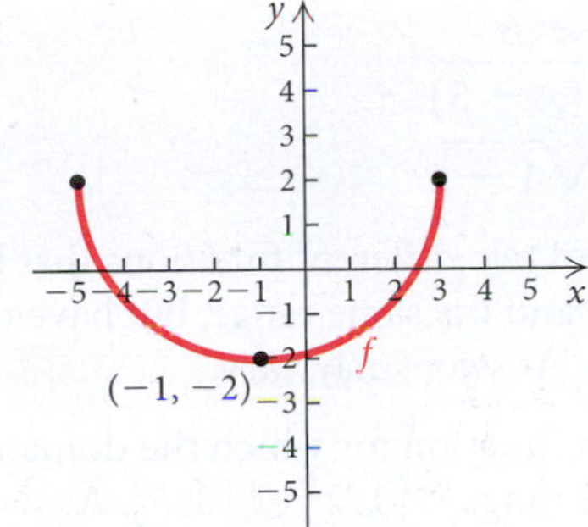

72.

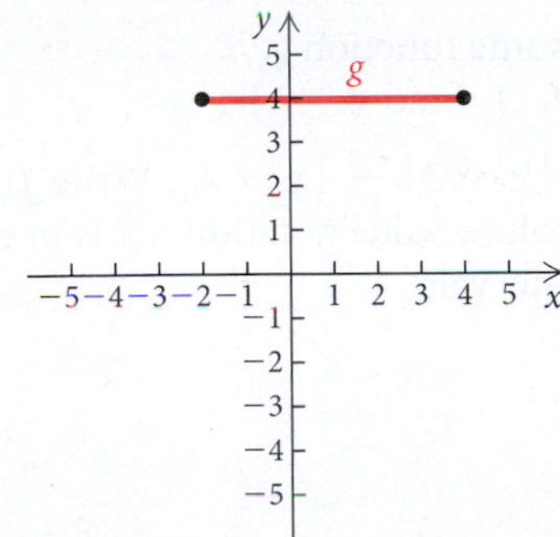

In Exercises 73–84, graph the given function. Then visually estimate the domain and the range.

73. $f(x) = |x|$

74. $f(x) = |x| - 2$

75. $f(x) = 3x - 2$

76. $f(x) = 5 - 3x$

77. $f(x) = \frac{1}{x - 3}$

78. $f(x) = \frac{1}{x + 1}$

79. $f(x) = (x - 1)^3 + 2$

80. $f(x) = (x - 2)^4 + 1$

81. $f(x) = \sqrt{7 - x}$

82. $f(x) = \sqrt{x + 8}$

83. $f(x) = -x^2 + 4x - 1$

84. $f(x) = 2x^2 - x^4 + 5$

85. *Decreasing Value of the Dollar.* In 2014, it took \$23.63 to equal the value of \$1 in 1913. In 2000, it took only \$17.39 to equal the value of \$1 in 1913. The amount that it takes to equal the value of \$1 in 1913 can be estimated by the linear function V given by

$$V(x) = 0.4306x + 11.0043,$$

where x is the number of years since 1985. Thus, $V(10)$ gives the amount that it took in 1995 to equal the value of \$1 in 1913.

Source: usinflationcalculator.com

a) Use this function to predict the amount that it will take in 2018 and in 2025 to equal the value of \$1 in 1913.

b) When will it take approximately \$32 to equal the value of \$1 in 1913?

86. *Population of the United States.* The population P of the United States in 1960 was 179,323,175. In 2010, the population was 308,745,538. The population of the United States can be estimated by the linear function P given by

$$P(x) = 2{,}578{,}409x + 151{,}116{,}864,$$

where x is the number of years after 1950. Thus, $P(20)$ gives the population in 1970.

a) Use this function to estimate the population in 1980 and in 2018.

b) When will the population be approximately 400,000,000?

87. *Boiling Point and Elevation.* The elevation E, in meters, above sea level at which the boiling point of water is t degrees Celsius is given by the function

$$E(t) = 1000(100 - t) + 580(100 - t)^2.$$

At what elevation is the boiling point 99.5°? 100°?

88. *Windmill Power.* Under certain conditions, the power P, in watts per hour, generated by a windmill with winds blowing v miles per hour is given by

$$P(v) = 0.015v^3.$$

Find the power generated by 15-mph winds and 35-mph winds.

Skill Maintenance

To the student and the instructor: *The Skill Maintenance exercises review skills covered previously in the text. You can expect such exercises in every exercise set. They provide excellent review for a final examination. Answers to all skill maintenance exercises, along with section references, appear in the answer section at the back of the book.*

Use substitution to determine whether the given ordered pairs are solutions of the given equation.

89. $(-3, -2), (2, -3)$; $y^2 - x^2 = -5$ [1.1]

90. $(0, -7), (8, 11)$; $y = 0.5x + 7$ [1.1]

91. $\left(\frac{4}{5}, -2\right), \left(\frac{11}{5}, \frac{1}{10}\right)$; $15x - 10y = 32$ [1.1]

Graph the equation. [1.1]

92. $y = (x - 1)^2$

93. $y = \frac{1}{3}x - 6$

94. $-2x - 5y = 10$

95. $(x - 3)^2 + y^2 = 4$

Synthesis

Find the domain of the function.

96. $f(x) = \sqrt[4]{2x + 5} + 3$

97. $f(x) = \dfrac{\sqrt{x + 1}}{x}$

98. $f(x) = \dfrac{\sqrt{x + 6}}{(x + 2)(x - 3)}$

99. $f(x) = \sqrt{x} - \sqrt{4 - x}$

100. Give an example of two different functions that have the same domain and the same range, but have no pairs in common. Answers may vary.

101. Draw a graph of a function for which the domain is $[-4, 4]$ and the range is $[1, 2] \cup [3, 5]$. Answers may vary.

102. Suppose that for some function g, $g(x + 3) = 2x + 1$. Find $g(-1)$.

103. Suppose $f(x) = |x + 3| - |x - 4|$. Write $f(x)$ without using absolute-value notation if x is in each of the following intervals.

a) $(-\infty, -3)$
b) $[-3, 4)$
c) $[4, \infty)$

1.3 Linear Functions, Slope, and Applications

- Determine the slope of a line given two points on the line.
- Solve applied problems involving slope, or average rate of change.
- Find the slope and the y-intercept of a line given the equation $y = mx + b$, or $f(x) = mx + b$.
- Graph a linear equation using the slope and the y-intercept.
- Solve applied problems involving linear functions.

In real-life situations, we often need to make decisions on the basis of limited information. When the given information is used to formulate an equation or an inequality that at least approximates the situation mathematically, we have created a **model**. One of the most frequently used mathematical models is *linear*. The graph of a linear model is a straight line.

Linear Functions

Let's examine the connections among equations, functions, and graphs that are *straight lines*. First, examine the graphs of linear functions and nonlinear functions shown here. Note that the graphs of the two types of functions are quite different.

Linear Functions **Nonlinear Functions**

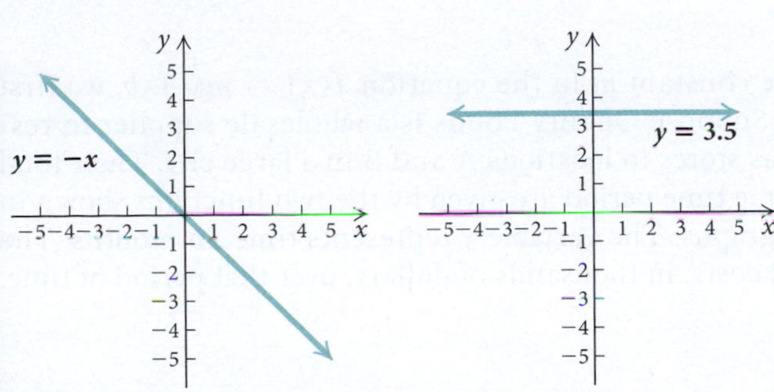

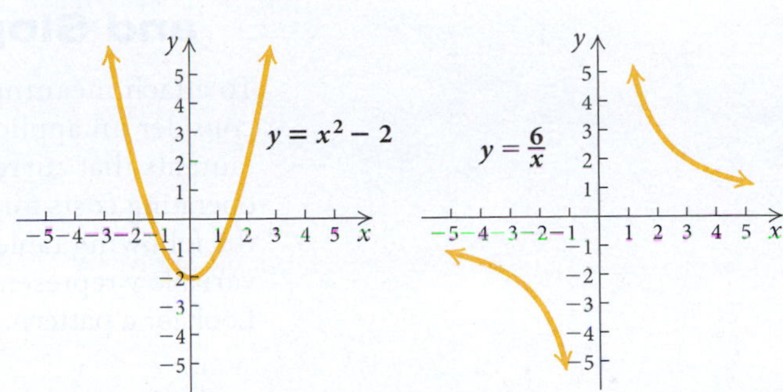

We begin with the definition of a linear function and related terminology, which are illustrated with graphs below.

LINEAR FUNCTIONS

A function f is a **linear function** if it can be written as

$$f(x) = mx + b,$$

where m and b are constants.

If $m = 0$, the function is a **constant function** $f(x) = b$. If $m = 1$ and $b = 0$, the function is the **identity function** $f(x) = x$.

Linear function:
$y = mx + b$

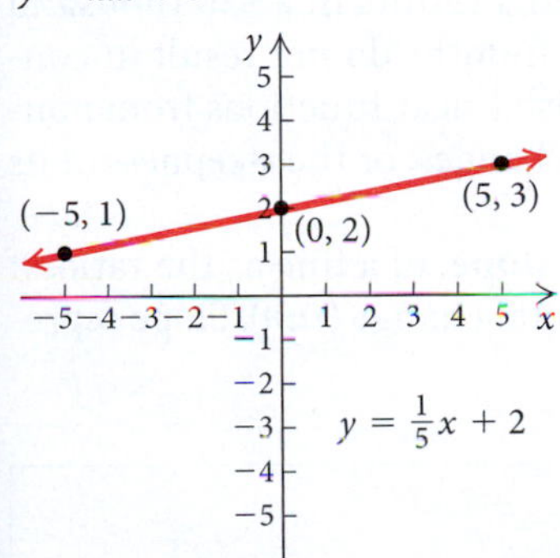

Identity function:
$y = 1 \cdot x + 0$, **or** $y = x$

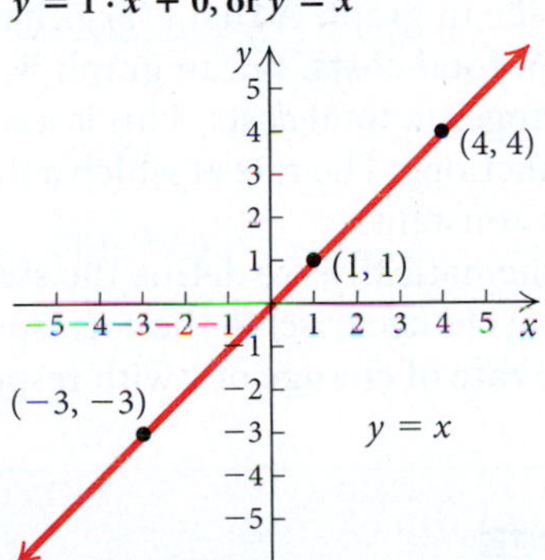

Constant function:
$y = 0 \cdot x + b$, **or** $y = b$ **(Horizontal line)**

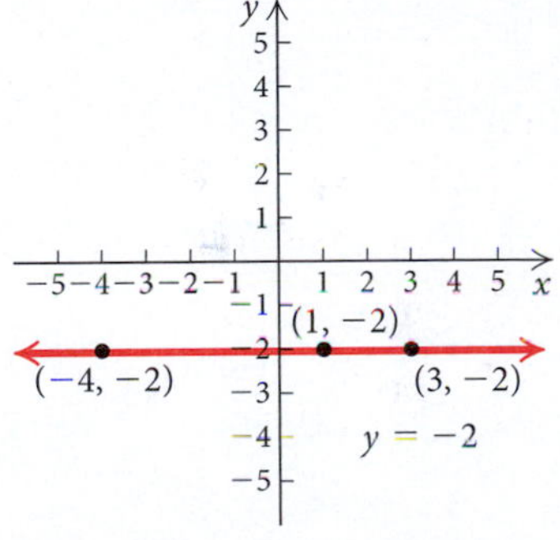

Vertical line: $x = a$
(*not* a function)

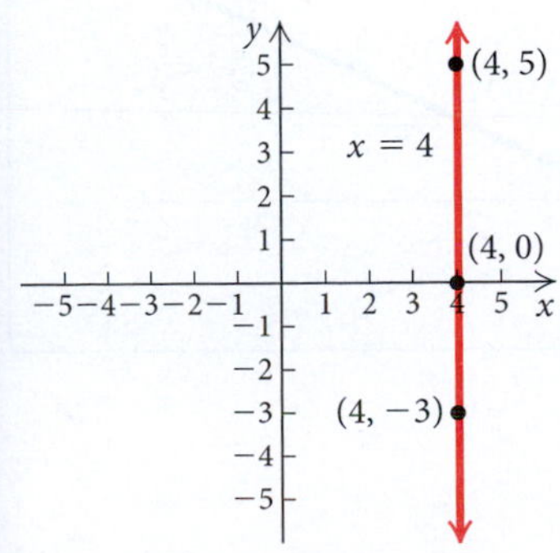

HORIZONTAL LINES AND VERTICAL LINES

Horizontal lines are given by equations of the type $y = b$ or $f(x) = b$. (They are functions.)

Vertical lines are given by equations of the type $x = a$. (They are *not* functions.)

The Linear Function $f(x) = mx + b$ and Slope

To attach meaning to the constant m in the equation $f(x) = mx + b$, we first consider an application. Suppose Quality Foods is a wholesale supplier to restaurants that currently has stores in locations A and B in a large city. Their total operating costs for the same time period are given by the two functions shown in the following tables and graphs. The variable x represents time, in months. The variable y represents total costs, in thousands of dollars, over that period of time. Look for a pattern.

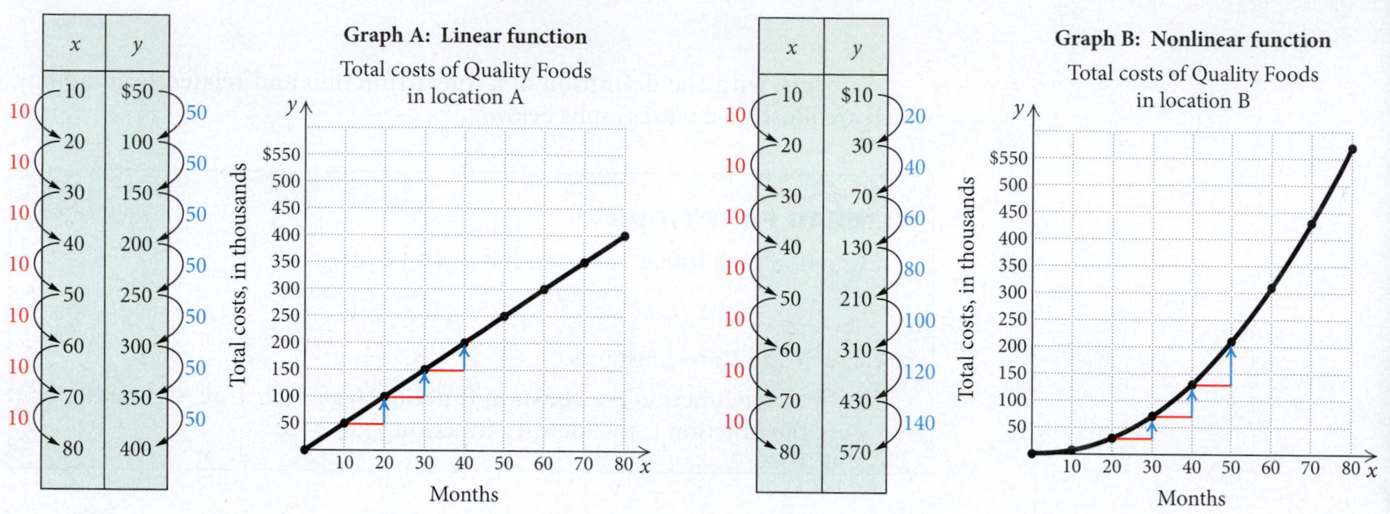

x	y
10	\$50
20	100
30	150
40	200
50	250
60	300
70	350
80	400

x	y
10	\$10
20	30
30	70
40	130
50	210
60	310
70	430
80	570

We see in graph A that *every* change of 10 months results in a \$50 thousand change in total costs. But in graph B, changes of 10 months do *not* result in constant changes in total costs. This is a way to distinguish linear functions from nonlinear functions. The rate at which a linear function changes, or the steepness of its graph, is constant.

Mathematically, we define the steepness, or the **slope**, of a line as the ratio of its vertical change (*rise*) to the corresponding horizontal change (*run*). Slope represents the **rate of change** of y with respect to x.

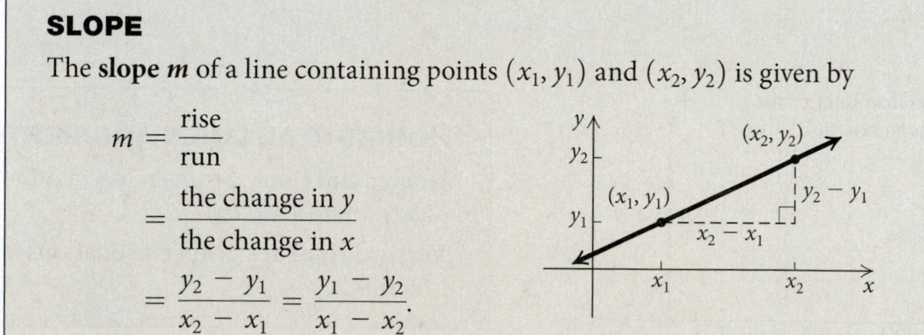

SLOPE

The **slope m** of a line containing points (x_1, y_1) and (x_2, y_2) is given by

$$m = \frac{\text{rise}}{\text{run}} = \frac{\text{the change in } y}{\text{the change in } x} = \frac{y_2 - y_1}{x_2 - x_1} = \frac{y_1 - y_2}{x_1 - x_2}.$$

EXAMPLE 1 Graph the function $f(x) = -\frac{2}{3}x + 1$ and determine its slope.

Solution Since the equation for f is in the form $f(x) = mx + b$, we know that it is a linear function. We can graph it by connecting two points on the graph with a straight line. We calculate two ordered pairs, plot the points, graph the function, and determine the slope:

$$f(3) = -\frac{2}{3} \cdot 3 + 1 = -2 + 1 = -1;$$

$$f(9) = -\frac{2}{3} \cdot 9 + 1 = -6 + 1 = -5;$$

Pairs: $(3, -1), (9, -5)$;

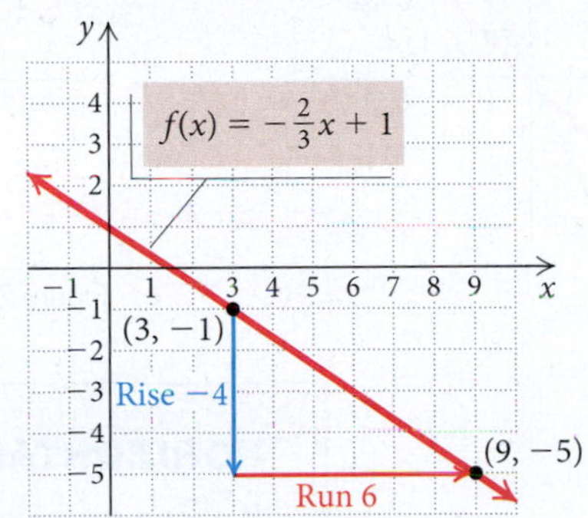

$$\text{Slope} = m = \frac{f(x_2) - f(x_1)}{x_2 - x_1} = \frac{y_2 - y_1}{x_2 - x_1}$$

$$= \frac{-5 - (-1)}{9 - 3} = \frac{-4}{6} = -\frac{2}{3}.$$

The slope is the same for any two points on a line. Thus, to check our work, we note that $f(6) = -\frac{2}{3} \cdot 6 + 1 = -4 + 1 = -3$. Using the points $(6, -3)$ and $(3, -1)$, we have

$$m = \frac{-1 - (-3)}{3 - 6} = \frac{2}{-3} = -\frac{2}{3}.$$

We can also use the points in the opposite order when computing slope:

$$m = \frac{-3 - (-1)}{6 - 3} = \frac{-2}{3} = -\frac{2}{3}.$$

Note too that the slope of the line is the number m in the equation for the function, $f(x) = -\frac{2}{3}x + 1$.

Now Try Exercise 7.

> The *slope* of the line given by $f(x) = mx + b$ is m.

Technology Connection

We can animate the effect of the slope m in linear functions of the type $f(x) = mx$ with a graphing calculator. Graph the equations

$y_1 = x, \quad y_2 = 2x,$

$y_3 = 5x, \quad$ and $\quad y_4 = 10x$

by entering them as $y_1 = \{1, 2, 5, 10\}x$. What do you think the graph of $y = 128x$ will look like?

Clear the screen and graph the equations

$y_1 = -x, \quad y_2 = -2x,$

$y_3 = -4x, \quad$ and $\quad y_4 = -10x.$

What do you think the graph of $y = -200x$ will look like?

If a line slants up from left to right, the change in x and the change in y have the same sign, so the line has a positive slope. The larger the slope, the steeper the line, as shown in Fig. 1. If a line slants down from left to right, the change in x and the change in y are of opposite signs, so the line has a negative slope. The larger the absolute value of the slope, the steeper the line, as shown in Fig. 2. Considering $y = mx$ when $m = 0$, we have $y = 0x$, or $y = 0$. Note that this horizontal line is the x-axis, as shown in Fig. 3.

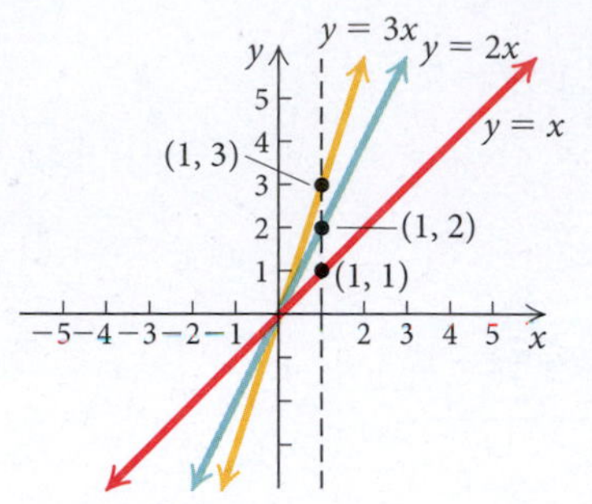

Figure 1.

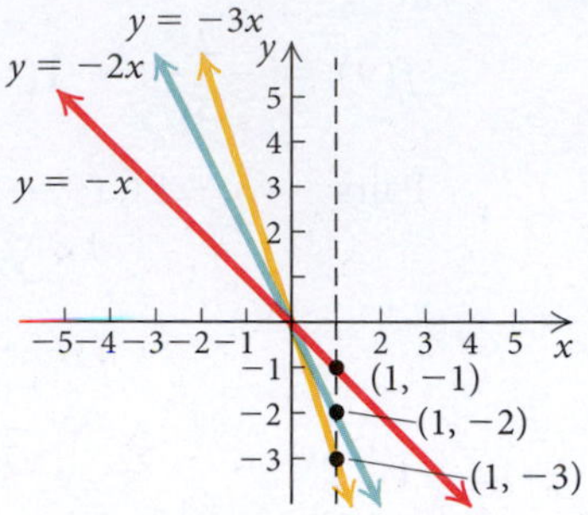

Figure 2.

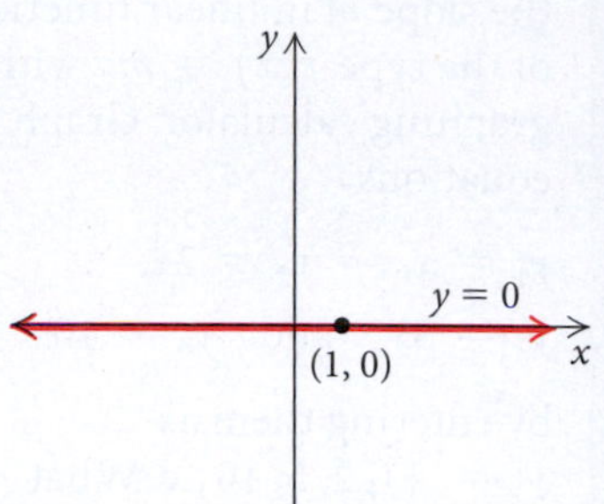

Figure 3.

HORIZONTAL LINES AND VERTICAL LINES

If a line is horizontal, the change in y for any two points is 0 and the change in x is nonzero. Thus a horizontal line has slope 0. (See Fig. 4.)

If a line is vertical, the change in x is 0. Thus the slope is *not defined* because we cannot divide by 0. (See Fig. 5.)

Horizontal lines

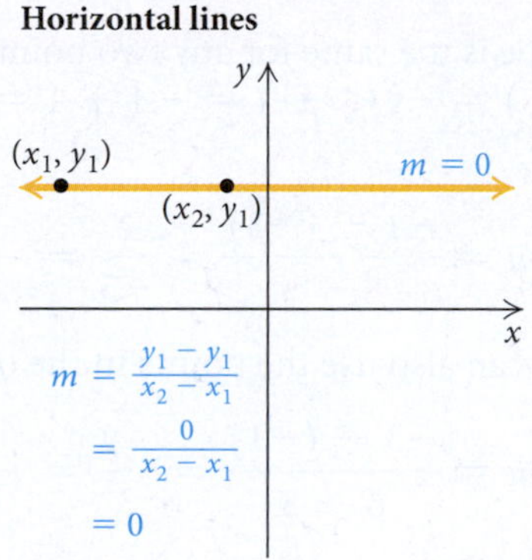

Figure 4.

Vertical lines

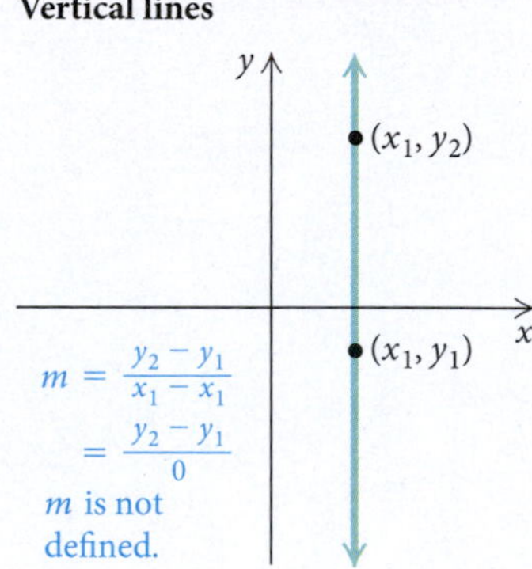

Figure 5.

Note that zero slope and an undefined slope are two very different concepts.

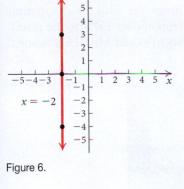

EXAMPLE 2 Graph each linear equation and determine its slope.

a) $x = -2$ **b)** $y = \frac{5}{2}$

Solution

a) Since y is missing in $x = -2$, any value for y will do.

x	y
-2	0
-2	3
-2	-4

Choose any number for y; x must be -2.

Figure 6.

The graph is a *vertical line* 2 units to the left of the y-axis. (See Fig. 6.) The slope is not defined. The graph is *not* the graph of a function.

b) Since x is missing in $y = \frac{5}{2}$, any value for x will do.

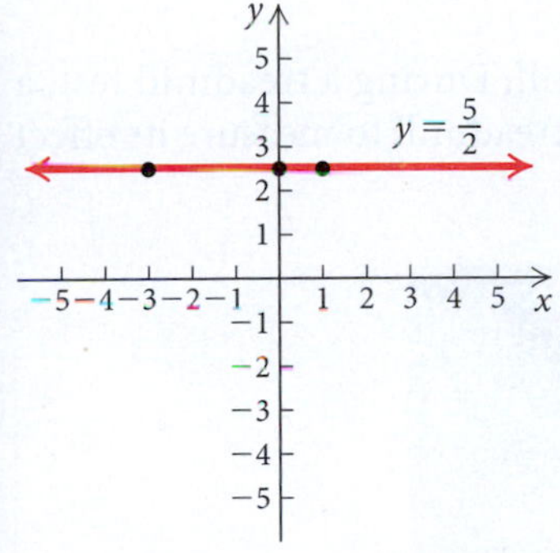

Figure 7.

x	y
0	$\frac{5}{2}$
-3	$\frac{5}{2}$
1	$\frac{5}{2}$

Choose any number for x; y must be $\frac{5}{2}$.

The graph is a *horizontal line* $\frac{5}{2}$, or $2\frac{1}{2}$, units above the x-axis. (See Fig. 7.) The slope is 0. The graph is the graph of a constant function.

Now Try Exercises 17 and 23.

Applications of Slope

Slope has many real-world applications. Numbers like 2%, 4%, and 7% are often used to represent the **grade** of a road. Such a number is meant to tell how steep a road is on a hill or a mountain. For example, a 4% grade means that the road rises (or falls) 4 ft for every horizontal distance of 100 ft.

The 2014 Olympic downhill course at Rosa Khutor Alpine Resort, located 40 km from Sochi, Russia, has the largest vertical drop ever built for an Olympic event. With a run of nearly 3500 m and a vertical drop of over 1075 m, the resulting grade, or slope, is approximately 31%. (*Source*: "Sochi's Gold Medal Ski Resort," by Brian Pinella, sochimagazine.com)

The concept of grade is also used with a treadmill. During a treadmill test, a cardiologist might change the slope, or grade, of the treadmill to measure its effect on heart rate.

Another example occurs in hydrology. The strength or force of a river depends on how far the river falls vertically compared to how far it flows horizontally.

EXAMPLE 3 ***Curb Ramps.*** Curb ramps provide independent access to sidewalks for those who use wheelchairs. Guidelines for the grade of a curb ramp suggest a grade between 5.9% and 8.3%. A federal law states that every vertical rise of 1 ft requires a horizontal run of at least 12 ft. (*Source*: Federal Highway Administration, Office of Planning, Environment, and Realty) Find the grade of the curb ramp shown in the following figure.

Solution The grade, or slope, is given by $m = \dfrac{3\text{ in.}}{42\text{ in.}} = \dfrac{1}{14} \approx 7.1\%$.

> **AVERAGE RATE OF CHANGE**
>
> Slope can also be considered as an **average rate of change**. To find the average rate of change between any two data points on a graph, we determine the slope of the line that passes through the two points.

EXAMPLE 4 ***Food Stamp Program.*** The number of people participating in the federal Supplemental Nutrition Assistance Program has increased from 17.2 million in 2000 to 47.6 million in 2013. The following graph illustrates this upward trend. Find the average rate of change in the number of people using food stamps from 2000 to 2013.

Enrollment in the Federal Supplemental Nutrition Assistance Program

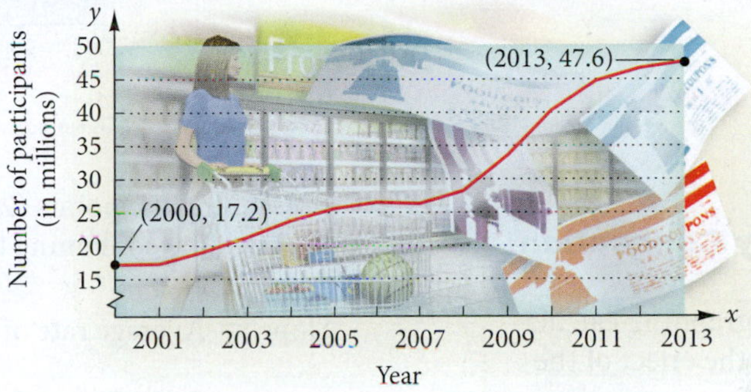

Source: U.S. Department of Agriculture

Solution We use the coordinates of two points on the graph. In this case, we use (2000, 17.2) and (2013, 47.6). Then we compute the slope, or average rate of change, as follows:

$$\begin{aligned}\text{Slope} &= \text{Average rate of change}\\ &= \frac{\text{Change in } y}{\text{Change in } x}\\ &= \frac{47.6 - 17.2}{2013 - 2000}\\ &= \frac{30.4}{13}\\ &\approx 2.3.\end{aligned}$$

The result tells us that each year from 2000 to 2013, the number of participants in the federal Supplemental Nutrition Assistance Program increased an average of 2.3 million. The average rate of change over this 13-year period was an increase of 2.3 million participants per year.

Now Try Exercise 41.

EXAMPLE 5 ***Oil Imports.*** Increased oil production in the United States has resulted in decreased imports of crude oil. The total number of barrels imported in 2008 was 3,590,000. This number had decreased to 2,810,000 barrels in 2013. (*Source*: U.S. Census Bureau) Find the average rate of change in crude oil imports from 2008 to 2013.

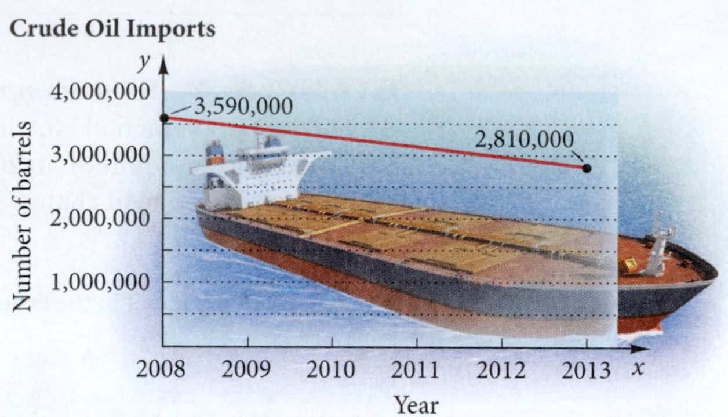

Source: U.S. Census Bureau

Solution Using the points (2008, 3,590,000) and (2013, 2,810,000), we compute the slope of the line containing these two points:

$$\text{Slope} = \text{Average rate of change} = \frac{\text{Change in } y}{\text{Change in } x}$$

$$= \frac{2,810,000 - 3,590,000}{2013 - 2008} = \frac{-780,000}{5} = -156,000.$$

The result tells us that each year from 2008 to 2013, the number of barrels of imported crude oil decreased on average 156,000 barrels. The average rate of change over the 5-year period was a decrease of 156,000 barrels per year.

Now Try Exercise 47.

Technology Connection

We can use a graphing calculator to explore the effect of the constant b in linear equations of the type $f(x) = mx + b$. Begin with the graph of $y = x$. Now graph the lines $y = x + 3$ and $y = x - 4$ in the same viewing window. Try entering these equations as $y = x + \{0, 3, -4\}$ and compare the graphs. How do the last two lines differ from $y = x$? What do you think the line $y = x - 6$ will look like?

Clear the first set of equations and graph $y = -0.5x$, $y = -0.5x + 3$, and $y = -0.5x - 4$ in the same viewing window. Describe what happens to the graph of $y = -0.5x$ when a number b is added.

Slope-Intercept Equations of Lines

Compare the graphs of the equations

$$y = 3x \quad \text{and} \quad y = 3x - 2.$$

Note that the graph of $y = 3x - 2$ is a shift of the graph of $y = 3x$ down 2 units and that $y = 3x - 2$ has y-intercept $(0, -2)$. That is, the graph is parallel to $y = 3x$ and it crosses the y-axis at $(0, -2)$. The point $(0, -2)$ is the ***y*-intercept** of the graph.

y-INTERCEPT
REVIEW SECTION 1.1

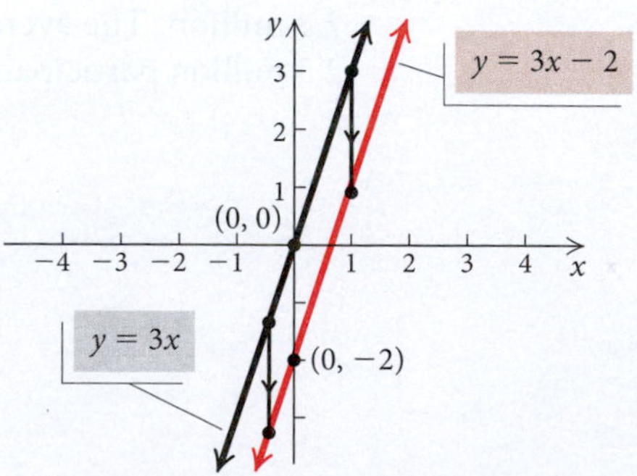

THE SLOPE–INTERCEPT EQUATION

The linear function f given by

$$f(x) = mx + b$$

is written in slope–intercept form. The graph of an equation in this form is a straight line parallel to $f(x) = mx$. The constant m is called the slope, and the y-intercept is $(0, b)$.

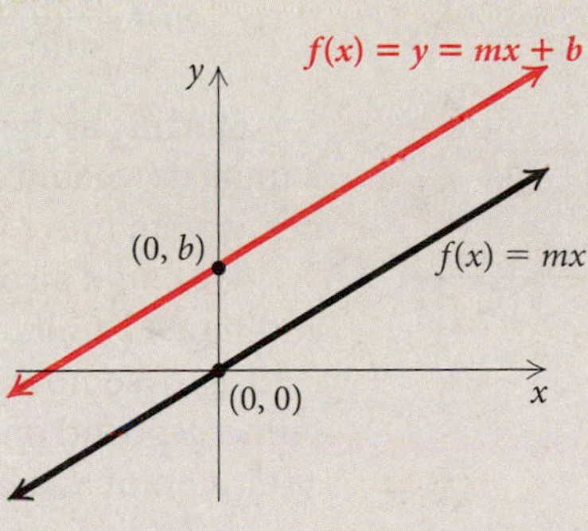

We can read the slope m and the y-intercept $(0, b)$ directly from the equation of a line written in slope–intercept form $y = mx + b$.

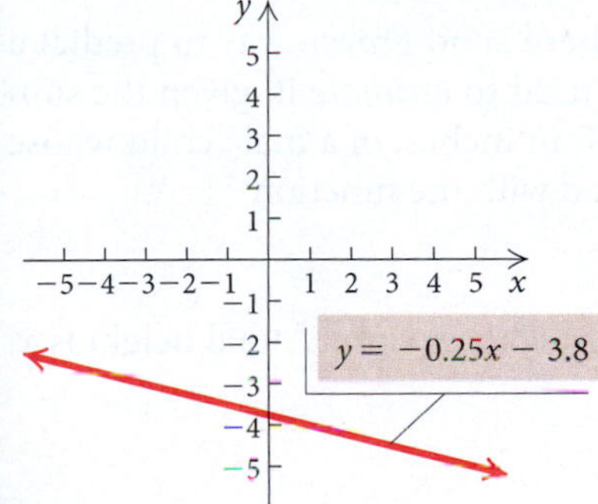

EXAMPLE 6 Find the slope and the y-intercept of the line with equation $y = -0.25x - 3.8$.

Solution

$$y = \underbrace{-0.25\,x}\ \underbrace{-\ 3.8}$$

Slope $= -0.25$; y-intercept $= (0, -3.8)$

Now Try Exercise 49.

Any equation whose graph is a straight line is a **linear equation**. To find the slope and the y-intercept of the graph of a nonvertical linear equation, we can solve for y, and then read the information from the equation.

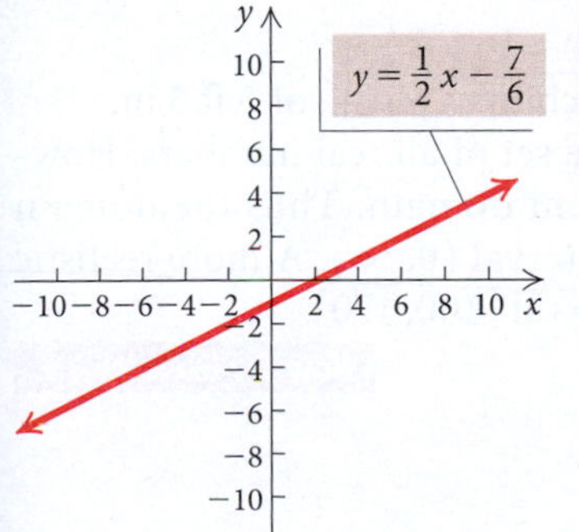

JUST IN TIME 2

EXAMPLE 7 Find the slope and the y-intercept of the line with equation $3x - 6y - 7 = 0$.

Solution We solve for y:

$$3x - 6y - 7 = 0$$
$$-6y = -3x + 7 \quad \text{Adding } -3x \text{ and } 7 \text{ on both sides}$$
$$-\tfrac{1}{6}(-6y) = -\tfrac{1}{6}(-3x + 7) \quad \text{Multiplying by } -\tfrac{1}{6}$$
$$y = \tfrac{1}{2}x - \tfrac{7}{6}. \quad \text{Using a distributive law}$$

Thus the slope is $\frac{1}{2}$, and the y-intercept is $\left(0, -\frac{7}{6}\right)$.

Now Try Exercise 61.

▶ Graphing $f(x) = mx + b$ Using m and b

We can also graph a linear equation using its slope and y-intercept.

EXAMPLE 8 Graph: $y = -\frac{2}{3}x + 4$.

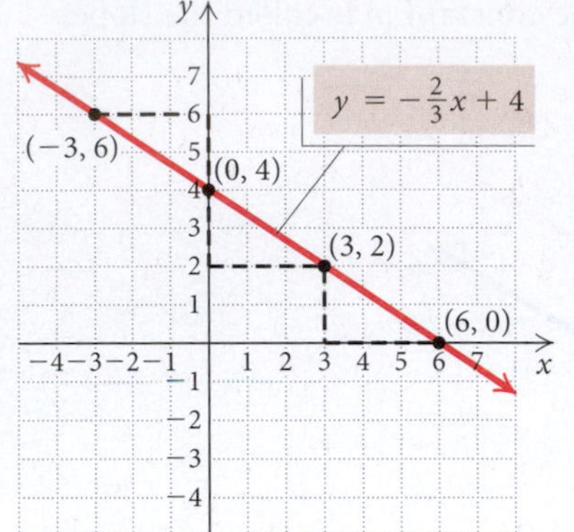

Solution This equation is in slope–intercept form, $y = mx + b$. The y-intercept is $(0, 4)$. We plot this point. We can think of the slope $\left(m = -\frac{2}{3}\right)$ as $\frac{-2}{3}$.

$$m = \frac{\text{rise}}{\text{run}} = \frac{\text{change in } y}{\text{change in } x} = \frac{-2}{3} \begin{array}{l} \leftarrow \text{Move 2 units down.} \\ \leftarrow \text{Move 3 units to the right.} \end{array}$$

Starting at the y-intercept and using the slope, we find another point by moving 2 units down and 3 units to the right. We get a new point, $(3, 2)$. In a similar manner, we can move from $(3, 2)$ to find another point, $(6, 0)$.

We could also think of the slope $\left(m = -\frac{2}{3}\right)$ as $\frac{2}{-3}$. Then we can start at $(0, 4)$ and move 2 units up and 3 units to the left. We get to another point on the graph, $(-3, 6)$. We now plot the points and draw the line. Note that we need only the y-intercept and one other point in order to graph the line, but it's a good idea to find a third point as a check that the first two points are correct.

Now Try Exercise 63.

▶ Applications of Linear Functions

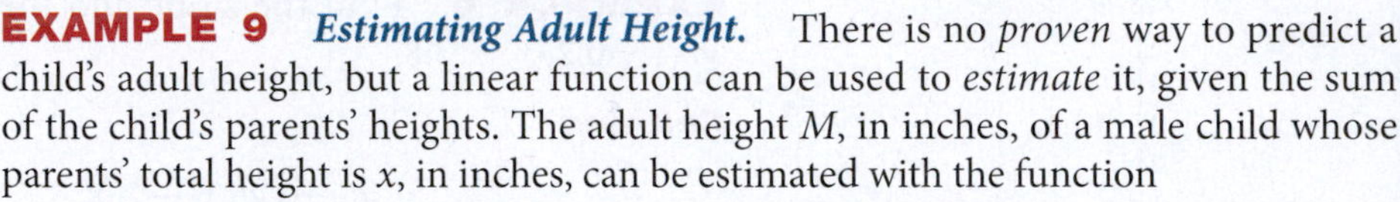

We now consider an application of linear functions.

EXAMPLE 9 ***Estimating Adult Height.*** There is no *proven* way to predict a child's adult height, but a linear function can be used to *estimate* it, given the sum of the child's parents' heights. The adult height M, in inches, of a male child whose parents' total height is x, in inches, can be estimated with the function

$$M(x) = 0.5x + 2.5.$$

The adult height F, in inches, of a female child whose parents' total height is x, in inches, can be estimated with the function

$$F(x) = 0.5x - 2.5.$$

(*Source*: Jay L. Hoecker, M.D., MayoClinic.com) Estimate the height of a female child whose parents' total height is 135 in. What is the domain of this function?

Solution We substitute in the function:

$$F(135) = 0.5(135) - 2.5 = 65.$$

Thus we can estimate the adult height of the female child as 65 in., or 5 ft 5 in.

Theoretically, the domain of the function is the set of all real numbers. However, the context of the problem dictates a different domain. Thus the domain consists of all positive real numbers—that is, the interval $(0, \infty)$. A more realistic domain might be 100 in. to 170 in.—that is, the interval $[100, 170]$.

Now Try Exercise 73.

Visualizing the Graph

Match the equation with its graph.

1. $y = 20$

2. $5y = 2x + 15$

3. $y = -\frac{1}{3}x - 4$

4. $x = \frac{5}{3}$

5. $y = -x - 2$

6. $y = 2x$

7. $y = -3$

8. $3y = -4x$

9. $x = -10$

10. $y = x + \frac{7}{2}$

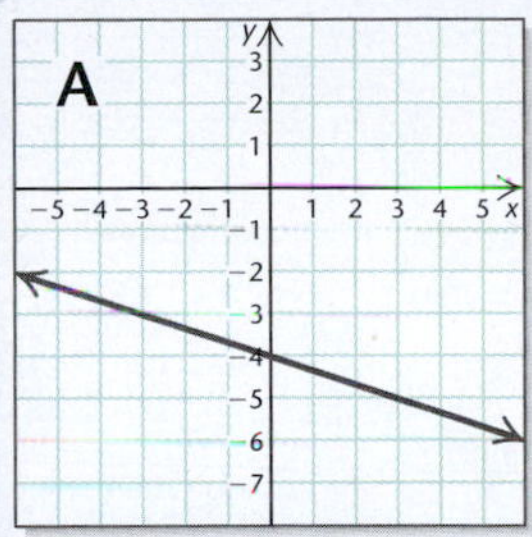

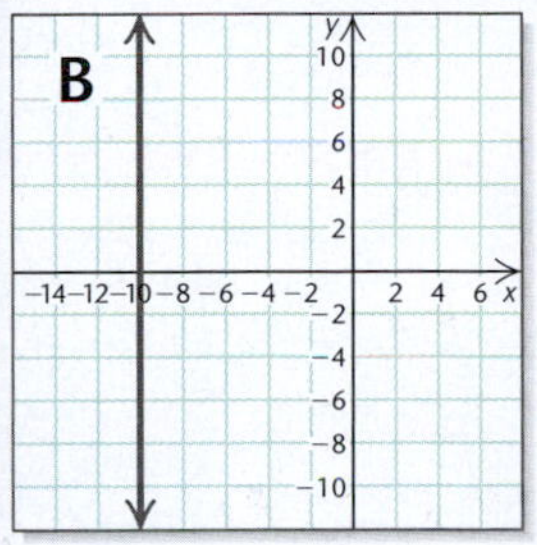

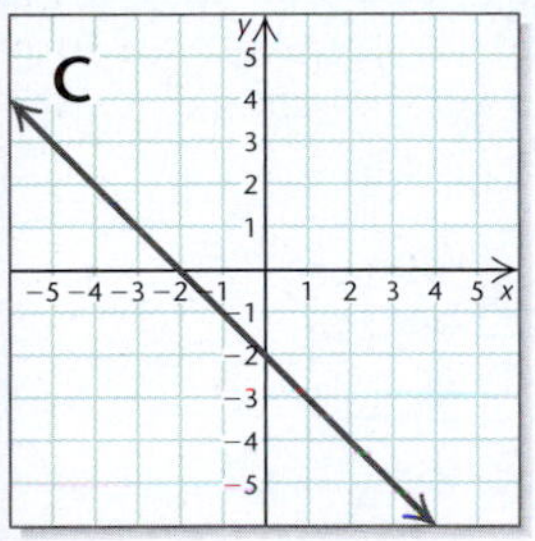

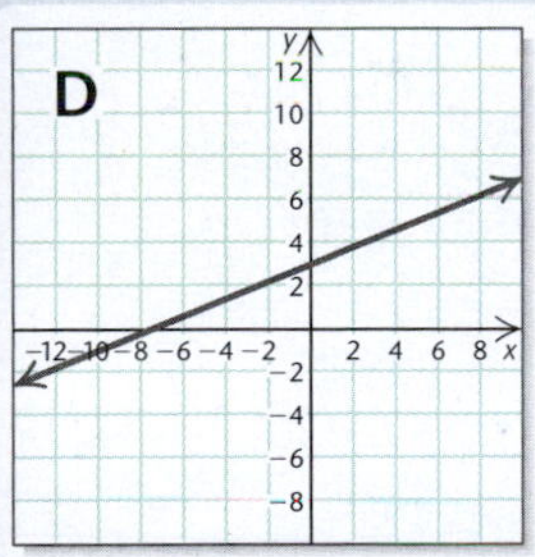

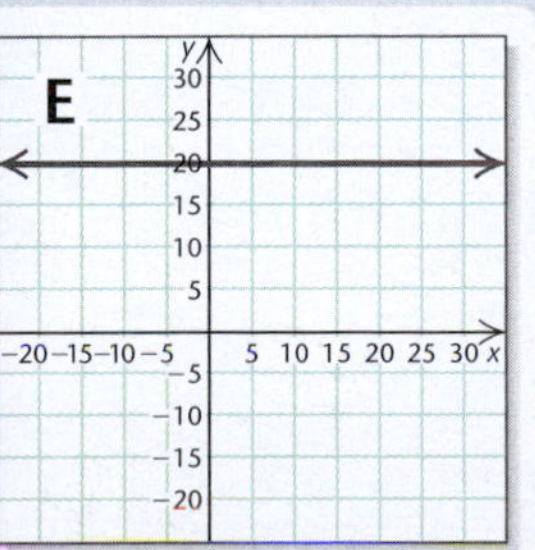

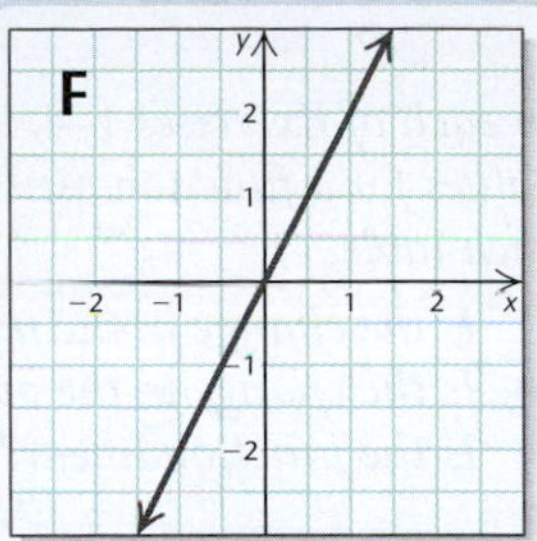

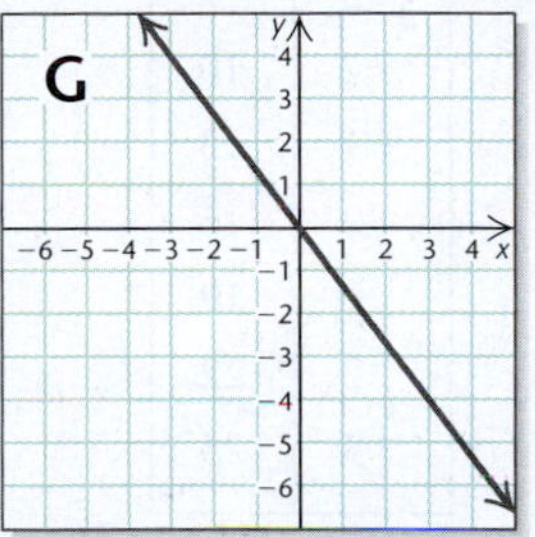

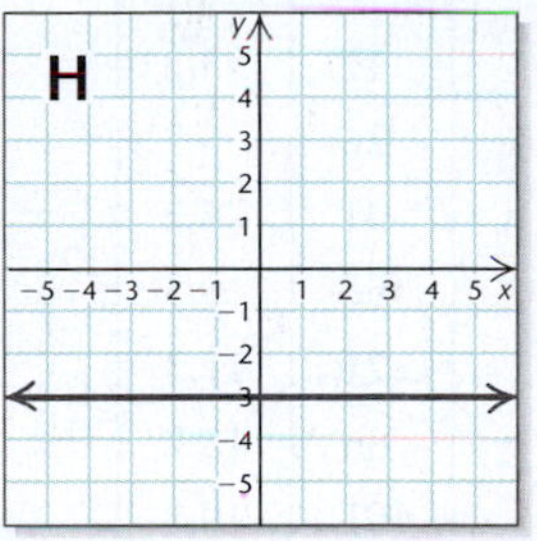

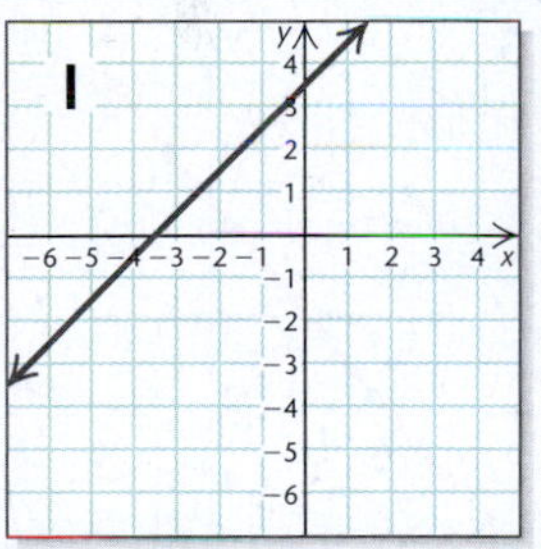

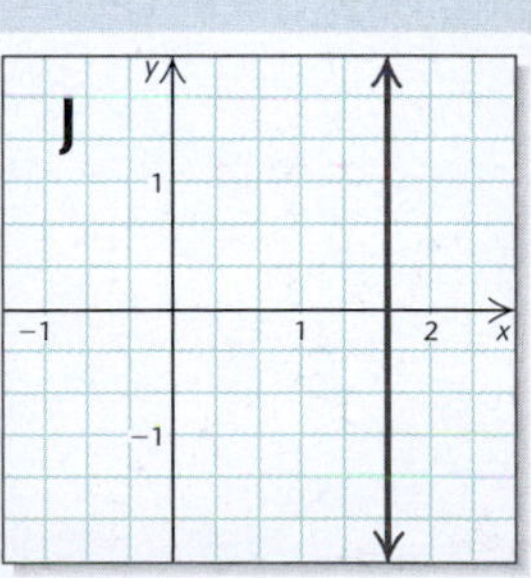

Answers on page A-3

1.3 Exercise Set

In each of Exercises 1–4, the table of data lists input–output values for a function. Answer the following questions for each table.

a) *Is the change in the inputs x the same?*
b) *Is the change in the outputs y the same?*
c) *Is the function linear?*

1.

x	y
−3	7
−2	10
−1	13
0	16
1	19
2	22
3	25

2.

x	y
20	12.4
30	24.8
40	49.6
50	99.2
60	198.4
70	396.8
80	793.6

3.

x	y
11	3.2
26	5.7
41	8.2
56	9.3
71	11.3
86	13.7
101	19.1

4.

x	y
2	−8
4	−12
6	−16
8	−20
10	−24
12	−28
14	−32

Find the slope of the line containing the given points.

5.

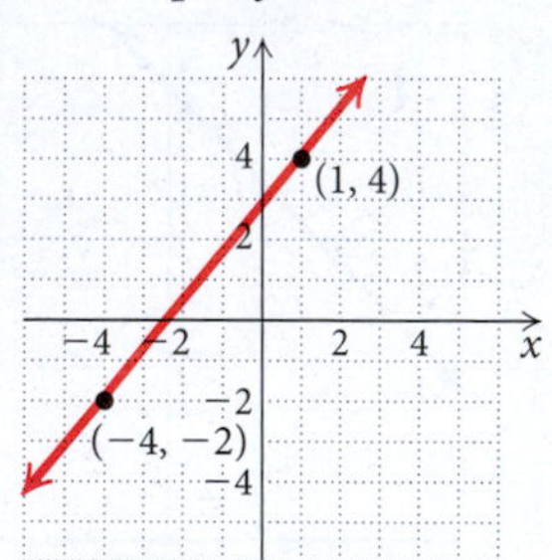

6.

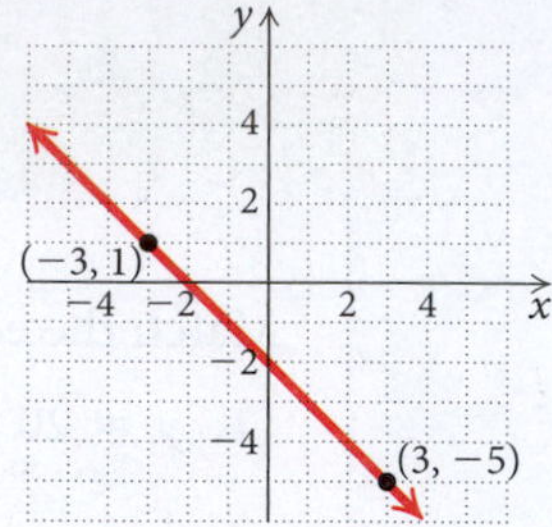

7.

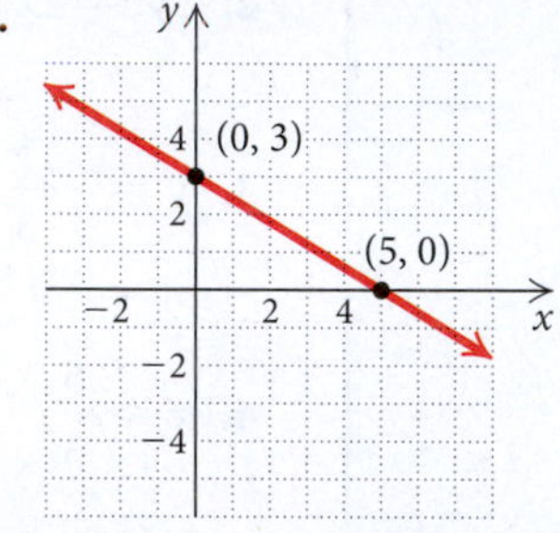

8.

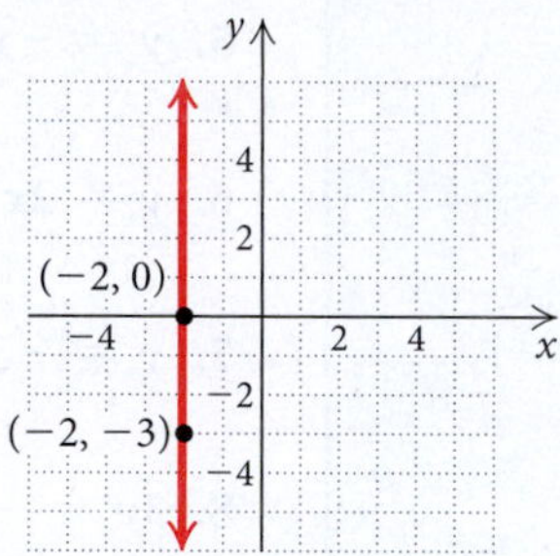

9.

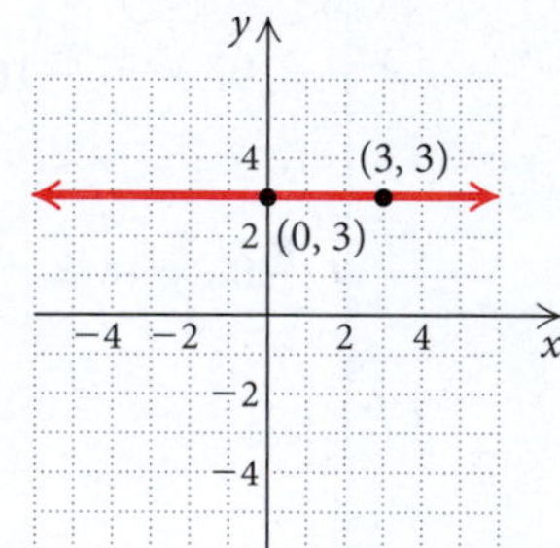

10.

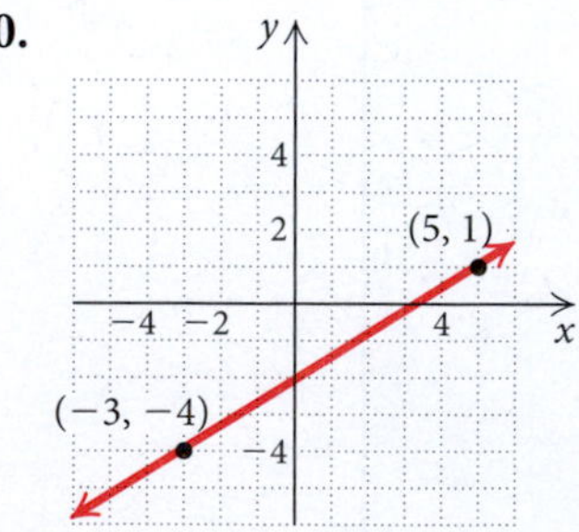

11. $(9, 4)$ and $(-1, 2)$
12. $(-3, 7)$ and $(5, -1)$
13. $(4, -9)$ and $(4, 6)$
14. $(-6, -1)$ and $(2, -13)$
15. $(0.7, -0.1)$ and $(-0.3, -0.4)$
16. $\left(-\frac{3}{4}, -\frac{1}{4}\right)$ and $\left(\frac{2}{7}, -\frac{5}{7}\right)$
17. $(2, -2)$ and $(4, -2)$
18. $(-9, 8)$ and $(7, -6)$
19. $\left(\frac{1}{2}, -\frac{3}{5}\right)$ and $\left(-\frac{1}{2}, \frac{3}{5}\right)$
20. $(-8.26, 4.04)$ and $(3.14, -2.16)$
21. $(16, -13)$ and $(-8, -5)$
22. $(\pi, -3)$ and $(\pi, 2)$
23. $(-10, -7)$ and $(-10, 7)$
24. $(\sqrt{2}, -4)$ and $(0.56, -4)$
25. $f(4) = 3$ and $f(-2) = 15$
26. $f(-4) = -5$ and $f(4) = 1$
27. $f\left(\frac{1}{5}\right) = \frac{1}{2}$ and $f(-1) = -\frac{11}{2}$
28. $f(8) = -1$ and $f\left(-\frac{2}{3}\right) = \frac{10}{3}$
29. $f(-6) = \frac{4}{5}$ and $f(0) = \frac{4}{5}$
30. $g\left(-\frac{9}{2}\right) = \frac{2}{9}$ and $g\left(\frac{2}{5}\right) = -\frac{5}{2}$

Determine the slope, if it exists, of the graph of the given linear equation.

31. $y = 1.3x - 5$
32. $y = -\frac{2}{5}x + 7$
33. $x = -2$
34. $f(x) = 4x - \frac{1}{4}$
35. $f(x) = -\frac{1}{2}x + 3$
36. $y = \frac{3}{4}$
37. $y = 9 - x$
38. $x = 8$
39. $y = 0.7$
40. $y = \frac{4}{5} - 2x$

41. *Electric Bicycle Sales.* Worldwide electric bicycle sales, with 92% of the market in China, totaled \$8.4 billion in 2013. By 2020, total sales are expected to reach \$10.8 billion. (*Source*: "Electric Bicycle Sales Expected to Climb," by Catherine Green, *Los Angeles Times*, July 27, 2013) Find the expected average rate of change in worldwide sales of electric bicycles from 2013 to 2020.

42. *Population Loss.* The population of Detroit, Michigan, decreased from 1,027,974 in 1990 to 701,475 in 2012 (*Source*: U.S. Census Bureau). Find the average rate of change in the population of Detroit, Michigan, over the 22-year period.

43. *Population Loss.* The population of Cleveland, Ohio, decreased from 478,403 in 2000 to 390,928 in 2012 (*Source*: U.S. Census Bureau). Find the average rate of change in the population of Cleveland, Ohio, over the 12-year period.

44. *Fireworks Revenue.* Revenue from display fireworks in the United States increased from \$141 million in 1998 to \$320 million in 2012 (*Source*: American Pyrotechnics Association). Find the average rate of change in the revenue from fireworks in the United States from 1998 to 2012.

45. *Growing Almonds.* In 2003, 550,000 acres of farmland in California were devoted to growing almonds. By 2012, the number of acres used to grow almonds had increased to 810,000. (*Source*: USDA National Agricultural Statistics Service) Find the average rate of change in the number of acres in California used to grow almonds from 2003 to 2012.

46. *Chicken Consumption.* The annual per-capita consumption of chicken in the United States was 42.5 lb in 1990. By 2011, this amount had increased to 58.4 lb. (*Source*: Economic Research Service, U.S. Department of Agriculture) Find the average rate of change in per-capita consumption of chicken from 1990 to 2011.

47. *Whole-Milk Consumption.* The annual per-capita consumption of whole milk in the United States was 25.3 gal in 1970. By 2011, this amount had decreased to 5.5 gal. Find the average rate of change in per-capita consumption of whole milk from 1970 to 2011.

48. *Minimum Wage.* In 1938, the minimum wage in the United States was $0.25. By 2009, the minimum wage had increased to $7.25. (*Source*: U.S. Department of Labor) Find the average rate of change in the minimum wage from 1938 to 2009.

Find the slope and the y-intercept of the line with the given equation.

49. $y = \frac{3}{5}x - 7$

50. $f(x) = -2x + 3$

51. $x = -\frac{2}{5}$

52. $y = \frac{4}{7}$

53. $f(x) = 5 - \frac{1}{2}x$

54. $y = 2 + \frac{3}{7}x$

55. $3x + 2y = 10$

56. $2x - 3y = 12$

57. $y = -6$

58. $x = 10$

59. $5y - 4x = 8$

60. $5x - 2y + 9 = 0$

61. $4y - x + 2 = 0$

62. $f(x) = 0.3 + x$

Graph the equation using the slope and the y-intercept.

63. $y = -\frac{1}{2}x - 3$

64. $y = \frac{3}{2}x + 1$

65. $f(x) = 3x - 1$

66. $f(x) = -2x + 5$

67. $3x - 4y = 20$

68. $2x + 3y = 15$

69. $x + 3y = 18$

70. $5y - 2x = -20$

71. *Whales and Pressure at Sea Depth.* Whales can withstand extreme atmospheric pressure changes because their bodies are flexible. Their rib cages and lungs can collapse safely under pressure. Sperm whales can hunt for squid at depths of 7000 ft or more. (*Sources*: National Ocean Service, National Oceanic and Atmospheric Administration)

The function P, given by

$$P(d) = \frac{1}{33}d + 1,$$

gives the pressure, in atmospheres (atm), at a given depth d, in feet, under the sea. Find $P(0)$, $P(33)$, $P(1000)$, $P(5000)$, and $P(7000)$.

72. *Stopping Distance on Glare Ice.* The stopping distance (at some fixed speed) of regular tires on glare ice is a function of the air temperature F, in degrees Fahrenheit. This function is estimated by

$$D(F) = 2F + 115,$$

where $D(F)$ is the stopping distance, in feet, when the air temperature is F, in degrees Fahrenheit.

a) Find $D(0°)$, $D(-20°)$, $D(10°)$, and $D(32°)$.
b) Explain why the domain should be restricted to $[-57.5°, 32°]$.

73. *Reaction Time.* Suppose that while driving a car, you suddenly see a deer standing in the road. Your brain registers the information and sends a signal to your foot to hit the brake. The car travels a distance D, in feet, during this time, where D is a function of the speed r, in miles per hour, of the car when you see the deer. That reaction distance is a linear function given by

$$D(r) = \frac{11}{10}r + \frac{1}{2}.$$

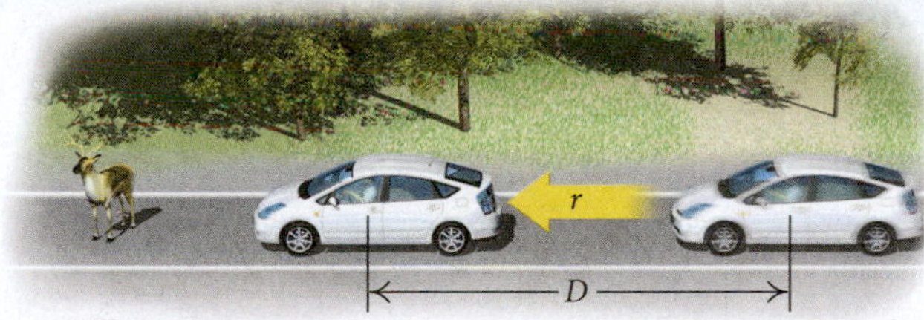

a) Find the slope of this line and interpret its meaning in this application.
b) Find $D(5)$, $D(10)$, $D(20)$, $D(50)$, and $D(65)$.
c) What is the domain of this function? Explain.

74. *Straight-Line Depreciation.* A contractor buys a new truck for \$38,000. The truck is purchased on January 1 and is expected to last 5 years, at the end of which time its *trade-in*, or *salvage, value* will be \$16,500. If the company figures the decline or depreciation in value to be the same each year, then the salvage value V, after t years, is given by the linear function

$$V(t) = \$38{,}000 - \$4300t, \quad \text{for } 0 \le t \le 5.$$

a) Find $V(0)$, $V(1)$, $V(2)$, $V(3)$, and $V(5)$.
b) Find the domain and the range of this function.

75. *Total Cost.* Richard is considering relocating to an assisted living facility. He learns that there is an initial community fee of \$2250 and a monthly charge of \$3380 for level-one care. Write an equation that can be used to determine the total cost $C(t)$ for t months of level-one care. Then find the total cost for 20 months.

76. *Total Cost.* Superior Cable Television charges a \$95 installation fee and \$125 per month for the Star plan. Write an equation that can be used to determine the total cost $C(t)$ for t months of the Star plan. Then find the total cost for 18 months of service.

In Exercises 77 and 78, the term ***fixed costs*** *refers to the start-up costs of operating a business. This includes machinery and building costs. The term* ***variable costs*** *refers to what it costs a business to produce or service one item.*

77. Max's Custom Lacrosse Stringing experienced fixed costs of \$750 and variable costs of \$15 for each lacrosse stick that was restrung. Write an equation that can be used to determine the total cost when x sticks are restrung. Then determine the total cost of restringing 32 lacrosse sticks.

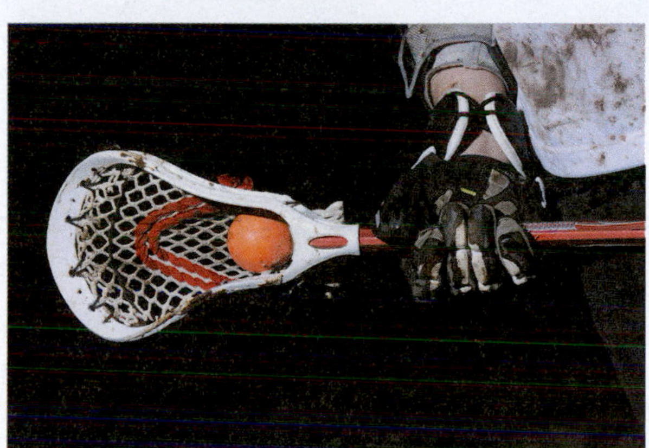

78. Soosie's Cookie Company had fixed costs of \$1250 and variable costs of \$4.25 per dozen gourmet cookies that were baked and packaged for sale. Write an equation that can be used to determine the total cost when x dozens of cookies are baked and sold. Then determine the total cost of baking and selling 85 dozen gourmet cookies.

Skill Maintenance

If $f(x) = x^2 - 3x$, find each of the following.

79. $f\left(\frac{1}{2}\right)$ [1.2]

80. $f(5)$ [1.2]

81. $f(-5)$ [1.2]

82. $f(-a)$ [1.2]

83. $f(a + h)$ [1.2]

Synthesis

84. *Grade of Treadmills.* A treadmill is 5 ft long and is set at an 8% grade. How high is the end of the treadmill?

Find the slope of the line containing the given points.

85. (a, a^2) and $(a + h, (a + h)^2)$

86. $(r, s + t)$ and (r, s)

Suppose that f is a linear function. Determine whether each of the following statements is true or false.

87. $f(c - d) = f(c) - f(d)$

88. $f(kx) = kf(x)$

Let $f(x) = mx + b$. Find a formula for $f(x)$ given each of the following.

89. $f(x + 2) = f(x) + 2$

90. $f(3x) = 3f(x)$

Mid-Chapter Mixed Review

Determine whether each of the following statements is true or false.

1. The x-intercept of the line that passes through $\left(-\frac{2}{3}, \frac{3}{2}\right)$ and the origin is $\left(-\frac{2}{3}, 0\right)$. [1.1]

2. All functions are relations, but not all relations are functions. [1.2]

3. The line parallel to the y-axis that passes through $(-5, 25)$ is $y = -5$. [1.3]

4. Find the intercepts of the graph of the line $-8x + 5y = -40$. [1.1]

For each pair of points, find the distance between the points and the midpoint of the segment having the points as endpoints. [1.1]

5. $(-8, -15)$ and $(3, 7)$

6. $\left(-\frac{3}{4}, \frac{1}{5}\right)$ and $\left(\frac{1}{4}, -\frac{4}{5}\right)$

7. Find an equation for a circle having center $(-5, 2)$ and radius 13. [1.1]

8. Find the center and the radius of the circle $(x - 3)^2 + (y + 1)^2 = 4$. [1.1]

Graph the equation.

9. $3x - 6y = 6$ [1.1]

10. $y = -\frac{1}{2}x + 3$ [1.3]

11. $y = 2 - x^2$ [1.1]

12. $(x + 4)^2 + y^2 = 4$ [1.1]

13. Given that $f(x) = x - 2x^2$, find $f(-4)$, $f(0)$, and $f(1)$. [1.2]

14. Given that $g(x) = \frac{x + 6}{x - 3}$, find $g(-6)$, $g(0)$, and $g(3)$. [1.2]

Find the domain of the function. [1.2]

15. $g(x) = x + 9$

16. $f(x) = \frac{-5}{x + 5}$

17. $h(x) = \frac{1}{x^2 + 2x - 3}$

Graph the function. [1.2]

18. $f(x) = -2x$

19. $g(x) = x^2 - 1$

20. Determine the domain and the range of the function shown in the following figure. [1.2]

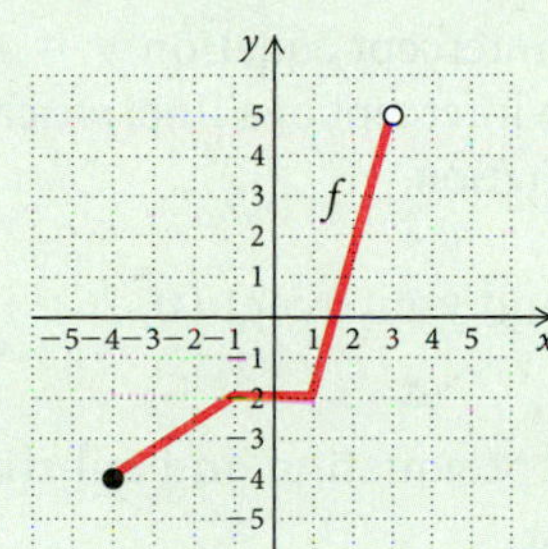

Find the slope of the line containing the given points. [1.3]

21. $(-2, 13)$ and $(-2, -5)$

22. $(10, -1)$ and $(-6, 3)$

23. $\left(\frac{5}{7}, \frac{1}{3}\right)$ and $\left(\frac{2}{7}, \frac{1}{3}\right)$

Determine the slope, if it exists, and the y-intercept of the line with the given equation. [1.3]

24. $f(x) = -\frac{1}{9}x + 12$

25. $y = -6$

26. $x = 2$

27. $3x - 16y + 1 = 0$

Collaborative Discussion and Writing

To the student and the instructor: *The Collaborative Discussion and Writing exercises are meant to be answered with one or more sentences. They can be discussed and answered collaboratively by the entire class or by small groups.*

28. Explain as you would to a fellow student how the numerical value of the slope of a line can be used to describe the slant and the steepness of that line. [1.3]

29. Discuss why the graph of a vertical line $x = a$ cannot represent a function. [1.3]

30. Explain in your own words the difference between the domain of a function and the range of a function. [1.2]

31. Explain how you could find the coordinates of a point $\frac{7}{8}$ of the way from point A to point B. [1.1]

1.4 Equations of Lines and Modeling

- Determine equations of lines.
- Given the equations of two lines, determine whether their graphs are parallel or perpendicular.
- Model a set of data with a linear function.

Slope–Intercept Equations of Lines

In Section 1.3, we developed the slope–intercept equation $y = mx + b$, or $f(x) = mx + b$. If we know the slope and the y-intercept of a line, we can find an equation of the line using the slope–intercept equation.

EXAMPLE 1 A line has slope $-\frac{7}{9}$ and y-intercept $(0, 16)$. Find an equation of the line.

Solution We use the slope–intercept equation and substitute $-\frac{7}{9}$ for m and 16 for b:

$$y = mx + b$$
$$y = -\tfrac{7}{9}x + 16, \quad \text{or}$$
$$f(x) = -\tfrac{7}{9}x + 16.$$

Now Try Exercise 7.

EXAMPLE 2 A line has slope $-\frac{2}{3}$ and contains the point $(-3, 6)$. Find an equation of the line.

Solution We use the slope–intercept equation, $y = mx + b$, and substitute $-\frac{2}{3}$ for m: $y = -\frac{2}{3}x + b$. Using the point $(-3, 6)$, we substitute -3 for x and 6 for y in $y = -\frac{2}{3}x + b$. Then we solve for b.

$$y = mx + b$$
$$y = -\tfrac{2}{3}x + b \qquad \text{Substituting } -\tfrac{2}{3} \text{ for } m$$
$$6 = -\tfrac{2}{3}(-3) + b \qquad \text{Substituting } -3 \text{ for } x \text{ and } 6 \text{ for } y$$
$$6 = 2 + b$$
$$4 = b \qquad \text{Solving for } b. \text{ The } y\text{-intercept is } (0, b).$$

The equation of the line is $y = -\frac{2}{3}x + 4$, or $f(x) = -\frac{2}{3}x + 4$.

Now Try Exercise 13.

Point–Slope Equations of Lines

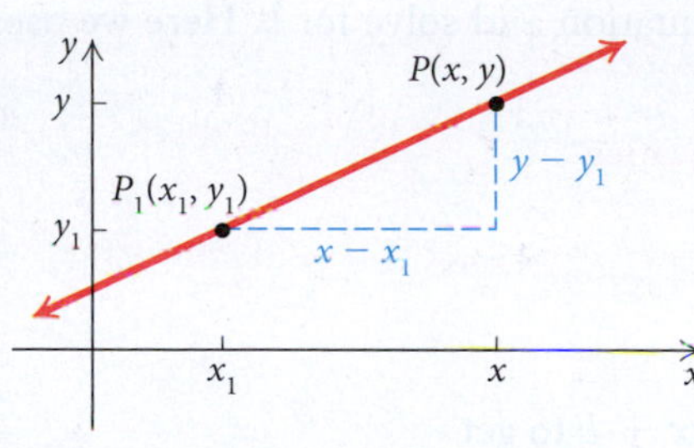

Another formula that can be used to determine an equation of a line is the *point–slope equation*. Suppose that we have a nonvertical line and that the coordinates of point P_1 on the line are (x_1, y_1). We can think of P_1 as fixed and imagine another point P on the line with coordinates (x, y). Thus the slope is given by

$$\frac{y - y_1}{x - x_1} = m.$$

Multiplying by $x - x_1$ on both sides, we get the *point–slope equation* of the line:

$$(x - x_1) \cdot \frac{y - y_1}{x - x_1} = m \cdot (x - x_1)$$

$$y - y_1 = m(x - x_1).$$

POINT–SLOPE EQUATION

The **point–slope equation** of the line with slope m passing through (x_1, y_1) is

$$y - y_1 = m(x - x_1).$$

If we know the slope of a line and the coordinates of one point on the line, we can find an equation of the line using either the point–slope equation,

$$y - y_1 = m(x - x_1),$$

or the slope–intercept equation,

$$y = mx + b.$$

EXAMPLE 3 Find an equation of the line containing the points $(2, 3)$ and $(1, -4)$.

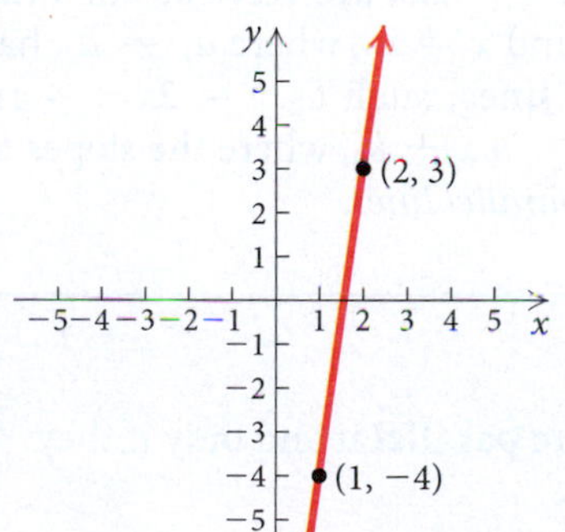

Solution We first determine the slope:

$$m = \frac{-4 - 3}{1 - 2} = \frac{-7}{-1} = 7.$$

Using the Point–Slope Equation: We substitute 7 for m and either of the points $(2, 3)$ or $(1, -4)$ for (x_1, y_1) in the point–slope equation. In this case, we use $(2, 3)$.

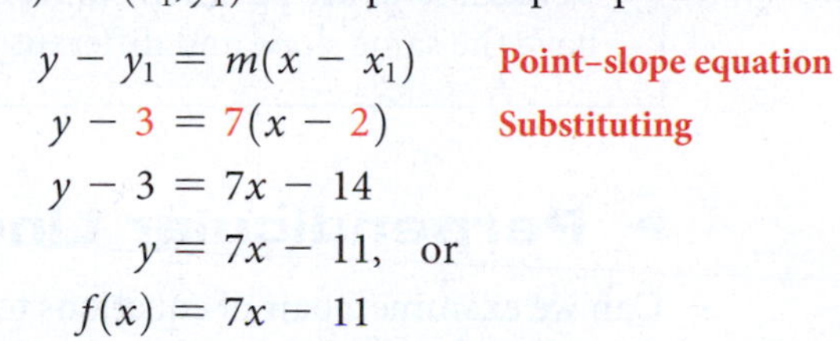

$$y - y_1 = m(x - x_1) \quad \textbf{Point–slope equation}$$

$$y - 3 = 7(x - 2) \quad \textbf{Substituting}$$

$$y - 3 = 7x - 14$$

$$y = 7x - 11, \text{ or}$$

$$f(x) = 7x - 11$$

Using the Slope–Intercept Equation: We substitute 7 for m and either of the points $(2, 3)$ or $(1, -4)$ for (x, y) in the slope–intercept equation and solve for b. Here we use $(1, -4)$.

$$\begin{aligned} y &= mx + b && \textbf{Slope–intercept equation} \\ -4 &= 7 \cdot 1 + b && \textbf{Substituting} \\ -4 &= 7 + b && \\ -11 &= b && \textbf{Solving for } b \end{aligned}$$

We substitute 7 for m and -11 for b in $y = mx + b$ to get

$$y = 7x - 11, \quad \text{or}$$
$$f(x) = 7x - 11.$$

Now Try Exercise 19.

Parallel Lines

Can we determine whether the graphs of two linear equations are parallel without graphing them? Let's look at three pairs of equations and their graphs.

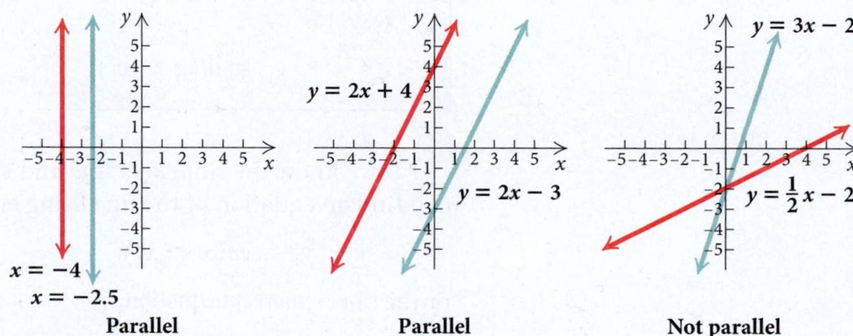

If two different lines, such as $x = -4$ and $x = -2.5$, are vertical, then they are parallel. Thus two equations such as $x = a_1$ and $x = a_2$, where $a_1 \neq a_2$, have graphs that are *parallel lines*. Two nonvertical lines, such as $y = 2x + 4$ and $y = 2x - 3$, or, in general, $y = mx + b_1$ and $y = mx + b_2$, where the slopes are the *same* and $b_1 \neq b_2$, also have graphs that are *parallel lines*.

PARALLEL LINES

Vertical lines are **parallel**. Nonvertical lines are **parallel** if and only if they have the same slope and different y-intercepts.

Perpendicular Lines

Can we examine a pair of equations to determine whether their graphs are perpendicular without graphing the equations? Let's look at the following pairs of equations and their graphs.

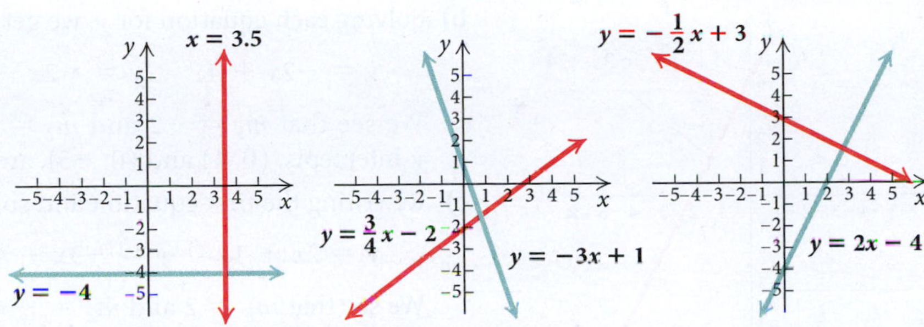

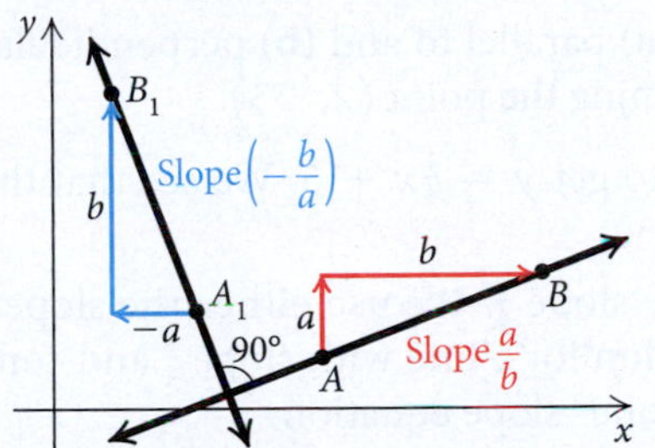

If one line is vertical and another is horizontal, they are perpendicular. For example, the lines $x = 5$ and $y = -3$ are perpendicular. Otherwise, how can we tell whether two lines are perpendicular? Consider a line $\overleftrightarrow{AB}$, as shown in the figure at left, with slope a/b. Then think of rotating the line 90° to get a line $\overleftrightarrow{A_1B_1}$ perpendicular to $\overleftrightarrow{AB}$. For the new line, the rise and the run are interchanged, but the run is now negative. Thus the slope of the new line is $-b/a$, which is the opposite of the reciprocal of the slope of the first line. Also note that when we multiply the slopes, we get

$$\frac{a}{b}\left(-\frac{b}{a}\right) = -1.$$

This is the condition under which lines will be perpendicular.

PERPENDICULAR LINES

Two lines with slopes m_1 and m_2 are **perpendicular** if and only if the product of their slopes is -1:

$$m_1 m_2 = -1.$$

Lines are also **perpendicular** if one is vertical ($x = a$) and the other is horizontal ($y = b$).

If a line has slope m_1, the slope m_2 of a line perpendicular to it is $-1/m_1$. The slope of one line is the *opposite of the reciprocal* of the other:

$$m_2 = -\frac{1}{m_1}, \quad \text{or} \quad m_1 = -\frac{1}{m_2}.$$

EXAMPLE 4 Determine whether each of the following pairs of lines is parallel, perpendicular, or neither.

a) $y + 2 = 5x,\ 5y + x = -15$

b) $2y + 4x = 8,\ 5 + 2x = -y$

c) $2x + 1 = y,\ y + 3x = 4$

Solution We use the slopes of the lines to determine whether the lines are parallel or perpendicular.

a) We solve each equation for y:

$$y = 5x - 2, \qquad y = -\tfrac{1}{5}x - 3.$$

The slopes are 5 and $-\frac{1}{5}$. Their product is -1, so the lines are perpendicular. (See Fig. 1.)

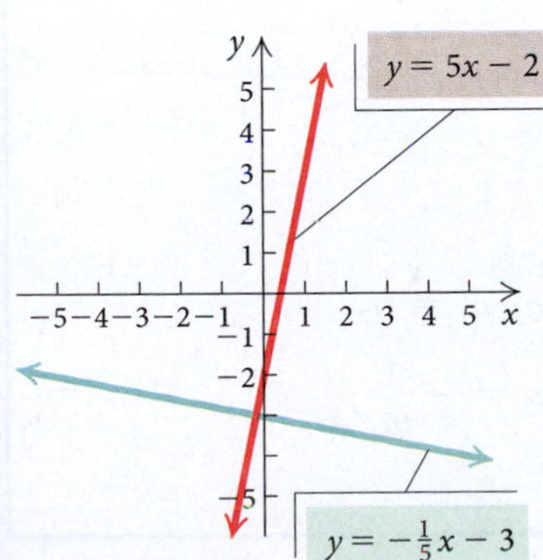

Figure 1.

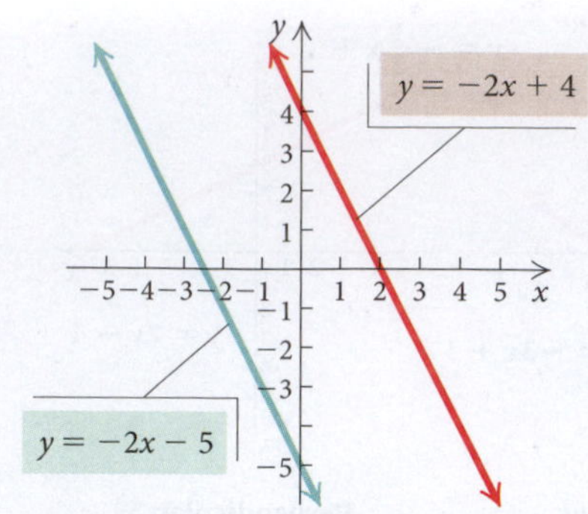

Figure 2.

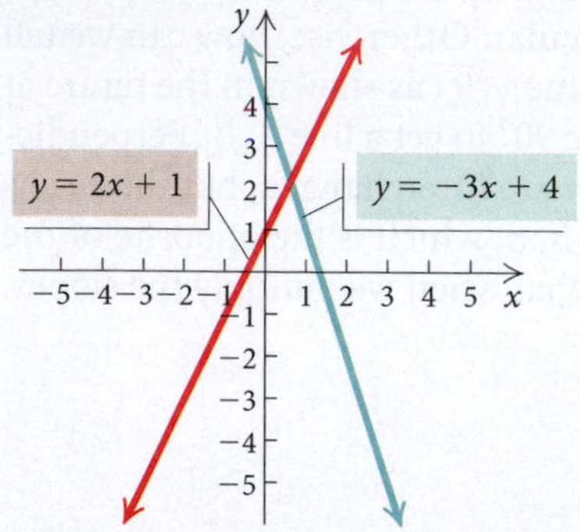

Figure 3.

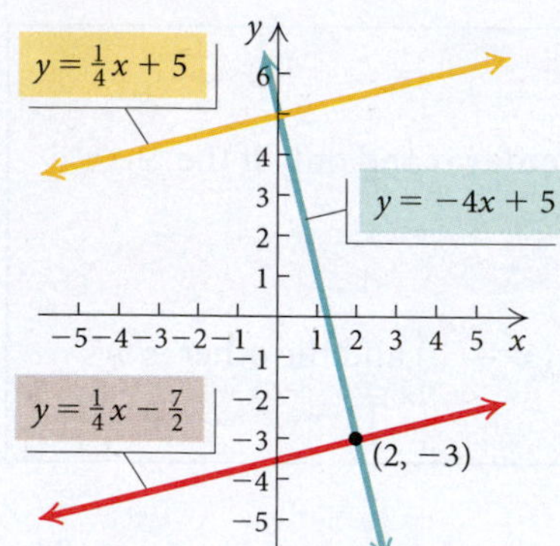

b) Solving each equation for y, we get

$$y = -2x + 4, \qquad y = -2x - 5.$$

We see that $m_1 = -2$ and $m_2 = -2$. Since the slopes are the same and the y-intercepts, $(0, 4)$ and $(0, -5)$, are different, the lines are parallel. (See Fig. 2.)

c) Rewriting the first equation and solving the second equation for y, we have

$$y = 2x + 1, \qquad y = -3x + 4.$$

We see that $m_1 = 2$ and $m_2 = -3$. Since the slopes are not the same and their product is not -1, it follows that the lines are neither parallel nor perpendicular. (See Fig. 3.)

Now Try Exercises 35 and 39.

EXAMPLE 5 Write equations of the lines **(a)** parallel to and **(b)** perpendicular to the graph of the line $4y - x = 20$ and containing the point $(2, -3)$.

Solution We first solve $4y - x = 20$ for y to get $y = \frac{1}{4}x + 5$. We see that the slope of the given line is $\frac{1}{4}$.

a) The line parallel to the given line will have slope $\frac{1}{4}$. We use either the slope–intercept equation or the point–slope equation for a line with slope $\frac{1}{4}$ and containing the point $(2, -3)$. Here we use the point–slope equation:

$$\begin{aligned} y - y_1 &= m(x - x_1) \\ y - (-3) &= \tfrac{1}{4}(x - 2) \\ y + 3 &= \tfrac{1}{4}x - \tfrac{1}{2} \\ y &= \tfrac{1}{4}x - \tfrac{7}{2}. \end{aligned}$$

b) The slope of the perpendicular line is the opposite of the reciprocal of $\frac{1}{4}$, or -4. Again we use the point–slope equation to write an equation for a line with slope -4 and containing the point $(2, -3)$:

$$\begin{aligned} y - y_1 &= m(x - x_1) \\ y - (-3) &= -4(x - 2) \\ y + 3 &= -4x + 8 \\ y &= -4x + 5. \end{aligned}$$

Now Try Exercise 43.

Summary of Terminology about Lines

TERMINOLOGY	MATHEMATICAL INTERPRETATION
Slope	$m = \dfrac{y_2 - y_1}{x_2 - x_1}$, or $\dfrac{y_1 - y_2}{x_1 - x_2}$
Slope–intercept equation	$y = mx + b$
Point–slope equation	$y - y_1 = m(x - x_1)$
Horizontal line	$y = b$
Vertical line	$x = a$
Parallel lines	$m_1 = m_2, b_1 \neq b_2$; also $x = a_1, x = a_2, a_1 \neq a_2$
Perpendicular lines	$m_1 m_2 = -1$, or $m_2 = -\dfrac{1}{m_1}$; also $x = a, y = b$

Creating a Mathematical Model

1. Recognize real-world problem.

2. Collect data.

3. Analyze data.

4. Construct model.

5. Test and refine model.

6. Explain and predict.

Mathematical Models

When a real-world situation can be described in mathematical language, we have a **mathematical model**. For example, the natural numbers constitute a mathematical model for situations in which counting is essential. Situations in which algebra can be brought to bear often require the use of functions as models.

Mathematical models are abstracted from real-world situations. The mathematical model gives results that allow one to predict what will happen in that real-world situation. If the predictions are inaccurate or the results of experimentation do not conform to the model, the model must be changed or discarded.

Mathematical modeling can be an ongoing process. For example, finding a mathematical model that will provide an accurate prediction of population growth is not a simple problem. Any population model that one might devise would need to be reshaped as further information is acquired.

Curve Fitting

We will develop and use many kinds of mathematical models in this text. In this chapter, we have used *linear* functions as models. Other types of functions, such as quadratic, cubic, and exponential functions, can also model data. These functions are *nonlinear*.

Quadratic function:
$y = ax^2 + bx + c, a > 0$

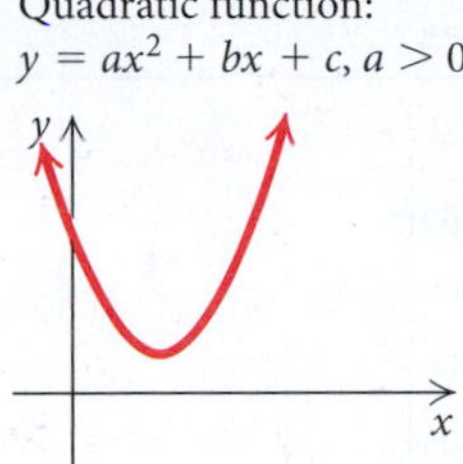

Cubic function:
$y = ax^3 + bx^2 + cx + d, a > 0$

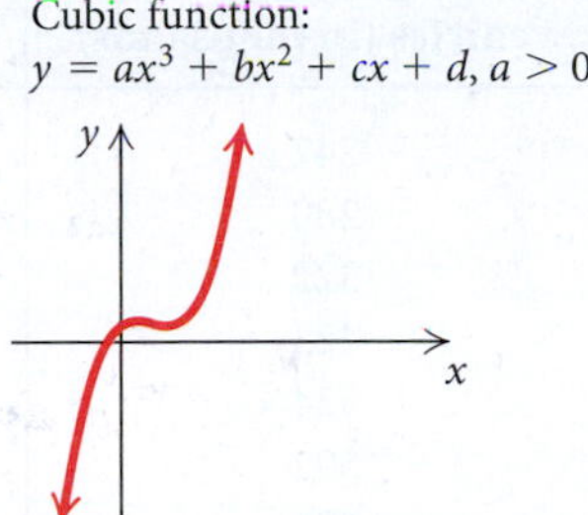

Exponential function:
$y = ab^x, a, b > 0, b \neq 1$

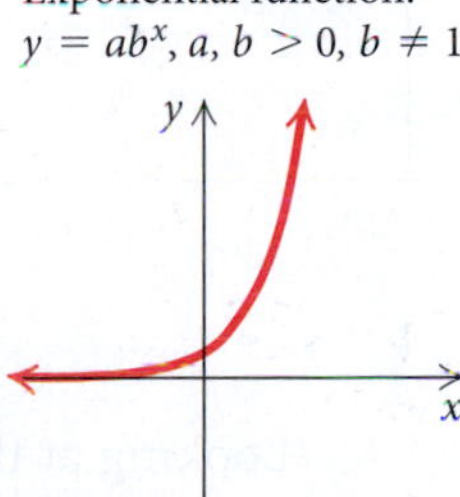

In general, we try to find a function that fits, as well as possible, observations (data), theoretical reasoning, and common sense. We call this **curve fitting**; it is one aspect of mathematical modeling.

Let's look at some data and related graphs, or **scatterplots**, and determine whether a linear function seems to fit the set of data.

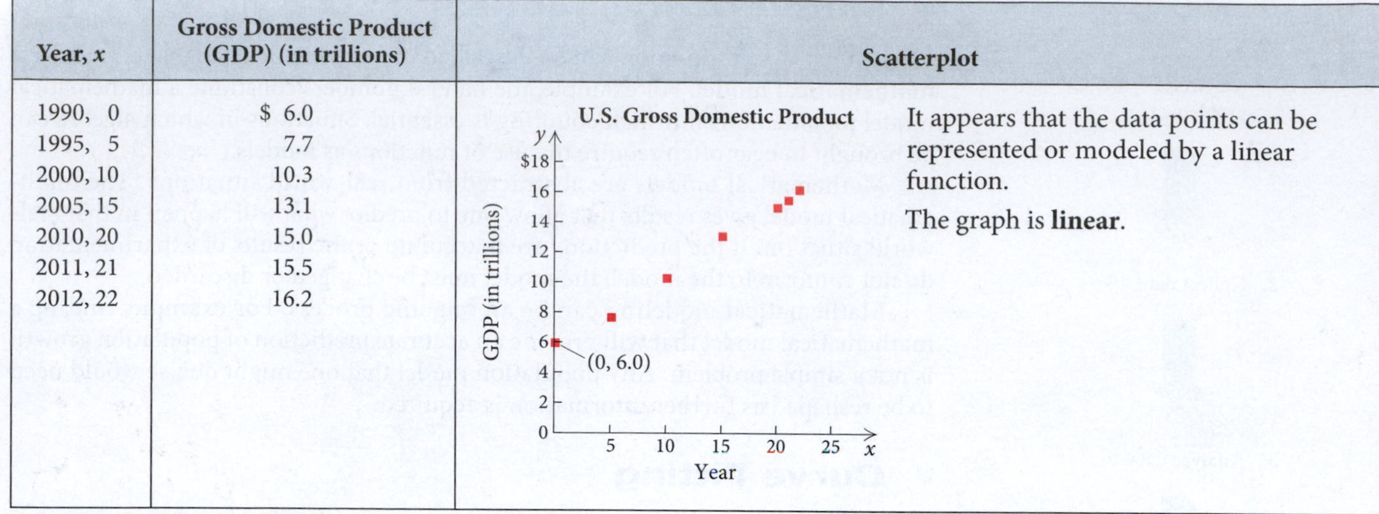

Year, x	Gross Domestic Product (GDP) (in trillions)	Scatterplot
1990, 0	\$ 6.0	It appears that the data points can be represented or modeled by a linear function. The graph is **linear**.
1995, 5	7.7	
2000, 10	10.3	
2005, 15	13.1	
2010, 20	15.0	
2011, 21	15.5	
2012, 22	16.2	

Sources: Bureau of Economic Analysis, U.S. Department of Commerce

Year, x	Estimated Number of Alternative-Fueled Vehicles (in thousands)	Scatterplot
1995, 0	247	It appears that the data points cannot be modeled accurately by a linear function. The graph is **nonlinear**.
1997, 2	280	
1999, 4	322	
2001, 6	425	
2003, 8	534	
2005, 10	592	
2007, 12	696	
2009, 14	826	
2011, 16	1192	

Source: U.S. Energy Information Administration

Looking at the scatterplots, we see that the data on gross domestic product seem to be rising in a manner to suggest that a *linear function* might fit, although a "perfect" straight line cannot be drawn through the data points. A linear function does not seem to fit the data on alternative-fueled vehicles.

EXAMPLE 6 ***U.S. Gross Domestic Product.*** The **gross domestic product** (GDP) of a country is the market value of final goods and services produced. Market value depends on the quantity of goods and services and their price. Model the data in the table above on the U.S. Gross Domestic Product with a linear function. Then estimate the GDP in 2018.

Solution We can choose any two of the data points to determine an equation. Note that the first coordinate is the number of years since 1990 and the second coordinate is the corresponding GDP in trillions of dollars. Let's use $(5, 7.7)$ and $(21, 15.5)$.

We first determine the slope of the line:

$$m = \frac{15.5 - 7.7}{21 - 5} = \frac{7.8}{16} = 0.4875.$$

Then we substitute 0.4875 for m and either of the points $(5, 7.7)$ or $(21, 15.5)$ for (x_1, y_1) in the point–slope equation. In this case, we use $(5, 7.7)$. We get

$$y - y_1 = m(x - x_1) \quad \textbf{Point–slope equation}$$
$$y - 7.7 = 0.4875(x - 5), \quad \textbf{Substituting}$$

which simplifies to

$$y = 0.4875x + 5.2625,$$

where x is the number of years after 1990 and y is in trillions of dollars.

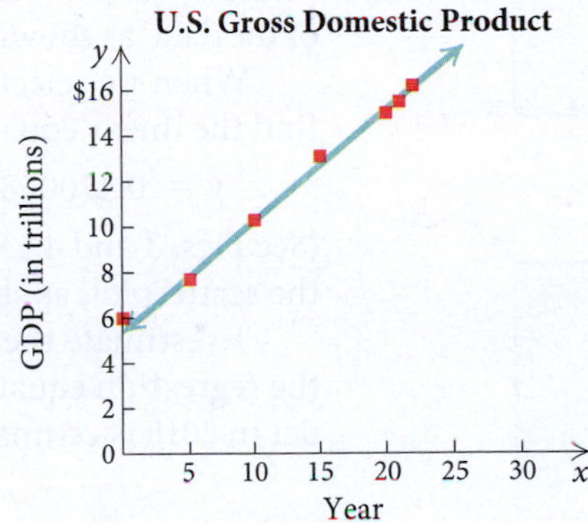

Next, we estimate the GDP in 2018 by substituting 28 $(2018 - 1990 = 28)$ for x in the model:

$$y = 0.4875x + 5.2625 \quad \textbf{Model}$$
$$= 0.4875(28) + 5.2625 \quad \textbf{Substituting}$$
$$= 18.9125 \approx 18.91.$$

We estimate that the gross domestic product will be \$18.91 trillion in 2018.

Now Try Exercise 61.

In Example 6, if we were to use the data points $(0, 6.0)$ and $(20, 15.0)$, our model would be

$$y = 0.45x + 6.0,$$

and our estimate for the GDP in 2018 would be \$18.60 trillion, about \$0.31 trillion less than the estimate provided by the first model. This illustrates that a model and the estimates that it produces are dependent on the data points used.

Models that consider all the data points, not just two, are generally better models. The model that best fits the data can be found using a graphing calculator and a procedure called **linear regression**. This procedure is explained in the following Technology Connection.

Technology Connection

L1	L2	L3 2
0	6	-----
5	7.7	
10	10.3	
15	13.1	
20	15	
21	15.5	
22	16.2	

L2(7) = 16.2

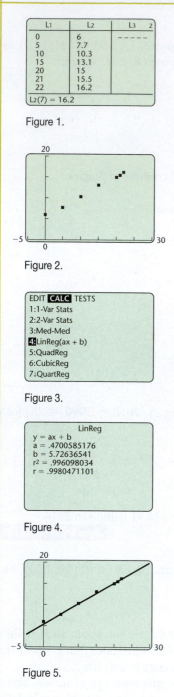

Figure 1.

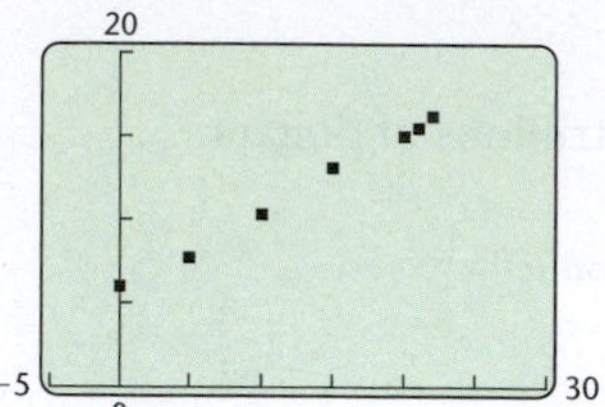

Figure 2.

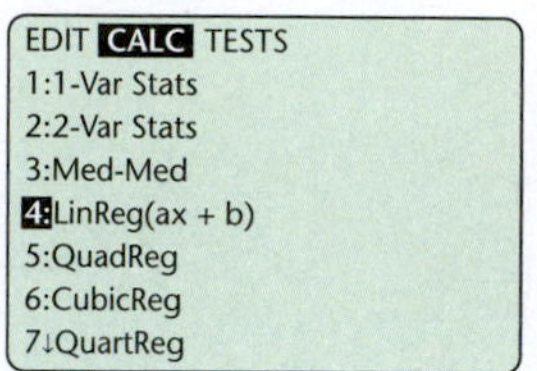

Figure 3.

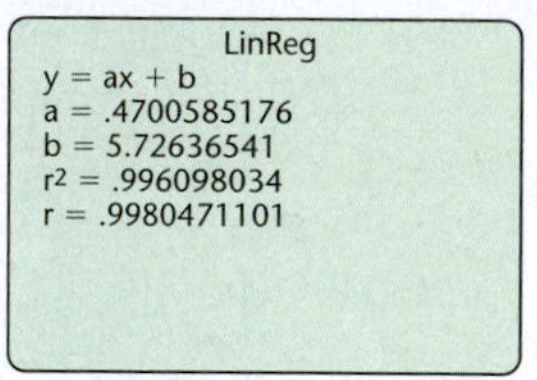

Figure 4.

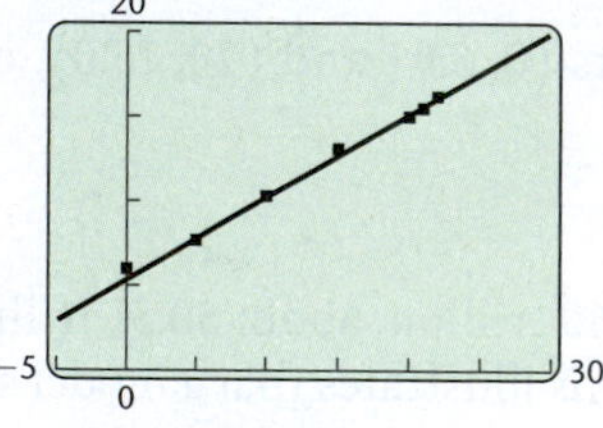

Figure 5.

We now consider **linear regression**, a procedure that can be used to model a set of data using a linear function. Although discussion leading to a complete understanding of this method belongs in a statistics course, we present the procedure here because we can carry it out easily using technology. The graphing calculator gives us the powerful capability to find linear models and to make predictions using them.

Consider the data presented before Example 6 on the gross domestic product. We can fit a regression line of the form $y = mx + b$ to the data using the LINEAR REGRESSION feature on a graphing calculator.

First, we enter the data in lists on the calculator. We enter the values of the independent variable x in list L1 and the corresponding values of the dependent variable y in L2. (See Fig. 1.) The graphing calculator can then create a scatterplot of the data, as shown at left in Fig. 2.

When we select the LINEAR REGRESSION feature from the STAT CALC menu, we find the linear equation that best models the data. It is

$$y = 0.4700585176x + 5.72636541. \qquad \text{Regression line}$$

(See Figs. 3 and 4.) We can then graph the regression line on the same graph as the scatterplot, as shown in Fig. 5.

To estimate the gross domestic product in 2012, we substitute 28 for x in the regression equation. Using this model, we see that the gross domestic product in 2018 is estimated to be about \$18.89 trillion. (See Fig. 6.)

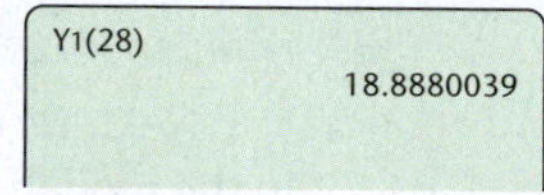

Figure 6.

Note that \$18.89 trillion is closer to the value \$18.91 trillion found with the data points $(5, 7.7)$ and $(21, 15.5)$ in Example 6 than to the value \$18.60 trillion found with the data points $(0, 6.0)$ and $(20, 15.0)$ following Example 6.

The Correlation Coefficient

On some graphing calculators with the DIAGNOSTIC feature turned on, a constant r between -1 and 1, called the **coefficient of linear correlation,** appears with the equation of the regression line. Though we cannot develop a formula for calculating r in this text, keep in mind that it is used to describe the strength of the linear relationship between x and y. The closer $|r|$ is to 1, the better the correlation. A positive value of r also indicates that the regression line has a positive slope, and a negative value of r indicates that the regression line has a negative slope. As shown in Fig. 4, for the data on gross domestic product just discussed, $r = 0.9980471101$, which indicates a good linear correlation.

The following scatterplots summarize the interpretation of a correlation coefficient.

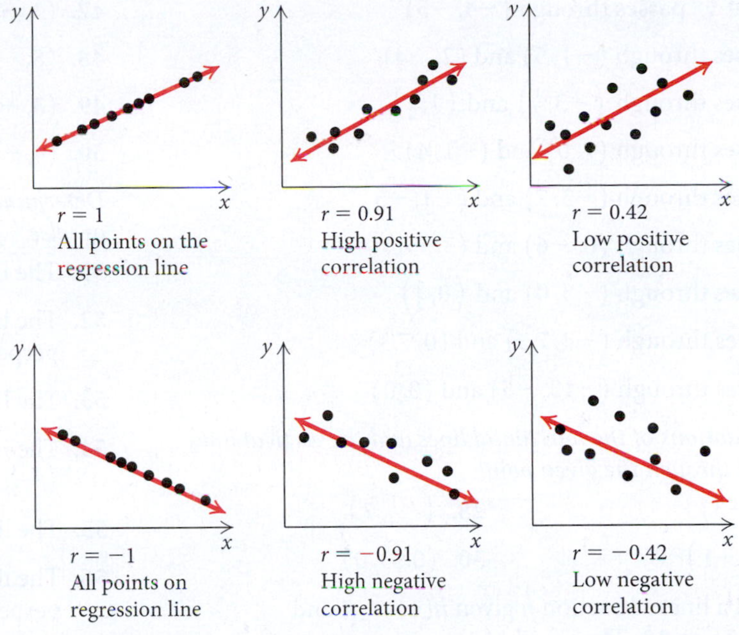

1.4 Exercise Set

Find the slope and the y-intercept of the graph of the linear equation. Then write the equation of the line in slope–intercept form.

1.

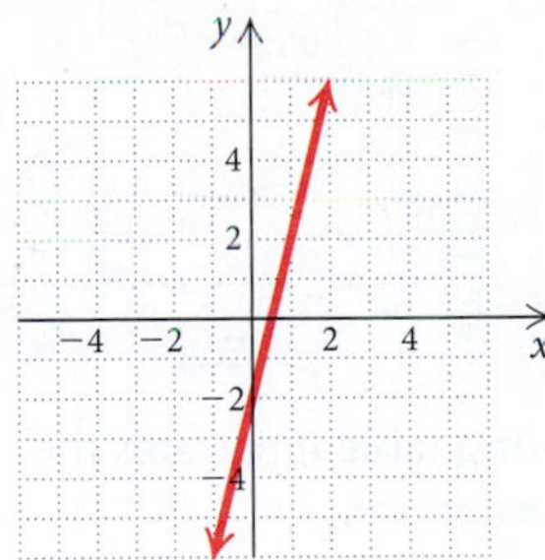

2.

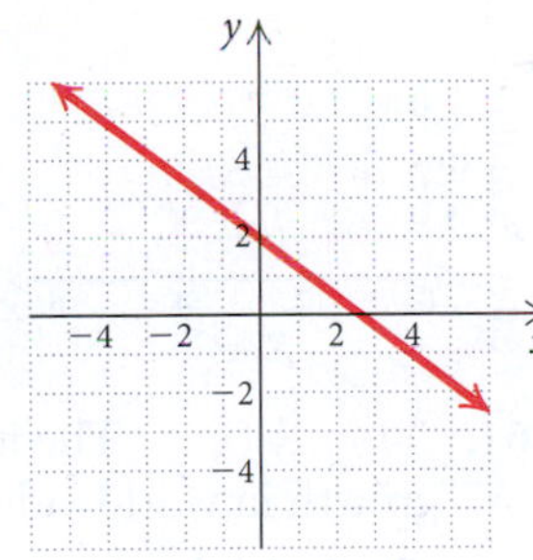

3.

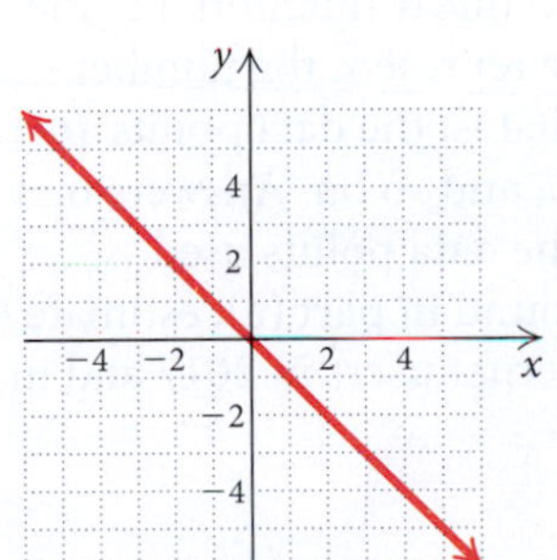

4.

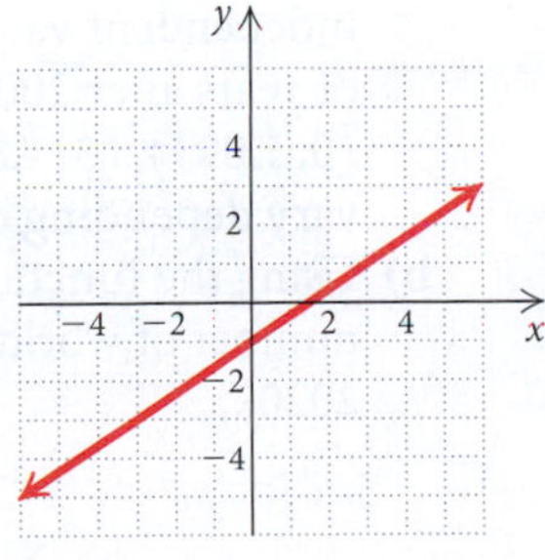

5.

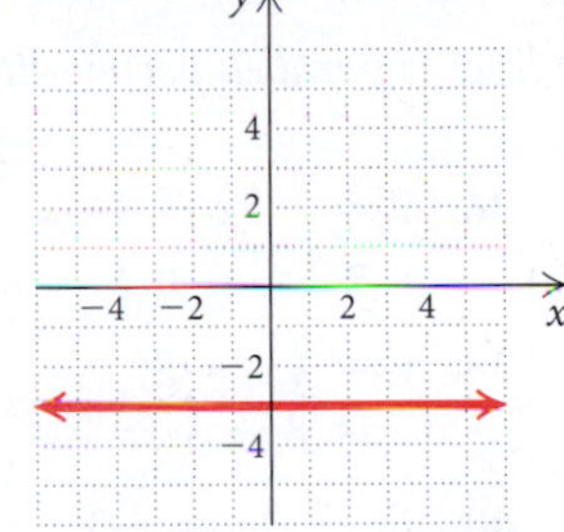

6.

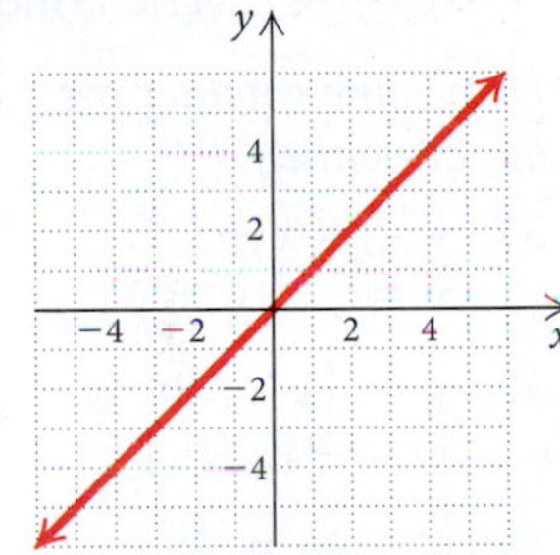

Write a slope–intercept equation for a line with the given characteristics.

7. $m = \frac{2}{9}$, y-intercept $(0, 4)$

8. $m = -\frac{3}{8}$, y-intercept $(0, 5)$

9. $m = -4$, y-intercept $(0, -7)$

10. $m = \frac{2}{7}$, y-intercept $(0, -6)$

11. $m = -4.2$, y-intercept $\left(0, \frac{3}{4}\right)$

12. $m = -4$, y-intercept $\left(0, -\frac{3}{2}\right)$

13. $m = \frac{2}{9}$, passes through $(3, 7)$

14. $m = -\frac{3}{8}$, passes through $(5, 6)$

15. $m = 0$, passes through $(-2, 8)$

16. $m = -2$, passes through $(-5, 1)$
17. $m = -\frac{3}{5}$, passes through $(-4, -1)$
18. $m = \frac{2}{3}$, passes through $(-4, -5)$
19. Passes through $(-1, 5)$ and $(2, -4)$
20. Passes through $\left(-3, \frac{1}{2}\right)$ and $\left(1, \frac{1}{2}\right)$
21. Passes through $(7, 0)$ and $(-1, 4)$
22. Passes through $(-3, 7)$ and $(-1, -5)$
23. Passes through $(0, -6)$ and $(3, -4)$
24. Passes through $(-5, 0)$ and $\left(0, \frac{4}{5}\right)$
25. Passes through $(-4, 7.3)$ and $(0, 7.3)$
26. Passes through $(-13, -5)$ and $(0, 0)$

Write equations of the horizontal lines and the vertical lines that pass through the given point.

27. $(0, -3)$
28. $\left(-\frac{1}{4}, 7\right)$
29. $\left(\frac{2}{11}, -1\right)$
30. $(0.03, 0)$
31. Find a linear function h given $h(1) = 4$ and $h(-2) = 13$. Then find $h(2)$.
32. Find a linear function g given $g\left(-\frac{1}{4}\right) = -6$ and $g(2) = 3$. Then find $g(-3)$.
33. Find a linear function f given $f(5) = 1$ and $f(-5) = -3$. Then find $f(0)$.
34. Find a linear function h given $h(-3) = 3$ and $h(0) = 2$. Then find $h(-6)$.

Determine whether the pair of lines is parallel, perpendicular, or neither.

35. $y = \frac{26}{3}x - 11$, $y = -\frac{3}{26}x - 11$
36. $y = -3x + 1$, $y = -\frac{1}{3}x + 1$
37. $y = \frac{2}{5}x - 4$, $y = -\frac{2}{5}x + 4$
38. $y = \frac{3}{2}x - 8$, $y = 8 + 1.5x$
39. $x + 2y = 5$, $2x + 4y = 8$
40. $2x - 5y = -3$, $2x + 5y = 4$
41. $y = 4x - 5$, $4y = 8 - x$
42. $y = 7 - x$, $y = x + 3$

Write a slope–intercept equation for a line passing through the given point that is parallel to the given line. Then write a second equation for a line passing through the given point that is perpendicular to the given line.

43. $(3, 5)$, $y = \frac{2}{7}x + 1$
44. $(-1, 6)$, $f(x) = 2x + 9$
45. $(-7, 0)$, $y = -0.3x + 4.3$
46. $(-4, -5)$, $2x + y = -4$
47. $(3, -2)$, $3x + 4y = 5$
48. $(8, -2)$, $y = 4.2(x - 3) + 1$
49. $(3, -3)$, $x = -1$
50. $(4, -5)$, $y = -1$

Determine whether each of the following statements is true or false.

51. The lines $x = -3$ and $y = 5$ are perpendicular.
52. The lines $y = 2x - 3$ and $y = -2x - 3$ are perpendicular.
53. The lines $y = \frac{2}{5}x + 4$ and $y = \frac{2}{5}x - 4$ are parallel.
54. The intersection of the lines $y = 2$ and $x = -\frac{3}{4}$ is $\left(-\frac{3}{4}, 2\right)$.
55. The lines $x = -1$ and $x = 1$ are perpendicular.
56. The lines $2x + 3y = 4$ and $3x - 2y = 4$ are perpendicular.

In Exercises 57–60, determine whether a linear model might fit the data.

57.

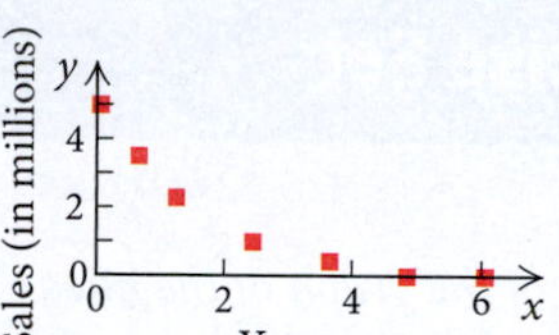

58.

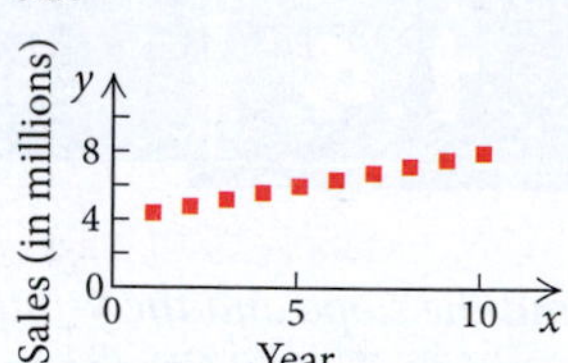

59.

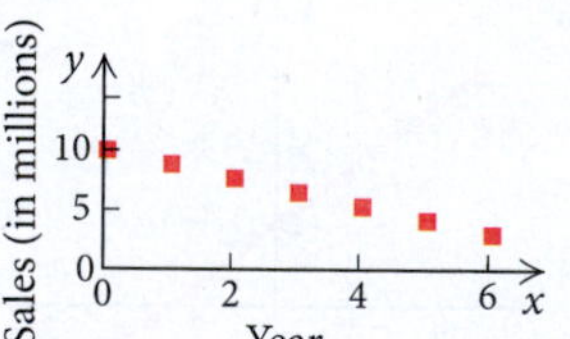

60.

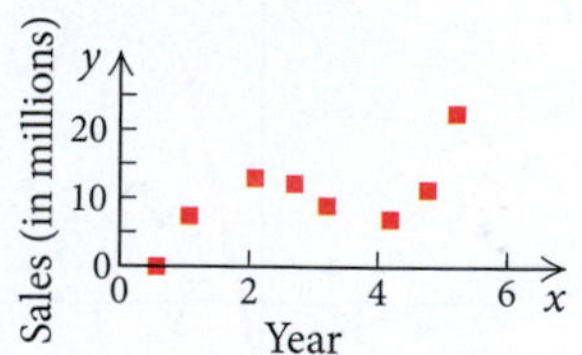

61. Internet Use. The following table illustrates the growth in worldwide Internet use.
    a) Model the data with a linear function. Let the independent variable represent the number of years after 2006; that is, the data points are $(0, 1.093)$, $(3, 1.802)$, and so on. Answers may vary depending on the data points used.
    b) Using the function found in part (a), estimate the number of world Internet users in 2017 and in 2020.

Year, x	Number of World Internet Users, y (in billions)
2006, 0	1.093
2007, 1	1.319
2008, 2	1.574
2009, 3	1.802
2010, 4	1.971
2011, 5	2.267
2012, 6	2.497
2013, 7	2.749

Source: Internet and Facebook World Stats

62. *Cremations.* The following table illustrates the upward trend in America to choose cremation.

a) Model the data with a linear function. Let the independent variable represent the number of years after 2005. Answers may vary depending on the data points used.

b) Using the function found in part (a), estimate the percentage of deaths followed by cremation in 2011 and in 2016.

Year, x	Percentage of Deaths Followed by Cremation, y
2005, 0	32.2%
2006, 1	33.5
2007, 2	34.3
2008, 3	35.3
2009, 4	36.7
2010, 5	40.6

Source: Cremation Association of North America

63. *Electricity Use.* Data on the average annual household use of electricity, in kilowatt-hours, are listed in the following table. Model the data with a linear function and predict the average annual household electricity use in 2019. Answers may vary depending on the data points used.

Year, x	Annual Electricity Use (in kilowatt-hours)
2010, 0	11,504
2011, 1	11,280
2012, 2	10,837
2013, 3	10,819

Source: Energy Information Administration

64. *Median Household Income.* Data on the median household income in the United States (adjusted for inflation) are listed in the following table. Model the data with a linear function, estimate the median household income in 2009, and predict the median household income in 2017. Answers may vary depending on the data points used.

Year, x	Median Household Income in the United States
2006, 0	$54,892
2008, 2	53,644
2010, 4	51,893
2012, 6	51,017

Source: U.S. Census Bureau

65. *Bottled Water.* Data on the per-capita consumption, in gallons, of bottled water in the United States are given in the following table. Model the data with a linear function and predict the per-capita consumption of bottled water in 2017. Answers may vary depending on the data points used.

Year, x	Per-Capita Consumption of Bottled Water (in gallons)
2009, 0	27.6
2010, 1	28.3
2011, 2	29.2
2012, 3	30.8
2013, 4	32.0

Source: Beverage Marketing Corporation

66. *Accessing the Internet by Smartphone.* Data on the percentage of adults who access the Internet by smartphone are given in the following table. Model the data with a linear function and predict the percentage of adults in 2016 who will access the Internet using a smartphone.

Year, x	Percentage of Adults Who Access the Internet with a Smartphone
2009, 0	31%
2010, 1	43
2011, 2	47
2012, 3	55
2013, 4	63

Source: Pew Internet and American Life Project

Skill Maintenance

Find the slope of the line containing the given points.

67. $(5, 7)$ and $(5, -7)$ [1.3]

68. $(2, -8)$ and $(-5, -1)$ [1.3]

Find an equation for a circle satisfying the given conditions.

69. Center $(0, 3)$, diameter of length 5 [1.1]

70. Center $(-7, -1)$, radius of length $\frac{9}{5}$ [1.1]

Synthesis

71. Find k so that the line containing the points $(-3, k)$ and $(4, 8)$ is parallel to the line containing the points $(5, 3)$ and $(1, -6)$.

72. Find an equation of the line passing through the point $(4, 5)$ and perpendicular to the line passing through the points $(-1, 3)$ and $(2, 9)$.

73. *Road Grade.* Using the following figure, find the road grade and an equation giving the height y as a function of the horizontal distance x.

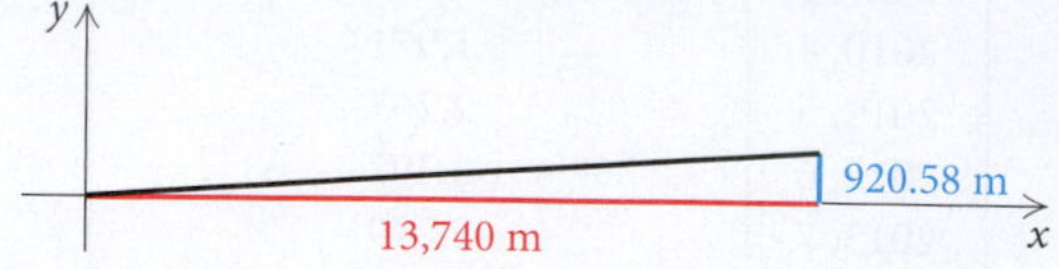

1.5 Linear Equations, Functions, Zeros, and Applications

- Solve linear equations.
- Solve applied problems using linear models.
- Find zeros of linear functions.

An **equation** is a statement that two expressions are equal. To **solve** an equation in one variable is to find all the values of the variable that make the equation true. Each of these values is a **solution** of the equation. The set of all solutions of an equation is its **solution set**. Some examples of **equations in one variable** are

$$2x + 3 = 5, \quad 3(x - 1) = 4x + 5,$$

$$x^2 - 3x + 2 = 0, \quad \text{and} \quad \frac{x - 3}{x + 4} = 1.$$

Linear Equations

The first two equations above are *linear equations* in one variable. We define such equations as follows.

> A **linear equation in one variable** is an equation that can be expressed in the form $mx + b = 0$, where m and b are real numbers and $m \neq 0$.

Equations that have the same solution set are **equivalent equations**. For example, $2x + 3 = 5$ and $x = 1$ are equivalent equations because 1 is the solution of each equation. On the other hand, $x^2 - 3x + 2 = 0$ and $x = 1$ are not equivalent equations because 1 and 2 are both solutions of $x^2 - 3x + 2 = 0$ but 2 is not a solution of $x = 1$.

To solve a linear equation, we find an equivalent equation in which the variable is isolated. The following principles allow us to solve linear equations.

JUST IN TIME 14

EQUATION-SOLVING PRINCIPLES

For any real numbers a, b, and c:

The Addition Principle: If $a = b$ is true, then $a + c = b + c$ is true.

The Multiplication Principle: If $a = b$ is true, then $ac = bc$ is true.

Note to the student and the instructor: We assume that students come to a College Algebra course with some equation-solving skills from their study of Intermediate Algebra. Thus a portion of the material in this section might be considered by some to be review in nature. We present this material here in order to use linear functions, with which students are familiar, to lay the groundwork for zeros of higher-order polynomial functions and their connection to solutions of equations and x-intercepts of graphs.

EXAMPLE 1 Solve: $\frac{3}{4}x - 1 = \frac{7}{5}$.

Solution When we have an equation that contains fractions, it is often convenient to multiply both sides of the equation by the least common denominator (LCD) of the fractions in order to clear the equation of fractions. We have

$$\frac{3}{4}x - 1 = \frac{7}{5}$$ The LCD is $4 \cdot 5$, or 20.

$$20\left(\frac{3}{4}x - 1\right) = 20 \cdot \frac{7}{5}$$ Multiplying by the LCD on both sides to clear fractions

$$20 \cdot \frac{3}{4}x - 20 \cdot 1 = 28$$

$$15x - 20 = 28$$

$$15x - 20 + 20 = 28 + 20$$ Using the addition principle to add 20 on both sides

$$15x = 48$$

$$\frac{15x}{15} = \frac{48}{15}$$ Using the multiplication principle to multiply by $\frac{1}{15}$, or divide by 15, on both sides

$$x = \frac{48}{15}$$

$$x = \frac{16}{5}.$$ Simplifying. Note that $\frac{3}{4}x - 1 = \frac{7}{5}$ and $x = \frac{16}{5}$ are equivalent equations.

Check:

$$\begin{array}{r|l} \frac{3}{4}x - 1 = & \frac{7}{5} \\ \hline \frac{3}{4}\cdot\frac{16}{5} - 1 \ ? & \frac{7}{5} \\ \frac{12}{5} - \frac{5}{5} & \\ \frac{7}{5} & \frac{7}{5} \quad \text{TRUE} \end{array}$$

Substituting $\frac{16}{5}$ for x

The solution is $\frac{16}{5}$.

Now Try Exercise 15.

Technology Connection

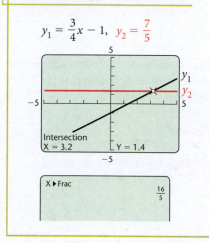

$y_1 = \frac{3}{4}x - 1, \; y_2 = \frac{7}{5}$

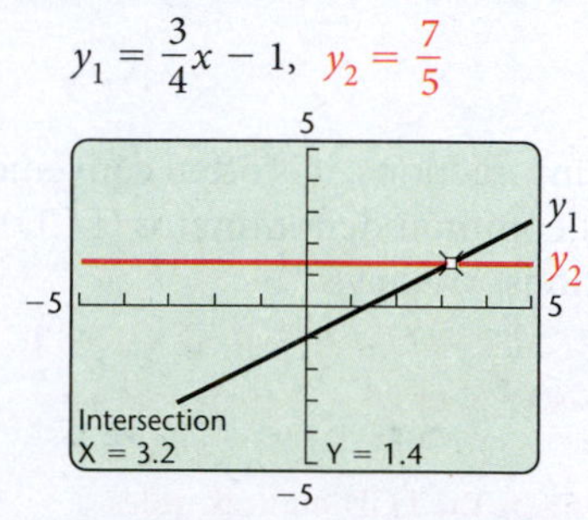

X▶Frac
16/5

We can use the INTERSECT feature on a graphing calculator to solve equations. We call this the **Intersect method**. To use the Intersect method to solve the equation in Example 1, for instance, we graph $y_1 = \frac{3}{4}x - 1$ and $y_2 = \frac{7}{5}$. The value of x for which $y_1 = y_2$ is the solution of the equation $\frac{3}{4}x - 1 = \frac{7}{5}$. This value of x is the first coordinate of the point of intersection of the graphs of y_1 and y_2. Using the INTERSECT feature, we find that the first coordinate of this point is 3.2. We can find fraction notation for the solution by using the ▶ FRAC feature. The solution is 3.2, or $\frac{16}{5}$.

EXAMPLE 2 Solve: $2(5 - 3x) = 8 - 3(x + 2)$.

Solution We have

$$\begin{aligned} 2(5 - 3x) &= 8 - 3(x + 2) \\ 10 - 6x &= 8 - 3x - 6 && \text{Using the distributive property} \\ 10 - 6x &= 2 - 3x && \text{Collecting like terms} \\ 10 - 6x + 6x &= 2 - 3x + 6x && \text{Using the addition principle to add } 6x \text{ on both sides} \\ 10 &= 2 + 3x \\ 10 - 2 &= 2 + 3x - 2 && \text{Using the addition principle to add } -2\text{, or subtract 2, on both sides} \\ 8 &= 3x \\ \frac{8}{3} &= \frac{3x}{3} && \text{Using the multiplication principle to multiply by } \tfrac{1}{3}\text{, or divide by 3, on both sides} \\ \frac{8}{3} &= x. \end{aligned}$$

Technology Connection

We can use the TABLE feature on a graphing calculator, set in ASK mode, to check the solutions of equations. In Example 2, for instance, let $y_1 = 2(5 - 3x)$ and $y_2 = 8 - 3(x + 2)$. When $\frac{8}{3}$ is entered for x, we see that $y_1 = y_2$, or $2(5 - 3x) = 8 - 3(x + 2)$. Thus, $\frac{8}{3}$ is the solution of the equation. (Note that the calculator converts $\frac{8}{3}$ to decimal notation in the table.)

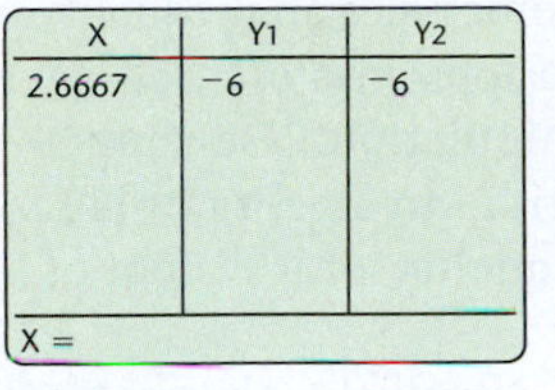

Check:

$2(5 - 3x)$	$= 8 - 3(x + 2)$	
$2\left(5 - 3 \cdot \frac{8}{3}\right)$	? $8 - 3\left(\frac{8}{3} + 2\right)$	Substituting $\frac{8}{3}$ for x
$2(5 - 8)$	$8 - 3\left(\frac{14}{3}\right)$	
$2(-3)$	$8 - 14$	
-6	-6	TRUE

The solution is $\frac{8}{3}$.

Now Try Exercise 27.

Special Cases

Some equations have *no* solution.

EXAMPLE 3 Solve: $-24x + 7 = 17 - 24x$.

Solution We have

$$-24x + 7 = 17 - 24x$$
$$24x - 24x + 7 = 24x + 17 - 24x \qquad \text{Adding } 24x$$
$$7 = 17. \qquad \text{We get a false equation.}$$

No matter what number we substitute for x, we get a false sentence. Thus the equation has *no* solution.

Now Try Exercise 11.

There are some equations for which *any* real number is a solution.

EXAMPLE 4 Solve: $3 - \frac{1}{3}x = -\frac{1}{3}x + 3$.

Solution We have

$$3 - \frac{1}{3}x = -\frac{1}{3}x + 3$$
$$\frac{1}{3}x + 3 - \frac{1}{3}x = \frac{1}{3}x - \frac{1}{3}x + 3 \qquad \text{Adding } \tfrac{1}{3}x$$
$$3 = 3. \qquad \text{We get a true equation.}$$

Replacing x with any real number gives a true sentence. Thus *any* real number is a solution. This equation has *infinitely* many solutions. The solution set is the set of real numbers, $\{x \mid x \text{ is a real number}\}$, or $(-\infty, \infty)$.

Now Try Exercise 3.

Applications Using Linear Models

Mathematical techniques are used to answer questions arising from real-world situations. Linear equations and functions *model* many of these situations.

The following strategy is of great assistance in problem solving.

FIVE STEPS FOR PROBLEM SOLVING

1. **Familiarize** yourself with the problem situation. If the problem is presented in words, this means to read carefully. Some or all of the following can also be helpful.
   a) Make a drawing, if it makes sense to do so.
   b) Make a written list of the known facts and a list of what you wish to find out.
   c) Assign variables to represent unknown quantities.
   d) Organize the information in a chart or a table.
   e) Find further information. Look up a formula, consult a reference book or an expert in the field, or do research on the Internet.
   f) Guess or estimate the answer and check your guess or estimate.
2. **Translate** the problem situation to mathematical language or symbolism. For most of the problems you will encounter in algebra, this means to write one or more equations, but sometimes an inequality or some other mathematical symbolism may be appropriate.
3. **Carry out** some type of mathematical manipulation. Use your mathematical skills to find a possible solution. In algebra, this usually means to solve an equation, an inequality, or a system of equations or inequalities.
4. **Check** to see whether your possible solution actually fits the problem situation and is thus really a solution of the problem. Although you may have solved an equation, the solution(s) of the equation might not be solution(s) of the original problem.
5. **State** the answer clearly using a complete sentence.

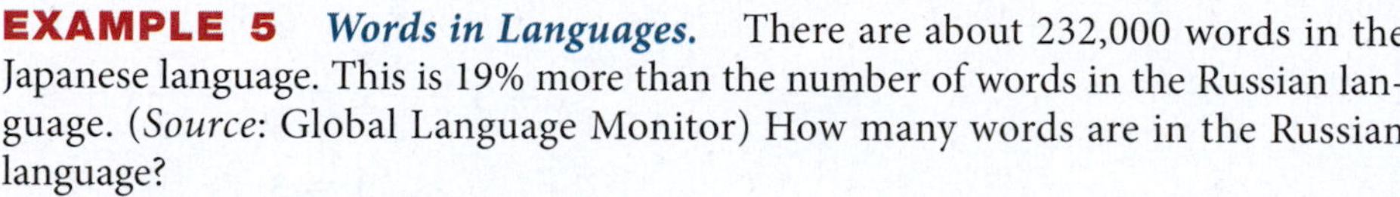

EXAMPLE 5 ***Words in Languages.*** There are about 232,000 words in the Japanese language. This is 19% more than the number of words in the Russian language. (*Source*: Global Language Monitor) How many words are in the Russian language?

Solution

1. **Familiarize.** Let's estimate that there are 200,000 words in the Russian language. Then the number of words in the Japanese language would be

$$200{,}000 + 19\% \cdot 200{,}000 = 1(200{,}000) + 0.19(200{,}000)$$
$$= 1.19(200{,}000) = 238{,}000.$$

Since we know that there are actually 232,000 words in the Japanese language, our estimate of 200,000 is too high. Nevertheless, the calculations performed indicate how we can translate the problem to an equation. We let $x =$ the number of words in the Russian language. Then $x + 19\%x$, or $1 \cdot x + 0.19x$, or $1.19x$, is the number of words in the Japanese language.

2. **Translate.** We translate to an equation:

$$\underbrace{\text{Number of words in Japanese language}} \quad \text{is} \quad 232{,}000.$$

$$1.19x = 232{,}000.$$

3. **Carry out.** We solve the equation, as follows:

$$1.19x = 232{,}000$$
$$x = \frac{232{,}000}{1.19} \quad \text{Dividing by 1.19 on both sides}$$
$$x \approx 195{,}000.$$

4. **Check.** 19% of 195,000 is 37,050, and 195,000 + 37,050 = 232,050. Since 232,050 ≈ 232,000, the answer checks. (Remember: We rounded the value of x.)

5. **State.** There are about 195,000 words in the Russian language.

Now Try Exercise 33.

EXAMPLE 6 ***Convenience Stores.*** In 2014, there were 151,282 convenience stores in the United States. The total number of convenience stores in Texas and in California was 26,379. There were 4003 more convenience stores in Texas than in California. (*Source*: 2014 NACS/Nielsen Convenience Industry store count) Find the number of convenience stores in Texas and in California.

Solution

1. **Familiarize.** The number of convenience stores in Texas is described in terms of the number in California, so we let x = the number of convenience stores in California. Then $x + 4003$ = the number of convenience stores in Texas.

2. **Translate.** We translate to an equation:

$$\underbrace{\text{Number of convenience stores in California}} \quad \text{plus} \quad \underbrace{\text{Number of convenience stores in Texas}} \quad \text{is} \quad 26{,}379.$$

$$x \quad + \quad x + 4003 \quad = \quad 26{,}379.$$

3. **Carry out.** We solve the equation, as follows:

$$x + x + 4003 = 26{,}379$$
$$2x + 4003 = 26{,}379 \quad \text{Collecting like terms}$$
$$2x = 22{,}376 \quad \text{Subtracting 4003 on both sides}$$
$$x = 11{,}188. \quad \text{Dividing by 2 on both sides}$$

If $x = 11{,}188$, then $x + 4003 = 11{,}188 + 4003 = 15{,}191$.

4. **Check.** If there were 15,191 convenience stores in Texas and 11,188 in California, then the total number of convenience stores in Texas and in California was 15,191 + 11,188, or 26,379. Also, 15,191 is 4003 more than 11,188. The answer checks.

5. **State.** In 2014, there were 15,191 convenience stores in Texas and 11,188 convenience stores in California.

Now Try Exercise 57.

In some applications, we need to use a formula that describes the relationships among variables. When a situation involves distance, rate (also called speed or velocity), and time, for example, we use the following formula.

THE MOTION FORMULA

The distance d traveled by an object moving at rate r in time t is given by

$$d = r \cdot t.$$

EXAMPLE 7 *Airplane Speed.* Delta Airlines' fleet includes B737/800's, each with a cruising speed of 517 mph, and Saab 340B's, each with a cruising speed of 290 mph (*Source*: Delta Airlines). Suppose that a Saab 340B takes off and travels at its cruising speed. One hour later, a B737/800 takes off and follows the same route, traveling at its cruising speed. How long will it take the B737/800 to overtake the Saab 340B?

Solution

1. **Familiarize.** We make a drawing showing both the known information and the unknown information. We let $t =$ the time, in hours, that the B737/800 travels before it overtakes the Saab 340B. Since the Saab 340B takes off 1 hr before the 737, it will travel for $t + 1$ hr before being overtaken. The planes will have traveled the same distance, d, when one overtakes the other.

We can also organize the information in a table, as follows.

	Distance (d)	Rate (r)	Time (t)	
B737/800	d	517	t	→ $d = 517t$
Saab 340B	d	290	$t + 1$	→ $d = 290(t + 1)$

2. **Translate.** Using the formula $d = rt$ in each row of the table, we get two expressions for d:

$$d = 517t \quad \text{and} \quad d = 290(t + 1).$$

Since the distances are the same, we have the following equation:

$$517t = 290(t + 1).$$

3. **Carry out.** We solve the equation, as follows:

$$517t = 290(t + 1)$$

$$517t = 290t + 290 \quad \text{Using the distributive property}$$

$$227t = 290 \quad \text{Subtracting } 290t \text{ on both sides}$$

$$t \approx 1.28. \quad \text{Dividing by 227 on both sides and rounding to the nearest hundredth}$$

4. **Check.** If the B737/800 travels for about 1.28 hr, then the Saab 340B travels for about 1.28 + 1, or 2.28 hr. In 2.28 hr, the Saab 340B travels 290(2.28), or 661.2 mi; and in 1.28 hr, the B737/800 travels 517(1.28), or 661.76 mi. Since 661.76 mi $\approx$ 661.2 mi, the answer checks. (Remember: We rounded the value of t.)

5. **State.** About 1.28 hr after the B737/800 has taken off, it will overtake the Saab 340B.

Now Try Exercise 41.

For some applications, we need to use a formula to find the amount of interest earned by an investment or the amount of interest due on a loan.

THE SIMPLE-INTEREST FORMULA

The **simple interest** I on a principal of P dollars at interest rate r for t years is given by

$$I = Prt.$$

EXAMPLE 8 ***Student Loans.*** Damarion's two student loans total $28,000. One loan is at 5% simple interest and the other is at 3% simple interest. After 1 year, Damarion owes $1040 in interest. What is the amount of each loan?

Solution

1. **Familiarize.** We let x = the amount borrowed at 5% interest. Then the remainder of the $28,000, or 28,000 − x, is borrowed at 3%. We organize the information in a table, keeping in mind the formula $I = Prt$.

	Amount Borrowed	Interest Rate	Time	Amount of Interest
5% Loan	x	5%, or 0.05	1 year	$x(0.05)(1)$, or $0.05x$
3% Loan	$28{,}000 - x$	3%, or 0.03	1 year	$(28{,}000 - x)(0.03)(1)$, or $0.03(28{,}000 - x)$
Total	28,000			1040

2. **Translate.** The total amount of interest on the two loans is \$1040. Thus we can translate to the following equation:

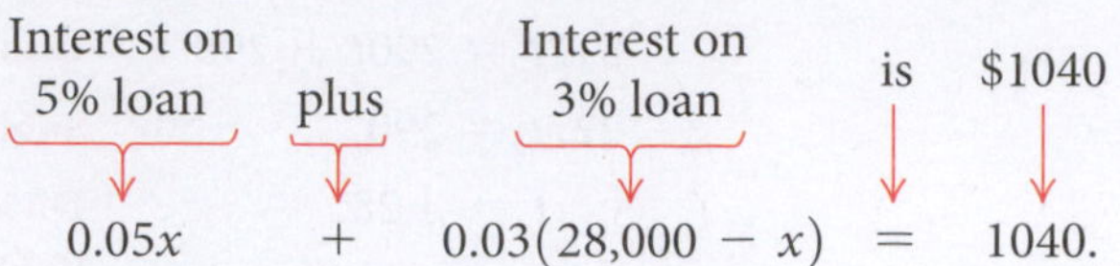

$$0.05x + 0.03(28{,}000 - x) = 1040.$$

3. **Carry out.** We solve the equation, as follows:

$$
\begin{aligned}
0.05x + 0.03(28{,}000 - x) &= 1040 \\
0.05x + 840 - 0.03x &= 1040 && \text{Using the distributive property} \\
0.02x + 840 &= 1040 && \text{Collecting like terms} \\
0.02x &= 200 && \text{Subtracting 840 on both sides} \\
x &= 10{,}000. && \text{Dividing by 0.02 on both sides}
\end{aligned}
$$

If $x = 10{,}000$, then $28{,}000 - x = 28{,}000 - 10{,}000 = 18{,}000$.

4. **Check.** The interest on \$10,000 at 5% for 1 year is \$10,000(0.05)(1), or \$500. The interest on \$18,000 at 3% for 1 year is \$18,000(0.03)(1), or \$540. Since \$500 + \$540 = \$1040, the answer checks.

5. **State.** Damarion borrowed \$10,000 at 5% interest and \$18,000 at 3% interest.

Now Try Exercise 55.

Sometimes we use formulas from geometry in solving applied problems. In the following example, we use the formula for the perimeter P of a rectangle with length l and width w: $P = 2l + 2w$.

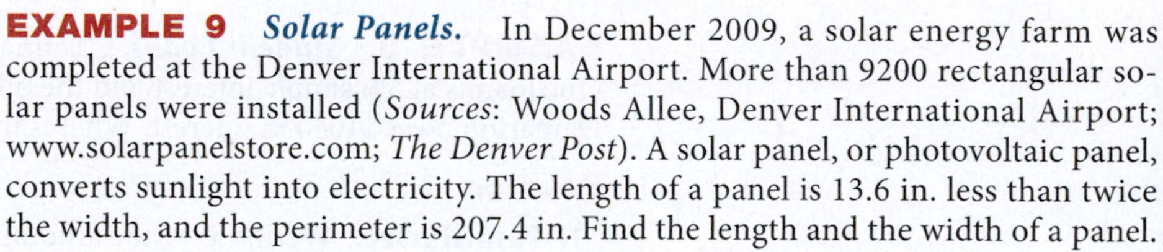

EXAMPLE 9 *Solar Panels.* In December 2009, a solar energy farm was completed at the Denver International Airport. More than 9200 rectangular solar panels were installed (*Sources*: Woods Allee, Denver International Airport; www.solarpanelstore.com; *The Denver Post*). A solar panel, or photovoltaic panel, converts sunlight into electricity. The length of a panel is 13.6 in. less than twice the width, and the perimeter is 207.4 in. Find the length and the width of a panel.

Solution

1. **Familiarize.** We first make a drawing. Since the length of the panel is described in terms of the width, we let w = the width, in inches. Then $2w - 13.6$ = the length, in inches.

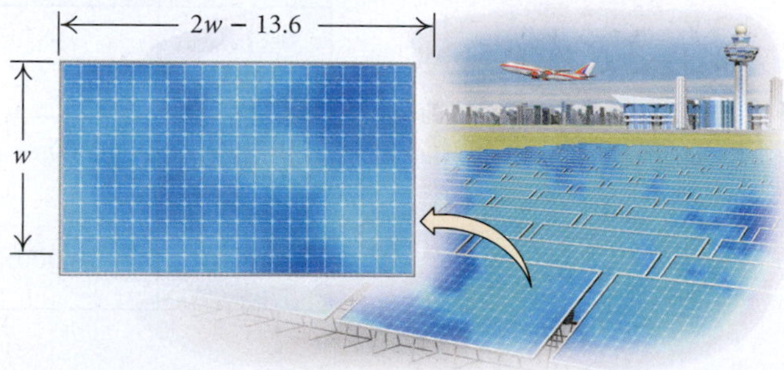

2. **Translate.** We use the formula for the perimeter of a rectangle:

$$
\begin{aligned}
P &= 2l + 2w \\
207.4 &= 2(2w - 13.6) + 2w. && \text{Substituting 207.4 for } P \text{ and } 2w - 13.6 \text{ for } l
\end{aligned}
$$

3. **Carry out.** We solve the equation:

$$207.4 = 2(2w - 13.6) + 2w$$
$$207.4 = 4w - 27.2 + 2w \quad \text{Using the distributive property}$$
$$207.4 = 6w - 27.2 \quad \text{Collecting like terms}$$
$$234.6 = 6w \quad \text{Adding 27.2 on both sides}$$
$$39.1 = w. \quad \text{Dividing by 6 on both sides}$$

If $w = 39.1$, then $2w - 13.6 = 2(39.1) - 13.6 = 78.2 - 13.6 = 64.6$.

4. **Check.** The length, 64.6 in., is 13.6 in. less than twice the width, 39.1 in. Also,

$$2 \cdot 64.6 \text{ in.} + 2 \cdot 39.1 \text{ in.} = 129.2 \text{ in.} + 78.2 \text{ in.} = 207.4 \text{ in.}$$

The answer checks.

5. **State.** The length of the solar panel is 64.6 in., and the width is 39.1 in.

Now Try Exercise 49.

EXAMPLE 10 *Cab Fare.* Metro Taxi charges a \$2.50 pickup fee and \$2 per mile traveled. Grayson's cab fare from the airport to his hotel is \$32.50. How many miles did he travel in the cab?

Solution

1. **Familiarize.** Let's guess that Grayson traveled 12 mi in the cab. Then his fare would be

$$\$2.50 + \$2 \cdot 12 = \$2.50 + \$24 = \$26.50.$$

We see that our guess is low, but the calculation shows us how to translate the problem to an equation. We let $m =$ the number of miles that Grayson traveled in the cab.

2. **Translate.** We translate to an equation:

Pickup fee	plus	Cost per mile	times	Number of miles traveled	is	Total charge
↓	↓	↓	↓	↓	↓	↓
2.50	+	2	·	m	=	32.50.

3. **Carry out.** We solve the equation:

$$2.50 + 2 \cdot m = 32.50$$
$$2m = 30 \quad \text{Subtracting 2.50 on both sides}$$
$$m = 15. \quad \text{Dividing by 2 on both sides}$$

4. **Check.** If Grayson travels 15 mi in the cab, the mileage charge is \$2 · 15, or \$30. Then, with the \$2.50 pickup fee included, his total charge is \$2.50 + \$30, or \$32.50. The answer checks.

5. **State.** Grayson traveled 15 mi in the cab.

Now Try Exercise 65.

Zeros of Linear Functions

An input for which a function's output is 0 is called a **zero** of the function. We will restrict our attention in this section to zeros of linear functions. This allows us to become familiar with the concept of a zero, and it lays the groundwork for working with zeros of other types of functions in succeeding chapters.

> **ZEROS OF FUNCTIONS**
>
> An input c of a function f is called a **zero** of the function if the output for the function is 0 when the input is c. That is, c is a zero of f if $f(c) = 0$.

LINEAR FUNCTIONS

REVIEW SECTION 1.3

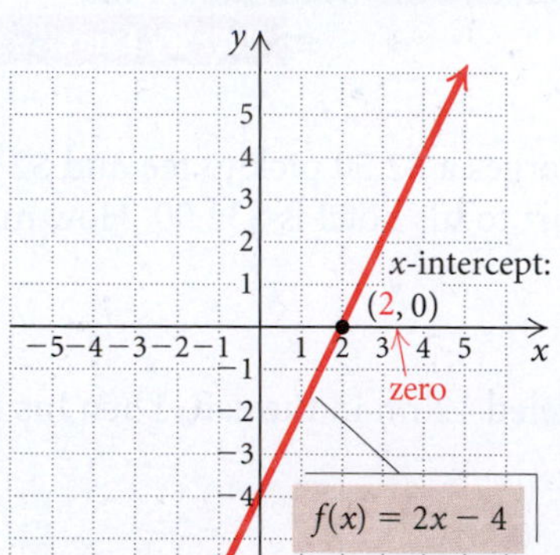

Recall that a linear function is given by $f(x) = mx + b$, where m and b are constants. For the linear function $f(x) = 2x - 4$, we have $f(2) = 2 \cdot 2 - 4 = 0$, so 2 is a **zero** of the function. In fact, 2 is the *only* zero of this function. In general, a **linear function $f(x) = mx + b$, with $m \neq 0$, has exactly one zero.**

Consider the graph of $f(x) = 2x - 4$, shown at left. We see from the graph that the zero, 2, is the first coordinate of the *x-intercept* of the graph. Thus when we find the zero of a linear function, we are also finding the first coordinate of the x-intercept of the graph of the function.

For every linear function $f(x) = mx + b$, there is an associated linear equation $mx + b = 0$. When we find the zero of a function $f(x) = mx + b$, we are also finding the solution of the equation $mx + b = 0$.

EXAMPLE 11 Find the zero of $f(x) = 5x - 9$.

Algebraic Solution

We find the value of x for which $f(x) = 0$:

$$5x - 9 = 0 \quad \text{Setting } f(x) = 0$$
$$5x = 9 \quad \text{Adding 9 on both sides}$$
$$x = \frac{9}{5}, \text{ or } 1.8. \quad \text{Dividing by 5 on both sides}$$

The zero is $\frac{9}{5}$, or 1.8. This means that $f\left(\frac{9}{5}\right) = 0$, or $f(1.8) = 0$. Note that the *zero* of the function $f(x) = 5x - 9$ is the *solution* of the equation $5x - 9 = 0$.

Visualizing the Solution

We graph $f(x) = 5x - 9$.

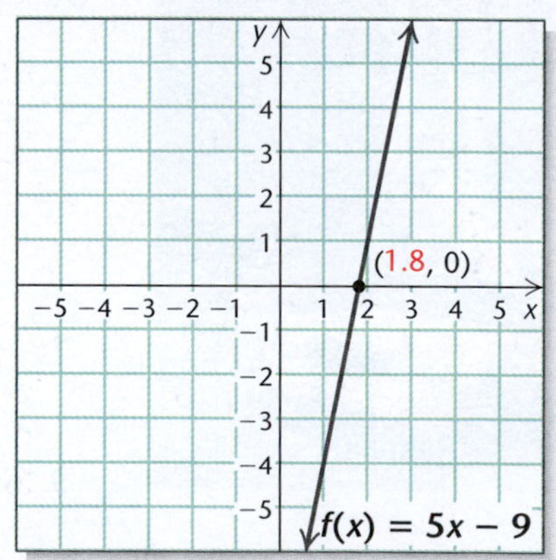

The x-intercept of the graph is $\left(\frac{9}{5}, 0\right)$, or (1.8, 0). Thus, $\frac{9}{5}$, or 1.8, is the zero of the function.

Now Try Exercise 73.

Technology Connection

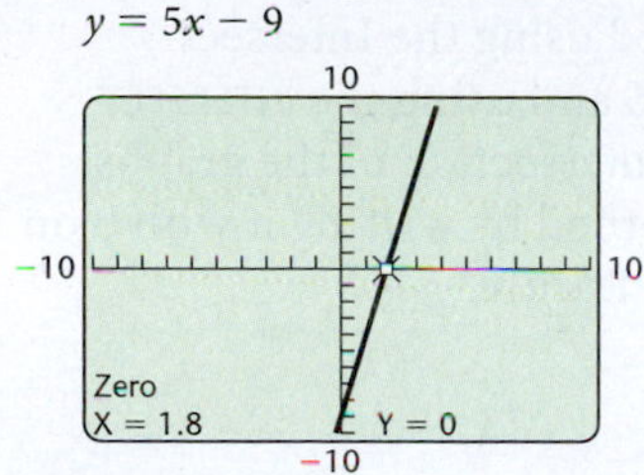

We can use the ZERO feature on a graphing calculator to find the zeros of a function $f(x)$ and to solve the corresponding equation $f(x) = 0$. We call this the **Zero method**. To use the Zero method in Example 11, for instance, we graph $y = 5x - 9$ and use the ZERO feature to find the coordinates of the x-intercept of the graph. Note that the x-intercept must appear in the window when the ZERO feature is used. We see that the zero of the function is 1.8.

CONNECTING THE CONCEPTS

Zeros, Solutions, and Intercepts

The zero of a linear function $f(x) = mx + b$, with $m \neq 0$, is the solution of the linear equation $mx + b = 0$ and is the first coordinate of the x-intercept of the graph of $f(x) = mx + b$. To find the zero of $f(x) = mx + b$, we solve $f(x) = 0$, or $mx + b = 0$.

FUNCTION	ZERO OF THE FUNCTION; SOLUTION OF THE EQUATION	ZERO OF THE FUNCTION; x-INTERCEPT OF THE GRAPH
Linear Function $f(x) = 2x - 4$, or $y = 2x - 4$	To find the **zero** of $f(x)$, we solve $f(x) = 0$: $2x - 4 = 0$ $2x = 4$ $x = 2$. The **solution** of $2x - 4 = 0$ is 2. This is the zero of the function $f(x) = 2x - 4$. That is, $f(2) = 0$.	The zero of $f(x)$ is the first coordinate of the ***x*-intercept** of the graph of $y = f(x)$. 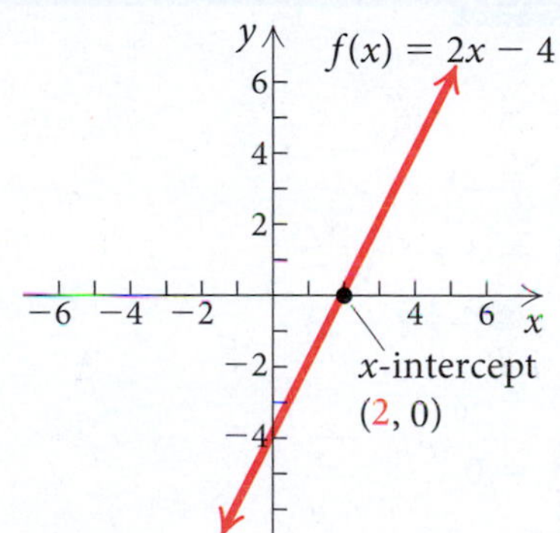

Technology Connection

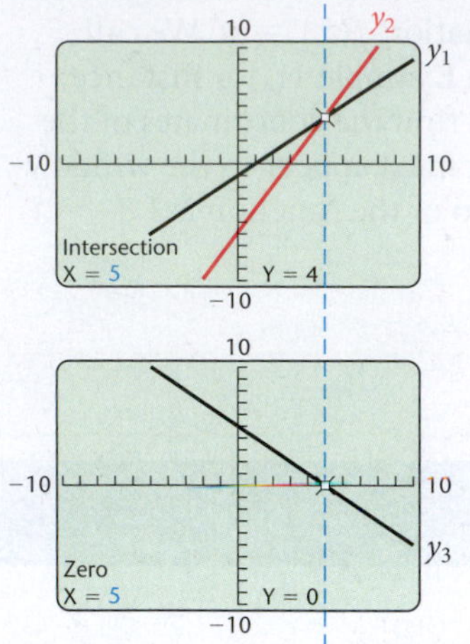

An equation such as $x - 1 = 2x - 6$ can be solved using the Intersect method by graphing $y_1 = x - 1$ and $y_2 = 2x - 6$ and using the INTERSECT feature to find the first coordinate of the point of intersection of the graphs. The equation can also be solved using the Zero method by writing it with 0 on one side of the equals sign and then using the ZERO feature.

Solve: $x - 1 = 2x - 6$.

The Intersect Method

Graph $y_1 = x - 1$ and $y_2 = 2x - 6$.
Point of intersection: $(5, 4)$
Solution: 5

The Zero Method

First, add $-2x$ and 6 on both sides of the equation to get 0 on one side:

$$x - 1 = 2x - 6$$

$$x - 1 - 2x + 6 = 0.$$

Graph $y_3 = x - 1 - 2x + 6$.
Zero: 5
Solution: 5

1.5 Exercise Set

Solve.

1. $4x + 5 = 21$
2. $2y - 1 = 3$
3. $23 - \frac{2}{5}x = -\frac{2}{5}x + 23$
4. $\frac{6}{5}y + 3 = \frac{3}{10}$
5. $4x + 3 = 0$
6. $3x - 16 = 0$
7. $3 - x = 12$
8. $4 - x = -5$
9. $3 - \frac{1}{4}x = \frac{3}{2}$
10. $10x - 3 = 8 + 10x$
11. $\frac{2}{11} - 4x = -4x + \frac{9}{11}$
12. $8 - \frac{2}{9}x = \frac{5}{6}$
13. $8 = 5x - 3$
14. $9 = 4x - 8$
15. $\frac{2}{5}y - 2 = \frac{1}{3}$
16. $-x + 1 = 1 - x$
17. $y + 1 = 2y - 7$
18. $5 - 4x = x - 13$
19. $2x + 7 = x + 3$
20. $5x - 4 = 2x + 5$
21. $3x - 5 = 2x + 1$
22. $4x + 3 = 2x - 7$
23. $4x - 5 = 7x - 2$
24. $5x + 1 = 9x - 7$
25. $5x - 2 + 3x = 2x + 6 - 4x$
26. $5x - 17 - 2x = 6x - 1 - x$
27. $7(3x + 6) = 11 - (x + 2)$
28. $4(5y + 3) = 3(2y - 5)$
29. $3(x + 1) = 5 - 2(3x + 4)$
30. $4(3x + 2) - 7 = 3(x - 2)$
31. $2(x - 4) = 3 - 5(2x + 1)$
32. $3(2x - 5) + 4 = 2(4x + 3)$

33. *New Words in the English Language.* During the nineteenth century, 75,029 new words entered the English language. This is about 46.9% more than the number of new words in the seventeenth century. (*Source*: Philip Durkin and Katherine Martin, Oxford University Press; "English by the Book," *National Geographic*, December 2013) Find the number of new words that appeared in the English language in the seventeenth century.

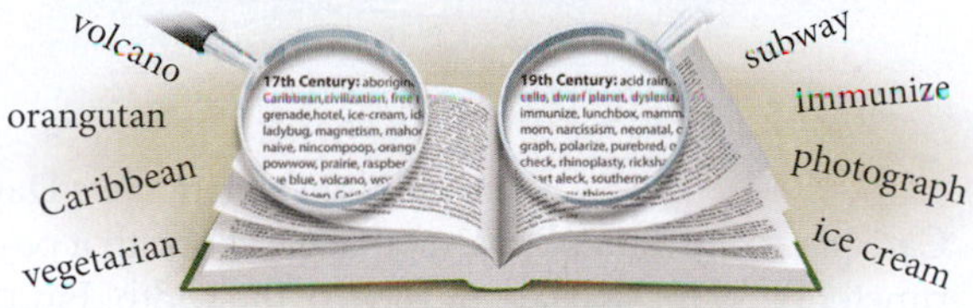

34. *Calorie Intake.* The average worldwide daily calorie intake per person has increased from 2200 to 2800 calories since the early 1960s. The average daily calorie intake per person in the United States is 3688. This is about 86.4% more than the average daily calorie intake per person in Haiti. (*Sources:* UN Food and Agriculture Organization; World Health Organization) Find the average daily calorie intake per person in Haiti.

35. *Amount Borrowed.* Kea borrowed money from her father at 5% simple interest to help pay her tuition at Wellington Community College. At the end of 1 year, she owed a total of $1365 in principal and interest. How much did she borrow?

36. *Amount of an Investment.* Khalid makes an investment at 4% simple interest. At the end of 1 year, the total value of the investment is $1560. How much was originally invested?

37. *Angle Measure.* In triangle *ABC*, angle *B* is five times as large as angle *A*. The measure of angle *C* is 2° less than that of angle *A*. Find the measures of the angles. (*Hint*: The sum of the angle measures is 180°.)

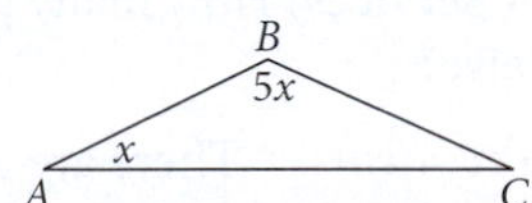

38. *Angle Measure.* In triangle *ABC*, angle *B* is twice as large as angle *A*. Angle *C* measures 20° more than angle *A*. Find the measures of the angles.

39. *Clothing Trade Deficit.* Imports of clothing to the United States totaled $84.916 billion in 2012. This amount was $1.459 billion less than twenty-five times the clothing exports that year. (*Source*: Bureau of Economic Analysis, U.S. Department of Commerce) Find the amount of clothing exports from the United States in 2012.

40. *Foreign Trade.* In 2012, the total value of exports from the United States was $2,210,585,000,000. That year, exports were $837,965,000,000 more than half of the U.S. imports. (*Source*: U.S. Bureau of Economic Analysis, U.S. Department of Commerce) Find the value of imports to the United States in 2012.

41. *Train Speeds.* A Central Railway freight train leaves a station and travels due north at a speed of 60 mph. One hour later, an Amtrak passenger train leaves the same station and travels due north on a parallel track at a speed of 80 mph. How long will it take the passenger train to overtake the freight train?

42. *Distance Traveled.* A private airplane leaves Midway Airport and flies due east at a speed of 180 km/h. Two hours later, a jet leaves Midway and flies due east at a speed of 900 km/h. How far from the airport will the jet overtake the private plane?

43. *Income Taxes.* In 2010, 40.9% of federal tax returns had zero or negative tax liability. This amount is 15.7% more than the percentage of filers who had zero or negative tax liability in 2000. (*Source*: The Tax Foundation) Find the percentage of tax filers in 2000 who had zero or negative tax liability.

44. *Salary Comparison.* The average annual salary of a restaurant manager is 24.8% less than the average annual salary of an office manager. The average annual salary of a restaurant manager is $48,533. (*Source*: www.salary.com) Find the average annual salary of an office manager.

45. *Commission vs. Salary.* Juliet has a choice between receiving a monthly salary of $1800 from Furniture by Design or a base salary of $1600 and a 4% commission on the amount of furniture she sells during the month. For what amount of sales will the two choices be equal?

46. *Sales Commission.* Edward, a consumer electronics salesperson, earns a base salary of $1270 per month and a commission of 6% on the amount of sales he makes. One month Edward received a paycheck for $3154. Find the amount of his sales for the month.

47. *Studying Abroad.* In the 2012–2013 school year, approximately 820,000 foreign students studied in the United States. The number of U.S. students who studied abroad that same year was about seven-twentieths of the number of foreign students who studied in the United States. (*Source*: Pew Research Center) Find the number of U.S. students who studied abroad during the 2012–2013 school year.

48. *Population Density.* The population density in China is 365.3 persons per square mile. The population density in the United States is approximately one-fourth of the density in China. (*Source*: *The World Almanac* 2014) Find the population density in the United States.

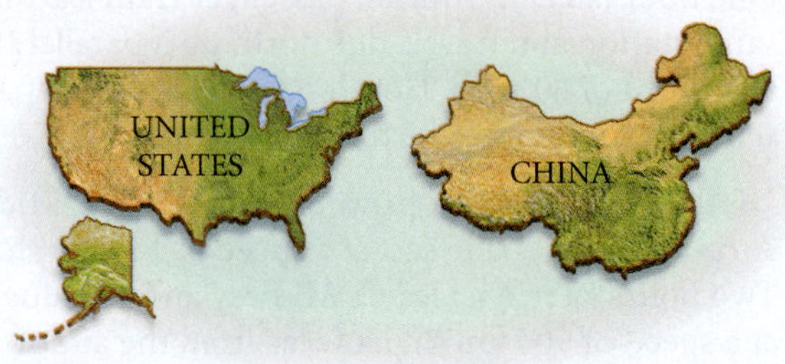

49. *Soccer-Field Dimensions.* The width of the soccer field recommended for players under the age of 12 is 35 yd less than the length. The perimeter of the field is 330 yd. (*Source*: U.S. Youth Soccer) Find the dimensions of the field.

50. *Poster Dimensions.* Marissa is designing a poster to promote the Talbot Street Art Fair. The width of the poster will be two-thirds of its height, and its perimeter will be 100 in. Find the dimensions of the poster.

51. *Test-Plot Dimensions.* Morgan's Seeds has a rectangular test plot with a perimeter of 322 m. The length is 25 m more than the width. Find the dimensions of the plot.

52. *Garden Dimensions.* The children at Tiny Tots Day Care plant a rectangular vegetable garden with a perimeter of 39 m. The length is twice the width. Find the dimensions of the garden.

53. *Flying into a Headwind.* An airplane that travels 450 mph in still air encounters a 30-mph headwind. How long will it take the plane to travel 1050 mi into the wind?

54. *Flying with a Tailwind.* An airplane that can travel 375 mph in still air is flying with a 25-mph tailwind. How long will it take the plane to travel 700 mi with the wind?

55. *Investment Income.* Katie invested a total of $5000, part at 3% simple interest and part at 4% simple interest. At the end of 1 year, the investments had earned $176 interest. How much was invested at each rate?

56. *Student Loans.* Anton's two student loans total $9000. One loan is at 5% simple interest and the other is at 6% simple interest. At the end of 1 year, Anton owes $492 in interest. What is the amount of each loan?

57. *Patents.* In 2013, IBM (International Business Machines) received 2133 more patents than Samsung. Together, they received 11,485 patents. (*Source:* IFI Claims Patent Services) How many patents did each company receive?

58. *Books about Presidents.* There are 5493 print and e-books written about both George Washington and Abraham Lincoln. There are 1675 more books about Lincoln than about Washington. (*Source*: Bowker Books in Print) How many books have been written about each president?

59. *Ocean Depth.* The average depth of the Pacific Ocean is 14,040 ft, and its depth is 8890 ft less than the sum of the average depths of the Atlantic and Indian Oceans. The average depth of the Indian Ocean is 272 ft less than four-fifths of the average depth of the Atlantic Ocean. (*Source*: *Time Almanac* 2010) Find the average depth of the Indian Ocean.

60. *Calcium Content of Foods.* Together, one 8-oz serving of plain nonfat yogurt and one 1-oz serving of Swiss cheese contain 676 mg of calcium. The yogurt contains 4 mg more than twice the calcium in the cheese. (*Source*: U.S. Department of Agriculture) Find the calcium content of each food.

61. *Water Weight.* Water accounts for 55% of a woman's weight (*Source*: ga.water.usgs.gov/edu). Lily weighs 135 lb. How much of her body weight is water?

62. *Water Weight.* Water accounts for 60% of a man's weight (*Source*: ga.water.usgs.gov/edu). Jake weighs 186 lb. How much of his body weight is water?

63. *Traveling Upstream.* A kayak moves at a rate of 12 mph in still water. If the river's current flows at a rate of 4 mph, how long does it take the boat to travel 36 mi upstream?

64. *Traveling Downstream.* Angelo's kayak travels 14 km/h in still water. If the river's current flows at a rate of 2 km/h, how long will it take him to travel 20 km downstream?

65. *Hourly Wage.* Rosalyn worked 48 hr one week and earned a $1066 paycheck. She earns time and a half (1.5 times her regular hourly wage) for the number of hours she works in excess of 40. What is Rosalyn's regular hourly wage?

66. *Cab Fare.* City Cabs charges a $1.75 pickup fee and $1.50 per mile traveled. Diego's fare for a cross-town cab ride is $19.75. How far did he travel in the cab?

67. *Olive Oil.* Together, Italy, Spain, and the United States consume 58% of the world's olive oil. The percentage consumed in Italy is $3\frac{3}{4}$ times the percentage consumed in the United States. The percentage consumed in Spain is $\frac{2}{3}$ of the percentage consumed in Italy. (*Source*: www.OliveOilEmporium.com) Find the percent of the world's olive oil consumed in each country.

68. *NFL Stadium Elevation.* The elevations of the 31 NFL stadiums range from 3 ft at Mercedes-Benz Superdome, New Orleans, Louisiana, to 5280 ft at Sports Authority Field at Mile High, Denver, Colorado. The elevation of Sports Authority Field at Mile High is 275 ft higher than seven times the elevation of Lucas Oil Stadium in Indianapolis, Indiana. What is the elevation of Lucas Oil Stadium?

Find the zero of the linear function.

69. $f(x) = x + 5$

70. $f(x) = 5x + 20$

71. $f(x) = -2x + 11$

72. $f(x) = 8 + x$

73. $f(x) = 16 - x$

74. $f(x) = -2x + 7$

75. $f(x) = x + 12$

76. $f(x) = 8x + 2$

77. $f(x) = -x + 6$

78. $f(x) = 4 + x$

79. $f(x) = 20 - x$

80. $f(x) = -3x + 13$

81. $f(x) = \frac{2}{5}x - 10$

82. $f(x) = 3x - 9$

83. $f(x) = -x + 15$

84. $f(x) = 4 - x$

In each of Exercises 85–90, use the given graph to find each of the following: **(a)** *the x-intercept and* **(b)** *the zero of the function.*

85.

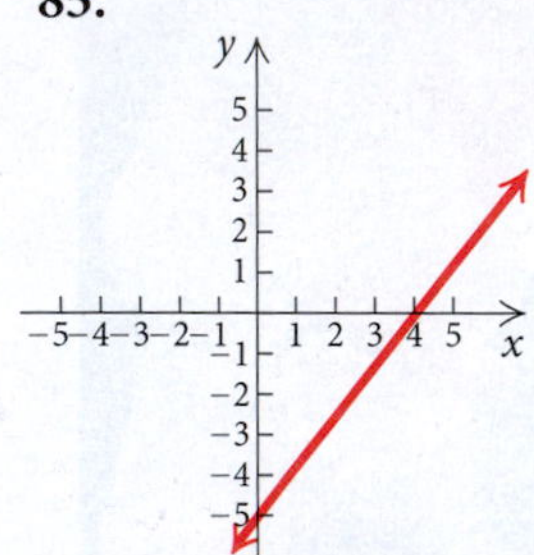

86.

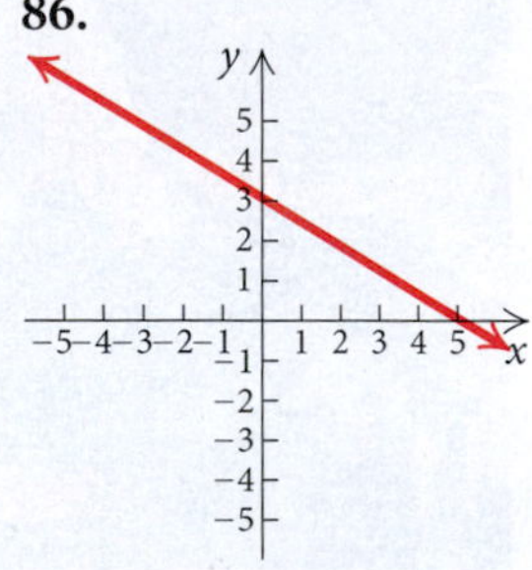

87.

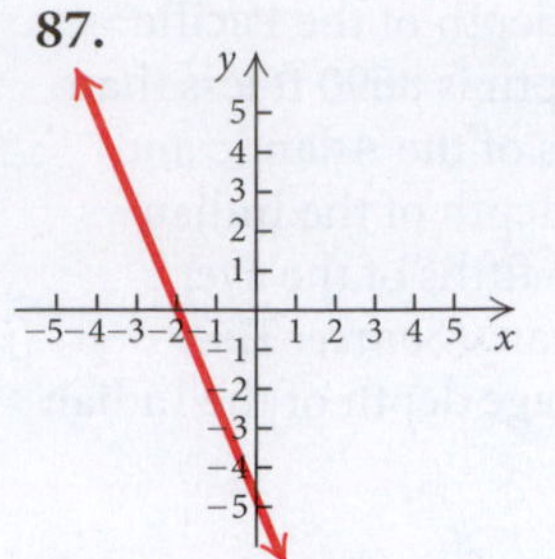

88.

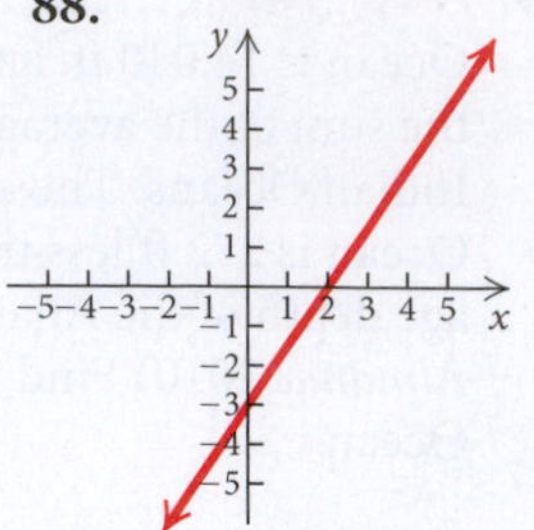

89.

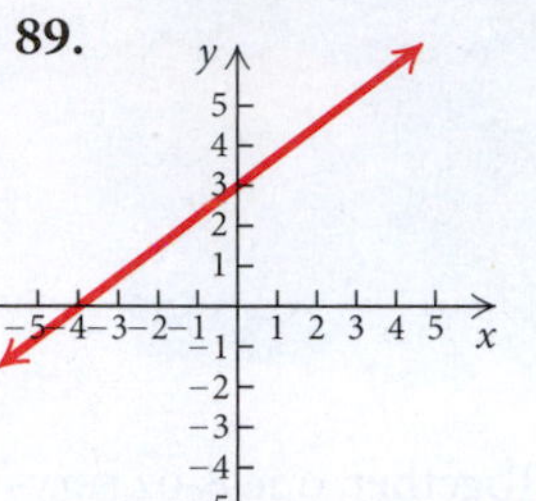

90.

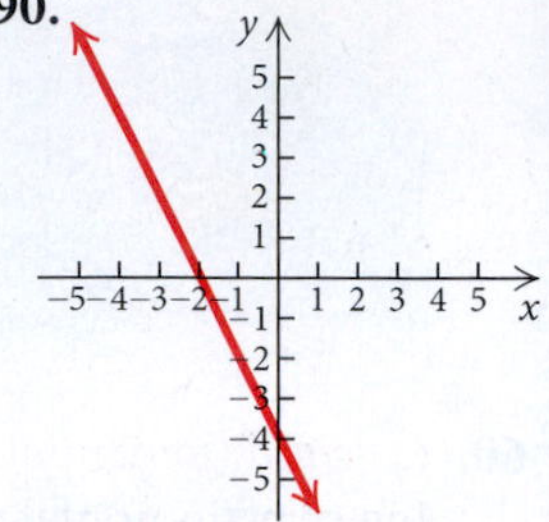

▶ Skill Maintenance

91. Write a slope–intercept equation for the line containing the point $(-1, 4)$ and parallel to the line $3x + 4y = 7$. [1.4]

92. Write an equation of the line containing the points $(-5, 4)$ and $(3, -2)$. [1.4]

93. Find the distance between $(2, 2)$ and $(-3, -10)$. [1.1]

94. Find the midpoint of the segment with endpoints $\left(-\frac{1}{2}, \frac{2}{5}\right)$ and $\left(-\frac{3}{2}, \frac{3}{5}\right)$. [1.1]

95. Given that $f(x) = \dfrac{x}{x - 3}$, find $f(-3)$, $f(0)$, and $f(3)$. [1.2]

96. Find the slope and the y-intercept of the line with the equation $7x - y = \frac{1}{2}$. [1.3]

▶ Synthesis

State whether each of the following is a linear function.

97. $f(x) = 7 - \dfrac{3}{2}x$

98. $f(x) = \dfrac{3}{2x} + 5$

99. $f(x) = x^2 + 1$

100. $f(x) = \frac{3}{4}x - (2.4)^2$

Solve.

101. $2x - \{x - [3x - (6x + 5)]\} = 4x - 1$

102. $14 - 2[3 + 5(x - 1)] = 3\{x - 4[1 + 6(2 - x)]\}$

103. *Packaging and Price.* Dannon recently replaced its 8-oz cup of yogurt with a 6-oz cup and reduced the suggested retail price from 89 cents to 71 cents (*Source*: IRI). Was the price per ounce reduced by the same percent as the size of the cup? If not, find the price difference per ounce in terms of a percent.

104. *Bestsellers.* One week 10 copies of the novel *The Last Song* by Nicholas Sparks were sold for every 7.9 copies of David Baldacci's *Deliver Us from Evil* that were sold (*Source*: *USA Today* Best-Selling Books). If a total of 10,919 copies of the two books were sold, how many copies of each were sold?

105. *Running vs. Walking.* A 150-lb person who runs at 6 mph for 1 hr burns about 720 calories. The same person, walking at 4 mph for 90 min, burns about 480 calories. (*Source*: FitSmart, *USA Weekend*, July 19–21, 2002) Suppose a 150-lb person runs at 6 mph for 75 min. How far must the person walk at 4 mph in order to burn the same number of calories burned running?

1.6 Solving Linear Inequalities

- Solve linear inequalities.
- Solve compound inequalities.
- Solve applied problems using inequalities.

An **inequality** is a sentence with $<$, $>$, $\leq$, or $\geq$ as its verb. An example is $3x - 5 < 6 - 2x$. To **solve** an inequality is to find all values of the variable that make the inequality true. Each of these values is a **solution** of the inequality, and the set of all such solutions is its **solution set**. Inequalities that have the same solution set are called **equivalent inequalities**.

JUST IN TIME 15

Linear Inequalities

The principles for solving inequalities are similar to those for solving equations.

> **PRINCIPLES FOR SOLVING INEQUALITIES**
>
> For any real numbers a, b, and c:
>
> ***The Addition Principle for Inequalities:***
> If $a < b$ is true, then $a + c < b + c$ is true.
>
> ***The Multiplication Principle for Inequalities:***
> **a)** If $a < b$ and $c > 0$ are true, then $ac < bc$ is true.
> **b)** If $a < b$ and $c < 0$ are true, then $ac > bc$ is true.
> (When both sides of an inequality are multiplied by a negative number, the inequality sign must be reversed.)
>
> Similar statements hold for $a \le b$.

First-degree inequalities with one variable, like those in Example 1 below, are **linear inequalities**.

JUST IN TIME 5, 6

EXAMPLE 1 Solve each of the following. Then graph the solution set.

a) $3x - 5 < 6 - 2x$ **b)** $13 - 7x \ge 10x - 4$

Solution

a) $3x - 5 < 6 - 2x$

$5x - 5 < 6$ Using the addition principle for inequalities; adding $2x$

$5x < 11$ Using the addition principle for inequalities; adding 5

$x < \frac{11}{5}$ Using the multiplication principle for inequalities; multiplying by $\frac{1}{5}$, or dividing by 5

Any number less than $\frac{11}{5}$ is a solution. The solution set is $\{x | x < \frac{11}{5}\}$, or $(-\infty, \frac{11}{5})$. The graph of the solution set is shown below.

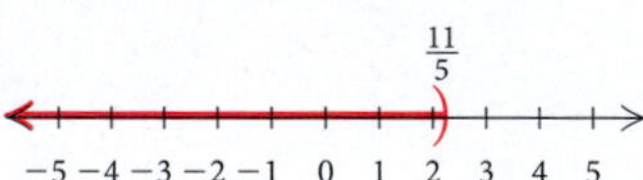

Technology Connection

To check Example 1(a), we can graph $y_1 = 3x - 5$ and $y_2 = 6 - 2x$. The graph shows that for $x < 2.2$, or $x < \frac{11}{5}$, the graph of y_1 lies below the graph of y_2, or $y_1 < y_2$.

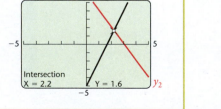

$y_1 = 3x - 5$, $y_2 = 6 - 2x$

Intersection X = 2.2 Y = 1.6

b) $13 - 7x \ge 10x - 4$

$13 - 17x \ge -4$ Subtracting $10x$

$-17x \ge -17$ Subtracting 13

$x \le 1$ Dividing by -17 and reversing the inequality sign

The solution set is $\{x | x \le 1\}$, or $(-\infty, 1]$. The graph of the solution set is shown below.

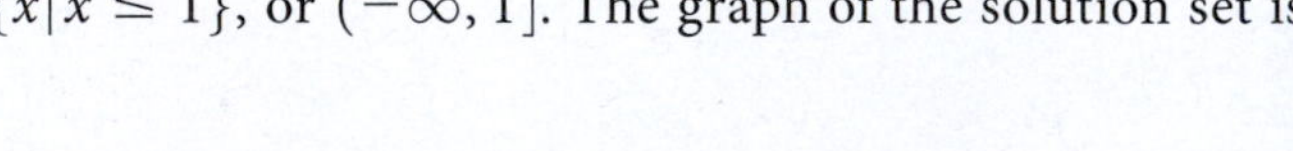

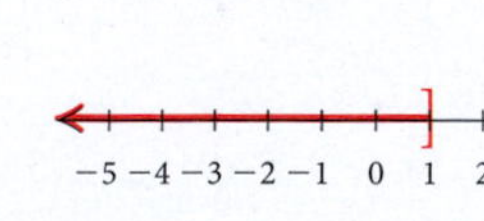

Now Try Exercises 1 and 3.

EXAMPLE 2 Find the domain of the function.

a) $f(x) = \sqrt{x - 6}$ b) $h(x) = \dfrac{x}{\sqrt{3 - x}}$

Solution

a) The radicand, $x - 6$, must be greater than or equal to 0. We solve the inequality $x - 6 \geq 0$:

$$x - 6 \geq 0$$
$$x \geq 6.$$

The domain is $\{x | x \geq 6\}$, or $[6, \infty)$.

b) Any real number can be an input for x in the numerator, but inputs for x must be restricted in the denominator. We must have $3 - x \geq 0$ and $\sqrt{3 - x} \neq 0$. Thus, $3 - x > 0$. We solve for x:

$$3 - x > 0$$
$$-x > -3 \quad \text{Subtracting 3}$$
$$x < 3. \quad \text{Multiplying by } -1 \text{ and reversing the inequality sign}$$

The domain is $\{x | x < 3\}$, or $(-\infty, 3)$.

Now Try Exercises 17 and 21.

Compound Inequalities

When two inequalities are joined by the word *and* or the word *or*, a **compound inequality** is formed. A compound inequality like

$$-3 < 2x + 5 \quad \textit{and} \quad 2x + 5 \leq 7$$

is called a **conjunction**, because it uses the word *and*. The sentence

$$-3 < 2x + 5 \leq 7$$

is an abbreviation for the preceding conjunction.

Compound inequalities can be solved using the addition and multiplication principles for inequalities.

EXAMPLE 3 Solve $-3 < 2x + 5 \leq 7$. Then graph the solution set.

Solution We have

$$-3 < 2x + 5 \leq 7$$
$$-8 < 2x \leq 2 \quad \text{Subtracting 5}$$
$$-4 < x \leq 1. \quad \text{Dividing by 2}$$

The solution set is $\{x | -4 < x \leq 1\}$, or $(-4, 1]$. The graph of the solution set is shown below.

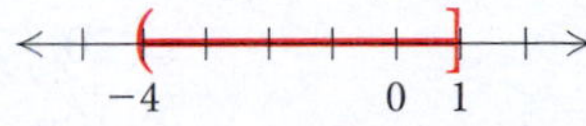

Now Try Exercise 23.

Technology Connection

To check the solution to Example 4, we graph $y_1 = 2x - 5$, $y_2 = -7$, and $y_3 = 1$. Note that for $\{x | x \le -1 \text{ or } x > 3\}$, $y_1 \le y_2$ or $y_1 > y_3$.

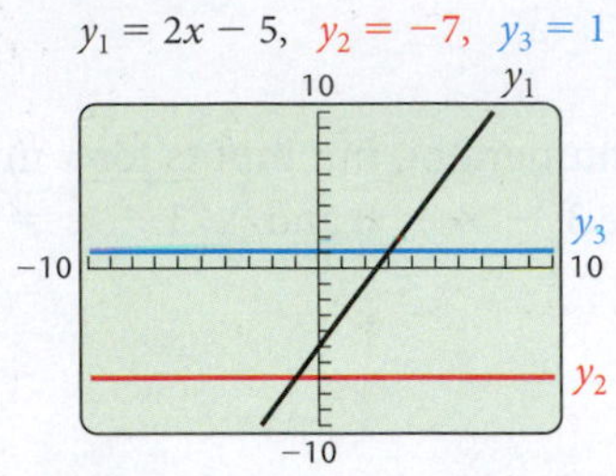

A compound inequality like $2x - 5 \le -7$ *or* $2x - 5 > 1$ is called a **disjunction**, because it contains the word *or*. Unlike some conjunctions, it cannot be abbreviated; that is, it cannot be written without the word *or*.

EXAMPLE 4 Solve $2x - 5 \le -7$ *or* $2x - 5 > 1$. Then graph the solution set.

Solution We have

$$\begin{aligned} 2x - 5 \le -7 \quad &or \quad 2x - 5 > 1 \\ 2x \le -2 \quad &or \quad 2x > 6 \qquad \text{Adding 5} \\ x \le -1 \quad &or \quad x > 3. \qquad \text{Dividing by 2} \end{aligned}$$

The solution set is $\{x | x \le -1 \text{ or } x > 3\}$. We can also write the solution set using interval notation and the symbol $\cup$ for the **union**, or inclusion, of both sets: $(-\infty, -1] \cup (3, \infty)$. The graph of the solution set is shown below.

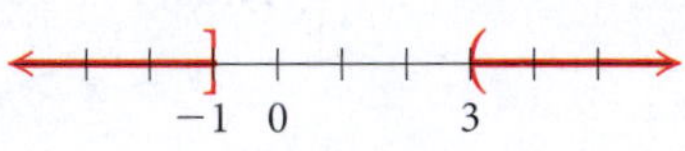

Now Try Exercise 35.

An Application

EXAMPLE 5 ***Income Plans.*** For her interior decorating job, Natália can be paid in one of two ways:

Plan A: \$250 plus \$10 per hour;

Plan B: \$20 per hour.

Suppose that a job takes n hours. For what values of n is plan B better for Natália?

Solution

1. **Familiarize.** Suppose that a job takes 20 hr. Then $n = 20$, and under plan A, Natália would earn $\$250 + \$10 \cdot 20$, or $\$250 + \$200$, or \$450. Her earnings under plan B would be $\$20 \cdot 20$, or \$400. This shows that plan A is better for Natália if a job takes 20 hr. If a job takes 30 hr, then $n = 30$, and under plan A, Natália would earn $\$250 + \$10 \cdot 30$, or $\$250 + \$300$, or \$550. Under plan B, he would earn $\$20 \cdot 30$, or \$600, so plan B is better in this case. To determine *all* values of n for which plan B is better for Natália, we solve an inequality. Our work in this step helps us write the inequality.

2. **Translate.** We translate to an inequality:

Income from plan B	is greater than	Income from plan A
$20n$	$>$	$250 + 10n.$

3. **Carry out.** We solve the inequality:

$$\begin{aligned} 20n &> 250 + 10n \\ 10n &> 250 \qquad \text{Subtracting } 10n \text{ on both sides} \\ n &> 25. \qquad \text{Dividing by 10 on both sides} \end{aligned}$$

4. **Check.** For $n = 25$, the income from plan A is $\$250 + \$10 \cdot 25$, or $\$250 + \$250$, or $500, and the income from plan B is $\$20 \cdot 25$, or $500. This shows that for a job that takes 25 hr to complete, the income is the same under either plan. In the *Familiarize* step, we saw that plan B pays more for a 30-hr job. Since $30 > 25$, this provides a partial check of the result. We cannot check all values of n.

5. **State.** For values of n greater than 25 hr, plan B is better for Natália.

Now Try Exercise 45.

1.6 Exercise Set

Solve and graph the solution set.

1. $4x - 3 > 2x + 7$
2. $8x + 1 \geq 5x - 5$
3. $x + 6 < 5x - 6$
4. $3 - x < 4x + 7$
5. $4 - 2x \leq 2x + 16$
6. $3x - 1 > 6x + 5$
7. $14 - 5y \leq 8y - 8$
8. $8x - 7 < 6x + 3$
9. $7x - 7 > 5x + 5$
10. $12 - 8y \geq 10y - 6$
11. $3x - 3 + 2x \geq 1 - 7x - 9$
12. $5y - 5 + y \leq 2 - 6y - 8$
13. $-\frac{3}{4}x \geq -\frac{5}{8} + \frac{2}{3}x$
14. $-\frac{5}{6}x \leq \frac{3}{4} + \frac{8}{3}x$
15. $4x(x - 2) < 2(2x - 1)(x - 3)$
16. $(x + 1)(x + 2) > x(x + 1)$

Find the domain of the function.

17. $h(x) = \sqrt{x - 7}$
18. $g(x) = \sqrt{x + 8}$
19. $f(x) = \sqrt{1 - 5x} + 2$
20. $f(x) = \sqrt{2x + 3} - 4$
21. $g(x) = \dfrac{5}{\sqrt{4 + x}}$
22. $h(x) = \dfrac{x}{\sqrt{8 - x}}$

Solve and write interval notation for the solution set. Then graph the solution set.

23. $-2 \leq x + 1 < 4$
24. $-3 < x + 2 \leq 5$
25. $5 \leq x - 3 \leq 7$
26. $-1 < x - 4 < 7$
27. $-3 \leq x + 4 \leq 3$
28. $-5 < x + 2 < 15$
29. $-2 < 2x + 1 < 5$
30. $-3 \leq 5x + 1 \leq 3$
31. $-4 \leq 6 - 2x < 4$
32. $-3 < 1 - 2x \leq 3$
33. $-5 < \frac{1}{2}(3x + 1) < 7$
34. $\frac{2}{3} \leq -\frac{4}{5}(x - 3) < 1$
35. $3x \leq -6$ *or* $x - 1 > 0$
36. $2x < 8$ *or* $x + 3 \geq 10$
37. $2x + 3 \leq -4$ *or* $2x + 3 \geq 4$
38. $3x - 1 < -5$ *or* $3x - 1 > 5$
39. $2x - 20 < -0.8$ *or* $2x - 20 > 0.8$
40. $5x + 11 \leq -4$ *or* $5x + 11 \geq 4$
41. $x + 14 \leq -\frac{1}{4}$ *or* $x + 14 \geq \frac{1}{4}$
42. $x - 9 < -\frac{1}{2}$ *or* $x - 9 > \frac{1}{2}$
43. *World Rice Production.* The three countries with the most rice production are China, India, and Indonesia. The equation $y = 9.06x + 410.81$ provides a good estimate of world rice production in millions of metric tons, where x is the number of years after 1980. (*Source*: www.geohive.com) For what years will world rice production exceed 820 million metric tons?

44. *Social Security Disability.* The equation $y = 0.326x + 7.148$ can be used to estimate the number of people, in millions, collecting Social Security disability payments, where x is the number of years after 2007 (*Source*: Social Security Administration). For what years will the number of people collecting disability payments be more than 12 million?

45. *Moving Costs.* Acme Movers charges \$200 plus \$45 per hour to move a household across town. Leo's Movers charges \$65 per hour. For what lengths of time does it cost less to hire Leo's Movers?

46. *Investment Income.* Jalyn plans to invest \$12,000, part at 4% simple interest and the rest at 6% simple interest. What is the most that she can invest at 4% and still be guaranteed at least \$650 in interest per year?

47. *Investment Income.* Dillon plans to invest \$7500, part at 4% simple interest and the rest at 5% simple interest. What is the most that he can invest at 4% and still be guaranteed at least \$325 in interest per year?

48. *Investment Income.* A foundation invests \$150,000 at simple interest, part at 7%, twice that amount at 4%, and the rest at 5.5%. What is the most that the foundation can invest at 4% and be guaranteed at least \$7575 in interest per year?

49. *Investment Income.* A university invests \$1,400,000 at simple interest, part at 5%, half that amount at 3.5%, and the rest at 5.5%. What is the most that the university can invest at 3.5% and be guaranteed at least \$68,000 in interest per year?

50. *Income Plans.* Karen can be paid in one of two ways for selling insurance policies:

    Plan A: A salary of \$750 per month, plus a commission of 10% of sales;
    Plan B: A salary of \$1000 per month, plus a commission of 8% of sales in excess of \$2000.

    For what amount of monthly sales is plan A better than plan B if we can assume that Karen's sales are always more than \$2000?

51. *Income Plans.* Curt can be paid in one of two ways for selling furniture:

    Plan A: A salary of \$900 per month, plus a commission of 10% of sales;
    Plan B: A salary of \$1200 per month, plus a commission of 15% of sales in excess of \$8000.

    For what amount of monthly sales is plan B better than plan A if we can assume that Curt's sales are always more than \$8000?

52. *Income Plans.* Jeanette can be paid in one of two ways for painting a house:

    Plan A: \$200 plus \$12 per hour;
    Plan B: \$20 per hour.

    Suppose a job takes n hours to complete. For what values of n is plan A better for Jeanette?

Skill Maintenance

Vocabulary Reinforcement

In each of Exercises 47–50, fill in the blank(s) with the correct term(s). Some of the given choices will not be used; others will be used more than once.

constant	domain
function	distance formula
any	exactly one
midpoint formula	identity
y-intercept	x-intercept
range	

53. A(n) ____________ is a correspondence between a first set, called the ____________, and a second set, called the ____________, such that each member of the ____________ corresponds to ____________ member of the ____________. [1.2]

54. The ____________ is $\left(\frac{x_1 + x_2}{2}, \frac{y_1 + y_2}{2}\right)$. [1.1]

55. A(n) ____________ is a point $(a, 0)$. [1.1]

56. A function f is a linear function if it can be written as $f(x) = mx + b$, where m and b are constants. If $m = 0$, the function is a(n) ____________ function $f(x) = b$. If $m = 1$ and $b = 0$, the function is the ____________ function $f(x) = x$. [1.3]

Synthesis

Solve.

57. $2x \leq 5 - 7x < 7 + x$

58. $x \leq 3x - 2 \leq 2 - x$

59. $3y < 4 - 5y < 5 + 3y$

60. $y - 10 < 5y + 6 \leq y + 10$

Chapter 1 Summary and Review

STUDY GUIDE

KEY TERMS AND CONCEPTS	EXAMPLES

SECTION 1.1: INTRODUCTION TO GRAPHING

Graphing Equations

To **graph** an equation is to make a drawing that represents the solutions of that equation. We can graph an equation by selecting values for one variable and finding the corresponding values for the other variable. We list the solutions (ordered pairs) in a table, plot the points, and draw the graph.

Graph: $y = 4 - x^2$.

x	$y = 4 - x^2$	(x, y)
0	4	$(0, 4)$
-1	3	$(-1, 3)$
1	3	$(1, 3)$
-2	0	$(-2, 0)$
2	0	$(2, 0)$

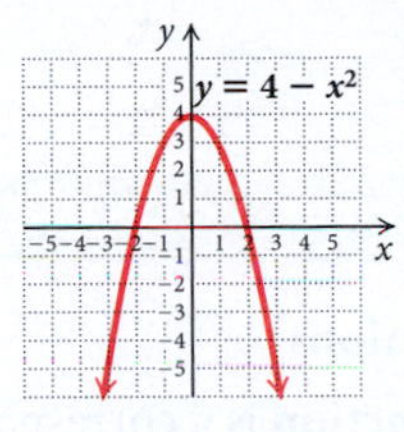

Intercepts

An ***x*-intercept** is a point $(a, 0)$. To find a, let $y = 0$ and solve for x. A ***y*-intercept** is a point $(0, b)$. To find b, let $x = 0$ and solve for y.

We can graph a straight line by plotting the intercepts and drawing the line containing them.

Graph using intercepts: $2x - y = 4$.

Let $y = 0$:

$$2x - 0 = 4$$
$$2x = 4$$
$$x = 2.$$

The x-intercept is $(2, 0)$.

Let $x = 0$:

$$2 \cdot 0 - y = 4$$
$$-y = 4$$
$$y = -4.$$

The y-intercept is $(0, -4)$.

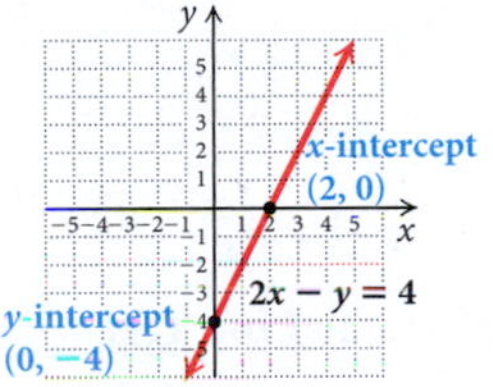

Distance Formula

The **distance** d between any two points (x_1, y_1) and (x_2, y_2) is given by

$$d = \sqrt{(x_2 - x_1)^2 + (y_2 - y_1)^2}.$$

Find the distance between $(-5, 7)$ and $(2, -3)$.

$$\begin{aligned} d &= \sqrt{[2 - (-5)]^2 + (-3 - 7)^2} \\ &= \sqrt{7^2 + (-10)^2} \\ &= \sqrt{49 + 100} \\ &= \sqrt{149} \approx 12.2 \end{aligned}$$

Midpoint Formula

If the endpoints of a segment are (x_1, y_1) and (x_2, y_2), then the coordinates of the **midpoint** of the segment are

$$\left(\frac{x_1 + x_2}{2}, \frac{y_1 + y_2}{2}\right).$$

Find the midpoint of the segment whose endpoints are $(-10, 4)$ and $(3, 8)$.

$$\begin{aligned} \left(\frac{x_1 + x_2}{2}, \frac{y_1 + y_2}{2}\right) &= \left(\frac{-10 + 3}{2}, \frac{4 + 8}{2}\right) \\ &= \left(-\frac{7}{2}, 6\right) \end{aligned}$$

Circles

The **standard form** of the equation of a circle with center (h, k) and radius r is

$$(x - h)^2 + (y - k)^2 = r^2.$$

Find an equation of a circle with center $(1, -6)$ and radius 8.

$$(x - h)^2 + (y - k)^2 = r^2$$
$$(x - 1)^2 + [y - (-6)]^2 = 8^2$$
$$(x - 1)^2 + (y + 6)^2 = 64$$

Given the circle

$$(x + 9)^2 + (y - 2)^2 = 121,$$

determine the center and the radius.

Writing in standard form, we have

$$[x - (-9)]^2 + (y - 2)^2 = 11^2.$$

The center is $(-9, 2)$, and the radius is 11.

SECTION 1.2: FUNCTIONS AND GRAPHS

Functions

A **function** is a correspondence between a first set, called the **domain**, and a second set, called the **range**, such that each member of the domain corresponds to *exactly one* member of the range.

Consider the function given by

$$g(x) = |x| - 1.$$
$$g(-3) = |-3| - 1$$
$$= 3 - 1$$
$$= 2$$

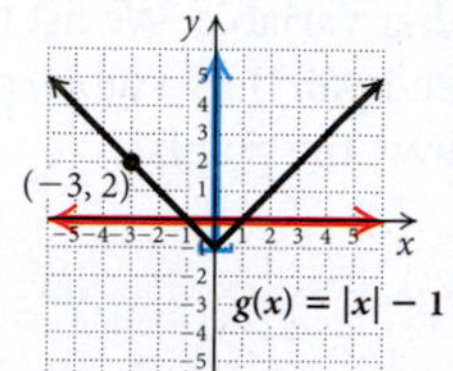

For the input -3, the output is 2:

$$f(-3) = 2.$$

The point $(-3, 2)$ is on the graph.

Domain: Set of all inputs $= \mathbb{R}$, or $(-\infty, \infty)$.

Range: Set of all outputs: $\{y \mid y \geq -1\}$, or $[-1, \infty)$.

The Vertical-Line Test

If it is possible for a vertical line to cross a graph more than once, then the graph *is not* the graph of a function.

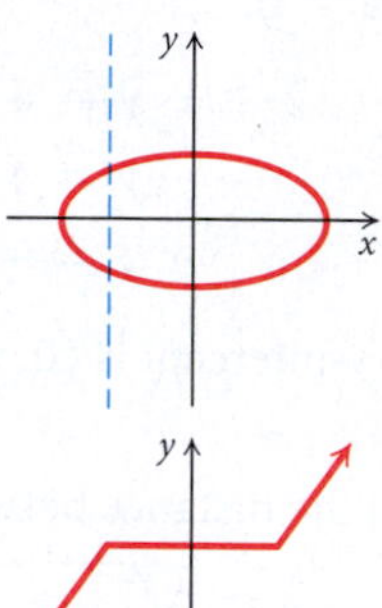

This *is not* the graph of a function because a vertical line can cross it more than once, as shown.

This *is* the graph of a function because no vertical line can cross it more than once.

Domain

When a function f whose inputs and outputs are real numbers is given by a formula, the **domain** is the set of all inputs for which the expression is defined as a real number.

Find the domain of the function given by

$$h(x) = \frac{x - 1}{(x + 5)(x - 10)}.$$

Division by 0 is not defined. Since $x + 5 = 0$ when $x = -5$ and $x - 10 = 0$ when $x = 10$, the domain of h is

$\{x \mid x$ is a real number *and* $x \neq -5$ *and* $x \neq 10\}$,
or $(-\infty, -5) \cup (-5, 10) \cup (10, \infty)$.

SECTION 1.3: LINEAR FUNCTIONS, SLOPE, AND APPLICATIONS

Slope $$m = \frac{\text{rise}}{\text{run}} = \frac{y_2 - y_1}{x_2 - x_1} = \frac{y_1 - y_2}{x_1 - x_2}$$ Slope can also be considered as an **average rate of change**. To find the average rate of change between two data points on a graph, determine the slope of the line that passes through the points.	The slope of the line containing the points $(3, -10)$ and $(-2, 6)$ is $$m = \frac{y_2 - y_1}{x_2 - x_1} = \frac{6 - (-10)}{-2 - 3} = \frac{16}{-5} = -\frac{16}{5}.$$ In 2000, the population of Flint, Michigan, was 124,943. By 2012, the population had decreased to 100,515. Find the average rate of change in population from 2000 to 2012. $$\text{Average rate of change} = m = \frac{100{,}515 - 124{,}943}{2012 - 2000} = \frac{-24{,}428}{12} \approx -2036$$ The average rate of change in population over the 12-year period was a decrease of about 2036 people per year.
Slope–Intercept Form of an Equation $$f(x) = mx + b$$ The slope of the line is m. The y-intercept of the line is $(0, b)$.	Determine the slope and the y-intercept of the line given by $5x - 7y = 14$. We first find the slope–intercept form: $5x - 7y = 14$ $-7y = -5x + 14$ **Adding $-5x$** $y = \frac{5}{7}x - 2.$ **Multiplying by $-\frac{1}{7}$** The slope is $\frac{5}{7}$, and the y-intercept is $(0, -2)$.
To graph an equation written in slope–intercept form, plot the y-intercept and use the slope to find another point. Then draw the line.	Graph: $f(x) = -\frac{2}{3}x + 4$. We plot the y-intercept, $(0, 4)$. Think of the slope as $\frac{-2}{3}$. From the y-intercept, we find another point by moving 2 units down and 3 units to the right to the point $(3, 2)$. We then draw the graph. 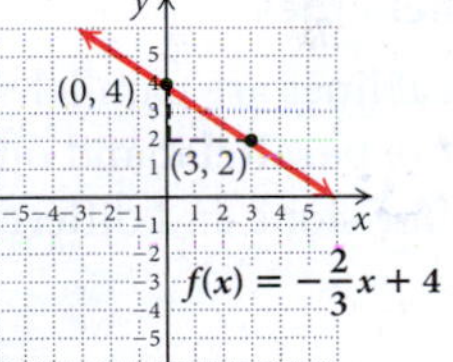
Horizontal Lines The graph of $y = b$, or $f(x) = b$, is a horizontal line with y-intercept $(0, b)$. The slope of a horizontal line is 0. **Vertical Lines** The graph of $x = a$ is a vertical line with x-intercept $(a, 0)$. The slope of a vertical line is *not* defined.	Graph $y = -4$ and determine its slope. 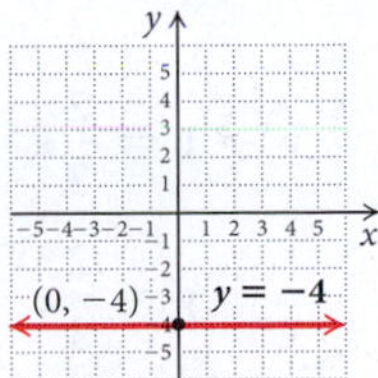 The slope is 0. Graph $x = 3$ and determine its slope. 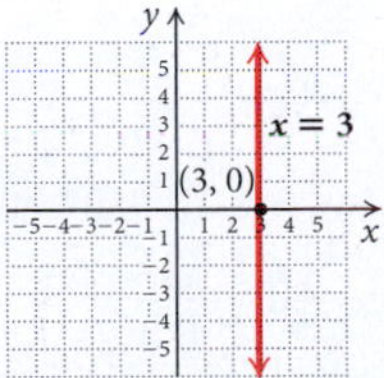 The slope is not defined.

SECTION 1.4: EQUATIONS OF LINES AND MODELING

Slope–Intercept Form of an Equation $y = mx + b$, or $f(x) = mx + b$ The slope of the line is m. The y-intercept of the line is $(0, b)$. **Point–Slope Form of an Equation** $y - y_1 = m(x - x_1)$ The slope of the line is m. The line passes through (x_1, y_1).	Write the slope–intercept equation for a line with slope $-\frac{2}{9}$ and y-intercept $(0, 4)$. $y = mx + b$ **Using the slope–intercept form** $y = -\frac{2}{9}x + 4$ **Substituting $-\frac{2}{9}$ for m and 4 for b** Write the slope–intercept equation for a line that passes through $(-5, 7)$ and $(3, -9)$. We first determine the slope: $m = \frac{-9 - 7}{3 - (-5)} = \frac{-16}{8} = -2.$ *Using the slope–intercept form:* We substitute -2 for m and either $(-5, 7)$ or $(3, -9)$ for (x, y) and solve for b: $y = mx + b$ $7 = -2 \cdot (-5) + b$ **Using $(-5, 7)$** $7 = 10 + b$ $-3 = b.$ The slope–intercept equation is $y = -2x - 3$. *Using the point–slope equation:* We substitute -2 for m and either $(-5, 7)$ or $(3, -9)$ for (x_1, y_1): $y - y_1 = m(x - x_1)$ $y - (-9) = -2(x - 3)$ **Using $(3, -9)$** $y + 9 = -2x + 6$ $y = -2x - 3.$ The slope–intercept equation is $y = -2x - 3$.
Parallel Lines Vertical lines are parallel. Nonvertical lines are **parallel** if and only if they have the same slope and different y-intercepts.	Write the slope–intercept equation for a line passing through $(-3, 1)$ that is parallel to the line $y = \frac{2}{3}x + 5$. The slope of $y = \frac{2}{3}x + 5$ is $\frac{2}{3}$, so the slope of a line parallel to this line is also $\frac{2}{3}$. We use either the slope–intercept equation or the point–slope equation for a line with slope $\frac{2}{3}$ and containing the point $(-3, 1)$. Here we use the point–slope equation and substitute $\frac{2}{3}$ for m, -3 for x_1, and 1 for y_1. $y - y_1 = m(x - x_1)$ $y - 1 = \frac{2}{3}[x - (-3)]$ $y - 1 = \frac{2}{3}x + 2$ $y = \frac{2}{3}x + 3$ **Slope–intercept form**

Perpendicular Lines

Two lines are **perpendicular** if and only if the product of their slopes is -1 or if one line is vertical $(x = a)$ and the other is horizontal $(y = b)$.

Write the slope–intercept equation for a line that passes through $(-3, 1)$ and is perpendicular to the line $y = \frac{2}{3}x + 5$.

The slope of $y = \frac{2}{3}x + 5$ is $\frac{2}{3}$, so the slope of a line perpendicular to this line is the opposite of the reciprocal of $\frac{2}{3}$, or $-\frac{3}{2}$. Here we use the point–slope equation and substitute $-\frac{3}{2}$ for m, -3 for x_1, and 1 for y_1.

$$y - y_1 = m(x - x_1)$$
$$y - 1 = -\frac{3}{2}[x - (-3)]$$
$$y - 1 = -\frac{3}{2}(x + 3)$$
$$y - 1 = -\frac{3}{2}x - \frac{9}{2}$$
$$y = -\frac{3}{2}x - \frac{7}{2}$$

SECTION 1.5: LINEAR EQUATIONS, FUNCTIONS, ZEROS, AND APPLICATIONS

Equation–Solving Principles

Addition Principle: If $a = b$ is true, then $a + c = b + c$ is true.

Multiplication Principle: If $a = b$ is true, then $ac = bc$ is true.

Solve: $2(3x - 7) = 15 - (x + 1)$.

$$2(3x - 7) = 15 - (x + 1)$$
$$6x - 14 = 15 - x - 1 \quad \text{Using the distributive property}$$
$$6x - 14 = 14 - x \quad \text{Collecting like terms}$$
$$6x - 14 + x = 14 - x + x \quad \text{Adding } x \text{ on both sides}$$
$$7x - 14 = 14$$
$$7x - 14 + 14 = 14 + 14 \quad \text{Adding 14 on both sides}$$
$$7x = 28$$
$$\frac{7x}{7} = \frac{28}{7} \quad \text{Dividing by 7 on both sides}$$
$$x = 4$$

Check:

$2(3x - 7) = 15 - (x + 1)$	
$2(3 \cdot 4 - 7)$?	$15 - (4 + 1)$
$2(12 - 7)$	$15 - 5$
$2 \cdot 5$	10
10	10 TRUE

The solution is 4.

Special Cases

Some equations have *no* solution.

Solve: $2 + 17x = 17x - 9$.

$$2 + 17x = 17x - 9$$
$$2 + 17x - 17x = 17x - 9 - 17x \quad \text{Subtracting } 17x \text{ on both sides}$$
$$2 = -9 \quad \text{False equation}$$

We get a false equation; thus the equation has *no* solution.

(*continued*)

There are some equations for which *any* real number is a solution.

Solve: $5 - \frac{1}{2}x = -\frac{1}{2}x + 5$.

$$5 - \frac{1}{2}x = -\frac{1}{2}x + 5$$

$$5 - \frac{1}{2}x + \frac{1}{2}x = -\frac{1}{2}x + 5 + \frac{1}{2}x \quad \text{Adding } \tfrac{1}{2}x \text{ on both sides}$$

$$5 = 5 \quad \text{True equation}$$

We get a true equation. Thus any real number is a solution. The solution set is

$$\{x \mid x \text{ is a real number}\}, \quad \text{or} \quad (-\infty, \infty).$$

Zeros of Functions

An input c of a function f is called a **zero** of the function if the output for the function is 0 when the input is c. That is,

c is a zero of f if $f(c) = 0$.

A linear function $f(x) = mx + b$, with $m \neq 0$, has exactly one zero.

Find the zero of the linear function

$$f(x) = \frac{5}{8}x - 40.$$

We find the value of x for which $f(x) = 0$:

$$\frac{5}{8}x - 40 = 0 \quad \text{Setting } f(x) = 0$$

$$\frac{5}{8}x = 40 \quad \text{Adding 40 on both sides}$$

$$\frac{8}{5} \cdot \frac{5}{8}x = \frac{8}{5} \cdot 40 \quad \text{Multiplying by } \tfrac{8}{5} \text{ on both sides}$$

$$x = 64.$$

The zero of $f(x) = \frac{5}{8}x - 40$ is 64.

SECTION 1.6: SOLVING LINEAR INEQUALITIES

Principles for Solving Linear Inequalities

Addition Principle:
If $a < b$ is true, then $a + c < b + c$ is true.

Multiplication Principle:
If $a < b$ and $c > 0$ are true, then $ac < bc$ is true.

If $a < b$ and $c < 0$ are true, then $ac > bc$ is true.

Similar statements hold for $a \leq b$.

Solve $3x - 2 \leq 22 - 5x$ and graph the solution set.

$$3x - 2 \leq 22 - 5x$$

$$3x - 2 + 5x \leq 22 - 5x + 5x \quad \text{Adding } 5x \text{ on both sides}$$

$$8x - 2 \leq 22$$

$$8x - 2 + 2 \leq 22 + 2 \quad \text{Adding 2 on both sides}$$

$$8x \leq 24$$

$$\frac{8x}{8} \leq \frac{24}{8} \quad \text{Dividing by 8 on both sides}$$

$$x \leq 3$$

The solution set is

$$\{x \mid x \leq 3\}, \quad \text{or} \quad (-\infty, 3].$$

The graph of the solution set is as follows.

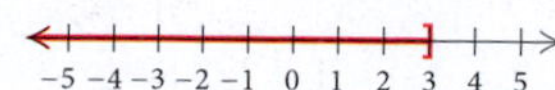

Compound Inequalities When two inequalities are joined by the word *and* or the word *or*, a compound inequality is formed. *A Conjunction:* $1 < 3x - 20$ *and* $3x - 20 \le 40$, or $1 < 3x - 20 \le 40$ *A Disjunction:* $8x - 1 \le -17$ *or* $8x - 1 > 7$	Solve: $1 < 3x - 20 \le 40$. $1 < 3x - 20 \le 40$ $21 < 3x \le 60$ **Adding 20** $7 < x \le 20$ **Dividing by 3** The solution set is $\{x \mid 7 < x \le 20\}$, or $(7, 20]$. Solve: $8x - 1 \le -17$ *or* $8x - 1 > 7$. $8x - 1 \le -17$ *or* $8x - 1 > 7$ $8x \le -16$ *or* $8x > 8$ **Adding 1** $x \le -2$ *or* $x > 1$ **Dividing by 8** The solution set is $\{x \mid x \le -2 \text{ or } x > 1\}$, or $(-\infty, -2] \cup (1, \infty)$.

REVIEW EXERCISES

Answers to all of the review exercises appear in the answer section at the back of the book. If you get an incorrect answer, restudy the objective indicated in red next to the exercise or the direction line that precedes it.

Determine whether each of the following statements is true or false.

1. If the line $ax + y = c$ is perpendicular to the line $x - by = d$, then $\frac{a}{b} = 1$. [1.4]

2. The intersection of the lines $y = \frac{1}{2}$ and $x = -5$ is $\left(-5, \frac{1}{2}\right)$. [1.3]

3. The domain of the function $f(x) = \frac{\sqrt{3 - x}}{x}$ does not contain -3 and 0. [1.2]

4. The line parallel to the x-axis that passes through $\left(-\frac{1}{4}, 7\right)$ is $x = -\frac{1}{4}$. [1.3]

5. The zero of a linear function f is the first coordinate of the x-intercept of the graph of $y = f(x)$. [1.5]

6. If $a < b$ is true and $c \neq 0$, then $ac < bc$ is true. [1.6]

Use substitution to determine whether the given ordered pairs are solutions of the given equation. [1.1]

7. $\left(3, \frac{24}{9}\right), (0, -9)$; $2x - 9y = -18$

8. $(0, 7), (7, 1)$; $y = 7$

Find the intercepts and then graph the line. [1.1]

9. $2x - 3y = 6$

10. $10 - 5x = 2y$

Graph the equation. [1.1]

11. $y = -\frac{2}{3}x + 1$

12. $2x - 4y = 8$

13. $y = 2 - x^2$

14. Find the distance between $(3, 7)$ and $(-2, 4)$. [1.1]

15. Find the midpoint of the segment with endpoints $(3, 7)$ and $(-2, 4)$. [1.1]

16. Find the center and the radius of the circle with equation $(x + 1)^2 + (y - 3)^2 = 9$. Then graph the circle. [1.1]

Find an equation for a circle satisfying the given conditions. [1.1]

17. Center: $(0, -4)$, radius of length $\frac{3}{2}$

18. Center: $(-2, 6)$, radius of length $\sqrt{13}$

19. Diameter with endpoints $(-3, 5)$ and $(7, 3)$

Determine whether the correspondence is a function. [1.2]

20. $-6 \to 1$, $-1 \to 3$, $2 \to 10$, $2 \to 3$, $7 \to 12$

21. h → r, i → s, j → s, k → t

Determine whether the relation is a function. Identify the domain and the range. [1.2]

22. $\{(3, 1), (5, 3), (7, 7), (3, 5)\}$

23. $\{(2, 7), (-2, -7), (7, -2), (0, 2), (1, -4)\}$

24. Given that $f(x) = x^2 - x - 3$, find each of the following. [1.2]

a) $f(0)$ b) $f(-3)$
c) $f(a - 1)$ d) $f(-x)$

25. Given that $f(x) = \dfrac{x - 7}{x + 5}$, find each of the following. [1.2]

a) $f(7)$ b) $f(x + 1)$
c) $f(-5)$ d) $f\left(-\frac{1}{2}\right)$

26. A graph of a function is shown. Find $f(2)$, $f(-4)$, and $f(0)$. [1.2]

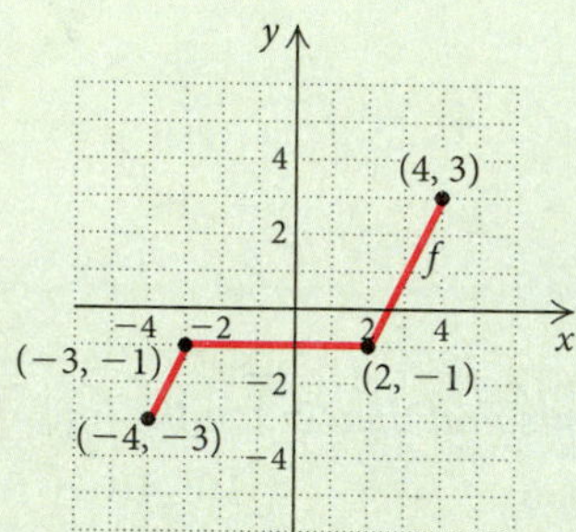

Determine whether the graph is that of a function. [1.2]

27.

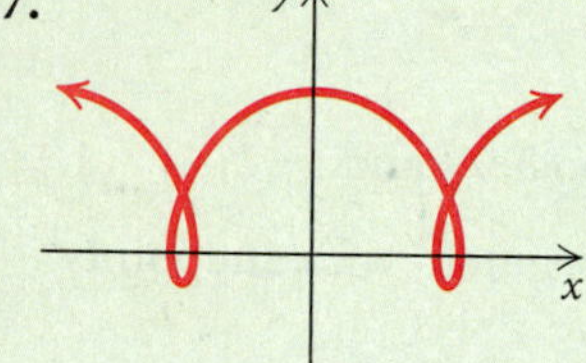

28.

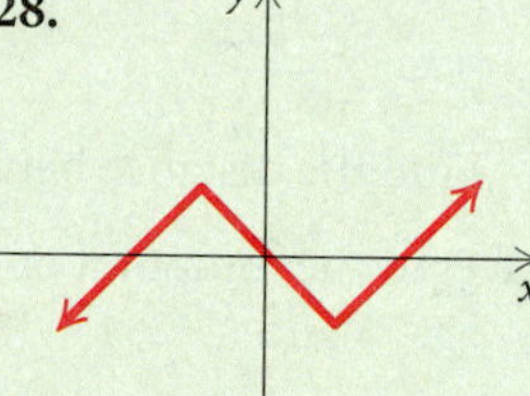

29.

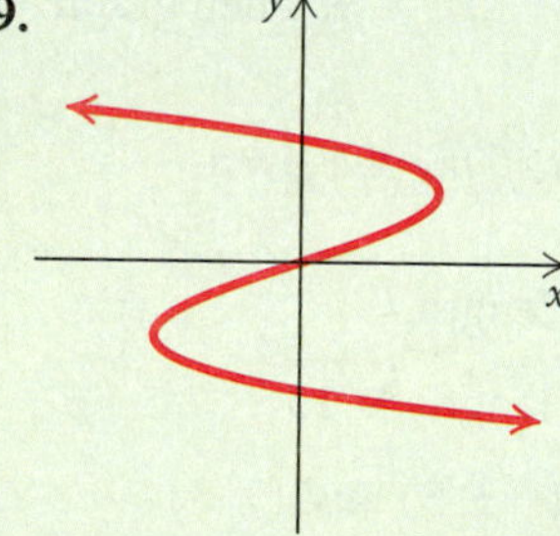

30.

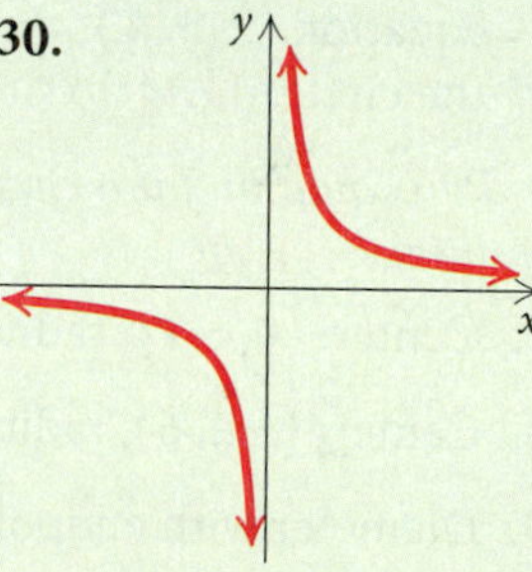

Find the domain of the function. [1.2]

31. $f(x) = 4 - 5x + x^2$

32. $f(x) = \dfrac{3}{x} + 2$

33. $f(x) = \dfrac{1}{x^2 - 6x + 5}$

34. $f(x) = \dfrac{-5x}{|16 - x^2|}$

Graph the function. Then visually estimate the domain and the range. [1.2]

35. $f(x) = \sqrt{16 - x^2}$

36. $g(x) = |x - 5|$

37. $f(x) = x^3 - 7$

38. $h(x) = x^4 + x^2$

In Exercises 39 and 40, the table of data contains input–output values for a function. Answer the following questions. [1.3]

a) *Is the change in the inputs, x, the same?*
b) *Is the change in the outputs, y, the same?*
c) *Is the function linear?*

39.

x	y
−3	8
−2	11
−1	14
0	17
1	20
2	22
3	26

40.

x	y
20	11.8
30	24.2
40	36.6
50	49.0
60	61.4
70	73.8
80	86.2

Find the slope of the line containing the given points. [1.3]

41. $(2, -11), (5, -6)$

42. $(5, 4), (-3, 4)$

43. $\left(\frac{1}{2}, 3\right), \left(\frac{1}{2}, 0\right)$

44. *Coffee Consumption.* The U.S. annual per-capita consumption of coffee was 26.8 gal in 1990. By 2011, this amount had decreased to 24.7 gal. (*Source:* Economic Research Service, U.S. Department of Agriculture) Find the average rate of change in per-capita coffee consumption from 1990 to 2011. [1.3]

Find the slope and the y-intercept of the line with the given equation. [1.3]

45. $y = -\frac{7}{11}x - 6$

46. $-2x - y = 7$

47. Graph $y = -\frac{1}{4}x + 3$ using the slope and the y-intercept. [1.3]

48. *Total Cost.* Clear County Cable Television charges a $110 installation fee and $85 per month for basic service. Write an equation that can be used to determine the total cost $C(t)$ of t months of basic cable television service. Find the total cost of 1 year of service. [1.3]

49. *Temperature and Depth of the Earth.* The function T given by $T(d) = 10d + 20$ can be used to determine the temperature T, in degrees Celsius, at a depth d, in kilometers, inside the earth. [1.3]

a) Find $T(5)$, $T(20)$, and $T(1000)$.

b) The radius of the earth is about 5600 km. Use this fact to determine the domain of the function.

Write a slope–intercept equation for a line with the following characteristics. [1.4]

50. $m = -\frac{2}{3}$, y-intercept $(0, -4)$

51. $m = 3$, passes through $(-2, -1)$

52. Passes through $(4, 1)$ and $(-2, -1)$

53. Write equations of the horizontal line and the vertical line that pass through $\left(-4, \frac{2}{5}\right)$. [1.4]

54. Find a linear function h given $h(-2) = -9$ and $h(4) = 3$. Then find $h(0)$. [1.4]

Determine whether the lines are parallel, perpendicular, or neither. [1.4]

55. $3x - 2y = 8,$
$6x - 4y = 2$

56. $y - 2x = 4,$
$2y - 3x = -7$

57. $y = \frac{3}{2}x + 7,$
$y = -\frac{2}{3}x - 4$

Given the point $(1, -1)$ *and the line* $2x + 3y = 4$:

58. Find an equation of the line containing the given point and parallel to the given line. [1.4]

59. Find an equation of the line containing the given point and perpendicular to the given line. [1.4]

60. *Female Medical School Graduates.* Data in the following table show the number of female medical school graduates for years 2005–2011. Model the data with a linear function, where the number of female medical school graduates W is a function of the year x and where x is the number of years after 2005. Then using this function, estimate the number of female graduates in 2008 and predict the number of female graduates in 2018. Answers may vary depending on the data points used. [1.4]

Year, x	Female Medical School Graduates in the United States, W
2005, 0	7412
2007, 2	7925
2009, 4	8036
2011, 6	8396

Source: The Kaiser Foundation

Solve. [1.5]

61. $4y - 5 = 1$

62. $3x - 4 = 5x + 8$

63. $5(3x + 1) = 2(x - 4)$

64. $2(n - 3) = 3(n + 5)$

65. $\frac{3}{5}y - 2 = \frac{3}{8}$

66. $5 - 2x = -2x + 3$

67. $x - 13 = -13 + x$

68. *Production of Quarters.* In 2013, the U.S. Mint produced 1455 million quarters. This was a 156% increase over the number of quarters produced in 2012. (*Source*: U.S. Mint) How many quarters were produced in 2012? [1.5]

69. *Amount of Investment.* James makes an investment at 5.2% simple interest. At the end of 1 year, the total value of the investment is $2419.60. How much was originally invested? [1.5]

70. *Flying into a Headwind.* An airplane that can travel 550 mph in still air encounters a 20-mph headwind. How long will it take the plane to travel 1802 mi? [1.5]

Find the zero of the function. [1.5]

71. $f(x) = 6x - 18$

72. $f(x) = x - 4$

73. $f(x) = 2 - 10x$

74. $f(x) = 8 - 2x$

Solve and write interval notation for the solution set. Then graph the solution set. [1.6]

75. $2x - 5 < x + 7$

76. $3x + 1 \geq 5x + 9$

77. $-3 \leq 3x + 1 \leq 5$

78. $-2 < 5x - 4 \leq 6$

79. $2x < -1$ *or* $x - 3 > 0$

80. $3x + 7 \leq 2$ *or* $2x + 3 \geq 5$

81. *Homeschooled Children in the United States.* The equation $y = 0.073x + 0.848$ can be used to estimate the number of homeschooled children in the United States, in millions, where x is the number of years after 1999 (*Source*: Department of Education's National Center for Education Statistics). For what years will the number of homeschooled children exceed 2.3 million? [1.6]

82. *Temperature Conversion.* The formula $C = \frac{5}{9}(F - 32)$ can be used to convert Fahrenheit temperatures F to Celsius temperatures C. For what Fahrenheit temperatures is the Celsius temperature lower than 45°C? [1.6]

83. The domain of the function

$$f(x) = \frac{x + 3}{8 - 4x}$$

is which of the following? [1.2]

A. $(-3, 2)$
B. $(-\infty, 2) \cup (2, \infty)$
C. $(-\infty, -3) \cup (-3, 2) \cup (2, \infty)$
D. $(-\infty, -3) \cup (-3, \infty)$

84. The center of the circle described by the equation $(x - 1)^2 + y^2 = 9$ is which of the following? [1.1]

A. $(-1, 0)$
B. $(1, 0)$
C. $(0, -3)$
D. $(-1, 3)$

85. The graph of $f(x) = -\frac{1}{2}x - 2$ is which of the following? [1.3]

A.

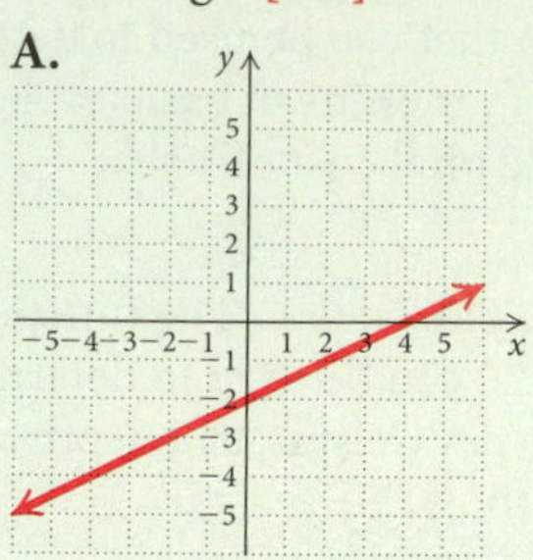

B.

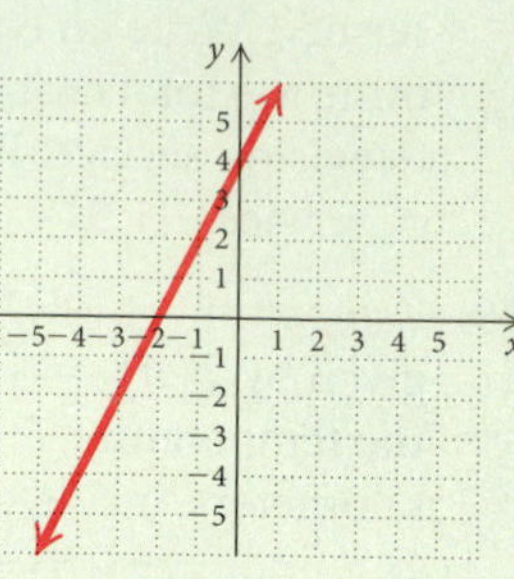

C.

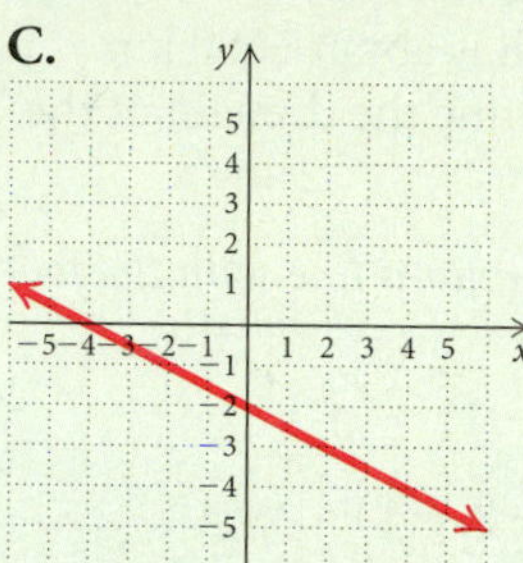

D.

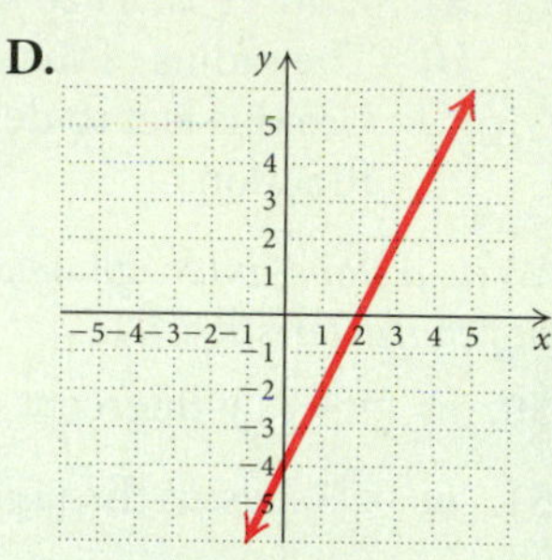

Synthesis

86. Find the point on the x-axis that is equidistant from the points $(1, 3)$ and $(4, -3)$. [1.1]

Find the domain. [1.2]

87. $f(x) = \dfrac{\sqrt{1 - x}}{x - |x|}$

88. $f(x) = (x - 9x^{-1})^{-1}$

Collaborative Discussion and Writing

89. Discuss why the graph of $f(x) = -\frac{3}{5}x + 4$ is steeper than the graph of $g(x) = \frac{1}{2}x - 6$. [1.3]

90. As the first step in solving

$$3x - 1 = 8,$$

Tenia multiplies by $\frac{1}{3}$ on both sides. What advice would you give her about the procedure for solving equations? [1.5]

91. Is it possible for a disjunction to have no solution? Why or why not? [1.6]

92. Explain in your own words why a linear function $f(x) = mx + b$, with $m \neq 0$, has exactly one zero. [1.5]

93. Why can the conjunction $3 < x$ *and* $x < 4$ be written as $3 < x < 4$, but the disjunction $x < 3$ *or* $x > 4$ cannot be written as $3 > x > 4$? [1.6]

94. Explain in your own words what a function is. [1.2]

1 Chapter Test

1. Determine whether the ordered pair $\left(\frac{1}{2}, \frac{9}{10}\right)$ is a solution of the equation $5y - 4 = x$.

2. Find the intercepts of $5x - 2y = -10$ and graph the line.

3. Find the distance between $(5, 8)$ and $(-1, 5)$.

4. Find the midpoint of the segment with endpoints $(-2, 6)$ and $(-4, 3)$.

5. Find the center and the radius of the circle
$$(x + 4)^2 + (y - 5)^2 = 36.$$

6. Find an equation of the circle with center $(-1, 2)$ and radius $\sqrt{5}$.

7. a) Determine whether the relation
$$\{(-4, 7), (3, 0), (1, 5), (0, 7)\}$$
is a function. Answer yes or no.
b) Find the domain of the relation.
c) Find the range of the relation.

8. Given that $f(x) = 2x^2 - x + 5$, find each of the following.
a) $f(-1)$ b) $f(a + 2)$

9. Given that $f(x) = \dfrac{1 - x}{x}$, find each of the following.
a) $f(0)$ b) $f(1)$

10. Using the graph below, find $f(-3)$.

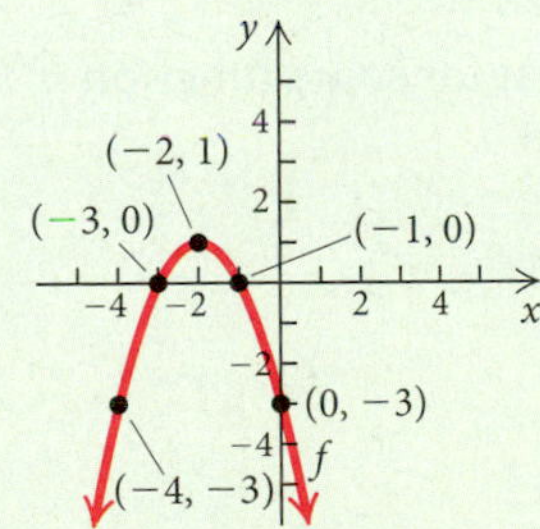

11. Determine whether each graph is that of a function. Answer yes or no.

a) 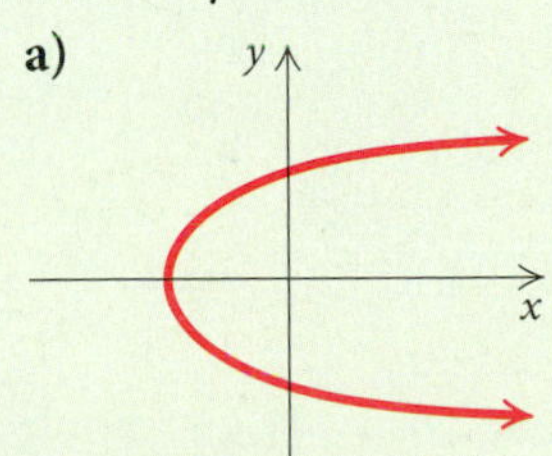

b) 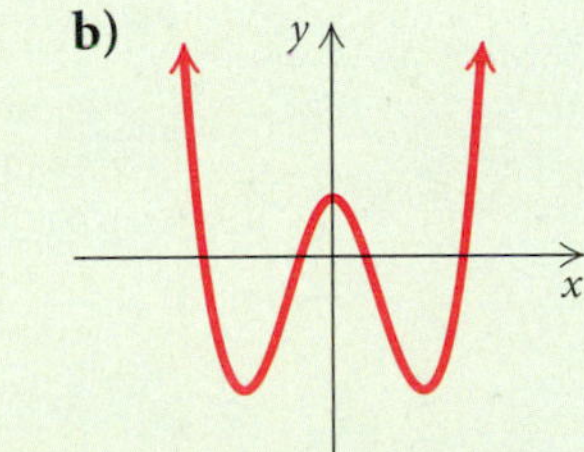

Find the domain of the function.

12. $f(x) = \dfrac{1}{x - 4}$

13. $g(x) = x^3 + 2$

14. $h(x) = \sqrt{25 - x^2}$

15. a) Graph: $f(x) = |x - 2| + 3$.
b) Visually estimate the domain of $f(x)$.
c) Visually estimate the range of $f(x)$.

Find the slope of the line containing the given points.

16. $\left(-2, \frac{2}{3}\right), (-2, 5)$

17. $(4, -10), (-8, 12)$

18. $(-5, 6), \left(\frac{3}{4}, 6\right)$

19. *Declining Number of Those Who Smoke.* Daily use of cigarettes by U.S. 12th graders is declining. In 1995, 21.6% of 12th graders smoked daily. This number had decreased to 9.3% in 2012. (*Source*: *Monitoring the Future*, University of Michigan Institute for Social Research and National Institute on Drug Abuse) Find the average rate of change in the percent of 12th graders who smoke daily from 1995 to 2012.

20. Find the slope and the y-intercept of the line with equation $-3x + 2y = 5$.

21. *Total Cost.* An electrician charges a basic rate of \$65 for a service call plus \$48 per hour for labor. Write an equation that can be used to determine the cost $C(t)$ of hiring an electrician to do repair work. Then find the total cost, not including parts, if the repair work takes 2.25 hr.

22. Write an equation for the line with $m = -\frac{5}{8}$ and y-intercept $(0, -5)$.

23. Write an equation for the line that passes through $(-5, 4)$ and $(3, -2)$.

24. Write the equation of the vertical line that passes through $\left(-\frac{3}{8}, 11\right)$.

25. Determine whether the lines are parallel, perpendicular, or neither.
$$2x + 3y = -12,$$
$$2y - 3x = 8$$

26. Find an equation of the line containing the point $(-1, 3)$ and parallel to the line $x + 2y = -6$.

27. Find an equation of the line containing the point $(-1, 3)$ and perpendicular to the line $x + 2y = -6$.

28. *Weekly Earnings.* Data in the following table show an increase in the average weekly earnings of U.S. production workers from 2000 to 2012. Model the data with a linear function and using this function, predict the average weekly earnings of U.S. production workers in 2016 and in 2020. Answers may vary depending on the data points used.

Year, x	Average Weekly Earnings of U.S. Production Workers
2000, 0	\$481.36
2002, 2	507.03
2004, 4	529.23
2006, 6	567.89
2008, 8	608.11
2010, 10	637.18
2012, 12	666.99

Source: Bureau of Labor Statistics, U.S. Department of Labor

Solve.

29. $6x + 7 = 1$

30. $2.5 - x = -x + 2.5$

31. $\frac{3}{2}y - 4 = \frac{5}{3}y + 6$

32. $2(4x + 1) = 8 - 3(x - 5)$

33. *Parking-Lot Dimensions.* The parking lot behind Kai's Kafé has a perimeter of 210 m. The width is three-fourths of the length. What are the dimensions of the parking lot?

34. *Pricing.* Kokona's Juice Bar prices its bottled juices by raising the wholesale price 50% and then adding 25¢. What is the wholesale price of a bottle of juice that sells for \$2.95?

35. Find the zero of the function $f(x) = 3x + 9$.

Solve and write interval notation for the solution set. Then graph the solution set.

36. $5 - x \geq 4x + 20$

37. $-7 < 2x + 3 < 9$

38. $2x - 1 \leq 3$ *or* $5x + 6 \geq 26$

39. *Moving Costs.* Morgan Movers charges \$200 plus \$40 per hour to move households across town. McKinley Movers charges \$75 per hour for crosstown moves. For what lengths of time does it cost less to hire Morgan Movers?

40. The graph of $g(x) = 1 - \frac{1}{2}x$ is which of the following?

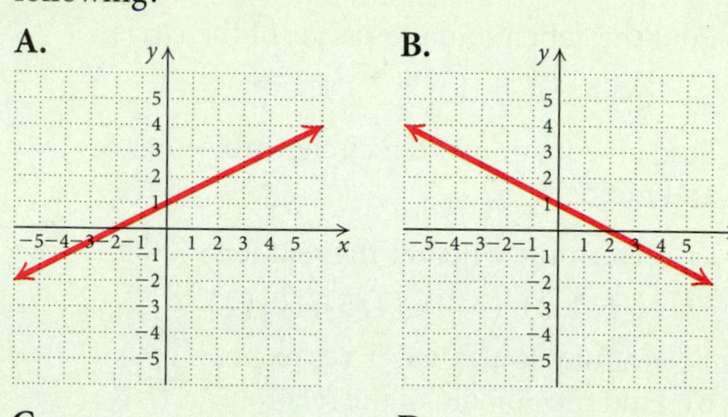

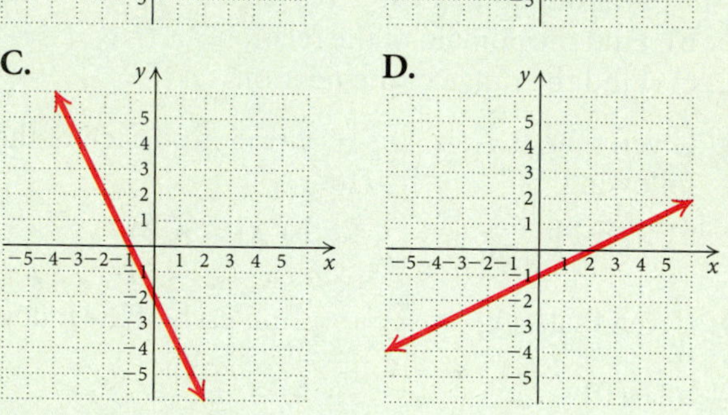

Synthesis

41. Suppose that for some function h, $h(x + 2) = \frac{1}{2}x$. Find $h(-2)$.

CHAPTER

2

More on Functions

PPLICATION This problem appears as Exercise 53 in Section 2.3.

A blouse that is size x in Japan is size $s(x)$ in the United States, where $s(x) = x - 3$. A blouse that is size x in the United States is size $t(x)$ in Australia, where $t(x) = x + 4$. (*Source:* www.onlineconversion.com) Find a function that will convert blouse sizes in Japan to blouse sizes in Australia.

2.1 Increasing, Decreasing, and Piecewise Functions; Applications

- Graph functions, looking for intervals on which the function is increasing, decreasing, or constant, and estimate relative maxima and minima.
- Given an application, find a function that models the application. Find the domain of the function and function values.
- Graph functions defined piecewise.

Because functions occur in so many real-world situations, it is important to be able to analyze them carefully.

Increasing, Decreasing, and Constant Functions

On a given interval, if the graph of a function rises from left to right, it is said to be **increasing** on that interval. If the graph drops from left to right, it is said to be **decreasing**. If the function values stay the same on the interval, the function is said to be **constant**.

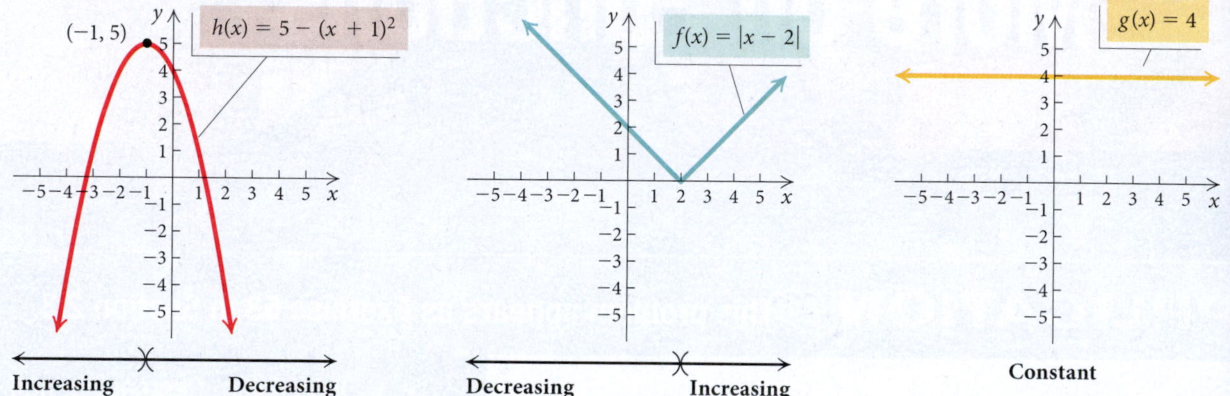

We are led to the following definitions.

INCREASING, DECREASING, AND CONSTANT FUNCTIONS

A function f is said to be **increasing** on an *open* interval I, if for all a and b in that interval, $a < b$ implies $f(a) < f(b)$. (See Fig. 1 on the following page.)

A function f is said to be **decreasing** on an *open* interval I, if for all a and b in that interval, $a < b$ implies $f(a) > f(b)$. (See Fig. 2.)

A function f is said to be **constant** on an *open* interval I, if for all a and b in that interval, $f(a) = f(b)$. (See Fig. 3.)

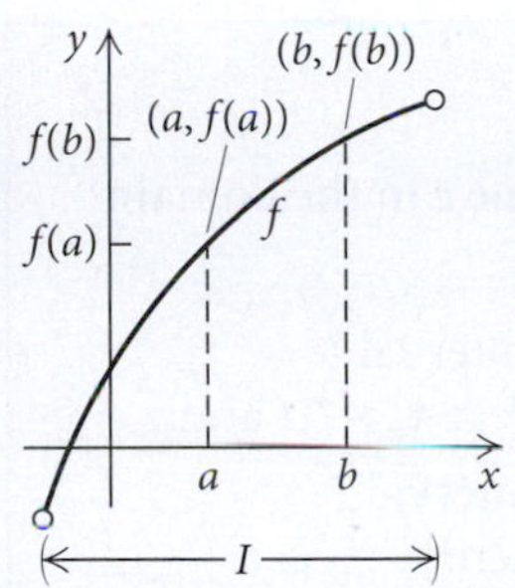

For $a < b$ in I, $f(a) < f(b)$;
f is **increasing** on I.

Figure 1.

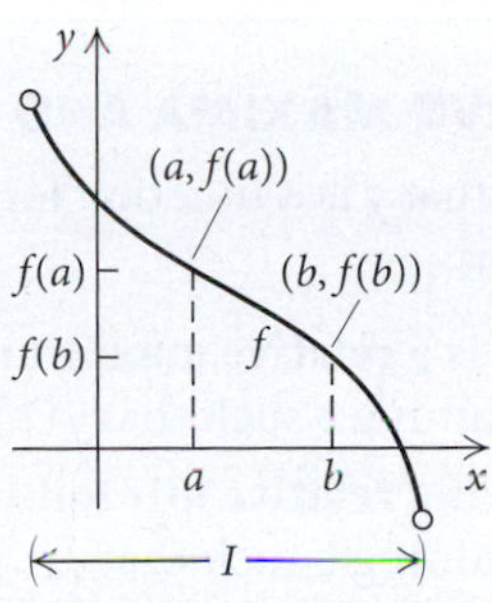

For $a < b$ in I, $f(a) > f(b)$;
f is **decreasing** on I.

Figure 2.

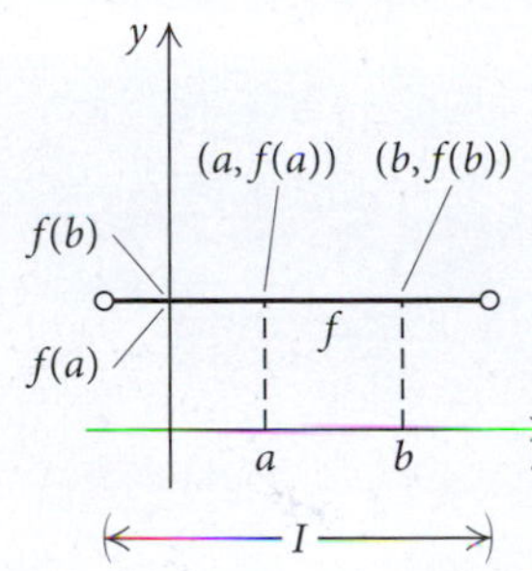

For all a and b in I, $f(a) = f(b)$;
f is **constant** on I.

Figure 3.

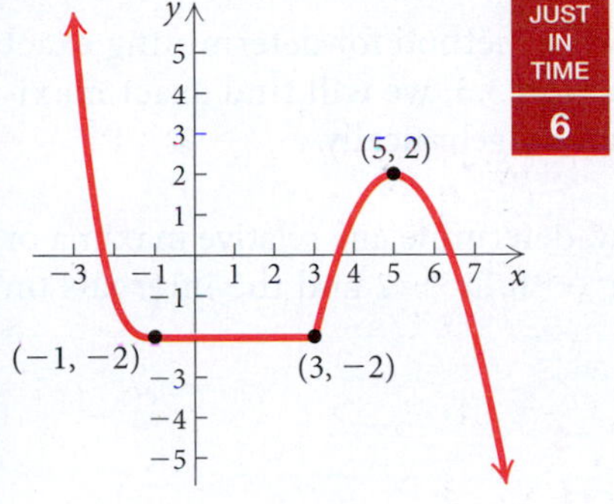

JUST IN TIME 6

EXAMPLE 1 Determine the intervals on which the function in the figure at left is **(a)** increasing; **(b)** decreasing; **(c)** constant.

Solution When expressing interval(s) on which a function is increasing, decreasing, or constant, we consider only values in the *domain* of the function. Since the domain of this function is $(-\infty, \infty)$, we consider all real values of x.

a) As x-values (that is, values in the domain) increase from $x = 3$ to $x = 5$, the y-values (that is, values in the range) increase from -2 to 2. Thus the function is increasing on the interval $(3, 5)$.

b) As x-values increase from negative infinity to -1, y-values decrease; y-values also decrease as x-values increase from 5 to positive infinity. Thus the function is decreasing on the intervals $(-\infty, -1)$ and $(5, \infty)$.

c) As x-values increase from -1 to 3, y remains -2. The function is constant on the interval $(-1, 3)$.

Now Try Exercise 5.

In calculus, the slope of a line tangent to the graph of a function at a particular point is used to determine whether the function is increasing, decreasing, or neither. If the slope is positive, the function is increasing; if the slope is negative, the function is decreasing; if the slope is 0, the function is constant. Since slope cannot be both positive and negative at the same point, a function cannot be both increasing and decreasing at a specific point. For this reason, increasing, decreasing, and constant intervals are expressed in *open-interval* notation. In Example 1, if $[3, 5]$ had been used for the increasing interval and $[5, \infty)$ for a decreasing interval, the function would be both increasing and decreasing at $x = 5$. This is not possible.

Relative Maximum and Minimum Values

Consider the graph shown below. Note the "peaks" and "valleys" at the x-values c_1, c_2, and c_3. The function value $f(c_2)$ is called a **relative maximum** (plural, **maxima**). Each of the function values $f(c_1)$ and $f(c_3)$ is called a **relative minimum** (plural, **minima**).

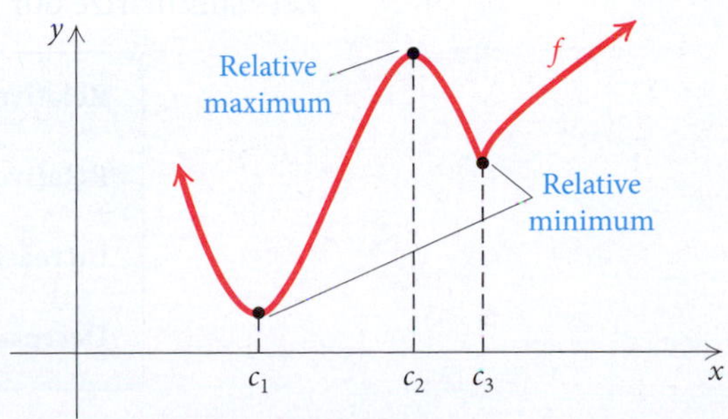

RELATIVE MAXIMA AND MINIMA

Suppose that f is a function for which $f(c)$ exists for some c in the domain of f. Then:

> $f(c)$ is a **relative maximum** if there exists an *open* interval I containing c such that $f(c) > f(x)$, for all x in I where $x \neq c$; and
>
> $f(c)$ is a **relative minimum** if there exists an *open* interval I containing c such that $f(c) < f(x)$, for all x in I where $x \neq c$.

Simply stated, $f(c)$ is a *relative maximum* if $(c, f(c))$ is the highest point in some *open* interval, and $f(c)$ is a *relative minimum* if $(c, f(c))$ is the lowest point in some *open* interval.

If you take a calculus course, you will learn a method for determining exact values of relative maxima and minima. In Section 3.3, we will find exact maximum and minimum values of quadratic functions algebraically.

Technology Connection

We can approximate relative maximum and minimum values with the MAXIMUM and MINIMUM features from the CALC menu on a graphing calculator. See the online *Graphing Calculator Manual* for more information on this procedure.

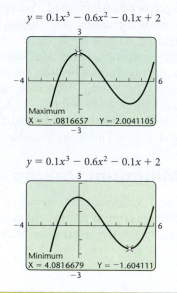

$y = 0.1x^3 - 0.6x^2 - 0.1x + 2$

3

−4 6

Maximum
X = −.0816657 Y = 2.0041105

−3

$y = 0.1x^3 - 0.6x^2 - 0.1x + 2$

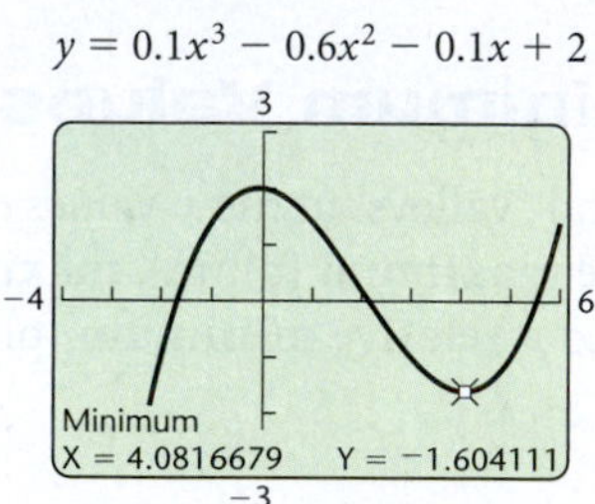

EXAMPLE 2 Using the graph shown below, determine any relative maxima or minima of the function $f(x) = 0.1x^3 - 0.6x^2 - 0.1x + 2$ and the intervals on which the function is increasing or decreasing.

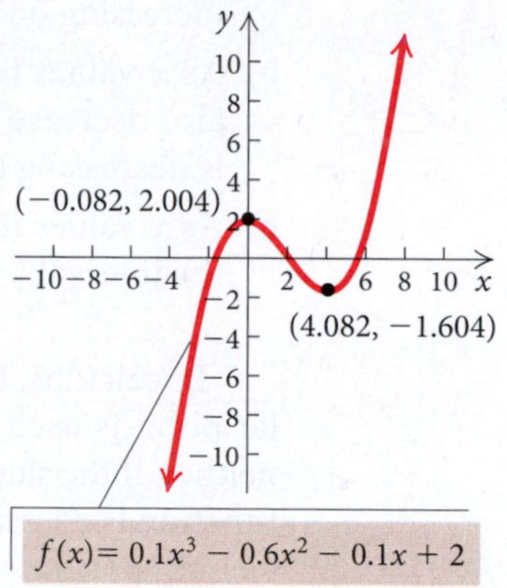

$f(x) = 0.1x^3 - 0.6x^2 - 0.1x + 2$

Solution We see that the *relative maximum* value of the function is 2.004. It occurs when $x = -0.082$. We also see the *relative minimum*: -1.604 at $x = 4.082$.

We note that the graph starts rising, or increasing, from the left and stops increasing at the relative maximum. From this point, the graph decreases to the relative minimum and then begins to rise again. The function is *increasing* on the intervals

$$(-\infty, -0.082) \quad \text{and} \quad (4.082, \infty)$$

and *decreasing* on the interval

$$(-0.082, 4.082).$$

Let's summarize our results.

Relative Maximum	2.004 at $x = -0.082$
Relative Minimum	-1.604 at $x = 4.082$
Increasing	$(-\infty, -0.082)$, $(4.082, \infty)$
Decreasing	$(-0.082, 4.082)$

Now Try Exercise 15.

Applications of Functions

Many real-world situations can be modeled by functions.

EXAMPLE 3 ***Car Distance.*** Two nurses, Kiara and Matias, drive away from a hospital at right angles to each other. Kiara's speed is 35 mph and Matias's is 40 mph.

a) Express the distance between the cars as a function of time, $d(t)$.

b) Find the domain of the function.

Solution

a) Suppose 1 hr goes by. At that time, Kiara has traveled 35 mi and Matias has traveled 40 mi. We can use the Pythagorean theorem to find the distance between them. This distance would be the length of the hypotenuse of a right triangle with legs measuring 35 mi and 40 mi. After 2 hr, the triangle's legs would measure $2 \cdot 35$, or 70 mi, and $2 \cdot 40$, or 80 mi. Noting that the distances will always be changing, we make a drawing and let $t =$ the time, in hours, that Kiara and Matias have been driving since leaving the hospital.

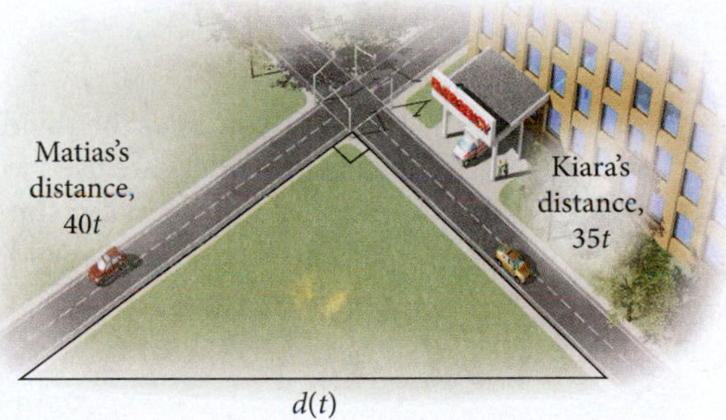

After t hours, Kiara has traveled $35t$ miles and Matias $40t$ miles. We now use the Pythagorean theorem:

$$[d(t)]^2 = (35t)^2 + (40t)^2.$$

Because distance must be nonnegative, we need consider only the positive square root when solving for $d(t)$:

$$\begin{aligned} d(t) &= \sqrt{(35t)^2 + (40t)^2} \\ &= \sqrt{1225t^2 + 1600t^2} \\ &= \sqrt{2825t^2} \\ &\approx 53.15|t| && \text{Approximating the root to two decimal places} \\ &\approx 53.15t. && \text{Since } t \geq 0, |t| = t. \end{aligned}$$

Thus, $d(t) = 53.15t$, $t \geq 0$.

b) Since the time traveled, t, must be nonnegative, the domain is the set of nonnegative real numbers $[0, \infty)$.

Now Try Exercise 25.

EXAMPLE 4 ***Area of Office Space.*** A community college has 30 ft of dividers with which to set off a rectangular area for a student testing center. If a corner of the math lab is used for the testing center, the partition need only form two sides of a rectangle.

a) Express the floor area of the office space as a function of the length of the partition.

b) Find the domain of the function.

c) Using the graph shown below, determine the dimensions that maximize the area of the floor.

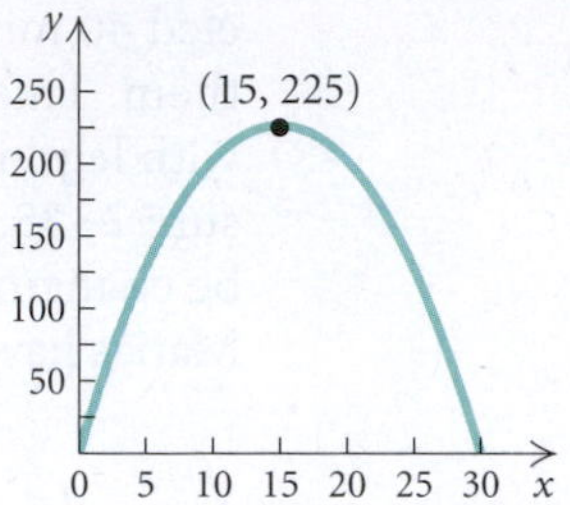

Solution

a) Note that the dividers will form two sides of a rectangle. If, for example, 14 ft of dividers are used for the length of the rectangle, that would leave $30 - 14$, or 16 ft of dividers for the width. Thus if $x =$ the length, in feet, of the rectangle, then $30 - x =$ the width. We represent this information in a drawing, as shown below.

The area, $A(x)$, is given by

$$A(x) = x(30 - x) \quad \text{Area = length} \cdot \text{width.}$$
$$= 30x - x^2.$$

The function $A(x) = 30x - x^2$ can be used to express the rectangle's area as a function of the length.

b) Because the rectangle's length and width must be positive and only 30 ft of dividers are available, we restrict the domain of A to $\{x \mid 0 < x < 30\}$—that is, the interval $(0, 30)$.

c) On the graph of the function shown at the top of the page, the maximum value of the area on the interval $(0, 30)$ appears to be 225 when $x = 15$. Thus the dimensions that maximize the area are

$$\text{Length} = x = 15 \text{ ft} \quad \text{and}$$
$$\text{Width} = 30 - x = 30 - 15 = 15 \text{ ft}.$$

Now Try Exercise 31.

JUST IN TIME 4

Functions Defined Piecewise

Sometimes functions are defined **piecewise** using different output formulas for different pieces, or parts, of the domain.

EXAMPLE 5 For the function defined as

$$f(x) = \begin{cases} x + 1, & \text{for } x < -2, \\ 5, & \text{for } -2 \le x \le 3, \\ x^2, & \text{for } x > 3, \end{cases}$$

find $f(-5), f(-3), f(0), f(3), f(4)$, and $f(10)$.

Solution First, we determine which part of the domain contains the given input. Then we use the corresponding formula to find the output.

Since $-5 < -2$, we use the formula $f(x) = x + 1$:

$$f(-5) = -5 + 1 = -4.$$

Since $-3 < -2$, we use the formula $f(x) = x + 1$ again:

$$f(-3) = -3 + 1 = -2.$$

Since $-2 \le 0 \le 3$, we use the formula $f(x) = 5$:

$$f(0) = 5.$$

Since $-2 \le 3 \le 3$, we use the formula $f(x) = 5$ a second time:

$$f(3) = 5.$$

Since $4 > 3$, we use the formula $f(x) = x^2$:

$$f(4) = 4^2 = 16.$$

Since $10 > 3$, we once again use the formula $f(x) = x^2$:

$$f(10) = 10^2 = 100.$$

Now Try Exercise 35.

EXAMPLE 6 Graph the function defined as

$$g(x) = \begin{cases} \frac{1}{3}x + 3, & \text{for } x < 3, \\ -x, & \text{for } x \ge 3. \end{cases}$$

Solution Since the function is defined in two pieces, or parts, we create the graph in two parts.

a) We graph $g(x) = \frac{1}{3}x + 3$ *only* for inputs x less than 3. That is, we use $g(x) = \frac{1}{3}x + 3$ only for x-values in the interval $(-\infty, 3)$. Some ordered pairs that are solutions of this piece of the function are shown in Table 1.

TABLE 1.

x $(x < 3)$	$g(x) = \frac{1}{3}x + 3$
-3	2
0	3
2	$3\frac{2}{3}$

TABLE 2.

x $(x \geq 3)$	$g(x) = -x$
3	−3
4	−4
6	−6

b) We graph $g(x) = -x$ *only* for inputs x greater than or equal to 3. That is, we use $g(x) = -x$ only for x-values in the interval $[3, \infty)$. Some ordered pairs that are solutions of this piece of the function are shown in Table 2.

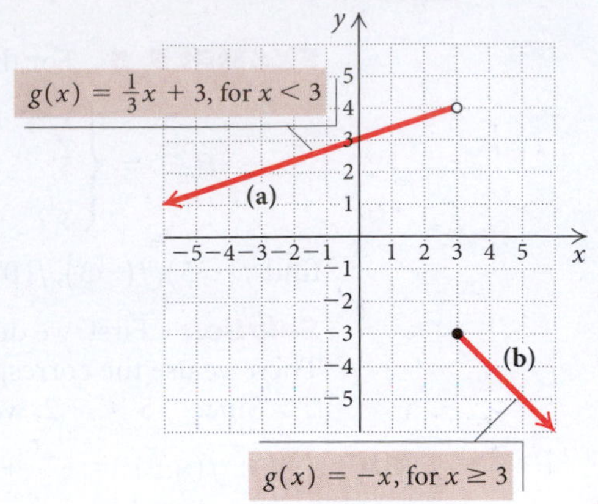

Now Try Exercise 39.

TABLE 3.

x $(x \leq 0)$	$f(x) = 4$
−5	4
−2	4
0	4

TABLE 4.

x $(0 < x \leq 2)$	$f(x) = 4 - x^2$
$\frac{1}{2}$	$3\frac{3}{4}$
1	3
2	0

TABLE 5.

x $(x > 2)$	$f(x) = 2x - 6$
$2\frac{1}{2}$	−1
3	0
5	4

EXAMPLE 7 Graph the function defined as

$$f(x) = \begin{cases} 4, & \text{for } x \leq 0, \\ 4 - x^2, & \text{for } 0 < x \leq 2, \\ 2x - 6, & \text{for } x > 2. \end{cases}$$

Solution We create the graph in three pieces, or parts.

a) We graph $f(x) = 4$ *only* for inputs x less than or equal to 0. That is, we use $f(x) = 4$ only for x-values in the interval $(-\infty, 0]$. Some ordered pairs that are solutions of this piece of the function are shown in Table 3.

b) We graph $f(x) = 4 - x^2$ *only* for inputs x greater than 0 and less than or equal to 2. That is, we use $f(x) = 4 - x^2$ only for x-values in the interval $(0, 2]$. Some ordered pairs that are solutions of this piece of the function are shown in Table 4.

c) We graph $f(x) = 2x - 6$ *only* for inputs x greater than 2. That is, we use $f(x) = 2x - 6$ only for x-values in the interval $(2, \infty)$. Some ordered pairs that are solutions of this piece of the function are shown in Table 5.

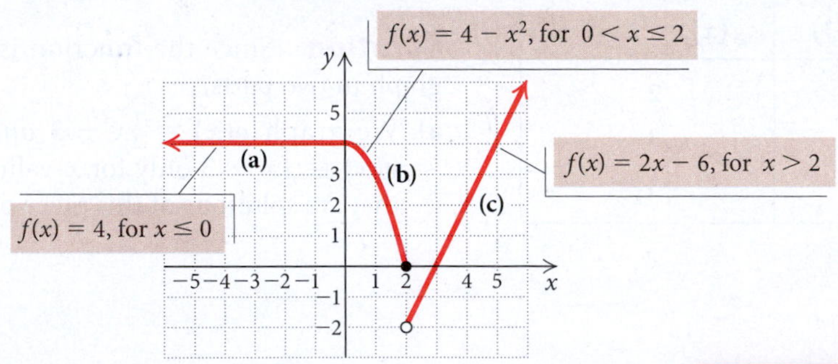

Now Try Exercise 43.

JUST IN TIME 18

EXAMPLE 8 Graph the function defined as

$$f(x) = \begin{cases} \dfrac{x^2 - 4}{x + 2}, & \text{for } x \neq -2, \\ 3, & \text{for } x = -2. \end{cases}$$

Solution When $x \neq -2$, the denominator of $(x^2 - 4)/(x + 2)$ is nonzero, so we can simplify:

$$\frac{x^2 - 4}{x + 2} = \frac{(x + 2)(x - 2)}{x + 2} = x - 2.$$

Thus,

$$f(x) = x - 2, \quad \text{for } x \neq -2.$$

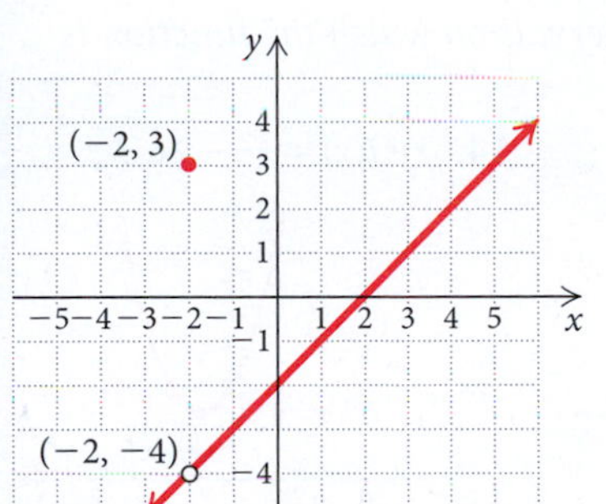

The graph of this part of the function consists of a line with a "hole" at the point $(-2, -4)$, indicated by the open circle. The hole occurs because the piece of the function represented by $(x^2 - 4)/(x + 2)$ is not defined for $x = -2$. By the definition of the function, we see that $f(-2) = 3$, so we plot the point $(-2, 3)$ above the open circle.

Now Try Exercise 47.

A piecewise function with importance in calculus and computer programming is the **greatest integer function**, f, denoted $f(x) = [\![x]\!]$, or $\text{int}(x)$.

GREATEST INTEGER FUNCTION

$f(x) = [\![x]\!] =$ the greatest integer *less than or equal to x*.

Technology Connection

To graph the greatest integer function

$$f(x) = [\![x]\!]$$

on a graphing calculator, select the greatest integer function from the MATH NUM menu. Notice that the graph does not show the open dots at the endpoints of segments.

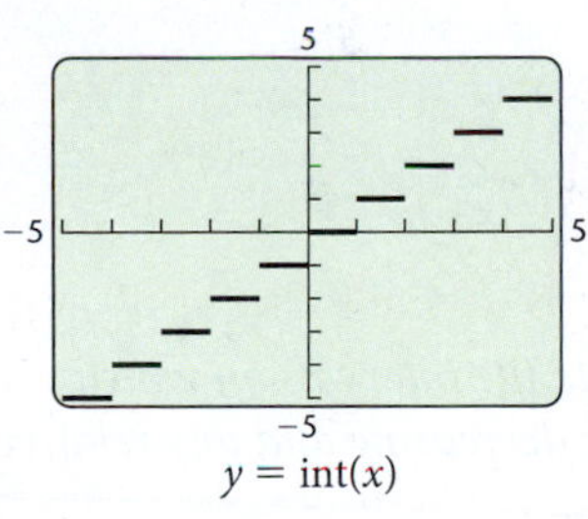

$y = \text{int}(x)$

The greatest integer function pairs each input with the greatest integer *less than or equal to* that input. Thus x-values 1, $1\frac{1}{2}$, and 1.8 are all paired with the y-value 1. Other pairings are shown below.

$-4, -3.6, -3\frac{1}{4} \to -4$; $-1, -\frac{7}{8}, -0.25 \to -1$; $0, \frac{1}{10}, 0.99 \to 0$; $2, 2.1, 2\frac{3}{4} \to 2$

EXAMPLE 9 Graph $f(x) = [\![x]\!]$ and determine its domain and range.

Solution The greatest integer function can also be defined as a piecewise function with an infinite number of statements.

JUST IN TIME 5

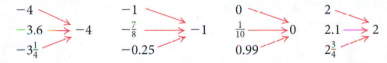

$$f(x) = [\![x]\!] = \begin{cases} \vdots \\ -3, & \text{for } -3 \le x < -2, \\ -2, & \text{for } -2 \le x < -1, \\ -1, & \text{for } -1 \le x < 0, \\ 0, & \text{for } 0 \le x < 1, \\ 1, & \text{for } 1 \le x < 2, \\ 2, & \text{for } 2 \le x < 3, \\ 3, & \text{for } 3 \le x < 4, \\ \vdots \end{cases}$$

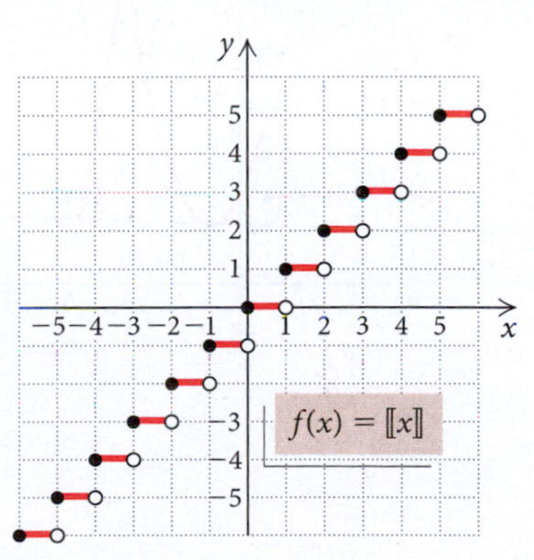

We see that the domain of this function is the set of all real numbers, $(-\infty, \infty)$, and the range is the set of all integers, $\{\ldots, -3, -2, -1, 0, 1, 2, 3, \ldots\}$.

Now Try Exercise 51.

2.1 Exercise Set

Determine the intervals on which the function is **(a)** *increasing;* **(b)** *decreasing;* **(c)** *constant.*

1.

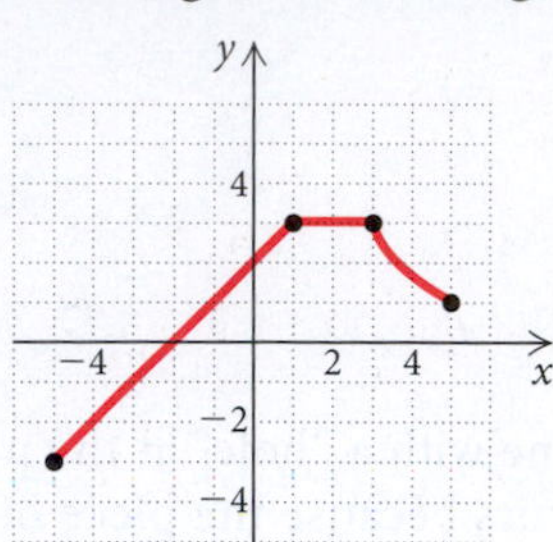

2.

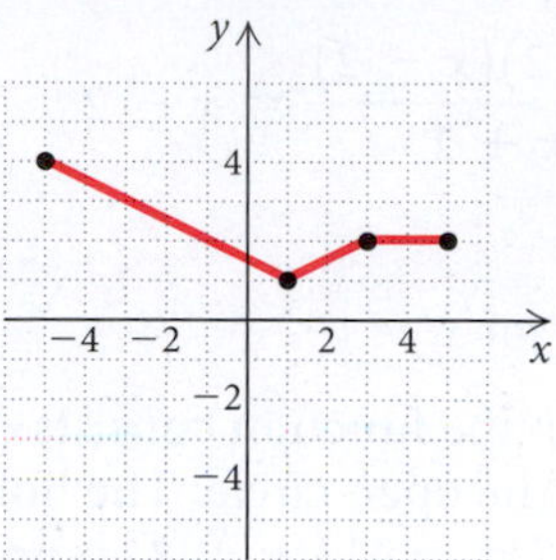

3.

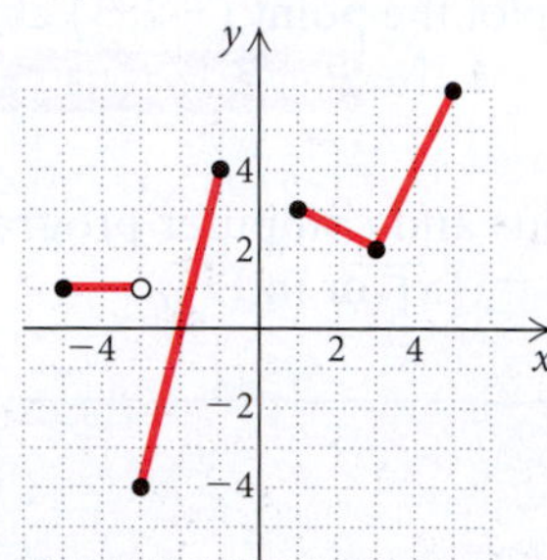

4.

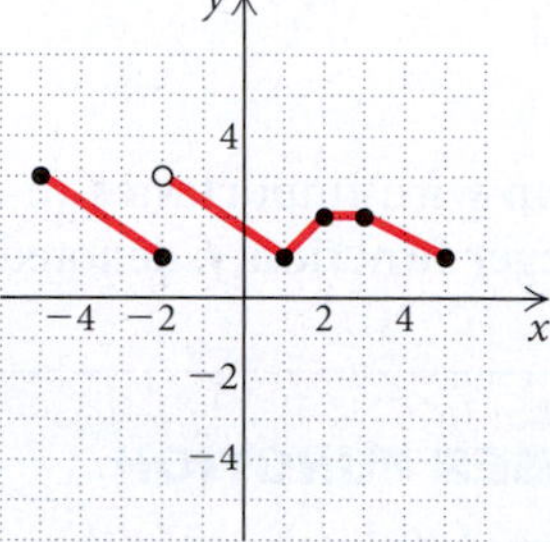

5.

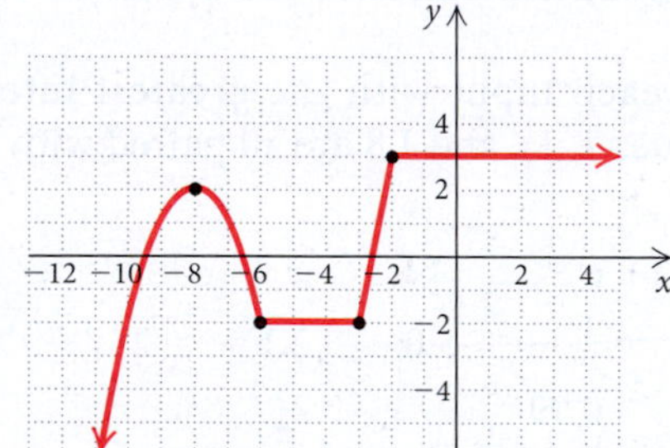

6.

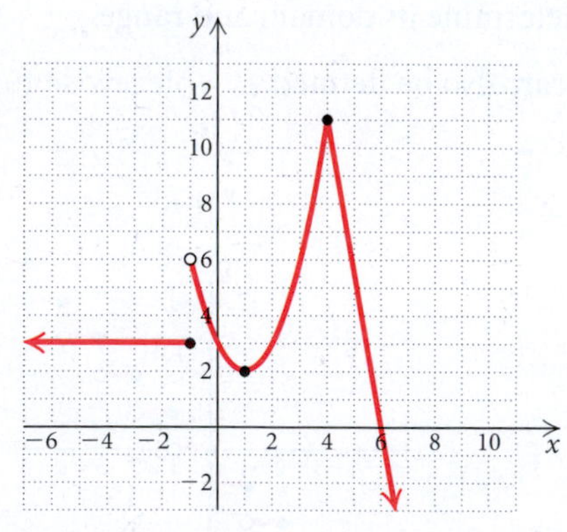

7.–12. Determine the domain and the range of each of the functions graphed in Exercises 1–6.

Using the graph, determine any relative maxima or minima of the function and the intervals on which the function is increasing or decreasing.

13. $f(x) = -x^2 + 5x - 3$

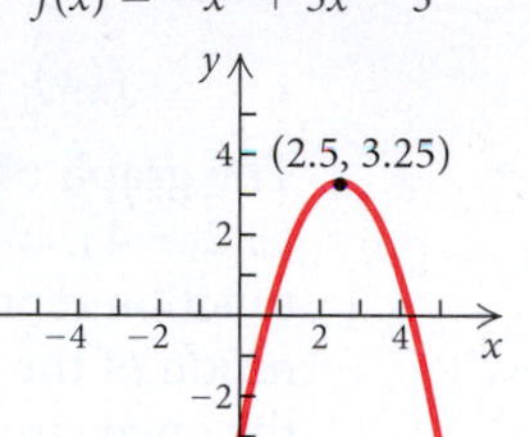

14. $f(x) = x^2 - 2x + 3$

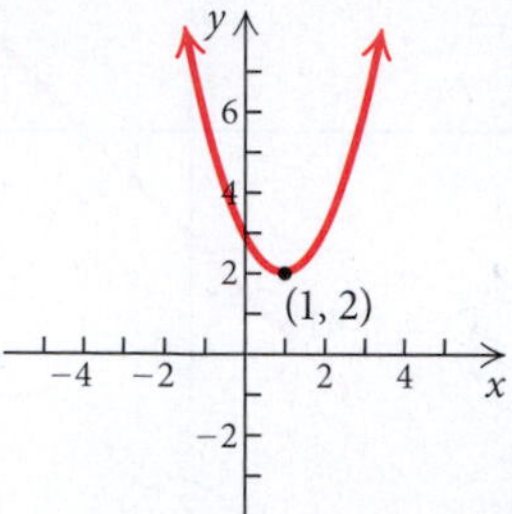

15. $f(x) = \frac{1}{4}x^3 - \frac{1}{2}x^2 - x + 2$

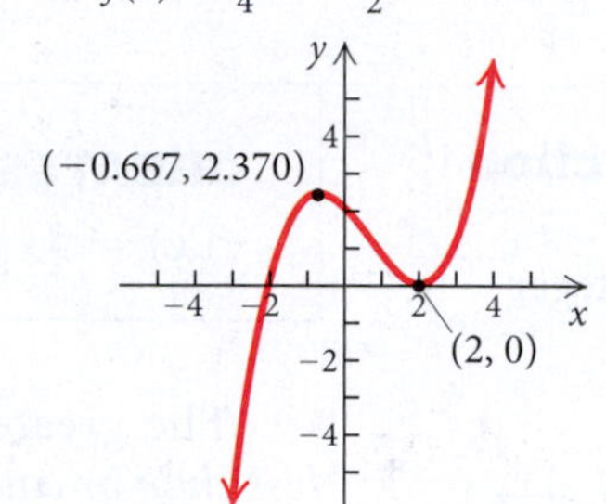

16. $f(x) = -0.09x^3 + 0.5x^2 - 0.1x + 1$

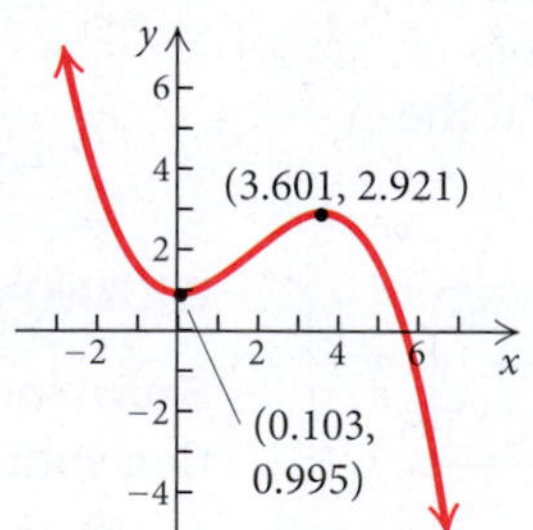

Graph the function. Estimate the intervals on which the function is increasing or decreasing and any relative maxima or minima.

17. $f(x) = x^2$

18. $f(x) = 4 - x^2$

19. $f(x) = 5 - |x|$

20. $f(x) = |x + 3| - 5$

21. $f(x) = x^2 - 6x + 10$

22. $f(x) = -x^2 - 8x - 9$

23. *Lumberyard.* Rick's lumberyard has 480 yd of fencing with which to enclose a rectangular area. If the enclosed area is x yards long, express its area as a function of its length.

24. *Triangular Flag.* A seamstress is designing a triangular flag so that the length of the base of the triangle, in inches, is 7 less than twice the height h. Express the area of the flag as a function of the height.

25. *Blimp Distance.* The Goodyear Blimp can be seen flying at an altitude of 3500 ft above the Motor Speedway during the Indianapolis 500 race. The slanted distance directly to the Pagoda at the start–finish line is d feet. Express the horizontal distance h as a function of d.

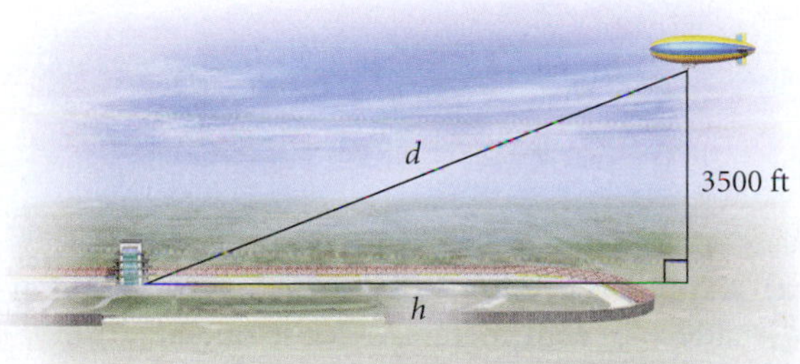

26. *Rising Balloon.* A hot-air balloon rises straight up from the ground at a rate of 120 ft/min. The balloon is tracked from a rangefinder on the ground at point P, which is 400 ft from the release point Q of the basket. Let $d =$ the distance from the balloon to the rangefinder and $t =$ the time, in minutes, since the balloon was released. Express d as a function of t.

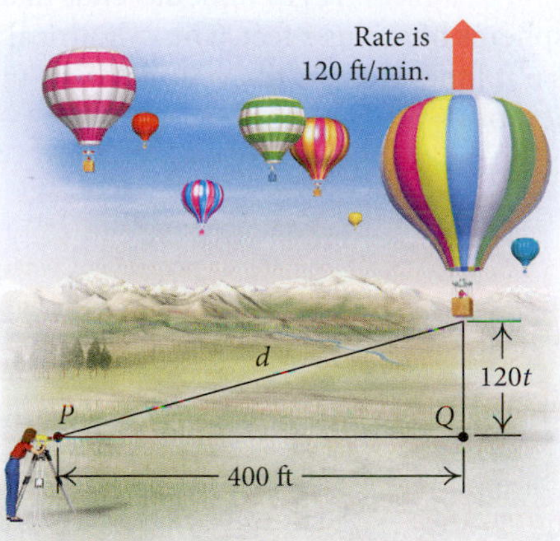

27. *Inscribed Rhombus.* A rhombus is inscribed in a rectangle that is w meters wide with a perimeter of 40 m. Each vertex of the rhombus is a midpoint of a side of the rectangle. Express the area of the rhombus as a function of the width of the rectangle.

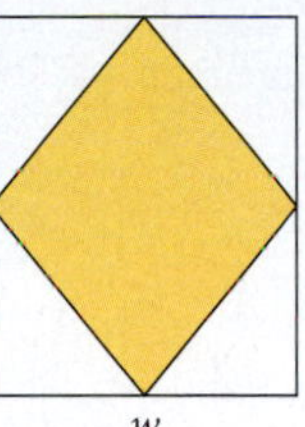

28. *Carpet Area.* A carpet installer uses 46 ft of linen tape to bind the edges of a rectangular hall runner. If the runner is w feet wide, express its area as a function of the width.

29. *Golf Distance Finder.* A device used in golf to estimate the distance d, in yards, to a hole measures the size s, in inches, that the 7-ft pin appears to be in a viewfinder. Express the distance d as a function of s.

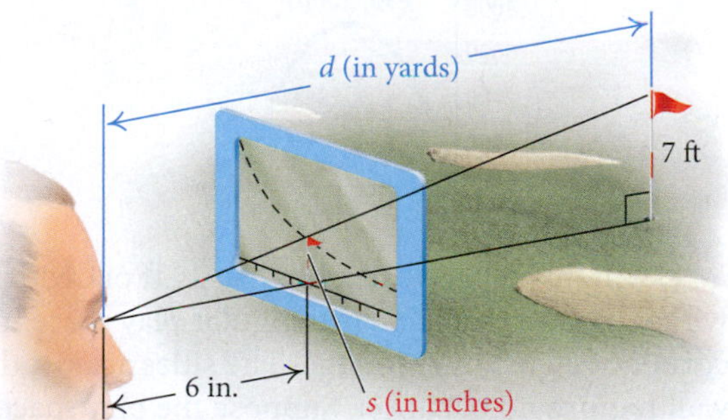

30. *Gas Tank Volume.* A gas tank has ends that are hemispheres of radius r feet. The cylindrical midsection is 6 ft long. Express the volume of the tank as a function of r.

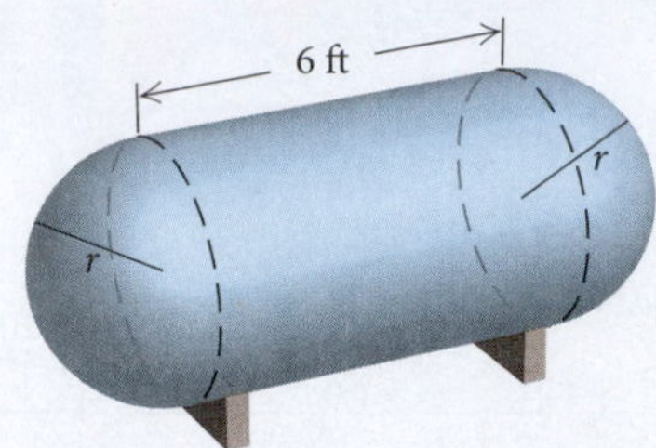

31. *Swimming Areas.* A summer camp has 240 ft of float line with which to rope off three adjacent rectangular areas of a lake for swimming lessons, one for each of three levels of swimming ability. A beach forms one side of the swimming areas. Suppose the width of each area is x yards.

a) Express the total area of the three swimming areas as a function of x.
b) Find the domain of the function.
c) Using the graph of the function shown below, determine the dimensions that yield the maximum area.

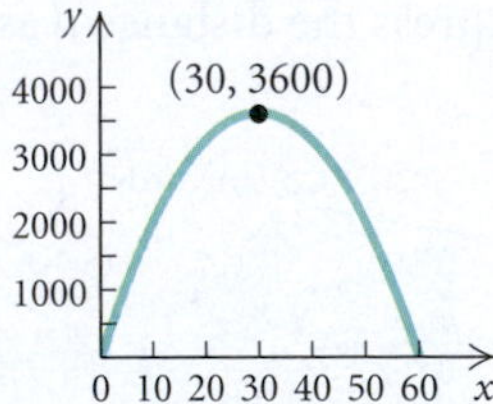

32. *Play Space.* A car dealership has 24 ft of dividers with which to enclose a rectangular play space in a corner of a customer lounge. The sides against the wall require no partition. Suppose the play space is x feet long.

a) Express the area of the play space as a function of x.
b) Find the domain of the function.
c) Using the graph shown below, determine the dimensions that yield the maximum area.

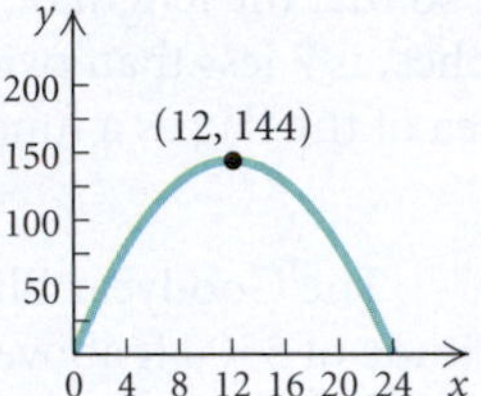

33. *Volume of a Box.* From a 12-cm by 12-cm piece of cardboard, square corners are cut out so that the sides can be folded up to make a box.

a) Express the volume of the box as a function of the side x, in centimeters, of a cut-out square.
b) Find the domain of the function.
c) Using the graph of the function shown below, determine the dimensions that yield the maximum volume.

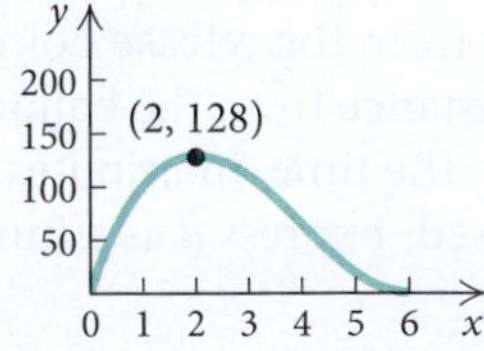

34. *Office File.* Designs Unlimited plans to produce a one-component vertical file by bending the long side of an 8-in. by 14-in. sheet of plastic along two lines to form a ⊔ shape.

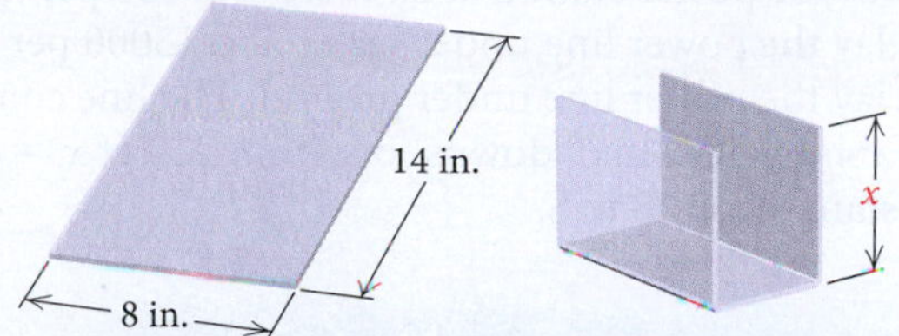

a) Express the volume of the file as a function of the height x, in inches, of the file.
b) Find the domain of the function.
c) Using the graph of the function shown below, determine how tall the file should be in order to maximize the volume that the file can hold.

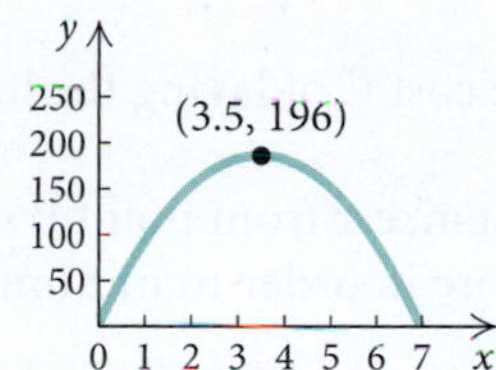

For each piecewise function, find the specified function values.

35. $g(x) = \begin{cases} x + 4, & \text{for } x \le 1, \\ 8 - x, & \text{for } x > 1 \end{cases}$

$g(-4), g(0), g(1)$, and $g(3)$

36. $f(x) = \begin{cases} 3, & \text{for } x \le -2, \\ \frac{1}{2}x + 6, & \text{for } x > -2 \end{cases}$

$f(-5), f(-2), f(0)$, and $f(2)$

37. $h(x) = \begin{cases} -3x - 18, & \text{for } x < -5, \\ 1, & \text{for } -5 \le x < 1, \\ x + 2, & \text{for } x \ge 1 \end{cases}$

$h(-5), h(0), h(1)$, and $h(4)$

38. $f(x) = \begin{cases} -5x - 8, & \text{for } x < -2, \\ \frac{1}{2}x + 5, & \text{for } -2 \le x \le 4, \\ 10 - 2x, & \text{for } x > 4 \end{cases}$

$f(-4), f(-2), f(4)$, and $f(6)$

Graph each of the following.

39. $f(x) = \begin{cases} \frac{1}{2}x, & \text{for } x < 0, \\ x + 3, & \text{for } x \ge 0 \end{cases}$

40. $f(x) = \begin{cases} -\frac{1}{3}x + 2, & \text{for } x \le 0, \\ x - 5, & \text{for } x > 0 \end{cases}$

41. $f(x) = \begin{cases} -\frac{3}{4}x + 2, & \text{for } x < 4, \\ -1, & \text{for } x \ge 4 \end{cases}$

42. $h(x) = \begin{cases} 2x - 1, & \text{for } x < 2, \\ 2 - x, & \text{for } x \ge 2 \end{cases}$

43. $f(x) = \begin{cases} x + 1, & \text{for } x \le -3, \\ -1, & \text{for } -3 < x < 4, \\ \frac{1}{2}x, & \text{for } x \ge 4 \end{cases}$

44. $f(x) = \begin{cases} 4, & \text{for } x \le -2, \\ x + 1, & \text{for } -2 < x < 3, \\ -x, & \text{for } x \ge 3 \end{cases}$

45. $g(x) = \begin{cases} \frac{1}{2}x - 1, & \text{for } x < 0, \\ 3, & \text{for } 0 \le x \le 1, \\ -2x, & \text{for } x > 1 \end{cases}$

46. $f(x) = \begin{cases} \dfrac{x^2 - 9}{x + 3}, & \text{for } x \ne -3, \\ 5, & \text{for } x = -3 \end{cases}$

47. $f(x) = \begin{cases} 2, & \text{for } x = 5, \\ \dfrac{x^2 - 25}{x - 5}, & \text{for } x \ne 5 \end{cases}$

48. $f(x) = \begin{cases} \dfrac{x^2 + 3x + 2}{x + 1}, & \text{for } x \ne -1, \\ 7, & \text{for } x = -1 \end{cases}$

49. $f(x) = [\![x]\!]$

50. $f(x) = 2[\![x]\!]$

51. $g(x) = 1 + [\![x]\!]$

52. $h(x) = \frac{1}{2}[\![x]\!] - 2$

53.–58. Find the domain and the range of each of the functions defined in Exercises 39–44.

Determine the domain and the range of the piecewise function. Then write an equation for the function.

59.

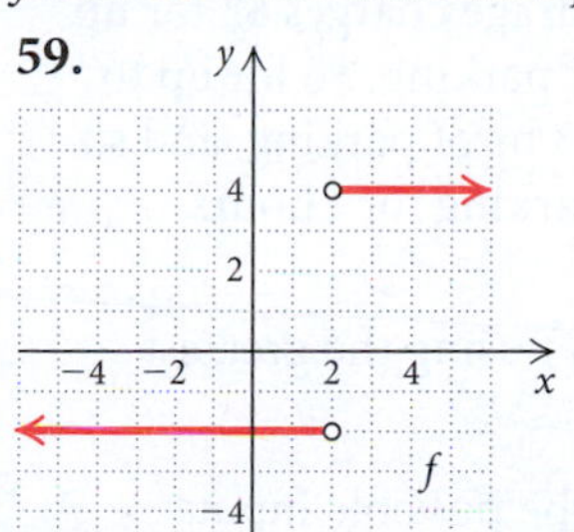

60.

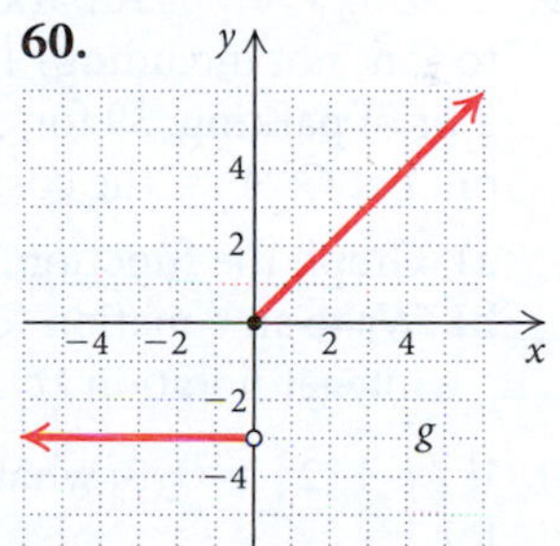

61.

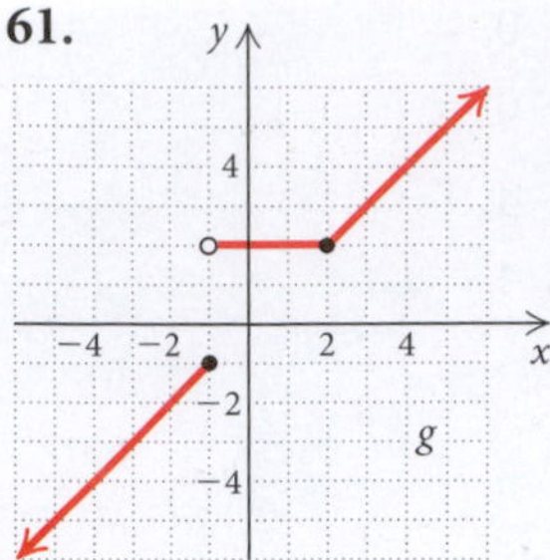

62.

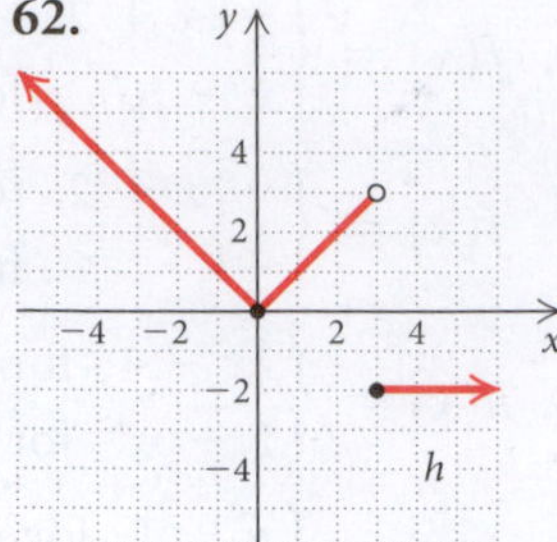

63.

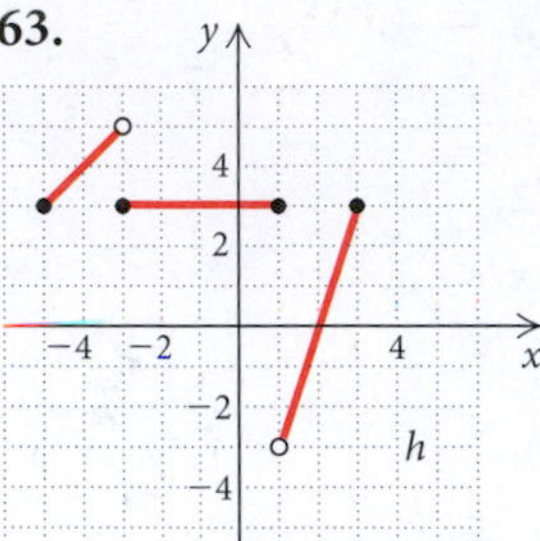

64.

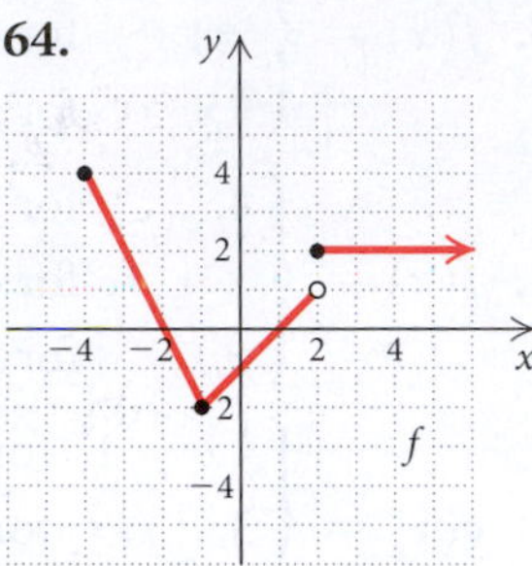

▶ Skill Maintenance

65. Given $f(x) = 5x^2 - 7$, find each of the following. [1.2]

a) $f(-3)$
b) $f(3)$
c) $f(a)$
d) $f(-a)$

66. Given $f(x) = 4x^3 - 5x$, find each of the following. [1.2]

a) $f(2)$
b) $f(-2)$
c) $f(a)$
d) $f(-a)$

67. Write an equation of the line perpendicular to the graph of the line $8x - y = 10$ and containing the point $(-1, 1)$. [1.4]

68. Find the slope and the y-intercept of the line with equation $2x - 9y + 1 = 0$. [1.4]

▶ Synthesis

69. *Parking Costs.* A parking garage charges $3 for up to (but not including) 1 hr of parking, $6 for up to 2 hr of parking, $9 for up to 3 hr of parking, and so on. Let $C(t) =$ the cost of parking for t hours.

a) Graph the function.
b) Write an equation for $C(t)$ using the greatest integer notation $[\![t]\!]$.

70. If $[\![x + 2]\!] = -3$, what are the possible inputs for x?

71. If $([\![x]\!])^2 = 25$, what are the possible inputs for x?

72. *Minimizing Power Line Costs.* A power line is constructed from a power station at point A to an island at point I, which is 1 mi directly out in the water from a point B on the shore. Point B is 4 mi downshore from the power station at A. It costs $5000 per mile to lay the power line under water and $3000 per mile to lay the power line under ground. The line comes to the shore at point S downshore from A. Let $x =$ the distance from B to S.

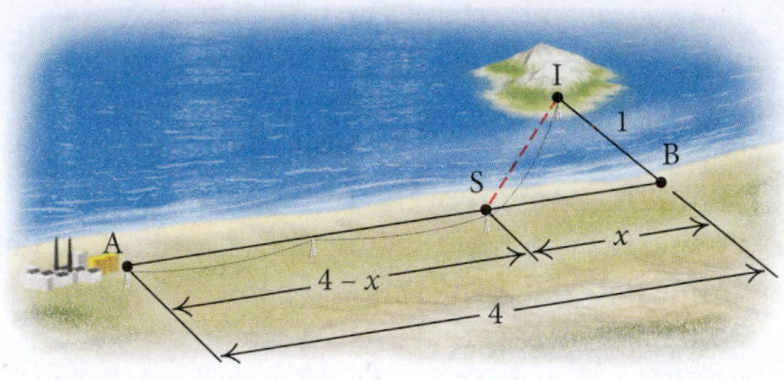

a) Express the cost C of laying the line as a function of x.
b) At what distance x from point B should the line come to shore in order to minimize cost?

73. *Volume of an Inscribed Cylinder.* A right circular cylinder of height h and radius r is inscribed in a right circular cone with a height of 10 ft and a base with radius 6 ft.

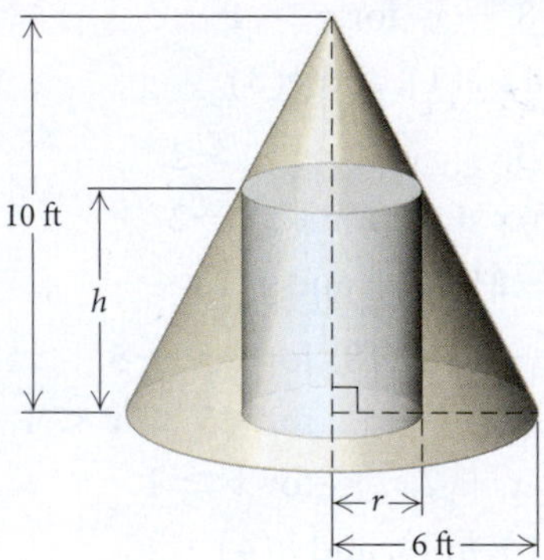

a) Express the height h of the cylinder as a function of r.
b) Express the volume V of the cylinder as a function of r.
c) Express the volume V of the cylinder as a function of h.

2.2 The Algebra of Functions

- Find the sum, the difference, the product, and the quotient of two functions, and determine the domains of the resulting functions.
- Find the difference quotient for a function.

The Algebra of Functions: Sums, Differences, Products, and Quotients

We now use addition, subtraction, multiplication, and division to combine functions and obtain new functions.

Consider the following two functions f and g:

$$f(x) = x + 2 \quad \text{and} \quad g(x) = x^2 + 1.$$

Since $f(3) = 3 + 2 = 5$ and $g(3) = 3^2 + 1 = 10$, we have

$$f(3) + g(3) = 5 + 10 = 15,$$
$$f(3) - g(3) = 5 - 10 = -5,$$
$$f(3) \cdot g(3) = 5 \cdot 10 = 50,$$

and

$$\frac{f(3)}{g(3)} = \frac{5}{10} = \frac{1}{2}.$$

In fact, so long as x is in the domain of *both* f and g, we can easily compute $f(x) + g(x)$, $f(x) - g(x)$, $f(x) \cdot g(x)$, and, assuming $g(x) \neq 0$, $f(x)/g(x)$. We use the notation shown below.

SUMS, DIFFERENCES, PRODUCTS, AND QUOTIENTS OF FUNCTIONS

If f and g are functions and x is in the domain of each function, then:

$$(f + g)(x) = f(x) + g(x),$$
$$(f - g)(x) = f(x) - g(x),$$
$$(fg)(x) = f(x) \cdot g(x),$$
$$(f/g)(x) = f(x)/g(x), \text{ provided } g(x) \neq 0.$$

EXAMPLE 1 Given that $f(x) = x + 1$ and $g(x) = \sqrt{x + 3}$, find each of the following.

a) $(f + g)(x)$ b) $(f + g)(6)$ c) $(f + g)(-4)$

Solution

a) $(f + g)(x) = f(x) + g(x)$
$= x + 1 + \sqrt{x + 3}$ This cannot be simplified.

b) We can find $(f + g)(6)$ provided 6 is in the domain of *each* function. The domain of f is all real numbers. The domain of g is all real numbers x for which $x + 3 \geq 0$, or $x \geq -3$. This is the interval $[-3, \infty)$. We see that 6 is in both domains, so we have

$$f(6) = 6 + 1 = 7, \qquad g(6) = \sqrt{6 + 3} = \sqrt{9} = 3, \quad \text{and}$$
$$(f + g)(6) = f(6) + g(6) = 7 + 3 = 10.$$

Another method is to use the formula found in part (a):

$$(f + g)(6) = 6 + 1 + \sqrt{6 + 3} = 7 + \sqrt{9} = 7 + 3 = 10.$$

c) To find $(f + g)(-4)$, we must first determine whether -4 is in the domain of both functions. We note that -4 is not in the domain of g, $[-3, \infty)$. That is, $\sqrt{-4 + 3}$ is not a real number. Thus, $(f + g)(-4)$ does not exist.

Now Try Exercise 15.

It is useful to view the concept of the sum of two functions graphically. In the graph below, we see the graphs of two functions f and g and their sum, $f + g$. Consider finding $(f + g)(4)$, or $f(4) + g(4)$. We can locate $g(4)$ on the graph of g and measure it. Then we add that length on top of $f(4)$ on the graph of f. The sum gives us $(f + g)(4)$.

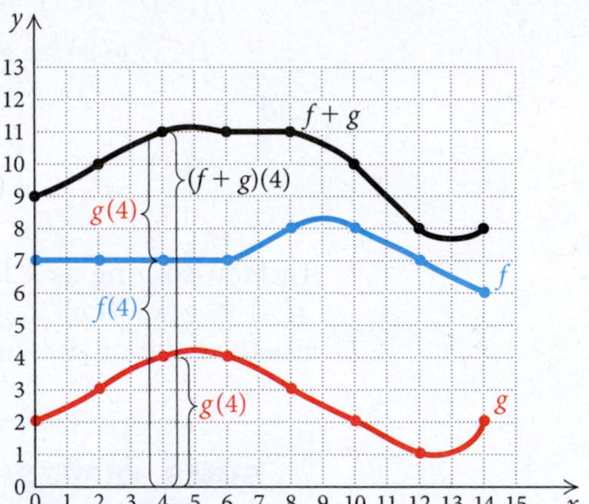
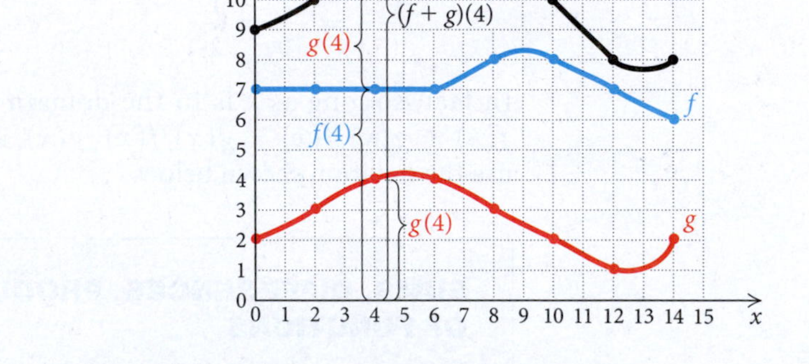

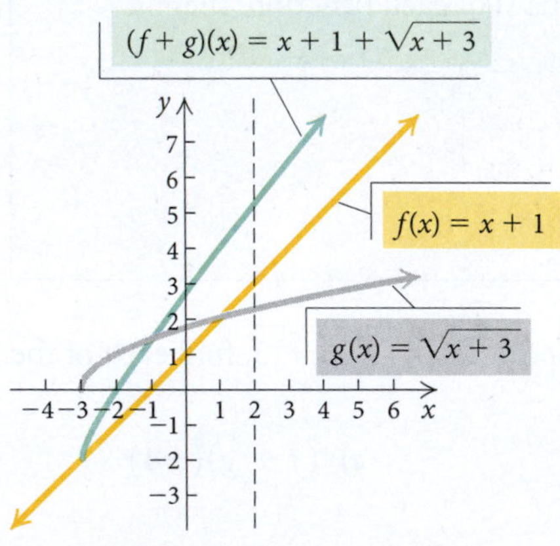

With this in mind, let's view Example 1 from a graphical perspective. Let's look at the graphs of

$$f(x) = x + 1, \qquad g(x) = \sqrt{x + 3}, \quad \text{and}$$
$$(f + g)(x) = x + 1 + \sqrt{x + 3}.$$

See the graph at left. Note that the domain of f is the set of all real numbers. The domain of g is $[-3, \infty)$. The domain of $f + g$ is the set of numbers in the intersection of the domains. This is the set of numbers in both domains.

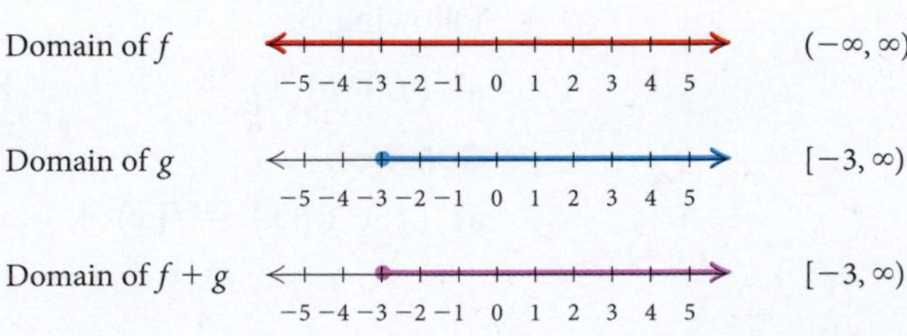

Thus the domain of $f + g$ is $[-3, \infty)$.

We can confirm that the y-coordinates of the graph of $(f + g)(x)$ are the sums of the corresponding y-coordinates of the graphs of $f(x)$ and $g(x)$. Here we confirm it for $x = 2$.

$$f(x) = x + 1 \qquad g(x) = \sqrt{x + 3}$$
$$f(2) = 2 + 1 = 3; \qquad g(2) = \sqrt{2 + 3} = \sqrt{5};$$
$$(f + g)(x) = x + 1 + \sqrt{x + 3}$$
$$(f + g)(2) = 2 + 1 + \sqrt{2 + 3}$$
$$= 3 + \sqrt{5} = f(2) + g(2)$$

DOMAIN

REVIEW SECTION 1.2

Let's also examine the domains of $f - g$, fg, and f/g for the functions $f(x) = x + 1$ and $g(x) = \sqrt{x + 3}$ of Example 1. The domains of $f - g$ and fg are the same as the domain of $f + g$, $[-3, \infty)$, because numbers in this interval are in the domains of *both* functions. For f/g, $g(x)$ cannot be 0. Since $\sqrt{x + 3} = 0$ when $x = -3$, we must exclude -3 and the domain of f/g is $(-3, \infty)$.

DOMAINS OF $f + g$, $f - g$, fg, and f/g

If f and g are functions, then the domain of the functions $f + g$, $f - g$, and fg is the intersection of the domain of f and the domain of g. The domain of f/g is also the intersection of the domains of f and g with the exclusion of any x-values for which $g(x) = 0$.

JUST IN TIME 11,12

EXAMPLE 2 Given that $f(x) = x^2 - 4$ and $g(x) = x + 2$, find each of the following.

a) The domain of $f + g$, $f - g$, fg, and f/g

b) $(f + g)(x)$

c) $(f - g)(x)$

d) $(fg)(x)$

e) $(f/g)(x)$

f) $(gg)(x)$

Solution

a) The domain of f is the set of all real numbers. The domain of g is also the set of all real numbers. The domain of $f + g$, $f - g$, and fg is the set of numbers in the intersection of the domains—that is, the set of numbers in both domains, which is again the set of real numbers. For f/g, we must exclude -2, since $g(-2) = 0$. Thus the domain of f/g is the set of real numbers excluding -2, or $(-\infty, -2) \cup (-2, \infty)$.

b) $(f + g)(x) = f(x) + g(x) = (x^2 - 4) + (x + 2) = x^2 + x - 2$

c) $(f - g)(x) = f(x) - g(x) = (x^2 - 4) - (x + 2) = x^2 - x - 6$

d) $(fg)(x) = f(x) \cdot g(x) = (x^2 - 4)(x + 2) = x^3 + 2x^2 - 4x - 8$

e) $(f/g)(x) = \dfrac{f(x)}{g(x)}$

$= \dfrac{x^2 - 4}{x + 2}$ — **Note that $g(x) = 0$ when $x = -2$, so $(f/g)(x)$ is not defined when $x = -2$.**

$= \dfrac{(x + 2)(x - 2)}{x + 2}$ — **Factoring**

$= x - 2$ — **Removing a factor of 1: $\dfrac{x + 2}{x + 2} = 1$**

Thus, $(f/g)(x) = x - 2$ with the added stipulation that $x \neq -2$ since -2 is not in the domain of $(f/g)(x)$.

f) $(gg)(x) = g(x) \cdot g(x) = [g(x)]^2 = (x + 2)^2 = x^2 + 4x + 4$

Now Try Exercise 21.

Difference Quotients

In Section 1.3, we learned that the slope of a line can be considered as an *average rate of change.* Here let's consider a nonlinear function f and draw a line through two points $(x, f(x))$ and $(x + h, f(x + h))$ as shown below.

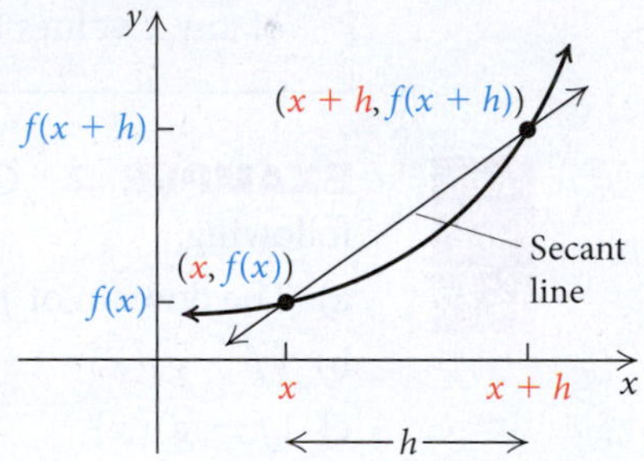

The slope of the line, called a **secant line**, is

$$\frac{f(x + h) - f(x)}{x + h - x},$$

which simplifies to

$$\frac{f(x + h) - f(x)}{h}. \quad \textbf{Difference quotient}$$

This ratio is called the **difference quotient**, or the **average rate of change**. In calculus, it is important to be able to find and simplify difference quotients.

EXAMPLE 3 For the function f given by $f(x) = 2x - 3$, find and simplify the difference quotient

$$\frac{f(x + h) - f(x)}{h}.$$

Solution

$$\frac{f(x+h)-f(x)}{h} = \frac{2(x+h)-3-(2x-3)}{h} \quad \text{Substituting}$$

$$= \frac{2x+2h-3-2x+3}{h} \quad \text{Removing parentheses}$$

$$= \frac{2h}{h}$$

$$= 2 \quad \text{Simplifying}$$

Now Try Exercise 47.

JUST IN TIME 21

EXAMPLE 4 For the function f given by $f(x) = \frac{1}{x}$, find and simplify the difference quotient

$$\frac{f(x+h)-f(x)}{h}.$$

Solution

$$\frac{f(x+h)-f(x)}{h} = \frac{\frac{1}{x+h}-\frac{1}{x}}{h} \quad \text{Substituting}$$

$$= \frac{\frac{1}{x+h}\cdot\frac{x}{x}-\frac{1}{x}\cdot\frac{x+h}{x+h}}{h} \quad \text{The LCD of } \frac{1}{x+h} \text{ and } \frac{1}{x} \text{ is } x(x+h).$$

$$= \frac{\frac{x}{x(x+h)}-\frac{x+h}{x(x+h)}}{h}$$

$$= \frac{\frac{x-(x+h)}{x(x+h)}}{h} \quad \text{Subtracting in the numerator}$$

$$= \frac{\frac{x-x-h}{x(x+h)}}{h} \quad \text{Removing parentheses}$$

$$= \frac{\frac{-h}{x(x+h)}}{h} \quad \text{Simplifying the numerator}$$

$$= \frac{-h}{x(x+h)}\cdot\frac{1}{h} \quad \text{Multiplying by the reciprocal of the divisor}$$

$$= \frac{-h\cdot 1}{x\cdot(x+h)\cdot h}$$

$$= \frac{-1\cdot h}{x\cdot(x+h)\cdot h} \quad \text{Rewriting } -h\cdot 1 \text{ as } -1\cdot h$$

$$= \frac{-1}{x(x+h)}, \text{ or } -\frac{1}{x(x+h)}$$

Now Try Exercise 55.

EXAMPLE 5 For the function f given by $f(x) = 2x^2 - x - 3$, find and simplify the difference quotient

$$\frac{f(x + h) - f(x)}{h}.$$

Solution We first find $f(x + h)$:

$$f(x + h) = 2(x + h)^2 - (x + h) - 3 \quad \text{Substituting } x + h \text{ for } x \text{ in } f(x) = 2x^2 - x - 3$$

$$= 2[x^2 + 2xh + h^2] - (x + h) - 3$$

$$= 2x^2 + 4xh + 2h^2 - x - h - 3.$$

Then we have

$$\frac{f(x + h) - f(x)}{h} = \frac{[2x^2 + 4xh + 2h^2 - x - h - 3] - [2x^2 - x - 3]}{h}$$

$$= \frac{2x^2 + 4xh + 2h^2 - x - h - 3 - 2x^2 + x + 3}{h}$$

$$= \frac{4xh + 2h^2 - h}{h}$$

$$= \frac{h(4x + 2h - 1)}{h \cdot 1} = \frac{4x + 2h - 1}{1} = 4x + 2h - 1.$$

Now Try Exercise 63.

2.2 Exercise Set

Given that $f(x) = x^2 - 3$ and $g(x) = 2x + 1$, find each of the following, if it exists.

1. $(f + g)(5)$
2. $(fg)(0)$
3. $(f - g)(-1)$
4. $(fg)(2)$
5. $(f/g)\left(-\frac{1}{2}\right)$
6. $(f - g)(0)$
7. $(fg)\left(-\frac{1}{2}\right)$
8. $(f/g)\left(-\sqrt{3}\right)$
9. $(g - f)(-1)$
10. $(g/f)\left(-\frac{1}{2}\right)$

Given that $h(x) = x + 4$ and $g(x) = \sqrt{x - 1}$, find each of the following, if it exists.

11. $(h - g)(-4)$
12. $(gh)(10)$
13. $(g/h)(1)$
14. $(h/g)(1)$
15. $(g + h)(1)$
16. $(hg)(3)$

For each pair of functions in Exercises 17–34:

a) *Find the domain of f, g, $f + g$, $f - g$, fg, ff, f/g, and g/f.*

b) *Find $(f + g)(x)$, $(f - g)(x)$, $(fg)(x)$, $(ff)(x)$, $(f/g)(x)$, and $(g/f)(x)$.*

17. $f(x) = 2x + 3$, $g(x) = 3 - 5x$
18. $f(x) = -x + 1$, $g(x) = 4x - 2$
19. $f(x) = x - 3$, $g(x) = \sqrt{x + 4}$
20. $f(x) = x + 2$, $g(x) = \sqrt{x - 1}$
21. $f(x) = 2x - 1$, $g(x) = -2x^2$
22. $f(x) = x^2 - 1$, $g(x) = 2x + 5$
23. $f(x) = \sqrt{x - 3}$, $g(x) = \sqrt{x + 3}$
24. $f(x) = \sqrt{x}$, $g(x) = \sqrt{2 - x}$
25. $f(x) = x + 1$, $g(x) = |x|$
26. $f(x) = 4|x|$, $g(x) = 1 - x$
27. $f(x) = x^3$, $g(x) = 2x^2 + 5x - 3$
28. $f(x) = x^2 - 4$, $g(x) = x^3$
29. $f(x) = \dfrac{4}{x + 1}$, $g(x) = \dfrac{1}{6 - x}$
30. $f(x) = 2x^2$, $g(x) = \dfrac{2}{x - 5}$

31. $f(x) = \frac{1}{x}$, $g(x) = x - 3$

32. $f(x) = \sqrt{x+6}$, $g(x) = \frac{1}{x}$

33. $f(x) = \frac{3}{x-2}$, $g(x) = \sqrt{x-1}$

34. $f(x) = \frac{2}{4-x}$, $g(x) = \frac{5}{x-1}$

In Exercises 35–40, consider the functions F and G as shown in the following graph.

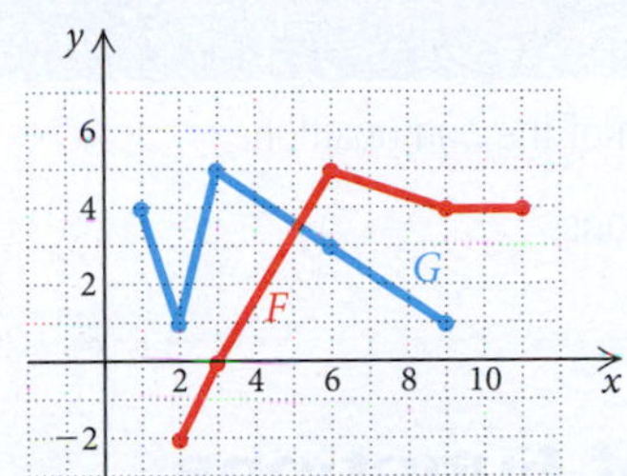

35. Find the domain of F, the domain of G, and the domain of $F + G$.

36. Find the domain of $F - G$, FG, and F/G.

37. Find the domain of G/F.

38. Graph $F + G$.

39. Graph $G - F$.

40. Graph $F - G$.

In Exercises 41–46, consider the functions F and G as shown in the following graph.

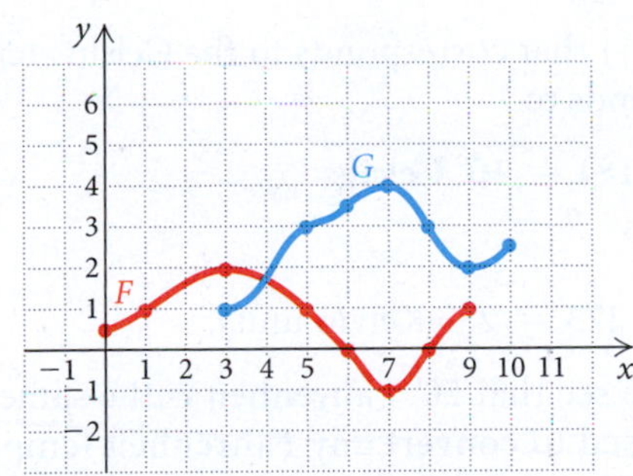

41. Find the domain of F, the domain of G, and the domain of $F + G$.

42. Find the domain of $F - G$, FG, and F/G.

43. Find the domain of G/F.

44. Graph $F + G$.

45. Graph $G - F$.

46. Graph $F - G$.

47. *Total Cost, Revenue, and Profit.* In economics, functions that involve revenue, cost, and profit are used. For example, suppose that $R(x)$ and $C(x)$ denote the total revenue and the total cost, respectively, of producing a new grocery cart for Ogata Wholesalers. Then the difference

$$P(x) = R(x) - C(x)$$

represents the total profit for producing x tools. Given

$$R(x) = 60x - 0.4x^2 \quad \text{and} \quad C(x) = 3x + 13,$$

find each of the following.

a) $P(x)$
b) $R(100)$, $C(100)$, and $P(100)$

48. *Total Cost, Revenue, and Profit.* Given that

$$R(x) = 200x - x^2 \quad \text{and} \quad C(x) = 5000 + 8x$$

for a new tablet produced by Visual Communications, find each of the following. (See Exercise 47.)

a) $P(x)$
b) $R(175)$, $C(175)$, and $P(175)$

For each function f, construct and simplify the difference quotient

$$\frac{f(x+h) - f(x)}{h}.$$

49. $f(x) = 3x - 5$

50. $f(x) = 4x - 1$

51. $f(x) = 6x + 2$

52. $f(x) = 5x + 3$

53. $f(x) = \frac{1}{3}x + 1$

54. $f(x) = -\frac{1}{2}x + 7$

55. $f(x) = \frac{1}{3x}$

56. $f(x) = \frac{1}{2x}$

57. $f(x) = -\frac{1}{4x}$

58. $f(x) = -\frac{1}{x}$

59. $f(x) = x^2 + 1$

60. $f(x) = x^2 - 3$

61. $f(x) = 4 - x^2$

62. $f(x) = 2 - x^2$

63. $f(x) = 3x^2 - 2x + 1$

64. $f(x) = 5x^2 + 4x$

65. $f(x) = 4 + 5|x|$

66. $f(x) = 2|x| + 3x$

67. $f(x) = x^3$

68. $f(x) = x^3 - 2x$

69. $f(x) = \frac{x-4}{x+3}$

70. $f(x) = \frac{x}{2-x}$

Skill Maintenance

Graph the equation. [1.1], [1.3]

71. $y = 3x - 1$

72. $2x + y = 4$

73. $x - 3y = 3$

74. $y = x^2 + 1$

▶ Synthesis

75. Write equations for two functions f and g such that the domain of $f - g$ is

$$\{x \mid x \neq -7 \text{ and } x \neq 3\}.$$

76. For functions h and f, find the domain of $h + f$, $h - f$, hf, and h/f if:

$$h = \left\{(-4, 13), (-1, 7), (0, 5), \left(\tfrac{5}{2}, 0\right), (3, -5)\right\}, \text{ and}$$
$$f = \left\{(-4, -7), (-2, -5), (0, -3), (3, 0), (5, 2), (9, 6)\right\}.$$

77. Find the domain of $(h/g)(x)$ given that

$$h(x) = \frac{5x}{3x - 7} \quad \text{and} \quad g(x) = \frac{x^4 - 1}{5x - 15}.$$

2.3 The Composition of Functions

▶ Find the composition of two functions and the domain of the composition.

▶ Decompose a function as a composition of two functions.

▶ The Composition of Functions

In real-world situations, it is not uncommon for the output of a function to depend on some input that is itself an output of another function. For instance, the amount that a person pays as state income tax usually depends on the amount of adjusted gross income on the person's federal tax return, which, in turn, depends on his or her annual earnings. Such functions are called **composite functions**.

To see how composite functions work, suppose a chemistry student needs a formula to convert Fahrenheit temperatures to Kelvin units. The formula

$$c(t) = \tfrac{5}{9}(t - 32)$$

gives the Celsius temperature $c(t)$ that corresponds to the Fahrenheit temperature t. The formula

$$k(c(t)) = c(t) + 273$$

gives the Kelvin temperature $k(c(t))$ that corresponds to the Celsius temperature $c(t)$. Thus, 50° Fahrenheit corresponds to

$$c(50) = \tfrac{5}{9}(50 - 32) = \tfrac{5}{9}(18) = 10^\circ \text{ Celsius}$$

and 10° Celsius corresponds to

$$k(c(50)) = k(10) = 10 + 273 = 283 \text{ Kelvin units,}$$

which is usually written 283 K. We see that 50° Fahrenheit is the same as 283 K. This two-step procedure can be used to convert any Fahrenheit temperature to Kelvin units.

	°F Fahrenheit	°C Celsius	K Kelvin
Boiling point of water	212°	100°	373 K
	50° →	10° →	283 K
Freezing point of water	32°	0°	273 K
Absolute zero	−460°	−273°	0 K

Technology Connection

With the TABLE feature, we can convert Fahrenheit temperatures, x, to Celsius temperatures, y_1, using

$$y_1 = \tfrac{5}{9}(x - 32).$$

We can also convert Celsius temperatures, y_1, to Kelvin units, y_2, using

$$y_2 = y_1 + 273.$$

$$y_1 = \frac{5}{9}(x - 32), \quad y_2 = y_1 + 273$$

X	Y1	Y2
50	10	283
59	15	288
68	20	293
77	25	298
86	30	303
95	35	308
104	40	313

X = 50

A student making numerous conversions might look for a formula that converts directly from Fahrenheit to Kelvin. Such a formula can be found by substitution:

$$\begin{aligned} k(c(t)) &= c(t) + 273 \\ &= \frac{5}{9}(t - 32) + 273 && \text{Substituting } \tfrac{5}{9}(t-32) \text{ for } c(t) \\ &= \frac{5}{9}t - \frac{160}{9} + 273 \\ &= \frac{5}{9}t - \frac{160}{9} + \frac{2457}{9} \\ &= \frac{5t + 2297}{9}. && \text{Simplifying} \end{aligned}$$

Since the formula found above expresses the Kelvin temperature as a new function K of the Fahrenheit temperature t, we can write

$$K(t) = \frac{5t + 2297}{9},$$

where $K(t)$ is the Kelvin temperature corresponding to the Fahrenheit temperature, t. Here we have $K(t) = k(c(t))$. The new function K is called the **composition** of k and c and can be denoted $k \circ c$ (read "k composed with c," "the composition of k and c," or "k circle c").

COMPOSITION OF FUNCTIONS

The **composite function** $f \circ g$, the **composition** of f and g, is defined as

$$(f \circ g)(x) = f(g(x)),$$

where x is in the domain of g and $g(x)$ is in the domain of f.

EXAMPLE 1 Given that $f(x) = 2x - 5$ and $g(x) = x^2 - 3x + 8$, find each of the following.

a) $(f \circ g)(x)$ and $(g \circ f)(x)$

b) $(f \circ g)(7)$ and $(g \circ f)(7)$

c) $(g \circ g)(1)$

d) $(f \circ f)(x)$

Solution Consider each function separately:

$$f(x) = 2x - 5 \qquad \text{This function multiplies each input by 2 and then subtracts 5.}$$

and

$$g(x) = x^2 - 3x + 8. \qquad \text{This function squares an input, subtracts three times the input from the result, and then adds 8.}$$

a) To find $(f \circ g)(x)$, we substitute $g(x)$ for x in the equation for $f(x)$:

$$\begin{aligned} (f \circ g)(x) = f(g(x)) &= f(x^2 - 3x + 8) && x^2 - 3x + 8 \text{ is the input for } f. \\ &= 2(x^2 - 3x + 8) - 5 && f \text{ multiplies the input by 2 and then subtracts 5.} \\ &= 2x^2 - 6x + 16 - 5 \\ &= 2x^2 - 6x + 11. \end{aligned}$$

> **Technology Connection**
>
> We can check our work in Example 1(b) using a graphing calculator. We enter the following on the equation-editor screen:
>
> $y_1 = 2x - 5$
>
> and
>
> $y_2 = x^2 - 3x + 8.$
>
> Then, on the home screen, we find $(f \circ g)(7)$ and $(g \circ f)(7)$ using the function notations Y1(Y2(7)) and Y2(Y1(7)), respectively.
>
> $y_1 = 2x - 5,\ y_2 = x^2 - 3x + 8$
>
> | | |
> |---|---|
> | Y1(Y2(7)) | |
> | | 67 |
> | Y2(Y1(7)) | |
> | | 62 |

To find $(g \circ f)(x)$, we substitute $f(x)$ for x in the equation for $g(x)$:

$$(g \circ f)(x) = g(f(x)) = g(2x - 5)$$

$2x - 5$ is the input for g.

$$= (2x - 5)^2 - 3(2x - 5) + 8$$

g squares the input, subtracts three times the input, and then adds 8.

$$= 4x^2 - 20x + 25 - 6x + 15 + 8$$
$$= 4x^2 - 26x + 48.$$

b) To find $(f \circ g)(7)$, we first find $g(7)$. Then we use $g(7)$ as an input for f:

$$(f \circ g)(7) = f(g(7)) = f(7^2 - 3 \cdot 7 + 8)$$
$$= f(36) = 2 \cdot 36 - 5$$
$$= 72 - 5 = 67.$$

To find $(g \circ f)(7)$, we first find $f(7)$. Then we use $f(7)$ as an input for g:

$$(g \circ f)(7) = g(f(7)) = g(2 \cdot 7 - 5)$$
$$= g(9) = 9^2 - 3 \cdot 9 + 8$$
$$= 81 - 27 + 8 = 62.$$

We could also find $(f \circ g)(7)$ and $(g \circ f)(7)$ by substituting 7 for x in the equations that we found in part (a):

$$(f \circ g)(x) = 2x^2 - 6x + 11$$
$$(f \circ g)(7) = 2 \cdot 7^2 - 6 \cdot 7 + 11 = 67;$$
$$(g \circ f)(x) = 4x^2 - 26x + 48$$
$$(g \circ f)(7) = 4 \cdot 7^2 - 26 \cdot 7 + 48 = 62.$$

c) $(g \circ g)(1) = g(g(1)) = g(1^2 - 3 \cdot 1 + 8)$
$$= g(1 - 3 + 8) = g(6)$$
$$= 6^2 - 3 \cdot 6 + 8$$
$$= 36 - 18 + 8 = 26$$

d) $(f \circ f)(x) = f(f(x)) = f(2x - 5)$
$$= 2(2x - 5) - 5$$
$$= 4x - 10 - 5 = 4x - 15$$

Now Try Exercises 1 and 15.

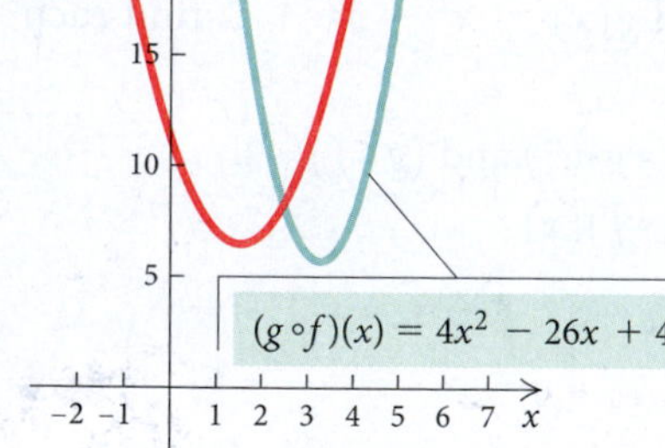

Example 1 illustrates that, as a rule, $(f \circ g)(x) \neq (g \circ f)(x)$. We can see this graphically, as shown in the graphs at left.

EXAMPLE 2 Given that $f(x) = \sqrt{x}$ and $g(x) = x - 3$:

a) Find $f \circ g$ and $g \circ f$.

b) Find the domain of $f \circ g$ and the domain of $g \circ f$.

Solution

a) $(f \circ g)(x) = f(g(x)) = f(x - 3) = \sqrt{x - 3}$

$(g \circ f)(x) = g(f(x)) = g(\sqrt{x}) = \sqrt{x} - 3$

b) Since $f(x)$ is not defined for negative radicands, the domain of $f(x)$ is $\{x|x \geq 0\}$, or $[0, \infty)$. Any real number can be an input for $g(x)$, so the domain of $g(x)$ is $(-\infty, \infty)$.

Since the inputs of $f \circ g$ are outputs of g, the domain of $f \circ g$ consists of the values of x in the domain of g, $(-\infty, \infty)$, for which $g(x)$ is nonnegative. (Recall that the inputs of $f(x)$ must be nonnegative.) Thus we have

$$\begin{aligned} g(x) &\geq 0 \\ x - 3 &\geq 0 \qquad \textbf{Substituting } x - 3 \textbf{ for } g(x) \\ x &\geq 3. \end{aligned}$$

We see that the domain of $f \circ g$ is $\{x|x \geq 3\}$, or $[3, \infty)$.

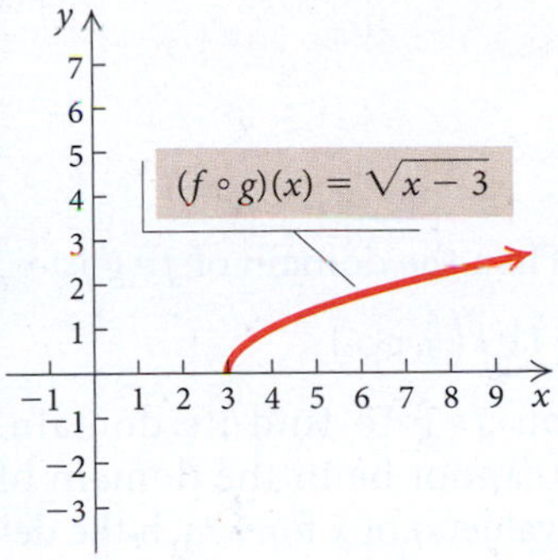

Figure 1.

We can also find the domain of $f \circ g$ by examining the composite function itself, $(f \circ g)(x) = \sqrt{x - 3}$. Since any real number can be an input for g, the only restriction on $f \circ g$ is that the radicand must be nonnegative. We have

$$\begin{aligned} x - 3 &\geq 0 \\ x &\geq 3. \end{aligned}$$

Again, we see that the domain of $f \circ g$ is $\{x|x \geq 3\}$, or $[3, \infty)$. The graph in Fig. 1 confirms this.

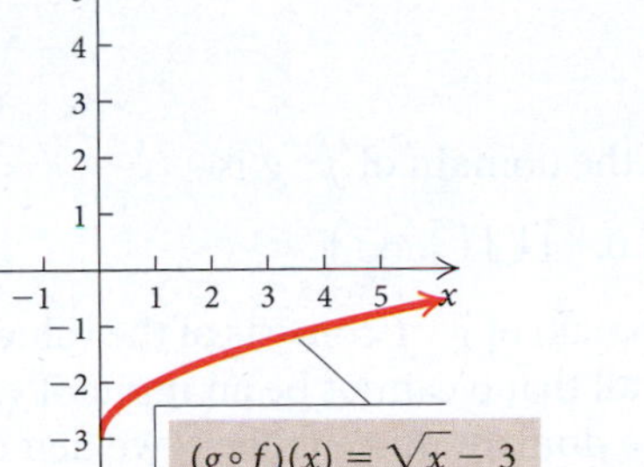

Figure 2.

The inputs of $g \circ f$ are outputs of f, so the domain of $g \circ f$ consists of the values of x in the domain of f, $[0, \infty)$, for which $g(x)$ is defined. Since g can accept *any* real number as an input, any output from f is acceptable, so the entire domain of f is the domain of $g \circ f$. That is, the domain of $g \circ f$ is $\{x|x \geq 0\}$, or $[0, \infty)$.

We can also examine the composite function itself to find its domain. First, recall that the domain of f is $\{x|x \geq 0\}$, or $[0, \infty)$. Then consider $(g \circ f)(x) = \sqrt{x} - 3$. The radicand cannot be negative, so we have $x \geq 0$. As above, we see that the domain of $g \circ f$ is the domain of f, $\{x|x \geq 0\}$, or $[0, \infty)$. The graph in Fig. 2 confirms this.

Now Try Exercise 27.

EXAMPLE 3 Given that $f(x) = \dfrac{1}{x - 2}$ and $g(x) = \dfrac{5}{x}$, find $f \circ g$ and $g \circ f$ and the domain of each.

Solution We have

$$(f \circ g)(x) = f(g(x)) = f\left(\frac{5}{x}\right) = \frac{1}{\frac{5}{x} - 2} = \frac{1}{\frac{5 - 2x}{x}} = \frac{x}{5 - 2x};$$

$$(g \circ f)(x) = g(f(x)) = g\left(\frac{1}{x - 2}\right) = \frac{5}{\frac{1}{x - 2}} = 5(x - 2).$$

Values of x that make the denominator 0 are not in the domains of these functions. Since $x - 2 = 0$ when $x = 2$, the domain of f is $\{x|x \neq 2\}$. The denominator of g is x, so the domain of g is $\{x|x \neq 0\}$.

The inputs of $f \circ g$ are outputs of g, so the domain of $f \circ g$ consists of the values of x in the domain of g for which $g(x) \neq 2$. (Recall that 2 cannot be an input of f.)

Since the domain of g is $\{x|x \neq 0\}$, 0 is not in the domain of $f \circ g$. In addition, we must find the value(s) of x for which $g(x) = 2$. We have

$$g(x) = 2$$
$$\frac{5}{x} = 2 \quad \text{Substituting } \frac{5}{x} \text{ for } g(x)$$
$$5 = 2x$$
$$\frac{5}{2} = x.$$

This tells us that $\frac{5}{2}$ is also *not* in the domain of $f \circ g$. Then the domain of $f \circ g$ is

$$\left\{x|x \neq 0 \text{ and } x \neq \tfrac{5}{2}\right\}, \text{ or } (-\infty, 0) \cup \left(0, \tfrac{5}{2}\right) \cup \left(\tfrac{5}{2}, \infty\right).$$

We can also examine the composite function $f \circ g$ to find its domain. First, recall that 0 is not in the domain of g, so it cannot be in the domain of $(f \circ g)(x) = x/(5 - 2x)$. We must also exclude the value(s) of x for which the denominator of $f \circ g$ is 0. We have

$$5 - 2x = 0$$
$$5 = 2x$$
$$\frac{5}{2} = x.$$

Again, we see that $\frac{5}{2}$ is also not in the domain, so the domain of $f \circ g$ is

$$\left\{x|x \neq 0 \text{ and } x \neq \tfrac{5}{2}\right\}, \text{ or } (-\infty, 0) \cup \left(0, \tfrac{5}{2}\right) \cup \left(\tfrac{5}{2}, \infty\right).$$

Since the inputs of $g \circ f$ are outputs of f, the domain of $g \circ f$ consists of the values of x in the domain of f for which $f(x) \neq 0$. (Recall that 0 cannot be an input of g.) The domain of f is $\{x|x \neq 2\}$, so 2 is not in the domain of $g \circ f$. Next, we determine whether there are values of x for which $f(x) = 0$:

$$f(x) = 0$$
$$\frac{1}{x - 2} = 0 \quad \text{Substituting } \frac{1}{x-2} \text{ for } f(x)$$
$$(x - 2) \cdot \frac{1}{x - 2} = (x - 2) \cdot 0 \quad \text{Multiplying by } x - 2$$
$$1 = 0. \quad \text{False equation}$$

We see that there are no values of x for which $f(x) = 0$, so there are no additional restrictions on the domain of $g \circ f$. Thus the domain of $g \circ f$ is

$$\{x|x \neq 2\}, \text{ or } (-\infty, 2) \cup (2, \infty).$$

We can also examine $g \circ f$ to find its domain. First, recall that 2 is not in the domain of f, so it cannot be in the domain of $(g \circ f)(x) = 5(x - 2)$. Since $5(x - 2)$ is defined for all real numbers, there are no additional restrictions on the domain of $g \circ f$. The domain is

$$\{x|x \neq 2\}, \text{ or } (-\infty, 2) \cup (2, \infty).$$

Now Try Exercise 23.

Decomposing a Function as a Composition

In calculus, one often needs to recognize how a function can be expressed as the composition of two functions. In this way, we are "decomposing" the function.

EXAMPLE 4 If $h(x) = (2x - 3)^5$, find $f(x)$ and $g(x)$ such that $h(x) = (f \circ g)(x)$.

Solution The function $h(x)$ raises $(2x - 3)$ to the 5th power. Two functions that can be used for the composition are

$$f(x) = x^5 \quad \text{and} \quad g(x) = 2x - 3.$$

We can check by forming the composition:

$$h(x) = (f \circ g)(x) = f(g(x)) = f(2x - 3) = (2x - 3)^5.$$

This is the most "obvious" solution. There can be other less obvious solutions. For example, if

$$f(x) = (x + 7)^5 \quad \text{and} \quad g(x) = 2x - 10,$$

then

$$\begin{aligned} h(x) &= (f \circ g)(x) = f(g(x)) \\ &= f(2x - 10) \\ &= [2x - 10 + 7]^5 = (2x - 3)^5. \end{aligned}$$

Now Try Exercise 39.

EXAMPLE 5 If $h(x) = \dfrac{1}{(x + 3)^3}$, find $f(x)$ and $g(x)$ such that $h(x) = (f \circ g)(x)$.

Solution Two functions that can be used are

$$f(x) = \frac{1}{x} \quad \text{and} \quad g(x) = (x + 3)^3.$$

We check by forming the composition:

$$h(x) = (f \circ g)(x) = f(g(x)) = f((x + 3)^3) = \frac{1}{(x + 3)^3}.$$

There are other functions that can be used as well. For example, if

$$f(x) = \frac{1}{x^3} \quad \text{and} \quad g(x) = x + 3,$$

then

$$h(x) = (f \circ g)(x) = f(g(x)) = f(x + 3) = \frac{1}{(x + 3)^3}.$$

Now Try Exercise 41.

2.3 Exercise Set

Given that $f(x) = 3x + 1$, $g(x) = x^2 - 2x - 6$, and $h(x) = x^3$, find each of the following.

1. $(f \circ g)(-1)$
2. $(g \circ f)(-2)$
3. $(h \circ f)(1)$
4. $(g \circ h)\left(\frac{1}{2}\right)$
5. $(g \circ f)(5)$
6. $(f \circ g)\left(\frac{1}{3}\right)$
7. $(f \circ h)(-3)$
8. $(h \circ g)(3)$
9. $(g \circ g)(-2)$
10. $(g \circ g)(3)$
11. $(h \circ h)(2)$
12. $(h \circ h)(-1)$
13. $(f \circ f)(-4)$
14. $(f \circ f)(1)$
15. $(h \circ h)(x)$
16. $(f \circ f)(x)$

Find $(f \circ g)(x)$ and $(g \circ f)(x)$ and the domain of each.

17. $f(x) = x + 3,\ g(x) = x - 3$

18. $f(x) = \frac{4}{5}x,\ g(x) = \frac{5}{4}x$

19. $f(x) = x + 1,\ g(x) = 3x^2 - 2x - 1$

20. $f(x) = 3x - 2,\ g(x) = x^2 + 5$

21. $f(x) = x^2 - 3,\ g(x) = 4x - 3$

22. $f(x) = 4x^2 - x + 10,\ g(x) = 2x - 7$

23. $f(x) = \dfrac{4}{1 - 5x},\ g(x) = \dfrac{1}{x}$

24. $f(x) = \dfrac{6}{x},\ g(x) = \dfrac{1}{2x + 1}$

25. $f(x) = 3x - 7,\ g(x) = \dfrac{x + 7}{3}$

26. $f(x) = \frac{2}{3}x - \frac{4}{5},\ g(x) = 1.5x + 1.2$

27. $f(x) = 2x + 1,\ g(x) = \sqrt{x}$

28. $f(x) = \sqrt{x},\ g(x) = 2 - 3x$

29. $f(x) = 20,\ g(x) = 0.05$

30. $f(x) = x^4,\ g(x) = \sqrt[4]{x}$

31. $f(x) = \sqrt{x + 5},\ g(x) = x^2 - 5$

32. $f(x) = x^5 - 2,\ g(x) = \sqrt[5]{x + 2}$

33. $f(x) = x^2 + 2,\ g(x) = \sqrt{3 - x}$

34. $f(x) = 1 - x^2,\ g(x) = \sqrt{x^2 - 25}$

35. $f(x) = \dfrac{1 - x}{x},\ g(x) = \dfrac{1}{1 + x}$

36. $f(x) = \dfrac{1}{x - 2},\ g(x) = \dfrac{x + 2}{x}$

37. $f(x) = x^3 - 5x^2 + 3x + 7,\ g(x) = x + 1$

38. $f(x) = x - 1,\ g(x) = x^3 + 2x^2 - 3x - 9$

Find $f(x)$ and $g(x)$ such that $h(x) = (f \circ g)(x)$. Answers may vary.

39. $h(x) = (4 + 3x)^5$

40. $h(x) = \sqrt[3]{x^2 - 8}$

41. $h(x) = \dfrac{1}{(x - 2)^4}$

42. $h(x) = \dfrac{1}{\sqrt{3x + 7}}$

43. $h(x) = \dfrac{x^3 - 1}{x^3 + 1}$

44. $h(x) = |9x^2 - 4|$

45. $h(x) = \left(\dfrac{2 + x^3}{2 - x^3}\right)^6$

46. $h(x) = (\sqrt{x} - 3)^4$

47. $h(x) = \sqrt{\dfrac{x - 5}{x + 2}}$

48. $h(x) = \sqrt{1 + \sqrt{1 + x}}$

49. $h(x) = (x + 2)^3 - 5(x + 2)^2 + 3(x + 2) - 1$

50. $h(x) = 2(x - 1)^{5/3} + 5(x - 1)^{2/3}$

51. *Ripple Spread.* A stone is thrown into a pond, creating a circular ripple that spreads over the pond in such a way that the radius is increasing at a rate of 3 ft/sec.

a) Find a function $r(t)$ for the radius in terms of t.

b) Find a function $A(r)$ for the area of the ripple in terms of the radius r.

c) Find $(A \circ r)(t)$. Explain the meaning of this function.

52. The surface area S of a right circular cylinder is given by the formula $S = 2\pi rh + 2\pi r^2$. If the height is twice the radius, find each of the following.

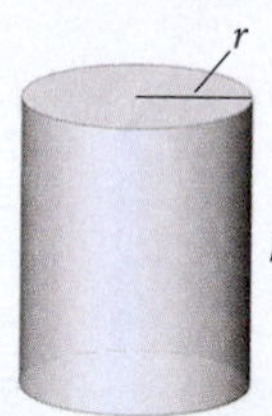

a) A function $S(r)$ for the surface area as a function of r

b) A function $S(h)$ for the surface area as a function of h

53. *Blouse Sizes.* A blouse that is size x in Japan is size $s(x)$ in the United States, where $s(x) = x - 3$. A blouse that is size x in the United States is size $t(x)$ in Australia, where $t(x) = x + 4$. (*Source*: www.onlineconversion.com) Find a function that will convert blouse sizes in Japan to blouse sizes in Australia.

54. A manufacturer of tools, selling rechargeable drills to a chain of home improvement stores, charges \$6 more per drill than its manufacturing cost m. The stores then sell each drill for 150% of the price that it paid the manufacturer. Find a function $P(m)$ for the price at the home improvement stores.

▶Skill Maintenance

Consider the following linear equations. Without graphing them, answer the questions in Exercises 55–62. [1.3], [1.4]

a) $y = x$ **b)** $y = -5x + 4$
c) $y = \frac{2}{3}x + 1$ **d)** $y = -0.1x + 6$
e) $y = 3x - 5$ **f)** $y = -x - 1$
g) $2x - 3y = 6$ **h)** $6x + 3y = 9$

55. Which, if any, have y-intercept $(0, 1)$?

56. Which, if any, have the same y-intercept?

57. Which slope down from left to right?

58. Which has the steepest slope?

59. Which pass(es) through the origin?

60. Which, if any, have the same slope?

61. Which, if any, are parallel?

62. Which, if any, are perpendicular?

▶Synthesis

63. Let $p(a)$ represent the number of pounds of grass seed required to seed a lawn with area a. Let $c(s)$ represent the cost of s pounds of grass seed. Which composition makes sense: $(c \circ p)(a)$ or $(p \circ c)(s)$? What does it represent?

64. Write equations of two functions f and g such that $f \circ g = g \circ f = x$. (In Section 5.1, we will study inverse functions. If $f \circ g = g \circ f = x$, functions f and g are *inverses* of each other.)

Mid-Chapter Mixed Review

Determine whether the statement is true or false.

1. $f(c)$ is a relative maximum if $(c, f(c))$ is the highest point in some open interval containing c. [2.1]

2. If f and g are functions, then the domain of the functions $f + g$, $f - g$, fg, and f/g is the intersection of the domain of f and the domain of g. [2.2]

3. In general, $(f \circ g)(x) \neq (g \circ f)(x)$. [2.3]

4. Determine the intervals on which the function is (a) increasing; (b) decreasing; (c) constant. [2.1]

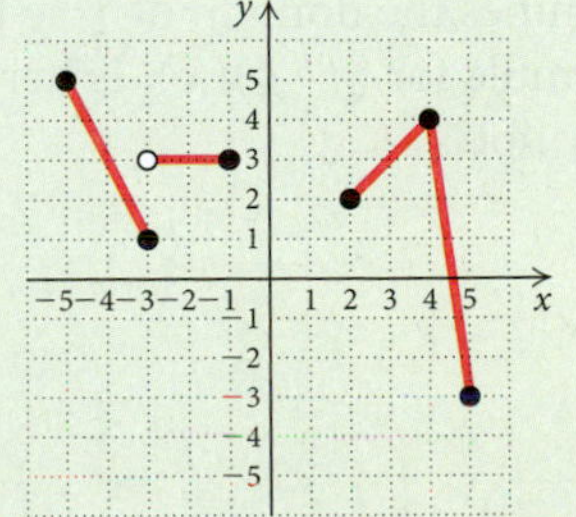

5. Using the graph shown below, determine any relative maxima or minima of the function and the intervals on which the function is increasing or decreasing. [2.1]

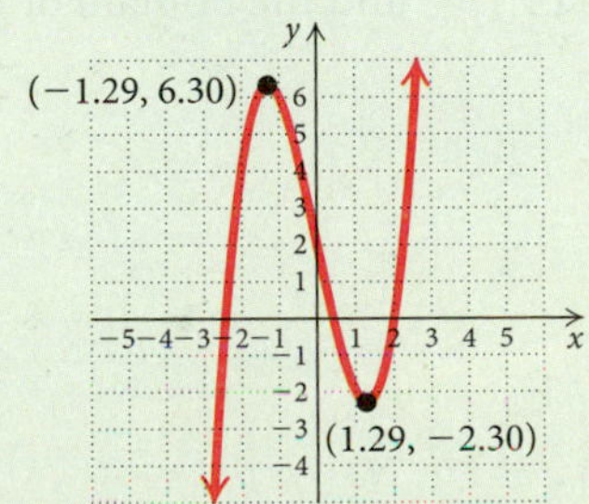

6. Determine the domain and the range of the function graphed in Exercise 4. [2.1]

7. *Window Design.* Lucas is designing a window for the peak of an A-frame house. The base is 4 ft more than the height h. Express the area of the window as a function of the height. [2.1]

8. For the function defined as

$$f(x) = \begin{cases} x - 5, & \text{for } x \le -3, \\ 2x + 3, & \text{for } -3 < x \le 0, \\ \frac{1}{2}x, & \text{for } x > 0, \end{cases}$$

$f(-5), f(-3), f(-1)$, and $f(6)$. [2.1]

9. Graph the function defined as

$$g(x) = \begin{cases} x + 2, & \text{for } x < -4, \\ -x, & \text{for } x \ge -4. \end{cases} \quad [2.1]$$

Given that $f(x) = 3x - 1$ and $g(x) = x^2 + 4$, find each of the following, if it exists. [2.2]

10. $(f + g)(-1)$

11. $(fg)(0)$

12. $(g - f)(3)$

13. $(g/f)\left(\frac{1}{3}\right)$

For each pair of functions in Exercises 14 and 15:

a) *Find the domains of $f, g, f + g, f - g, fg, ff, f/g$, and g/f.*

b) *Find $(f + g)(x)$, $(f - g)(x)$, $(fg)(x)$, $(ff)(x)$, $(f/g)(x)$, and $(g/f)(x)$.* [2.2]

14. $f(x) = 2x + 5,\ g(x) = -x - 4$

15. $f(x) = x - 1,\ g(x) = \sqrt{x + 2}$

For each function f, construct and simplify the difference quotient

$$\frac{f(x + h) - f(x)}{h}. \quad [2.2]$$

16. $f(x) = 4x - 3$

17. $f(x) = 6 - x^2$

Given that $f(x) = 5x - 4$, $g(x) = x^3 + 1$, and $h(x) = x^2 - 2x + 3$, find each of the following. [2.3]

18. $(f \circ g)(1)$

19. $(g \circ h)(2)$

20. $(f \circ f)(0)$

21. $(h \circ f)(-1)$

Find $(f \circ g)(x)$ and $(g \circ f)(x)$ and the domain of each. [2.3]

22. $f(x) = \frac{1}{2}x,\ g(x) = 6x + 4$

23. $f(x) = 3x + 2,\ g(x) = \sqrt{x}$

Collaborative Discussion and Writing

24. If $g(x) = b$, where b is a positive constant, describe how the graphs of $y = h(x)$ and $y = (h - g)(x)$ will differ. [2.2]

25. If the domain of a function f is the set of real numbers and the domain of a function g is also the set of real numbers, under what circumstances do $(f + g)(x)$ and $(f/g)(x)$ have different domains? [2.2]

26. If f and g are linear functions, what can you say about the domain of $f \circ g$ and the domain of $g \circ f$? [2.3]

27. Nora determines the domain of $f \circ g$ by examining only the formula for $(f \circ g)(x)$. Is her approach valid? Why or why not? [2.3]

2.4 Symmetry

- Determine whether a graph is symmetric with respect to the x-axis, the y-axis, and the origin.
- Determine whether a function is even, odd, or neither even nor odd.

Symmetry

Symmetry occurs often in nature and in art. For example, when viewed from the front, the bodies of most animals are at least approximately symmetric. This means that each eye is the same distance from the center of the bridge of the nose, each shoulder is the same distance from the center of the chest, and so on. Architects have used symmetry for thousands of years to enhance the beauty of buildings.

A knowledge of symmetry in mathematics helps us graph and analyze equations and functions.

Consider the points $(4, 2)$ and $(4, -2)$ that appear on the graph of $x = y^2$, as shown below. Points like these have the same x-value but opposite y-values and are **reflections** of each other across the x-axis. If, for any point (x, y) on a graph, the point $(x, -y)$ is also on the graph, then the graph is said to be **symmetric with respect to the x-axis**. If we fold the graph on the x-axis, the parts above and below the x-axis will coincide.

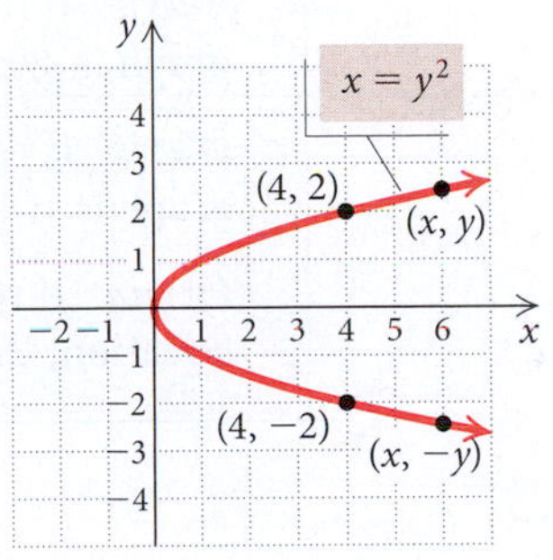

Consider the points $(3, 4)$ and $(-3, 4)$ that appear on the graph of $y = x^2 - 5$, as shown below. Points like these have the same y-value but opposite x-values and are **reflections** of each other across the y-axis. If, for any point (x, y) on a graph, the point $(-x, y)$ is also on the graph, then the graph is said to be **symmetric with respect to the y-axis.** If we fold the graph on the y-axis, the parts to the left and right of the y-axis will coincide.

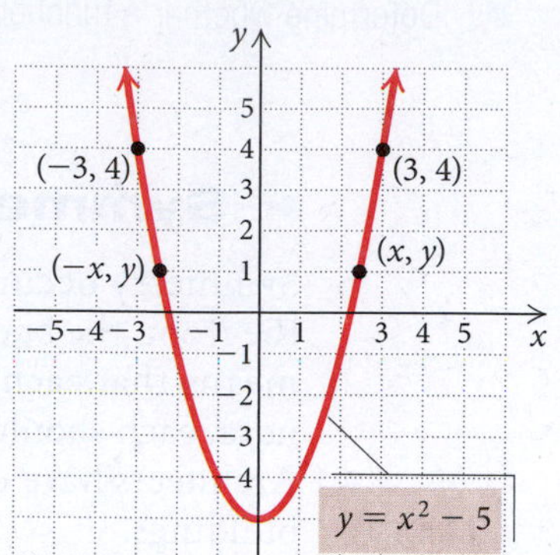

Consider the points $(-3, \sqrt{7})$ and $(3, -\sqrt{7})$ that appear on the graph of $x^2 = y^2 + 2$, as shown below. Note that if we take the opposites of the coordinates of one pair, we get the other pair. If, for any point (x, y) on a graph, the point $(-x, -y)$ is also on the graph, then the graph is said to be **symmetric with respect to the origin**. Visually, if we rotate the graph 180° about the origin, the resulting figure coincides with the original.

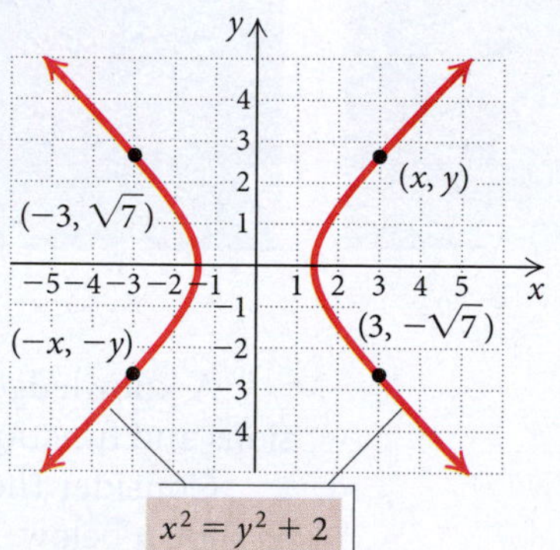

ALGEBRAIC TESTS OF SYMMETRY

x-axis: If replacing y with $-y$ produces an equivalent equation, then the graph is *symmetric with respect to the x-axis.*

y-axis: If replacing x with $-x$ produces an equivalent equation, then the graph is *symmetric with respect to the y-axis.*

Origin: If replacing x with $-x$ and y with $-y$ produces an equivalent equation, then the graph is *symmetric with respect to the origin.*

EXAMPLE 1 Test $y = x^2 + 2$ for symmetry with respect to the x-axis, the y-axis, and the origin.

Algebraic Solution

***x*-Axis:**
We replace y with $-y$:

$$y = x^2 + 2$$
$$-y = x^2 + 2$$
$$y = -x^2 - 2. \quad \text{Multiplying by } -1 \text{ on both sides}$$

The resulting equation *is not* equivalent to the original equation, so the graph *is not* symmetric with respect to the x-axis.

***y*-Axis:**
We replace x with $-x$:

$$y = x^2 + 2$$
$$y = (-x)^2 + 2$$
$$y = x^2 + 2. \quad \text{Simplifying}$$

The resulting equation *is* equivalent to the original equation, so the graph *is* symmetric with respect to the y-axis.

Origin:
We replace x with $-x$ and y with $-y$:

$$y = x^2 + 2$$
$$-y = (-x)^2 + 2$$
$$-y = x^2 + 2 \quad \text{Simplifying}$$
$$y = -x^2 - 2.$$

The resulting equation *is not* equivalent to the original equation, so the graph *is not* symmetric with respect to the origin.

Visualizing the Solution

Let's look at the graph of $y = x^2 + 2$.

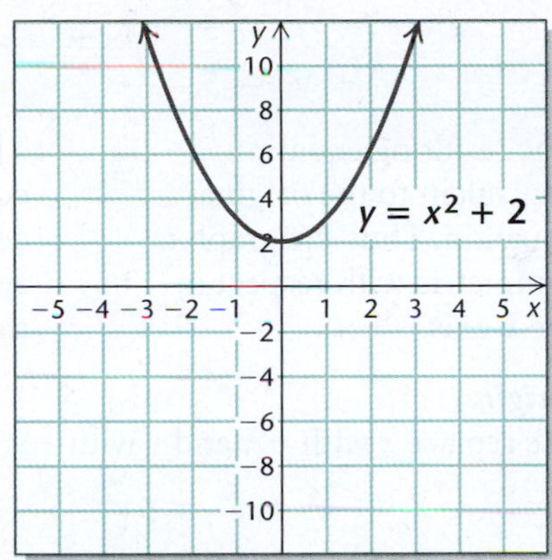

Note that if the graph were folded on the x-axis, the parts above and below the x-axis would not coincide. If it were folded on the y-axis, the parts to the left and right of the y-axis would coincide. If we rotated it 180° about the origin, the resulting graph would not coincide with the original graph.

Thus we see that the graph *is not* symmetric with respect to the x-axis or the origin. The graph *is* symmetric with respect to the y-axis.

Now Try Exercise 11.

EXAMPLE 2 Test $x^2 + y^4 = 5$ for symmetry with respect to the x-axis, the y-axis, and the origin.

Algebraic Solution

***x*-Axis:**
We replace y with $-y$:

$$x^2 + y^4 = 5$$
$$x^2 + (-y)^4 = 5$$
$$x^2 + y^4 = 5.$$

The resulting equation *is* equivalent to the original equation. Thus the graph *is* symmetric with respect to the x-axis.

***y*-Axis:**
We replace x with $-x$:

$$x^2 + y^4 = 5$$
$$(-x)^2 + y^4 = 5$$
$$x^2 + y^4 = 5.$$

The resulting equation *is* equivalent to the original equation, so the graph *is* symmetric with respect to the y-axis.

Origin:
We replace x with $-x$ and y with $-y$:

$$x^2 + y^4 = 5$$
$$(-x)^2 + (-y)^4 = 5$$
$$x^2 + y^4 = 5.$$

The resulting equation *is* equivalent to the original equation, so the graph *is* symmetric with respect to the origin.

Visualizing the Solution

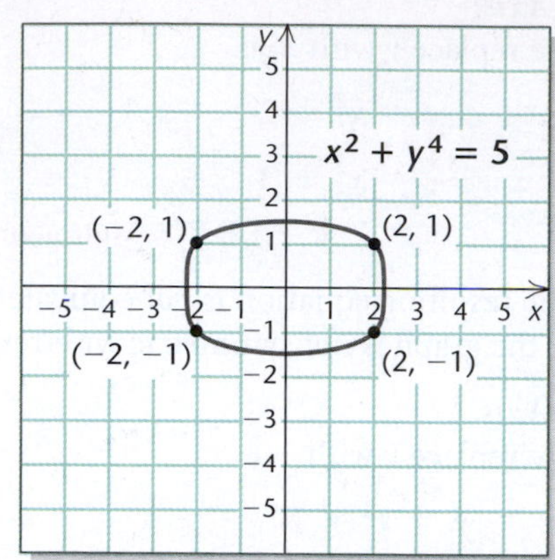

From the graph of the equation, we see symmetry with respect to both axes and with respect to the origin.

Now Try Exercise 21.

Even Functions and Odd Functions

Now we relate symmetry to graphs of functions.

EVEN FUNCTIONS AND ODD FUNCTIONS

If the graph of a function f is symmetric with respect to the y-axis, we say that it is an **even function**. That is, for each x in the domain of f, $f(x) = f(-x)$.

If the graph of a function f is symmetric with respect to the origin, we say that it is an **odd function**. That is, for each x in the domain of f, $f(-x) = -f(x)$.

Algebraic Procedure for Determining Even Functions and Odd Functions

Given the function $f(x)$:

1. Find $f(-x)$ and simplify. If $f(x) = f(-x)$, then f is even.
2. Find $-f(x)$, simplify, and compare with $f(-x)$ from step (1). If $f(-x) = -f(x)$, then f is odd.

Except for the function $f(x) = 0$, a function cannot be *both* even and odd. Thus if $f(x) \neq 0$ and we see in step (1) that $f(x) = f(-x)$ (that is, f is even), we need not continue.

An algebraic procedure for determining even functions and odd functions is shown at left. Below we show an even function and an odd function. Many functions are neither even nor odd.

EXAMPLE 3 Determine whether each of the following functions is even, odd, or neither.

a) $f(x) = 5x^7 - 6x^3 - 2x$

b) $h(x) = 5x^6 - 3x^2 - 7$

a) Algebraic Solution

$f(x) = 5x^7 - 6x^3 - 2x$

1. $f(-x) = 5(-x)^7 - 6(-x)^3 - 2(-x)$

$= 5(-x^7) - 6(-x^3) + 2x$

$(-x)^7 = (-1 \cdot x)^7 = (-1)^7 x^7 = -x^7; (-x)^3 = -x^3$

$= -5x^7 + 6x^3 + 2x$

We see that $f(x) \neq f(-x)$. Thus, f is not even.

2. $-f(x) = -(5x^7 - 6x^3 - 2x)$

$= -5x^7 + 6x^3 + 2x$

We see that $f(-x) = -f(x)$. Thus, f is odd.

Visualizing the Solution

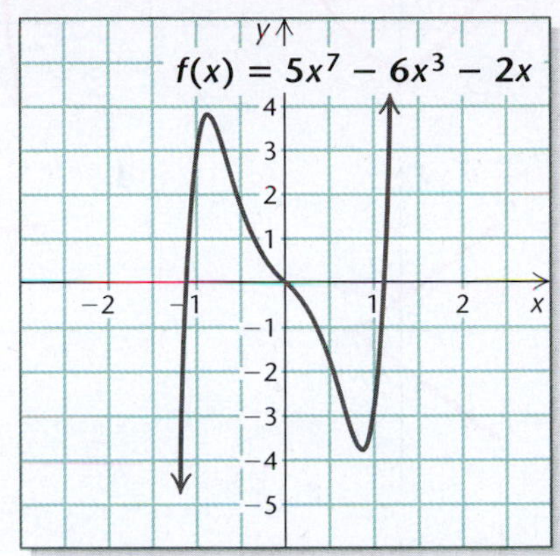

We see that the graph appears to be symmetric with respect to the origin. The function is odd.

b) Algebraic Solution

$h(x) = 5x^6 - 3x^2 - 7$

1. $h(-x) = 5(-x)^6 - 3(-x)^2 - 7$

$= 5x^6 - 3x^2 - 7$

We see that $h(x) = h(-x)$. Thus the function is even.

Visualizing the Solution

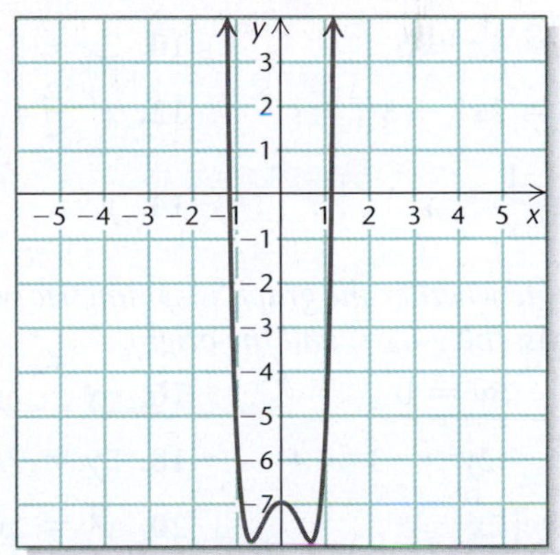

$h(x) = 5x^6 - 3x^2 - 7$

We see that the graph appears to be symmetric with respect to the y-axis. The function is even.

Now Try Exercises 39 and 41.

2.4 Exercise Set

Determine visually whether the graph is symmetric with respect to the x-axis, the y-axis, and the origin.

1.

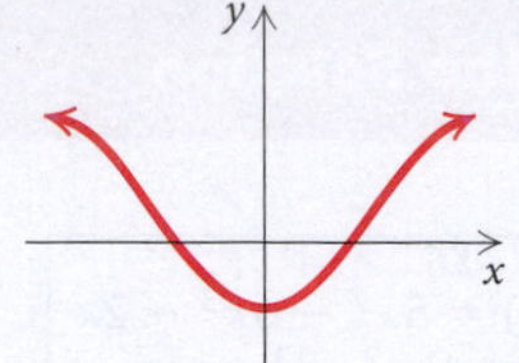

2.

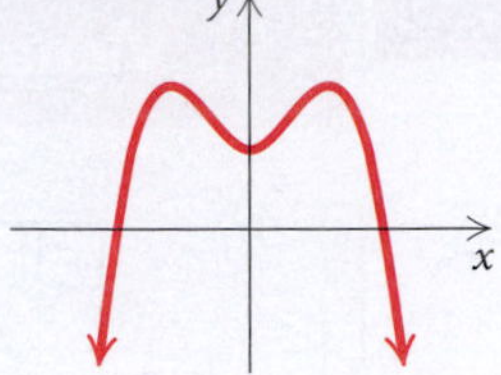

3.

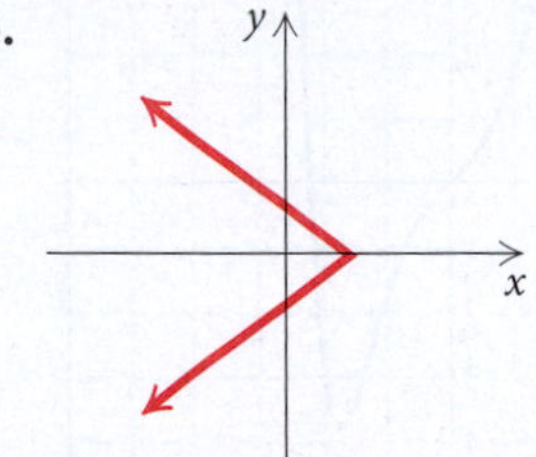

4.

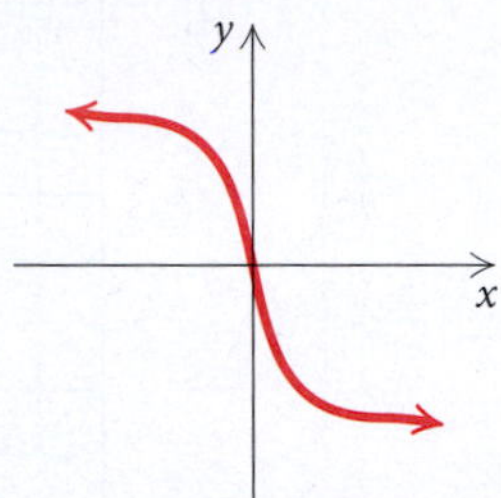

5.

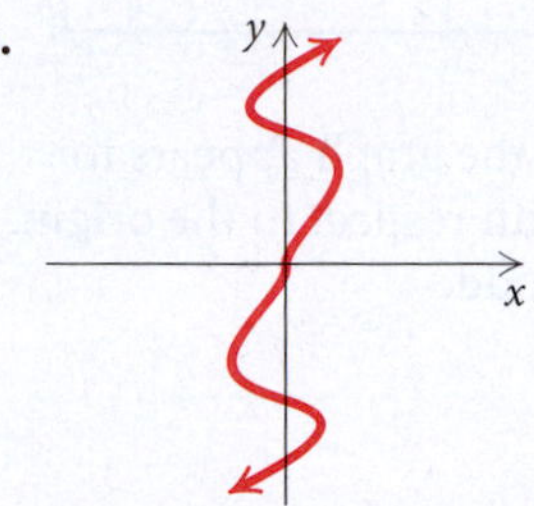

6.

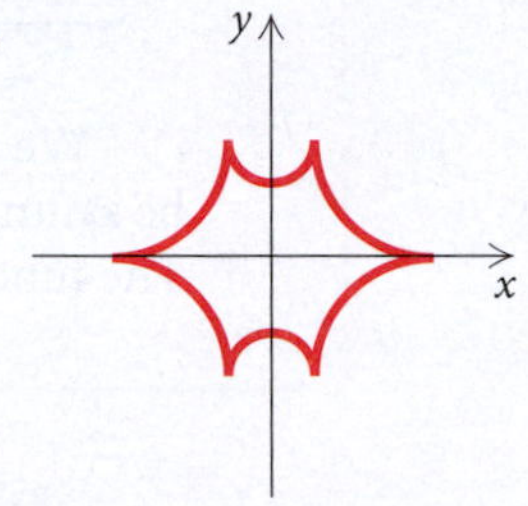

First, graph the equation and determine visually whether it is symmetric with respect to the x-axis, the y-axis, and the origin. Then verify your assertion algebraically.

7. $y = |x| - 2$

8. $y = |x + 5|$

9. $5y = 4x + 5$

10. $2x - 5 = 3y$

11. $5y = 2x^2 - 3$

12. $x^2 + 4 = 3y$

13. $y = \dfrac{1}{x}$

14. $y = -\dfrac{4}{x}$

Determine whether the graph is symmetric with respect to the x-axis, the y-axis, and the origin.

15. $5x - 5y = 0$

16. $6x + 7y = 0$

17. $3x^2 - 2y^2 = 3$

18. $5y = 7x^2 - 2x$

19. $y = |2x|$

20. $y^3 = 2x^2$

21. $2x^4 + 3 = y^2$

22. $2y^2 = 5x^2 + 12$

23. $3y^3 = 4x^3 + 2$

24. $3x = |y|$

25. $xy = 12$

26. $xy - x^2 = 3$

Find the point that is symmetric to the given point with respect to the x-axis, the y-axis, and the origin.

27. $(-5, 6)$

28. $\left(\frac{7}{2}, 0\right)$

29. $(-10, -7)$

30. $\left(1, \frac{3}{8}\right)$

31. $(0, -4)$

32. $(8, -3)$

Determine visually whether the function is even, odd, or neither even nor odd.

33.

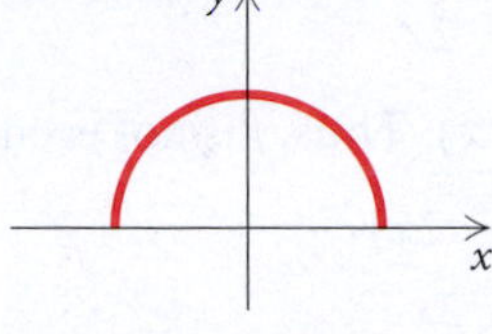

34.

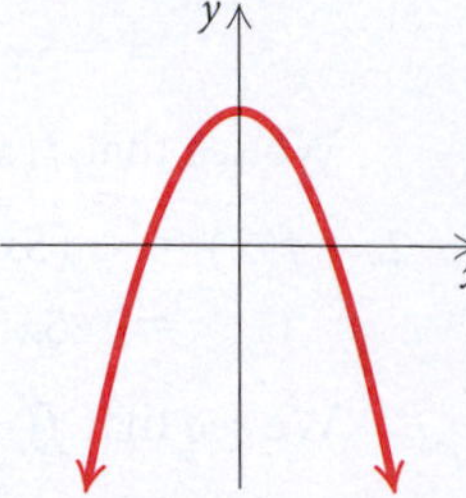

35.

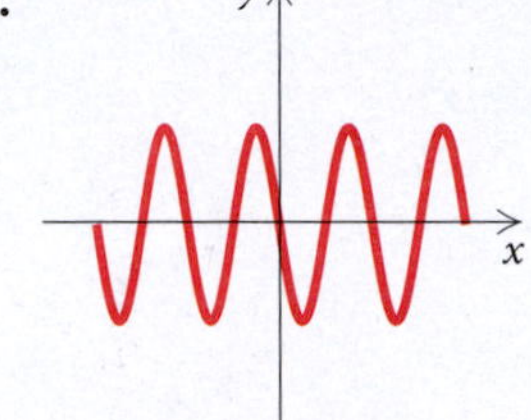

36.

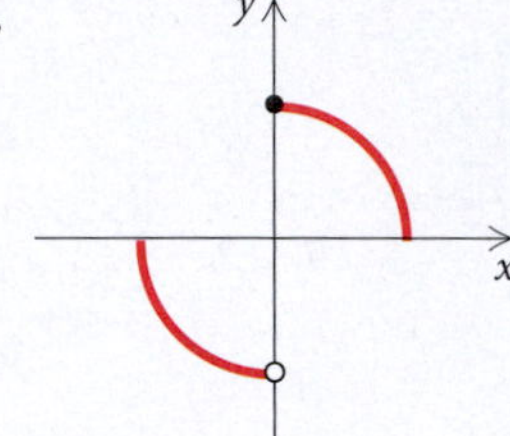

37.

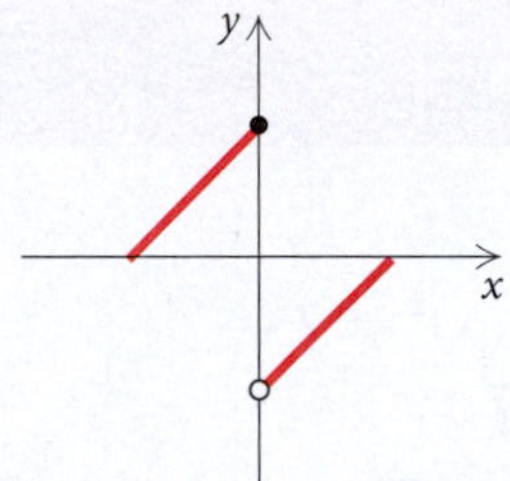

38.

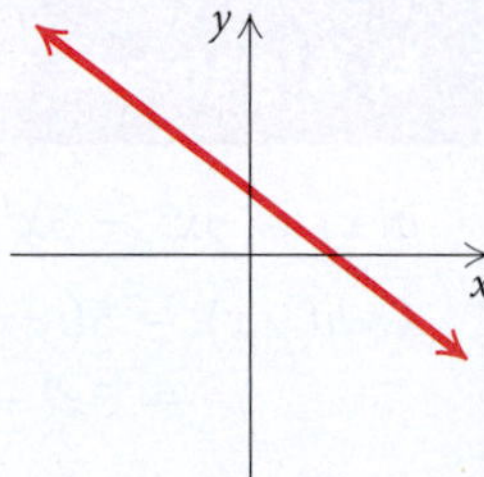

Determine whether the function is even, odd, or neither even nor odd.

39. $f(x) = -3x^3 + 2x$

40. $f(x) = 7x^3 + 4x - 2$

41. $f(x) = 5x^2 + 2x^4 - 1$

42. $f(x) = x + \dfrac{1}{x}$

43. $f(x) = x^{17}$

44. $f(x) = \sqrt[3]{x}$

45. $f(x) = x - |x|$

46. $f(x) = \dfrac{1}{x^2}$

47. $f(x) = 8$

48. $f(x) = \sqrt{x^2 + 1}$

Skill Maintenance

49. Graph: $f(x) = \begin{cases} x - 2, & \text{for } x \leq -1, \\ 3, & \text{for } -1 < x \leq 2, \\ x, & \text{for } x > 2. \end{cases}$ [2.1]

50. *Peace Corps Volunteers.* Since 1961, there has been a total of 6688 Peace Corps volunteers from the University of California–Berkeley and the University of Wisconsin–Madison. The number of volunteers from the University of California–Berkeley is 464 more than the number of volunteers from the University of Wisconsin–Madison. (*Source:* Peace Corps 2014) Find the number of Peace Corps volunteers from each university. [1.5]

Synthesis

Determine whether the function is even, odd, or neither even nor odd.

51. $f(x) = x\sqrt{10 - x^2}$

52. $f(x) = \dfrac{x^2 + 1}{x^3 - 1}$

Determine whether the graph is symmetric with respect to the x-axis, the y-axis, and the origin.

53. $x^3 = y^2(2 - x)$

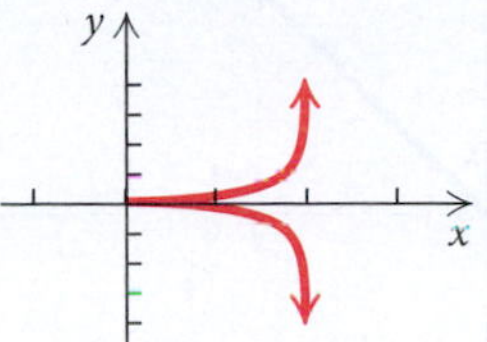

54. $(x^2 + y^2)^2 = 2xy$

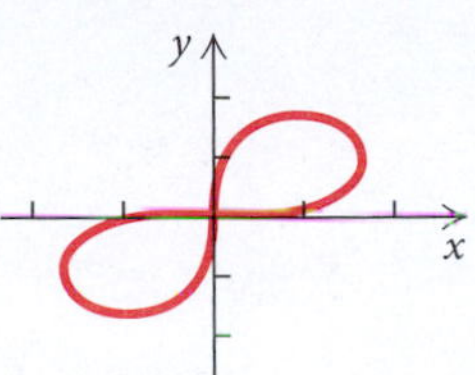

55. Show that if f is *any* function, then the function E defined by

$$E(x) = \frac{f(x) + f(-x)}{2}$$

is even.

56. Show that if f is *any* function, then the function O defined by

$$O(x) = \frac{f(x) - f(-x)}{2}$$

is odd.

57. Consider the functions E and O of Exercises 55 and 56.
   a) Show that $f(x) = E(x) + O(x)$. This means that every function can be expressed as the sum of an even function and an odd function.
   b) Let $f(x) = 4x^3 - 11x^2 + \sqrt{x} - 10$. Express f as a sum of an even function and an odd function.

Determine whether the statement is true or false.

58. The product of two odd functions is odd.

59. The sum of two even functions is even.

60. The product of an even function and an odd function is odd.

2.5 Transformations

- Given the graph of a function, graph its transformation under translations, reflections, stretchings, and shrinkings.

Transformations of Functions

The graphs of some basic functions are shown on the following page. Others can be seen on the inside back cover.

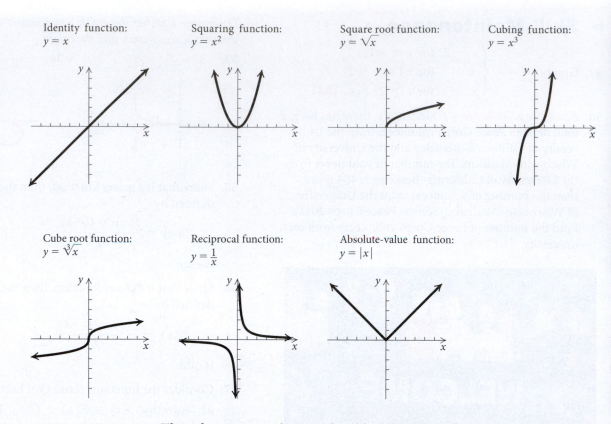

Identity function:
$y = x$

Squaring function:
$y = x^2$

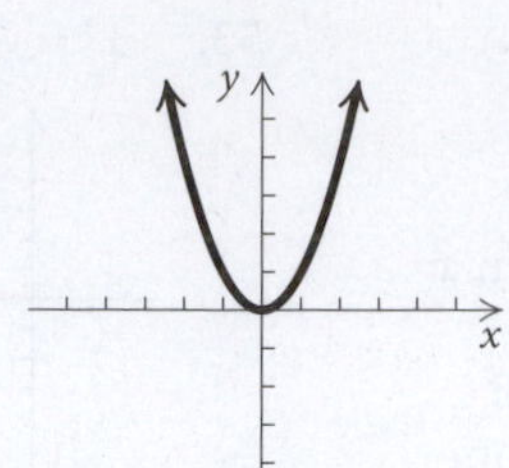

Square root function:
$y = \sqrt{x}$

Cubing function:
$y = x^3$

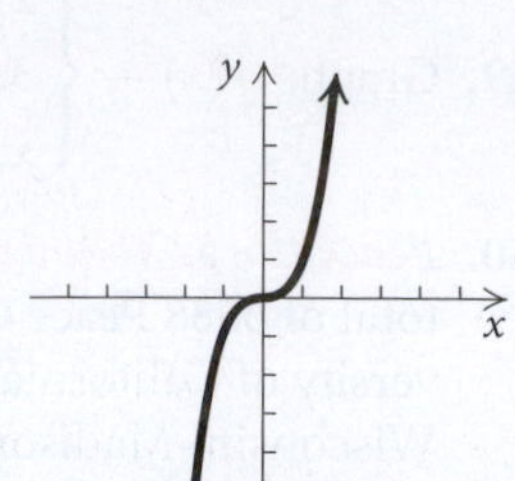

Cube root function:
$y = \sqrt[3]{x}$

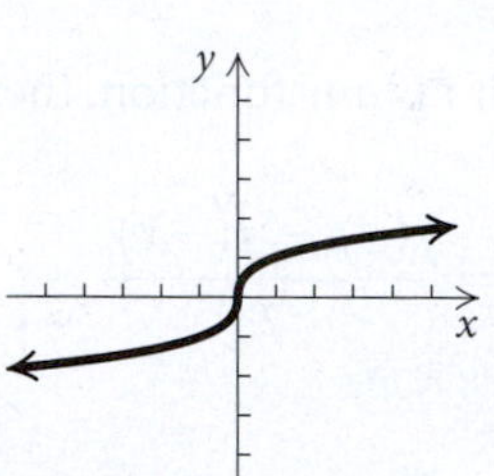

Reciprocal function:
$y = \frac{1}{x}$

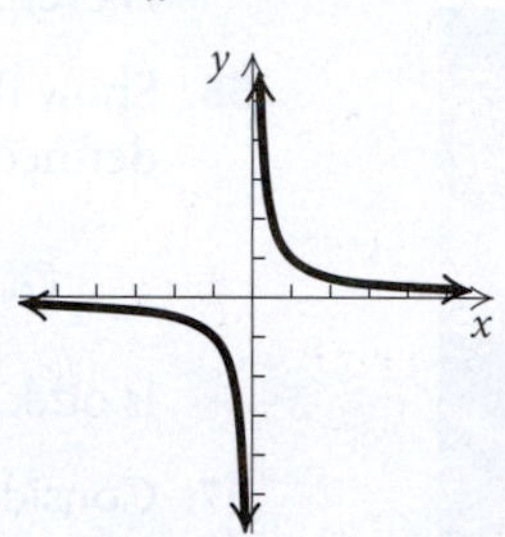

Absolute-value function:
$y = |x|$

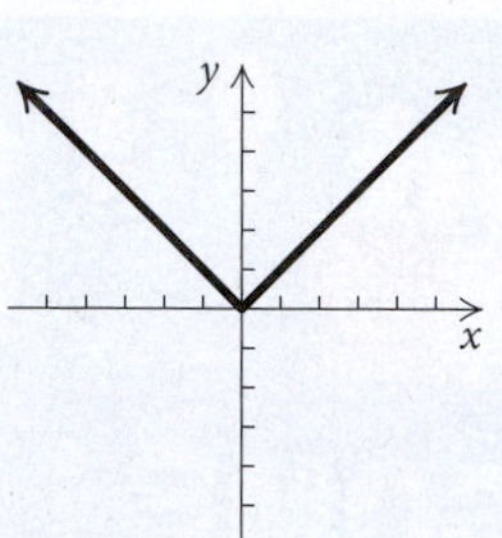

These functions can be considered building blocks for many other functions. We can create graphs of new functions by shifting them horizontally or vertically, stretching or shrinking them, and reflecting them across an axis. We now consider these **transformations**.

Vertical Translations and Horizontal Translations

Suppose that we have a function given by $y = f(x)$. Let's explore the graphs of the new functions $y = f(x) + b$ and $y = f(x) - b$, for $b > 0$.

Consider the functions $y = \frac{1}{5}x^4$, $y = \frac{1}{5}x^4 + 5$, and $y = \frac{1}{5}x^4 - 3$ and compare their graphs. What pattern do you see? Test it with some other functions.

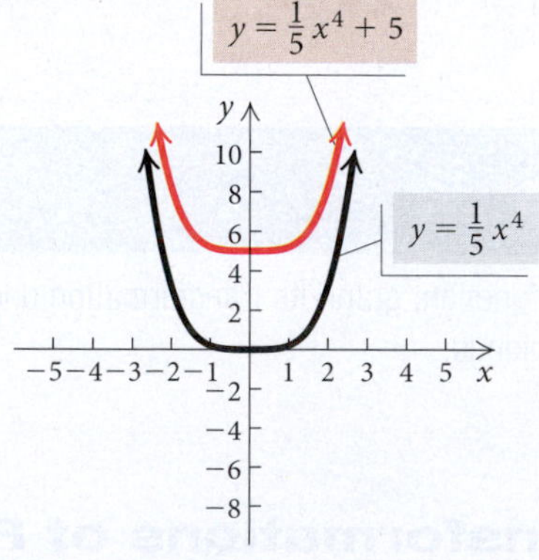

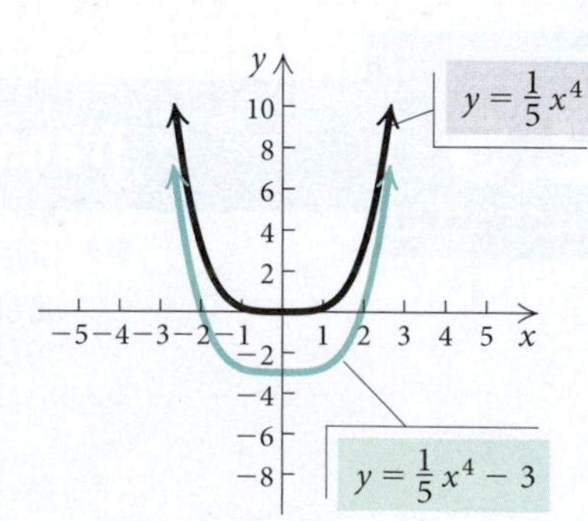

The effect of adding a constant to or subtracting a constant from $f(x)$ in $y = f(x)$ is a shift of the graph of $f(x)$ up or down. Such a shift is called a **vertical translation**.

VERTICAL TRANSLATION

For $b > 0$:

the graph of $y = f(x) + b$ is the graph of $y = f(x)$ shifted *up* b units;
the graph of $y = f(x) - b$ is the graph of $y = f(x)$ shifted *down* b units.

Suppose that we have a function given by $y = f(x)$. Let's explore the graphs of the new functions $y = f(x - d)$ and $y = f(x + d)$, for $d > 0$.

Consider the functions $y = \frac{1}{5}x^4$, $y = \frac{1}{5}(x - 3)^4$, and $y = \frac{1}{5}(x + 7)^4$ and compare their graphs. What pattern do you observe? Test it with some other functions.

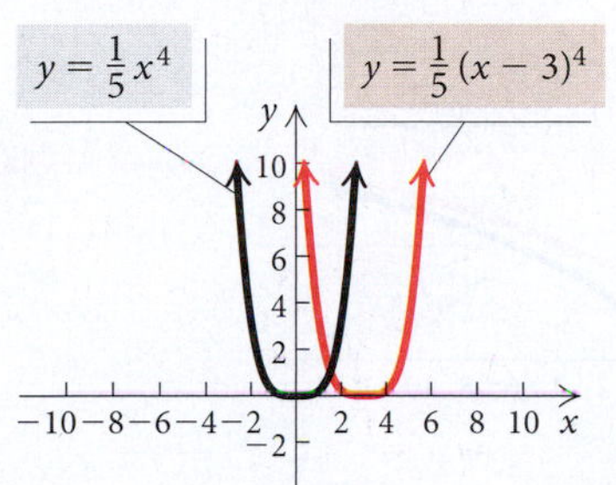

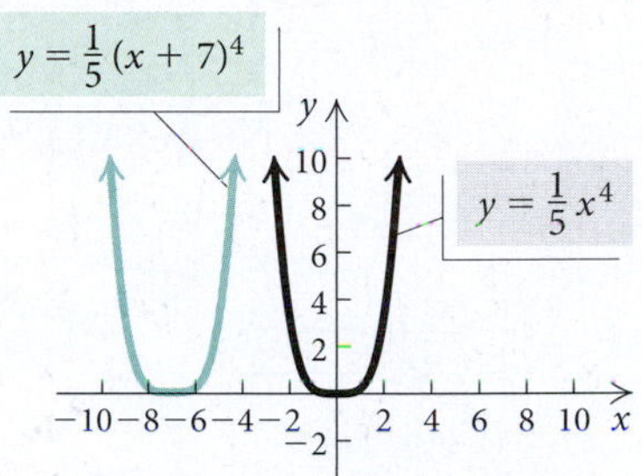

The effect of subtracting a constant from the x-value or adding a constant to the x-value in $y = f(x)$ is a shift of the graph of $f(x)$ to the right or to the left. Such a shift is called a **horizontal translation**.

HORIZONTAL TRANSLATION

For $d > 0$:

the graph of $y = f(x - d)$ is the graph of $y = f(x)$ shifted *right* d units;
the graph of $y = f(x + d)$ is the graph of $y = f(x)$ shifted *left* d units.

EXAMPLE 1 Graph each of the following. Before doing so, describe how each graph can be obtained from one of the basic graphs shown on the preceding pages.

a) $g(x) = x^2 - 6$
b) $h(x) = |x - 4|$
c) $g(x) = \sqrt{x + 2}$
d) $h(x) = \sqrt{x + 2} - 3$

Solution

a) To graph $g(x) = x^2 - 6$, think of the graph of $f(x) = x^2$. Since $g(x) = f(x) - 6$, the graph of $g(x) = x^2 - 6$ is the graph of $f(x) = x^2$, shifted, or translated, *down* 6 units. (See Fig. 1.)

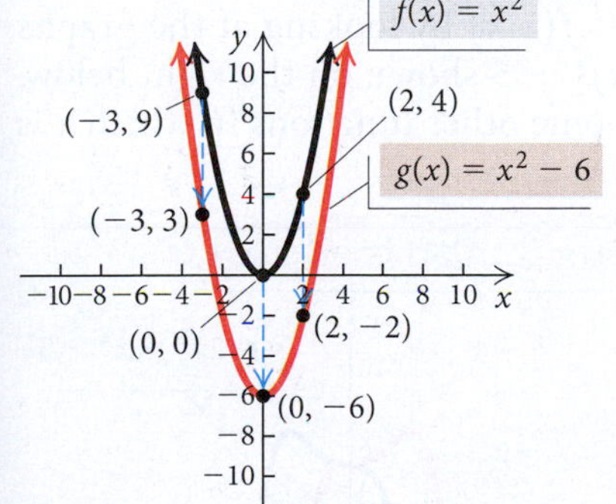

Figure 1.

Let's compare some points on the graphs of f and g.

Points on f:	$(-3, 9)$,	$(0, 0)$,	$(2, 4)$
	↓	↓	↓
Corresponding points on g:	$(-3, 3)$,	$(0, -6)$,	$(2, -2)$

We note that the y-coordinate of a point on the graph of g is 6 less than the corresponding y-coordinate on the graph of f.

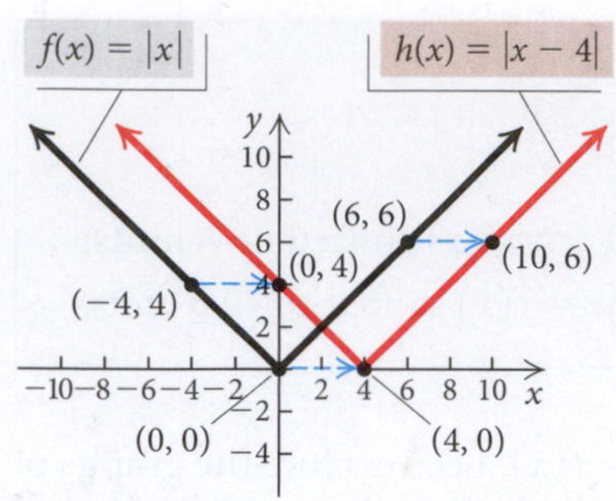

Figure 2.

b) To graph $h(x) = |x - 4|$, think of the graph of $f(x) = |x|$. Since $h(x) = f(x - 4)$, the graph of $h(x) = |x - 4|$ is the graph of $f(x) = |x|$ shifted *right* 4 units. (See Fig. 2.)

Let's again compare points on the two graphs.

Points on f:	$(-4, 4)$,	$(0, 0)$,	$(6, 6)$
	↓	↓	↓
Corresponding points on h:	$(0, 4)$,	$(4, 0)$,	$(10, 6)$

Noting points on f and h, we see that the x-coordinate of a point on the graph of h is 4 more than the x-coordinate of the corresponding point on f.

c) To graph $g(x) = \sqrt{x + 2}$, think of the graph of $f(x) = \sqrt{x}$. Since $g(x) = f(x + 2)$, the graph of $g(x) = \sqrt{x + 2}$ is the graph of $f(x) = \sqrt{x}$, shifted *left* 2 units. (See Fig. 3.)

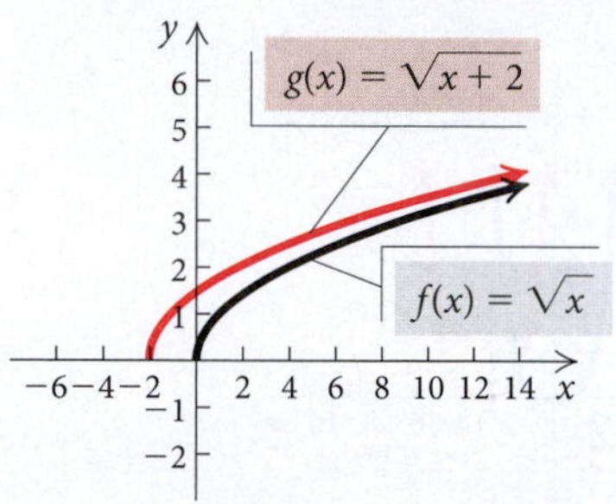

Figure 3.

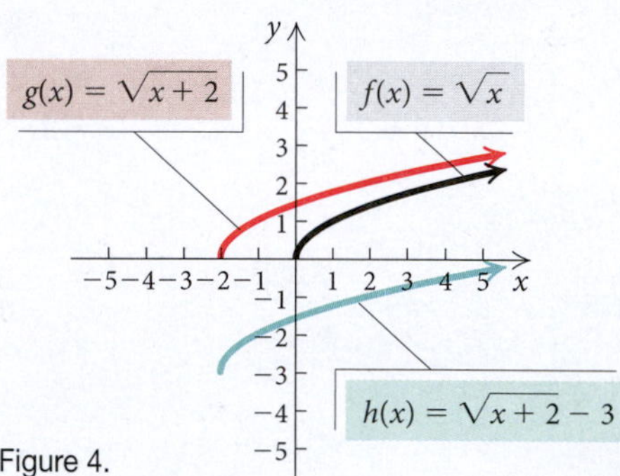

Figure 4.

d) To graph $h(x) = \sqrt{x + 2} - 3$, think of the graph of $f(x) = \sqrt{x}$. In part (c), we found that the graph of $g(x) = \sqrt{x + 2}$ is the graph of $f(x) = \sqrt{x}$ shifted *left* 2 units. Since $h(x) = g(x) - 3$, we shift the graph of $g(x) = \sqrt{x + 2}$ *down* 3 units. Together, the graph of $f(x) = \sqrt{x}$ is shifted *left* 2 units and *down* 3 units. (See Fig. 4.)

Now Try Exercises 3 and 15.

Reflections

Suppose that we have a function given by $y = f(x)$. Let's explore the graphs of the new functions $y = -f(x)$ and $y = f(-x)$.

Compare the functions $y = f(x)$ and $y = -f(x)$ by looking at the graphs of $y = \frac{1}{5}x^4$ and $y = -\frac{1}{5}x^4$ shown on the left below. What do you see? Test your observation with some other functions y_1 and y_2 where $y_2 = -y_1$.

Compare the functions $y = f(x)$ and $y = f(-x)$ by looking at the graphs of $y = 2x^3 - x^4 + 5$ and $y = 2(-x)^3 - (-x)^4 + 5$ shown on the right below. What do you see? Test your observation with some other functions in which x is replaced with $-x$.

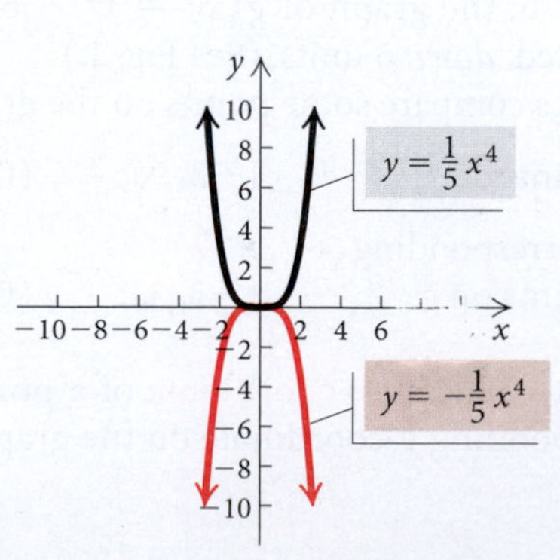

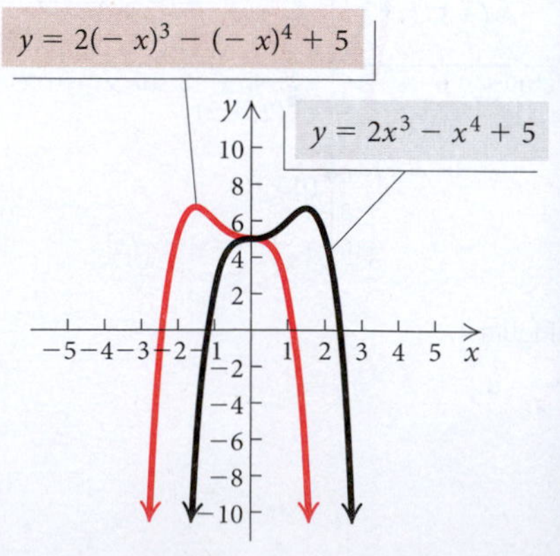

Given the graph of $y = f(x)$, we can reflect each point *across the x-axis* to obtain the graph of $y = -f(x)$. We can reflect each point of $y = f(x)$ *across the y-axis* to obtain the graph of $y = f(-x)$. The new graphs are called **reflections** of $y = f(x)$.

The following photographs illustrate reflection.

REFLECTIONS

The graph of $y = -f(x)$ is the **reflection** of the graph of $y = f(x)$ across the x-axis.

The graph of $y = f(-x)$ is the **reflection** of the graph of $y = f(x)$ across the y-axis.

If a point (x, y) is on the graph of $y = f(x)$, then $(x, -y)$ is on the graph of $y = -f(x)$, and $(-x, y)$ is on the graph of $y = f(-x)$.

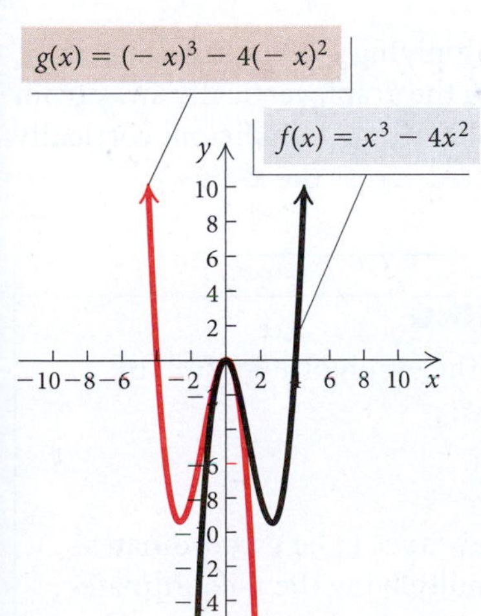

EXAMPLE 2 Graph each of the following. Before doing so, describe how each graph can be obtained from the graph of $f(x) = x^3 - 4x^2$.

a) $g(x) = (-x)^3 - 4(-x)^2$ **b)** $h(x) = 4x^2 - x^3$

Solution

a) We first note that

$$f(-x) = (-x)^3 - 4(-x)^2 = g(x).$$

Thus the graph of g is a *reflection* of the graph of f across the y-axis. (See the figure at left.) If (x, y) is on the graph of f, then $(-x, y)$ is on the graph of g. For example, $(2, -8)$ is on f and $(-2, -8)$ is on g.

b) We first note that

$$\begin{aligned} -f(x) &= -(x^3 - 4x^2) \\ &= -x^3 + 4x^2 \\ &= 4x^2 - x^3 \\ &= h(x). \end{aligned}$$

Thus the graph of h is a reflection of the graph of f across the x-axis. (See the figure at right.) If (x, y) is on the graph of f, then $(x, -y)$ is on the graph of h. For example, $(2, -8)$ is on f and $(2, 8)$ is on h.

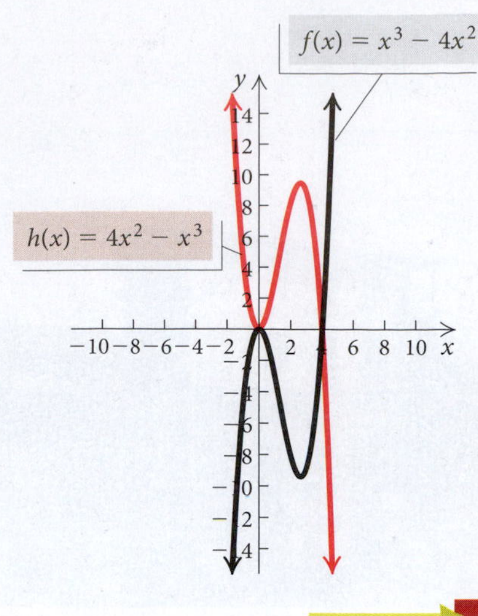

Vertical and Horizontal Stretchings and Shrinkings

Suppose that we have a function given by $y = f(x)$. Let's explore the graphs of the new functions $y = af(x)$ and $y = f(cx)$.

Consider the functions $y = f(x) = x^3 - x$, $y = \frac{1}{10}(x^3 - x) = \frac{1}{10}f(x)$, $y = 2(x^3 - x) = 2f(x)$, and $y = -2(x^3 - x) = -2f(x)$ and compare their graphs. What pattern do you observe? Test it with some other functions.

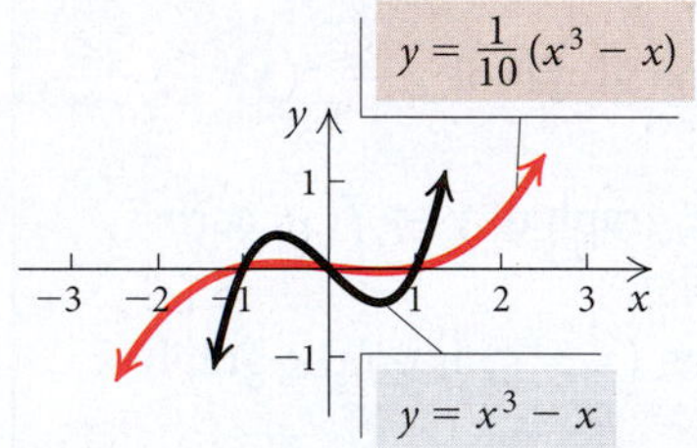

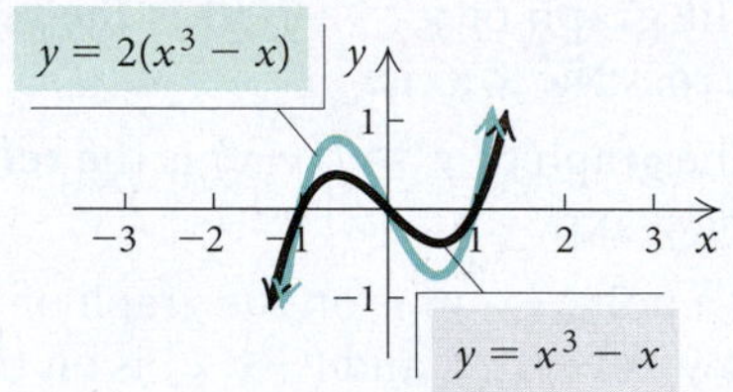

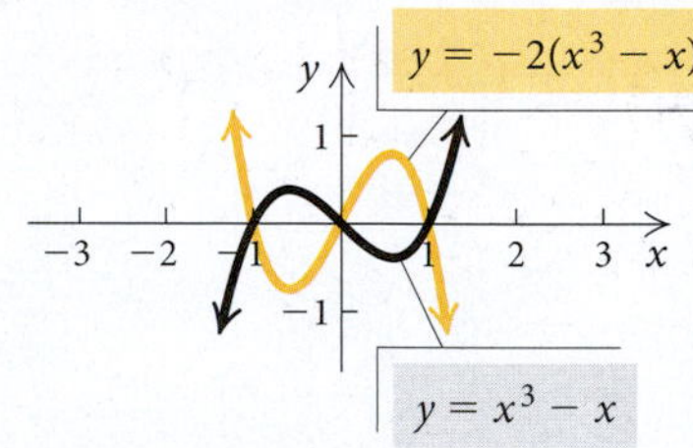

Consider any function f given by $y = f(x)$. Multiplying $f(x)$ by any constant a, where $|a| > 1$, to obtain $g(x) = af(x)$ will *stretch* the graph vertically away from the x-axis. If $0 < |a| < 1$, then the graph will be flattened or *shrunk* vertically toward the x-axis. If $a < 0$, the graph is also reflected across the x-axis.

JUST IN TIME 3

VERTICAL STRETCHING AND SHRINKING

The graph of $y = af(x)$ can be obtained from the graph of $y = f(x)$ by

stretching vertically for $|a| > 1$, or
shrinking vertically for $0 < |a| < 1$.

For $a < 0$, the graph is also reflected across the x-axis. (The y-coordinates of the graph of $y = af(x)$ can be obtained by multiplying the y-coordinates of $y = f(x)$ by a.)

Consider the functions $y = f(x) = x^3 - x$, $y = (2x)^3 - (2x) = f(2x)$, $y = \left(\frac{1}{2}x\right)^3 - \left(\frac{1}{2}x\right) = f\left(\frac{1}{2}x\right)$, and $y = \left(-\frac{1}{2}x\right)^3 - \left(-\frac{1}{2}x\right) = f\left(-\frac{1}{2}x\right)$ and compare their graphs. What pattern do you observe? Test it with some other functions.

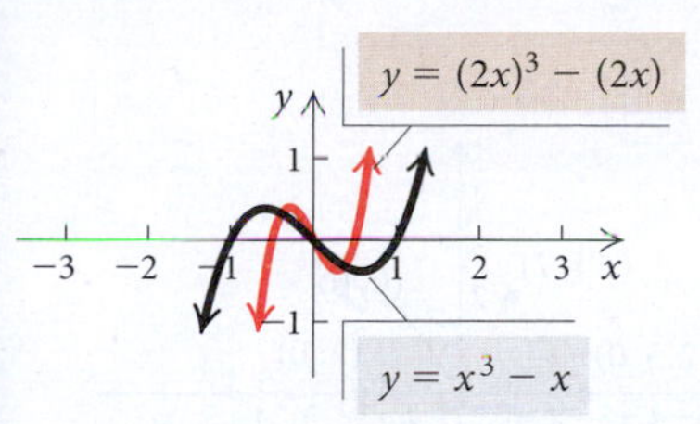

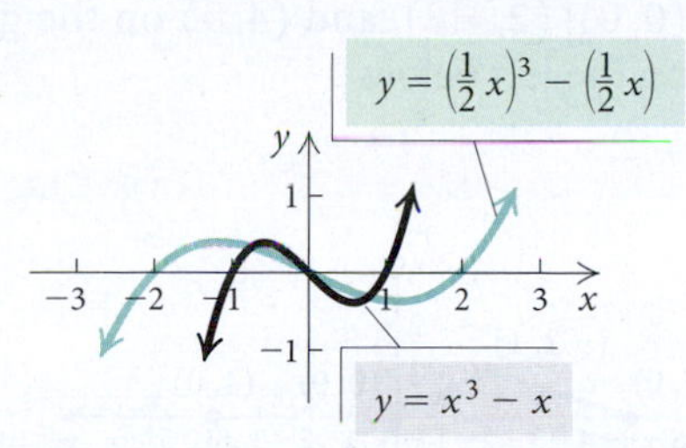

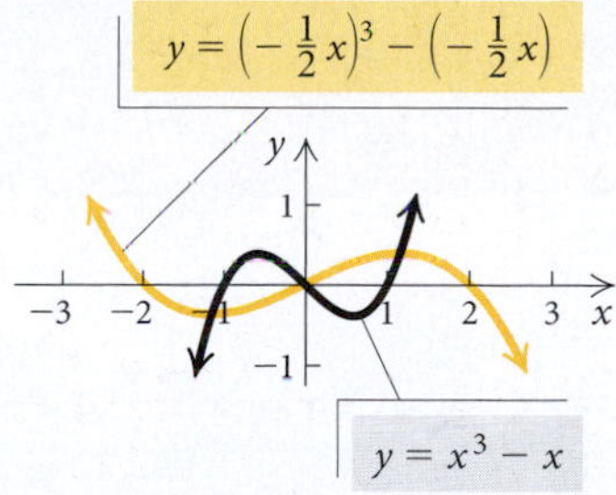

The constant c in the equation $g(x) = f(cx)$ will *shrink* the graph of $y = f(x)$ horizontally toward the y-axis if $|c| > 1$. If $0 < |c| < 1$, the graph will be *stretched* horizontally away from the y-axis. If $c < 0$, the graph is also reflected across the y-axis.

HORIZONTAL STRETCHING AND SHRINKING

The graph of $y = f(cx)$ can be obtained from the graph of $y = f(x)$ by

shrinking horizontally for $|c| > 1$, or

stretching horizontally for $0 < |c| < 1$.

For $c < 0$, the graph is also reflected across the y-axis. (The x-coordinates of the graph of $y = f(cx)$ can be obtained by dividing the x-coordinates of the graph of $y = f(x)$ by c.)

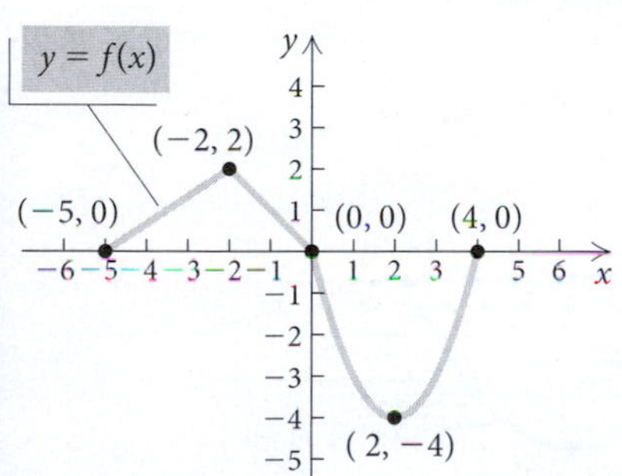

EXAMPLE 3 Shown at left is a graph of $y = f(x)$ for some function f. No formula for f is given. Graph each of the following.

a) $g(x) = 2f(x)$ **b)** $h(x) = \frac{1}{2}f(x)$ **c)** $r(x) = f(2x)$

d) $s(x) = f\left(\frac{1}{2}x\right)$ **e)** $t(x) = f\left(-\frac{1}{2}x\right)$

Solution

a) Since $|2| > 1$, the graph of $g(x) = 2f(x)$ is a vertical stretching of the graph of $y = f(x)$ by a factor of 2. We can consider the key points $(-5, 0)$, $(-2, 2)$, $(0, 0)$, $(2, -4)$, and $(4, 0)$ on the graph of $y = f(x)$. The transformation multiplies each y-coordinate by 2 to obtain the key points $(-5, 0)$, $(-2, 4)$, $(0, 0)$, $(2, -8)$, and $(4, 0)$ on the graph of $g(x) = 2f(x)$, as shown below.

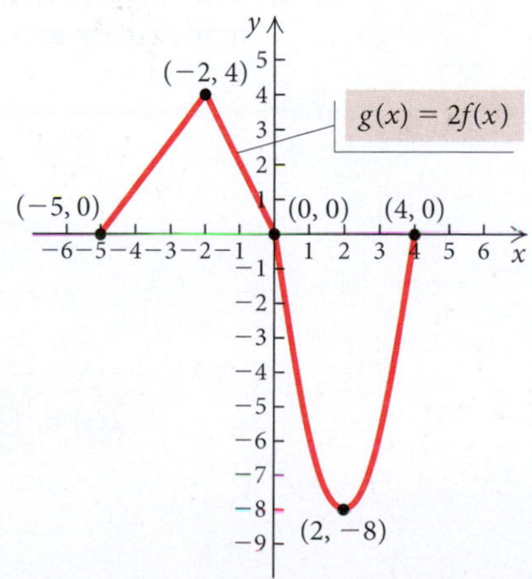

b) Since $\left|\frac{1}{2}\right| < 1$, the graph of $h(x) = \frac{1}{2}f(x)$ is a vertical shrinking of the graph of $y = f(x)$ by a factor of $\frac{1}{2}$. We again consider the key points $(-5, 0)$, $(-2, 2)$, $(0, 0)$, $(2, -4)$, and $(4, 0)$ on the graph of $y = f(x)$. The transformation multiplies each y-coordinate by $\frac{1}{2}$ to obtain the key points $(-5, 0)$, $(-2, 1)$, $(0, 0)$, $(2, -2)$, and $(4, 0)$ on the graph of $h(x) = \frac{1}{2}f(x)$. The graph is shown on the left below.

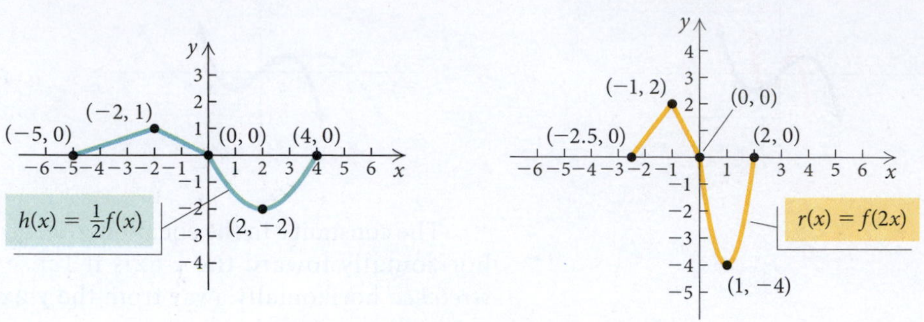

c) Since $|2| > 1$, the graph of $r(x) = f(2x)$ is a horizontal shrinking of the graph of $y = f(x)$. We consider the key points $(-5, 0)$, $(-2, 2)$, $(0, 0)$, $(2, -4)$, and $(4, 0)$ on the graph of $y = f(x)$. The transformation divides each x-coordinate by 2 to obtain the key points $(-2.5, 0)$, $(-1, 2)$, $(0, 0)$, $(1, -4)$, and $(2, 0)$ on the graph of $r(x) = f(2x)$. The graph is shown on the right above.

d) Since $\left|\frac{1}{2}\right| < 1$, the graph of $s(x) = f\left(\frac{1}{2}x\right)$ is a horizontal stretching of the graph of $y = f(x)$. We consider the key points $(-5, 0)$, $(-2, 2)$, $(0, 0)$, $(2, -4)$, and $(4, 0)$ on the graph of $y = f(x)$. The transformation divides each x-coordinate by $\frac{1}{2}$ (which is the same as multiplying by 2) to obtain the key points $(-10, 0)$, $(-4, 2)$, $(0, 0)$, $(4, -4)$, and $(8, 0)$ on the graph of $s(x) = f\left(\frac{1}{2}x\right)$. The graph is shown below.

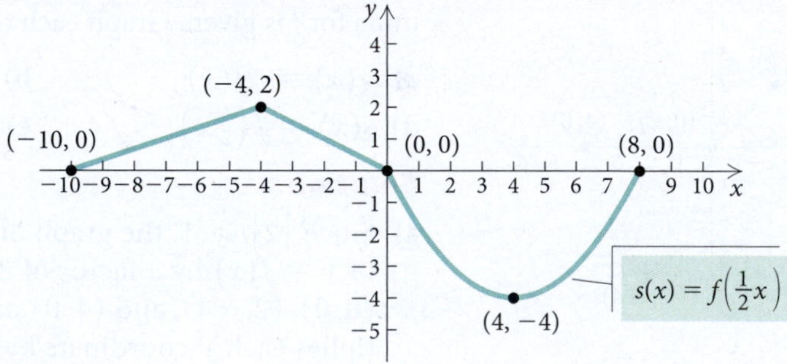

e) The graph of $t(x) = f\left(-\frac{1}{2}x\right)$ can be obtained by reflecting the graph in part (d) across the y-axis.

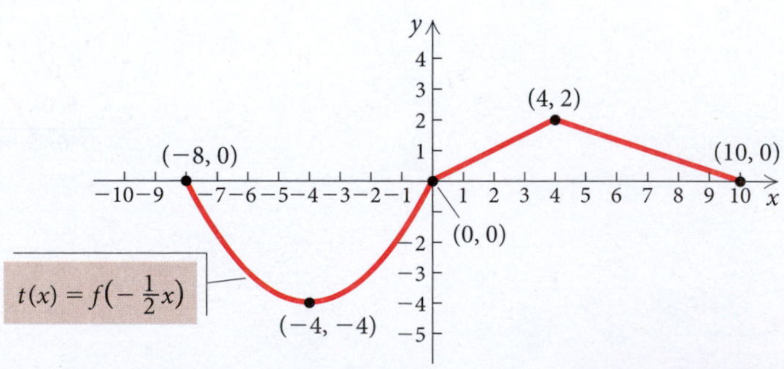

Now Try Exercises 59 and 61.

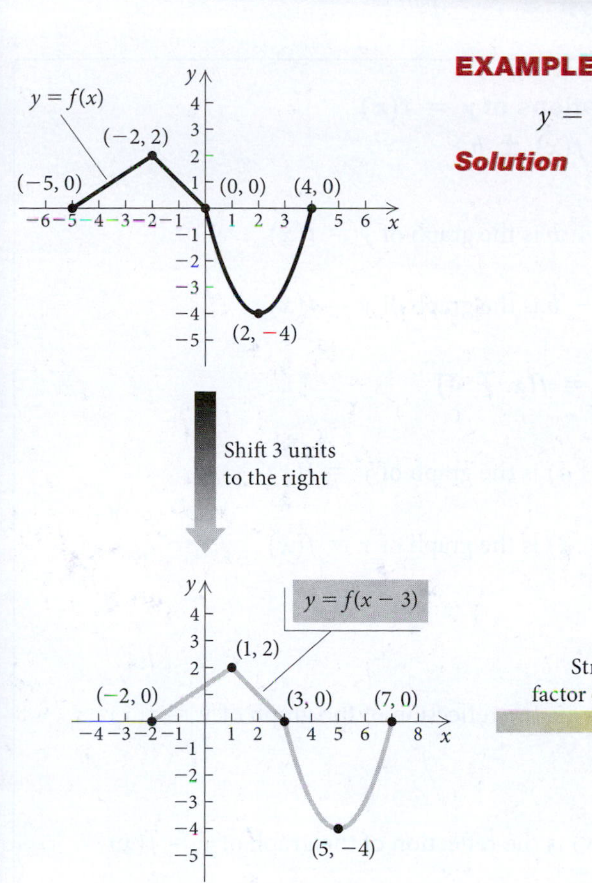

EXAMPLE 4 Use the graph of $y = f(x)$ shown at left to graph

$$y = -2f(x - 3) + 1.$$

Solution

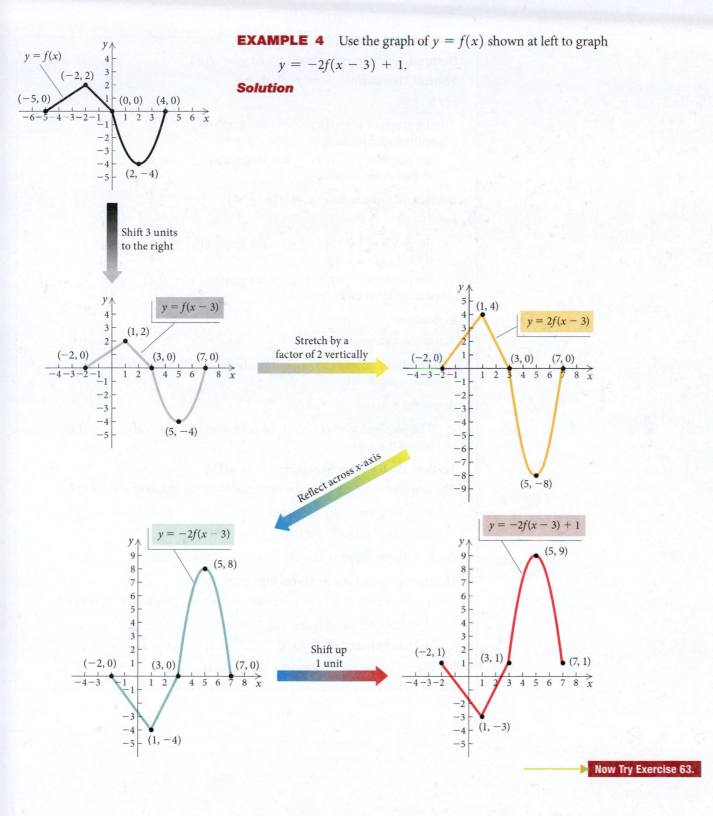

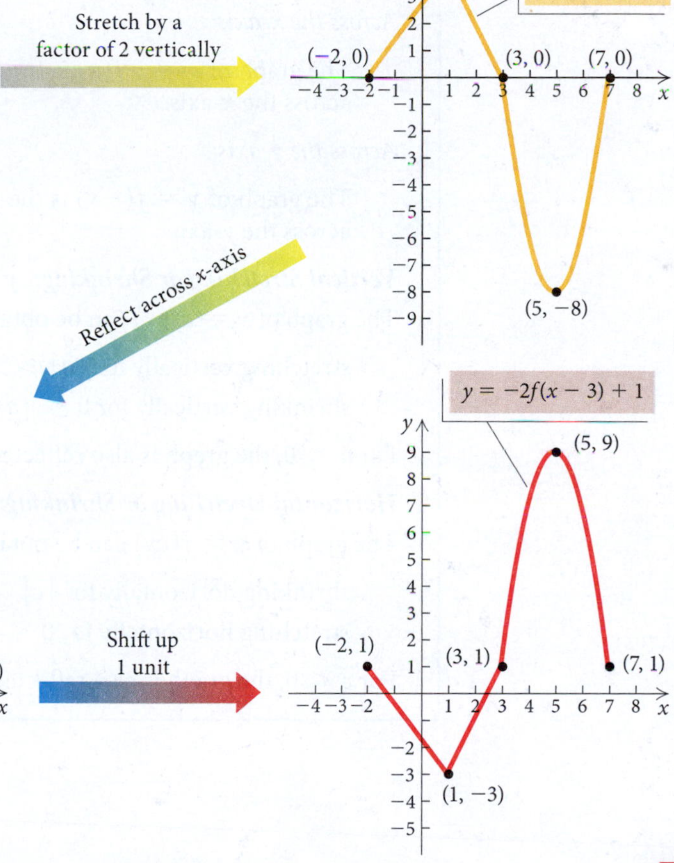

Now Try Exercise 63.

Summary of Transformations of $y = f(x)$

Vertical Translation: $y = f(x) \pm b$

For $b > 0$:

the graph of $y = f(x) + b$ is the graph of $y = f(x)$ shifted *up* b units;

the graph of $y = f(x) - b$ is the graph of $y = f(x)$ shifted *down* b units.

Horizontal Translation: $y = f(x \mp d)$

For $d > 0$:

the graph of $y = f(x - d)$ is the graph of $y = f(x)$ shifted *right* d units;

the graph of $y = f(x + d)$ is the graph of $y = f(x)$ shifted *left* d units.

Reflections

Across the x-axis:

The graph of $y = -f(x)$ is the reflection of the graph of $y = f(x)$ across the x-axis.

Across the y-axis:

The graph of $y = f(-x)$ is the reflection of the graph of $y = f(x)$ across the y-axis.

Vertical Stretching or Shrinking: $y = af(x)$

The graph of $y = af(x)$ can be obtained from the graph of $y = f(x)$ by

stretching vertically for $|a| > 1$, or
shrinking vertically for $0 < |a| < 1$.

For $a < 0$, the graph is also reflected across the x-axis.

Horizontal Stretching or Shrinking: $y = f(cx)$

The graph of $y = f(cx)$ can be obtained from the graph of $y = f(x)$ by

shrinking horizontally for $|c| > 1$, or
stretching horizontally for $0 < |c| < 1$.

For $c < 0$, the graph is also reflected across the y-axis.

Visualizing the Graph

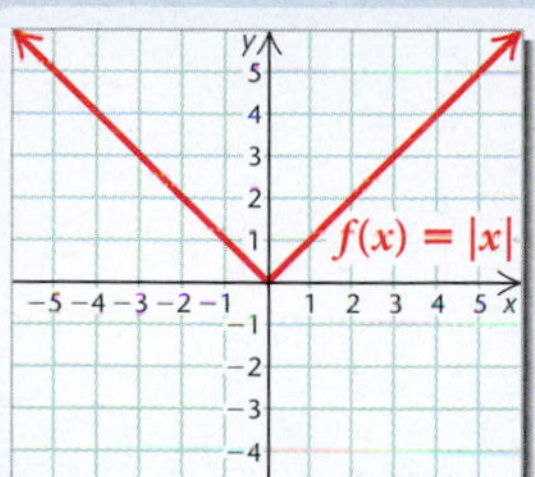

Match the function with its graph. Use transformation graphing techniques to obtain the graph of g from the basic function $f(x) = |x|$ shown at top left.

1. $g(x) = -2|x|$
2. $g(x) = |x - 1| + 1$
3. $g(x) = -\left|\frac{1}{3}x\right|$
4. $g(x) = |2x|$
5. $g(x) = |x + 2|$
6. $g(x) = |x| + 3$
7. $g(x) = -\frac{1}{2}|x - 4|$
8. $g(x) = \frac{1}{2}|x| - 3$
9. $g(x) = -|x| - 2$

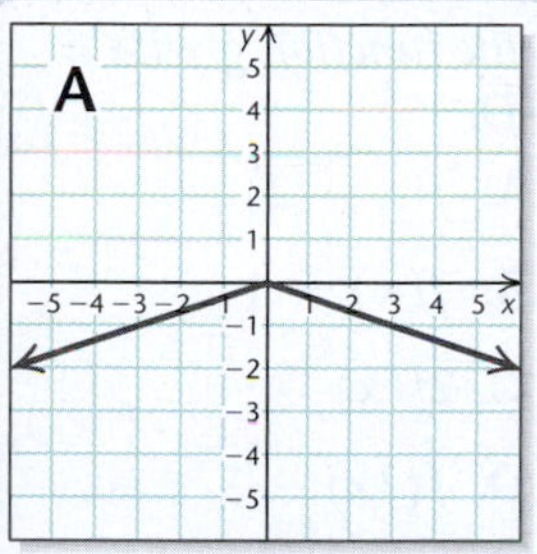

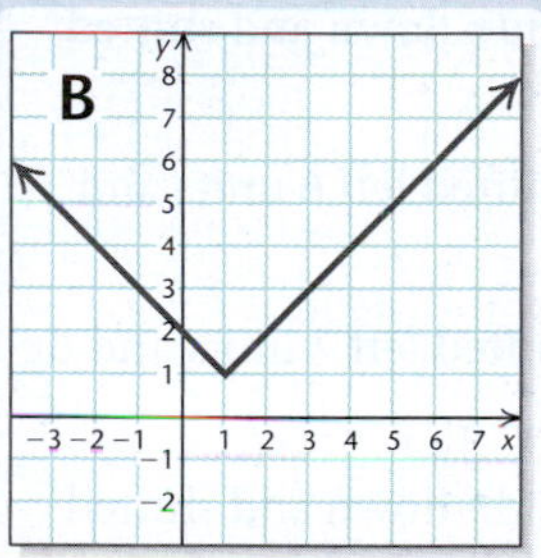

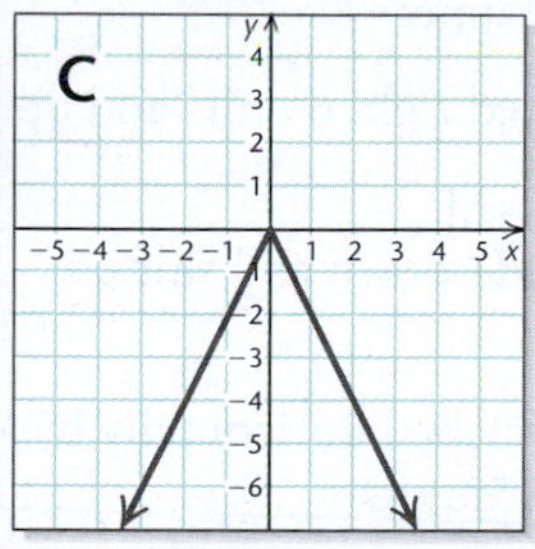

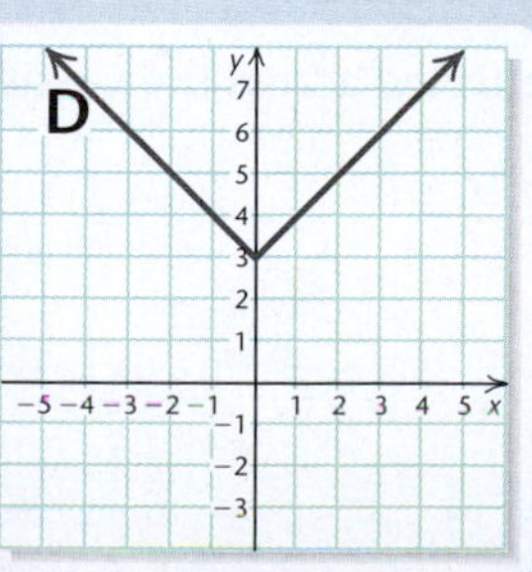

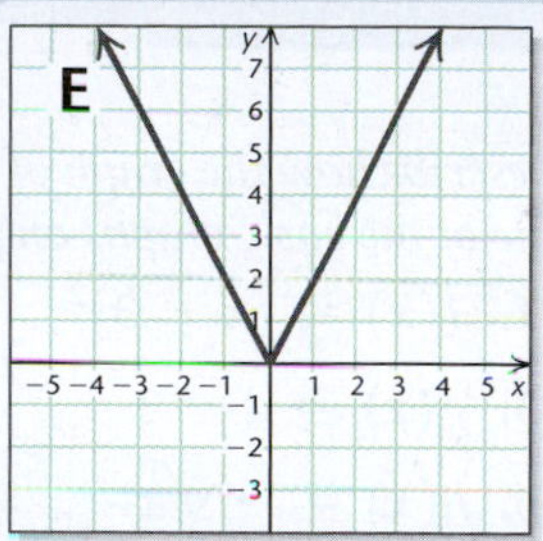

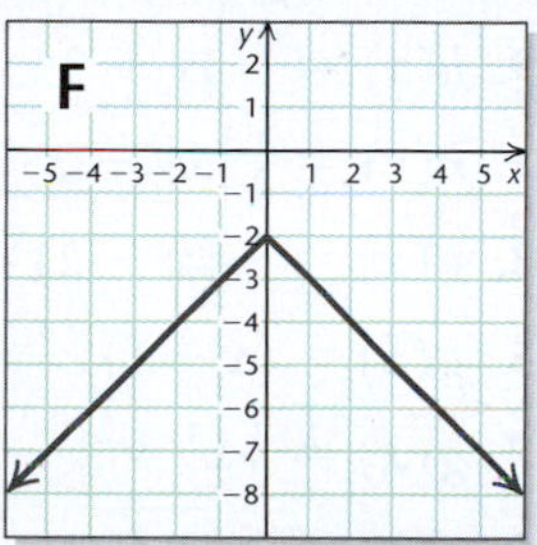

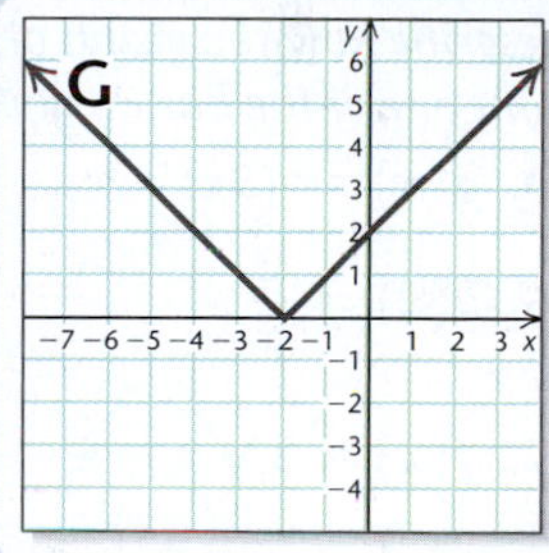

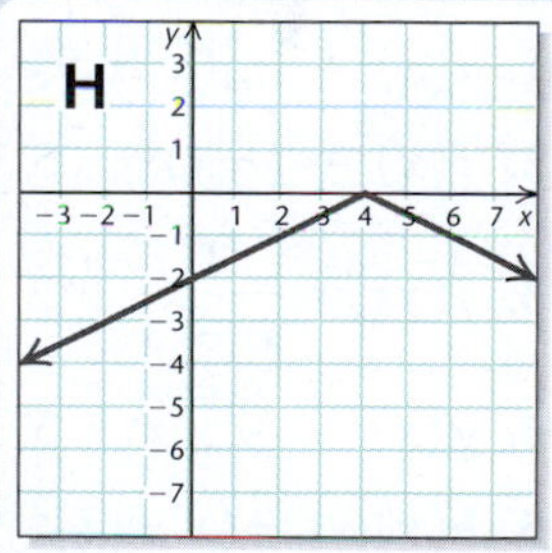

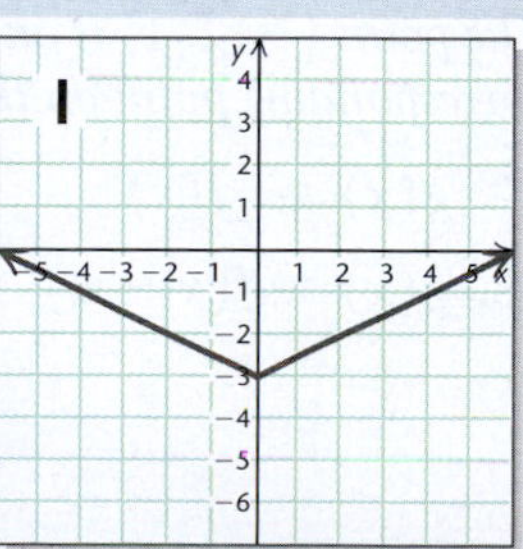

Answers on page A-10

2.5 Exercise Set

Describe how the graph of the function can be obtained from one of the basic graphs on p. 134. Then graph the function.

1. $f(x) = (x - 3)^2$
2. $g(x) = x^2 + \frac{1}{2}$
3. $g(x) = x - 3$
4. $g(x) = -x - 2$
5. $h(x) = -\sqrt{x}$
6. $g(x) = \sqrt{x - 1}$
7. $h(x) = \dfrac{1}{x} + 4$
8. $g(x) = \dfrac{1}{x - 2}$
9. $h(x) = -3x + 3$
10. $f(x) = 2x + 1$
11. $h(x) = \frac{1}{2}|x| - 2$
12. $g(x) = -|x| + 2$
13. $g(x) = -(x - 2)^3$
14. $f(x) = (x + 1)^3$
15. $g(x) = (x + 1)^2 - 1$
16. $h(x) = -x^2 - 4$
17. $g(x) = \frac{1}{3}x^3 + 2$
18. $h(x) = (-x)^3$
19. $f(x) = \sqrt{x + 2}$
20. $f(x) = -\frac{1}{2}\sqrt{x - 1}$
21. $f(x) = \sqrt[3]{x} - 2$
22. $h(x) = \sqrt[3]{x + 1}$

Describe how the graph of the function can be obtained from one of the basic graphs on p. 134.

23. $g(x) = |3x|$
24. $f(x) = \frac{1}{2}\sqrt[3]{x}$
25. $h(x) = \dfrac{2}{x}$
26. $f(x) = |x - 3| - 4$
27. $f(x) = 3\sqrt{x} - 5$
28. $f(x) = 5 - \dfrac{1}{x}$
29. $g(x) = |\frac{1}{3}x| - 4$
30. $f(x) = \frac{2}{3}x^3 - 4$
31. $f(x) = -\frac{1}{4}(x - 5)^2$
32. $f(x) = (-x)^3 - 5$
33. $f(x) = \dfrac{1}{x + 3} + 2$
34. $g(x) = \sqrt{-x} + 5$
35. $h(x) = -(x - 3)^2 + 5$
36. $f(x) = 3(x + 4)^2 - 3$

The point $(-12, 4)$ is on the graph of $y = f(x)$. Find the corresponding point on the graph of $y = g(x)$.

37. $g(x) = \frac{1}{2}f(x)$
38. $g(x) = f(x - 2)$
39. $g(x) = f(-x)$
40. $g(x) = f(4x)$
41. $g(x) = f(x) - 2$
42. $g(x) = f(\frac{1}{2}x)$
43. $g(x) = 4f(x)$
44. $g(x) = -f(x)$

Given that $f(x) = x^2 + 3$, match the function g with a transformation of f from one of A–D.

45. $g(x) = x^2 + 4$ — A. $f(x - 2)$
46. $g(x) = 9x^2 + 3$ — B. $f(x) + 1$
47. $g(x) = (x - 2)^2 + 3$ — C. $2f(x)$
48. $g(x) = 2x^2 + 6$ — D. $f(3x)$

Write an equation for a function that has a graph with the given characteristics.

49. The shape of $y = x^2$, but upside-down and shifted right 8 units
50. The shape of $y = \sqrt{x}$, but shifted left 6 units and down 5 units
51. The shape of $y = |x|$, but shifted left 7 units and up 2 units
52. The shape of $y = x^3$, but upside-down and shifted right 5 units
53. The shape of $y = 1/x$, but shrunk horizontally by a factor of 2 and shifted down 3 units
54. The shape of $y = x^2$, but shifted right 6 units and up 2 units
55. The shape of $y = x^2$, but upside-down and shifted right 3 units and up 4 units
56. The shape of $y = |x|$, but stretched horizontally by a factor of 2 and shifted down 5 units
57. The shape of $y = \sqrt{x}$, but reflected across the y-axis and shifted left 2 units and down 1 unit
58. The shape of $y = 1/x$, but reflected across the x-axis and shifted up 1 unit

A graph of $y = f(x)$ follows. No formula for f is given. In Exercises 59–66, graph the given equation.

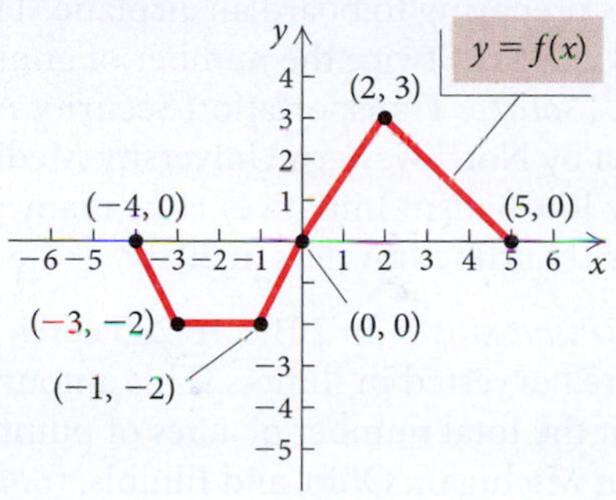

59. $g(x) = -2f(x)$

60. $g(x) = \frac{1}{2}f(x)$

61. $g(x) = f\left(-\frac{1}{2}x\right)$

62. $g(x) = f(2x)$

63. $g(x) = -\frac{1}{2}f(x - 1) + 3$

64. $g(x) = -3f(x + 1) - 4$

65. $g(x) = f(-x)$

66. $g(x) = -f(x)$

A graph of $y = g(x)$ follows. No formula for g is given. In Exercises 67–70, graph the given equation.

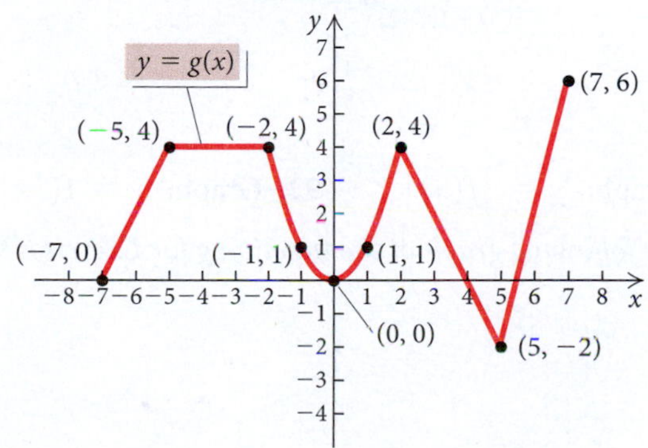

67. $h(x) = -g(x + 2) + 1$

68. $h(x) = \frac{1}{2}g(-x)$

69. $h(x) = g(2x)$

70. $h(x) = 2g(x - 1) - 3$

The graph of the function f is shown in figure (a) below. In each of Exercises 71–78, match the function g with one of the graphs (a)–(h) that follow. Some graphs may be used more than once and some may not be used at all.

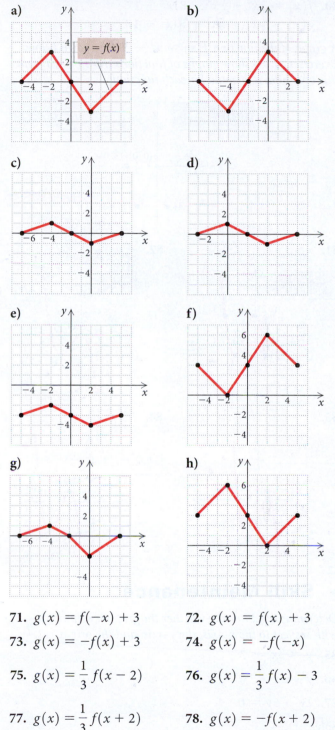

a)

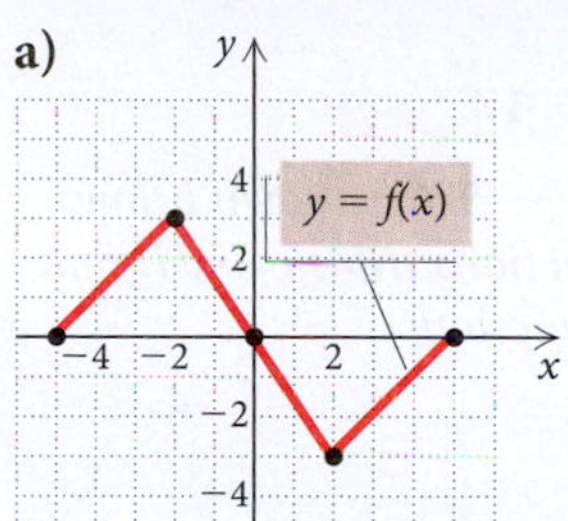

b)

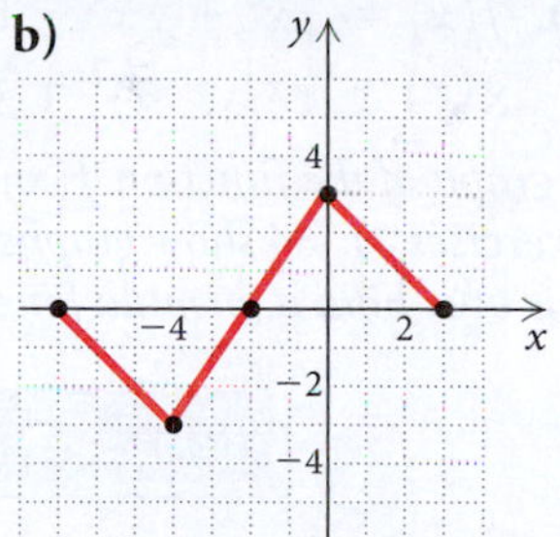

c)

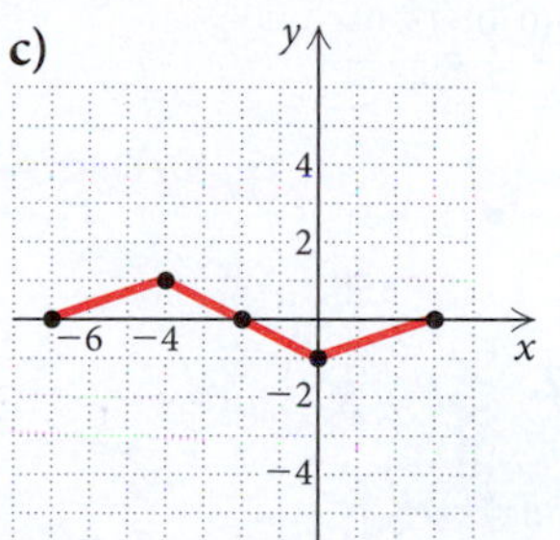

d)

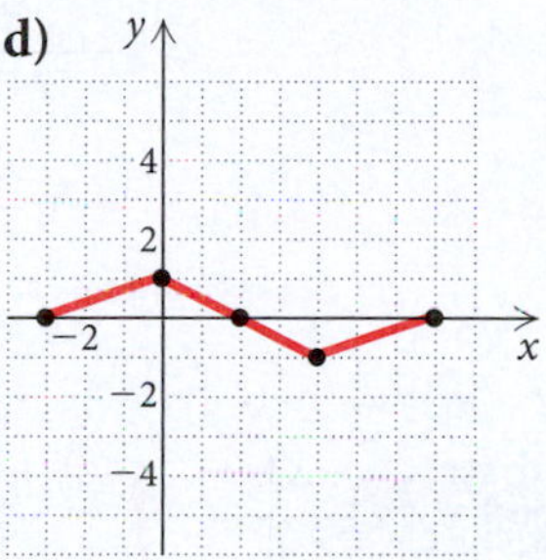

e)

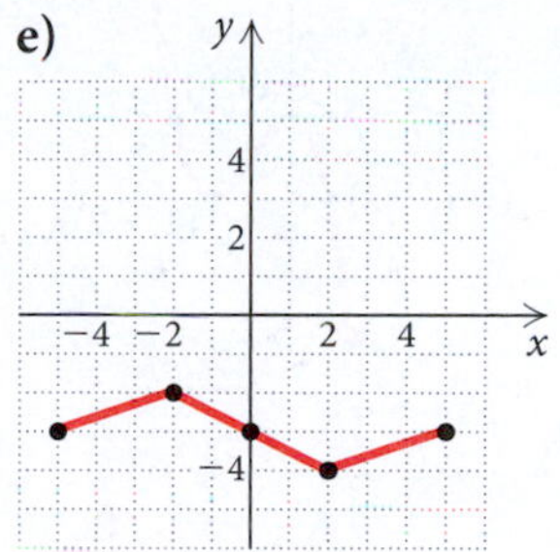

f)

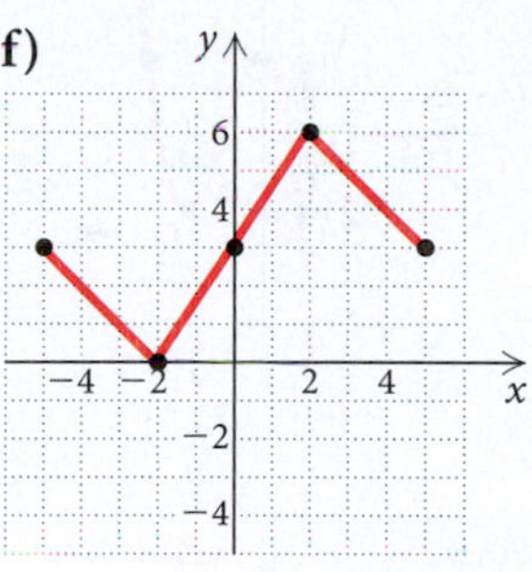

g)

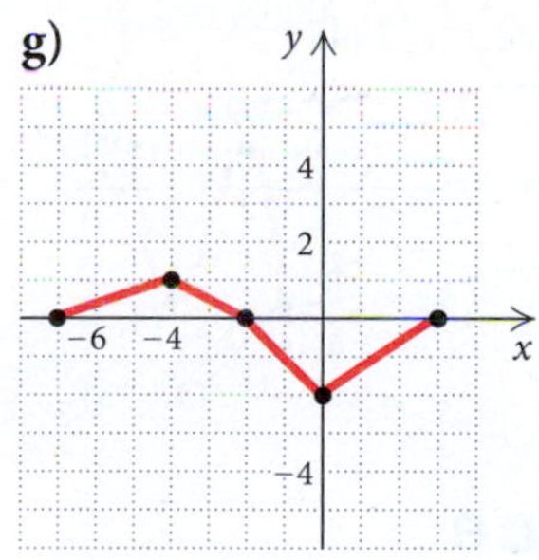

h)

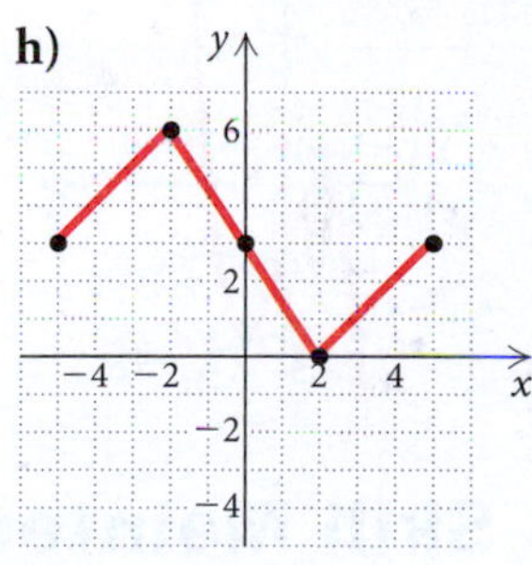

71. $g(x) = f(-x) + 3$

72. $g(x) = f(x) + 3$

73. $g(x) = -f(x) + 3$

74. $g(x) = -f(-x)$

75. $g(x) = \dfrac{1}{3}f(x - 2)$

76. $g(x) = \dfrac{1}{3}f(x) - 3$

77. $g(x) = \dfrac{1}{3}f(x + 2)$

78. $g(x) = -f(x + 2)$

For each pair of functions, determine if $g(x) = f(-x)$.

79. $f(x) = 2x^4 - 35x^3 + 3x - 5$,
$g(x) = 2x^4 + 35x^3 - 3x - 5$

80. $f(x) = \frac{1}{4}x^4 + \frac{1}{5}x^3 - 81x^2 - 17$,
$g(x) = \frac{1}{4}x^4 + \frac{1}{5}x^3 + 81x^2 - 17$

A graph of the function $f(x) = x^3 - 3x^2$ is shown below. Exercises 81–84 show graphs of functions transformed from this one. Find a formula for each function.

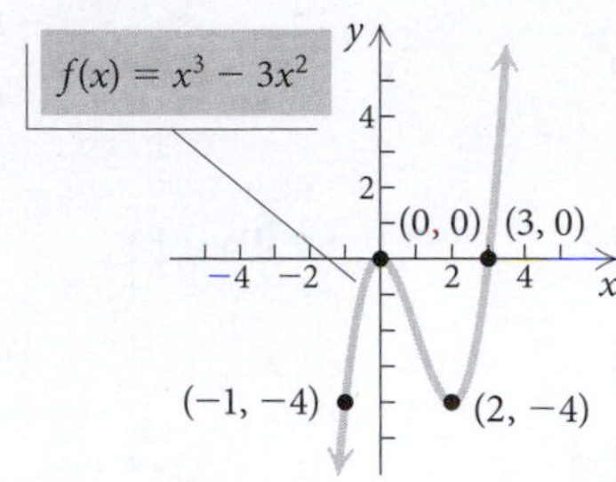

81.

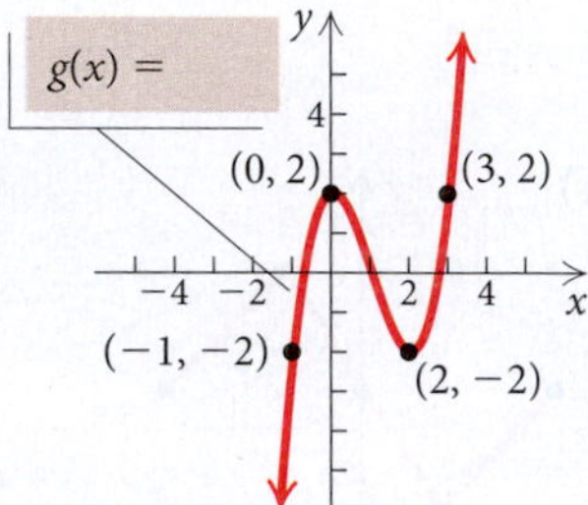

82.

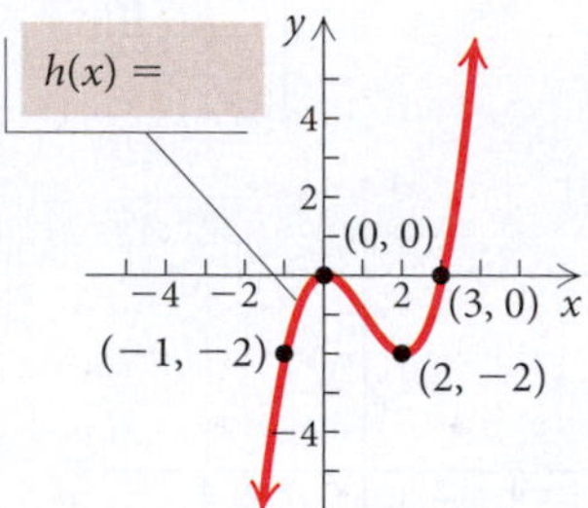

83.

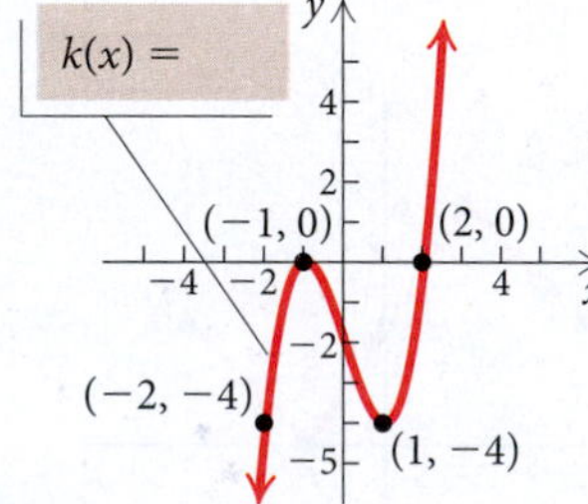

84.

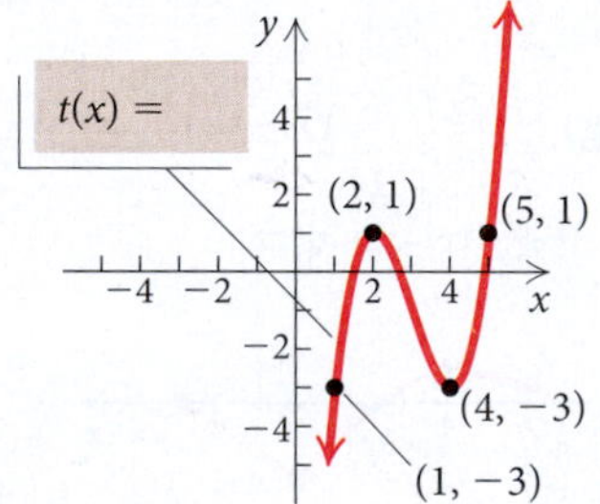

Skill Maintenance

Determine algebraically whether the graph is symmetric with respect to the x-axis, the y-axis, and the origin. [2.4]

85. $y = 3x^4 - 3$

86. $y^2 = x$

87. $2x - 5y = 0$

Solve. [1.5]

88. *Federal Tax Rules.* The number of pages of U.S. federal tax rules that explain the tax code and regulations totaled 74,608 in 2014 (for tax year 2013). This number was an increase of 84.2% over the number of pages in 1995 (for tax year 1994). (*Source:* Wolters Kluwer, CCH: 2014) Find the number of pages of federal tax rules in 1995.

89. *Guns with Airline Passengers.* In 2013, the Transportation Security Administration found 1828 guns with travelers preparing to board an airplane. This number was 418 less than twice the number of guns discovered in 2010. (*Source:* Transportation Security Administration data by Northwestern University Medill National Security Journalism Initiative) How many guns were found with airline travelers in 2010?

90. *Acres of Pumpkins.* In 2012, 16,200 acres of pumpkins were harvested in Illinois. This amount was about 54.5% of the total number of acres of pumpkins harvested in Michigan, Ohio, and Illinois. (*Source:* U.S. Department of Agriculture) Find the total number of acres of pumpkins harvested in Michigan, Ohio, and Illinois.

Synthesis

Use the following graph of the function f for Exercises 91 and 92.

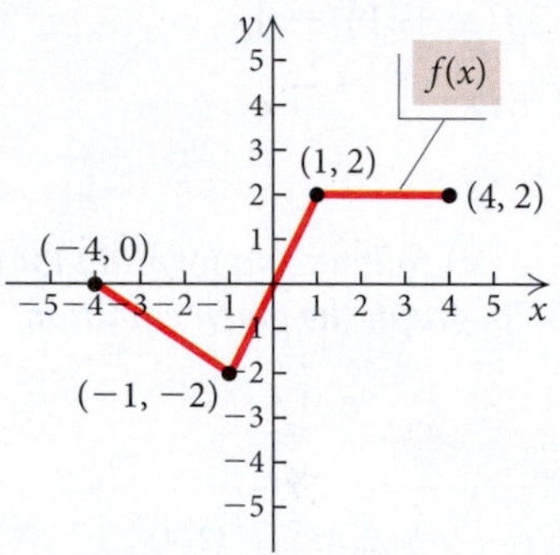

91. Graph: $y = |f(x)|$.

92. Graph: $y = f(|x|)$.

Use the following graph of the function g for Exercises 93 and 94.

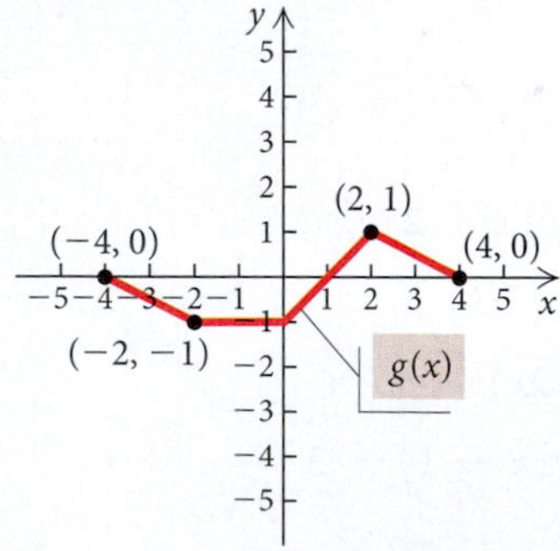

93. Graph: $y = g(|x|)$.

94. Graph: $y = |g(x)|$.

95. If $(-1, 5)$ is a point on the graph of $y = f(x)$, find b such that $(2, b)$ is on the graph of $y = f(x - 3)$.

96. The graph of $f(x) = |x|$ passes through the points $(-3, 3)$, $(0, 0)$, and $(3, 3)$. Transform this function to one whose graph passes through the points $(5, 1)$, $(8, 4)$, and $(11, 1)$.

2.6 Variation and Applications

- Find equations of direct variation, inverse variation, and combined variation given values of the variables.
- Solve applied problems involving variation.

We now consider applications involving variation.

Direct Variation

The median hourly wage for an elevator and escalator installer/repairer is \$35 per hour (*Source:* U.S. Bureau of Labor Statistics). In 1 hr, \$35 is earned; in 2 hr, \$70 is earned; in 3 hr, \$105 is earned; and so on. This gives rise to a set of ordered pairs:

$$(1, 35),\quad (2, 70),\quad (3, 105),\quad (4, 140),\quad \text{and so on.}$$

Note that the ratio of the second coordinate to the first coordinate is the same number for each pair:

$$\frac{35}{1} = 35,\quad \frac{70}{2} = 35,\quad \frac{105}{3} = 35,\quad \frac{140}{4} = 35,\quad \text{and so on.}$$

Earnings for Elevator and Escalator Installer/Repairer

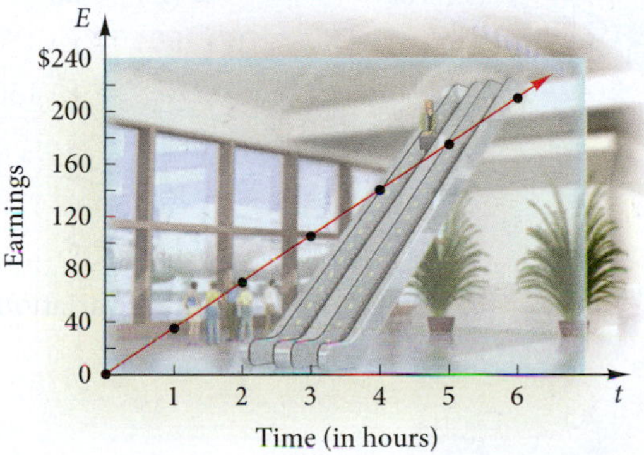

The graph of $y = kx$, $k > 0$, always goes through the origin and rises from left to right. Note that as x increases, y increases. That is, the function is increasing on the interval $(0, \infty)$. The constant k is also the slope of the line.

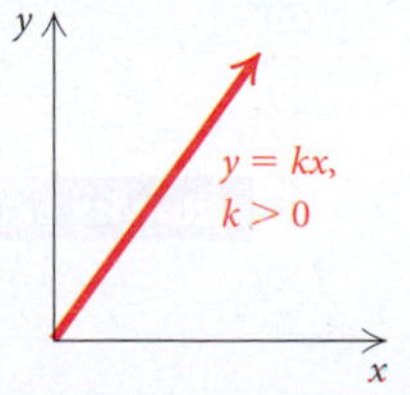

Whenever a situation produces pairs of numbers in which the *ratio is constant*, we say that there is **direct variation**. In this case, the amount earned E varies directly as the time worked t:

$$\frac{E}{t} = 35 \text{ (a constant)}, \quad \text{or} \quad E = 35t,$$

or, if we use function notation, $E(t) = 35t$. This equation is an equation of **direct variation**. The coefficient, 35, is called the **variation constant**. In this case, it is the rate of change of earnings with respect to time.

DIRECT VARIATION

If a situation gives rise to a linear function $f(x) = kx$, or $y = kx$, where k is a positive constant, we say that we have **direct variation**, or that y **varies directly as** x, or that y **is directly proportional to** x. The number k is called the **variation constant**, or the **constant of proportionality**.

EXAMPLE 1 Find the variation constant and an equation of variation in which y varies directly as x, and $y = 32$ when $x = 2$.

Solution We know that $(2, 32)$ is a solution of $y = kx$. Thus,

$$y = kx$$
$$32 = k \cdot 2 \quad \text{Substituting}$$
$$\frac{32}{2} = k \quad \text{Solving for } k$$
$$16 = k. \quad \text{Simplifying}$$

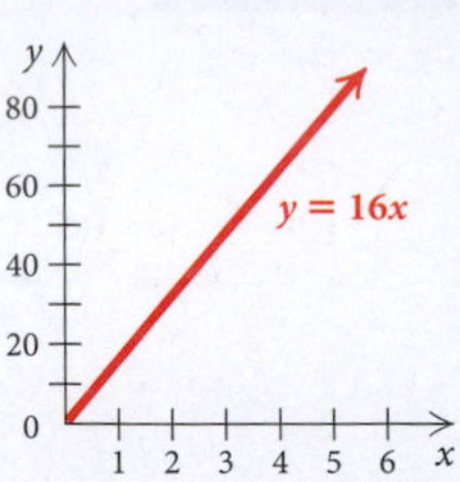

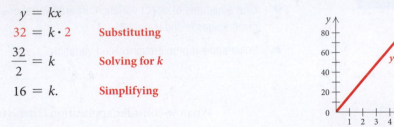

The variation constant, 16, is the rate of change of y with respect to x. The equation of variation is $y = 16x$.

Now Try Exercise 1.

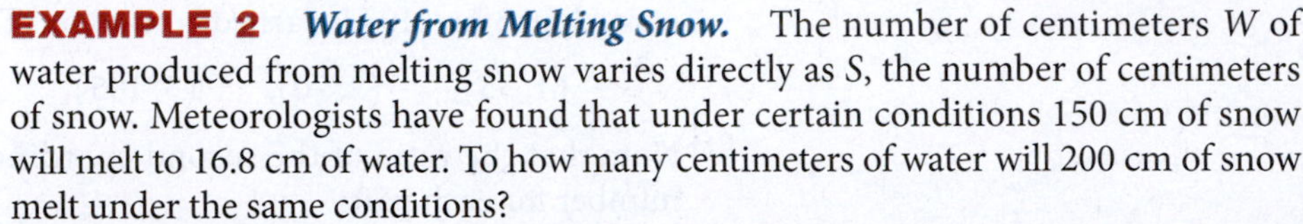

EXAMPLE 2 ***Water from Melting Snow.*** The number of centimeters W of water produced from melting snow varies directly as S, the number of centimeters of snow. Meteorologists have found that under certain conditions 150 cm of snow will melt to 16.8 cm of water. To how many centimeters of water will 200 cm of snow melt under the same conditions?

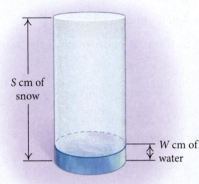

Solution We can express the amount of water as a function of the amount of snow. Thus, $W(S) = kS$, where k is the variation constant. We first find k using the given data and then find an equation of variation:

$$W(S) = kS \quad W \text{ varies directly as } S.$$
$$W(150) = k \cdot 150 \quad \text{Substituting 150 for } S$$
$$16.8 = k \cdot 150 \quad \text{Replacing } W(150) \text{ with 16.8}$$
$$\frac{16.8}{150} = k \quad \text{Solving for } k$$
$$0.112 = k. \quad \text{This is the variation constant.}$$

The equation of variation is $W(S) = 0.112S$.

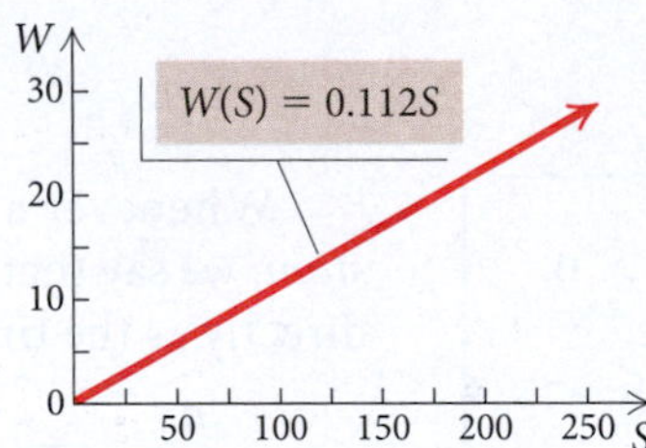

Next, we use the equation to find how many centimeters of water will result from melting 200 cm of snow:

$$W(S) = 0.112S$$
$$W(200) = 0.112(200) \quad \text{Substituting}$$
$$= 22.4.$$

Thus, 200 cm of snow will melt to 22.4 cm of water.

Now Try Exercise 17.

Inverse Variation

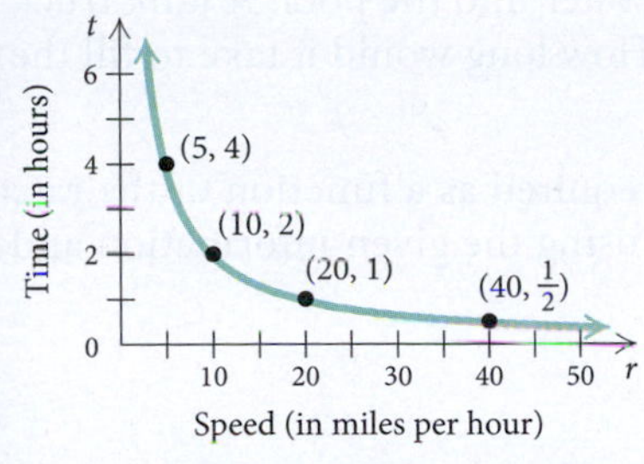

Suppose a bus is traveling a distance of 20 mi. At a speed of 5 mph, the trip will take 4 hr; at 10 mph, it will take 2 hr; at 20 mph, it will take 1 hr; at 40 mph, it will take $\frac{1}{2}$ hr; and so on. We plot this information on a graph, using speed as the first coordinate and time as the second coordinate to determine a set of ordered pairs:

$$(5, 4), \quad (10, 2), \quad (20, 1), \quad \left(40, \tfrac{1}{2}\right), \quad \text{and so on.}$$

Note that the products of the coordinates are all the same number:

$$5 \cdot 4 = 20, \quad 10 \cdot 2 = 20, \quad 20 \cdot 1 = 20, \quad 40 \cdot \tfrac{1}{2} = 20, \quad \text{and so on.}$$

Whenever a situation produces pairs of numbers in which the *product is constant*, we say that there is **inverse variation**. In this case, the time varies inversely as the speed, or rate:

$$rt = 20 \text{ (a constant)}, \quad \text{or} \quad t = \frac{20}{r},$$

or, if we use function notation, $t(r) = 20/r$. This equation is an equation of **inverse variation**. The coefficient, 20, is called the **variation constant**. Note that as the first number increases, the second number decreases.

The graph of $y = k/x$, $k > 0$, is like the one shown below. Note that as x increases, y decreases. That is, the function is decreasing on the interval $(0, \infty)$.

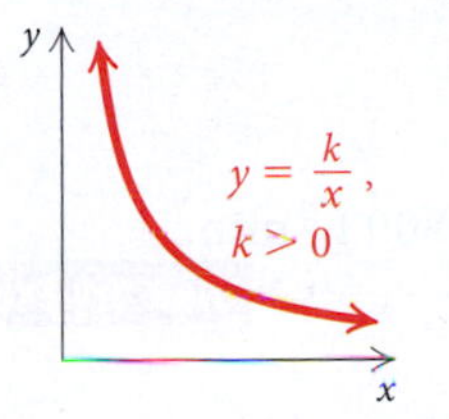

INVERSE VARIATION

If a situation gives rise to a function $f(x) = k/x$, or $y = k/x$, where k is a positive constant, we say that we have **inverse variation**, or that **y varies inversely as x**, or that **y is inversely proportional to x**. The number k is called the **variation constant**, or the **constant of proportionality**.

EXAMPLE 3 Find the variation constant and an equation of variation in which y varies inversely as x, and $y = 16$ when $x = 0.3$.

Solution We know that $(0.3, 16)$ is a solution of $y = k/x$. We substitute:

$$y = \frac{k}{x}$$

$$16 = \frac{k}{0.3} \quad \text{Substituting}$$

$$(0.3)16 = k \quad \text{Solving for } k$$

$$4.8 = k.$$

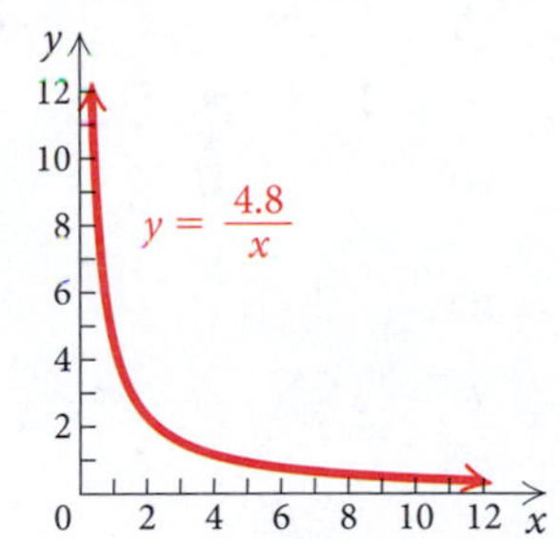

The variation constant is 4.8. The equation of variation is $y = 4.8/x$.

Now Try Exercise 3.

There are many real-world problems that translate to an equation of inverse variation.

EXAMPLE 4 ***Filling a Swimming Pool.*** The time t required to fill a swimming pool varies inversely as the rate of flow r of water into the pool. A tank truck can fill a pool in 90 min at a rate of 1500 L/min. How long would it take to fill the pool at a rate of 1800 L/min?

Solution We can express the amount of time required as a function of the rate of flow. Thus we have $t(r) = k/r$. We first find k using the given information and then find an equation of variation:

$$t(r) = \frac{k}{r} \qquad t \text{ varies inversely as } r.$$

$$t(1500) = \frac{k}{1500} \qquad \text{Substituting 1500 for } r$$

$$90 = \frac{k}{1500} \qquad \text{Replacing } t(1500) \text{ with 90}$$

$$90 \cdot 1500 = k \qquad \text{Solving for } k$$

$$135{,}000 = k. \qquad \text{This is the variation constant.}$$

The equation of variation is

$$t(r) = \frac{135{,}000}{r}.$$

Next, we use the equation to find the time that it would take to fill the pool at a rate of 1800 L/min:

$$t(r) = \frac{135{,}000}{r}$$

$$t(1800) = \frac{135{,}000}{1800} \qquad \text{Substituting}$$

$$t = 75.$$

Thus it would take 75 min to fill the pool at a rate of 1800 L/min.

Now Try Exercise 15.

Let's summarize the procedure for solving variation problems.

SOLVING VARIATION PROBLEMS

1. Determine whether direct variation or inverse variation applies.
2. Write an equation of the form $y = kx$ (for direct variation) or $y = k/x$ (for inverse variation), substitute the known values, and solve for k.
3. Write the equation of variation, and use it to find the unknown value(s) in the problem.

Combined Variation

We now look at other kinds of variation.

y varies **directly as the nth power of x** if there is some positive constant k such that

$$y = kx^n.$$

y varies **inversely as the nth power of x** if there is some positive constant k such that

$$y = \frac{k}{x^n}.$$

y varies **jointly as x and z** if there is some positive constant k such that

$$y = kxz.$$

There are other types of combined variation as well. Consider the formula for the volume of a right circular cylinder, $V = \pi r^2 h$, in which V, r, and h are variables and π is a constant. We say that V varies jointly as h and the square of r. In this formula, π is the variation constant.

EXAMPLE 5 Find an equation of variation in which y varies directly as the square of x, and $y = 12$ when $x = 2$.

Solution We write an equation of variation and find k:

$$\begin{aligned} y &= kx^2 \\ 12 &= k \cdot 2^2 \quad \text{Substituting} \\ 12 &= k \cdot 4 \\ 3 &= k. \end{aligned}$$

Thus, $y = 3x^2$.

Now Try Exercise 27.

EXAMPLE 6 Find an equation of variation in which y varies jointly as x and z, and $y = 42$ when $x = 2$ and $z = 3$.

Solution We have

$$\begin{aligned} y &= kxz \\ 42 &= k \cdot 2 \cdot 3 \quad \text{Substituting} \\ 42 &= k \cdot 6 \\ 7 &= k. \end{aligned}$$

Thus, $y = 7xz$.

Now Try Exercise 29.

EXAMPLE 7 Find an equation of variation in which y varies jointly as x and z and inversely as the square of w, and $y = 105$ when $x = 3$, $z = 20$, and $w = 2$.

Solution We have

$$y = k \cdot \frac{xz}{w^2}$$

$$105 = k \cdot \frac{3 \cdot 20}{2^2} \quad \text{Substituting}$$

$$105 = k \cdot 15$$

$$7 = k.$$

Thus, $y = 7\frac{xz}{w^2}$, or $y = \frac{7xz}{w^2}$.

Now Try Exercise 33.

EXAMPLE 8 ***Volume of a Tree.*** The volume of wood V in a tree varies jointly as the height h and the square of the girth g. (Girth is distance around.) If the volume of a redwood tree is 216 m^3 when the height is 30 m and the girth is 1.5 m, what is the height of a tree whose volume is 344 m^3 and whose girth is 1.6 m?

Solution We first find k using the first set of data. Then we solve for h using the second set of data.

$$V = khg^2$$

$$216 = k \cdot 30 \cdot 1.5^2$$

$$216 = k \cdot 30 \cdot 2.25$$

$$216 = k \cdot 67.5$$

$$3.2 = k$$

Thus the equation of variation is $V = 3.2hg^2$. We substitute the second set of data into the equation:

$$344 = 3.2 \cdot h \cdot 1.6^2$$

$$344 = 3.2 \cdot h \cdot 2.56$$

$$344 = 8.192 \cdot h$$

$$42 \approx h.$$

The height of the tree is about 42 m.

Now Try Exercise 35.

2.6 Exercise Set

Find the variation constant and an equation of variation for the given situation.

1. y varies directly as x, and $y = 54$ when $x = 12$
2. y varies directly as x, and $y = 0.1$ when $x = 0.2$
3. y varies inversely as x, and $y = 3$ when $x = 12$
4. y varies inversely as x, and $y = 12$ when $x = 5$
5. y varies directly as x, and $y = 1$ when $x = \frac{1}{4}$
6. y varies inversely as x, and $y = 0.1$ when $x = 0.5$
7. y varies inversely as x, and $y = 32$ when $x = \frac{1}{8}$
8. y varies directly as x, and $y = 3$ when $x = 33$
9. y varies directly as x, and $y = \frac{3}{4}$ when $x = 2$
10. y varies inversely as x, and $y = \frac{1}{5}$ when $x = 35$
11. y varies inversely as x, and $y = 1.8$ when $x = 0.3$
12. y varies directly as x, and $y = 0.9$ when $x = 0.4$

13. *Child's Allowance.* The Harrisons decide to give their children a weekly allowance that is directly proportional to each child's age. Their 6-year-old daughter receives an allowance of \$5.50. What is their 9-year-old son's allowance?

14. *Sales Tax.* The amount of sales tax paid on a product is directly proportional to the purchase price. In Iowa, the sales tax on a Nook Glowlight™ that sells for \$119 is \$7.14. What is the sales tax on an e-book that sells for \$21?

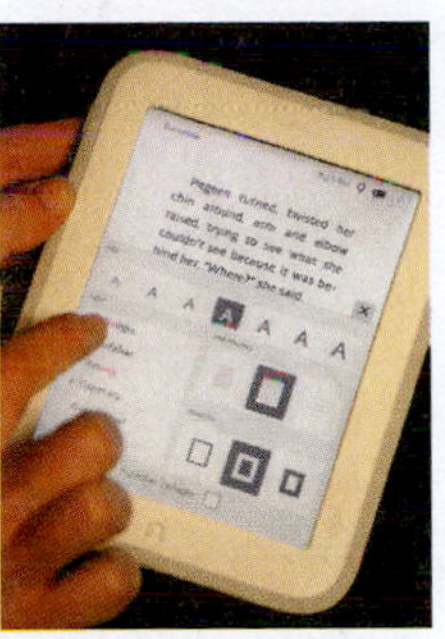

15. *Rate of Travel.* The time t required to drive a fixed distance varies inversely as the speed r. It takes 5 hr at a speed of 80 km/h to drive a fixed distance. How long will it take to drive the same distance at a speed of 70 km/h?

16. *Beam Weight.* The weight W that a horizontal beam can support varies inversely as the length L of the beam. Suppose an 8-m beam can support 1200 kg. How many kilograms can a 14-m beam support?

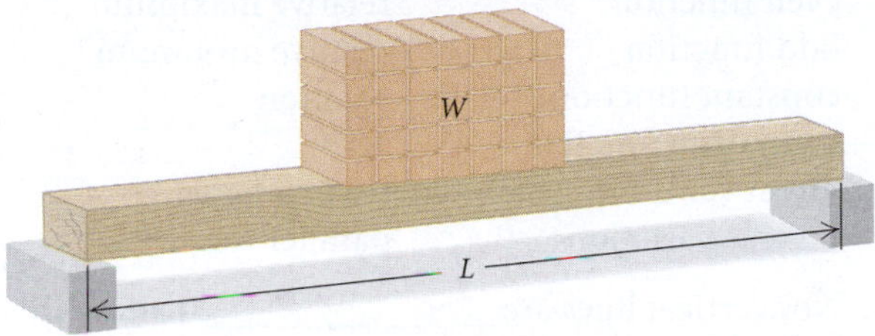

17. *Fat Intake.* The maximum number of grams of fat that should be in a diet varies directly as a person's weight. A person weighing 120 lb should have no more than 60 g of fat per day. What is the maximum daily fat intake for a person weighing 180 lb?

18. *House of Representatives.* The number of representatives N that each state has varies directly as the number of people P living in the state. If California, with 38,040,000 residents, has 53 representatives, how many representatives does Texas, with a population of 26,060,000, have?

19. *Work Rate.* The time T required to do a job varies inversely as the number of people P working. It takes 5 hr for 7 bricklayers to build a park wall. How long will it take 10 bricklayers to complete the job?

20. *Pumping Rate.* The time t required to empty a tank varies inversely as the rate r of pumping. If a pump can empty a tank in 45 min at the rate of 600 kL/min, how long will it take the pump to empty the same tank at the rate of 1000 kL/min?

21. *Hooke's Law.* Hooke's law states that the distance d that a spring will stretch varies directly as the mass m of an object hanging from the spring. If a 3-kg mass stretches a spring 40 cm, how far will a 5-kg mass stretch the spring?

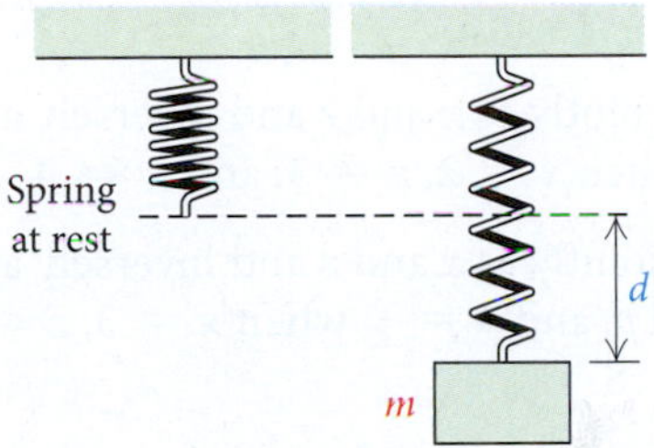

22. *Relative Aperture.* The relative aperture, or f-stop, of a 23.5-mm diameter lens is directly proportional to the focal length F of the lens. If a 150-mm focal length has an f-stop of 6.3, find the f-stop of a 23.5-mm diameter lens with a focal length of 80 mm.

23. *Musical Pitch.* The pitch P of a musical tone varies inversely as its wavelength W. One tone has a pitch of 330 vibrations per second and a wavelength of 3.2 ft. Find the wavelength of another tone that has a pitch of 550 vibrations per second.

24. *Weight on Mars.* The weight M of an object on Mars varies directly as its weight E on Earth. A person who weighs 95 lb on Earth weighs 35.9 lb on Mars. How much would a 100-lb person weigh on Mars?

Find an equation of variation for the given situation.

25. y varies inversely as the square of x, and $y = 0.15$ when $x = 0.1$

26. y varies inversely as the square of x, and $y = 6$ when $x = 3$

27. y varies directly as the square of x, and $y = 0.15$ when $x = 0.1$

28. y varies directly as the square of x, and $y = 6$ when $x = 3$

29. y varies jointly as x and z, and $y = 56$ when $x = 7$ and $z = 8$

30. y varies directly as x and inversely as z, and $y = 4$ when $x = 12$ and $z = 15$

31. y varies jointly as x and the square of z, and $y = 105$ when $x = 14$ and $z = 5$

32. y varies jointly as x and z and inversely as w, and $y = \frac{3}{2}$ when $x = 2, z = 3$, and $w = 4$

33. y varies jointly as x and z and inversely as the product of w and p, and $y = \frac{3}{28}$ when $x = 3, z = 10, w = 7$, and $p = 8$

34. y varies jointly as x and z and inversely as the square of w, and $y = \frac{12}{5}$ when $x = 16, z = 3$, and $w = 5$

35. *Intensity of Light.* The intensity I of light from a light bulb varies inversely as the square of the distance d from the bulb. Suppose that I is 90 W/m^2 (watts per square meter) when the distance is 5 m. How much *farther* would it be to a point where the intensity is 40 W/m^2?

36. *Atmospheric Drag.* Wind resistance, or atmospheric drag, tends to slow down moving objects. Atmospheric drag varies jointly as an object's surface area A and velocity v. If a car traveling at a speed of 40 mph with a surface area of 37.8 ft^2 experiences a drag of 222 N (Newtons), how fast must a car with 51 ft^2 of surface area travel in order to experience a drag force of 430 N?

37. *Stopping Distance of a Car.* The stopping distance d of a car after the brakes have been applied varies directly as the square of the speed r. If a car traveling 60 mph can stop in 200 ft, how fast can a car travel and still stop in 72 ft?

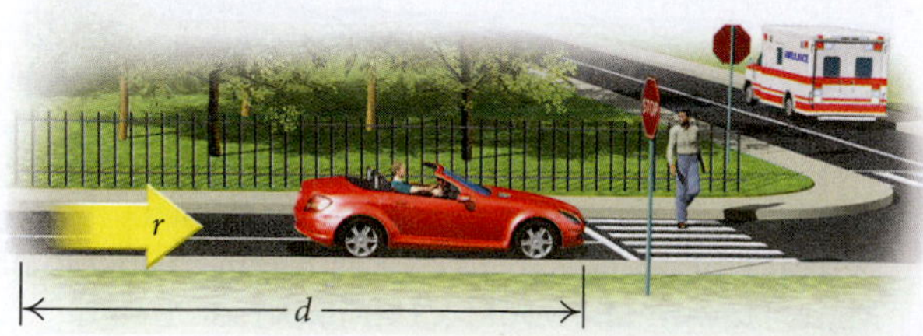

38. *Weight of an Astronaut.* The weight W of an object varies inversely as the square of the distance d from the center of the earth. At sea level (3978 mi from the center of the earth), an astronaut weighs 220 lb. Find his weight when he is 200 mi above the surface of the earth.

39. *Earned-Run Average.* A pitcher's earned-run average E varies directly as the number R of earned runs allowed and inversely as the number I of innings pitched. In 2013, Jon Lester of the Boston Red Sox had an earned-run average of 3.75. He gave up 89 earned runs in 213.1 innings. How many earned runs would he have given up had he pitched 235 innings with the same average? Round to the nearest whole number.

40. *Boyle's Law.* The volume V of a given mass of a gas varies directly as the temperature T and inversely as the pressure P. If $V = 231\text{ cm}^3$ when $T = 42°$ and $P = 20\text{ kg/cm}^2$, what is the volume when $T = 30°$ and $P = 15\text{ kg/cm}^2$?

Skill Maintenance

Vocabulary Reinforcement

In each of Exercises 41–45, fill in the blank with the correct term. Some of the given choices will not be used.

even function	relative maximum
odd function	relative minimum
constant function	solution
composite function	zero
direct variation	perpendicular
inverse variation	parallel

41. Nonvertical lines are ______________ if and only if they have the same slope and different y-intercepts. [1.4]

42. An input c of a function f is a(n) ______________ of the function if $f(c) = 0$. [1.5]

43. For a function f for which $f(c)$ exists, $f(c)$ is a(n) ______________ if $f(c)$ is the lowest point in some open interval. [2.1]

44. If the graph of a function is symmetric with respect to the origin, then f is a(n) ______________. [2.4]

45. An equation $y = k/x$ is an equation of ______________. [2.6]

Synthesis

46. In each of the following equations, state whether y varies directly as x, inversely as x, or neither directly nor inversely as x.

a) $7xy = 14$

b) $x - 2y = 12$

c) $-2x + 3y = 0$

d) $x = \frac{3}{4}y$

e) $\dfrac{x}{y} = 2$

47. *Volume and Cost.* An 18-oz jar of peanut butter in the shape of a right circular cylinder is 5 in. high and 3 in. in diameter and sells for \$2.89. In the same store, a 22-oz jar of the same brand is $5\frac{1}{4}$ in. high and $3\frac{1}{4}$ in. in diameter. If the cost is directly proportional to volume, what should the price of the larger jar be? If the cost is directly proportional to weight, what should the price of the larger jar be?

48. Describe in words the variation given by the equation

$$Q = \frac{kp^2}{q^3}.$$

49. *Area of a Circle.* The area of a circle varies directly as the square of the length of a diameter. What is the variation constant?

Chapter 2 Summary and Review

STUDY GUIDE

KEY TERMS AND CONCEPTS	EXAMPLES
SECTION 2.1: INCREASING, DECREASING, AND PIECEWISE FUNCTIONS; APPLICATIONS	
Increasing, Decreasing, and Constant Functions A function f is said to be **increasing** on an *open* interval I if for all a and b in that interval, $a < b$ implies $f(a) < f(b)$. A function f is said to be **decreasing** on an *open* interval I if for all a and b in that interval, $a < b$ implies $f(a) > f(b)$. A function f is said to be **constant** on an *open* interval I if for all a and b in that interval, $f(a) = f(b)$.	Determine the intervals on which the function is **(a)** increasing; **(b)** decreasing; **(c)** constant. 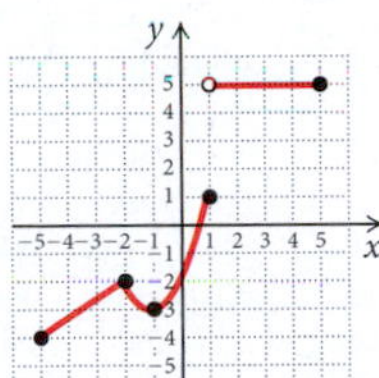 **a)** As x-values increase from -5 to -2, y-values increase from -4 to -2; y-values also increase as x-values increase from -1 to 1. Thus the function is increasing on the intervals $(-5, -2)$ and $(-1, 1)$. **b)** As x-values increase from -2 to -1, y-values decrease from -2 to -3, so the function is decreasing on the interval $(-2, -1)$. **c)** As x-values increase from 1 to 5, y remains 5, so the function is constant on the interval (1, 5).
Relative Maxima and Minima Suppose that f is a function for which $f(c)$ exists for some c in the domain of f. Then: $f(c)$ is a **relative maximum** if there exists an *open* interval I containing c such that $f(c) > f(x)$ for all x in I, where $x \neq c$; and $f(c)$ is a **relative minimum** if there exists an *open* interval I containing c such that $f(c) < f(x)$ for all x in I, where $x \neq c$.	Determine any relative maxima or minima of the function. 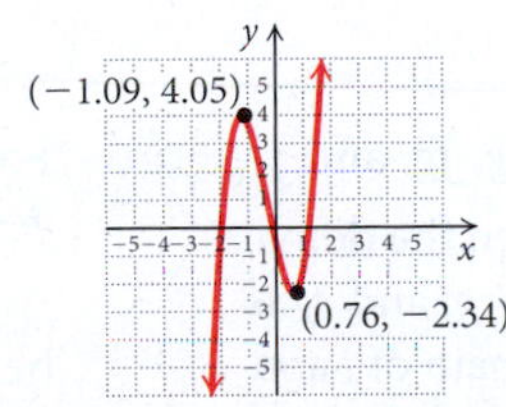 We see from the graph that the function has one relative maximum, 4.05. It occurs when $x = -1.09$. We also see that there is one relative minimum, -2.34. It occurs when $x = 0.76$.

Some applied problems can be modeled by functions.	See Examples 3 and 4 on pages 101 and 102.
To graph a function that is defined **piecewise**, graph the function in parts as defined by its output formulas.	Graph the function defined as $$f(x) = \begin{cases} 2x - 3, & \text{for } x < 1, \\ x + 1, & \text{for } x \geq 1. \end{cases}$$ We create the graph in two parts. First, we graph $f(x) = 2x - 3$ for inputs x less than 1. Then we graph $f(x) = x + 1$ for inputs x greater than or equal to 1. 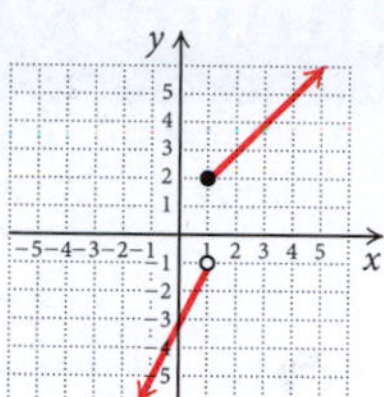
Greatest Integer Function $f(x) = [\![x]\!] =$ the greatest integer less than or equal to x.	The graph of the greatest integer function is shown below. Each input is paired with the greatest integer less than or equal to that input. 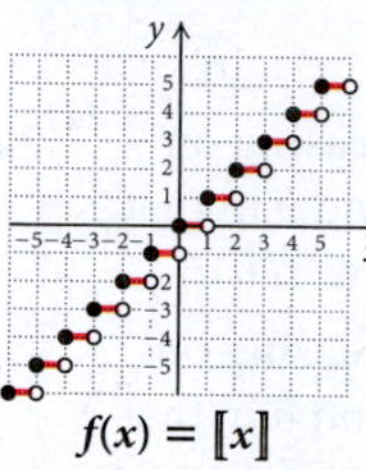 $f(x) = [\![x]\!]$

SECTION 2.2: THE ALGEBRA OF FUNCTIONS

Sums, Differences, Products, and Quotients of Functions If f and g are functions and x is in the domain of each function, then: $(f + g)(x) = f(x) + g(x)$, $(f - g)(x) = f(x) - g(x)$, $(fg)(x) = f(x) \cdot g(x)$, $(f/g)(x) = f(x)/g(x)$, provided $g(x) \neq 0$.	Given that $f(x) = x - 4$ and $g(x) = \sqrt{x + 5}$, find each of the following. **a)** $(f + g)(x)$ **b)** $(f - g)(x)$ **c)** $(fg)(x)$ **d)** $(f/g)(x)$ **a)** $(f + g)(x) = f(x) + g(x) = x - 4 + \sqrt{x + 5}$ **b)** $(f - g)(x) = f(x) - g(x) = x - 4 - \sqrt{x + 5}$ **c)** $(fg)(x) = f(x) \cdot g(x) = (x - 4)\sqrt{x + 5}$ **d)** $(f/g)(x) = f(x)/g(x) = \dfrac{x - 4}{\sqrt{x + 5}}, x \neq -5$
Domains of $f + g$, $f - g$, fg, and f/g If f and g are functions, then the domain of the functions $f + g, f - g$, and fg is the intersection of the domain of f and the domain of g. The domain of f/g is also the intersection of the domain of f and the domain of g, with the exclusion of any x-values for which $g(x) = 0$.	For the functions f and g above, find the domains of $f + g, f - g, fg$, and f/g. The domain of $f(x) = x - 4$ is the set of all real numbers. The domain of $g(x) = \sqrt{x + 5}$ is the set of all real numbers for which $x + 5 \geq 0$, or $x \geq -5$, or $[-5, \infty)$. Then the domain of $f + g, f - g$, and fg is the set of numbers in the intersection of these domains, or $[-5, \infty)$. Since $g(-5) = 0$, we must exclude -5. Thus the domain of f/g is $[-5, \infty)$ excluding -5, or $(-5, \infty)$.

The **difference quotient** for a function $f(x)$ is the ratio $$\frac{f(x+h)-f(x)}{h}.$$	For the function $f(x) = x^2 - 4$, construct and simplify the difference quotient. $$\frac{f(x+h)-f(x)}{h} = \frac{[(x+h)^2-4]-(x^2-4)}{h} = \frac{x^2+2xh+h^2-4-x^2+4}{h} = \frac{2xh+h^2}{h} = \frac{h(2x+h)}{h} = 2x+h$$

SECTION 2.3: THE COMPOSITION OF FUNCTIONS

The **composition of functions**, $f \circ g$, is defined as $$(f \circ g)(x) = f(g(x)),$$ where x is in the domain of g and $g(x)$ is in the domain of f.	Given that $f(x) = 2x - 1$ and $g(x) = \sqrt{x}$, find each of the following. **a)** $(f \circ g)(4)$ **b)** $(g \circ g)(625)$ **c)** $(f \circ g)(x)$ **d)** $(g \circ f)(x)$ **e)** The domain of $f \circ g$ and the domain of $g \circ f$ **a)** $(f \circ g)(4) = f(g(4)) = f(\sqrt{4}) = f(2) = 2 \cdot 2 - 1 = 4 - 1 = 3$ **b)** $(g \circ g)(625) = g(g(625)) = g(\sqrt{625}) = g(25) = \sqrt{25} = 5$ **c)** $(f \circ g)(x) = f(g(x)) = f(\sqrt{x}) = 2\sqrt{x} - 1$ **d)** $(g \circ f)(x) = g(f(x)) = g(2x-1) = \sqrt{2x-1}$ **e)** The domain and the range of $f(x)$ are both $(-\infty, \infty)$, and the domain and the range of $g(x)$ are both $[0, \infty)$. Since the inputs of $f \circ g$ are outputs of g and since f can accept any real number as an input, the domain of $f \circ g$ consists of all real numbers that are outputs of g, or $[0, \infty)$. The inputs of $g \circ f$ consist of all real numbers that are in the domain of g. Thus we must have $2x - 1 \geq 0$, or $x \geq \frac{1}{2}$, so the domain of $g \circ f$ is $\left[\frac{1}{2}, \infty\right)$.
When we **decompose** a function, we write it as the composition of two functions.	If $h(x) = \sqrt{3x+7}$, find $f(x)$ and $g(x)$ such that $h(x) = (f \circ g)(x)$. This function finds the square root of $3x + 7$, so one decomposition is $f(x) = \sqrt{x}$ and $g(x) = 3x + 7$. There are other correct answers, but this one is probably the most obvious.

SECTION 2.4: SYMMETRY

Algebraic Tests of Symmetry

x-axis: If replacing y with $-y$ produces an equivalent equation, then the graph is *symmetric with respect to the x-axis.*

y-axis: If replacing x with $-x$ produces an equivalent equation, then the graph is *symmetric with respect to the y-axis.*

Origin: If replacing x with $-x$ and y with $-y$ produces an equivalent equation, then the graph is *symmetric with respect to the origin.*

Test $y = 2x^3$ for symmetry with respect to the x-axis, the y-axis, and the origin.

x-axis: We replace y with $-y$:

$$-y = 2x^3$$
$$y = -2x^3. \quad \text{Multiplying by } -1$$

The resulting equation *is not* equivalent to the original equation, so the graph *is not* symmetric with respect to the x-axis.

y-axis: We replace x with $-x$:

$$y = 2(-x)^3$$
$$y = -2x^3.$$

The resulting equation *is not* equivalent to the original equation, so the graph *is not* symmetric with respect to the y-axis.

Origin: We replace x with $-x$ and y with $-y$:

$$-y = 2(-x)^3$$
$$-y = -2x^3$$
$$y = 2x^3.$$

The resulting equation *is* equivalent to the original equation, so the graph *is* symmetric with respect to the origin.

Even Functions and Odd Functions

If the graph of a function is symmetric with respect to the y-axis, we say that it is an **even function**. That is, for each x in the domain of f, $f(x) = f(-x)$.

If the graph of a function is symmetric with respect to the origin, we say that it is an **odd function**. That is, for each x in the domain of f, $f(-x) = -f(x)$.

Determine whether each function is even, odd, or neither.

a) $g(x) = 2x^2 - 4$ **b)** $h(x) = x^5 - 3x^3 - x$

a) We first find $g(-x)$ and simplify:

$$g(-x) = 2(-x)^2 - 4$$
$$= 2x^2 - 4.$$

$g(x) = g(-x)$, so g *is* even. Since a function other than $f(x) = 0$ cannot be *both* even and odd and g is even, we need not test to see if it is an odd function.

b) We first find $h(-x)$ and simplify:

$$h(-x) = (-x)^5 - 3(-x)^3 - (-x)$$
$$= -x^5 + 3x^3 + x.$$

$h(x) \neq h(-x)$, so h *is not* even.

Next, we find $-h(x)$ and simplify:

$$-h(x) = -(x^5 - 3x^3 - x)$$
$$= -x^5 + 3x^3 + x.$$

$h(-x) = -h(x)$, so h *is* odd.

SECTION 2.5: TRANSFORMATIONS

Vertical Translation

For $b > 0$:

the graph of $y = f(x) + b$ is the graph of $y = f(x)$ shifted *up* b units;

the graph of $y = f(x) - b$ is the graph of $y = f(x)$ shifted *down* b units.

Horizontal Translation

For $d > 0$:

the graph of $y = f(x - d)$ is the graph of $y = f(x)$ shifted *right* d units;

the graph of $y = f(x + d)$ is the graph of $y = f(x)$ shifted *left* d units.

Graph $g(x) = (x - 2)^2 + 1$. Before doing so, describe how the graph can be obtained from the graph of $f(x) = x^2$.

First, note that the graph of $h(x) = (x - 2)^2$ is the graph of $f(x) = x^2$ shifted right 2 units. Then the graph of $g(x) = (x - 2)^2 + 1$ is the graph of $h(x) = (x - 2)^2$ shifted up 1 unit. Thus the graph of g is obtained by shifting the graph of $f(x) = x^2$ right 2 units and up 1 unit.

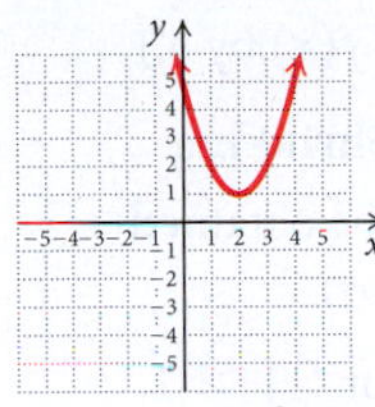

$g(x) = (x - 2)^2 + 1$

Reflections

The graph of $y = -f(x)$ is the **reflection** of $y = f(x)$ across the x-axis.

The graph of $y = f(-x)$ is the **reflection** of $y = f(x)$ across the y-axis.

If a point (x, y) is on the graph of $y = f(x)$, then $(x, -y)$ is on the graph of $y = -f(x)$, and $(-x, y)$ is on the graph of $y = f(-x)$.

Graph each of the following. Before doing so, describe how each graph can be obtained from the graph of $f(x) = x^2 - x$.

a) $g(x) = x - x^2$ **b)** $h(x) = (-x)^2 - (-x)$

a) Note that

$$\begin{aligned} -f(x) &= -(x^2 - x) \\ &= -x^2 + x \\ &= x - x^2 \\ &= g(x). \end{aligned}$$

Thus the graph is a reflection of the graph of $f(x) = x^2 - x$ across the x-axis.

$f(x) = x^2 - x$

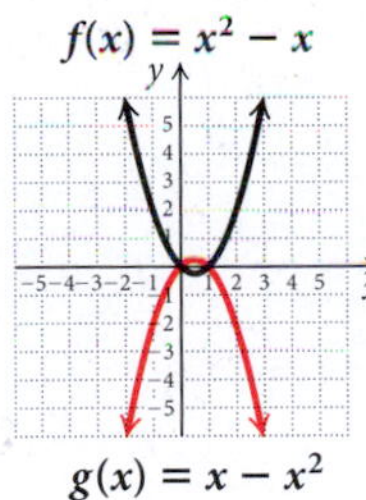

$g(x) = x - x^2$

b) Note that

$$f(-x) = (-x)^2 - (-x) = h(x).$$

Thus the graph of $h(x) = (-x)^2 - (-x)$ is a reflection of the graph of $f(x) = x^2 - x$ across the y-axis.

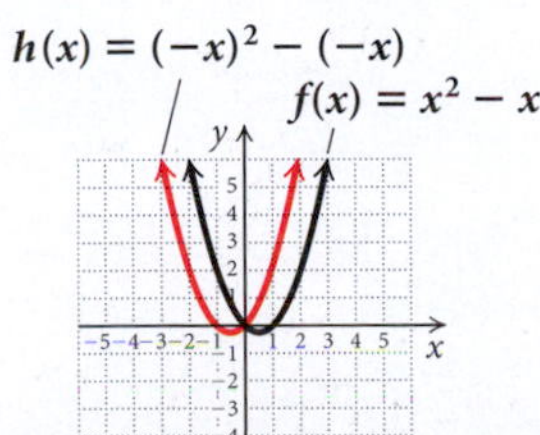

Vertical Stretching and Shrinking

The graph of $y = af(x)$ can be obtained from the graph of $y = f(x)$ by:

stretching vertically for $|a| > 1$, or

shrinking vertically for $0 < |a| < 1$.

For $a < 0$, the graph is also reflected across the x-axis.

(The y-coordinates of the graph of $y = af(x)$ can be obtained by multiplying the y-coordinates of $y = f(x)$ by a.)

Horizontal Stretching and Shrinking

The graph of $y = f(cx)$ can be obtained from the graph of $y = f(x)$ by:

shrinking horizontally for $|c| > 1$, or

stretching horizontally for $0 < |c| < 1$.

For $c < 0$, the graph is also reflected across the y-axis.

(The x-coordinates of the graph of $y = f(cx)$ can be obtained by dividing the x-coordinates of $y = f(x)$ by c.)

A graph of $y = g(x)$ is shown below. Use this graph to graph each of the given equations.

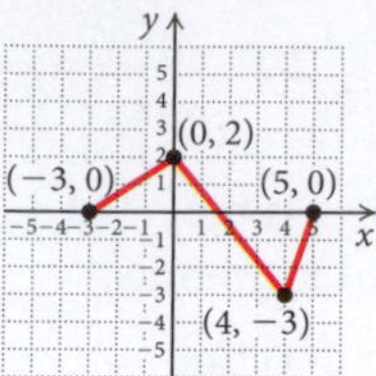

a) $f(x) = g(2x)$　　**b)** $f(x) = -2g(x)$

c) $f(x) = \frac{1}{2}g(x)$　　**d)** $f(x) = g\left(\frac{1}{2}x\right)$

a) Since $|2| > 1$, the graph of $f(x) = g(2x)$ is a horizontal shrinking of the graph of $y = g(x)$. The transformation divides each x-coordinate of g by 2.

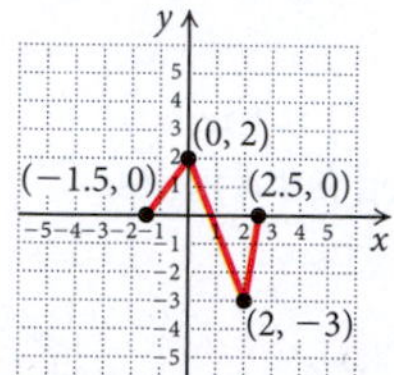

b) Since $|-2| > 1$, the graph of $f(x) = -2g(x)$ is a vertical stretching of the graph of $y = g(x)$. The transformation multiplies each y-coordinate of g by 2. Since $-2 < 0$, the graph is also reflected across the x-axis.

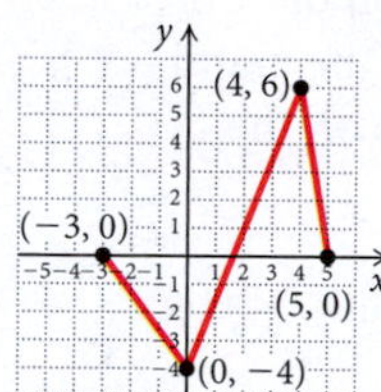

c) Since $\left|\frac{1}{2}\right| < 1$, the graph of $f(x) = \frac{1}{2}g(x)$ is a vertical shrinking of the graph of $y = g(x)$. The transformation multiplies each y-coordinate of g by $\frac{1}{2}$.

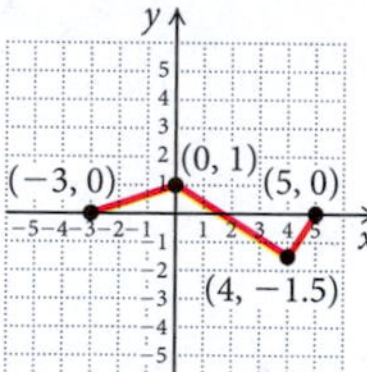

(continued)

d) Since $\left|\frac{1}{2}\right| < 1$, the graph of $f(x) = g\left(\frac{1}{2}x\right)$ is a horizontal stretching of the graph of $y = g(x)$. The transformation divides each x-coordinate of g by $\frac{1}{2}$ (which is the same as multiplying by 2).

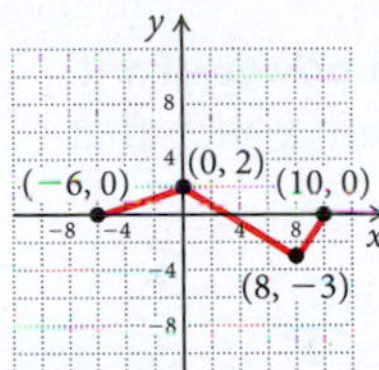

SECTION 2.6: VARIATION AND APPLICATIONS

Direct Variation

If a situation gives rise to a linear function $f(x) = kx$, or $y = kx$, where k is a positive constant, we say that we have **direct variation**, or that y **varies directly as** x, or that y **is directly proportional to** x. The number k is called the **variation constant**, or the **constant of proportionality**.

Find an equation of variation in which y varies directly as x, and $y = 24$ when $x = 8$. Then find the value of y when $x = 5$.

First, we have

$$y = kx \quad \textit{y varies directly as x.}$$
$$24 = k \cdot 8 \quad \text{Substituting}$$
$$3 = k \quad \text{Variation constant}$$

The equation of variation is $y = 3x$. Now we use the equation to find the value of y when $x = 5$:

$$y = 3x$$
$$= 3 \cdot 5 \quad \text{Substituting}$$
$$= 15.$$

When $x = 5$, the value of y is 15.

Inverse Variation

If a situation gives rise to a function $f(x) = k/x$, or $y = k/x$, where k is a positive constant, we say that we have **inverse variation**, or that y **varies inversely as** x, or that y **is inversely proportional to** x. The number k is called the **variation constant**, or the **constant of proportionality**.

Find an equation of variation in which y varies inversely as x, and $y = 5$ when $x = 0.1$. Then find the value of y when $x = 10$.

First, we have

$$y = \frac{k}{x} \quad \textit{y varies inversely as x.}$$
$$5 = \frac{k}{0.1} \quad \text{Substituting}$$
$$0.5 = k. \quad \text{Variation constant}$$

The equation of variation is $y = \dfrac{0.5}{x}$. Now we use the equation to find the value of y when $x = 10$:

$$y = \frac{0.5}{x}$$
$$= \frac{0.5}{10} \quad \text{Substituting}$$
$$= 0.05.$$

When $x = 10$, the value of y is 0.05.

Combined Variation y varies **directly as the nth power of x** if there is some positive constant k such that $y = kx^n$. y varies **inversely as the nth power of x** if there is some positive constant k such that $y = \frac{k}{x^n}$. y varies **jointly as x and z** if there is some positive constant k such that $y = kxz$.	Find an equation of variation in which y varies jointly as w and the square of x and inversely as z, and $y = 8$ when $w = 3, x = 2$, and $z = 6$. First, we have $y = k \cdot \frac{wx^2}{z}$ $8 = k \cdot \frac{3 \cdot 2^2}{6}$ **Substituting** $8 = k \cdot \frac{3 \cdot 4}{6}$ $8 = 2k$ $4 = k$. **Variation constant** The equation of variation is $y = 4\frac{wx^2}{z}$, or $y = \frac{4wx^2}{z}$.

REVIEW EXERCISES

Determine whether the statement is true or false.

1. The greatest integer function pairs each input with the greatest integer less than or equal to that input. [2.1]
2. In general, for functions f and g, the domain of $f \circ g =$ the domain of $g \circ f$. [2.3]
3. The graph of $y = (x - 2)^2$ is the graph of $y = x^2$ shifted right 2 units. [2.5]
4. The graph of $y = -x^2$ is the reflection of the graph of $y = x^2$ across the x-axis. [2.5]

Determine the intervals on which the function is **(a)** *increasing,* **(b)** *decreasing, and* **(c)** *constant.* [2.1]

5. 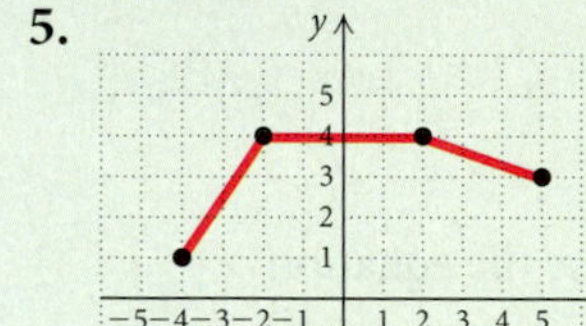

6. 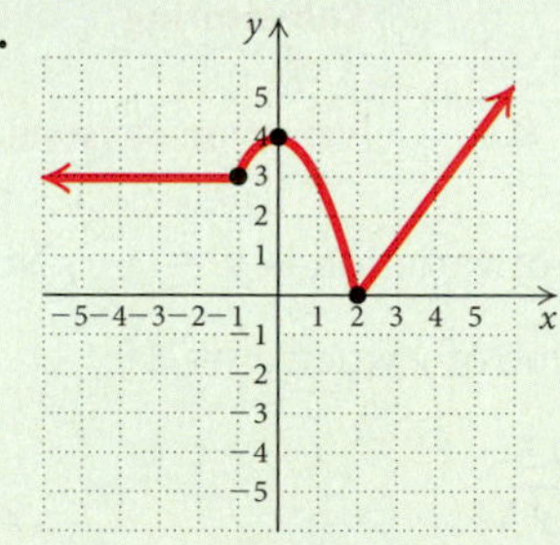

Graph the function. Estimate the intervals on which the function is increasing or decreasing and estimate any relative maxima or minima. [2.1]

7. $f(x) = x^2 - 1$
8. $f(x) = 2 - |x|$
9. *Fenced Patio.* Syd has 48 ft of rolled bamboo fence to enclose a rectangular patio. The house forms one side of the patio. Suppose two sides of the patio are each x feet. Express the area of the patio as a function of x. [2.1]
10. *Inscribed Rectangle.* A rectangle is inscribed in a semicircle of radius 2, as shown. The variable $x =$ half the length of the rectangle. Express the area of the rectangle as a function of x. [2.1]

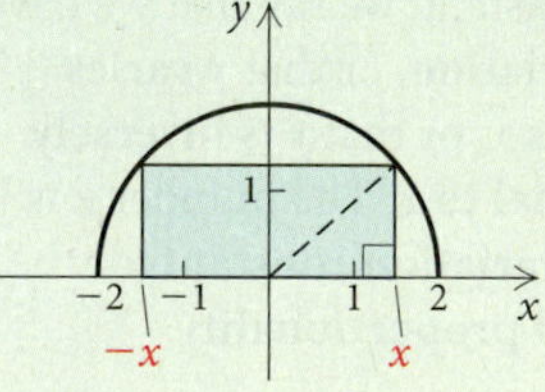

11. *Minimizing Surface Area.* A container firm is designing an open-top rectangular box, with a square base, that will hold 108 in^3. Let $x =$ the length of a side of the base.

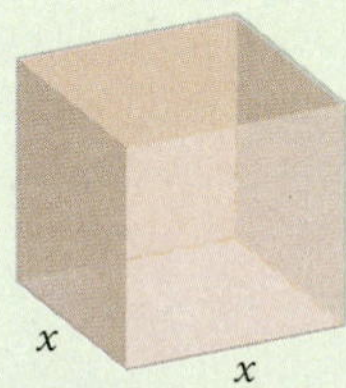

a) Express the surface area as a function of x. [2.1]
b) Find the domain of the function. [2.1]

c) Using the following graph, determine the dimensions that will minimize the surface area of the box. [2.1]

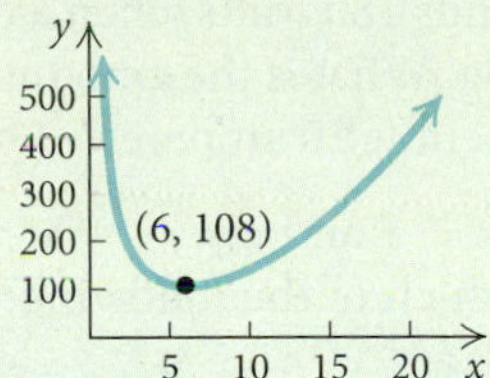

Graph each of the following. [2.1]

12. $f(x) = \begin{cases} -x, & \text{for } x \leq -4, \\ \frac{1}{2}x + 1, & \text{for } x > -4 \end{cases}$

13. $f(x) = \begin{cases} x^3, & \text{for } x < -2, \\ |x|, & \text{for } -2 \leq x \leq 2, \\ \sqrt{x-1}, & \text{for } x > 2 \end{cases}$

14. $f(x) = \begin{cases} \dfrac{x^2 - 1}{x + 1}, & \text{for } x \neq -1, \\ 3, & \text{for } x = -1 \end{cases}$

15. $f(x) = [\![x]\!]$

16. $f(x) = [\![x - 3]\!]$

17. For the function in Exercise 13, find $f(-1), f(5), f(-2)$, and $f(-3)$. [2.1]

18. For the function in Exercise 14, find $f(-2), f(-1), f(0)$, and $f(4)$. [2.1]

Given that $f(x) = \sqrt{x - 2}$ and $g(x) = x^2 - 1$, find each of the following if it exists. [2.2]

19. $(f - g)(6)$

20. $(fg)(2)$

21. $(f + g)(-1)$

For each pair of functions in Exercises 22 and 23:

a) *Find the domains of f, g, $f + g$, $f - g$, fg, and f/g.* [2.2]

b) *Find $(f + g)(x)$, $(f - g)(x)$, $(fg)(x)$, and $(f/g)(x)$.* [2.2]

22. $f(x) = \dfrac{4}{x^2}$, $g(x) = 3 - 2x$

23. $f(x) = 3x^2 + 4x$, $g(x) = 2x - 1$

24. Given the total-revenue and total-cost functions $R(x) = 120x - 0.5x^2$ and $C(x) = 15x + 6$, find the total-profit function $P(x)$. [2.2]

For each function f, construct and simplify the difference quotient. [2.2]

25. $f(x) = 2x + 7$

26. $f(x) = 3 - x^2$

27. $f(x) = \dfrac{4}{x}$

Given that $f(x) = 2x - 1$, $g(x) = x^2 + 4$, and $h(x) = 3 - x^3$, find each of the following. [2.3]

28. $(f \circ g)(1)$

29. $(g \circ f)(1)$

30. $(h \circ f)(-2)$

31. $(g \circ h)(3)$

32. $(f \circ h)(-1)$

33. $(h \circ g)(2)$

34. $(f \circ f)(x)$

35. $(h \circ h)(x)$

For each pair of functions in Exercises 36 and 37:

a) *Find $(f \circ g)(x)$ and $(g \circ f)(x)$.* [2.3]

b) *Find the domain of $f \circ g$ and the domain of $g \circ f$.* [2.3]

36. $f(x) = \dfrac{4}{x^2}$, $g(x) = 3 - 2x$

37. $f(x) = 3x^2 + 4x$, $g(x) = 2x - 1$

Find $f(x)$ and $g(x)$ such that $h(x) = (f \circ g)(x)$. [2.3]

38. $h(x) = \sqrt{5x + 2}$

39. $h(x) = 4(5x - 1)^2 + 9$

Graph the given equation and determine visually whether it is symmetric with respect to the x-axis, the y-axis, and the origin. Then verify your assertion algebraically. [2.4]

40. $x^2 + y^2 = 4$

41. $y^2 = x^2 + 3$

42. $x + y = 3$

43. $y = x^2$

44. $y = x^3$

45. $y = x^4 - x^2$

Determine visually whether the function is even, odd, or neither even nor odd. [2.4]

46.

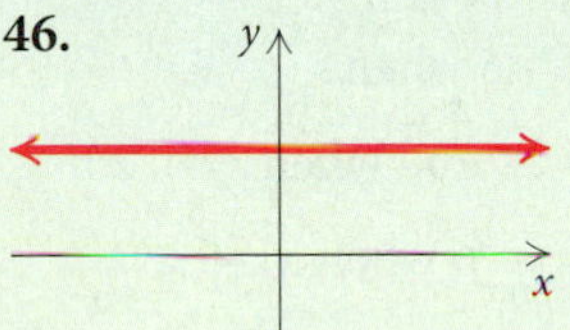

47.

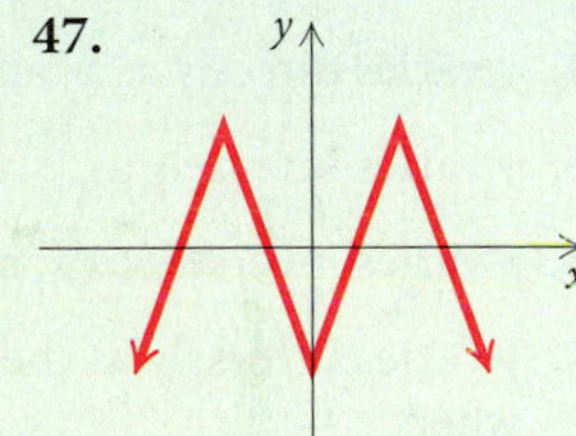

48.

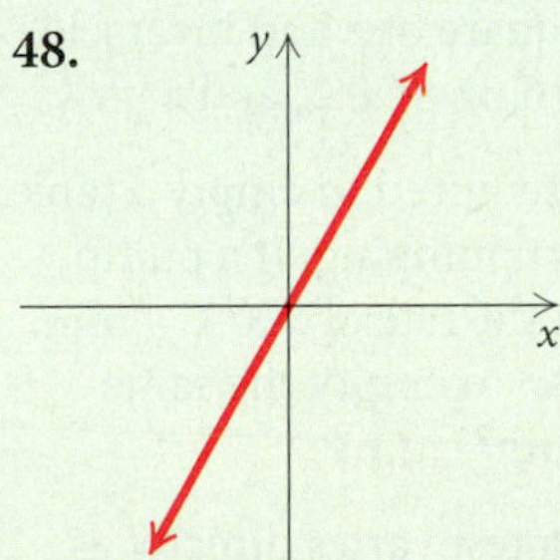

49.

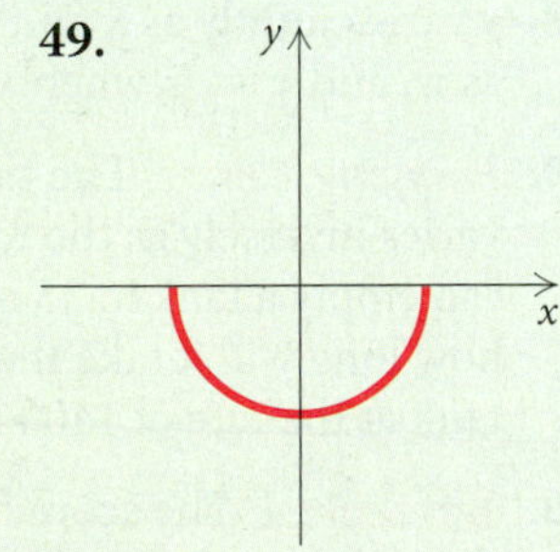

In Exercises 50–55, test whether the function is even, odd, or neither even nor odd. [2.4]

50. $f(x) = 9 - x^2$

51. $f(x) = x^3 - 2x + 4$

52. $f(x) = x^7 - x^5$

53. $f(x) = |x|$

54. $f(x) = \sqrt{16 - x^2}$

55. $f(x) = \dfrac{10x}{x^2 + 1}$

Write an equation for a function that has a graph with the given characteristics. [2.5]

56. The shape of $y = x^2$, but shifted left 3 units

57. The shape of $y = \sqrt{x}$, but upside down and shifted right 3 units and up 4 units

58. The shape of $y = |x|$, but stretched vertically by a factor of 2 and shifted right 3 units

A graph of $y = f(x)$ is shown below. No formula for f is given. Graph each of the following. [2.5]

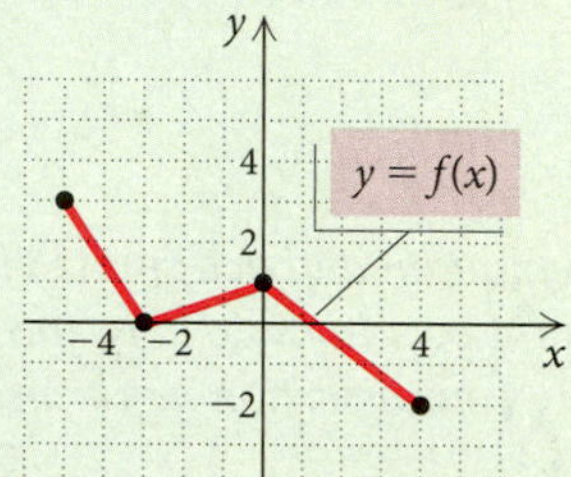

59. $y = f(x - 1)$

60. $y = f(2x)$

61. $y = -2f(x)$

62. $y = 3 + f(x)$

Find an equation of variation for the given situation. [2.6]

63. y varies directly as x, and $y = 100$ when $x = 25$.

64. y varies directly as x, and $y = 6$ when $x = 9$.

65. y varies inversely as x, and $y = 100$ when $x = 25$.

66. y varies inversely as x, and $y = 6$ when $x = 9$.

67. y varies inversely as the square of x, and $y = 12$ when $x = 2$.

68. y varies jointly as x and the square of z and inversely as w, and $y = 2$ when $x = 16$, $w = 0.2$, and $z = \frac{1}{2}$.

69. *Pumping Time.* The time t required to empty a tank varies inversely as the rate r of pumping. If a pump can empty a tank in 35 min at the rate of 800 kL/min, how long will it take the pump to empty the same tank at the rate of 1400 kL/min? [2.6]

70. *Test Score.* The score N on a test varies directly as the number of correct responses a. Sam answers 29 questions correctly and earns a score of 87. What would Sam's score have been if he had answered 25 questions correctly? [2.6]

71. *Power of Electric Current.* The power P expended by heat in an electric circuit of fixed resistance varies directly as the square of the current C in the circuit. A circuit expends 180 watts when a current of 6 amperes is flowing. What is the amount of heat expended when the current is 10 amperes? [2.6]

72. For $f(x) = x + 1$ and $g(x) = \sqrt{x}$, the domain of $(g \circ f)(x)$ is which of the following? [2.3]

A. $[-1, \infty)$ **B.** $[-1, 0)$
C. $[0, \infty)$ **D.** $(-\infty, \infty)$

73. For $b > 0$, the graph of $y = f(x) + b$ is the graph of $y = f(x)$ shifted in which of the following ways? [2.5]

A. Right b units **B.** Left b units
C. Up b units **D.** Down b units

74. The graph of the function f is shown below.

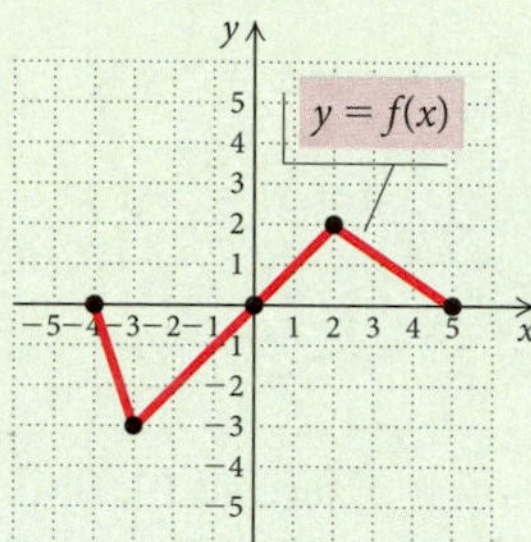

The graph of $g(x) = -\frac{1}{2}f(x) + 1$ is which of the following? [2.5]

A.

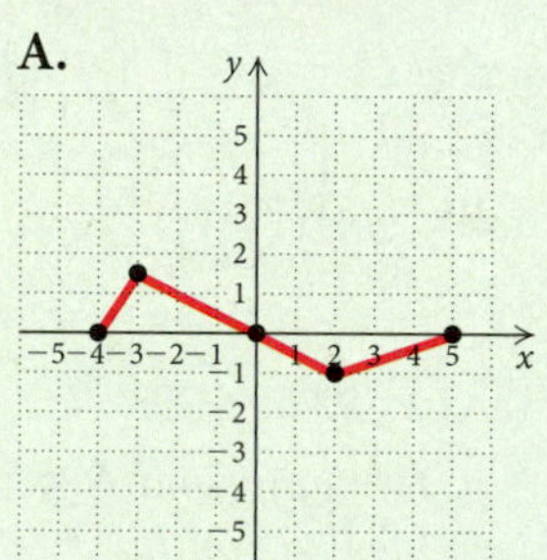

B.

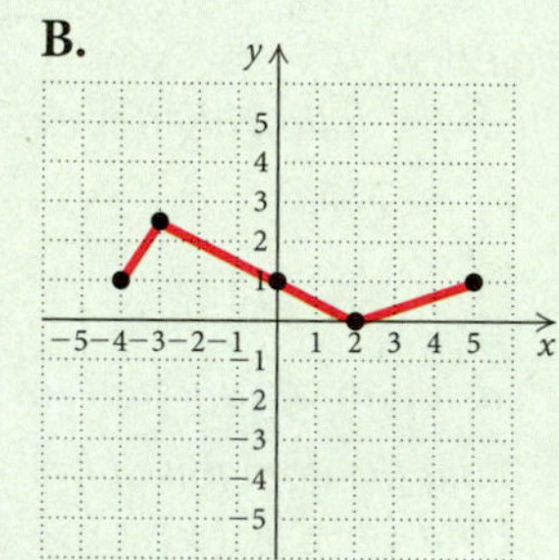

C.

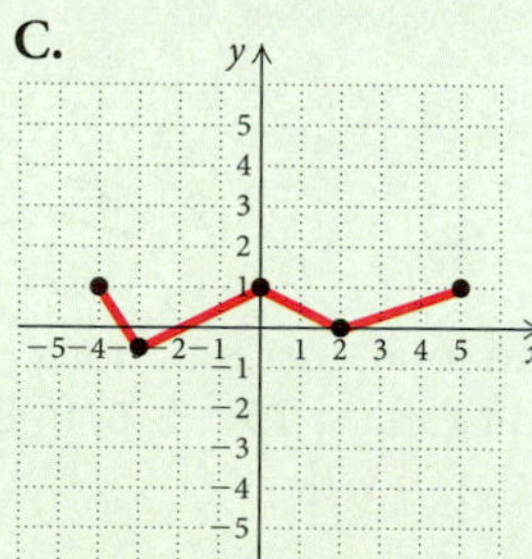

D.

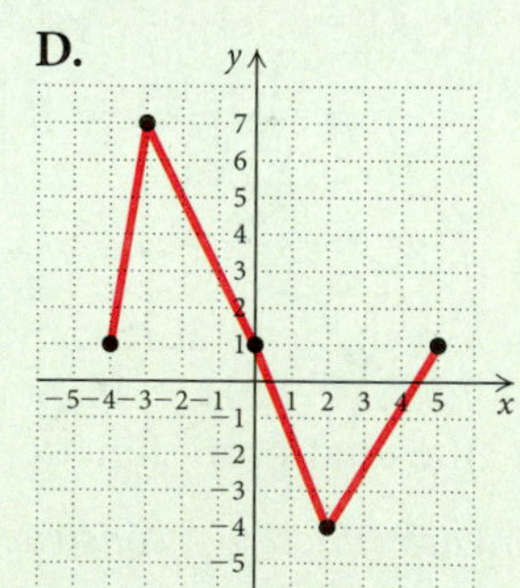

Synthesis

75. Prove that the sum of two odd functions is odd. [2.2], [2.4]

76. Describe how the graph of $y = -f(-x)$ is obtained from the graph of $y = f(x)$. [2.5]

Collaborative Discussion and Writing

77. Given that $f(x) = 4x^3 - 2x + 7$, find each of the following. Then discuss how each expression differs from the other. [1.2], [2.5]

a) $f(x) + 2$
b) $f(x + 2)$
c) $f(x) + f(2)$

78. Given the graph of $y = f(x)$, explain and contrast the effect of the constant c on the graphs of $y = f(cx)$ and $y = cf(x)$. [2.5]

79. Consider the constant function $f(x) = 0$. Determine whether the graph of this function is symmetric with respect to the x-axis, the y-axis, and/or the origin. Determine whether this function is even or odd. [2.4]

80. Describe conditions under which you would know whether a polynomial function

$$f(x) = a_nx^n + a_{n-1}x^{n-1} + \cdots + a_2x^2 + a_1x + a_0$$

is even or odd without using an algebraic procedure. Explain. [2.4]

81. If y varies directly as x^2, explain why doubling x would not cause y to be doubled as well. [2.6]

82. If y varies directly as x and x varies inversely as z, how does y vary with regard to z? Why? [2.6]

2 Chapter Test

1. Determine the intervals on which the function is **(a)** increasing; **(b)** decreasing; **(c)** constant.

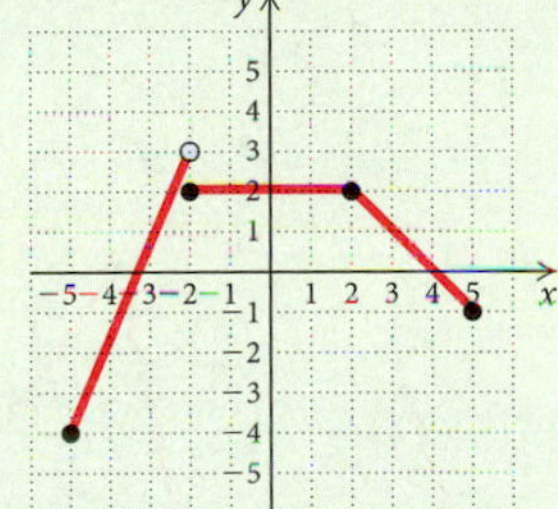

2. Graph the function $f(x) = 2 - x^2$. Estimate the intervals on which the function is increasing or decreasing and estimate any relative maxima or minima.

3. *Triangular Pennant.* A softball team is designing a triangular pennant such that the height is 6 in. less than four times the length of the base b. Express the area of the pennant as a function of b.

4. Graph:

$$f(x) = \begin{cases} x^2, & \text{for } x < -1, \\ |x|, & \text{for } -1 \le x \le 1, \\ \sqrt{x - 1}, & \text{for } x > 1. \end{cases}$$

5. For the function in Exercise 4, find $f\left(-\frac{7}{8}\right), f(5)$, and $f(-4)$.

Given that $f(x) = x^2 - 4x + 3$ and $g(x) = \sqrt{3 - x}$, find each of the following, if it exists.

6. $(f + g)(-6)$

7. $(f - g)(-1)$

8. $(fg)(2)$

9. $(f/g)(1)$

For $f(x) = x^2$ and $g(x) = \sqrt{x - 3}$, find each of the following.

10. The domain of f

11. The domain of g

12. The domain of $f + g$

13. The domain of $f - g$

14. The domain of fg

15. The domain of f/g

16. $(f + g)(x)$

17. $(f - g)(x)$

18. $(fg)(x)$

19. $(f/g)(x)$

For each function, construct and simplify the difference quotient.

20. $f(x) = \frac{1}{2}x + 4$

21. $f(x) = 2x^2 - x + 3$

Given that $f(x) = x^2 - 1$, $g(x) = 4x + 3$, *and* $h(x) = 3x^2 + 2x + 4$, *find each of the following.*

22. $(g \circ h)(2)$

23. $(f \circ g)(-1)$

24. $(h \circ f)(1)$

25. $(g \circ g)(x)$

For $f(x) = \sqrt{x - 5}$ *and* $g(x) = x^2 + 1$:

26. Find $(f \circ g)(x)$ and $(g \circ f)(x)$.

27. Find the domain of $(f \circ g)(x)$ and the domain of $(g \circ f)(x)$.

28. Find $f(x)$ and $g(x)$ such that
$$h(x) = (f \circ g)(x) = (2x - 7)^4.$$

29. Determine whether the graph of $y = x^4 - 2x^2$ is symmetric with respect to the x-axis, the y-axis, and the origin.

30. Test whether the function
$$f(x) = \frac{2x}{x^2 + 1}$$
is even, odd, or neither even nor odd. Show your work.

31. Write an equation for a function that has the shape of $y = x^2$, but shifted right 2 units and down 1 unit.

32. Write an equation for a function that has the shape of $y = x^2$, but shifted left 2 units and down 3 units.

33. The graph of a function $y = f(x)$ is shown below. No formula for f is given. Graph $y = -\frac{1}{2}f(x)$.

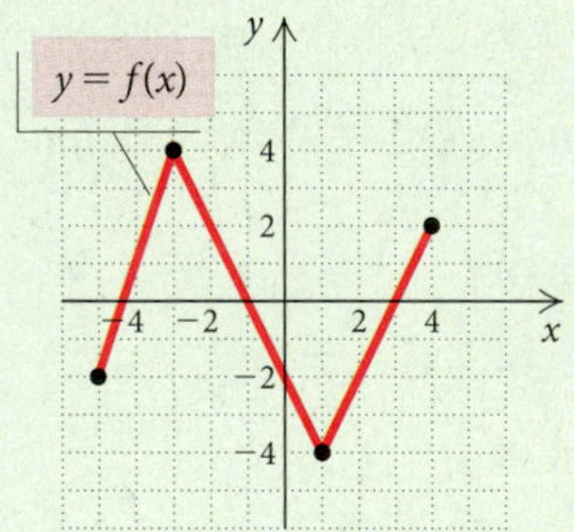

34. Find an equation of variation in which y varies inversely as x, and $y = 5$ when $x = 6$.

35. Find an equation of variation in which y varies directly as x, and $y = 60$ when $x = 12$.

36. Find an equation of variation where y varies jointly as x and the square of z and inversely as w, and $y = 100$ when $x = 0.1$, $z = 10$, and $w = 5$.

37. The stopping distance d of a car after the brakes have been applied varies directly as the square of the speed r. If a car traveling 60 mph can stop in 200 ft, how long will it take a car traveling 30 mph to stop?

38. The graph of the function f is shown below.

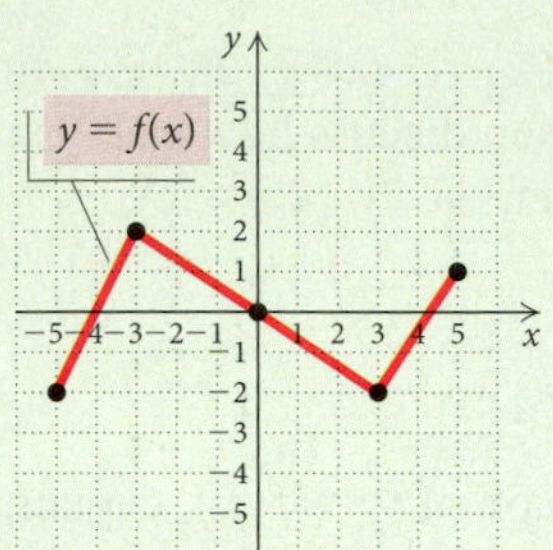

The graph of $g(x) = 2f(x) - 1$ is which of the following?

A.

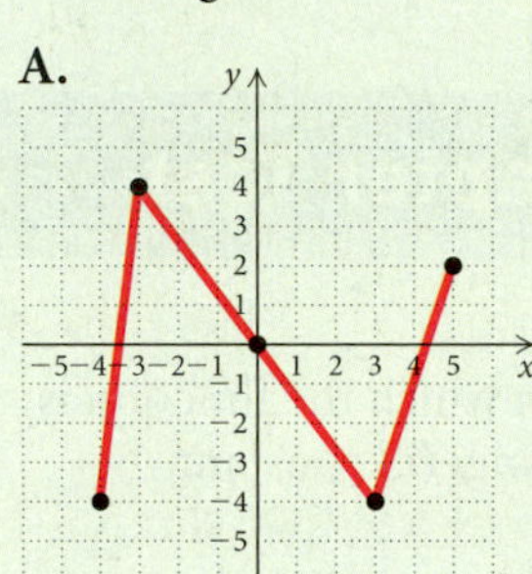

B.

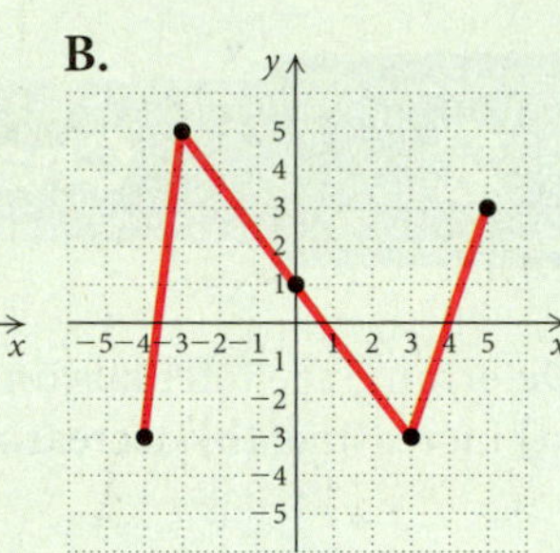

C.

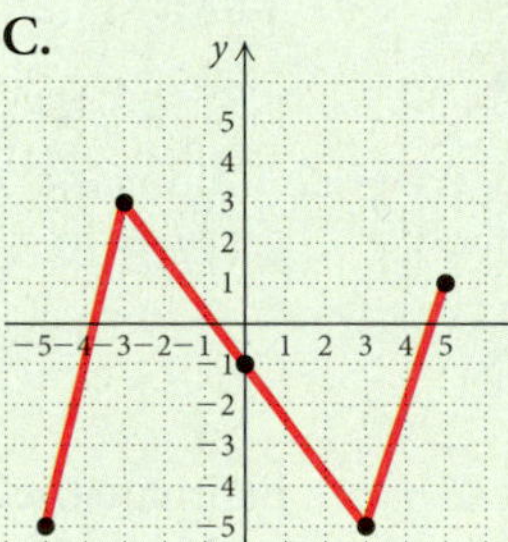

D.

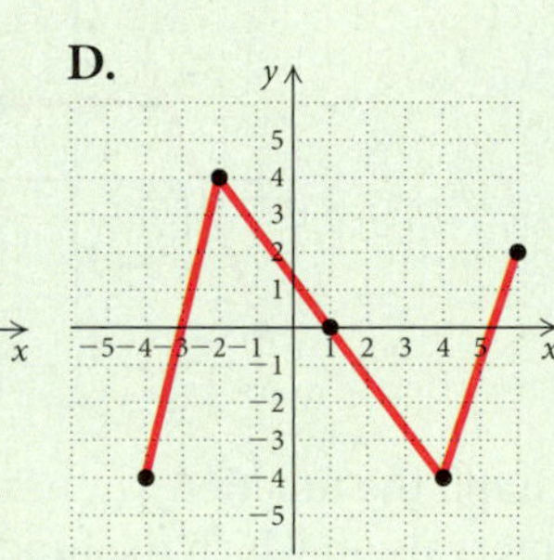

▸ Synthesis

39. If $(-3, 1)$ is a point on the graph of $y = f(x)$, what point do you know is on the graph of $y = f(3x)$?

Quadratic Functions and Equations; Inequalities

PPLICATION This problem appears as Exercises 95 and 96 in Section 3.2.

The number of U.S. forces in Afghanistan decreased to approximately 34,000 in 2014 from a high of about 100,000 in 2010. The amount of U.S. funding for Afghan security forces also decreased during this period. The function

$$f(x) = -1.321x^2 + 5.156x + 5.517$$

can be used to estimate the amount of U.S. funding for Afghan security forces, in billions of dollars, x years after 2009 (*Sources:* U.S. Department of Defense; Brookings Institution; International Security Assistance Force; ESRI). Find the amount of U.S. funding for Afghan security forces in 2011 and in 2013.

3.1 The Complex Numbers

► Perform computations involving complex numbers.

Some functions have zeros that are not real numbers. In order to find the zeros of such functions, we must consider the **complex-number system**.

► The Complex-Number System

We know that the square root of a negative number is not a real number. For example, $\sqrt{-1}$ is not a real number because there is no real number x such that $x^2 = -1$. This means that certain equations, like $x^2 = -1$ or $x^2 + 1 = 0$, do not have real-number solutions, and certain functions, like $f(x) = x^2 + 1$, do not have real-number zeros. Consider the graph of $f(x) = x^2 + 1$.

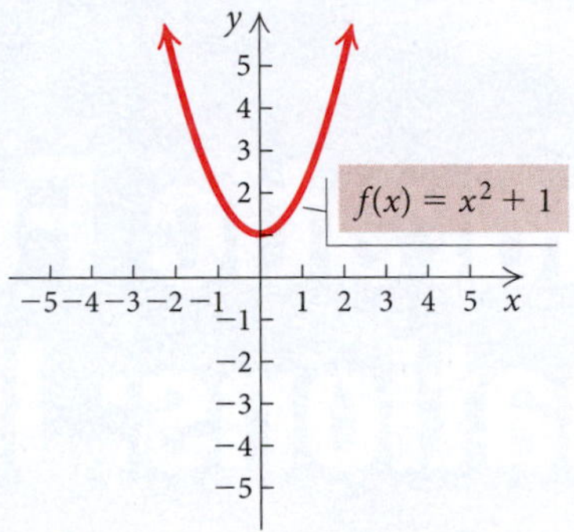

We see that the graph does not cross the x-axis and thus has no x-intercepts. This illustrates that the function $f(x) = x^2 + 1$ has no real-number zeros. Thus there are no real-number solutions of the corresponding equation $x^2 + 1 = 0$.

We can define a nonreal number that is a solution of the equation $x^2 + 1 = 0$.

> **THE NUMBER *i***
>
> The number i is defined such that
>
> $$i = \sqrt{-1} \quad \text{and} \quad i^2 = -1.$$

To express roots of negative numbers in terms of i, we can use the fact that

$$\sqrt{-p} = \sqrt{-1 \cdot p} = \sqrt{-1} \cdot \sqrt{p} = i\sqrt{p}$$

when p is a positive real number.

EXAMPLE 1 Express each number in terms of i.

a) $\sqrt{-7}$ b) $\sqrt{-16}$ c) $-\sqrt{-13}$

d) $-\sqrt{-64}$ e) $\sqrt{-48}$

Solution

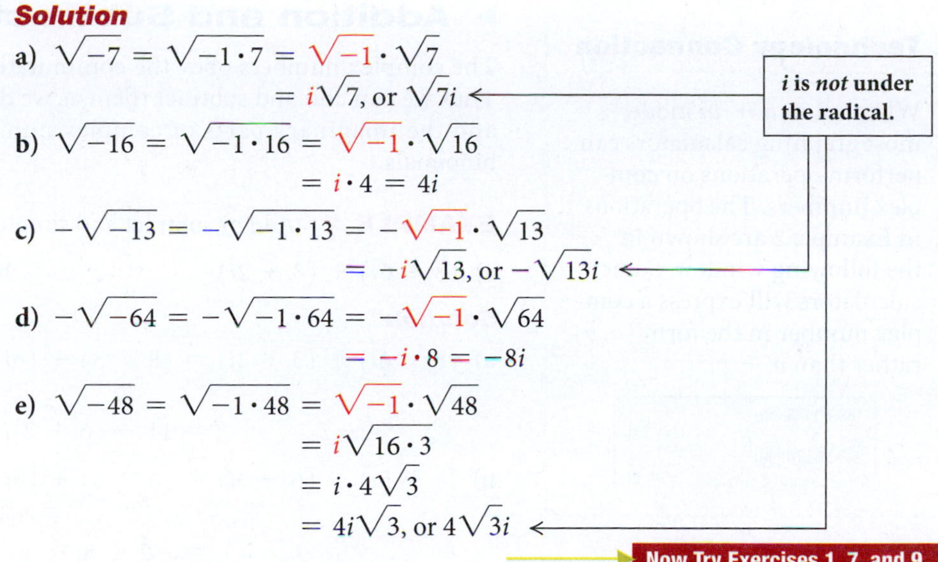

a) $\sqrt{-7} = \sqrt{-1 \cdot 7} = \sqrt{-1} \cdot \sqrt{7}$
$= i\sqrt{7}$, or $\sqrt{7}i$

b) $\sqrt{-16} = \sqrt{-1 \cdot 16} = \sqrt{-1} \cdot \sqrt{16}$
$= i \cdot 4 = 4i$

c) $-\sqrt{-13} = -\sqrt{-1 \cdot 13} = -\sqrt{-1} \cdot \sqrt{13}$
$= -i\sqrt{13}$, or $-\sqrt{13}i$

d) $-\sqrt{-64} = -\sqrt{-1 \cdot 64} = -\sqrt{-1} \cdot \sqrt{64}$
$= -i \cdot 8 = -8i$

e) $\sqrt{-48} = \sqrt{-1 \cdot 48} = \sqrt{-1} \cdot \sqrt{48}$
$= i\sqrt{16 \cdot 3}$
$= i \cdot 4\sqrt{3}$
$= 4i\sqrt{3}$, or $4\sqrt{3}i$

Now Try Exercises 1, 7, and 9.

The complex numbers are formed by adding real numbers and multiples of i.

COMPLEX NUMBERS

A **complex number** is a number of the form $a + bi$, where a and b are real numbers. The number a is said to be the **real part** of $a + bi$ and the number b is said to be the **imaginary part** of $a + bi$.*

Note that either a or b or both can be 0. When $b = 0$, $a + bi = a + 0i = a$, so every real number is a complex number. A complex number like $3 + 4i$ or $17i$, in which $b \neq 0$, is called an **imaginary number**. A complex number like $17i$ or $-4i$, in which $a = 0$ and $b \neq 0$, is sometimes called a **pure imaginary number**. The relationships among various types of complex numbers are shown in the following figure.

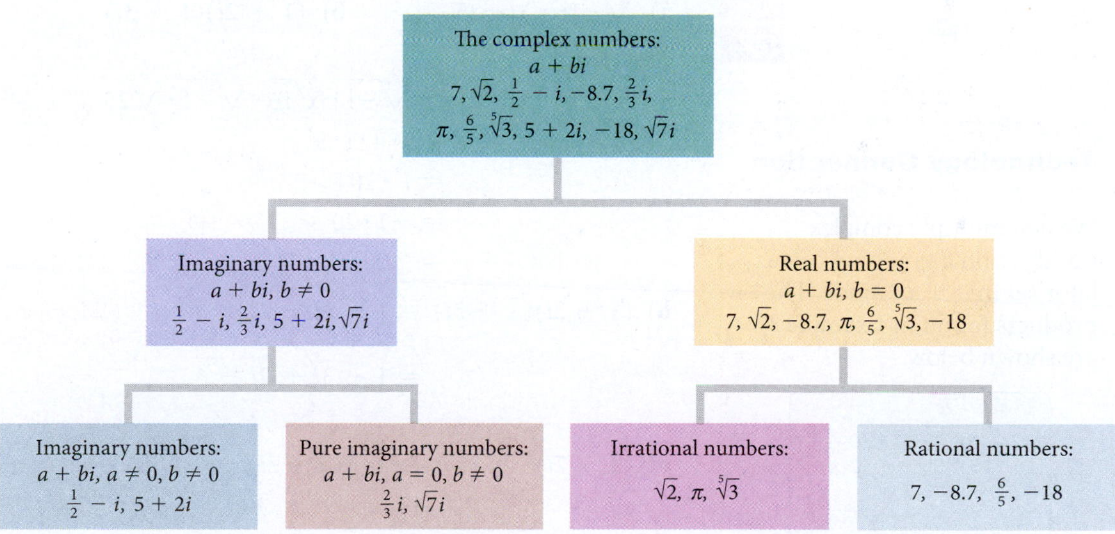

*Sometimes bi is considered to be the imaginary part.

Technology Connection

When set in $a + bi$ mode, most graphing calculators can perform operations on complex numbers. The operations in Example 2 are shown in the following window. Some calculators will express a complex number in the form (a, b) rather than $a + bi$.

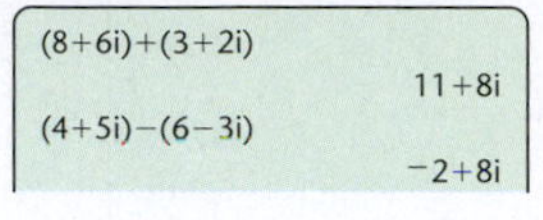

Addition and Subtraction

The complex numbers obey the commutative, associative, and distributive laws. Thus we can add and subtract them as we do binomials. We collect the real parts and the imaginary parts of complex numbers just as we collect like terms in binomials.

EXAMPLE 2 Add or subtract and simplify each of the following.

a) $(8 + 6i) + (3 + 2i)$ **b)** $(4 + 5i) - (6 - 3i)$

Solution

a) $(8 + 6i) + (3 + 2i) = (8 + 3) + (6i + 2i)$

Collecting the real parts and the imaginary parts

$= 11 + (6 + 2)i = 11 + 8i$

b) $(4 + 5i) - (6 - 3i) = (4 - 6) + [5i - (-3i)]$

Note that 6 and $-3i$ are both being subtracted.

$= -2 + 8i$

Now Try Exercises 11 and 21.

Multiplication

When $\sqrt{a}$ and $\sqrt{b}$ are real numbers, $\sqrt{a} \cdot \sqrt{b} = \sqrt{ab}$, but this is not true when $\sqrt{a}$ and $\sqrt{b}$ are not real numbers. Thus,

$$\sqrt{-2} \cdot \sqrt{-5} = \sqrt{-1} \cdot \sqrt{2} \cdot \sqrt{-1} \cdot \sqrt{5}$$
$$= i\sqrt{2} \cdot i\sqrt{5}$$
$$= i^2\sqrt{10} = -1\sqrt{10} = -\sqrt{10} \quad \text{is correct!}$$

But

$$\sqrt{-2} \cdot \sqrt{-5} = \sqrt{(-2)(-5)} = \sqrt{10} \quad \text{is wrong!}$$

Keeping this and the fact that $i^2 = -1$ in mind, we multiply with imaginary numbers in much the same way that we do with real numbers.

EXAMPLE 3 Multiply and simplify each of the following.

a) $\sqrt{-16} \cdot \sqrt{-25}$ **b)** $(1 + 2i)(1 + 3i)$ **c)** $(3 - 7i)^2$

Solution

a) $\sqrt{-16} \cdot \sqrt{-25} = \sqrt{-1} \cdot \sqrt{16} \cdot \sqrt{-1} \cdot \sqrt{25}$
$= i \cdot 4 \cdot i \cdot 5$
$= i^2 \cdot 20$
$= -1 \cdot 20$ $\quad i^2 = -1$
$= -20$

b) $(1 + 2i)(1 + 3i) = 1 + 3i + 2i + 6i^2$ Multiplying each term of one number by every term of the other (FOIL)
$= 1 + 3i + 2i - 6$ $\quad i^2 = -1$
$= -5 + 5i$ Collecting like terms

c) $(3 - 7i)^2 = 3^2 - 2 \cdot 3 \cdot 7i + (7i)^2$ Recall that $(A - B)^2 = A^2 - 2AB + B^2$.
$= 9 - 42i + 49i^2$
$= 9 - 42i - 49$ $\quad i^2 = -1$
$= -40 - 42i$

Now Try Exercises 31, 39, and 55.

Technology Connection

We can multiply complex numbers on a graphing calculator set in $a + bi$ mode. The products found in Example 3 are shown below.

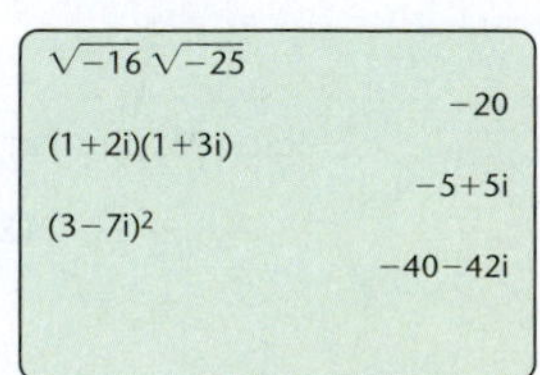

JUST IN TIME 7

Recall that -1 raised to an *even* power is 1, and -1 raised to an *odd* power is -1. Simplifying powers of i can then be done by using the fact that $i^2 = -1$ and expressing the given power of i in terms of i^2. Consider the following:

$$\begin{aligned}
i &= \sqrt{-1},\\
i^2 &= -1,\\
i^3 &= i^2 \cdot i = (-1)i = -i,\\
i^4 &= (i^2)^2 = (-1)^2 = 1,\\
i^5 &= i^4 \cdot i = (i^2)^2 \cdot i = (-1)^2 \cdot i = 1 \cdot i = i,\\
i^6 &= (i^2)^3 = (-1)^3 = -1,\\
i^7 &= i^6 \cdot i = (i^2)^3 \cdot i = (-1)^3 \cdot i = -1 \cdot i = -i,\\
i^8 &= (i^2)^4 = (-1)^4 = 1.
\end{aligned}$$

Note that the powers of i cycle through the values i, -1, $-i$, and 1.

EXAMPLE 4 Simplify each of the following.

a) i^{37} **b)** i^{58}

c) i^{75} **d)** i^{80}

Solution

a) $i^{37} = i^{36} \cdot i = (i^2)^{18} \cdot i = (-1)^{18} \cdot i = 1 \cdot i = i$

b) $i^{58} = (i^2)^{29} = (-1)^{29} = -1$

c) $i^{75} = i^{74} \cdot i = (i^2)^{37} \cdot i = (-1)^{37} \cdot i = -1 \cdot i = -i$

d) $i^{80} = (i^2)^{40} = (-1)^{40} = 1$

Now Try Exercises 79 and 83.

These powers of i can also be simplified in terms of i^4 rather than i^2. Consider i^{37} in Example 4(a), for instance. When we divide 37 by 4, we get 9 with a remainder of 1. Then $37 = 4 \cdot 9 + 1$, so

$$i^{37} = (i^4)^9 \cdot i = 1^9 \cdot i = 1 \cdot i = i.$$

The other examples shown above can be done in a similar manner.

Conjugates and Division

Conjugates of complex numbers are defined as follows.

CONJUGATE OF A COMPLEX NUMBER

The **conjugate** of a complex number $a + bi$ is $a - bi$. The numbers $a + bi$ and $a - bi$ are **complex conjugates**.

Each of the following pairs of numbers are complex conjugates:

$$-3 + 7i \text{ and } -3 - 7i; \quad 14 - 5i \text{ and } 14 + 5i; \quad \text{and} \quad 8i \text{ and } -8i.$$

The product of a complex number and its conjugate is a real number.

EXAMPLE 5 Multiply each of the following.

a) $(5 + 7i)(5 - 7i)$

b) $(8i)(-8i)$

Solution

a) $$\begin{aligned}(5 + 7i)(5 - 7i) &= 5^2 - (7i)^2 \quad \text{Using } (A + B)(A - B) = A^2 - B^2\\ &= 25 - 49i^2\\ &= 25 - 49(-1)\\ &= 25 + 49\\ &= 74\end{aligned}$$

b) $$\begin{aligned}(8i)(-8i) &= -64i^2\\ &= -64(-1)\\ &= 64\end{aligned}$$

Now Try Exercise 49.

Conjugates are used when we divide complex numbers.

EXAMPLE 6 Divide $2 - 5i$ by $1 - 6i$.

Solution We write fraction notation and then multiply by 1, using the conjugate of the denominator to form the symbol for 1.

$$\begin{aligned}\frac{2 - 5i}{1 - 6i} &= \frac{2 - 5i}{1 - 6i} \cdot \frac{1 + 6i}{1 + 6i} \quad \text{Note that } 1 + 6i \text{ is the conjugate of the divisor, } 1 - 6i.\\ &= \frac{(2 - 5i)(1 + 6i)}{(1 - 6i)(1 + 6i)}\\ &= \frac{2 + 12i - 5i - 30i^2}{1 - 36i^2}\\ &= \frac{2 + 7i + 30}{1 + 36} \quad i^2 = -1\\ &= \frac{32 + 7i}{37}\\ &= \frac{32}{37} + \frac{7}{37}i. \quad \text{Writing the quotient in the form } a + bi\end{aligned}$$

Now Try Exercise 69.

Technology Connection

With a graphing calculator set in $a + bi$ mode, we can divide complex numbers and express the real and imaginary parts in fraction form, just as we did in Example 6.

(2−5i)/(1−6i)▸Frac
32/37 + 7/37 i

3.1 Exercise Set

Express the number in terms of i.

1. $\sqrt{-3}$
2. $\sqrt{-21}$
3. $\sqrt{-25}$
4. $\sqrt{-100}$
5. $-\sqrt{-33}$
6. $-\sqrt{-59}$
7. $-\sqrt{-81}$
8. $-\sqrt{-9}$
9. $\sqrt{-98}$
10. $\sqrt{-28}$

Simplify. Write answers in the form $a + bi$, where a and b are real numbers.

11. $(-5 + 3i) + (7 + 8i)$
12. $(-6 - 5i) + (9 + 2i)$
13. $(4 - 9i) + (1 - 3i)$
14. $(7 - 2i) + (4 - 5i)$

15. $(12 + 3i) + (-8 + 5i)$

16. $(-11 + 4i) + (6 + 8i)$

17. $(-1 - i) + (-3 - i)$

18. $(-5 - i) + (6 + 2i)$

19. $(3 + \sqrt{-16}) + (2 + \sqrt{-25})$

20. $(7 - \sqrt{-36}) + (2 + \sqrt{-9})$

21. $(10 + 7i) - (5 + 3i)$

22. $(-3 - 4i) - (8 - i)$

23. $(13 + 9i) - (8 + 2i)$

24. $(-7 + 12i) - (3 - 6i)$

25. $(6 - 4i) - (-5 + i)$

26. $(8 - 3i) - (9 - i)$

27. $(-5 + 2i) - (-4 - 3i)$

28. $(-6 + 7i) - (-5 - 2i)$

29. $(4 - 9i) - (2 + 3i)$

30. $(10 - 4i) - (8 + 2i)$

31. $\sqrt{-4} \cdot \sqrt{-36}$

32. $\sqrt{-49} \cdot \sqrt{-9}$

33. $\sqrt{-81} \cdot \sqrt{-25}$

34. $\sqrt{-16} \cdot \sqrt{-100}$

35. $7i(2 - 5i)$

36. $3i(6 + 4i)$

37. $-2i(-8 + 3i)$

38. $-6i(-5 + i)$

39. $(1 + 3i)(1 - 4i)$

40. $(1 - 2i)(1 + 3i)$

41. $(2 + 3i)(2 + 5i)$

42. $(3 - 5i)(8 - 2i)$

43. $(-4 + i)(3 - 2i)$

44. $(5 - 2i)(-1 + i)$

45. $(8 - 3i)(-2 - 5i)$

46. $(7 - 4i)(-3 - 3i)$

47. $(3 + \sqrt{-16})(2 + \sqrt{-25})$

48. $(7 - \sqrt{-16})(2 + \sqrt{-9})$

49. $(5 - 4i)(5 + 4i)$

50. $(5 + 9i)(5 - 9i)$

51. $(3 + 2i)(3 - 2i)$

52. $(8 + i)(8 - i)$

53. $(7 - 5i)(7 + 5i)$

54. $(6 - 8i)(6 + 8i)$

55. $(4 + 2i)^2$

56. $(5 - 4i)^2$

57. $(-2 + 7i)^2$

58. $(-3 + 2i)^2$

59. $(1 - 3i)^2$

60. $(2 - 5i)^2$

61. $(-1 - i)^2$

62. $(-4 - 2i)^2$

63. $(3 + 4i)^2$

64. $(6 + 5i)^2$

65. $\dfrac{3}{5 - 11i}$

66. $\dfrac{i}{2 + i}$

67. $\dfrac{5}{2 + 3i}$

68. $\dfrac{-3}{4 - 5i}$

69. $\dfrac{4 + i}{-3 - 2i}$

70. $\dfrac{5 - i}{-7 + 2i}$

71. $\dfrac{5 - 3i}{4 + 3i}$

72. $\dfrac{6 + 5i}{3 - 4i}$

73. $\dfrac{2 + \sqrt{3}i}{5 - 4i}$

74. $\dfrac{\sqrt{5} + 3i}{1 - i}$

75. $\dfrac{1 + i}{(1 - i)^2}$

76. $\dfrac{1 - i}{(1 + i)^2}$

77. $\dfrac{4 - 2i}{1 + i} + \dfrac{2 - 5i}{1 + i}$

78. $\dfrac{3 + 2i}{1 - i} + \dfrac{6 + 2i}{1 - i}$

Simplify.

79. i^{11}

80. i^7

81. i^{35}

82. i^{24}

83. i^{64}

84. i^{42}

85. $(-i)^{71}$

86. $(-i)^6$

87. $(5i)^4$

88. $(2i)^5$

Skill Maintenance

89. Write a slope–intercept equation for the line containing the point $(3, -5)$ and perpendicular to the line $3x - 6y = 7$. [1.4]

Given that $f(x) = x^2 + 4$ *and* $g(x) = 3x + 5$, *find each of the following.* [2.2]

90. The domain of $f - g$

91. The domain of f/g

92. $(f - g)(x)$

93. $(f/g)(2)$

94. For the function $f(x) = x^2 - 3x + 4$, construct and simplify the difference quotient
$$\frac{f(x+h) - f(x)}{h}. \text{ [2.2]}$$

Synthesis

Determine whether the statement is true or false.

95. The sum of two numbers that are conjugates of each other is always a real number.

96. The conjugate of a sum is the sum of the conjugates of the individual complex numbers.

97. The conjugate of a product is the product of the conjugates of the individual complex numbers.

Let $z = a + bi$ and $\overline{z} = a - bi$.

98. Find a general expression for $1/z$.

99. Find a general expression for $z\overline{z}$.

100. Solve $z + 6\overline{z} = 7$ for z.

101. Multiply and simplify:
$$[x - (3 + 4i)][x - (3 - 4i)].$$

3.2 Quadratic Equations, Functions, Zeros, and Models

- Find zeros of quadratic functions and solve quadratic equations by using the principle of zero products, by using the principle of square roots, by completing the square, and by using the quadratic formula.
- Solve equations that are reducible to quadratic.
- Solve applied problems using quadratic equations.

Quadratic Equations and Quadratic Functions

In this section, we will explore the relationship between the solutions of quadratic equations and the zeros of quadratic functions. We define quadratic equations and quadratic functions as follows.

QUADRATIC EQUATIONS

A **quadratic equation** is an equation that can be written in the form
$$ax^2 + bx + c = 0, \quad a \neq 0,$$
where a, b, and c are real numbers.

QUADRATIC FUNCTIONS

A **quadratic function** f is a function that can be written in the form
$$f(x) = ax^2 + bx + c, \quad a \neq 0,$$
where a, b, and c are real numbers.

A quadratic equation written in the form $ax^2 + bx + c = 0$ is said to be in **standard form**.

ZEROS OF A FUNCTION

REVIEW SECTION 1.5

The *zeros* of a quadratic function $f(x) = ax^2 + bx + c$ are the *solutions* of the associated quadratic equation $ax^2 + bx + c = 0$. (These solutions are sometimes called *roots* of the equation.) Quadratic functions can have real-number or imaginary-number zeros and quadratic equations can have real-number or imaginary-number solutions. If the zeros or solutions are real numbers, they are also the first coordinates of the x-intercepts of the graph of the quadratic function.

The following principles allow us to solve many quadratic equations.

EQUATION-SOLVING PRINCIPLES

The Principle of Zero Products: If $ab = 0$ is true, then $a = 0$ or $b = 0$, and if $a = 0$ *or* $b = 0$, then $ab = 0$.

The Principle of Square Roots: If $x^2 = k$, then $x = \sqrt{k}$ *or* $x = -\sqrt{k}$.

JUST IN TIME 13, 16

EXAMPLE 1 Solve: $2x^2 - x = 3$.

Algebraic Solution

We have

$$2x^2 - x = 3$$

$$2x^2 - x - 3 = 0 \quad \text{Subtracting 3 on both sides}$$

Factoring

$$(x + 1)(2x - 3) = 0$$

$$x + 1 = 0 \quad or \quad 2x - 3 = 0 \quad \text{Using the principle of zero products}$$

$$x = -1 \quad or \quad 2x = 3$$

$$x = -1 \quad or \quad x = \tfrac{3}{2}.$$

Check: For $x = -1$:

$$\begin{array}{c|c} 2x^2 - x & = 3 \\ \hline 2(-1)^2 - (-1) & ?\ 3 \\ 2 \cdot 1 + 1 & \\ 2 + 1 & \\ 3 & 3 \quad \text{TRUE} \end{array}$$

For $x = \frac{3}{2}$:

$$\begin{array}{c|c} 2x^2 - x & = 3 \\ \hline 2\left(\frac{3}{2}\right)^2 - \frac{3}{2} & ?\ 3 \\ 2 \cdot \frac{9}{4} - \frac{3}{2} & \\ \frac{9}{2} - \frac{3}{2} & \\ \frac{6}{2} & \\ 3 & 3 \quad \text{TRUE} \end{array}$$

The solutions are -1 and $\frac{3}{2}$.

Visualizing the Solution

The solutions of the equation $2x^2 - x = 3$, or the equivalent equation $2x^2 - x - 3 = 0$, are the zeros of the function $f(x) = 2x^2 - x - 3$. They are also the first coordinates of the x-intercepts of the graph of $f(x) = 2x^2 - x - 3$.

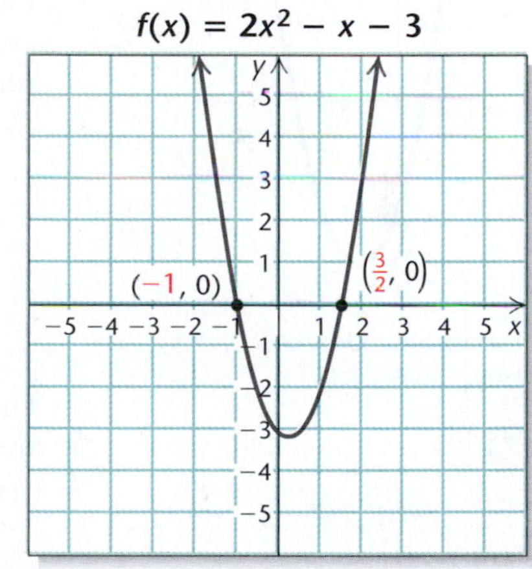

The solutions are -1 and $\frac{3}{2}$.

Now Try Exercise 3.

JUST IN TIME 17

EXAMPLE 2 Solve: $2x^2 - 10 = 0$.

Algebraic Solution

We have

$$2x^2 - 10 = 0$$
$$2x^2 = 10 \quad \text{Adding 10 on both sides}$$
$$x^2 = 5 \quad \text{Dividing by 2 on both sides}$$
$$x = \sqrt{5} \quad or \quad x = -\sqrt{5}. \quad \text{Using the principle of square roots}$$

Check:

$$\begin{array}{c|l} 2x^2 - 10 & = 0 \\ \hline 2(\pm\sqrt{5})^2 - 10 & ?\ 0 \\ 2 \cdot 5 - 10 & \\ 10 - 10 & \\ 0 & 0 \quad \text{TRUE} \end{array}$$

We can check both solutions at once.

The solutions are $\sqrt{5}$ and $-\sqrt{5}$, or $\pm\sqrt{5}$.

Visualizing the Solution

The solutions of the equation $2x^2 - 10 = 0$ are the zeros of the function $f(x) = 2x^2 - 10$. Note that they are also the first coordinates of the x-intercepts of the graph of $f(x) = 2x^2 - 10$.

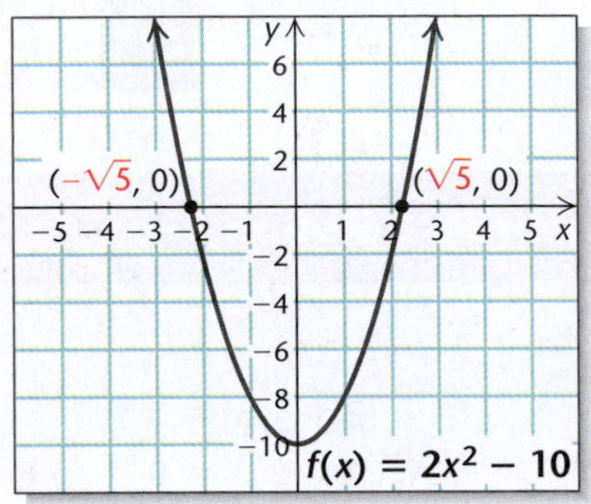

The solutions are $-\sqrt{5}$ and $\sqrt{5}$.

Now Try Exercise 7.

We have seen that some quadratic equations can be solved by factoring and using the principle of zero products. For example, consider the equation $x^2 - 3x - 4 = 0$:

$$x^2 - 3x - 4 = 0$$
$$(x + 1)(x - 4) = 0 \quad \text{Factoring}$$
$$x + 1 = 0 \quad or \quad x - 4 = 0 \quad \text{Using the principle of zero products}$$
$$x = -1 \quad or \quad x = 4.$$

The equation $x^2 - 3x - 4 = 0$ has *two real-number* solutions, -1 and 4. These are the zeros of the associated quadratic function $f(x) = x^2 - 3x - 4$ and the first coordinates of the x-intercepts of the graph of this function. (See Fig. 1.)

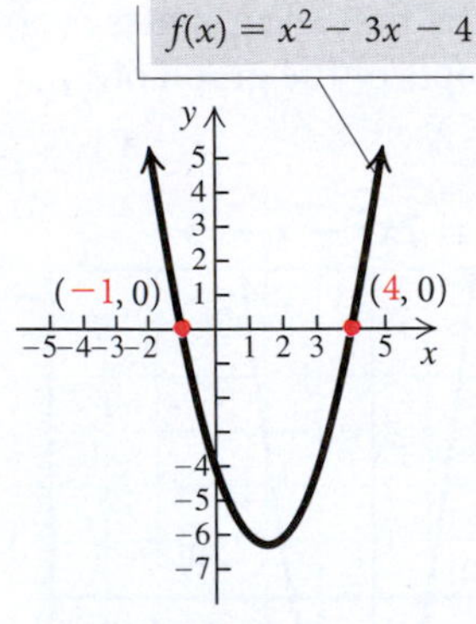

Two real-number zeros
Two x-intercepts

Figure 1.

Next, consider the equation $x^2 - 6x + 9 = 0$. Again, we factor and use the principle of zero products:

$$x^2 - 6x + 9 = 0$$
$$(x - 3)(x - 3) = 0 \quad \text{Factoring}$$
$$x - 3 = 0 \quad or \quad x - 3 = 0 \quad \text{Using the principle of zero products}$$
$$x = 3 \quad or \quad x = 3.$$

The equation $x^2 - 6x + 9 = 0$ has *one real-number* solution, 3. It is the zero of the quadratic function $g(x) = x^2 - 6x + 9$ and the first coordinate of the x-intercept of the graph of this function. (See Fig. 2.)

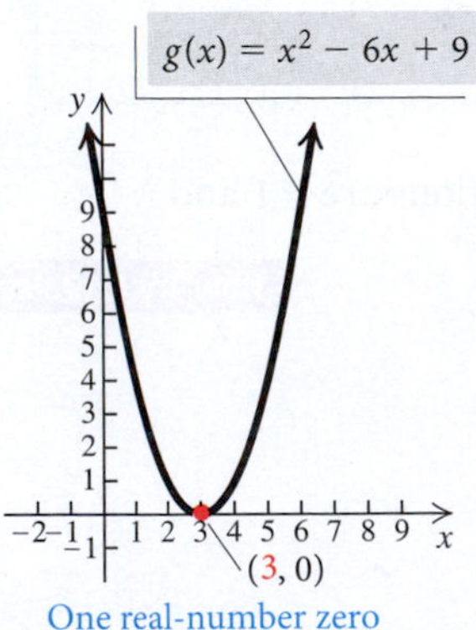

One real-number zero
One x-intercept

Figure 2.

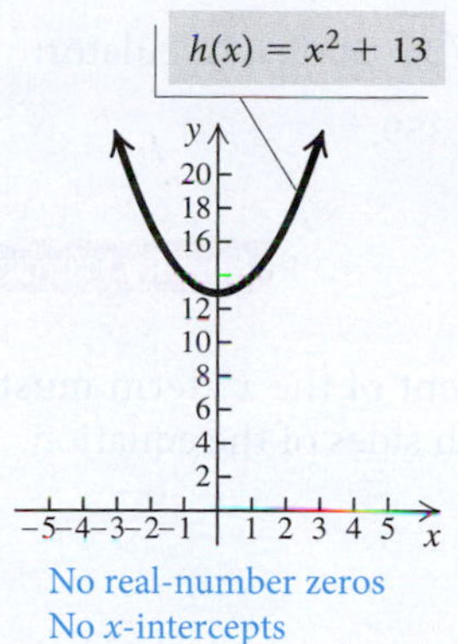

Figure 3.

The principle of square roots can be used to solve quadratic equations like $x^2 + 13 = 0$:

$$x^2 + 13 = 0$$
$$x^2 = -13$$
$$x = \pm\sqrt{-13} \quad \text{Using the principle of square roots}$$
$$x = \pm\sqrt{13}i. \quad \sqrt{-13} = \sqrt{-1}\cdot\sqrt{13} = i\cdot\sqrt{13} = \sqrt{13}i$$

The equation has *two imaginary-number* solutions, $-\sqrt{13}i$ and $\sqrt{13}i$. These are the zeros of the associated quadratic function $h(x) = x^2 + 13$. Since the zeros are not real numbers, the graph of the function has no x-intercepts. (See Fig. 3.)

Completing the Square

Neither the principle of zero products nor the principle of square roots would yield the *exact* zeros of a function like $f(x) = x^2 - 6x - 10$ or the *exact* solutions of the associated equation $x^2 - 6x - 10 = 0$. If we wish to find exact zeros or solutions, we can use a procedure called **completing the square** and then use the principle of square roots.

EXAMPLE 3 Find the zeros of $f(x) = x^2 - 6x - 10$ by completing the square.

Solution We find the values of x for which $f(x) = 0$; that is, we solve the associated equation $x^2 - 6x - 10 = 0$. Our goal is to find an equivalent equation of the form $x^2 + bx + c = d$ in which $x^2 + bx + c$ is a perfect square. Since

$$x^2 + bx + \left(\frac{b}{2}\right)^2 = \left(x + \frac{b}{2}\right)^2,$$

the number c is found by taking half the coefficient of the x-term and squaring it. Thus for the equation $x^2 - 6x - 10 = 0$, we have

$$x^2 - 6x - 10 = 0$$
$$x^2 - 6x = 10 \quad \text{Adding 10}$$
$$x^2 - 6x + 9 = 10 + 9 \quad \text{Adding 9 to complete the square: } \left(\frac{b}{2}\right)^2 = \left(\frac{-6}{2}\right)^2 = (-3)^2 = 9$$
$$x^2 - 6x + 9 = 19.$$

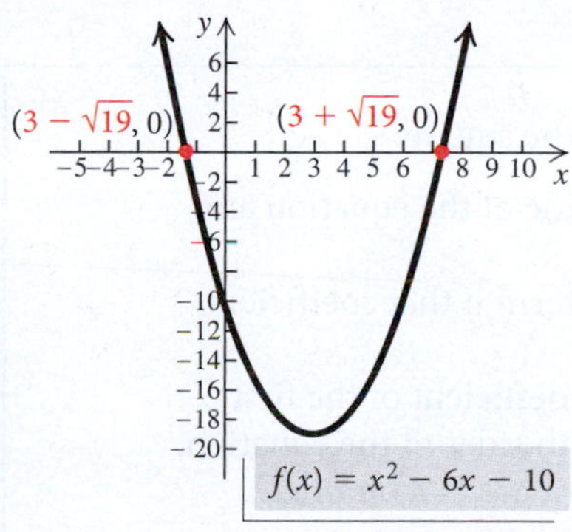

Because $x^2 - 6x + 9$ is a perfect square, we are able to write it as $(x - 3)^2$, the square of a binomial. We can then use the principle of square roots to finish the solution:

$$(x - 3)^2 = 19 \quad \text{Factoring}$$
$$x - 3 = \pm\sqrt{19} \quad \text{Using the principle of square roots}$$
$$x = 3 \pm \sqrt{19}. \quad \text{Adding 3}$$

Therefore, the solutions of the equation are $3 + \sqrt{19}$ and $3 - \sqrt{19}$, or simply $3 \pm \sqrt{19}$. The zeros of $f(x) = x^2 - 6x - 10$ are also $3 + \sqrt{19}$ and $3 - \sqrt{19}$, or $3 \pm \sqrt{19}$.

We can find decimal approximations for $3 \pm \sqrt{19}$ using a calculator:

$$3 + \sqrt{19} \approx 7.359 \quad \text{and} \quad 3 - \sqrt{19} \approx -1.359.$$

The zeros are approximately 7.359 and -1.359.

Now Try Exercise 31.

Technology Connection

Approximations for the zeros of the quadratic function $f(x) = x^2 - 6x - 10$ in Example 3 can be found using the Zero method.

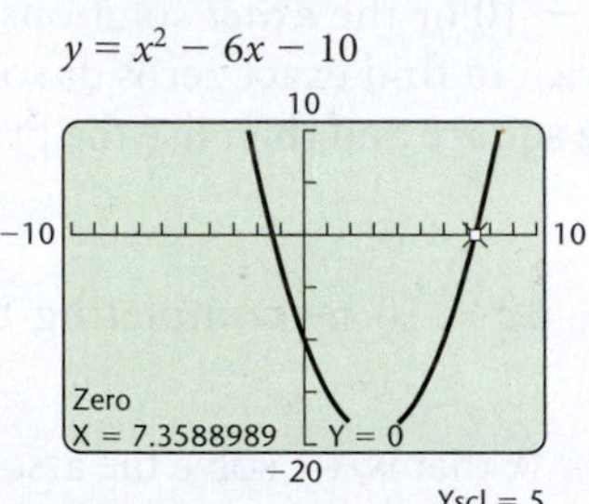

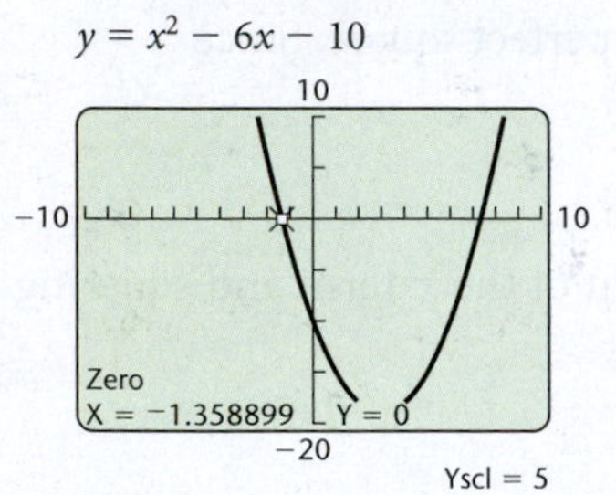

Before we can complete the square, the coefficient of the x^2-term must be 1. When it is not, we divide by the x^2-coefficient on both sides of the equation.

EXAMPLE 4 Solve: $2x^2 - 1 = 3x$.

Solution We have

$$2x^2 - 1 = 3x$$

$$2x^2 - 3x - 1 = 0 \qquad \text{Subtracting } 3x\text{. We are unable to factor the result.}$$

$$2x^2 - 3x = 1 \qquad \text{Adding 1}$$

$$x^2 - \frac{3}{2}x = \frac{1}{2} \qquad \text{Dividing by 2 to make the } x^2\text{-coefficient 1}$$

$$x^2 - \frac{3}{2}x + \frac{9}{16} = \frac{1}{2} + \frac{9}{16} \qquad \text{Completing the square: } \tfrac{1}{2}\left(-\tfrac{3}{2}\right) = -\tfrac{3}{4} \text{ and } \left(-\tfrac{3}{4}\right)^2 = \tfrac{9}{16}\text{; adding } \tfrac{9}{16}$$

$$\left(x - \frac{3}{4}\right)^2 = \frac{17}{16} \qquad \text{Factoring and simplifying}$$

$$x - \frac{3}{4} = \pm\frac{\sqrt{17}}{4} \qquad \text{Using the principle of square roots and the quotient rule for radicals}$$

$$x = \frac{3}{4} \pm \frac{\sqrt{17}}{4} \qquad \text{Adding } \tfrac{3}{4}$$

$$x = \frac{3 \pm \sqrt{17}}{4}.$$

The solutions are

$$\frac{3 + \sqrt{17}}{4} \quad \text{and} \quad \frac{3 - \sqrt{17}}{4}, \quad \text{or} \quad \frac{3 \pm \sqrt{17}}{4}.$$

Now Try Exercise 35.

To solve a quadratic equation by completing the square:

1. Isolate the terms with variables on one side of the equation and arrange them in descending order.
2. Divide by the coefficient of the squared term if that coefficient is not 1.
3. Complete the square by taking half the coefficient of the first-degree term and adding its square on both sides of the equation.
4. Express one side of the equation as the square of a binomial.
5. Use the principle of square roots.
6. Solve for the variable.

Using the Quadratic Formula

Because completing the square works for *any* quadratic equation, it can be used to solve the general quadratic equation $ax^2 + bx + c = 0$ for x. The result will be a formula that can be used to solve any quadratic equation quickly.

Consider any quadratic equation in standard form:

$$ax^2 + bx + c = 0, \quad a \neq 0.$$

For now, we assume that $a > 0$ and solve by completing the square. As the steps are carried out, compare them with those of Example 4.

$$ax^2 + bx + c = 0 \qquad \textbf{Standard form}$$

$$ax^2 + bx = -c \qquad \textbf{Adding } -c$$

$$x^2 + \frac{b}{a}x = -\frac{c}{a} \qquad \textbf{Dividing by } a$$

Half of $\frac{b}{a}$ is $\frac{b}{2a}$, and $\left(\frac{b}{2a}\right)^2 = \frac{b^2}{4a^2}$. Thus we add $\frac{b^2}{4a^2}$:

$$x^2 + \frac{b}{a}x + \frac{b^2}{4a^2} = -\frac{c}{a} + \frac{b^2}{4a^2} \qquad \textbf{Adding } \frac{b^2}{4a^2} \textbf{ to complete the square}$$

$$\left(x + \frac{b}{2a}\right)^2 = -\frac{4ac}{4a^2} + \frac{b^2}{4a^2}$$

Factoring on the left; finding a common denominator on the right: $-\frac{c}{a} = -\frac{c}{a} \cdot \frac{4a}{4a} = -\frac{4ac}{4a^2}$

$$\left(x + \frac{b}{2a}\right)^2 = \frac{b^2 - 4ac}{4a^2}$$

$$x + \frac{b}{2a} = \pm \frac{\sqrt{b^2 - 4ac}}{2a}$$

Using the principle of square roots and the quotient rule for radicals. Since $a > 0$, $\sqrt{4a^2} = 2a$.

$$x = -\frac{b}{2a} \pm \frac{\sqrt{b^2 - 4ac}}{2a} \qquad \textbf{Adding } -\frac{b}{2a}$$

$$x = \frac{-b \pm \sqrt{b^2 - 4ac}}{2a}.$$

It can also be shown that this result holds if $a < 0$.

THE QUADRATIC FORMULA

The solutions of $ax^2 + bx + c = 0$, $a \neq 0$, are given by

$$x = \frac{-b \pm \sqrt{b^2 - 4ac}}{2a}.$$

EXAMPLE 5 Solve $3x^2 + 2x = 7$. Find exact solutions and approximate solutions rounded to three decimal places.

Solution After writing the equation in standard form, we are unable to factor, so we identify a, b, and c in order to use the quadratic formula:

$$3x^2 + 2x = 7$$
$$3x^2 + 2x - 7 = 0;$$
$$a = 3, \quad b = 2, \quad c = -7.$$

We then use the quadratic formula:

$$x = \frac{-b \pm \sqrt{b^2 - 4ac}}{2a}$$
$$= \frac{-2 \pm \sqrt{2^2 - 4(3)(-7)}}{2(3)} \quad \text{Substituting}$$
$$= \frac{-2 \pm \sqrt{4 + 84}}{6} = \frac{-2 \pm \sqrt{88}}{6}$$
$$= \frac{-2 \pm \sqrt{4 \cdot 22}}{6} = \frac{-2 \pm 2\sqrt{22}}{6} = \frac{2(-1 \pm \sqrt{22})}{2 \cdot 3}$$
$$= \frac{2}{2} \cdot \frac{-1 \pm \sqrt{22}}{3} = \frac{-1 \pm \sqrt{22}}{3}.$$

The exact solutions are

$$\frac{-1 - \sqrt{22}}{3} \quad \text{and} \quad \frac{-1 + \sqrt{22}}{3}.$$

Using a calculator, we approximate the solutions to be -1.897 and 1.230.

Now Try Exercise 41.

EXAMPLE 6 Solve: $x^2 + 5x + 8 = 0$.

Algebraic Solution

To find the solutions, we use the quadratic formula. For $x^2 + 5x + 8 = 0$, we have

$$a = 1, \quad b = 5, \quad c = 8;$$
$$x = \frac{-b \pm \sqrt{b^2 - 4ac}}{2a}$$
$$= \frac{-5 \pm \sqrt{5^2 - 4(1)(8)}}{2 \cdot 1} \quad \text{Substituting}$$
$$= \frac{-5 \pm \sqrt{25 - 32}}{2}$$
$$= \frac{-5 \pm \sqrt{-7}}{2} = \frac{-5 \pm \sqrt{7}i}{2}.$$

The solutions are $-\frac{5}{2} - \frac{\sqrt{7}}{2}i$ and $-\frac{5}{2} + \frac{\sqrt{7}}{2}i$.

Visualizing the Solution

The graph of the function $f(x) = x^2 + 5x + 8$ has no x-intercepts.

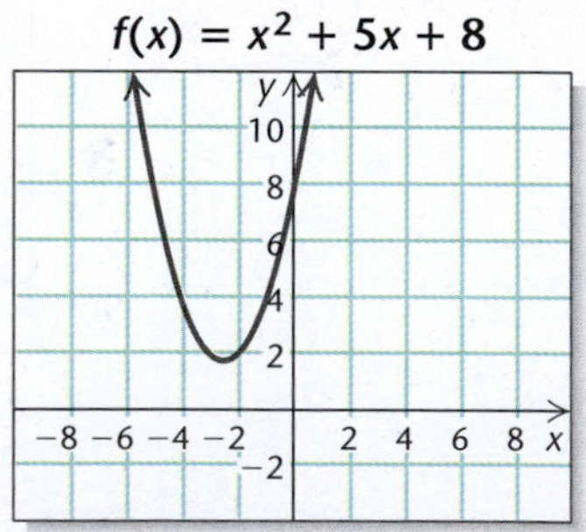

Thus the function has no real-number zeros and there are no real-number solutions of the associated equation $x^2 + 5x + 8 = 0$.

Now Try Exercise 47.

Technology Connection

We can solve the equation $3x^2 + 2x = 7$ in Example 5 using the Intersect method. We graph $y_1 = 3x^2 + 2x$ and $y_2 = 7$ and use the INTERSECT feature to find the coordinates of the points of intersection. The first coordinates of these points are the solutions of the equation $y_1 = y_2$, or $3x^2 + 2x = 7$.

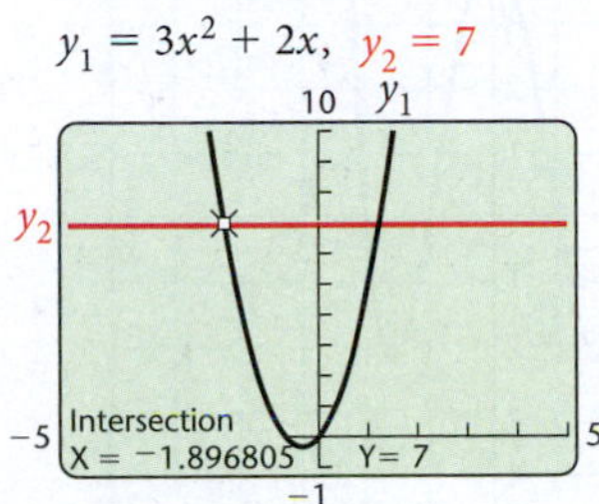

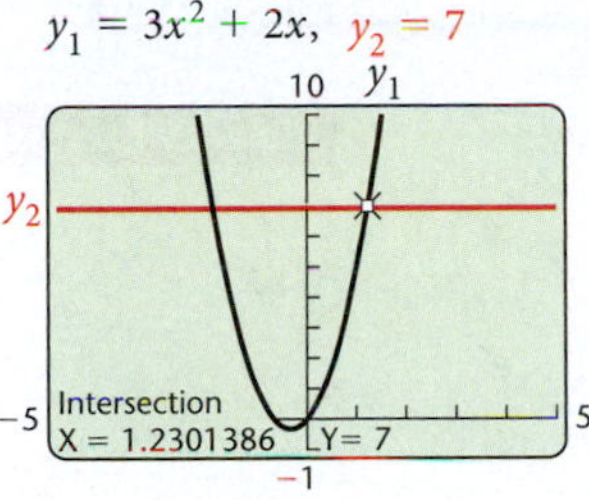

The solutions are approximately −1.897 and 1.230. We could also write the equation in standard form, $3x^2 + 2x - 7 = 0$, and use the Zero method.

The Discriminant

From the quadratic formula, we know that the solutions x_1 and x_2 of a quadratic equation are given by

$$x_1 = \frac{-b + \sqrt{b^2 - 4ac}}{2a} \quad \text{and} \quad x_2 = \frac{-b - \sqrt{b^2 - 4ac}}{2a}.$$

The expression $b^2 - 4ac$ shows the nature of the solutions. This expression is called the **discriminant**. If it is 0, then it makes no difference whether we choose the plus sign or the minus sign in the formula. That is, $x_1 = -\frac{b}{2a} = x_2$, so there is just one solution. In this case, we sometimes say that there is one repeated real solution. If the discriminant is positive, there will be two real solutions. If it is negative, we will be taking the square root of a negative number; hence there will be two imaginary-number solutions, and they will be complex conjugates.

DISCRIMINANT

For $ax^2 + bx + c = 0$, where a, b, and c are real numbers:

$b^2 - 4ac = 0 \longrightarrow$ One real-number solution;
$b^2 - 4ac > 0 \longrightarrow$ Two different real-number solutions;
$b^2 - 4ac < 0 \longrightarrow$ Two different imaginary-number solutions, complex conjugates.

In Example 5, the discriminant, 88, is positive, indicating that there are two different real-number solutions. The negative discriminant, −7, in Example 6 indicates that there are two different imaginary-number solutions.

Equations Reducible to Quadratic

Some equations can be treated as quadratic, provided we make a suitable substitution. For example, consider the following:

$$x^4 - 5x^2 + 4 = 0$$
$$(x^2)^2 - 5x^2 + 4 = 0 \qquad x^4 = (x^2)^2$$
$$u^2 - 5u + 4 = 0. \qquad \text{Substituting } u \text{ for } x^2$$

The equation $u^2 - 5u + 4 = 0$ can be solved for u by factoring or using the quadratic formula. Then we can reverse the substitution, replacing u with x^2, and solve for x. Equations like the one above are said to be **reducible to quadratic**, or **quadratic in form**.

EXAMPLE 7 Solve: $x^4 - 5x^2 + 4 = 0$.

Algebraic Solution

We let $u = x^2$ and substitute:

$$u^2 - 5u + 4 = 0 \quad \text{Substituting } u \text{ for } x^2$$
$$(u - 1)(u - 4) = 0 \quad \text{Factoring}$$
$$u - 1 = 0 \quad or \quad u - 4 = 0 \quad \text{Using the principle of zero products}$$
$$u = 1 \quad or \quad u = 4.$$

Don't stop here! We must solve for the original variable. We substitute x^2 for u and solve for x:

$$x^2 = 1 \quad or \quad x^2 = 4$$
$$x = \pm 1 \quad or \quad x = \pm 2. \quad \text{Using the principle of square roots}$$

The solutions are -1, 1, -2, and 2.

Visualizing the Solution

The solutions of the given equation are the zeros of $f(x) = x^4 - 5x^2 + 4$. Note that the zeros occur at the x-values -2, -1, 1, and 2.

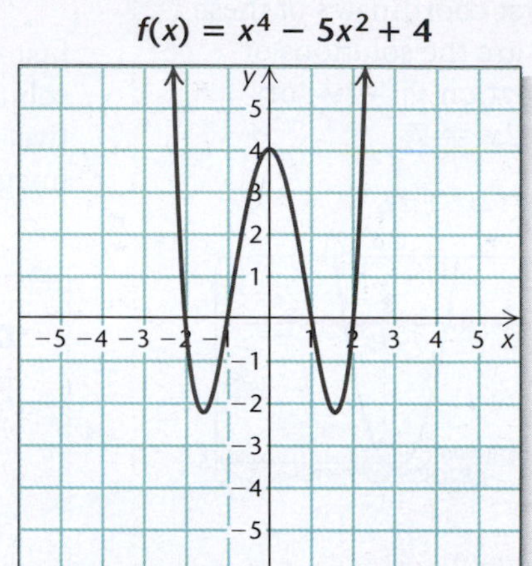

Now Try Exercise 79.

JUST IN TIME 24

EXAMPLE 8 Solve: $t^{2/3} - 2t^{1/3} - 3 = 0$.

Solution We let $u = t^{1/3}$ and substitute:

$$t^{2/3} - 2t^{1/3} - 3 = 0$$
$$(t^{1/3})^2 - 2t^{1/3} - 3 = 0$$
$$u^2 - 2u - 3 = 0 \quad \text{Substituting } u \text{ for } t^{1/3}$$
$$(u + 1)(u - 3) = 0 \quad \text{Factoring}$$
$$u + 1 = 0 \quad or \quad u - 3 = 0 \quad \text{Using the principle of zero products}$$
$$u = -1 \quad or \quad u = 3.$$

Now we must solve for the original variable, t. We substitute $t^{1/3}$ for u and solve for t:

$$t^{1/3} = -1 \quad or \quad t^{1/3} = 3$$
$$(t^{1/3})^3 = (-1)^3 \quad or \quad (t^{1/3})^3 = 3^3 \quad \text{Cubing on both sides}$$
$$t = -1 \quad or \quad t = 27.$$

The solutions are -1 and 27.

Now Try Exercise 87.

Applications

EXAMPLE 9 ***Museums in China.*** The number of museums in China increased from approximately 2000 in the year 2000 to over 3500 by the end of 2012. In 2012, a record 451 new museums opened. For comparison, in the United States, only 20–40 new museums were opened per year from 2000 to 2008. The function

$$h(x) = 30.992x^2 + 4.108x + 2294.594$$

can be used to estimate the number of museums in China, x years after 2005. (*Source:* The Economist/www.economist.com) Use this function to answer the following.

a) Estimate the number of museums that will be in China in 2017 if the number of new museums that open per year continues at the same rate.

b) In what year was the number of museums in China 2600?

Solution

a) For 2017, $x = 2017 - 2005 = 12$. We substitute 12 for x and find $h(12)$:

$$h(x) = 30.992x^2 + 4.108x + 2294.594$$
$$h(12) = 30.992(12)^2 + 4.108(12) + 2294.594$$
$$h(12) = 4462.848 + 49.296 + 2294.594 \approx 6807.$$

In 2017, there will be approximately 6807 museums in China.

b) We substitute 2600 for $h(x)$ and solve for x:

$$h(x) = 30.992x^2 + 4.108x + 2294.594$$
$$2600 = 30.992x^2 + 4.108x + 2294.594$$
$$0 = 30.992x^2 + 4.108x - 305.406.$$

We then use the quadratic formula, with $a = 30.992$, $b = 4.108$, and $c = -305.406$:

$$x = \frac{-b \pm \sqrt{b^2 - 4ac}}{2a}$$
$$x = \frac{-4.108 \pm \sqrt{(4.108)^2 - 4(30.992)(-305.406)}}{2(30.992)}$$
$$x = \frac{-4.108 \pm \sqrt{37{,}877.44667}}{61.984}$$
$$x = 3.074 \quad or \quad x = -3.206.$$

Because we are looking for a year after 2005, we use the positive solution. Thus there were about 2600 museums in China 3 years after 2005, or in 2008.

Now Try Exercise 97.

EXAMPLE 10 ***Sales of New Homes.*** Sales of new homes have increased in recent years. The function

$$h(x) = 22.1x^2 - 72.2x + 371.9$$

can be used to estimate the sales of new homes, in thousands, in the United States, where x is the number of years after 2009 (*Source*: IHS Global Insight). In what year were the number of sales of new homes about 563,400, or 563.4 thousands?

Solution We substitute 563.4 for $h(x)$ and solve for x:

$$563.4 = 22.1x^2 - 72.2x + 371.9$$
$$0 = 22.1x^2 - 72.2x - 191.5.$$

We then use the quadratic formula, with $a = 22.1$, $b = -72.2$, and $c = -191.5$:

$$x = \frac{-(-72.2) \pm \sqrt{(-72.2)^2 - 4(22.1)(-191.5)}}{2(22.1)} \qquad \text{Substituting}$$
$$x = \frac{72.2 \pm \sqrt{22{,}141.44}}{44.2}$$
$$x = 5 \quad or \quad x = -1.7$$

Because we are looking for a year after 2009, we use the positive solution. Thus there were about 563,400 sales of new homes 5 years after 2009, or in 2014.

Now Try Exercise 95.

EXAMPLE 11 ***Train Speeds.*** Two trains leave a station at the same time. One train travels due west, and the other travels due south. The train traveling west travels 20 km/h faster than the train traveling south. After 2 hr, the trains are 200 km apart. Find the speed of each train.

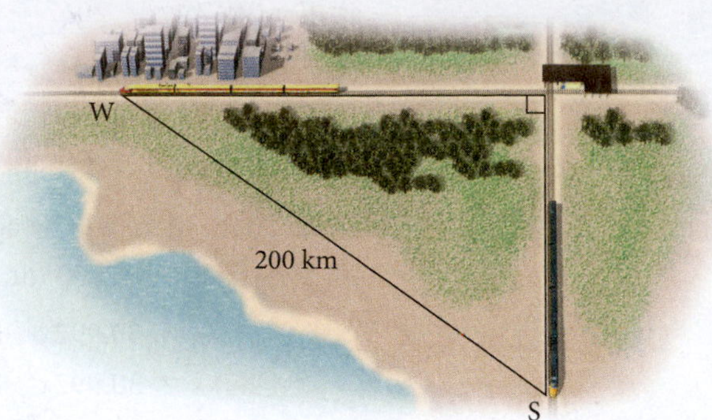

Solution

1. **Familiarize.** First, we make a drawing. We let $r =$ the speed of the train traveling south, in kilometers per hour. Then $r + 20 =$ the speed of the train traveling west, in kilometers per hour. We use the motion formula $d = rt$, where d is the distance, r is the rate (or speed), and t is the time. After 2 hr, the train traveling south has traveled $2r$ kilometers, and the train traveling west has traveled $2(r + 20)$ kilometers. We add these distances to the drawing.

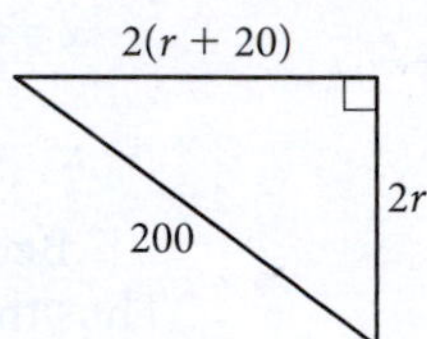

2. **Translate.** We use the Pythagorean theorem, $a^2 + b^2 = c^2$, where a and b are the lengths of the legs of a right triangle and c is the length of the hypotenuse:

$$[2(r + 20)]^2 + (2r)^2 = 200^2.$$

3. **Carry out.** We solve the equation:

$$[2(r + 20)]^2 + (2r)^2 = 200^2$$
$$4(r^2 + 40r + 400) + 4r^2 = 40{,}000$$
$$4r^2 + 160r + 1600 + 4r^2 = 40{,}000$$
$$8r^2 + 160r + 1600 = 40{,}000 \quad \text{Collecting like terms}$$
$$8r^2 + 160r - 38{,}400 = 0 \quad \text{Subtracting 40,000}$$
$$r^2 + 20r - 4800 = 0 \quad \text{Dividing by 8}$$
$$(r + 80)(r - 60) = 0 \quad \text{Factoring}$$
$$r + 80 = 0 \quad or \quad r - 60 = 0 \quad \text{Principle of zero products}$$
$$r = -80 \quad or \quad r = 60.$$

4. **Check.** Since speed cannot be negative, we need check only 60. If the speed of the train traveling south is 60 km/h, then the speed of the train traveling west is $60 + 20$, or 80 km/h. In 2 hr, the train heading south travels $60 \cdot 2$, or 120 km, and the train heading west travels $80 \cdot 2$, or 160 km. Then they are $\sqrt{120^2 + 160^2}$, or $\sqrt{40{,}000}$, or 200 km apart. The answer checks.

5. **State.** The speed of the train heading south is 60 km/h, and the speed of the train heading west is 80 km/h.

Now Try Exercise 101.

CONNECTING THE CONCEPTS

Zeros, Solutions, and Intercepts

The zeros of a function $y = f(x)$ are also the solutions of the equation $f(x) = 0$, and the real-number zeros are the first coordinates of the x-intercepts of the graph of the function.

FUNCTION	ZERO OF THE FUNCTION; SOLUTION OF THE EQUATION	x-INTERCEPTS OF THE GRAPH
Linear Function $f(x) = 2x - 4$, or $y = 2x - 4$	To find the **zero** of $f(x)$, we solve $f(x) = 0$: $2x - 4 = 0$; $2x = 4$; $x = 2$. The **solution** of the equation $2x - 4 = 0$ is 2. This is the zero of the function $f(x) = 2x - 4$; that is, $f(2) = 0$.	The zero of $f(x)$ is the first coordinate of the **x-intercept** of the graph of $y = f(x)$. 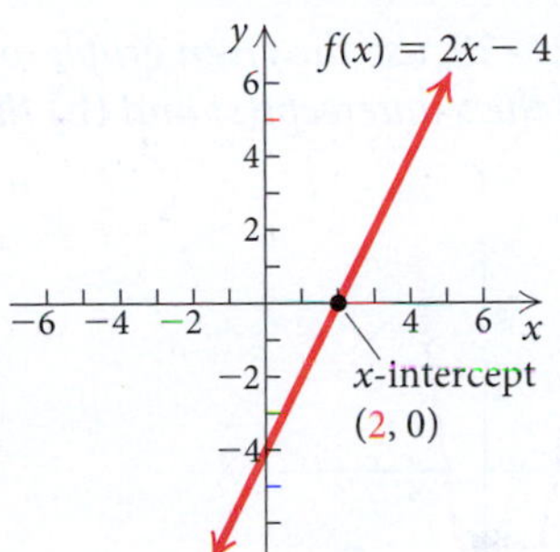
Quadratic Function $g(x) = x^2 - 3x - 4$, or $y = x^2 - 3x - 4$	To find the **zeros** of $g(x)$, we solve $g(x) = 0$: $x^2 - 3x - 4 = 0$; $(x + 1)(x - 4) = 0$; $x + 1 = 0$ *or* $x - 4 = 0$; $x = -1$ *or* $x = 4$. The **solutions** of the equation $x^2 - 3x - 4 = 0$ are -1 and 4. They are the zeros of the function $g(x)$; that is, $g(-1) = 0$ and $g(4) = 0$.	The real-number zeros of $g(x)$ are the first coordinates of the **x-intercepts** of the graph of $y = g(x)$. 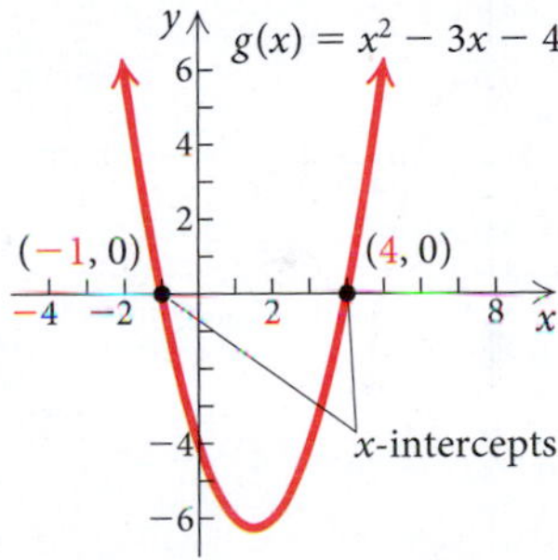

3.2 Exercise Set

Solve.

1. $(2x - 3)(3x - 2) = 0$
2. $(2x + 3)(5x - 2) = 0$
3. $x^2 - 8x - 20 = 0$
4. $x^2 + 6x + 8 = 0$
5. $3x^2 + x - 2 = 0$
6. $10x^2 - 16x + 6 = 0$
7. $4x^2 - 12 = 0$
8. $6x^2 = 36$
9. $3x^2 = 21$
10. $2x^2 - 20 = 0$
11. $5x^2 + 10 = 0$

12. $4x^2 + 12 = 0$

13. $x^2 + 16 = 0$

14. $x^2 + 25 = 0$

15. $2x^2 = 6x$

16. $18x + 9x^2 = 0$

17. $3y^3 - 5y^2 - 2y = 0$

18. $3t^3 + 2t = 5t^2$

19. $7x^3 + x^2 - 7x - 1 = 0$
(*Hint:* Factor by grouping.)

20. $3x^3 + x^2 - 12x - 4 = 0$
(*Hint:* Factor by grouping.)

In Exercises 21–28, use the given graph to find each of the following: **(a)** *the x-intercept(s) and* **(b)** *the zero(s) of the function.*

21.

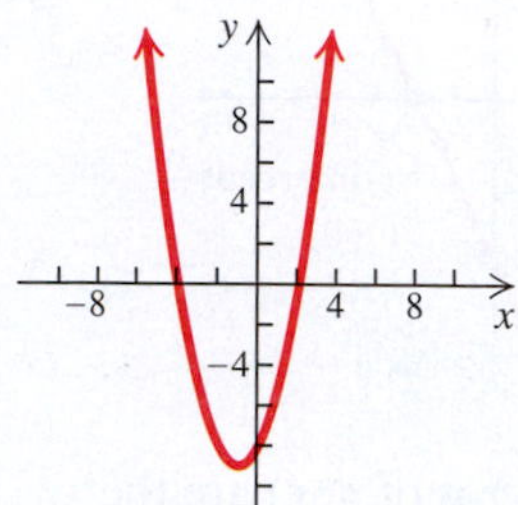

22.

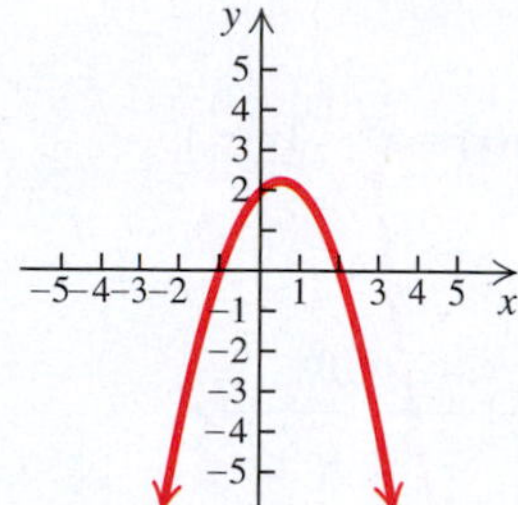

23.

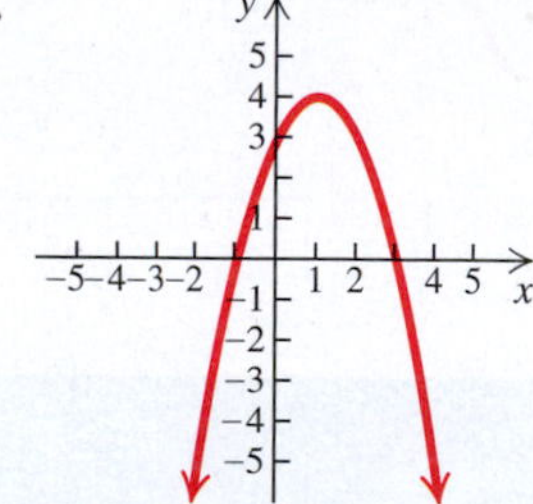

24.

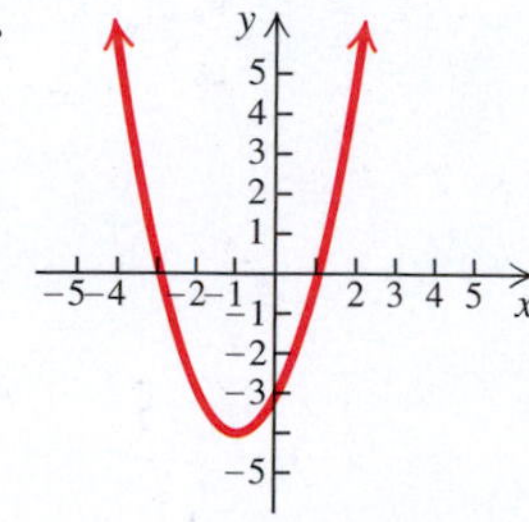

25.

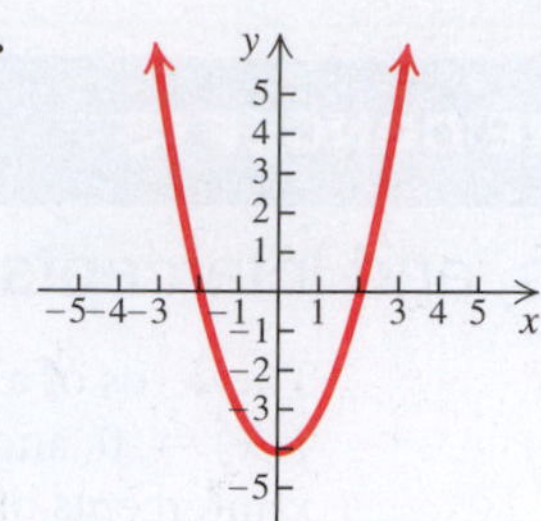

26.

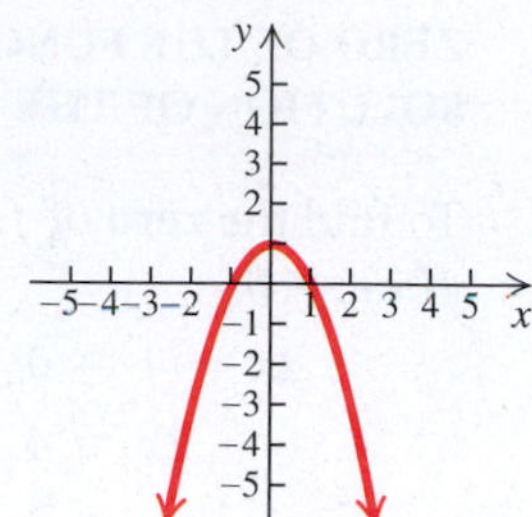

27.

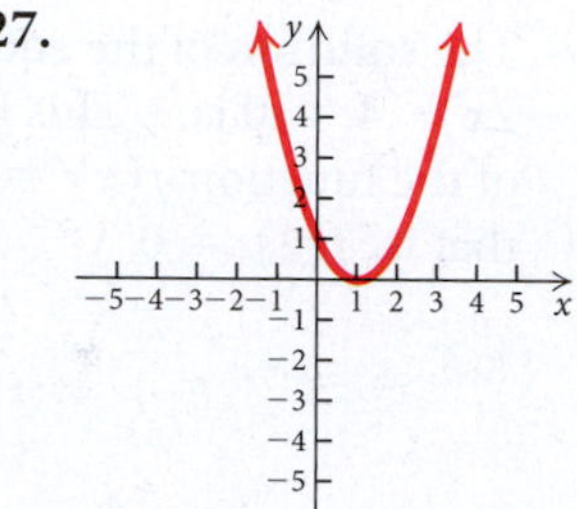

28.

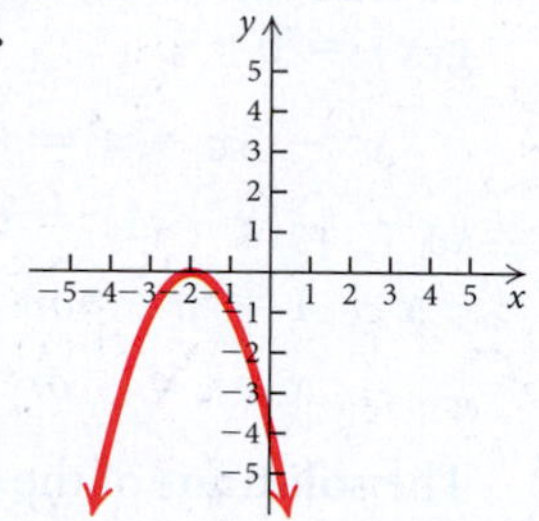

Solve by completing the square to obtain exact solutions.

29. $x^2 + 6x = 7$

30. $x^2 + 8x = -15$

31. $x^2 = 8x - 9$

32. $x^2 = 22 + 10x$

33. $x^2 + 8x + 25 = 0$

34. $x^2 + 6x + 13 = 0$

35. $3x^2 + 5x - 2 = 0$

36. $2x^2 - 5x - 3 = 0$

Use the quadratic formula to find exact solutions.

37. $x^2 - 2x = 15$

38. $x^2 + 4x = 5$

39. $5m^2 + 3m = 2$

40. $2y^2 - 3y - 2 = 0$

41. $3x^2 + 6 = 10x$

42. $3t^2 + 8t + 3 = 0$

43. $x^2 + x + 2 = 0$

44. $x^2 + 1 = x$

45. $5t^2 - 8t = 3$

46. $5x^2 + 2 = x$

47. $3x^2 + 4 = 5x$

48. $2t^2 - 5t = 1$

49. $x^2 - 8x + 5 = 0$

50. $x^2 - 6x + 3 = 0$

51. $3x^2 + x = 5$

52. $5x^2 + 3x = 1$

53. $2x^2 + 1 = 5x$

54. $4x^2 + 3 = x$

55. $5x^2 + 2x = -2$

56. $3x^2 + 3x = -4$

For each of the following, find the discriminant, $b^2 - 4ac$, and then determine whether one real-number solution, two different real-number solutions, or two different imaginary-number solutions exist.

57. $4x^2 = 8x + 5$

58. $4x^2 - 12x + 9 = 0$

59. $x^2 + 3x + 4 = 0$

60. $x^2 - 2x + 4 = 0$

61. $5t^2 - 7t = 0$

62. $5t^2 - 4t = 11$

Find the zeros of the function. Give exact answers and approximate solutions rounded to three decimal places when possible.

63. $f(x) = x^2 + 6x + 5$

64. $f(x) = x^2 - x - 2$

65. $f(x) = x^2 - 3x - 3$

66. $f(x) = 3x^2 + 8x + 2$

67. $f(x) = x^2 - 5x + 1$

68. $f(x) = x^2 - 3x - 7$

69. $f(x) = x^2 + 2x - 5$

70. $f(x) = x^2 - x - 4$

71. $f(x) = 2x^2 - x + 4$

72. $f(x) = 2x^2 + 3x + 2$

73. $f(x) = 3x^2 - x - 1$

74. $f(x) = 3x^2 + 5x + 1$

75. $f(x) = 5x^2 - 2x - 1$

76. $f(x) = 4x^2 - 4x - 5$

77. $f(x) = 4x^2 + 3x - 3$

78. $f(x) = x^2 + 6x - 3$

Solve.

79. $x^4 - 3x^2 + 2 = 0$

80. $x^4 + 3 = 4x^2$

81. $x^4 + 3x^2 = 10$

82. $x^4 - 8x^2 = 9$

83. $y^4 + 4y^2 - 5 = 0$

84. $y^4 - 15y^2 - 16 = 0$

85. $x - 3\sqrt{x} - 4 = 0$
$\left(\textit{Hint}\text{: Let } u = \sqrt{x}.\right)$

86. $2x - 9\sqrt{x} + 4 = 0$

87. $m^{2/3} - 2m^{1/3} - 8 = 0$
(*Hint*: Let $u = m^{1/3}$.)

88. $t^{2/3} + t^{1/3} - 6 = 0$

89. $x^{1/2} - 3x^{1/4} + 2 = 0$

90. $x^{1/2} - 4x^{1/4} = -3$

91. $(2x - 3)^2 - 5(2x - 3) + 6 = 0$
(*Hint*: Let $u = 2x - 3$.)

92. $(3x + 2)^2 + 7(3x + 2) - 8 = 0$

93. $(2t^2 + t)^2 - 4(2t^2 + t) + 3 = 0$

94. $12 = (m^2 - 5m)^2 + (m^2 - 5m)$

Funding for Afghan Security. *The number of U.S. forces in Afghanistan decreased to approximately 34,000 in 2014 from a high of about 100,000 in 2010. The amount of U.S. funding for Afghan security forces also decreased during this period. The function*

$$f(x) = -1.321x^2 + 5.156x + 5.517$$

can be used to estimate the amount of U.S. funding for Afghan security forces, in billions of dollars, x years after 2009 (Source: U.S. Department of Defense; Brookings Institution; International Security Assistance Force; ESRI). Use this function for Exercises 95 and 96.

95. In what year was the amount of U.S. funding for Afghan security forces about $10.5 billion?

96. In what year was the amount of U.S. funding for Afghan security forces about $5.0 billion?

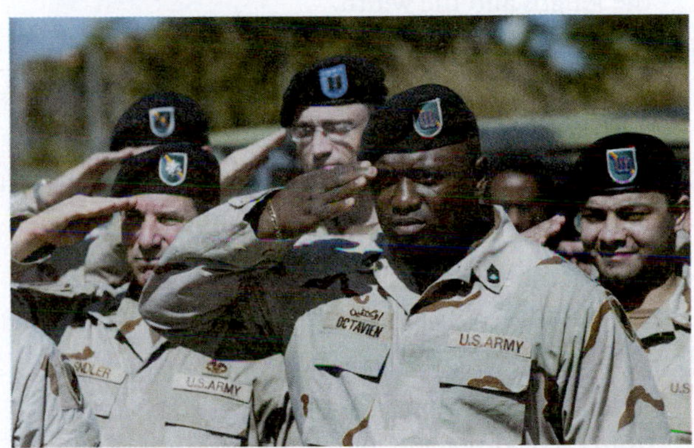

Multigenerational Households. *After declining between 1940 and 1980, the number of multigenerational American households has been increasing since 1980. The function*

$$h(x) = 0.012x^2 - 0.583x + 35.727$$

can be used to estimate the number of multigenerational households in the United States, in millions, x years after 1940 (Source: Pew Research Center). Use this function for Exercises 97 and 98.

97. In what year were there 40 million multigenerational households?

98. In what year were there 55 million multigenerational households?

Time of a Free Fall. *The formula $s = 16t^2$ is used to approximate the distance s, in feet, that an object falls freely from rest in t seconds. Use this formula for Exercises 99 and 100.*

99. The Taipei 101 Tower, also known as the Taipei Financial Center, in Taipei, Taiwan, is 1670 ft tall. How long would it take an object dropped from the top to reach the ground?

100. At 630 ft, the Gateway Arch in St. Louis is the tallest man-made monument in the United States. How long would it take an object dropped from the top to reach the ground?

101. The length of a rectangular poster is 1 ft more than the width, and a diagonal of the poster is 5 ft. Find the length and the width.

102. One leg of a right triangle is 7 cm less than the length of the other leg. The length of the hypotenuse is 13 cm. Find the lengths of the legs.

103. One number is 5 greater than another. The product of the numbers is 36. Find the numbers.

104. One number is 6 less than another. The product of the numbers is 72. Find the numbers.

105. *Box Construction.* An open box is made from a 10-cm by 20-cm piece of tin by cutting a square from each corner and folding up the edges. The area of the resulting base is 96 cm². What is the length of the sides of the squares?

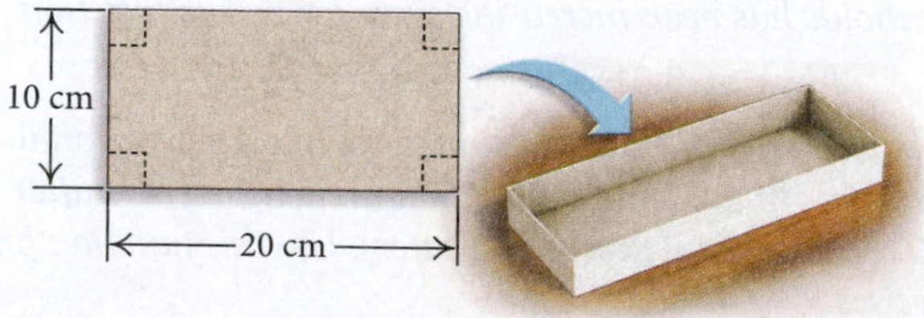

106. *Petting Zoo Dimensions.* At the Glen Island Zoo, 170 m of fencing was used to enclose a rectangular petting area of 1750 m². Find the dimensions of the petting area.

107. *Dimensions of a Rug.* Find the dimensions of a rectangular Persian rug whose perimeter is 28 ft and whose area is 48 ft².

108. *Picture Frame Dimensions.* The rectangular frame on a picture is 8 in. by 10 in. outside and is of uniform width. What is the width of the frame if 48 in² of the picture shows?

State whether the function is linear or quadratic.

109. $f(x) = 4 - 5x$

110. $f(x) = 4 - 5x^2$

111. $f(x) = 7x^2$

112. $f(x) = 23x + 6$

113. $f(x) = 1.2x - (3.6)^2$

114. $f(x) = 2 - x - x^2$

▶ Skill Maintenance

Cost of a Super Bowl Ad. *The cost of a 30-sec Super Bowl ad has increased more than 70% since 2004. The function*

$$C(x) = 0.17x + 2.25$$

can be used to estimate the cost of a 30-sec ad, in millions of dollars, x years after 2004 (Source: Katar Media). Use this function for Exercises 115 and 116. **[1.2]**

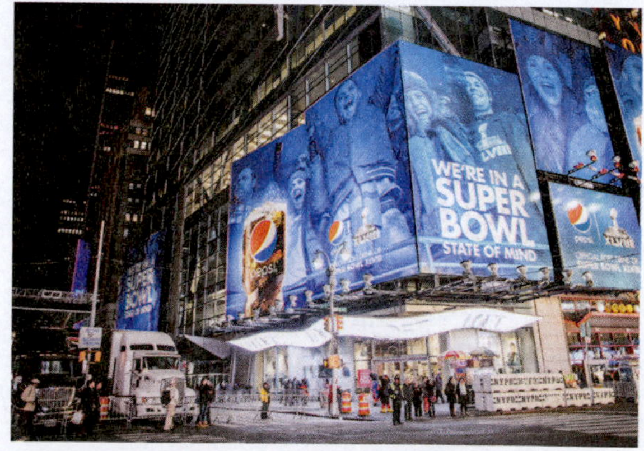

115. Estimate the cost of a 30-sec Super Bowl ad in 2014.

116. When will the cost of a 30-sec Super Bowl ad reach $5.0 million?

Determine whether the graph is symmetric with respect to the x-axis, the y-axis, and the origin. **[2.4]**

117. $3x^2 + 4y^2 = 5$

118. $y^3 = 6x^2$

Determine whether the function is even, odd, or neither even nor odd. [2.4]

119. $f(x) = 2x^3 - x$

120. $f(x) = 4x^2 + 2x - 3$

Synthesis

For each equation in Exercises 121–124, under the given condition: **(a)** *find k and* **(b)** *find a second solution.*

121. $kx^2 - 17x + 33 = 0$; one solution is 3

122. $kx^2 - 2x + k = 0$; one solution is -3

123. $x^2 - kx + 2 = 0$; one solution is $1 + i$

124. $x^2 - (6 + 3i)x + k = 0$; one solution is 3

Solve.

125. $(x - 2)^3 = x^3 - 2$

126. $(x + 1)^3 = (x - 1)^3 + 26$

127. $(6x^3 + 7x^2 - 3x)(x^2 - 7) = 0$

128. $\left(x - \frac{1}{5}\right)\left(x^2 - \frac{1}{4}\right) + \left(x - \frac{1}{5}\right)\left(x^2 + \frac{1}{8}\right) = 0$

129. $x^2 + x - \sqrt{2} = 0$

130. $x^2 + \sqrt{5}x - \sqrt{3} = 0$

131. $2t^2 + (t - 4)^2 = 5t(t - 4) + 24$

132. $9t(t + 2) - 3t(t - 2) = 2(t + 4)(t + 6)$

133. $\sqrt{x - 3} - \sqrt[4]{x - 3} = 2$

134. $x^2 + 3x + 1 - \sqrt{x^2 + 3x + 1} = 8$

135. $\left(y + \frac{2}{y}\right)^2 + 3y + \frac{6}{y} = 4$

136. Solve $\frac{1}{2}at^2 + v_0t + x_0 = 0$ for t.

3.3 Analyzing Graphs of Quadratic Functions

- Find the vertex, the axis of symmetry, and the maximum or minimum value of a quadratic function using the method of completing the square.
- Graph quadratic functions.
- Solve applied problems involving maximum and minimum function values.

Graphing Quadratic Functions of the Type $f(x) = a(x - h)^2 + k$

The graph of a quadratic function is called a **parabola**. The graph of every parabola evolves from the graph of the squaring function $f(x) = x^2$ using transformations.

TRANSFORMATIONS

REVIEW SECTION 2.5

We get the graph of $f(x) = a(x - h)^2 + k$ from the graph of $f(x) = x^2$ as follows:

$f(x) = x^2$
↓
$f(x) = ax^2$ — **Vertical stretching or shrinking with a reflection across the x-axis if $a < 0$**
↓
$f(x) = a(x - h)^2$ — **Horizontal translation**
↓
$f(x) = a(x - h)^2 + k.$ — **Vertical translation**

Consider the following graphs of the form $f(x) = a(x - h)^2 + k$. The point (h, k) at which the graph turns is called the **vertex**. The maximum or minimum value of $f(x)$ occurs at the vertex. Each graph has a line $x = h$ that is called the **axis of symmetry**.

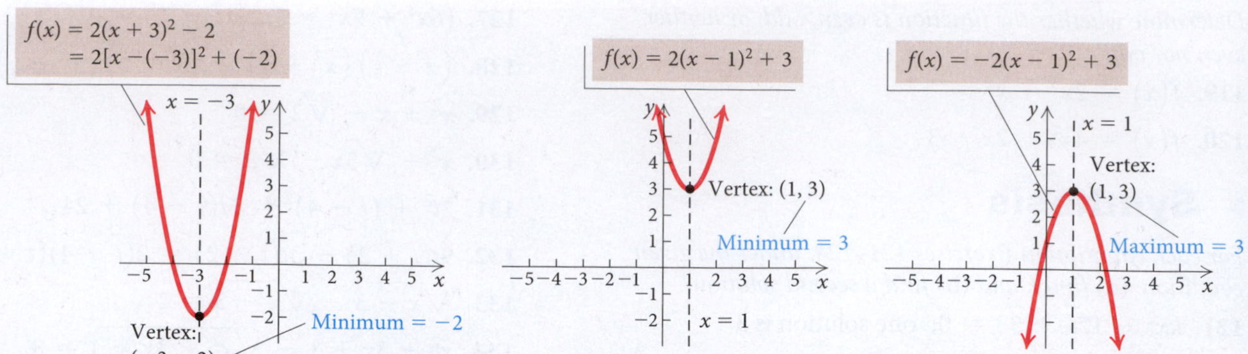

Graphing Quadratic Functions

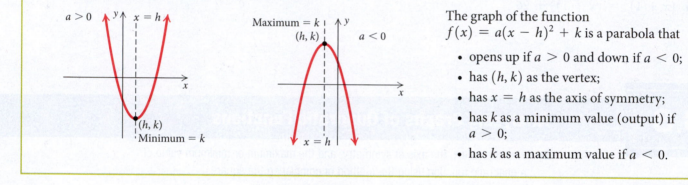

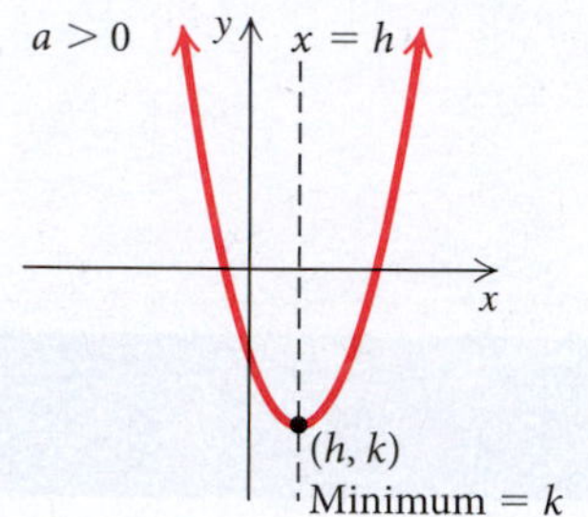

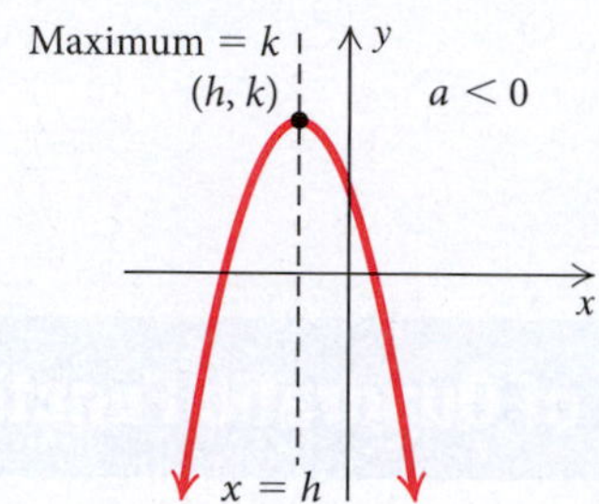

The graph of the function $f(x) = a(x - h)^2 + k$ is a parabola that

- opens up if $a > 0$ and down if $a < 0$;
- has (h, k) as the vertex;
- has $x = h$ as the axis of symmetry;
- has k as a minimum value (output) if $a > 0$;
- has k as a maximum value if $a < 0$.

As we saw in Section 2.5, the constant a serves to stretch or shrink the graph vertically. As a parabola is stretched vertically, it becomes narrower, and as it is shrunk vertically, it becomes wider. That is, as $|a|$ increases, the graph becomes narrower, and as $|a|$ gets close to 0, the graph becomes wider.

If the equation is in the form $f(x) = a(x - h)^2 + k$, we can learn a great deal about the graph without actually graphing the function.

Function	$f(x) = 3\left(x - \frac{1}{4}\right)^2 - 2$ $= 3\left(x - \frac{1}{4}\right)^2 + (-2)$	$g(x) = -3(x + 5)^2 + 7$ $= -3[x - (-5)]^2 + 7$
Vertex	$\left(\frac{1}{4}, -2\right)$	$(-5, 7)$
Axis of Symmetry	$x = \frac{1}{4}$	$x = -5$
Maximum	None ($3 > 0$, so the graph opens up.)	7 ($-3 < 0$, so the graph opens down.)
Minimum	-2 ($3 > 0$, so the graph opens up.)	None ($-3 < 0$, so the graph opens down.)

Note that the vertex (h, k) is used to find the maximum or minimum value of the function. The maximum or minimum value is the number k, *not* the ordered pair (h, k).

Graphing Quadratic Functions of the Type $f(x) = ax^2 + bx + c,\ a \neq 0$

We now use a modification of the method of completing the square as an aid in graphing and analyzing quadratic functions of the form $f(x) = ax^2 + bx + c$, $a \neq 0$.

EXAMPLE 1 Find the vertex, the axis of symmetry, and the maximum or minimum value of $f(x) = x^2 + 10x + 23$. Then graph the function.

Solution To express $f(x) = x^2 + 10x + 23$ in the form $f(x) = a(x - h)^2 + k$, we complete the square on the terms involving x. To do so, we take half the coefficient of x and square it, obtaining $(10/2)^2$, or 25. We now add and subtract that number on the *right side*:

$$f(x) = x^2 + 10x + 23 = x^2 + 10x + 25 - 25 + 23.$$

Since $25 - 25 = 0$, the new expression for the function is equivalent to the original expression. Note that this process differs from the one we used to complete the square in order to solve a quadratic equation, where we added the same number on both sides of the equation to obtain an equivalent equation. Instead, when we complete the square to write a function in the form $f(x) = a(x - h)^2 + k$, we add and subtract the same number on the one side. The entire process is shown below:

$$f(x) = x^2 + 10x + 23$$ Note that 25 completes the square for $x^2 + 10x$.

$$= x^2 + 10x + 25 - 25 + 23$$ Adding $25 - 25$, or 0, on the right side

$$= (x^2 + 10x + 25) - 25 + 23$$ Regrouping

$$= (x + 5)^2 - 2$$ Factoring and simplifying

$$= [x - (-5)]^2 + (-2).$$ Writing in the form $f(x) = a(x - h)^2 + k$

Keeping in mind that this function will have a minimum value since $a > 0$ $(a = 1)$, from this form of the function we know the following:

Vertex: $(-5, -2)$;

Axis of symmetry: $x = -5$;

Minimum value of the function: -2.

To graph the function, we first plot the vertex and find several points on either side of it. Then we plot these points and connect them with a smooth curve. We see that the points $(-4, -1)$ and $(-3, 2)$ are reflections of the points $(-6, -1)$ and $(-7, 2)$, respectively, across the axis of symmetry, $x = -5$.

x	$f(x)$
-5	-2 ← Vertex
-4	-1
-3	2
-6	-1
-7	2

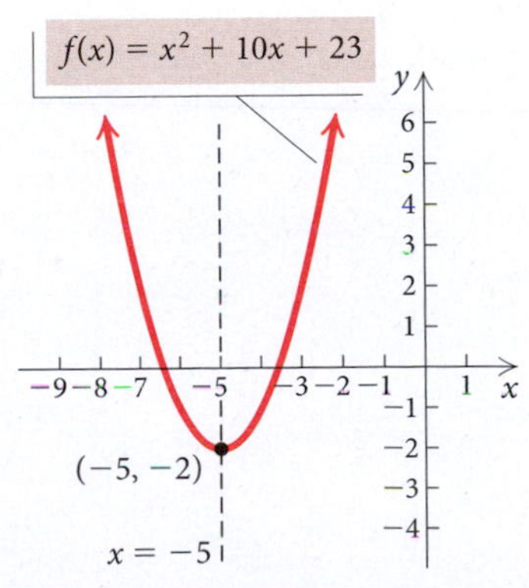

The graph of $f(x) = x^2 + 10x + 23$, or $f(x) = [x - (-5)]^2 + (-2)$, shown above, is a shift of the graph of $y = x^2$ left 5 units and down 2 units.

Now Try Exercise 3.

Keep in mind that the axis of symmetry is not part of the graph; it is a characteristic of the graph. If you fold the graph on its axis of symmetry, the two halves of the graph will coincide.

EXAMPLE 2 Find the vertex, the axis of symmetry, and the maximum or minimum value of $g(x) = x^2/2 - 4x + 8$. Then graph the function.

Solution We complete the square in order to write the function in the form $g(x) = a(x - h)^2 + k$. First, we factor $\frac{1}{2}$ out of the first two terms. This makes the coefficient of x^2 within the parentheses 1:

$$g(x) = \frac{x^2}{2} - 4x + 8$$

$$= \frac{1}{2}(x^2 - 8x) + 8. \qquad \text{Factoring } \tfrac{1}{2} \text{ out of the first two terms: } \frac{x^2}{2} - 4x = \frac{1}{2}\cdot x^2 - \frac{1}{2}\cdot 8x$$

Next, we complete the square inside the parentheses: Half of -8 is -4, and $(-4)^2 = 16$. We add and subtract 16 inside the parentheses:

$$g(x) = \tfrac{1}{2}(x^2 - 8x + 16 - 16) + 8$$

$$= \tfrac{1}{2}(x^2 - 8x + 16) - \tfrac{1}{2}\cdot 16 + 8 \qquad \text{Using the distributive law to remove } -16 \text{ from within the parentheses}$$

$$= \tfrac{1}{2}(x^2 - 8x + 16) - 8 + 8$$

$$= \tfrac{1}{2}(x - 4)^2 + 0, \text{ or } \tfrac{1}{2}(x - 4)^2. \qquad \text{Factoring and simplifying}$$

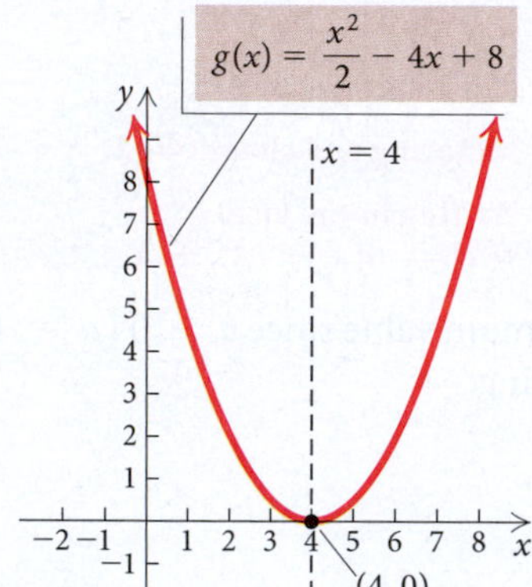

We know the following:

Vertex: $(4, 0)$;
Axis of symmetry: $x = 4$;
Minimum value of the function: 0.

Finally, we plot the vertex and several points on either side of it and draw the graph of the function. The graph of g is a vertical shrinking of the graph of $y = x^2$ along with a shift right 4 units.

Now Try Exercise 9.

EXAMPLE 3 Find the vertex, the axis of symmetry, and the maximum or minimum value of $f(x) = -2x^2 + 10x - \frac{23}{2}$. Then graph the function.

Solution We have

$$f(x) = -2x^2 + 10x - \tfrac{23}{2}$$

$$= -2(x^2 - 5x) - \tfrac{23}{2} \qquad \text{Factoring } -2 \text{ out of the first two terms}$$

$$= -2\left(x^2 - 5x + \tfrac{25}{4} - \tfrac{25}{4}\right) - \tfrac{23}{2} \qquad \text{Completing the square inside the parentheses}$$

$$= -2\left(x^2 - 5x + \tfrac{25}{4}\right) - 2\left(-\tfrac{25}{4}\right) - \tfrac{23}{2} \qquad \text{Using the distributive law to remove } -\tfrac{25}{4} \text{ from within the parentheses}$$

$$= -2\left(x^2 - 5x + \tfrac{25}{4}\right) + \tfrac{25}{2} - \tfrac{23}{2}$$

$$= -2\left(x - \tfrac{5}{2}\right)^2 + 1.$$

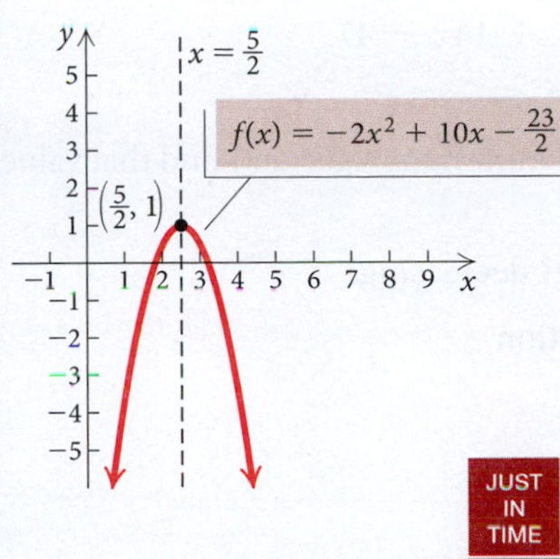

This form of the function yields the following:

Vertex: $\left(\frac{5}{2}, 1\right)$;

Axis of symmetry: $x = \frac{5}{2}$;

Maximum value of the function: 1.

The graph is found by shifting the graph of $f(x) = x^2$ right $\frac{5}{2}$ units, reflecting it across the x-axis, stretching it vertically, and shifting it up 1 unit.

Now Try Exercise 13.

JUST IN TIME 21

In many situations, we want to use a formula to find the coordinates of the vertex directly from the equation $f(x) = ax^2 + bx + c$. One way to develop such a formula is to observe that the x-coordinate of the vertex is centered between the x-intercepts, or zeros, of the function. By averaging the two solutions of $ax^2 + bx + c = 0$, we find a formula for the x-coordinate of the vertex:

$$x\text{-coordinate of vertex} = \frac{\dfrac{-b - \sqrt{b^2 - 4ac}}{2a} + \dfrac{-b + \sqrt{b^2 - 4ac}}{2a}}{2}$$

$$= \frac{\dfrac{-2b}{2a}}{2} = \frac{-\dfrac{b}{a}}{2}$$

$$= -\frac{b}{a} \cdot \frac{1}{2} = -\frac{b}{2a}.$$

We use this value of x to find the y-coordinate of the vertex, $f\left(-\dfrac{b}{2a}\right)$.

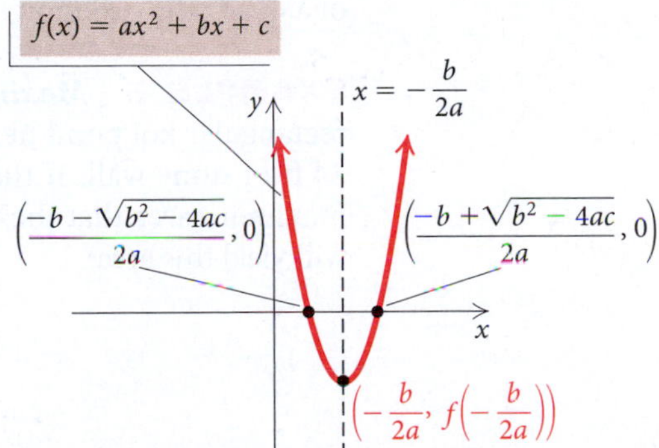

THE VERTEX OF A PARABOLA

The **vertex** of the graph of $f(x) = ax^2 + bx + c$ is

$$\left(-\frac{b}{2a}, f\left(-\frac{b}{2a}\right)\right).$$

We calculate the x-coordinate. We substitute to find the y-coordinate.

Technology Connection

We can use a graphing calculator to do Example 4. Once we have graphed $y = -x^2 + 14x - 47$, we see that the graph opens down and thus has a maximum value. We can use the MAXIMUM feature to find the coordinates of the vertex. Using these coordinates, we can then find the maximum value and the range of the function along with the intervals on which the function is increasing or decreasing.

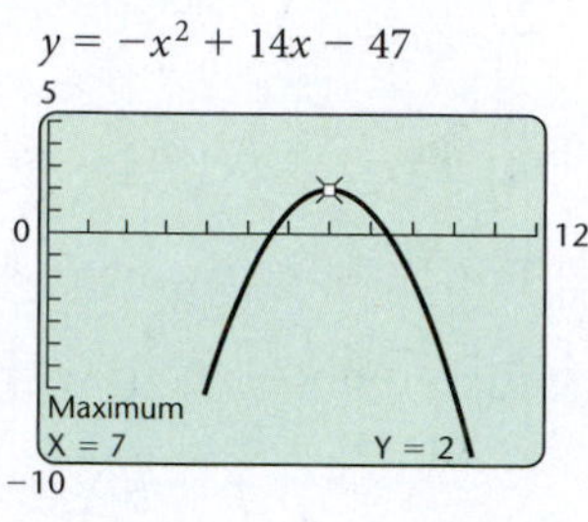

EXAMPLE 4 For the function $f(x) = -x^2 + 14x - 47$:

a) Find the vertex.

b) Determine whether there is a maximum or a minimum value and find that value.

c) Find the range.

d) On what intervals is the function increasing? decreasing?

Solution There is no need to graph the function.

a) The x-coordinate of the vertex is

$$-\frac{b}{2a} = -\frac{14}{2(-1)} = -\frac{14}{-2} = 7.$$

Since

$$f(7) = -7^2 + 14 \cdot 7 - 47 = -49 + 98 - 47 = 2,$$

the vertex is $(7, 2)$.

b) Since a is negative $(a = -1)$, the graph opens down so the second coordinate of the vertex, 2, is the maximum value of the function.

c) The range is $(-\infty, 2]$.

d) Since the graph opens down, function values increase as we approach the vertex from the left and decrease as we move to the right from the vertex. Thus the function is increasing on the interval $(-\infty, 7)$ and decreasing on $(7, \infty)$.

Now Try Exercise 31.

Applications

Many real-world situations involve finding the maximum or the minimum value of a quadratic function.

EXAMPLE 5 ***Maximizing Area.*** A landscaper has enough stone to enclose a rectangular koi pond next to an existing garden wall of the Englemans' house with 24 ft of stone wall. If the garden wall forms one side of the rectangle, what is the maximum area that the landscaper can enclose? What dimensions of the koi pond will yield this area?

PROBLEM-SOLVING STRATEGY
REVIEW SECTION 1.5

Solution We will use the five-step problem-solving strategy.

1. **Familiarize.** We first make a drawing of the situation, using w to represent the width of the koi pond, in feet. Then $(24 - 2w)$ feet of stone is available for the length. Suppose the koi pond were 1 ft wide. Then its length would be $24 - 2 \cdot 1 = 22$ ft, and its area would be $(22 \text{ ft})(1 \text{ ft}) = 22 \text{ ft}^2$. If the koi pond were 2 ft wide, its length would be $24 - 2 \cdot 2 = 20$ ft, and its area would be $(20 \text{ ft})(2 \text{ ft}) = 40 \text{ ft}^2$. This is larger than the first area we found, but we do not know if it is the maximum possible area. To find the maximum area, we will find a function that represents the area and then determine its maximum value.

2. **Translate.** We write a function for the area of the koi pond. We have

$$\begin{aligned} A(w) &= (24 - 2w)w \qquad A = lw; l = 24 - 2w \\ &= -2w^2 + 24w, \end{aligned}$$

where $A(w)$ is the area of the koi pond, in square feet, as a function of the width w.

3. **Carry out.** To solve this problem, we need to determine the maximum value of $A(w)$ and find the dimensions for which that maximum occurs. Since A is a quadratic function and w^2 has a negative coefficient, we know that the function has a maximum value that occurs at the vertex of the graph of the function. The first coordinate of the vertex, $(w, A(w))$, is

$$w = -\frac{b}{2a} = -\frac{24}{2(-2)} = -\frac{24}{-4} = 6.$$

Thus, if $w = 6$ ft, then the length $l = 24 - 2 \cdot 6 = 12$ ft, and the area is $(12 \text{ ft})(6 \text{ ft}) = 72 \text{ ft}^2$.

4. **Check.** As a partial check, we note that $72 \text{ ft}^2 > 40 \text{ ft}^2$, the larger of the two areas that we found in the *Familiarize* step. We could also complete the square to write the function in the form $A(w) = a(w - h)^2 + k$, and then use this form of the function to determine the coordinates of the vertex. We get

$$A(w) = -2(w - 6)^2 + 72.$$

This confirms that the vertex is $(6, 72)$, so the answer checks.

5. **State.** The maximum possible area is 72 ft^2 when the koi pond is 6 ft wide and 12 ft long.

Now Try Exercise 45.

EXAMPLE 6 ***Height of a Rocket.*** A model rocket is launched with an initial velocity of 100 ft/sec from the top of a hill that is 20 ft high. Its height, in feet, t seconds after it has been launched is given by the function $s(t) = -16t^2 + 100t + 20$. Determine the time at which the rocket reaches its maximum height and find the maximum height.

Solution

1., 2. Familiarize and **Translate.** We are given the function in the statement of the problem: $s(t) = -16t^2 + 100t + 20$.

3. Carry out. We need to find the maximum value of the function and the value of t for which it occurs. Since $s(t)$ is a quadratic function and t^2 has a negative coefficient, we know that the maximum value of the function occurs at the vertex of the graph of the function. The first coordinate of the vertex gives the time t at which the rocket reaches its maximum height. It is

$$t = -\frac{b}{2a} = -\frac{100}{2(-16)} = -\frac{100}{-32} = 3.125.$$

The second coordinate of the vertex gives the maximum height of the rocket. We substitute in the function to find it:

$$s(3.125) = -16(3.125)^2 + 100(3.125) + 20 = 176.25.$$

4. Check. As a check, we can complete the square to write the function in the form $s(t) = a(t - h)^2 + k$ and determine the coordinates of the vertex from this form of the function. We get

$$s(t) = -16(t - 3.125)^2 + 176.25.$$

This confirms that the vertex is $(3.125, 176.25)$, so the answer checks.

5. State. The rocket reaches a maximum height of 176.25 ft. This occurs 3.125 sec after it has been launched.

Now Try Exercise 41.

EXAMPLE 7 ***Determining the Height of an Elevator Shaft.*** Jared drops a screwdriver from the top of an elevator shaft. Exactly 5 sec later, he hears the sound of the screwdriver hitting the bottom of the shaft. The speed of sound is 1100 ft/sec. How tall is the elevator shaft?

Solution

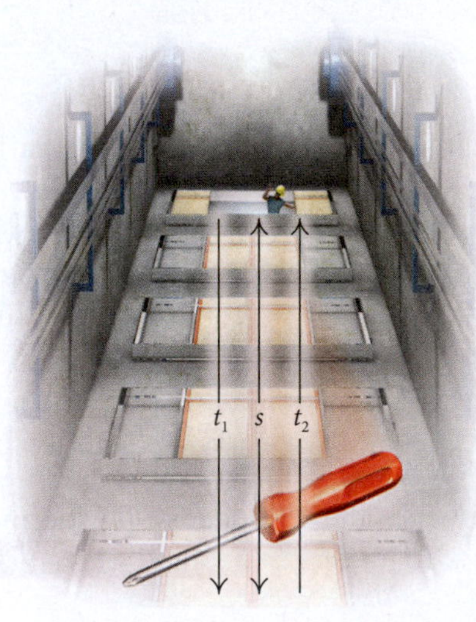

1. Familiarize. We first make a drawing and label it with known and unknown information. We let s = the height of the elevator shaft, in feet, t_1 = the time, in seconds, that it takes for the screwdriver to hit the bottom of the elevator shaft, and t_2 = the time, in seconds, that it takes for the sound to reach the top of the elevator shaft. This gives us the equation

$$t_1 + t_2 = 5. \qquad (1)$$

2. Translate. Can we find any relationship between the two times and the distance s? Often in problem solving you may need to look up related formulas in a physics book, another mathematics book, or on the Internet. We find that the formula

$$s = 16t^2$$

gives the distance, in feet, that a dropped object falls in t seconds. The time t_1 that it takes the screwdriver to hit the bottom of the elevator shaft can be found as follows:

$$s = 16t_1^2, \quad \text{or} \quad \frac{s}{16} = t_1^2, \quad \text{so} \quad t_1 = \frac{\sqrt{s}}{4}. \qquad (2)$$

Taking the positive square root

To find an expression for t_2, the time that it takes the sound to travel to the top of the well, recall that *Distance* = *Rate* · *Time*. Thus,

$$s = 1100t_2, \quad \text{or} \quad t_2 = \frac{s}{1100}. \tag{3}$$

We now have expressions for t_1 and t_2, both in terms of s. Substituting into equation (1), we obtain

$$t_1 + t_2 = 5, \quad \text{or} \quad \frac{\sqrt{s}}{4} + \frac{s}{1100} = 5. \tag{4}$$

3. **Carry out.** We solve equation (4) for s. Multiplying by 1100, we get

$$275\sqrt{s} + s = 5500, \quad \text{or} \quad s + 275\sqrt{s} - 5500 = 0.$$

This equation is reducible to quadratic with $u = \sqrt{s}$. Substituting, we get

$$u^2 + 275u - 5500 = 0.$$

Using the quadratic formula, we can solve for u:

$$\begin{aligned} u &= \frac{-b \pm \sqrt{b^2 - 4ac}}{2a} \\ &= \frac{-275 + \sqrt{275^2 - 4 \cdot 1 \cdot (-5500)}}{2 \cdot 1} \qquad \text{We want only the positive solution.} \\ &= \frac{-275 + \sqrt{97{,}625}}{2} \\ &\approx 18.725. \end{aligned}$$

Since $u \approx 18.725$, we have

$$\sqrt{s} \approx 18.725$$

$$s \approx 350.6. \qquad \text{Squaring both sides and rounding to the nearest tenth}$$

4. **Check.** To check, we can substitute 350.6 for s in equation (4) and see that $t_1 + t_2 \approx 5$. We leave the computation to the student.

5. **State.** The height of the elevator shaft is about 350.6 ft.

Now Try Exercise 55.

Technology Connection

We can solve the equation in Example 7 graphically using the Intersect method. It will probably require some trial and error to determine an appropriate window.

$$y_1 = \frac{\sqrt{x}}{4} + \frac{x}{1100}, \quad y_2 = 5$$

6

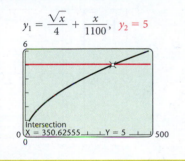

Intersection

X = 350.62555 Y = 5

0 0 500

Visualizing the Graph

Match the equation with its graph.

1. $y = 3x$

2. $y = -(x - 1)^2 + 3$

3. $(x + 2)^2 + (y - 2)^2 = 9$

4. $y = 3$

5. $2x - 3y = 6$

6. $(x - 1)^2 + (y + 3)^2 = 4$

7. $y = -2x + 1$

8. $y = 2x^2 - x - 4$

9. $x = -2$

10. $y = -3x^2 + 6x - 2$

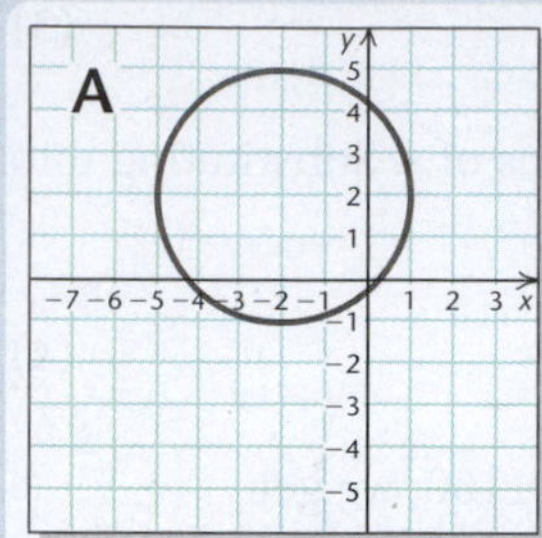

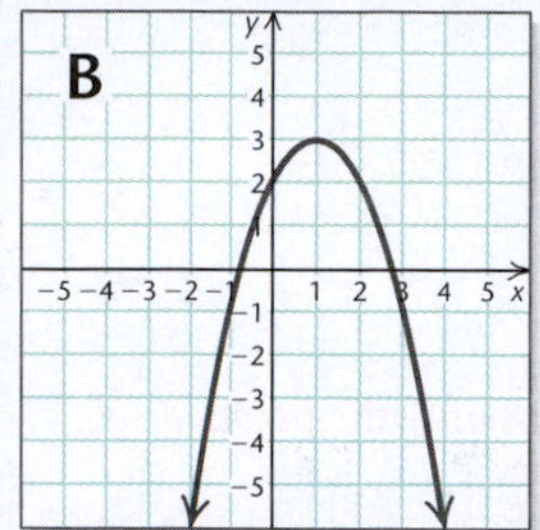

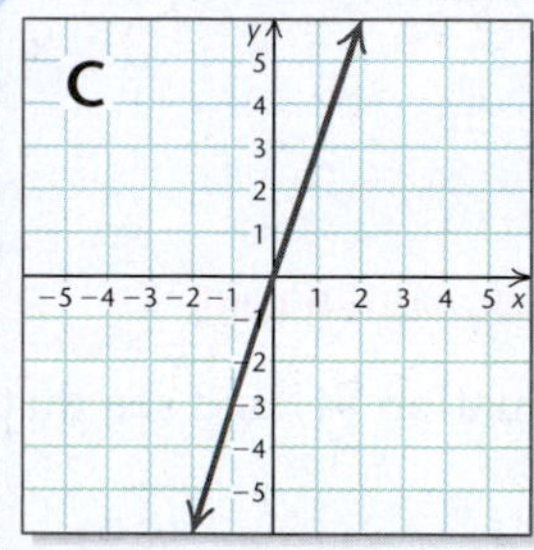

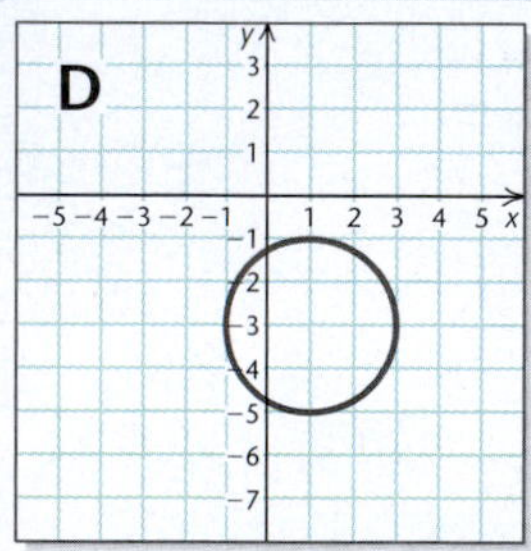

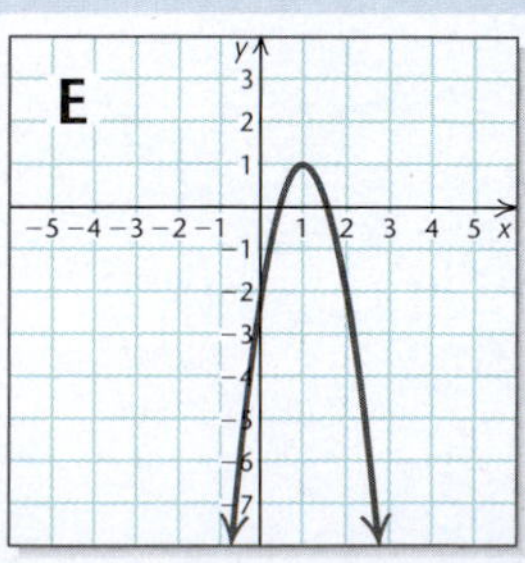

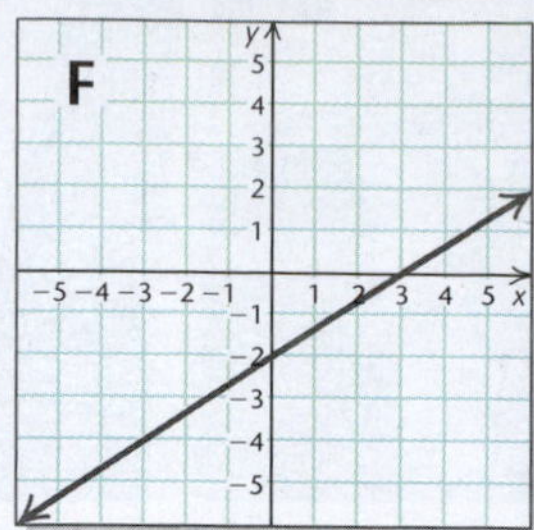

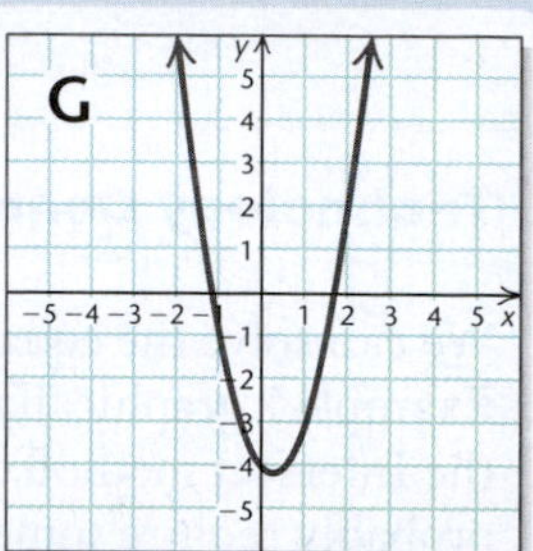

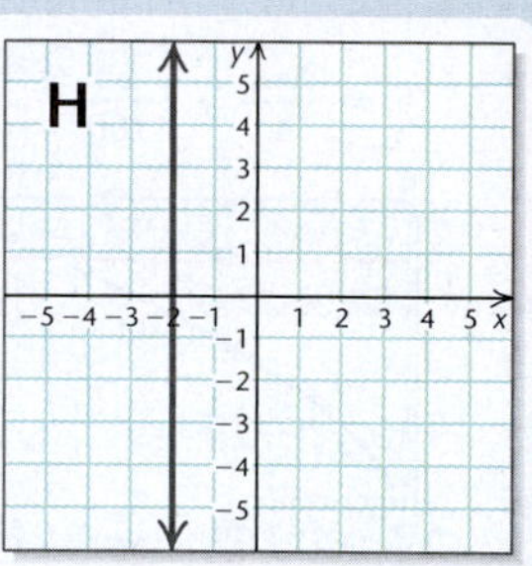

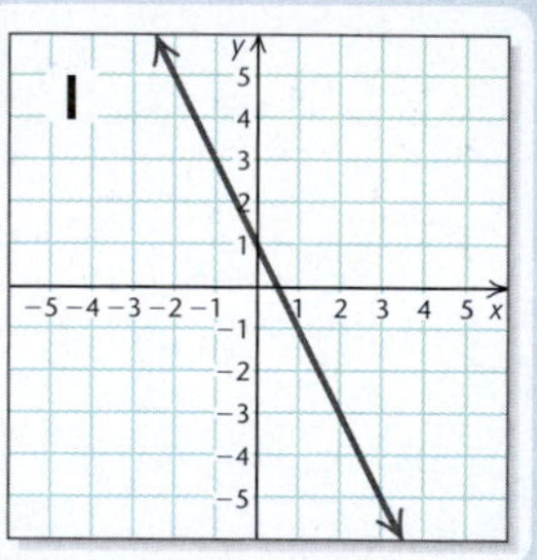

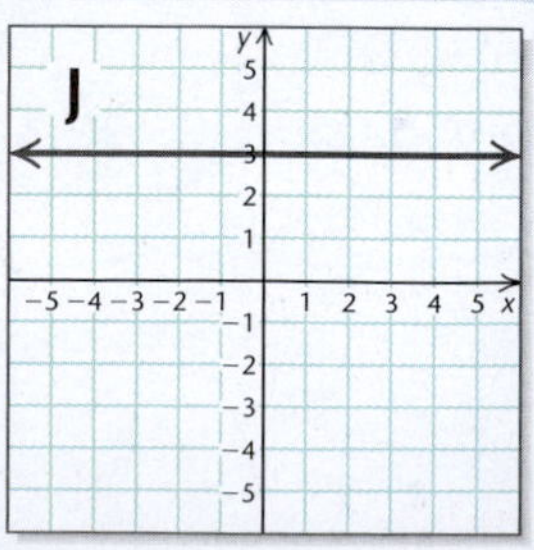

Answers on page A-14

3.3 Exercise Set

In Exercises 1 and 2, use the given graph to find each of the following: **(a)** *the vertex;* **(b)** *the axis of symmetry; and* **(c)** *the maximum or the minimum value of the function.*

1.

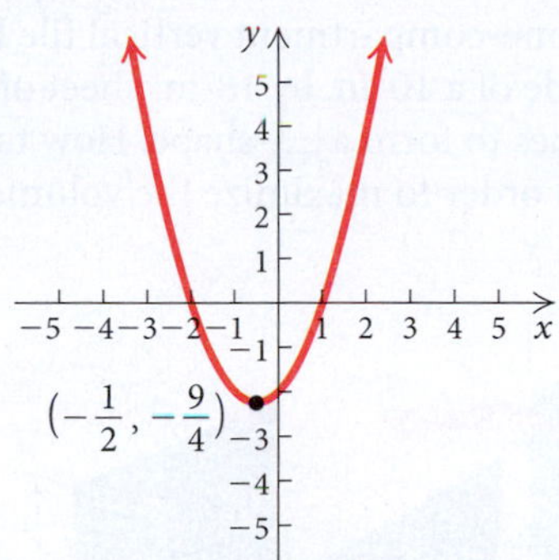

2.

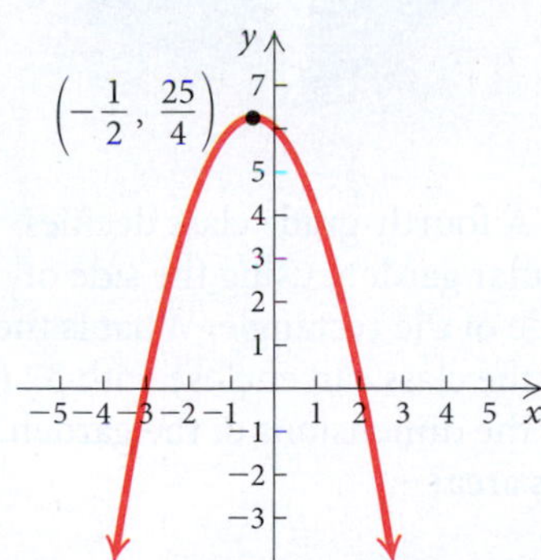

In Exercises 3–16, **(a)** *find the vertex;* **(b)** *find the axis of symmetry;* **(c)** *determine whether there is a maximum or a minimum value, and find that value; and* **(d)** *graph the function.*

3. $f(x) = x^2 - 8x + 12$

4. $g(x) = x^2 + 7x - 8$

5. $f(x) = x^2 - 7x + 12$

6. $g(x) = x^2 - 5x + 6$

7. $f(x) = x^2 + 4x + 5$

8. $f(x) = x^2 + 2x + 6$

9. $g(x) = \dfrac{x^2}{2} + 4x + 6$

10. $g(x) = \dfrac{x^2}{3} - 2x + 1$

11. $g(x) = 2x^2 + 6x + 8$

12. $f(x) = 2x^2 - 10x + 14$

13. $f(x) = -x^2 - 6x + 3$

14. $f(x) = -x^2 - 8x + 5$

15. $g(x) = -2x^2 + 2x + 1$

16. $f(x) = -3x^2 - 3x + 1$

In Exercises 17–24, match the equation with one of the graphs (a)–(h) that follow.

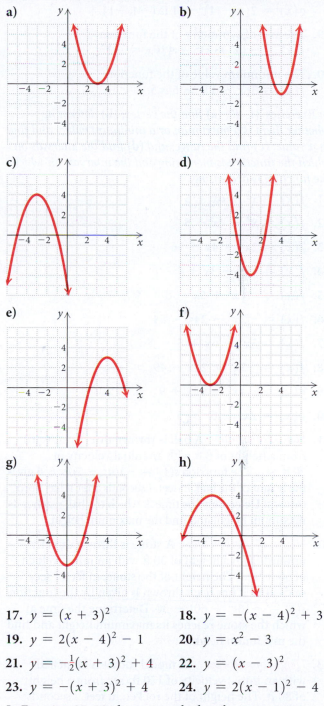

a)

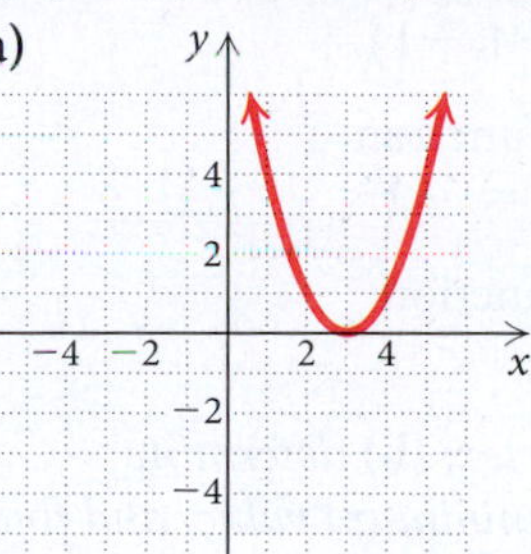

b)

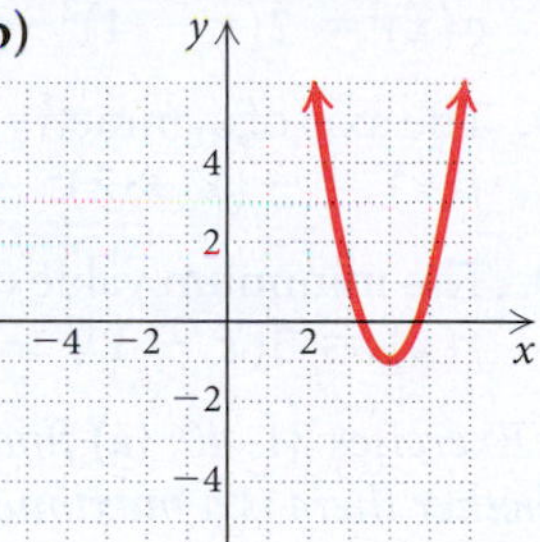

c)

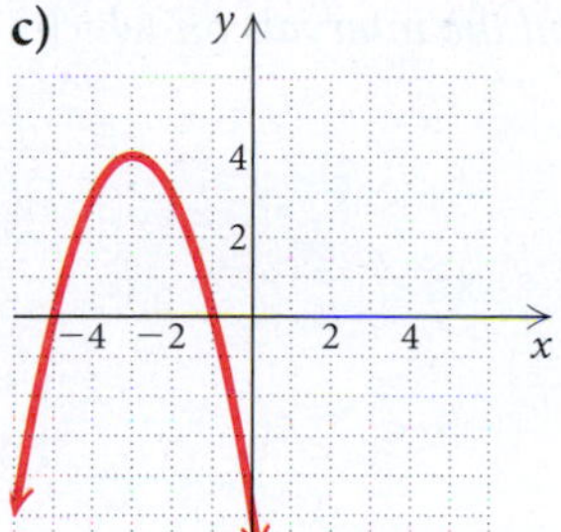

d)

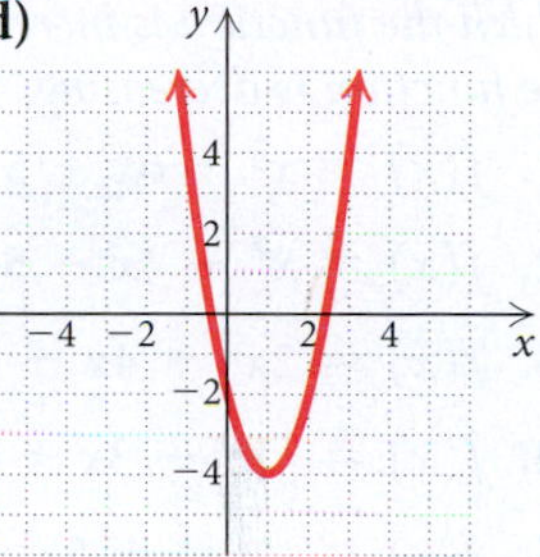

e)

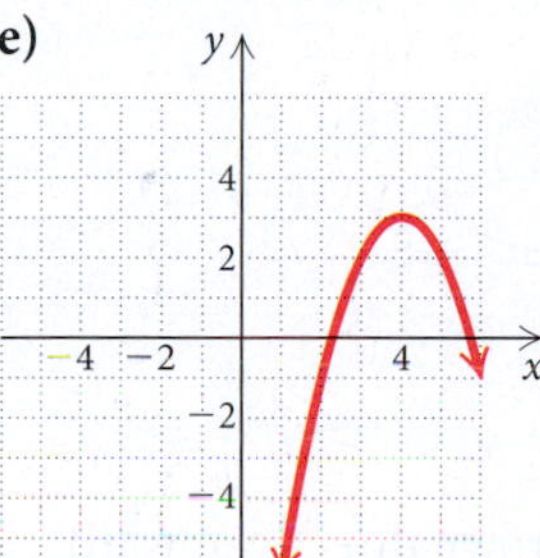

f)

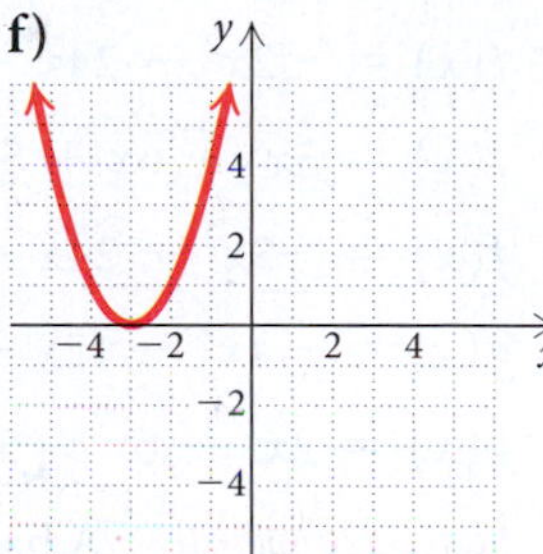

g)

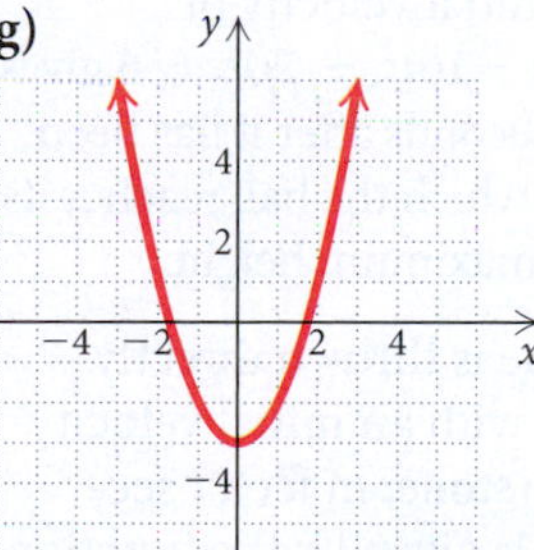

h)

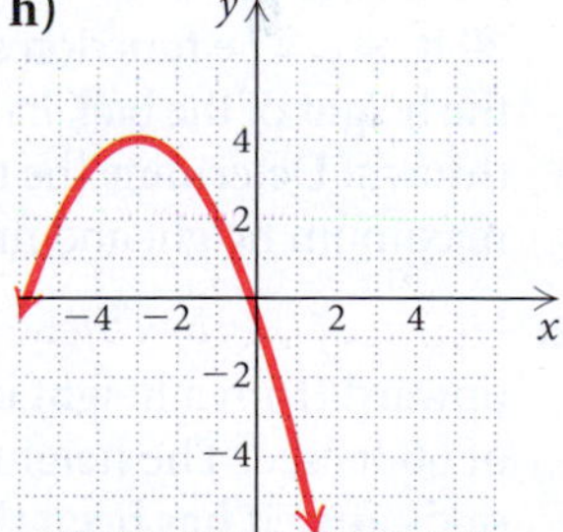

17. $y = (x + 3)^2$

18. $y = -(x - 4)^2 + 3$

19. $y = 2(x - 4)^2 - 1$

20. $y = x^2 - 3$

21. $y = -\frac{1}{2}(x + 3)^2 + 4$

22. $y = (x - 3)^2$

23. $y = -(x + 3)^2 + 4$

24. $y = 2(x - 1)^2 - 4$

In Exercises 25–30, determine whether the statement is true or false.

25. The function $f(x) = -3x^2 + 2x + 5$ has a maximum value.

26. The vertex of the graph of $f(x) = ax^2 + bx + c$ is $-\frac{b}{2a}$.

27. The graph of $h(x) = (x + 2)^2$ can be obtained by translating the graph of $h(x) = x^2$ right 2 units.

28. The vertex of the graph of the function $g(x) = 2(x - 4)^2 - 1$ is $(-4, -1)$.

29. The axis of symmetry of the function $f(x) = -(x + 2)^2 - 4$ is $x = -2$.

30. The minimum value of the function $f(x) = 3(x - 1)^2 + 5$ is 5.

In Exercises 31–40; **(a)** *find the vertex;* **(b)** *determine whether there is a maximum or a minimum value, and find that value;* **(c)** *find the range; and* **(d)** *find the intervals on which the function is increasing and the intervals on which the function is decreasing.*

31. $f(x) = x^2 - 6x + 5$

32. $f(x) = x^2 + 4x - 5$

33. $f(x) = 2x^2 + 4x - 16$

34. $f(x) = \frac{1}{2}x^2 - 3x + \frac{5}{2}$

35. $f(x) = -\frac{1}{2}x^2 + 5x - 8$

36. $f(x) = -2x^2 - 24x - 64$

37. $f(x) = 3x^2 + 6x + 5$

38. $f(x) = -3x^2 + 24x - 49$

39. $g(x) = -4x^2 - 12x + 9$

40. $g(x) = 2x^2 - 6x + 5$

41. *Height of a Ball.* A ball is thrown directly upward from a height of 6 ft with an initial velocity of 20 ft/sec. The function $s(t) = -16t^2 + 20t + 6$ gives the height of the ball, in feet, t seconds after it has been thrown. Determine the time at which the ball reaches its maximum height and find the maximum height.

42. *Height of a Projectile.* A stone is thrown directly upward from a height of 30 ft with an initial velocity of 60 ft/sec. The height of the stone, in feet, t seconds after it has been thrown is given by the function $s(t) = -16t^2 + 60t + 30$. Determine the time at which the stone reaches its maximum height and find the maximum height.

43. *Height of a Rocket.* A model rocket is launched with an initial velocity of 120 ft/sec from a height of 80 ft. The height of the rocket, in feet, t seconds after it has been launched is given by the function $s(t) = -16t^2 + 120t + 80$. Determine the time at which the rocket reaches its maximum height and find the maximum height.

44. *Height of a Rocket.* A model rocket is launched with an initial velocity of 150 ft/sec from a height of 40 ft. The function $s(t) = -16t^2 + 150t + 40$ gives the height of the rocket, in feet, t seconds after it has been launched. Determine the time at which the rocket reaches its maximum height and find the maximum height.

45. *Maximizing Volume.* Mendoza Manufacturing plans to produce a one-compartment vertical file by bending the long side of a 10-in. by 18-in. sheet of plastic along two lines to form a ⊔-shape. How tall should the file be in order to maximize the volume that it can hold?

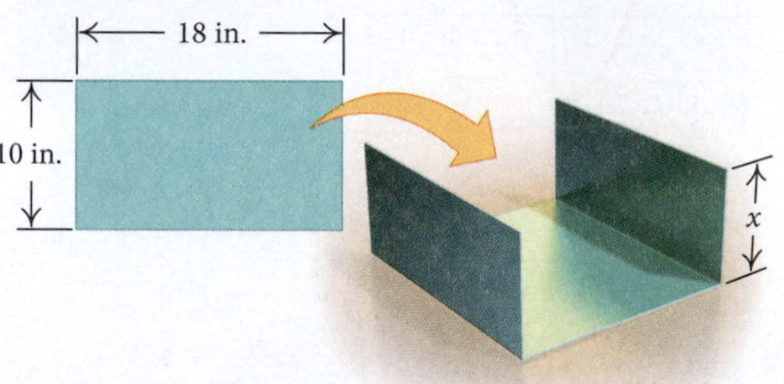

46. *Maximizing Area.* A fourth-grade class decides to enclose a rectangular garden, using the side of the school as one side of the rectangle. What is the maximum area that the class can enclose with 32 ft of fence? What should the dimensions of the garden be in order to yield this area?

47. *Maximizing Area.* The sum of the base and the height of a triangle is 20 cm. Find the dimensions for which the area is a maximum.

48. *Maximizing Area.* The sum of the base and the height of a parallelogram is 69 cm. Find the dimensions for which the area is a maximum.

49. *Minimizing Cost.* Designs for #1 Canines has determined that when x hundred wooden doghouses are built, the average cost per doghouse is given by

$$C(x) = 0.1x^2 - 0.7x + 1.625,$$

where $C(x)$ is in hundreds of dollars. How many doghouses should be built in order to minimize the average cost per doghouse?

Maximizing Profit. In business, profit is the difference between revenue and cost; that is,

Total profit = Total revenue − Total cost,

$$P(x) = R(x) - C(x),$$

where x is the number of units sold. Find the maximum profit and the number of units that must be sold in order to yield the maximum profit for each of the following.

50. $R(x) = 5x,\ C(x) = 0.001x^2 + 1.2x + 60$

51. $R(x) = 50x - 0.5x^2,\ C(x) = 10x + 3$

52. $R(x) = 20x - 0.1x^2,\ C(x) = 4x + 2$

53. *Maximizing Area.* A berry farmer needs to separate and enclose two adjacent rectangular fields, one for blueberries and one for strawberries. If a lake forms one side of the fields and 240 yd of fencing is available, what is the largest total area that can be enclosed?

54. *Norman Window.* A Norman window is a rectangle with a semicircle on top. Sky Blue Windows is designing a Norman window that will require 24 ft of trim on the outer edges. What dimensions will allow the maximum amount of light to enter a house?

55. *Finding the Depth of a Well.* Two seconds after a chlorine tablet has been dropped into a well, a splash is heard. The speed of sound is 1100 ft/sec. How far is the top of the well from the water? (*Hint*: See Example 7.)

56. *Finding the Height of a Cliff.* A water balloon is dropped from a cliff. Exactly 3 sec later, the sound of the balloon hitting the ground reaches the top of the cliff. How high is the cliff? (*Hint*: See Example 7.)

▶ Skill Maintenance

For each function f, construct and simplify the difference quotient

$$\frac{f(x+h) - f(x)}{h}.\ [2.2]$$

57. $f(x) = 3x - 7$

58. $f(x) = 2x^2 - x + 4$

A graph of $y = f(x)$ follows. No formula is given for f. Graph each of the following. [2.5]

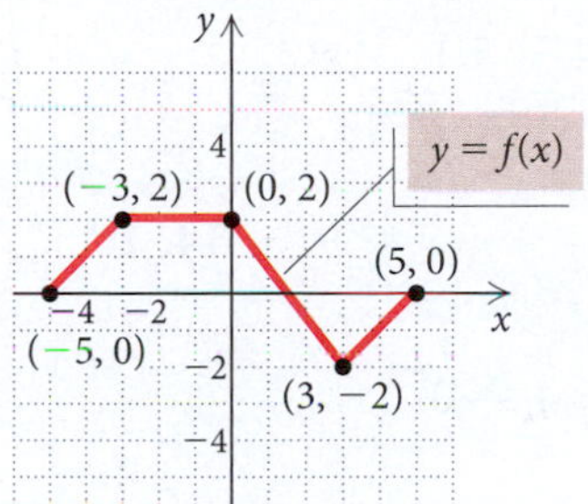

59. $g(x) = -2f(x)$

60. $g(x) = f(2x)$

▶ Synthesis

61. Find c such that

$$f(x) = -0.2x^2 - 3x + c$$

has a maximum value of -225.

62. Find b such that

$$f(x) = -4x^2 + bx + 3$$

has a maximum value of 50.

63. Graph: $f(x) = (|x| - 5)^2 - 3$.

64. Find a quadratic function with vertex $(4, -5)$ and containing the point $(-3, 1)$.

65. *Minimizing Area.* A 24-in. piece of string is cut into two pieces. One piece is used to form a circle while the other is used to form a square. How should the string be cut so that the sum of the areas is a minimum?

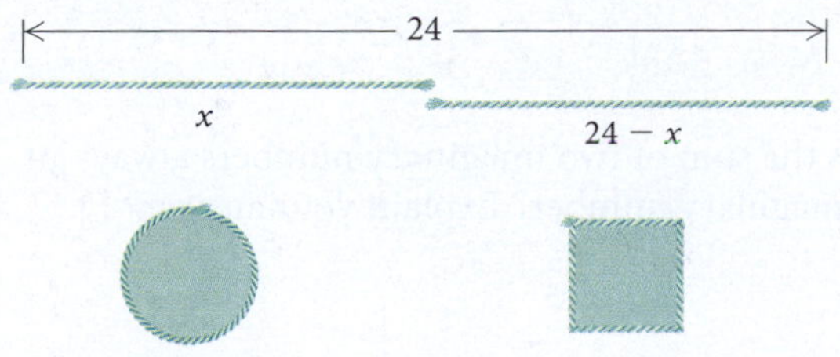

Mid-Chapter Mixed Review

Determine whether the statement is true or false.

1. The product of a complex number and its conjugate is a real number. [3.1]
2. Every quadratic equation has at least one x-intercept. [3.2]
3. If a quadratic equation has two different real-number solutions, then its discriminant is positive. [3.2]
4. The vertex of the graph of the function $f(x) = 3(x + 4)^2 + 5$ is $(4, 5)$. [3.3]

Express the number in terms of i. [3.1]

5. $\sqrt{-36}$
6. $\sqrt{-5}$
7. $-\sqrt{-16}$
8. $\sqrt{-32}$

Simplify. Write answers in the form a + bi, where a and b are real numbers. [3.1]

9. $(3 - 2i) + (-4 + 3i)$
10. $(-5 + i) - (2 - 4i)$
11. $(2 + 3i)(4 - 5i)$
12. $\dfrac{3 + i}{-2 + 5i}$

Simplify. [3.1]

13. i^{13}
14. i^{44}
15. $(-i)^5$
16. $(2i)^6$

Solve. [3.2]

17. $x^2 + 3x - 4 = 0$
18. $2x^2 + 6 = -7x$
19. $4x^2 = 24$
20. $x^2 + 100 = 0$
21. Find the zeros of $f(x) = 4x^2 - 8x - 3$ by completing the square. Show your work. [3.2]

In Exercises 22–24, **(a)** *find the discriminant $b^2 - 4ac$, and then determine whether one real-number solution, two different real-number solutions, or two different imaginary-number solutions exist; and* **(b)** *solve the equation, finding exact solutions and approximate solutions rounded to three decimal places, where appropriate.* [3.2]

22. $x^2 - 3x - 5 = 0$
23. $4x^2 - 12x + 9 = 0$
24. $3x^2 + 2x = -1$

Solve. [3.2]

25. $x^4 + 5x^2 - 6 = 0$
26. $2x - 5\sqrt{x} + 2 = 0$
27. One number is 2 more than another. The product of the numbers is 35. Find the numbers. [3.2]

In Exercises 28 and 29, **(a)** *find the vertex;* **(b)** *find the axis of symmetry;* **(c)** *determine whether there is a maximum or a minimum value, and find that value;* **(d)** *find the range;* **(e)** *find the intervals on which the function is increasing or decreasing; and* **(f)** *graph the function.* [3.3]

28. $f(x) = x^2 - 6x + 7$
29. $f(x) = -2x^2 - 4x - 5$
30. The sum of the base and the height of a triangle is 16 in. Find the dimensions for which the area is a maximum. [3.3]

Collaborative Discussion and Writing

31. Is the sum of two imaginary numbers always an imaginary number? Explain your answer. [3.1]
32. The graph of a quadratic function can have 0, 1, or 2 x-intercepts. How can you predict the number of x-intercepts without drawing the graph or (completely) solving an equation? [3.2]
33. Discuss two ways in which we used completing the square in this chapter. [3.2], [3.3]
34. Suppose that the graph of $f(x) = ax^2 + bx + c$ has x-intercepts $(x_1, 0)$ and $(x_2, 0)$. What are the x-intercepts of $g(x) = -ax^2 - bx - c$? Explain. [3.3]

3.4 Solving Rational Equations and Radical Equations

- Solve rational equations.
- Solve radical equations.

Rational Equations

Equations containing rational expressions are called **rational equations**. Solving such equations involves multiplying on both sides by the least common denominator (LCD) of all the rational expressions to *clear the equation of fractions.*

EXAMPLE 1 Solve: $\frac{x-8}{3}+\frac{x-3}{2}=0$.

Algebraic Solution

We have

$$\frac{x-8}{3}+\frac{x-3}{2}=0$$

The LCD is $3 \cdot 2$, or 6.

$$6\left(\frac{x-8}{3}+\frac{x-3}{2}\right)=6 \cdot 0$$

Multiplying by the LCD on both sides to clear fractions

$$6\left(\frac{x-8}{3}\right)+6\left(\frac{x-3}{2}\right)=0$$

$$2(x-8)+3(x-3)=0$$

$$2x-16+3x-9=0$$

$$5x-25=0$$

$$5x=25$$

$$x=5.$$

The possible solution is 5.

Check:

$$\frac{x-8}{3}+\frac{x-3}{2}=0$$

$$\frac{5-8}{3}+\frac{5-3}{2} \stackrel{?}{=} 0$$

$$\frac{-3}{3}+\frac{2}{2}$$

$$-1+1$$

$$0 \quad | \quad 0 \quad \text{TRUE}$$

The solution is 5.

Visualizing the Solution

The solution of the given equation is the zero of the function

$$f(x)=\frac{x-8}{3}+\frac{x-3}{2}.$$

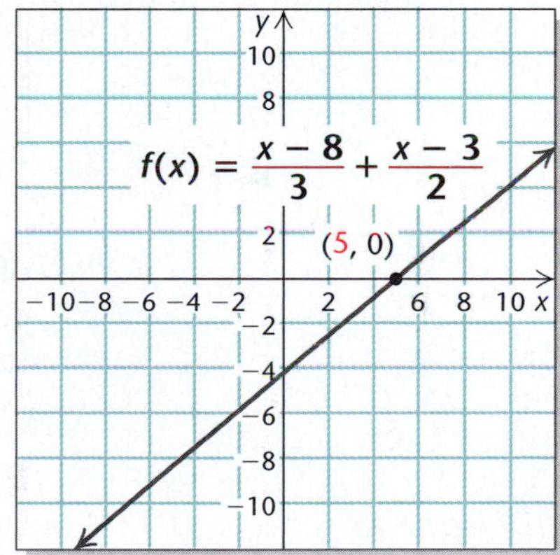

The zero of the function is 5. Thus the solution of the equation is 5.

Now Try Exercise 3.

Technology Connection

$y = \frac{x-8}{3} + \frac{x-3}{2}$

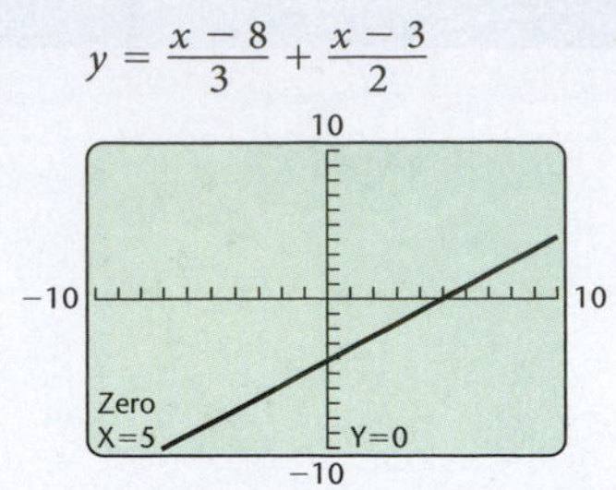

We can use the Zero method to solve the equation in Example 1. We find the zero of the function

$$f(x) = \frac{x-8}{3} + \frac{x-3}{2}.$$

The zero of the function is 5. Thus the solution of the equation is 5.

CAUTION! Clearing fractions is a valid procedure when solving rational equations but not when adding, subtracting, multiplying, or dividing rational expressions. A rational expression may have operation signs but it will have no equals sign. A rational equation *always* has an equals sign. For example, $\frac{x-8}{3} + \frac{x-3}{2}$ is a rational expression but $\frac{x-8}{3} + \frac{x-3}{2} = 0$ is a rational equation.

To *simplify* the rational *expression* $\frac{x-8}{3} + \frac{x-3}{2}$, we first find the LCD and write each fraction with that denominator. The final result is usually a rational expression.

To *solve* the rational *equation* $\frac{x-8}{3} + \frac{x-3}{2} = 0$, we first multiply by the LCD on both sides to clear fractions. The final result is one or more numbers. As we will see in Example 2, these numbers must be checked in the original equation.

When we use the multiplication principle to multiply (or divide) on both sides of an equation by an expression with a variable, we might not obtain an equivalent equation. We must check the possible solutions obtained in this manner by substituting them in the original equation. The next example illustrates this.

EXAMPLE 2 Solve: $\frac{x^2}{x-3} = \frac{9}{x-3}$.

Solution The LCD is $x - 3$.

$$(x-3) \cdot \frac{x^2}{x-3} = (x-3) \cdot \frac{9}{x-3}$$

$$x^2 = 9$$

$$x = -3 \quad or \quad x = 3 \qquad \text{Using the principle of square roots}$$

The possible solutions are -3 and 3. We check.

Check: For -3:

$$\begin{array}{c|c} \frac{x^2}{x-3} & \frac{9}{x-3} \\ \hline \frac{(-3)^2}{-3-3} \ ? & \frac{9}{-3-3} \\ \frac{9}{-6} & \frac{9}{-6} \end{array} \quad \text{TRUE}$$

For 3:

$$\begin{array}{c|c} \frac{x^2}{x-3} & \frac{9}{x-3} \\ \hline \frac{(3)^2}{3-3} \ ? & \frac{9}{3-3} \\ \frac{9}{0} & \frac{9}{0} \end{array} \quad \text{NOT DEFINED}$$

The number -3 checks, so it is a solution. Since division by 0 is not defined, 3 is not a solution. Note that 3 is not in the domain of either $x^2/(x-3)$ or $9/(x-3)$.

Now Try Exercise 9.

Technology Connection

We can use a table on a graphing calculator to check the possible solutions in Example 2. We enter

$$y_1 = \frac{x^2}{x-3}$$

and

$$y_2 = \frac{9}{x-3}.$$

$y_1 = \frac{x^2}{x-3}$, $y_2 = \frac{9}{x-3}$

X	Y1	Y2
−3	−1.5	−1.5
3	ERROR	ERROR

X =

When $x = -3$, we see that $y_1 = -1.5 = y_2$, so -3 is a solution. When $x = 3$, we get ERROR messages. This indicates that 3 is not in the domain of y_1 or y_2 and thus is not a solution.

EXAMPLE 3 Solve: $\frac{2}{3x+6} + \frac{1}{x^2-4} = \frac{4}{x-2}$.

Solution We first factor the denominators in order to determine the LCD:

$$\frac{2}{3(x+2)} + \frac{1}{(x+2)(x-2)} = \frac{4}{x-2} \qquad \text{The LCD is } 3(x+2)(x-2).$$

$$3(x+2)(x-2)\left(\frac{2}{3(x+2)} + \frac{1}{(x+2)(x-2)}\right) = 3(x+2)(x-2)\cdot\frac{4}{x-2}$$

Multiplying by the LCD to clear fractions

$$\begin{aligned} 2(x-2) + 3 &= 3\cdot 4(x+2) \\ 2x - 4 + 3 &= 12x + 24 \\ 2x - 1 &= 12x + 24 \\ -10x &= 25 \\ x &= -\frac{5}{2}. \end{aligned}$$

The possible solution is $-\frac{5}{2}$. This number checks. It is the solution.

Now Try Exercise 21.

Technology Connection

We can use a table on a graphing calculator to check the possible solution in Example 3. We enter

$$y_1 = \frac{2}{3x+6} + \frac{1}{x^2-4}$$

and

$$y_2 = \frac{4}{x-2}.$$

$y_1 = \frac{2}{3x+6} + \frac{1}{x^2-4}$, $y_2 = \frac{4}{x-2}$

X	Y1	Y2
−2.5	−.8889	−.8889

X =

We see that $y_1 = y_2$ when $x = -\frac{5}{2}$, or -2.5, so $-\frac{5}{2}$ is the solution.

Radical Equations

A **radical equation** is an equation in which variables appear in one or more radicands. For example,

$$\sqrt{2x-5} - \sqrt{x-3} = 1$$

is a radical equation. The following principle is used to solve such equations.

THE PRINCIPLE OF POWERS

For any positive integer n:

If $a = b$ is true, then $a^n = b^n$ is true.

EXAMPLE 4 Solve: $\sqrt{3x+1} = 4$.

Solution We have

$$\sqrt{3x+1} = 4$$
$$(\sqrt{3x+1})^2 = 4^2 \quad \text{Using the principle of powers; squaring both sides}$$
$$3x + 1 = 16$$
$$3x = 15$$
$$x = 5.$$

Check:

$\sqrt{3x+1} = 4$	
$\sqrt{3 \cdot 5 + 1}$?	4
$\sqrt{15+1}$	
$\sqrt{16}$	
4	4 TRUE

The solution is 5.

Now Try Exercise 31.

In Example 4, the radical was isolated on one side of the equation. If this had not been the case, our first step would have been to isolate the radical. We do so in the next example.

Technology Connection

We can use the Intersect method to solve the equation in Example 4. We graph $y_1 = \sqrt{3x+1}$ and $y_2 = 4$ and then use the INTERSECT feature. We see that the solution is 5.

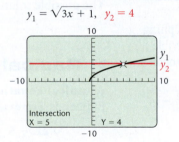

$y_1 = \sqrt{3x+1},\ y_2 = 4$

EXAMPLE 5 Solve: $5 + \sqrt{x + 7} = x$.

Algebraic Solution

We first isolate the radical and then use the principle of powers.

$$5 + \sqrt{x + 7} = x$$

$$\sqrt{x + 7} = x - 5 \quad \text{Subtracting 5 on both sides}$$

$$(\sqrt{x + 7})^2 = (x - 5)^2 \quad \text{Using the principle of powers; squaring both sides}$$

$$x + 7 = x^2 - 10x + 25$$

$$0 = x^2 - 11x + 18 \quad \text{Subtracting } x \text{ and 7}$$

$$0 = (x - 9)(x - 2) \quad \text{Factoring}$$

$$x - 9 = 0 \quad or \quad x - 2 = 0$$

$$x = 9 \quad or \quad x = 2$$

The possible solutions are 9 and 2.

Check: For 9:

$$\begin{array}{c|c} 5 + \sqrt{x + 7} & x \\ \hline 5 + \sqrt{9 + 7} \; ? & 9 \\ 5 + \sqrt{16} & \\ 5 + 4 & \\ 9 & 9 \quad \text{TRUE} \end{array}$$

For 2:

$$\begin{array}{c|c} 5 + \sqrt{x + 7} & x \\ \hline 5 + \sqrt{2 + 7} \; ? & 2 \\ 5 + \sqrt{9} & \\ 5 + 3 & \\ 8 & 2 \quad \text{FALSE} \end{array}$$

Since 9 checks but 2 does not, the only solution is 9.

Visualizing the Solution

When we graph $y = 5 + \sqrt{x + 7}$ and $y = x$, we find that the first coordinate of the point of intersection of the graphs is 9. Thus the solution of $5 + \sqrt{x + 7} = x$ is 9.

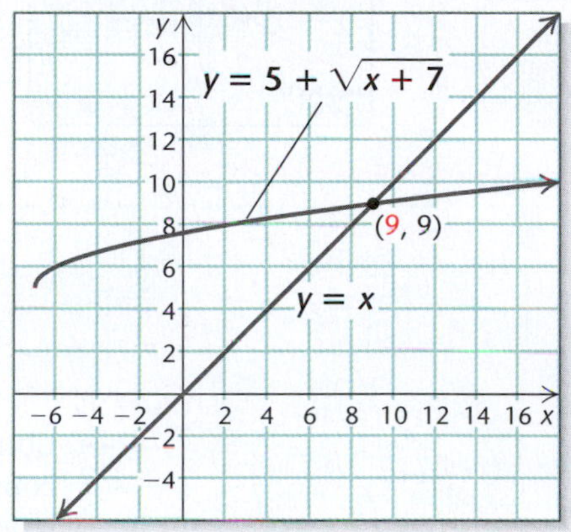

Note that the graphs show that the equation has only one solution.

Now Try Exercise 55.

Technology Connection

We can use a graphing calculator to solve the equation in Example 5. We can graph $y_1 = 5 + \sqrt{x + 7}$ and $y_2 = x$. Using the INTERSECT feature, we see in the window below that the solution is 9.

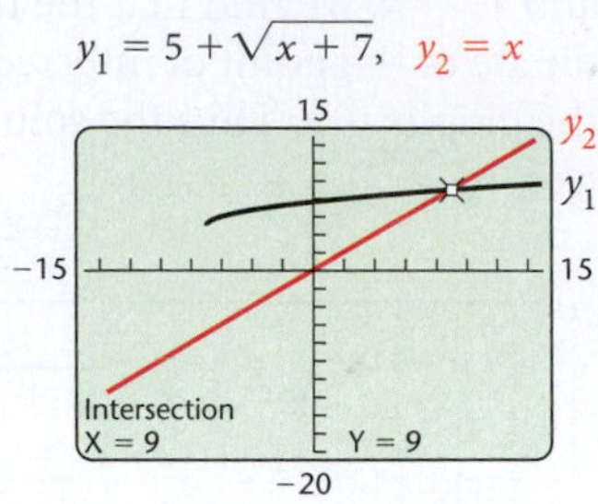

We can also use the ZERO feature to get this result, as shown in the window below. To do so, we first write the equivalent equation $5 + \sqrt{x + 7} - x = 0$. The zero of the function $f(x) = 5 + \sqrt{x + 7} - x$ is 9, so the solution of the original equation is 9.

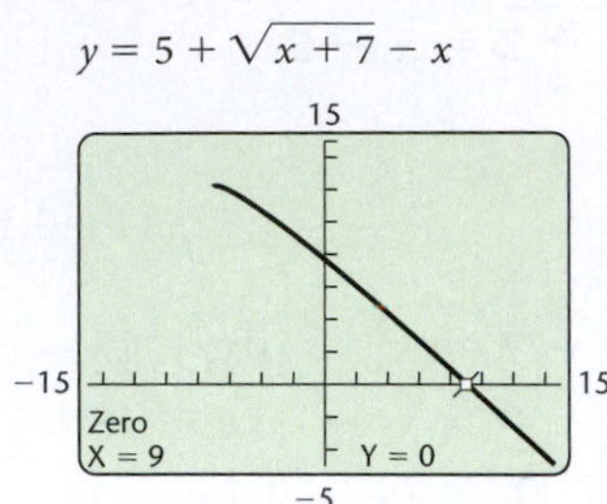

When we raise both sides of an equation to an even power, the resulting equation can have solutions that the original equation does not. This is because the converse of the principle of powers is not necessarily true. That is, if $a^n = b^n$ is true, we do not know that $a = b$ is true. For example, $(-2)^2 = 2^2$, but $-2 \neq 2$. Thus, as we saw in Example 5, it is necessary to check the possible solutions in the original equation when the principle of powers is used to raise both sides of an equation to an even power.

When a radical equation has two radical terms on one side, we isolate one of them and then use the principle of powers. If, after doing so, a radical term remains, we repeat these steps.

EXAMPLE 6 Solve: $\sqrt{x - 3} + \sqrt{x + 5} = 4$.

Solution We have

$$\sqrt{x - 3} = 4 - \sqrt{x + 5}$$ Isolating one radical

$$\left(\sqrt{x - 3}\right)^2 = \left(4 - \sqrt{x + 5}\right)^2$$ Using the principle of powers; squaring both sides

$$x - 3 = 16 - 8\sqrt{x + 5} + (x + 5)$$

$$x - 3 = 21 - 8\sqrt{x + 5} + x$$ Combining like terms

$$-24 = -8\sqrt{x + 5}$$ Isolating the remaining radical; subtracting x and 21 on both sides

$$3 = \sqrt{x + 5}$$ Dividing by −8 on both sides

$$3^2 = \left(\sqrt{x + 5}\right)^2$$ Using the principle of powers; squaring both sides

$$9 = x + 5$$

$$4 = x.$$ Subtracting 5 on both sides

The number 4 checks. It is the solution.

Now Try Exercise 65.

Technology Connection

We check the possible solution in Example 6 on a graphing calculator.

$y_1 = \sqrt{x - 3} + \sqrt{x + 5}$, $y_2 = 4$

X	Y1	Y2
4	4	4

X =

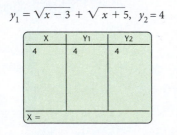

Since $y_1 = y_2$ when $x = 4$, the number 4 checks. It is the solution.

3.4 Exercise Set

Solve.

1. $\frac{1}{4} + \frac{1}{5} = \frac{1}{t}$

2. $\frac{1}{3} - \frac{5}{6} = \frac{1}{x}$

3. $\frac{x+2}{4} - \frac{x-1}{5} = 15$

4. $\frac{t+1}{3} - \frac{t-1}{2} = 1$

5. $\frac{1}{2} + \frac{2}{x} = \frac{1}{3} + \frac{3}{x}$

6. $\frac{1}{t} + \frac{1}{2t} + \frac{1}{3t} = 5$

7. $\frac{5}{3x+2} = \frac{3}{2x}$

8. $\frac{2}{x-1} = \frac{3}{x+2}$

9. $\frac{y^2}{y+4} = \frac{16}{y+4}$

10. $\frac{49}{w-7} = \frac{w^2}{w-7}$

11. $x + \frac{6}{x} = 5$

12. $x - \frac{12}{x} = 1$

13. $\frac{6}{y+3} + \frac{2}{y} = \frac{5y-3}{y^2-9}$

14. $\frac{3}{m+2} + \frac{2}{m} = \frac{4m-4}{m^2-4}$

15. $\frac{2x}{x-1} = \frac{5}{x-3}$

16. $\frac{2x}{x+7} = \frac{5}{x+1}$

17. $\frac{2}{x+5} + \frac{1}{x-5} = \frac{16}{x^2-25}$

18. $\frac{2}{x^2-9} + \frac{5}{x-3} = \frac{3}{x+3}$

19. $\frac{3x}{x+2} + \frac{6}{x} = \frac{12}{x^2+2x}$

20. $\frac{3y+5}{y^2+5y} + \frac{y+4}{y+5} = \frac{y+1}{y}$

21. $\frac{1}{5x+20} - \frac{1}{x^2-16} = \frac{3}{x-4}$

22. $\frac{1}{4x+12} - \frac{1}{x^2-9} = \frac{5}{x-3}$

23. $\frac{2}{5x+5} - \frac{3}{x^2-1} = \frac{4}{x-1}$

24. $\frac{1}{3x+6} - \frac{1}{x^2-4} = \frac{3}{x-2}$

25. $\frac{8}{x^2-2x+4} = \frac{x}{x+2} + \frac{24}{x^3+8}$

26. $\frac{18}{x^2-3x+9} - \frac{x}{x+3} = \frac{81}{x^3+27}$

27. $\frac{x}{x-4} - \frac{4}{x+4} = \frac{32}{x^2-16}$

28. $\frac{x}{x-1} - \frac{1}{x+1} = \frac{2}{x^2-1}$

29. $\frac{1}{x-6} - \frac{1}{x} = \frac{6}{x^2-6x}$

30. $\frac{1}{x-15} - \frac{1}{x} = \frac{15}{x^2-15x}$

31. $\sqrt{3x-4} = 1$

32. $\sqrt{4x+1} = 3$

33. $\sqrt{2x-5} = 2$

34. $\sqrt{3x+2} = 6$

35. $\sqrt{7-x} = 2$

36. $\sqrt{5-x} = 1$

37. $\sqrt{1-2x} = 3$

38. $\sqrt{2-7x} = 2$

39. $\sqrt[3]{5x-2} = -3$

40. $\sqrt[3]{2x+1} = -5$

41. $\sqrt[4]{x^2-1} = 1$

42. $\sqrt[5]{3x+4} = 2$

43. $\sqrt{y-1} + 4 = 0$

44. $\sqrt{m+1} - 5 = 8$

45. $\sqrt{b+3} - 2 = 1$

46. $\sqrt{x-4} + 1 = 5$

47. $\sqrt{z+2} + 3 = 4$

48. $\sqrt{y-5} - 2 = 3$

49. $\sqrt{2x+1} - 3 = 3$

50. $\sqrt{3x-1} + 2 = 7$

51. $\sqrt{2-x} - 4 = 6$

52. $\sqrt{5-x} + 2 = 8$

53. $\sqrt[3]{6x+9} + 8 = 5$

54. $\sqrt[5]{2x-3} - 1 = 1$

55. $\sqrt{x+4} + 2 = x$

56. $\sqrt{x+1} + 1 = x$

57. $\sqrt{x-3} + 5 = x$

58. $\sqrt{x+3} - 1 = x$

59. $\sqrt{x+7} = x+1$

60. $\sqrt{6x+7} = x+2$

61. $\sqrt{3x+3} = x+1$

62. $\sqrt{2x+5} = x-5$

63. $\sqrt{5x+1} = x-1$

64. $\sqrt{7x+4} = x+2$

65. $\sqrt{x-3} + \sqrt{x+2} = 5$

66. $\sqrt{x} - \sqrt{x-5} = 1$

67. $\sqrt{3x-5} + \sqrt{2x+3} + 1 = 0$

68. $\sqrt{2m-3} = \sqrt{m+7} - 2$

69. $\sqrt{x} - \sqrt{3x - 3} = 1$

70. $\sqrt{2x + 1} - \sqrt{x} = 1$

71. $\sqrt{2y - 5} - \sqrt{y - 3} = 1$

72. $\sqrt{4p + 5} + \sqrt{p + 5} = 3$

73. $\sqrt{y + 4} - \sqrt{y - 1} = 1$

74. $\sqrt{y + 7} + \sqrt{y + 16} = 9$

75. $\sqrt{x + 5} + \sqrt{x + 2} = 3$

76. $\sqrt{6x + 6} = 5 + \sqrt{21 - 4x}$

77. $x^{1/3} = -2$

78. $t^{1/5} = 2$

79. $t^{1/4} = 3$

80. $m^{1/2} = -7$

Solve.

81. $\dfrac{P_1V_1}{T_1} = \dfrac{P_2V_2}{T_2}$, for T_1
(A chemistry formula for gases)

82. $\dfrac{1}{F} = \dfrac{1}{m} + \dfrac{1}{p}$, for F
(A formula from optics)

83. $W = \sqrt{\dfrac{1}{LC}}$, for C
(An electricity formula)

84. $s = \sqrt{\dfrac{A}{6}}$, for A
(A geometry formula)

85. $\dfrac{1}{R} = \dfrac{1}{R_1} + \dfrac{1}{R_2}$, for R_2
(A formula for resistance)

86. $\dfrac{1}{t} = \dfrac{1}{a} + \dfrac{1}{b}$, for t
(A formula for work rate)

87. $I = \sqrt{\dfrac{A}{P}} - 1$, for P
(A compound-interest formula)

88. $T = 2\pi\sqrt{\dfrac{1}{g}}$, for g
(A pendulum formula)

89. $\dfrac{1}{F} = \dfrac{1}{m} + \dfrac{1}{p}$, for p
(A formula from optics)

90. $\dfrac{V^2}{R^2} = \dfrac{2g}{R + h}$, for h
(A formula for escape velocity)

Skill Maintenance

Find the zero of the function. [1.5]

91. $f(x) = 15 - 2x$

92. $f(x) = -3x + 9$

Solve. [1.5]

93. *Pork Production.* Together, China and the United States, the top two pork producers worldwide, produced 64,308,000 metric tons of pork in 2013. China produced 1,260,000 metric tons more than five times the number of metric tons produced by the United States. (*Source:* United Nations Food and Agriculture Organization) How many metric tons of pork did each country produce in 2013?

94. *Sports Injuries.* In 2012 in the United States, there were 172,470 injuries among soccer players ages 19 and under. This was about 44% more than the number of injuries among baseball players ages 19 and under. (*Source:* Safe Kids Worldwide, based on hospital ER reports, 2012) How many baseball players ages 19 and under were injured in 2012?

Synthesis

Solve.

95. $(x - 3)^{2/3} = 2$

96. $\dfrac{x + 3}{x + 2} - \dfrac{x + 4}{x + 3} = \dfrac{x + 5}{x + 4} - \dfrac{x + 6}{x + 5}$

97. $\sqrt{x + 5} + 1 = \dfrac{6}{\sqrt{x + 5}}$

98. $\sqrt{15 + \sqrt{2x + 80}} = 5$

99. $x^{2/3} = x$

3.5 Solving Equations and Inequalities with Absolute Value

- Solve equations with absolute value.
- Solve inequalities with absolute value.

Equations with Absolute Value

Recall that the absolute value of a number is its distance from 0 on the number line. We use this concept to solve equations with absolute value.

For $a > 0$ and an algebraic expression X:

$$|X| = a \quad \text{is equivalent to} \quad X = -a \quad \text{or} \quad X = a.$$

EXAMPLE 1 Solve: $|x| = 5$.

Algebraic Solution

We have

$$|x| = 5$$

$x = -5$ *or* $x = 5$. **Writing an equivalent statement**

The solutions are -5 and 5.

To check, note that -5 and 5 are both 5 units from 0 on the number line.

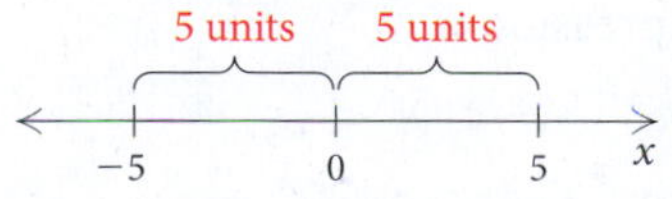

Visualizing the Solution

The first coordinates of the points of intersection of the graphs of $y = |x|$ and $y = 5$ are -5 and 5. These are the solutions of the equation $|x| = 5$.

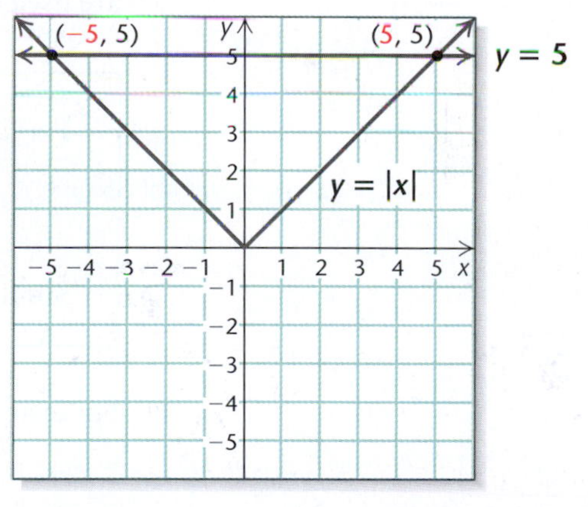

Now Try Exercise 1.

EXAMPLE 2 Solve: $|x - 3| - 1 = 4$.

Solution First, we add 1 on both sides to get an expression of the form $|X| = a$:

$$|x - 3| - 1 = 4$$

$$|x - 3| = 5$$

$x - 3 = -5$ *or* $x - 3 = 5$ **$|X| = a$ is equivalent to $X = -a$ or $X = a$.**

$x = -2$ *or* $x = 8$. **Adding 3**

Check: For -2:

$$\begin{array}{c|l} |x-3|-1 = 4 & \\ \hline |-2-3|-1 & ?\ 4 \\ |-5|-1 & \\ 5-1 & \\ 4 & 4 \quad \text{TRUE} \end{array}$$

For 8:

$$\begin{array}{c|l} |x-3|-1 = 4 & \\ \hline |8-3|-1 & ?\ 4 \\ |5|-1 & \\ 5-1 & \\ 4 & 4 \quad \text{TRUE} \end{array}$$

The solutions are -2 and 8.

Now Try Exercise 21.

When $a = 0$, $|X| = a$ is equivalent to $X = 0$. Note that for $a < 0, |X| = a$ has no solution, because the absolute value of an expression is never negative. We can use a graph to illustrate the last statement for a specific value of a. For example, if we let $a = -3$ and graph $y = |x|$ and $y = -3$, we see that the graphs do not intersect, as shown below. Thus the equation $|x| = -3$ has no solution. The solution set is the **empty set**, denoted $\varnothing$.

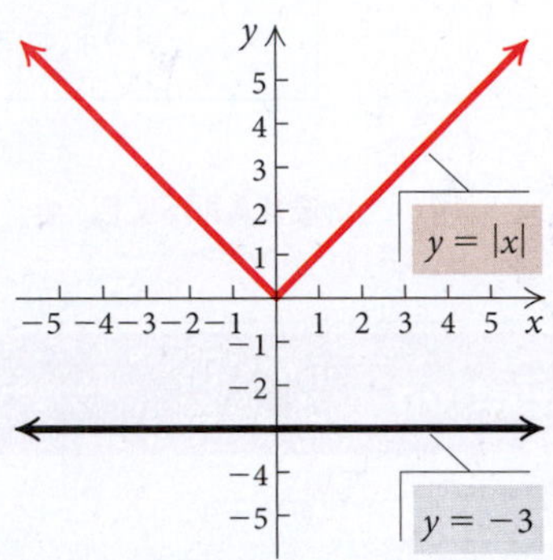

▶ Inequalities with Absolute Value

Inequalities sometimes contain absolute-value notation. The following properties are used to solve them.

> For $a > 0$ and an algebraic expression X:
>
> $|X| < a$ is equivalent to $-a < X < a$.
>
> $|X| > a$ is equivalent to $X < -a$ *or* $X > a$.
>
> Similar statements hold for $|X| \le a$ and $|X| \ge a$.

For example,

$|x| < 3$ is equivalent to $-3 < x < 3$;

$|y| \ge 1$ is equivalent to $y \le -1$ *or* $y \ge 1$; and

$|2x + 3| \le 4$ is equivalent to $-4 \le 2x + 3 \le 4$.

EXAMPLE 3 Solve and graph the solution set: $|3x + 2| < 5$.

Solution We have

$$|3x + 2| < 5$$

$-5 < 3x + 2 < 5$ Writing an equivalent inequality

$-7 < 3x < 3$ Subtracting 2

$-\frac{7}{3} < x < 1.$ Dividing by 3

The solution set is $\left\{x \mid -\frac{7}{3} < x < 1\right\}$, or $\left(-\frac{7}{3}, 1\right)$. The graph of the solution set is shown below.

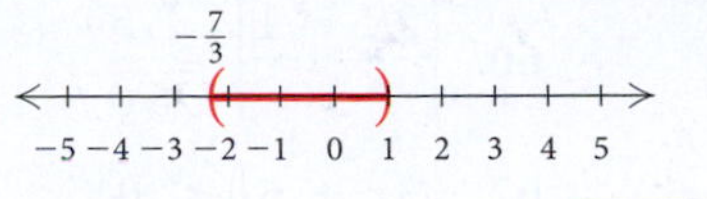

Now Try Exercise 45.

EXAMPLE 4 Solve and graph the solution set: $|5 - 2x| \geq 1$.

Solution We have

$$|5 - 2x| \geq 1$$

$$5 - 2x \leq -1 \quad or \quad 5 - 2x \geq 1 \qquad \text{Writing an equivalent inequality}$$

$$-2x \leq -6 \quad or \quad -2x \geq -4 \qquad \text{Subtracting 5}$$

$$x \geq 3 \quad or \quad x \leq 2. \qquad \text{Dividing by } -2 \text{ and reversing the inequality signs}$$

The solution set is $\{x \mid x \leq 2 \text{ or } x \geq 3\}$, or $(-\infty, 2] \cup [3, \infty)$. The graph of the solution set is shown below.

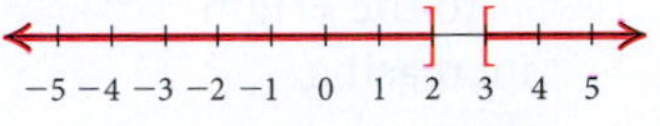

Now Try Exercise 47.

3.5 Exercise Set

Solve.

1. $|x| = 7$
2. $|x| = 4.5$
3. $|x| = 0$
4. $|x| = \frac{3}{2}$
5. $|x| = \frac{5}{6}$
6. $|x| = -\frac{3}{5}$
7. $|x| = -10.7$
8. $|x| = 12$
9. $|3x| = 1$
10. $|5x| = 4$
11. $|8x| = 24$
12. $|6x| = 0$
13. $|x - 1| = 4$
14. $|x - 7| = 5$
15. $|x + 2| = 6$
16. $|x + 5| = 1$
17. $|3x + 2| = 1$
18. $|7x - 4| = 8$
19. $\left|\frac{1}{2}x - 5\right| = 17$
20. $\left|\frac{1}{3}x - 4\right| = 13$
21. $|x - 1| + 3 = 6$
22. $|x + 2| - 5 = 9$
23. $|x + 3| - 2 = 8$
24. $|x - 4| + 3 = 9$
25. $|3x + 1| - 4 = -1$
26. $|2x - 1| - 5 = -3$
27. $|4x - 3| + 1 = 7$
28. $|5x + 4| + 2 = 5$
29. $12 - |x + 6| = 5$
30. $9 - |x - 2| = 7$
31. $7 - |2x - 1| = 6$
32. $5 - |4x + 3| = 2$

Solve and write interval notation for the solution set. Then graph the solution set.

33. $|x| < 7$
34. $|x| \leq 4.5$
35. $|x| \leq 2$
36. $|x| < 3$
37. $|x| \geq 4.5$
38. $|x| > 7$
39. $|x| > 3$
40. $|x| \geq 2$
41. $|3x| < 1$
42. $|5x| \leq 4$
43. $|2x| \geq 6$
44. $|4x| > 20$
45. $|x + 8| < 9$
46. $|x + 6| \leq 10$
47. $|x + 8| \geq 9$
48. $|x + 6| > 10$
49. $\left|x - \frac{1}{4}\right| < \frac{1}{2}$
50. $|x - 0.5| \leq 0.2$
51. $|2x + 3| \leq 9$
52. $|3x + 4| < 13$
53. $|x - 5| > 0.1$
54. $|x - 7| \geq 0.4$

55. $|6 - 4x| \geq 8$

56. $|5 - 2x| > 10$

57. $\left|x + \frac{2}{3}\right| \leq \frac{5}{3}$

58. $\left|x + \frac{3}{4}\right| < \frac{1}{4}$

59. $\left|\frac{2x + 1}{3}\right| > 5$

60. $\left|\frac{2x - 1}{3}\right| \geq \frac{5}{6}$

61. $|2x - 4| < -5$

62. $|3x + 5| < 0$

63. $|7 - x| \geq -4$

64. $|2x + 1| > -\frac{1}{2}$

Skill Maintenance

Vocabulary Reinforcement

In each of Exercises 65–72, fill in the blank with the correct term. Some of the given choices will not be used.

distance formula
midpoint formula
function
relation
x-intercept
y-intercept
perpendicular
parallel
horizontal lines
vertical lines
symmetric with respect to the x-axis
symmetric with respect to the y-axis
symmetric with respect to the origin
increasing
decreasing
constant

65. A(n) ______________ is a point $(0, b)$. [1.1]

66. The ____________ is $d = \sqrt{(x_2 - x_1)^2 + (y_2 - y_1)^2}$. [1.1]

67. A(n) ______________ is a correspondence such that each member of the domain corresponds to at least one member of the range. [1.2]

68. A(n) ______________ is a correspondence such that each member of the domain corresponds to exactly one member of the range. [1.2]

69. ____________ are given by equations of the type $y = b$, or $f(x) = b$. [1.3]

70. Nonvertical lines are ____________ if and only if they have the same slope and different y-intercepts. [1.4]

71. A function f is said to be ______________ on an open interval I if, for all a and b in that interval, $a < b$ implies $f(a) > f(b)$. [2.1]

72. For an equation $y = f(x)$, if replacing x with $-x$ produces an equivalent equation, then the graph is ______________. [2.4]

Synthesis

Solve.

73. $|3x - 1| > 5x - 2$

74. $|x + 2| \leq |x - 5|$

75. $|p - 4| + |p + 4| < 8$

76. $|x| + |x + 1| < 10$

77. $|x - 3| + |2x + 5| > 6$

Chapter 3 Summary and Review

STUDY GUIDE

KEY TERMS AND CONCEPTS	EXAMPLE
SECTION 3.1: THE COMPLEX NUMBERS	
The number i is defined such that $i = \sqrt{-1}$ and $i^2 = -1$.	Express each number in terms of i. $\sqrt{-5} = \sqrt{-1 \cdot 5} = \sqrt{-1} \cdot \sqrt{5} = i\sqrt{5}$, or $\sqrt{5}i$; $-\sqrt{-36} = -\sqrt{-1 \cdot 36} = -\sqrt{-1} \cdot \sqrt{36} = -i \cdot 6 = -6i$

A **complex number** is a number of the form $a + bi$, where a and b are real numbers. The number a is said to be the **real part** of $a + bi$, and the number b is said to be the **imaginary part** of $a + bi$. To **add** or **subtract complex numbers**, we add or subtract the real parts, and we add or subtract the imaginary parts.	Add or subtract. $(-3 + 4i) + (5 - 8i) = (-3 + 5) + (4i - 8i) = 2 - 4i;$ $(6 - 7i) - (10 + 3i) = (6 - 10) + (-7i - 3i) = -4 - 10i$
When we **multiply complex numbers**, we must keep in mind the fact that $i^2 = -1$. Note that $\sqrt{a} \cdot \sqrt{b} \neq \sqrt{ab}$ when $\sqrt{a}$ and $\sqrt{b}$ are not real numbers.	Multiply. $\sqrt{-4} \cdot \sqrt{-100} = \sqrt{-1} \cdot \sqrt{4} \cdot \sqrt{-1} \cdot \sqrt{100}$ $= i \cdot 2 \cdot i \cdot 10$ $= i^2 \cdot 20$ $= -1 \cdot 20$ $i^2 = -1$ $= -20;$ $(2 - 5i)(3 + i) = 6 + 2i - 15i - 5i^2$ $= 6 - 13i - 5(-1)$ $= 6 - 13i + 5$ $= 11 - 13i$
The **conjugate of a complex number** $a + bi$ is $a - bi$. The numbers $a + bi$ and $a - bi$ are **complex conjugates**. Conjugates are used when we **divide complex numbers**.	Divide. $\frac{5 - 2i}{3 + i} = \frac{5 - 2i}{3 + i} \cdot \frac{3 - i}{3 - i}$ **$3 - i$ is the conjugate of the divisor, $3 + i$.** $= \frac{15 - 5i - 6i + 2i^2}{9 - i^2}$ $= \frac{15 - 11i - 2}{9 + 1}$ $i^2 = -1$ $= \frac{13 - 11i}{10}$ $= \frac{13}{10} - \frac{11}{10}i$

SECTION 3.2: QUADRATIC EQUATIONS, FUNCTIONS, ZEROS, AND MODELS

A **quadratic equation** is an equation that can be written in the form $ax^2 + bx + c = 0, \quad a \neq 0,$ where a, b, and c are real numbers. A **quadratic function** f is a function that can be written in the form $f(x) = ax^2 + bx + c, \quad a \neq 0,$ where a, b, and c are real numbers. The **zeros** of a quadratic function $f(x) = ax^2 + bx + c$ are the *solutions* of the associated quadratic equation $ax^2 + bx + c = 0$.	$3x^2 - 2x + 4 = 0$ and $5 - 4x = x^2$ are examples of quadratic equations. The equation $3x^2 - 2x + 4 = 0$ is written in **standard form.** The functions $f(x) = 2x^2 + x + 1$ and $f(x) = 5x^2 - 4$ are examples of quadratic functions.

The Principle of Zero Products If $ab = 0$ is true, then $a = 0$ *or* $b = 0$, and if $a = 0$ *or* $b = 0$, then $ab = 0$.	Solve: $3x^2 - 4 = 11x$. $3x^2 - 4 = 11x$ $3x^2 - 11x - 4 = 0$ **Subtracting $11x$ on both sides to get 0 on one side of the equation** $(3x + 1)(x - 4) = 0$ **Factoring** $3x + 1 = 0$ *or* $x - 4 = 0$ **Using the principle of zero products** $3x = -1$ *or* $x = 4$ $x = -\frac{1}{3}$ *or* $x = 4$ The solutions are $-\frac{1}{3}$ and 4.
The Principle of Square Roots If $x^2 = k$, then $x = \sqrt{k}$ *or* $x = -\sqrt{k}$.	Solve: $3x^2 - 18 = 0$. $3x^2 - 18 = 0$ $3x^2 = 18$ **Adding 18 on both sides** $x^2 = 6$ **Dividing by 3 on both sides** $x = \sqrt{6}$ *or* $x = -\sqrt{6}$ **Using the principle of square roots** The solutions are $\sqrt{6}$ and $-\sqrt{6}$, or $\pm\sqrt{6}$.
To solve a quadratic equation by **completing the square**: **1.** Isolate the terms with variables on one side of the equation and arrange them in descending order. **2.** Divide by the coefficient of the squared term if that coefficient is not 1. **3.** Complete the square by taking half the coefficient of the first-degree term and adding its square on both sides of the equation. **4.** Express one side of the equation as the square of a binomial. **5.** Use the principle of square roots. **6.** Solve for the variable.	Solve: $2x^2 - 3 = 6x$. $2x^2 - 3 = 6x$ $2x^2 - 6x - 3 = 0$ **Subtracting $6x$** $2x^2 - 6x = 3$ **Adding 3** $x^2 - 3x = \frac{3}{2}$ **Dividing by 2 to make the x^2-coefficient 1** $x^2 - 3x + \frac{9}{4} = \frac{3}{2} + \frac{9}{4}$ **Completing the square: $\frac{1}{2}(-3) = -\frac{3}{2}$ and $(-\frac{3}{2})^2 = \frac{9}{4}$; adding $\frac{9}{4}$** $\left(x - \frac{3}{2}\right)^2 = \frac{15}{4}$ **Factoring and simplifying** $x - \frac{3}{2} = \pm\frac{\sqrt{15}}{2}$ **Using the principle of square roots and the quotient rule for radicals** $x = \frac{3}{2} \pm \frac{\sqrt{15}}{2} = \frac{3 \pm \sqrt{15}}{2}$ The solutions are $\frac{3 + \sqrt{15}}{2}$ and $\frac{3 - \sqrt{15}}{2}$, or $\frac{3 \pm \sqrt{15}}{2}$.

The solutions of $ax^2 + bx + c = 0$, $a \neq 0$, can be found using the **quadratic formula**: $$x = \frac{-b \pm \sqrt{b^2 - 4ac}}{2a}.$$	Solve: $x^2 - 6 = 3x$. $$x^2 - 6 = 3x$$ $$x^2 - 3x - 6 = 0 \quad \textbf{Standard form}$$ $$a = 1, b = -3, c = -6$$ $$x = \frac{-b \pm \sqrt{b^2 - 4ac}}{2a}$$ $$= \frac{-(-3) \pm \sqrt{(-3)^2 - 4(1)(-6)}}{2 \cdot 1}$$ $$= \frac{3 \pm \sqrt{9 + 24}}{2}$$ $$= \frac{3 \pm \sqrt{33}}{2} \quad \textbf{Exact solutions}$$ Using a calculator, we approximate the solutions to be 4.372 and -1.372.
Discriminant For $ax^2 + bx + c = 0$, where a, b, and c are real numbers: $b^2 - 4ac = 0 \rightarrow$ One real-number solution; $b^2 - 4ac > 0 \rightarrow$ Two different real-number solutions; $b^2 - 4ac < 0 \rightarrow$ Two different imaginary-number solutions, complex conjugates.	For the equation above, $x^2 - 6 = 3x$, we see that $b^2 - 4ac$ is 33. Since 33 is positive, there are two different real-number solutions. For $2x^2 - x + 4 = 0$, with $a = 2$, $b = -1$, and $c = 4$, the discriminant, $(-1)^2 - 4 \cdot 2 \cdot 4 = 1 - 32 = -31$, is negative, so there are two different imaginary-number (or nonreal) solutions. For $x^2 - 6x + 9 = 0$, with $a = 1$, $b = -6$, and $c = 9$, the discriminant, $(-6)^2 - 4 \cdot 1 \cdot 9 = 36 - 36 = 0$, is 0 so there is one real-number solution.
Equations **reducible to quadratic,** or **quadratic in form,** can be treated as quadratic equations if a suitable substitution is made.	Solve: $x^4 - x^2 - 12 = 0$. $$x^4 - x^2 - 12 = 0 \quad \textbf{Let } u = x^2\textbf{. Then } u^2 = (x^2)^2 = x^4.$$ $$u^2 - u - 12 = 0 \quad \textbf{Substituting}$$ $$(u - 4)(u + 3) = 0$$ $$u - 4 = 0 \quad or \quad u + 3 = 0$$ $$u = 4 \quad or \quad u = -3 \quad \textbf{Solving for } u$$ $$x^2 = 4 \quad or \quad x^2 = -3$$ $$x = \pm 2 \quad or \quad x = \pm\sqrt{3}i \quad \textbf{Solving for } x$$ The solutions are 2, -2, $\sqrt{3}i$, and $-\sqrt{3}i$.

SECTION 3.3: ANALYZING GRAPHS OF QUADRATIC FUNCTIONS

Graphing Quadratic Equations

The graph of the function $f(x) = a(x - h)^2 + k$ is a parabola that:

- opens up if $a > 0$ and down if $a < 0$;
- has (h, k) as the vertex;
- has $x = h$ as the axis of symmetry;
- has k as a minimum value (output) if $a > 0$;
- has k as a maximum value if $a < 0$.

We can use a modification of the technique of completing the square as an aid in analyzing and graphing quadratic functions.

Find the vertex, the axis of symmetry, and the maximum or minimum value of $f(x) = 2x^2 + 12x + 12$.

$f(x) = 2x^2 + 12x + 12$

$= 2(x^2 + 6x) + 12$ — **Note that 9 completes the square for $x^2 + 6x$.**

$= 2(x^2 + 6x + 9 - 9) + 12$ — **Adding $9 - 9$, or 0, inside the parentheses**

$= 2(x^2 + 6x + 9) - 2 \cdot 9 + 12$ — **Using the distributive law to remove -9 from within the parentheses**

$= 2(x + 3)^2 - 6$

$= 2[x - (-3)]^2 + (-6)$

The function is now written in the form $f(x) = a(x - h)^2 + k$ with $a = 2$, $h = -3$, and $k = -6$. Because $a > 0$, we know the graph opens up and thus the function has a minimum value. We also know the following:

Vertex (h, k): $(-3, -6)$;

Axis of symmetry $x = h$: $x = -3$;

Minimum value of the function k: -6.

To graph the function, we first plot the vertex and then find several points on either side of it. We plot these points and connect them with a smooth curve.

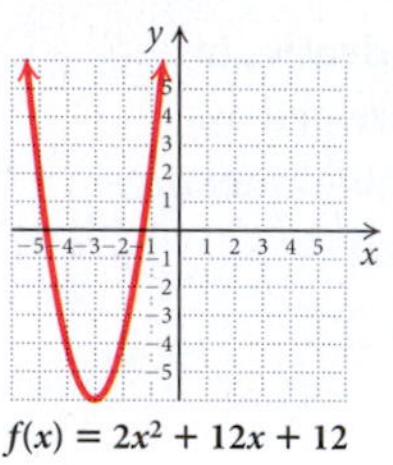

$f(x) = 2x^2 + 12x + 12$

The Vertex of a Parabola

The **vertex** of the graph of $f(x) = ax^2 + bx + c$ is

$$\left(-\frac{b}{2a}, f\left(-\frac{b}{2a}\right)\right).$$

We calculate the x-coordinate. We substitute to find the y-coordinate.

Find the vertex of the function $f(x) = -3x^2 + 6x + 1$.

$$-\frac{b}{2a} = -\frac{6}{2(-3)} = 1$$

$$f(1) = -3 \cdot 1^2 + 6 \cdot 1 + 1 = 4$$

The vertex is $(1, 4)$.

Some applied problems can be solved by finding the maximum or minimum value of a quadratic function.

See Examples 5–7 on pp. 194–197.

SECTION 3.4: SOLVING RATIONAL EQUATIONS AND RADICAL EQUATIONS

A **rational equation** is an equation containing one or more rational expressions. When we solve a rational equation, we usually first multiply by the least common denominator (LCD) of all the rational expressions to clear the fractions.

CAUTION! When we multiply by an expression containing a variable, we might not obtain an equation equivalent to the original equation, so we *must* check the possible solutions obtained by substituting them in the original equation.

Solve.

$$\frac{5}{x+2} - \frac{4}{x^2-4} = \frac{x-3}{x-2}$$

$$\frac{5}{x+2} - \frac{4}{(x+2)(x-2)} = \frac{x-3}{x-2} \quad \text{The LCD is } (x+2)(x-2).$$

$$(x+2)(x-2)\left(\frac{5}{x+2} - \frac{4}{(x+2)(x-2)}\right) = (x+2)(x-2)\cdot\frac{x-3}{x-2}$$

$$5(x-2) - 4 = (x+2)(x-3)$$

$$5x - 10 - 4 = x^2 - x - 6$$

$$5x - 14 = x^2 - x - 6$$

$$0 = x^2 - 6x + 8$$

$$0 = (x-2)(x-4)$$

$$x - 2 = 0 \quad or \quad x - 4 = 0$$

$$x = 2 \quad or \quad x = 4$$

The number 2 does not check, but 4 does. The solution is 4.

A **radical equation** is an equation that contains one or more radicals. We use the **principle of powers** to solve radical equations.

For any positive integer n:

If $a = b$ is true, then $a^n = b^n$ is true.

CAUTION! If $a^n = b^n$ is true, it is not necessarily true that $a = b$, so we *must* check the possible solutions obtained by substituting them in the original equation.

Solve: $\sqrt{x+2} + \sqrt{x-1} = 3$.

$$\sqrt{x+2} + \sqrt{x-1} = 3$$

$$\sqrt{x+2} = 3 - \sqrt{x-1} \quad \text{Isolating one radical}$$

$$\left(\sqrt{x+2}\right)^2 = \left(3 - \sqrt{x-1}\right)^2$$

$$x + 2 = 9 - 6\sqrt{x-1} + (x-1)$$

$$x + 2 = 8 - 6\sqrt{x-1} + x$$

$$-6 = -6\sqrt{x-1} \quad \text{Isolating the remaining radical}$$

$$1 = \sqrt{x-1} \quad \text{Dividing by } -6$$

$$1^2 = \left(\sqrt{x-1}\right)^2$$

$$1 = x - 1$$

$$2 = x$$

The number 2 checks. It is the solution.

SECTION 3.5: SOLVING EQUATIONS AND INEQUALITIES WITH ABSOLUTE VALUE

We use the following property to **solve equations with absolute value.** For $a > 0$ and an algebraic expression X: $\lvert X \rvert = a$ is equivalent to $X = -a$ or $X = a$	Solve: $\lvert x + 1 \rvert = 4$. $\lvert x + 1 \rvert = 4$ $x + 1 = -4$ *or* $x + 1 = 4$ $x = -5$ *or* $x = 3$ Both numbers check. The solutions are -5 and 3.
The following properties are used to **solve inequalities with absolute value.** For $a > 0$ and an algebraic expression X: $\lvert X \rvert < a$ is equivalent to $-a < X < a$. $\lvert X \rvert > a$ is equivalent to $X < -a$ or $X > a$. Similar statements hold for $\lvert X \rvert \le a$ and $\lvert X \rvert \ge a$.	Solve: $\lvert x - 2 \rvert < 3$. $\lvert x - 2 \rvert < 3$ $-3 < x - 2 < 3$ $-1 < x < 5$ **Adding 2** The solution set is $\{x \mid -1 < x < 5\}$, or $(-1, 5)$. $\lvert 3x \rvert \ge 6$ $3x \le -6$ *or* $3x \ge 6$ $x \le -2$ *or* $x \ge 2$ **Dividing by 3** The solution set is $\{x \mid x \le -2 \text{ or } x \ge 2\}$, or $(-\infty, -2] \cup [2, \infty)$.

REVIEW EXERCISES

Determine whether the statement is true or false.

1. We can use the quadratic formula to solve any quadratic equation. [3.2]
2. The function $f(x) = -3(x + 4)^2 - 1$ has a maximum value. [3.3]
3. For any positive integer n, if $a^n = b^n$ is true, then $a = b$ is true. [3.4]
4. An equation with absolute value cannot have two negative-number solutions. [3.5]

Solve. [3.2]

5. $(2y + 5)(3y - 1) = 0$
6. $x^2 + 4x - 5 = 0$
7. $3x^2 + 2x = 8$
8. $5x^2 = 15$
9. $x^2 + 10 = 0$

Find the zero(s) of the function. [3.2]

10. $f(x) = x^2 - 2x + 1$
11. $f(x) = x^2 + 2x - 15$
12. $f(x) = 2x^2 - x - 5$
13. $f(x) = 3x^2 + 2x + 3$

Solve.

14. $\dfrac{5}{2x + 3} + \dfrac{1}{x - 6} = 0$ [3.4]
15. $\dfrac{3}{8x + 1} + \dfrac{8}{2x + 5} = 1$ [3.4]
16. $\sqrt{5x + 1} - 1 = \sqrt{3x}$ [3.4]
17. $\sqrt{x - 1} - \sqrt{x - 4} = 1$ [3.4]
18. $\lvert x - 4 \rvert = 3$ [3.5]
19. $\lvert 2y + 7 \rvert = 9$ [3.5]

Solve and write interval notation for the solution set. Then graph the solution set. [3.5]

20. $|5x| \geq 15$

21. $|3x + 4| < 10$

22. $|1 - 6x| < 5$

23. $|x + 4| \geq 2$

24. Solve $\frac{1}{M} + \frac{1}{N} = \frac{1}{P}$ for P. [3.4]

Express in terms of i. [3.1]

25. $-\sqrt{-40}$

26. $\sqrt{-12} \cdot \sqrt{-20}$

27. $\frac{\sqrt{-49}}{-\sqrt{-64}}$

Simplify each of the following. Write the answer in the form $a + bi$, where a and b are real numbers. [3.1]

28. $(6 + 2i) + (-4 - 3i)$

29. $(3 - 5i) - (2 - i)$

30. $(6 + 2i)(-4 - 3i)$

31. $\frac{2 - 3i}{1 - 3i}$

32. i^{23}

Solve by completing the square to obtain exact solutions. Show your work. [3.2]

33. $x^2 - 3x = 18$

34. $3x^2 - 12x - 6 = 0$

Solve. Give exact solutions. [3.2]

35. $3x^2 + 10x = 8$

36. $r^2 - 2r + 10 = 0$

37. $x^2 = 10 + 3x$

38. $x = 2\sqrt{x} - 1$

39. $y^4 - 3y^2 + 1 = 0$

40. $(x^2 - 1)^2 - (x^2 - 1) - 2 = 0$

41. $(p + 2)(3p + 2)(p - 3) = 0$

42. $x^3 + 5x^2 - 4x - 20 = 0$

In Exercises 43 and 44, complete the square to **(a)** *find the vertex;* **(b)** *find the axis of symmetry;* **(c)** *determine whether there is a maximum or minimum value and find that value;* **(d)** *find the range; and* **(e)** *graph the function.* [3.3]

43. $f(x) = -4x^2 + 3x - 1$

44. $f(x) = 5x^2 - 10x + 3$

In Exercises 45–48, match the equation with one of the figures (a)–(d) that follow. [3.3]

a)

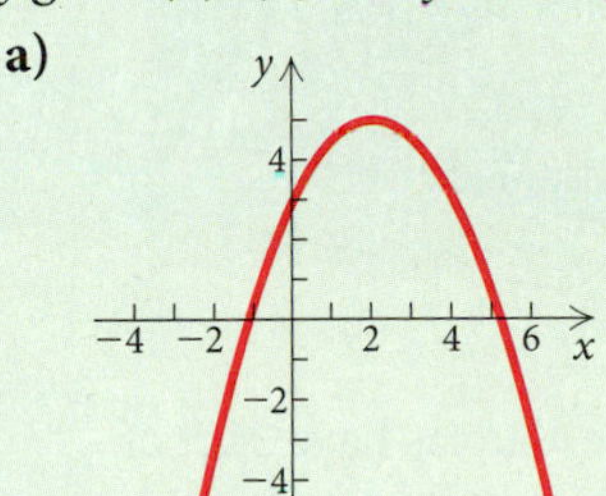

b)

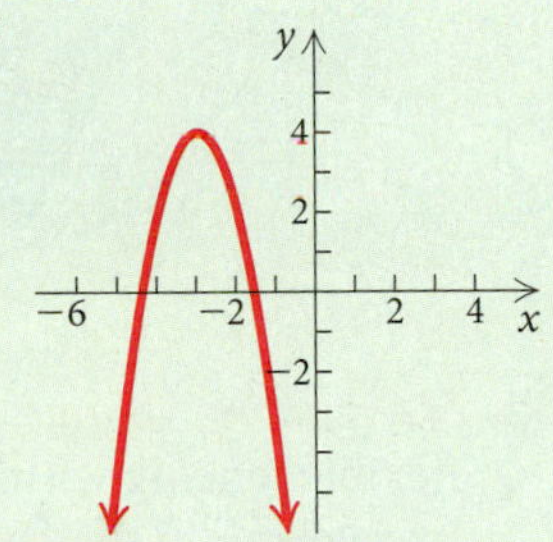

c)

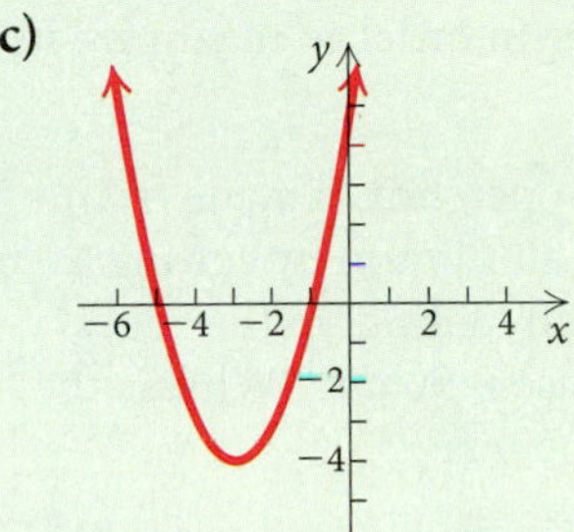

d)

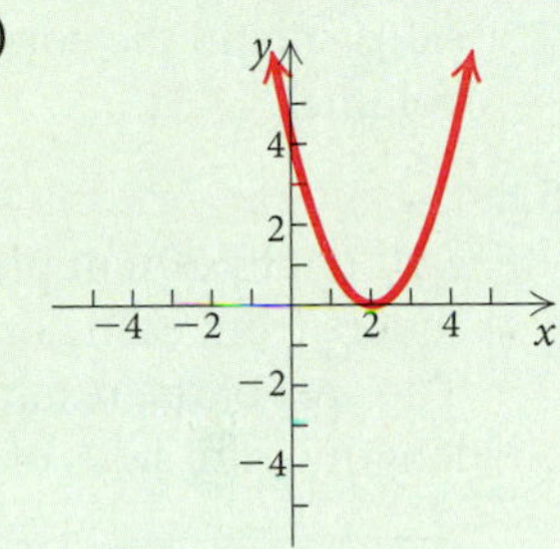

45. $y = (x - 2)^2$

46. $y = (x + 3)^2 - 4$

47. $y = -2(x + 3)^2 + 4$

48. $y = -\frac{1}{2}(x - 2)^2 + 5$

49. *Legs of a Right Triangle.* The hypotenuse of a right triangle is 50 ft. One leg is 10 ft longer than the other. What are the lengths of the legs? [3.2]

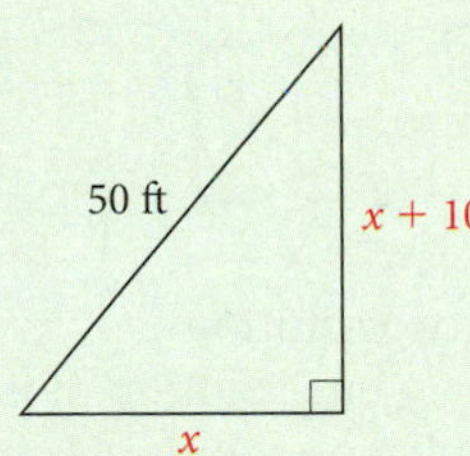

50. *Bicycling Speed.* Harry and Rebecca leave a campsite, Harry biking due north and Rebecca biking due east. Harry bikes 7 km/h slower than Rebecca. After 4 hr, they are 68 km apart. Find the speed of each bicyclist. [3.2]

51. *Sidewalk Width.* A 60-ft by 80-ft parking lot is torn up to install a sidewalk of uniform width around its perimeter. The new area of the parking lot is two-thirds of the old area. How wide is the sidewalk? [3.2]

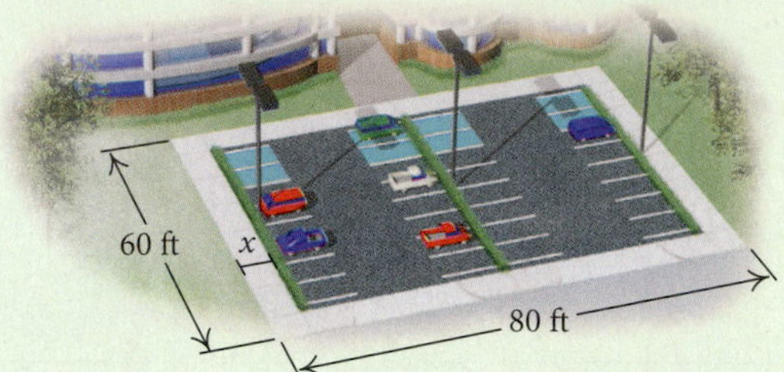

52. *Maximizing Volume.* The Garcias have 24 ft of flexible fencing with which to build a rectangular "toy corral." If the fencing is 2 ft high, what dimensions should the corral have in order to maximize its volume? [3.3]

53. *Dimensions of a Box.* An open box is made from a 10-cm by 20-cm piece of aluminum by cutting a square from each corner and folding up the edges. The area of the resulting base is 90 cm². What is the length of the sides of the squares? [3.2]

54. Find the zeros of $f(x) = 2x^2 - 5x + 1$. [3.2]

A. $\dfrac{5 \pm \sqrt{17}}{2}$ B. $\dfrac{5 \pm \sqrt{17}}{4}$

C. $\dfrac{5 \pm \sqrt{33}}{4}$ D. $\dfrac{-5 \pm \sqrt{17}}{4}$

55. Solve: $\sqrt{4x + 1} + \sqrt{2x} = 1$. [3.4]

A. There are two solutions.
B. There is only one solution. It is less than 1.
C. There is only one solution. It is greater than 1.
D. There is no solution.

56. The graph of $f(x) = (x - 2)^2 - 3$ is which of the following? [3.3]

A.

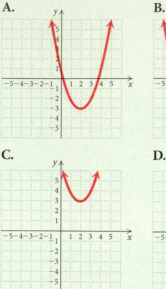

B.

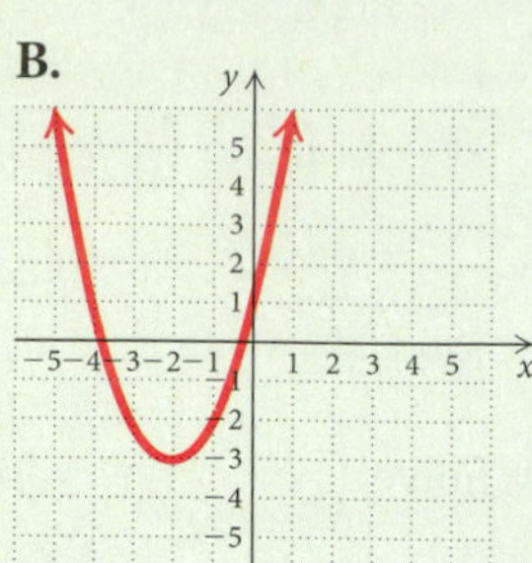

C.

D.

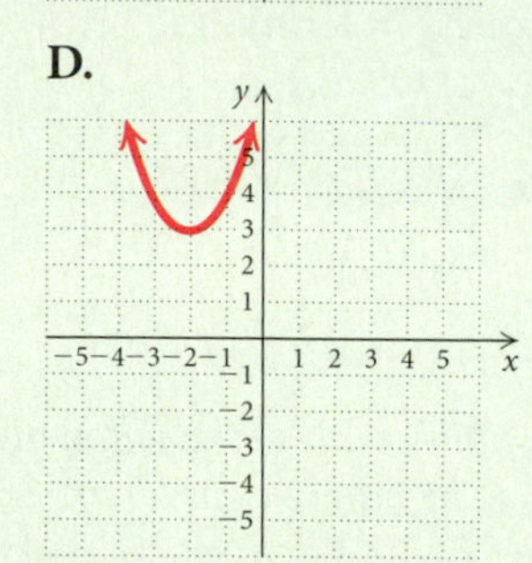

Synthesis

Solve.

57. $\sqrt{\sqrt{\sqrt{x}}} = 2$ [3.4]

58. $(t - 4)^{4/5} = 3$ [3.4]

59. $(x - 1)^{2/3} = 4$ [3.4]

60. $(2y - 2)^2 + y - 1 = 5$ [3.2]

61. $\sqrt{x + 2} + \sqrt[4]{x + 2} - 2 = 0$ [3.2]

62. At the beginning of the year, \$3500 was deposited in a savings account. One year later, \$4000 was deposited in another account. The interest rate was the same for both accounts. At the end of the second year, there was a total of \$8518.35 in the accounts. What was the annual interest rate? [3.2]

63. Find b such that $f(x) = -3x^2 + bx - 1$ has a maximum value of 2. [3.3]

Collaborative Discussion and Writing

64. Is the product of two imaginary numbers always an imaginary number? Explain your answer. [3.1]

65. Is it possible for a quadratic function to have one real zero and one imaginary zero? Why or why not? [3.2]

66. If the graphs of

$$f(x) = a_1(x - h_1)^2 + k_1$$

and

$$g(x) = a_2(x - h_2)^2 + k_2$$

have the same shape, what, if anything, can you conclude about the a's, the h's, and the k's? Explain your answer. [3.3]

67. Explain why it is necessary to check the possible solutions of a rational equation. [3.4]

68. Explain in your own words why it is necessary to check the possible solutions when the principle of powers is used to solve an equation. [3.4]

69. Explain why $|x| < p$ has no solution for $p \leq 0$. [3.5]

70. Explain why all real numbers are solutions of $|x| > p$, for $p < 0$. [3.5]

3 Chapter Test

Solve. Find exact solutions.

1. $(2x - 1)(x + 5) = 0$
2. $6x^2 - 36 = 0$
3. $x^2 + 4 = 0$
4. $x^2 - 2x - 3 = 0$
5. $x^2 - 5x + 3 = 0$
6. $2t^2 - 3t + 4 = 0$
7. $x + 5\sqrt{x} - 36 = 0$
8. $\dfrac{3}{3x + 4} + \dfrac{2}{x - 1} = 2$
9. $\sqrt{x + 4} - 2 = 1$
10. $\sqrt{x + 4} - \sqrt{x - 4} = 2$
11. $|x + 4| = 7$
12. $|4y - 3| = 5$

Solve and write interval notation for the solution set. Then graph the solution set.

13. $|x + 3| \leq 4$
14. $|2x - 1| < 5$
15. $|x + 5| > 2$
16. $|3 - 2x| \geq 7$
17. Solve $\dfrac{1}{A} + \dfrac{1}{B} = \dfrac{1}{C}$ for B.
18. Solve $R = \sqrt{3np}$ for n.
19. Solve $x^2 + 4x = 1$ by completing the square. Find the exact solutions. Show your work.
20. The tallest structure in the United States, at 2063 ft, is the KTHI-TV tower in Blanchard, North Dakota (*Source: The Cambridge Fact Finder*). How long would it take an object falling freely from the top to reach the ground? (Use the formula $s = 16t^2$, where s is the distance, in feet, that an object falls freely from rest in t seconds.)

Express in terms of i.

21. $\sqrt{-43}$
22. $-\sqrt{-25}$

Simplify.

23. $(5 - 2i) - (2 + 3i)$
24. $(3 + 4i)(2 - i)$
25. $\dfrac{1 - i}{6 + 2i}$
26. i^{33}

Find the zeros of each function.

27. $f(x) = 4x^2 - 11x - 3$
28. $f(x) = 2x^2 - x - 7$
29. For the graph of the function $f(x) = -x^2 + 2x + 8$; **(a)** find the vertex; **(b)** find the axis of symmetry; **(c)** state whether there is a maximum or a minimum value and find that value; **(d)** find the range; and **(e)** graph the function.
30. *Maximizing Area.* A homeowner wants to fence a rectangular play yard using 80 ft of fencing. The side of the house will be used as one side of the rectangle. Find the dimensions for which the area is a maximum.

31. The graph of $f(x) = (x - 1)^2 - 2$ is which of the following?

A.

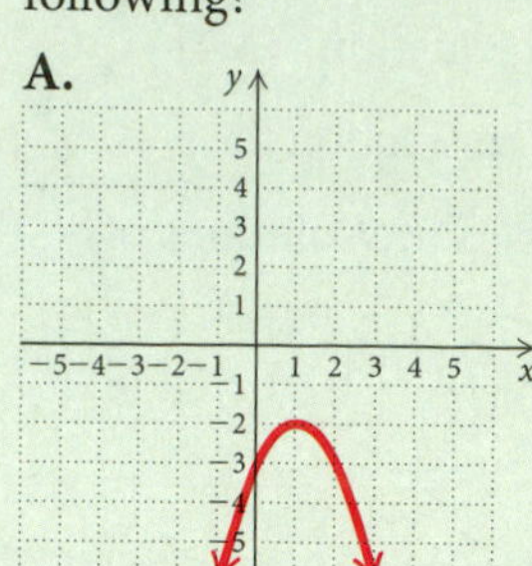

B.

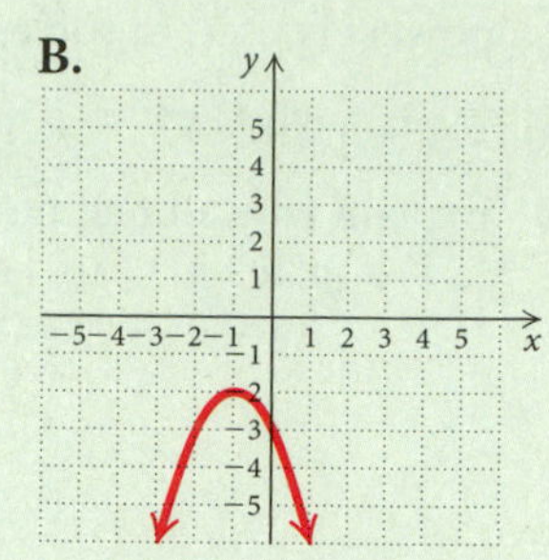

C.

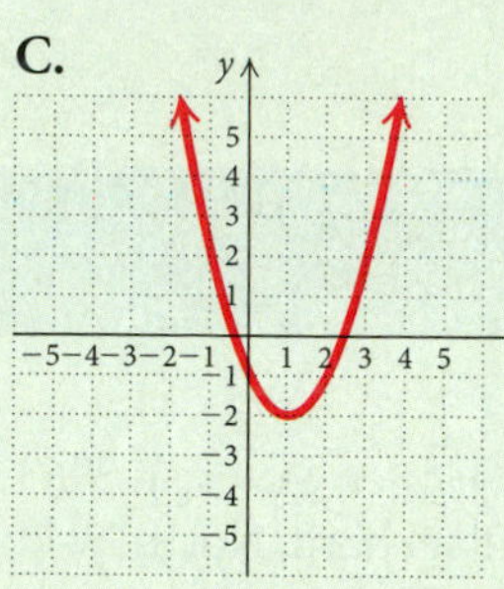

D.

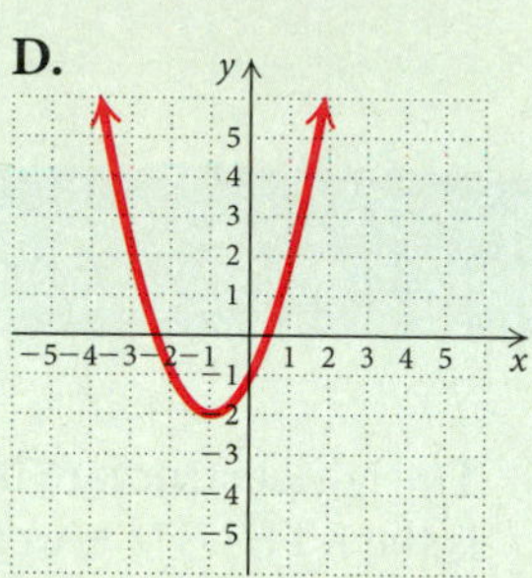

Synthesis

32. Find a such that $f(x) = ax^2 - 4x + 3$ has a maximum value of 12.

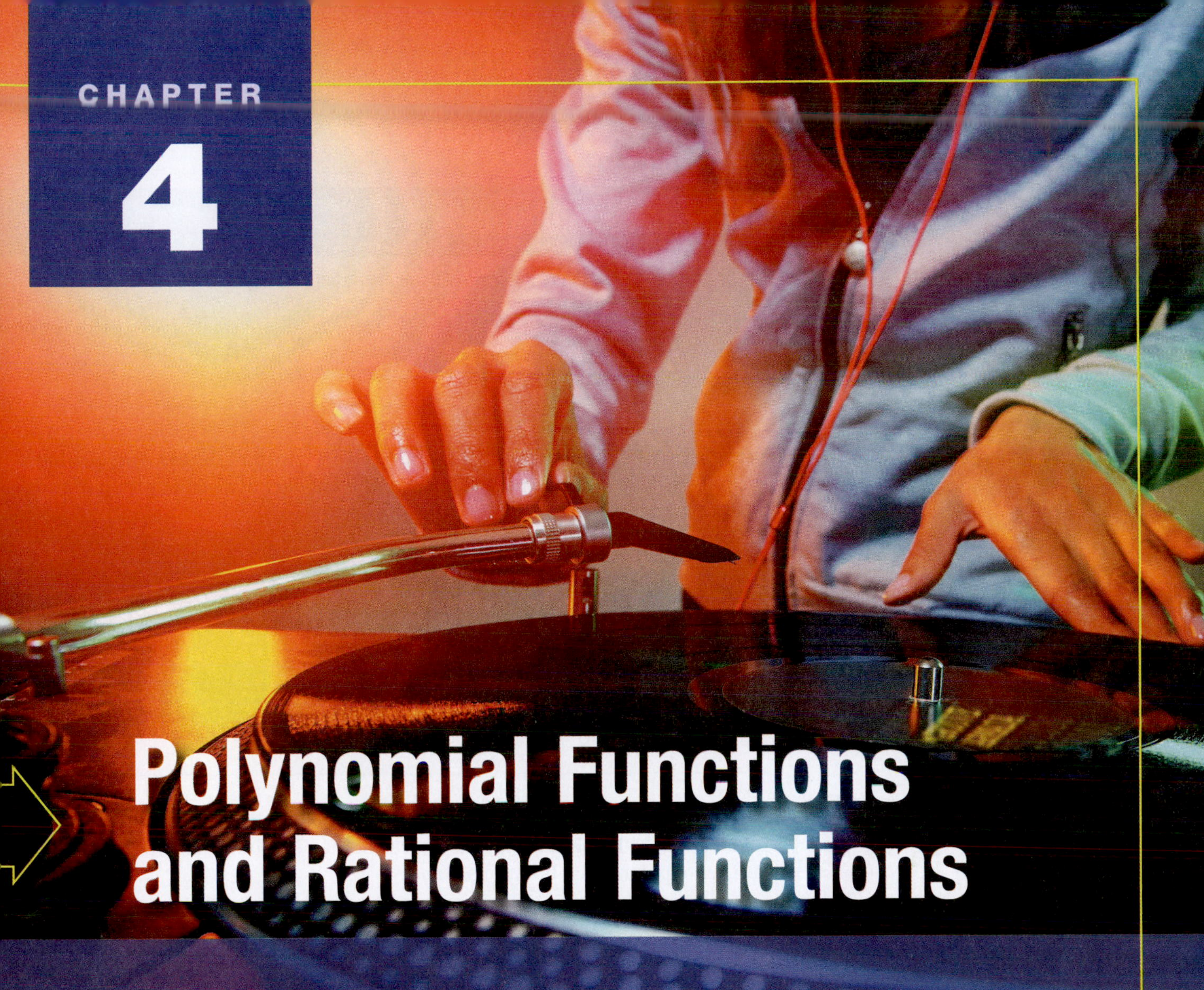

CHAPTER 4

Polynomial Functions and Rational Functions

APPLICATION This problem appears as Exercise 47 in Section 4.1.

Vinyl record albums are making a comeback. Sales of vinyl albums rose 32% from 2012 to 2013. The sales data over the years 2001 to 2013 are modeled by the quartic function

$$f(x) = -0.000913x^4 + 0.248x^3 - 0.1515x^2 + 0.2136x + 1.2779,$$

where x is the number of years after 2001 and $f(x)$ is the number of albums in millions (*Source:* Nielsen SoundScan). Find the number of vinyl albums sold in 2008 and in 2012, and estimate the number sold in 2016.

4.1 Polynomial Functions and Models

- Determine the behavior of the graph of a polynomial function using the leading-term test.
- Factor polynomial functions and find their zeros and their multiplicities.
- Solve applied problems using polynomial models.

There are many different kinds of functions. The constant, linear, and quadratic functions that we studied in Chapters 1 and 3 are part of a larger group of functions called *polynomial functions.*

POLYNOMIAL FUNCTION

A **polynomial function** P is given by

$$P(x) = a_n x^n + a_{n-1}x^{n-1} + a_{n-2}x^{n-2} + \cdots + a_1 x + a_0,$$

where the coefficients $a_n, a_{n-1}, \ldots, a_1, a_0$ are real numbers and the exponents are whole numbers.

The first nonzero coefficient, a_n, is called the **leading coefficient**. The term $a_n x^n$ is called the **leading term**. The **degree** of the polynomial function is n. Some examples of polynomial functions follow.

POLYNOMIAL FUNCTION	EXAMPLE	DEGREE	LEADING TERM	LEADING COEFFICIENT
Constant	$f(x) = 3 \quad (f(x) = 3 = 3x^0)$	0	3	3
Linear	$f(x) = \frac{2}{3}x + 5 \quad (f(x) = \frac{2}{3}x + 5 = \frac{2}{3}x^1 + 5)$	1	$\frac{2}{3}x$	$\frac{2}{3}$
Quadratic	$f(x) = 4x^2 - x + 3$	2	$4x^2$	4
Cubic	$f(x) = x^3 + 2x^2 + x - 5$	3	x^3	1
Quartic	$f(x) = -x^4 - 1.1x^3 + 0.3x^2 - 2.8x - 1.7$	4	$-x^4$	-1

The function $f(x) = 0$ can be described in many ways:

$$f(x) = 0 = 0x^2 = 0x^{15} = 0x^{48},$$

and so on. For this reason, we say that the constant function $f(x) = 0$ has no degree.

Functions such as

$$f(x) = \frac{2}{x} + 5, \text{ or } 2x^{-1} + 5, \quad \text{and} \quad g(x) = \sqrt{x} - 6, \text{ or } x^{1/2} - 6,$$

are *not* polynomial functions because the exponents -1 and $\frac{1}{2}$ are *not* whole numbers.

From our study of functions in Chapters 1–3, we know how to find or at least estimate many characteristics of a polynomial function. Let's consider two examples for review.

Quadratic Function

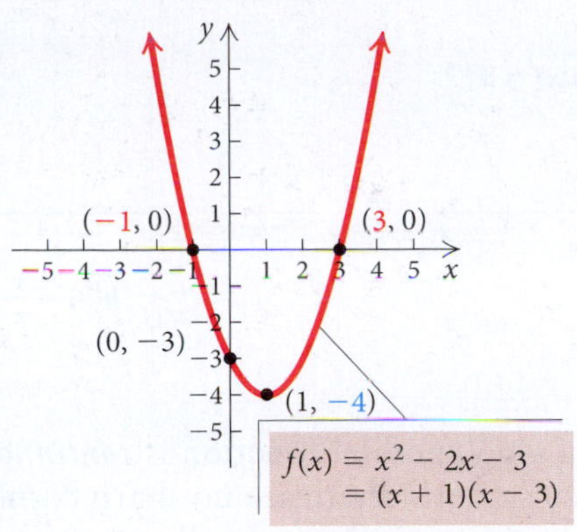

Function: $f(x) = x^2 - 2x - 3$
$= (x + 1)(x - 3)$

Zeros: $-1, 3$

x-intercepts: $(-1, 0), (3, 0)$

y-intercept: $(0, -3)$

Minimum: -4 at $x = 1$

Maximum: None

Domain: All real numbers, $(-\infty, \infty)$

Range: $[-4, \infty)$

Cubic Function

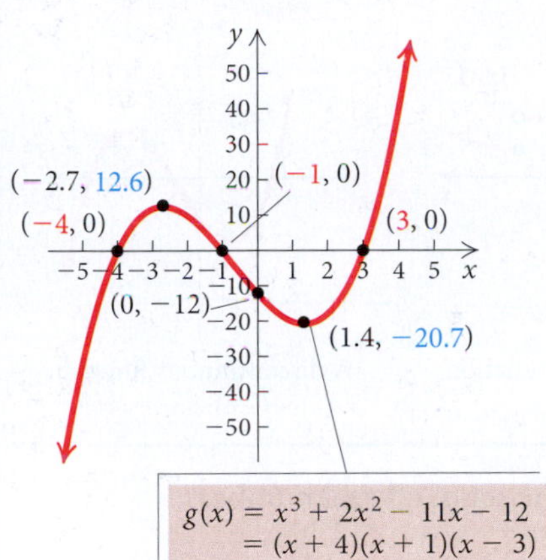

Function: $g(x) = x^3 + 2x^2 - 11x - 12$
$= (x + 4)(x + 1)(x - 3)$

Zeros: $-4, -1, 3$

x-intercepts: $(-4, 0), (-1, 0), (3, 0)$

y-intercept: $(0, -12)$

Relative minimum: -20.7 at $x = 1.4$

Relative maximum: 12.6 at $x = -2.7$

Domain: All real numbers, $(-\infty, \infty)$

Range: All real numbers, $(-\infty, \infty)$

All graphs of polynomial functions have some characteristics in common. Compare the following graphs. How do the graphs of polynomial functions differ from the graphs of nonpolynomial functions? Describe some characteristics of the graphs of polynomial functions that you observe.

Polynomial Functions

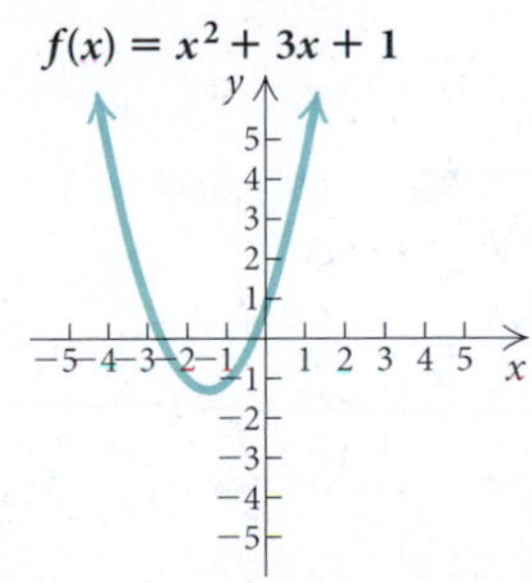

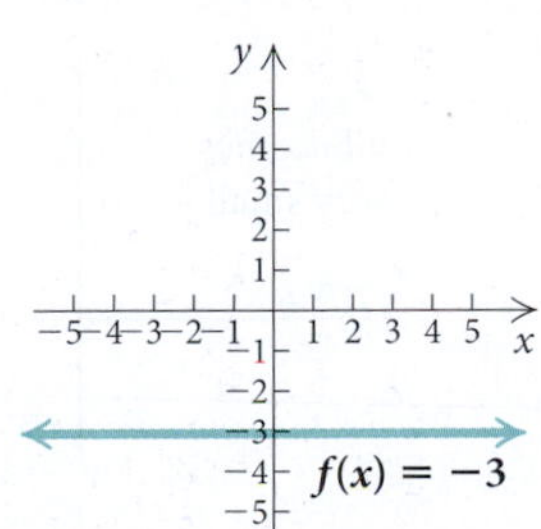

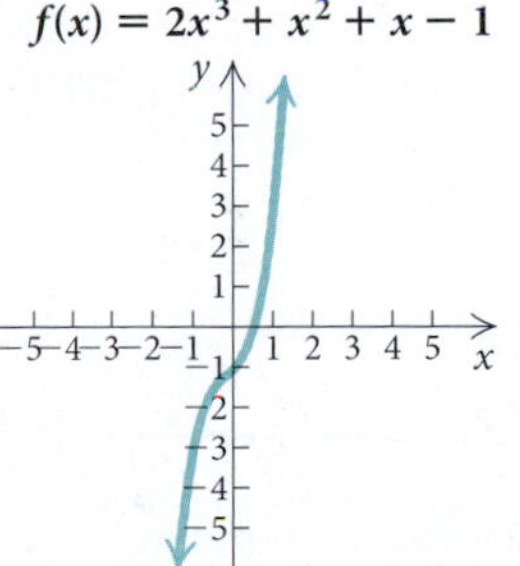

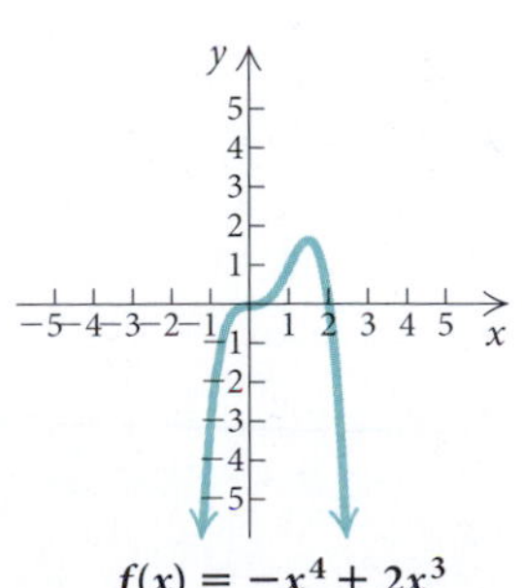

Nonpolynomial Functions

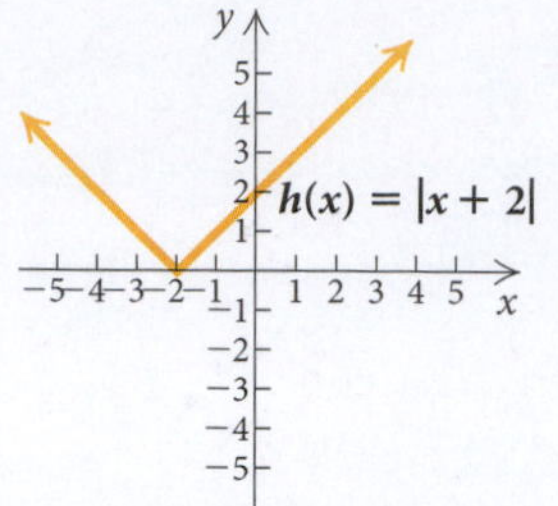

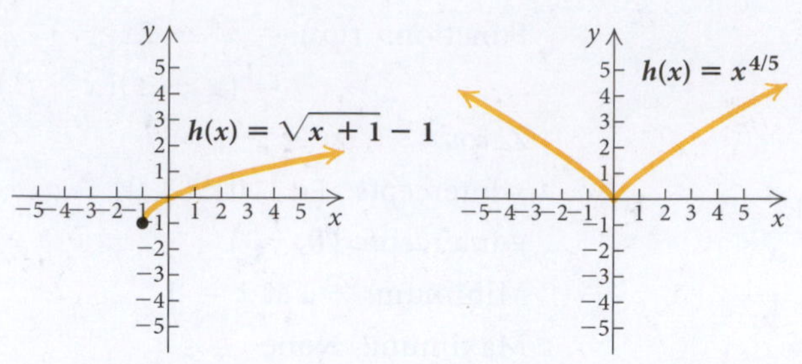

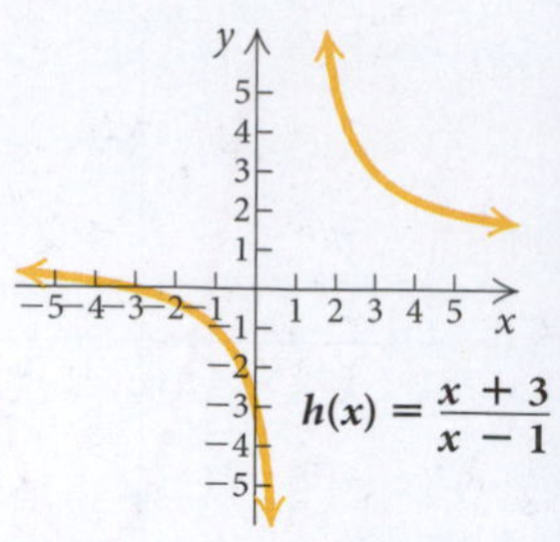

DOMAIN OF A FUNCTION

REVIEW SECTION 1.2

You probably noted that the graph of a polynomial function is *continuous*; that is, it has no holes or breaks. It is also smooth; there are no sharp corners. Furthermore, the *domain* of a polynomial function is the set of all real numbers, $(-\infty, \infty)$.

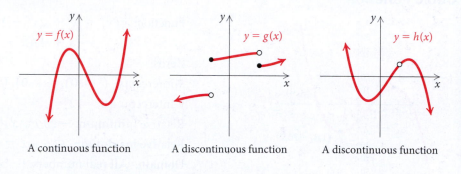

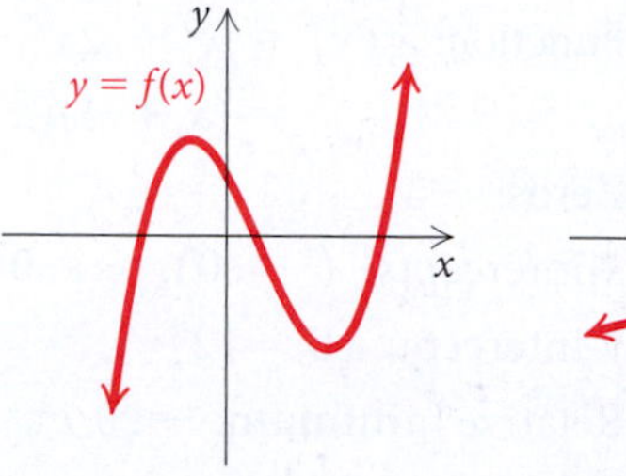

A continuous function

A discontinuous function

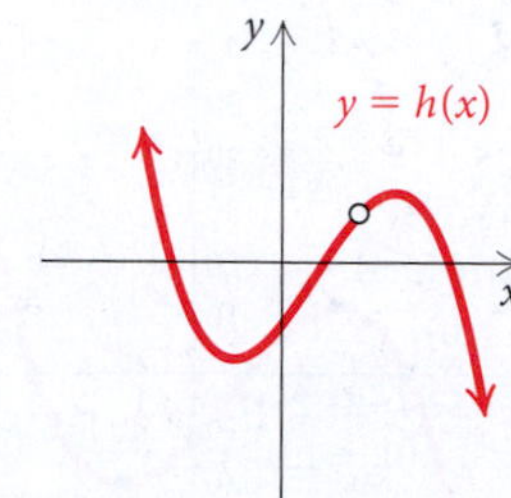

A discontinuous function

> The *domain* of a polynomial function is the set of all real numbers, $(-\infty, \infty)$.

▶ The Leading-Term Test

The behavior of the graph of a polynomial function as x becomes very large $(x \to \infty)$ or very small $(x \to -\infty)$ is referred to as the end behavior of the graph. The leading term of a polynomial function determines its end behavior.

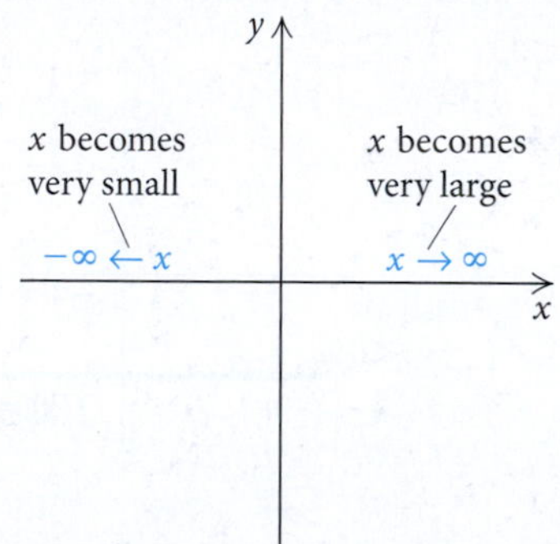

Using the graphs shown on the following page, let's see if we can discover some general patterns by comparing the end behavior of even- and odd-degree functions. We also note the effect of positive and negative leading coefficients.

Even Degree

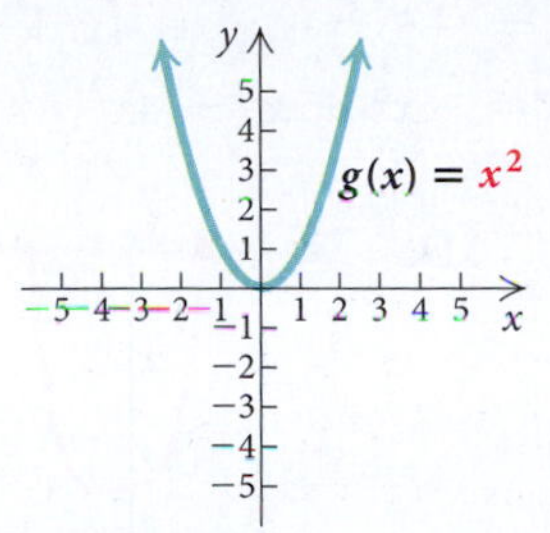

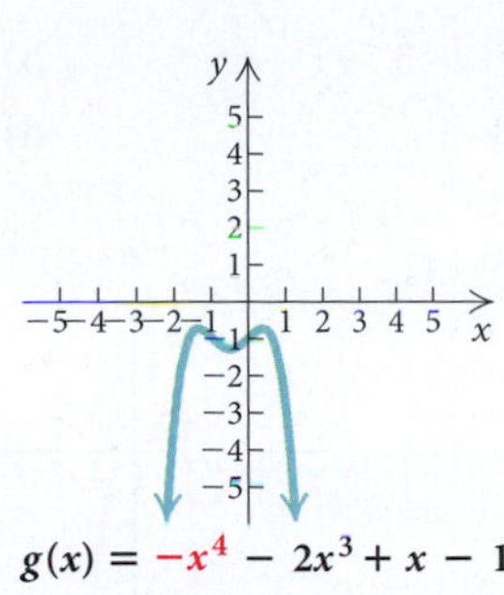

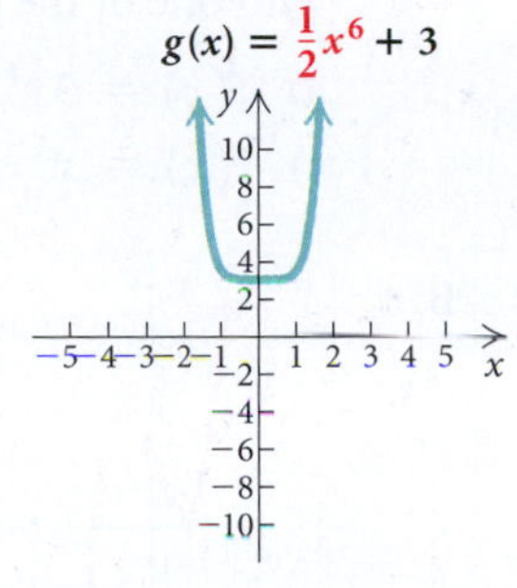

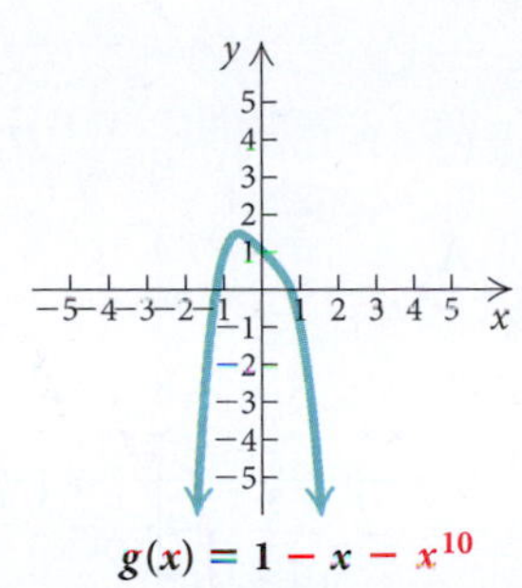

Odd Degree

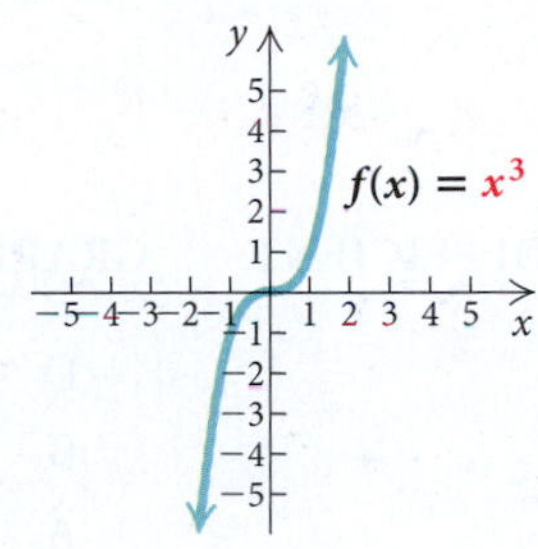

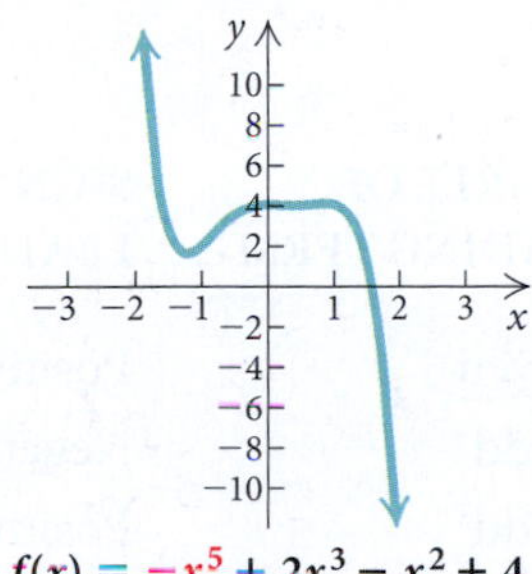

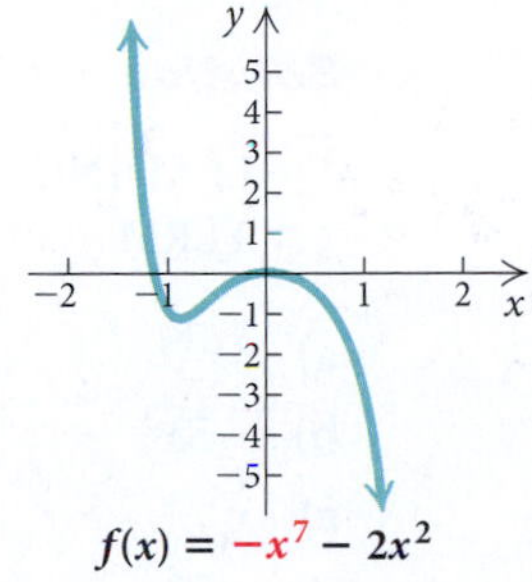

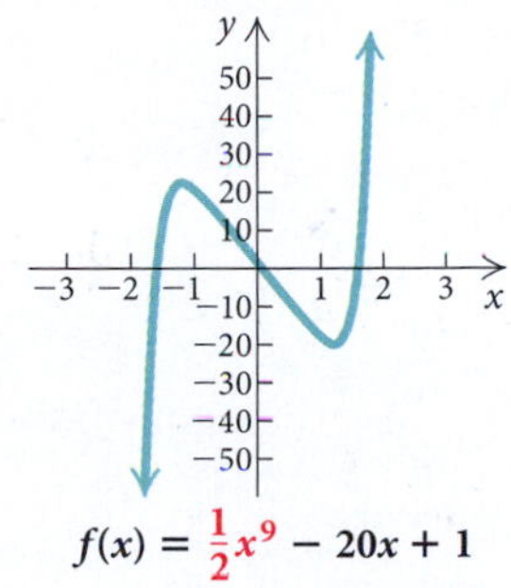

We can summarize our observations as follows.

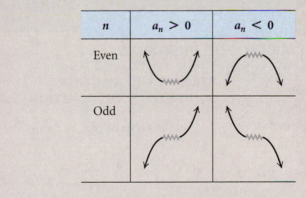

THE LEADING-TERM TEST

If $a_n x^n$ is the leading term of a polynomial function, then the behavior of the graph as $x \to \infty$ or as $x \to -\infty$ can be described in one of the four following ways.

n	$a_n > 0$	$a_n < 0$
Even		
Odd		

The portion of the graph is not determined by this test.

EXAMPLE 1 Using the leading-term test, match each of the following functions with one of the graphs A–D that follow.

a) $f(x) = 3x^4 - 2x^3 + 3$ **b)** $f(x) = -5x^3 - x^2 + 4x + 2$

c) $f(x) = x^5 + \frac{1}{4}x + 1$ **d)** $f(x) = -x^6 + x^5 - 4x^3$

A.
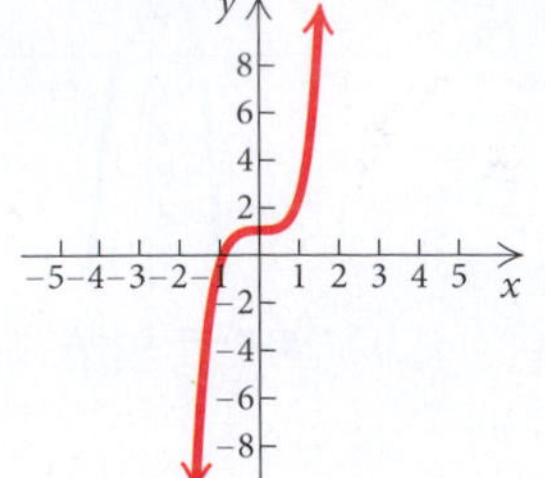

B.
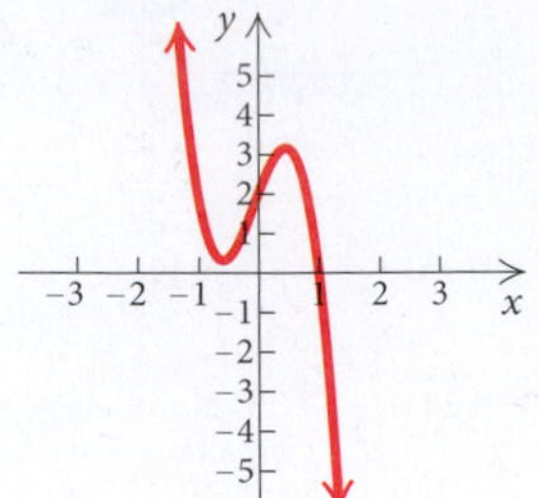

C.
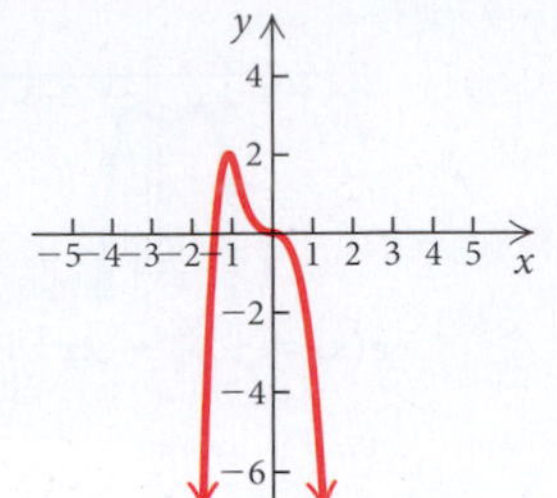

D.
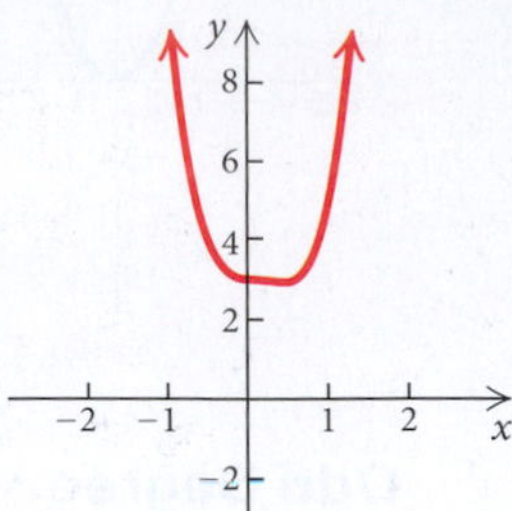

Solution

	LEADING TERM	DEGREE OF LEADING TERM	SIGN OF LEADING COEFFICIENT	GRAPH
a)	$3x^4$	4, even	Positive	D
b)	$-5x^3$	3, odd	Negative	B
c)	x^5	5, odd	Positive	A
d)	$-x^6$	6, even	Negative	C

Now Try Exercise 19.

Finding Zeros of Polynomial Functions

Let's review the meaning of the real zeros of a function and their connection to the x-intercepts of the function's graph.

CONNECTING THE CONCEPTS

Zeros, Solutions, and Intercepts

FUNCTION	ZEROS OF THE FUNCTION; SOLUTIONS OF THE EQUATION	ZEROS OF THE FUNCTION; x-INTERCEPTS OF THE GRAPH
Quadratic Polynomial $g(x) = x^2 - 2x - 8 = (x+2)(x-4)$, or $y = (x+2)(x-4)$	To find the **zeros** of $g(x)$, we solve $g(x) = 0$: $x^2 - 2x - 8 = 0$; $(x+2)(x-4) = 0$; $x + 2 = 0$ *or* $x - 4 = 0$; $x = -2$ *or* $x = 4$. The **solutions** of $x^2 - 2x - 8 = 0$ are -2 and 4. They are the zeros of the function $g(x)$; that is, $g(-2) = 0$ and $g(4) = 0$.	The real-number zeros of $g(x)$ are the x-coordinates of the ***x*-intercepts** of the graph of $y = g(x)$. 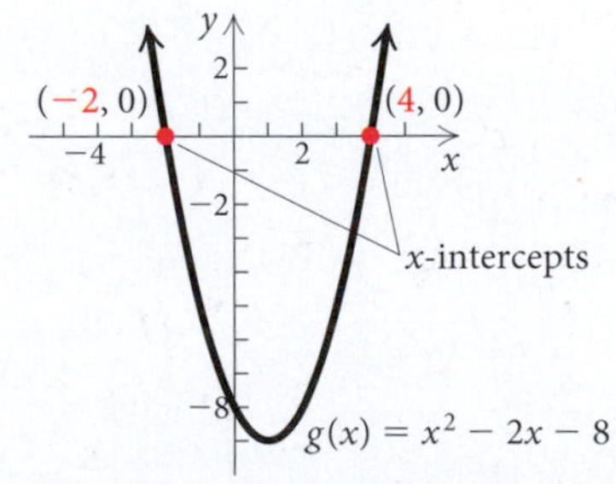

(continued)

Cubic Polynomial

$h(x)$
$= x^3 + 2x^2 - 5x - 6$
$= (x + 3)(x + 1)(x - 2),$

or

$y = (x + 3)(x + 1)(x - 2)$

To find the **zeros** of $h(x)$, we solve $h(x) = 0$:

$$x^3 + 2x^2 - 5x - 6 = 0$$
$$(x + 3)(x + 1)(x - 2) = 0$$
$$x + 3 = 0 \quad or \quad x + 1 = 0 \quad or \quad x - 2 = 0$$
$$x = -3 \quad or \quad x = -1 \quad or \quad x = 2.$$

The **solutions** of $x^3 + 2x^2 - 5x - 6 = 0$ are -3, -1, and 2. They are the zeros of the function $h(x)$; that is,

$$h(-3) = 0,$$
$$h(-1) = 0, \quad \text{and}$$
$$h(2) = 0.$$

The real-number zeros of $h(x)$ are the x-coordinates of the **x-intercepts** of the graph of $y = h(x)$.

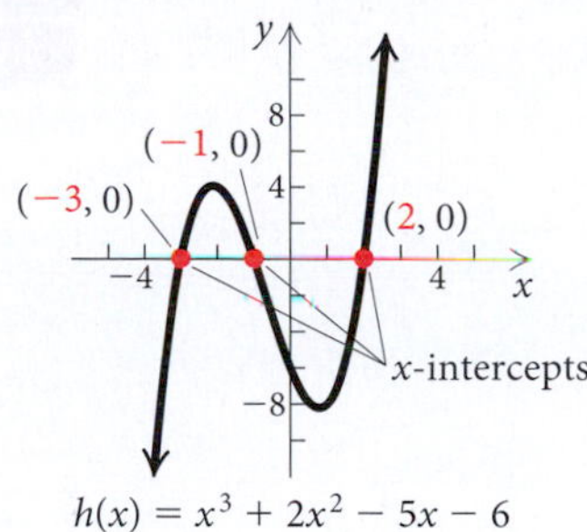

The connection between the real-number zeros of a function and the x-intercepts of the graph of the function is easily seen in the preceding examples. If c is a real zero of a function (that is, $f(c) = 0$), then $(c, 0)$ is an x-intercept of the graph of the function.

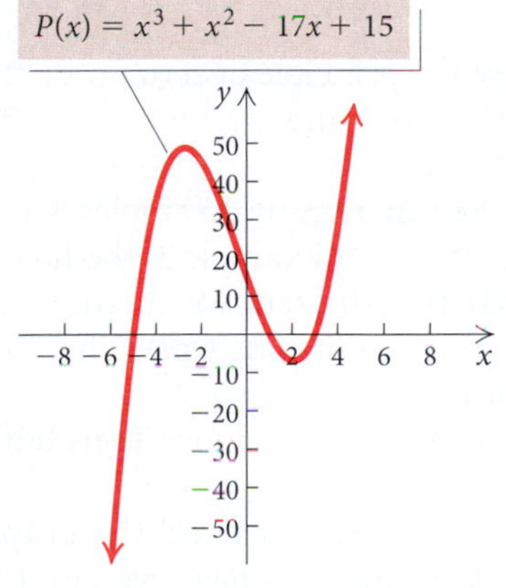

EXAMPLE 2 Consider $P(x) = x^3 + x^2 - 17x + 15$. Determine whether each of the numbers 2 and -5 is a zero of $P(x)$.

Solution We first evaluate $P(2)$:

$$P(2) = (2)^3 + (2)^2 - 17(2) + 15 = -7. \qquad \text{Substituting 2 into the polynomial}$$

Since $P(2) \neq 0$, we know that 2 is *not* a zero of the polynomial function.

We then evaluate $P(-5)$:

$$P(-5) = (-5)^3 + (-5)^2 - 17(-5) + 15 = 0. \qquad \text{Substituting } -5 \text{ into the polynomial}$$

Since $P(-5) = 0$, we know that -5 is a zero of $P(x)$.

Now Try Exercise 23.

Let's take a closer look at the polynomial function

$$h(x) = x^3 + 2x^2 - 5x - 6$$

(see Connecting the Concepts above). The factors of $h(x)$ are

$$x + 3, \quad x + 1, \quad \text{and} \quad x - 2,$$

and the zeros are

$$-3, \quad -1, \quad \text{and} \quad 2.$$

We note that when the polynomial is expressed as a product of linear factors, each factor determines a zero of the function. Thus if we know the linear factors of a polynomial function $f(x)$, we can easily find the zeros of $f(x)$ by solving the equation $f(x) = 0$ using the principle of zero products.

EXAMPLE 3 Find the zeros of

$$\begin{aligned} f(x) &= 5(x-2)(x-2)(x-2)(x+1) \\ &= 5(x-2)^3(x+1). \end{aligned}$$

Solution To solve the equation $f(x) = 0$, we use the principle of zero products, solving $x - 2 = 0$ and $x + 1 = 0$. The zeros of $f(x)$ are 2 and -1.

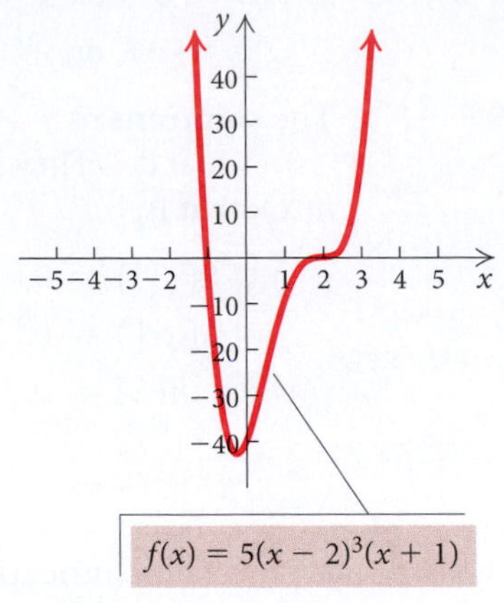

EXAMPLE 4 Find the zeros of

$$\begin{aligned} g(x) &= -(x-1)(x-1)(x+2)(x+2) \\ &= -(x-1)^2(x+2)^2. \end{aligned}$$

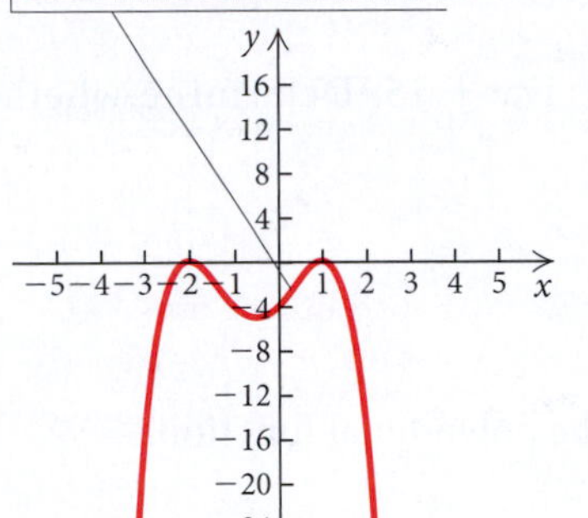

Solution To solve the equation $g(x) = 0$, we use the principle of zero products, solving $x - 1 = 0$ and $x + 2 = 0$. The zeros of $g(x)$ are 1 and -2.

Let's consider the occurrences of the zeros in the functions in Examples 3 and 4 and their relationship to the graphs of those functions. In Example 3, the factor $x - 2$ occurs three times. In a case like this, we say that the zero we obtain from this factor, 2, has a **multiplicity** of 3. The factor $x + 1$ occurs one time. The zero we obtain from this factor, -1, has a *multiplicity* of 1.

In Example 4, the factors $x - 1$ and $x + 2$ each occur two times. Thus both zeros, 1 and -2, have a *multiplicity* of 2.

Note, in Example 3, that the zeros have odd multiplicities and the graph crosses the x-axis at both -1 and 2. But in Example 4, the zeros have even multiplicities and the graph is tangent to (touches but does not cross) the x-axis at -2 and 1. This leads us to the following generalization.

EVEN MULTIPLICITY AND ODD MULTIPLICITY

If $(x - c)^k$, $k \geq 1$, is a factor of a polynomial function $P(x)$ and $(x - c)^{k+1}$ is not a factor and:

- if k is odd, then the graph crosses the x-axis at $(c, 0)$;
- if k is even, then the graph is tangent to the x-axis at $(c, 0)$.

Some polynomials can be factored by grouping. Then we use the principle of zero products to find their zeros.

EXAMPLE 5 Find the zeros of

$$f(x) = x^3 - 2x^2 - 9x + 18.$$

JUST IN TIME 13

Solution We factor by grouping, as follows:

$$\begin{aligned} f(x) &= x^3 - 2x^2 - 9x + 18 \\ &= x^2(x - 2) - 9(x - 2) && \text{Grouping } x^3 \text{ with } -2x^2 \text{ and } -9x \text{ with } 18 \text{ and factoring each group} \\ &= (x - 2)(x^2 - 9) && \text{Factoring out } x - 2 \\ &= (x - 2)(x + 3)(x - 3). && \text{Factoring } x^2 - 9 \end{aligned}$$

Then, by the principle of zero products, the solutions of the equation $f(x) = 0$ are 2, -3, and 3. These are the zeros of $f(x)$.

Now Try Exercise 39.

Technology Connection

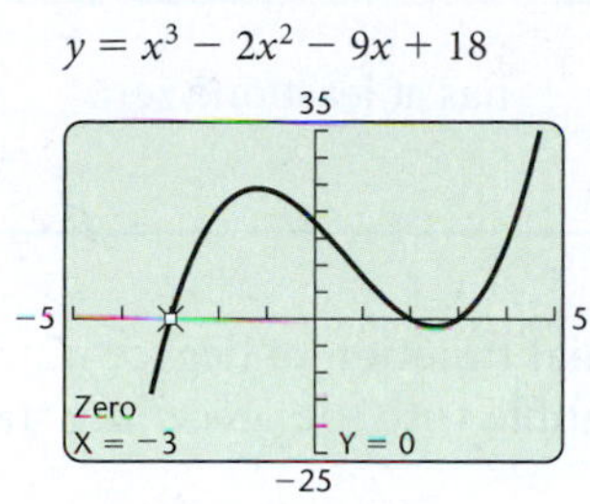

Using the Zero method, we can determine the zeros of the function in Example 5. The window at left shows the calculator display when we find the leftmost zero. The other zeros, 2 and 3, can be found in the same manner.

Find the real zeros of the function f given by $f(x) = 0.1x^3 - 0.6x^2 - 0.1x + 2$. Approximate the zeros to three decimal places.

We use a graphing calculator to create a graph that clearly shows the curvature. It appears that there are three zeros, one near -2, one near 2, and one near 6. We use the ZERO feature to find them. The zeros are approximately -1.680, 2.154, and 5.526.

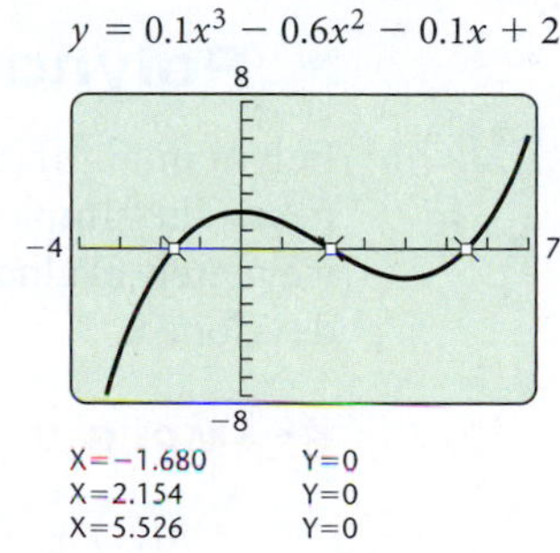

Other factoring techniques can also be used.

EXAMPLE 6 Find the zeros of

$$f(x) = x^4 + 4x^2 - 45.$$

Solution We factor as follows:

$$f(x) = x^4 + 4x^2 - 45 = (x^2 - 5)(x^2 + 9).$$

We now solve the equation $f(x) = 0$ to determine the zeros. We use the principle of zero products:

$$\begin{aligned} &(x^2 - 5)(x^2 + 9) = 0 \\ &x^2 - 5 = 0 \quad \text{or} \quad x^2 + 9 = 0 \\ &x^2 = 5 \quad \text{or} \quad x^2 = -9 \\ &x = \pm\sqrt{5} \quad \text{or} \quad x = \pm\sqrt{-9} = \pm 3i. \end{aligned}$$

The solutions are $\pm\sqrt{5}$ and $\pm 3i$. These are the zeros of $f(x)$.

Now Try Exercise 37.

Only the real-number zeros of a function correspond to the x-intercepts of its graph. For instance, the real-number zeros of the function in Example 6, $-\sqrt{5}$ and $\sqrt{5}$, can be seen on the graph of the function below, but the nonreal zeros, $-3i$ and $3i$, cannot.

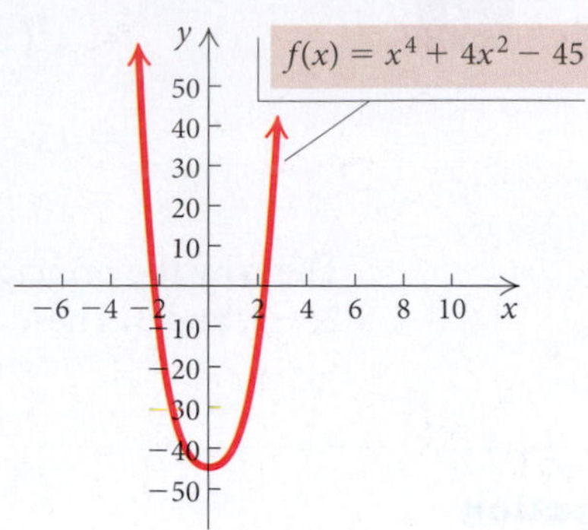

> Every polynomial function of degree n, with $n \geq 1$, has at least one zero and at most n zeros.

This is often stated as follows: "Every polynomial function of degree n, with $n \geq 1$, has *exactly* n zeros." This statement is compatible with the preceding statement, if one takes multiplicities into account.

Polynomial Models

Polynomial functions have many uses as models in science, engineering, and business. The simplest use of polynomial functions in applied problems occurs when we merely evaluate a polynomial function. In such cases, a model has already been developed.

EXAMPLE 7 ***Ibuprofen in the Bloodstream.*** The polynomial function

$$M(t) = 0.5t^4 + 3.45t^3 - 96.65t^2 + 347.7t$$

can be used to estimate the number of milligrams of the pain relief medication ibuprofen in the bloodstream t hours after 400 mg of the medication has been taken. Find the number of milligrams in the bloodstream at $t = 0, 0.5, 1, 1.5$, and so on, up to 6 hr. Round the function values to the nearest tenth.

Solution Using a calculator, we compute function values:

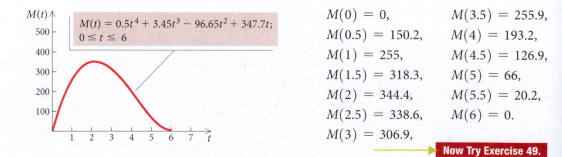

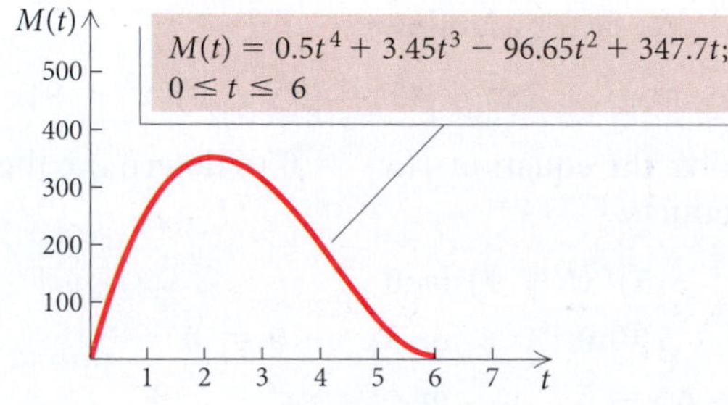

$M(0) = 0$, $M(0.5) = 150.2$, $M(1) = 255$, $M(1.5) = 318.3$, $M(2) = 344.4$, $M(2.5) = 338.6$, $M(3) = 306.9$, $M(3.5) = 255.9$, $M(4) = 193.2$, $M(4.5) = 126.9$, $M(5) = 66$, $M(5.5) = 20.2$, $M(6) = 0$.

Now Try Exercise 49.

Recall that the domain of a polynomial function, unless restricted by a statement of the function, is $(-\infty, \infty)$. The implications of the application in Example 7 restrict the domain of the function. If we assume that a patient had not taken any of the medication before, it seems reasonable that $M(0) = 0$; that is, at time 0, there is 0 mg of the medication in the bloodstream. After the medication has been taken, $M(t)$ will be positive for a period of time and eventually decrease back to 0 when $t = 6$ and not increase again (unless another dose is taken). Thus the restricted domain is $[0, 6]$.

Technology Connection

X	Y1
0	0
.5	150.15
1	255
1.5	318.26
2	344.4
2.5	338.63
3	306.9

X =

We can evaluate the function in Example 7 with the TABLE feature of a graphing calculator set in AUTO mode. We start at 0 and use a step-value of 0.5.

As discussed above, the domain of $M(t)$ is $[0, 6]$. To determine the range, we find the relative maximum value of the function using the MAXIMUM feature.

The maximum is about 345.76 mg. It occurs approximately 2.15 hr, or 2 hr 9 min, after the initial dose has been taken. The range is about $[0, 345.76]$.

$y = 0.5x^4 + 3.45x^3 - 96.65x^2 + 347.7x$

500, 0, 6, −200

Maximum
X = 2.1485504 Y = 345.76165

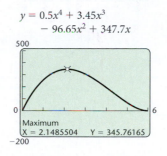

4.1 Exercise Set

Determine the leading term, the leading coefficient, and the degree of the polynomial. Then classify the polynomial function as constant, linear, quadratic, cubic, or quartic.

1. $g(x) = \frac{1}{2}x^3 - 10x + 8$
2. $f(x) = 15x^2 - 10 + 0.11x^4 - 7x^3$
3. $h(x) = 0.9x - 0.13$
4. $f(x) = -6$
5. $g(x) = 305x^4 + 4021$
6. $h(x) = 2.4x^3 + 5x^2 - x + \frac{7}{8}$
7. $h(x) = -5x^2 + 7x^3 + x^4$
8. $f(x) = 2 - x^2$
9. $g(x) = 4x^3 - \frac{1}{2}x^2 + 8$
10. $f(x) = 12 + x$

In Exercises 11–18, select one of the following four sketches to describe the end behavior of the graph of the function.

a) 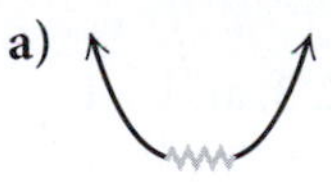

b) 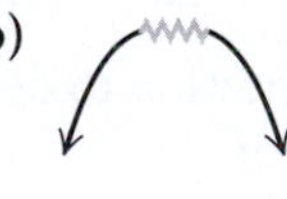

c) 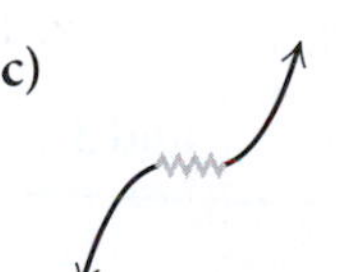

d) 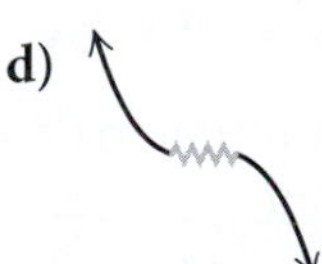

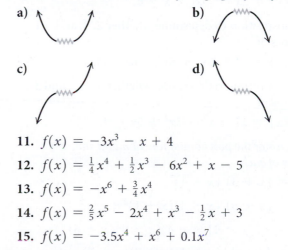

11. $f(x) = -3x^3 - x + 4$
12. $f(x) = \frac{1}{4}x^4 + \frac{1}{2}x^3 - 6x^2 + x - 5$
13. $f(x) = -x^6 + \frac{3}{4}x^4$
14. $f(x) = \frac{2}{5}x^5 - 2x^4 + x^3 - \frac{1}{2}x + 3$
15. $f(x) = -3.5x^4 + x^6 + 0.1x^7$

16. $f(x) = -x^3 + x^5 - 0.5x^6$

17. $f(x) = 10 + \frac{1}{10}x^4 - \frac{2}{5}x^3$

18. $f(x) = 2x + x^3 - x^5$

In Exercises 19–22, use the leading-term test to match the function with one of the graphs (a)–(d) that follow.

a)

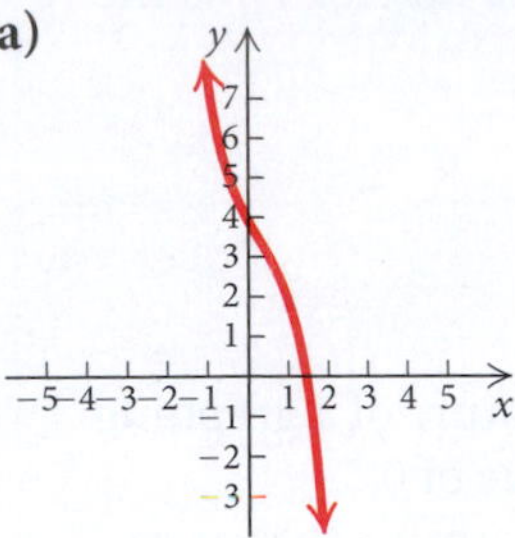

b)

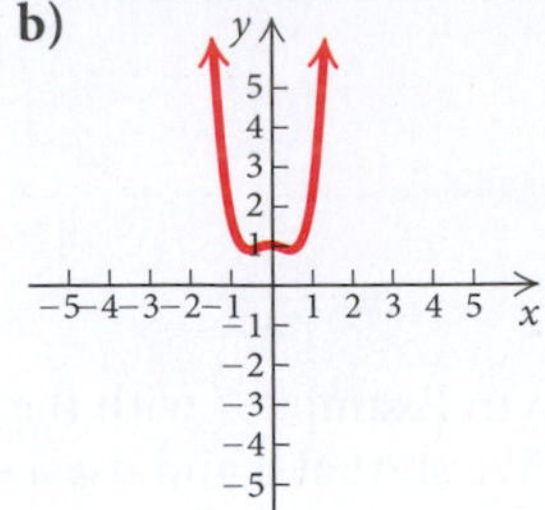

c)

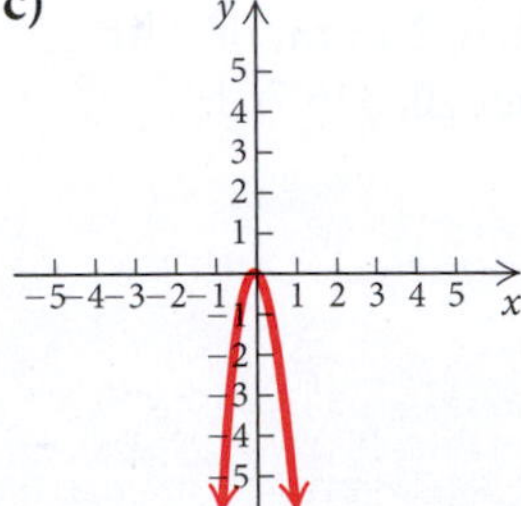

d)

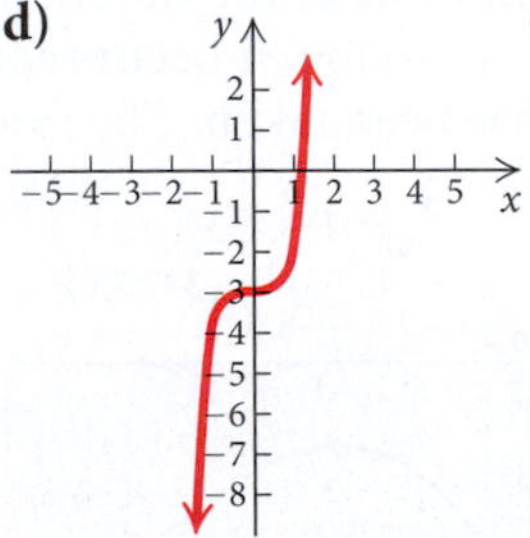

19. $f(x) = -x^6 + 2x^5 - 7x^2$

20. $f(x) = 2x^4 - x^2 + 1$

21. $f(x) = x^5 + \frac{1}{10}x - 3$

22. $f(x) = -x^3 + x^2 - 2x + 4$

23. Use substitution to determine whether 4, 5, and -2 are zeros of
$$f(x) = x^3 - 9x^2 + 14x + 24.$$

24. Use substitution to determine whether 2, 3, and -1 are zeros of
$$f(x) = 2x^3 - 3x^2 + x + 6.$$

25. Use substitution to determine whether 2, 3, and -1 are zeros of
$$g(x) = x^4 - 6x^3 + 8x^2 + 6x - 9.$$

26. Use substitution to determine whether 1, -2, and 3 are zeros of
$$g(x) = x^4 - x^3 - 3x^2 + 5x - 2.$$

Find the zeros of the polynomial function and state the multiplicity of each.

27. $f(x) = (x + 3)^2(x - 1)$

28. $f(x) = (x + 5)^3(x - 4)(x + 1)^2$

29. $f(x) = -2(x - 4)(x - 4)(x - 4)(x + 6)$

30. $f(x) = \left(x + \frac{1}{2}\right)(x + 7)(x + 7)(x + 5)$

31. $f(x) = (x^2 - 9)^3$

32. $f(x) = (x^2 - 4)^2$

33. $f(x) = x^3(x - 1)^2(x + 4)$

34. $f(x) = x^2(x + 3)^2(x - 4)(x + 1)^4$

35. $f(x) = -8(x - 3)^2(x + 4)^3x^4$

36. $f(x) = (x^2 - 5x + 6)^2$

37. $f(x) = x^4 - 4x^2 + 3$

38. $f(x) = x^4 - 10x^2 + 9$

39. $f(x) = x^3 + 3x^2 - x - 3$

40. $f(x) = x^3 - x^2 - 2x + 2$

41. $f(x) = 2x^3 - x^2 - 8x + 4$

42. $f(x) = 3x^3 + x^2 - 48x - 16$

Determine whether the statement is true or false.

43. If $P(x) = (x - 3)^4(x + 1)^3$, then the graph of the polynomial function $y = P(x)$ crosses the x-axis at $(3, 0)$.

44. If $P(x) = (x + 2)^2\left(x - \frac{1}{4}\right)^5$, then the graph of the polynomial function $y = P(x)$ crosses the x-axis at $\left(\frac{1}{4}, 0\right)$.

45. If $P(x) = (x - 2)^3(x + 5)^6$, then the graph of $y = P(x)$ is tangent to the x-axis at $(-5, 0)$.

46. If $P(x) = (x + 4)^2(x - 1)^2$, then the graph of $y = P(x)$ is tangent to the x-axis at $(4, 0)$.

47. *Vinyl Album Sales.* Vinyl record albums are making a comeback. Sales of vinyl albums rose 32% from 2012 to 2013. The sales data over the years 2001 to 2013 are modeled by the quartic function
$$f(x) = -0.000913x^4 + 0.0248x^3 - 0.1515x^2 + 0.2136x + 1.2779,$$
where x is the number of years after 2001 and $f(x)$ is the number of albums in millions (*Source:* Nielsen SoundScan). Find the number of vinyl albums sold in 2008, in 2012, and in 2016.

48. *Railroad Miles.* The greatest combined length of U.S.-owned operating railroad track existed in 1916, when industrial activity increased during World War I. The total length has decreased ever since. The data over the years 1900 to 2011 are modeled by the quartic function

$$f(x) = -0.002391x^4 + 0.949686x^3 - 123.648199x^2 + 4729.3635x + 198{,}846.4097,$$

where x is the number of years after 1900 and $f(x)$ is in miles (*Source:* Association of American Railroads). Find the number of miles of operating railroad track in the United States in 1916, in 1960, in 2000, and in 2016.

49. *Dog Years.* A dog's life span is typically much shorter than that of a human. The cubic function

$$d(x) = 0.010255x^3 - 0.340119x^2 + 7.397499x + 6.618361,$$

where x is the dog's age, in years, approximates the equivalent human age in years. Estimate the equivalent human age for dogs that are 3, 12, and 16 years old.

50. *Threshold Weight.* In a study performed by Alvin Shemesh, it was found that the **threshold weight W**, defined as the weight above which the risk of death rises dramatically, is given by

$$W(h) = \left(\frac{h}{12.3}\right)^3,$$

where W is in pounds and h is a person's height, in inches. Find the threshold weight of a person who is 5 ft 7 in. tall.

51. *Projectile Motion.* A stone thrown downward with an initial velocity of 34.3 m/sec will travel a distance of s meters, where

$$s(t) = 4.9t^2 + 34.3t$$

and t is in seconds. If a stone is thrown downward at 34.3 m/sec from a height of 294 m, how long will it take the stone to hit the ground?

52. *Games in a Sports League.* If there are x teams in a sports league and all the teams play each other twice, a total of $N(x)$ games are played, where

$$N(x) = x^2 - x.$$

A softball league has 9 teams, each of which plays the others twice. If the league pays \$110 per game for the field and the umpires, how much will it cost to play the entire schedule?

53. *Prison Admissions.* Since 2006, total admissions to state and federal prisons have been declining (*Source:* Bureau of Justice Statistics). The quartic function

$$p(x) = 6.213x^4 - 432.347x^3 + 1922.987x^2 + 20{,}503.912x + 638{,}684.984,$$

where x is the number of years after 2001, can be used to estimate the number of admissions to state and federal prisons from 2001 to 2012. Estimate the number of prison admissions in 2003, in 2006, and in 2011.

54. *Obesity.* The percentage of adults who are obese is rising (*Source:* Gallup–Healthways Well-Being Index). The cubic function

$$f(x) = 0.102x^3 - 0.764x^2 + 1.595x + 25.494,$$

where x is the number of years after 2008, can be used to estimate the percentage of adults who are obese. Using this function, estimate the percentage of adults who were obese in 2009 and in 2013.

55. *Interest Compounded Annually.* When P dollars is invested at interest rate i, compounded annually, for t years, the investment grows to A dollars, where

$$A = P(1 + i)^t.$$

Trevor's parents deposit \$8000 in a savings account when Trevor is 16 years old. The principal plus interest is to be used for a truck when Trevor is 18 years old. Find the interest rate i if the \$8000 grows to \$9039.75 in 2 years.

56. *Interest Compounded Annually.* When P dollars is invested at interest rate i, compounded annually, for t years, the investment grows to A dollars, where

$$A = P(1 + i)^t.$$

When Sara enters the 11th grade, her grandparents deposit $10,000 in a college savings account. Find the interest rate i if the $10,000 grows to $11,193.64 in 2 years.

Skill Maintenance

Find the distance between the pair of points. [1.1]

57. $(3, -5)$ and $(0, -1)$

58. $(4, 2)$ and $(-2, -4)$

59. Find the center and the radius of the circle $(x - 3)^2 + (y + 5)^2 = 49$. [1.1]

60. The diameter of a circle connects the points $(-6, 5)$ and $(-2, 1)$ on the circle. Find the coordinates of the center of the circle and the length of the radius. [1.1]

Solve.

61. $2y - 3 \geq 1 - y + 5$ [1.6]

62. $(x - 2)(x + 5) > x(x - 3)$ [1.6]

63. $|x + 6| \geq 7$ [3.5]

64. $|x + \frac{1}{4}| \leq \frac{2}{3}$ [3.5]

Synthesis

Determine the degree and the leading term of the polynomial function.

65. $f(x) = (x^5 - 1)^2(x^2 + 2)^3$

66. $f(x) = (10 - 3x^5)^2(5 - x^4)^3(x + 4)$

4.2 Graphing Polynomial Functions

- Graph polynomial functions.
- Use the intermediate value theorem to determine whether a function has a real zero between two given real numbers.

Graphing Polynomial Functions

In addition to using the leading-term test and finding the zeros of the function, it is helpful to consider the following facts when graphing a polynomial function.

> If $P(x)$ is a polynomial function of degree n, the graph of the function has:
>
> - at most n real zeros, and thus at most n x-intercepts;
> - at most $n - 1$ turning points.
>
> (Turning points on a graph, also called relative maxima and minima, occur when the function changes from decreasing to increasing or from increasing to decreasing.)

EXAMPLE 1 Graph the polynomial function $h(x) = -2x^4 + 3x^3$.

Solution

1. First, we use the leading-term test to determine the end behavior of the graph. The leading term is $-2x^4$. The degree, 4, is even, and the coefficient, -2, is negative. Thus the end behavior of the graph as $x \to \infty$ and as $x \to -\infty$ can be sketched as follows.

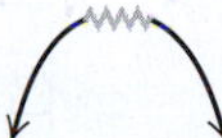

2. The zeros of the function are the first coordinates of the x-intercepts of the graph. To find the zeros, we solve $h(x) = 0$ by factoring and using the principle of zero products.

$$-2x^4 + 3x^3 = 0$$

$$-x^3(2x - 3) = 0 \qquad \text{Factoring}$$

$$-x^3 = 0 \quad or \quad 2x - 3 = 0 \qquad \text{Using the principle of zero products}$$

$$x = 0 \quad or \quad x = \frac{3}{2}.$$

The zeros of the function are 0 and $\frac{3}{2}$. Note that the multiplicity of 0 is 3 and the multiplicity of $\frac{3}{2}$ is 1. The x-intercepts are $(0, 0)$ and $\left(\frac{3}{2}, 0\right)$.

3. The zeros divide the x-axis into three intervals:

$$(-\infty, 0), \quad \left(0, \frac{3}{2}\right), \quad \text{and} \quad \left(\frac{3}{2}, \infty\right).$$

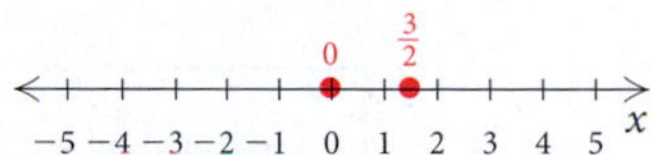

The sign of $h(x)$ is the same for all values of x in each of the three intervals. That is, **$h(x)$ is positive for all x-values in an interval or $h(x)$ is negative for all x-values in an interval**. To determine which, we choose a test value for x from each interval and find $h(x)$.

Interval	$(-\infty, 0)$	$\left(0, \frac{3}{2}\right)$	$\left(\frac{3}{2}, \infty\right)$
Test Value, x	-1	1	2
Function Value, $h(x)$	-5	1	-8
Sign of $h(x)$	$-$	$+$	$-$
Location of Points on Graph	Below x-axis	Above x-axis	Below x-axis

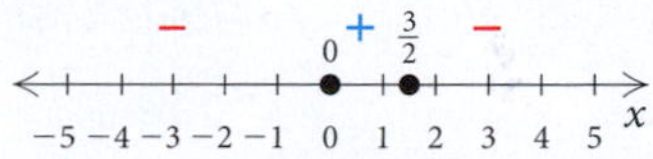

This test-point procedure also gives us three points to plot. In this case, we have $(-1, -5)$, $(1, 1)$, and $(2, -8)$.

4. To determine the y-intercept, we find $h(0)$:

$$h(x) = -2x^4 + 3x^3$$
$$h(0) = -2 \cdot 0^4 + 3 \cdot 0^3 = 0.$$

The y-intercept is $(0, 0)$.

5. A few additional points are helpful when completing the graph.

x	$h(x)$
-1.5	-20.25
-0.5	-0.5
0.5	0.25
2.5	-31.25

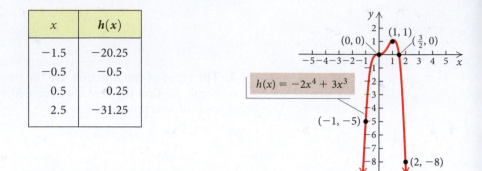

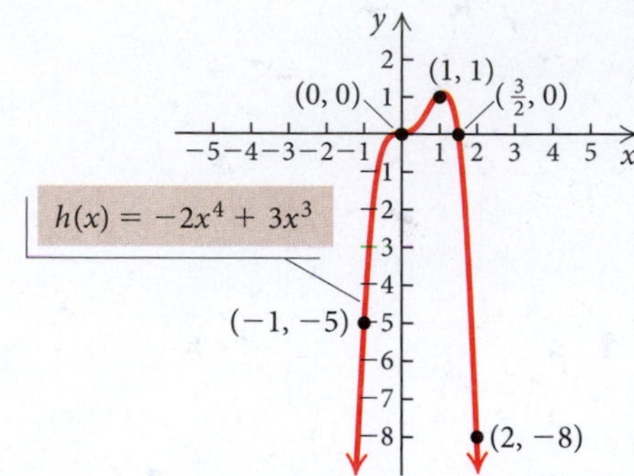

6. The degree of h is 4. The graph of h can have at most 4 x-intercepts and at most 3 turning points. In fact, it has 2 x-intercepts and 1 turning point. The zeros, 0 and $\frac{3}{2}$, each have odd multiplicities: 3 for 0 and 1 for $\frac{3}{2}$. Since the multiplicities are odd, the graph crosses the x-axis at 0 and $\frac{3}{2}$. The end behavior of the graph is what we described in step (1). As $x \to \infty$ and also as $x \to -\infty$, $h(x) \to -\infty$. The graph appears to be correct.

Now Try Exercise 23.

The following is a procedure for graphing polynomial functions.

To graph a polynomial function:

1. Use the leading-term test to determine the end behavior.
2. Find the zeros of the function by solving $f(x) = 0$. Any real zeros are the first coordinates of the x-intercepts.
3. Use the x-intercepts (zeros) to divide the x-axis into intervals and choose a test point in each interval to determine the sign of all function values in that interval.
4. Find $f(0)$. This gives the y-intercept of the function.
5. If necessary, find additional function values to determine the general shape of the graph and then draw the graph.
6. As a partial check, use the facts that the graph has at most n x-intercepts and at most $n - 1$ turning points. Multiplicity of zeros can also be considered in order to check where the graph crosses or is tangent to the x-axis.

EXAMPLE 2 Graph the polynomial function

$$f(x) = 2x^3 + x^2 - 8x - 4.$$

Solution

1. The leading term is $2x^3$. The degree, 3, is odd, and the coefficient, 2, is positive. Thus the end behavior of the graph will appear as follows.

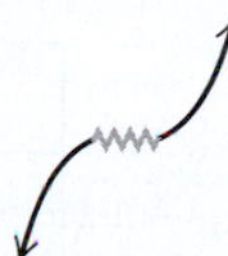

2. To find the zeros, we solve $f(x) = 0$. Here we can use factoring by grouping.

$$2x^3 + x^2 - 8x - 4 = 0$$
$$x^2(2x + 1) - 4(2x + 1) = 0 \qquad \text{Factoring by grouping}$$
$$(2x + 1)(x^2 - 4) = 0$$
$$(2x + 1)(x + 2)(x - 2) = 0 \qquad \text{Factoring a difference of squares}$$

The zeros are $-\frac{1}{2}$, -2, and 2. Each is of multiplicity 1. The x-intercepts are $(-2, 0)$, $\left(-\frac{1}{2}, 0\right)$, and $(2, 0)$.

3. The zeros divide the x-axis into four intervals:

$$(-\infty, -2), \quad \left(-2, -\frac{1}{2}\right), \quad \left(-\frac{1}{2}, 2\right), \quad \text{and} \quad (2, \infty).$$

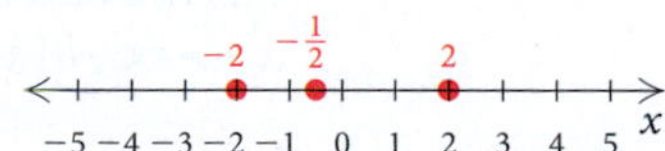

We choose a test value for x from each interval and find $f(x)$.

Interval	$(-\infty, -2)$	$\left(-2, -\frac{1}{2}\right)$	$\left(-\frac{1}{2}, 2\right)$	$(2, \infty)$
Test Value, x	-3	-1	1	3
Function Value, $f(x)$	-25	3	-9	35
Sign of $f(x)$	$-$	$+$	$-$	$+$
Location of Points on Graph	Below x-axis	Above x-axis	Below x-axis	Above x-axis

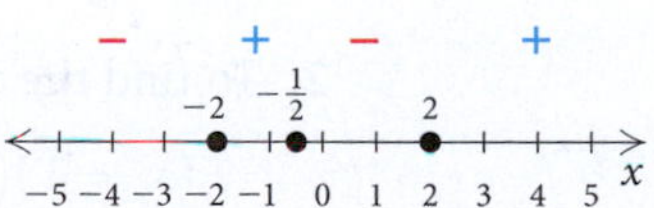

The test values and corresponding function values also give us four points on the graph: $(-3, -25)$, $(-1, 3)$, $(1, -9)$, and $(3, 35)$.

4. To determine the y-intercept, we find $f(0)$:

$$f(x) = 2x^3 + x^2 - 8x - 4$$
$$f(0) = 2 \cdot 0^3 + 0^2 - 8 \cdot 0 - 4 = -4.$$

The y-intercept is $(0, -4)$.

5. We find a few additional points and complete the graph.

x	$f(x)$
−2.5	−9
−1.5	3.5
0.5	−7.5
1.5	−7

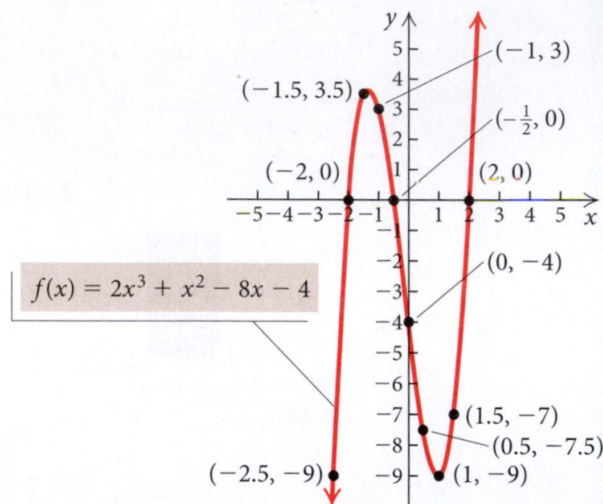

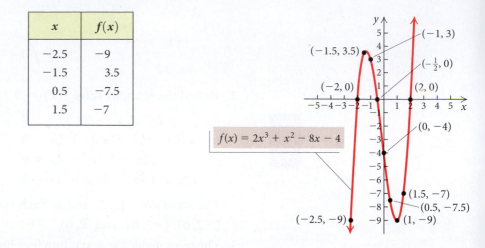

6. The degree of f is 3. The graph of f can have at most 3 x-intercepts and at most 2 turning points. It has 3 x-intercepts and 2 turning points. Each zero has a multiplicity of 1; thus the graph crosses the x-axis at -2, $-\frac{1}{2}$, and 2. The graph has the end behavior described in step (1). As $x \to -\infty$, $h(x) \to -\infty$, and as $x \to \infty$, $h(x) \to \infty$. The graph appears to be correct.

Now Try Exercise 33.

Some polynomials are difficult to factor. In the next example, the polynomial is given in factored form. In Sections 4.3 and 4.4, we will learn methods that facilitate determining factors of such polynomials.

EXAMPLE 3 Graph the polynomial function

$$g(x) = x^4 - 7x^3 + 12x^2 + 4x - 16$$
$$= (x + 1)(x - 2)^2(x - 4).$$

Solution

1. The leading term is x^4. The degree, 4, is even, and the coefficient, 1, is positive. The sketch below shows the end behavior.

2. To find the zeros, we solve $g(x) = 0$:

$$(x + 1)(x - 2)^2(x - 4) = 0.$$

The zeros are -1, 2, and 4; 2 is of multiplicity 2; the others are of multiplicity 1. The x-intercepts are $(-1, 0)$, $(2, 0)$, and $(4, 0)$.

3. The zeros divide the x-axis into four intervals:

$$(-\infty, -1), \quad (-1, 2), \quad (2, 4), \quad \text{and} \quad (4, \infty).$$

We choose a test value for x from each interval and find $g(x)$.

Interval	$(-\infty, -1)$	$(-1, 2)$	$(2, 4)$	$(4, \infty)$
Test Value, x	-1.25	1	3	4.25
Function Value, $g(x)$	≈ 13.9	-6	-4	≈ 6.6
Sign of $g(x)$	+	−	−	+
Location of Points on Graph	Above x-axis	Below x-axis	Below x-axis	Above x-axis

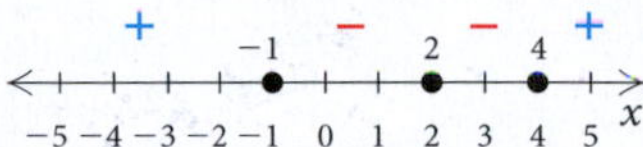

The test values and the corresponding function values also give us four points on the graph: $(-1.25, 13.9)$, $(1, -6)$, $(3, -4)$, and $(4.25, 6.6)$.

4. To determine the y-intercept, we find $g(0)$:

$$g(x) = x^4 - 7x^3 + 12x^2 + 4x - 16$$
$$g(0) = 0^4 - 7 \cdot 0^3 + 12 \cdot 0^2 + 4 \cdot 0 - 16 = -16.$$

The y-intercept is $(0, -16)$.

5. We find a few additional points and draw the graph.

x	$g(x)$
-0.5	-14.1
0.5	-11.8
1.5	-1.6
2.5	-1.3
3.5	-5.1

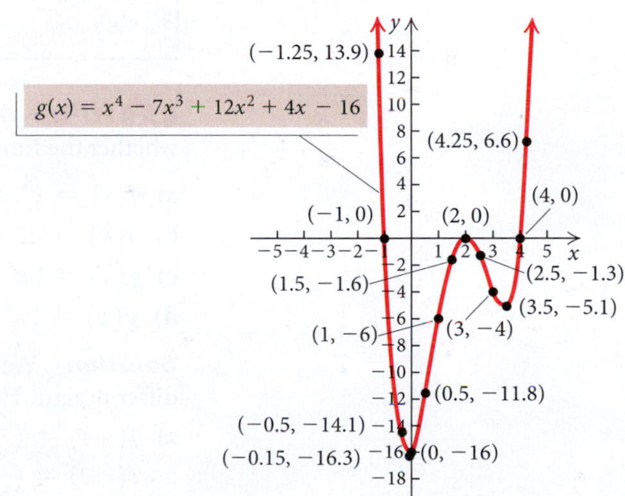

6. The degree of g is 4. The graph of g can have at most 4 x-intercepts and at most 3 turning points. It has 3 x-intercepts and 3 turning points. One of the zeros, 2, has a multiplicity of 2, so the graph is tangent to the x-axis at 2. The other zeros, -1 and 4, each have a multiplicity of 1 so the graph crosses the x-axis at -1 and 4. The graph has the end behavior described in step (1). As $x \to \infty$ and as $x \to -\infty$, $g(x) \to \infty$. The graph appears to be correct.

Now Try Exercise 19.

The Intermediate Value Theorem

Polynomial functions are continuous, hence their graphs are unbroken. The domain of a polynomial function, unless restricted by the statement of the function, is $(-\infty, \infty)$. Suppose two polynomial function values $P(a)$ and $P(b)$ have opposite signs. Since P is continuous, its graph must be a curve from $(a, P(a))$ to $(b, P(b))$ without a break. Then it follows that the curve must cross the x-axis at at least one point c between a and b; that is, the function has a zero at c between a and b.

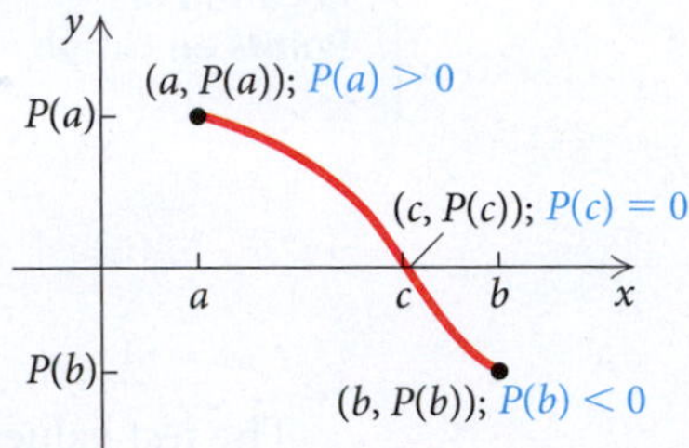

THE INTERMEDIATE VALUE THEOREM

For any polynomial function $P(x)$ with real coefficients, suppose that for $a \neq b$, $P(a)$ and $P(b)$ are of opposite signs. Then the function has at least one real zero between a and b.

The intermediate value theorem *cannot* be used to determine whether there is a real zero between a and b when $P(a)$ and $P(b)$ have the *same* sign.

EXAMPLE 4 Using the intermediate value theorem, determine, if possible, whether the function has at least one real zero between a and b.

a) $f(x) = x^3 + x^2 - 6x$; $a = -4, b = -2$

b) $f(x) = x^3 + x^2 - 6x$; $a = -1, b = 3$

c) $g(x) = \frac{1}{3}x^4 - x^3$; $a = -\frac{1}{2}, b = \frac{1}{2}$

d) $g(x) = \frac{1}{3}x^4 - x^3$; $a = 1, b = 2$

Solution We find $f(a)$ and $f(b)$ or $g(a)$ and $g(b)$ and determine whether they differ in sign. The graphs of $f(x)$ and $g(x)$ provide visual checks of the conclusions.

a) $f(-4) = (-4)^3 + (-4)^2 - 6(-4) = -24$,

$f(-2) = (-2)^3 + (-2)^2 - 6(-2) = 8$

Note that $f(-4)$ is negative and $f(-2)$ is positive. By the intermediate value theorem, since $f(-4)$ and $f(-2)$ have opposite signs, then $f(x)$ has at least one zero between -4 and -2. The following graph confirms this.

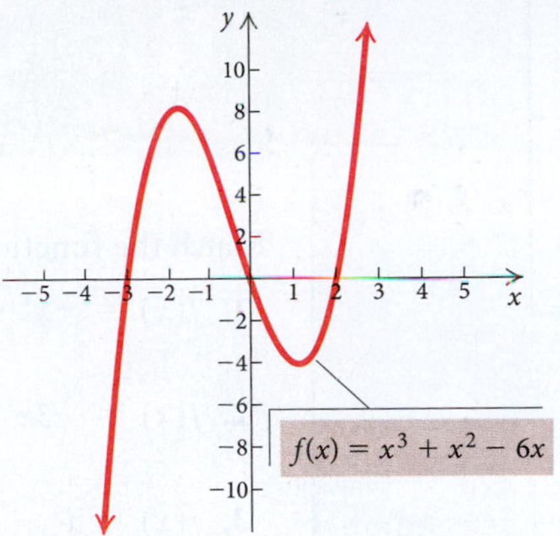

b) $f(-1) = (-1)^3 + (-1)^2 - 6(-1) = 6,$
$f(3) = 3^3 + 3^2 - 6(3) = 18$

Both $f(-1)$ and $f(3)$ are positive. Thus the intermediate value theorem *does not allow* us to determine whether there is a real zero between -1 and 3. Note that the graph of $f(x)$ above shows that there are two zeros between -1 and 3.

c) $g\left(-\frac{1}{2}\right) = \frac{1}{3}\left(-\frac{1}{2}\right)^4 - \left(-\frac{1}{2}\right)^3 = \frac{7}{48},$
$g\left(\frac{1}{2}\right) = \frac{1}{3}\left(\frac{1}{2}\right)^4 - \left(\frac{1}{2}\right)^3 = -\frac{5}{48}$

Since $g\left(-\frac{1}{2}\right)$ and $g\left(\frac{1}{2}\right)$ have opposite signs, $g(x)$ has at least one zero between $-\frac{1}{2}$ and $\frac{1}{2}$. The following graph confirms this.

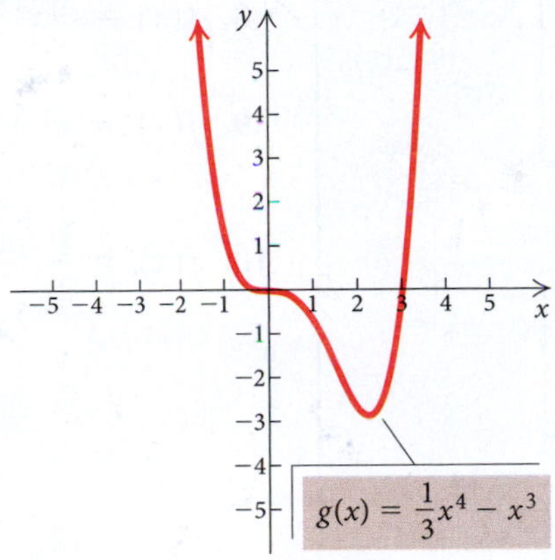

d) $g(1) = \frac{1}{3}(1)^4 - 1^3 = -\frac{2}{3},$
$g(2) = \frac{1}{3}(2)^4 - 2^3 = -\frac{8}{3}$

Both $g(1)$ and $g(2)$ are negative. Thus the intermediate value theorem *does not allow* us to determine whether there is a real zero between 1 and 2. Note that the graph of $g(x)$ above shows that there are no zeros between 1 and 2.

Now Try Exercises 39 and 43.

Visualizing the Graph

Match the function with its graph.

1. $f(x) = -x^4 - x + 5$

2. $f(x) = -3x^2 + 6x - 3$

3. $f(x) = x^4 - 4x^3 + 3x^2 + 4x - 4$

4. $f(x) = -\frac{2}{5}x + 4$

5. $f(x) = x^3 - 4x^2$

6. $f(x) = x^6 - 9x^4$

7. $f(x) = x^5 - 3x^3 + 2$

8. $f(x) = -x^3 - x - 1$

9. $f(x) = x^2 + 7x + 6$

10. $f(x) = \frac{7}{2}$

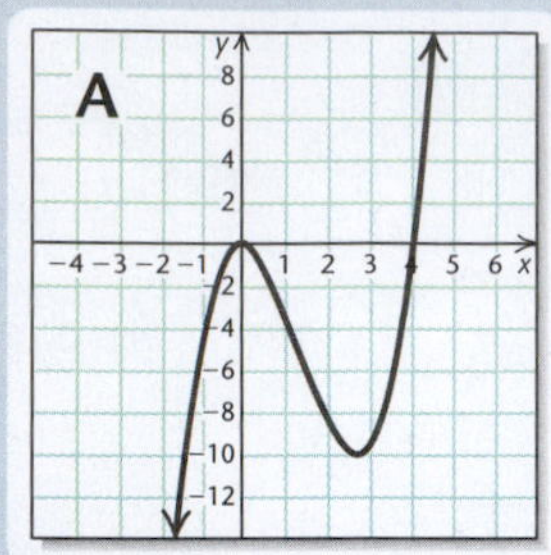

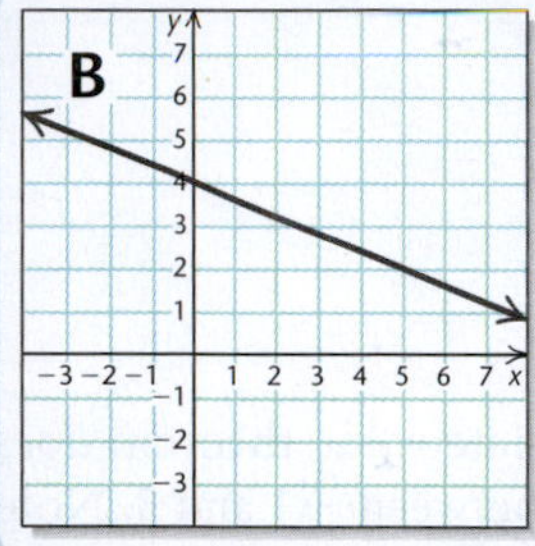

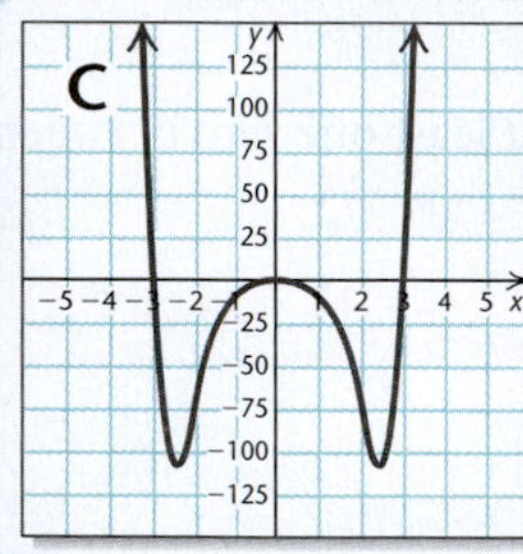

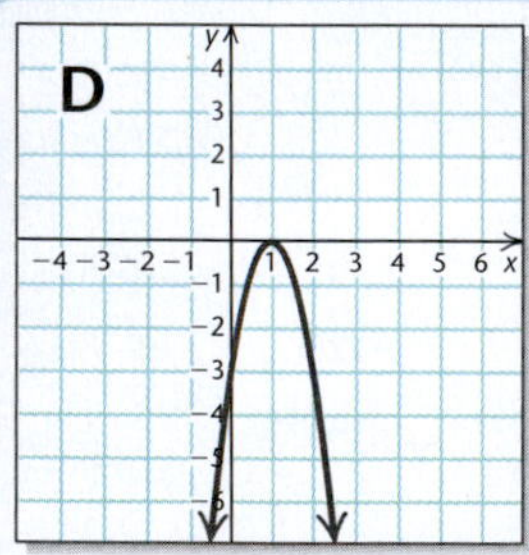

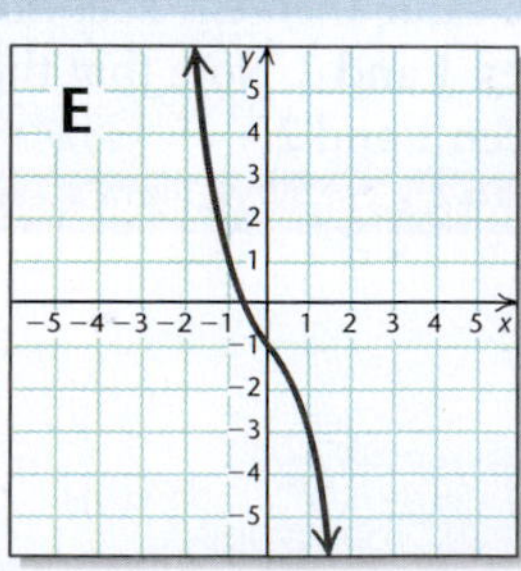

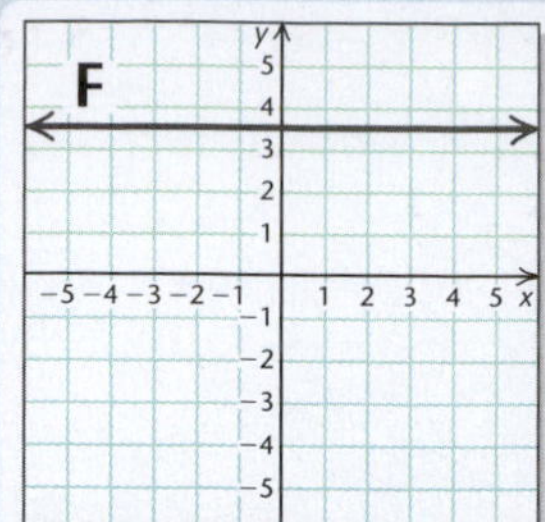

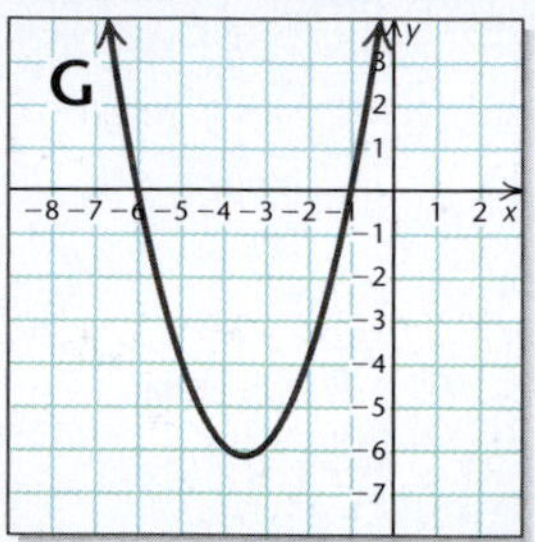

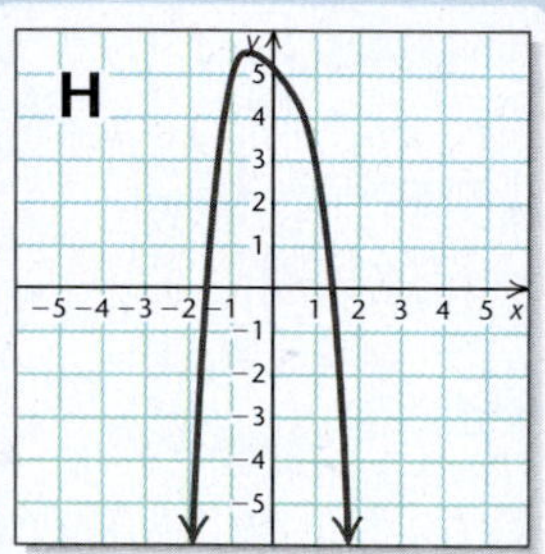

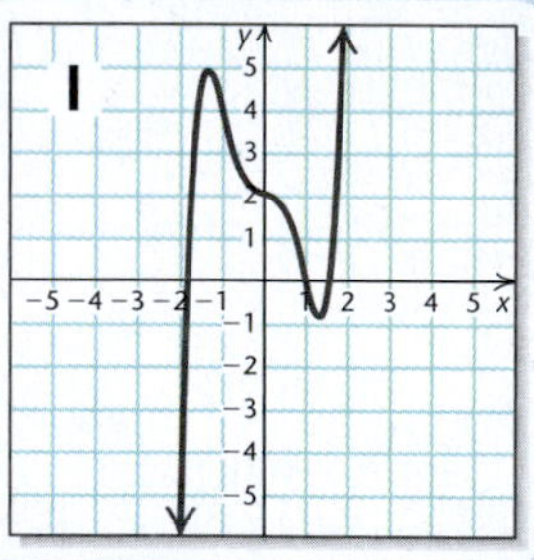

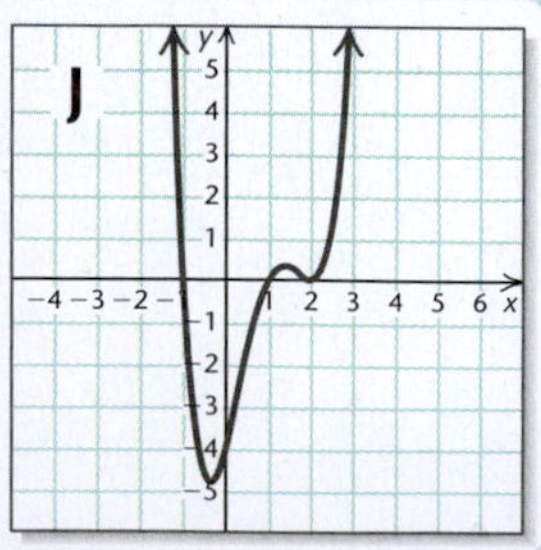

Answers on page A-17

4.2 Exercise Set

For each function in Exercises 1–6, state:

a) *the maximum number of real zeros that the function can have;*

b) *the maximum number of x-intercepts that the graph of the function can have; and*

c) *the maximum number of turning points that the graph of the function can have.*

1. $f(x) = x^5 - x^2 + 6$

2. $f(x) = -x^2 + x^4 - x^6 + 3$

3. $f(x) = x^{10} - 2x^5 + 4x - 2$

4. $f(x) = \frac{1}{4}x^3 + 2x^2$

5. $f(x) = -x - x^3$

6. $f(x) = -3x^4 + 2x^3 - x - 4$

In Exercises 7–12, use the leading-term test and your knowledge of y-intercepts to match the function with one of the graphs (a)–(f) that follow.

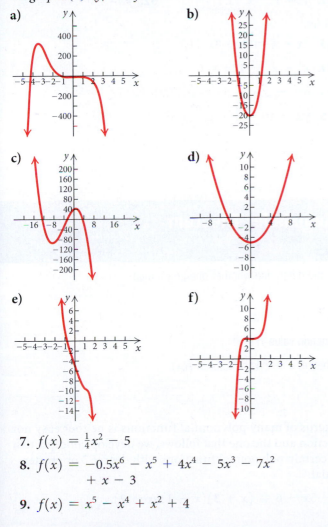

a)

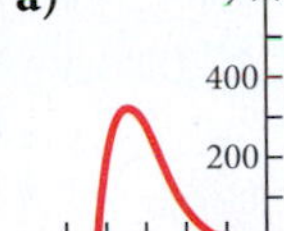

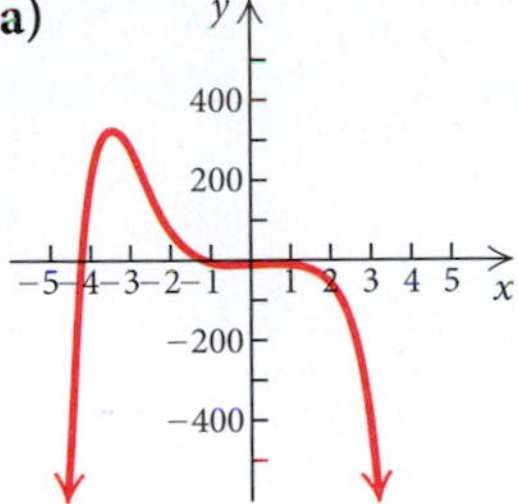

b)

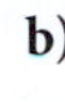

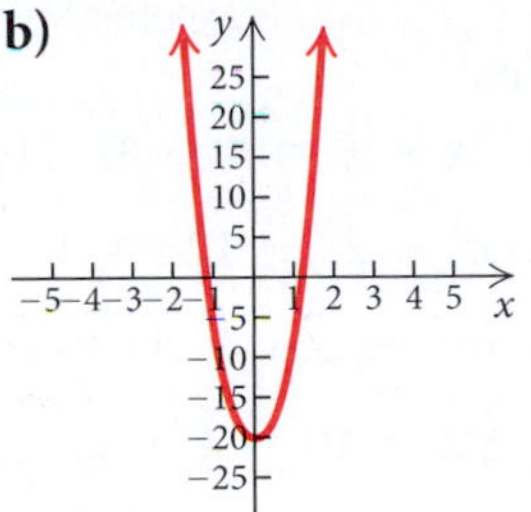

c)

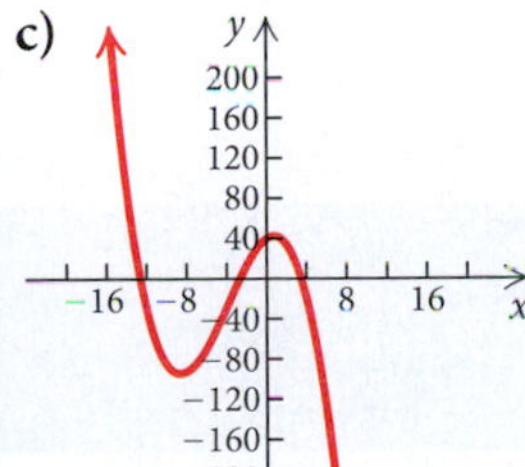

d)

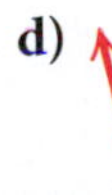

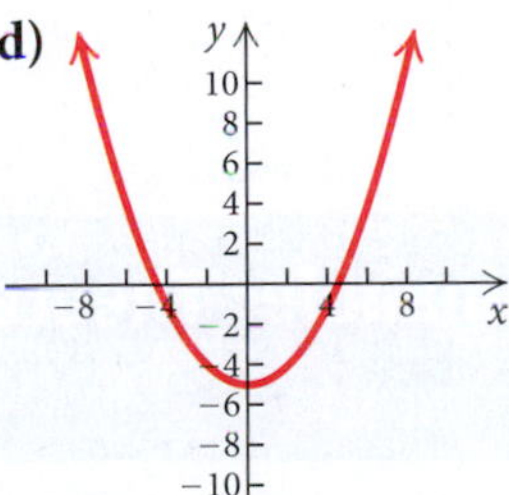

e)

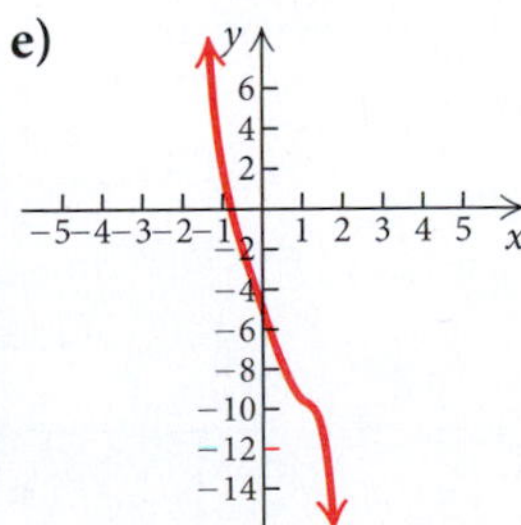

f)

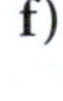

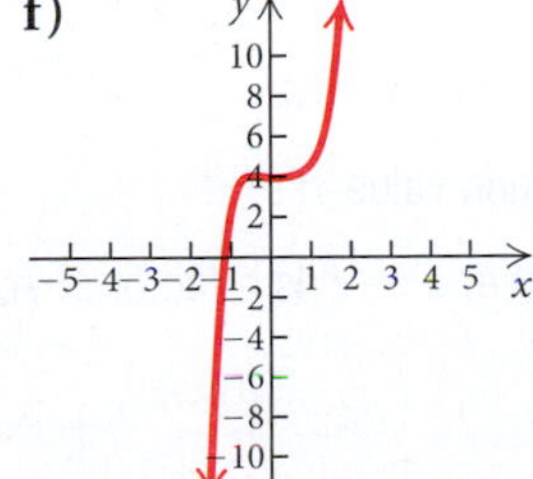

7. $f(x) = \frac{1}{4}x^2 - 5$

8. $f(x) = -0.5x^6 - x^5 + 4x^4 - 5x^3 - 7x^2 + x - 3$

9. $f(x) = x^5 - x^4 + x^2 + 4$

10. $f(x) = -\frac{1}{3}x^3 - 4x^2 + 6x + 42$

11. $f(x) = x^4 - 2x^3 + 12x^2 + x - 20$

12. $f(x) = -0.3x^7 + 0.11x^6 - 0.25x^5 + x^4 + x^3 - 6x - 5$

Graph the polynomial function. Follow the steps outlined in the procedure on p. 240.

13. $f(x) = -x^3 - 2x^2$

14. $g(x) = x^4 - 4x^3 + 3x^2$

15. $h(x) = x^2 + 2x - 3$

16. $f(x) = x^2 - 5x + 4$

17. $h(x) = x^5 - 4x^3$

18. $f(x) = x^3 - x$

19. $h(x) = x(x - 4)(x + 1)(x - 2)$

20. $f(x) = x(x - 1)(x + 3)(x + 5)$

21. $g(x) = -\frac{1}{4}x^3 - \frac{3}{4}x^2$

22. $f(x) = \frac{1}{2}x^3 + \frac{5}{2}x^2$

23. $g(x) = -x^4 - 2x^3$

24. $h(x) = x^3 - 3x^2$

25. $f(x) = -\frac{1}{2}(x - 2)(x + 1)^2(x - 1)$

26. $g(x) = (x - 2)^3(x + 3)$

27. $g(x) = -x(x - 1)^2(x + 4)^2$

28. $h(x) = -x(x - 3)(x - 3)(x + 2)$

29. $f(x) = (x - 2)^2(x + 1)^4$

30. $g(x) = x^4 - 9x^2$

31. $g(x) = -(x - 1)^4$

32. $h(x) = (x + 2)^3$

33. $h(x) = x^3 + 3x^2 - x - 3$

34. $g(x) = -x^3 + 2x^2 + 4x - 8$

35. $f(x) = 6x^3 - 8x^2 - 54x + 72$

36. $h(x) = x^5 - 5x^3 + 4x$

Graph each piecewise function.

37. $g(x) = \begin{cases} -x + 3, & \text{for } x \leq -2, \\ 4, & \text{for } -2 < x < 1, \\ \frac{1}{2}x^3, & \text{for } x \geq 1 \end{cases}$

38. $h(x) = \begin{cases} -x^2, & \text{for } x < -2, \\ x + 1, & \text{for } -2 \leq x < 0, \\ x^3 - 1, & \text{for } x \geq 0 \end{cases}$

Using the intermediate value theorem, determine, if possible, whether the function f has at least one real zero between a and b.

39. $f(x) = x^3 + 3x^2 - 9x - 13;\ a = -5, b = -4$

40. $f(x) = x^3 + 3x^2 - 9x - 13;\ a = 1, b = 2$

41. $f(x) = 3x^2 - 2x - 11;\ a = -3, b = -2$

42. $f(x) = 3x^2 - 2x - 11;\ a = 2, b = 3$

43. $f(x) = x^4 - 2x^2 - 6;\ a = 2, b = 3$

44. $f(x) = 2x^5 - 7x + 1;\ a = 1, b = 2$

45. $f(x) = x^3 - 5x^2 + 4;\ a = 4, b = 5$

46. $f(x) = x^4 - 3x^2 + x - 1;\ a = -3, b = -2$

Skill Maintenance

In Exercises 47–52, match the equation with one of the graphs (a)–(f) that follow.

a)

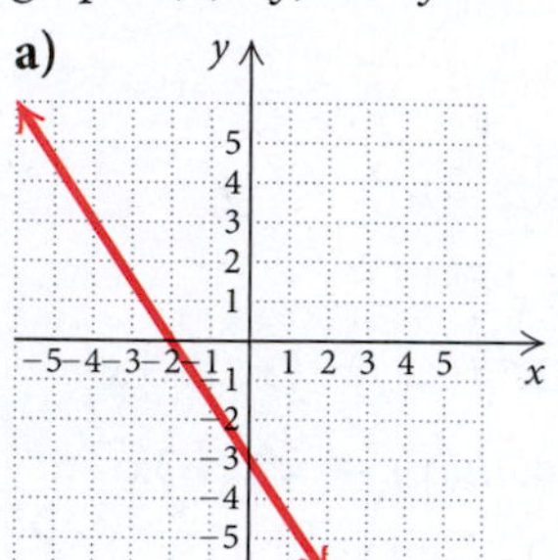

b)

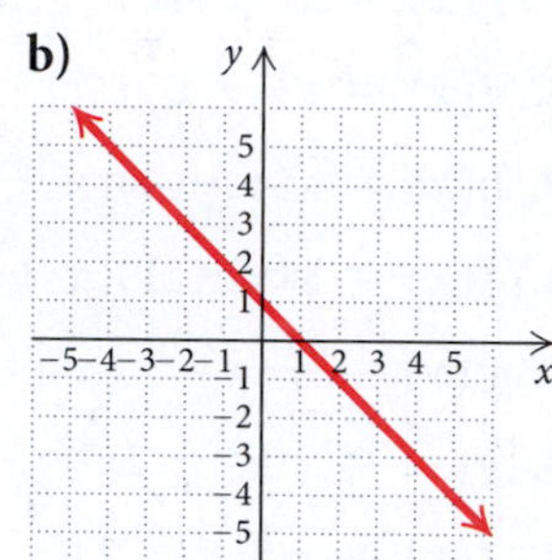

c)

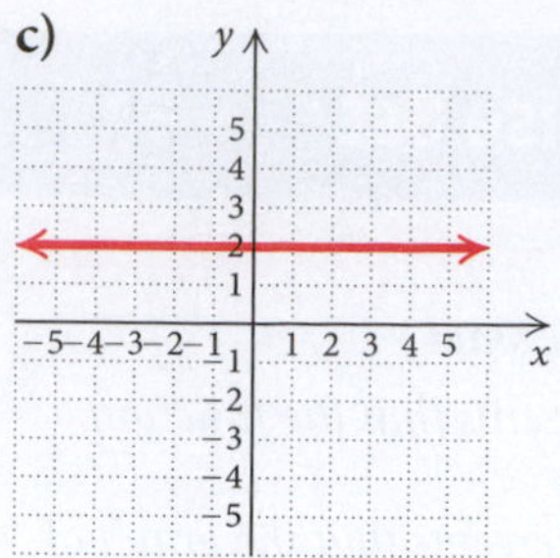

d)

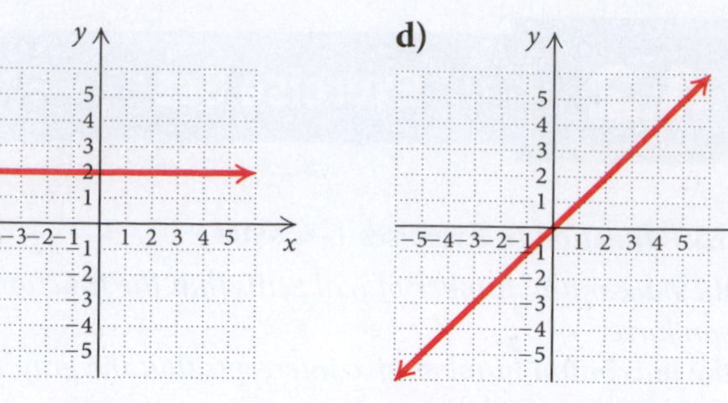

e)

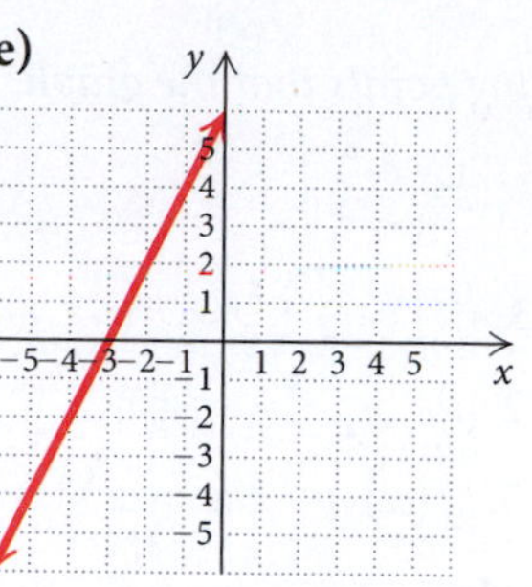

f)

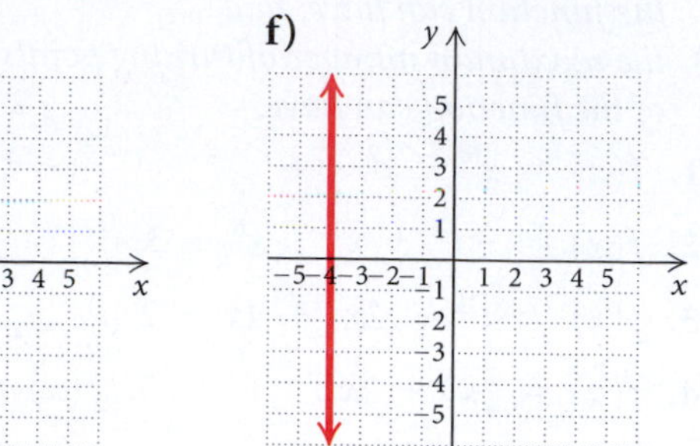

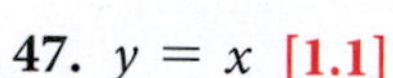

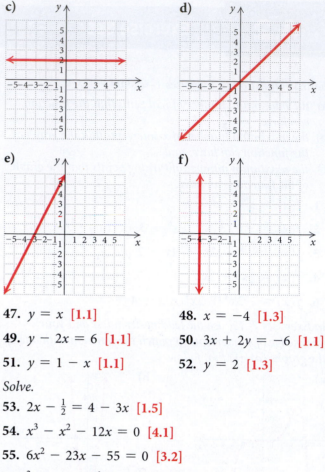

47. $y = x$ [1.1]

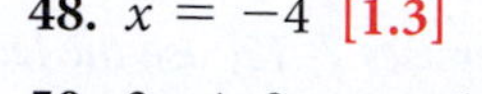

48. $x = -4$ [1.3]

49. $y - 2x = 6$ [1.1]

50. $3x + 2y = -6$ [1.1]

51. $y = 1 - x$ [1.1]

52. $y = 2$ [1.3]

Solve.

53. $2x - \frac{1}{2} = 4 - 3x$ [1.5]

54. $x^3 - x^2 - 12x = 0$ [4.1]

55. $6x^2 - 23x - 55 = 0$ [3.2]

56. $\frac{3}{4}x + 10 = \frac{1}{5} + 2x$ [1.5]

4.3 Polynomial Division; The Remainder Theorem and the Factor Theorem

- Perform long division with polynomials and determine whether one polynomial is a factor of another.
- Use synthetic division to divide a polynomial by $x - c$.
- Use the remainder theorem to find a function value $f(c)$.
- Use the factor theorem to determine whether $x - c$ is a factor of $f(x)$.

In general, finding exact zeros of many polynomial functions is neither easy nor straightforward. In this section and the one that follows, we develop concepts that help us find exact zeros of certain polynomial functions with degree 3 or greater.

Consider the polynomial

$$h(x) = x^3 + 2x^2 - 5x - 6 = (x + 3)(x + 1)(x - 2).$$

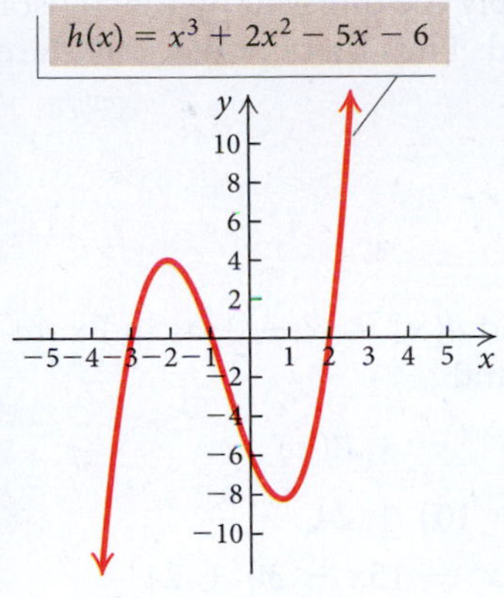

The factors are

$$x + 3, \quad x + 1, \quad \text{and} \quad x - 2,$$

and the zeros are

$$-3, \qquad -1, \quad \text{and} \quad 2.$$

When a polynomial is expressed in factored form, each factor determines a zero of the function. Thus if we know the factors of a polynomial, we can easily find the zeros. The "reverse" is also true: If we know the zeros of a polynomial function, we can find the factors of the polynomial.

Division and Factors

When we divide one polynomial by another, we obtain a quotient and a remainder. If the remainder is 0, then the divisor is a factor of the dividend.

EXAMPLE 1 Divide to determine whether $x + 1$ and $x - 3$ are factors of

$$x^3 + 2x^2 - 5x - 6.$$

Solution We divide $x^3 + 2x^2 - 5x - 6$ by $x + 1$.

$$\begin{array}{r l}
 & \overbrace{x^2 + x - 6}^{\text{Quotient}} \\
\underbrace{x + 1}_{\text{Divisor}} \,\big) & \overline{x^3 + 2x^2 - 5x - 6} \leftarrow \text{Dividend} \\
 & \underline{x^3 + x^2} \\
 & \quad x^2 - 5x \\
 & \quad \underline{x^2 + x} \\
 & \qquad -6x - 6 \\
 & \qquad \underline{-6x - 6} \\
 & \qquad\qquad 0 \leftarrow \text{Remainder}
\end{array}$$

Since the remainder is 0, we know that $x + 1$ is a factor of $x^3 + 2x^2 - 5x - 6$. In fact, we know that

$$x^3 + 2x^2 - 5x - 6 = (x + 1)(x^2 + x - 6).$$

We divide $x^3 + 2x^2 - 5x - 6$ by $x - 3$.

$$\begin{array}{r l}
 & x^2 + 5x + 10 \\
x - 3 \,\big) & \overline{x^3 + 2x^2 - 5x - 6} \\
 & \underline{x^3 - 3x^2} \\
 & \quad 5x^2 - 5x \\
 & \quad \underline{5x^2 - 15x} \\
 & \qquad 10x - 6 \\
 & \qquad \underline{10x - 30} \\
 & \qquad\qquad 24 \leftarrow \text{Remainder}
\end{array}$$

Since the remainder, 24, is not 0, we know that $x - 3$ is *not* a factor of $x^3 + 2x^2 - 5x - 6$.

Now Try Exercise 3.

When we divide a polynomial $P(x)$ by a divisor $d(x)$, a polynomial $Q(x)$ is the quotient and a polynomial $R(x)$ is the remainder. The quotient $Q(x)$ must have degree less than that of the dividend $P(x)$. The remainder $R(x)$ must either be 0 or have degree less than that of the divisor $d(x)$.

As in arithmetic, to check a division, we multiply the quotient by the divisor and add the remainder, to see if we get the dividend. Thus these polynomials are related as follows:

$$P(x) = d(x) \cdot Q(x) + R(x).$$

Dividend **Divisor** **Quotient** **Remainder**

For instance, if $P(x) = x^3 + 2x^2 - 5x - 6$ and $d(x) = x - 3$, as in Example 1, then $Q(x) = x^2 + 5x + 10$ and $R(x) = 24$, and

$$\begin{aligned} P(x) \qquad &= \quad d(x) \quad \cdot \quad Q(x) \qquad + R(x) \\ x^3 + 2x^2 - 5x - 6 &= (x - 3) \cdot (x^2 + 5x + 10) + 24 \\ &= x^3 + 5x^2 + 10x - 3x^2 - 15x - 30 + 24 \\ &= x^3 + 2x^2 - 5x - 6. \end{aligned}$$

The Remainder Theorem and Synthetic Division

Consider the function

$$h(x) = x^3 + 2x^2 - 5x - 6.$$

When we divided $h(x)$ by $x + 1$ and $x - 3$ in Example 1, the remainders were 0 and 24, respectively. Let's now find the function values $h(-1)$ and $h(3)$:

$$\begin{aligned} h(-1) &= (-1)^3 + 2(-1)^2 - 5(-1) - 6 = 0; \\ h(3) &= (3)^3 + 2(3)^2 - 5(3) - 6 = 24. \end{aligned}$$

Note that the function values are the same as the remainders. This suggests the following theorem.

THE REMAINDER THEOREM

If a number c is substituted for x in the polynomial $f(x)$, then the result $f(c)$ is the remainder that would be obtained by dividing $f(x)$ by $x - c$. That is, if $f(x) = (x - c) \cdot Q(x) + R$, then $f(c) = R$.

Proof (Optional). The equation $f(x) = d(x) \cdot Q(x) + R(x)$, where $d(x) = x - c$, is the basis of this proof. If we divide $f(x)$ by $x - c$, we obtain a quotient $Q(x)$ and a remainder $R(x)$ related as follows:

$$f(x) = (x - c) \cdot Q(x) + R(x).$$

The remainder $R(x)$ must either be 0 or have degree less than $x - c$. Thus, $R(x)$ must be a constant. Let's call this constant R. The equation above is true for any replacement of x, so we replace x with c. We get

$$\begin{aligned} f(c) &= (c - c) \cdot Q(c) + R \\ &= 0 \cdot Q(c) + R \\ &= R. \end{aligned}$$

Thus the function value $f(c)$ is the remainder obtained when we divide $f(x)$ by $x - c$. ■

The remainder theorem motivates us to find a rapid way of dividing by $x - c$ in order to find function values. To streamline division, we can arrange the work so that duplicate and unnecessary writing is avoided. Consider the following:

$$(4x^3 - 3x^2 + x + 7) \div (x - 2).$$

A.

$$\begin{array}{rl} & 4x^2 + 5x + 11 \\ x - 2 \overline{)} & 4x^3 - 3x^2 + x + 7 \\ & \underline{4x^3 - 8x^2} \\ & 5x^2 + x \\ & \underline{5x^2 - 10x} \\ & 11x + 7 \\ & \underline{11x - 22} \\ & 29 \end{array}$$

B.

$$\begin{array}{rl} & 4 \quad 5 \quad 11 \\ 1 - 2 \overline{)} & 4 - 3 + 1 + 7 \\ & \underline{4 - 8} \\ & 5 + 1 \\ & \underline{5 - 10} \\ & 11 + 7 \\ & \underline{11 - 22} \\ & 29 \end{array}$$

The division in (B) is the same as that in (A), but we wrote only the coefficients. The red numerals are duplicated, so we look for an arrangement in which they are not duplicated. In place of the divisor in the form $x - c$, we can simply use c and then add rather than subtract. When the procedure is "collapsed," we have the algorithm known as **synthetic division**.

C. *Synthetic Division*

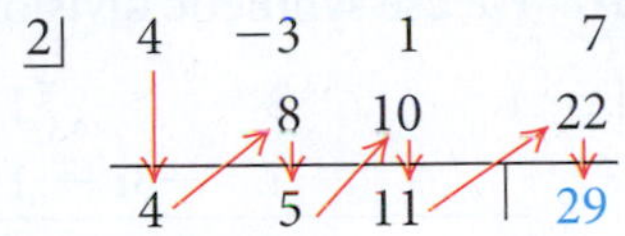

The divisor is $x - 2$; thus we use 2 in synthetic division.

We "bring down" the 4. Then we multiply it by the 2 to get 8 and add to get 5. We then multiply 5 by 2 to get 10, add, and so on. The last number, 29, is the remainder. The others, 4, 5, and 11, are the coefficients of the quotient, $4x^2 + 5x + 11$. (Note that the degree of the quotient is 1 less than the degree of the dividend when the degree of the divisor is 1.)

When using synthetic division, we write a 0 for a missing term in the dividend.

JUST IN TIME 4

EXAMPLE 2 Use synthetic division to find the quotient and the remainder:

$$(2x^3 + 7x^2 - 5) \div (x + 3).$$

Solution First, we note that $x + 3 = x - (-3)$.

$$\begin{array}{r|rrrr} -3 & 2 & 7 & 0 & -5 \\ & & -6 & -3 & 9 \\ \hline & 2 & 1 & -3 & 4 \end{array}$$

Note: **We must write a 0 for the missing x-term.**

The quotient is $2x^2 + x - 3$. The remainder is 4. **Now Try Exercise 13.**

We can now use synthetic division to find polynomial function values.

EXAMPLE 3 Given that $f(x) = 2x^5 - 3x^4 + x^3 - 2x^2 + x - 8$, find $f(10)$.

Solution By the remainder theorem, $f(10)$ is the remainder when $f(x)$ is divided by $x - 10$. We use synthetic division to find that remainder.

$$\begin{array}{r|rrrrrr} 10 & 2 & -3 & 1 & -2 & 1 & -8 \\ & & 20 & 170 & 1710 & 17{,}080 & 170{,}810 \\ \hline & 2 & 17 & 171 & 1708 & 17{,}081 & 170{,}802 \end{array}$$

Thus, $f(10) = 170{,}802$. **Now Try Exercise 25.**

Technology Connection

The computations in synthetic division are less complicated than those involved in substituting. In Example 3, the easiest way to find $f(10)$ is to use one of the methods for evaluating a function on a graphing calculator. For example, we can enter $y_1 = 2x^5 - 3x^4 + x^3 - 2x^2 + x - 8$ and then use function notation on the home screen.

Compare the computations in Example 3 with those in a direct substitution:

$$\begin{aligned} f(10) &= 2(10)^5 - 3(10)^4 + (10)^3 - 2(10)^2 + 10 - 8 \\ &= 2 \cdot 100{,}000 - 3 \cdot 10{,}000 + 1000 - 2 \cdot 100 + 10 - 8 \\ &= 200{,}000 - 30{,}000 + 1000 - 200 + 10 - 8 \\ &= 170{,}802. \end{aligned}$$

EXAMPLE 4 Determine whether 5 is a zero of $g(x)$, where

$$g(x) = x^4 - 26x^2 + 25.$$

Solution We use synthetic division and the remainder theorem to find $g(5)$.

$$\begin{array}{r|rrrrr} 5 & 1 & 0 & -26 & 0 & 25 \\ & & 5 & 25 & -5 & -25 \\ \hline & 1 & 5 & -1 & -5 & 0 \end{array}$$

Writing 0's for missing terms: $x^4 + 0x^3 - 26x^2 + 0x + 25$

Since $g(5) = 0$, the number 5 is a zero of $g(x)$. **Now Try Exercise 31.**

EXAMPLE 5 Determine whether i is a zero of $f(x)$, where

$$f(x) = x^3 - 3x^2 + x - 3.$$

Solution We use synthetic division and the remainder theorem to find $f(i)$.

$$\begin{array}{r|rrrr} i & 1 & -3 & 1 & -3 \\ & & i & -3i - 1 & 3 \\ \hline & 1 & -3 + i & -3i & 0 \end{array}$$

$i(-3 + i) = -3i + i^2 = -3i - 1$

$i(-3i) = -3i^2 = 3$

Since $f(i) = 0$, the number i is a zero of $f(x)$. **Now Try Exercise 35.**

Finding Factors of Polynomials

We now consider a useful result that follows from the remainder theorem.

THE FACTOR THEOREM

For a polynomial $f(x)$, if $f(c) = 0$, then $x - c$ is a factor of $f(x)$.

Proof (Optional). If we divide $f(x)$ by $x - c$, we obtain a quotient and a remainder, related as follows:

$$f(x) = (x - c) \cdot Q(x) + f(c).$$

Then if $f(c) = 0$, we have

$$f(x) = (x - c) \cdot Q(x),$$

so $x - c$ is a factor of $f(x)$. ■

The factor theorem is very useful in factoring polynomials and hence in solving polynomial equations and finding zeros of polynomial functions. If we know a zero of a polynomial function, we know a factor.

EXAMPLE 6 Let $f(x) = x^3 - 3x^2 - 6x + 8$. Factor $f(x)$ and solve the equation $f(x) = 0$.

Solution We look for linear factors of the form $x - c$. Let's try $x + 1$, or $x - (-1)$. (In the next section, we will learn a method for choosing the numbers to try for c.) We use synthetic division to determine whether $f(-1) = 0$.

$$\begin{array}{r|rrrr} -1 & 1 & -3 & -6 & 8 \\ & & -1 & 4 & 2 \\ \hline & 1 & -4 & -2 & 10 \end{array}$$

Since $f(-1) \neq 0$, we know that $x + 1$ *is not a factor* of $f(x)$. We now try $x - 1$.

$$\begin{array}{r|rrrr} 1 & 1 & -3 & -6 & 8 \\ & & 1 & -2 & -8 \\ \hline & 1 & -2 & -8 & 0 \end{array}$$

Since $f(1) = 0$, we know that $x - 1$ *is one factor* of $f(x)$ and the quotient, $x^2 - 2x - 8$, is another. Thus,

$$f(x) = (x - 1)(x^2 - 2x - 8).$$

The trinomial $x^2 - 2x - 8$ is easily factored, so we have

$$f(x) = (x - 1)(x - 4)(x + 2).$$

Our goal is to solve the equation $f(x) = 0$. To do so, we use the principle of zero products:

$$(x - 1)(x - 4)(x + 2) = 0$$

$$x - 1 = 0 \quad or \quad x - 4 = 0 \quad or \quad x + 2 = 0$$

$$x = 1 \quad or \quad x = 4 \quad or \quad x = -2.$$

The solutions of the equation $x^3 - 3x^2 - 6x + 8 = 0$ are -2, 1, and 4. They are also the zeros of the function $f(x) = x^3 - 3x^2 - 6x + 8$.

Now Try Exercise 41.

Technology Connection

In Example 6, we can use a table set in ASK mode to check the solutions of the equation

$$x^3 - 3x^2 - 6x + 8 = 0.$$

We check the solutions, -2, 1, and 4, of $f(x) = 0$ by evaluating $f(-2)$, $f(1)$, and $f(4)$.

X	Y1
−2	0
1	0
4	0

X =

CONNECTING THE CONCEPTS

Consider the function

$$f(x) = (x - 2)(x + 3)(x + 1), \quad \text{or} \quad f(x) = x^3 + 2x^2 - 5x - 6,$$

and its graph.

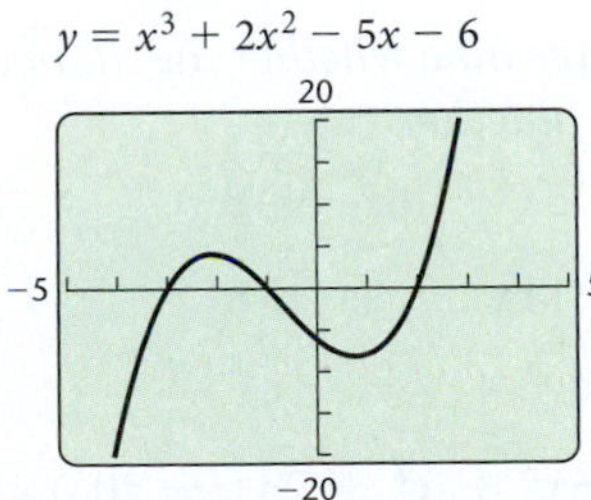

We can make the following statements:

- -3 is a zero of f.
- $f(-3) = 0$.
- -3 is a solution of $f(x) = 0$.
- $(-3, 0)$ is an x-intercept of the graph of f.
- 0 is the remainder when $f(x)$ is divided by $x - (-3)$.
- $x - (-3)$ is a factor of f.

Similar statements are also true for -1 and 2.

4.3 Exercise Set

1. For the function

$$f(x) = x^4 - 6x^3 + x^2 + 24x - 20,$$

use long division to determine whether each of the following is a factor of $f(x)$.

a) $x + 1$ b) $x - 2$ c) $x + 5$

2. For the function

$$h(x) = x^3 - x^2 - 17x - 15,$$

use long division to determine whether each of the following is a factor of $h(x)$.

a) $x + 5$ b) $x + 1$ c) $x + 3$

3. For the function

$$g(x) = x^3 - 2x^2 - 11x + 12,$$

use long division to determine whether each of the following is a factor of $g(x)$.

a) $x - 4$ b) $x - 3$ c) $x - 1$

4. For the function

$$f(x) = x^4 + 8x^3 + 5x^2 - 38x + 24,$$

use long division to determine whether each of the following is a factor of $f(x)$.

a) $x + 6$ b) $x + 1$ c) $x - 4$

In each of the following, a polynomial $P(x)$ and a divisor $d(x)$ are given. Use long division to find the quotient $Q(x)$ and the remainder $R(x)$ when $P(x)$ is divided by $d(x)$. Express $P(x)$ in the form $d(x) \cdot Q(x) + R(x)$.

5. $P(x) = x^3 - 8$, $d(x) = x + 2$

6. $P(x) = 2x^3 - 3x^2 + x - 1$, $d(x) = x - 3$

7. $P(x) = x^3 + 6x^2 - 25x + 18$, $d(x) = x + 9$

8. $P(x) = x^3 - 9x^2 + 15x + 25$, $d(x) = x - 5$

9. $P(x) = x^4 - 2x^2 + 3$, $d(x) = x + 2$

10. $P(x) = x^4 + 6x^3$, $d(x) = x - 1$

Use synthetic division to find the quotient and the remainder.

11. $(2x^4 + 7x^3 + x - 12) \div (x + 3)$

12. $(x^3 - 7x^2 + 13x + 3) \div (x - 2)$

13. $(x^3 - 2x^2 - 8) \div (x + 2)$

14. $(x^3 - 3x + 10) \div (x - 2)$

15. $(3x^3 - x^2 + 4x - 10) \div (x + 1)$

16. $(4x^4 - 2x + 5) \div (x + 3)$

17. $(x^5 + x^3 - x) \div (x - 3)$

18. $(x^7 - x^6 + x^5 - x^4 + 2) \div (x + 1)$

19. $(x^4 - 1) \div (x - 1)$

20. $(x^5 + 32) \div (x + 2)$

21. $(2x^4 + 3x^2 - 1) \div \left(x - \frac{1}{2}\right)$

22. $(3x^4 - 2x^2 + 2) \div \left(x - \frac{1}{4}\right)$

Use synthetic division to find the function values. Then check your work using a graphing calculator.

23. $f(x) = x^3 - 6x^2 + 11x - 6$; find $f(1)$, $f(-2)$, and $f(3)$.

24. $f(x) = x^3 + 7x^2 - 12x - 3$; find $f(-3)$, $f(-2)$, and $f(1)$.

25. $f(x) = x^4 - 3x^3 + 2x + 8$; find $f(-1)$, $f(4)$, and $f(-5)$.

26. $f(x) = 2x^4 + x^2 - 10x + 1$; find $f(-10)$, $f(2)$, and $f(3)$.

27. $f(x) = 2x^5 - 3x^4 + 2x^3 - x + 8$; find $f(20)$ and $f(-3)$.

28. $f(x) = x^5 - 10x^4 + 20x^3 - 5x - 100$; find $f(-10)$ and $f(5)$.

29. $f(x) = x^4 - 16$; find $f(2)$, $f(-2)$, $f(3)$, and $f\left(1 - \sqrt{2}\right)$.

30. $f(x) = x^5 + 32$; find $f(2)$, $f(-2)$, $f(3)$, and $f(2 + 3i)$.

Using synthetic division, determine whether the numbers are zeros of the polynomial function.

31. $-3, 2$; $f(x) = 3x^3 + 5x^2 - 6x + 18$

32. $-4, 2$; $f(x) = 3x^3 + 11x^2 - 2x + 8$

33. $-3, 1$; $h(x) = x^4 + 4x^3 + 2x^2 - 4x - 3$

34. $2, -1$; $g(x) = x^4 - 6x^3 + x^2 + 24x - 20$

35. $i, -2i$; $g(x) = x^3 - 4x^2 + 4x - 16$

36. $\frac{1}{3}, 2$; $h(x) = x^3 - x^2 - \frac{1}{9}x + \frac{1}{9}$

37. $-3, \frac{1}{2}$; $f(x) = x^3 - \frac{7}{2}x^2 + x - \frac{3}{2}$

38. $i, -i, -2$; $f(x) = x^3 + 2x^2 + x + 2$

Factor the polynomial function $f(x)$. Then solve the equation $f(x) = 0$.

39. $f(x) = x^3 + 4x^2 + x - 6$

40. $f(x) = x^3 + 5x^2 - 2x - 24$

41. $f(x) = x^3 - 6x^2 + 3x + 10$

42. $f(x) = x^3 + 2x^2 - 13x + 10$

43. $f(x) = x^3 - x^2 - 14x + 24$

44. $f(x) = x^3 - 3x^2 - 10x + 24$

45. $f(x) = x^4 - 7x^3 + 9x^2 + 27x - 54$

46. $f(x) = x^4 - 4x^3 - 7x^2 + 34x - 24$

47. $f(x) = x^4 - x^3 - 19x^2 + 49x - 30$

48. $f(x) = x^4 + 11x^3 + 41x^2 + 61x + 30$

Sketch the graph of the polynomial function. Follow the procedure outlined on p. 240. Use synthetic division and the remainder theorem to find the zeros.

49. $f(x) = x^4 - x^3 - 7x^2 + x + 6$

50. $f(x) = x^4 + x^3 - 3x^2 - 5x - 2$

51. $f(x) = x^3 - 7x + 6$

52. $f(x) = x^3 - 12x + 16$

53. $f(x) = -x^3 + 3x^2 + 6x - 8$

54. $f(x) = -x^4 + 2x^3 + 3x^2 - 4x - 4$

▸ Skill Maintenance

Solve. Find exact solutions. [3.2]

55. $2x^2 + 12 = 5x$

56. $7x^2 + 4x = 3$

Consider the function

$$g(x) = x^2 + 5x - 14$$

in Exercises 57–59.

57. What are the inputs if the output is -14? [3.2]

58. What is the output if the input is 3? [1.2]

59. Given an output of -20, find the corresponding inputs. [3.2]

60. *Movie Ticket Price.* The average price of a movie ticket has increased linearly over the years, rising from \$2.69 in 1980 to \$8.38 in 2013 (*Source:* Motion Picture Association of America). Using these two data points, find a linear function, $f(x) = mx + b$, that models the data. Let x represent the number of years after 1980. Then use this function to estimate the average price of a movie ticket in 1995 and in 2018. [1.4]

61. The sum of the base and the height of a triangle is 30 in. Find the dimensions for which the area is a maximum. [3.2]

▸ Synthesis

In Exercises 62 and 63, a graph of a polynomial function is given. On the basis of the graph:

a) *Find as many factors of the polynomial as you can.*

b) *Construct a polynomial function with the zeros shown in the graph.*

c) *Can you find any other polynomial functions with the given zeros?*

d) *Can you find any other polynomial functions with the given zeros and the same graph?*

62.

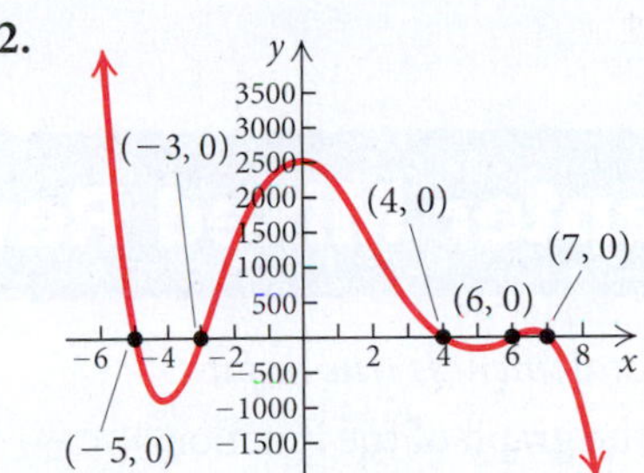

63.

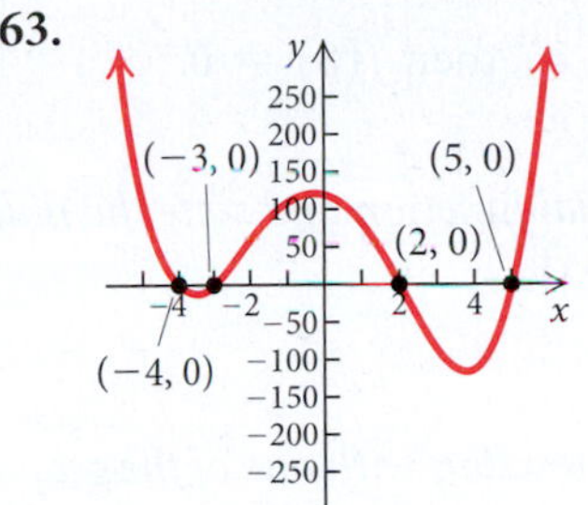

64. For what values of k will the remainder be the same when $x^2 + kx + 4$ is divided by $x - 1$ and by $x + 1$?

65. Find k such that $x + 2$ is a factor of $x^3 - kx^2 + 3x + 7k$.

66. *Beam Deflection.* A beam rests at two points A and B and has a concentrated load applied to its center, as shown below. Let $y =$ the deflection, in feet, of the beam at a distance of x feet from A. Under certain conditions, this deflection is given by

$$y = \frac{1}{13}x^3 - \frac{1}{14}x.$$

Find the zeros of the polynomial in the interval $[0, 2]$.

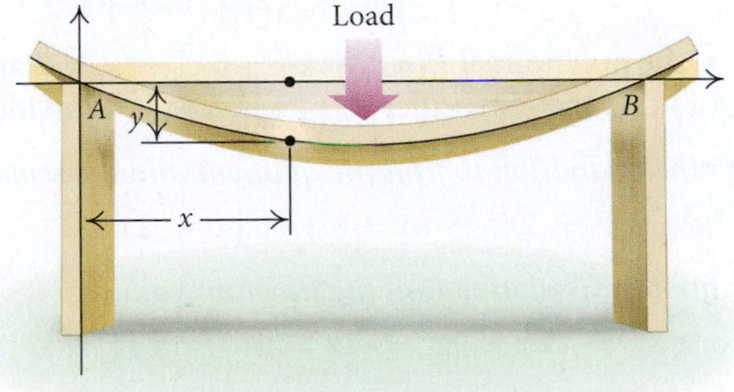

Solve.

67. $\frac{2x^2}{x^2 - 1} + \frac{4}{x + 3} = \frac{12x - 4}{x^3 + 3x^2 - x - 3}$

68. $\frac{6x^2}{x^2 + 11} + \frac{60}{x^3 - 7x^2 + 11x - 77} = \frac{1}{x - 7}$

69. Find a 15th-degree polynomial for which $x - 1$ is a factor. Answers may vary.

Use synthetic division to divide.

70. $(x^4 - y^4) \div (x - y)$

71. $(x^3 + 3ix^2 - 4ix - 2) \div (x + i)$

72. $(x^2 - 4x - 2) \div [x - (3 + 2i)]$

73. $(x^2 - 3x + 7) \div (x - i)$

Mid-Chapter Mixed Review

Determine whether the statement is true or false.

1. The y-intercept of the graph of the function $P(x) = 5 - 2x^3$ is $(5, 0)$. [4.2]

2. The degree of the polynomial $x - \frac{1}{2}x^4 - 3x^6 + x^5$ is 6. [4.1]

3. If $f(x) = (x + 7)(x - 8)$, then $f(8) = 0$. [4.3]

4. If $f(12) = 0$, then $x + 12$ is a factor of $f(x)$. [4.3]

Find the zeros of the polynomial function and state the multiplicity of each. [4.1]

5. $f(x) = (x^2 - 10x + 25)^3$

6. $h(x) = 2x^3 + x^2 - 50x - 25$

7. $g(x) = x^4 - 3x^2 + 2$

8. $f(x) = -6(x - 3)^2(x + 4)$

In Exercises 9–12, match the function with one of the graphs (a)–(d) that follow. [4.2]

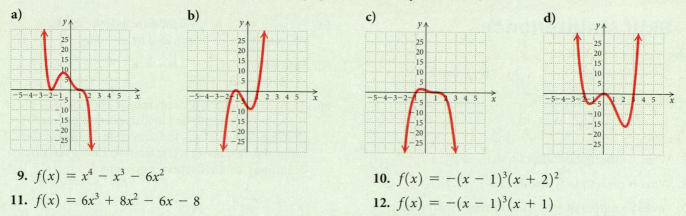

a)

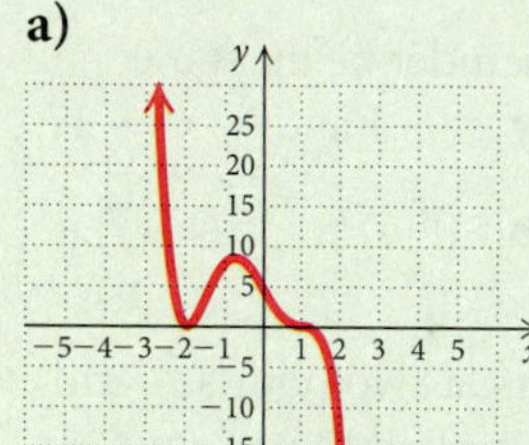

b)

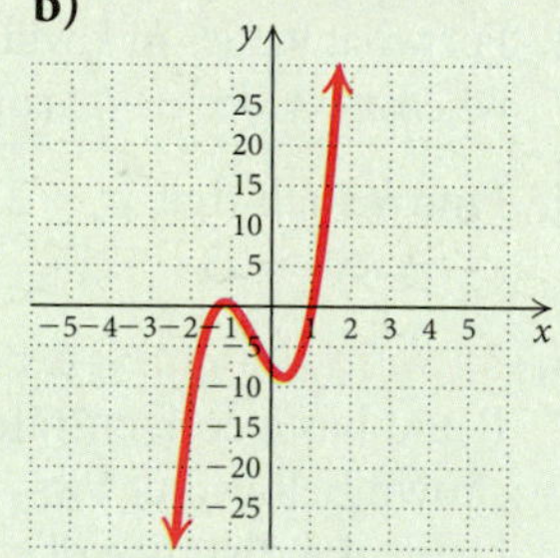

c)

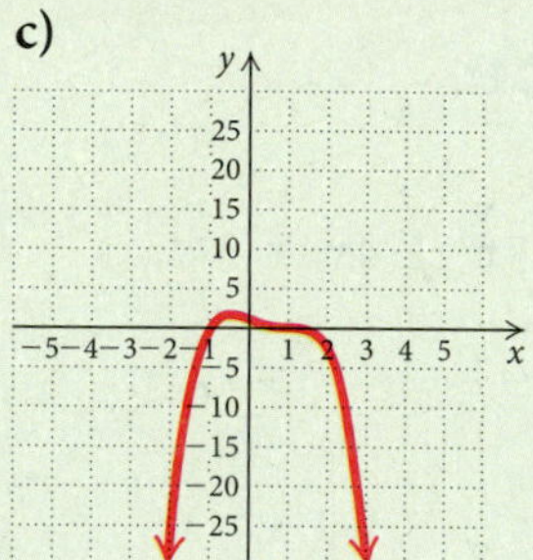

d)

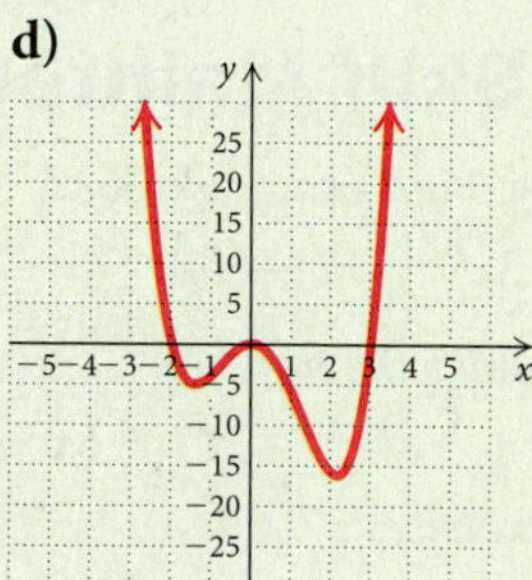

9. $f(x) = x^4 - x^3 - 6x^2$

10. $f(x) = -(x - 1)^3(x + 2)^2$

11. $f(x) = 6x^3 + 8x^2 - 6x - 8$

12. $f(x) = -(x - 1)^3(x + 1)$

Using the intermediate value theorem, determine, if possible, whether the function has at least one real zero between a and b. [4.2]

13. $f(x) = x^3 - 2x^2 + 3;\ a = -2, b = 0$

14. $f(x) = x^3 - 2x^2 + 3;\ a = -\frac{1}{2}, b = 1$

15. For the polynomial $P(x) = x^4 - 6x^3 + x - 2$ and the divisor $d(x) = x - 1$, use long division to find the quotient $Q(x)$ and the remainder $R(x)$ when $P(x)$ is divided by $d(x)$. Express $P(x)$ in the form $d(x) \cdot Q(x) + R(x)$. [4.3]

Use synthetic division to find the quotient and the remainder. [4.3]

16. $(3x^4 - x^3 + 2x^2 - 6x + 6) \div (x - 2)$

17. $(x^5 - 5) \div (x + 1)$

Use synthetic division to find the function values. [4.3]

18. For $g(x) = x^3 - 9x^2 + 4x - 10$, find $g(-5)$.

19. For $f(x) = 20x^2 - 40x$, find $f\left(\frac{1}{2}\right)$.

20. For $f(x) = 5x^4 + x^3 - x$, find $f(-\sqrt{2})$.

Using synthetic division, determine whether the numbers are zeros of the polynomial function. [4.3]

21. $-3i, 3;\ f(x) = x^3 - 4x^2 + 9x - 36$

22. $-1, 5;\ f(x) = x^6 - 35x^4 + 259x^2 - 225$

Factor the polynomial function $f(x)$. Then solve the equation $f(x) = 0$. [4.3]

23. $h(x) = x^3 - 2x^2 - 55x + 56$

24. $g(x) = x^4 - 2x^3 - 13x^2 + 14x + 24$

Collaborative Discussion and Writing

25. How is the range of a polynomial function related to the degree of the polynomial? [4.1]

26. Is it possible for the graph of a polynomial function to have no y-intercept? no x-intercepts? Explain your answer. [4.2]

27. Explain why values of a function must be all positive or all negative between consecutive zeros. [4.2]

28. In synthetic division, why is the degree of the quotient 1 less than that of the dividend? [4.3]

4.4 Theorems about Zeros of Polynomial Functions

- Find a polynomial with specified zeros.
- For a polynomial function with integer coefficients, find the rational zeros and the other zeros, if possible.
- Use Descartes' rule of signs to find information about the number of real zeros of a polynomial function with real coefficients.

We will now allow the coefficients of a polynomial to be complex numbers. In certain cases, we will restrict the coefficients to be real numbers, rational numbers, or integers, as shown in the following examples.

Polynomial	Type of Coefficient
$5x^3 - 3x^2 + (2 + 4i)x + i$	Complex
$5x^3 - 3x^2 + \sqrt{2}x - \pi$	Real
$5x^3 - 3x^2 + \frac{2}{3}x - \frac{7}{4}$	Rational
$5x^3 - 3x^2 + 8x - 11$	Integer

The Fundamental Theorem of Algebra

A linear, or first-degree, polynomial function $f(x) = mx + b$ (where $m \neq 0$) has just one zero, $-b/m$. It can be shown that any quadratic polynomial function $f(x) = ax^2 + bx + c$ with complex numbers for coefficients has at least one, and at most two, complex zeros. The following theorem is a generalization. No proof is given in this text.

THE FUNDAMENTAL THEOREM OF ALGEBRA

Every polynomial function of degree n, with $n \geq 1$, has at least one zero in the set of complex numbers.

Note that although the fundamental theorem of algebra guarantees that a zero exists, it does not tell how to find it. Recall that the zeros of a polynomial function $f(x)$ are the solutions of the polynomial equation $f(x) = 0$. We now develop some concepts that can help in finding zeros. First, we consider one of the results of the fundamental theorem of algebra.

Every polynomial function f of degree n, with $n \geq 1$, can be factored into n linear factors (not necessarily unique); that is,

$$f(x) = a_n(x - c_1)(x - c_2) \cdots (x - c_n).$$

Finding Polynomials with Given Zeros

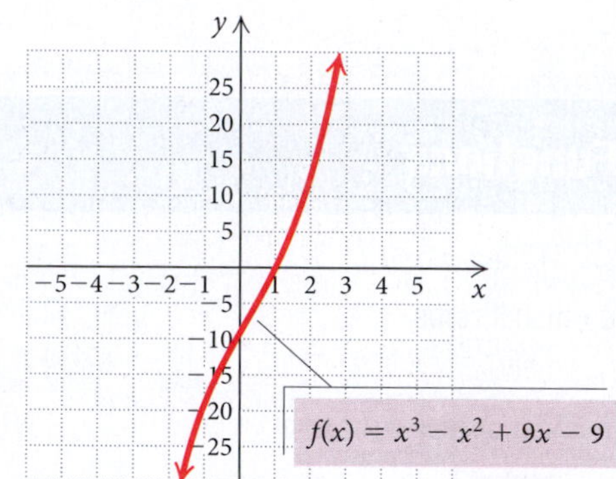

Given several numbers, we can find a polynomial function with those numbers as its zeros.

EXAMPLE 1 Find a polynomial function of degree 3, having the zeros 1, $3i$, and $-3i$.

Solution Such a function has factors $x - 1$, $x - 3i$, and $x - (-3i)$, or $x + 3i$, so we have

$$f(x) = a_n(x - 1)(x - 3i)(x + 3i).$$

JUST IN TIME 12

The number a_n can be any nonzero number. The simplest polynomial function will be obtained if we let it be 1. If we then multiply the factors, we obtain

$$\begin{aligned} f(x) &= (x - 1)(x^2 - 9i^2) && \text{Multiplying } (x - 3i)(x + 3i) \\ &= (x - 1)(x^2 + 9) && -9i^2 = -9(-1) = 9 \\ &= x^3 - x^2 + 9x - 9. \end{aligned}$$

Now Try Exercise 3.

EXAMPLE 2 Find a polynomial function of degree 5 with -1 as a zero of multiplicity 3, 4 as a zero of multiplicity 1, and 0 as a zero of multiplicity 1.

Solution Proceeding as in Example 1, letting $a_n = 1$, we obtain

$$\begin{aligned} f(x) &= [x - (-1)]^3(x - 4)(x - 0) \\ &= (x + 1)^3(x - 4)x \\ &= (x^3 + 3x^2 + 3x + 1)(x^2 - 4x) \\ &= x^5 - x^4 - 9x^3 - 11x^2 - 4x. \end{aligned}$$

Now Try Exercise 13.

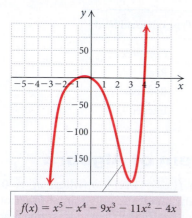

$f(x) = x^5 - x^4 - 9x^3 - 11x^2 - 4x$

Zeros of Polynomial Functions with Real Coefficients

Consider the quadratic equation $x^2 - 2x + 2 = 0$, with real coefficients. Its solutions are $1 + i$ and $1 - i$. Note that they are complex conjugates. This generalizes to any polynomial equation with real coefficients.

NONREAL ZEROS: $a + bi$ and $a - bi$, $b \neq 0$

If a complex number $a + bi$, $b \neq 0$, is a zero of a polynomial function $f(x)$ with *real* coefficients, then its conjugate, $a - bi$, is also a zero. For example, if $2 + 7i$ is a zero of a polynomial function $f(x)$ with real coefficients, then its conjugate, $2 - 7i$, is also a zero. (Nonreal zeros occur in conjugate pairs.)

In order for the preceding to be true, it is essential that the coefficients be *real* numbers.

Rational Coefficients

When a polynomial function has *rational* numbers for coefficients, certain irrational zeros also occur in pairs, as described in the following theorem.

IRRATIONAL ZEROS: $a + c\sqrt{b}$ and $a - c\sqrt{b}$, b IS NOT A PERFECT SQUARE

If $a + c\sqrt{b}$, where a, b, and c are rational and b is not a perfect square, is a zero of a polynomial function $f(x)$ with *rational* coefficients, then its conjugate, $a - c\sqrt{b}$, is also a zero. For example, if $-3 + 5\sqrt{2}$ is a zero of a polynomial function $f(x)$ with rational coefficients, then its conjugate, $-3 - 5\sqrt{2}$, is also a zero. (Irrational zeros occur in conjugate pairs.)

EXAMPLE 3 Suppose that a polynomial function of degree 6 with rational coefficients has

$$-2 + 5i, \quad -2i, \quad \text{and} \quad 1 - \sqrt{3}$$

as three of its zeros. Find the other zeros.

Solution Since the coefficients are rational, the other zeros are the conjugates of the given zeros:

$$-2 - 5i, \quad 2i, \quad \text{and} \quad 1 + \sqrt{3}.$$

There are no other zeros because a polynomial function of degree 6 can have at most 6 zeros.

Now Try Exercise 19.

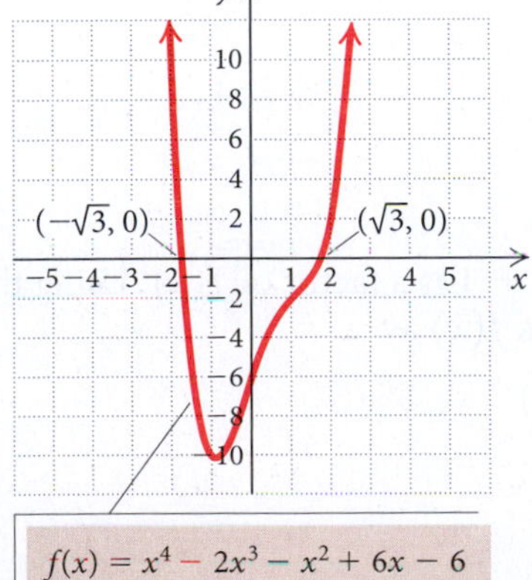

EXAMPLE 4 Find a polynomial function of lowest degree with rational coefficients that has $-\sqrt{3}$ and $1 + i$ as two of its zeros.

Solution The function must also have the zeros $\sqrt{3}$ and $1 - i$. Because we want to find the polynomial function of lowest degree with the given zeros, we will not include additional zeros; that is, we will write a polynomial function of degree 4. Thus if we let $a_n = 1$, the polynomial function is

$$\begin{aligned} f(x) &= [x - (-\sqrt{3})][x - \sqrt{3}][x - (1 + i)][x - (1 - i)] \\ &= (x + \sqrt{3})(x - \sqrt{3})[(x - 1) - i][(x - 1) + i] \\ &= (x^2 - 3)[(x - 1)^2 - i^2] \\ &= (x^2 - 3)[x^2 - 2x + 1 + 1] \\ &= (x^2 - 3)(x^2 - 2x + 2) \\ &= x^4 - 2x^3 - x^2 + 6x - 6. \end{aligned}$$

Now Try Exercise 39.

▶ Integer Coefficients and the Rational Zeros Theorem

It is not always easy to find the zeros of a polynomial function. However, if a polynomial function has integer coefficients, there is a procedure that will yield all the rational zeros.

> **THE RATIONAL ZEROS THEOREM**
>
> Let
>
> $$P(x) = a_n x^n + a_{n-1}x^{n-1} + \cdots + a_1 x + a_0,$$
>
> where all the coefficients are integers. Consider a rational number denoted by p/q, where p and q are relatively prime (having no common factor besides -1 and 1). If p/q is a zero of $P(x)$, then p is a factor of a_0 and q is a factor of a_n.

Technology Connection

We can narrow the list of possibilities for the rational zeros of a function by examining its graph.

$y = 3x^4 - 11x^3 + 10x - 4$

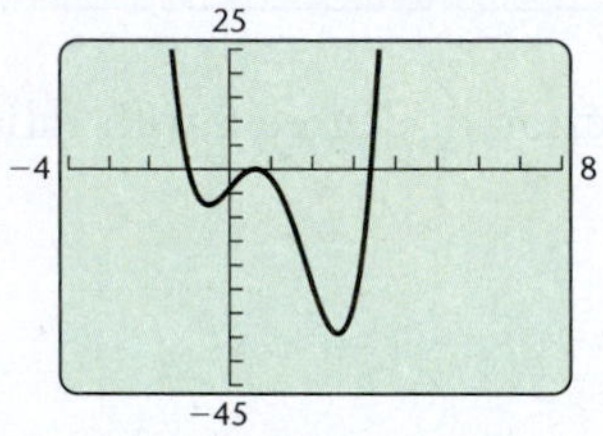

From the graph of the function in Example 5, we see that of the possibilities for rational zeros, only the numbers -1, $\frac{1}{3}$, and $\frac{2}{3}$ might actually be rational zeros. We can check these with the TABLE feature. Note that the graphing calculator converts $\frac{1}{3}$ and $\frac{2}{3}$ to decimal notation.

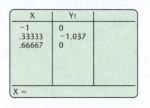

X	Y1
-1	0
.33333	-1.037
.66667	0

X =

Since $f(-1) = 0$ and $f\left(\frac{2}{3}\right) = 0$, -1 and $\frac{2}{3}$ are zeros. Since $f\left(\frac{1}{3}\right) \neq 0$, $\frac{1}{3}$ is not a zero.

EXAMPLE 5 Given $f(x) = 3x^4 - 11x^3 + 10x - 4$:

a) Find the rational zeros and then the other zeros; that is, solve $f(x) = 0$.

b) Factor $f(x)$ into linear factors.

Solution

a) Because the degree of $f(x)$ is 4, there are at most 4 distinct zeros. The rational zeros theorem says that if a rational number p/q is a zero of $f(x)$, then p must be a factor of -4 and q must be a factor of 3. Thus the possibilities for p/q are

$$\frac{\textit{Possibilities for } p}{\textit{Possibilities for } q}: \quad \frac{\pm 1, \pm 2, \pm 4}{\pm 1, \pm 3};$$

Possibilities for p/q: $1, -1, 2, -2, 4, -4, \frac{1}{3}, -\frac{1}{3}, \frac{2}{3}, -\frac{2}{3}, \frac{4}{3}, -\frac{4}{3}.$

To find which are zeros, we could use substitution, but synthetic division is usually more efficient. It is easier to consider the integers first. Then we consider the fractions, if the integers do not produce all the zeros.

We try 1:

$$\begin{array}{r|rrrrr} 1 & 3 & -11 & 0 & 10 & -4 \\ & & 3 & -8 & -8 & 2 \\ \hline & 3 & -8 & -8 & 2 & -2. \end{array}$$

Since $f(1) = -2$, 1 is not a zero.

We try -1:

$$\begin{array}{r|rrrrr} -1 & 3 & -11 & 0 & 10 & -4 \\ & & -3 & 14 & -14 & 4 \\ \hline & 3 & -14 & 14 & -4 & 0. \end{array}$$

We have $f(-1) = 0$, so -1 is a zero. Thus, $x + 1$ is a factor of $f(x)$. Using the results of the synthetic division, we can express $f(x)$ as

$$f(x) = (x + 1)(3x^3 - 14x^2 + 14x - 4).$$

We now consider the factor $3x^3 - 14x^2 + 14x - 4$ and check the other possible zeros. We use synthetic division again, to determine whether -1 is a zero of multiplicity 2 of $f(x)$:

$$\begin{array}{r|rrrr} -1 & 3 & -14 & 14 & -4 \\ & & -3 & 17 & -31 \\ \hline & 3 & -17 & 31 & -35. \end{array}$$

We see that -1 is not a double zero. We leave it to the student to verify that 2, -2, 4, and -4 are not zeros. There are no other zeros that are integers, so we start checking the fractions. Let's try $\frac{2}{3}$.

$$\begin{array}{r|rrrr} 2/3 & 3 & -14 & 14 & -4 \\ & & 2 & -8 & 4 \\ \hline & 3 & -12 & 6 & 0 \end{array}$$

Since the remainder is 0, we know that $x - \frac{2}{3}$ is a factor of $3x^3 - 14x^2 + 14x - 4$ and is also a factor of $f(x)$. Thus, $\frac{2}{3}$ is a zero of $f(x)$.

Using the results of the synthetic division, we can factor further:

$$f(x) = (x+1)\left(x - \tfrac{2}{3}\right)(3x^2 - 12x + 6) \quad \text{Using the results of the last synthetic division}$$

$$= (x+1)\left(x - \tfrac{2}{3}\right) \cdot 3 \cdot (x^2 - 4x + 2). \quad \text{Removing a factor of 3}$$

The quadratic formula can be used to find the values of x for which $x^2 - 4x + 2 = 0$. Those values are also zeros of $f(x)$:

$$x = \frac{-b \pm \sqrt{b^2 - 4ac}}{2a}$$

$$= \frac{-(-4) \pm \sqrt{(-4)^2 - 4 \cdot 1 \cdot 2}}{2 \cdot 1} \quad a = 1, b = -4, \text{ and } c = 2$$

$$= \frac{4 \pm \sqrt{8}}{2} = \frac{4 \pm 2\sqrt{2}}{2} = \frac{2(2 \pm \sqrt{2})}{2} = 2 \pm \sqrt{2}.$$

The rational zeros are -1 and $\frac{2}{3}$. The other zeros are $2 \pm \sqrt{2}$.

b) The complete factorization of $f(x)$ is

$$f(x) = 3(x+1)\left(x - \tfrac{2}{3}\right)\left[x - (2 - \sqrt{2})\right]\left[x - (2 + \sqrt{2})\right], \text{ or}$$

$$(x+1)(3x-2)\left[x - (2 - \sqrt{2})\right]\left[x - (2 + \sqrt{2})\right].$$

Replacing $3\left(x - \frac{2}{3}\right)$ with $(3x - 2)$

Now Try Exercise 55.

EXAMPLE 6 Given $f(x) = 2x^5 - x^4 - 4x^3 + 2x^2 - 30x + 15$:

a) Find the rational zeros and then the other zeros; that is, solve $f(x) = 0$.

b) Factor $f(x)$ into linear factors.

Solution

a) Because the degree of $f(x)$ is 5, there are at most 5 distinct zeros. According to the rational zeros theorem, any rational zero of f must be of the form p/q, where p is a factor of 15 and q is a factor of 2. The possibilities are

$$\frac{\textit{Possibilities for } p}{\textit{Possibilities for } q}: \quad \frac{\pm 1, \pm 3, \pm 5, \pm 15}{\pm 1, \pm 2};$$

Possibilities for p/q: $1, -1, 3, -3, 5, -5, 15, -15, \frac{1}{2}, -\frac{1}{2}, \frac{3}{2}, -\frac{3}{2}, \frac{5}{2}, -\frac{5}{2}, \frac{15}{2}, -\frac{15}{2}$.

> **Technology Connection**
>
> We can inspect the graph of the function in Example 6 for zeros that appear to be near any of the possible rational zeros.
>
> $y = 2x^5 - x^4 - 4x^3 + 2x^2 - 30x + 15$
>
> 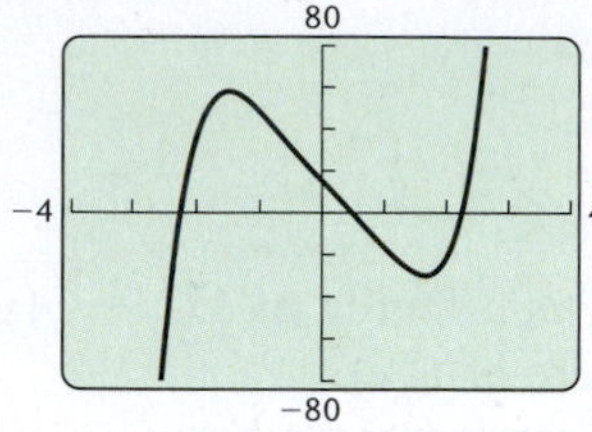
>
>
> We see that, of the possibilities in the list, only the numbers
>
> $-\frac{5}{2}$, $\frac{1}{2}$, and $\frac{5}{2}$
>
> might be rational zeros. With the TABLE feature, we can determine that only $\frac{1}{2}$ is actually a rational zero.
>
> | X | Y1 | |
> |---|---|---|
> | −2.5 | −69.38 | |
> | .5 | 0 | |
> | 2.5 | 46.25 | |
>
> X =

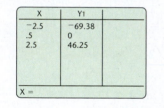

We use synthetic division to check each of the possibilities. Let's try 1 and −1.

$$\begin{array}{r|rrrrrr} 1 & 2 & -1 & -4 & 2 & -30 & 15 \\ & & 2 & 1 & -3 & -1 & -31 \\ \hline & 2 & 1 & -3 & -1 & -31 & \vert\ -16 \end{array}$$

$$\begin{array}{r|rrrrrr} -1 & 2 & -1 & -4 & 2 & -30 & 15 \\ & & -2 & 3 & 1 & -3 & 33 \\ \hline & 2 & -3 & -1 & 3 & -33 & \vert\ 48 \end{array}$$

Since $f(1) = -16$ and $f(-1) = 48$, neither 1 nor −1 is a zero of the function. We leave it to the student to verify that the other integer possibilities are not zeros. We now check $\frac{1}{2}$.

$$\begin{array}{r|rrrrrr} 1/2 & 2 & -1 & -4 & 2 & -30 & 15 \\ & & 1 & 0 & -2 & 0 & -15 \\ \hline & 2 & 0 & -4 & 0 & -30 & \vert\ 0 \end{array}$$

This means that $x - \frac{1}{2}$ is a factor of $f(x)$. We write the factorization and try to factor further:

$$\begin{aligned} f(x) &= \left(x - \tfrac{1}{2}\right)\left(2x^4 - 4x^2 - 30\right) \\ &= \left(x - \tfrac{1}{2}\right) \cdot 2 \cdot (x^4 - 2x^2 - 15) && \text{Factoring out the 2} \\ &= \left(x - \tfrac{1}{2}\right) \cdot 2 \cdot (x^2 - 5)(x^2 + 3). && \text{Factoring the trinomial} \end{aligned}$$

We now solve the equation $f(x) = 0$ to determine the zeros. We use the principle of zero products:

$$\left(x - \tfrac{1}{2}\right) \cdot 2 \cdot (x^2 - 5)(x^2 + 3) = 0$$

$$\begin{array}{lclcl} x - \frac{1}{2} = 0 & or & x^2 - 5 = 0 & or & x^2 + 3 = 0 \\ x = \frac{1}{2} & or & x^2 = 5 & or & x^2 = -3 \\ x = \frac{1}{2} & or & x = \pm\sqrt{5} & or & x = \pm\sqrt{3}i. \end{array}$$

There is only one rational zero, $\frac{1}{2}$. The other zeros are $\pm\sqrt{5}$ and $\pm\sqrt{3}i$.

b) The factorization into linear factors is

$$\begin{aligned} f(x) &= 2\left(x - \tfrac{1}{2}\right)\left(x + \sqrt{5}\right)\left(x - \sqrt{5}\right)\left(x + \sqrt{3}i\right)\left(x - \sqrt{3}i\right), \quad \text{or} \\ &\quad (2x - 1)\left(x + \sqrt{5}\right)\left(x - \sqrt{5}\right)\left(x + \sqrt{3}i\right)\left(x - \sqrt{3}i\right). \end{aligned}$$

Replacing $2\left(x - \frac{1}{2}\right)$ with $(2x - 1)$

Now Try Exercise 61.

Descartes' Rule of Signs

The development of a rule that helps determine the number of positive real zeros and the number of negative real zeros of a polynomial function is credited to the French mathematician René Descartes. To use the rule, we must have the polynomial arranged in descending order or ascending order, with no zero terms written in and the constant term not 0. Then we determine the number of *variations of sign*, that is, the number of times, in reading through the polynomial, that successive coefficients are of different signs.

EXAMPLE 7 Determine the number of variations of sign in the polynomial function $P(x) = 2x^5 - 3x^2 + x + 4$.

Solution We have

$$P(x) = 2x^5 - 3x^2 + x + 4$$

From positive to negative; a variation

From negative to positive; a variation

Both positive; no variation

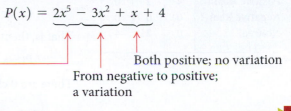

The number of variations of sign is 2.

JUST IN TIME 7

Note the following:

$$\begin{aligned} P(-x) &= 2(-x)^5 - 3(-x)^2 + (-x) + 4 \\ &= -2x^5 - 3x^2 - x + 4. \end{aligned}$$

We see that the number of variations of sign in $P(-x)$ is 1. It occurs as we go from $-x$ to 4.

We now state Descartes' rule, without proof.

DESCARTES' RULE OF SIGNS

Let $P(x)$, written in descending order or ascending order, be a polynomial function with real coefficients and a nonzero constant term. The number of positive real zeros of $P(x)$ is either:

1. The same as the number of variations of sign in $P(x)$, or
2. Less than the number of variations of sign in $P(x)$ by a positive even integer.

The number of negative real zeros of $P(x)$ is either:

3. The same as the number of variations of sign in $P(-x)$, or
4. Less than the number of variations of sign in $P(-x)$ by a positive even integer.

A zero of multiplicity m must be counted m times.

In each of Examples 8–10, what does Descartes' rule of signs tell you about the number of positive real zeros and the number of negative real zeros?

EXAMPLE 8 $P(x) = 2x^5 - 5x^2 - 3x + 6$

Solution The number of variations of sign in $P(x)$ is 2. Therefore, the number of positive real zeros is either 2 or less than 2 by 2, 4, 6, and so on. Thus the number of positive real zeros is either 2 or 0, since a negative number of zeros has no meaning.

$$P(-x) = -2x^5 - 5x^2 + 3x + 6$$

The number of variations of sign in $P(-x)$ is 1. Thus there is exactly 1 negative real zero. Since nonreal, complex conjugates occur in pairs, we also know the possible ways in which nonreal zeros might occur. The table shown at left summarizes all the possibilities for real zeros and nonreal zeros of $P(x)$.

Total Number of Zeros	5	
Positive Real	2	0
Negative Real	1	1
Nonreal	2	4

Now Try Exercise 93.

Total Number of Zeros	4		
Positive Real	4	2	0
Negative Real	0	0	0
Nonreal	0	2	4

EXAMPLE 9 $P(x) = 5x^4 - 3x^3 + 7x^2 - 12x + 4$

Solution There are 4 variations of sign. Thus the number of positive real zeros is either

$$4 \text{ or } 4 - 2 \text{ or } 4 - 4.$$

That is, the number of positive real zeros is 4, 2, or 0.

$$P(-x) = 5x^4 + 3x^3 + 7x^2 + 12x + 4$$

There are 0 changes in sign, so there are no negative real zeros.

Now Try Exercise 81.

EXAMPLE 10 $P(x) = 6x^6 - 2x^2 - 5x$

Solution As written, the polynomial does not satisfy the conditions of Descartes' rule of signs because the constant term is 0. But because x is a factor of every term, we know that the polynomial has 0 as a zero. We can then factor as follows:

$$P(x) = x(6x^5 - 2x - 5).$$

Now we analyze $Q(x) = 6x^5 - 2x - 5$ and $Q(-x) = -6x^5 + 2x - 5$. The number of variations of sign in $Q(x)$ is 1. Therefore, there is exactly 1 positive real zero. The number of variations of sign in $Q(-x)$ is 2. Thus the number of negative real zeros is 2 or 0. The same results apply to $P(x)$. Since nonreal, complex conjugates occur in pairs, we know the possible ways in which nonreal zeros might occur. The table at left summarizes all the possibilities for real zeros and nonreal zeros of $P(x)$.

Total Number of Zeros	6	
0 as a Zero	1	1
Positive Real	1	1
Negative Real	2	0
Nonreal	2	4

Now Try Exercise 95.

4.4 Exercise Set

Find a polynomial function of degree 3 with the given numbers as zeros.

1. $-2, 3, 5$
2. $-1, 0, 4$
3. $-3, 2i, -2i$
4. $2, i, -i$
5. $\sqrt{2}, -\sqrt{2}, 3$
6. $-5, \sqrt{3}, -\sqrt{3}$
7. $1 - \sqrt{3}, 1 + \sqrt{3}, -2$
8. $-4, 1 - \sqrt{5}, 1 + \sqrt{5}$
9. $1 + 6i, 1 - 6i, -4$
10. $1 + 4i, 1 - 4i, -1$
11. $-\frac{1}{3}, 0, 2$
12. $-3, 0, \frac{1}{2}$
13. Find a polynomial function of degree 5 with -1 as a zero of multiplicity 3, 0 as a zero of multiplicity 1, and 1 as a zero of multiplicity 1.
14. Find a polynomial function of degree 4 with -2 as a zero of multiplicity 1, 3 as a zero of multiplicity 2, and -1 as a zero of multiplicity 1.
15. Find a polynomial function of degree 4 with $a_4 = 1$ and with -1 as a zero of multiplicity 3 and 0 as a zero of multiplicity 1.
16. Find a polynomial function of degree 5 with $a_5 = 1$ and with $-\frac{1}{2}$ as a zero of multiplicity 2, 0 as a zero of multiplicity 1, and 1 as a zero of multiplicity 2.

Suppose that a polynomial function of degree 4 with rational coefficients has the given numbers as zeros. Find the other zero(s).

17. $-1, \sqrt{3}, \frac{11}{3}$
18. $-\sqrt{2}, -1, \frac{4}{5}$
19. $-i, 2 - \sqrt{5}$
20. $i, -3 + \sqrt{3}$

21. $3i, 0, -5$

22. $3, 0, -2i$

23. $-4 - 3i, 2 - \sqrt{3}$

24. $6 - 5i, -1 + \sqrt{7}$

Suppose that a polynomial function of degree 5 with rational coefficients has the given numbers as zeros. Find the other zero(s).

25. $-\frac{1}{2}, \sqrt{5}, -4i$

26. $\frac{3}{4}, -\sqrt{3}, 2i$

27. $-5, 0, 2 - i, 4$

28. $-2, 3, 4, 1 - i$

29. $6, -3 + 4i, 4 - \sqrt{5}$

30. $-3 - 3i, 2 + \sqrt{13}, 6$

31. $-\frac{3}{4}, \frac{3}{4}, 0, 4 - i$

32. $-0.6, 0, 0.6, -3 + \sqrt{2}$

Find a polynomial function of lowest degree with rational coefficients that has the given numbers as some of its zeros.

33. $1 + i, 2$

34. $2 - i, -1$

35. $4i$

36. $-5i$

37. $-4i, 5$

38. $3, -i$

39. $1 - i, -\sqrt{5}$

40. $2 - \sqrt{3}, 1 + i$

41. $\sqrt{5}, -3i$

42. $-\sqrt{2}, 4i$

Given that the polynomial function has the given zero, find the other zeros.

43. $f(x) = x^3 + 5x^2 - 2x - 10;\ -5$

44. $f(x) = x^3 - x^2 + x - 1;\ 1$

45. $f(x) = x^4 - 5x^3 + 7x^2 - 5x + 6;\ -i$

46. $f(x) = x^4 - 16;\ 2i$

47. $f(x) = x^3 - 6x^2 + 13x - 20;\ 4$

48. $f(x) = x^3 - 8;\ 2$

List all possible rational zeros of the function.

49. $f(x) = x^5 - 3x^2 + 1$

50. $f(x) = x^7 + 37x^5 - 6x^2 + 12$

51. $f(x) = 2x^4 - 3x^3 - x + 8$

52. $f(x) = 3x^3 - x^2 + 6x - 9$

53. $f(x) = 15x^6 + 47x^2 + 2$

54. $f(x) = 10x^{25} + 3x^{17} - 35x + 6$

For each polynomial function, **(a)** *find the rational zeros and then the other zeros; that is, solve $f(x) = 0$; and* **(b)** *factor $f(x)$ into linear factors.*

55. $f(x) = x^3 + 3x^2 - 2x - 6$

56. $f(x) = x^3 - x^2 - 3x + 3$

57. $f(x) = 3x^3 - x^2 - 15x + 5$

58. $f(x) = 4x^3 - 4x^2 - 3x + 3$

59. $f(x) = x^3 - 3x + 2$

60. $f(x) = x^3 - 2x + 4$

61. $f(x) = 2x^3 + 3x^2 + 18x + 27$

62. $f(x) = 2x^3 + 7x^2 + 2x - 8$

63. $f(x) = 5x^4 - 4x^3 + 19x^2 - 16x - 4$

64. $f(x) = 3x^4 - 4x^3 + x^2 + 6x - 2$

65. $f(x) = x^4 - 3x^3 - 20x^2 - 24x - 8$

66. $f(x) = x^4 + 5x^3 - 27x^2 + 31x - 10$

67. $f(x) = x^3 - 4x^2 + 2x + 4$

68. $f(x) = x^3 - 8x^2 + 17x - 4$

69. $f(x) = x^3 + 8$

70. $f(x) = x^3 - 8$

71. $f(x) = \frac{1}{3}x^3 - \frac{1}{2}x^2 - \frac{1}{6}x + \frac{1}{6}$

72. $f(x) = \frac{2}{3}x^3 - \frac{1}{2}x^2 + \frac{2}{3}x - \frac{1}{2}$

Find only the rational zeros of the function.

73. $f(x) = x^4 + 2x^3 - 5x^2 - 4x + 6$

74. $f(x) = x^4 - 3x^3 - 9x^2 - 3x - 10$

75. $f(x) = x^3 - x^2 - 4x + 3$

76. $f(x) = 2x^3 + 3x^2 + 2x + 3$

77. $f(x) = x^4 + 2x^3 + 2x^2 - 4x - 8$

78. $f(x) = x^4 + 6x^3 + 17x^2 + 36x + 66$

79. $f(x) = x^5 - 5x^4 + 5x^3 + 15x^2 - 36x + 20$

80. $f(x) = x^5 - 3x^4 - 3x^3 + 9x^2 - 4x + 12$

What does Descartes' rule of signs tell you about the number of positive real zeros and the number of negative real zeros of the function?

81. $f(x) = 3x^5 - 2x^2 + x - 1$

82. $g(x) = 5x^6 - 3x^3 + x^2 - x$

83. $h(x) = 6x^7 + 2x^2 + 5x + 4$

84. $P(x) = -3x^5 - 7x^3 - 4x - 5$

85. $F(p) = 3p^{18} + 2p^4 - 5p^2 + p + 3$

86. $H(t) = 5t^{12} - 7t^4 + 3t^2 + t + 1$

87. $C(x) = 7x^6 + 3x^4 - x - 10$

88. $g(z) = -z^{10} + 8z^7 + z^3 + 6z - 1$

89. $h(t) = -4t^5 - t^3 + 2t^2 + 1$

90. $P(x) = x^6 + 2x^4 - 9x^3 - 4$

91. $f(y) = y^4 + 13y^3 - y + 5$

92. $Q(x) = x^4 - 2x^2 + 12x - 8$

93. $r(x) = x^4 - 6x^2 + 20x - 24$

94. $f(x) = x^5 - 2x^3 - 8x$

95. $R(x) = 3x^5 - 5x^3 - 4x$

96. $f(x) = x^4 - 9x^2 - 6x + 4$

Sketch the graph of the polynomial function. Follow the procedure outlined on p. 240. Use the rational zeros theorem when finding the zeros.

97. $f(x) = 4x^3 + x^2 - 8x - 2$

98. $f(x) = 3x^3 - 4x^2 - 5x + 2$

99. $f(x) = 2x^4 - 3x^3 - 2x^2 + 3x$

100. $f(x) = 4x^4 - 37x^2 + 9$

Skill Maintenance

For Exercises 101 and 102, complete the square to **(a)** *find the vertex;* **(b)** *find the axis of symmetry; and* **(c)** *determine whether there is a maximum or a minimum function value and find that value.*

101. $f(x) = x^2 - 8x + 10$ [3.3]

102. $f(x) = 3x^2 - 6x - 1$ [3.3]

Find the zeros of the function.

103. $f(x) = -\frac{4}{5}x + 8$ [1.5]

104. $g(x) = x^2 - 8x - 33$ [3.2]

Determine the leading term, the leading coefficient, and the degree of the polynomial. Then describe the end behavior of the function's graph and classify the polynomial function as constant, linear, quadratic, cubic, or quartic.

105. $g(x) = -x^3 - 2x^2$ [4.1]

106. $f(x) = -x^2 - 3x + 6$ [3.3]

107. $f(x) = -\frac{4}{9}$ [1.3]

108. $h(x) = x - 2$ [1.3]

109. $g(x) = x^4 - 2x^3 + x^2 - x + 2$ [4.1]

110. $h(x) = x^3 + \frac{1}{2}x^2 - 4x - 3$ [4.1]

Synthesis

111. Consider $f(x) = 2x^3 - 5x^2 - 4x + 3$. Find the solutions of each equation.
   a) $f(x) = 0$
   b) $f(x - 1) = 0$
   c) $f(x + 2) = 0$
   d) $f(2x) = 0$

112. Use the rational zeros theorem and the equation $x^4 - 12 = 0$ to show that $\sqrt[4]{12}$ is irrational.

Find the rational zeros of the function.

113. $P(x) = 2x^5 - 33x^4 - 84x^3 + 2203x^2 - 3348x - 10{,}080$

114. $P(x) = x^6 - 6x^5 - 72x^4 - 81x^2 + 486x + 5832$

4.5 Rational Functions

- For a rational function, find the domain and graph the function, identifying all of the asymptotes.
- Solve applied problems involving rational functions.

Now we turn our attention to functions that represent the quotient of two polynomials. Whereas the sum, the difference, or the product of two polynomials is a polynomial, in general the quotient of two polynomials is *not* itself a polynomial.

A *rational number* can be expressed as the quotient of two integers, p/q, where $q \neq 0$. A *rational function* is formed by the quotient of two polynomials, $p(x)/q(x)$, where $q(x) \neq 0$. Here are some examples of rational functions and their graphs.

$f(x) = \frac{1}{x}$

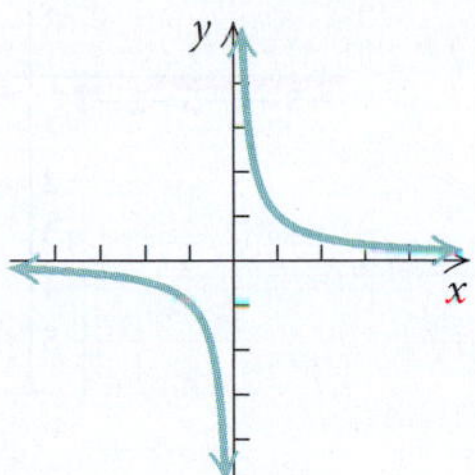

$f(x) = \frac{1}{x^2}$

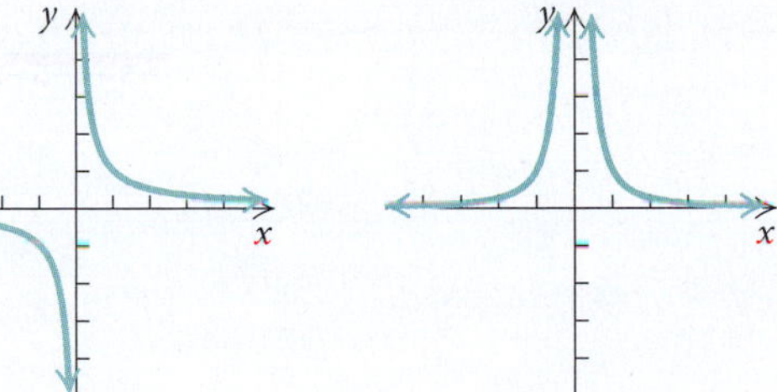

$f(x) = \frac{x-3}{x^2+x-2}$

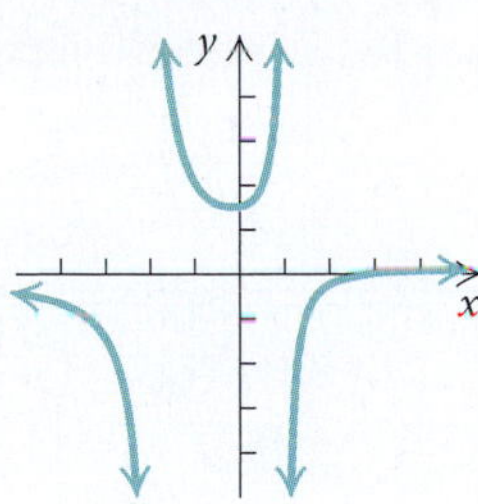

$f(x) = \frac{2x+5}{2x-6}$

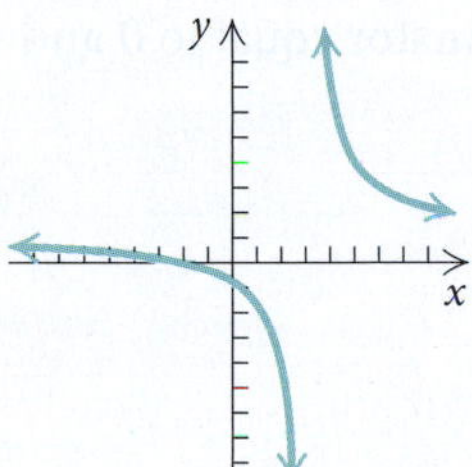

$f(x) = \frac{x^2+2x-3}{x^2-x-2}$

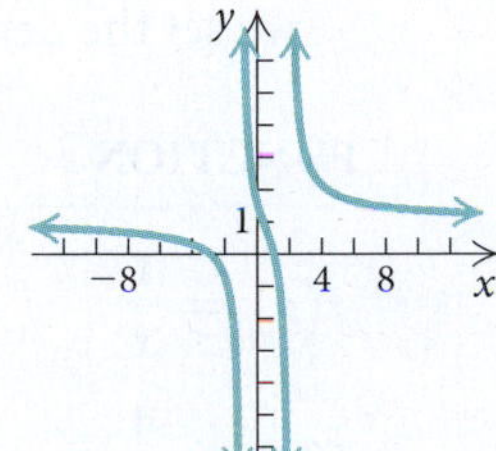

$f(x) = \frac{-x^2}{x+1}$

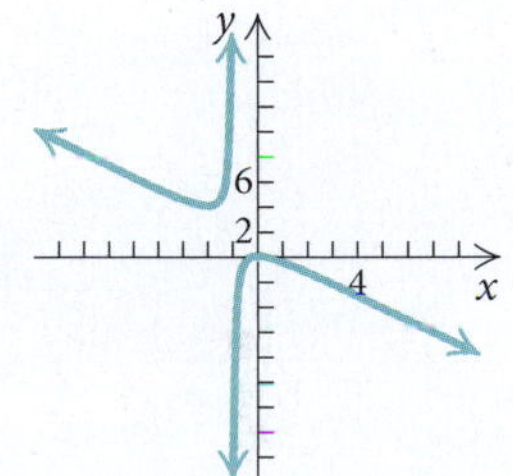

RATIONAL FUNCTION

A **rational function** is a function f that is a quotient of two polynomials. That is,

$$f(x) = \frac{p(x)}{q(x)},$$

where $p(x)$ and $q(x)$ are polynomials and where $q(x)$ is not the zero polynomial. The domain of f consists of all inputs x for which $q(x) \neq 0$.

The Domain of a Rational Function

EXAMPLE 1 Consider

$$f(x) = \frac{1}{x-3}.$$

Find the domain and graph f.

DOMAINS OF FUNCTIONS
REVIEW SECTION 1.2

Solution When the denominator $x - 3$ is 0, we have $x = 3$, so the only input that results in a denominator of 0 is 3. Thus the domain is

$$\{x \mid x \neq 3\}, \text{ or } (-\infty, 3) \cup (3, \infty).$$

The graph of this function is the graph of $y = 1/x$ translated right 3 units.

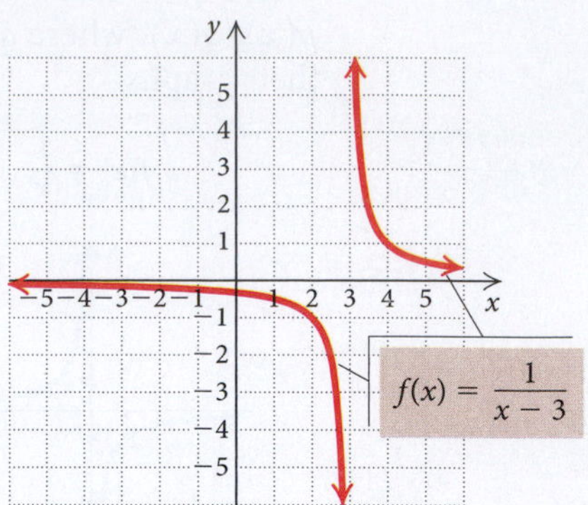

JUST IN TIME 6

EXAMPLE 2 Determine the domain of each of the functions illustrated at the beginning of this section.

Solution The domain of each rational function will be the set of all real numbers except those values that make the denominator 0. To determine those exceptions, we set the denominator equal to 0 and solve for x.

FUNCTION	DOMAIN
$f(x) = \frac{1}{x}$	$\{x \mid x \neq 0\}$, or $(-\infty, 0) \cup (0, \infty)$
$f(x) = \frac{1}{x^2}$	$\{x \mid x \neq 0\}$, or $(-\infty, 0) \cup (0, \infty)$
$f(x) = \frac{x-3}{x^2+x-2} = \frac{x-3}{(x+2)(x-1)}$	$\{x \mid x \neq -2 \text{ and } x \neq 1\}$, or $(-\infty, -2) \cup (-2, 1) \cup (1, \infty)$
$f(x) = \frac{2x+5}{2x-6} = \frac{2x+5}{2(x-3)}$	$\{x \mid x \neq 3\}$, or $(-\infty, 3) \cup (3, \infty)$
$f(x) = \frac{x^2+2x-3}{x^2-x-2} = \frac{x^2+2x-3}{(x+1)(x-2)}$	$\{x \mid x \neq -1 \text{ and } x \neq 2\}$, or $(-\infty, -1) \cup (-1, 2) \cup (2, \infty)$
$f(x) = \frac{-x^2}{x+1}$	$\{x \mid x \neq -1\}$, or $(-\infty, -1) \cup (-1, \infty)$

As a partial check of the domains, we can observe the discontinuities (breaks) in the graphs of these functions. (See p. 267.)

Asymptotes

Vertical Asymptotes

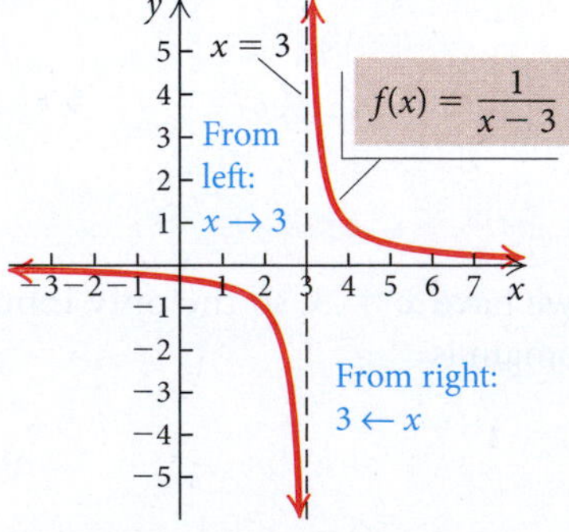

Vertical asymptote: $x = 3$

Look at the graph of $f(x) = 1/(x - 3)$, shown at left. (Also see Example 1.) Let's explore what happens as x-values get closer and closer to 3 from the left. We then explore what happens as x-values get closer and closer to 3 from the right.

From the left:

x	2	$2\frac{1}{2}$	$2\frac{99}{100}$	$2\frac{9999}{10{,}000}$	$2\frac{999{,}999}{1{,}000{,}000}$	$\to 3$
$f(x)$	-1	-2	-100	$-10{,}000$	$-1{,}000{,}000$	$\to -\infty$

From the right:

x	4	$3\frac{1}{2}$	$3\frac{1}{100}$	$3\frac{1}{10,000}$	$3\frac{1}{1,000,000}$	$\rightarrow 3$
$f(x)$	1	2	100	10,000	1,000,000	$\rightarrow \infty$

We see that as x-values get closer and closer to 3 from the left, the function values (y-values) decrease without bound (that is, they approach negative infinity, $-\infty$). Similarly, as the x-values approach 3 from the right, the function values increase without bound (that is, they approach positive infinity, ∞). We write this as

$$f(x) \rightarrow -\infty \text{ as } x \rightarrow 3^- \quad \text{and} \quad f(x) \rightarrow \infty \text{ as } x \rightarrow 3^+.$$

We read "$f(x) \rightarrow -\infty$ as $x \rightarrow 3^-$" as "$f(x)$ decreases without bound as x approaches 3 from the left." We read "$f(x) \rightarrow \infty$ as $x \rightarrow 3^+$" as "$f(x)$ increases without bound as x approaches 3 from the right." The notation $x \rightarrow 3$ means that x gets as close to 3 as possible without being equal to 3. The vertical line $x = 3$ is said to be a *vertical asymptote* for this curve.

In general, the line $x = a$ is a **vertical asymptote** for the graph of f if any of the following is true:

$$f(x) \rightarrow \infty \text{ as } x \rightarrow a^-, \quad \text{or} \quad f(x) \rightarrow -\infty \text{ as } x \rightarrow a^-, \quad \text{or}$$
$$f(x) \rightarrow \infty \text{ as } x \rightarrow a^+, \quad \text{or} \quad f(x) \rightarrow -\infty \text{ as } x \rightarrow a^+.$$

The following figures show the four ways in which a vertical asymptote can occur.

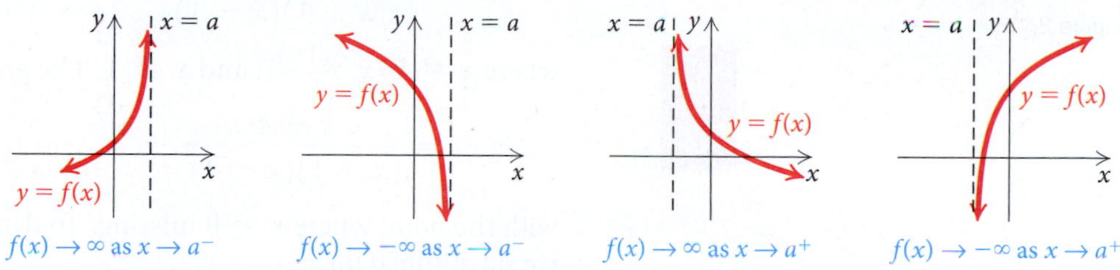

The vertical asymptotes of a rational function $f(x) = p(x)/q(x)$ are found by determining the zeros of $q(x)$ that are not also zeros of $p(x)$. If $p(x)$ and $q(x)$ are polynomials with no common factors other than constants, we need determine only the zeros of the denominator $q(x)$.

DETERMINING VERTICAL ASYMPTOTES

For a rational function $f(x) = p(x)/q(x)$, where $p(x)$ and $q(x)$ are polynomials with *no common factors* other than constants, if a is a zero of the denominator, then the line $x = a$ is a vertical asymptote for the graph of the function.

EXAMPLE 3 Determine the vertical asymptotes for the graph of each of the following functions.

a) $f(x) = \dfrac{2x - 11}{x^2 + 2x - 8}$

b) $h(x) = \dfrac{x^2 - 4x}{x^3 - x}$

c) $g(x) = \dfrac{x - 2}{x^3 - 5x}$

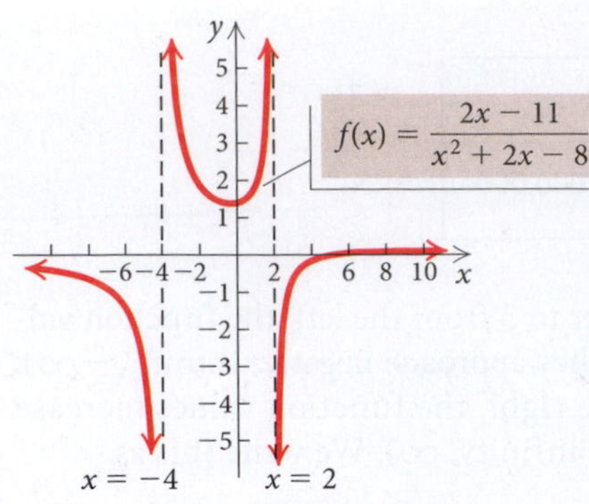

Figure 1.

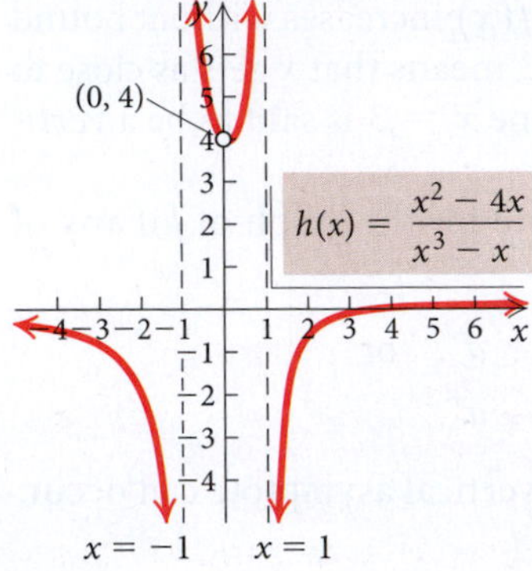

Figure 2.

Solution

a) First, we factor the denominator:

$$f(x) = \frac{2x - 11}{x^2 + 2x - 8} = \frac{2x - 11}{(x + 4)(x - 2)}.$$

The numerator and the denominator have no common factors. The zeros of the denominator are -4 and 2. Thus the vertical asymptotes for the graph of $f(x)$ are the lines $x = -4$ and $x = 2$. (See Fig. 1.)

b) We factor the numerator and the denominator:

$$h(x) = \frac{x^2 - 4x}{x^3 - x} = \frac{x(x - 4)}{x(x^2 - 1)} = \frac{x(x - 4)}{x(x + 1)(x - 1)}.$$

The domain of the function is $\{x \mid x \neq -1 \text{ and } x \neq 0 \text{ and } x \neq 1\}$, or $(-\infty, -1) \cup (-1, 0) \cup (0, 1) \cup (1, \infty)$. Note that the numerator and the denominator share a common factor, x. The vertical asymptotes of $h(x)$ are found by determining the zeros of the denominator, $x(x + 1)(x - 1)$, that are *not* also zeros of the numerator, $x(x - 4)$. The zeros of $x(x + 1)(x - 1)$ are 0, -1, and 1. The zeros of $x(x - 4)$ are 0 and 4. Thus, although the denominator has three zeros, the graph of $h(x)$ has only two vertical asymptotes, $x = -1$ and $x = 1$. (See Fig. 2.)

The rational expression $[x(x - 4)]/[x(x + 1)(x - 1)]$ can be simplified. Thus,

$$h(x) = \frac{x(x - 4)}{x(x + 1)(x - 1)} = \frac{x - 4}{(x + 1)(x - 1)},$$

where $x \neq 0$, $x \neq -1$, and $x \neq 1$. The graph of $h(x)$ is the graph of

$$h(x) = \frac{x - 4}{(x + 1)(x - 1)}$$

with the point where $x = 0$ missing. To determine the y-coordinate of the hole, we substitute 0 for x:

$$h(0) = \frac{0 - 4}{(0 + 1)(0 - 1)} = \frac{-4}{1 \cdot (-1)} = 4.$$

Thus the hole is located at $(0, 4)$.

c) We factor the denominator:

$$g(x) = \frac{x - 2}{x^3 - 5x} = \frac{x - 2}{x(x^2 - 5)}.$$

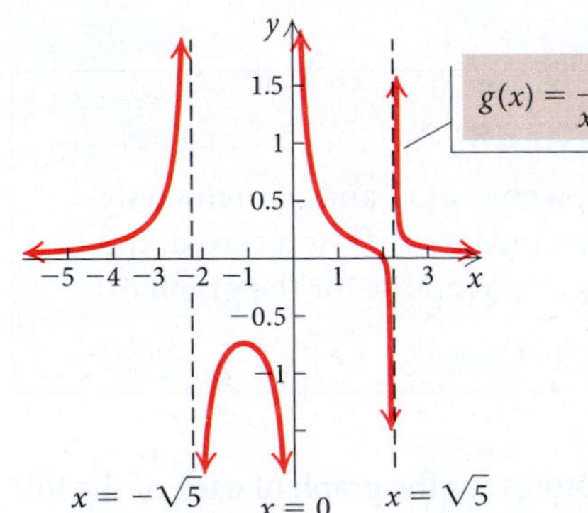

Figure 3.

The numerator and the denominator have no common factors. We find the zeros of the denominator, $x(x^2 - 5)$. Solving $x(x^2 - 5) = 0$, we get

$$x = 0 \quad or \quad x^2 - 5 = 0$$
$$x = 0 \quad or \quad x^2 = 5$$
$$x = 0 \quad or \quad x = \pm\sqrt{5}.$$

The zeros of the denominator are 0, $\sqrt{5}$, and $-\sqrt{5}$. Thus the vertical asymptotes are the lines $x = 0$, $x = \sqrt{5}$, and $x = -\sqrt{5}$. (See Fig. 3.)

Now Try Exercises 15 and 19.

Horizontal Asymptotes

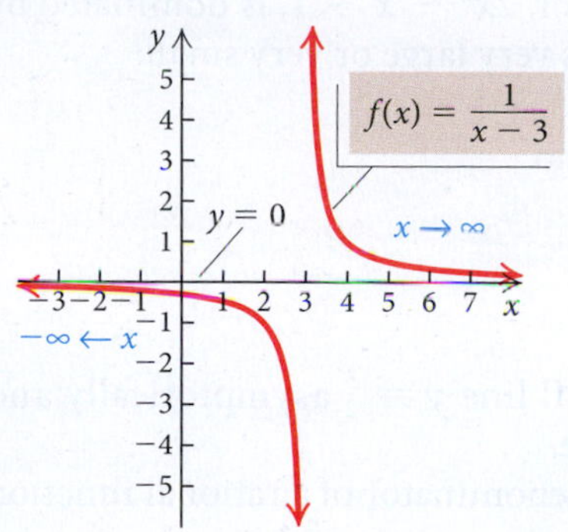

Horizontal asymptote: $y = 0$

Looking again at the graph of $f(x) = 1/(x - 3)$, shown at left (also see Example 1), let's explore what happens to $f(x) = 1/(x - 3)$ as x increases without bound (approaches positive infinity, ∞) and as x decreases without bound (approaches negative infinity, $-\infty$).

x increases without bound:

x	100	5000	1,000,000	$\to \infty$
$f(x)$	≈ 0.0103	≈ 0.0002	≈ 0.000001	$\to 0$

x decreases without bound:

x	-300	-8000	$-1{,}000{,}000$	$\to -\infty$
$f(x)$	≈ -0.0033	≈ -0.0001	≈ -0.000001	$\to 0$

We see that

$$\frac{1}{x-3} \to 0 \text{ as } x \to \infty \quad \text{and} \quad \frac{1}{x-3} \to 0 \text{ as } x \to -\infty.$$

Since $y = 0$ is the equation of the x-axis, we say that the curve approaches the x-axis asymptotically and that the x-axis is a *horizontal asymptote* for the curve.

In general, the line $y = b$ is a **horizontal asymptote** for the graph of f if either or both of the following are true:

$$f(x) \to b \text{ as } x \to \infty \quad \text{or} \quad f(x) \to b \text{ as } x \to -\infty.$$

The following figures illustrate four ways in which horizontal asymptotes can occur. In each case, the curve gets close to the line $y = b$ either as $x \to \infty$ or as $x \to -\infty$. Keep in mind that the symbols ∞ and $-\infty$ convey the idea of increasing without bound and decreasing without bound, respectively.

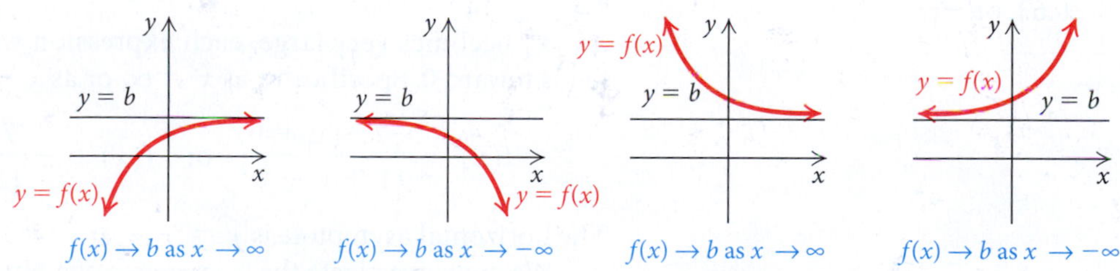

How can we determine a horizontal asymptote? As x gets very large or very small, the value of a polynomial function $p(x)$ is dominated by the function's leading term. Because of this, if $p(x)$ and $q(x)$ have the *same* degree, the value of $p(x)/q(x)$ as $x \to \infty$ or as $x \to -\infty$ is dominated by the ratio of the numerator's leading coefficient to the denominator's leading coefficient.

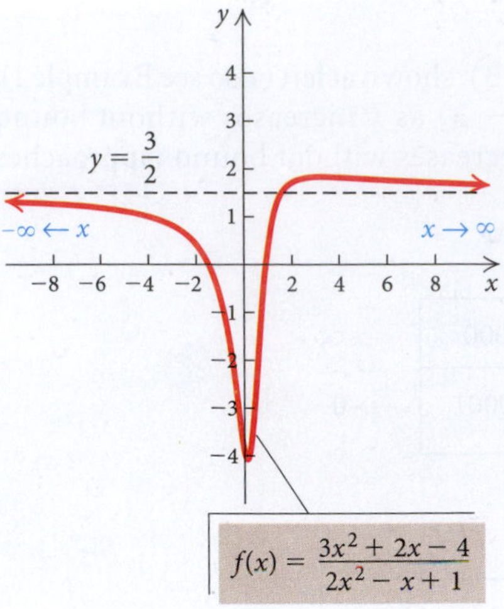

For $f(x) = (3x^2 + 2x - 4)/(2x^2 - x + 1)$, we see that the numerator, $3x^2 + 2x - 4$, is dominated by $3x^2$ and the denominator, $2x^2 - x + 1$, is dominated by $2x^2$, so $f(x)$ approaches $3x^2/2x^2$, or $3/2$, as x gets very large or very small:

$$\frac{3x^2 + 2x - 4}{2x^2 - x + 1} \rightarrow \frac{3}{2}, \text{ or } 1.5, \text{ as } x \rightarrow \infty, \quad \text{and}$$

$$\frac{3x^2 + 2x - 4}{2x^2 - x + 1} \rightarrow \frac{3}{2}, \text{ or } 1.5, \text{ as } x \rightarrow -\infty.$$

We say that the curve approaches the horizontal line $y = \frac{3}{2}$ asymptotically and that $y = \frac{3}{2}$ is a *horizontal asymptote* for the curve.

It follows that when the numerator and the denominator of a rational function have the same degree, the line $y = a/b$ is the horizontal asymptote, where a and b are the leading coefficients of the numerator and the denominator, respectively.

EXAMPLE 4 Find the horizontal asymptote: $f(x) = \dfrac{-7x^4 - 10x^2 + 1}{11x^4 + x - 2}$.

Solution The numerator and the denominator have the same degree. The ratio of the leading coefficients is $-\frac{7}{11}$, so the line $y = -\frac{7}{11}$, or $-0.\overline{63}$, is the horizontal asymptote.

Now Try Exercise 21.

Technology Connection

We can check Example 4 by using a graphing calculator to evaluate the function for a very large and a very small value of x.

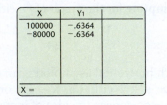

X	Y1
100000	−.6364
−80000	−.6364
X =	

For each number, the function value is close to $-0.\overline{63}$, or $-\frac{7}{11}$.

To check Example 4, we could evaluate the function for a very large and a very small value of x. Another check, one that is useful in calculus, is to multiply by 1, using $(1/x^4)/(1/x^4)$:

$$f(x) = \frac{-7x^4 - 10x^2 + 1}{11x^4 + x - 2} \cdot \frac{\frac{1}{x^4}}{\frac{1}{x^4}} = \frac{\frac{-7x^4}{x^4} - \frac{10x^2}{x^4} + \frac{1}{x^4}}{\frac{11x^4}{x^4} + \frac{x}{x^4} - \frac{2}{x^4}}$$

$$= \frac{-7 - \frac{10}{x^2} + \frac{1}{x^4}}{11 + \frac{1}{x^3} - \frac{2}{x^4}}.$$

As $|x|$ becomes very large, each expression whose denominator is a power of x tends toward 0. Specifically, as $x \rightarrow \infty$ or as $x \rightarrow -\infty$, we have

$$f(x) \rightarrow \frac{-7 - 0 + 0}{11 + 0 - 0}, \quad \text{or} \quad f(x) \rightarrow -\frac{7}{11}.$$

The horizontal asymptote is $y = -\frac{7}{11}$, or $-0.\overline{63}$.

We now investigate the occurrence of a horizontal asymptote when the degree of the numerator is less than the degree of the denominator.

EXAMPLE 5 Find the horizontal asymptote: $f(x) = \dfrac{2x + 3}{x^3 - 2x^2 + 4}$.

Solution Let $p(x) = 2x + 3$, $q(x) = x^3 - 2x^2 + 4$, and $f(x) = p(x)/q(x)$. Note that as $x \rightarrow \infty$, the value of $q(x)$ grows much faster than the value of $p(x)$. Because of this, the ratio $p(x)/q(x)$ shrinks toward 0. As $x \rightarrow -\infty$, the ratio $p(x)/q(x)$ behaves in a similar manner. The horizontal asymptote is $y = 0$, the x-axis. This is the case for all rational functions for which the degree of the numerator is less than the degree of the denominator. Note in Example 1 that $y = 0$, the x-axis, is the horizontal asymptote of $f(x) = 1/(x - 3)$.

Now Try Exercise 23.

The following statements describe the two ways in which a horizontal asymptote occurs.

DETERMINING A HORIZONTAL ASYMPTOTE

- When the numerator and the denominator of a rational function have the same degree, the line $y = a/b$ is the horizontal asymptote, where a and b are the leading coefficients of the numerator and the denominator, respectively.
- When the degree of the numerator of a rational function is less than the degree of the denominator, the x-axis, or $y = 0$, is the horizontal asymptote.
- When the degree of the numerator of a rational function is greater than the degree of the denominator, there is no horizontal asymptote.

EXAMPLE 6 Graph

$$g(x) = \frac{2x^2 + 1}{x^2}.$$

Include and label all asymptotes.

Solution Since 0 is the zero of the denominator and is not a zero of the numerator, the y-axis, $x = 0$, is the vertical asymptote. Note also that the degree of the numerator is the same as the degree of the denominator. Thus, $y = 2/1$, or 2, is the horizontal asymptote.

To draw the graph, we first draw the asymptotes with dashed lines. Then we compute and plot some ordered pairs and draw the two branches of the curve.

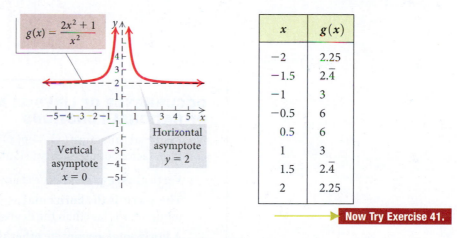

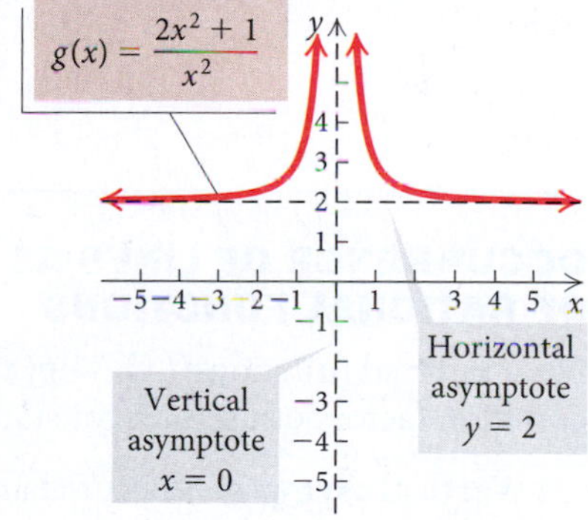

x	$g(x)$
−2	2.25
−1.5	$2.\overline{4}$
−1	3
−0.5	6
0.5	6
1	3
1.5	$2.\overline{4}$
2	2.25

Now Try Exercise 41.

Oblique Asymptotes

Sometimes a line that is neither horizontal nor vertical is an asymptote. Such a line is called an **oblique asymptote**, or a **slant asymptote**.

EXAMPLE 7 Find all the asymptotes of

$$f(x) = \frac{2x^2 - 3x - 1}{x - 2}.$$

Solution The line $x = 2$ is the vertical asymptote because 2 is the zero of the denominator and is not a zero of the numerator. There is no horizontal asymptote because the degree of the numerator is greater than the degree of the denominator. When the degree of the numerator is 1 greater than the degree of the denominator, we divide to find an equivalent expression:

$$\frac{2x^2 - 3x - 1}{x - 2} = (2x + 1) + \frac{1}{x - 2}.$$

$$\begin{array}{r} 2x + 1 \\ x - 2 \overline{)\, 2x^2 - 3x - 1} \\ \underline{2x^2 - 4x} \\ x - 1 \\ \underline{x - 2} \\ 1 \end{array}$$

Now we see that when $x \to \infty$ or $x \to -\infty$, $1/(x - 2) \to 0$ and the value of $f(x) \to 2x + 1$. This means that as $|x|$ becomes very large, the graph of $f(x)$ gets very close to the graph of $y = 2x + 1$. Thus the line $y = 2x + 1$ is the oblique asymptote.

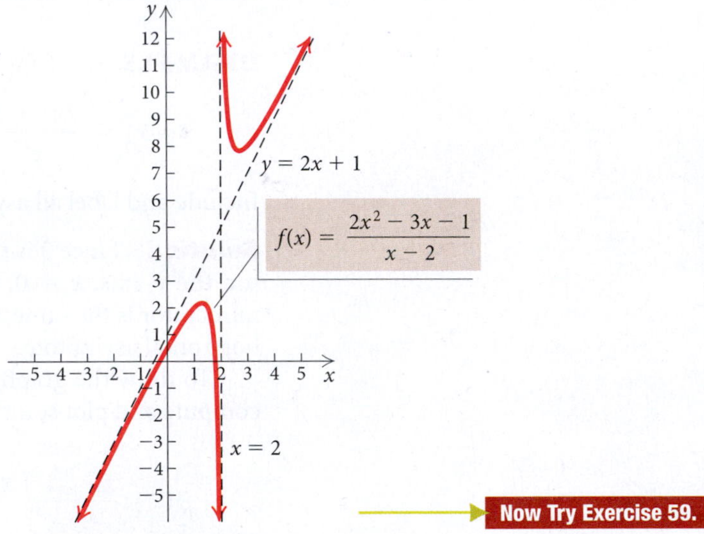

Now Try Exercise 59.

OCCURRENCE OF LINES AS ASYMPTOTES OF RATIONAL FUNCTIONS

For a rational function $f(x) = p(x)/q(x)$, where $p(x)$ and $q(x)$ have no common factors other than constants:

Vertical asymptotes occur at any x-values that make the denominator 0.

The x-axis is the horizontal asymptote when the degree of the numerator is less than the degree of the denominator.

A horizontal asymptote other than the x-axis occurs when the numerator and the denominator have the same degree.

An oblique asymptote occurs when the degree of the numerator is 1 greater than the degree of the denominator.

There can be only one horizontal asymptote or one oblique asymptote and never both.

An asymptote is *not* part of the graph of the function.

The following statements are also true.

Crossing an Asymptote

- The graph of a rational function *never crosses* a vertical asymptote.
- The graph of a rational function *might cross* a horizontal asymptote but does not necessarily do so.

Shown below is an outline of a procedure that we can follow to create accurate graphs of rational functions.

To graph a rational function $f(x) = p(x)/q(x)$, where $p(x)$ and $q(x)$ have no common factor other than constants:

1. Find any real zeros of the denominator. Determine the domain of the function and sketch any vertical asymptotes.
2. Find the horizontal asymptote or the oblique asymptote, if there is one, and sketch it.
3. Find any zeros of the function. The zeros are found by determining the zeros of the numerator. These are the first coordinates of the x-intercepts of the graph.
4. Find $f(0)$. This gives the y-intercept, $(0, f(0))$, of the function.
5. Find other function values to determine the general shape. Then draw the graph.

EXAMPLE 8 Graph: $f(x) = \dfrac{2x + 3}{3x^2 + 7x - 6}$.

Solution

1. We find the zeros of the denominator by solving $3x^2 + 7x - 6 = 0$. Since

   $$3x^2 + 7x - 6 = (3x - 2)(x + 3),$$

   the zeros are $\frac{2}{3}$ and -3. Thus the domain excludes $\frac{2}{3}$ and -3 and is

   $$(-\infty, -3) \cup \left(-3, \tfrac{2}{3}\right) \cup \left(\tfrac{2}{3}, \infty\right).$$

   Since neither zero of the denominator is a zero of the numerator, the graph has vertical asymptotes $x = -3$ and $x = \frac{2}{3}$. We sketch these as dashed lines.
2. Because the degree of the numerator is less than the degree of the denominator, the x-axis, $y = 0$, is the horizontal asymptote.
3. To find the zeros of the numerator, we solve $2x + 3 = 0$ and get $x = -\frac{3}{2}$. Thus, $-\frac{3}{2}$ is the zero of the function, and the pair $\left(-\frac{3}{2}, 0\right)$ is the x-intercept.
4. We find $f(0)$:

   $$f(0) = \frac{2 \cdot 0 + 3}{3 \cdot 0^2 + 7 \cdot 0 - 6} = \frac{3}{-6} = -\frac{1}{2}.$$

   The point $\left(0, -\frac{1}{2}\right)$ is the y-intercept.

5. We find other function values to determine the general shape. We choose values in each interval of the domain as shown in the table below and then draw the graph. Note that the graph of this function crosses its horizontal asymptote at $x = -\frac{3}{2}$.

x	y
−4.5	−0.26
−3.25	−1.19
−2.5	0.42
−0.5	−0.23
0.5	−2.29
0.75	4.8
1.5	0.53
3.5	0.18

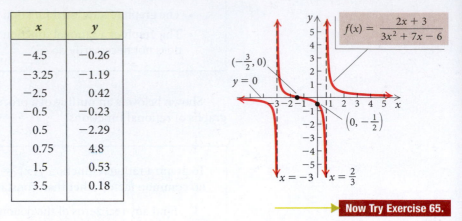

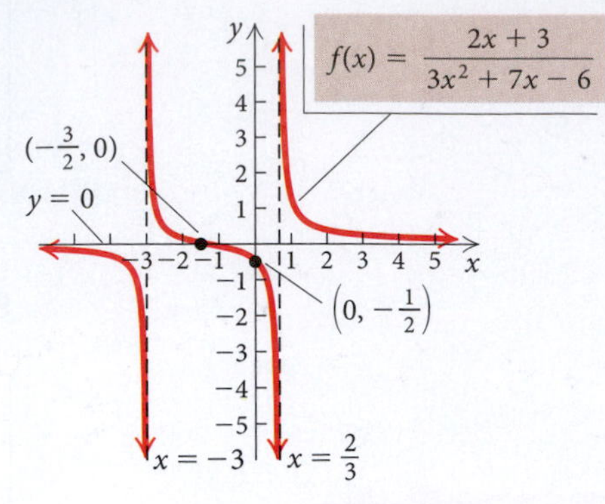

Now Try Exercise 65.

EXAMPLE 9 Graph: $g(x) = \dfrac{x^2 - 1}{x^2 + x - 6}$.

Solution

1. We find the zeros of the denominator by solving $x^2 + x - 6 = 0$. Since

$$x^2 + x - 6 = (x + 3)(x - 2),$$

the zeros are -3 and 2. Thus the domain excludes the x-values -3 and 2 and is

$$(-\infty, -3) \cup (-3, 2) \cup (2, \infty).$$

Since neither zero of the denominator is a zero of the numerator, the graph has vertical asymptotes $x = -3$ and $x = 2$. We sketch these as dashed lines.

2. The numerator and the denominator have the same degree, so the horizontal asymptote is determined by the ratio of the leading coefficients: $1/1$, or 1. Thus, $y = 1$ is the horizontal asymptote. We sketch it with a dashed line.

3. To find the zeros of the numerator, we solve $x^2 - 1 = 0$. The solutions are -1 and 1. Thus, -1 and 1 are the zeros of the function and the pairs $(-1, 0)$ and $(1, 0)$ are the x-intercepts.

4. We find $g(0)$:

$$g(0) = \frac{0^2 - 1}{0^2 + 0 - 6} = \frac{-1}{-6} = \frac{1}{6}.$$

Thus, $(0, \frac{1}{6})$ is the y-intercept.

5. We find other function values to determine the general shape and then draw the graph.

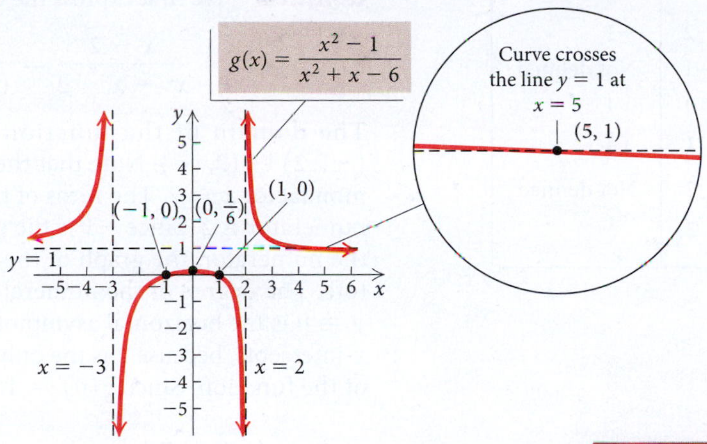

Now Try Exercise 77.

The magnified portion of the graph in Example 9 above shows another situation in which a graph can cross its horizontal asymptote. The point where $g(x)$ crosses $y = 1$ can be found by setting $g(x) = 1$ and solving for x:

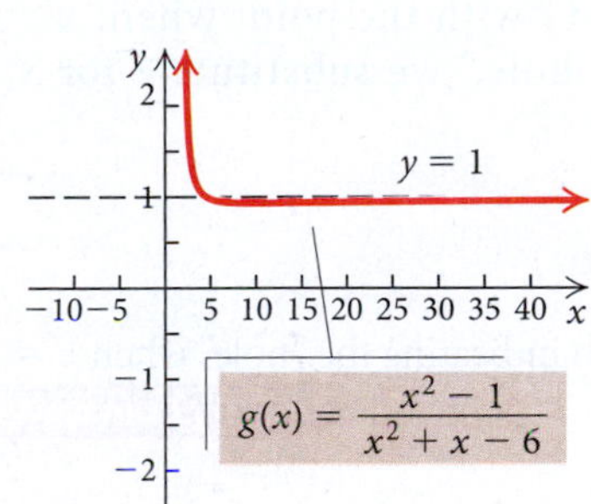

$$\frac{x^2 - 1}{x^2 + x - 6} = 1$$

$$x^2 - 1 = x^2 + x - 6$$

$$-1 = x - 6 \quad \text{Subtracting } x^2$$

$$5 = x. \quad \text{Adding 6}$$

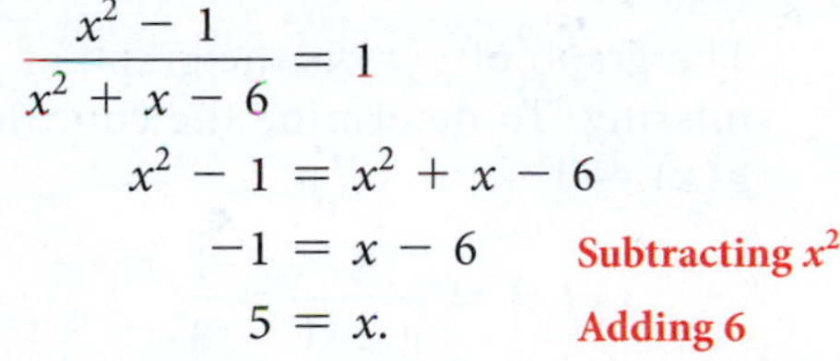

The point of intersection is $(5, 1)$. Note the behavior of the curve after it crosses the horizontal asymptote at $x = 5$. (See the graph at left.) It continues to decrease for a short interval and then begins to increase, getting closer and closer to $y = 1$ as $x \to \infty$.

Graphs of rational functions can also cross an oblique asymptote. The graph of

$$f(x) = \frac{2x^3}{x^2 + 1}$$

shown below crosses its oblique asymptote $y = 2x$. **Remember, graphs can cross horizontal asymptotes or oblique asymptotes, but they cannot cross vertical asymptotes.**

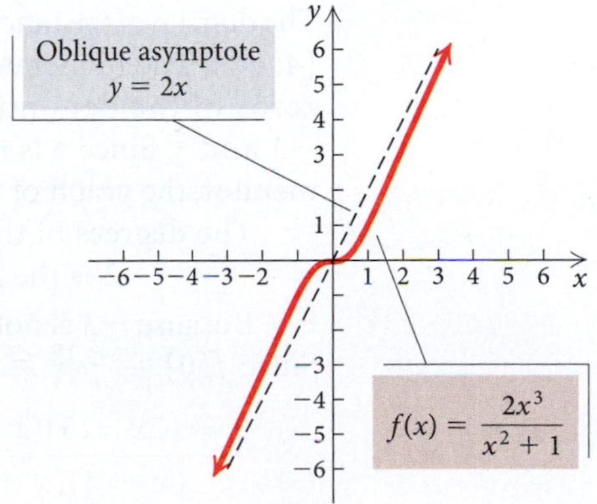

Let's now graph a rational function $f(x) = p(x)/q(x)$, where $p(x)$ and $q(x)$ have a common factor, $x - c$. The graph of such a function has a "hole" in it. We first saw this situation in Example 3(b), where the common factor was x.

x	y
-3	$-\frac{1}{2}$
-2	-1
-1	Not defined
0	1
1	$\frac{1}{2}$
2	Not defined
3	$\frac{1}{4}$

EXAMPLE 10 Graph: $g(x) = \dfrac{x - 2}{x^2 - x - 2}$.

Solution We first express the denominator in factored form:

$$g(x) = \frac{x - 2}{x^2 - x - 2} = \frac{x - 2}{(x + 1)(x - 2)}.$$

The domain of the function is $\{x | x \neq -1 \textit{ and } x \neq 2\}$, or $(-\infty, -1) \cup (-1, 2) \cup (2, \infty)$. Note that the numerator and the denominator have the common factor $x - 2$. The zeros of the denominator are -1 and 2, and the zero of the numerator is 2. Since -1 is the only zero of the denominator that is *not* a zero of the numerator, the graph of the function has $x = -1$ as its *only* vertical asymptote. The degree of the numerator is less than the degree of the denominator, so $y = 0$ is the horizontal asymptote. There are no zeros of the function and thus no x-intercepts, because 2 is the only zero of the numerator and 2 is not in the domain of the function. Since $g(0) = 1$, $(0, 1)$ is the y-intercept. The rational expression

$$\frac{x - 2}{(x + 1)(x - 2)}$$

can be simplified. Thus,

$$g(x) = \frac{x - 2}{(x + 1)(x - 2)} = \frac{1}{x + 1}, \quad \text{where } x \neq -1 \text{ and } x \neq 2.$$

$x = -1$

$g(x) = \dfrac{x - 2}{(x + 1)(x - 2)}$

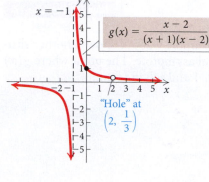

"Hole" at $\left(2, \frac{1}{3}\right)$

The graph of $g(x)$ is the graph of $y = 1/(x + 1)$ with the point where $x = 2$ missing. To determine the coordinates of the "hole," we substitute 2 for x in $g(x) = 1/(x + 1)$:

$$g(2) = \frac{1}{2 + 1} = \frac{1}{3}.$$

Thus the "hole" is located at $\left(2, \frac{1}{3}\right)$. We draw the graph indicating the "hole" when $x = 2$ with an open circle.

Now Try Exercise 49.

JUST IN TIME 18

EXAMPLE 11 Graph: $f(x) = \dfrac{-2x^2 - x + 15}{x^2 - x - 12}$.

Solution We first express the numerator and the denominator in factored form:

$$f(x) = \frac{-2x^2 - x + 15}{x^2 - x - 12} = \frac{-(2x^2 + x - 15)}{x^2 - x - 12} = \frac{-(2x - 5)(x + 3)}{(x - 4)(x + 3)}.$$

The domain of the function is $\{x | x \neq -3 \textit{ and } x \neq 4\}$, or $(-\infty, -3) \cup (-3, 4) \cup (4, \infty)$. The numerator and the denominator have the common factor $x + 3$. The zeros of the denominator are -3 and 4, and the zeros of the numerator are -3 and $\frac{5}{2}$. Since 4 is the only zero of the denominator that is *not* a zero of the numerator, the graph of the function has $x = 4$ as its *only* vertical asymptote.

The degrees of the numerator and the denominator are the same, so the line $y = \frac{-2}{1} = -2$ is the horizontal asymptote. The zeros of the numerator are $\frac{5}{2}$ and -3. Because -3 is not in the domain of the function, the only x-intercept is $\left(\frac{5}{2}, 0\right)$. Since $f(0) = -\frac{15}{12} = -\frac{5}{4}$, then $\left(0, -\frac{5}{4}\right)$ is the y-intercept. The rational function

$$\frac{-(2x - 5)(x + 3)}{(x - 4)(x + 3)}$$

can be simplified. Thus,

$$f(x) = \frac{-(2x - 5)(x + 3)}{(x - 4)(x + 3)} = \frac{-(2x - 5)}{x - 4}, \quad \text{where } x \neq -3 \text{ and } x \neq 4.$$

The graph of $f(x)$ is the graph of $y = -(2x - 5)/(x - 4)$ with the point where $x = -3$ missing. To determine the coordinates of the hole, we substitute -3 for x in $f(x) = (-2x + 5)/(x - 4)$:

$$f(-3) = \frac{-[2(-3) - 5]}{-3 - 4} = \frac{-[-11]}{-7} = \frac{11}{-7} = -\frac{11}{7}.$$

Thus the hole is located at $\left(-3, -\frac{11}{7}\right)$. We draw the graph indicating the hole when $x = -3$ with an open circle.

x	y
−5	−1.67
−4	−1.63
−3	Not defined
−2	−1.5
−1	−1.4
0	−1.25
1	−1
2	−0.5

x	y
3	1
3.5	4
4	Not defined
5	−5
6	−3.5
7	−3
8	−2.75

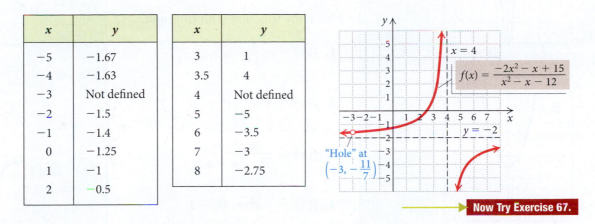

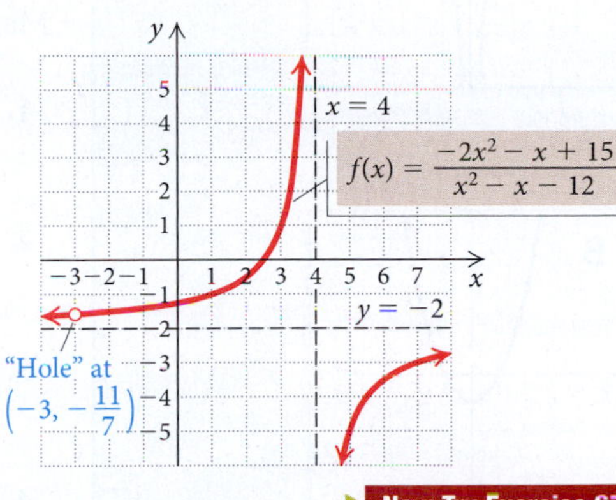

Now Try Exercise 67.

▶ Applications

EXAMPLE 12 ***Temperature During an Illness.*** A person's temperature T, in degrees Fahrenheit, during an illness is given by the function

$$T(t) = \frac{4t}{t^2 + 1} + 98.6,$$

where time t is given in hours since the onset of the illness. The graph of this function is shown at left.

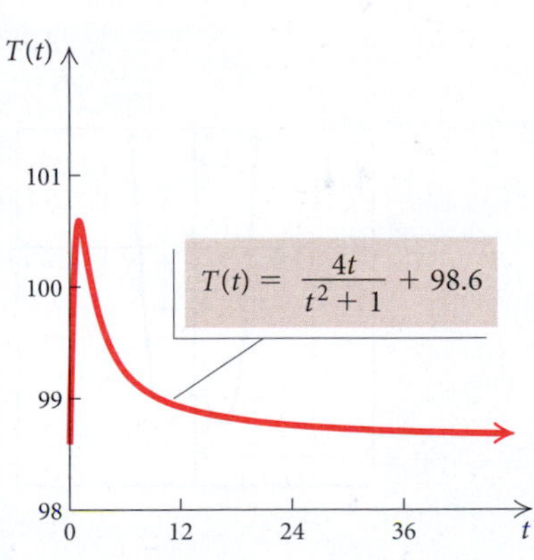

a) Find the temperature at $t = 0, 1, 2, 5, 12$, and 24.

b) Find the horizontal asymptote of the graph of $T(t)$. Complete:

$T(t) \to \square$ as $t \to \infty$.

c) Give the meaning of the answer to part (b) in terms of the application.

Solution

a) We have

$$T(0) = 98.6, \quad T(1) = 100.6, \quad T(2) = 100.2,$$
$$T(5) \approx 99.369, \quad T(12) \approx 98.931, \quad \text{and} \quad T(24) \approx 98.766.$$

b) Since

$$T(t) = \frac{4t}{t^2 + 1} + 98.6 = \frac{98.6t^2 + 4t + 98.6}{t^2 + 1},$$

the horizontal asymptote is $y = 98.6/1$, or 98.6. Then it follows that $T(t) \to 98.6$ as $t \to \infty$.

c) As time goes on, the temperature returns to "normal," which is 98.6°F.

Now Try Exercise 83.

Visualizing the Graph

Match the function with its graph.

1. $f(x) = -\dfrac{1}{x^2}$

2. $f(x) = x^3 - 3x^2 + 2x + 3$

3. $f(x) = \dfrac{x^2 - 4}{x^2 - x - 6}$

4. $f(x) = -x^2 + 4x - 1$

5. $f(x) = \dfrac{x - 3}{x^2 + x - 6}$

6. $f(x) = \dfrac{3}{4}x + 2$

7. $f(x) = x^2 - 1$

8. $f(x) = x^4 - 2x^2 - 5$

9. $f(x) = \dfrac{8x - 4}{3x + 6}$

10. $f(x) = 2x^2 - 4x - 1$

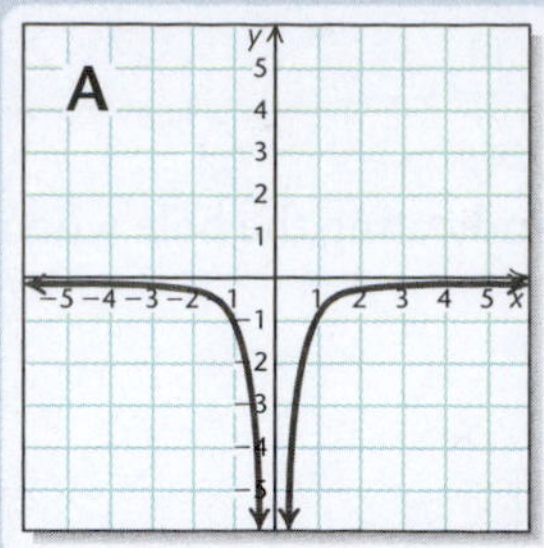

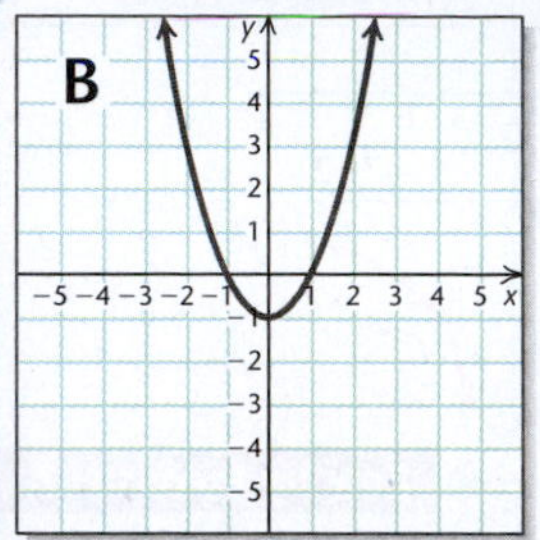

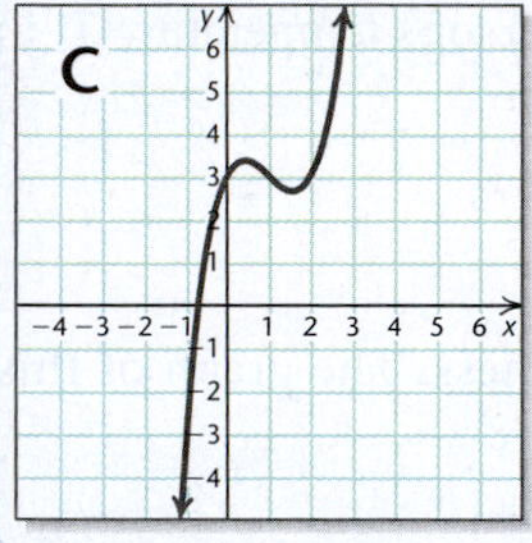

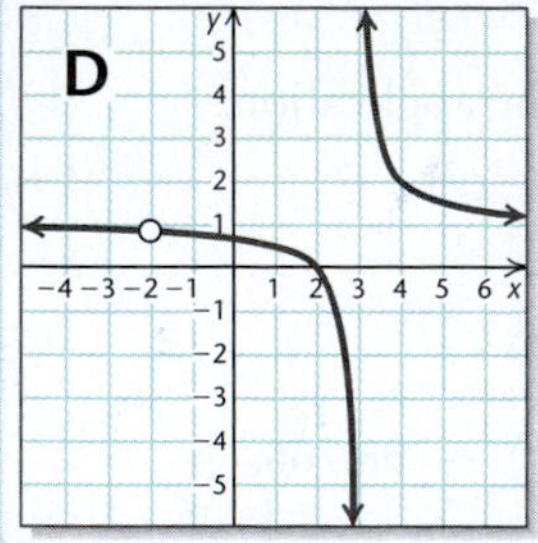

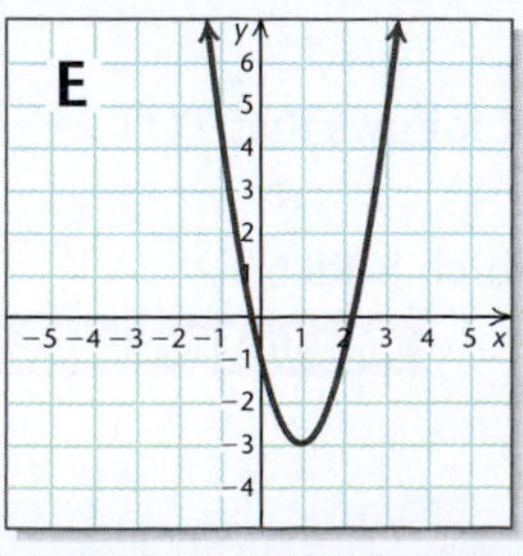

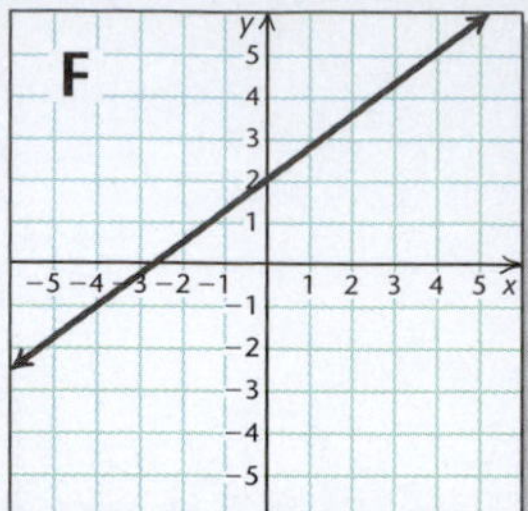

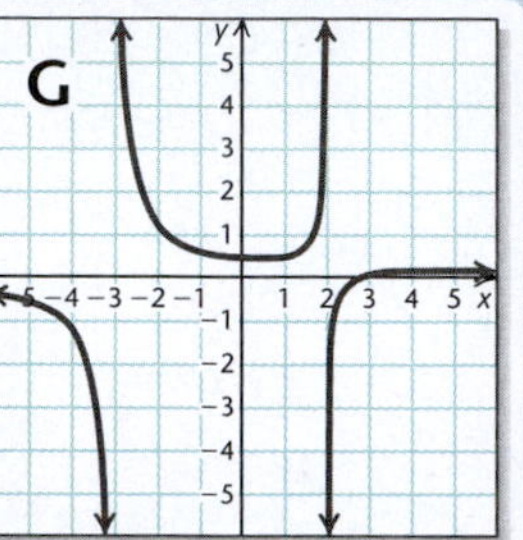

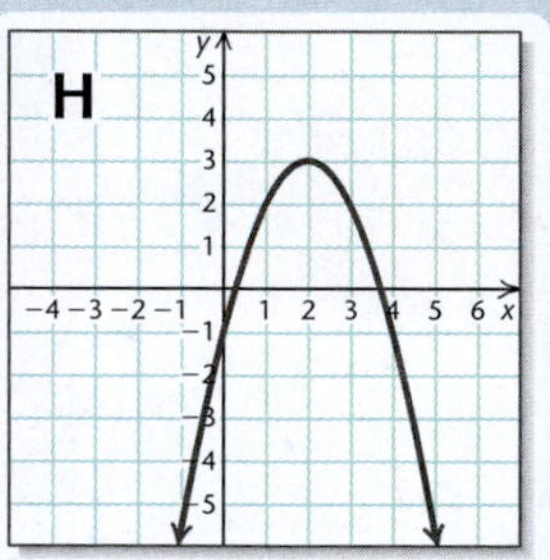

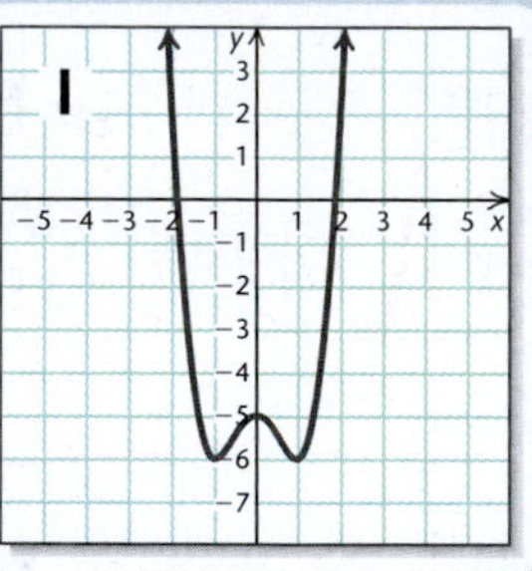

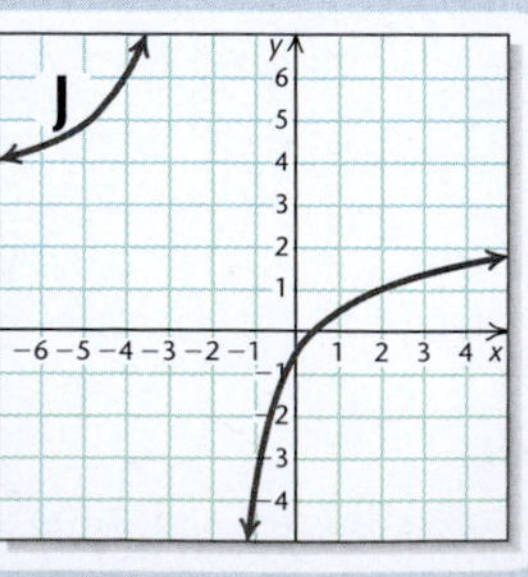

Answers on page A-19

4.5 Exercise Set

Determine the domain of the function.

1. $f(x) = \frac{x^2}{2 - x}$

2. $f(x) = \frac{1}{x^3}$

3. $f(x) = \frac{x + 1}{x^2 - 6x + 5}$

4. $f(x) = \frac{(x + 4)^2}{4x - 3}$

5. $f(x) = \frac{3x - 4}{3x + 15}$

6. $f(x) = \frac{x^2 + 3x - 10}{x^2 + 2x}$

In Exercises 7–12, use your knowledge of asymptotes and intercepts to match the equation with one of the graphs (a)–(f) that follow. List all asymptotes.

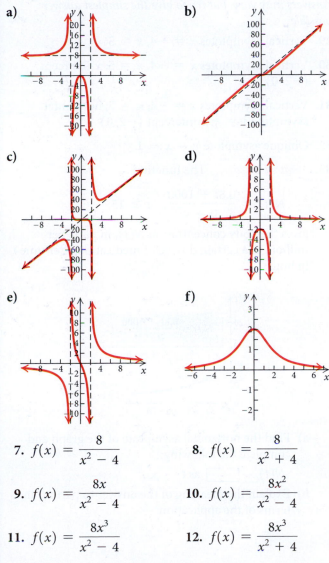

a)

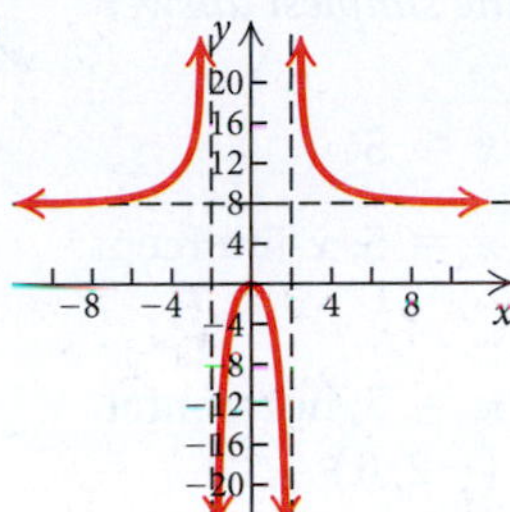

b)

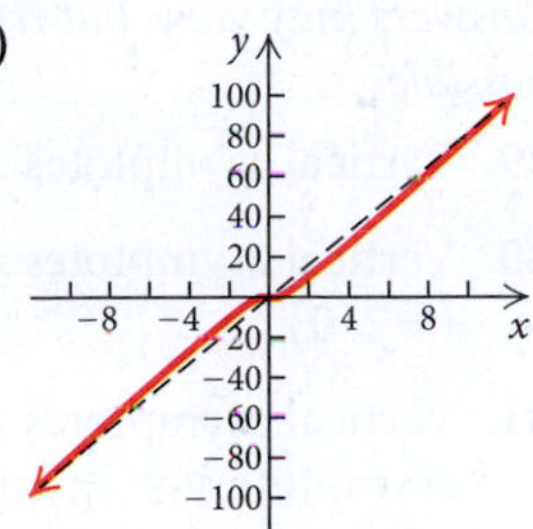

c)

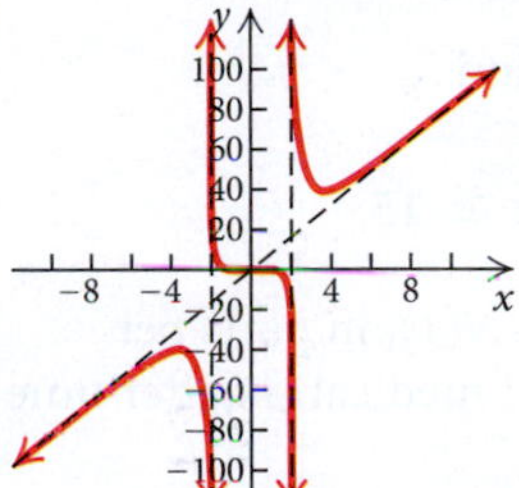

d)

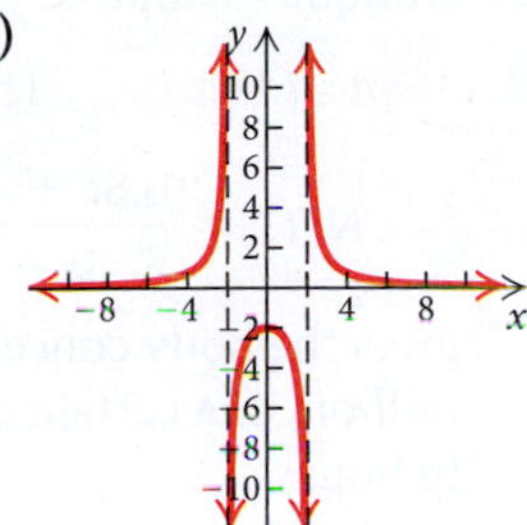

e)

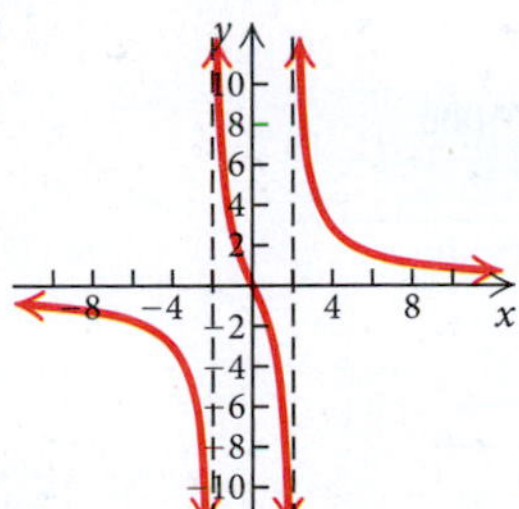

f)

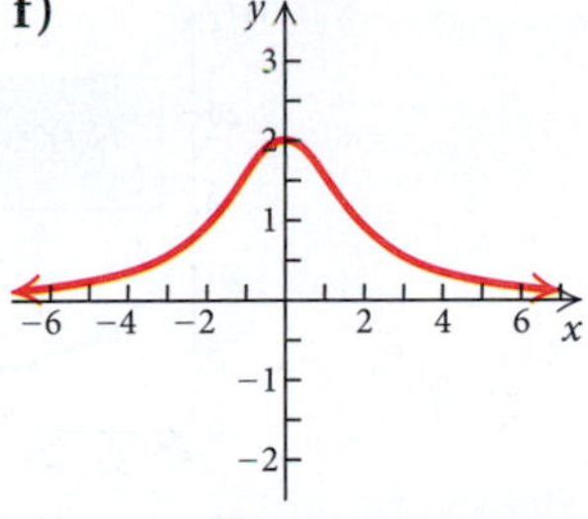

7. $f(x) = \frac{8}{x^2 - 4}$

8. $f(x) = \frac{8}{x^2 + 4}$

9. $f(x) = \frac{8x}{x^2 - 4}$

10. $f(x) = \frac{8x^2}{x^2 - 4}$

11. $f(x) = \frac{8x^3}{x^2 - 4}$

12. $f(x) = \frac{8x^3}{x^2 + 4}$

Determine the vertical asymptotes of the graph of the function.

13. $g(x) = \frac{1}{x^2}$

14. $f(x) = \frac{4x}{x^2 + 10x}$

15. $h(x) = \frac{x + 7}{2 - x}$

16. $g(x) = \frac{x^4 + 2}{x}$

17. $f(x) = \frac{3 - x}{(x - 4)(x + 6)}$

18. $h(x) = \frac{x^2 - 4}{x(x + 5)(x - 2)}$

19. $g(x) = \frac{x^3}{2x^3 - x^2 - 3x}$

20. $f(x) = \frac{x + 5}{x^2 + 4x - 32}$

Determine the horizontal asymptote of the graph of the function.

21. $f(x) = \frac{3x^2 + 5}{4x^2 - 3}$

22. $g(x) = \frac{x + 6}{x^3 + 2x^2}$

23. $h(x) = \frac{x^2 - 4}{2x^4 + 3}$

24. $f(x) = \frac{x^5}{x^5 + x}$

25. $g(x) = \frac{x^3 - 2x^2 + x - 1}{x^2 - 16}$

26. $h(x) = \frac{8x^4 + x - 2}{2x^4 - 10}$

Determine the oblique asymptote of the graph of the function.

27. $g(x) = \frac{x^2 + 4x - 1}{x + 3}$

28. $f(x) = \frac{x^2 - 6x}{x - 5}$

29. $h(x) = \frac{x^4 - 2}{x^3 + 1}$

30. $g(x) = \frac{12x^3 - x}{6x^2 + 4}$

31. $f(x) = \frac{x^3 - x^2 + x - 4}{x^2 + 2x - 1}$

32. $h(x) = \frac{5x^3 - x^2 + x - 1}{x^2 - x + 2}$

Graph the function. Be sure to label all the asymptotes. List the domain and the x- and y-intercepts.

33. $f(x) = \frac{1}{x}$

34. $g(x) = \frac{1}{x^2}$

35. $h(x) = -\frac{4}{x^2}$

36. $f(x) = -\frac{6}{x}$

37. $g(x) = \frac{x^2 - 4x + 3}{x + 1}$

38. $h(x) = \frac{2x^2 - x - 3}{x - 1}$

39. $f(x) = \frac{-2}{x - 5}$

40. $f(x) = \frac{1}{x - 5}$

41. $f(x) = \frac{2x + 1}{x}$

42. $f(x) = \frac{3x - 1}{x}$

43. $f(x) = \frac{x + 3}{x^2 - 9}$

44. $f(x) = \frac{x - 1}{x^2 - 1}$

45. $f(x) = \frac{x}{x^2 + 3x}$

46. $f(x) = \frac{3x}{3x - x^2}$

47. $f(x) = \frac{1}{(x - 2)^2}$

48. $f(x) = \frac{-2}{(x - 3)^2}$

49. $f(x) = \frac{x^2 + 2x - 3}{x^2 + 4x + 3}$

50. $f(x) = \frac{x^2 - x - 2}{x^2 - 5x - 6}$

51. $f(x) = \frac{1}{x^2 + 3}$

52. $f(x) = \frac{-1}{x^2 + 2}$

53. $f(x) = \frac{x^2 - 4}{x - 2}$

54. $f(x) = \frac{x^2 - 9}{x + 3}$

55. $f(x) = \frac{x - 1}{x + 2}$

56. $f(x) = \frac{x - 2}{x + 1}$

57. $f(x) = \frac{x^2 + 3x}{2x^3 - 5x^2 - 3x}$

58. $f(x) = \frac{3x}{x^2 + 5x + 4}$

59. $f(x) = \frac{x^2 - 9}{x + 1}$

60. $f(x) = \frac{x^3 - 4x}{x^2 - x}$

61. $f(x) = \frac{x^2 + x - 2}{2x^2 + 1}$

62. $f(x) = \frac{x^2 - 2x - 3}{3x^2 + 2}$

63. $g(x) = \frac{3x^2 - x - 2}{x - 1}$

64. $f(x) = \frac{2x^2 - 5x - 3}{2x + 1}$

65. $f(x) = \frac{x - 1}{x^2 - 2x - 3}$

66. $f(x) = \frac{x + 2}{x^2 + 2x - 15}$

67. $f(x) = \frac{3x^2 + 11x - 4}{x^2 + 2x - 8}$

68. $f(x) = \frac{2x^2 - 3x - 9}{x^2 - 2x - 3}$

69. $f(x) = \frac{x - 3}{(x + 1)^3}$

70. $f(x) = \frac{x + 2}{(x - 1)^3}$

71. $f(x) = \frac{x^3 + 1}{x}$

72. $f(x) = \frac{x^3 - 1}{x}$

73. $f(x) = \frac{x^3 + 2x^2 - 15x}{x^2 - 5x - 14}$

74. $f(x) = \frac{x^3 + 2x^2 - 3x}{x^2 - 25}$

75. $f(x) = \frac{5x^4}{x^4 + 1}$

76. $f(x) = \frac{x + 1}{x^2 + x - 6}$

77. $f(x) = \frac{x^2}{x^2 - x - 2}$

78. $f(x) = \frac{x^2 - x - 2}{x + 2}$

Find a rational function that satisfies the given conditions. Answers may vary, but try to give the simplest answer possible.

79. Vertical asymptotes $x = -4$, $x = 5$

80. Vertical asymptotes $x = -4$, $x = 5$; x-intercept $(-2, 0)$

81. Vertical asymptotes $x = -4$, $x = 5$; horizontal asymptote $y = \frac{3}{2}$; x-intercept $(-2, 0)$

82. Oblique asymptote $y = x - 1$

83. *Medical Dosage.* The function

$$N(t) = \frac{0.8t + 1000}{5t + 4}, \quad t \geq 15,$$

gives the body concentration $N(t)$, in parts per million, of a certain dosage of medication after time t, in hours.

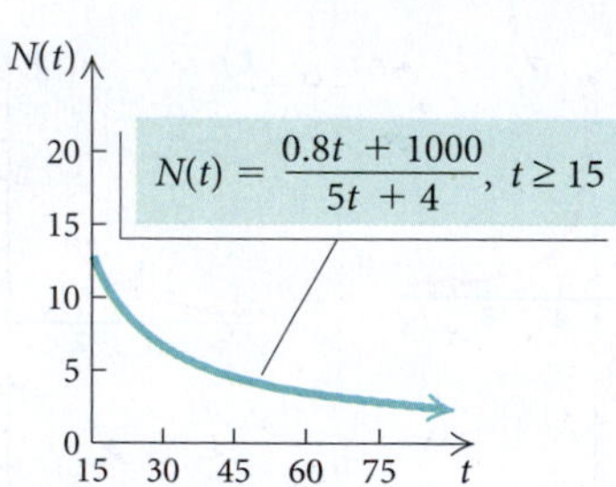

a) Find the horizontal asymptote of the graph and complete the following:

$N(t) \rightarrow \square$ as $t \rightarrow \infty$.

b) Explain the meaning of the answer to part (a) in terms of the application.

84. *Average Cost.* The average cost per light, in dollars, for a company to produce x roadside emergency lights is given by the function

$$A(x) = \frac{2x + 100}{x}, \quad x > 0.$$

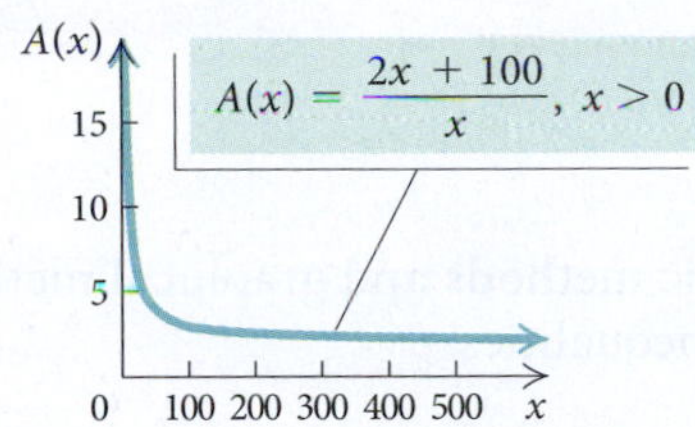

a) Find the horizontal asymptote of the graph and complete the following:

$A(x) \to \square$ as $x \to \infty$.

b) Explain the meaning of the answer to part (a) in terms of the application.

85. *Population Growth.* The population P, in thousands, of a resort community is given by

$$P(t) = \frac{500t}{2t^2 + 9},$$

where t is the time, in months, since the city council raised the property taxes.

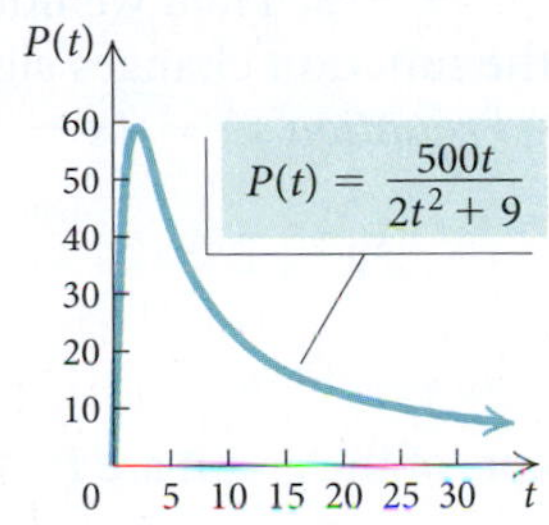

a) Find the population at $t = 0, 1, 3$, and 8 months.

b) Find the horizontal asymptote of the graph and complete the following:

$P(t) \to \square$ as $t \to \infty$.

c) Explain the meaning of the answer to part (b) in terms of the application.

Skill Maintenance

Vocabulary Reinforcement

In each of Exercises 86–94, fill in the blank with the correct term. Some of the given choices will not be used. Others will be used more than once.

x-intercept	even function
y-intercept	domain
odd function	range
slope	slope–intercept equation
distance formula	difference quotient
midpoint formula	$f(x) = f(-x)$
horizontal lines	$f(-x) = -f(x)$
vertical lines	
point–slope equation	

86. A function is a correspondence between a first set, called the __________, and a second set, called the __________, such that each member of the __________ corresponds to exactly one member of the __________. [1.2]

87. The __________ of a line containing (x_1, y_1) and (x_2, y_2) is given by $(y_2 - y_1)/(x_2 - x_1)$. [1.3]

88. The __________ of the line with slope m and y-intercept $(0, b)$ is $y = mx + b$. [1.3]

89. The __________ of the line with slope m passing through (x_1, y_1) is $y - y_1 = m(x - x_1)$. [1.4]

90. A(n) __________ is a point $(a, 0)$. [1.1]

91. For each x in the domain of an odd function f, __________. [2.4]

92. __________ are given by equations of the type $x = a$. [1.3]

93. The __________ is $\left(\frac{x_1 + x_2}{2}, \frac{y_1 + y_2}{2}\right)$. [1.1]

94. A(n) __________ is a point $(0, b)$. [1.1]

Synthesis

Find the nonlinear asymptote of the function.

95. $f(x) = \dfrac{x^5 + 2x^3 + 4x^2}{x^2 + 2}$

96. $f(x) = \dfrac{x^4 + 3x^2}{x^2 + 1}$

Graph the function.

97. $f(x) = \dfrac{2x^3 + x^2 - 8x - 4}{x^3 + x^2 - 9x - 9}$

98. $f(x) = \dfrac{x^3 + 4x^2 + x - 6}{x^2 - x - 2}$

4.6 Polynomial Inequalities and Rational Inequalities

- Solve polynomial inequalities.
- Solve rational inequalities.

We will use a combination of algebraic methods and graphical methods to solve polynomial inequalities and rational inequalities.

Polynomial Inequalities

Just as a quadratic equation can be written in the form $ax^2 + bx + c = 0$, a **quadratic inequality** can be written in the form $ax^2 + bx + c \;\square\; 0$, where $\square$ is $<, >, \leq,$ or $\geq$. Here are some examples of quadratic inequalities:

$$x^2 - 4x - 5 < 0 \quad \text{and} \quad -\tfrac{1}{2}x^2 + 4x - 7 \geq 0.$$

When the inequality symbol in a polynomial inequality is replaced with an equals sign, a **related equation** is formed. Polynomial inequalities can be solved once the related equation has been solved.

EXAMPLE 1 Solve: $x^2 - 4x - 5 > 0$.

Solution We are asked to find all x-values for which $x^2 - 4x - 5 > 0$. To locate these values, we graph $f(x) = x^2 - 4x - 5$. Then we note that whenever its graph passes through an x-intercept, the function changes sign. Thus to solve $x^2 - 4x - 5 > 0$, we first solve the *related equation* $x^2 - 4x - 5 = 0$ to find all zeros of the function:

$$x^2 - 4x - 5 = 0$$
$$(x + 1)(x - 5) = 0.$$

The zeros are -1 and 5. Thus the x-intercepts of the graph are $(-1, 0)$ and $(5, 0)$, as shown below.

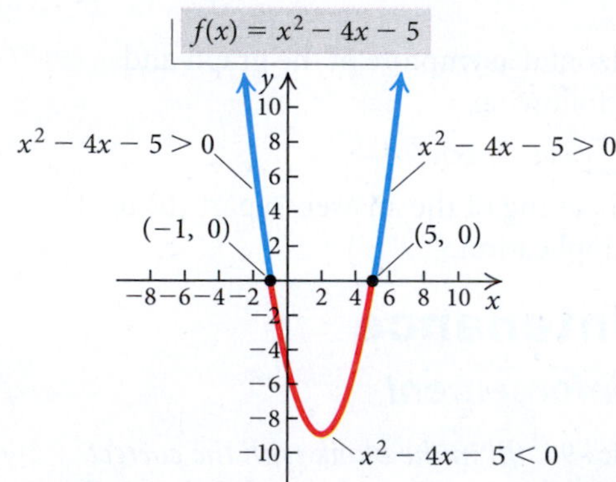

The zeros divide the x-axis into three intervals:

$$(-\infty, -1), \quad (-1, 5), \quad \text{and} \quad (5, \infty).$$

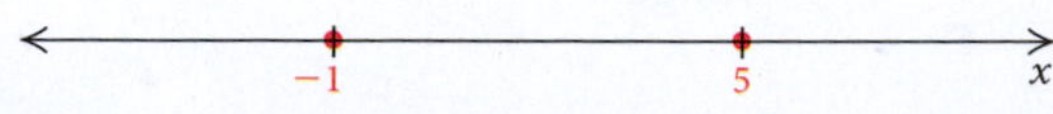

The sign of $x^2 - 4x - 5$ is the same for all values of x in a given interval. Thus we choose a test value for x from each interval and find $f(x)$. We can also determine the sign of $f(x)$ in each interval by simply looking at the graph of the function.

Interval	$(-\infty, -1)$	$(-1, 5)$	$(5, \infty)$
Test Value	$f(-2) = 7$	$f(0) = -5$	$f(7) = 16$
Sign of $f(x)$	Positive	Negative	Positive

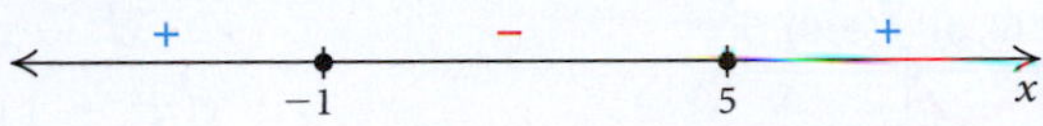

JUST IN TIME 5

Since we are solving $x^2 - 4x - 5 > 0$, the solution set consists of only two of the three intervals, those in which the sign of $f(x)$ is positive. Since the inequality sign is $>$, we do not include the endpoints of the intervals in the solution set. The solution set is $(-\infty, -1) \cup (5, \infty)$, or $\{x \mid x < -1 \text{ or } x > 5\}$.

Now Try Exercise 27.

EXAMPLE 2 Solve: $x^2 + 3x - 5 \leq x + 3$.

Solution By subtracting $x + 3$, we form an equivalent inequality:

$$x^2 + 3x - 5 - x - 3 \leq 0$$
$$x^2 + 2x - 8 \leq 0.$$

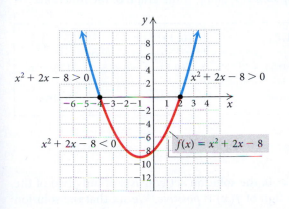

We need to find all x-values for which $x^2 + 2x - 8 \leq 0$. To visualize these values, we first graph $f(x) = x^2 + 2x - 8$ and then determine the zeros of the function. To find the zeros, we solve the related equation $x^2 + 2x - 8 = 0$:

$$x^2 + 2x - 8 = 0$$
$$(x + 4)(x - 2) = 0.$$

The zeros are -4 and 2. Thus the x-intercepts of the graph are $(-4, 0)$ and $(2, 0)$, as shown in the figure at left.

The zeros divide the x-axis into three intervals:

$$(-\infty, -4), \quad (-4, 2), \quad \text{and} \quad (2, \infty).$$

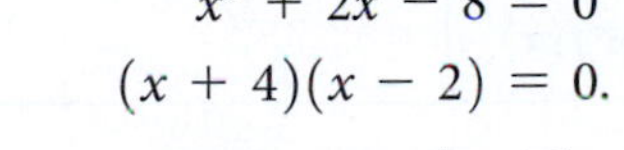

We choose a test value for x from each interval and find $f(x)$. The sign of $x^2 + 2x - 8$ is the same for all values of x in a given interval.

Interval	$(-\infty, -4)$	$(-4, 2)$	$(2, \infty)$
Test Value	$f(-5) = 7$	$f(0) = -8$	$f(4) = 16$
Sign of $f(x)$	Positive	Negative	Positive

Function values are negative on the interval $(-4, 2)$. We can also see from the graph where the function values are negative. Since the inequality symbol is $\leq$, we include the endpoints of the interval in the solution set. The solution set of $x^2 + 3x - 5 \leq x + 3$ is $[-4, 2]$, or $\{x \mid -4 \leq x \leq 2\}$.

Now Try Exercise 29.

Quadratic inequalities are one type of **polynomial inequality**. Other examples of polynomial inequalities are

$$-2x^4 + x^2 - 3 < 7, \quad \tfrac{2}{3}x + 4 \geq 0, \quad \text{and} \quad 4x^3 - 2x^2 > 5x + 7.$$

EXAMPLE 3 Solve: $x^3 - x > 0$.

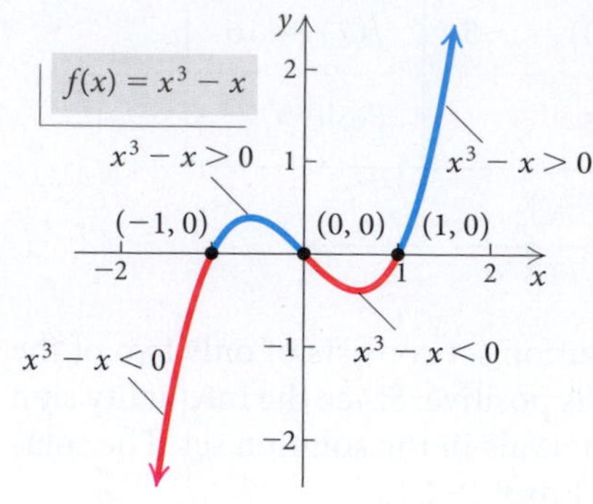

Solution We are asked to find all x-values for which $x^3 - x > 0$. To locate these values, we graph $f(x) = x^3 - x$. Then we note that whenever the function changes sign, its graph passes through an x-intercept. Thus to solve $x^3 - x > 0$, we first solve the related equation $x^3 - x = 0$ to find all zeros of the function:

$$x^3 - x = 0$$
$$x(x^2 - 1) = 0$$
$$x(x + 1)(x - 1) = 0.$$

The zeros are -1, 0, and 1. Thus the x-intercepts of the graph are $(-1, 0)$, $(0, 0)$, and $(1, 0)$, as shown in the figure at left. The zeros divide the x-axis into four intervals:

$$(-\infty, -1), \quad (-1, 0), \quad (0, 1), \quad \text{and} \quad (1, \infty).$$

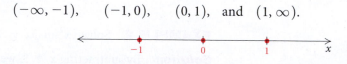

The sign of $x^3 - x$ is the same for all values of x in a given interval. Thus we choose a test value for x from each interval and find $f(x)$. We can also determine the sign of $f(x)$ in each interval by simply looking at the graph of the function.

Interval	$(-\infty, -1)$	$(-1, 0)$	$(0, 1)$	$(1, \infty)$
Test Value	$f(-2) = -6$	$f(-0.5) = 0.375$	$f(0.5) = -0.375$	$f(2) = 6$
Sign of $f(x)$	Negative	Positive	Negative	Positive

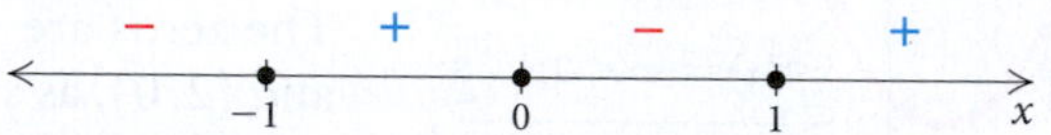

Since we are solving $x^3 - x > 0$, the solution set consists of only two of the four intervals, those in which the sign of $f(x)$ is *positive*. We see that the solution set is $(-1, 0) \cup (1, \infty)$, or $\{x \mid -1 < x < 0 \text{ or } x > 1\}$.

Now Try Exercise 39.

To solve a polynomial inequality:

1. Find an equivalent inequality with $P(x)$ on one side and 0 on the other.
2. Change the inequality symbol to an equals sign and solve the related equation; that is, solve $P(x) = 0$.
3. Use the solutions to divide the x-axis into intervals. Then select a test value from each interval and determine the polynomial's sign on the interval.
4. Determine the intervals for which the inequality is satisfied and write interval notation or set-builder notation for the solution set. Include the endpoints of the intervals in the solution set if the inequality symbol is $\leq$ or $\geq$.

EXAMPLE 4 Solve: $3x^4 + 10x \leq 11x^3 + 4$.

Solution By subtracting $11x^3 + 4$, we form the equivalent inequality

$$3x^4 - 11x^3 + 10x - 4 \leq 0.$$

Algebraic Solution

To solve the related equation

$$3x^4 - 11x^3 + 10x - 4 = 0,$$

we need to use the theorems of Section 4.4. We solved this equation in Example 5 in Section 4.4. The solutions are

$$-1, \quad 2 - \sqrt{2}, \quad \tfrac{2}{3}, \quad \text{and} \quad 2 + \sqrt{2},$$

or approximately

$$-1, \quad 0.586, \quad 0.667, \quad \text{and} \quad 3.414.$$

These numbers divide the x-axis into five intervals:

$(-\infty, -1)$, $(-1, 2 - \sqrt{2})$, $(2 - \sqrt{2}, \frac{2}{3})$, $(\frac{2}{3}, 2 + \sqrt{2})$, and $(2 + \sqrt{2}, \infty)$.

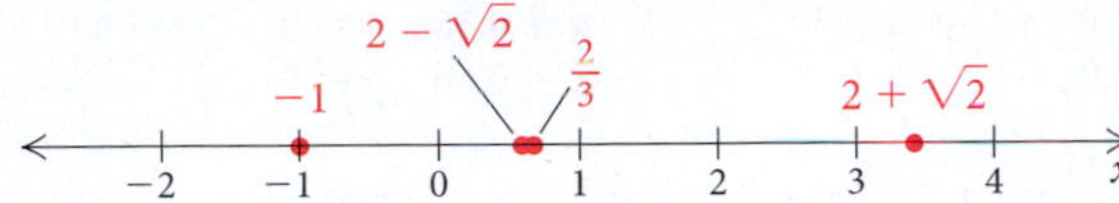

We then let $f(x) = 3x^4 - 11x^3 + 10x - 4$ and, using test values for $f(x)$, determine the sign of $f(x)$ in each interval.

INTERVAL	TEST VALUE	SIGN OF $f(x)$
$(-\infty, -1)$	$f(-2) = 112$	+
$(-1, 2 - \sqrt{2})$	$f(0) = -4$	−
$(2 - \sqrt{2}, 2/3)$	$f(0.6) = 0.0128$	+
$(2/3, 2 + \sqrt{2})$	$f(1) = -2$	−
$(2 + \sqrt{2}, \infty)$	$f(4) = 100$	+

Function values are negative in the intervals $(-1, 2 - \sqrt{2})$ and $(\frac{2}{3}, 2 + \sqrt{2})$. Since the inequality sign is $\leq$, we include the endpoints of the intervals in the solution set. The solution set is

$$[-1, 2 - \sqrt{2}] \cup [\tfrac{2}{3}, 2 + \sqrt{2}], \quad \text{or}$$
$$\{x \mid -1 \leq x \leq 2 - \sqrt{2} \text{ or } \tfrac{2}{3} \leq x \leq 2 + \sqrt{2}\}.$$

Visualizing the Solution

Observing the graph of the function

$$f(x) = 3x^4 - 11x^3 + 10x - 4$$

and a closeup view of the graph on the interval $(0, 1)$, we see the intervals on which $f(x) \leq 0$. The values of $f(x)$ are less than or equal to 0 in two intervals.

$f(x) = 3x^4 - 11x^3 + 10x - 4$

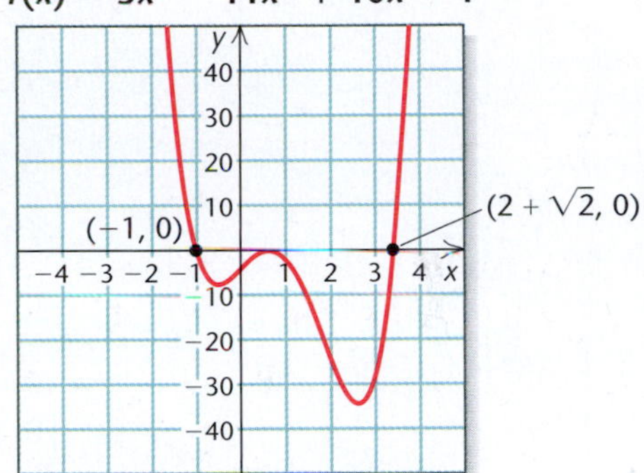

$f(x) = 3x^4 - 11x^3 + 10x - 4$

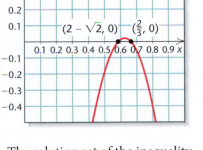

The solution set of the inequality

$$3x^4 - 11x^3 + 10x - 4 \leq 0$$

is

$$[-1, 2 - \sqrt{2}] \cup [\tfrac{2}{3}, 2 + \sqrt{2}].$$

Now Try Exercise 45.

Technology Connection

Polynomial inequalities can be solved quickly with a graphing calculator. Consider the inequality in Example 4:

$$3x^4 + 10x \le 11x^3 + 4, \quad \text{or}$$
$$3x^4 - 11x^3 + 10x - 4 \le 0.$$

We graph the related equation

$$y = 3x^4 - 11x^3 + 10x - 4$$

and use the ZERO feature. We see in the window on the left below that two of the zeros are -1 and approximately 3.414 $(2 + \sqrt{2} \approx 3.414)$. However, this window leaves us uncertain about the number of zeros of the function on the interval $[0, 1]$.

$y = 3x^4 - 11x^3 + 10x - 4$

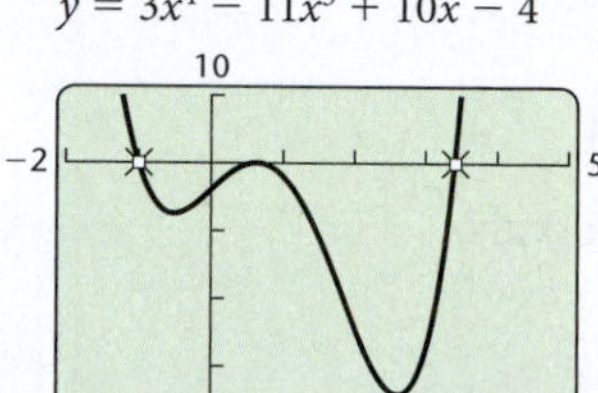

$x = -1, \quad y = 0$
$x = 3.414, \quad y = 0$

$y = 3x^4 - 11x^3 + 10x - 4$

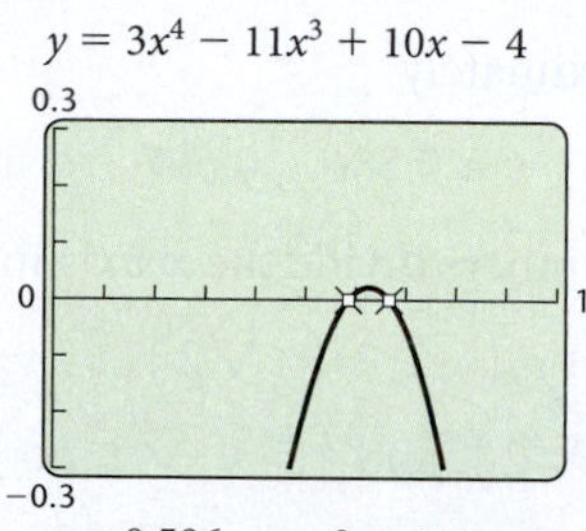

$x = 0.586, \quad y = 0$
$x = 0.667, \quad y = 0$

Xscl = 0.1
Yscl = 0.1

The window on the right above shows another view of the zeros on the interval $[0, 1]$. Those zeros are about 0.586 and 0.667 $(2 - \sqrt{2} \approx 0.586; \frac{2}{3} \approx 0.667)$. The intervals to be considered are $(-\infty, -1)$, $(-1, 0.586)$, $(0.586, 0.667)$, $(0.667, 3.414)$, and $(3.414, \infty)$. We note on the graph where the function is negative. Then, including appropriate endpoints, we find that the solution set is approximately

$$[-1, 0.586] \cup [0.667, 3.414], \quad \text{or}$$
$$\{x \mid -1 \le x \le 0.586 \text{ } or \text{ } 0.667 \le x \le 3.414\}.$$

Rational Inequalities

Some inequalities involve rational expressions and functions. These are called **rational inequalities**. To solve rational inequalities, we must make some adjustments to the preceding method.

EXAMPLE 5 Solve: $\dfrac{3x}{x + 6} < 0.$

Solution We look for all values of x for which the related function

$$f(x) = \frac{3x}{x + 6}$$

is not defined or is 0. These are called **critical values**.

The denominator tells us that $f(x)$ is not defined when $x = -6$. Next, we solve $f(x) = 0$:

$$\frac{3x}{x+6} = 0$$

$$(x+6)\cdot\frac{3x}{x+6} = (x+6)\cdot 0 \quad \text{Multiplying by } x+6$$

$$3x = 0$$

$$x = 0.$$

The critical values are -6 and 0. These values divide the x-axis into three intervals:

$$(-\infty, -6), \qquad (-6, 0), \quad \text{and} \quad (0, \infty).$$

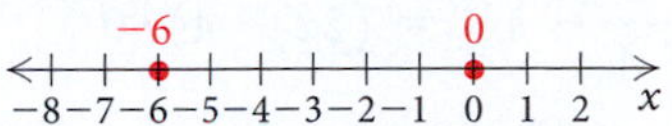

We then use a test value to determine the sign of $f(x)$ in each interval.

Interval	$(-\infty, -6)$	$(-6, 0)$	$(0, \infty)$
Test Value	$f(-8) = 12$	$f(-2) = -\frac{3}{2}$	$f(3) = 1$
Sign of $f(x)$	Positive	Negative	Positive

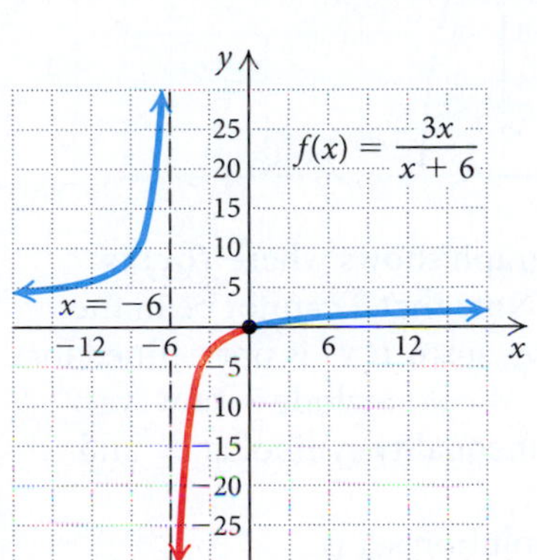

Function values are negative on only the interval $(-6, 0)$. Although $f(0) = 0$, the inequality symbol is $<$, so we know that 0 is not included in the solution set. Note that since -6 is not in the domain of f, -6 cannot be part of the solution set. The solution set is

$$(-6, 0), \quad \text{or} \quad \{x \mid -6 < x < 0\}.$$

The graph of $f(x)$ shows where $f(x)$ is positive and where it is negative.

Now Try Exercise 57.

EXAMPLE 6 Solve: $\dfrac{x+1}{2x-4} \leq 1$.

Solution We first subtract 1 on both sides in order to find an equivalent inequality with 0 on one side:

$$\frac{x+1}{2x-4} - 1 \leq 0.$$

Algebraic Solution

We look for all values of x for which the related function

$$f(x) = \frac{x+1}{2x-4} - 1$$

is not defined or is 0. These are called **critical values.**

A look at the denominator shows that $f(x)$ is not defined for $x = 2$. Next, we solve $f(x) = 0$:

$$\frac{x+1}{2x-4} - 1 = 0$$

$$(2x-4)\left(\frac{x+1}{2x-4} - 1\right) = (2x-4)\cdot 0$$

$$x + 1 - (2x-4)\cdot 1 = 0$$

$$x + 1 - 2x + 4 = 0$$

$$-x = -5$$

$$x = 5.$$

The critical values are 2 and 5. These values divide the x-axis into three intervals:

$$(-\infty, 2) \qquad (2, 5), \quad \text{and} \quad (5, \infty).$$

We then use a test value to determine the sign of $f(x)$ in each interval.

INTERVAL	TEST VALUE	SIGN OF $f(x)$
$(-\infty, 2)$	$f(0) = -\frac{5}{4}$	$-$
$(2, 5)$	$f(3) = 1$	$+$
$(5, \infty)$	$f(6) = -\frac{1}{8}$	$-$

Function values are negative on the intervals $(-\infty, 2)$ and $(5, \infty)$. Since $f(5) = 0$ and the inequality symbol is $\leq$, we know that 5 is in the solution set. Note that since 2 is not in the domain of f, it cannot be part of the solution set.

The solution set is $(-\infty, 2) \cup [5, \infty)$.

Visualizing the Solution

The graph of the related function

$$f(x) = \frac{x+1}{2x-4} - 1$$

confirms the two critical values found algebraically: 2 where $f(x)$ is not defined and 5 where $f(x) = 0$.

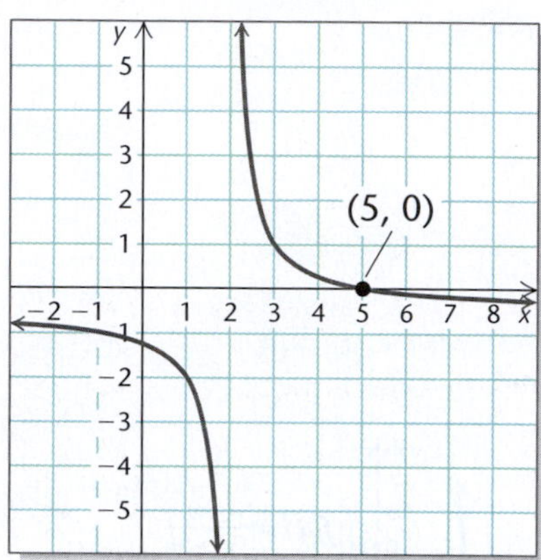

The graph shows where $f(x)$ is negative. Note that 2 cannot be in the solution set since $f(x)$ is not defined for this value. We do include 5, however, since the inequality symbol is $\leq$ and $f(5) = 0$.

The solution set is

$$(-\infty, 2) \cup [5, \infty).$$

Now Try Exercise 59.

EXAMPLE 7 Solve: $\dfrac{x-3}{x+4} \geq \dfrac{x+2}{x-5}$.

Solution We first subtract $(x+2)/(x-5)$ on both sides in order to find an equivalent inequality with 0 on one side:

$$\frac{x-3}{x+4} - \frac{x+2}{x-5} \geq 0.$$

Algebraic Solution

We look for all values of x for which the related function

$$f(x) = \frac{x-3}{x+4} - \frac{x+2}{x-5}$$

is not defined or is 0. These are called **critical values**.

A look at the denominators shows that $f(x)$ is not defined for $x = -4$ and $x = 5$. Next, we solve $f(x) = 0$:

$$\frac{x-3}{x+4} - \frac{x+2}{x-5} = 0$$

$$(x+4)(x-5)\left(\frac{x-3}{x+4} - \frac{x+2}{x-5}\right) = (x+4)(x-5)\cdot 0$$

$$(x-5)(x-3) - (x+4)(x+2) = 0$$

$$(x^2 - 8x + 15) - (x^2 + 6x + 8) = 0$$

$$-14x + 7 = 0$$

$$x = \tfrac{1}{2}.$$

The critical values are -4, $\frac{1}{2}$, and 5. These values divide the x-axis into four intervals:

$$(-\infty, -4), \quad \left(-4, \tfrac{1}{2}\right), \quad \left(\tfrac{1}{2}, 5\right), \quad \text{and} \quad (5, \infty).$$

-4, 1/2, 5
−5 −4 −3 −2 −1 0 1 2 3 4 5 6 x

We then use a test value to determine the sign of $f(x)$ in each interval.

INTERVAL	TEST VALUE	SIGN OF $f(x)$
$(-\infty, -4)$	$f(-5) = 7.7$	+
$\left(-4, \frac{1}{2}\right)$	$f(-2) = -2.5$	−
$\left(\frac{1}{2}, 5\right)$	$f(3) = 2.5$	+
$(5, \infty)$	$f(6) = -7.7$	−

+ −4 − 1/2 + 5 −
−5 −4 −3 −2 −1 0 1 2 3 4 5 6 x

Function values are positive in the intervals $(-\infty, -4)$ and $\left(\frac{1}{2}, 5\right)$. Since $f\left(\frac{1}{2}\right) = 0$ and the inequality symbol is $\geq$, we know that $\frac{1}{2}$ must be in the solution set. Note that since neither -4 nor 5 is in the domain of f, they cannot be part of the solution set.

The solution set is $(-\infty, -4) \cup \left[\frac{1}{2}, 5\right)$.

Visualizing the Solution

The graph of the related function

$$f(x) = \frac{x-3}{x+4} - \frac{x+2}{x-5}$$

confirms the three critical values found algebraically: -4 and 5 where $f(x)$ is not defined and $\frac{1}{2}$ where $f(x) = 0$.

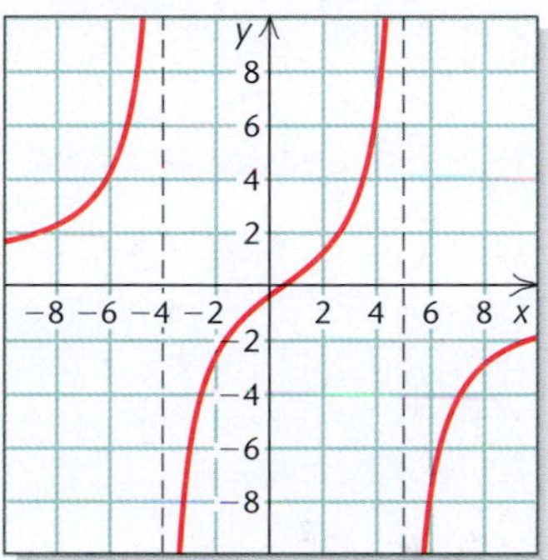

The graph shows where $f(x)$ is positive and where it is negative. Note that -4 and 5 cannot be in the solution set since $f(x)$ is not defined for these values. We do include $\frac{1}{2}$, however, since the inequality symbol is $\geq$ and $f\left(\frac{1}{2}\right) = 0$.

The solution set is

$$(-\infty, -4) \cup \left[\tfrac{1}{2}, 5\right).$$

Now Try Exercise 61.

The following is a method for solving rational inequalities.

To solve a rational inequality:

1. Find an equivalent inequality with 0 on one side.
2. Change the inequality symbol to an equals sign and solve the related equation.
3. Find values of the variable for which the related rational function is not defined.
4. The numbers found in steps (2) and (3) are called critical values. Use the critical values to divide the x-axis into intervals. Then determine the function's sign in each interval using an x-value from the interval or using the graph of the equation.
5. Select the intervals for which the inequality is satisfied and write interval notation or set-builder notation for the solution set. If the inequality symbol is $\leq$ or $\geq$, then the solutions to step (2) should be included in the solution set. The x-values found in step (3) are never included in the solution set.

It works well to use a combination of algebraic methods and graphical methods to solve polynomial inequalities and rational inequalities. The algebraic methods give exact numbers for the critical values, and the graphical methods usually allow us to see easily what intervals satisfy the inequality.

4.6 Exercise Set

For the function $f(x) = x^2 + 2x - 15$, solve each of the following.

1. $f(x) = 0$

2. $f(x) < 0$

3. $f(x) \leq 0$

4. $f(x) > 0$

5. $f(x) \geq 0$

For the function $g(x) = \dfrac{x - 2}{x + 4}$, solve each of the following.

6. $g(x) = 0$

7. $g(x) > 0$

8. $g(x) \leq 0$

9. $g(x) \geq 0$

10. $g(x) < 0$

For the function

$$h(x) = \frac{7x}{(x - 1)(x + 5)},$$

solve each of the following.

11. $h(x) = 0$

12. $h(x) \leq 0$

13. $h(x) \geq 0$

14. $h(x) > 0$

15. $h(x) < 0$

For the function $g(x) = x^5 - 9x^3$, solve each of the following.

16. $g(x) = 0$

17. $g(x) < 0$

18. $g(x) \leq 0$

19. $g(x) > 0$

20. $g(x) \geq 0$

In Exercises 21–24, a related function is graphed. Solve the given inequality.

21. $x^3 + 6x^2 < x + 30$

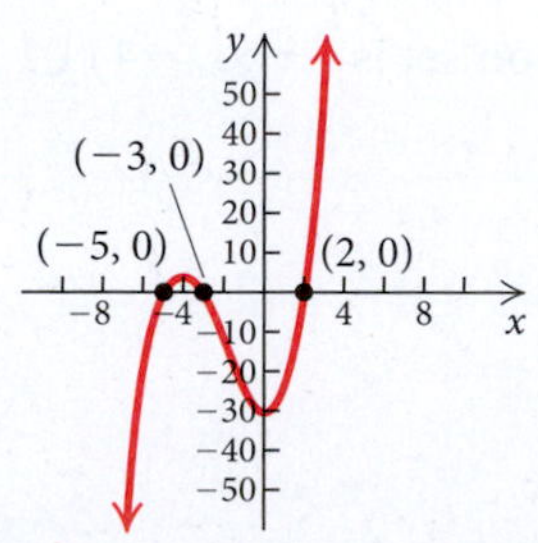

22. $x^4 - 27x^2 - 14x + 120 \geq 0$

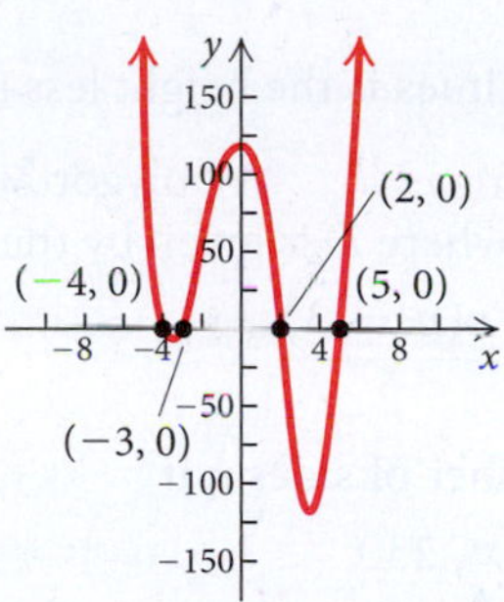

23. $\dfrac{8x}{x^2 - 4} \geq 0$

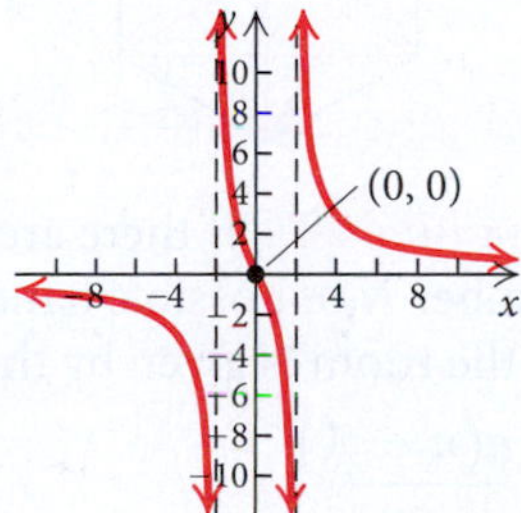

24. $\dfrac{8}{x^2 - 4} < 0$

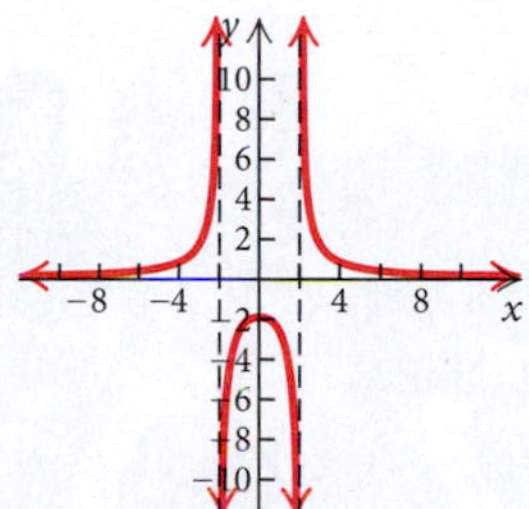

Solve.

25. $(x - 1)(x + 4) < 0$

26. $(x + 3)(x - 5) < 0$

27. $x^2 + x - 2 > 0$

28. $x^2 - x - 6 > 0$

29. $x^2 - x - 5 \geq x - 2$

30. $x^2 + 4x + 7 \geq 5x + 9$

31. $x^2 > 25$

32. $x^2 \leq 1$

33. $4 - x^2 \leq 0$

34. $11 - x^2 \geq 0$

35. $6x - 9 - x^2 < 0$

36. $x^2 + 2x + 1 \leq 0$

37. $x^2 + 12 < 4x$

38. $x^2 - 8 > 6x$

39. $4x^3 - 7x^2 \leq 15x$

40. $2x^3 - x^2 < 5x$

41. $x^3 + 3x^2 - x - 3 \geq 0$

42. $x^3 + x^2 - 4x - 4 \geq 0$

43. $x^3 - 2x^2 < 5x - 6$

44. $x^3 + x \leq 6 - 4x^2$

45. $x^5 + x^2 \geq 2x^3 + 2$

46. $x^5 + 24 > 3x^3 + 8x^2$

47. $2x^3 + 6 \leq 5x^2 + x$

48. $2x^3 + x^2 < 10 + 11x$

49. $x^3 + 5x^2 - 25x \leq 125$

50. $x^3 - 9x + 27 \geq 3x^2$

51. $0.1x^3 - 0.6x^2 - 0.1x + 2 < 0$

52. $19.2x^3 + 12.8x^2 + 144 \geq 172.8x + 3.2x^4$

List the critical values of the related function. Then solve the inequality.

53. $\dfrac{1}{x + 4} > 0$

54. $\dfrac{1}{x - 3} \leq 0$

55. $\dfrac{-4}{2x + 5} < 0$

56. $\dfrac{-2}{5 - x} \geq 0$

57. $\dfrac{2x}{x - 4} \geq 0$

58. $\dfrac{5x}{x + 1} < 0$

59. $\dfrac{x + 1}{x - 2} \geq 3$

60. $\dfrac{x}{x - 5} < 2$

61. $\dfrac{x - 4}{x + 3} - \dfrac{x + 2}{x - 1} \leq 0$

62. $\dfrac{x + 1}{x - 2} - \dfrac{x - 3}{x - 1} < 0$

63. $\dfrac{x + 6}{x - 2} > \dfrac{x - 8}{x - 5}$

64. $\dfrac{x - 7}{x + 2} \geq \dfrac{x - 9}{x + 3}$

65. $x - 2 > \dfrac{1}{x}$

66. $4 \geq \dfrac{4}{x} + x$

67. $\dfrac{2}{x^2 - 4x + 3} \leq \dfrac{5}{x^2 - 9}$

68. $\dfrac{3}{x^2 - 4} \leq \dfrac{5}{x^2 + 7x + 10}$

69. $\dfrac{3}{x^2 + 1} \geq \dfrac{6}{5x^2 + 2}$

70. $\dfrac{4}{x^2 - 9} < \dfrac{3}{x^2 - 25}$

71. $\dfrac{5}{x^2 + 3x} < \dfrac{3}{2x + 1}$

72. $\dfrac{2}{x^2 + 3} > \dfrac{3}{5 + 4x^2}$

73. $\dfrac{5x}{7x - 2} > \dfrac{x}{x + 1}$

74. $\dfrac{x^2 - x - 2}{x^2 + 5x + 6} < 0$

75. $\dfrac{x}{x^2 + 4x - 5} + \dfrac{3}{x^2 - 25} \leq \dfrac{2x}{x^2 - 6x + 5}$

76. $\dfrac{2x}{x^2 - 9} + \dfrac{x}{x^2 + x - 12} \geq \dfrac{3x}{x^2 + 7x + 12}$

77. *Temperature During an Illness.* A person's temperature T, in degrees Fahrenheit, during an illness is given by the function

$$T(t) = \frac{4t}{t^2 + 1} + 98.6,$$

where t is the time since the onset of the illness, in hours. Find the interval on which the temperature was over 100°F. (See Example 12 in Section 4.5.)

78. *Population Growth.* The population P, in thousands, of a resort community is given by

$$P(t) = \frac{500t}{2t^2 + 9},$$

where t is the time, in months, since the city council raised the property taxes. Find the interval on which the population was 40,000 or greater. (See Exercise 85 in Exercise Set 4.5.)

79. *Total Profit.* Flexl, Inc., determines that its total profit is given by the function

$$P(x) = -3x^2 + 630x - 6000.$$

a) Flexl makes a profit for those nonnegative values of x for which $P(x) > 0$. Find the values of x for which Flexl makes a profit.

b) Flexl loses money for those nonnegative values of x for which $P(x) < 0$. Find the values of x for which Flexl loses money.

80. *Height of a Thrown Object.* The function

$$S(t) = -16t^2 + 32t + 1920$$

gives the height S, in feet, of an object thrown upward from a cliff that is 1920 ft high. Here t is the time, in seconds, that the object is in the air.

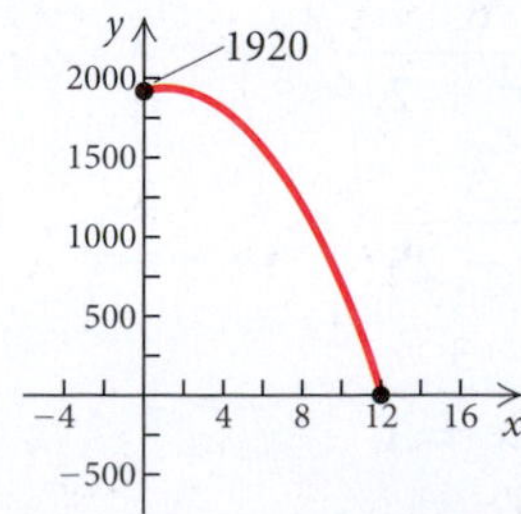

a) For what times is the height greater than 1920 ft?

b) For what times is the height less than 640 ft?

81. *Number of Diagonals.* A polygon with n sides has D diagonals, where D is given by the function

$$D(n) = \frac{n(n - 3)}{2}.$$

Find the number of sides n if

$$27 \leq D \leq 230.$$

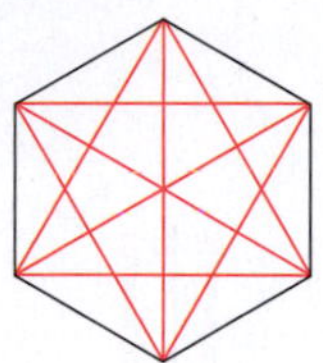

82. *Number of Handshakes.* If there are n people in a room, the number N of possible handshakes by all the people in the room is given by the function

$$N(n) = \frac{n(n - 1)}{2}.$$

For what number n of people is

$$66 \leq N \leq 300?$$

Skill Maintenance

Find an equation for a circle satisfying the given conditions. [1.1]

83. Center: $(-2, 4)$; radius of length 3

84. Center: $(0, -3)$; diameter of length $\frac{7}{2}$

In Exercises 85 and 86, **(a)** *find the vertex;* **(b)** *determine whether there is a maximum or a minimum value and find that value; and* **(c)** *find the range.*

85. $h(x) = -2x^2 + 3x - 8$ [3.3]

86. $g(x) = x^2 - 10x + 2$ [3.3]

Synthesis

Solve.

87. $|x^2 - 5| = 5 - x^2$

88. $x^4 - 6x^2 + 5 > 0$

89. $2|x|^2 - |x| + 2 \le 5$

90. $(7 - x)^{-2} < 0$

91. $\left|1 + \frac{1}{x}\right| < 3$

92. $\left|2 - \frac{1}{x}\right| \le 2 + \left|\frac{1}{x}\right|$

93. Write a quadratic inequality for which the solution set is $(-4, 3)$.

94. Write a polynomial inequality for which the solution set is $[-4, 3] \cup [7, \infty)$.

Find the domain of the function.

95. $f(x) = \sqrt{\dfrac{72}{x^2 - 4x - 21}}$

96. $f(x) = \sqrt{x^2 - 4x - 21}$

Chapter 4 Summary and Review

STUDY GUIDE

KEY TERMS AND CONCEPTS	EXAMPLES

SECTION 4.1: POLYNOMIAL FUNCTIONS AND MODELS

Polynomial Function

$$P(x) = a_n x^n + a_{n-1}x^{n-1} + a_{n-2}x^{n-2} + \cdots + a_1 x + a_0,$$

where the coefficients $a_n, a_{n-1}, \ldots, a_1, a_0$ are real numbers and the exponents are whole numbers.

The first nonzero coefficient, a_n, is called the **leading coefficient**. The term $a_n x^n$ is called the **leading term**. The **degree** of the polynomial function is n.

Classifying polynomial functions by degree:

Type	Degree
Constant	0
Linear	1
Quadratic	2
Cubic	3
Quartic	4

Consider the polynomial

$$P(x) = \frac{1}{3}x^2 + x - 4x^5 + 2.$$

Leading term: $-4x^5$

Leading coefficient: -4

Degree of polynomial: 5

Classify the following polynomial functions:

Function	Type
$f(x) = -2$	Constant
$f(x) = 0.6x - 11$	Linear
$f(x) = 5x^2 + x - 4$	Quadratic
$f(x) = 5x^3 - x + 10$	Cubic
$f(x) = -x^4 + 8x^3 + x$	Quartic

The Leading-Term Test

If $a_n x^n$ is the leading term of a polynomial function, then the behavior of the graph as $x \to \infty$ and as $x \to -\infty$ can be described in one of the following four ways.

a) If n is even, and $a_n > 0$:

b) If n is even, and $a_n < 0$:

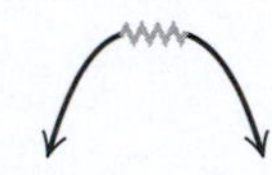

c) If n is odd, and $a_n > 0$:

d) If n is odd, and $a_n < 0$:

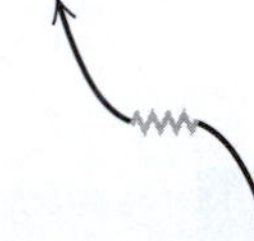

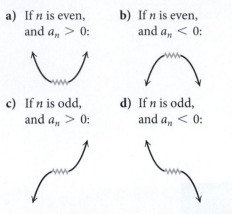

Using the leading-term test, describe the end behavior of the graph of each function by selecting one of (a)–(d) shown at left.

$$h(x) = -2x^6 + x^4 - 3x^2 + x$$

The leading term $a_n x^n$ is $-2x^6$. Since 6 is even and $-2 < 0$, the shape is shown in (b).

$$g(x) = 4x^3 - 8x + 1$$

The leading term, $a_n x^n$, is $4x^3$. Since 3 is odd and $4 > 0$, the shape is shown in (c).

Zeros of Functions

If c is a real zero of a function $f(x)$ (that is, $f(c) = 0$), then $x - c$ is a factor of $f(x)$ and $(c, 0)$ is an x-intercept of the graph of the function.

If we know the linear factors of a polynomial function $f(x)$, we can find the zeros of $f(x)$ by solving the equation $f(x) = 0$ using the principle of zero products.

Every function of degree n, with $n \geq 1$, has at least one zero and at most n zeros.

To find the zeros of

$$f(x) = -2(x - 3)(x + 8)^2,$$

solve $-2(x - 3)(x + 8)^2 = 0$ using the principle of zero products:

$$x - 3 = 0 \quad or \quad x + 8 = 0$$
$$x = 3 \quad or \quad x = -8.$$

The zeros of $f(x)$ are 3 and -8.

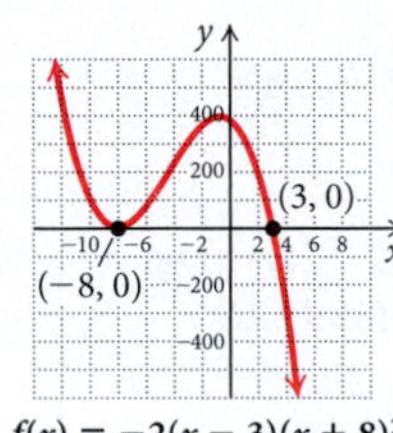

$f(x) = -2(x - 3)(x + 8)^2$

To find the zeros of

$$h(x) = x^4 - 12x^2 - 64,$$

solve $h(x) = 0$:

$$x^4 - 12x^2 - 64 = 0$$
$$(x^2 - 16)(x^2 + 4) = 0$$
$$(x + 4)(x - 4)(x^2 + 4) = 0$$
$$x + 4 = 0 \quad or \quad x - 4 = 0 \quad or \quad x^2 + 4 = 0$$
$$x = -4 \quad or \quad x = 4 \quad or \quad x^2 = -4$$
$$= \pm\sqrt{-4}$$
$$= \pm 2i.$$

The zeros of $h(x)$ are -4, 4, and $\pm 2i$.

Even and Odd Multiplicity

If $(x - c)^k$, $k \geq 1$, is a factor of a polynomial function $P(x)$ and $(x - c)^{k+1}$ is not a factor and:

- k is odd, then the graph crosses the x-axis at $(c, 0)$;
- k is even, then the graph is tangent to the x-axis at $(c, 0)$.

For $f(x) = -2(x - 3)(x + 8)^2$ graphed above, note that for the factor $x - 3$, or $(x - 3)^1$, the exponent 1 is odd and the graph crosses the x-axis at $(3, 0)$. For the factor $(x + 8)^2$, the exponent 2 is even and the graph is tangent to the x-axis at $(-8, 0)$.

SECTION 4.2: GRAPHING POLYNOMIAL FUNCTIONS

If $P(x)$ is a polynomial function of degree n, the graph of the function has:

- at most n real zeros, and thus at most n x-intercepts, and
- at most $n - 1$ turning points.

To Graph a Polynomial Function

1. Use the leading-term test to determine the end behavior.
2. Find the zeros of the function by solving $f(x) = 0$. Any real zeros are the first coordinates of the x-intercepts.
3. Use the x-intercepts (zeros) to divide the x-axis into intervals and choose a test point in each interval to determine the sign of all function values in that interval. For all x-values in an interval, $f(x)$ is either always positive for all values or always negative for all values.
4. Find $f(0)$. This gives the y-intercept of the function.
5. If necessary, find additional function values to determine the general shape of the graph and then draw the graph.

Graph: $h(x) = x^4 - 12x^2 - 16x = x(x - 4)(x + 2)^2$.

1. The leading term is x^4. Since 4 is even and $1 > 0$, the end behavior of the graph can be sketched as follows.

2. Solve $x(x - 4)(x + 2)^2 = 0$. The solutions are 0, 4, and -2. The zeros of $h(x)$ are 0, 4, and -2. The x-intercepts are $(0, 0)$, $(4, 0)$, and $(-2, 0)$. The multiplicity of 0 and 4 is 1. The graph will cross the x-axis at 0 and 4. The multiplicity of -2 is 2. The graph is tangent to the x-axis at -2.
3. The zeros divide the x-axis into four intervals.

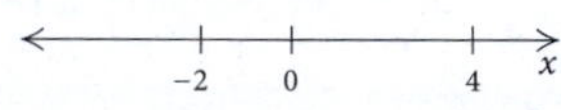

Interval	$(-\infty, -2)$	$(-2, 0)$	$(0, 4)$	$(4, \infty)$
Test Value	-3	-1	1	5
Function Value, $h(x)$	21	5	-27	245
Sign of $h(x)$	+	+	−	+
Location of Points on Graph	Above x-axis	Above x-axis	Below x-axis	Above x-axis

Four points on the graph are $(-3, 21)$, $(-1, 5)$, $(1, -27)$, and $(5, 245)$.

4. Find $h(0)$:

$$h(0) = 0(0 - 4)(0 + 2)^2 = 0.$$

The y-intercept is $(0, 0)$.

(*continued*)

5. Find additional points and draw the graph.

x	$h(x)$
-2.5	4.1
-1.5	2.1
-0.5	5.1
0.5	-10.9
2	-64
3	-75

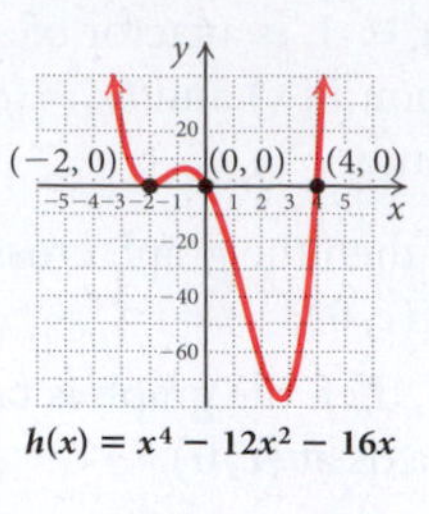

The Intermediate Value Theorem

For any polynomial function $P(x)$ with real coefficients, suppose that for $a \neq b$, $P(a)$ and $P(b)$ are of opposite signs. Then the function has at least one real zero between a and b.

The intermediate value theorem *cannot* be used to determine whether there is a real zero between a and b when $P(a)$ and $P(b)$ have the *same* sign.

Use the intermediate value theorem to determine, if possible, whether the function has a zero between a and b.

$$f(x) = 2x^3 - 5x^2 + x - 2, \quad a = 2, b = 3;$$
$$f(2) = -4; \; f(3) = 10$$

Since $f(2)$ and $f(3)$ have opposite signs, $f(x)$ has at least one zero between 2 and 3.

$$f(x) = 2x^3 - 5x^2 + x - 2, \quad a = -2, b = -1;$$
$$f(-2) = -40; \; f(-1) = -10$$

Both $f(-2)$ and $f(-1)$ are negative. Thus the intermediate value theorem does not allow us to determine whether there is a real zero between -2 and -1.

SECTION 4.3: POLYNOMIAL DIVISION; THE REMAINDER THEOREM AND THE FACTOR THEOREM

Polynomial Division

$$P(x) = d(x) \cdot Q(x) + R(x)$$

Dividend Divisor Quotient Remainder

When we divide a polynomial $P(x)$ by a divisor $d(x)$, a polynomial $Q(x)$ is the quotient and a polynomial $R(x)$ is the remainder. The quotient $Q(x)$ must have degree less than that of the dividend $P(x)$. The remainder $R(x)$ must either be 0 or have degree less than that of the divisor $d(x)$. If $R(x) = 0$, then the divisor $d(x)$ is a factor of the dividend.

Given $P(x) = x^4 - 6x^3 + 9x^2 + 4x - 12$ and $d(x) = x + 2$, use long division to find the quotient and the remainder when $P(x)$ is divided by $d(x)$. Express $P(x)$ in the form $d(x) \cdot Q(x) + R(x)$.

$$\begin{array}{r} x^3 - 8x^2 + 25x - 46 \\ x + 2 \overline{) x^4 - 6x^3 + 9x^2 + 4x - 12} \\ \underline{x^4 + 2x^3} \qquad\qquad\qquad\quad \\ -8x^3 + 9x^2 \qquad\qquad \\ \underline{-8x^3 - 16x^2} \qquad\qquad \\ 25x^2 + 4x \qquad \\ \underline{25x^2 + 50x} \qquad \\ -46x - 12 \\ \underline{-46x - 92} \\ 80 \end{array}$$

$Q(x) = x^3 - 8x^2 + 25x - 46$ and $R(x) = 80$. Thus, $P(x) = (x + 2)(x^3 - 8x^2 + 25x - 46) + 80$. Since $R(x) \neq 0$, $x + 2$ is not a factor of $P(x)$.

The Remainder Theorem

If a number c is substituted for x in the polynomial $f(x)$, then the result $f(c)$ is the remainder that would be obtained by dividing $f(x)$ by $x - c$. That is, if $f(x) = (x - c) \cdot Q(x) + R$, then $f(c) = R$.

The long-division process can be streamlined with synthetic division. Synthetic division can also be used to find polynomial function values.

Repeat the division shown above using synthetic division. Note that the divisor $x + 2 = x - (-2)$.

$$
\begin{array}{r|rrrrr}
-2 & 1 & -6 & 9 & 4 & -12 \\
 & & -2 & 16 & -50 & 92 \\
\hline
 & 1 & -8 & 25 & -46 & \big| \; 80
\end{array}
$$

Again, note that $Q(x) = x^3 - 8x^2 + 25x - 46$ and $R(x) = 80$. Since $R(x) \neq 0$, $x - (-2)$, or $x + 2$, is not a factor of $P(x)$.

Now divide $P(x)$ by $x - 3$.

$$
\begin{array}{r|rrrrr}
3 & 1 & -6 & 9 & 4 & -12 \\
 & & 3 & -9 & 0 & 12 \\
\hline
 & 1 & -3 & 0 & 4 & \big| \; 0
\end{array}
$$

$Q(x) = x^3 - 3x^2 + 4$ and $R(x) = 0$. Since $R(x) = 0$, $x - 3$ is a factor of $P(x)$.

For $f(x) = 2x^5 - x^3 - 3x^2 - 4x + 15$, find $f(-2)$.

$$
\begin{array}{r|rrrrrr}
-2 & 2 & 0 & -1 & -3 & -4 & 15 \\
 & & -4 & 8 & -14 & 34 & -60 \\
\hline
 & 2 & -4 & 7 & -17 & 30 & \big| \; -45
\end{array}
$$

Thus, $f(-2) = -45$.

The Factor Theorem

For a polynomial $f(x)$, if $f(c) = 0$, then $x - c$ is a factor of $f(x)$.

Let $g(x) = x^4 + 8x^3 + 6x^2 - 40x + 25$. Factor $g(x)$ and solve $g(x) = 0$.

Use synthetic division to look for factors of the form $x - c$. Let's try $x + 5$.

$$
\begin{array}{r|rrrrr}
-5 & 1 & 8 & 6 & -40 & 25 \\
 & & -5 & -15 & 45 & -25 \\
\hline
 & 1 & 3 & -9 & 5 & \big| \; 0
\end{array}
$$

Since $g(-5) = 0$, the number -5 is a zero of $g(x)$ and $x - (-5)$, or $x + 5$, is a factor of $g(x)$. This gives us

$$g(x) = (x + 5)(x^3 + 3x^2 - 9x + 5).$$

Let's try $x + 5$ again with the factor $x^3 + 3x^2 - 9x + 5$.

$$
\begin{array}{r|rrrr}
-5 & 1 & 3 & -9 & 5 \\
 & & -5 & 10 & -5 \\
\hline
 & 1 & -2 & 1 & \big| \; 0
\end{array}
$$

Now we have

$$g(x) = (x + 5)^2(x^2 - 2x + 1).$$

The trinomial $x^2 - 2x + 1$ easily factors, so

$$g(x) = (x + 5)^2(x - 1)^2.$$

Solve $g(x) = 0$. The solutions of $(x + 5)^2(x - 1)^2 = 0$ are -5 and 1. They are also the zeros of $g(x)$.

SECTION 4.4: THEOREMS ABOUT ZEROS OF POLYNOMIAL FUNCTIONS

The Fundamental Theorem of Algebra

Every polynomial function of degree n, $n \geq 1$, with complex coefficients has at least one zero in the system of complex numbers.

Every polynomial function f of degree n, with $n \geq 1$, can be factored into n linear factors (not necessarily unique); that is,

$$f(x) = a_n(x - c_1)(x - c_2) \cdots (x - c_n).$$

Find a polynomial function of degree 5 with -4 and 2 as zeros of multiplicity 1, and -1 as a zero of multiplicity 3.

$$\begin{aligned} f(x) &= [x - (-4)][x - 2][x - (-1)]^3 \\ &= (x + 4)(x - 2)(x + 1)^3 \\ &= x^5 + 5x^4 + x^3 - 17x^2 - 22x - 8 \end{aligned}$$

Nonreal Zeros

$\boldsymbol{a + bi}$ **and** $\boldsymbol{a - bi,\ b \neq 0}$

If a complex number $a + bi$, $b \neq 0$, is a zero of a polynomial function $f(x)$ with *real* coefficients, then its conjugate, $a - bi$, is also a zero. (Nonreal zeros occur in conjugate pairs.)

Irrational Zeros

$\boldsymbol{a + c\sqrt{b}}$, **and** $\boldsymbol{a - c\sqrt{b}}$,
b not a perfect square

If $a + c\sqrt{b}$, where a, b, and c are rational and b is not a perfect square, is a zero of a polynomial function $f(x)$ with *rational* coefficients, then its conjugate, $a - c\sqrt{b}$, is also a zero. (Irrational zeros occur in conjugate pairs.)

Find a polynomial function with rational coefficients of lowest degree with $1 - i$ and $\sqrt{7}$ as two of its zeros.

If $1 - i$ is a zero, then $1 + i$ is also a zero. If $\sqrt{7}$ is a zero, then $-\sqrt{7}$ is also a zero.

$$\begin{aligned} f(x) &= [x - (1 - i)][x - (1 + i)][x - \sqrt{7}] \times \\ &\qquad [x - (-\sqrt{7})] \\ &= [(x - 1) + i][(x - 1) - i](x - \sqrt{7})(x + \sqrt{7}) \\ &= [(x - 1)^2 - i^2](x^2 - 7) \\ &= (x^2 - 2x + 1 + 1)(x^2 - 7) \\ &= (x^2 - 2x + 2)(x^2 - 7) \\ &= x^4 - 2x^3 - 5x^2 + 14x - 14 \end{aligned}$$

The Rational Zeros Theorem

Consider the polynomial function

$$P(x) = a_n x^n + a_{n-1}x^{n-1} + a_{n-2}x^{n-2} + \cdots + a_1x + a_0,$$

where all the coefficients are integers and $n \geq 1$. Also, consider a rational number p/q, where p and q have no common factor other than -1 and 1. If p/q is a zero of $P(x)$, then p is a factor of a_0 and q is a factor of a_n.

For $f(x) = 2x^4 - 9x^3 - 16x^2 - 9x - 18$, solve $f(x) = 0$ and factor $f(x)$ into linear factors.

There are at most 4 distinct zeros. Any rational zeros of f must be of the form p/q, where p is a factor of -18 and q is a factor of 2.

Possibilities for p: $\pm 1, \pm 2, \pm 3, \pm 6, \pm 9, \pm 18$
Possibilities for q: $\pm 1, \pm 2$
Possibilities for p/q: $1, -1, 2, -2, 3, -3, 6, -6, 9, -9, 18, -18, \frac{1}{2}, -\frac{1}{2}, \frac{3}{2}, -\frac{3}{2}, \frac{9}{2}, -\frac{9}{2}$

Use synthetic division to check the possibilities. We leave it to the student to verify that ± 1, ± 2, and ± 3 are not zeros. Let's try 6.

$$\begin{array}{r|rrrrr} 6 & 2 & -9 & -16 & -9 & -18 \\ & & 12 & 18 & 12 & 18 \\ \hline & 2 & 3 & 2 & 3 & 0 \end{array}$$

(continued)

Since $f(6) = 0$, 6 is a zero and $x - 6$ is a factor of $f(x)$. Now express $f(x)$ as

$$f(x) = (x - 6)(2x^3 + 3x^2 + 2x + 3).$$

Consider the factor $2x^3 + 3x^2 + 2x + 3$ and check the other possibilities. Let's try $-\frac{3}{2}$.

$$\begin{array}{r|rrrr} -\frac{3}{2} & 2 & 3 & 2 & 3 \\ & & -3 & 0 & -3 \\ \hline & 2 & 0 & 2 & 0 \end{array}$$

Since $f\left(-\frac{3}{2}\right) = 0$, $-\frac{3}{2}$ is also a zero and $x + \frac{3}{2}$ is a factor of $f(x)$. We express $f(x)$ as

$$\begin{aligned} f(x) &= (x - 6)\left(x + \tfrac{3}{2}\right)(2x^2 + 2) \\ &= 2(x - 6)\left(x + \tfrac{3}{2}\right)(x^2 + 1). \end{aligned}$$

Now solve the equation $f(x) = 0$ to determine the zeros. We see that the only rational zeros are 6 and $-\frac{3}{2}$. The other zeros are $\pm i$.

The factorization into linear factors is

$$\begin{aligned} f(x) &= 2(x - 6)\left(x + \tfrac{3}{2}\right)(x - i)(x + i), \quad \text{or} \\ &\quad (x - 6)(2x + 3)(x - i)(x + i). \end{aligned}$$

Descartes' Rule of Signs

Let $P(x)$, written in descending or ascending order, be a polynomial function with real coefficients and a nonzero constant term. The number of positive real zeros of $P(x)$ is either:

1. The same as the number of variations of sign in $P(x)$, or
2. Less than the number of variations of sign in $P(x)$ by a positive even integer.

The number of negative real zeros of $P(x)$ is either:

3. The same as the number of variations of sign in $P(-x)$, or
4. Less than the number of variations of sign in $P(-x)$ by a positive even integer.

A zero of multiplicity m must be counted m times.

Determine the number of positive real zeros and the number of negative real zeros of

$$P(x) = 4x^5 - x^4 - 2x^3 + 8x - 10.$$

There are 3 variations of sign in $P(x)$. Thus the number of positive real zeros is 3 or 1.

$$P(-x) = -4x^5 - x^4 + 2x^3 - 8x - 10$$

There are 2 variations of sign in $P(-x)$. Thus the number of negative real zeros is 2 or 0.

SECTION 4.5: RATIONAL FUNCTIONS

Rational Function

$$f(x) = \frac{p(x)}{q(x)},$$

where $p(x)$ and $q(x)$ are polynomials and $q(x)$ is not the zero polynomial. The domain of $f(x)$ consists of all x for which $q(x) \neq 0$.

Determine the domain of each function.

FUNCTION	DOMAIN
$f(x) = \frac{1}{x^5}$	$(-\infty, 0) \cup (0, \infty)$
$f(x) = \frac{x+6}{x^2+2x-8} = \frac{x+6}{(x-2)(x+4)}$	$(-\infty, -4) \cup (-4, 2) \cup (2, \infty)$

Vertical Asymptotes

For a rational function $f(x) = p(x)/q(x)$, where $p(x)$ and $q(x)$ are polynomials with *no common factors* other than constants, if a is a zero of the denominator, then the line $x = a$ is a vertical asymptote for the graph of the function.

Horizontal Asymptotes

When the numerator and the denominator have the same degree, the line $y = a/b$ is the horizontal asymptote, where a and b are the leading coefficients of the numerator and the denominator, respectively.

When the degree of the numerator is less than the degree of the denominator, the x-axis, or $y = 0$, is the horizontal asymptote.

When the degree of the numerator is greater than the degree of the denominator, there is *no* horizontal asymptote.

Oblique Asymptotes

When the degree of the numerator is 1 greater than the degree of the denominator, there is an oblique asymptote.

Determine the vertical, horizontal, and oblique asymptotes of the graph of the function.

FUNCTION	ASYMPTOTES
$f(x) = \frac{x^2-2}{x-1}$	Vertical: $x = 1$ Horizontal: None Oblique: $y = x + 1$
$f(x) = \frac{3x-4}{x^2+6x-7}$	Vertical: $x = -7; x = 1$ Horizontal: $y = 0$ Oblique: None
$f(x) = \frac{2x^2+9x-5}{3x^2+13x+12}$	Vertical: $x = -\frac{4}{3}; x = -3$ Horizontal: $y = \frac{2}{3}$ Oblique: None

To Graph a Rational Function

$f(x) = p(x)/q(x)$, where $p(x)$ and $q(x)$ have no common factor other than constants:

1. Find any real zeros of the denominator. Determine the domain of the function and sketch any vertical asymptotes.
2. Find the horizontal asymptote or the oblique asymptote, if there is one, and sketch it.

Graph: $g(x) = \frac{x^2-4}{x^2+4x-5}$.

Domain: The zeros of the denominator are -5 and 1. The domain is $(-\infty, -5) \cup (-5, 1) \cup (1, \infty)$.

Vertical asymptotes: Since neither zero of the denominator is a zero of the numerator, the graph has vertical asymptotes at $x = -5$ and $x = 1$.

Horizontal asymptote: The degree of the numerator is the same as the degree of the denominator, so the horizontal asymptote is determined by the ratio of the leading coefficients: $1/1$, or 1. The horizontal asymptote is $y = 1$.

(continued)

3. Find any zeros of the function. The zeros are found by determining the zeros of the numerator. These are the first coordinates of the x-intercepts of the graph.
4. Find $f(0)$. This gives the y-intercept, $(0, f(0))$, of the graph.
5. Find other function values to determine the general shape. Then draw the graph.

Crossing an Asymptote

The graph of a rational function never crosses a vertical asymptote.

The graph of a rational function might cross a horizontal asymptote but does not necessarily do so.

Oblique asymptote: None

Zeros of g: Solving $g(x) = 0$ gives us -2 and 2, so the zeros are -2 and 2.

x-intercepts: $(-2, 0)$ and $(2, 0)$

y-intercept: $\left(0, \frac{4}{5}\right)$, because $g(0) = \frac{4}{5}$

Other values:

x	y
-8	2.22
-6	4.57
-4	-2.4
-3	-0.63
-1	0.38
0.5	1.36
1.5	-0.54
3	0.31
4	0.44

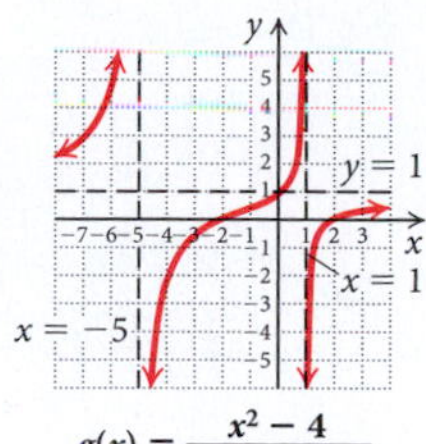

Graph: $f(x) = \frac{x + 3}{x^2 - x - 12}$.

Domain: The zeros of the denominator are -3 and 4. The domain is $(-\infty, -3) \cup (-3, 4) \cup (4, \infty)$.

Vertical asymptote: Since 4 is the only zero of the denominator that is not a zero of the numerator, the only vertical asymptote is $x = 4$.

Horizontal asymptote: Because the degree of the numerator is less than the degree of the denominator, the x-axis, $y = 0$, is the horizontal asymptote.

Oblique asymptote: None

Zeros of f: The equation $f(x) = 0$ has no solutions, so there are no zeros of f.

x-intercepts: None

y-intercept: $\left(0, -\frac{1}{4}\right)$ because $f(0) = -\frac{1}{4}$

Hole in the graph:

$$f(x) = \frac{x + 3}{(x + 3)(x - 4)} = \frac{1}{x - 4},$$

where $x \neq -3$ and $x \neq 4$.

(continued)

To determine the coordinates of the hole, substitute -3 for x in $f(x) = 1/(x - 4)$:

$$f(-3) = \frac{1}{-3 - 4} = -\frac{1}{7}.$$

The hole is located at $\left(-3, -\frac{1}{7}\right)$.

Other values:

x	y
-4	-0.13
-2	-0.17
1	-0.33
3	-1
5	1
7	0.33

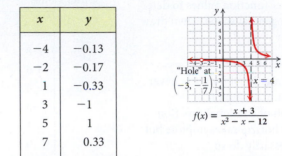

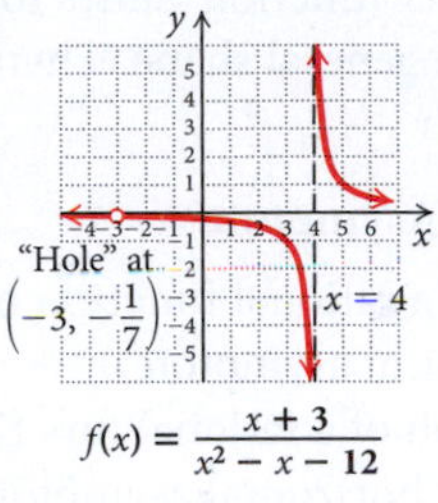

SECTION 4.6: POLYNOMIAL INEQUALITIES AND RATIONAL INEQUALITIES

To Solve a Polynomial Inequality

1. Find an equivalent inequality with 0 on one side.
2. Change the inequality symbol to an equals sign and solve the related equation.
3. Use the solutions to divide the x-axis into intervals. Then select a test value from each interval and determine the polynomial's sign on the interval.
4. Determine the intervals for which the inequality is satisfied and write interval notation or set-builder notation for the solution set. Include the endpoints of the intervals in the solution set if the inequality symbol is $\leq$ or $\geq$.

Solve: $x^3 - 3x^2 \leq 6x - 8$.

Equivalent inequality: $x^3 - 3x^2 - 6x + 8 \leq 0$.

First, solve the related equation:

$$x^3 - 3x^2 - 6x + 8 = 0.$$

The solutions are -2, 1, and 4. The numbers divide the x-axis into 4 intervals. Next, let $f(x) = x^3 - 3x^2 - 6x + 8$ and, using test values for $f(x)$, determine the sign of $f(x)$ in each interval.

INTERVAL	TEST VALUE	SIGN OF $f(x)$
$(-\infty, -2)$	$f(-3) = -28$	$-$
$(-2, 1)$	$f(0) = 8$	$+$
$(1, 4)$	$f(2) = -8$	$-$
$(4, \infty)$	$f(6) = 80$	$+$

$-\quad+\quad-\quad+$

$-2 \quad 1 \quad 4 \quad x$

Test values are negative in the intervals $(-\infty, -2)$ and $(1, 4)$. Since the inequality sign is $\leq$, include the endpoints of the intervals in the solution set.

The solution set is

$$(-\infty, -2] \cup [1, 4].$$

To Solve a Rational Inequality

1. Find an equivalent inequality with 0 on one side.
2. Change the inequality symbol to an equals sign and solve the related equation.
3. Find values of the variable for which the related rational function is not defined.
4. The numbers found in steps (2) and (3) are called *critical values*. Use the critical values to divide the x-axis into intervals. Then determine the function's sign in each interval using an x-value from the interval or using the graph of the equation.
5. Select the intervals for which the inequality is satisfied and write interval notation or set-builder notation for the solution set. If the inequality symbol is $\le$ or $\ge$, then the solutions to step (2) should be included in the solution set. The x-values found in step (3) are never included in the solution set.

Solve: $\dfrac{x-1}{x+5} > \dfrac{x+3}{x-2}$.

Equivalent inequality: $\dfrac{x-1}{x+5} - \dfrac{x+3}{x-2} > 0$

Related function: $f(x) = \dfrac{x-1}{x+5} - \dfrac{x+3}{x-2}$

The function is not defined for $x = -5$ and $x = 2$. Solving $f(x) = 0$, we get $x = -\frac{13}{11}$. The critical values are -5, $-\frac{13}{11}$, and 2. These divide the x-axis into four intervals.

INTERVAL	TEST VALUE	SIGN OF $f(x)$
$(-\infty, -5)$	$f(-6) = 6.63$	+
$\left(-5, -\frac{13}{11}\right)$	$f(-2) = -0.75$	−
$\left(-\frac{13}{11}, 2\right)$	$f(0) = 1.3$	+
$(2, \infty)$	$f(3) = -5.75$	−

+ − + −
−5 $-\frac{13}{11}$ 2 x

Test values are positive in the intervals $(-\infty, -5)$ and $\left(-\frac{13}{11}, 2\right)$. Since $f\left(-\frac{13}{11}\right) = 0$ and -5 and 2 are not in the domain of f, -5, $-\frac{13}{11}$, and 2 cannot be part of the solution set.

The solution set is

$$(-\infty, -5) \cup \left(-\tfrac{13}{11}, 2\right).$$

REVIEW EXERCISES

Determine whether the statement is true or false.

1. If $f(x) = (x + a)(x + b)(x - c)$, then $f(-b) = 0$. [4.3]
2. The graph of a rational function never crosses a vertical asymptote. [4.5]
3. For the function $g(x) = x^4 - 8x^2 - 9$, the only possible rational zeros are 1, -1, 3, and -3. [4.4]
4. The graph of $P(x) = x^6 - x^8$ has at most 6 x-intercepts. [4.2]
5. The domain of the function
$$f(x) = \frac{x-4}{(x+2)(x-3)}$$
is $(-\infty, -2) \cup (3, \infty)$. [4.5]

Determine the leading term, the leading coefficient, and the degree of the polynomial. Then classify the polynomial function as constant, linear, quadratic, cubic, or quartic. [4.1]

6. $f(x) = 7x^2 - 5 + 0.45x^4 - 3x^3$

7. $h(x) = -25$

8. $g(x) = 6 - 0.5x$

9. $f(x) = \frac{1}{3}x^3 - 2x + 3$

Use the leading-term test to describe the end behavior of the graph of the function. [4.1]

10. $f(x) = -\frac{1}{2}x^4 + 3x^2 + x - 6$

11. $f(x) = x^5 + 2x^3 - x^2 + 5x + 4$

Find the zeros of the polynomial function and state the multiplicity of each. [4.1]

12. $g(x) = \left(x - \frac{2}{3}\right)(x + 2)^3(x - 5)^2$

13. $f(x) = x^4 - 26x^2 + 25$

14. $h(x) = x^3 + 4x^2 - 9x - 36$

15. *Interest Compounded Annually.* When P dollars is invested at interest rate i, compounded annually, for t years, the investment grows to A dollars, where

$A = P(1 + i)^t$.

a) Find the interest rate i if \$6250 grows to \$6760 in 2 years. [4.1]

b) Find the interest rate i if \$1,000,000 grows to \$1,215,506.25 in 4 years. [4.1]

Sketch the graph of the polynomial function.

16. $f(x) = -x^4 + 2x^3$ [4.2]

17. $g(x) = (x - 1)^3(x + 2)^2$ [4.2]

18. $h(x) = x^3 + 3x^2 - x - 3$ [4.2]

19. $f(x) = x^4 - 5x^3 + 6x^2 + 4x - 8$ [4.2], [4.3], [4.4]

20. $g(x) = 2x^3 + 7x^2 - 14x + 5$ [4.2], [4.4]

Using the intermediate value theorem, determine, if possible, whether the function f has a zero between a and b. [4.2]

21. $f(x) = 4x^2 - 5x - 3$; $a = 1, b = 2$

22. $f(x) = x^3 - 4x^2 + \frac{1}{2}x + 2$; $a = -1, b = 1$

In each of the following, a polynomial $P(x)$ and a divisor $d(x)$ are given. Use long division to find the quotient $Q(x)$ and the remainder $R(x)$ when $P(x)$ is divided by $d(x)$. Express $P(x)$ in the form $d(x) \cdot Q(x) + R(x)$. [4.3]

23. $P(x) = 6x^3 - 2x^2 + 4x - 1$,
$d(x) = x - 3$

24. $P(x) = x^4 - 2x^3 + x + 5$,
$d(x) = x + 1$

Use synthetic division to find the quotient and the remainder. [4.3]

25. $(x^3 + 2x^2 - 13x + 10) \div (x - 5)$

26. $(x^4 + 3x^3 + 3x^2 + 3x + 2) \div (x + 2)$

27. $(x^5 - 2x) \div (x + 1)$

Use synthetic division to find the indicated function value. [4.3]

28. $f(x) = x^3 + 2x^2 - 13x + 10$; $f(-2)$

29. $f(x) = x^4 - 16$; $f(-2)$

30. $f(x) = x^5 - 4x^4 + x^3 - x^2 + 2x - 100$; $f(-10)$

Using synthetic division, determine whether the given numbers are zeros of the polynomial function. [4.3]

31. $-i, -5$; $f(x) = x^3 - 5x^2 + x - 5$

32. $-1, -2$; $f(x) = x^4 - 4x^3 - 3x^2 + 14x - 8$

33. $\frac{1}{3}, 1$; $f(x) = x^3 - \frac{4}{3}x^2 - \frac{5}{3}x + \frac{2}{3}$

34. $2, -\sqrt{3}$; $f(x) = x^4 - 5x^2 + 6$

Factor the polynomial $f(x)$. Then solve the equation $f(x) = 0$. [4.3], [4.4]

35. $f(x) = x^3 + 2x^2 - 7x + 4$

36. $f(x) = x^3 + 4x^2 - 3x - 18$

37. $f(x) = x^4 - 4x^3 - 21x^2 + 100x - 100$

38. $f(x) = x^4 - 3x^2 + 2$

Find a polynomial function of degree 3 with the given numbers as zeros. [4.4]

39. $-4, -1, 2$

40. $-3, 1 - i, 1 + i$

41. $\frac{1}{2}, 1 - \sqrt{2}, 1 + \sqrt{2}$

42. Find a polynomial function of degree 4 with -5 as a zero of multiplicity 3 and $\frac{1}{2}$ as a zero of multiplicity 1. [4.4]

43. Find a polynomial function of degree 5 with -3 as a zero of multiplicity 2, 2 as a zero of multiplicity 1, and 0 as a zero of multiplicity 2. [4.4]

Suppose that a polynomial function of degree 5 with rational coefficients has the given zeros. Find the other zero(s). [4.4]

44. $-\frac{2}{3}, \sqrt{5}, i$

45. $0, 1 + \sqrt{3}, -\sqrt{3}$

46. $-\sqrt{2}, \frac{1}{2}, 1, 2$

Find a polynomial function of lowest degree with rational coefficients and the following as some of its zeros. [4.4]

47. $\sqrt{11}$

48. $-i, 6$

49. $-1, 4, 1 + i$

50. $\sqrt{5}, -2i$

51. $\frac{1}{3}, 0, -3$

List all possible rational zeros. [4.4]

52. $h(x) = 4x^5 - 2x^3 + 6x - 12$

53. $g(x) = 3x^4 - x^3 + 5x^2 - x + 1$

54. $f(x) = x^3 - 2x^2 + x - 24$

For each polynomial function, **(a)** *find the rational zeros and then the other zeros; that is, solve* $f(x) = 0$*; and* **(b)** *factor* $f(x)$ *into linear factors.* [4.4]

55. $f(x) = 3x^5 + 2x^4 - 25x^3 - 28x^2 + 12x$

56. $f(x) = x^3 - 2x^2 - 3x + 6$

57. $f(x) = x^4 - 6x^3 + 9x^2 + 6x - 10$

58. $f(x) = x^3 + 3x^2 - 11x - 5$

59. $f(x) = 3x^3 - 8x^2 + 7x - 2$

60. $f(x) = x^5 - 8x^4 + 20x^3 - 8x^2 - 32x + 32$

61. $f(x) = x^6 + x^5 - 28x^4 - 16x^3 + 192x^2$

62. $f(x) = 2x^5 - 13x^4 + 32x^3 - 38x^2 + 22x - 5$

What does Descartes' rule of signs tell you about the number of positive real zeros and the number of negative real zeros of each of the following polynomial functions? [4.4]

63. $f(x) = 2x^6 - 7x^3 + x^2 - x$

64. $h(x) = -x^8 + 6x^5 - x^3 + 2x - 2$

65. $g(x) = 5x^5 - 4x^2 + x - 1$

Graph the function. Be sure to label all the asymptotes. List the domain and the x- and y-intercepts. [4.5]

66. $f(x) = \dfrac{x^2 - 5}{x + 2}$

67. $f(x) = \dfrac{5}{(x - 2)^2}$

68. $f(x) = \dfrac{x^2 + x - 6}{x^2 - x - 20}$

69. $f(x) = \dfrac{x - 2}{x^2 - 2x - 15}$

In Exercises 70 and 71, find a rational function that satisfies the given conditions. Answers may vary, but try to give the simplest answer possible. [4.5]

70. Vertical asymptotes $x = -2, x = 3$

71. Vertical asymptotes $x = -2, x = 3$; horizontal asymptote $y = 4$; x-intercept $(-3, 0)$

72. *Medical Dosage.* The function

$$N(t) = \frac{0.7t + 2000}{8t + 9}, \quad t \geq 5,$$

gives the body concentration $N(t)$, in parts per million, of a certain dosage of medication after time t, in hours.

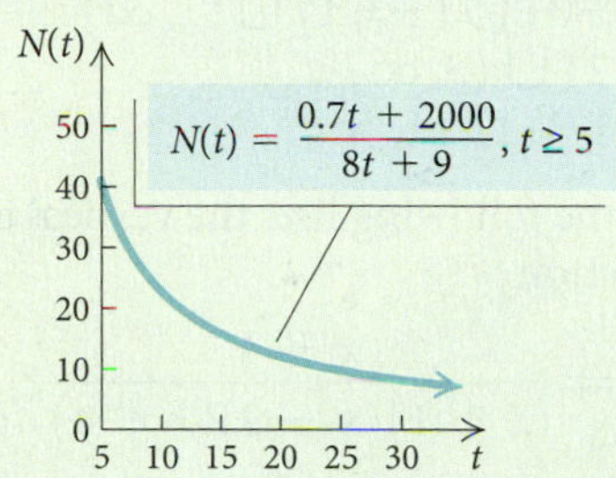

a) Find the horizontal asymptote of the graph and complete the following:

$N(t) \to \square$ as $t \to \infty$. [4.5]

b) Explain the meaning of the answer to part (a) in terms of the application. [4.5]

Solve. [4.6]

73. $x^2 - 9 < 0$

74. $2x^2 > 3x + 2$

75. $(1 - x)(x + 4)(x - 2) \leq 0$

76. $\dfrac{x - 2}{x + 3} < 4$

77. *Height of a Rocket.* The function

$$S(t) = -16t^2 + 80t + 224$$

gives the height S, in feet, of a model rocket launched with a velocity of 80 ft/sec from a hill that is 224 ft high, where t is the time, in seconds.

a) Determine when the rocket reaches the ground. [4.1]

b) On what interval is the height greater than 320 ft? [4.1], [4.6]

78. *Population Growth.* The population P, in thousands, of Novi is given by

$$P(t) = \frac{8000t}{4t^2 + 10},$$

where t is the time, in months. Find the interval on which the population was 400,000 or greater. [4.6]

79. Which of the following is the domain of the function

$$g(x) = \frac{x^2 + 2x - 3}{x^2 - 5x + 6}? \text{ [4.5]}$$

A. $(-\infty, 2) \cup (2, 3) \cup (3, \infty)$
B. $(-\infty, -3) \cup (-3, 1) \cup (1, \infty)$
C. $(-\infty, 2) \cup (3, \infty)$
D. $(-\infty, -3) \cup (1, \infty)$

80. Which of the following lists the vertical asymptotes of the function

$$f(x) = \frac{x - 4}{(x + 1)(x - 2)(x + 4)}? \text{ [4.5]}$$

A. $x = 1, x = -2$, and $x = 4$
B. $x = -1, x = 2, x = -4$, and $x = 4$
C. $x = -1, x = 2$, and $x = -4$
D. $x = 4$

81. The graph of $f(x) = -\frac{1}{2}x^4 + x^3 + 1$ is which of the following? [4.2]

A.

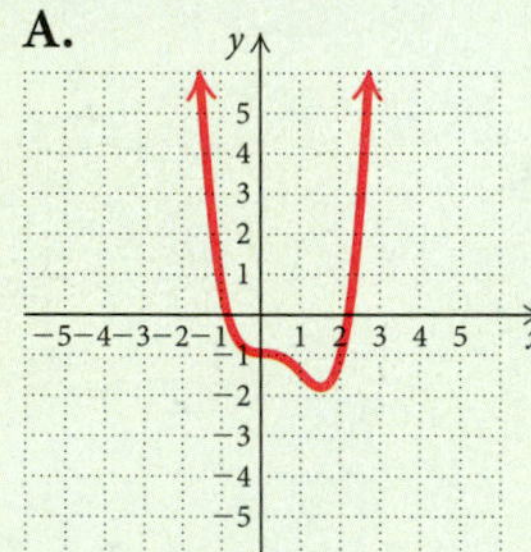

B.

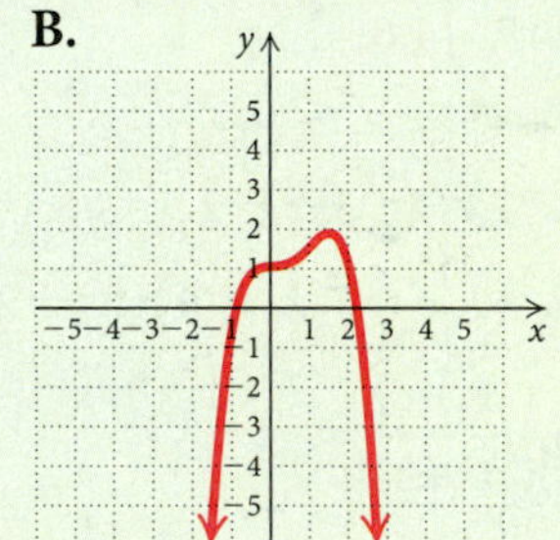

C.

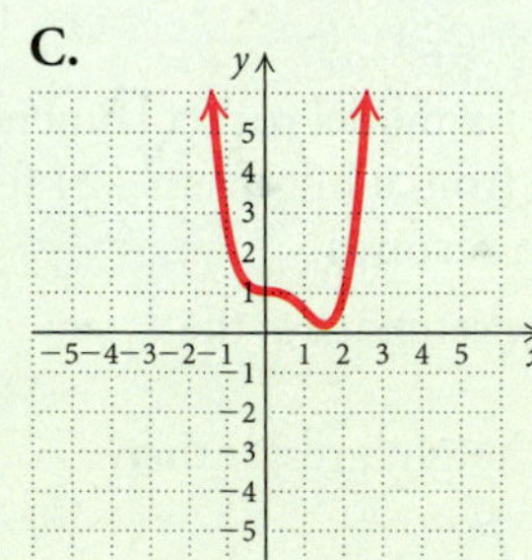

D.

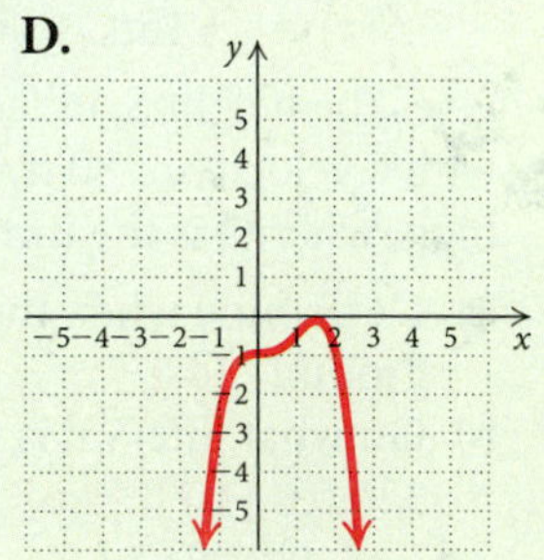

Synthesis

Solve.

82. $x^2 \geq 5 - 2x$ [4.6]

83. $\left|1 - \frac{1}{x^2}\right| < 3$ [4.6]

84. $x^4 - 2x^3 + 3x^2 - 2x + 2 = 0$ [4.4]

85. $(x - 2)^{-3} < 0$ [4.6]

86. Express $x^3 - 1$ as a product of linear factors. [4.4]

87. Find k such that $x + 3$ is a factor of $x^3 + kx^2 + kx - 15$. [4.3]

88. When $x^2 - 4x + 3k$ is divided by $x + 5$, the remainder is 33. Find the value of k. [4.3]

Find the domain of the function. [4.5]

89. $f(x) = \sqrt{x^2 + 3x - 10}$

90. $f(x) = \sqrt{x^2 - 3.1x + 2.2} + 1.75$

91. $f(x) = \frac{1}{\sqrt{5 - |7x + 2|}}$

Collaborative Discussion and Writing

92. Explain the difference between a polynomial function and a rational function. [4.1], [4.5]

93. Is it possible for a third-degree polynomial with rational coefficients to have no real zeros? Why or why not? [4.4]

94. Explain and contrast the three types of asymptotes considered for rational functions. [4.5]

95. If $P(x)$ is an even function, and by Descartes' rule of signs, $P(x)$ has one positive real zero, how many negative real zeros does $P(x)$ have? Explain. [4.4]

96. Explain why the graph of a rational function cannot have both a horizontal asymptote and an oblique asymptote. [4.5]

97. Under what circumstances would a quadratic inequality have a solution set that is a closed interval? [4.6]

4 Chapter Test

Determine the leading term, the leading coefficient, and the degree of the polynomial. Then classify the polynomial as constant, linear, quadratic, cubic, or quartic.

1. $f(x) = 2x^3 + 6x^2 - x^4 + 11$
2. $h(x) = -4.7x + 29$
3. Find the zeros of the polynomial function and state the multiplicity of each:
$$f(x) = x(3x - 5)(x - 3)^2(x + 1)^3.$$
4. *Hybrid Automobiles.* In 2004, only 84,199 hybrid automobiles were sold, while in 2012, 431,798 were sold (*Source:* U.S. Department of Transportation). The quartic function
$$f(x) = 897.690x^4 - 10{,}349.487x^3 + 19{,}202.137x^2 + 91{,}597.838x + 88{,}209.580,$$
where x is the number of years after 2004, can be used to estimate the number of hybrid automobiles sold in years 2004 to 2012. Use this function to estimate the number of hybrid automobiles sold in 2008 and in 2011.

Sketch the graph of the polynomial function.

5. $f(x) = x^3 - 5x^2 + 2x + 8$
6. $f(x) = -2x^4 + x^3 + 11x^2 - 4x - 12$

Using the intermediate value theorem, determine, if possible, whether the function has a zero between a and b.

7. $f(x) = -5x^2 + 3;\ a = 0, b = 2$
8. $g(x) = 2x^3 + 6x^2 - 3;\ a = -2, b = -1$
9. Use long division to find the quotient $Q(x)$ and the remainder $R(x)$ when $P(x)$ is divided by $d(x)$. Express $P(x)$ in the form $d(x) \cdot Q(x) + R(x)$. Show your work.
$$P(x) = x^4 + 3x^3 + 2x - 5,$$
$$d(x) = x - 1$$
10. Use synthetic division to find the quotient and the remainder. Show your work.
$$(3x^3 - 12x + 7) \div (x - 5)$$
11. Use synthetic division to find $P(-3)$ for $P(x) = 2x^3 - 6x^2 + x - 4$. Show your work.
12. Use synthetic division to determine whether -2 is a zero of $f(x) = x^3 + 4x^2 + x - 6$. Answer yes or no. Show your work.
13. Find a polynomial function of degree 4 with -3 as a zero of multiplicity 2 and 0 and 6 as zeros of multiplicity 1.
14. Suppose that a polynomial function of degree 5 with rational coefficients has 1, $\sqrt{3}$, and $2 - i$ as zeros. Find the other zeros.

Find a polynomial function of lowest degree with rational coefficients and the following as some of its zeros.

15. $-10, 3i$
16. $0, -\sqrt{3}, 1 - i$

List all possible rational zeros.

17. $f(x) = 2x^3 + x^2 - 2x + 12$
18. $h(x) = 10x^4 - x^3 + 2x - 5$

For each polynomial function, **(a)** *find the rational zeros and then the other zeros; that is, solve $f(x) = 0$; and* **(b)** *factor $f(x)$ into linear factors.*

19. $f(x) = x^3 + x^2 - 5x - 5$
20. $f(x) = 2x^4 - 11x^3 + 16x^2 - x - 6$
21. $f(x) = x^3 + 4x^2 + 4x + 16$
22. $f(x) = 3x^4 - 11x^3 + 15x^2 - 9x + 2$
23. What does Descartes' rule of signs tell you about the number of positive real zeros and the number of negative real zeros of the following function?
$$g(x) = -x^8 + 2x^6 - 4x^3 - 1$$

Graph the function. Be sure to label all the asymptotes. List the domain and the x- and y-intercepts.

24. $f(x) = \dfrac{2}{(x - 3)^2}$
25. $f(x) = \dfrac{x + 3}{x^2 - 3x - 4}$
26. Find a rational function that has vertical asymptotes $x = -1$ and $x = 2$ and x-intercept $(-4, 0)$.

Solve.

27. $2x^2 > 5x + 3$
28. $\dfrac{x + 1}{x - 4} \le 3$

29. The function $S(t) = -16t^2 + 64t + 192$ gives the height S, in feet, of a model rocket launched with a velocity of 64 ft/sec from a hill that is 192 ft high.

a) Determine how long it will take the rocket to reach the ground.

b) Find the interval on which the height of the rocket is greater than 240 ft.

30. The graph of $f(x) = x^3 - x^2 - 2$ is which of the following?

A.

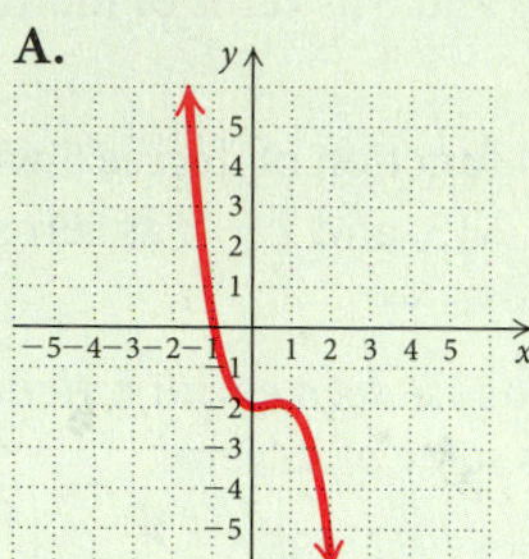

B.

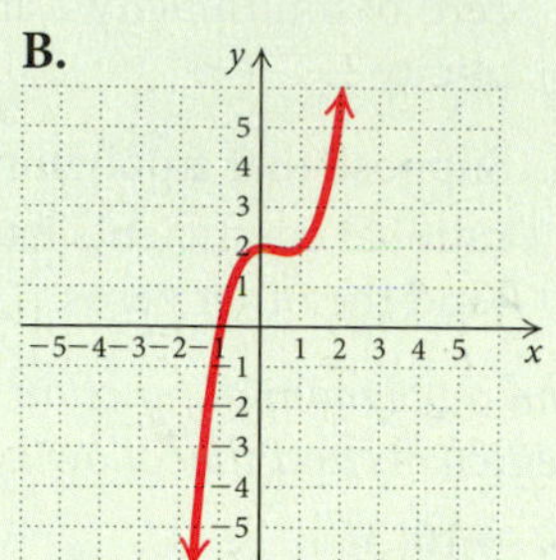

C.

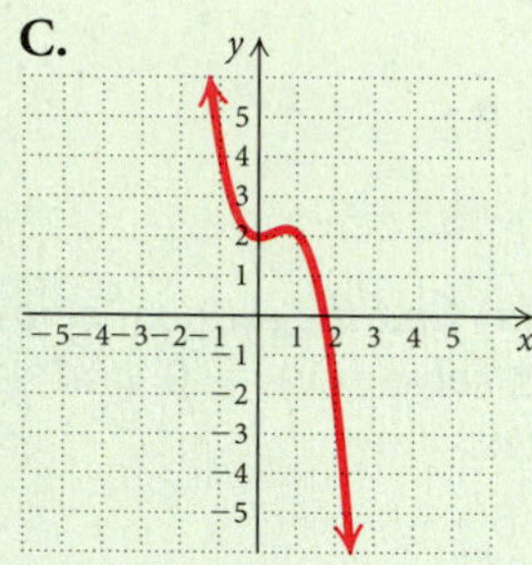

D.

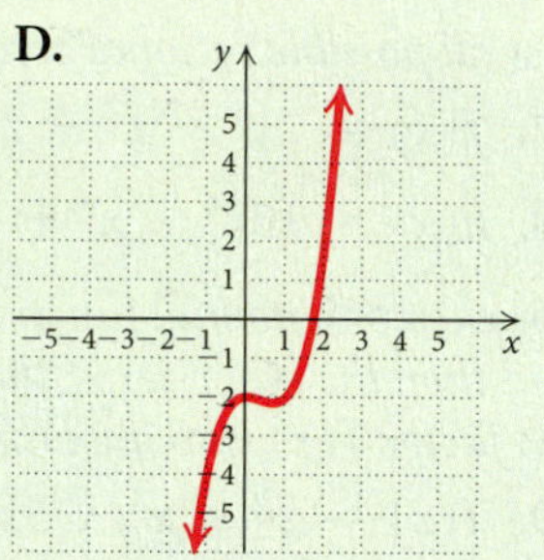

Synthesis

31. Find the domain of $f(x) = \sqrt{x^2 + x - 12}$.

Exponential Functions and Logarithmic Functions

APPLICATION This problem appears as Exercise 69 in Section 5.2.

The centenarian population in the United States has grown over 65% in the last 30 years. In 1980, there were only 32,194 residents ages 100 and over. This number had grown to 53,364 by 2010. (*Sources:* Population Projections Program; U.S. Census Bureau; U.S. Department of Commerce; "What People Who Live to 100 Have in Common," by Emily Brandon, *U.S. News and World Report*, January 7, 2013) The exponential function

$$H(t) = 80{,}040.68(1.0481)^t,$$

where t is the number of years after 2015, can be used to project the number of centenarians, in thousands. Use this function to project the centenarian population in 2020 and in 2050.

5.1 Inverse Functions

- Determine whether a function is one-to-one, and if it is, find a formula for its inverse.
- Simplify expressions of the type $(f \circ f^{-1})(x)$ and $(f^{-1} \circ f)(x)$.

Inverses

When we go from an output of a function back to its input or inputs, we get an inverse relation. When that relation is a function, we have an inverse function.

Consider the relation h given as follows:

$$h = \{(-8, 5), (4, -2), (-7, 1), (3.8, 6.2)\}.$$

RELATIONS
REVIEW SECTION 1.2

Suppose we *interchange* the first and second coordinates. The relation we obtain is called the **inverse** of the relation h and is given as follows:

$$\text{Inverse of } h = \{(5, -8), (-2, 4), (1, -7), (6.2, 3.8)\}.$$

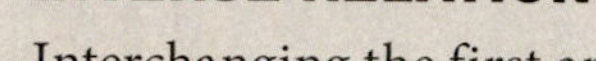

INVERSE RELATION

Interchanging the first and second coordinates of each ordered pair in a relation produces the **inverse relation**.

EXAMPLE 1 Consider the relation g given by

$$g = \{(2, 4), (-1, 3), (-2, 0)\}.$$

Graph the relation in blue. Find the inverse and graph it in red.

Solution The relation g is shown in blue in the figure at left. The inverse of the relation is

$$\{(4, 2), (3, -1), (0, -2)\}$$

and is shown in red. The pairs in the inverse are reflections of the pairs in g across the line $y = x$.

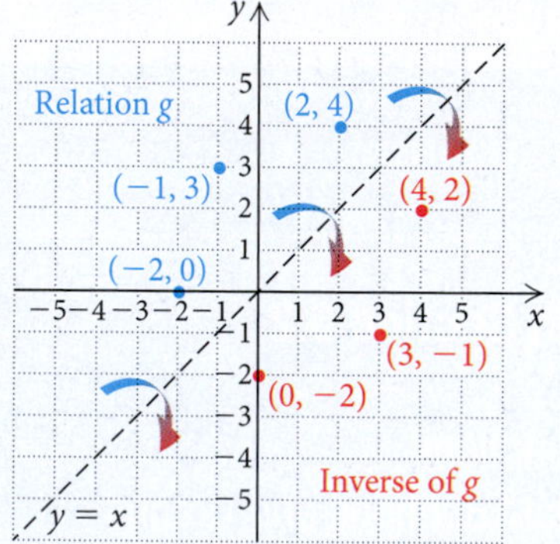

Now Try Exercise 1.

INVERSE RELATION

If a relation is defined by an equation, interchanging the variables produces an equation of the **inverse relation**.

EXAMPLE 2 Find an equation for the inverse of the relation

$$y = x^2 - 5x.$$

Solution We interchange x and y and obtain an equation of the inverse:

$$x = y^2 - 5y.$$

Now Try Exercise 9.

If a relation is given by an equation, then the solutions of the inverse can be found from those of the original equation by interchanging the first and second coordinates of each ordered pair. Thus the graphs of a relation and its inverse are always reflections of each other across the line $y = x$. This is illustrated with the equations of Example 2 in the tables and graph below. We will explore inverses and their graphs later in this section.

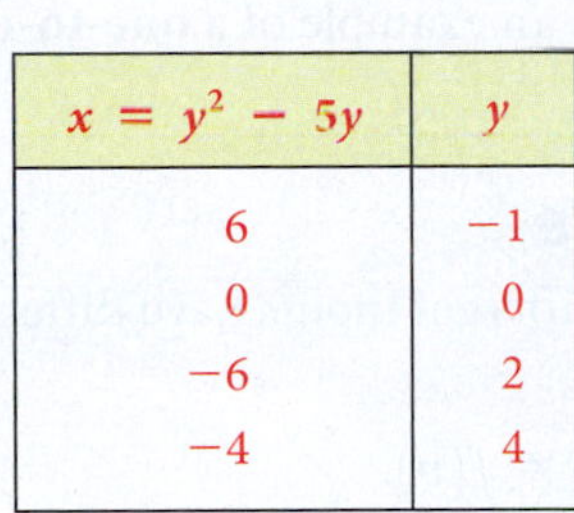

$x = y^2 - 5y$	y
6	−1
0	0
−6	2
−4	4

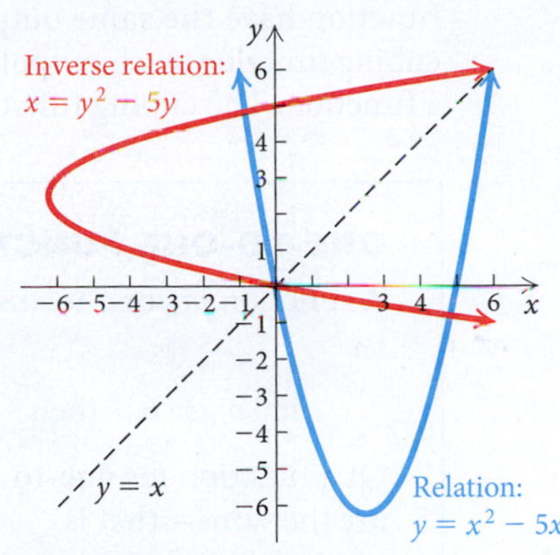

x	$y = x^2 - 5x$
−1	6
0	0
2	−6
4	−4

Inverses and One-to-One Functions

Let's consider the following two functions.

Year (domain)	First-Class Postage Cost, in cents (range)
2006 →	39
2007 →	41
2008 →	42
2009 →	44
2010 →	44
2011 →	44
2012 →	45
2013 →	46
2014 →	49

Source: U.S. Postal Service

Number (domain)	Cube (range)
−3 →	−27
−2 →	−8
−1 →	−1
0 →	0
1 →	1
2 →	8
3 →	27

Suppose we reverse the arrows. Are these inverse relations functions?

Year (range)	First-Class Postage Cost, in cents (domain)
2006	← 39
2007	← 41
2008	← 42
2009	← 44
2010	← 44
2011	← 44
2012	← 45
2013	← 46
2014	← 49

Source: U.S. Postal Service

Number (range)	Cube (domain)
−3	← −27
−2	← −8
−1	← −1
0	← 0
1	← 1
2	← 8
3	← 27

We see that the inverse of the postage function is not a function. Like all functions, each input in the postage function has exactly one output. However, the output for 2009, 2010, and 2011 is 44. Thus in the inverse of the postage function, the input 44 has three outputs, 2009, 2010, and 2011. When two or more inputs of a function have the same output, the inverse relation cannot be a function. In the cubing function, each output corresponds to exactly one input, so its inverse is also a function. The cubing function is an example of a **one-to-one function**.

ONE-TO-ONE FUNCTIONS

A function f is **one-to-one** if different inputs have different outputs—that is,

$$\text{if } a \neq b, \text{ then } f(a) \neq f(b).$$

Or a function f is **one-to-one** if when the outputs are the same, the inputs are the same—that is,

$$\text{if } f(a) = f(b), \text{ then } a = b.$$

If the inverse of a function f is also a function, it is named f^{-1} (read "f-inverse").

The -1 in f^{-1} is *not* an exponent!

Do *not* misinterpret the -1 in f^{-1} as a negative exponent: f^{-1} does *not* mean the reciprocal of f and $f^{-1}(x)$ is *not* equal to $\frac{1}{f(x)}$.

ONE-TO-ONE FUNCTIONS AND INVERSES

- If a function f is one-to-one, then its inverse f^{-1} is a function.
- The domain of a one-to-one function f is the range of the inverse f^{-1}.
- The range of a one-to-one function f is the domain of the inverse f^{-1}.
- A function that is increasing over its entire domain or is decreasing over its entire domain is a one-to-one function.

D_f $D_{f^{-1}}$
R_f $R_{f^{-1}}$

EXAMPLE 3 Given the function f described by $f(x) = 2x - 3$, prove that f is one-to-one (that is, it has an inverse that is a function).

Solution To show that f is one-to-one, we show that if $f(a) = f(b)$, then $a = b$. Assume that $f(a) = f(b)$ for a and b in the domain of f. Since $f(a) = 2a - 3$ and $f(b) = 2b - 3$, we have

$$2a - 3 = 2b - 3$$
$$2a = 2b \quad \text{Adding 3}$$
$$a = b. \quad \text{Dividing by 2}$$

Thus, if $f(a) = f(b)$, then $a = b$. This shows that f is one-to-one.

Now Try Exercise 17.

EXAMPLE 4 Given the function g described by $g(x) = x^2$, prove that g is not one-to-one.

Solution We can prove that g is not one-to-one by finding two numbers a and b for which $a \neq b$ and $g(a) = g(b)$. Two such numbers are -3 and 3, because $-3 \neq 3$ and $g(-3) = g(3) = 9$. Thus g is not one-to-one.

Now Try Exercise 21.

The following graphs show a function, in blue, and its inverse, in red. To determine whether the inverse is a function, we can apply the vertical-line test to its graph. By reflecting each such vertical line across the line $y = x$, we obtain an equivalent **horizontal-line test** for the original function.

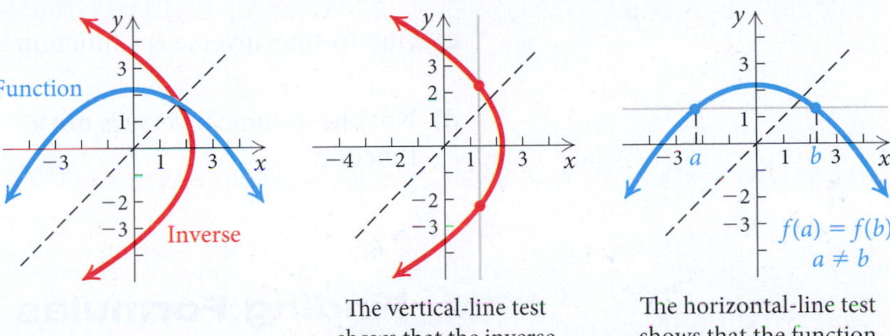

The vertical-line test shows that the inverse is not a function.

The horizontal-line test shows that the function is not one-to-one.

> **HORIZONTAL-LINE TEST**
>
> If it is possible for a horizontal line to intersect the graph of a function more than once, then the function is *not* one-to-one and its inverse is *not* a function.

EXAMPLE 5 From the graphs shown, determine whether each function is one-to-one and thus has an inverse that is a function.

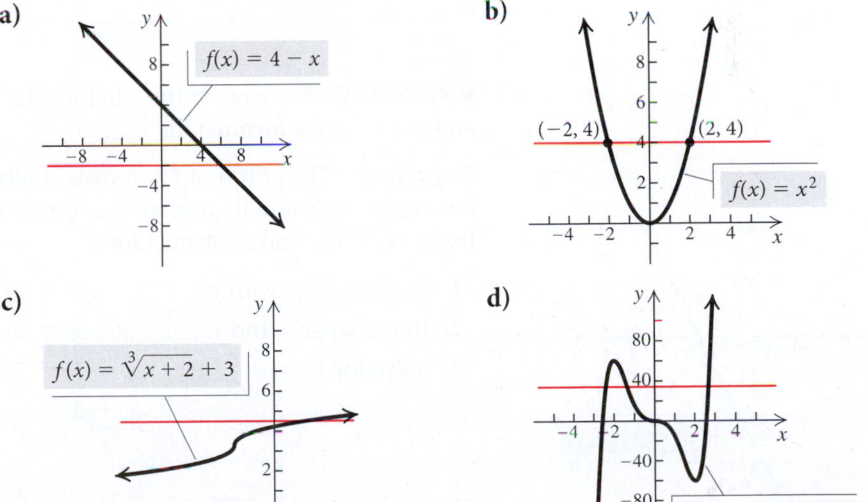

Solution For each function, we apply the horizontal-line test.

RESULT	REASON
a) One-to-one; inverse is a function	No horizontal line intersects the graph more than once.
b) Not one-to-one; inverse is not a function	There are many horizontal lines that intersect the graph more than once. Note that where the line $y = 4$ intersects the graph, the first coordinates are -2 and 2. Although these are different inputs, they have the same output, 4.
c) One-to-one; inverse is a function	No horizontal line intersects the graph more than once.
d) Not one-to-one; inverse is not a function	There are many horizontal lines that intersect the graph more than once.

Now Try Exercises 25 and 27.

Finding Formulas for Inverses

Suppose that a function is described by a formula. If it has an inverse that is a function, we proceed as follows to find a formula for f^{-1}.

> **Obtaining a Formula for an Inverse**
>
> If a function f is one-to-one, a formula for its inverse can generally be found as follows:
>
> 1. Replace $f(x)$ with y.
> 2. Interchange x and y.
> 3. Solve for y.
> 4. Replace y with $f^{-1}(x)$.

EXAMPLE 6 Determine whether the function $f(x) = 2x - 3$ is one-to-one, and if it is, find a formula for $f^{-1}(x)$.

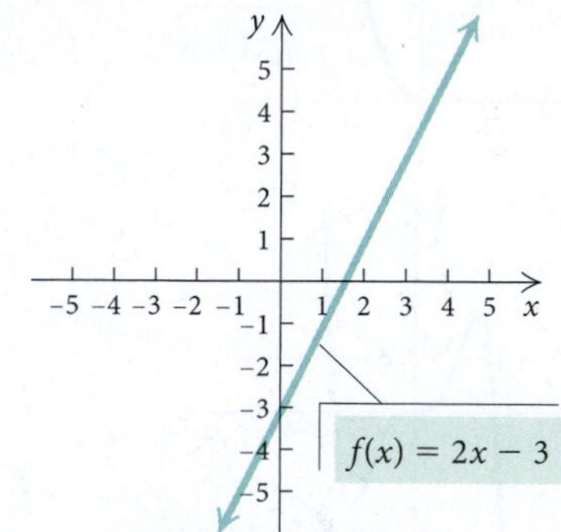

Solution The graph of f is shown at left. It passes the horizontal-line test. Thus it is one-to-one and its inverse is a function. We also proved that f is one-to-one in Example 3. We find a formula for $f^{-1}(x)$.

1. Replace $f(x)$ with y: $\quad y = 2x - 3$

2. Interchange x and y: $\quad x = 2y - 3$

3. Solve for y: $\quad x + 3 = 2y$

$$\frac{x + 3}{2} = y$$

4. Replace y with $f^{-1}(x)$: $\quad f^{-1}(x) = \dfrac{x + 3}{2}$.

Now Try Exercise 47.

Consider

$$f(x) = 2x - 3 \quad \text{and} \quad f^{-1}(x) = \frac{x + 3}{2}$$

from Example 6. For the input 5, we have

$$f(5) = 2 \cdot 5 - 3 = 10 - 3 = 7.$$

The output is 7. Now we use 7 for the input in the inverse:

$$f^{-1}(7) = \frac{7 + 3}{2} = \frac{10}{2} = 5.$$

The function f takes the number 5 to 7. The inverse function f^{-1} takes the number 7 back to 5.

EXAMPLE 7 Graph

$$f(x) = 2x - 3 \quad \text{and} \quad f^{-1}(x) = \frac{x + 3}{2}$$

using the same set of axes. Then compare the two graphs.

Solution The graphs of f and f^{-1} are shown at left. The solutions of the inverse function can be found from those of the original function by interchanging the first and second coordinates of each ordered pair.

$f^{-1}(x) = \frac{x+3}{2}$

$f(x) = 2x - 3$

$y = x$

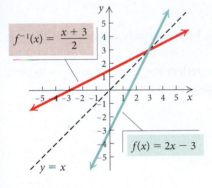

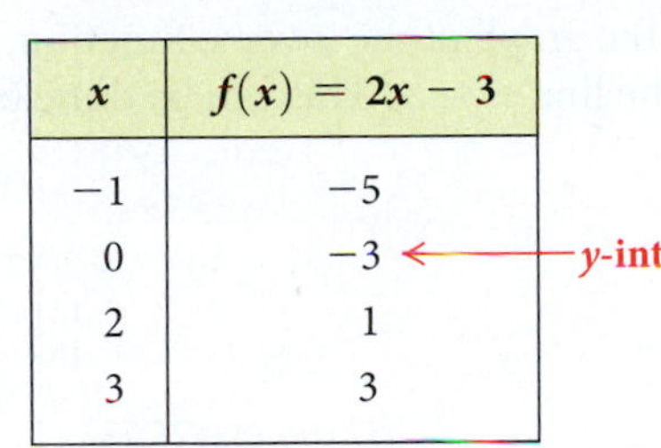

x	$f(x) = 2x - 3$
−1	−5
0	−3 ← *y*-intercept
2	1
3	3

x	$f^{-1}(x) = \frac{x + 3}{2}$
−5	−1
−3	0 ← *x*-intercept
1	2
3	3

When we interchange x and y in finding a formula for the inverse of $f(x) = 2x - 3$, we are in effect reflecting the graph of that function across the line $y = x$. For example, when the coordinates of the y-intercept, $(0, -3)$, of the graph of f are reversed, we get the x-intercept, $(-3, 0)$, of the graph of f^{-1}. If we were to graph $f(x) = 2x - 3$ in wet ink and fold along the line $y = x$, the graph of $f^{-1}(x) = (x + 3)/2$ would be formed by the ink transferred from f. ■

> The graph of f^{-1} is a reflection of the graph of f across the line $y = x$.

Technology Connection

On some graphing calculators, we can graph the inverse of a function after graphing the function itself by accessing a drawing feature. Consult your user's manual or the online *Graphing Calculator Manual* that accompanies this text for the procedure.

EXAMPLE 8 Consider $g(x) = x^3 + 2$.

a) Determine whether the function is one-to-one.

b) If it is one-to-one, find a formula for its inverse.

c) Graph the function and its inverse.

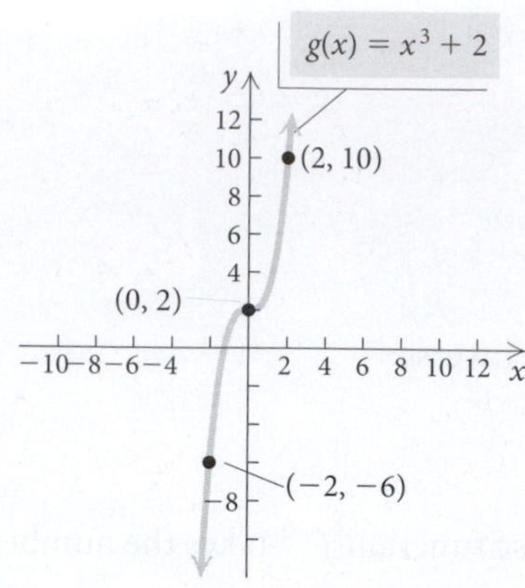

Solution

a) The graph of $g(x) = x^3 + 2$ is shown at left. It passes the horizontal-line test and thus has an inverse that is a function. We also know that $g(x)$ is one-to-one because it is an increasing function over its entire domain.

b) We follow the procedure for finding an inverse.

1. Replace $g(x)$ with y: $\quad y = x^3 + 2$
2. Interchange x and y: $\quad x = y^3 + 2$
3. Solve for y: $\quad x - 2 = y^3$
   $\sqrt[3]{x-2} = y$
4. Replace y with $g^{-1}(x)$: $\quad g^{-1}(x) = \sqrt[3]{x-2}$.

We can test a point as a partial check:

$$g(x) = x^3 + 2$$
$$g(3) = 3^3 + 2 = 27 + 2 = 29.$$

Will $g^{-1}(29) = 3$? We have

$$g^{-1}(x) = \sqrt[3]{x-2}$$
$$g^{-1}(29) = \sqrt[3]{29-2} = \sqrt[3]{27} = 3.$$

Since $g(3) = 29$ and $g^{-1}(29) = 3$, we can be reasonably certain that the formula for $g^{-1}(x)$ is correct.

c) To find the graph of the inverse function, we reflect the graph of $g(x) = x^3 + 2$ across the line $y = x$. This can be done by plotting points.

x	$g(x)$
−2	−6
−1	1
0	2
1	3
2	10

x	$g^{-1}(x)$
−6	−2
1	−1
2	0
3	1
10	2

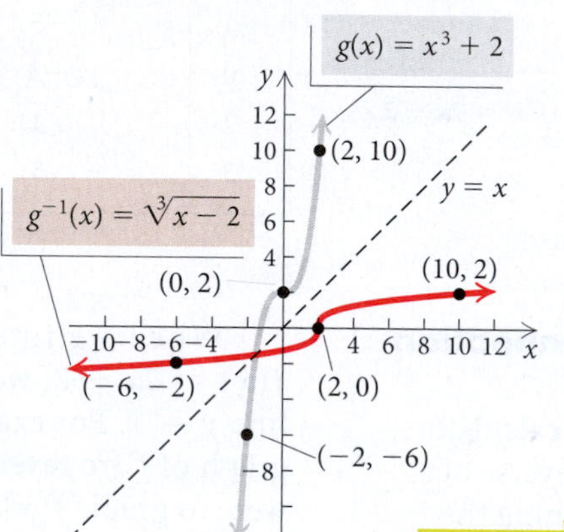

Now Try Exercise 69.

Inverse Functions and Composition

Suppose that we were to use some input a for a one-to-one function f and find its output, $f(a)$. The function f^{-1} would then take that output back to a. Similarly, if we began with an input b for the function f^{-1} and found its output, $f^{-1}(b)$, the original function f would then take that output back to b. This is summarized as follows.

If a function f is one-to-one, then f^{-1} is the unique function such that each of the following holds:

$(f^{-1} \circ f)(x) = f^{-1}(f(x)) = x$, for each x in the domain of f, and
$(f \circ f^{-1})(x) = f(f^{-1}(x)) = x$, for each x in the domain of f^{-1}.

EXAMPLE 9 Given that $f(x) = 5x + 8$, use composition of functions to show that

$$f^{-1}(x) = \frac{x - 8}{5}.$$

Solution We find $(f^{-1} \circ f)(x)$ and $(f \circ f^{-1})(x)$ and check to see that each is x:

$$(f^{-1} \circ f)(x) = f^{-1}(f(x))$$
$$= f^{-1}(5x + 8) = \frac{(5x + 8) - 8}{5} = \frac{5x}{5} = x;$$
$$(f \circ f^{-1})(x) = f(f^{-1}(x))$$
$$= f\left(\frac{x - 8}{5}\right) = 5\left(\frac{x - 8}{5}\right) + 8 = x - 8 + 8 = x.$$

Now Try Exercise 77.

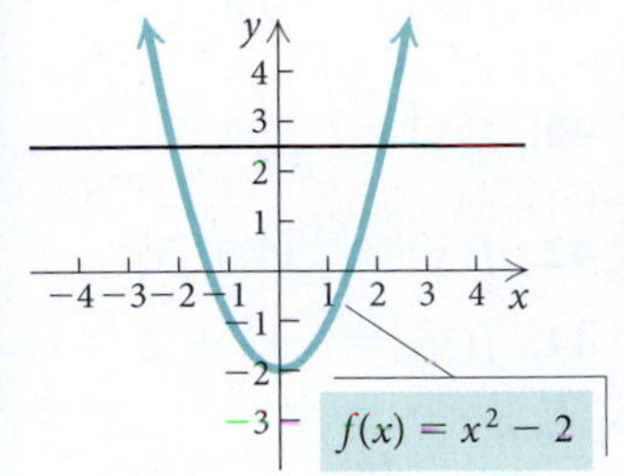

► Restricting a Domain

In the case in which the inverse of a function is not a function, the domain of the function can be restricted to allow the inverse to be a function. Let's consider the function $f(x) = x^2 - 2$. It is not one-to-one. The graph is shown at left.

Suppose that we had tried to find a formula for the inverse as follows:

$$y = x^2 - 2 \quad \text{Replacing } f(x) \text{ with } y$$
$$x = y^2 - 2 \quad \text{Interchanging } x \text{ and } y$$
$$x + 2 = y^2$$
$$\pm\sqrt{x + 2} = y. \quad \text{Solving for } y$$

This is not the equation of a function. An input of, say, 2 would yield two outputs, -2 and 2. In such cases, it is convenient to consider "part" of the function by restricting the domain of $f(x)$. For example, if we restrict the domain of $f(x) = x^2 - 2$ to nonnegative numbers, then its inverse is a function, as shown at left by the graphs of $f(x) = x^2 - 2, x \geq 0$, and $f^{-1}(x) = \sqrt{x + 2}$.

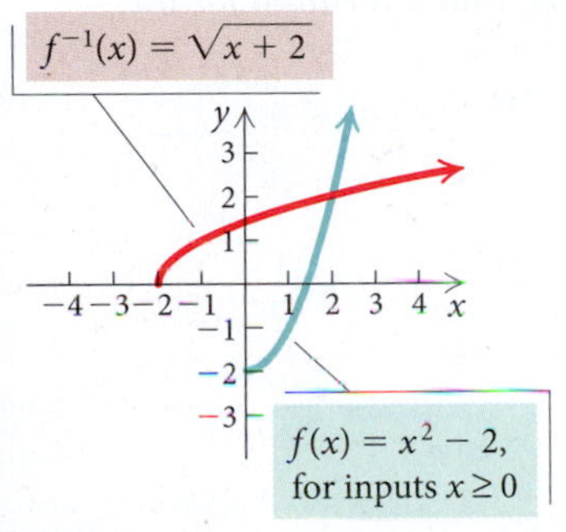

5.1 Exercise Set

Find the inverse of the relation.

1. $\{(7, 8), (-2, 8), (3, -4), (8, -8)\}$
2. $\{(0, 1), (5, 6), (-2, -4)\}$
3. $\{(-1, -1), (-3, 4)\}$
4. $\{(-1, 3), (2, 5), (-3, 5), (2, 0)\}$

Find an equation of the inverse relation.

5. $y = 4x - 5$
6. $2x^2 + 5y^2 = 4$
7. $x^3y = -5$
8. $y = 3x^2 - 5x + 9$

9. $x = y^2 - 2y$

10. $x = \frac{1}{2}y + 4$

Graph the equation by substituting and plotting points. Then reflect the graph across the line $y = x$ to obtain the graph of its inverse.

11. $x = y^2 - 3$ **12.** $y = x^2 + 1$

13. $y = 3x - 2$ **14.** $x = -y + 4$

15. $y = |x|$ **16.** $x + 2 = |y|$

Given the function f, prove that f is one-to-one using the definition of a one-to-one function on p. 314.

17. $f(x) = \frac{1}{3}x - 6$ **18.** $f(x) = 4 - 2x$

19. $f(x) = x^3 + \frac{1}{2}$ **20.** $f(x) = \sqrt[3]{x}$

Given the function g, prove that g is not one-to-one using the definition of a one-to-one function on p. 314.

21. $g(x) = 1 - x^2$ **22.** $g(x) = 3x^2 + 1$

23. $g(x) = x^4 - x^2$ **24.** $g(x) = \dfrac{1}{x^6}$

Using the horizontal-line test, determine whether the function is one-to-one.

25. $f(x) = 2.7^x$

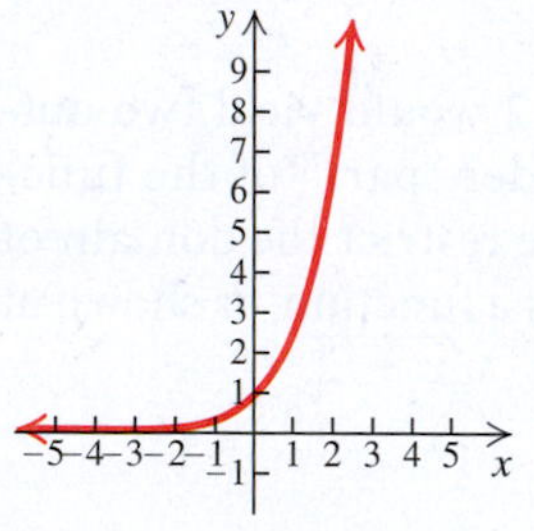

26. $f(x) = 2^{-x}$

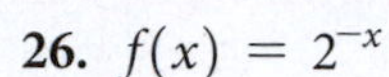

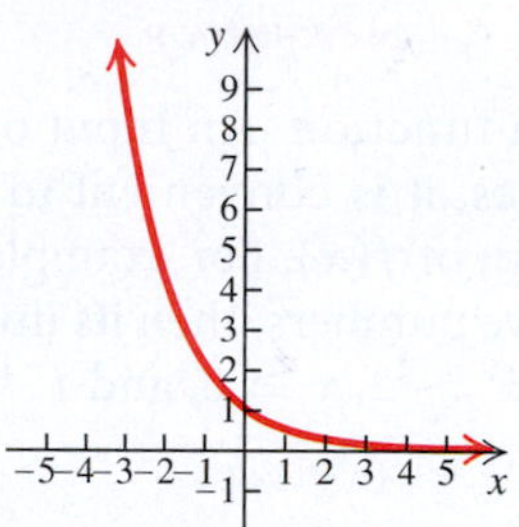

27. $f(x) = 4 - x^2$

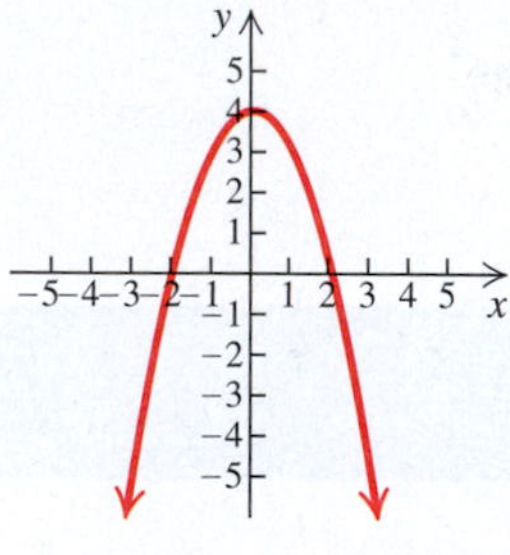

28. $f(x) = x^3 - 3x + 1$

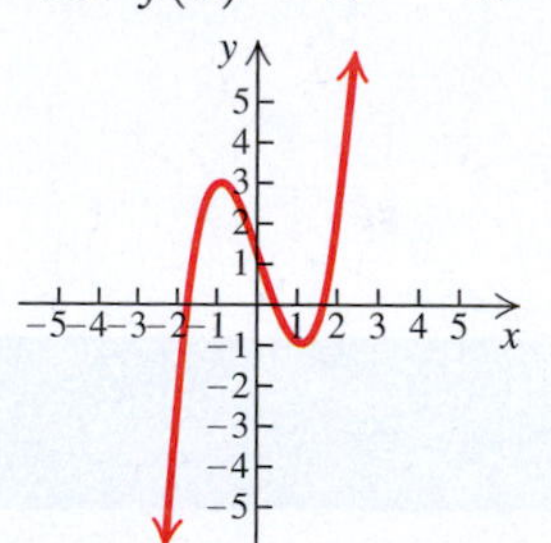

29. $f(x) = \dfrac{8}{x^2 - 4}$

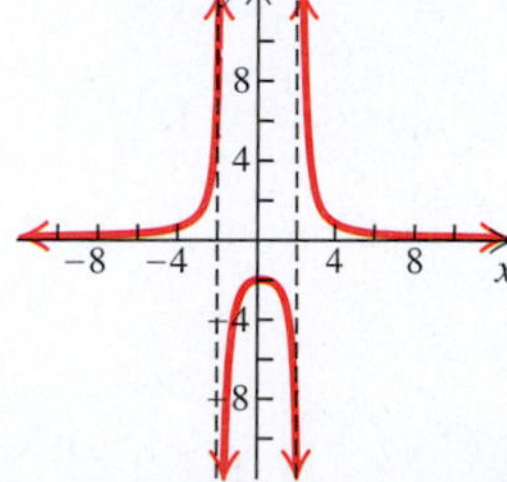

30. $f(x) = \sqrt{\dfrac{10}{4 + x}}$

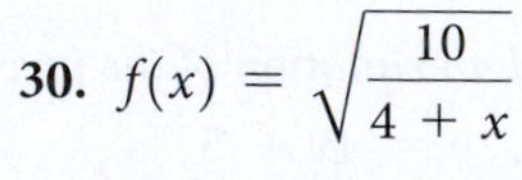

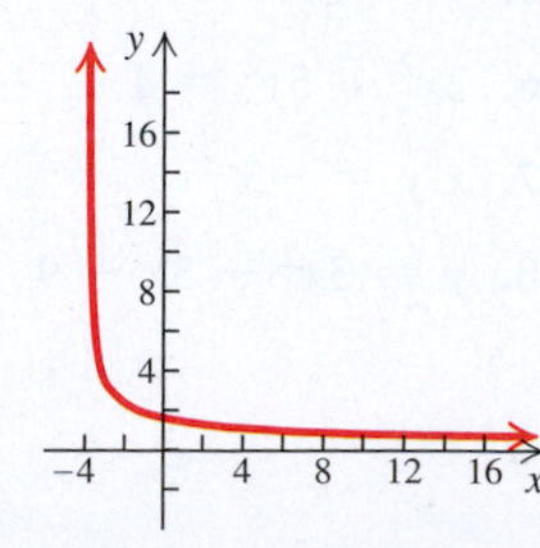

31. $f(x) = \sqrt[3]{x + 2} - 2$

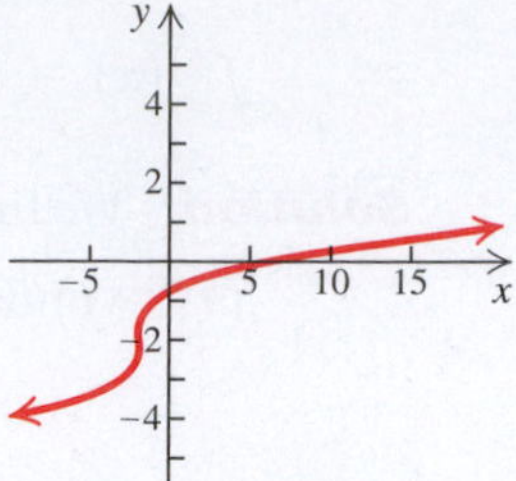

32. $f(x) = \dfrac{8}{x}$

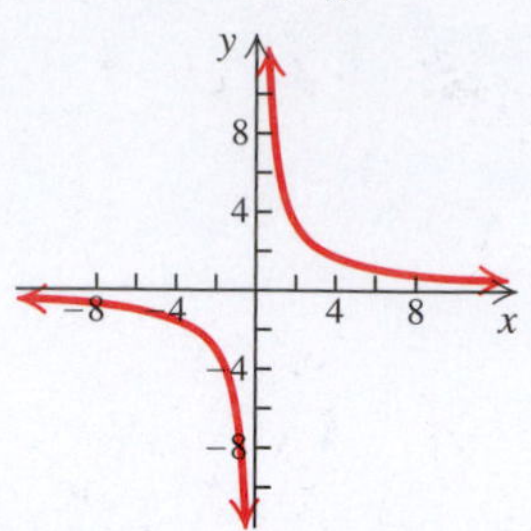

Graph the function and determine whether the function is one-to-one using the horizontal-line test.

33. $f(x) = 5x - 8$ **34.** $f(x) = 3 + 4x$

35. $f(x) = 1 - x^2$ **36.** $f(x) = |x| - 2$

37. $f(x) = |x + 2|$ **38.** $f(x) = -0.8$

39. $f(x) = -\dfrac{4}{x}$ **40.** $f(x) = \dfrac{2}{x + 3}$

41. $f(x) = \frac{2}{3}$ **42.** $f(x) = \frac{1}{2}x^2 + 3$

43. $f(x) = \sqrt{25 - x^2}$ **44.** $f(x) = -x^3 + 2$

In Exercises 45–60, for each function:

a) *Determine whether it is one-to-one.*

b) *If the function is one-to-one, find a formula for the inverse.*

45. $f(x) = x + 4$

46. $f(x) = 7 - x$

47. $f(x) = 2x - 1$

48. $f(x) = 5x + 8$

49. $f(x) = \dfrac{4}{x + 7}$

50. $f(x) = -\dfrac{3}{x}$

51. $f(x) = \dfrac{x + 4}{x - 3}$

52. $f(x) = \dfrac{5x - 3}{2x + 1}$

53. $f(x) = x^3 - 1$

54. $f(x) = (x + 5)^3$

55. $f(x) = x\sqrt{4 - x^2}$

56. $f(x) = 2x^2 - x - 1$

57. $f(x) = 5x^2 - 2, \ x \geq 0$

58. $f(x) = 4x^2 + 3, \ x \geq 0$

59. $f(x) = \sqrt{x + 1}$

60. $f(x) = \sqrt[3]{x - 8}$

Find the inverse by thinking about the operations of the function and then reversing, or undoing, them. Check your work algebraically.

FUNCTION	INVERSE
61. $f(x) = 3x$	$f^{-1}(x) =$
62. $f(x) = \frac{1}{4}x + 7$	$f^{-1}(x) =$
63. $f(x) = -x$	$f^{-1}(x) =$
64. $f(x) = \sqrt[3]{x} - 5$	$f^{-1}(x) =$
65. $f(x) = \sqrt[3]{x - 5}$	$f^{-1}(x) =$
66. $f(x) = x^{-1}$	$f^{-1}(x) =$

Each graph in Exercises 67–72 is the graph of a one-to-one function f. Sketch the graph of the inverse function f^{-1}.

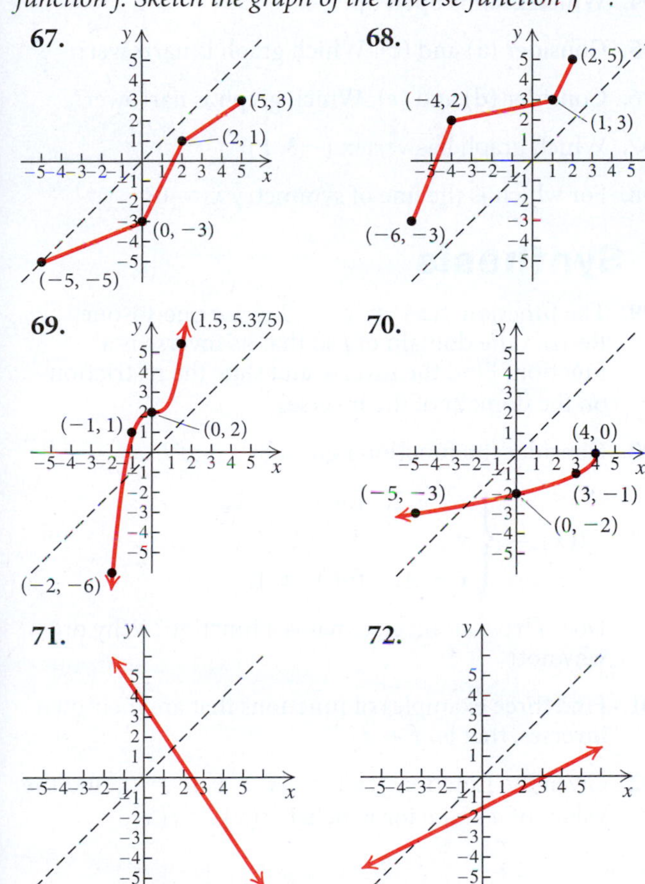

For the function f, use composition of functions to show that f^{-1} is as given.

73. $f(x) = \frac{7}{8}x,\ f^{-1}(x) = \frac{8}{7}x$

74. $f(x) = \dfrac{x + 5}{4},\ f^{-1}(x) = 4x - 5$

75. $f(x) = \dfrac{1 - x}{x},\ f^{-1}(x) = \dfrac{1}{x + 1}$

76. $f(x) = \sqrt[3]{x + 4},\ f^{-1}(x) = x^3 - 4$

77. $f(x) = \dfrac{2}{5}x + 1,\ f^{-1}(x) = \dfrac{5x - 5}{2}$

78. $f(x) = \dfrac{x + 6}{3x - 4},\ f^{-1}(x) = \dfrac{4x + 6}{3x - 1}$

Find the inverse of the given one-to-one function f. Give the domain and the range of f and of f^{-1}, and then graph both f and f^{-1} on the same set of axes.

79. $f(x) = 5x - 3$

80. $f(x) = 2 - x$

81. $f(x) = \dfrac{2}{x}$

82. $f(x) = -\dfrac{3}{x + 1}$

83. $f(x) = \dfrac{1}{3}x^3 - 2$

84. $f(x) = \sqrt[3]{x} - 1$

85. $f(x) = \dfrac{x + 1}{x - 3}$

86. $f(x) = \dfrac{x - 1}{x + 2}$

87. Find $f(f^{-1}(5))$ and $f^{-1}(f(a))$:

$$f(x) = x^3 - 4.$$

88. Find $(f^{-1}(f(p))$ and $f(f^{-1}(1253))$:

$$f(x) = \sqrt[5]{\frac{2x - 7}{3x + 4}}.$$

89. *Hitting Lessons.* A summer little-league baseball team determines that the cost per player of a group hitting lesson is given by the formula

$$C(x) = \frac{72 + 2x}{x},$$

where x is the number of players in the group and $C(x)$ is in dollars.

a) Determine the cost per player of a group hitting lesson when there are 2, 5, and 8 players in the group.

b) Find a formula for the inverse of the function and explain what it represents.

c) Use the inverse function to determine the number of players in the group lesson when the cost per player is \$74, \$20, and \$11.

90. *Women's Shoe Sizes.* A function that will convert women's shoe sizes in the United States to those in Australia is

$$s(x) = \frac{2x - 3}{2}$$

(*Source:* OnlineConversion.com).

a) Determine the women's shoe sizes in Australia that correspond to sizes 5, $7\frac{1}{2}$, and 8 in the United States.

b) Find a formula for the inverse of the function and explain what it represents.

c) Use the inverse function to determine the women's shoe sizes in the United States that correspond to sizes 3, $5\frac{1}{2}$, and 7 in Australia.

91. *E-Commerce Holiday Sales.* Retail e-commerce holiday season sales (November and December), in billions of dollars, x years after 2008 is given by the function

$$H(x) = 6.58x + 27.7$$

(*Source:* statista.com).

a) Determine the total amount of e-commerce holiday sales in 2010 and in 2013.

b) Find $H^{-1}(x)$ and explain what it represents.

92. *Converting Temperatures.* The following formula can be used to convert Fahrenheit temperatures x to Celsius temperatures $T(x)$:

$$T(x) = \frac{5}{9}(x - 32).$$

a) Find $T(-13°)$ and $T(86°)$.

b) Find $T^{-1}(x)$ and explain what it represents.

Skill Maintenance

Consider the quadratic functions (a)–(h) that follow. Without graphing them, answer the questions below. [3.3]

a) $f(x) = 2x^2$

b) $f(x) = -x^2$

c) $f(x) = \frac{1}{4}x^2$

d) $f(x) = -5x^2 + 3$

e) $f(x) = \frac{2}{3}(x - 1)^2 - 3$

f) $f(x) = -2(x + 3)^2 + 1$

g) $f(x) = (x - 3)^2 + 1$

h) $f(x) = -4(x + 1)^2 - 3$

93. Which functions have a maximum value?

94. Which graphs open up?

95. Consider (a) and (c). Which graph is narrower?

96. Consider (d) and (e). Which graph is narrower?

97. Which graph has vertex $(-3, 1)$?

98. For which is the line of symmetry $x = 0$?

Synthesis

99. The function $f(x) = x^2 - 3$ is not one-to-one. Restrict the domain of f so that its inverse is a function. Find the inverse and state the restriction on the domain of the inverse.

100. Consider the function f given by

$$f(x) = \begin{cases} x^3 + 2, & \text{for } x \leq -1, \\ x^2, & \text{for } -1 < x < 1, \\ x + 1, & \text{for } x \geq 1. \end{cases}$$

Does f have an inverse that is a function? Why or why not?

101. Find three examples of functions that are their own inverses; that is, $f = f^{-1}$.

102. Given the function $f(x) = ax + b$, $a \neq 0$, find the values of a and b for which $f^{-1}(x) = f(x)$.

5.2 Exponential Functions and Graphs

- Graph exponential equations and exponential functions.
- Solve applied problems involving exponential functions and their graphs.

We now turn our attention to the study of a set of functions that are very rich in application. Consider the following graphs. Each one illustrates an *exponential function*. In this section, we consider such functions and some important applications.

Skype Users Online at Same Time

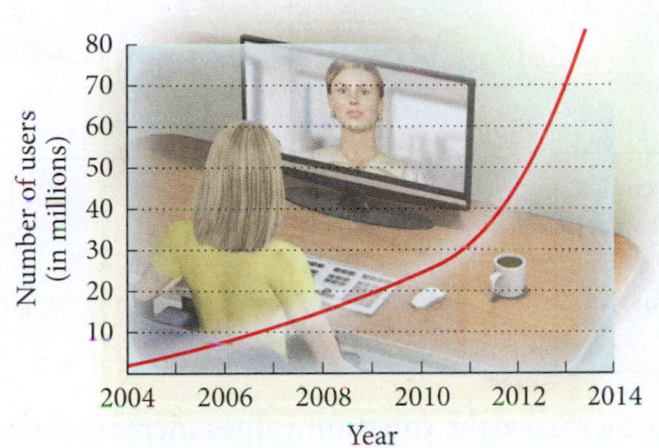

Source: Skype Numerology Blog

Postseason Bowl Games

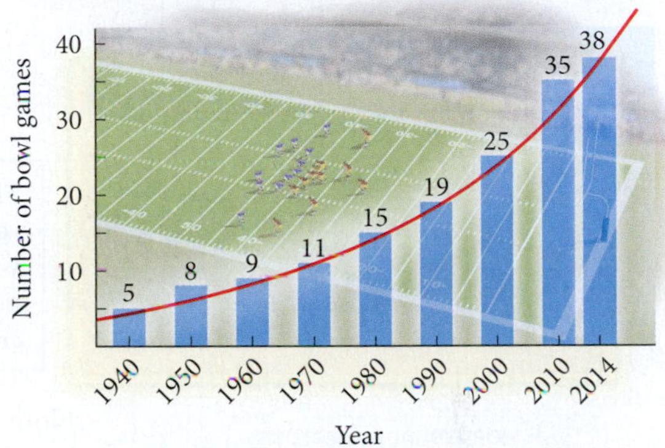

Source: USA TODAY research, College Football Data Warehouse

Graphing Exponential Functions

We now define exponential functions. We assume that a^x has meaning for any real number x and any positive real number a and that the laws of exponents still hold, though we will not prove them here.

> **EXPONENTIAL FUNCTION**
>
> The function $f(x) = a^x$, where x is a real number, $a > 0$ and $a \neq 1$, is called the **exponential function, base *a*.**

We require the **base** to be positive in order to avoid the imaginary numbers that would occur by taking even roots of negative numbers—an example is $(-1)^{1/2}$, the square root of -1, which is not a real number. The restriction $a \neq 1$ is made to exclude the constant function $f(x) = 1^x = 1$, which does not have an inverse that is a function because it is not one-to-one.

The following are examples of exponential functions:

$$f(x) = 2^x, \qquad f(x) = \left(\frac{1}{2}\right)^x, \qquad f(x) = (3.57)^x.$$

Note that, in contrast to functions like $f(x) = x^5$ and $f(x) = x^{1/2}$ in which the variable is the base of an exponential expression, the variable in an exponential function is *in the exponent*.

Let's now consider graphs of exponential functions.

JUST IN TIME 7

EXAMPLE 1 Graph the exponential function $y = f(x) = 2^x$.

Solution We compute some function values and list the results in a table.

$$f(0) = 2^0 = 1; \quad f(-1) = 2^{-1} = \frac{1}{2^1} = \frac{1}{2};$$
$$f(1) = 2^1 = 2;$$
$$f(2) = 2^2 = 4; \quad f(-2) = 2^{-2} = \frac{1}{2^2} = \frac{1}{4};$$
$$f(3) = 2^3 = 8;$$
$$f(-3) = 2^{-3} = \frac{1}{2^3} = \frac{1}{8}.$$

x	y $y = f(x) = 2^x$	(x, y)
0	1	$(0, 1)$
1	2	$(1, 2)$
2	4	$(2, 4)$
3	8	$(3, 8)$
-1	$\frac{1}{2}$	$\left(-1, \frac{1}{2}\right)$
-2	$\frac{1}{4}$	$\left(-2, \frac{1}{4}\right)$
-3	$\frac{1}{8}$	$\left(-3, \frac{1}{8}\right)$

Next, we plot these points and connect them with a smooth curve. Be sure to plot enough points to determine how steeply the curve rises.

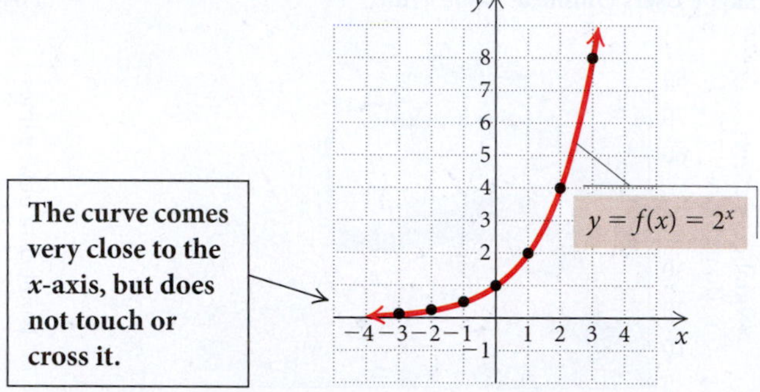

HORIZONTAL ASYMPTOTES

REVIEW SECTION 4.5

Note that as x increases, the function values increase without bound. As x decreases, the function values decrease, getting close to 0. That is, as $x \rightarrow -\infty$, $y \rightarrow 0$. Thus the x-axis, or the line $y = 0$, is a horizontal asymptote. As the x-inputs decrease, the curve gets closer and closer to this line, but does not cross it.

Now Try Exercise 11.

EXAMPLE 2 Graph the exponential function $y = f(x) = \left(\frac{1}{2}\right)^x$.

Solution Before we plot points and draw the curve, note that

$$y = f(x) = \left(\frac{1}{2}\right)^x = (2^{-1})^x = 2^{-x}.$$

This tells us that this graph is a reflection of the graph of $y = 2^x$ across the y-axis. For example, if $(3, 8)$ is a point of the graph of $g(x) = 2^x$, then $(-3, 8)$ is a point of the graph of $f(x) = 2^{-x}$. Selected points are listed in the table at left.

Points of $g(x) = 2^x$	Points of $f(x) = \left(\frac{1}{2}\right)^x = 2^{-x}$
$(0, 1)$	$(0, 1)$
$(1, 2)$	$(-1, 2)$
$(2, 4)$	$(-2, 4)$
$(3, 8)$	$(-3, 8)$
$\left(-1, \frac{1}{2}\right)$	$\left(1, \frac{1}{2}\right)$
$\left(-2, \frac{1}{4}\right)$	$\left(2, \frac{1}{4}\right)$
$\left(-3, \frac{1}{8}\right)$	$\left(3, \frac{1}{8}\right)$

Next, we plot these points and connect them with a smooth curve.

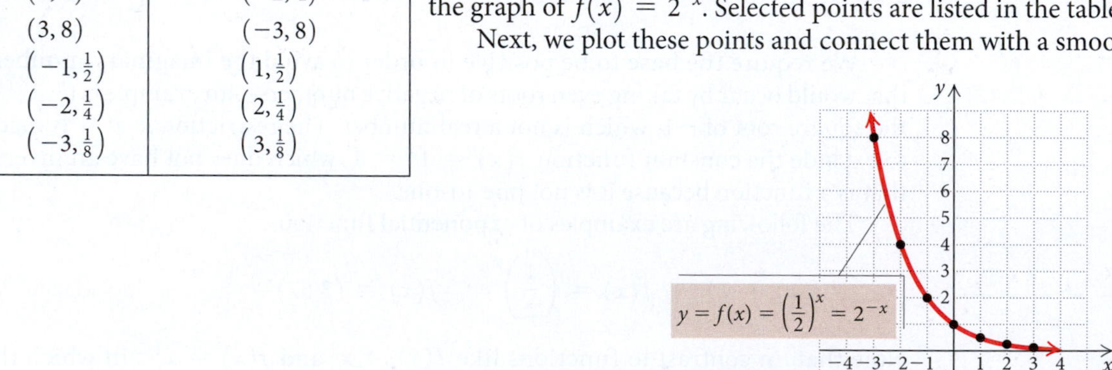

Note that as x increases, the function values decrease, getting close to 0. The x-axis, $y = 0$, is the horizontal asymptote. As x decreases, the function values increase without bound.

Now Try Exercise 15.

CONNECTING THE CONCEPTS

Properties of Exponential Functions

Let's list and compare some characteristics of exponential functions, keeping in mind that the definition of an exponential function, $f(x) = a^x$, requires that a be positive and different from 1.

$f(x) = a^x, a > 0, a \neq 1$
Continuous
One-to-one
Domain: $(-\infty, \infty)$
Range: $(0, \infty)$
Increasing if $a > 1$
Decreasing if $0 < a < 1$
Horizontal asymptote is x-axis
y-intercept: $(0, 1)$

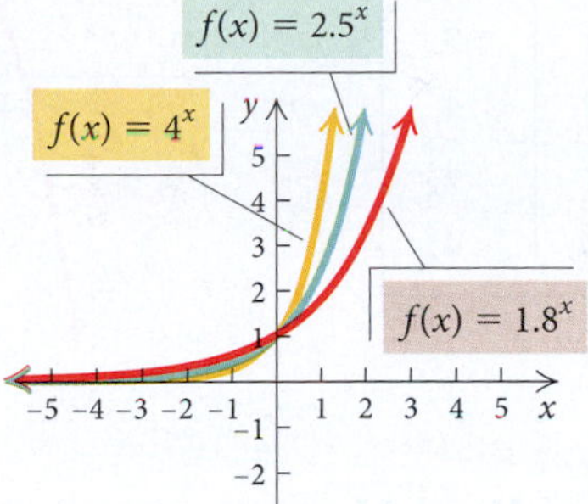

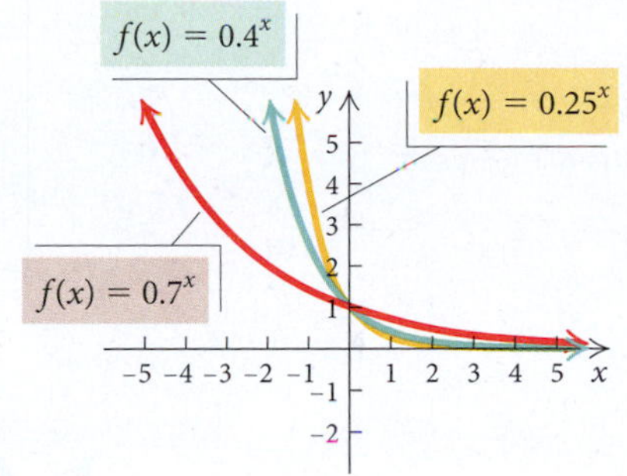

TRANSFORMATIONS
REVIEW SECTION 2.5

To graph other types of exponential functions, keep in mind the ideas of translation, stretching, and reflection. All these concepts allow us to visualize the graph before drawing it.

EXAMPLE 3 Graph each of the following. Before doing so, describe how each graph can be obtained from the graph of $f(x) = 2^x$.

a) $f(x) = 2^{x-2}$ **b)** $f(x) = 2^x - 4$ **c)** $f(x) = 5 - 0.5^x$

Solution

a) The graph of $f(x) = 2^{x-2}$ is the graph of $y = 2^x$ shifted *right* 2 units.

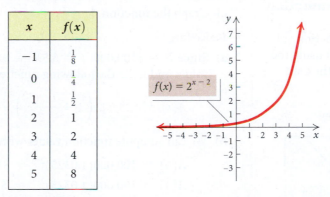

x	$f(x)$
-1	$\frac{1}{8}$
0	$\frac{1}{4}$
1	$\frac{1}{2}$
2	1
3	2
4	4
5	8

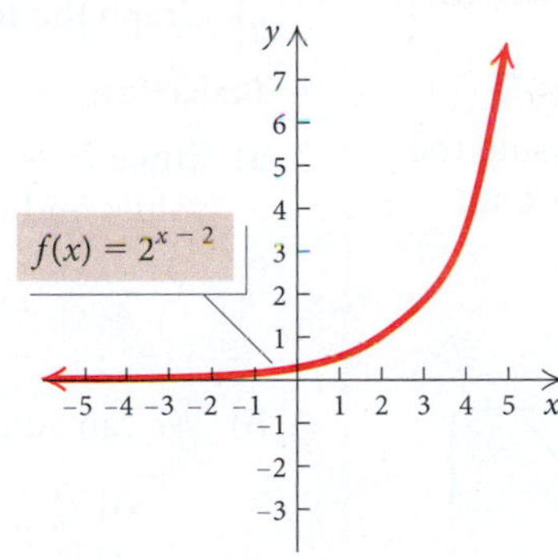

b) The graph of $f(x) = 2^x - 4$ is the graph of $y = 2^x$ shifted *down* 4 units.

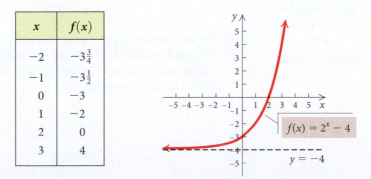

x	$f(x)$
-2	$-3\frac{3}{4}$
-1	$-3\frac{1}{2}$
0	-3
1	-2
2	0
3	4

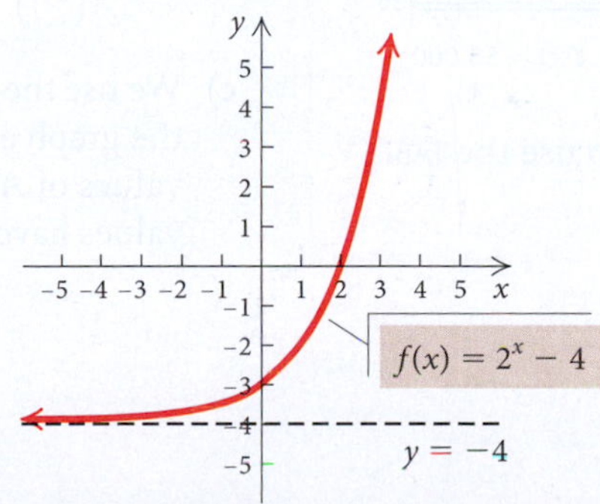

c) The graph of $f(x) = 5 - 0.5^x = 5 - \left(\frac{1}{2}\right)^x = 5 - 2^{-x}$ is a reflection of the graph of $y = 2^x$ across the y-axis, followed by a reflection across the x-axis and then a shift *up* 5 units.

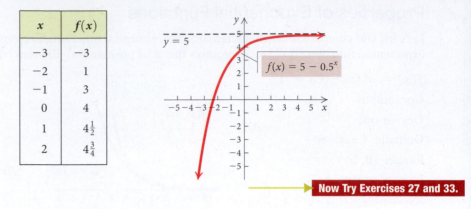

x	$f(x)$
−3	−3
−2	1
−1	3
0	4
1	$4\frac{1}{2}$
2	$4\frac{3}{4}$

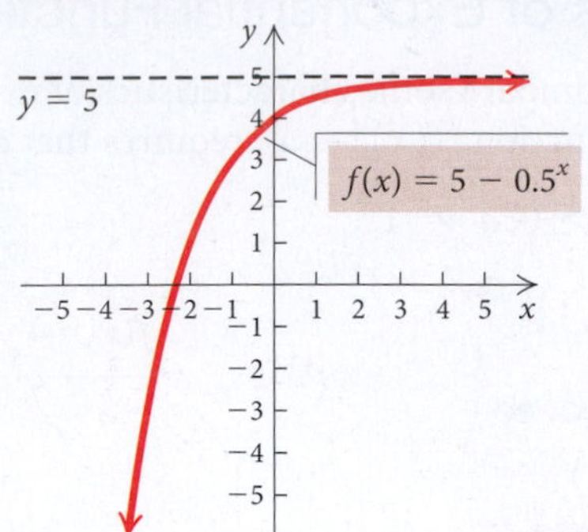

Now Try Exercises 27 and 33.

Applications

One of the most frequent applications of exponential functions occurs with compound interest.

EXAMPLE 4 ***Compound Interest.*** The amount of money A to which a principal P will grow after t years at interest rate r (in decimal form), compounded n times per year, is given by the formula

$$A = P\left(1 + \frac{r}{n}\right)^{nt}.$$

Suppose that \$100,000 is invested at 6.5% interest, compounded semiannually.

a) Find a function for the amount to which the investment grows after t years.

b) Find the amount of money in the account at $t = 0, 4, 8$, and 10 years.

c) Graph the function.

> **Technology Connection**
>
> We can find the function values in Example 4(b) using the VALUE feature from the CALC menu.
>
> $y = 100{,}000(1.0325)^{2x}$
>
> 500,000
>
> 0 0 30
>
> Xscl = 5, Yscl = 50,000
>
> We could also use the TABLE feature.

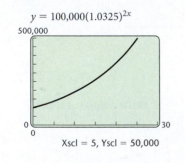

Solution

a) Since $P = \$100{,}000$, $r = 6.5\% = 0.065$, and $n = 2$, we can substitute these values and write the following function:

$$A(t) = 100{,}000\left(1 + \frac{0.065}{2}\right)^{2 \cdot t} = \$100{,}000(1.0325)^{2t}.$$

b) We can compute function values with a calculator:

$$A(0) = 100{,}000(1.0325)^{2 \cdot 0} = \$100{,}000;$$
$$A(4) = 100{,}000(1.0325)^{2 \cdot 4} \approx \$129{,}157.75;$$
$$A(8) = 100{,}000(1.0325)^{2 \cdot 8} \approx \$166{,}817.25;$$
$$A(10) = 100{,}000(1.0325)^{2 \cdot 10} \approx \$189{,}583.79.$$

c) We use the function values computed in part (b) and others if we wish, and draw the graph as follows. Note that the axes are scaled differently because of the large values of A and that t is restricted to nonnegative values, because negative time values have no meaning here.

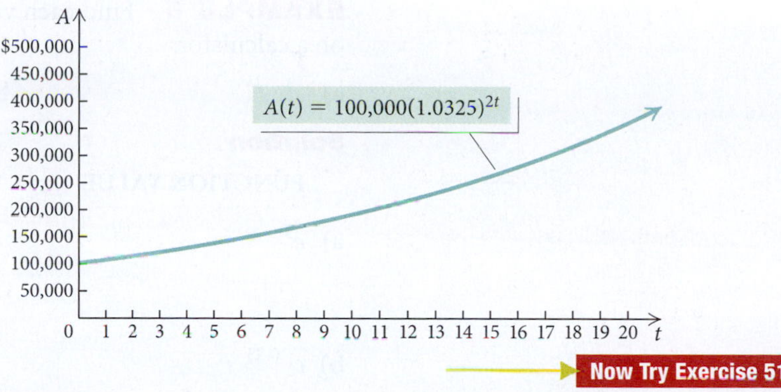

Now Try Exercise 51.

The Number e

We now consider a very special number in mathematics. In 1741, Leonhard Euler named this number e. Though you may not have encountered it before, you will see here and in future mathematics courses that it has many important applications. To explain this number, we use the compound interest formula $A = P(1 + r/n)^{nt}$ discussed in Example 4. Suppose that \$1 is invested at 100% interest for 1 year. Since $P = 1$, $r = 100\% = 1$, and $t = 1$, the formula above becomes a function A defined in terms of the number of compounding periods n:

$$A = P\left(1 + \frac{r}{n}\right)^{nt} = 1\left(1 + \frac{1}{n}\right)^{n \cdot 1} = \left(1 + \frac{1}{n}\right)^{n}.$$

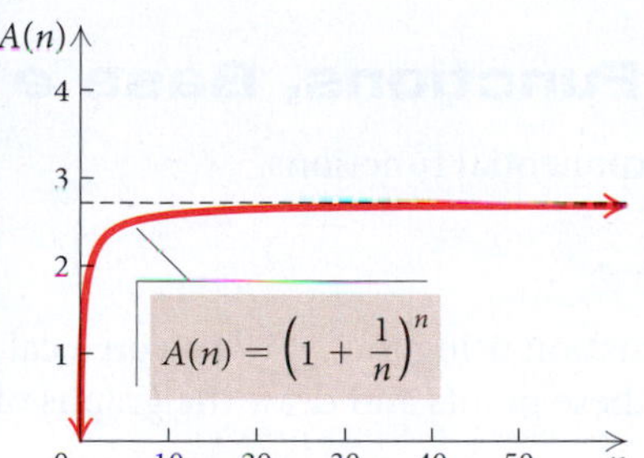

Let's visualize this function using its graph, shown at left, and explore the values of $A(n)$ as $n \to \infty$. Consider the graph for larger and larger values of n. Does this function have a horizontal asymptote?

Let's find some function values using a calculator.

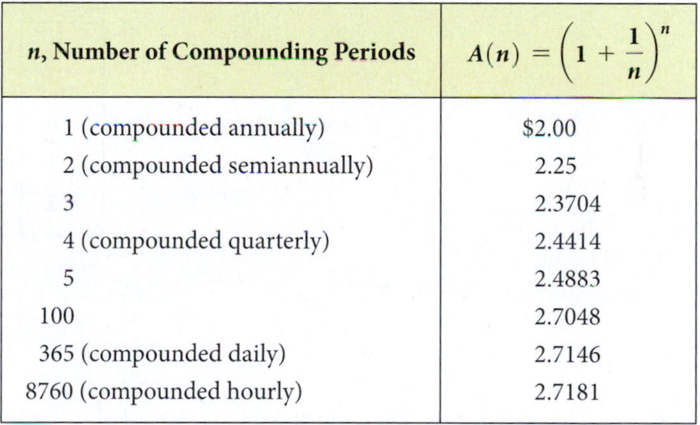

n, Number of Compounding Periods	$A(n) = \left(1 + \frac{1}{n}\right)^n$
1 (compounded annually)	\$2.00
2 (compounded semiannually)	2.25
3	2.3704
4 (compounded quarterly)	2.4414
5	2.4883
100	2.7048
365 (compounded daily)	2.7146
8760 (compounded hourly)	2.7181

It appears from these values that the graph does have a horizontal asymptote, $y \approx 2.7$. As the values of n get larger and larger, the function values get closer and closer to the number Euler named e. Its decimal representation does not terminate or repeat; it is irrational.

$$e = 2.7182818284\ldots$$

EXAMPLE 5 Find each value of e^x, to four decimal places, using the e^x key on a calculator.

a) e^3 b) $e^{-0.23}$ c) e^0

Solution

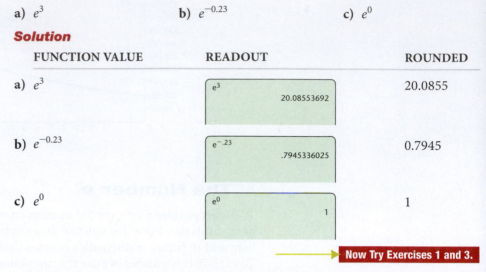

FUNCTION VALUE	READOUT	ROUNDED
a) e^3	e3 20.08553692	20.0855
b) $e^{-0.23}$	e-.23 .7945336025	0.7945
c) e^0	e0 1	1

Now Try Exercises 1 and 3.

Graphs of Exponential Functions, Base *e*

We now demonstrate ways in which to graph exponential functions.

EXAMPLE 6 Graph $f(x) = e^x$ and $g(x) = e^{-x}$.

Solution We can compute points for each equation using the e^x key on a calculator. (See the following table.) Then we plot these points and draw the graphs of the functions.

x	$f(x) = e^x$	$g(x) = e^{-x}$
−2	0.135	7.389
−1	0.368	2.718
0	1	1
1	2.718	0.368
2	7.389	0.135

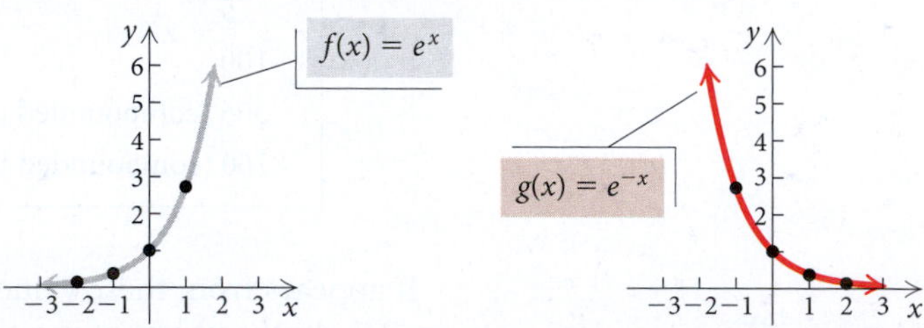

Note that the graphs are reflections of each other across the y-axis.

Now Try Exercise 23.

TRANSFORMATIONS

REVIEW SECTION 2.5

EXAMPLE 7 Graph each of the following. Before doing so, describe how each graph can be obtained from the graph of $y = e^x$.

a) $f(x) = e^{x+3}$ b) $f(x) = e^{-0.5x}$ c) $f(x) = 1 - e^{-2x}$

Solution

a) The graph of $f(x) = e^{x+3}$ is a translation of the graph of $y = e^x$ left 3 units.

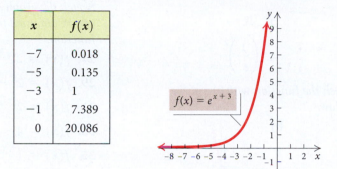

x	$f(x)$
−7	0.018
−5	0.135
−3	1
−1	7.389
0	20.086

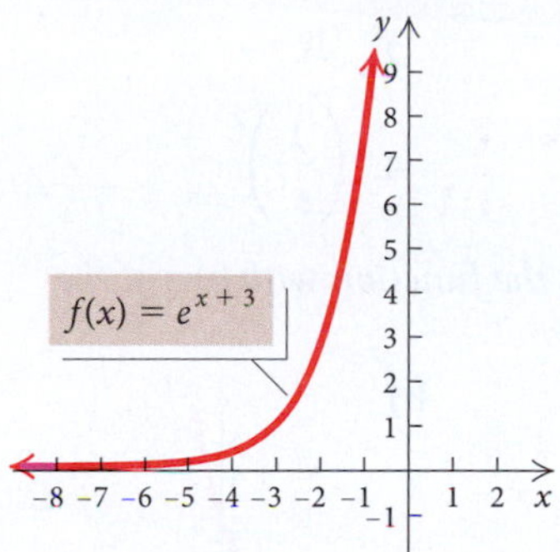

b) We note that the graph of $f(x) = e^{-0.5x}$ is a horizontal stretching of the graph of $y = e^x$ followed by a reflection across the y-axis.

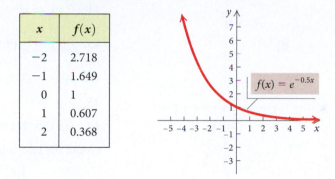

x	$f(x)$
−2	2.718
−1	1.649
0	1
1	0.607
2	0.368

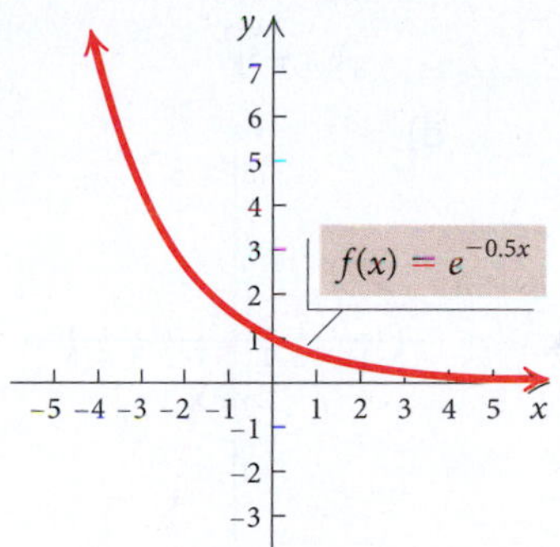

c) The graph of $f(x) = 1 - e^{-2x}$ is a horizontal shrinking of the graph of $y = e^x$, followed by a reflection across the y-axis, then across the x-axis, and followed by a translation up 1 unit.

x	$f(x)$
−1	−6.389
0	0
1	0.865
2	0.982
3	0.998

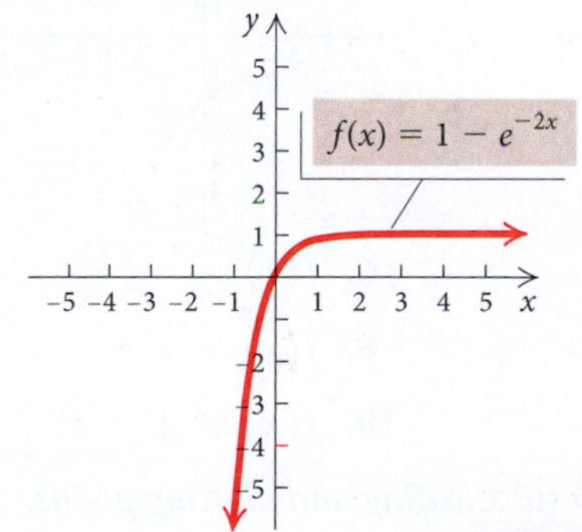

Now Try Exercises 41 and 47.

5.2 Exercise Set

Find each of the following, to four decimal places, using a calculator.

1. e^4
2. e^{10}
3. $e^{-2.458}$
4. $\left(\dfrac{1}{e^3}\right)^2$

In Exercises 5–10, match the function with one of the graphs (a)–(f) that follow.

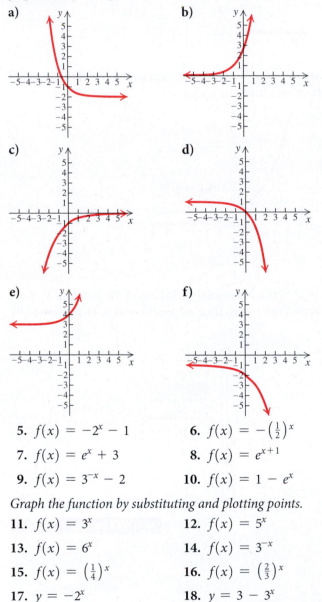

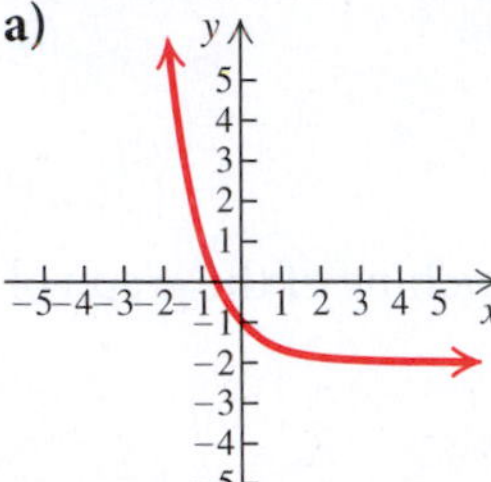

b)

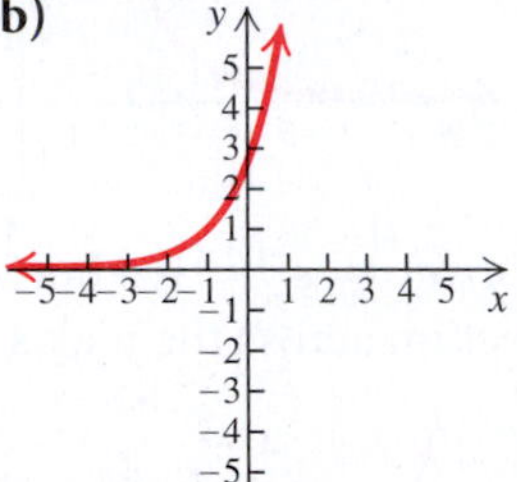

c)

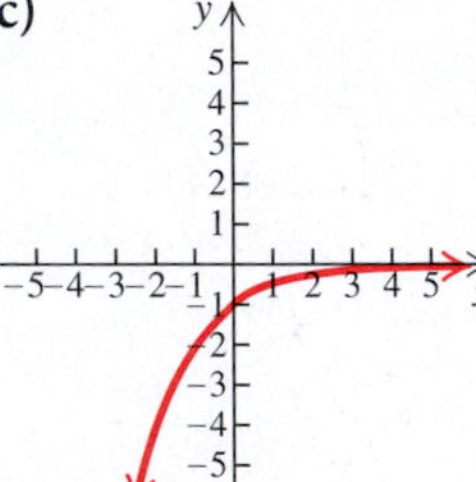

d)

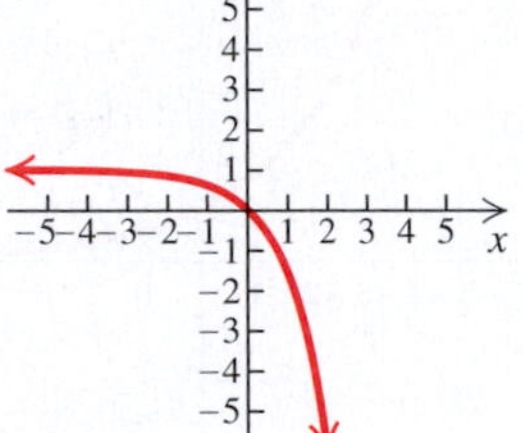

e)

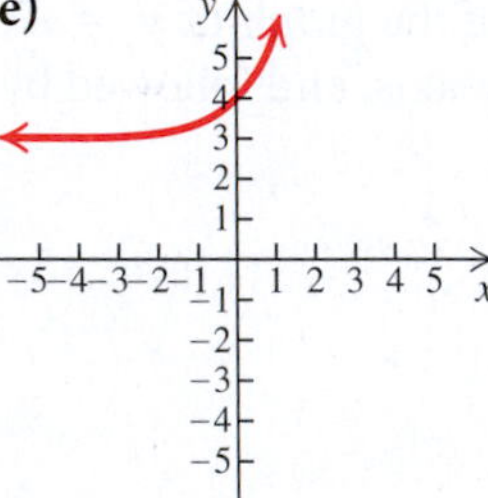

f)

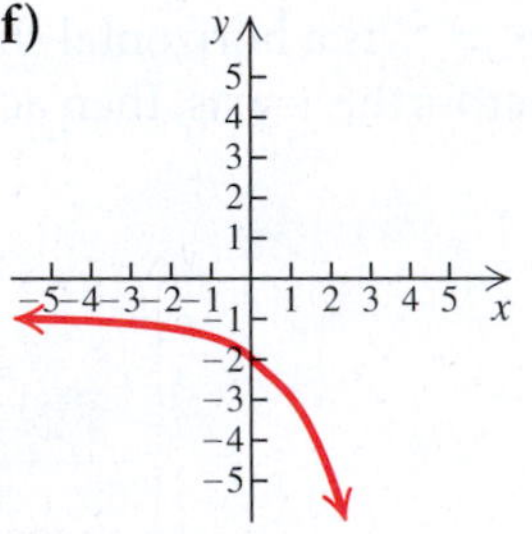

5. $f(x) = -2^x - 1$
6. $f(x) = -\left(\frac{1}{2}\right)^x$
7. $f(x) = e^x + 3$
8. $f(x) = e^{x+1}$
9. $f(x) = 3^{-x} - 2$
10. $f(x) = 1 - e^x$

Graph the function by substituting and plotting points.

11. $f(x) = 3^x$
12. $f(x) = 5^x$
13. $f(x) = 6^x$
14. $f(x) = 3^{-x}$
15. $f(x) = \left(\frac{1}{4}\right)^x$
16. $f(x) = \left(\frac{2}{3}\right)^x$
17. $y = -2^x$
18. $y = 3 - 3^x$
19. $f(x) = -0.25^x + 4$
20. $f(x) = 0.6^x - 3$
21. $f(x) = 1 + e^{-x}$
22. $f(x) = 2 - e^{-x}$
23. $y = \frac{1}{4}e^x$
24. $y = 2e^{-x}$
25. $f(x) = 1 - e^{-x}$
26. $f(x) = e^x - 2$

Sketch the graph of the function. Describe how each graph can be obtained from the graph of a basic exponential function.

27. $f(x) = 2^{x+1}$
28. $f(x) = 2^{x-1}$
29. $f(x) = 2^x - 3$
30. $f(x) = 2^x + 1$
31. $f(x) = 2^{1-x} + 2$
32. $f(x) = 5 - 2^{-x}$
33. $f(x) = 4 - 3^{-x}$
34. $f(x) = 2^{x-1} - 3$
35. $f(x) = \left(\frac{3}{2}\right)^{x-1}$
36. $f(x) = 3^{4-x}$
37. $f(x) = 2^{x+3} - 5$
38. $f(x) = -3^{x-2}$
39. $f(x) = 3 \cdot 2^{x-1} + 1$
40. $f(x) = 2 \cdot 3^{x+1} - 2$
41. $f(x) = e^{2x}$
42. $f(x) = e^{-0.2x}$
43. $f(x) = \dfrac{1}{2}(1 - e^x)$
44. $f(x) = 3(1 + e^x) - 2$
45. $y = e^{-x+1}$
46. $y = e^{2x} + 1$
47. $f(x) = 2(1 - e^{-x})$
48. $f(x) = 1 - e^{-0.01x}$

Graph the piecewise function.

49. $f(x) = \begin{cases} e^{-x} - 4, & \text{for } x < -2, \\ x + 3, & \text{for } -2 \le x < 1, \\ x^2, & \text{for } x \ge 1 \end{cases}$

50. $g(x) = \begin{cases} 4, & \text{for } x \le -3, \\ x^2 - 6, & \text{for } -3 < x < 0, \\ e^x, & \text{for } x \ge 0 \end{cases}$

51. *Compound Interest.* Suppose that \$82,000 is invested at $4\frac{1}{2}\%$ interest, compounded quarterly.
    a) Find the function for the amount to which the investment grows after t years.
    b) Find the amount of money in the account at $t = 0, 2, 5,$ and 10 years.

52. *Compound Interest.* Suppose that \$750 is invested at 7% interest, compounded semiannually.
    a) Find the function for the amount to which the investment grows after t years.
    b) Find the amount of money in the account at $t = 1, 6, 10, 15,$ and 25 years.

53. *Interest on a CD.* On Elizabeth's sixth birthday, her grandparents present her with a \$3000 certificate of deposit (CD) that earns 5% interest, compounded quarterly. If the CD matures on her sixteenth birthday, what amount will be available then?

54. *Interest in a College Trust Fund.* Following the birth of his child, Benjamin deposits $10,000 in a college trust fund where interest is 3.9%, compounded semiannually.

a) Find a function for the amount in the account after t years.

b) Find the amount of money in the account at $t = 0$, 4, 8, 10, 18, and 21 years.

In Exercises 55–62, use the compound-interest formula to find the account balance A with the given conditions:

P = principal,
r = interest rate,
n = number of compounding periods per year,
t = time, in years,
A = account balance.

	P	r	Compounded	n	t	A
55.	$3,000	4%	Semiannually		2	
56.	$12,500	3%	Quarterly		3	
57.	$120,000	2.5%	Annually		10	
58.	$120,000	2.5%	Quarterly		10	
59.	$53,500	$5\frac{1}{2}$%	Quarterly		$6\frac{1}{2}$	
60.	$6,250	$6\frac{3}{4}$%	Semiannually		$4\frac{1}{2}$	
61.	$17,400	8.1%	Daily		5	
62.	$900	7.3%	Daily		$7\frac{1}{4}$	

63. *Alternative-Fuel Vehicles.* The sales of alternative-fuel vehicles have more than tripled since 1995 (*Source*: Energy Information Administration). The exponential function

$$A(x) = 246{,}855(1.0931)^x,$$

where x is the number of years after 1995, can be used to estimate the number of alternative-fuel vehicles in a given year. Find the number of alternative-fuel vehicles in 1998 and in 2010. Then project the number of alternative-fuel vehicles in 2018.

64. *Increasing CPU Power.* The central processing unit (CPU) power in computers has increased significantly over the years. The CPU power in Macintosh computers has grown exponentially from 8 MHz in 1984 to 3400 MHz in 2013 (*Source:* Apple). The exponential function

$$M(t) = 7.91477(1.26698)^t,$$

where t is the number of years after 1984, can be used to estimate the CPU power in a Macintosh computer in a given year. Find the CPU power of a Macintosh Performa 5320CD in 1995 and of an iMac G6 in 2009. Round to the nearest one MHz.

65. *E-Cigarette Sales.* The electronic cigarette was launched in 2007, and since then sales have increased from about $20 million in 2008 to about $500 million in 2012 (*Sources:* UBS; forbes.com). The exponential function

$$S(x) = 20.913(2.236)^x,$$

where x is the number of years after 2008, models the sales, in millions of dollars. Use this function to estimate the sales of e-cigarettes in 2011 and in 2015. Round to the nearest million dollars.

66. *Foreign High School Students.* The number of foreign students studying in high schools in the United States grew from 6541 in 2007 to 65,452 in 2012 (*Source*: Council on Standards for International Educational Travel). The increase can be modeled by the exponential function

$$E(x) = 6541(1.5851)^x,$$

where x is the number of years after 2007. Find the number of foreign students enrolled in U.S. high schools in 2009 and in 2011. Then use this function to project the number of foreign students enrolled in 2016.

67. *Skype Concurrent Users.* The number of concurrent users of Skype has increased dramatically since 2004 (*Source*: Skype NumerologyBlog). By 2013, Skype could connect concurrently 70 million users online. The exponential function

$$P(t) = 2.307(1.483)^t,$$

where t is the number of years after 2004, models this increase in millions of users. Estimate the number of

Skype users that could be online concurrently in 2005, in 2009, and in 2012. Round to the nearest million users.

68. *U.S. Imports.* The amount of imports to the United States has increased exponentially since 1980 (*Sources*: U.S. Census Bureau; U.S. Bureau of Economic Analysis; U.S. Department of Commerce). The exponential function

$$P(x) = 307.368(1.072)^x,$$

where x is the number of years after 1980, can be used to estimate the total amount of U.S. imports, in billions of dollars. Find the total amount of imports to the United States in 1990, in 2000, and in 2012. Round to nearest billion dollars.

69. *Centenarian Population.* The centenarian population in the United States has grown over 65% in the last 30 years.

In 1980, there were only 32,194 residents ages 100 and over. This number had grown to 53,364 by 2010. (*Sources:* Population Projections Program; U.S. Census Bureau; U.S. Department of Commerce; "What People Who Live to 100 Have in Common," by Emily Brandon, *U.S. News and World Report*, January 7, 2013) The exponential function

$$H(t) = 80{,}040.68(1.0481)^t,$$

where t is the number of years after 2015, can be used to project the number of centenarians. Use this function to project the centenarian population in 2020 and in 2050.

70. *Bachelor's Degrees Earned.* The exponential function

$$D(t) = 347(1.024)^t$$

gives the number of bachelor's degrees, in thousands, earned in the United States t years after 1970 (*Sources:* National Center for Educational Statistics; U.S. Department of Education). Find the number of bachelor's degrees earned in 1985, in 2000, and in 2014. Then estimate the number of bachelor's degrees that will be earned in 2020. Round to the nearest thousand degrees.

71. *Charitable Giving.* Over the last four decades, the amount of charitable giving in the United States has grown exponentially from approximately \$20.7 billion in 1969 to approximately \$316.2 billion in 2012 (*Sources:* Giving USA Foundation; Volunteering in America by the Corporation for National & Community Service; National Philanthropic Trust; School of Philanthropy, Indiana University Purdue University Indianapolis). The exponential function

$$G(x) = 20.7(1.066)^x,$$

where x is the number of years after 1969, can be used to estimate the amount of charitable giving, in billions of dollars, in a given year. Find the amount of charitable giving in 1982, in 1995, and in 2010. Round to the nearest billion dollars.

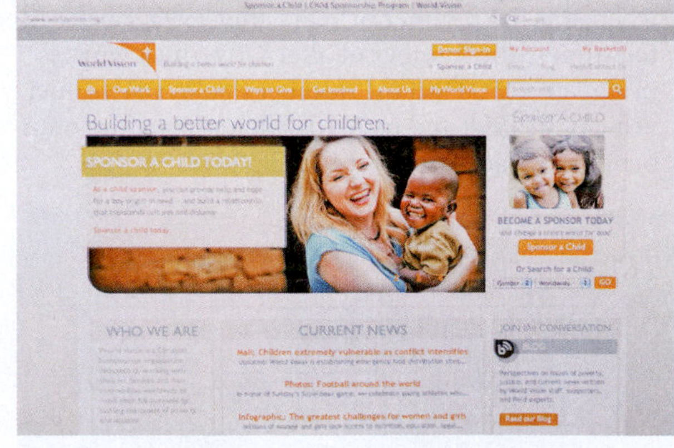

72. *Tennis Participation.* In recent years, U.S. tennis participation has increased. The exponential function

$$T(x) = 23.7624(1.0752)^x,$$

where $T(x)$ is in millions and x is the number of years after 2006, models the number of people who had played tennis at least once in a given year (*Source*: USTA/Coyne Public Relations). Estimate the number who had played tennis at least once in 2010. Then use this function to project participation in tennis in 2018. Round to the nearest million participants.

73. *Salvage Value.* A restaurant purchased a 72-in. range with six burners for \$6982. The value of the range each year is 85% of the value of the preceding year. After t years, its value, in dollars, is given by the exponential function

$$V(t) = 6982(0.85)^t.$$

Find the value of the range after 0, 1, 2, 5, and 8 years. Round to the nearest dollar.

74. *Salvage Value.* A landscape company purchased a backhoe for \$56,395. The value of the backhoe each year is 90% of the value of the preceding year. After t years, its value, in dollars, is given by the exponential function

$$V(t) = 56{,}395(0.9)^t.$$

Find the value of the backhoe after 0, 1, 3, 6, and 10 years. Round to the nearest dollar.

75. *Advertising.* A company begins an Internet advertising campaign to market a new telephone. The percentage of the target market that buys a product is generally a function of the length of the advertising campaign. The estimated percentage is given by the exponential function

$$f(t) = 100(1 - e^{-0.04t}),$$

where t is the number of days of the campaign. Find $f(25)$, the percentage of the target market that has bought the product after a 25-day advertising campaign.

76. *Growth of a Stock.* The value of a stock is given by the function

$$V(t) = 58(1 - e^{-1.1t}) + 20,$$

where V is the value of the stock after time t, in months. Find $V(1)$, $V(2)$, $V(4)$, $V(6)$, and $V(12)$.

Skill Maintenance

Simplify. [3.1]

77. $(1 - 4i)(7 + 6i)$

78. $\dfrac{2 - i}{3 + i}$

Find the x-intercepts and the zeros of the function.

79. $f(x) = 2x^2 - 13x - 7$ [3.2]

80. $h(x) = x^3 - 3x^2 + 3x - 1$ [4.4]

81. $h(x) = x^4 - x^2$ [4.1]

82. $g(x) = x^3 + x^2 - 12x$ [4.1]

Solve.

83. $x^3 + 6x^2 - 16x = 0$ [4.1]

84. $3x^2 - 6 = 5x$ [3.2]

Synthesis

85. Which is larger, 7^π or π^7? 70^{80} or 80^{70}?

86. For the function f, construct and simplify the difference quotient.

$$f(x) = 2e^x - 3$$

5.3 Logarithmic Functions and Graphs

- Find common logarithms and natural logarithms with and without a calculator.
- Convert between exponential equations and logarithmic equations.
- Change logarithmic bases.
- Graph logarithmic functions.
- Solve applied problems involving logarithmic functions.

We now consider *logarithmic*, or *logarithm, functions*. These functions are inverses of exponential functions and have many applications.

Logarithmic Functions

We have noted that every exponential function (with $a > 0$ and $a \neq 1$) is one-to-one. Thus such a function has an inverse that is a function. In this section, we will name these inverse functions logarithmic functions and use them in applications. We can draw the graph of the inverse of an exponential function by interchanging x and y.

EXAMPLE 1 Graph: $x = 2^y$.

Solution Note that x is alone on one side of the equation. We can find ordered pairs that are solutions by choosing values for y and then computing the corresponding x-values.

For $y = 0$, $x = 2^0 = 1$.
For $y = 1$, $x = 2^1 = 2$.
For $y = 2$, $x = 2^2 = 4$.
For $y = 3$, $x = 2^3 = 8$.
For $y = -1$, $x = 2^{-1} = \frac{1}{2^1} = \frac{1}{2}$.
For $y = -2$, $x = 2^{-2} = \frac{1}{2^2} = \frac{1}{4}$.
For $y = -3$, $x = 2^{-3} = \frac{1}{2^3} = \frac{1}{8}$.

x		
$x = 2^y$	y	(x, y)
1	0	$(1, 0)$
2	1	$(2, 1)$
4	2	$(4, 2)$
8	3	$(8, 3)$
$\frac{1}{2}$	-1	$\left(\frac{1}{2}, -1\right)$
$\frac{1}{4}$	-2	$\left(\frac{1}{4}, -2\right)$
$\frac{1}{8}$	-3	$\left(\frac{1}{8}, -3\right)$

(1) Choose values for y.
(2) Compute values for x.

We plot the points and connect them with a smooth curve. Note that the curve does not touch or cross the y-axis. The y-axis is a vertical asymptote.

Note too that this curve is the graph of $y = 2^x$ reflected across the line $y = x$, as we would expect for an inverse. The inverse of $y = 2^x$ is $x = 2^y$.

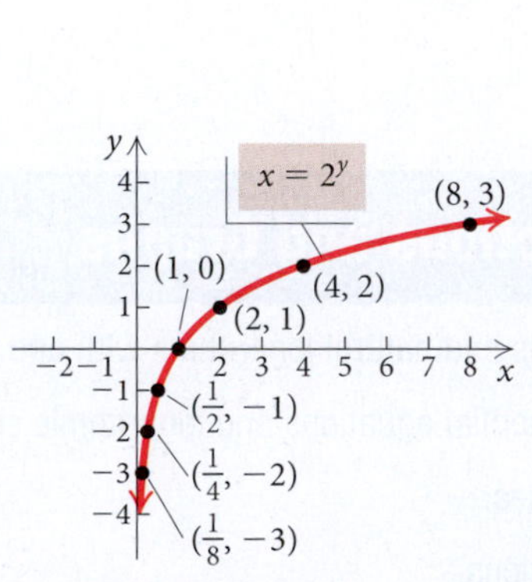

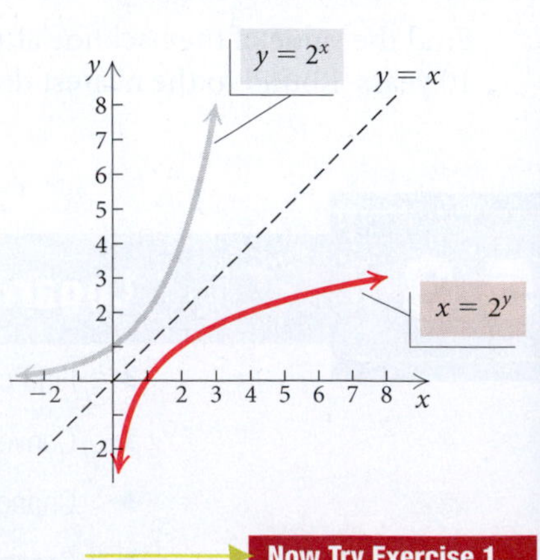

Now Try Exercise 1.

To find a formula for f^{-1} when $f(x) = 2^x$, we use the method discussed in Section 5.1:

1. Replace $f(x)$ with y: $\quad y = 2^x$
2. Interchange x and y: $\quad x = 2^y$
3. Solve for y: $\quad y =$ the power to which we raise 2 to get x.
4. Replace y with $f^{-1}(x)$: $\quad f^{-1}(x) =$ the power to which we raise 2 to get x.

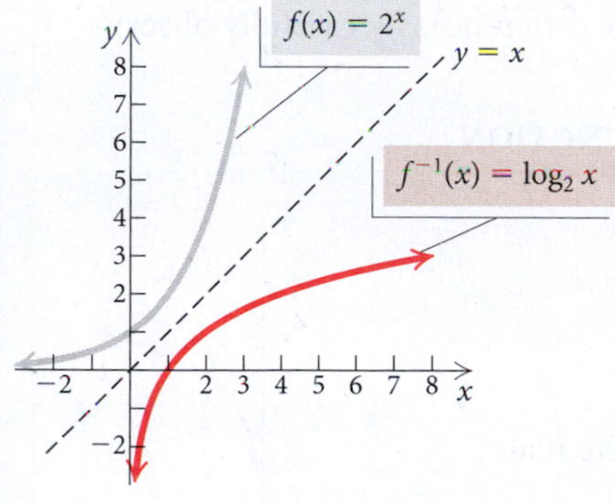

Mathematicians have defined a new symbol to replace the words "the power to which we raise 2 to get x." That symbol is "$\log_2 x$," read "the logarithm, base 2, of x."

Logarithmic Function, Base 2

"$\log_2 x$," read "the logarithm, base 2, of x," means "the power to which we raise 2 to get x."

Thus if $f(x) = 2^x$, then $f^{-1}(x) = \log_2 x$. For example, $f^{-1}(8) = \log_2 8 = 3$, because *3 is the power to which we raise 2 to get 8.* Similarly, $\log_2 13$ is the power to which we raise 2 to get 13. As yet, we have no simpler way to say this other than

"$\log_2 13$ is the power to which we raise 2 to get 13."

Later, however, we will learn how to approximate this expression using a calculator.

For any exponential function $f(x) = a^x$, its inverse is called a **logarithmic function, base *a*.** The graph of the inverse can be obtained by reflecting the graph of $y = a^x$ across the line $y = x$, to obtain $x = a^y$. Then $x = a^y$ is equivalent to $y = \log_a x$. We read $\log_a x$ as "the logarithm, base a, of x."

The inverse of $f(x) = a^x$ is given by $f^{-1}(x) = \log_a x$.

LOGARITHMIC FUNCTION, BASE *a*

We define $y = \log_a x$ as that number y such that $x = a^y$, where $x > 0$ and a is a positive constant other than 1.

Let's look at the graphs of $f(x) = a^x$ and $f^{-1}(x) = \log_a x$ for $a > 1$ and for $0 < a < 1$.

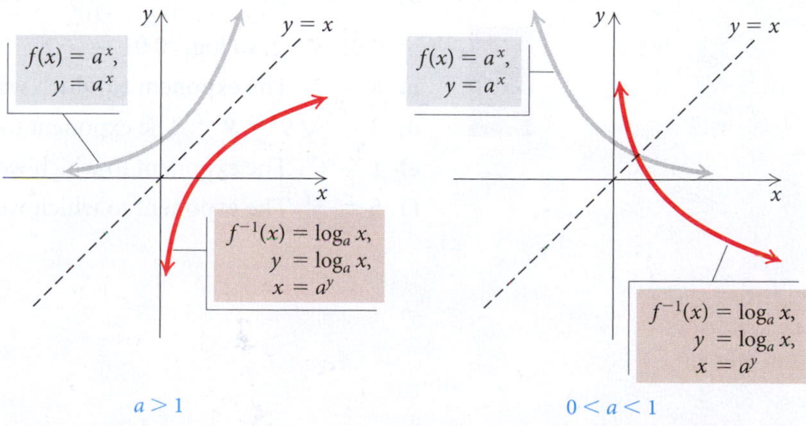

$a > 1$ $\qquad$ $0 < a < 1$

Note that the graphs of $f(x)$ and $f^{-1}(x)$ are reflections of each other across the line $y = x$.

CONNECTING THE CONCEPTS

Comparing Exponential Functions and Logarithmic Functions

In the following table, we compare exponential functions and logarithmic functions with bases a greater than 1. Similar statements could be made for a, where $0 < a < 1$. It is helpful to visualize the differences by carefully observing the graphs.

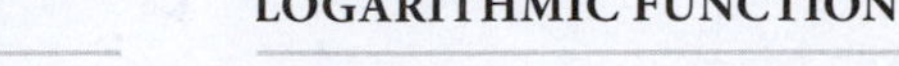

EXPONENTIAL FUNCTION	LOGARITHMIC FUNCTION
$y = a^x$	$x = a^y$
$f(x) = a^x$	$f^{-1}(x) = \log_a x$
$a > 1$	$a > 1$
Continuous	Continuous
One-to-one	One-to-one
Domain: All real numbers, $(-\infty, \infty)$	Domain: All positive real numbers, $(0, \infty)$
Range: All positive real numbers, $(0, \infty)$	Range: All real numbers, $(-\infty, \infty)$
Increasing	Increasing
Horizontal asymptote is x-axis: $(a^x \to 0 \text{ as } x \to -\infty)$	Vertical asymptote is y-axis: $(\log_a x \to -\infty \text{ as } x \to 0^+)$
y-intercept: $(0, 1)$	x-intercept: $(1, 0)$
There is no x-intercept.	There is no y-intercept.

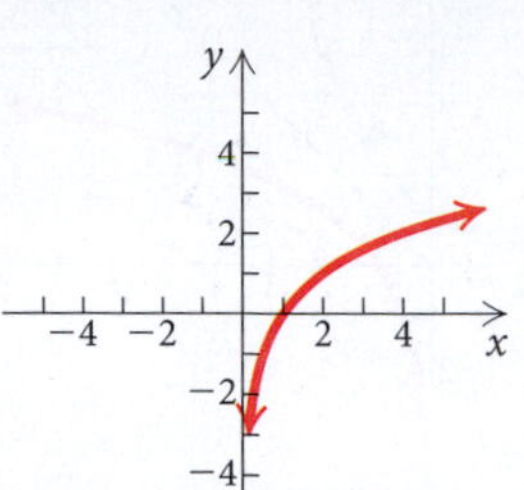

▶ Finding Certain Logarithms

Let's use the definition of logarithms to find some logarithmic values.

EXAMPLE 2 Find each of the following logarithms.

a) $\log_{10} 10{,}000$ **b)** $\log_{10} 0.01$ **c)** $\log_2 8$
d) $\log_9 3$ **e)** $\log_6 1$ **f)** $\log_8 8$

Solution

a) The exponent to which we raise 10 to obtain 10,000 is 4; thus $\log_{10} 10{,}000 = 4$.

b) We have $0.01 = \frac{1}{100} = \frac{1}{10^2} = 10^{-2}$. The exponent to which we raise 10 to get 0.01 is -2, so $\log_{10} 0.01 = -2$.

c) $8 = 2^3$. The exponent to which we raise 2 to get 8 is 3, so $\log_2 8 = 3$.

d) $3 = \sqrt{9} = 9^{1/2}$. The exponent to which we raise 9 to get 3 is $\frac{1}{2}$, so $\log_9 3 = \frac{1}{2}$.

e) $1 = 6^0$. The exponent to which we raise 6 to get 1 is 0, so $\log_6 1 = 0$.

f) $5 = 5^1$. The exponent to which we raise 5 to get 5 is 1, so $\log_5 5 = 1$.

Now Try Exercises 9 and 15.

Examples 2(e) and 2(f) illustrate two important properties of logarithms. The property $\log_a 1 = 0$ follows from the fact that $a^0 = 1$. Thus, $\log_5 1 = 0$, $\log_{10} 1 = 0$, and so on. The property $\log_a a = 1$ follows from the fact that $a^1 = a$. Thus, $\log_5 5 = 1$, $\log_{10} 10 = 1$, and so on.

$\log_a 1 = 0$ and $\log_a a = 1$, for any logarithmic base a.

Converting Between Exponential Equations and Logarithmic Equations

In dealing with logarithmic functions, it is helpful to remember that a logarithm of a number is an *exponent*. It is the exponent y in $x = a^y$. You might think to yourself, "the logarithm, base a, of a number x is the power to which a must be raised to get x."

We are led to the following. (The symbol $\longleftrightarrow$ means that the two statements are equivalent; that is, when one is true, the other is true. The words "if and only if" can be used in place of $\longleftrightarrow$.)

$\log_a x = y \longleftrightarrow x = a^y$ A logarithm is an exponent!

EXAMPLE 3 Convert each of the following to a logarithmic equation.

a) $16 = 2^x$ b) $10^{-3} = 0.001$ c) $e^t = 70$

Solution

The exponent is the logarithm.

a) $16 = 2^x \longrightarrow \log_2 16 = x$

The base remains the same.

b) $10^{-3} = 0.001 \rightarrow \log_{10} 0.001 = -3$

c) $e^t = 70 \rightarrow \log_e 70 = t$

Now Try Exercise 37.

EXAMPLE 4 Convert each of the following to an exponential equation.

a) $\log_2 32 = 5$ b) $\log_a Q = 8$ c) $x = \log_t M$

Solution

The logarithm is the exponent.

a) $\log_2 32 = 5 \qquad 2^5 = 32$

The base remains the same.

b) $\log_a Q = 8 \rightarrow a^8 = Q$

c) $x = \log_t M \rightarrow t^x = M$

Now Try Exercise 45.

▶ Finding Logarithms on a Calculator

Before calculators became so widely available, base-10 logarithms, or **common logarithms**, were used extensively to simplify complicated calculations. In fact, that is why logarithms were invented. The abbreviation **log**, with no base written, is used to represent common logarithms, or base-10 logarithms. Thus,

$$\textbf{log}\,x \quad \textbf{means} \quad \textbf{log}_{10}\,x.$$

For example, log 29 means $\log_{10} 29$. Let's compare log 29 with log 10 and log 100:

$$\left.\begin{aligned} \log 10 &= \log_{10} 10 = 1 \\ \log 29 &= \ ? \\ \log 100 &= \log_{10} 100 = 2 \end{aligned}\right\}$$ Since 29 is between 10 and 100, it seems reasonable that log 29 is between 1 and 2.

On a calculator, the key for common logarithms is generally marked LOG. Using that key, we find that

$$\log 29 \approx 1.462397998 \approx 1.4624$$

rounded to four decimal places. Since $1 < 1.4624 < 2$, our answer seems reasonable. This also tells us that $10^{1.4624} \approx 29$.

EXAMPLE 5 Find each of the following common logarithms on a calculator. If you are using a graphing calculator, set the calculator in REAL mode. Round to four decimal places.

a) log 645,778 **b)** log 0.0000239 **c)** log (-3)

Solution

FUNCTION VALUE	READOUT	ROUNDED
a) log 645,778	log(645778) 5.810083246	5.8101
b) log 0.0000239	log(0.0000239) −4.621602099	−4.6216
c) log (-3)	ERR:NONREAL ANS *	Does not exist as a real number

Since 5.810083246 is the power to which we raise 10 to get 645,778, we can check part (a) by finding $10^{5.810083246}$. We can check part (b) in a similar manner. In part (c), log (-3) does not exist as a real number because there is no real-number power to which we can raise 10 to get -3. The number 10 raised to any real-number power is positive. The common logarithm of a negative number does not exist as a real number. Recall that logarithmic functions are inverses of exponential functions, and since the range of an exponential function is $(0, \infty)$, the domain of $f(x) = \log_a x$ is $(0, \infty)$.

Now Try Exercises 57 and 61.

*If the graphing calculator is set in $a + bi$ mode, the readout is .4771212547 + 1.364376354i.

▶ Natural Logarithms

Logarithms, base e, are called **natural logarithms**. The abbreviation "ln" is generally used for natural logarithms. Thus,

$$\ln x \quad \text{means} \quad \log_e x.$$

For example, ln 53 means $\log_e 53$. On a calculator, the key for natural logarithms is generally marked LN. Using that key, we find that

$$\ln 53 \approx 3.970291914 \approx 3.9703$$

rounded to four decimal places. This also tells us that $e^{3.9703} \approx 53$.

EXAMPLE 6 Find each of the following natural logarithms on a calculator. If you are using a graphing calculator, set the calculator in REAL mode. Round to four decimal places.

a) ln 645,778 b) ln 0.0000239 c) ln (−5)
d) ln e e) ln 1

Solution

FUNCTION VALUE	READOUT	ROUNDED
a) ln 645,778	ln(645778) 13.37821107	13.3782
b) ln 0.0000239	ln(0.0000239) −10.6416321	−10.6416
c) ln (−5)	ERR:NONREAL ANS *	Does not exist
d) ln e	ln(e) 1	1
e) ln 1	ln(1) 0	0

Since 13.37821107 is the power to which we raise e to get 645,778, we can check part (a) by finding $e^{13.37821107}$. We can check parts (b), (d), and (e) in a similar manner. In parts (d) and (e), note that $\ln e = \log_e e = 1$ and $\ln 1 = \log_e 1 = 0$.

Now Try Exercises 65 and 67.

$\ln 1 = 0$ and $\ln e = 1$, for the logarithmic base e.

*If the graphing calculator is set in $a + bi$ mode, the readout is $1.609437912 + 3.141592654i$.

Changing Logarithmic Bases

Most calculators give the values of both common logarithms and natural logarithms. To find a logarithm with a base other than 10 or *e*, we can use the following conversion formula.

Technology Connection

With some calculators, it is possible to find a logarithm with any logarithmic base using the logBASE operation from the MATH MATH menu. The computation of $\log_4 15$ is shown in the following window.

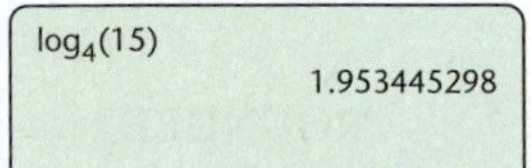

THE CHANGE-OF-BASE FORMULA

For any logarithmic bases *a* and *b*, and any positive number *M*,

$$\log_b M = \frac{\log_a M}{\log_a b}.$$

We will prove this result in the next section.

EXAMPLE 7 Find $\log_5 8$ using common logarithms.

Solution First, we let $a = 10$, $b = 5$, and $M = 8$. Then we substitute into the change-of-base formula:

$$\log_5 8 = \frac{\log_{10} 8}{\log_{10} 5} \qquad \text{Substituting}$$

$$\approx 1.2920. \qquad \text{Using a calculator}$$

Since $\log_5 8$ is the power to which we raise 5 to get 8, we would expect this power to be greater than 1 $(5^1 = 5)$ and less than 2 $(5^2 = 25)$, so the result is reasonable.

Now Try Exercise 69.

We can also use base *e* for a conversion.

EXAMPLE 8 Find $\log_5 8$ using natural logarithms.

Solution Substituting *e* for *a*, 5 for *b*, and 8 for *M*, we have

$$\log_5 8 = \frac{\log_e 8}{\log_e 5}$$

$$= \frac{\ln 8}{\ln 5} \approx 1.2920.$$

Note that we get the same value using base *e* for the conversion that we did using base 10 in Example 7.

Now Try Exercise 75.

Graphs of Logarithmic Functions

Let's now consider graphs of logarithmic functions.

EXAMPLE 9 Graph: $y = f(x) = \log_5 x$.

Solution The equation $y = \log_5 x$ is equivalent to $x = 5^y$. We can find ordered pairs that are solutions by choosing values for *y* and computing the corresponding *x*-values. We then plot points, remembering that *x* is still the first coordinate.

For $y = 0, x = 5^0 = 1$.
For $y = 1, x = 5^1 = 5$.
For $y = 2, x = 5^2 = 25$.
For $y = 3, x = 5^3 = 125$.
For $y = -1, x = 5^{-1} = \frac{1}{5}$.
For $y = -2, x = 5^{-2} = \frac{1}{25}$.

x, or 5^y	y
1	0
5	1
25	2
125	3
$\frac{1}{5}$	-1
$\frac{1}{25}$	-2

(1) Select y.
(2) Compute x.

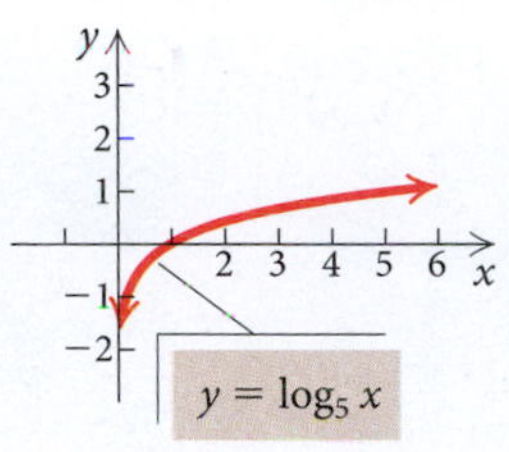

Now Try Exercise 5.

Technology Connection

To graph $y = \log_5 x$ in Example 9 with a graphing calculator, we must first change the base to 10 or e. Here we change from base 5 to base e:

$$y = \log_5 x = \frac{\ln x}{\ln 5}.$$

The graph is shown below.

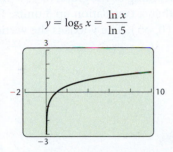

$y = \log_5 x = \frac{\ln x}{\ln 5}$

Some graphing calculators can graph inverses without the need to first find an equation of the inverse. If we begin with $y_1 = 5^x$, the graphs of both y_1 and its inverse, $y_2 = \log_5 x$, will be drawn as shown below.

$y_1 = 5^x$, $y_2 = \log_5 x$

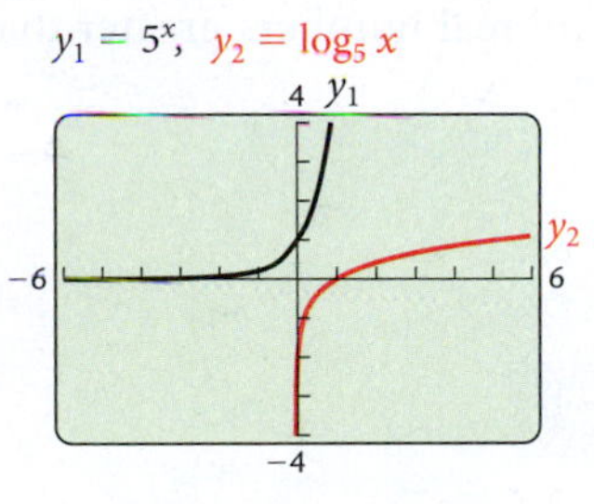

EXAMPLE 10 Graph: $g(x) = \ln x$.

Solution To graph $y = g(x) = \ln x$, we select values for x and use the LN key on a calculator to find the corresponding values of $\ln x$. We then plot points and draw the curve.

x	$g(x) = \ln x$
0.5	-0.7
1	0
2	0.7
3	1.1
4	1.4
5	1.6

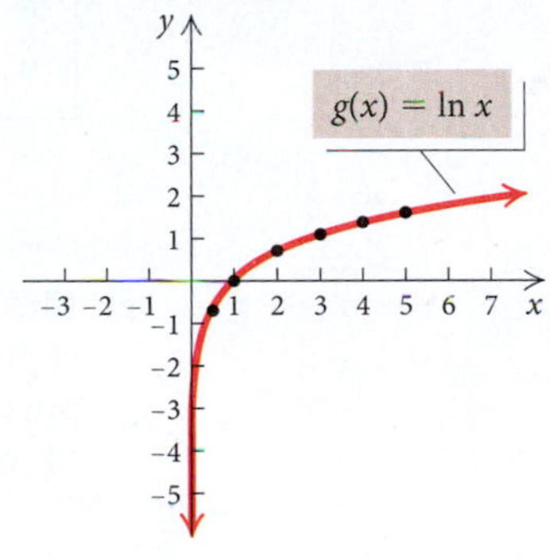

We could also write $g(x) = \ln x$, or $y = \ln x$, as $x = e^y$, select values for y, and use a calculator to find the corresponding values of x.

Now Try Exercise 7.

Recall that the graph of $f(x) = \log_a x$, for any base a, has the x-intercept $(1, 0)$. The domain is the set of positive real numbers, and the range is the set of all real numbers. The y-axis is the vertical asymptote.

EXAMPLE 11 Graph each of the following. Before doing so, describe how each graph can be obtained from the graph of $y = \ln x$. Give the domain and the vertical asymptote of each function.

a) $f(x) = \ln(x + 3)$
b) $f(x) = 3 - \frac{1}{2}\ln x$
c) $f(x) = |\ln(x - 1)|$

Solution

a) The graph of $f(x) = \ln(x + 3)$ is a shift of the graph of $y = \ln x$ left 3 units. The domain is the set of all real numbers greater than -3, $(-3, \infty)$. The line $x = -3$ is the vertical asymptote.

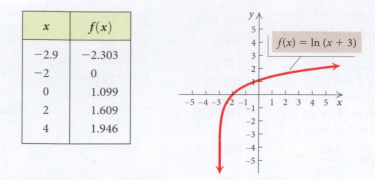

x	$f(x)$
−2.9	−2.303
−2	0
0	1.099
2	1.609
4	1.946

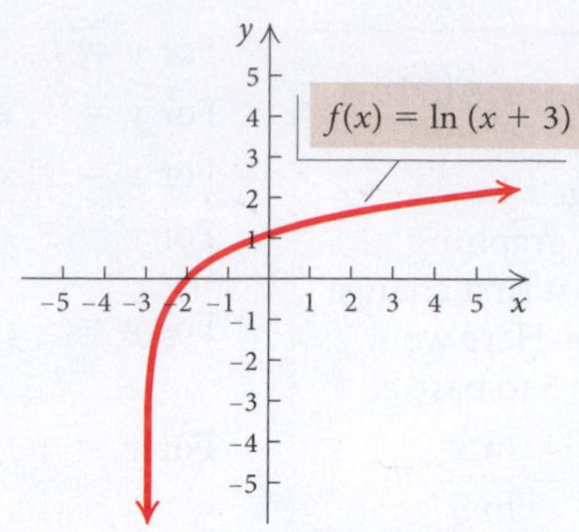

b) The graph of $f(x) = 3 - \frac{1}{2}\ln x$ is a vertical shrinking of the graph of $y = \ln x$, followed by a reflection across the x-axis, and then a translation up 3 units. The domain is the set of all positive real numbers, $(0, \infty)$. The y-axis is the vertical asymptote.

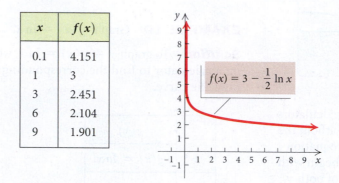

x	$f(x)$
0.1	4.151
1	3
3	2.451
6	2.104
9	1.901

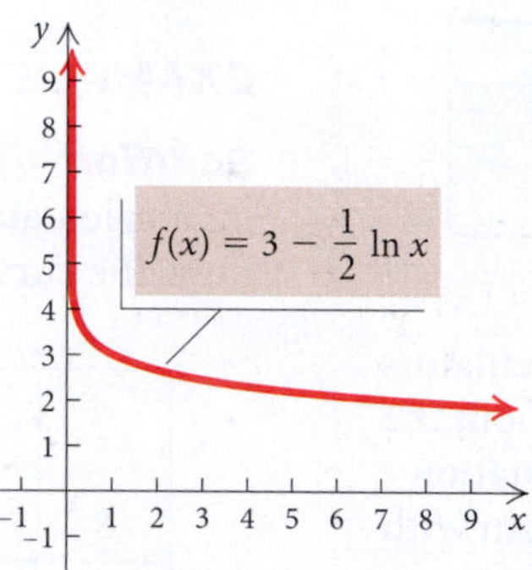

c) The graph of $f(x) = |\ln(x - 1)|$ is a translation of the graph of $y = \ln x$ right 1 unit. Then the absolute value has the effect of reflecting negative outputs across the x-axis. The domain is the set of all real numbers greater than 1, $(1, \infty)$. The line $x = 1$ is the vertical asymptote.

x	$f(x)$
1.1	2.303
2	0
4	1.099
6	1.609
8	1.946

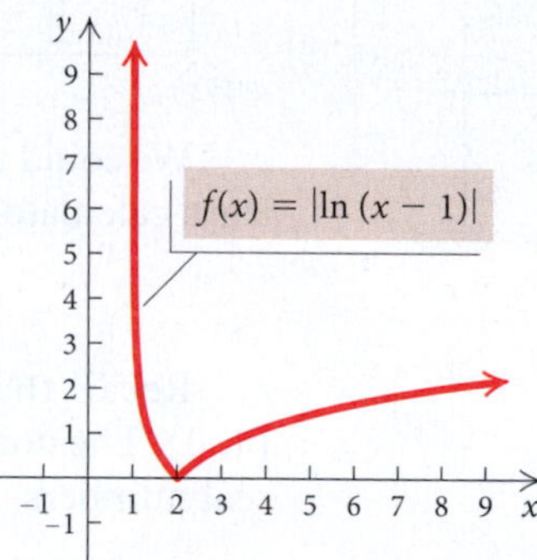

Now Try Exercise 89.

Applications

EXAMPLE 12 ***Walking Speed.*** In a study by psychologists Bornstein and Bornstein, it was found that the average walking speed w, in feet per second, of a person living in a city of population P, in thousands, is given by the function

$$w(P) = 0.37 \ln P + 0.05$$

(*Source: International Journal of Psychology*).

a) The population of Billings, Montana, is 106,954. Find the average walking speed of people living in Billings.

b) The population of Chicago, Illinois, is 2,714,856. Find the average walking speed of people living in Chicago.

Solution

a) Since P is in thousands and $106{,}954 = 106.954$ thousand, we substitute 106.954 for P:

$$w(106.954) = 0.37 \ln 106.954 + 0.05 \quad \text{Substituting}$$
$$\approx 1.8. \quad \text{Finding the natural logarithm and simplifying}$$

The average walking speed of people living in Billings is about 1.8 ft/sec.

b) We substitute 2714.856 for P:

$$w(2714.856) = 0.37 \ln 2714.856 + 0.05 \quad \text{Substituting}$$
$$\approx 3.0.$$

The average walking speed of people living in Chicago is about 3.0 ft/sec.

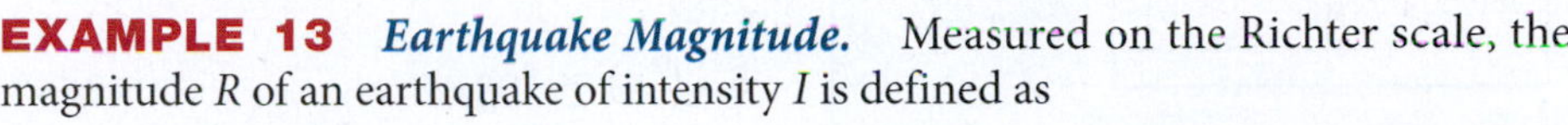

Now Try Exercise 95(d).

EXAMPLE 13 ***Earthquake Magnitude.*** Measured on the Richter scale, the magnitude R of an earthquake of intensity I is defined as

$$R = \log \frac{I}{I_0},$$

where I_0 is a minimum intensity used for comparison. We can think of I_0 as a threshold intensity that is the weakest earthquake that can be recorded on a seismograph. If one earthquake is 10 times as intense as another, its magnitude on the Richter scale is 1 greater than that of the other. If one earthquake is 100 times as intense as another, its magnitude on the Richter scale is 2 higher, and so on. Thus an earthquake whose magnitude is 7 on the Richter scale is 10 times as intense as an earthquake whose magnitude is 6. Earthquake intensities can be interpreted as multiples of the minimum intensity I_0.

The undersea Tohoku earthquake and tsunami, near the northeast coast of Honshu, Japan, on March 11, 2011, had an intensity of $10^{9.0} \cdot I_0$ (*Source*: earthquake.usgs.gov). They caused extensive loss of life and severe structural damage to buildings, railways, and roads. What was the magnitude on the Richter scale?

Solution We substitute into the formula:

$$R = \log \frac{I}{I_0} = \log \frac{10^{9.0} \cdot I_0}{I_0} = \log 10^{9.0} = 9.0.$$

The magnitude of the earthquake was 9.0 on the Richter scale.

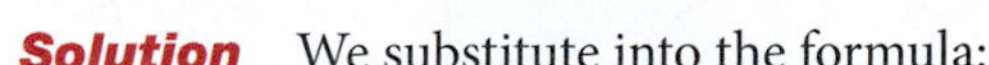

Now Try Exercise 97(a).

Visualizing the Graph

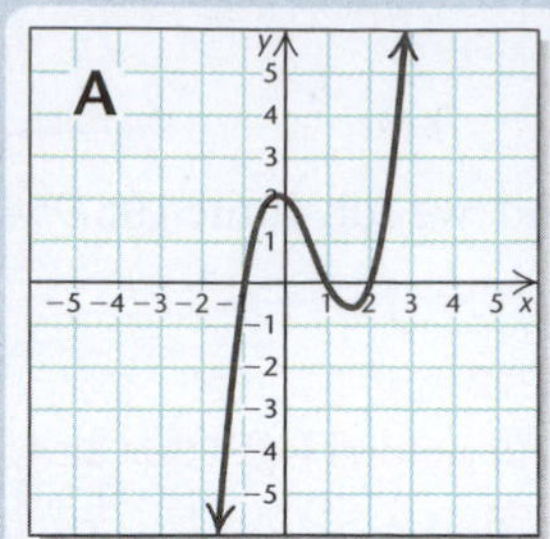

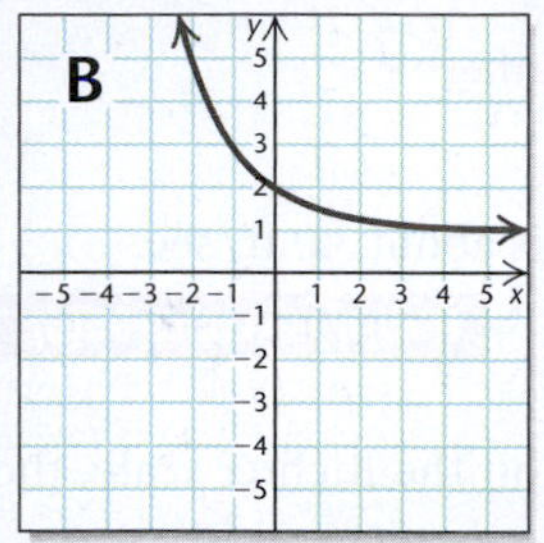

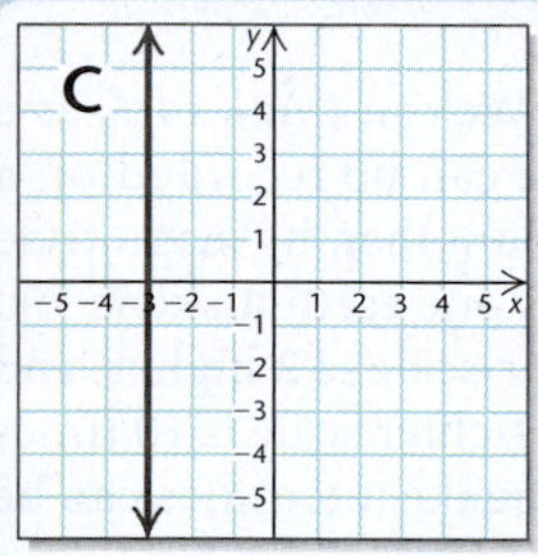

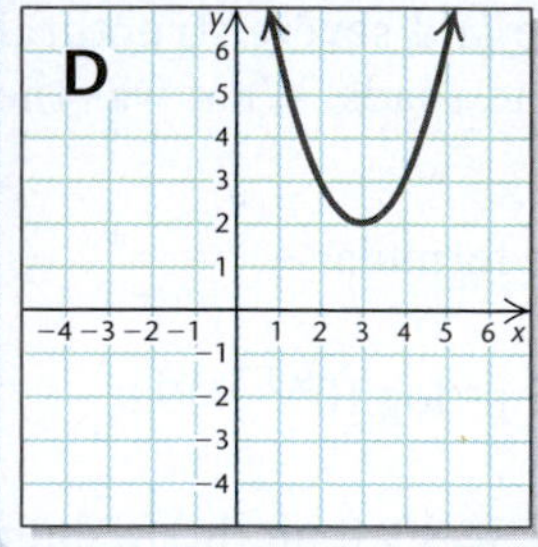

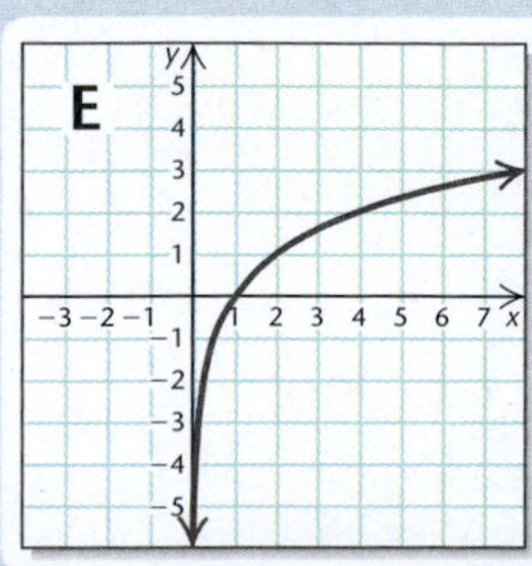

Match the equation or function with its graph.

1. $f(x) = 4^x$
2. $f(x) = \ln x - 3$
3. $(x + 3)^2 + y^2 = 9$
4. $f(x) = 2^{-x} + 1$
5. $f(x) = \log_2 x$
6. $f(x) = x^3 - 2x^2 - x + 2$
7. $x = -3$
8. $f(x) = e^x - 4$
9. $f(x) = (x - 3)^2 + 2$
10. $3x = 6 + y$

Answers on page A-27

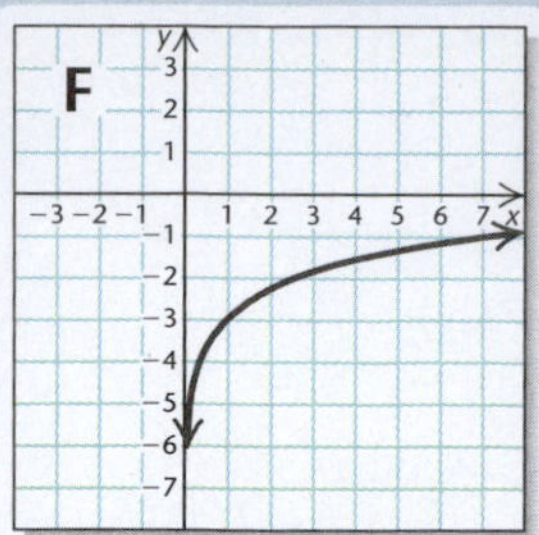

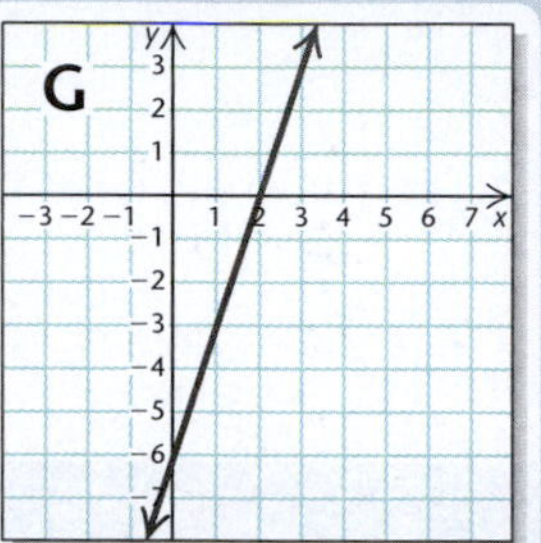

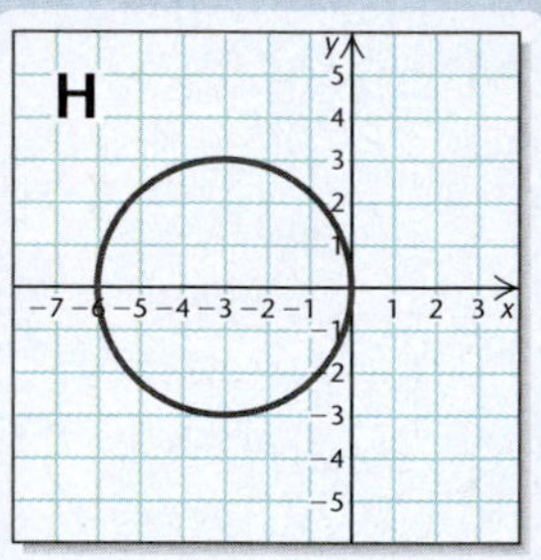

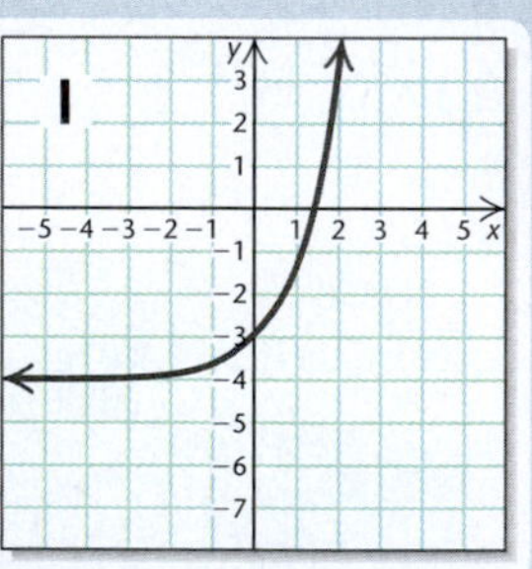

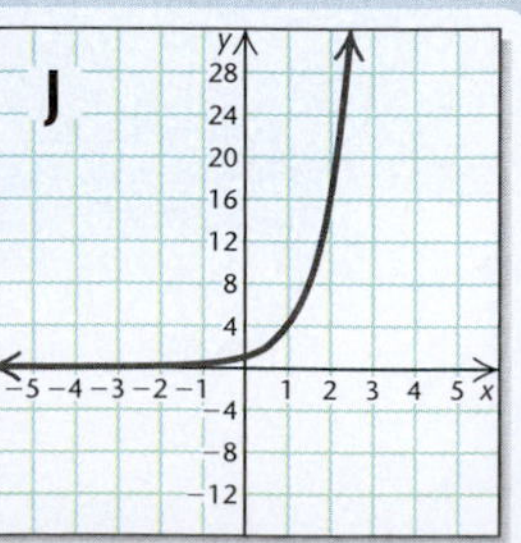

5.3 Exercise Set

Graph.

1. $x = 3^y$
2. $x = 4^y$
3. $x = \left(\frac{1}{2}\right)^y$
4. $x = \left(\frac{4}{3}\right)^y$
5. $y = \log_3 x$
6. $y = \log_4 x$
7. $f(x) = \log x$
8. $f(x) = \ln x$

Find each of the following. Do not use a calculator.

9. $\log_2 16$
10. $\log_3 9$
11. $\log_5 125$
12. $\log_2 64$
13. $\log 0.001$
14. $\log 100$
15. $\log_2 \frac{1}{4}$
16. $\log_8 2$
17. $\ln 1$
18. $\ln e$
19. $\log 10$
20. $\log 1$
21. $\log_5 5^4$
22. $\log \sqrt{10}$
23. $\log_3 \sqrt[4]{3}$
24. $\log 10^{8/5}$
25. $\log 10^{-7}$
26. $\log_5 1$
27. $\log_{49} 7$
28. $\log_3 3^{-2}$
29. $\ln e^{3/4}$
30. $\log_2 \sqrt{2}$
31. $\log_4 1$
32. $\ln e^{-5}$
33. $\ln \sqrt{e}$
34. $\log_{64} 4$

Convert to a logarithmic equation.

35. $10^3 = 1000$
36. $5^{-3} = \frac{1}{125}$
37. $8^{1/3} = 2$
38. $10^{0.3010} = 2$
39. $e^3 = t$
40. $Q^t = x$
41. $e^2 = 7.3891$
42. $e^{-1} = 0.3679$
43. $p^k = 3$
44. $e^{-t} = 4000$

Convert to an exponential equation.

45. $\log_5 5 = 1$
46. $t = \log_4 7$
47. $\log 0.01 = -2$
48. $\log 7 = 0.845$
49. $\ln 30 = 3.4012$
50. $\ln 0.38 = -0.9676$
51. $\log_a M = -x$
52. $\log_t Q = k$
53. $\log_a T^3 = x$
54. $\ln W^5 = t$

Find each of the following using a calculator. Round to four decimal places.

55. $\log 3$
56. $\log 8$
57. $\log 532$
58. $\log 93{,}100$
59. $\log 0.57$
60. $\log 0.082$
61. $\log (-2)$
62. $\ln 50$
63. $\ln 2$
64. $\ln (-4)$
65. $\ln 809.3$
66. $\ln 0.00037$
67. $\ln (-1.32)$
68. $\ln 0$

Find the logarithm using common logarithms and the change-of-base formula. Round to four decimal places.

69. $\log_4 100$
70. $\log_3 20$
71. $\log_{100} 0.3$
72. $\log_\pi 100$
73. $\log_{200} 50$
74. $\log_{5.3} 1700$

Find the logarithm using natural logarithms and the change-of-base formula. Round to four decimal places.

75. $\log_3 12$
76. $\log_4 25$
77. $\log_{100} 15$
78. $\log_9 100$

Graph the function and its inverse using the same set of axes. Use any method.

79. $f(x) = 3^x,\ f^{-1}(x) = \log_3 x$
80. $f(x) = \log_4 x,\ f^{-1}(x) = 4^x$
81. $f(x) = \log x,\ f^{-1}(x) = 10^x$
82. $f(x) = e^x,\ f^{-1}(x) = \ln x$

For each of the following functions, briefly describe how the graph can be obtained from the graph of a basic logarithmic function. Then graph the function. Give the domain and the vertical asymptote of each function.

83. $f(x) = \log_2 (x + 3)$
84. $f(x) = \log_3 (x - 2)$
85. $y = \log_3 x - 1$
86. $y = 3 + \log_2 x$
87. $f(x) = 4 \ln x$
88. $f(x) = \frac{1}{2} \ln x$
89. $y = 2 - \ln x$
90. $y = \ln (x + 1)$
91. $f(x) = \frac{1}{2} \log (x - 1) - 2$
92. $f(x) = 5 - 2 \log (x + 1)$

Graph the piecewise function.

93. $g(x) = \begin{cases} 5, & \text{for } x \le 0, \\ \log x + 1, & \text{for } x > 0 \end{cases}$

94. $f(x) = \begin{cases} 1 - x, & \text{for } x \le -1, \\ \ln (x + 1), & \text{for } x > -1 \end{cases}$

95. *Walking Speed.* Refer to Example 12. Various cities and their populations are given below. Find the average walking speed in each city. Round to the nearest tenth of a foot per second.

a) El Paso, Texas: 672,538
b) Phoenix, Arizona: 1,488,750
c) Birmingham, Alabama: 212,038
d) Milwaukee, Wisconsin: 598,916
e) Honolulu, Hawaii: 345,610
f) Charlotte, North Carolina: 775,202
g) Omaha, Nebraska: 421,570
h) Sydney, Australia: 3,908,643

96. *Forgetting.* Students in an accounting class took a final exam and then took equivalent forms of the exam at monthly intervals thereafter. The average score $S(t)$, as a percent, after t months was found to be given by the function

$$S(t) = 78 - 15 \log (t + 1), \quad t \geq 0.$$

a) What was the average score when the students initially took the test, $t = 0$?
b) What was the average score after 4 months? after 24 months?

97. *Earthquake Magnitude.* Refer to Example 13. Various locations of earthquakes and their intensities are given below. Find the magnitude of each earthquake on the Richter scale.

a) San Francisco, California, 1906: $10^{7.7} \cdot I_0$
b) Chile, 1960: $10^{9.5} \cdot I_0$
c) Iran, 2003: $10^{6.6} \cdot I_0$
d) Turkey, 1999: $10^{7.6} \cdot I_0$
e) Peru, 2007: $10^{8.0} \cdot I_0$
f) China, 2008: $10^{7.9} \cdot I_0$
g) Spain, 2011: $10^{5.1} \cdot I_0$
h) Sumatra, 2004: $10^{9.3} \cdot I_0$

98. *pH of Substances in Chemistry.* In chemistry, the pH of a substance is defined as

$$\text{pH} = -\log [\text{H}^+],$$

where H^+ is the hydrogen ion concentration, in moles per liter. Find the pH of each substance.

SUBSTANCE	HYDROGEN ION CONCENTRATION
a) Pineapple juice	1.6×10^{-4}
b) Hair conditioner	0.0013
c) Mouthwash	6.3×10^{-7}
d) Eggs	1.6×10^{-8}
e) Tomatoes	6.3×10^{-5}

99. Find the hydrogen ion concentration of each substance, given the pH. (See Exercise 98.) Express the answer in scientific notation.

SUBSTANCE	pH
a) Tap water	7
b) Rainwater	5.4
c) Orange juice	3.2
d) Wine	4.8

100. *Advertising.* A model for advertising response is given by the function

$$N(a) = 1000 + 200 \ln a, \quad a \geq 1,$$

where $N(a)$ is the number of units sold when a is the amount spent on advertising, in thousands of dollars.

a) How many units were sold after spending $1000 ($a = 1$) on advertising?
b) How many units were sold after spending $5000?

101. *Loudness of Sound.* The **loudness** L, in bels (after Alexander Graham Bell), of a sound of intensity I is defined to be

$$L = \log \frac{I}{I_0},$$

where I_0 is the minimum intensity detectable by the human ear (such as the tick of a watch at 20 ft under quiet conditions). If a sound is 10 times as intense as another, its loudness is 1 bel greater than that of the other. If a sound is 100 times as intense as another, its loudness is 2 bels greater, and so on. The bel is a large unit, so a subunit, the **decibel**, is generally used. For L, in decibels, the formula is

$$L = 10 \log \frac{I}{I_0}.$$

Find the loudness, in decibels, of each sound with the given intensity.

SOUND	INTENSITY
a) Jet engine at 100 ft	$10^{14} \cdot I_0$
b) Loud rock concert	$10^{11.5} \cdot I_0$
c) Bird calls	$10^{4} \cdot I_0$
d) Normal conversation	$10^{6.5} \cdot I_0$
e) Thunder	$10^{12} \cdot I_0$
f) Loudest sound possible	$10^{19.4} \cdot I_0$

▶ Skill Maintenance

Find the slope and the y-intercept of the line. [1.3]

102. $3x - 10y = 14$

103. $y = 6$

104. $x = -4$

Use synthetic division to find the function values. [4.3]

105. $g(x) = x^3 - 6x^2 + 3x + 10$; find $g(-5)$

106. $f(x) = x^4 - 2x^3 + x - 6$; find $f(-1)$

Find a polynomial function of degree 3 with the given numbers as zeros. Answers may vary. [4.4]

107. $\sqrt{7}, -\sqrt{7}, 0$

108. $4i, -4i, 1$

▶ Synthesis

Simplify.

109. $\dfrac{\log_5 8}{\log_5 2}$

110. $\dfrac{\log_3 64}{\log_3 16}$

Find the domain of the function.

111. $f(x) = \log_5 x^3$

112. $f(x) = \log_4 x^2$

113. $f(x) = \ln |x|$

114. $f(x) = \log (3x - 4)$

Solve.

115. $\log_2 (2x + 5) < 0$

116. $\log_2 (x - 3) \geq 4$

In Exercises 117–120, match the equation with one of the figures (a)–(d) that follow.

a)

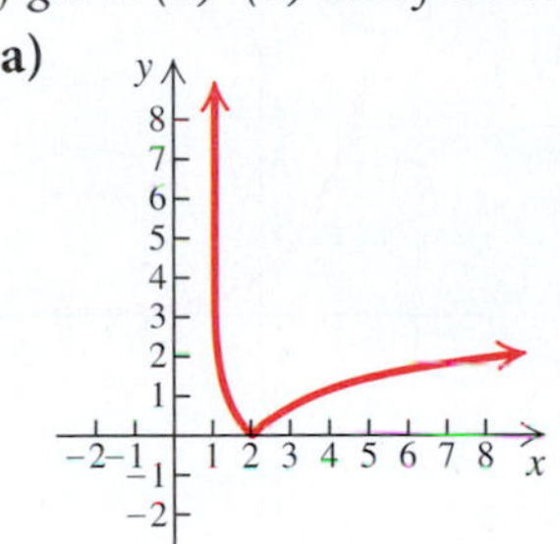

b)

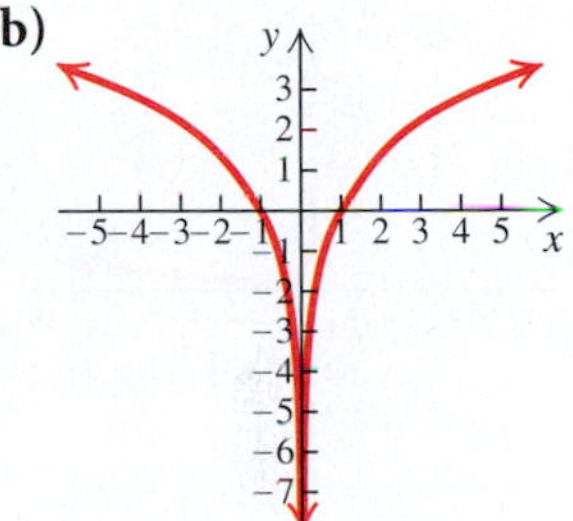

c)

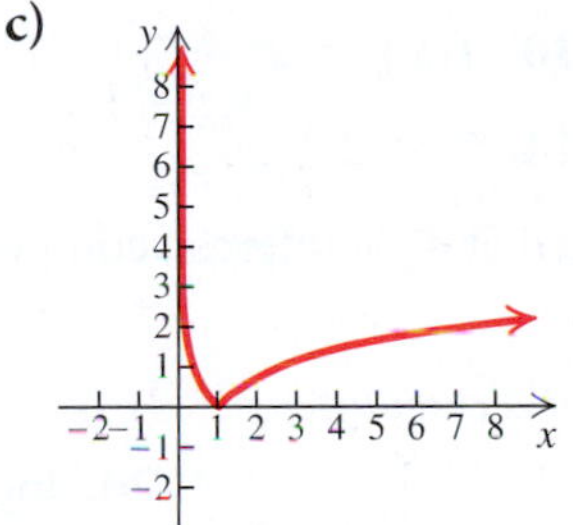

d)

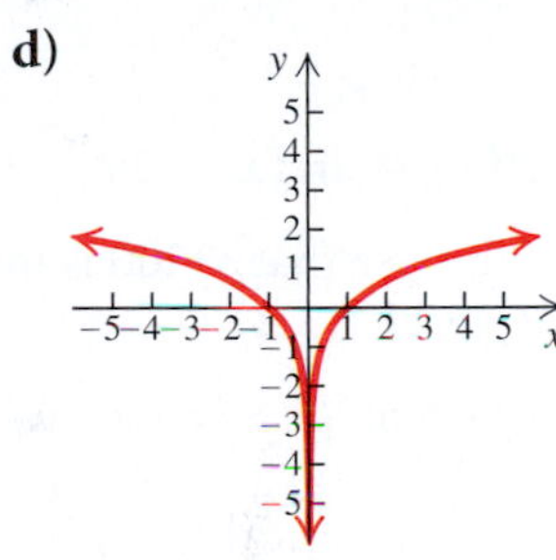

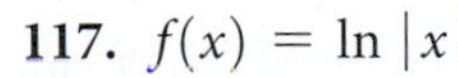

117. $f(x) = \ln |x|$

118. $f(x) = |\ln x|$

119. $f(x) = \ln x^2$

120. $g(x) = |\ln (x - 1)|$

Mid-Chapter Mixed Review

Determine whether the statement is true or false.

1. The domain of all logarithmic functions is $[1, \infty)$. [5.3]
2. The range of a one-to-one function f is the domain of its inverse f^{-1}. [5.1]
3. The y-intercept of $f(x) = e^{-x}$ is $(0, -1)$. [5.2]

For each function, determine whether it is one-to-one, and if the function is one-to-one, find a formula for its inverse. [5.1]

4. $f(x) = -\dfrac{2}{x}$

5. $f(x) = 3 + x^2$

6. $f(x) = \dfrac{5}{x - 2}$

7. Given the function $f(x) = \sqrt{x - 5}$, use composition of functions to show that $f^{-1}(x) = x^2 + 5$. [5.1]

8. Given the one-to-one function $f(x) = x^3 + 2$, find the inverse, give the domain and the range of f and f^{-1}, and graph both f and f^{-1} on the same set of axes. [5.1]

Match the function with one of the graphs (a)–(h) that follow. [5.2], [5.3]

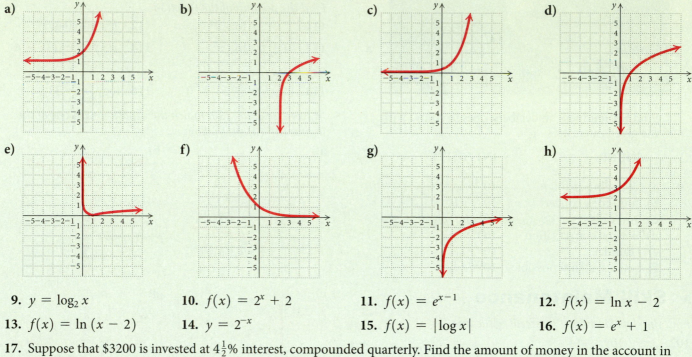

a)

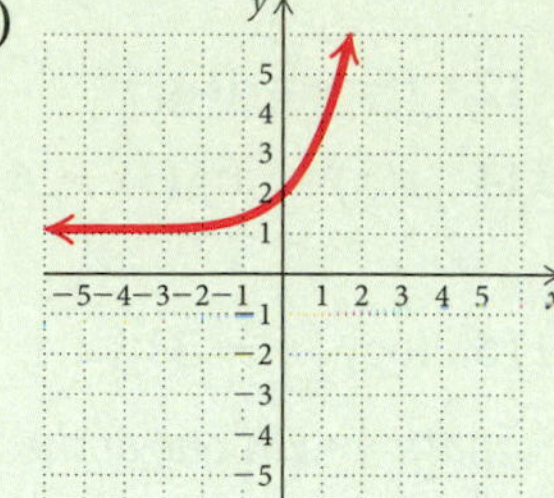

b)

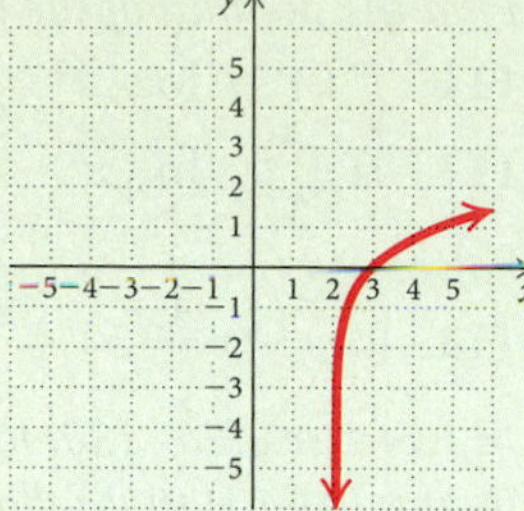

c)

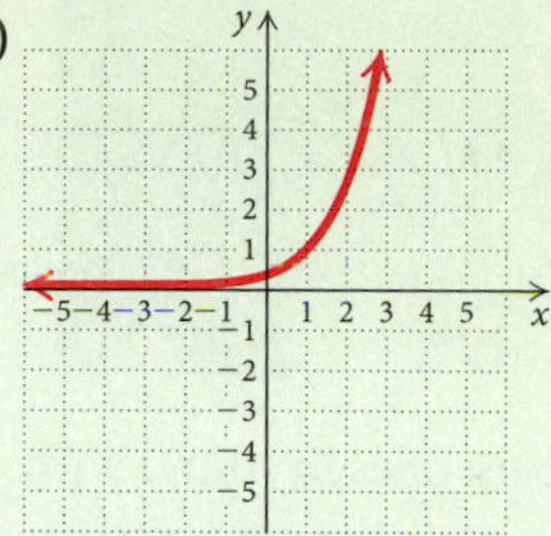

d)

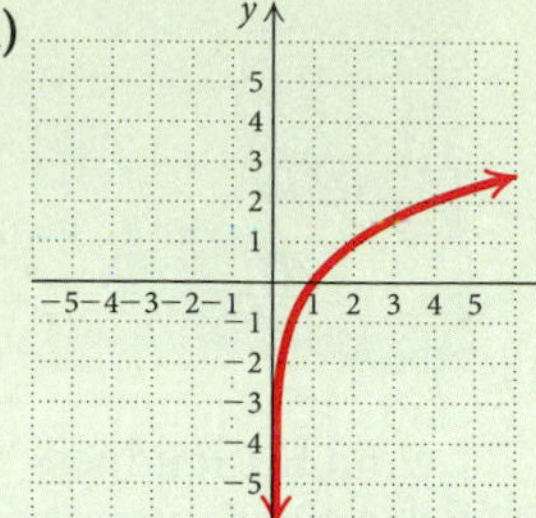

e)

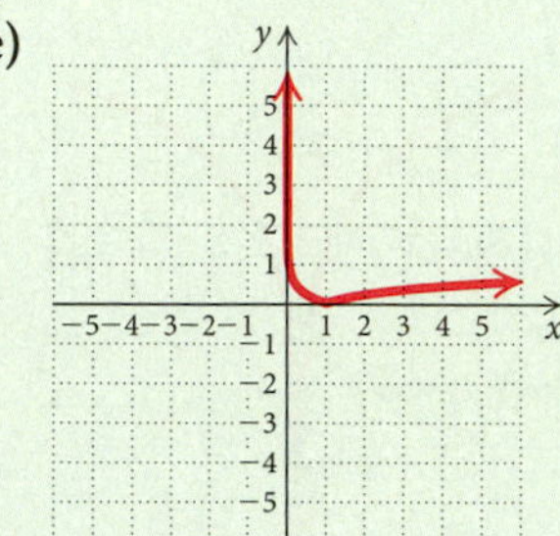

f)

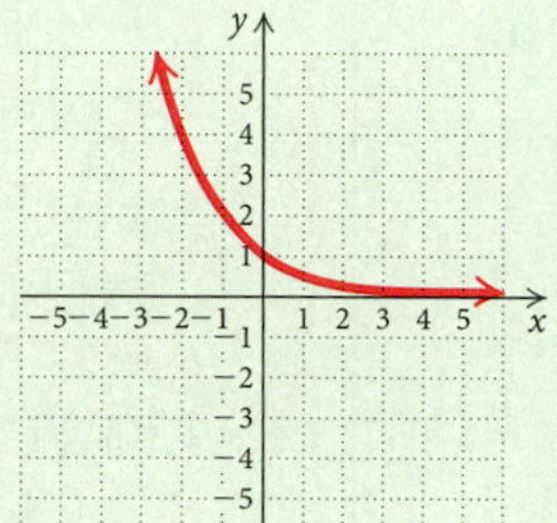

g)

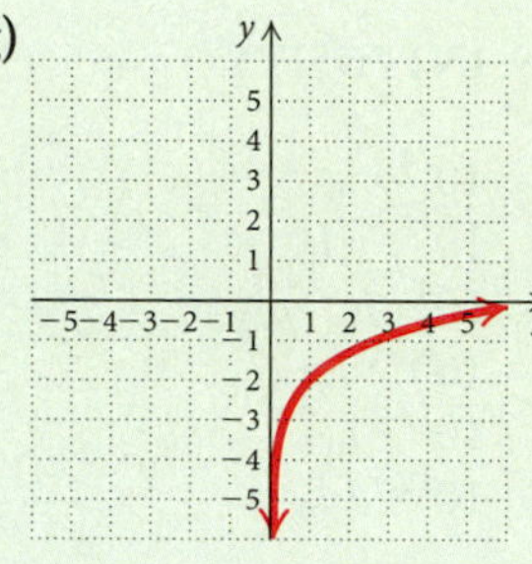

h)

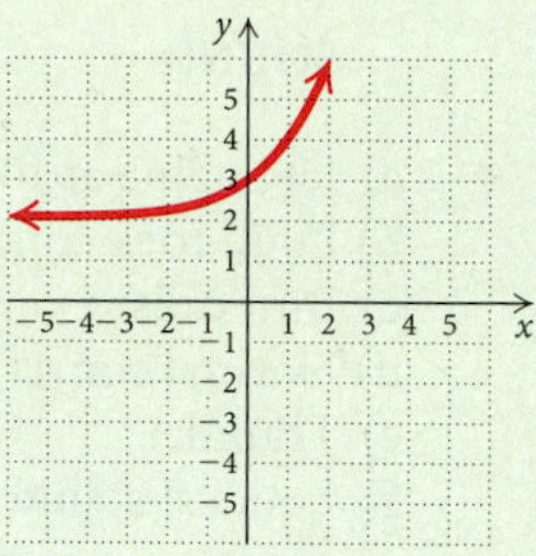

9. $y = \log_2 x$

10. $f(x) = 2^x + 2$

11. $f(x) = e^{x-1}$

12. $f(x) = \ln x - 2$

13. $f(x) = \ln (x - 2)$

14. $y = 2^{-x}$

15. $f(x) = |\log x|$

16. $f(x) = e^x + 1$

17. Suppose that \$3200 is invested at $4\frac{1}{2}\%$ interest, compounded quarterly. Find the amount of money in the account in 6 years. [5.2]

Find each of the following without a calculator. [5.3]

18. $\log_4 1$

19. $\ln e^{-4/5}$

20. $\log 0.01$

21. $\ln e^2$

22. $\ln 1$

23. $\log_2 \dfrac{1}{16}$

24. $\log 1$

25. $\log_3 27$

26. $\log \sqrt[4]{10}$

27. $\ln e$

28. Convert $e^{-6} = 0.0025$ to a logarithmic equation. [5.3]

29. Convert $\log T = r$ to an exponential equation. [5.3]

Find the logarithm using the change-of-base formula. [5.3]

30. $\log_3 20$

31. $\log_\pi 10$

Collaborative Discussion and Writing

32. Explain why an even function f does not have an inverse f^{-1} that is a function. [5.1]

33. Suppose that \$10,000 is invested for 8 years at 6.4% interest, compounded annually. In what year will the most interest be earned? Why? [5.2]

34. Describe the differences between the graph of $f(x) = x^3$ and the graph of $g(x) = 3^x$. [5.2]

35. If $\log b < 0$, what can you say about b? [5.3]

5.4 Properties of Logarithmic Functions

- Convert from logarithms of products, powers, and quotients to expressions in terms of individual logarithms, and conversely.
- Simplify expressions of the type $\log_a a^x$ and $a^{\log_a x}$.

We now establish some properties of logarithmic functions. These properties are based on the corresponding rules for exponents.

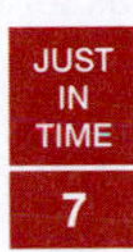

Logarithms of Products

The first property of logarithms corresponds to the product rule for exponents: $a^m \cdot a^n = a^{m+n}$.

> **THE PRODUCT RULE**
>
> For any positive numbers M and N and any logarithmic base a,
>
> $$\log_a MN = \log_a M + \log_a N.$$
>
> (The logarithm of a product is the sum of the logarithms of the factors.)

EXAMPLE 1 Express as a sum of logarithms: $\log_3 (9 \cdot 27)$.

Solution We have

$$\log_3 (9 \cdot 27) = \log_3 9 + \log_3 27. \quad \text{Using the product rule}$$

As a check, note that

$$\log_3 (9 \cdot 27) = \log_3 243 = 5 \quad 3^5 = 243$$

and $\log_3 9 + \log_3 27 = 2 + 3 = 5.$ $\quad 3^2 = 9;\ 3^3 = 27$

Now Try Exercise 1.

EXAMPLE 2 Express as a single logarithm: $\log_2 p^3 + \log_2 q$.

Solution We have

$$\log_2 p^3 + \log_2 q = \log_2 (p^3 q).$$

Now Try Exercise 35.

A Proof of the Product Rule. Let $\log_a M = x$ and $\log_a N = y$. Converting to exponential equations, we have $a^x = M$ and $a^y = N$. Then

$$MN = a^x \cdot a^y = a^{x+y}.$$

Converting back to a logarithmic equation, we get

$$\log_a MN = x + y.$$

Remembering what x and y represent, we know it follows that

$$\log_a MN = \log_a M + \log_a N.$$

■

Logarithms of Powers

The second property of logarithms corresponds to the power rule for exponents: $(a^m)^n = a^{mn}$.

THE POWER RULE

For any positive number M, any logarithmic base a, and any real number p,

$$\log_a M^p = p\log_a M.$$

(The logarithm of a power of M is the exponent times the logarithm of M.)

EXAMPLE 3 Express each of the following as a product.

a) $\log_a 11^{-3}$ b) $\log_a \sqrt[4]{7}$ c) $\ln x^6$

Solution

a) $\log_a 11^{-3} = -3\log_a 11$ Using the power rule

b) $\log_a \sqrt[4]{7} = \log_a 7^{1/4}$ Writing exponential notation

$= \frac{1}{4}\log_a 7$ Using the power rule

c) $\ln x^6 = 6\ln x$ Using the power rule

Now Try Exercises 13 and 15.

A Proof of the Power Rule. Let $x = \log_a M$. The equivalent exponential equation is $a^x = M$. Raising both sides to the power p, we obtain

$$(a^x)^p = M^p, \quad \text{or} \quad a^{xp} = M^p.$$

Converting back to a logarithmic equation, we get

$$\log_a M^p = xp.$$

But $x = \log_a M$, so substituting gives us

$$\log_a M^p = (\log_a M)p = p\log_a M.$$

■

Logarithms of Quotients

The third property of logarithms corresponds to the quotient rule for exponents: $a^m/a^n = a^{m-n}$.

THE QUOTIENT RULE

For any positive numbers M and N, and any logarithmic base a,

$$\log_a \frac{M}{N} = \log_a M - \log_a N.$$

(The logarithm of a quotient is the logarithm of the numerator minus the logarithm of the denominator.)

EXAMPLE 4 Express as a difference of logarithms: $\log_t \frac{8}{w}$.

Solution We have

$$\log_t \frac{8}{w} = \log_t 8 - \log_t w. \quad \text{Using the quotient rule}$$

Now Try Exercise 17.

EXAMPLE 5 Express as a single logarithm: $\log_b 64 - \log_b 16$.

Solution We have

$$\log_b 64 - \log_b 16 = \log_b \frac{64}{16} = \log_b 4.$$

Now Try Exercise 37.

A Proof of the Quotient Rule. The proof follows from both the product rule and the power rule:

$$\begin{aligned}\log_a \frac{M}{N} &= \log_a MN^{-1} \\ &= \log_a M + \log_a N^{-1} && \text{Using the product rule} \\ &= \log_a M + (-1)\log_a N && \text{Using the power rule} \\ &= \log_a M - \log_a N.\end{aligned}$$

■

Common Errors

$\log_a MN \neq (\log_a M)(\log_a N)$	The logarithm of a product is *not* the product of the logarithms.
$\log_a (M + N) \neq \log_a M + \log_a N$	The logarithm of a sum is *not* the sum of the logarithms.
$\log_a \frac{M}{N} \neq \frac{\log_a M}{\log_a N}$	The logarithm of a quotient is *not* the quotient of the logarithms.
$(\log_a M)^p \neq p \log_a M$	The power of a logarithm is *not* the exponent times the logarithm.

Applying the Properties

EXAMPLE 6 Express each of the following in terms of sums and differences of logarithms.

a) $\log_a \frac{x^2 y^5}{z^4}$ b) $\log_a \sqrt[3]{\frac{a^2 b}{c^5}}$ c) $\log_b \frac{ay^5}{m^3 n^4}$

Solution

a)
$$\begin{aligned}\log_a \frac{x^2 y^5}{z^4} &= \log_a (x^2 y^5) - \log_a z^4 && \text{Using the quotient rule} \\ &= \log_a x^2 + \log_a y^5 - \log_a z^4 && \text{Using the product rule} \\ &= 2 \log_a x + 5 \log_a y - 4 \log_a z && \text{Using the power rule}\end{aligned}$$

b) $\log_a \sqrt[3]{\dfrac{a^2b}{c^5}} = \log_a \left(\dfrac{a^2b}{c^5}\right)^{1/3}$ — Writing exponential notation

$= \dfrac{1}{3}\log_a \dfrac{a^2b}{c^5}$ — Using the power rule

$= \dfrac{1}{3}(\log_a a^2b - \log_a c^5)$ — Using the quotient rule. The parentheses are necessary.

$= \dfrac{1}{3}(2\log_a a + \log_a b - 5\log_a c)$ — Using the product rule and the power rule

$= \dfrac{1}{3}(2 + \log_a b - 5\log_a c)$ — $\log_a a = 1$

$= \dfrac{2}{3} + \dfrac{1}{3}\log_a b - \dfrac{5}{3}\log_a c$ — Multiplying to remove parentheses

c) $\log_b \dfrac{ay^5}{m^3n^4} = \log_b ay^5 - \log_b m^3n^4$ — Using the quotient rule

$= (\log_b a + \log_b y^5) - (\log_b m^3 + \log_b n^4)$ — Using the product rule

$= \log_b a + \log_b y^5 - \log_b m^3 - \log_b n^4$ — Removing parentheses

$= \log_b a + 5\log_b y - 3\log_b m - 4\log_b n$ — Using the power rule

Now Try Exercises 25 and 31.

EXAMPLE 7 Express as a single logarithm:

$$5\log_b x - \log_b y + \frac{1}{4}\log_b z.$$

Solution We have

$5\log_b x - \log_b y + \dfrac{1}{4}\log_b z = \log_b x^5 - \log_b y + \log_b z^{1/4}$ — Using the power rule

$= \log_b \dfrac{x^5}{y} + \log_b z^{1/4}$ — Using the quotient rule

$= \log_b \dfrac{x^5z^{1/4}}{y}$, or $\log_b \dfrac{x^5\sqrt[4]{z}}{y}$. — Using the product rule

Now Try Exercise 41.

EXAMPLE 8 Express as a single logarithm:

$$\ln(3x + 1) - \ln(3x^2 - 5x - 2).$$

Solution We have

$\ln(3x + 1) - \ln(3x^2 - 5x - 2)$

$= \ln \dfrac{3x + 1}{3x^2 - 5x - 2}$ — Using the quotient rule

$= \ln \dfrac{3x + 1}{(3x + 1)(x - 2)}$ — Factoring

$= \ln \dfrac{1}{x - 2}$. — Simplifying

Now Try Exercise 45.

EXAMPLE 9 Given that $\log_a 2 \approx 0.301$ and $\log_a 3 \approx 0.477$, find each of the following, if possible.

a) $\log_a 6$
b) $\log_a \frac{2}{3}$
c) $\log_a 81$
d) $\log_a \frac{1}{4}$
e) $\log_a 5$
f) $\frac{\log_a 3}{\log_a 2}$

Solution

a) $\log_a 6 = \log_a (2 \cdot 3) = \log_a 2 + \log_a 3$ — Using the product rule
$\approx 0.301 + 0.477$
≈ 0.778

b) $\log_a \frac{2}{3} = \log_a 2 - \log_a 3$ — Using the quotient rule
$\approx 0.301 - 0.477 \approx -0.176$

c) $\log_a 81 = \log_a 3^4 = 4 \log_a 3$ — Using the power rule
$\approx 4(0.477) \approx 1.908$

d) $\log_a \frac{1}{4} = \log_a 1 - \log_a 4$ — Using the quotient rule
$= 0 - \log_a 2^2$ — $\log_a 1 = 0$; $4 = 2^2$
$= -2 \log_a 2$ — Using the power rule
$\approx -2(0.301) \approx -0.602$

e) $\log_a 5$ *cannot* be found using these properties and the given information.
$\log_a 5 \neq \log_a 2 + \log_a 3$ — $\log_a 2 + \log_a 3 = \log_a (2 \cdot 3) = \log_a 6$

f) $\frac{\log_a 3}{\log_a 2} \approx \frac{0.477}{0.301} \approx 1.585$ — We simply divide, not using any of the properties.

Now Try Exercises 53 and 55.

Simplifying Expressions of the Type $\log_a a^x$ and $a^{\log_a x}$

We have two final properties of logarithms to consider. The first follows from the product rule: Since $\log_a a^x = x \log_a a = x \cdot 1 = x$, we have $\log_a a^x = x$. This property also follows from the definition of a logarithm: x is the power to which we raise a in order to get a^x.

THE LOGARITHM OF A BASE TO A POWER

For any base a and any real number x,

$$\log_a a^x = x.$$

(The logarithm, base a, of a to a power is the power.)

EXAMPLE 10 Simplify each of the following.

a) $\log_a a^8$
b) $\ln e^{-t}$
c) $\log 10^{3k}$

Solution

a) $\log_a a^8 = 8$ — 8 is the power to which we raise a in order to get a^8.
b) $\ln e^{-t} = \log_e e^{-t} = -t$ — $\ln e^x = x$
c) $\log 10^{3k} = \log_{10} 10^{3k} = 3k$

Now Try Exercises 65 and 73.

Let $M = \log_a x$. Then $a^M = x$. Substituting $\log_a x$ for M, we obtain $a^{\log_a x} = x$. This also follows from the definition of a logarithm: $\log_a x$ is the power to which a is raised in order to get x.

A BASE TO A LOGARITHMIC POWER

For any base a and any positive real number x,

$$a^{\log_a x} = x.$$

(The number a raised to the power $\log_a x$ is x.)

EXAMPLE 11 Simplify each of the following.

a) $4^{\log_4 k}$ **b)** $e^{\ln 5}$ **c)** $10^{\log 7t}$

Solution

a) $4^{\log_4 k} = k$

b) $e^{\ln 5} = e^{\log_e 5} = 5$

c) $10^{\log 7t} = 10^{\log_{10} 7t} = 7t$

Now Try Exercises 69 and 71.

A Proof of the Change-of-Base Formula. We close this section by proving the change-of-base formula and summarizing the properties of logarithms considered thus far in this chapter. In Section 5.3, we used the change-of-base formula,

$$\log_b M = \frac{\log_a M}{\log_a b},$$

to make base conversions in order to find logarithmic values using a calculator. Let $x = \log_b M$. Then

$$b^x = M \quad \text{Definition of logarithm}$$

$$\log_a b^x = \log_a M \quad \text{Taking the logarithm on both sides}$$

$$x \log_a b = \log_a M \quad \text{Using the power rule}$$

$$x = \frac{\log_a M}{\log_a b}, \quad \text{Dividing by } \log_a b$$

so $\quad x = \log_b M = \dfrac{\log_a M}{\log_a b}.$ ■

CHANGE-OF-BASE FORMULA

REVIEW SECTION 5.3

Following is a summary of the properties of logarithms.

Summary of the Properties of Logarithms

The Product Rule: $\log_a MN = \log_a M + \log_a N$

The Power Rule: $\log_a M^p = p \log_a M$

The Quotient Rule: $\log_a \dfrac{M}{N} = \log_a M - \log_a N$

The Change-of-Base Formula: $\log_b M = \dfrac{\log_a M}{\log_a b}$

Other Properties: $\log_a a = 1, \quad \log_a 1 = 0,$

$\log_a a^x = x, \quad a^{\log_a x} = x$

5.4 Exercise Set

Express as a sum of logarithms.

1. $\log_3 (81 \cdot 27)$
2. $\log_2 (8 \cdot 64)$
3. $\log_5 (5 \cdot 125)$
4. $\log_4 (64 \cdot 4)$
5. $\log_t 8Y$
6. $\log 0.2x$
7. $\ln xy$
8. $\ln ab$

Express as a product.

9. $\log_b t^3$
10. $\log_a x^4$
11. $\log y^8$
12. $\ln y^5$
13. $\log_c K^{-6}$
14. $\log_b Q^{-8}$
15. $\ln \sqrt[3]{4}$
16. $\ln \sqrt{a}$

Express as a difference of logarithms.

17. $\log_t \frac{M}{8}$
18. $\log_a \frac{76}{13}$
19. $\log \frac{x}{y}$
20. $\ln \frac{a}{b}$
21. $\ln \frac{r}{s}$
22. $\log_b \frac{3}{w}$

Express in terms of sums and differences of logarithms.

23. $\log_a 6xy^5z^4$
24. $\log_a x^3y^2z$
25. $\log_b \frac{p^2q^5}{m^4b^9}$
26. $\log_b \frac{x^2y}{b^3}$
27. $\ln \frac{2}{3x^3y}$
28. $\log \frac{5a}{4b^2}$
29. $\log \sqrt{r^3t}$
30. $\ln \sqrt[3]{5x^5}$
31. $\log_a \sqrt{\frac{x^6}{p^5q^8}}$
32. $\log_c \sqrt[3]{\frac{y^3z^2}{x^4}}$
33. $\log_a \sqrt[4]{\frac{m^8n^{12}}{a^3b^5}}$
34. $\log_a \sqrt{\frac{a^6b^8}{a^2b^5}}$

Express as a single logarithm and, if possible, simplify.

35. $\log_a 75 + \log_a 2$
36. $\log 0.01 + \log 1000$
37. $\log 10{,}000 - \log 100$
38. $\ln 54 - \ln 6$
39. $\frac{1}{2} \log n + 3 \log m$
40. $\frac{1}{2} \log a - \log 2$
41. $\frac{1}{2} \log_a x + 4 \log_a y - 3 \log_a x$
42. $\frac{2}{5} \log_a x - \frac{1}{3} \log_a y$
43. $\ln x^2 - 2 \ln \sqrt{x}$
44. $\ln 2x + 3(\ln x - \ln y)$
45. $\ln (x^2 - 4) - \ln (x + 2)$
46. $\log (x^3 - 8) - \log (x - 2)$
47. $\log (x^2 - 5x - 14) - \log (x^2 - 4)$
48. $\log_a \frac{a}{\sqrt{x}} - \log_a \sqrt{ax}$
49. $\ln x - 3[\ln (x - 5) + \ln (x + 5)]$
50. $\frac{2}{3}\left[\ln (x^2 - 9) - \ln (x + 3)\right] + \ln (x + y)$
51. $\frac{3}{2} \ln 4x^6 - \frac{4}{5} \ln 2y^{10}$
52. $120(\ln \sqrt[5]{x^3} + \ln \sqrt[3]{y^2} - \ln \sqrt[4]{16z^5})$

Given that $\log_a 2 \approx 0.301$, $\log_a 7 \approx 0.845$, and $\log_a 11 \approx 1.041$, find each of the following, if possible. Round the answer to the nearest thousandth.

53. $\log_a \frac{2}{11}$
54. $\log_a 14$
55. $\log_a 98$
56. $\log_a \frac{1}{7}$
57. $\frac{\log_a 2}{\log_a 7}$
58. $\log_a 9$

Given that $\log_b 2 \approx 0.693$, $\log_b 3 \approx 1.099$, and $\log_b 5 \approx 1.609$, find each of the following, if possible. Round the answer to the nearest thousandth.

59. $\log_b 125$
60. $\log_b \frac{5}{3}$
61. $\log_b \frac{1}{6}$
62. $\log_b 30$
63. $\log_b \frac{3}{b}$
64. $\log_b 15b$

Simplify.

65. $\log_p p^3$
66. $\log_t t^{2713}$
67. $\log_e e^{|x-4|}$
68. $\log_q q^{\sqrt{3}}$
69. $3^{\log_3 4x}$
70. $5^{\log_5 (4x-3)}$
71. $10^{\log w}$
72. $e^{\ln x^3}$
73. $\ln e^{8t}$
74. $\log 10^{-k}$
75. $\log_b \sqrt{b}$
76. $\log_b \sqrt{b^3}$

Skill Maintenance

In each of Exercises 77–86, classify the function as linear, quadratic, cubic, quartic, rational, exponential, or logarithmic.

77. $f(x) = 5 - x^2 + x^4$ [4.1]

78. $f(x) = 2^x$ [5.2]

79. $f(x) = -\frac{3}{4}$ [1.3]

80. $f(x) = 4^x - 8$ [5.2]

81. $f(x) = -\dfrac{3}{x}$ [4.5]

82. $f(x) = \log x + 6$ [5.3]

83. $f(x) = -\frac{1}{3}x^3 - 4x^2 + 6x + 42$ [4.1]

84. $f(x) = \dfrac{x^2 - 1}{x^2 + x - 6}$ [4.5]

85. $f(x) = \frac{1}{2}x + 3$ [1.3]

86. $f(x) = 2x^2 - 6x + 3$ [3.3]

Synthesis

Solve for x.

87. $5^{\log_5 8} = 2x$

88. $\ln e^{3x-5} = -8$

Express as a single logarithm and, if possible, simplify.

89. $\log_a (x^2 + xy + y^2) + \log_a (x - y)$

90. $\log_a (a^{10} - b^{10}) - \log_a (a + b)$

Express as a sum or a difference of logarithms.

91. $\log_a \dfrac{x - y}{\sqrt{x^2 - y^2}}$

92. $\log_a \sqrt{9 - x^2}$

93. Given that $\log_a x = 2$, $\log_a y = 3$, and $\log_a z = 4$, find

$$\log_a \frac{\sqrt[4]{y^2 z^5}}{\sqrt[4]{x^3 z^{-2}}}.$$

Determine whether each of the following is true. Assume that a, x, M, and N are positive.

94. $\log_a M + \log_a N = \log_a (M + N)$

95. $\log_a M - \log_a N = \log_a \dfrac{M}{N}$

96. $\dfrac{\log_a M}{\log_a N} = \log_a M - \log_a N$

97. $\dfrac{\log_a M}{x} = \log_a M^{1/x}$

98. $\log_a x^3 = 3 \log_a x$

99. $\log_a 8x = \log_a x + \log_a 8$

100. $\log_N (MN)^x = x \log_N M + x$

Suppose that $\log_a x = 2$. Find each of the following.

101. $\log_a \left(\dfrac{1}{x}\right)$

102. $\log_{1/a} x$

103. Simplify:

$\log_{10} 11 \cdot \log_{11} 12 \cdot \log_{12} 13 \cdots \log_{998} 999 \cdot \log_{999} 1000.$

Write each of the following without using logarithms.

104. $\log_a x + \log_a y - mz = 0$

105. $\ln a - \ln b + xy = 0$

Prove each of the following for any base a and any positive number x.

106. $\log_a \left(\dfrac{1}{x}\right) = -\log_a x = \log_{1/a} x$

107. $\log_a \left(\dfrac{x + \sqrt{x^2 - 5}}{5}\right) = -\log_a (x - \sqrt{x^2 - 5})$

5.5 Solving Exponential Equations and Logarithmic Equations

- Solve exponential equations.
- Solve logarithmic equations.

Solving Exponential Equations

Equations with variables in the exponents, such as

$$3^x = 20 \quad \text{and} \quad 2^{5x} = 64,$$

are called **exponential equations**.

Sometimes, as is the case with the equation $2^{5x} = 64$, we can write each side as a power of the same number:

$$2^{5x} = 2^6.$$

We can then set the exponents equal and solve:

$$5x = 6$$
$$x = \tfrac{6}{5}, \text{ or } 1.2.$$

We use the following property to solve exponential equations.

BASE-EXPONENT PROPERTY

For any $a > 0$, $a \neq 1$,

$$a^x = a^y \longleftrightarrow x = y.$$

ONE-TO-ONE FUNCTIONS
REVIEW SECTION 5.1

This property follows from the fact that for any $a > 0$, $a \neq 1$, $f(x) = a^x$ is a one-to-one function. If $a^x = a^y$, then $f(x) = f(y)$. Then since f is one-to-one, it follows that $x = y$. Conversely, if $x = y$, it follows that $a^x = a^y$, since we are raising a to the same power in each case.

EXAMPLE 1 Solve: $2^{3x-7} = 32$.

Algebraic Solution

Note that $32 = 2^5$. Thus we can write each side as a power of the same number:

$$2^{3x-7} = 2^5.$$

Since the bases are the same number, 2, we can use the base-exponent property and set the exponents equal:

$$3x - 7 = 5$$
$$3x = 12$$
$$x = 4.$$

Check:

2^{3x-7}	$= 32$
$2^{3(4)-7}$	? 32
2^{12-7}	
2^5	
32	32 TRUE

The solution is 4.

Visualizing the Solution

We graph $y = 2^{3x-7}$ and $y = 32$. The first coordinate of the point of intersection of the graphs is the value of x for which $2^{3x-7} = 32$ and is thus the solution of the equation.

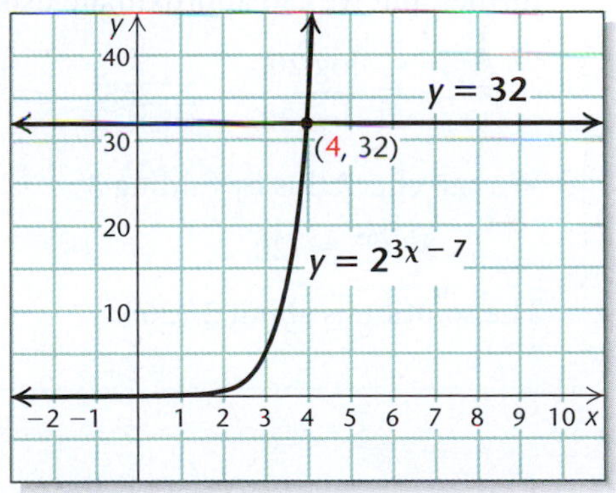

The solution of $2^{3x-7} = 32$ is 4.

Now Try Exercise 7.

Another property that is used when solving some exponential equations and logarithmic equations is as follows.

PROPERTY OF LOGARITHMIC EQUALITY

For any $M > 0, N > 0, a > 0$, and $a \neq 1$,

$$\log_a M = \log_a N \leftrightarrow M = N.$$

This property follows from the fact that for any $a > 0, a \neq 1, f(x) = \log_a x$ is a one-to-one function. If $\log_a x = \log_a y$, then $f(x) = f(y)$. Then since f is one-to-one, it follows that $x = y$. Conversely, if $x = y$, it follows that $\log_a x = \log_a y$, since we are taking the logarithm of the same number in each case.

When it does not seem possible to write each side as a power of the same base, we can use the property of logarithmic equality and take the logarithm with any base on each side and then use the power rule for logarithms.

EXAMPLE 2 Solve: $3^x = 20$.

Algebraic Solution

We have

$$3^x = 20$$

$$\log 3^x = \log 20 \quad \text{Taking the common logarithm on both sides}$$

$$x \log 3 = \log 20 \quad \text{Using the power rule}$$

$$x = \frac{\log 20}{\log 3}. \quad \text{Dividing by log 3}$$

This is an exact answer. We cannot simplify further, but we can approximate using a calculator:

$$x = \frac{\log 20}{\log 3} \approx 2.7268.$$

We can check this by finding $3^{2.7268}$:

$$3^{2.7268} \approx 20.$$

The solution is about 2.7268.

Visualizing the Solution

We graph $y = 3^x$ and $y = 20$. The first coordinate of the point of intersection of the graphs is the value of x for which $3^x = 20$ and is thus the solution of the equation.

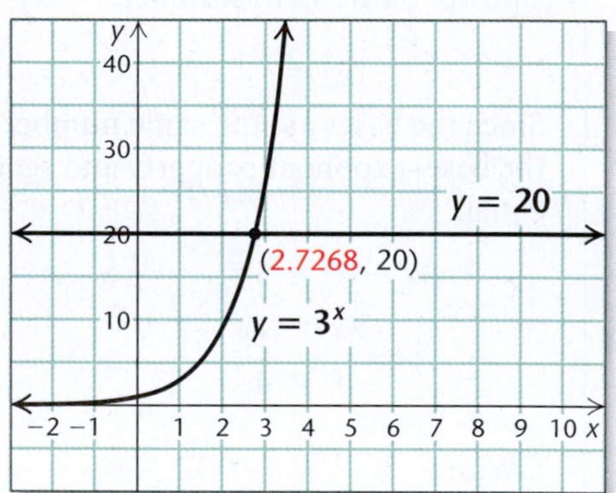

The solution is approximately 2.7268.

Now Try Exercise 11.

Technology Connection

With some calculators, it is possible to find a logarithm with any logarithmic base using the logBASE operation from the MATH MATH menu. The computation $\log_3 20$ is shown in the following window.

```
log3(20)
        2.726833028
```

In Example 2, we took the common logarithm on both sides of the equation. Any base will give the same result. Let's try base 3. We have

$$3^x = 20$$
$$\log_3 3^x = \log_3 20$$
$$x = \log_3 20 \qquad \log_a a^x = x$$
$$x = \frac{\log 20}{\log 3} \qquad \text{Using the change-of-base formula}$$
$$x \approx 2.7268.$$

Note that we must change the base in order to do the final calculation.

EXAMPLE 3 Solve: $100e^{0.08t} = 2500$.

Algebraic Solution

It will make our work easier if we take the natural logarithm when working with equations that have e as a base.

We have

$$100e^{0.08t} = 2500$$
$$e^{0.08t} = 25 \qquad \text{Dividing by 100}$$
$$\ln e^{0.08t} = \ln 25 \qquad \text{Taking the natural logarithm on both sides}$$
$$0.08t = \ln 25 \qquad \text{Finding the logarithm of a base to a power: } \log_a a^x = x$$
$$t = \frac{\ln 25}{0.08} \qquad \text{Dividing by 0.08}$$
$$t \approx 40.2.$$

The solution is about 40.2.

Visualizing the Solution

The first coordinate of the point of intersection of the graphs of $y = 100e^{0.08t}$ and $y = 2500$ is about 40.2. This is the solution of the equation.

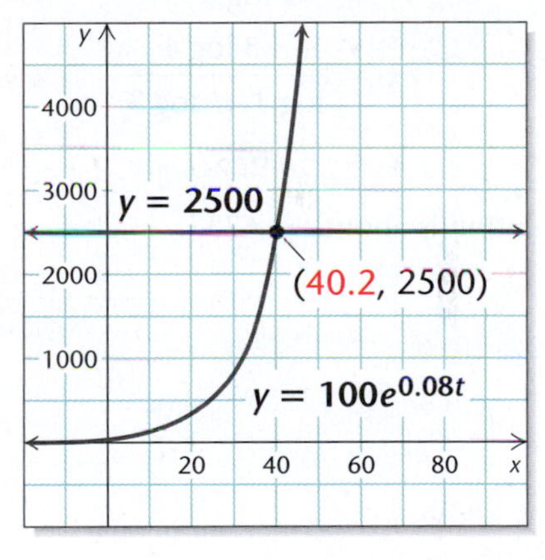

Now Try Exercise 19.

Technology Connection

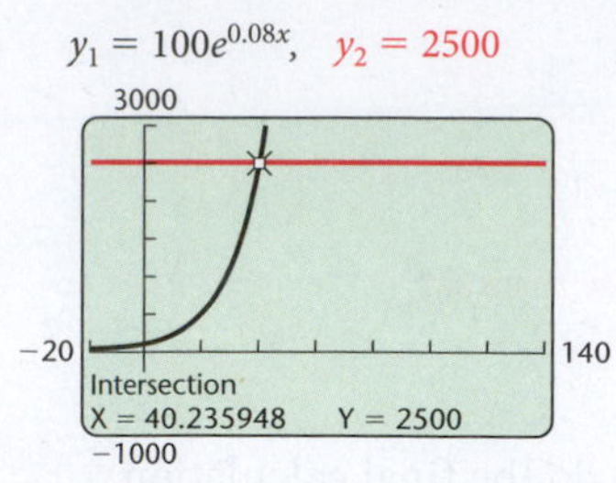

We can solve the equations in Examples 1–4 using the Intersect method. In Example 3, for instance, we graph $y = 100e^{0.08x}$ and $y = 2500$ and use the INTERSECT feature to find the coordinates of the point of intersection.

The first coordinate of the point of intersection is the solution of the equation $100e^{0.08x} = 2500$. The solution is about 40.2. We could also write the equation in the form $100e^{0.08x} - 2500 = 0$ and use the Zero method.

EXAMPLE 4 Solve: $4^{x+3} = 3^{-x}$.

Algebraic Solution

We have

$$4^{x+3} = 3^{-x}$$

$$\log 4^{x+3} = \log 3^{-x} \qquad \text{Taking the common logarithm on both sides}$$

$$(x + 3) \log 4 = -x \log 3 \qquad \text{Using the power rule}$$

$$x \log 4 + 3 \log 4 = -x \log 3 \qquad \text{Removing parentheses}$$

$$x \log 4 + x \log 3 = -3 \log 4 \qquad \text{Adding } x \log 3 \text{ and subtracting } 3 \log 4$$

$$x(\log 4 + \log 3) = -3 \log 4 \qquad \text{Factoring on the left}$$

$$x = \frac{-3 \log 4}{\log 4 + \log 3} \qquad \text{Dividing by } \log 4 + \log 3$$

$$x \approx -1.6737.$$

The solution is about -1.6737.

Visualizing the Solution

We graph $y = 4^{x+3}$ and $y = 3^{-x}$. The first coordinate of the point of intersection of the graphs is the value of x for which $4^{x+3} = 3^{-x}$ and is thus the solution of the equation.

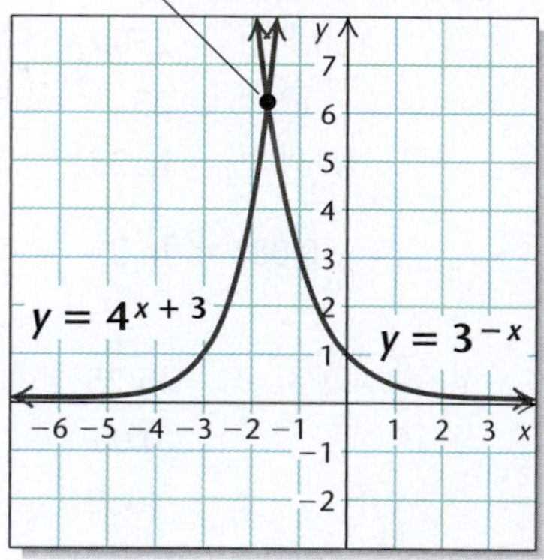

The solution is approximately -1.6737.

Now Try Exercise 21.

EQUATIONS REDUCIBLE TO QUADRATIC

REVIEW SECTION 3.2

EXAMPLE 5 Solve: $e^x + e^{-x} - 6 = 0$.

Algebraic Solution

In this case, we have more than one term with x in the exponent:

$$e^x + e^{-x} - 6 = 0$$

$$e^x + \frac{1}{e^x} - 6 = 0 \qquad \text{Rewriting } e^{-x} \text{ with a positive exponent}$$

$$e^{2x} + 1 - 6e^x = 0. \qquad \text{Multiplying by } e^x \text{ on both sides}$$

This equation is reducible to quadratic with $u = e^x$:

$$u^2 - 6u + 1 = 0.$$

We use the quadratic formula with $a = 1$, $b = -6$, and $c = 1$:

$$u = \frac{-b \pm \sqrt{b^2 - 4ac}}{2a}$$

$$u = \frac{-(-6) \pm \sqrt{(-6)^2 - 4 \cdot 1 \cdot 1}}{2 \cdot 1}$$

$$u = \frac{6 \pm \sqrt{32}}{2} = \frac{6 \pm 4\sqrt{2}}{2}$$

$$u = \frac{2(3 \pm 2\sqrt{2})}{2}$$

$$u = 3 \pm 2\sqrt{2}$$

$$e^x = 3 \pm 2\sqrt{2}. \qquad \text{Replacing } u \text{ with } e^x$$

We now take the natural logarithm on both sides:

$$\ln e^x = \ln(3 \pm 2\sqrt{2})$$

$$x = \ln(3 \pm 2\sqrt{2}). \qquad \text{Using } \ln e^x = x$$

Approximating each of the solutions, we obtain 1.76 and −1.76.

Visualizing the Solution

The solutions of the equation

$$e^x + e^{-x} - 6 = 0$$

are the zeros of the function

$$f(x) = e^x + e^{-x} - 6.$$

Note that the solutions are also the first coordinates of the x-intercepts of the graph of the function.

$f(x) = e^x + e^{-x} - 6$

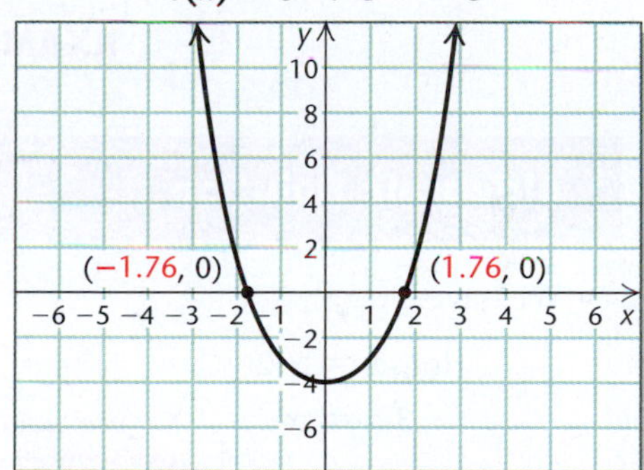

The leftmost zero is about −1.76. The zero on the right is about 1.76. The solutions of the equation are approximately −1.76 and 1.76.

Now Try Exercise 25.

Technology Connection

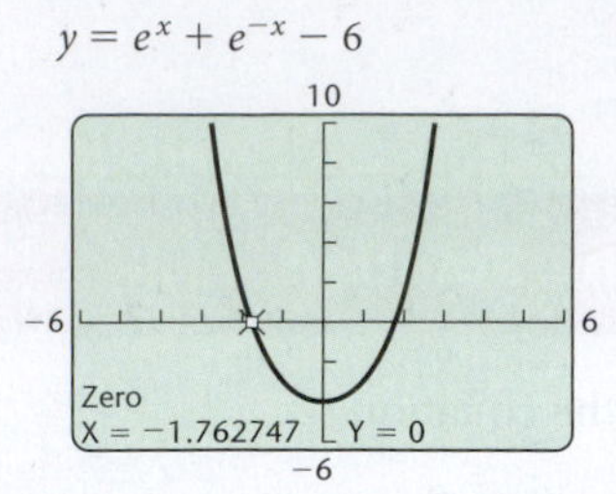

We can use the Zero method in Example 5 to solve the equation $e^x + e^{-x} - 6 = 0$. We graph the function $y = e^x + e^{-x} - 6$ and use the ZERO feature to find the zeros.

The leftmost zero is about -1.76. Using the ZERO feature one more time, we find that the other zero is about 1.76.

Solving Logarithmic Equations

Equations containing variables in logarithmic expressions, such as $\log_2 x = 4$ and $\log x + \log(x + 3) = 1$, are called **logarithmic equations**. To solve logarithmic equations algebraically, we first try to obtain a single logarithmic expression on one side and then write an equivalent exponential equation.

EXAMPLE 6 Solve: $\log_3 x = -2$.

Algebraic Solution

We have

$$\log_3 x = -2$$

$$3^{-2} = x \quad \text{Converting to an exponential equation}$$

$$\frac{1}{3^2} = x$$

$$\frac{1}{9} = x.$$

Check:

$\log_3 x = -2$	
$\log_3 \frac{1}{9}$?	-2
$\log_3 3^{-2}$	
-2	-2 TRUE

The solution is $\frac{1}{9}$.

Visualizing the Solution

When we graph $y = \log_3 x$ and $y = -2$, we find that the first coordinate of the point of intersection of the graphs is $\frac{1}{9}$.

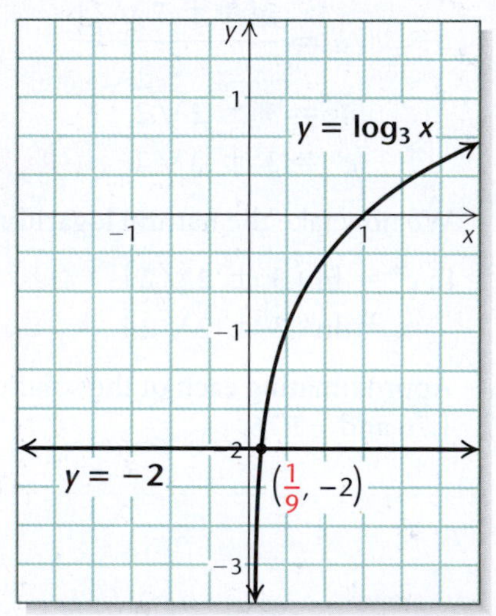

The solution of $\log_3 x = -2$ is $\frac{1}{9}$.

Now Try Exercise 33.

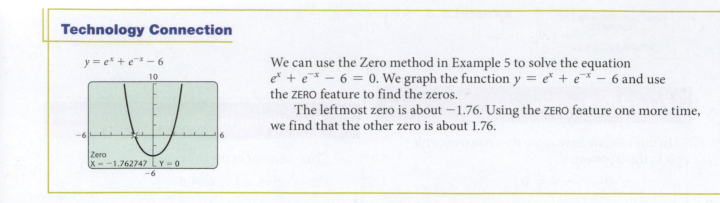

EXAMPLE 7 Solve: $\log x + \log(x + 3) = 1$.

Algebraic Solution

In this case, we have common logarithms. Writing the base of 10 will help us understand the problem:

$$\log_{10} x + \log_{10}(x + 3) = 1$$

$$\log_{10}[x(x + 3)] = 1 \quad \text{Using the product rule to obtain a single logarithm}$$

$$x(x + 3) = 10^1 \quad \text{Writing an equivalent exponential equation}$$

$$x^2 + 3x = 10$$

$$x^2 + 3x - 10 = 0$$

$$(x - 2)(x + 5) = 0 \quad \text{Factoring}$$

$$x - 2 = 0 \quad or \quad x + 5 = 0$$

$$x = 2 \quad or \quad x = -5.$$

Check: For 2:

$$\begin{array}{r|l} \log x + \log(x + 3) & 1 \\ \hline \log 2 + \log(2 + 3) & ?\ 1 \\ \log 2 + \log 5 & \\ \log(2 \cdot 5) & \\ \log 10 & \\ 1 & 1 \quad \text{TRUE} \end{array}$$

For -5:

$$\begin{array}{r|l} \log x + \log(x + 3) & 1 \\ \hline \log(-5) + \log(-5 + 3) & ?\ 1 \quad \text{FALSE} \end{array}$$

The number -5 is not a solution because negative numbers do not have real-number logarithms. The solution is 2.

Visualizing the Solution

The solution of the equation

$$\log x + \log(x + 3) = 1$$

is the zero of the function

$$f(x) = \log x + \log(x + 3) - 1.$$

The solution is also the first coordinate of the x-intercept of the graph of the function.

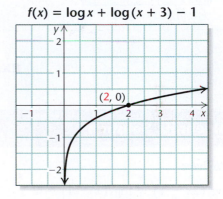

$f(x) = \log x + \log(x + 3) - 1$

The solution of the equation is 2. From the graph, we can easily see that there is only one solution.

Now Try Exercise 41.

Technology Connection

$y_1 = \log x + \log(x + 3)$, $y_2 = 1$

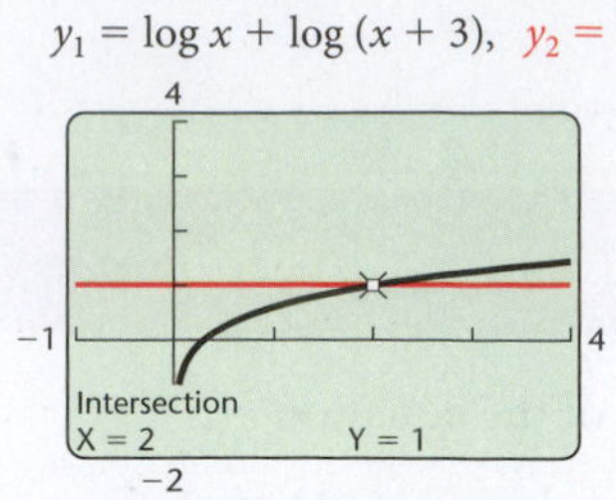

$y_1 = \log x + \log(x + 3) - 1$

4

−1 4

Zero

X = 2 Y = 0

−2

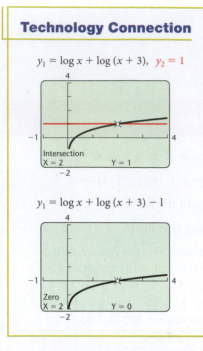

In Example 7, we can graph the equations

$$y_1 = \log x + \log(x + 3)$$

and

$$y_2 = 1$$

and use the Intersect method. The first coordinate of the point of intersection is the solution of the equation.

We could also graph the function

$$y = \log x + \log(x + 3) - 1$$

and use the Zero method. The zero of the function is the solution of the equation.

With either method, we see that the solution is 2. Note that the graphical solution gives only the one *true* solution.

EXAMPLE 8 Solve: $\log_3(2x - 1) - \log_3(x - 4) = 2$.

Algebraic Solution

We have

$$\log_3(2x - 1) - \log_3(x - 4) = 2$$

$$\log_3 \frac{2x - 1}{x - 4} = 2 \qquad \text{Using the quotient rule}$$

$$\frac{2x - 1}{x - 4} = 3^2 \qquad \text{Writing an equivalent exponential equation}$$

$$\frac{2x - 1}{x - 4} = 9$$

$$(x - 4) \cdot \frac{2x - 1}{x - 4} = 9(x - 4) \qquad \text{Multiplying by the LCD, } x - 4$$

$$2x - 1 = 9x - 36$$

$$35 = 7x$$

$$5 = x.$$

Check:

$$\log_3(2x - 1) - \log_3(x - 4) = 2$$

$$\log_3(2 \cdot 5 - 1) - \log_3(5 - 4) \stackrel{?}{=} 2$$

$$\log_3 9 - \log_3 1$$

$$2 - 0$$

$$2 \mid 2 \quad \text{TRUE}$$

The solution is 5.

Visualizing the Solution

We see that the first coordinate of the point of intersection of the graphs of

$$y = \log_3(2x - 1) - \log_3(x - 4)$$

and

$$y = 2$$

is 5.

$y = \log_3(2x - 1) - \log_3(x - 4)$

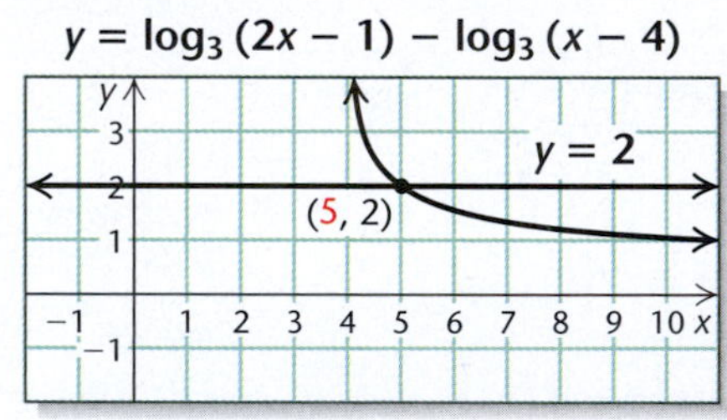

The solution is 5.

Now Try Exercise 45.

EXAMPLE 9 Solve: $\ln(4x + 6) - \ln(x + 5) = \ln x$.

Algebraic Solution

We have

$$\ln(4x + 6) - \ln(x + 5) = \ln x$$

$$\ln \frac{4x + 6}{x + 5} = \ln x \quad \text{Using the quotient rule}$$

$$\frac{4x + 6}{x + 5} = x \quad \text{Using the property of logarithmic equality}$$

$$(x + 5) \cdot \frac{4x + 6}{x + 5} = x(x + 5) \quad \text{Multiplying by } x + 5$$

$$4x + 6 = x^2 + 5x$$

$$0 = x^2 + x - 6$$

$$0 = (x + 3)(x - 2) \quad \text{Factoring}$$

$$x + 3 = 0 \quad or \quad x - 2 = 0$$

$$x = -3 \quad or \quad x = 2.$$

The number -3 is not a solution because $4(-3) + 6 = -6$ and $\ln(-6)$ is not a real number. The value 2 checks and is the solution.

Visualizing the Solution

The solution of the equation

$$\ln(4x + 6) - \ln(x + 5) = \ln x$$

is the zero of the function

$$f(x) = \ln(4x + 6) - \ln(x + 5) - \ln x.$$

The solution is also the first coordinate of the x-intercept of the graph of the function.

$f(x) = \ln(4x + 6) - \ln(x + 5) - \ln x$

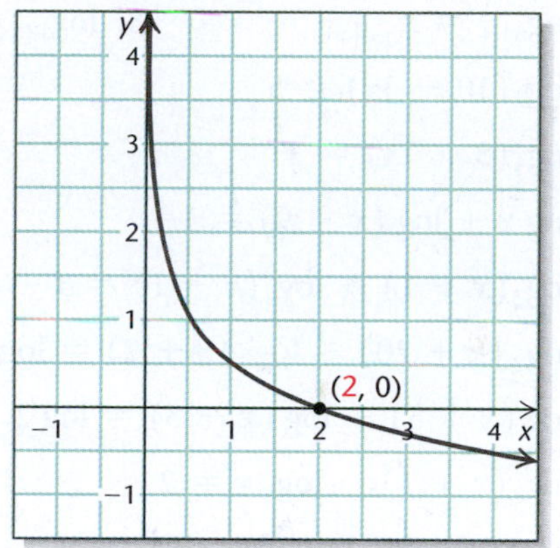

The solution of the equation is 2. From the graph, we can easily see that there is only one solution.

Now Try Exercise 43.

5.5 Exercise Set

Solve the exponential equation.

1. $3^x = 81$
2. $2^x = 32$
3. $2^{2x} = 8$
4. $3^{7x} = 27$
5. $2^x = 33$
6. $2^x = 40$
7. $5^{4x-7} = 125$
8. $4^{3x-5} = 16$
9. $27 = 3^{5x} \cdot 9^{x^2}$
10. $3^{x^2+4x} = \frac{1}{27}$
11. $84^x = 70$
12. $28^x = 10^{-3x}$
13. $10^{-x} = 5^{2x}$
14. $15^x = 30$
15. $e^{-c} = 5^{2c}$
16. $e^{4t} = 200$
17. $e^t = 1000$
18. $e^{-t} = 0.04$
19. $e^{-0.03t} = 0.08$
20. $1000e^{0.09t} = 5000$
21. $3^x = 2^{x-1}$
22. $5^{x+2} = 4^{1-x}$

23. $(3.9)^x = 48$

24. $250 - (1.87)^x = 0$

25. $e^x + e^{-x} = 5$

26. $e^x - 6e^{-x} = 1$

27. $3^{2x-1} = 5^x$

28. $2^{x+1} = 5^{2x}$

29. $2e^x = 5 - e^{-x}$

30. $e^x + e^{-x} = 4$

Solve the logarithmic equation.

31. $\log_5 x = 4$

32. $\log_2 x = -3$

33. $\log x = -4$

34. $\log x = 1$

35. $\ln x = 1$

36. $\ln x = -2$

37. $\log_{64} \frac{1}{4} = x$

38. $\log_{125} \frac{1}{25} = x$

39. $\log_2 (10 + 3x) = 5$

40. $\log_5 (8 - 7x) = 3$

41. $\log x + \log (x - 9) = 1$

42. $\log_2 (x + 1) + \log_2 (x - 1) = 3$

43. $\log_2 (x + 20) - \log_2 (x + 2) = \log_2 x$

44. $\log (x + 5) - \log (x - 3) = \log 2$

45. $\log_8 (x + 1) - \log_8 x = 2$

46. $\log x - \log (x + 3) = -1$

47. $\log x + \log (x + 4) = \log 12$

48. $\log_3 (x + 14) - \log_3 (x + 6) = \log_3 x$

49. $\log (x + 8) - \log (x + 1) = \log 6$

50. $\ln x - \ln (x - 4) = \ln 3$

51. $\log_4 (x + 3) + \log_4 (x - 3) = 2$

52. $\ln (x + 1) - \ln x = \ln 4$

53. $\log (2x + 1) - \log (x - 2) = 1$

54. $\log_5 (x + 4) + \log_5 (x - 4) = 2$

55. $\ln (x + 8) + \ln (x - 1) = 2 \ln x$

56. $\log_3 x + \log_3 (x + 1) = \log_3 2 + \log_3 (x + 3)$

Solve.

57. $\log_6 x = 1 - \log_6 (x - 5)$

58. $2^{x^2 - 9x} = \frac{1}{256}$

59. $9^{x-1} = 100(3^x)$

60. $2 \ln x - \ln 5 = \ln (x + 10)$

61. $e^x - 2 = -e^{-x}$

62. $2 \log 50 = 3 \log 25 + \log (x - 2)$

▶ Skill Maintenance

In Exercises 63–66:

a) *Find the vertex.*
b) *Find the axis of symmetry.*
c) *Determine whether there is a maximum or a minimum value and find that value.* [3.3]

63. $g(x) = x^2 - 6$

64. $f(x) = -x^2 + 6x - 8$

65. $G(x) = -2x^2 - 4x - 7$

66. $H(x) = 3x^2 - 12x + 16$

▶ Synthesis

Solve using any method.

67. $\dfrac{e^x + e^{-x}}{e^x - e^{-x}} = 3$

68. $\ln (\ln x) = 2$

69. $\sqrt{\ln x} = \ln \sqrt{x}$

70. $\ln \sqrt[4]{x} = \sqrt{\ln x}$

71. $(\log_3 x)^2 - \log_3 x^2 = 3$

72. $\log_3 (\log_4 x) = 0$

73. $\ln x^2 = (\ln x)^2$

74. $x \left(\ln \frac{1}{6}\right) = \ln 6$

75. $5^{2x} - 3 \cdot 5^x + 2 = 0$

76. $x^{\log x} = \dfrac{x^3}{100}$

77. $\ln x^{\ln x} = 4$

78. $|2^{x^2} - 8| = 3$

79. $\dfrac{\sqrt{(e^{2x} \cdot e^{-5x})^{-4}}}{e^x \div e^{-x}} = e^7$

80. Given that $a = (\log_{125} 5)^{\log_5 125}$, find the value of $\log_3 a$.

81. Given that $a = \log_8 225$ and $b = \log_2 15$, express a as a function of b.

82. Given that $f(x) = e^x - e^{-x}$, find $f^{-1}(x)$ if it exists.

5.6 Applications and Models: Growth and Decay; Compound Interest

- Solve applied problems involving exponential growth and decay.
- Solve applied problems involving compound interest.

Exponential functions and logarithmic functions with base e are rich in applications to many fields such as business, science, psychology, and sociology.

Population Growth

The function

$$P(t) = P_0e^{kt}, \quad k > 0,$$

is a model of many kinds of population growth, whether it be a population of people, bacteria, smartphones, or money. In this function, P_0 is the population at time 0, P is the population after time t, and k is called the **exponential growth rate**. The graph of such an equation is shown at right.

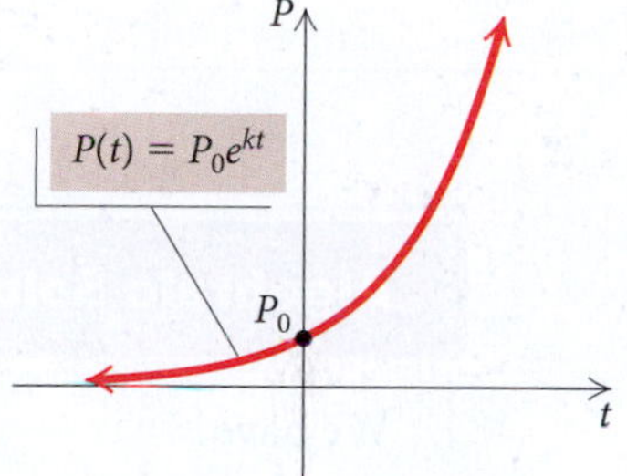

EXAMPLE 1 ***Population Growth of Ghana.*** In 2013, the population of Ghana, located on the west coast of Africa, was about 25.2 million, and the exponential growth rate was 2.19% per year (*Source: CIA World Factbook*, 2014).

a) Find the exponential growth function.

b) Estimate the population in 2018.

c) After how long will the population be double what it was in 2013?

d) At this growth rate, when will the population be 40 million?

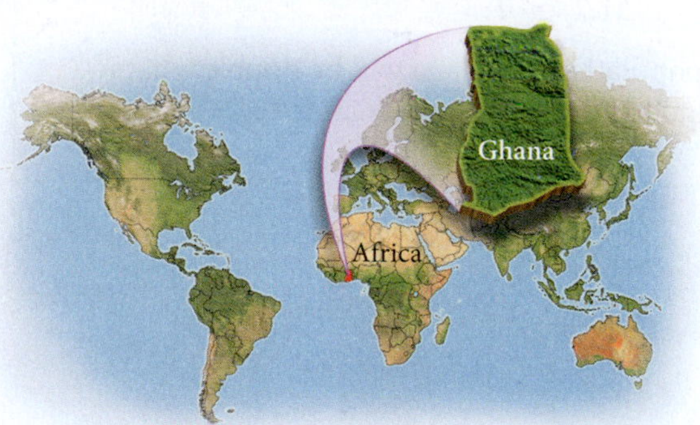

Technology Connection

We can find function values using a graphing calculator. Below, we find $P(5)$ from Example 1(b) with the VALUE feature from the CALC menu. We see that $P(5) \approx 28.1$.

$y_1 = 25.2e^{0.0219x}$

100

0 X = 5 Y = 28.116146 60

0

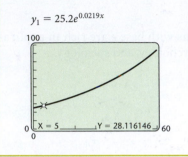

Solution

a) At $t = 0$ (2013), the population was 25.2 million, and the exponential growth rate was 2.19% per year. We substitute 25.2 for P_0 and 2.19%, or 0.0219, for k to obtain the exponential growth function

$$P(t) = 25.2e^{0.0219t},$$

where t is the number of years after 2013 and $P(t)$ is in millions.

b) In 2018, $t = 5$; that is, 5 years have passed since 2013. To find the population in 2018, we substitute 5 for t:

$$P(5) = 25.2e^{0.0219(5)} = 25.2e^{0.1095} \approx 28.1.$$

The population will be about 28.1 million, or 28,100,000, in 2018.

c) We are looking for the time T for which $P(T) = 2 \cdot 25.2$, or 50.4. The number T is called the **doubling time**. To find T, we solve the equation

$$50.4 = 25.2e^{0.0219T}.$$

Algebraic Solution

We have

$$50.4 = 25.2e^{0.0219T} \quad \text{Substituting 50.4 for } P(T)$$

$$2 = e^{0.0219T} \quad \text{Dividing by 25.2}$$

$$\ln 2 = \ln e^{0.0219T} \quad \text{Taking the natural logarithm on both sides}$$

$$\ln 2 = 0.0219T \quad \ln e^x = x$$

$$\frac{\ln 2}{0.0219} = T \quad \text{Dividing by 0.0219}$$

$$31.7 \approx T.$$

The population of Ghana will be double what it was in 2013 about 31.7 years after 2013.

Visualizing the Solution

From the graphs of $y = 50.4$ and $y = 25.2e^{0.0219T}$, we see that the first coordinate of their point of intersection is about 31.7.

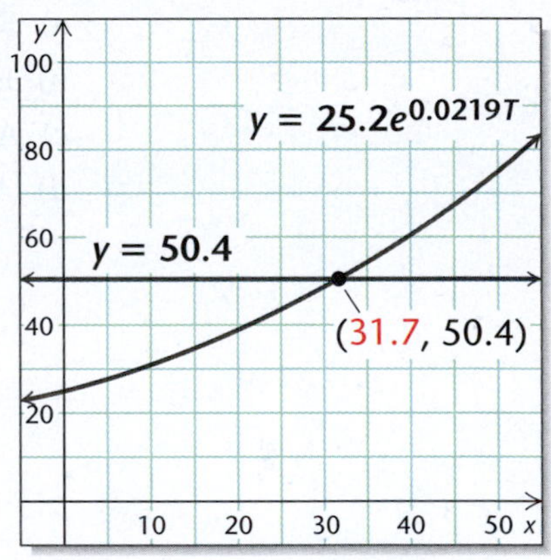

The solution of the equation is approximately 31.7.

d) To determine the time t for which $P(t) = 40$, we solve the following equation for t:

$$40 = 25.2e^{0.0219t} \quad \text{Substituting 40 for } P(t)$$

$$\frac{40}{25.2} = e^{0.0219t} \quad \text{Dividing by 25.2}$$

$$\ln\frac{40}{25.2} = \ln e^{0.0219t} \quad \text{Taking the natural logarithm on both sides}$$

$$\ln\frac{40}{25.2} = 0.0219t \quad \ln e^x = x$$

$$\frac{\ln\frac{40}{25.2}}{0.0219} = t \quad \text{Dividing by 0.0219}$$

$$21 \approx t.$$

The population of Ghana will be 40 million about 21 years after 2013.

Now Try Exercise 1.

Technology Connection

$y_1 = 25.2e^{0.0219x}$, $y_2 = 50.4$

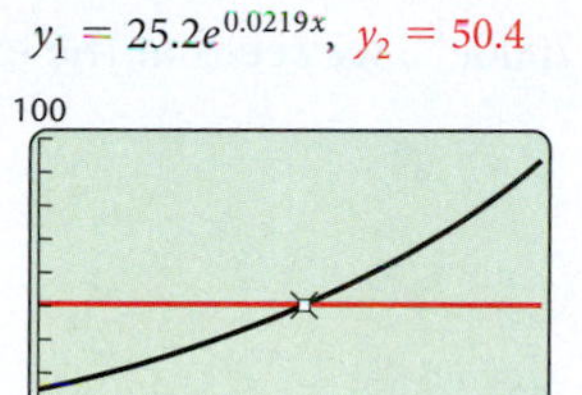

Using the Intersect method in Example 1(c), we graph the equations

$$y_1 = 25.2e^{0.0219x} \quad \text{and} \quad y_2 = 50.4$$

and find the first coordinate of their point of intersection. It is about 31.7, so the population of Ghana will be double that of 2013 about 31.7 years after 2013.

Interest Compounded Continuously

When interest is paid on interest, we call it **compound interest**. Suppose that an amount P_0 is invested in a savings account at interest rate k, **compounded continuously**. The amount $P(t)$ in the account after t years is given by the exponential function

$$P(t) = P_0e^{kt}.$$

EXAMPLE 2 ***Interest Compounded Continuously.*** Suppose that \$2000 is invested at interest rate k, compounded continuously, and grows to \$2504.65 in 5 years.

a) What is the interest rate?

b) Find the exponential growth function.

c) What will the balance be after 10 years?

d) After how long will the \$2000 have doubled?

Technology Connection

We can also find k in Example 2(a) by graphing the equations

$$y_1 = 2000e^{5x}$$

and

$$y_2 = 2504.65$$

and use the Intersect feature to approximate the first coordinate of the point of intersection.

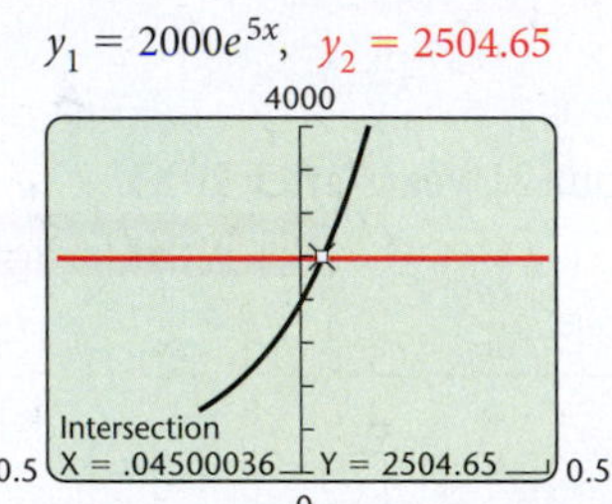

The interest rate is about 0.045, or 4.5%.

Solution

a) At $t = 0$, $P(0) = P_0 = \$2000$. Thus the exponential growth function is of the form

$$P(t) = 2000e^{kt}.$$

We know that $P(5) = \$2504.65$. We substitute and solve for k:

$$2504.65 = 2000e^{k(5)} \qquad \text{Substituting 2504.65 for } P(t) \text{ and 5 for } t$$

$$2504.65 = 2000e^{5k}$$

$$\frac{2504.65}{2000} = e^{5k} \qquad \text{Dividing by 2000}$$

$$\ln \frac{2504.65}{2000} = \ln e^{5k} \qquad \text{Taking the natural logarithm}$$

$$\ln \frac{2504.65}{2000} = 5k \qquad \text{Using } \ln e^x = x$$

$$\frac{\ln \frac{2504.65}{2000}}{5} = k \qquad \text{Dividing by 5}$$

$$0.045 \approx k.$$

The interest rate is about 0.045, or 4.5%.

b) Substituting 0.045 for k in the function $P(t) = 2000e^{kt}$, we see that the exponential growth function is

$$P(t) = 2000e^{0.045t}.$$

c) The balance after 10 years is

$$P(10) = 2000e^{0.045(10)} = 2000e^{0.45} \approx \$3136.62.$$

d) To find the doubling time T, we set $P(T) = 2 \cdot P_0 = 2 \cdot \$2000 = \$4000$ and solve for T. We solve

$$4000 = 2000e^{0.045T}.$$

Algebraic Solution

We have

$$\begin{aligned} 4000 &= 2000e^{0.045T} \\ 2 &= e^{0.045T} && \text{Dividing by 2000} \\ \ln 2 &= \ln e^{0.045T} && \text{Taking the natural logarithm} \\ \ln 2 &= 0.045T && \ln e^x = x \\ \frac{\ln 2}{0.045} &= T && \text{Dividing by 0.045} \\ 15.4 &\approx T. \end{aligned}$$

Thus the original investment of $2000 will double in about 15.4 years.

Visualizing the Solution

The solution of the equation

$$4000 = 2000e^{0.045T}$$

or

$$2000e^{0.045T} - 4000 = 0,$$

is the zero of the function

$$y = 2000e^{0.045T} - 4000.$$

Note the zero from the graph shown here.

$y = 2000e^{0.045T} - 4000$

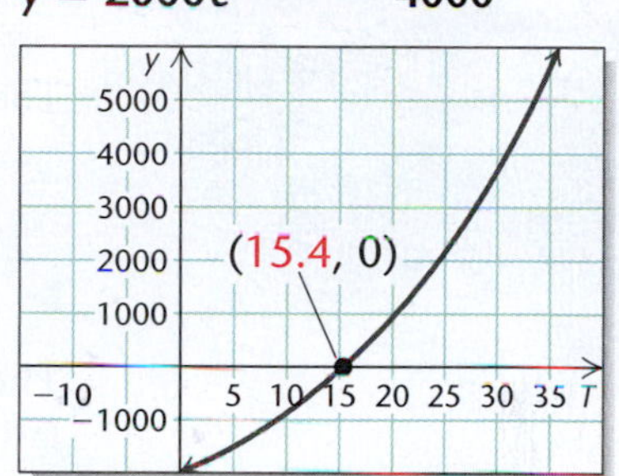

The zero is about 15.4. Thus the solution of the equation is approximately 15.4.

Now Try Exercise 7.

Technology Connection

$y = 2000e^{0.045x} - 4000$

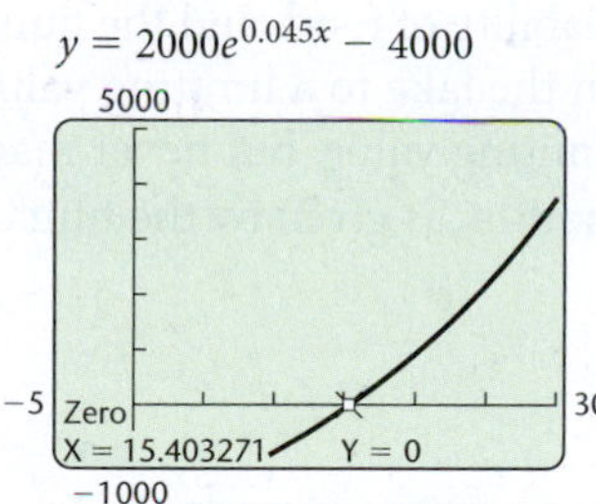

The amount of money in Example 2 will have doubled when $P(t) = 2 \cdot P_0 = 4000$, or when $2000e^{0.045t} = 4000$. We use the Zero method. We graph the equation

$$y = 2000e^{0.045x} - 4000$$

and find the zero of the function. The zero of the function is the solution of the equation. The zero is about 15.4, so the original investment of $2000 will double in about 15.4 years.

We can find a general expression relating the growth rate k and the doubling time T by solving the following equation:

$$\begin{aligned} 2P_0 &= P_0e^{kT} && \text{Substituting } 2P_0 \text{ for } P \text{ and } T \text{ for } t \\ 2 &= e^{kT} && \text{Dividing by } P_0 \\ \ln 2 &= \ln e^{kT} && \text{Taking the natural logarithm} \\ \ln 2 &= kT && \text{Using } \ln e^x = x \\ \frac{\ln 2}{k} &= T. \end{aligned}$$

> **GROWTH RATE AND DOUBLING TIME**
>
> The **growth rate k** and the **doubling time T** are related by
>
> $$kT = \ln 2, \quad \text{or} \quad k = \frac{\ln 2}{T}, \quad \text{or} \quad T = \frac{\ln 2}{k}.$$

Note that the relationship between k and T does not depend on P_0.

EXAMPLE 3 ***Population Growth.*** The population of the Philippines is now doubling every 37.7 years (*Source: CIA World Factbook*, 2014). What is the exponential growth rate?

Solution We have

$$k = \frac{\ln 2}{T} = \frac{\ln 2}{37.7} \approx 0.0184 \approx 1.84\%.$$

The growth rate of the population of the Philippines is about 1.84% per year.

Now Try Exercise 3(e).

Models of Limited Growth

The model $P(t) = P_0e^{kt}, k > 0$, has many applications involving unlimited population growth. However, in some populations, there can be factors that prevent a population from exceeding some limiting value—perhaps a limitation on food, living space, or other natural resources. One model of such growth is

$$P(t) = \frac{a}{1 + be^{-kt}}.$$

This is called a **logistic function**. This function increases toward a *limiting value a* as $t \longrightarrow \infty$. Thus, $y = a$ is the horizontal asymptote of the graph of $P(t)$.

EXAMPLE 4 ***Limited Population Growth in a Lake.*** A lake is stocked with 400 fish of a new variety. The size of the lake, the availability of food, and the number of other fish restrict the growth of that type of fish in the lake to a limiting value of 2500. The population gets closer and closer to this limiting value, but never reaches it. The population of fish in the lake after time t, in months, is given by the function

$$P(t) = \frac{2500}{1 + 5.25e^{-0.32t}}.$$

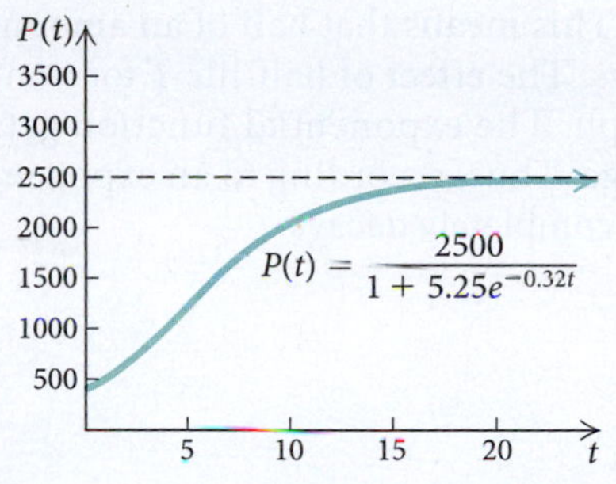

The graph of $P(t)$ is the curve shown at left. Note that this function increases toward a limiting value of 2500. The graph has $y = 2500$ as a horizontal asymptote. Find the population after 0, 1, 5, 10, 15, and 20 months.

Solution Using a calculator, we compute the function values. We find that

$$P(0) = 400, \quad P(10) \approx 2059,$$
$$P(1) \approx 520, \quad P(15) \approx 2396,$$
$$P(5) \approx 1214, \quad P(20) \approx 2478.$$

Thus the population will be about 400 after 0 months, 520 after 1 month, 1214 after 5 months, 2059 after 10 months, 2396 after 15 months, and 2478 after 20 months.

Now Try Exercise 17(b).

Another model of limited growth is provided by the function

$$P(t) = L(1 - e^{-kt}), \quad k > 0,$$

which is shown graphed below. This function also increases toward a limiting value L, as $t \rightarrow \infty$, so $y = L$ is the horizontal asymptote of the graph of $P(t)$.

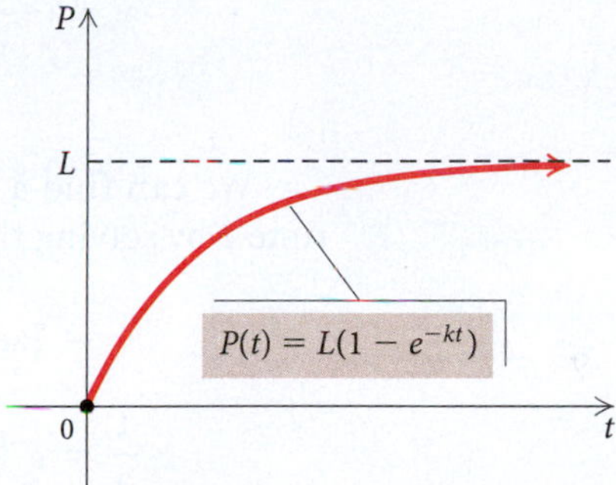

Exponential Decay

The function

$$P(t) = P_0 e^{-kt}, \quad k > 0,$$

is an effective model of the decline, or decay, of a population. An example is the decay of a radioactive substance. In this case, P_0 is the amount of the substance at time $t = 0$, and $P(t)$ is the amount of the substance left after time t, where k is a positive constant that depends on the situation. The constant k is called the **decay rate**.

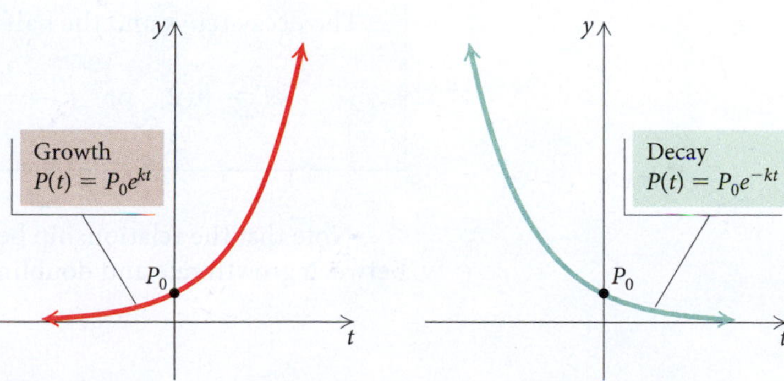

The **half-life** of bismuth (Bi-210) is 5 days. This means that half of an amount of bismuth will cease to be radioactive in 5 days. The effect of half-life T for nonnegative inputs is shown in the following graph. The exponential function gets close to 0, but never reaches 0, as t gets very large. Thus, according to an exponential decay model, a radioactive substance never completely decays.

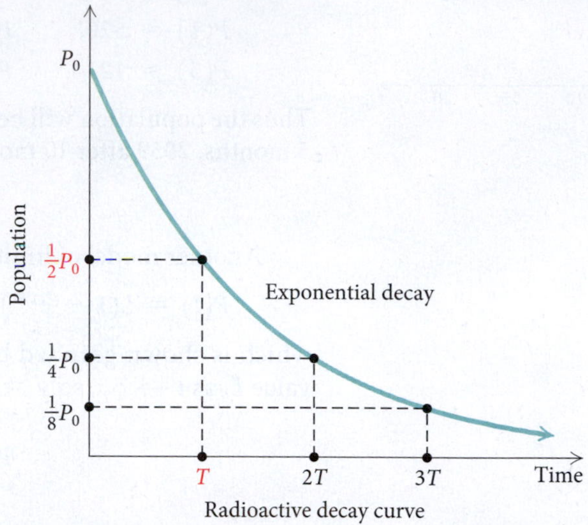

Radioactive decay curve

We can find a general expression relating the decay rate k and the half-life time T by solving the following equation:

$$\frac{1}{2}P_0 = P_0e^{-kT} \quad \text{Substituting } \frac{1}{2}P_0 \text{ for } P \text{ and } T \text{ for } t$$

$$\frac{1}{2} = e^{-kT} \quad \text{Dividing by } P_0$$

$$\ln\frac{1}{2} = \ln e^{-kT} \quad \text{Taking the natural logarithm}$$

$$\ln 2^{-1} = -kT \quad \frac{1}{2} = 2^{-1};\ \ln e^x = x$$

$$-\ln 2 = -kT \quad \text{Using the power rule}$$

$$\frac{\ln 2}{k} = T. \quad \text{Dividing by } -k$$

DECAY RATE AND HALF-LIFE

The decay rate k and the half-life T are related by

$$kT = \ln 2, \quad \text{or} \quad k = \frac{\ln 2}{T}, \quad \text{or} \quad T = \frac{\ln 2}{k}.$$

Note that the relationship between decay rate and half-life is the same as that between growth rate and doubling time.

How can scientists determine that an animal bone has lost 30% of its carbon-14? The assumption is that the percentage of carbon-14 in the atmosphere is the same as that in living plants and animals. When a plant or an animal dies, the amount of carbon-14 that it contains decays exponentially. A scientist can burn an animal bone and use a Geiger counter to determine the percentage of the smoke that is carbon-14. The amount by which this varies from the percentage in the atmosphere tells how much carbon-14 has been lost.

The process of carbon-14 dating was developed by the American chemist Willard E. Libby in 1952. It is known that the radioactivity in a living plant is 16 disintegrations per gram per minute. Since the half-life of carbon-14 is 5750 years, an object with an activity of 8 disintegrations per gram per minute is 5750 years old, one with an activity of 4 disintegrations per gram per minute is 11,500 years old, and so on. Carbon-14 dating can be used to measure the age of objects up to 40,000 years old. Beyond such an age, it is too difficult to measure the radioactivity and some other method would have to be used.

Carbon-14 dating was used to find the age of the Dead Sea Scrolls. It was also used to refute the authenticity of the Shroud of Turin, presumed to have covered the body of Christ.

EXAMPLE 5 ***Carbon Dating.*** The radioactive element carbon-14 has a half-life of 5750 years. The percentage of carbon-14 present in the remains of organic matter can be used to determine the age of that organic matter. Archaeologists discovered that the linen wrapping from one of the Dead Sea Scrolls had lost 22.3% of its carbon-14 at the time it was found. How old was the linen wrapping?

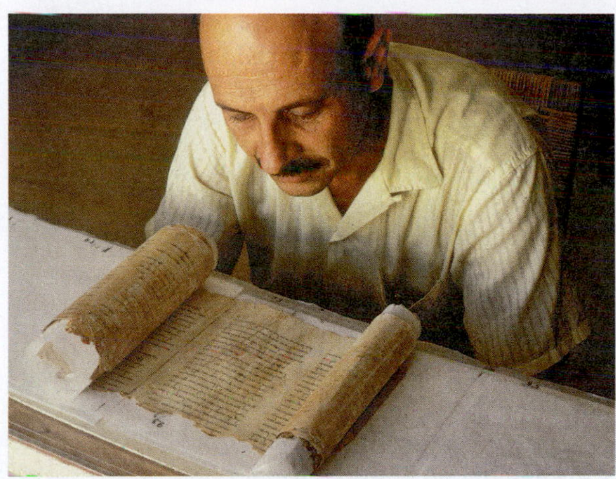

In 1947, a Bedouin youth looking for a stray goat climbed into a cave at Kirbet Qumran on the shores of the Dead Sea near Jericho and came upon earthenware jars containing an incalculable treasure of ancient manuscripts. Shown here are fragments of those Dead Sea Scrolls, a portion of some 600 or so texts found so far and which concern the Jewish books of the Bible. Officials date them before 70 A.D., making them the oldest Biblical manuscripts by 1000 years.

Solution We first find k when the half-life T is 5750 years:

$$k = \frac{\ln 2}{T}$$

$$k = \frac{\ln 2}{5750} \qquad \text{Substituting 5750 for } T$$

$$k = 0.00012.$$

Now we have the function

$$P(t) = P_0 e^{-0.00012t}.$$

(This function can be used for any subsequent carbon-dating problem.) If the linen wrapping has lost 22.3% of its carbon-14 from an initial amount P_0, then $77.7\% P_0$ is the amount present. To find the age t of the wrapping, we solve the following equation for t:

$$77.7\% P_0 = P_0 e^{-0.00012t} \qquad \text{Substituting } 77.7\% P_0 \text{ for } P$$

$$0.777 = e^{-0.00012t} \qquad \text{Dividing by } P_0 \text{ and writing 77.7\% as 0.777}$$

$$\ln 0.777 = \ln e^{-0.00012t} \qquad \text{Taking the natural logarithm on both sides}$$

$$\ln 0.777 = -0.00012t \qquad \ln e^x = x$$

$$\frac{\ln 0.777}{-0.00012} = t \qquad \text{Dividing by } -0.00012$$

$$2103 \approx t.$$

Thus the linen wrapping on the Dead Sea Scrolls was about 2103 years old when it was found.

Now Try Exercise 9.

5.6 Exercise Set

1. *Population Growth of Houston.* The Houston–Woodlands—Sugar Land metropolitan area is the fifth largest metropolitan area in the United States. In 2012, the population of this area was 6.18 million, and the exponential growth rate was 2.14% per year.

   a) Find the exponential growth function.
   b) Estimate the population of the Houston–Woodlands—Sugar Land metropolitan area in 2018.
   c) When will the population of this metropolitan area be 8 million?
   d) Find the doubling time.

2. *Population Growth of Rabbits.* Under ideal conditions, a population of rabbits has an exponential growth rate of 11.7% per day. Consider an initial population of 100 rabbits.
   a) Find the exponential growth function.
   b) What will the population be after 7 days? after 2 weeks?
   c) Find the doubling time.

3. *Population Growth.* Complete the following table.

Country	Growth Rate, k	Doubling Time, T
a) United States		77.0 years
b) Bolivia		42.5 years
c) Uganda	3.32%	
d) Australia	1.11%	
e) Sweden		385 years
f) Laos	2.32%	
g) India	1.28%	
h) China		150.7 years
i) Guinea		26.3 years
j) Hong Kong	0.39%	

4. *E-Book Sales.* The revenue from e-book sales (not including educational textbooks) accounted for only 0.5% of U.S. publishing sales in 2006. This percentage grew to 22.6% in 2012. (*Source:* Association of American Publishers) Assuming that the exponential growth model applies:
   a) Find the value of k and write the function.
   b) Estimate the percentage of U.S. publishing sales that were e-book sales in 2009 and in 2010. Round to the nearest tenth.

5. *Population Growth of Haiti.* The population of Haiti has a growth rate of 0.99% per year. In 2013, the population was 9,893,934, and the land area of Haiti is 32,961,561,600 square yards. (*Source:* U.S. Census Bureau) Assuming that this growth rate continues and is exponential, after how long will there be one person for every square yard of land?

6. *Picasso Painting.* In May 2010, a 1932 painting, *Nude, Green Leaves, and Bust,* by Pablo Picasso, sold at a New York City art auction for $106.5 million to an anonymous buyer. The painting had belonged to the estate of Sydney and Francis Brody, who bought it for $17,000 in 1952 from a New York art dealer, who had acquired it from Picasso in 1936. (*Sources:* Associated Press, "Picasso Painting Fetches World Record $106.5M at NYC Auction," by Ula Ilnytzky, May 4, 2010; Online Associated Newspapers, May 6, 2010)

Assuming that the value A_0 of the painting has grown exponentially:

a) Find the value of k, and determine the exponential growth function, assuming that $A_0 = 17{,}000$ and t is the number of years after 1952.

b) Estimate the value of the painting in 2020.

c) What is the doubling time for the value of the painting?

d) After how long will the value of the painting be $240 million, assuming that there is no change in the growth rate?

7. *Interest Compounded Continuously.* Suppose that $10,000 is invested at an interest rate of 5.4% per year, compounded continuously.

a) Find the exponential function that describes the amount in the account after time t, in years.

b) What is the balance after 1 year? 2 years? 5 years? 10 years?

c) What is the doubling time?

8. *Interest Compounded Continuously.* Complete the following table.

Initial Investment at $t = 0$, P_0	Interest Rate, k	Doubling Time, T	Amount After 5 Years
a) $35,000	3.2%		
b) $5000			$7,130.90
c)	5.6%		$9,923.47
d)		11 years	$17,539.32
e) $109,000			$136,503.18
f)		46.2 years	$19,552.82

9. *Carbon Dating.* In 1970, Amos Flora of Flora, Indiana, discovered teeth and jawbones while dredging a creek. Scientists determined that the bones were from a mastodon and had lost 77.2% of their carbon-14. How old were the bones at the time they were discovered? (*Sources:* "Farm Yields Bones Thousands of Years Old," by Dan McFeely, *Indianapolis Star,* October 20, 2008; Field Museum of Chicago, Bill Turnbull, anthropologist)

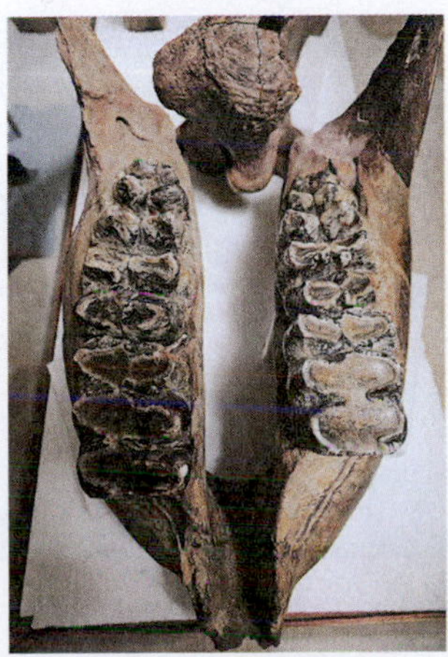

10. *Tomb in the Valley of the Kings.* In February 2006, in the Valley of the Kings in Egypt, a team of archaeologists uncovered the first tomb since King Tut's tomb was found in 1922. The tomb contained five wooden sarcophagi that contained mummies. The archaeologists believe that the mummies are from the 18th Dynasty, about 3300 to 3500 years ago. Determine the amount of carbon-14 that the mummies have lost.

11. *Radioactive Decay.* Complete the following table.

Radioactive Substance	Decay Rate, k	Half-Life T
a) Polonium (Po-218)		3.1 min
b) Lead (Pb-210)		22.3 years
c) Iodine (I-125)	1.15% per day	
d) Krypton (Kr-85)	6.5% per year	
e) Strontium (Sr-90)		29.1 years
f) Uranium (U-232)		70.0 years
g) Plutonium (Pu-239)		24,100 years

12. *Advertising Revenue.* The amount of advertising revenue in U.S. newspapers has declined continually since 2006. In 2006, the advertising revenue was $49.3 billion, and in 2013 that amount had decreased to $20.7 billion (*Source:* Newspaper Association of America). Assuming that the amount of newspaper advertising revenue decreased according to the exponential decay model:

a) Find the value of k, and write an exponential function that describes the advertising revenue after time t, in years, where t is the number of years after 2006.

b) Estimate the advertising revenue in 2008 and in 2012.

c) At this decay rate, when will the advertising revenue be $16 billion?

13. *Married Adults.* The data in the following table show that the percentage of adults in the United States who are currently married is declining.

Year	Percent of Adults Who Are Married
1960	72.2%
1980	62.3
2000	57.4
2010	51.4
2012	50.5

Sources: Pew Research Center; U.S. Census Bureau

Assuming that the percentage of adults who are married will continue to decrease according to the exponential decay model:

a) Use the data for 1960 and 2012 to find the value of k and to write an exponential function that describes the percent of adults married after time t, in years, where t is the number of years after 1960.

b) Estimate the percent of adults who are married in 2015 and in 2018.

c) At this decay rate, in which year will the percent of adults who are married be 40%?

14. *Lamborghini 350 GT.* The market value of the 1964–1965 Lamborghini 350 GT has had a recent upswing. In a decade, the car's value increased from $66,000 in 1999 to $220,000 in 2009 (*Source:* "1964–1965 Lamborghini 350 GT," by David LaChance, *Hemmings Motor News,* July, 2010, p. 28).

Assuming that the value V_0 of the car has grown exponentially:

a) Find the value of k, and determine the exponential growth function, assuming that $V_0 = 66{,}000$ and t is the number of years after 1999.

b) Estimate the value of the car in 2011.

c) After how long was the value of the car $300,000, assuming that there is no change in the growth rate?

15. *Oil Consumption.* In 1980, China consumed 1.85 million barrels of oil per day. By 2012, that consumption had grown to 10.28 million barrels per day. (*Sources:* U.S. Energy Information Administration; NextBigThingInvestor.com)

Assuming that the consumption of oil C_0 in China has grown exponentially:

a) Find the value of k, and determine the exponential growth function, assuming that $C_0 = 1.85$ and t is the number of years after 1980.

b) Estimate the consumption of oil in 2005.

c) What is the doubling time for the consumption of oil in China?

d) After how long will the consumption of oil in China be 13 million barrels per day, assuming that there is no change in the growth rate?

16. *T206 Wagner Baseball Card.* In 1909, the Pittsburgh Pirates shortstop Honus Wagner forced the American Tobacco Company to withdraw his baseball card because it was packaged with cigarettes. Fewer than 60 of the Wagner cards still exist. In 1971, a Wagner card sold for $1000; and in September 2007, a card in near-mint condition was purchased for a record $2.8 million (*Source: USA Today*, 9/6/07; Kathy Willens/AP).

Assuming that the value W_0 of the baseball card has grown exponentially:

a) Find the value of k, and determine the exponential growth function, assuming that $W_0 = 1000$ and t is the number of years since 1971.
b) Estimate the value of the Wagner card in 2011.
c) What is the doubling time for the value of the card?
d) After how long was the value of the Wagner card $3 million, assuming that there is no change in the growth rate?

17. *Spread of an Epidemic.* In a town whose population is 3500, a disease creates an epidemic. The number of people N infected t days after the disease has begun is given by the function

$$N(t) = \frac{3500}{1 + 19.9e^{-0.6t}}.$$

a) How many are initially infected with the disease $(t = 0)$?
b) Find the number infected after 2 days, 5 days, 8 days, 12 days, and 16 days.
c) Using this model, can you say whether all 3500 people will ever be infected? Explain.

18. *Limited Population Growth in a Lake.* A lake is stocked with 640 fish of a new variety. The size of the lake, the availability of food, and the number of other fish restrict the growth of that type of fish in the lake to a limiting value of 3040. The population of fish in the lake after time t, in months, is given by the function

$$P(t) = \frac{3040}{1 + 3.75e^{-0.32t}}.$$

Find the population after 0, 1, 5, 10, 15, and 20 months.

Newton's Law of Cooling. *Suppose that a body with temperature T_1 is placed in surroundings with temperature T_0 different from that of T_1. The body will either cool or warm to temperature $T(t)$ after time t, in minutes, where*

$$T(t) = T_0 + (T_1 - T_0)e^{-kt}.$$

Use this law in Exercises 19–22.

19. A cup of coffee with temperature 105°F is placed in a freezer with temperature 0°F. After 5 min, the temperature of the coffee is 70°F. What will its temperature be after 10 min?

20. A dish of lasagna baked at 375°F is taken out of the oven at 11:15 A.M. into a kitchen that is 72°F. After 3 min, the temperature of the lasagna is 365°F. What will the temperature of the lasagna be at 11:30 A.M.?

21. A chilled jello salad that has a temperature of 43°F is taken from the refrigerator and placed on the dining room table in a room that is 68°F. After 12 min, the temperature of the salad is 55°F. What will the temperature of the salad be after 20 min?

22. *When Was the Murder Committed?* The police discover the body of a murder victim. Critical to solving the crime is determining when the murder was committed. The coroner arrives at the murder scene at 12:00 P.M. She immediately takes the temperature of the body and finds it to be 94.6°F. She then takes the temperature 1 hr later and finds it to be 93.4°F. The temperature of the room is 70°F. When was the murder committed?

Skill Maintenance

Vocabulary Reinforcement

In Exercises 23–28, choose the correct name of the principle or the rule from the given choices.

principle of zero products
multiplication principle for equations
product rule
addition principle for inequalities
power rule
multiplication principle for inequalities
principle of square roots
quotient rule

23. For any real numbers a, b, and c: If $a < b$ and $c > 0$ are true, then $ac < bc$ is true. If $a < b$ and $c < 0$ are true, then $ac > bc$ is true. [1.6] ____________

24. For any positive numbers M and N and any logarithmic base a, $\log_a MN = \log_a M + \log_a N$. [5.4] ____________

25. If $ab = 0$ is true, then $a = 0$ or $b = 0$, and if $a = 0$ or $b = 0$, then $ab = 0$. [3.2] ____________

26. If $x^2 = k$, then $x = \sqrt{k}$ or $x = -\sqrt{k}$. [3.2] ____________

27. For any positive number M, any logarithmic base a, and any real number p, $\log_a M^p = p \log_a M$. [5.4] ____________

28. For any real numbers a, b, and c: If $a = b$ is true, then $ac = bc$ is true. [1.5] ____________

Synthesis

29. *Supply and Demand.* The supply function and the demand function for the sale of a certain type of DVD player are given by

$$S(p) = 150e^{0.004p} \quad \text{and} \quad D(p) = 480e^{-0.003p},$$

respectively, where $S(p)$ is the number of DVD players that the company is willing to sell at price p and $D(p)$ is the quantity that the public is willing to buy at price p. Find p such that $D(p) = S(p)$. This is called the **equilibrium price**.

30. *Carbon Dating.* Recently, while digging in Chaco Canyon, New Mexico, archaeologists found corn pollen that was 4000 years old (*Source: American Anthropologist*). This was evidence that Native Americans had been cultivating crops in the Southwest centuries earlier than scientists had thought. What percent of the carbon-14 had been lost from the pollen?

31. *Present Value.* Following the birth of a child, a grandparent wants to make an initial investment P_0 that will grow to \$50,000 for the child's education at age 18. Interest is compounded continuously at 5.2%. What should the initial investment be? Such an amount is called the **present value** of \$50,000 due 18 years from now.

32. *Present Value.*

a) Solve $P = P_0 e^{kt}$ for P_0.
b) Referring to Exercise 31, find the present value of \$50,000 due 18 years from now at interest rate 6.4%, compounded continuously.

33. *Electricity.* The formula

$$i = \frac{V}{R}\left[1 - e^{-(R/L)t}\right]$$

occurs in the theory of electricity. Solve for t.

34. *The Beer–Lambert Law.* A beam of light enters a medium such as water or smog with initial intensity I_0. Its intensity decreases depending on the thickness (or concentration) of the medium. The intensity I at a depth (or concentration) of x units is given by

$$I = I_0 e^{-\mu x}.$$

The constant μ (the Greek letter "mu") is called the **coefficient of absorption**, and it varies with the medium. For sea water, $\mu = 1.4$.

a) What percentage of light intensity I_0 remains in sea water at a depth of 1 m? 3 m? 5 m? 50 m?
b) Plant life cannot exist below 10 m. What percentage of I_0 remains at 10 m?

35. Given that $y = ae^x$, take the natural logarithm on both sides. Let $Y = \ln y$. Consider Y as a function of x. What kind of function is Y?

36. Given that $y = ax^b$, take the natural logarithm on both sides. Let $Y = \ln y$ and $X = \ln x$. Consider Y as a function of X. What kind of function is Y?

Chapter 5 Summary and Review

STUDY GUIDE

KEY TERMS AND CONCEPTS	EXAMPLES
SECTION 5.1: INVERSE FUNCTIONS	
Inverse Relation If a relation is defined by an equation, then interchanging the variables produces an equation of the inverse relation.	Given $y = -5x + 7$, find an equation of the inverse relation. $y = -5x + 7$ **Relation** $\downarrow \quad \downarrow$ $x = -5y + 7$ **Inverse relation**
One-to-One Functions A function f is one-to-one if different inputs have different outputs—that is, if $a \neq b$, then $f(a) \neq f(b)$. Or a function f is one-to-one if when the outputs are the same, the inputs are the same—that is, if $f(a) = f(b)$, then $a = b$.	Prove that $f(x) = 16 - 3x$ is one-to-one. Show that if $f(a) = f(b)$, then $a = b$. Assume $f(a) = f(b)$. Since $f(a) = 16 - 3a$ and $f(b) = 16 - 3b$, $16 - 3a = 16 - 3b$ $-3a = -3b$ $a = b$. Thus, if $f(a) = f(b)$, then $a = b$ and f is one-to-one.
Horizontal-Line Test If it is possible for a horizontal line to intersect the graph of a function more than once, then the function is *not* one-to-one and its inverse is *not* a function. **One-to-One Functions and Inverses** • If a function f is one-to-one, then its inverse f^{-1} is a function. • The domain of a one-to-one function f is the range of the inverse f^{-1}. • The range of a one-to-one function f is the domain of the inverse f^{-1}. • A function that is increasing over its entire domain or is decreasing over its entire domain is a one-to-one function. The -1 in f^{-1} is *not* an exponent.	Using its graph, determine whether each function is one-to-one. **a)** 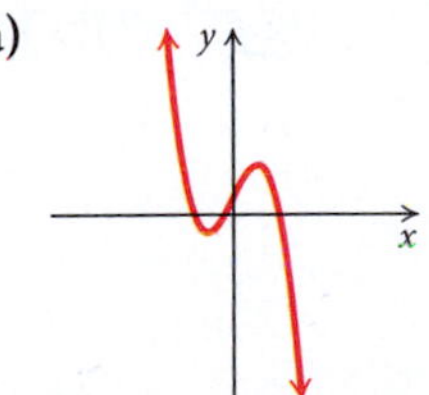 **b)** 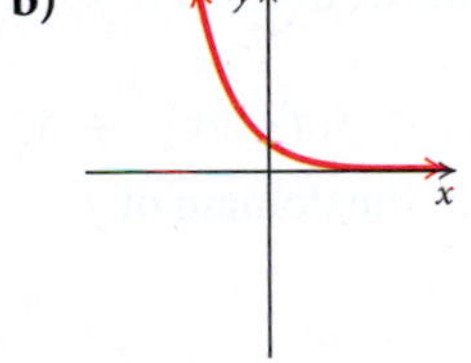 **a)** There are many horizontal lines that intersect the graph more than once. Thus the function is *not* one-to-one and its inverse is *not* a function. **b)** No horizontal line intersects the graph more than once. Thus the function is one-to-one and its inverse is a function.

Obtaining a Formula for an Inverse

If a function f is one-to-one, a formula for its inverse can generally be found as follows:

1. Replace $f(x)$ with y.
2. Interchange x and y.
3. Solve for y.
4. Replace y with $f^{-1}(x)$.

The graph of f^{-1} is a reflection of the graph of f across the line $y = x$.

Given the one-to-one function $f(x) = 2 - x^3$, find a formula for its inverse. Then graph the function and its inverse on the same set of axes.

$$f(x) = 2 - x^3$$

1. $y = 2 - x^3$ **Replacing $f(x)$ with y**
2. $x = 2 - y^3$ **Interchanging x and y**
3. Solve for y:

$$y^3 = 2 - x \quad \textbf{Adding } y^3 \textbf{ and subtracting } x$$

$$y = \sqrt[3]{2 - x}.$$

4. $f^{-1}(x) = \sqrt[3]{2 - x}$ **Replacing y with $f^{-1}(x)$**

$f(x) = 2 - x^3$ and $f^{-1}(x) = \sqrt[3]{2 - x}$

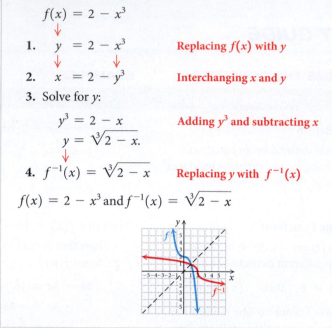

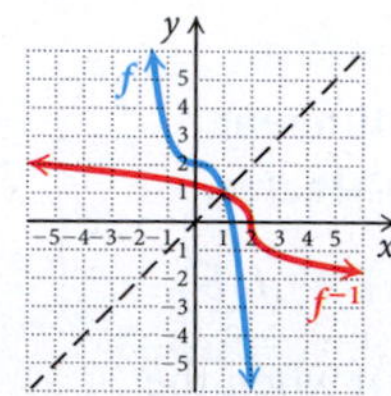

If a function f is one-to-one, then f^{-1} is the unique function such that each of the following holds:

$(f^{-1} \circ f)(x) = f^{-1}(f(x)) = x$, for each x in the domain of f, and

$(f \circ f^{-1})(x) = f(f^{-1}(x)) = x$, for each x in the domain of f^{-1}.

Given $f(x) = \dfrac{3 + x}{x}$, use composition of functions to show that $f^{-1}(x) = \dfrac{3}{x - 1}$.

$$(f^{-1} \circ f)(x) = f^{-1}(f(x))$$

$$= f^{-1}\left(\frac{3 + x}{x}\right) = \frac{3}{\dfrac{3 + x}{x} - 1}$$

$$= \frac{3}{\dfrac{3 + x - x}{x}} = \frac{3}{\dfrac{3}{x}} = 3 \cdot \frac{x}{3} = x;$$

$$(f \circ f^{-1})(x) = f(f^{-1}(x)) = f\left(\frac{3}{x - 1}\right)$$

$$= \frac{3 + \dfrac{3}{x - 1}}{\dfrac{3}{x - 1}} = \frac{3(x - 1) + 3}{x - 1} \cdot \frac{x - 1}{3}$$

$$= \frac{3x - 3 + 3}{3} = \frac{3x}{3} = x$$

SECTION 5.2: EXPONENTIAL FUNCTIONS AND GRAPHS

Exponential Function

$y = a^x$, or $f(x) = a^x$, $a > 0, a \neq 1$

Continuous
One-to-one
Domain: $(-\infty, \infty)$
Range: $(0, \infty)$
Increasing if $a > 1$
Decreasing if $0 < a < 1$
Horizontal asymptote is x-axis
y-intercept: $(0, 1)$

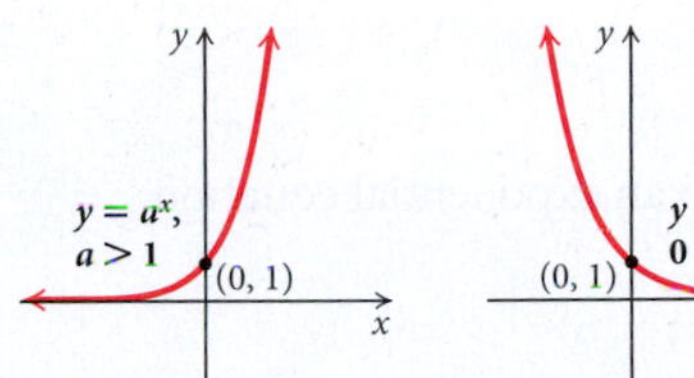

Graph: $f(x) = 2^x, g(x) = 2^{-x}, h(x) = 2^{x-1}$, and $t(x) = 2^x - 1$.

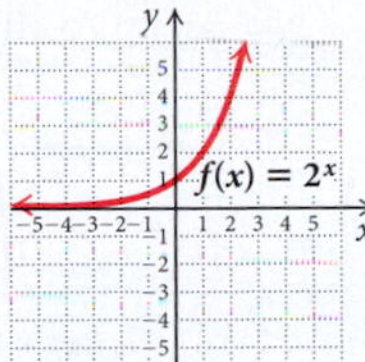

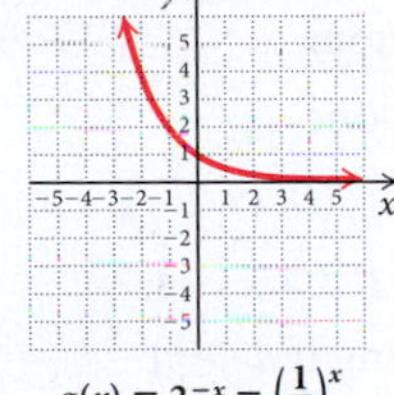

$g(x) = 2^{-x} = \left(\frac{1}{2}\right)^x$

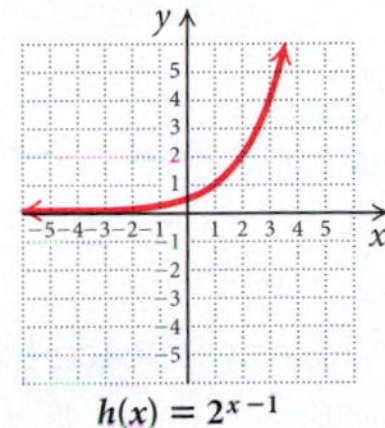

$h(x) = 2^{x-1}$

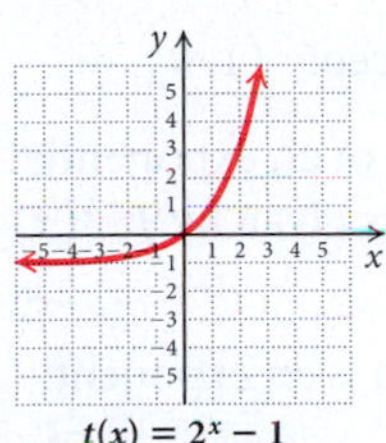

$t(x) = 2^x - 1$

Compound Interest

The amount of money A to which a principal P will grow after t years at interest rate r (in decimal form), compounded n times per year, is given by the formula

$$A = P\left(1 + \frac{r}{n}\right)^{nt}.$$

Suppose that \$5000 is invested at 3.5% interest, compounded quarterly. Find the money in the account after 3 years.

$$A = P\left(1 + \frac{r}{n}\right)^{nt} = 5000\left(1 + \frac{0.035}{4}\right)^{4 \cdot 3}$$
$$= \$5551.02$$

The Number e

$e = 2.7182818284\ldots$

Find each of the following, to four decimal places, using a calculator.

$e^{-3} = 0.0498;$
$e^{4.5} = 90.0171$

Graph: $f(x) = e^x$ and $g(x) = e^{-x+2} - 4$.

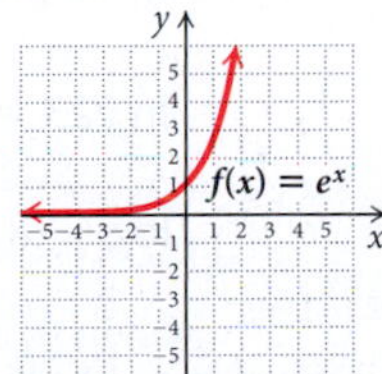

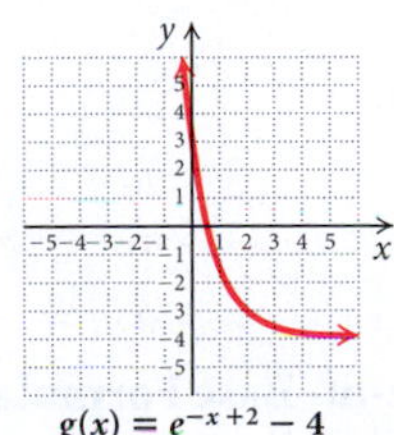

$g(x) = e^{-x+2} - 4$

SECTION 5.3: LOGARITHMIC FUNCTIONS AND GRAPHS

Logarithmic Function

$y = \log_a x, \quad x > 0, a > 0, a \neq 1$

Continuous

One-to-one

Domain: $(0, \infty)$

Range: $(-\infty, \infty)$

Increasing if $a > 1$

Vertical asymptote is y-axis

x-intercept: $(1, 0)$

The inverse of an exponential function $f(x) = a^x$ is given by $f^{-1}(x) = \log_a x$.

Graph: $f(x) = \log_2 x$ and $g(x) = \ln(x - 1) + 2$.

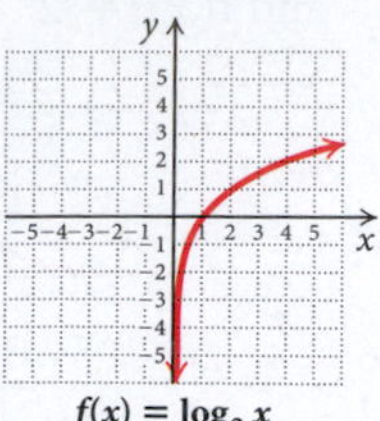

$f(x) = \log_2 x$

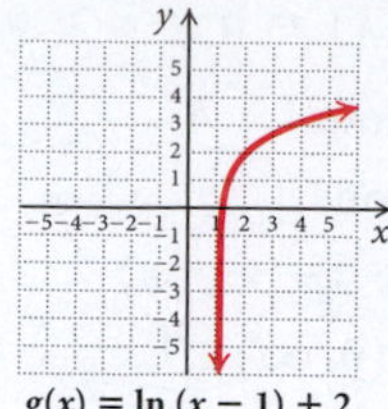

$g(x) = \ln(x - 1) + 2$

A logarithm is an exponent:

$$\log_a x = y \longleftrightarrow x = a^y.$$

Convert each logarithmic equation to an exponential equation.

$$\log_4 \frac{1}{16} = -2 \longleftrightarrow 4^{-2} = \frac{1}{16};$$

$$\ln R = 3 \longleftrightarrow e^3 = R$$

Convert each exponential equation to a logarithmic equation.

$$e^{-5} = 0.0067 \longleftrightarrow \ln 0.0067 = -5;$$

$$7^2 = 49 \longleftrightarrow \log_7 49 = 2$$

$\log x$ means $\log_{10} x$ **Common logarithms**

$\ln x$ means $\log_e x$ **Natural logarithms**

For any logarithm base a,

$$\log_a 1 = 0 \quad \text{and} \quad \log_a a = 1.$$

For the logarithm base e,

$$\ln 1 = 0 \quad \text{and} \quad \ln e = 1.$$

Find each of the following without using a calculator.

$\log 100 = 2;$ $\log 10^{-5} = -5;$

$\ln 1 = 0;$ $\log_9 9 = 1;$

$\ln \sqrt[3]{e} = \frac{1}{3};$ $\log_2 64 = 6;$

$\log_8 1 = 0;$ $\ln e = 1$

Find each of the following using a calculator and rounding to four decimal places.

$\ln 223 = 5.4072;$ $\log \frac{2}{9} = -0.6532;$

$\log(-8)$ Does not exist; $\ln 0.06 = -2.8134$

The Change-of-Base Formula

For any logarithmic bases a and b, and any positive number M,

$$\log_b M = \frac{\log_a M}{\log_a b}.$$

Find $\log_3 11$ using common logarithms:

$$\log_3 11 = \frac{\log 11}{\log 3} \approx 2.1827.$$

Find $\log_3 11$ using natural logarithms:

$$\log_3 11 = \frac{\ln 11}{\ln 3} \approx 2.1827.$$

Earthquake Magnitude The magnitude R, measured on the Richter scale, of an earthquake of intensity I is defined as $R = \log \frac{I}{I_0}$, where I_0 is a minimum intensity used for comparison.	What is the magnitude on the Richter scale of an earthquake of intensity $10^{6.8} \cdot I_0$? $R = \log \frac{I}{I_0} = \log \frac{10^{6.8} \cdot I_0}{I_0} = \log 10^{6.8} = 6.8$

SECTION 5.4: PROPERTIES OF LOGARITHMIC FUNCTIONS

The Product Rule For any positive numbers M and N, and any logarithmic base a, $\log_a MN = \log_a M + \log_a N$. **The Power Rule** For any positive number M, any logarithmic base a, and any real number p, $\log_a M^p = p \log_a M$. **The Quotient Rule** For any positive numbers M and N, and any logarithmic base a, $\log_a \frac{M}{N} = \log_a M - \log_a N$.	Express $\log_c \sqrt{\frac{c^2 r}{b^3}}$ in terms of sums and differences of logarithms. $\log_c \sqrt{\frac{c^2 r}{b^3}} = \log_c \left(\frac{c^2 r}{b^3}\right)^{1/2}$ $= \frac{1}{2} \log_c \left(\frac{c^2 r}{b^3}\right)$ $= \frac{1}{2}(\log_c c^2 r - \log_c b^3)$ $= \frac{1}{2}(\log_c c^2 + \log_c r - \log_c b^3)$ $= \frac{1}{2}(2 + \log_c r - 3 \log_c b)$ $= 1 + \frac{1}{2} \log_c r - \frac{3}{2} \log_c b$ Express $\ln (3x^2 + 5x - 2) - \ln (x + 2)$ as a single logarithm. $\ln (3x^2 + 5x - 2) - \ln (x + 2) = \ln \frac{3x^2 + 5x - 2}{x + 2}$ $= \ln \frac{(3x - 1)(x + 2)}{x + 2}$ $= \ln (3x - 1)$ Given $\log_a 7 \approx 0.8451$ and $\log_a 5 \approx 0.6990$, find $\log_a \frac{1}{7}$ and $\log_a 35$. $\log_a \frac{1}{7} = \log_a 1 - \log_a 7 \approx 0 - 0.8451 \approx -0.8451;$ $\log_a 35 = \log_a (7 \cdot 5) = \log_a 7 + \log_a 5$ $\approx 0.8451 + 0.6990$ ≈ 1.5441
For any base a and any real number x, $\log_a a^x = x$. For any base a and any positive real number x, $a^{\log_a x} = x$.	Simplify each of the following. $8^{\log_8 k} = k;$ $\log 10^{43} = 43;$ $\log_a a^4 = 4;$ $e^{\ln 2} = 2$

SECTION 5.5: SOLVING EXPONENTIAL EQUATIONS AND LOGARITHMIC EQUATIONS

The Base–Exponent Property For any $a > 0, a \neq 1$, $a^x = a^y \longleftrightarrow x = y.$	Solve: $3^{2x-3} = 81$. $3^{2x-3} = 3^4$ $81 = 3^4$ $2x - 3 = 4$ $2x = 7$ $x = \frac{7}{2}$ The solution is $\frac{7}{2}$.
The Property of Logarithmic Equality For any $M > 0, N > 0, a > 0$, and $a \neq 1$, $\log_a M = \log_a N \longleftrightarrow M = N.$	Solve: $6^{x-2} = 2^{-3x}$. $\log 6^{x-2} = \log 2^{-3x}$ $(x-2)\log 6 = -3x \log 2$ $x \log 6 - 2 \log 6 = -3x \log 2$ $x \log 6 + 3x \log 2 = 2 \log 6$ $x(\log 6 + 3 \log 2) = 2 \log 6$ $x = \dfrac{2 \log 6}{\log 6 + 3 \log 2}$ $x \approx 0.9257$
	Solve: $\log_3 (x-2) + \log_3 x = 1$. $\log_3 [x(x-2)] = 1$ $x(x-2) = 3^1$ $x^2 - 2x - 3 = 0$ $(x-3)(x+1) = 0$ $x - 3 = 0$ *or* $x + 1 = 0$ $x = 3$ *or* $x = -1$ The number -1 is not a solution because negative numbers do not have real-number logarithms. The value 3 checks and is the solution.
	Solve: $\ln (x+10) - \ln (x+4) = \ln x$. $\ln \dfrac{x+10}{x+4} = \ln x$ $\dfrac{x+10}{x+4} = x$ $x + 10 = x(x+4)$ $x + 10 = x^2 + 4x$ $0 = x^2 + 3x - 10$ $0 = (x+5)(x-2)$ $x + 5 = 0$ *or* $x - 2 = 0$ $x = -5$ *or* $x = 2$ The number -5 is not a solution because $-5 + 4 = -1$ and $\ln (-1)$ is not a real number. The value 2 checks and is the solution.

SECTION 5.6: APPLICATIONS AND MODELS: GROWTH AND DECAY; COMPOUND INTEREST

Exponential Growth Model $P(t) = P_0 e^{kt}, \quad k > 0$ **Doubling Time** $kT = \ln 2, \quad \text{or} \quad k = \dfrac{\ln 2}{T},$ $\text{or} \quad T = \dfrac{\ln 2}{k}$	In July 2013, the population of the United States was 316.7 million, and the exponential growth rate was 0.9% per year (*Source: CIA World Factbook* 2014). After how long will the population be double what it was in 2013? Estimate the population in 2020. With a population growth rate of 0.9%, or 0.009, the doubling time T is $T = \dfrac{\ln 2}{k} = \dfrac{\ln 2}{0.009} \approx 77.$ The population of the United States will be double what it was in 2013 in about 77 years. The exponential growth function is $P(t) = 316.7e^{0.009t},$ where t is the number of years after 2013 and $P(t)$ is in millions. Since in 2020, $t = 7$, we substitute 7 for t: $P(7) = 316.7e^{0.009 \cdot 7} = 316.7e^{0.063} \approx 337.3.$ The population will be about 337.3 million, or 337,300,000, in 2020.
Interest Compounded Continuously $P(t) = P_0 e^{kt}, \quad k > 0$	Suppose that \$20,000 is invested at interest rate k, compounded continuously, and grows to \$23,236.68 in 3 years. What is the interest rate? What will the balance be in 8 years? The exponential growth function is of the form $P(t) = 20{,}000e^{kt}$. Given that $P(3) = \$23{,}236.68$, substituting 3 for t and 23,236.68 for $P(t)$ gives $23{,}236.68 = 20{,}000e^{k(3)}$ to get $k \approx 0.05$, or 5%. We then substitute 0.05 for k and 8 for t and determine $P(8)$: $P(8) = 20{,}000e^{0.05(8)} = 20{,}000e^{0.4} \approx \$29{,}836.49.$
Exponential Decay Model $P(t) = P_0 e^{-kt}, \quad k > 0$ **Half-Life** $kT = \ln 2, \quad \text{or} \quad k = \dfrac{\ln 2}{T},$ $\text{or} \quad T = \dfrac{\ln 2}{k}$	Archaeologists discovered an animal bone that had lost 65.2% of its carbon-14 at the time it was found. How old was the bone? The decay rate for carbon-14 is 0.012%, or 0.00012. If the bone has lost 65.2% of its carbon-14 from an initial amount P_0, then 34.8% P_0 is the amount present. We substitute 34.8% P_0 for $P(t)$ and solve: $34.8\% P_0 = P_0 e^{-0.00012t}$ $0.348 = e^{-0.00012t}$ $\ln 0.348 = -0.00012t$ $\dfrac{\ln 0.348}{-0.00012} = t$ $8796 \approx t.$ The bone was about 8796 years old when it was found.

REVIEW EXERCISES

Determine whether the statement is true or false.

1. The domain of a one-to-one function f is the range of the inverse f^{-1}. [5.1]

2. The x-intercept of $f(x) = \log x$ is $(0, 1)$. [5.3]

3. The graph of f^{-1} is a reflection of the graph of f across $y = 0$. [5.1]

4. If it is not possible for a horizontal line to intersect the graph of a function more than once, then the function is one-to-one and its inverse is a function. [5.1]

5. The range of all exponential functions is $[0, \infty)$. [5.2]

6. The horizontal asymptote of $y = 2^x$ is $y = 0$. [5.2]

7. Find the inverse of the relation $\{(1.3, -2.7), (8, -3), (-5, 3), (6, -3), (7, -5)\}$. [5.1]

8. Find an equation of the inverse relation. [5.1]
   a) $y = -2x + 3$
   b) $y = 3x^2 + 2x - 1$
   c) $0.8x^3 - 5.4y^2 = 3x$

Graph the function and determine whether the function is one-to-one using the horizontal-line test. [5.1]

9. $f(x) = -|x| + 3$
10. $f(x) = x^2 + 1$
11. $f(x) = 2x - \dfrac{3}{4}$
12. $f(x) = -\dfrac{6}{x + 1}$

In Exercises 13–18, given the function:

a) *Sketch the graph and determine whether the function is one-to-one.* [5.1], [5.3]

b) *If it is one-to-one, find a formula for the inverse.* [5.1], [5.3]

13. $f(x) = 2 - 3x$
14. $f(x) = \dfrac{x + 2}{x - 1}$
15. $f(x) = \sqrt{x - 6}$
16. $f(x) = x^3 - 8$
17. $f(x) = 3x^2 + 2x - 1$
18. $f(x) = e^x$

For the function f, use composition of functions to show that f^{-1} is as given. [5.1]

19. $f(x) = 6x - 5$, $f^{-1}(x) = \dfrac{x + 5}{6}$

20. $f(x) = \dfrac{x + 1}{x}$, $f^{-1}(x) = \dfrac{1}{x - 1}$

Find the inverse of the given one-to-one function f. Give the domain and the range of f and of f^{-1} and then graph both f and f^{-1} on the same set of axes. [5.1]

21. $f(x) = 2 - 5x$
22. $f(x) = \dfrac{x - 3}{x + 2}$

23. Find $f(f^{-1}(657))$:
$$f(x) = \frac{4x^5 - 16x^{37}}{119x}, \quad x > 1. \text{ [5.1]}$$

24. Find $f(f^{-1}(a))$: $f(x) = \sqrt[3]{3x - 4}$. [5.1]

Graph the function.

25. $f(x) = \left(\frac{1}{3}\right)^x$ [5.2]
26. $f(x) = 1 + e^x$ [5.2]
27. $f(x) = -e^{-x}$ [5.2]
28. $f(x) = \log_2 x$ [5.3]
29. $f(x) = \frac{1}{2} \ln x$ [5.3]
30. $f(x) = \log x - 2$ [5.3]

In Exercises 31–36, match the equation with one of the figures (a)–(f) that follow.

a)
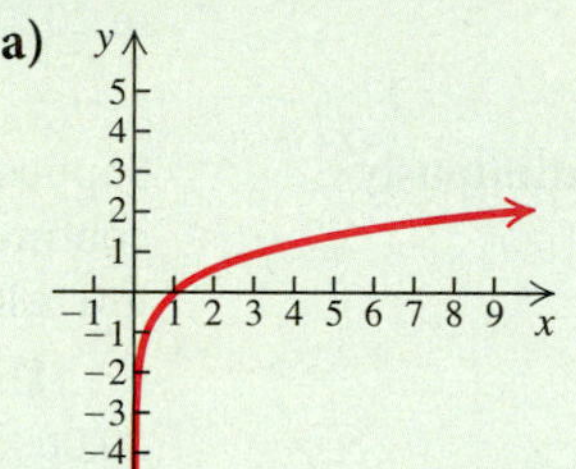

b)
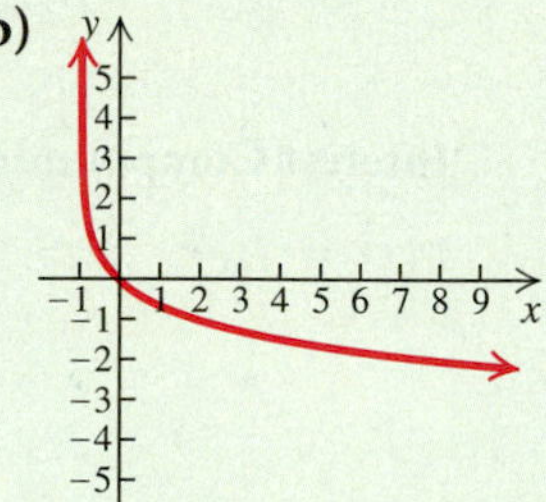

c)
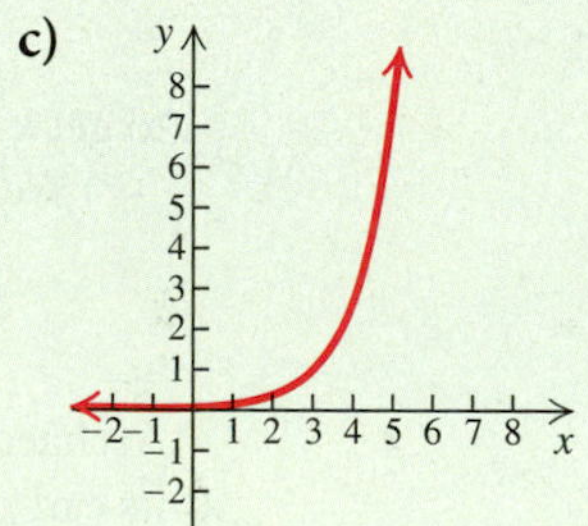

d)
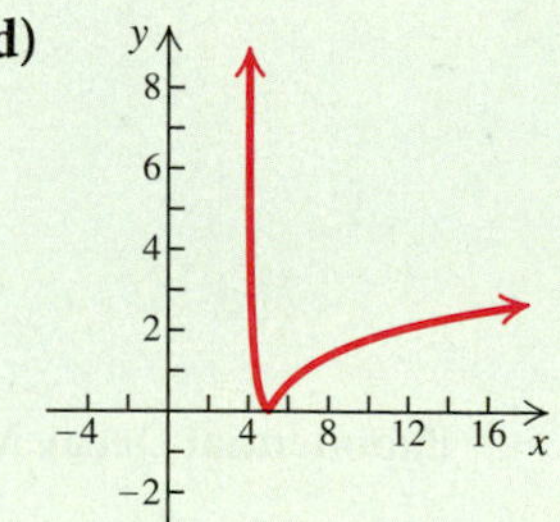

e)
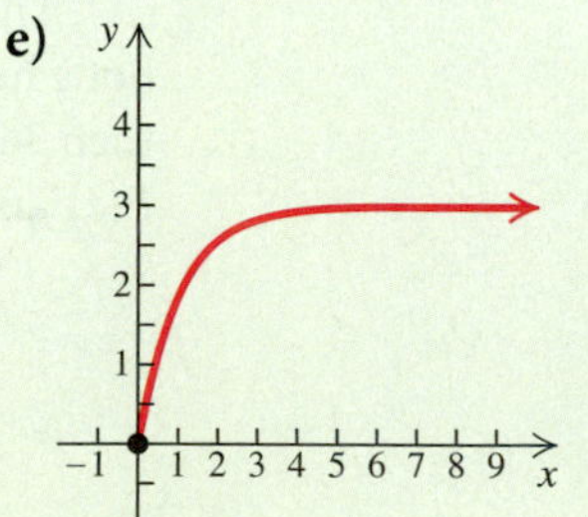

f)
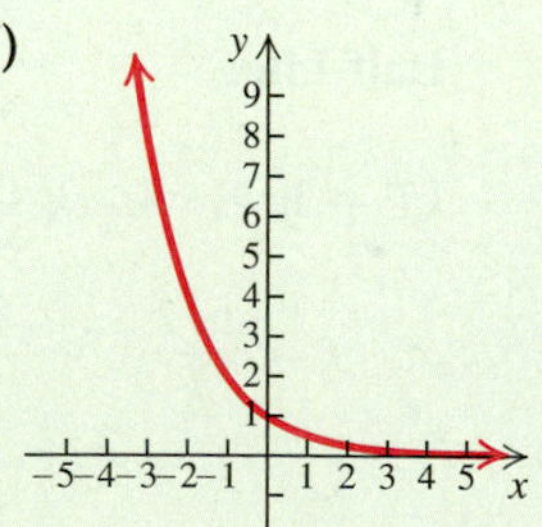

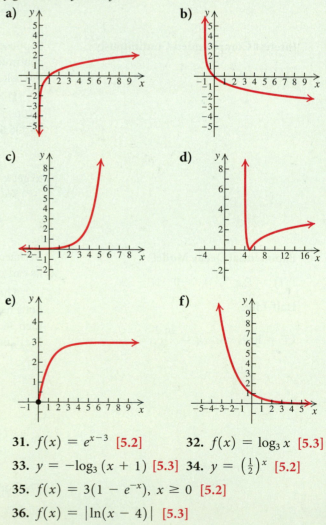

31. $f(x) = e^{x-3}$ [5.2]
32. $f(x) = \log_3 x$ [5.3]
33. $y = -\log_3 (x + 1)$ [5.3]
34. $y = \left(\frac{1}{2}\right)^x$ [5.2]
35. $f(x) = 3(1 - e^{-x})$, $x \geq 0$ [5.2]
36. $f(x) = |\ln(x - 4)|$ [5.3]

Find each of the following. Do not use a calculator. [5.3]

37. $\log_5 125$

38. $\log 100{,}000$

39. $\ln e$

40. $\ln 1$

41. $\log 10^{1/4}$

42. $\log_3 \sqrt{3}$

43. $\log 1$

44. $\log 10$

45. $\log_2 \sqrt[3]{2}$

46. $\log 0.01$

Convert to an exponential equation. [5.3]

47. $\log_4 x = 2$

48. $\log_a Q = k$

Convert to a logarithmic equation. [5.3]

49. $4^{-3} = \frac{1}{64}$

50. $e^x = 80$

Find each of the following using a calculator. Round to four decimal places. [5.3]

51. $\log 11$

52. $\log 0.234$

53. $\ln 3$

54. $\ln 0.027$

55. $\log(-3)$

56. $\ln 0$

Find the logarithm using the change-of-base formula. [5.3]

57. $\log_5 24$

58. $\log_8 3$

Express as a single logarithm and, if possible, simplify. [5.4]

59. $3\log_b x - 4\log_b y + \frac{1}{2}\log_b z$

60. $\ln(x^3 - 8) - \ln(x^2 + 2x + 4) + \ln(x + 2)$

Express in terms of sums and differences of logarithms. [5.4]

61. $\ln \sqrt[4]{wr^2}$

62. $\log \sqrt[3]{\frac{M^2}{N}}$

Given that $\log_a 2 = 0.301$, $\log_a 5 = 0.699$, *and* $\log_a 6 = 0.778$, *find each of the following.* [5.4]

63. $\log_a 3$

64. $\log_a 50$

65. $\log_a \frac{1}{5}$

66. $\log_a \sqrt[3]{5}$

Simplify. [5.4]

67. $\ln e^{-5k}$

68. $\log_5 5^{-6t}$

Solve. [5.5]

69. $\log_4 x = 2$

70. $3^{1-x} = 9^{2x}$

71. $e^x = 80$

72. $4^{2x-1} - 3 = 61$

73. $\log_{16} 4 = x$

74. $\log_x 125 = 3$

75. $\log_2 x + \log_2 (x - 2) = 3$

76. $\log(x^2 - 1) - \log(x - 1) = 1$

77. $\log x^2 = \log x$

78. $e^{-x} = 0.02$

79. *Saving for College.* Following the birth of triplets, the grandparents deposit \$30,000 in a college trust fund that earns 4.2% interest, compounded quarterly.

a) Find a function for the amount in the account after t years. [5.2]

b) Find the amount in the account at $t = 0, 6, 12,$ and 18 years. [5.2]

80. *Wind Power Capacity.* Global wind power capacity is increasing exponentially. The total capacity, in gigawatts (GW), can be estimated with the exponential function

$$W(t) = 29.9(1.26)^t,$$

where t is the number of years after 2002 (*Source:* REN21). Find the global wind power capacity in 2005 and in 2010. Then use this function to estimate the capacity in 2016. [5.2]

81. How long will it take an investment to double if it is invested at 4.5%, compounded continuously? [5.6]

82. The population of a metropolitan area consisting of 8 counties doubled in 26 years. What was the exponential growth rate? [5.6]

83. How old is a skeleton that has lost 27% of its carbon-14? [5.6]

84. The hydrogen ion concentration of milk is 2.3×10^{-6}. What is the pH? (See Exercise 98 in Exercise Set 5.3.) [5.3]

85. *Earthquake Magnitude.* The earthquake in Kashgar, China, on February 25, 2003, had an intensity of $10^{6.3} \cdot I_0$ (*Source:* U.S. Geological Survey). What is the magnitude on the Richter scale? [5.3]

86. What is the loudness, in decibels, of a sound whose intensity is $1000I_0$? (See Exercise 101 in Exercise Set 5.3.) [5.3]

87. *Walking Speed.* The average walking speed w, in feet per second, of a person living in a city of population P, in thousands, is given by the function

$$w(P) = 0.37 \ln P + 0.05.$$

a) The population of Wichita, Kansas, is 353,823. Find the average walking speed. [5.3]

b) A city's population has an average walking speed of 3.4 ft/sec. Find the population. [5.6]

88. *Social Security Distributions.* Cash Social Security distributions were \$35 million, or \$0.035 billion, in 1940. This amount has increased exponentially to \$786 billion in 2012. (*Source:* Pew Research Center) Assuming that the exponential growth model applies:

a) Find the exponential growth rate k. [5.6]

b) Find the exponential growth function. [5.6]

c) Estimate the total cash distributions in 1970, in 2000, and in 2015. [5.6]

d) In what year will the cash benefits reach \$2 trillion? [5.6]

89. *The Population of Cambodia.* The population of Cambodia was 15.2 million in 2013, and the exponential growth rate was 1.67% per year (*Source:* U.S. Census Bureau, World Population Profile).

a) Find the exponential growth function. [5.6]
b) What will the population be in 2017? in 2020? [5.6]
c) When will the population be 18 million? [5.6]
d) What is the doubling time? [5.6]

90. Which of the following is the horizontal asymptote of the graph of $f(x) = e^{x-3} + 2$? [5.2]

A. $y = -2$ B. $y = -3$
C. $y = 3$ D. $y = 2$

91. Which of the following is the domain of the logarithmic function $f(x) = \log(2x - 3)$? [5.3]

A. $\left(\frac{3}{2}, \infty\right)$ B. $\left(-\infty, \frac{3}{2}\right)$
C. $(3, \infty)$ D. $(-\infty, \infty)$

92. The graph of $f(x) = 2^{x-2}$ is which of the following? [5.2]

A.
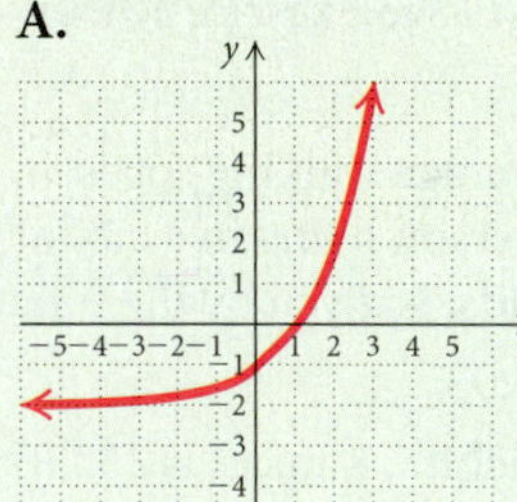

B.
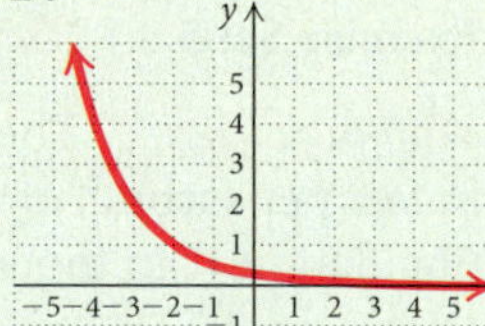

C.
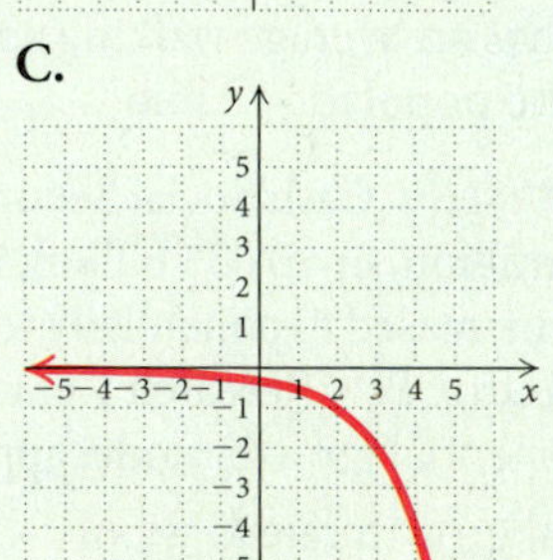

D.
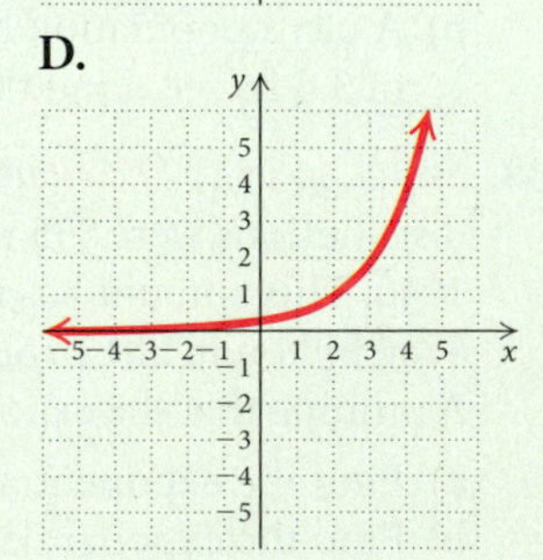

93. The graph of $f(x) = \log_2 x$ is which of the following? [5.3]

A. B.
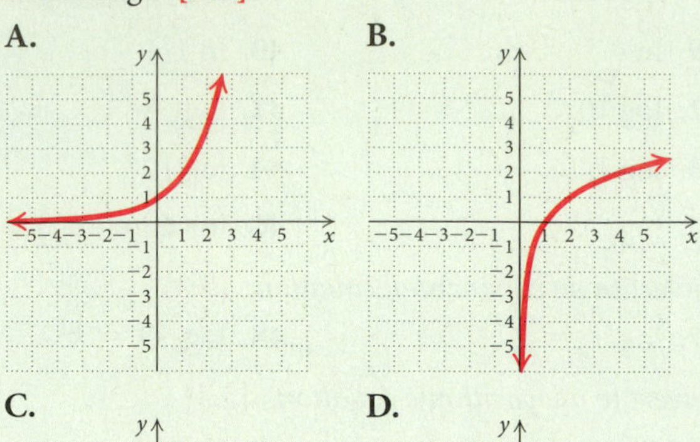

C. D.
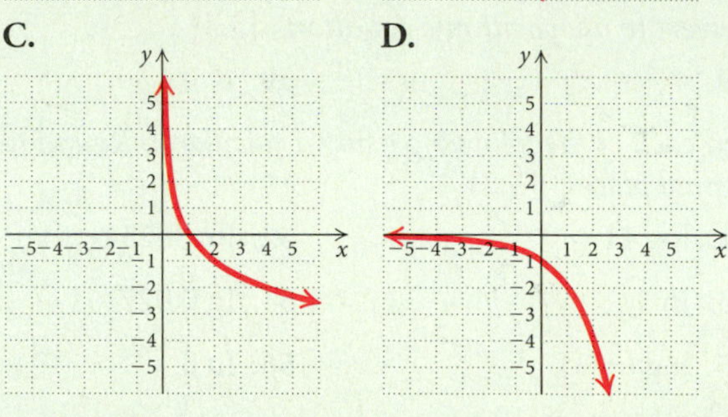

▶ Synthesis

Solve. [5.5]

94. $|\log_4 x| = 3$

95. $\log x = \ln x$

96. $5^{\sqrt{x}} = 625$

▶ Collaborative Discussion and Writing

97. *Atmospheric Pressure.* Atmospheric pressure P at an altitude a is given by

$$P = P_0 e^{-0.00005a},$$

where P_0 is the pressure at sea level, approximately 14.7 lb/in^2 (pounds per square inch). Explain how a barometer, or some device for measuring atmospheric pressure, can be used to find the height of a skyscraper. [5.6]

98. Explain how the graph of $f(x) = \ln x$ can be used to obtain the graph of $g(x) = e^{x-2}$. [5.3]

99. Describe the difference between $f^{-1}(x)$ and $[f(x)]^{-1}$. [5.1]

100. Explain the errors, if any, in the following:

$$\log_a ab^3 = (\log_a a)(\log_a b^3) = 3 \log_a b. \text{ [5.4]}$$

5 Chapter Test

1. Find the inverse of the relation
$\{(-2, 5), (4, 3), (0, -1), (-6, -3)\}$.

Determine whether the function is one-to-one. Answer yes or no.

2. 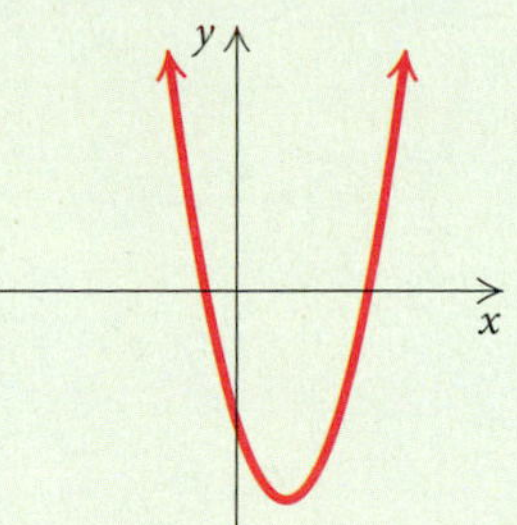

3. 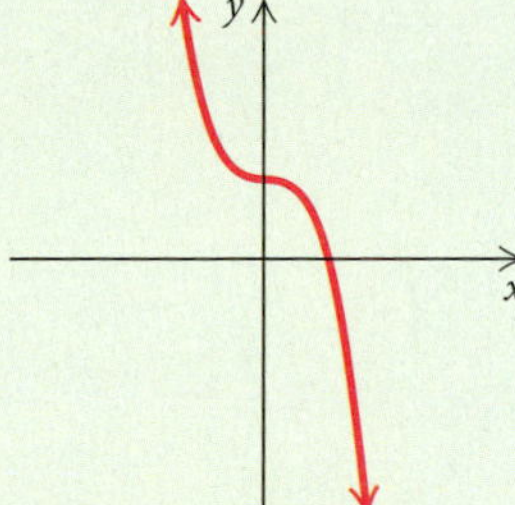

In Exercises 4–7, given the function:

a) *Sketch the graph and determine whether the function is one-to-one.*
b) *If it is one-to-one, find a formula for the inverse.*

4. $f(x) = x^3 + 1$
5. $f(x) = 1 - x$
6. $f(x) = \dfrac{x}{2 - x}$
7. $f(x) = x^2 + x - 3$

8. Use composition of functions to show that f^{-1} is as given:
$$f(x) = -4x + 3, \quad f^{-1}(x) = \frac{3 - x}{4}.$$

9. Find the inverse of the one-to-one function
$$f(x) = \frac{1}{x - 4}.$$
Give the domain and the range of f and of f^{-1} and then graph both f and f^{-1} on the same set of axes.

Graph the function.

10. $f(x) = 4^{-x}$
11. $f(x) = \log x$
12. $f(x) = e^x - 3$
13. $f(x) = \ln(x + 2)$

Find each of the following. Do not use a calculator.

14. $\log 0.00001$
15. $\ln e$
16. $\ln 1$
17. $\log_4 \sqrt[5]{4}$
18. Convert to an exponential equation: $\ln x = 4$.
19. Convert to a logarithmic equation: $3^x = 5.4$.

Find each of the following using a calculator. Round to four decimal places.

20. $\ln 16$
21. $\log 0.293$
22. Find $\log_6 10$ using the change-of-base formula.
23. Express as a single logarithm:
$$2 \log_a x - \log_a y + \tfrac{1}{2} \log_a z.$$
24. Express $\ln \sqrt[5]{x^2 y}$ in terms of sums and differences of logarithms.
25. Given that $\log_a 2 = 0.328$ and $\log_a 8 = 0.984$, find $\log_a 4$.
26. Simplify: $\ln e^{-4t}$.

Solve.

27. $\log_{25} 5 = x$
28. $\log_3 x + \log_3 (x + 8) = 2$
29. $3^{4-x} = 27^x$
30. $e^x = 65$
31. *Earthquake Magnitude.* The earthquake in Bam, in southeast Iran, on December 26, 2003, had an intensity of $10^{6.6} \cdot I_0$ (*Source:* U.S. Geological Survey). What was its magnitude on the Richter scale?
32. *Growth Rate.* A country's population doubled in 45 years. What was the exponential growth rate?
33. *Compound Interest.* Suppose \$1000 is invested at interest rate k, compounded continuously, and grows to \$1144.54 in 3 years.
    a) Find the interest rate.
    b) Find the exponential growth function.
    c) Find the balance after 8 years.
    d) Find the doubling time.

34. The graph of $f(x) = 2^{x-1} + 1$ is which of the following?

A.

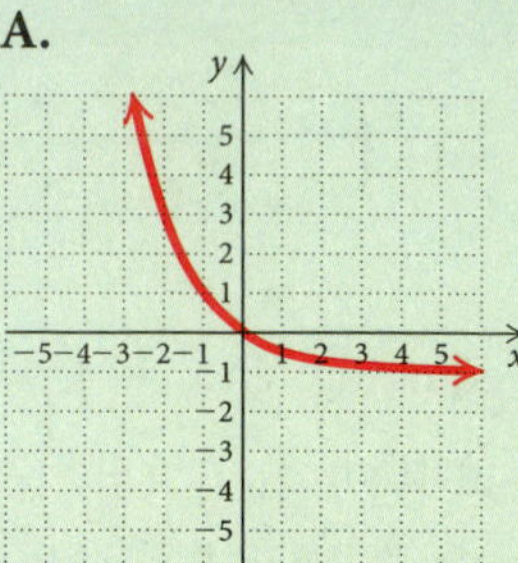

B.

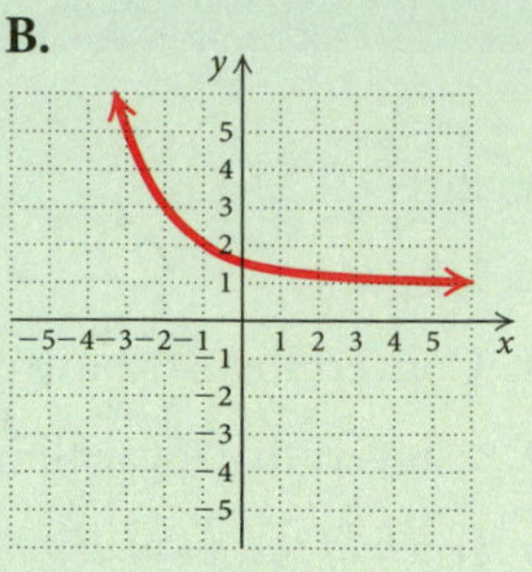

C.

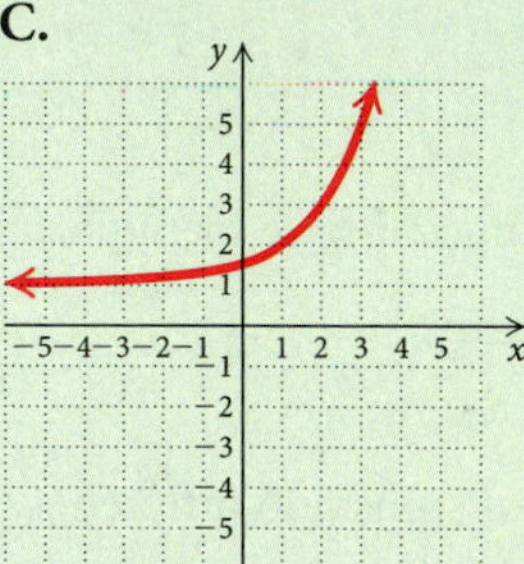

D.

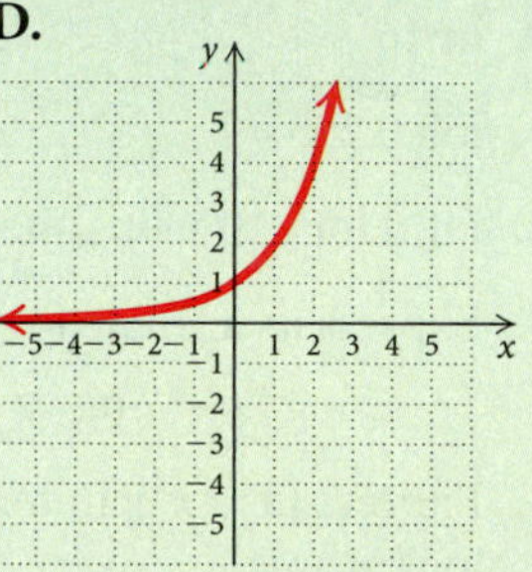

► Synthesis

35. Solve: $4^{\sqrt[3]{x}} = 8$.

CHAPTER

6

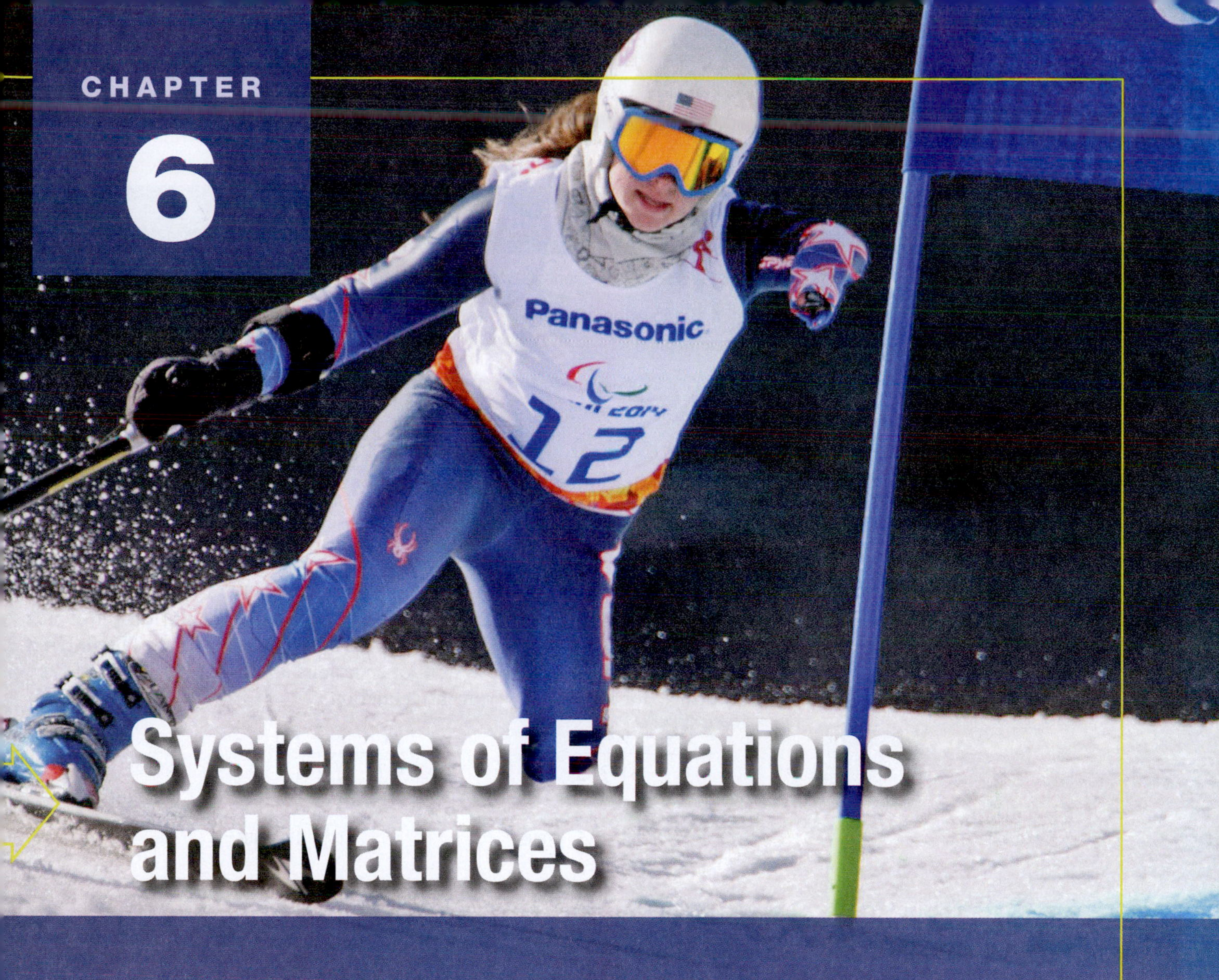

Systems of Equations and Matrices

APPLICATION This problem appears as Exercise 17 in Exercise Set 6.2.

At the 2014 Paralympic Games in Sochi, Russia, the top three countries—the Russian Federation, Ukraine, and the United States—won a total of 123 medals. Ukraine won 7 more medals than the United States. The Russian Federation won 37 more medals than the total amount won by Ukraine and the United States. (*Source*: International Paralympic Committee) How many medals did each of the top three countries win?

6.1 Systems of Equations in Two Variables

- Solve a system of two linear equations in two variables by graphing.
- Solve a system of two linear equations in two variables using the substitution method and the elimination method.
- Use systems of two linear equations to solve applied problems.

A **system of equations** is composed of two or more equations considered simultaneously. For example,

$$x - y = 5,$$
$$2x + y = 1$$

is a **system of two linear equations in two variables**. The solution set of this system consists of all ordered pairs that make *both* equations true. The ordered pair $(2, -3)$ is a solution of the system of equations above. We can verify this by substituting 2 for x and -3 for y in *each* equation.

$x - y = 5$			$2x + y = 1$		
$2 - (-3)$	? 5		$2 \cdot 2 + (-3)$	? 1	
$2 + 3$			$4 - 3$		
5	5	TRUE	1	1	TRUE

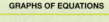

GRAPHS OF EQUATIONS

REVIEW SECTION 1.1

Solving Systems of Equations Graphically

Recall that the graph of a linear equation is a line that contains all the ordered pairs in the solution set of the equation. When we graph a system of linear equations, each point at which the graphs intersect is a solution of *both* equations and therefore a **solution of the system of equations**.

Technology Connection

To use a graphing calculator to solve the system of equations in Example 1, it might be necessary to write each equation in "$Y = \cdots$" form. If so, we would graph $y_1 = x - 5$ and $y_2 = -2x + 1$ and then use the INTERSECT feature. We see that the solution is $(2, -3)$.

$y_1 = x - 5, \quad y_2 = -2x + 1$

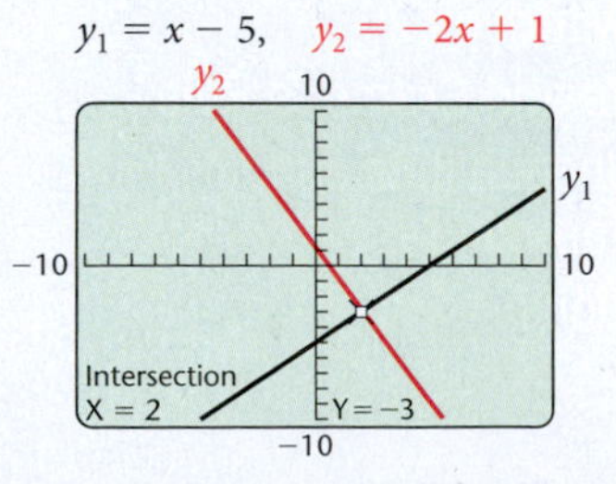

EXAMPLE 1 Solve the following system of equations graphically.

$$x - y = 5,$$
$$2x + y = 1$$

Solution We graph the equations on the same set of axes, as shown below.

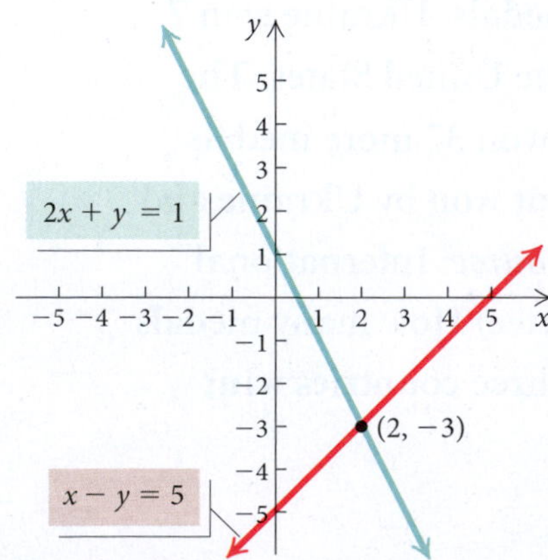

We see that the graphs intersect at a single point, $(2, -3)$, so $(2, -3)$ is the solution of the system of equations. To check this solution, we substitute 2 for x and -3 for y in both equations as we did above.

Now Try Exercise 7.

The graphs of most of the systems of equations that we use to model applications intersect at a single point, like the system above. However, it is possible that the graphs will have no points in common or infinitely many points in common. Each of these possibilities is illustrated below.

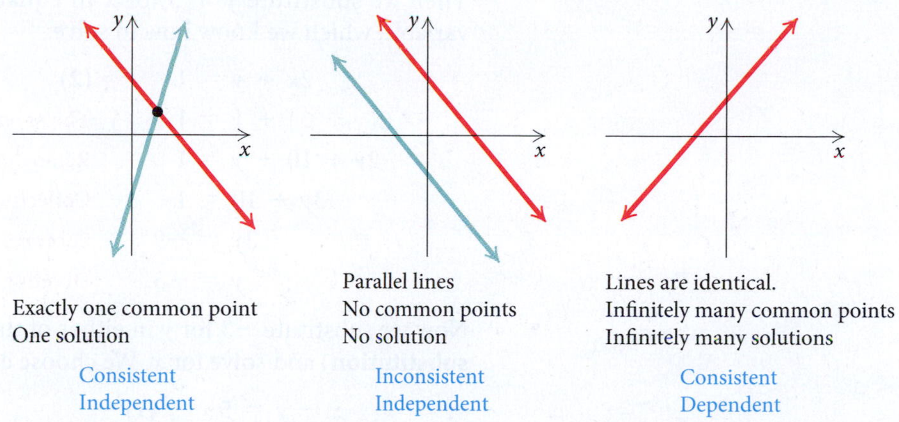

Exactly one common point
One solution
Consistent
Independent

Parallel lines
No common points
No solution
Inconsistent
Independent

Lines are identical.
Infinitely many common points
Infinitely many solutions
Consistent
Dependent

If a system of equations has at least one solution, it is **consistent**. If the system has no solutions, it is **inconsistent**. In addition, for a system of two linear equations in two variables, if one equation can be obtained by multiplying by a constant on both sides of the other equation, then the equations are **dependent**. Otherwise, they are **independent**. A system of two dependent linear equations in two variables has an infinite number of solutions.

The Substitution Method

Solving a system of equations graphically is not always accurate when the solutions are not integers. A solution like $\left(\frac{43}{27}, -\frac{19}{27}\right)$, for instance, will be difficult to determine from a hand-drawn graph.

Algebraic methods for solving systems of equations, when used correctly, always give accurate results. One such technique is the **substitution method**. It is used most often when a variable is alone on one side of an equation or when it is easy to solve for a variable. To apply the substitution method, we begin by using one of the equations to express one variable in terms of the other; then we substitute that expression in the other equation of the system.

EXAMPLE 2 Use the substitution method to solve the system

$$x - y = 5, \quad (1)$$
$$2x + y = 1. \quad (2)$$

Solution First, we solve equation (1) for x. (We could have solved for y instead.) We have

$$x - y = 5, \quad (1)$$
$$x = y + 5. \quad \text{Solving for } x$$

Then we substitute $y + 5$ for x in equation (2). This gives an equation in one variable, which we know how to solve:

$$2x + y = 1 \quad (2)$$
$$2(y + 5) + y = 1 \quad \text{The parentheses are necessary.}$$
$$2y + 10 + y = 1 \quad \text{Removing parentheses}$$
$$3y + 10 = 1 \quad \text{Collecting like terms on the left}$$
$$3y = -9 \quad \text{Subtracting 10 on both sides}$$
$$y = -3. \quad \text{Dividing by 3 on both sides}$$

Now we substitute -3 for y in either of the original equations (this is called **back-substitution**) and solve for x. We choose equation (1):

$$x - y = 5 \quad (1)$$
$$x - (-3) = 5 \quad \text{Substituting } -3 \text{ for } y$$
$$x + 3 = 5$$
$$x = 2. \quad \text{Subtracting 3 on both sides}$$

We have previously checked the pair $(2, -3)$ in both equations. The solution of the system of equations is $(2, -3)$. Since there is exactly one solution, the system of equations is consistent, and the equations are independent.

Now Try Exercise 17.

The Elimination Method

Another algebraic technique for solving systems of equations is the **elimination method.** With this method, we eliminate a variable by adding two equations. If the coefficients of a particular variable are opposites, we can eliminate that variable simply by adding the original equations. For example, if the x-coefficient is -3 in one equation and is 3 in the other equation, then the sum of the x-terms will be 0 and thus the variable x will be eliminated when we add the equations.

EXAMPLE 3 Use the elimination method to solve the system of equations

$$2x + y = 2, \quad (1)$$
$$x - y = 7. \quad (2)$$

Algebraic Solution

Since the y-coefficients, 1 and -1, are opposites, we can eliminate y by adding the equations:

$$\begin{aligned} 2x + y &= 2 \quad &(1)\\ x - y &= 7 \quad &(2)\\ \hline 3x \quad &= 9 \quad &\text{Adding}\\ x &= 3. \end{aligned}$$

We then back-substitute 3 for x in either equation and solve for y. We choose equation (1):

$$\begin{aligned} 2x + y &= 2 \quad &(1)\\ 2 \cdot 3 + y &= 2 \quad &\text{Substituting 3 for } x\\ 6 + y &= 2\\ y &= -4. \end{aligned}$$

We check the solution by substituting the pair $(3, -4)$ in both equations.

$$\begin{array}{c|l} 2x + y = 2 & \\ \hline 2 \cdot 3 + (-4) \ ? & 2\\ 6 - 4 & \\ 2 & 2 \quad \text{TRUE} \end{array} \qquad \begin{array}{c|l} x - y = 7 & \\ \hline 3 - (-4) \ ? & 7\\ 3 + 4 & \\ 7 & 7 \quad \text{TRUE} \end{array}$$

The solution is $(3, -4)$. Since there is exactly one solution, the system of equations is consistent, and the equations are independent.

Visualizing the Solution

We graph $2x + y = 2$ and $x - y = 7$.

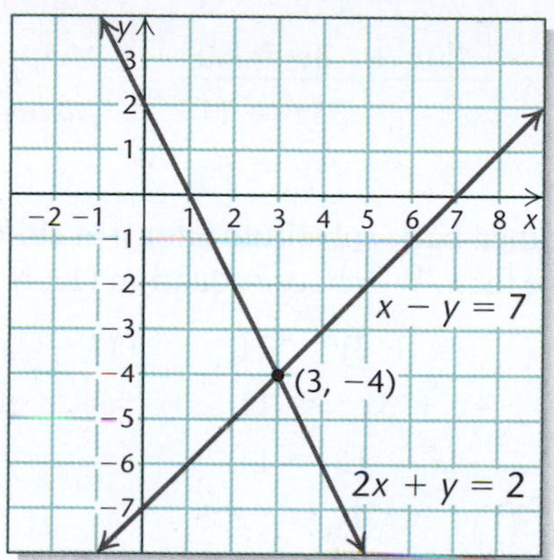

The graphs intersect at the point $(3, -4)$, so the solution of the system of equations is $(3, -4)$.

Now Try Exercise 31.

Before we add, it might be necessary to multiply one or both equations by suitable constants in order to find two equations in which the coefficients of a variable are opposites.

EXAMPLE 4 Use the elimination method to solve the system of equations

$$4x + 3y = 11, \quad (1)$$
$$-5x + 2y = 15. \quad (2)$$

Algebraic Solution

We can obtain x-coefficients that are opposites by multiplying the first equation by 5 and the second equation by 4:

$$\begin{aligned} 20x + 15y &= 55 && \text{Multiplying equation (1) by 5} \\ -20x + 8y &= 60 && \text{Multiplying equation (2) by 4} \\ 23y &= 115 && \text{Adding} \\ y &= 5. \end{aligned}$$

We then back-substitute 5 for y in either equation (1) or (2) and solve for x. We choose equation (1):

$$\begin{aligned} 4x + 3y &= 11 && (1) \\ 4x + 3 \cdot 5 &= 11 && \text{Substituting 5 for } y \\ 4x + 15 &= 11 \\ 4x &= -4 \\ x &= -1. \end{aligned}$$

We can check the pair $(-1, 5)$ by substituting in both equations. The solution is $(-1, 5)$. The system of equations is consistent, and the equations are independent.

Visualizing the Solution

We graph $4x + 3y = 11$ and $-5x + 2y = 15$.

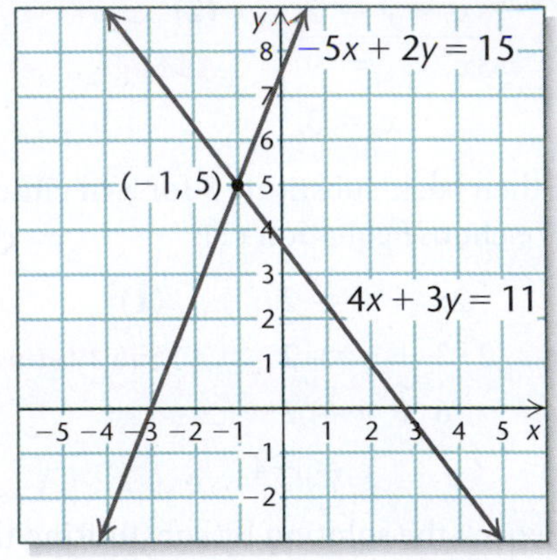

The graphs intersect at the point $(-1, 5)$, so the solution of the system of equations is $(-1, 5)$.

Now Try Exercise 33.

In Example 4, the two systems

$$\begin{aligned} 4x + 3y &= 11, \\ -5x + 2y &= 15 \end{aligned} \quad \text{and} \quad \begin{aligned} 20x + 15y &= 55, \\ -20x + 8y &= 60 \end{aligned}$$

are **equivalent** because they have exactly the same solutions. When we use the elimination method, we often multiply one or both equations by constants to find equivalent equations that allow us to eliminate a variable by adding.

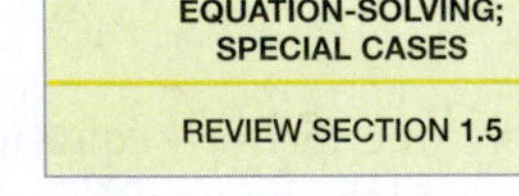

EXAMPLE 5 Solve each of the following systems using the elimination method.

a) $x - 3y = 1, \quad (1)$
$-2x + 6y = 5 \quad (2)$

b) $2x + 3y = 6, \quad (1)$
$4x + 6y = 12 \quad (2)$

Solution

a) We multiply equation (1) by 2 and add:

$$\begin{aligned} 2x - 6y &= 2 && \text{Multiplying equation (1) by 2} \\ -2x + 6y &= 5 && (2) \\ 0 &= 7. && \text{Adding} \end{aligned}$$

There are no values of x and y for which $0 = 7$ is true, so the system has *no solution*. The solution set is $\emptyset$. The system of equations is inconsistent and the equations are independent. The graphs of the equations are parallel lines, as shown in Fig. 1.

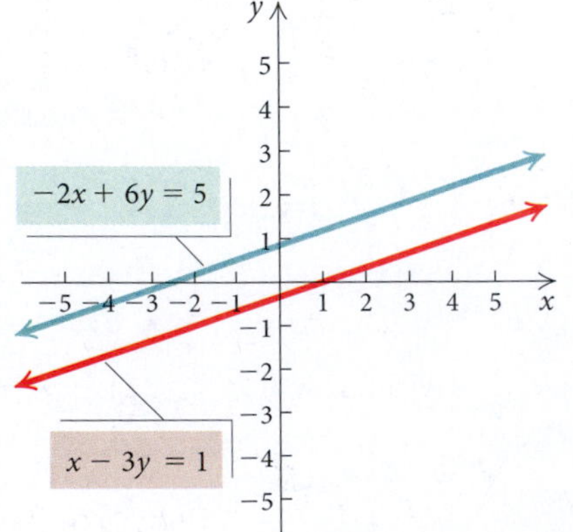

Figure 1.

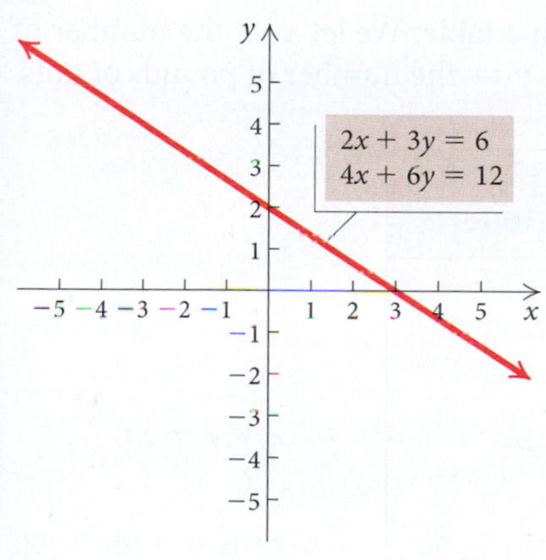

Figure 2.

b) We multiply equation (1) by -2 and add:

$$\begin{aligned} -4x - 6y &= -12 &\quad& \text{Multiplying equation (1) by } -2 \\ 4x + 6y &= 12 && (2) \\ \hline 0 &= 0. && \text{Adding} \end{aligned}$$

We obtain the equation $0 = 0$, which is true for all values of x and y. This tells us that the equations are dependent, so there are *infinitely many solutions.* That is, any solution of one equation of the system is also a solution of the other. The system of equations is consistent. The graphs of the equations are identical, as shown in Fig. 2.

Solving either equation for y, we have $y = -\frac{2}{3}x + 2$, so we can write the solutions of the system as ordered pairs (x, y), where y is expressed as $-\frac{2}{3}x + 2$. Thus the solutions can be written in the form $\left(x, -\frac{2}{3}x + 2\right)$. Any real value that we choose for x then gives us a value for y and thus an ordered pair in the solution set. For example,

$$\begin{aligned} &\text{if } x = -3, &&\text{then } -\tfrac{2}{3}x + 2 = -\tfrac{2}{3}(-3) + 2 = 4, \\ &\text{if } x = 0, &&\text{then } -\tfrac{2}{3}x + 2 = -\tfrac{2}{3}\cdot 0 + 2 = 2, \quad \text{and} \\ &\text{if } x = 6, &&\text{then } -\tfrac{2}{3}x + 2 = -\tfrac{2}{3}\cdot 6 + 2 = -2. \end{aligned}$$

x	$-\frac{2}{3}x + 2$
$-\frac{3}{2}y + 3$	y
-3	4
0	2
6	-2

Thus some of the solutions are $(-3, 4)$, $(0, 2)$, and $(6, -2)$.

Similarly, solving either equation for x, we have $x = -\frac{3}{2}y + 3$, so the solutions (x, y) can also be written, expressing x as $-\frac{3}{2}y + 3$, in the form $\left(-\frac{3}{2}y + 3, y\right)$.

Since the two forms of the solutions are equivalent, they yield the same solution set, as illustrated in the table at left. Note, for example, that when $y = 4$, we have the solution $(-3, 4)$; when $y = 2$, we have $(0, 2)$; and when $y = -2$, we have $(6, -2)$.

Now Try Exercises 35 and 37.

Applications

Frequently the most challenging and time-consuming step in the problem-solving process is translating a situation to mathematical language. However, in many cases, this task is made easier if we translate to more than one equation in more than one variable.

EXAMPLE 6 ***Snack Mixtures.*** At Max's Munchies, caramel corn worth \$2.50 per pound is mixed with honey-roasted mixed nuts worth \$7.50 per pound in order to get 20 lb of a mixture worth \$4.50 per pound. How much of each snack is used?

Solution We use the five-step problem-solving process.

1. **Familiarize.** Let's begin by making a guess. Suppose 16 lb of caramel corn and 4 lb of nuts are used. Then the total weight of the mixture would be 16 lb + 4 lb, or 20 lb, the desired weight. The total values of these amounts of ingredients are found by multiplying the price per pound by the number of pounds used:

   Caramel corn: $\$2.50(16) = \$40$
   Nuts: $\$7.50(4) = \underline{\$30}$
   Total value: $\$70.$

   The desired value of the mixture is \$4.50 per pound, so the value of 20 lb would be $\$4.50(20)$, or \$90. Thus we see that our guess, which led to a total of \$70, is incorrect. Nevertheless, these calculations will help us to translate.

2. **Translate.** We organize the information in a table. We let $x =$ the number of pounds of caramel corn in the mixture and $y =$ the number of pounds of nuts.

	Caramel Corn	Nuts	Mixture	
Price per Pound	$2.50	$7.50	$4.50	
Number of Pounds	x	y	20	$\rightarrow x + y = 20$
Value of Mixture	$2.50x$	$7.50y$	4.50(20), or 90	$\rightarrow 2.50x + 7.50y = 90$

From the second row of the table, we get one equation:

$$x + y = 20.$$

The last row of the table yields a second equation:

$$2.50x + 7.50y = 90, \quad \text{or} \quad 2.5x + 7.5y = 90.$$

We can multiply by 10 on both sides of the second equation to clear the decimals. This gives us the following system of equations:

$$x + y = 20, \quad (1)$$
$$25x + 75y = 900. \quad (2)$$

3. **Carry out.** We carry out the solution as follows.

Algebraic Solution

Using the elimination method, we multiply equation (1) by -25 and add it to equation (2):

$$\begin{aligned} -25x - 25y &= -500 \\ 25x + 75y &= 900 \\ \hline 50y &= 400 \\ y &= 8. \end{aligned}$$

Then we back-substitute to find x:

$$x + y = 20 \quad (1)$$
$$x + 8 = 20 \quad \textbf{Substituting 8 for } y$$
$$x = 12.$$

The solution is $(12, 8)$.

Visualizing the Solution

The solution of the system of equations is the point of intersection of the graphs of the equations.

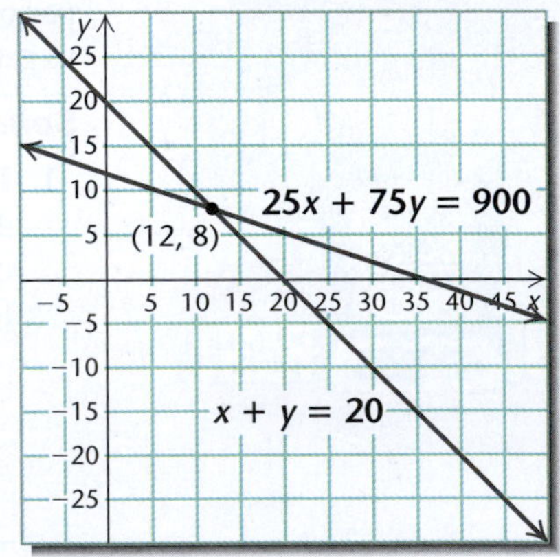

The graphs intersect at the point $(12, 8)$, so the solution of the system of equations is $(12, 8)$.

4. **Check.** If 12 lb of caramel corn and 8 lb of nuts are used, the mixture weighs $12 + 8$, or 20 lb. The value of the mixture is $\$2.50(12) + \$7.50(8)$, or $\$30 + \$60$, or \$90. Since the possible solution yields the desired weight and value of the mixture, our result checks.
5. **State.** The mixture should consist of 12 lb of caramel corn and 8 lb of honey-roasted mixed nuts.

Now Try Exercise 61.

EXAMPLE 7 ***Airplane Travel.*** An airplane flies the 3000-mi distance from Los Angeles to New York, with a tailwind, in 5 hr. The return trip, against the wind, takes 6 hr. Find the speed of the airplane in still air and the speed of the wind.

Solution

1. **Familiarize.** We first make a drawing, letting $p =$ the speed of the plane in still air, in miles per hour, and $w =$ the speed of the wind, also in miles per hour. When the plane is traveling with a tailwind, the wind increases the speed of the plane, so the speed with the tailwind is $p + w$. On the other hand, the headwind slows the plane down, so the speed with the headwind is $p - w$.

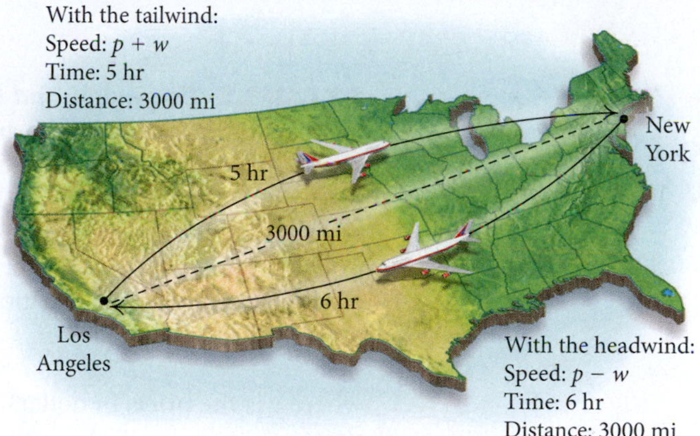

2. **Translate.** We organize the information in a table. Using the formula *Distance* = *Rate* (or *Speed*) · *Time*, we find that each row of the table yields an equation.

	Distance	Rate	Time	
With Tailwind	3000	$p + w$	5	→ $3000 = (p + w)5$
With Headwind	3000	$p - w$	6	→ $3000 = (p - w)6$

We now have a system of equations:

$3000 = (p + w)5,$ or $600 = p + w,$ (1) **Dividing by 5**

$3000 = (p - w)6,$ or $500 = p - w.$ (2) **Dividing by 6**

3. **Carry out.** We use the elimination method:

$$\begin{aligned} 600 &= p + w \quad (1) \\ 500 &= p - w \quad (2) \\ \hline 1100 &= 2p \quad \text{Adding} \\ 550 &= p. \quad \text{Dividing by 2 on both sides} \end{aligned}$$

Now we substitute in one of the equations to find w:

$$\begin{aligned} 600 &= p + w \quad (1) \\ 600 &= 550 + w \quad \text{Substituting 550 for } p \\ 50 &= w. \quad \text{Subtracting 550 on both sides} \end{aligned}$$

4. **Check.** If $p = 550$ and $w = 50$, then the speed of the plane with the tailwind is $550 + 50$, or 600 mph, and the speed with the headwind is $550 - 50$, or 500 mph. At 600 mph, the time it takes to travel 3000 mi is $3000/600$, or 5 hr. At 500 mph, the time it takes to travel 3000 mi is $3000/500$, or 6 hr. The times check, so the answer is correct.
5. **State.** The speed of the plane is 550 mph, and the speed of the wind is 50 mph.

EXAMPLE 8 ***Supply and Demand.*** Suppose that the price and the supply of the Star Station satellite radio are related by the equation

$$y = 90 + 30x,$$

where y is the price, in dollars, at which the seller is willing to supply x thousand units. Suppose too that the price and the demand for the same model of satellite radio are related by the equation

$$y = 200 - 25x,$$

where y is the price, in dollars, at which the consumer is willing to buy x thousand units.

The **equilibrium point** for this radio is the pair (x, y) that is a solution of both equations. The **equilibrium price** is the price at which the amount of the product that the seller is willing to supply is the same as the amount demanded by the consumer. Find the equilibrium point for this radio.

Solution

1., 2. **Familiarize** and **Translate.** We are given a system of equations in the statement of the problem, so no further translation is necessary.

$$\begin{aligned} y &= 90 + 30x, \quad (1) \\ y &= 200 - 25x \quad (2) \end{aligned}$$

We substitute some values for x in each equation to get an idea of the corresponding prices. When $x = 1$,

$$\begin{aligned} y &= 90 + 30 \cdot 1 = 120, \quad \text{Substituting in equation (1)} \\ y &= 200 - 25 \cdot 1 = 175. \quad \text{Substituting in equation (2)} \end{aligned}$$

This indicates that the price when 1 thousand units are supplied is lower than the price when 1 thousand units are demanded.

When $x = 4$,

$y = 90 + 30 \cdot 4 = 210,$ **Substituting in equation (1)**

$y = 200 - 25 \cdot 4 = 100.$ **Substituting in equation (2)**

In this case, the price related to supply is higher than the price related to demand. It would appear that the x-value we are looking for is between 1 and 4.

3. **Carry out.** We use the substitution method:

$y = 90 + 30x$ **Equation (1)**

$200 - 25x = 90 + 30x$ **Substituting $200 - 25x$ for y**

$110 = 55x$ **Adding $25x$ and subtracting 90 on both sides**

$2 = x.$ **Dividing by 55 on both sides**

We now back-substitute 2 for x in either equation and find y:

$y = 200 - 25x$ (2)

$= 200 - 25 \cdot 2$ **Substituting 2 for x**

$= 200 - 50$

$= 150.$

We can visualize the solution as the coordinates of the point of intersection of the graphs of the equations $y = 90 + 30x$ and $y = 200 - 25x$.

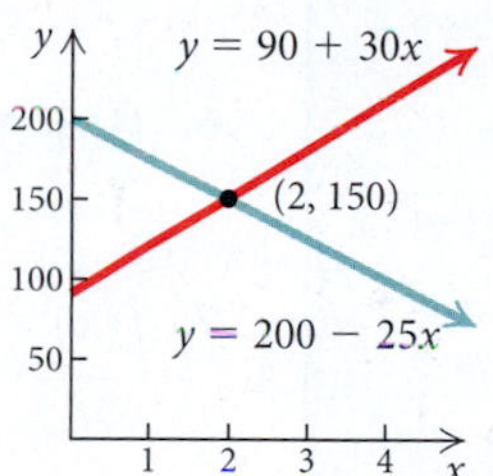

4. **Check.** We can check by substituting 2 for x and 150 for y in both equations. Also note that 2 is between 1 and 4, as we expected from the *Familiarize* and *Translate* steps.
5. **State.** The equilibrium point is (2, \$150). That is, the equilibrium quantity is 2 thousand units and the equilibrium price is \$150.

Now Try Exercise 69.

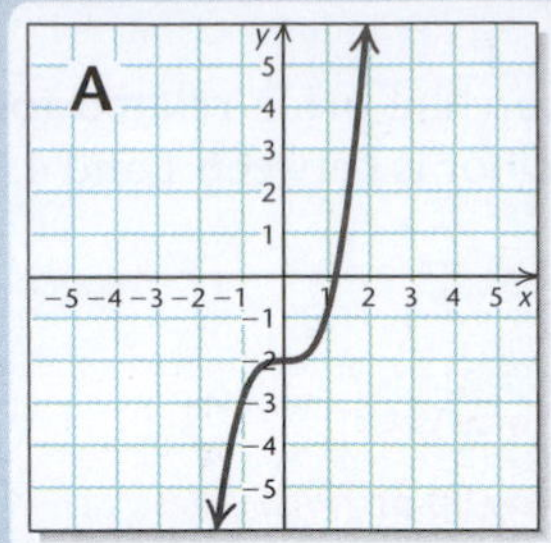

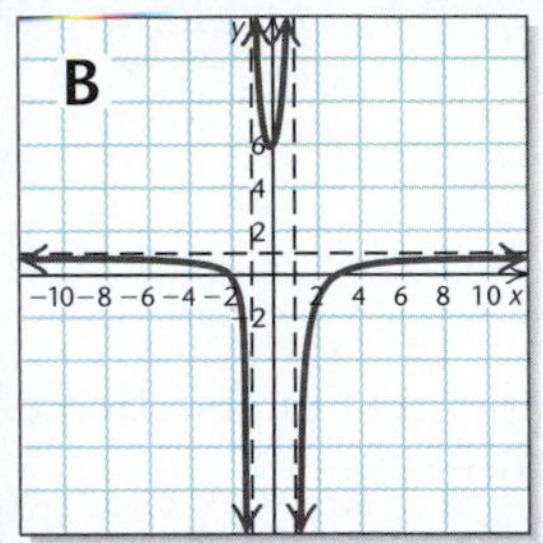

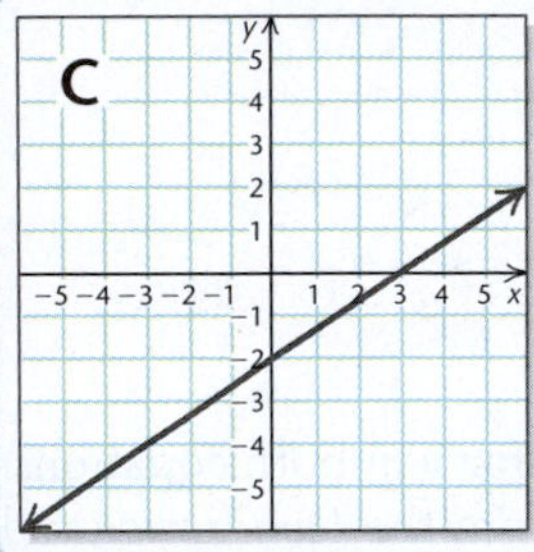

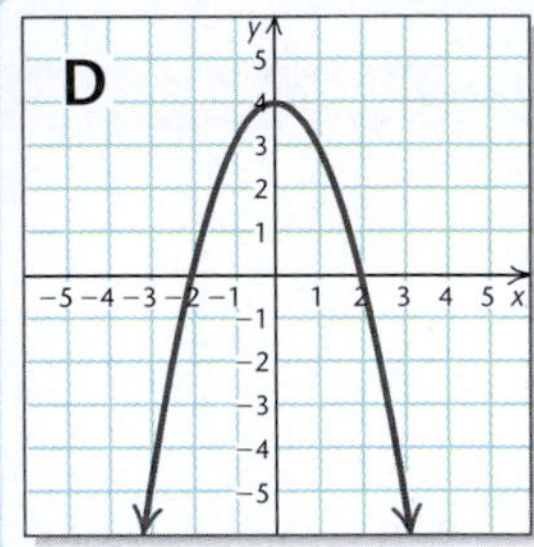

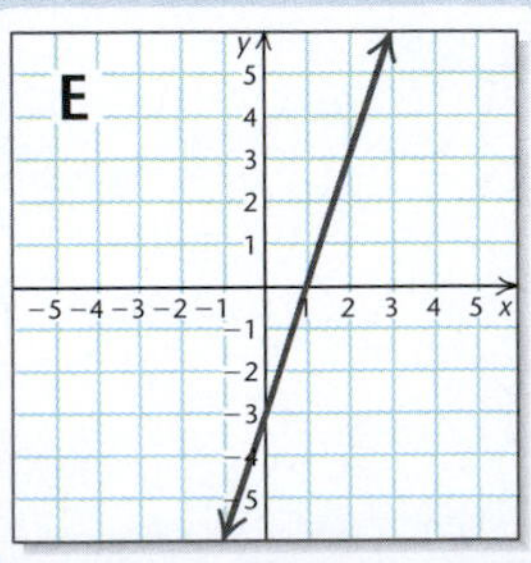

Visualizing the Graph

Match the equation or system of equations with its graph.

1. $2x - 3y = 6$
2. $f(x) = x^2 - 2x - 3$
3. $f(x) = -x^2 + 4$
4. $(x - 2)^2 + (y + 3)^2 = 9$
5. $f(x) = x^3 - 2$
6. $f(x) = -(x - 1)^2(x + 1)^2$
7. $f(x) = \dfrac{x - 1}{x^2 - 4}$
8. $f(x) = \dfrac{x^2 - x - 6}{x^2 - 1}$
9. $x - y = -1,$
$2x - y = 2$
10. $3x - y = 3,$
$2y = 6x - 6$

Answers on page A-31

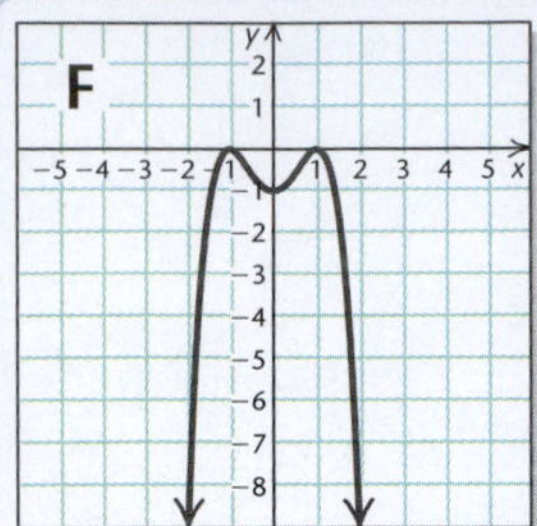

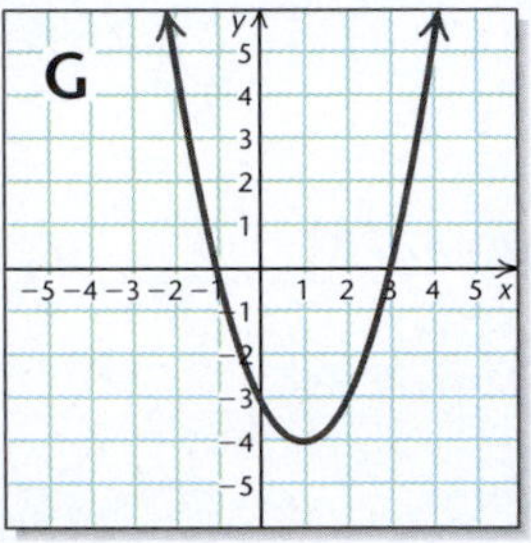

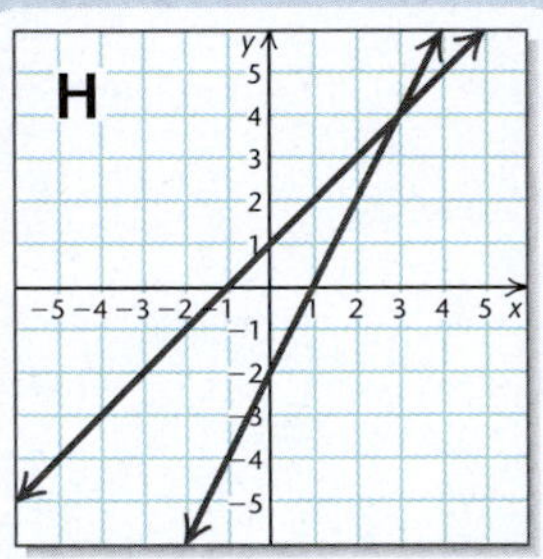

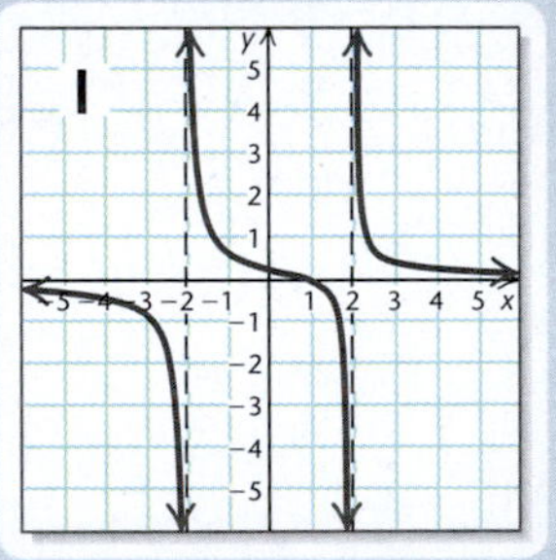

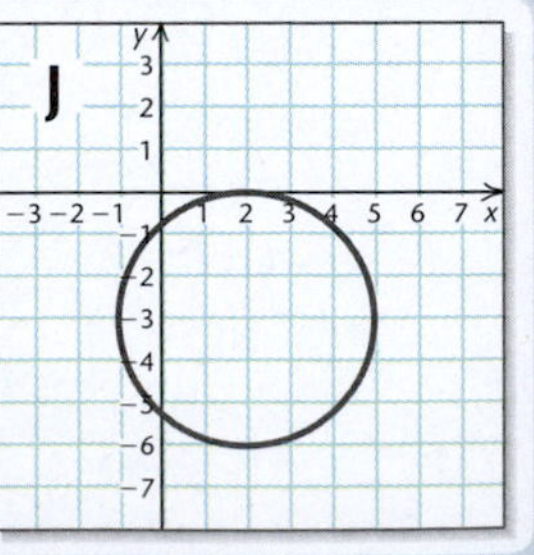

6.1 Exercise Set

In Exercises 1–6, match the system of equations with one of the graphs (a)–(f) that follow.

a)

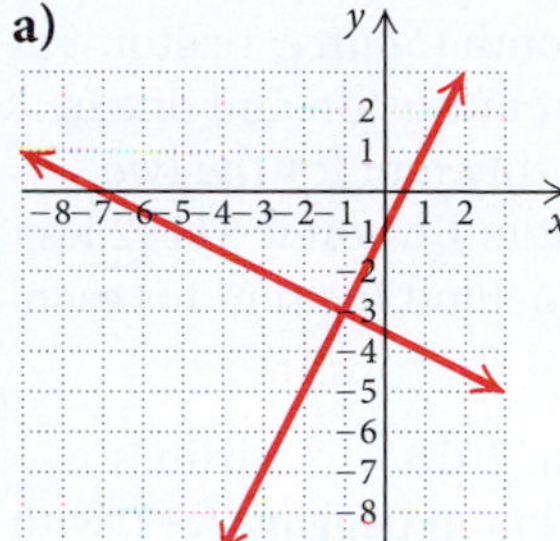

b)

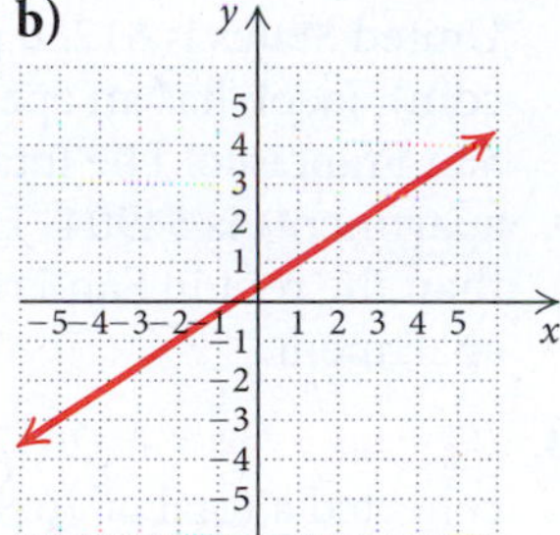

c)

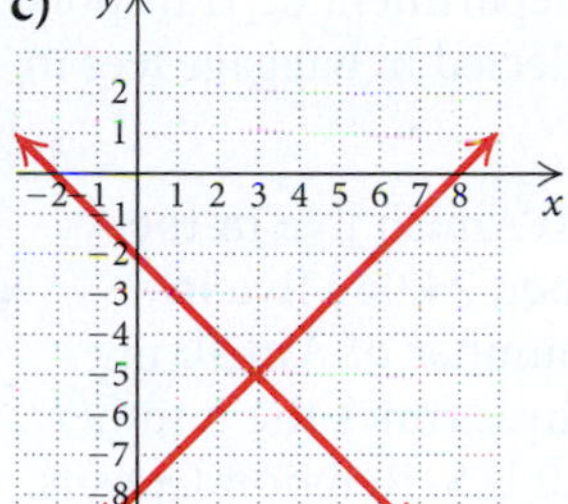

d)

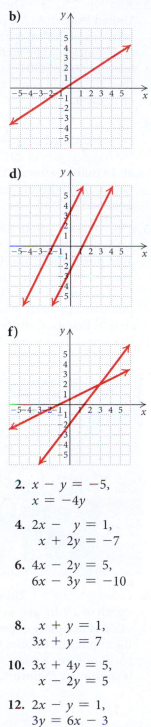

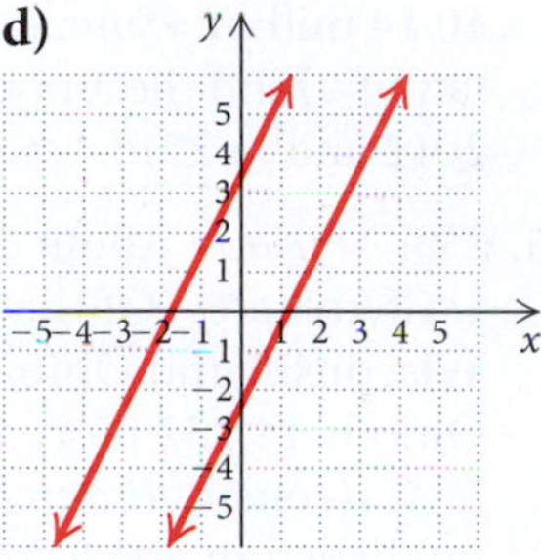

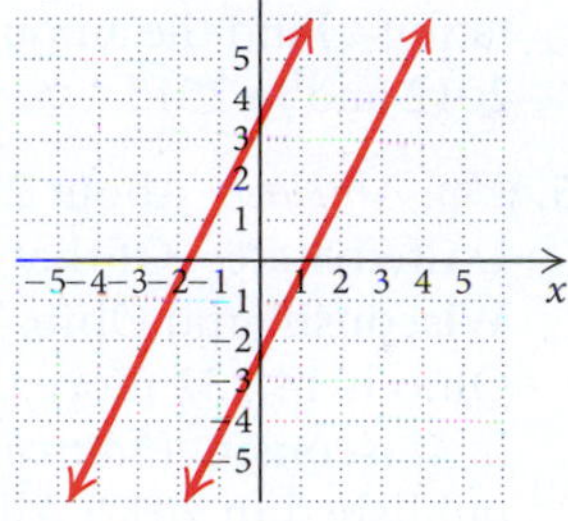

e)

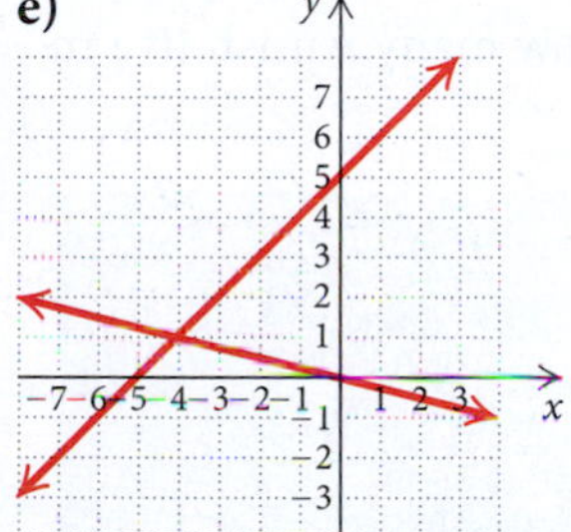

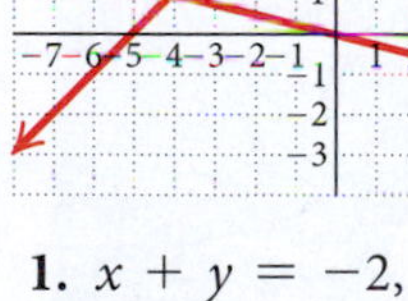

f)

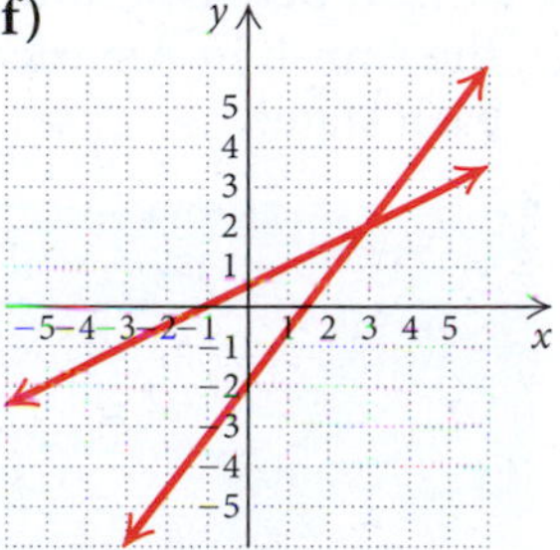

1. $x + y = -2,$
   $y = x - 8$

2. $x - y = -5,$
   $x = -4y$

3. $x - 2y = -1,$
   $4x - 3y = 6$

4. $2x - y = 1,$
   $x + 2y = -7$

5. $2x - 3y = -1,$
   $-4x + 6y = 2$

6. $4x - 2y = 5,$
   $6x - 3y = -10$

Solve graphically.

7. $x + y = 2,$
   $3x + y = 0$

8. $x + y = 1,$
   $3x + y = 7$

9. $x + 2y = 1,$
   $x + 4y = 3$

10. $3x + 4y = 5,$
    $x - 2y = 5$

11. $y + 1 = 2x,$
    $y - 1 = 2x$

12. $2x - y = 1,$
    $3y = 6x - 3$

13. $x - y = -6,$
    $y = -2x$

14. $2x + y = 5,$
    $x = -3y$

15. $2y = x - 1,$
    $3x = 6y + 3$

16. $y = 3x + 2,$
    $3x - y = -3$

Solve using the substitution method.

17. $x + y = 9,$
    $2x - 3y = -2$

18. $3x - y = 5,$
    $x + y = \frac{1}{2}$

19. $x - 2y = 7,$
    $x = y + 4$

20. $x + 4y = 6,$
    $x = -3y + 3$

21. $y = 2x - 6,$
    $5x - 3y = 16$

22. $3x + 5y = 2,$
    $2x - y = -3$

23. $x + y = 3,$
    $y = 4 - x$

24. $x - 2y = 3,$
    $2x = 4y + 6$

25. $x - 5y = 4,$
    $y = 7 - 2x$

26. $5x + 3y = -1,$
    $x + y = 1$

27. $x + 2y = 2,$
    $4x + 4y = 5$

28. $2x - y = 2,$
    $4x + y = 3$

29. $3x - y = 5,$
    $3y = 9x - 15$

30. $2x - y = 7,$
    $y = 2x - 5$

Solve using the elimination method. Also determine whether each system is consistent or inconsistent and whether the equations are dependent or independent.

31. $x + 2y = 7,$
    $x - 2y = -5$

32. $3x + 4y = -2,$
    $-3x - 5y = 1$

33. $x - 3y = 2,$
    $6x + 5y = -34$

34. $x + 3y = 0,$
    $20x - 15y = 75$

35. $3x - 12y = 6,$
    $2x - 8y = 4$

36. $2x + 6y = 7,$
    $3x + 9y = 10$

37. $4x - 2y = 3,$
    $2x - y = 4$

38. $6x + 9y = 12,$
    $4x + 6y = 8$

39. $2x = 5 - 3y,$
    $4x = 11 - 7y$

40. $7(x - y) = 14,$
    $2x = y + 5$

41. $0.3x - 0.2y = -0.9,$
    $0.2x - 0.3y = -0.6$
    (*Hint*: Since each coefficient has one decimal place, first multiply each equation by 10 to clear the decimals.)

42. $0.2x - 0.3y = 0.3,$
    $0.4x + 0.6y = -0.2$
    (*Hint*: Since each coefficient has one decimal place, first multiply each equation by 10 to clear the decimals.)

43. $\frac{1}{5}x + \frac{1}{2}y = 6,$
$\frac{3}{5}x - \frac{1}{2}y = 2$
(*Hint*: First multiply by the least common denominator to clear fractions.)

44. $\frac{2}{3}x + \frac{3}{5}y = -17,$
$\frac{1}{2}x - \frac{1}{3}y = -1$
(*Hint*: First multiply by the least common denominator to clear fractions.)

In Exercises 45–50, determine whether the statement is true or false.

45. If the graph of a system of equations is a pair of parallel lines, then the system of equations is inconsistent.

46. If we obtain the equation $0 = 0$ when using the elimination method to solve a system of equations, then the system has no solution.

47. If a system of two linear equations in two variables is consistent, then it has exactly one solution.

48. If a system of two linear equations in two variables is dependent, then it has infinitely many solutions.

49. It is possible for a system of two linear equations in two variables to be consistent and dependent.

50. It is possible for a system of two linear equations in two variables to be inconsistent and dependent.

51. *Cosmetic Survey.* Liposuction and breast augmentation are the two most popular cosmetic surgeries in the United States. Together, these two procedures accounted for 677,239 surgeries in 2013. The number of breast augmentation surgeries was 50,585 fewer than the number of liposuction surgeries. (*Source:* American Society for Aesthetic Plastic Surgery) Find the number of each type of surgery.

52. *Military Spending.* In 2013, the United States spent \$452 billion more on the military than China spent on its military. Together, China and the United States spent \$828 billion. (*Source:* Stockholm International Peace Research Institute) Find the amount spent on the military in China and in the United States.

53. *Apartment Rent.* The average apartment rent in the United States is \$1230 per month (*Source:* realtor.com). Jacob has an apartment in Boston and one in San Francisco. The total monthly rent for the two apartments is \$4904. The rent in Boston is \$1142 less than the rent in San Francisco. Find the rent for each apartment.

54. *Baggage Fees.* In 2012 and in 2013, U.S. airlines collected a total of \$6.84 million in baggage fees, with the baggage fees in 2012 exceeding those in 2013 by \$0.14 million (*Source:* U.S. Department of Transportation). Find the amount collected in baggage fees in 2012 and in 2013.

55. *The Amish.* About 270,000 Amish live in the United States. Of that number, 74,060 live in Wisconsin and Ohio. The number of Amish in Ohio is 14,232 more than three times the number in Wisconsin. (*Sources:* 2010 U.S. Religion Census, published in 2012; Albrecht Powell, "Amish 101–Amish Beliefs, Culture & Lifestyle, History of the Amish in America.") How many Amish live in each state?

56. *Calories in a Pie.* Using the calorie count of one-eighth of a 9-in. pie, we find that the number of calories in a piece of pecan pie is 221 less than twice the number of calories in a piece of lemon meringue pie. If one eats a piece of each, a total of 865 calories is consumed. (*Source: Good Housekeeping*, Good

Health, p. 41, November 2007) How many calories are there in each piece of pie?

57. *Mail-Order Business.* A mail-order gardening equipment business shipped 120 packages one day. Customers are charged \$6.50 for each standard-delivery package and \$10.00 for each express-delivery package. Total shipping charges for the day were \$934. How many of each kind of package were shipped?

58. *Concert Ticket Prices.* One evening 1500 concert tickets were sold for the Fairmont Summer Jazz Festival. Tickets cost \$25 for a covered pavilion seat and \$15 for a lawn seat. Total receipts were \$28,500. How many of each type of ticket were sold?

59. *Investment.* Charles inherited \$15,000 and invested it in two municipal bonds that pay 4% and 5% simple interest. The annual interest is \$690. Find the amount invested at each rate.

60. *Coffee Mixtures.* The owner of The Daily Grind coffee shop mixes French roast coffee worth \$12.00 per pound with Colombian coffee worth \$9.50 per pound in order to get 20 lb of a mixture worth \$10.50 per pound. How much of each type of coffee was used?

61. *Infinity Scarves.* During the holiday season, Brianna sold scarves at a kiosk in a shopping mall. Embroidered floral scarves cost \$24 each, and sheer chevron scarves cost \$18. One day she sold 39 scarves. Total receipts for the day were \$798. How many of each kind of scarf did she sell?

62. *Commissions.* Bentley's Office Solutions offers its sales representatives a choice between being paid a commission of 8% of sales or being paid a monthly salary of \$1500 plus a commission of 1% of sales. For what amount of monthly sales do the two plans pay the same?

63. *Nutrition.* A one-cup serving of spaghetti with meatballs contains 260 Cal (calories) and 32 g of carbohydrates. A one-cup serving of chopped iceberg lettuce contains 5 Cal and 1 g of carbohydrates. (*Source*: U.S. Department of Agriculture) How many servings of each would be required to obtain 400 Cal and 50 g of carbohydrates?

64. *Nutrition.* One serving of tomato soup contains 100 Cal and 18 g of carbohydrates. One slice of whole wheat bread contains 70 Cal and 13 g of carbohydrates. (*Source*: U.S. Department of Agriculture) How many servings of each would be required to obtain 230 Cal and 42 g of carbohydrates?

65. *Motion.* A Leisure Time Cruises riverboat travels 46 km downstream in 2 hr. It travels 51 km upstream in 3 hr. Find the speed of the boat and the speed of the stream.

66. *Motion.* A DC10 airplane travels 3000 km with a tailwind in 3 hr. It travels 3000 km with a headwind in 4 hr. Find the speed of the plane and the speed of the wind.

67. *Motion.* Two private airplanes travel toward each other from cities that are 780 km apart at speeds of 190 km/h and 200 km/h. They left at the same time. In how many hours will they meet?

68. *Motion.* Mackenzie's boat travels 45 mi downstream in 3 hr. The return trip upstream takes 5 hr. Find the speed of the boat in still water and the speed of the current.

69. *Supply and Demand.* The supply and demand for an all-terrain skateboard are related to price by the equations

$$y = 140 + 4x,$$
$$y = 275 - 5x,$$

respectively, where y is the price, in dollars, and x is the number of units, in thousands. Find the equilibrium point for this product.

70. *Supply and Demand.* The supply and demand for a particular model of treadmill are related to price by the equations

$$y = 240 + 40x,$$
$$y = 500 - 25x,$$

respectively, where y is the price, in dollars, and x is the number of units, in thousands. Find the equilibrium point for this product.

The point at which a company's costs equal its revenues is the **break-even point**. *In Exercises 71–74, C represents the production cost, in dollars, of x units of a product and R represents the revenue, in dollars, from the sale of x units. Find the number of units that must be produced and sold in order to break even. That is, find the value of x for which C = R.*

71. $C = 14x + 350,$
$R = 16.5x$

72. $C = 8.5x + 75,$
$R = 10x$

73. $C = 15x + 12{,}000,$
$R = 18x - 6000$

74. $C = 3x + 400,$
$R = 7x - 600$

Skill Maintenance

75. *Registered Snowmobiles.* There were 251,986 registered snowmobiles in Minnesota in 2013. This was 21,952 more than twice the number of registered snowmobiles in New York. (*Sources:* International Snowmobile Manufacturers Association; Maine Snowmobile Association) Find the number of registered snowmobiles in New York. [1.5]

76. *International Adoptions.* In 2013, the number of international adoptions in the United States was at its lowest level since 2004. The number of international adoptions in 2013 totaled 7094, a decrease of 69.1% from 2004 (*Source:* U.S. State Department). Find the number of international adoptions in 2004. Round to the nearest ten. [1.5]

Consider the function

$$f(x) = x^2 - 4x + 3$$

in Exercises 77–80.

77. What are the inputs if the output is 15? [3.2]

78. Given an output of 8, find the corresponding inputs. [3.2]

79. What is the output if the input is -2? [1.2]

80. Find the zeros of the function. [3.2]

Synthesis

81. *Gas Mileage.* The 2014 Volkswagen CC 2.0T gets 22 miles per gallon (mpg) in city driving and 31 mpg in highway driving (*Source: Motor Trend*, May 2014). The car is driven 314 mi on 11 gal of gasoline. How many miles were driven in the city and how many miles were driven on the highway?

82. *e-Commerce.* Shirts.com advertises a limited-time sale, offering 1 turtleneck for \$15 and 2 turtlenecks for \$25. A total of 1250 turtlenecks are sold and \$16,750 is taken in. How many customers ordered 2 turtlenecks?

83. *Motion.* A train leaves Union Station for Central Station, 216 km away, at 9 A.M. One hour later, a train leaves Central Station for Union Station. They meet at noon. If the second train had started at 9 A.M. and the first train at 10:30 A.M., they would still have met at noon. Find the speed of each train.

84. *Antifreeze Mixtures.* An automobile radiator contains 16 L of antifreeze and water. This mixture is 30% antifreeze. How much of this mixture should be drained and replaced with pure antifreeze so that the final mixture will be 50% antifreeze?

85. Two solutions of the equation $Ax + By = 1$ are $(3, -1)$ and $(-4, -2)$. Find A and B.

86. *Ticket Line.* You are in line at a ticket window. There are 2 more people ahead of you in line than there are behind you. In the entire line, there are three times as many people as there are behind you. How many people are ahead of you?

6.2 Systems of Equations in Three Variables

- Solve systems of linear equations in three variables.
- Use systems of three equations to solve applied problems.
- Model a situation using a quadratic function.

A **linear equation in three variables** is an equation equivalent to one of the form $Ax + By + Cz = D$, where A, B, C, and D are real numbers and none of A, B, and C is 0. A **solution of a system of three equations in three variables** is an ordered triple that makes all three equations true. For example, the triple $(2, -1, 0)$ is a solution of the system of equations

$$\begin{aligned} 4x + 2y + 5z &= 6, \\ 2x - y + z &= 5, \\ 3x + 2y - z &= 4. \end{aligned}$$

We can verify this by substituting 2 for x, -1 for y, and 0 for z in each equation.

Solving Systems of Equations in Three Variables

We will solve systems of equations in three variables using an algebraic method called **Gaussian elimination**, named for the German mathematician Karl Friedrich Gauss (1777–1855). Our goal is to transform the original system to an equivalent system (one with the same solution set) of the form

$$\begin{aligned} Ax + By + Cz &= D, \\ Ey + Fz &= G, \\ Hz &= K. \end{aligned}$$

Then we solve the third equation for z and back-substitute to find y and then x.

Each of the following operations can be used to transform the original system to an equivalent system in the desired form.

1. Interchange any two equations.
2. Multiply both sides of one of the equations by a nonzero constant.
3. Add a nonzero multiple of one equation to another equation.

EXAMPLE 1 Solve the following system:

$$\begin{aligned} x - 2y + 3z &= 11, \quad &(1) \\ 4x + 2y - 3z &= 4, \quad &(2) \\ 3x + 3y - z &= 4. \quad &(3) \end{aligned}$$

Solution First, we choose one of the variables to eliminate using two different pairs of equations. Let's eliminate x from equations (2) and (3). We multiply equation (1) by -4 and add it to equation (2). We also multiply equation (1) by -3 and add it to equation (3).

$$\begin{aligned} -4x + 8y - 12z &= -44 && \text{Multiplying (1) by } -4 \\ 4x + 2y - 3z &= 4 && (2) \\ \hline 10y - 15z &= -40; && (4) \end{aligned}$$

$$\begin{aligned} -3x + 6y - 9z &= -33 && \text{Multiplying (1) by } -3 \\ 3x + 3y - z &= 4 && (3) \\ \hline 9y - 10z &= -29. && (5) \end{aligned}$$

This gives us

$$\begin{aligned} x - 2y + 3z &= 11, && (1) \\ 10y - 15z &= -40, && (4) \\ 9y - 10z &= -29. && (5) \end{aligned}$$

Next, we multiply equation (5) by 10 to make the y-coefficient a multiple of the y-coefficient in the equation above it:

$$\begin{aligned} x - 2y + 3z &= 11, && (1) \\ 10y - 15z &= -40, && (4) \\ 90y - 100z &= -290. && (6) \end{aligned}$$

Next, we multiply equation (4) by -9 and add it to equation (6):

$$\begin{aligned} -90y + 135z &= 360 && \text{Multiplying (4) by } -9 \\ 90y - 100z &= -290 && (6) \\ \hline 35z &= 70. && (7) \end{aligned}$$

This gives us the system of equations

$$\begin{aligned} x - 2y + 3z &= 11, && (1) \\ 10y - 15z &= -40, && (4) \\ 35z &= 70. && (7) \end{aligned}$$

Now we solve equation (7) for z:

$$\begin{aligned} 35z &= 70 \\ z &= 2. \end{aligned}$$

Then we back-substitute 2 for z in equation (4) and solve for y:

$$\begin{aligned} 10y - 15 \cdot 2 &= -40 \\ 10y - 30 &= -40 \\ 10y &= -10 \\ y &= -1. \end{aligned}$$

Finally, we back-substitute -1 for y and 2 for z in equation (1) and solve for x:

$$\begin{aligned} x - 2(-1) + 3 \cdot 2 &= 11 \\ x + 2 + 6 &= 11 \\ x &= 3. \end{aligned}$$

We can check the triple $(3, -1, 2)$ in each of the three original equations. Since it makes all three equations true, the solution is $(3, -1, 2)$.

Now Try Exercise 1.

EXAMPLE 2 Solve the following system:

$$\begin{aligned} x + y + z &= 7, \quad &(1)\\ 3x - 2y + z &= 3, \quad &(2)\\ x + 6y + 3z &= 25. \quad &(3) \end{aligned}$$

Solution We multiply equation (1) by -3 and add it to equation (2). We also multiply equation (1) by -1 and add it to equation (3).

$$\begin{aligned} x + y + z &= 7, \quad &(1)\\ -5y - 2z &= -18, \quad &(4)\\ 5y + 2z &= 18 \quad &(5) \end{aligned}$$

Next, we add equation (4) to equation (5):

$$\begin{aligned} x + y + z &= 7, \quad &(1)\\ -5y - 2z &= -18, \quad &(4)\\ 0 &= 0. \end{aligned}$$

The equation $0 = 0$ tells us that equations (1), (2), and (3) are dependent. This means that the original system of three equations is equivalent to a system of two equations. One way to see this is to note that four times equation (1) minus equation (2) is equation (3). Thus removing equation (3) from the system does not affect the solution of the system. We can say that the original system is equivalent to

$$\begin{aligned} x + y + z &= 7, \quad &(1)\\ 3x - 2y + z &= 3. \quad &(2) \end{aligned}$$

In this particular case, the original system has infinitely many solutions. (In some cases, a system containing dependent equations is inconsistent.) To find an expression for these solutions, we first solve equation (4) for either y or z. We choose to solve for y:

$$\begin{aligned} -5y - 2z &= -18 \quad (4)\\ -5y &= 2z - 18\\ y &= -\tfrac{2}{5}z + \tfrac{18}{5}. \end{aligned}$$

Then we back-substitute in equation (1) to find an expression for x in terms of z:

$$\begin{aligned} x - \tfrac{2}{5}z + \tfrac{18}{5} + z &= 7 \qquad \text{Substituting } -\tfrac{2}{5}z + \tfrac{18}{5} \text{ for } y\\ x + \tfrac{3}{5}z + \tfrac{18}{5} &= 7\\ x + \tfrac{3}{5}z &= \tfrac{17}{5}\\ x &= -\tfrac{3}{5}z + \tfrac{17}{5}. \end{aligned}$$

The solutions of the system of equations are ordered triples of the form $\left(-\frac{3}{5}z + \frac{17}{5}, -\frac{2}{5}z + \frac{18}{5}, z\right)$, where z can be any real number. Any real number that we use for z then gives us values for x and y and thus an ordered triple in the solution set. For example, if we choose $z = 0$, we have the solution $\left(\frac{17}{5}, \frac{18}{5}, 0\right)$. If we choose $z = -1$, we have $(4, 4, -1)$.

Now Try Exercise 9.

If we get a false equation, such as $0 = -5$, at some stage of the elimination process, we conclude that the original system is *inconsistent*; that is, it has no solutions.

Although systems of three linear equations in three variables do not lend themselves well to graphical solutions, it is of interest to picture some possible solutions. The graph of a linear equation in three variables is a plane. Thus the solution set of such a system is the intersection of three planes. Some possibilities are shown below.

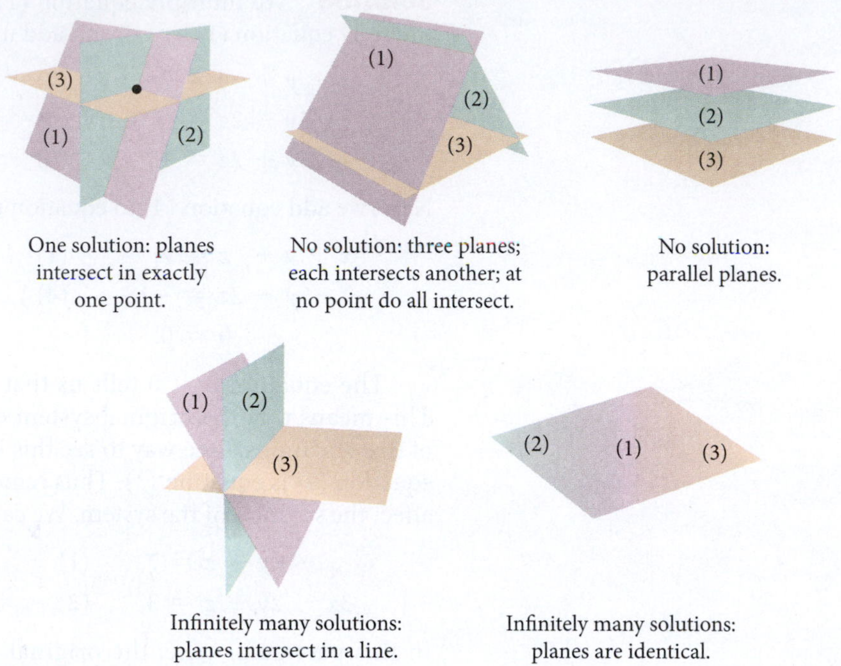

One solution: planes intersect in exactly one point.

No solution: three planes; each intersects another; at no point do all intersect.

No solution: parallel planes.

Infinitely many solutions: planes intersect in a line.

Infinitely many solutions: planes are identical.

Applications

Systems of equations in three or more variables allow us to solve many problems in fields such as business, the social and natural sciences, and engineering.

EXAMPLE 3 ***Investment.*** Luis inherited \$15,000 and invested part of it in a money market account, part in municipal bonds, and part in a mutual fund. After 1 year, he received a total of \$730 in simple interest from the three investments. The money market account paid 4% annually, the bonds paid 5% annually, and the mutual fund paid 6% annually. There was \$2000 more invested in the mutual fund than in bonds. Find the amount that Luis invested in each category.

Solution

1. **Familiarize.** We let x, y, and z represent the amounts invested in the money market account, the bonds, and the mutual fund, respectively. Then the amounts of income produced annually by each investment are given by $4\%x$, $5\%y$, and $6\%z$, or $0.04x$, $0.05y$, and $0.06z$.
2. **Translate.** The fact that a total of \$15,000 is invested gives us one equation:

$$x + y + z = 15{,}000.$$

Since the total amount of interest is \$730, we have a second equation:

$$0.04x + 0.05y + 0.06z = 730.$$

Another statement in the problem gives us a third equation.

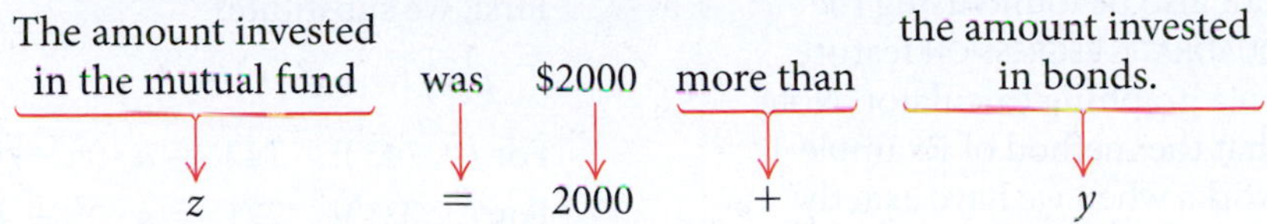
The amount invested in the mutual fund
was
$2000
more than
the amount invested in bonds.
z
=
2000
+
y

We now have a system of three equations:

$$\begin{aligned} x + y + z &= 15{,}000, \\ 0.04x + 0.05y + 0.06z &= 730, \\ z &= 2000 + y; \end{aligned} \quad \text{or} \quad \begin{aligned} x + y + z &= 15{,}000, \\ 4x + 5y + 6z &= 73{,}000, \\ -y + z &= 2000. \end{aligned}$$

3. **Carry out.** Solving the system of equations, we get

   $(7000, 3000, 5000)$.

4. **Check.** The sum of the numbers is 15,000. The income produced is

   $0.04(7000) + 0.05(3000) + 0.06(5000) = 280 + 150 + 300$, or \$730.

   Also, the amount invested in the mutual fund, \$5000, is \$2000 more than the amount invested in bonds, \$3000. Our solution checks in the original problem.

5. **State.** Luis invested \$7000 in a money market account, \$3000 in municipal bonds, and \$5000 in a mutual fund.

Now Try Exercise 29.

Mathematical Models and Applications

Recall that when we model a situation using a linear function $f(x) = mx + b$, we need to know two data points in order to determine m and b. For a quadratic model, $f(x) = ax^2 + bx + c$, we need three data points in order to determine a, b, and c.

EXAMPLE 4 ***Airline Passengers.*** The following table lists the number of scheduled airline passengers in three recent years. Use the data to find a quadratic function that gives the number of airline passengers as a function of the number of years after 2008. Then use the function to estimate the number of airline passengers in 2013.

Year x	Number of Airline Passengers (in millions)
2008, 0	743
2010, 2	721
2012, 4	737

Source: Airlines of America

Technology Connection

The function in Example 4 can also be found using the QUADRATIC REGRESSION feature on a graphing calculator. Note that the method of Example 4 works when we have exactly three data points, whereas the QUADRATIC REGRESSION feature on a graphing calculator can be used for *three or more* points.

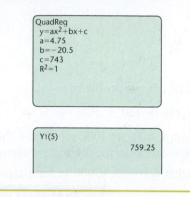

```
QuadReg
 y=ax²+bx+c
 a=4.75
 b=-20.5
 c=743
 R²=1
```

```
Y1(5)
                759.25
```

Solution We let $x =$ the number of years after 2008 and $f(x) =$ the number of airline passengers, in millions. Then $x = 0$ corresponds to 2008, $x = 2$ corresponds to 2010, and $x = 4$ corresponds to 2012. We use the three data points $(0, 743)$, $(2, 721)$, and $(4, 737)$ to find a, b, and c in the function $f(x) = ax^2 + bx + c$.

First, we substitute:

$$f(x) = ax^2 + bx + c.$$

$$\begin{aligned} &\text{For } (0, 743): \quad 743 = a \cdot 0^2 + b \cdot 0 + c; \\ &\text{For } (2, 721): \quad 721 = a \cdot 2^2 + b \cdot 2 + c; \\ &\text{For } (4, 737): \quad 737 = a \cdot 4^2 + b \cdot 4 + c. \end{aligned}$$

We now have a system of equations in the variables a, b, and c:

$$\begin{aligned} c &= 743, \\ 4a + 2b + c &= 721, \\ 16a + 4b + c &= 737. \end{aligned}$$

Solving this system, we get $(4.75, -20.5, 743)$. Thus,

$$f(x) = 4.75x^2 - 20.5x + 743.$$

To estimate the number of airline passengers in 2013, we find $f(5)$, since 2013 is 5 years after 2008:

$$\begin{aligned} f(5) &= 4.75 \cdot 5^2 - 20.5 \cdot 5 + 743 \\ &= 759.25. \end{aligned}$$

The number of airline passengers in 2013 is estimated to be 759.25 million.

Now Try Exercise 33.

6.2 Exercise Set

Solve the system of equations.

1. $x + y + z = 2,$
$6x - 4y + 5z = 31,$
$5x + 2y + 2z = 13$

2. $x + 6y + 3z = 4,$
$2x + y + 2z = 3,$
$3x - 2y + z = 0$

3. $x - y + 2z = -3,$
$x + 2y + 3z = 4,$
$2x + y + z = -3$

4. $x + y + z = 6,$
$2x - y - z = -3,$
$x - 2y + 3z = 6$

5. $x + 2y - z = 5,$
$2x - 4y + z = 0,$
$3x + 2y + 2z = 3$

6. $2x + 3y - z = 1,$
$x + 2y + 5z = 4,$
$3x - y - 8z = -7$

7. $x + 2y - z = -8,$
$2x - y + z = 4,$
$8x + y + z = 2$

8. $x + 2y - z = 4,$
$4x - 3y + z = 8,$
$5x - y = 12$

9. $2x + y - 3z = 1,$
$x - 4y + z = 6,$
$4x - 7y - z = 13$

10. $x + 3y + 4z = 1,$
$3x + 4y + 5z = 3,$
$x + 8y + 11z = 2$

11. $4a + 9b = 8,$
$8a + 6c = -1,$
$6b + 6c = -1$

12. $3p + 2r = 11,$
$q - 7r = 4,$
$p - 6q = 1$

13. $2x + z = 1,$
$3y - 2z = 6,$
$x - 2y = -9$

14. $3x + 4z = -11,$
$x - 2y = 5,$
$4y - z = -10$

15. $w + x + y + z = 2,$
$w + 2x + 2y + 4z = 1,$
$-w + x - y - z = -6,$
$-w + 3x + y - z = -2$

16. $w + x - y + z = 0,$
$-w + 2x + 2y + z = 5,$
$-w + 3x + y - z = -4,$
$-2w + x + y - 3z = -7$

17. *Paralympic Medals.* At the 2014 Paralympic Games in Sochi, Russia, the top three countries—the Russian Federation, Ukraine, and the United States—won a total of 123 medals. Ukraine won 7 more medals than

the United States. The Russian Federation won 37 more medals than the total amount won by Ukraine and the United States. (*Source:* International Paralympic Committee) How many medals did each of the top three countries win?

18. *Top Apple Growers.* The top three apple growers in the world—China, the United States, and Turkey—grew a total of about 74 billion lb of apples in a recent year. China produced 44 billion lb more than the combined production of the United States and Turkey. The United States produced twice as many pounds of apples as Turkey. (*Source:* U.S. Apple Association) Find the number of pounds of apples produced by each country.

19. *Restaurant Meals.* The total number of restaurant-purchased meals that the average person will eat in a restaurant, in a car, or at home in a year is 170. The total number of these meals eaten in a car or at home exceeds the number eaten in a restaurant by 14. Twenty more restaurant-purchased meals will be eaten in a restaurant than at home. (*Source:* The NPD Group) Find the number of restaurant-purchased meals eaten in a restaurant, the number eaten in a car, and the number eaten at home.

20. *Acres Planted in Crops.* The top three crops in the United States in 2013, in terms of the number of acres planted, were corn, soybeans, and wheat. The total number of acres planted in these crops was 227.9 million. The number of acres planted in wheat was 39.2 million fewer than the number of acres planted in corn. The number of acres planted in soybeans was 35.7 million fewer than twice the number of acres planted in wheat. (*Source:* U.S. Department of Agriculture) Find the number of acres planted in each crop.

21. *Top Art Auction Sales.* The top three prices for works of art sold at auction in 2013 totaled $306.2 million. These works included Francis Bacon's 1969 *Three Studies of Lucian Freud*, Andy Warhol's *Silver Car Crash* (*Double Disaster*), and Jeff Koons's *Balloon Dog* (*Orange*). The selling price of the Warhol art was $47 million more than that of the Koons art. Together, the Warhol art and the Koons art sold for $21.4 million more than the Bacon art. (*Source:* Bloomberg.com, Katya Kazakina and Scott Reyburn, December 24, 2013) What was the selling price of each work?

22. *Jolts of Caffeine.* One 8-oz serving each of brewed coffee, Red Bull energy drink, and Mountain Dew soda contains 197 mg of caffeine. One serving of brewed coffee has 6 mg more caffeine than two servings of Mountain Dew. One serving of Red Bull contains 37 mg less caffeine than one serving each of brewed coffee and Mountain Dew. (*Source:* Australian Institute of Sport) Find the amount of caffeine in one serving of each beverage.

23. *Pet Ownership.* Americans own a total of about 152.3 million dogs, cats, and birds. The number of dogs owned is 12.5 million less than the total number of cats and birds owned, and 4.2 million more cats are owned than dogs. (*Source:* American Veterinary Medical Association) How many of each type of pet do Americans own?

24. *Mail-Order Business.* Natural Fibers Clothing charges $6 to ship orders of $25 or less, $10 for orders from $25.01 to $75, and $12 for orders over $75. One week, shipping charges for 600 orders totaled $5480. Eighty more orders for $25 or less were shipped than orders for more than $75. Find the number of orders shipped at each rate.

25. *Foreign Aid.* In 2013, the top three donors of foreign aid were the United States, Great Britain, and Germany. The total amount of their aid was \$63.5 billion. Together, Great Britain and Germany donated \$0.5 billion more than the United States did. The United States donated \$17.4 billion more than Germany did. (*Source:* Organization for Economic Cooperation and Development) How much in foreign aid did each country donate?

26. *Spring Cleaning.* In a group of 100 adults, 70 say they are most likely to do spring housecleaning in March, April, or May. Of these 70, the number who clean in April is 14 more than the total number who clean in March and May. The total number who clean in April and May is 2 more than three times the number who clean in March. (*Source:* Zoomerang online survey) Find the number who clean in each month.

27. *Nutrition.* A hospital dietician must plan a lunch menu that provides 485 Cal, 41.5 g of carbohydrates, and 35 mg of calcium. A 3-oz serving of broiled ground beef contains 245 Cal, 0 g of carbohydrates, and 9 mg of calcium. One baked potato contains 145 Cal, 34 g of carbohydrates, and 8 mg of calcium. A one-cup serving of strawberries contains 45 Cal, 10 g of carbohydrates, and 21 mg of calcium. (*Source:* U.S. Department of Agriculture) How many servings of each are required to provide the desired nutritional values?

28. *Nutrition.* A diabetic patient wishes to prepare a meal consisting of roasted chicken breast, mashed potatoes, and peas. A 3-oz serving of roasted skinless chicken breast contains 140 Cal, 27 g of protein, and 64 mg of sodium. A one-cup serving of mashed potatoes contains 160 Cal, 4 g of protein, and 636 mg of sodium, and a one-cup serving of peas contains 125 Cal, 8 g of protein, and 139 mg of sodium. (*Source:* U.S. Department of Agriculture) How many servings of each should be used if the meal is to contain 415 Cal, 50.5 g of protein, and 553 mg of sodium?

29. *Investment.* Santiago receives \$126 per year in simple interest from three investments. Part is invested at 2%, part at 3%, and part at 4%. There is \$500 more invested at 3% than at 2%. The amount invested at 4% is three times the amount invested at 3%. Find the amount invested at each rate.

30. *Investment.* Walter earns a year-end bonus of \$5000 and puts it in 3 one-year investments that pay \$243 in simple interest. Part is invested at 3%, part at 4%, and part at 6%. There is \$1500 more invested at 6% than at 3%. Find the amount invested at each rate.

31. *Price Increases.* Orange juice, a raisin bagel, and a cup of coffee from Kelly's Koffee Kart cost a total of \$6.30. Kelly posts a notice announcing that, effective the following week, the price of orange juice will increase 25% and the price of bagels will increase 20%. After the increase, the same purchase will cost a total of \$7.30, and orange juice will cost 70¢ more than coffee. Find the price of each item before the increase.

32. *Cost of Snack Food.* Martin and Eva pool their loose change to buy snacks on their coffee break. One day, they spent \$6.75 on 1 carton of milk, 2 donuts, and 1 cup of coffee. The next day, they spent \$8.50 on 3 donuts and 2 cups of coffee. The third day, they bought 1 carton of milk, 1 donut, and 2 cups of coffee and spent \$7.25. On the fourth day, they have a total of \$6.45 left. Is this enough to buy 2 cartons of milk and 2 donuts?

33. *Hours Spent Studying.* The following table lists the percent of college freshmen who responded that they spent 6 or more hours per week studying during their high school senior year, which is represented in terms of the number of years after 1992.

Year, x	Percent Who Say They Studied 6 or More Hours per Week
1992, 0	43
2002, 10	33
2012, 20	38

Source: 2014 Brown Center Report on Education, compiled from "The American Freshman," UCLA Higher Education Research Institute

a) Use a system of equations to fit a quadratic function $f(x) = ax^2 + bx + c$ to the data.
b) Use the function to estimate the percent of college freshmen who say they spent 6 or more hours per week studying during their high school senior year in 2007 and in 2014.

34. *Existing Home Sales.* The following table lists the number of U.S. existing home sales, in millions, represented in terms of years after 2006.

Year, x	Existing Home Sales (in millions)
2006, 0	6.5
2009, 3	4.3
2012, 6	4.7

Source: National Association of Realtors from Haver Analytics

a) Use a system of equations to fit a quadratic function $f(x) = ax^2 + bx + c$ to the data.
b) Use the function to estimate the number of existing home sales in 2013.

35. *Deportations.* The following table lists the number of foreigners deported, in thousands, represented in terms of the number of years after 2007.

Year, x	U.S. Deportations (in thousands)
2007, 0	291
2010, 3	393
2013, 6	369

Source: U.S. Immigration and Customs Enforcement

a) Use a system of equations to fit a quadratic function $f(x) = ax^2 + bx + c$ to the data.
b) Use the function to estimate the number of deportations in 2014.

36. *Strength of Army.* The following table lists the number of active-duty personnel in the U.S. army, in thousands, represented in terms of the number of years after 2009.

Year, x	Active-Duty Army Personnel (in thousands)
2009, 0	553
2011, 2	565
2013, 4	541

Source: Department of the Army, U.S. Department of Defense

a) Use a system of equations to fit a quadratic function $f(x) = ax^2 + bx + c$ to the data.
b) Use the function to estimate the number of active-duty Army personnel in 2010 and in 2014.

Skill Maintenance

In each of Exercises 37–44, fill in the blank with the correct term. Some of the given choices will not be used.

Descartes' rule of signs	constant function
the leading-term test	horizontal asymptote
the intermediate value theorem	vertical asymptote
the fundamental theorem of algebra	oblique asymptote
polynomial function	direct variation
rational function	inverse variation
one-to-one function	horizontal line
	vertical line
	parallel
	perpendicular

37. Two lines with slopes m_1 and m_2 are ____________ if and only if the product of their slopes is -1. [1.4]

38. We can use ____________ to determine the behavior of the graph of a polynomial function as $x \longrightarrow \infty$ or as $x \longrightarrow -\infty$. [4.1]

39. If it is possible for a(n) ____________ to cross a graph more than once, then the graph is not the graph of a function. [1.2]

40. A function is a(n) ____________ if different inputs have different outputs. [5.1]

41. A(n) ____________ is a function that is a quotient of two polynomials. [4.5]

42. If a situation gives rise to a function $f(x) = k/x$, or $y = k/x$, where k is a positive constant, we say that we have ____________. [2.5]

43. A(n) ____________ of a rational function $p(x)/q(x)$, where $p(x)$ and $q(x)$ have no common factors other than constants, occurs at an x-value that makes the denominator 0. [4.5]

44. When the numerator and the denominator of a rational function have the same degree, the graph of the function has a(n) ____________. [4.5]

Synthesis

In Exercises 45 and 46, let u represent $1/x$, v represent $1/y$, and w represent $1/z$. Solve first for u, v, and w. Then solve the system of equations.

45.
$$\frac{2}{x} - \frac{1}{y} - \frac{3}{z} = -1,$$
$$\frac{2}{x} - \frac{1}{y} + \frac{1}{z} = -9,$$
$$\frac{1}{x} + \frac{2}{y} - \frac{4}{z} = 17$$

46.
$$\frac{2}{x} + \frac{2}{y} - \frac{3}{z} = 3,$$
$$\frac{1}{x} - \frac{2}{y} - \frac{3}{z} = 9,$$
$$\frac{7}{x} - \frac{2}{y} + \frac{9}{z} = -39$$

47. Find the sum of the angle measures at the tips of the star.

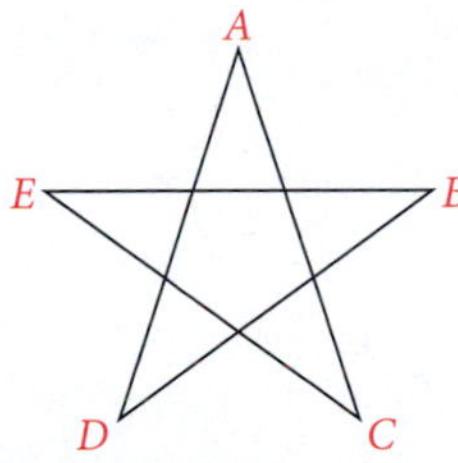

48. *Transcontinental Railroad.* Use the following facts to find the year in which the first U.S. transcontinental railroad was completed. The sum of the digits in the year is 24. The units digit is 1 more than the hundreds digit. Both the tens and the units digits are multiples of three.

In Exercises 49 and 50, three solutions of an equation are given. Use a system of three equations in three variables to find the constants and write the equation.

49. $Ax + By + Cz = 12;$
$\left(1, \frac{3}{4}, 3\right), \left(\frac{4}{3}, 1, 2\right)$, and $(2, 1, 1)$

50. $y = B - Mx - Nz;$
$(1, 1, 2), (3, 2, -6)$, and $\left(\frac{3}{2}, 1, 1\right)$

In Exercises 51 and 52, four solutions of the equation $y = ax^3 + bx^2 + cx + d$ are given. Use a system of four equations in four variables to find the constants a, b, c, and d and write the equation.

51. $(-2, 59), (-1, 13), (1, -1)$, and $(2, -17)$

52. $(-2, -39), (-1, -12), (1, -6)$, and $(3, 16)$

53. *Theater Attendance.* A performance at the Bingham Performing Arts Center was attended by 100 people. The audience consisted of adults, students, and children. The ticket prices were \$10 each for adults, \$3 each for students, and 50 cents each for children. The total amount of money taken in was \$100. How many adults, students, and children were in attendance? Does there seem to be some information missing? Do some careful reasoning.

6.3 Matrices and Systems of Equations

▶ Solve systems of equations using matrices.

▶ Matrices and Row-Equivalent Operations

In this section, we consider additional techniques for solving systems of equations. You have probably observed that when we solve a system of equations, we perform computations with the coefficients and the constants and continually rewrite the variables. We can streamline the solution process by omitting the variables until a solution is found. For example, the system

$$2x - 3y = 7,$$
$$x + 4y = -2$$

can be written more simply as

$$\left[\begin{array}{cc|c} 2 & -3 & 7 \\ 1 & 4 & -2 \end{array}\right].$$

The vertical line replaces the equals signs.

A rectangular array of numbers like the one above is called a **matrix** (pl., **matrices**). The matrix above is called an **augmented matrix** for the given system of equations, because it contains not only the coefficients but also the constant terms. The matrix

$$\begin{bmatrix} 2 & -3 \\ 1 & 4 \end{bmatrix}$$

is called the **coefficient matrix** of the system.

The **rows** of a matrix are horizontal, and the **columns** are vertical. The augmented matrix above has 2 rows and 3 columns, and the coefficient matrix has 2 rows and 2 columns. A matrix with m rows and n columns is said to be of **order** $m \times n$. Thus the order of the augmented matrix above is 2×3, and the order of the coefficient matrix is 2×2. When $m = n$, a matrix is said to be **square**. The coefficient matrix above is a square matrix. The numbers 2 and 4 lie on the **main diagonal** of the coefficient matrix. The numbers in a matrix are called **entries**, or **elements**.

Gaussian Elimination with Matrices

In Section 6.2, we described a series of operations that can be used to transform a system of equations to an equivalent system. Each of these operations corresponds to one that can be used to produce *row-equivalent matrices.*

ROW-EQUIVALENT OPERATIONS

1. Interchange any two rows.
2. Multiply each entry in a row by the same nonzero constant.
3. Add a nonzero multiple of one row to another row.

We can use these operations on the augmented matrix of a system of equations to solve the system.

EXAMPLE 1 Solve the following system:

$$\begin{aligned} 2x - y + 4z &= -3, \\ x - 2y - 10z &= -6, \\ 3x \quad\;\; + 4z &= 7. \end{aligned}$$

Solution First, we write the augmented matrix, writing 0 for the missing y-term in the last equation:

$$\left[\begin{array}{rrr|r} 2 & -1 & 4 & -3 \\ 1 & -2 & -10 & -6 \\ 3 & 0 & 4 & 7 \end{array}\right].$$

Our goal is to find a row-equivalent matrix of the form

$$\left[\begin{array}{ccc|c} 1 & a & b & c \\ 0 & 1 & d & e \\ 0 & 0 & 1 & f \end{array}\right].$$

The variables can then be reinserted to form equations from which we can complete the solution. This is done by working from the bottom equation to the top and using back-substitution.

The first step is to multiply and/or interchange rows so that each number in the first column below the first number is a multiple of that number. In this case, we interchange the first and second rows to obtain a 1 in the upper left-hand corner.

$$\left[\begin{array}{rrr|r} 1 & -2 & -10 & -6 \\ 2 & -1 & 4 & -3 \\ 3 & 0 & 4 & 7 \end{array}\right] \quad \begin{array}{l} \text{New row 1} = \text{row 2} \\ \text{New row 2} = \text{row 1} \\ \end{array}$$

JUST IN TIME 4

Next, we multiply the first row by -2 and add it to the second row. We also multiply the first row by -3 and add it to the third row.

$$\left[\begin{array}{rrr|r} 1 & -2 & -10 & -6 \\ 0 & 3 & 24 & 9 \\ 0 & 6 & 34 & 25 \end{array}\right] \quad \begin{array}{l} \text{Row 1 is unchanged.} \\ \text{New row 2} = -2(\text{row 1}) + \text{row 2} \\ \text{New row 3} = -3(\text{row 1}) + \text{row 3} \end{array}$$

Technology Connection

Row-equivalent operations can be performed on a graphing calculator. For example, to interchange the first and second rows of the augmented matrix, as we did in the first step in Example 1, we enter the matrix as matrix **A** and select "rowSwap" from the MATRIX MATH menu. Some graphing calculators will not automatically store the matrix produced using a row-equivalent operation, so when several operations are to be performed in succession, it is helpful to store the result of each operation as it is produced. In the window below, we see both the matrix produced by the rowSwap operation and the indication that this matrix is stored as matrix **B**.

```
rowSwap([A],1,2)→[B]
[[1  -2  -10  -6]
 [2  -1   4   -3]
 [3   0   4    7 ]]
```

Now we multiply the second row by $\frac{1}{3}$ to get a 1 in the second row, second column.

$$\left[\begin{array}{rrr|r} 1 & -2 & -10 & -6 \\ 0 & 1 & 8 & 3 \\ 0 & 6 & 34 & 25 \end{array}\right] \quad \text{New row 2} = \tfrac{1}{3}(\text{row 2})$$

Then we multiply the second row by -6 and add it to the third row.

$$\left[\begin{array}{rrr|r} 1 & -2 & -10 & -6 \\ 0 & 1 & 8 & 3 \\ 0 & 0 & -14 & 7 \end{array}\right] \quad \text{New row 3} = -6(\text{row 2}) + \text{row 3}$$

Finally, we multiply the third row by $-\frac{1}{14}$ to get a 1 in the third row, third column.

$$\left[\begin{array}{rrr|r} 1 & -2 & -10 & -6 \\ 0 & 1 & 8 & 3 \\ 0 & 0 & 1 & -\frac{1}{2} \end{array}\right] \quad \text{New row 3} = -\tfrac{1}{14}(\text{row 3})$$

Now we can write the system of equations that corresponds to the last matrix above:

$$\begin{aligned} x - 2y - 10z &= -6, && (1) \\ y + 8z &= 3, && (2) \\ z &= -\tfrac{1}{2}. && (3) \end{aligned}$$

We back-substitute $-\frac{1}{2}$ for z in equation (2) and solve for y:

$$\begin{aligned} y + 8\left(-\tfrac{1}{2}\right) &= 3 \\ y - 4 &= 3 \\ y &= 7. \end{aligned}$$

Next, we back-substitute 7 for y and $-\frac{1}{2}$ for z in equation (1) and solve for x:

$$\begin{aligned} x - 2 \cdot 7 - 10\left(-\tfrac{1}{2}\right) &= -6 \\ x - 14 + 5 &= -6 \\ x - 9 &= -6 \\ x &= 3. \end{aligned}$$

The triple $\left(3, 7, -\frac{1}{2}\right)$ checks in the original system of equations, so it is the solution.

Now Try Exercise 27.

The procedure followed in Example 1 is called **Gaussian elimination with matrices**. The last matrix in Example 1 is in **row-echelon form**. To be in this form, a matrix must have the following properties.

ROW-ECHELON FORM

1. If a row does not consist entirely of 0's, then the first nonzero element in the row is a 1 (called a **leading 1**).
2. For any two successive nonzero rows, the leading 1 in the lower row is farther to the right than the leading 1 in the higher row.
3. All the rows consisting entirely of 0's are at the bottom of the matrix.

If a fourth property is also satisfied, a matrix is said to be in **reduced row-echelon form**:

4. Each column that contains a leading 1 has 0's everywhere else.

EXAMPLE 2 Which of the following matrices are in row-echelon form? Which, if any, are in reduced row-echelon form?

a) $\left[\begin{array}{ccc|c} 1 & -3 & 5 & -2 \\ 0 & 1 & -4 & 3 \\ 0 & 0 & 1 & 10 \end{array}\right]$ b) $\left[\begin{array}{cc|c} 0 & -1 & 2 \\ 0 & 1 & 5 \end{array}\right]$ c) $\left[\begin{array}{cccc|c} 1 & -2 & -6 & 4 & 7 \\ 0 & 3 & 5 & -8 & -1 \\ 0 & 0 & 1 & 9 & 2 \end{array}\right]$

d) $\left[\begin{array}{ccc|c} 1 & 0 & 0 & -2.4 \\ 0 & 1 & 0 & 0.8 \\ 0 & 0 & 1 & 5.6 \end{array}\right]$ e) $\left[\begin{array}{cccc|c} 1 & 0 & 0 & 0 & \frac{2}{3} \\ 0 & 1 & 0 & 0 & -\frac{1}{4} \\ 0 & 0 & 1 & 0 & \frac{6}{7} \\ 0 & 0 & 0 & 0 & 0 \end{array}\right]$ f) $\left[\begin{array}{ccc|c} 1 & -4 & 2 & 5 \\ 0 & 0 & 0 & 0 \\ 0 & 1 & -3 & -8 \end{array}\right]$

Solution The matrices in (a), (d), and (e) satisfy the row-echelon criteria and, thus, are in row-echelon form. In (b) and (c), the first nonzero elements of the first and second rows, respectively, are not 1. In (f), the row consisting entirely of 0's is not at the bottom of the matrix. Thus the matrices in (b), (c), and (f) are not in row-echelon form. In (d) and (e), not only are the row-echelon criteria met but each column that contains a leading 1 also has 0's elsewhere, so these matrices are in reduced row-echelon form.

Gauss–Jordan Elimination

We have seen that with Gaussian elimination we perform row-equivalent operations on a matrix to obtain a row-equivalent matrix in row-echelon form. When we continue to apply these operations until we have a matrix in *reduced* row-echelon form, we are using **Gauss–Jordan elimination**. This method is named for Karl Friedrich Gauss and Wilhelm Jordan (1842–1899).

EXAMPLE 3 Use Gauss–Jordan elimination to solve the system of equations in Example 1.

Solution Using Gaussian elimination in Example 1, we obtained the matrix

$$\left[\begin{array}{ccc|c} 1 & -2 & -10 & -6 \\ 0 & 1 & 8 & 3 \\ 0 & 0 & 1 & -\frac{1}{2} \end{array}\right].$$

We continue to perform row-equivalent operations until we have a matrix in reduced row-echelon form. We multiply the third row by 10 and add it to the first row. We also multiply the third row by -8 and add it to the second row.

$$\left[\begin{array}{ccc|c} 1 & -2 & 0 & -11 \\ 0 & 1 & 0 & 7 \\ 0 & 0 & 1 & -\frac{1}{2} \end{array}\right] \qquad \begin{array}{l} \text{New row 1} = 10(\text{row 3}) + \text{row 1} \\ \text{New row 2} = -8(\text{row 3}) + \text{row 2} \end{array}$$

Next, we multiply the second row by 2 and add it to the first row.

$$\left[\begin{array}{ccc|c} 1 & 0 & 0 & 3 \\ 0 & 1 & 0 & 7 \\ 0 & 0 & 1 & -\frac{1}{2} \end{array}\right] \qquad \text{New row 1} = 2(\text{row 2}) + \text{row 1}$$

Writing the system of equations that corresponds to this matrix, we have

$$\begin{aligned} x \qquad &= 3, \\ y \quad &= 7, \\ z &= -\tfrac{1}{2}. \end{aligned}$$

We can actually read the solution, $\left(3, 7, -\frac{1}{2}\right)$, directly from the last column of the reduced row-echelon matrix.

Now Try Exercise 27.

Technology Connection

After an augmented matrix is entered in a graphing calculator, reduced row-echelon form can be found directly using the "rref" operation from the MATRIX MATH menu.

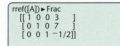

```
rref([A])►Frac
  [[1 0 0 3   ]
   [0 1 0 7   ]
   [0 0 1 −1/2]]
```

The application PolySmlt from the APPS menu can also be used to solve a system of equations.

EXAMPLE 4 Solve the following system:

$$
\begin{aligned}
3x - 4y - z &= 6,\\
2x - y + z &= -1,\\
4x - 7y - 3z &= 13.
\end{aligned}
$$

Solution First, we write the augmented matrix and use Gauss–Jordan elimination.

$$
\left[\begin{array}{ccc|c} 3 & -4 & -1 & 6 \\ 2 & -1 & 1 & -1 \\ 4 & -7 & -3 & 13 \end{array}\right]
$$

We then multiply the second and third rows by 3 so that each number in the first column below the first number, 3, is a multiple of that number.

$$
\left[\begin{array}{ccc|c} 3 & -4 & -1 & 6 \\ 6 & -3 & 3 & -3 \\ 12 & -21 & -9 & 39 \end{array}\right]
\quad
\begin{array}{l} \\ \text{New row 2} = 3(\text{row 2}) \\ \text{New row 3} = 3(\text{row 3}) \end{array}
$$

Next, we multiply the first row by -2 and add it to the second row. We also multiply the first row by -4 and add it to the third row.

$$
\left[\begin{array}{ccc|c} 3 & -4 & -1 & 6 \\ 0 & 5 & 5 & -15 \\ 0 & -5 & -5 & 15 \end{array}\right]
\quad
\begin{array}{l} \\ \text{New row 2} = -2(\text{row 1}) + \text{row 2} \\ \text{New row 3} = -4(\text{row 1}) + \text{row 3} \end{array}
$$

Now we add the second row to the third row.

$$
\left[\begin{array}{ccc|c} 3 & -4 & -1 & 6 \\ 0 & 5 & 5 & -15 \\ 0 & 0 & 0 & 0 \end{array}\right]
\quad
\begin{array}{l} \\ \\ \text{New row 3} = \text{row 2} + \text{row 3} \end{array}
$$

We can stop at this stage because we have a row consisting entirely of 0's. The last row of the matrix corresponds to the equation $0 = 0$, which is true for all values of x, y, and z. Therefore, the equations are dependent and the system is equivalent to

$$
\begin{aligned}
3x - 4y - z &= 6,\\
5y + 5z &= -15.
\end{aligned}
$$

This particular system has infinitely many solutions. (A system containing dependent equations could be inconsistent.)

Solving the second equation for y gives us

$$y = -z - 3.$$

Substituting $-z - 3$ for y in the first equation and solving for x, we get

$$
\begin{aligned}
3x - 4(-z - 3) - z &= 6\\
3x + 4z + 12 - z &= 6\\
3x + 3z + 12 &= 6\\
3x &= -3z - 6\\
x &= -z - 2.
\end{aligned}
$$

Then the solutions of this system are of the form

$$(-z - 2, -z - 3, z),$$

where z can be any real number.

Now Try Exercise 33.

Similarly, if we obtain a row whose only nonzero entry occurs in the last column, we have an inconsistent system of equations. For example, in the matrix

$$\left[\begin{array}{ccc|c} 1 & 0 & 3 & -2 \\ 0 & 1 & 5 & 4 \\ 0 & 0 & 0 & 6 \end{array}\right],$$

the last row corresponds to the false equation $0 = 6$, so we know the original system of equations has no solution.

6.3 Exercise Set

Determine the order of the matrix.

1. $\begin{bmatrix} 1 & -6 \\ -3 & 2 \\ 0 & 5 \end{bmatrix}$

2. $\begin{bmatrix} 7 \\ -5 \\ -1 \\ 3 \end{bmatrix}$

3. $\begin{bmatrix} 2 & -4 & 0 & 9 \end{bmatrix}$

4. $\begin{bmatrix} -8 \end{bmatrix}$

5. $\begin{bmatrix} 1 & -5 & -8 \\ 6 & 4 & -2 \\ -3 & 0 & 7 \end{bmatrix}$

6. $\begin{bmatrix} 13 & 2 & -6 & 4 \\ -1 & 18 & 5 & -12 \end{bmatrix}$

Write the augmented matrix for the system of equations.

7. $2x - y = 7,$
   $x + 4y = -5$

8. $3x + 2y = 8,$
   $2x - 3y = 15$

9. $x - 2y + 3z = 12,$
   $2x - 4z = 8,$
   $3y + z = 7$

10. $x + y - z = 7,$
    $3y + 2z = 1,$
    $-2x - 5y = 6$

Write the system of equations that corresponds to the augmented matrix.

11. $\left[\begin{array}{cc|c} 3 & -5 & 1 \\ 1 & 4 & -2 \end{array}\right]$

12. $\left[\begin{array}{cc|c} 1 & 2 & -6 \\ 4 & 1 & -3 \end{array}\right]$

13. $\left[\begin{array}{ccc|c} 2 & 1 & -4 & 12 \\ 3 & 0 & 5 & -1 \\ 1 & -1 & 1 & 2 \end{array}\right]$

14. $\left[\begin{array}{ccc|c} -1 & -2 & 3 & 6 \\ 0 & 4 & 1 & 2 \\ 2 & -1 & 0 & 9 \end{array}\right]$

Solve the system of equations using Gaussian elimination or Gauss–Jordan elimination.

15. $4x + 2y = 11,$
    $3x - y = 2$

16. $2x + y = 1,$
    $3x + 2y = -2$

17. $5x - 2y = -3,$
    $2x + 5y = -24$

18. $2x + y = 1,$
    $3x - 6y = 4$

19. $3x + 4y = 7,$
    $-5x + 2y = 10$

20. $5x - 3y = -2,$
    $4x + 2y = 5$

21. $3x + 2y = 6,$
    $2x - 3y = -9$

22. $x - 4y = 9,$
    $2x + 5y = 5$

23. $x - 3y = 8,$
    $-2x + 6y = 3$

24. $4x - 8y = 12,$
    $-x + 2y = -3$

25. $-2x + 6y = 4,$
    $3x - 9y = -6$

26. $6x + 2y = -10,$
    $-3x - y = 6$

27. $x + 2y - 3z = 9,$
    $2x - y + 2z = -8,$
    $3x - y - 4z = 3$

28. $x - y + 2z = 0,$
    $x - 2y + 3z = -1,$
    $2x - 2y + z = -3$

29. $4x - y - 3z = 1,$
    $8x + y - z = 5,$
    $2x + y + 2z = 5$

30. $3x + 2y + 2z = 3,$
    $x + 2y - z = 5,$
    $2x - 4y + z = 0$

31. $x - 2y + 3z = -4,$
    $3x + y - z = 0,$
    $2x + 3y - 5z = 1$

32. $2x - 3y + 2z = 2,$
    $x + 4y - z = 9,$
    $-3x + y - 5z = 5$

33. $2x - 4y - 3z = 3,$
    $x + 3y + z = -1,$
    $5x + y - 2z = 2$

34. $x + y - 3z = 4,$
    $4x + 5y + z = 1,$
    $2x + 3y + 7z = -7$

35. $p + q + r = 1,$
    $p + 2q + 3r = 4,$
    $4p + 5q + 6r = 7$

36. $m + n + t = 9,$
    $m - n - t = -15,$
    $3m + n + t = 2$

37. $a + b - c = 7,$
    $a - b + c = 5,$
    $3a + b - c = -1$

38. $a - b + c = 3,$
    $2a + b - 3c = 5,$
    $4a + b - c = 11$

39. $-2w + 2x + 2y - 2z = -10,$
$w + x + y + z = -5,$
$3w + x - y + 4z = -2,$
$w + 3x - 2y + 2z = -6$

40. $-w + 2x - 3y + z = -8,$
$-w + x + y - z = -4,$
$w + x + y + z = 22,$
$-w + x - y - z = -14$

Use Gaussian elimination or Gauss–Jordan elimination in Exercises 41–44.

41. *Borrowing.* Greenfield Manufacturing borrowed \$30,000 to buy a new piece of equipment. Part of the money was borrowed at 8%, part at 10%, and part at 12%. The annual interest was \$3040, and the total amount borrowed at 8% and at 10% was twice the amount borrowed at 12%. How much was borrowed at each rate?

42. *Time of Return.* The Patels pay their babysitter \$11 per hour before 11 P.M. and \$14.50 after 11 P.M. One evening, they went out for 6 hr and paid the sitter \$73. What time did they return?

43. *Stamp Purchase.* For her business, Olivia spent \$86.80 on both 49¢ and 21¢ stamps. She bought a total of 200 stamps. How many of each type did she buy?

44. *Advertising Expense.* eAuction.com spent a total of \$11 million on advertising in fiscal years 2010, 2011, and 2012. The amount spent in 2012 was three times the amount spent in 2010. The amount spent in 2011 was \$3 million less than the amount spent in 2012. How much was spent on advertising each year?

Skill Maintenance

In Exercises 45–52, classify the function as linear, quadratic, cubic, quartic, rational, exponential, or logarithmic.

45. $f(x) = 3^{x-1}$ [5.2]

46. $f(x) = 3x - 1$ [1.3]

47. $f(x) = \dfrac{3x - 1}{x^2 + 4}$ [4.5]

48. $f(x) = -\frac{3}{4}x^4 + \frac{9}{2}x^3 + 2x^2 - 4$ [4.1]

49. $f(x) = \ln(3x - 1)$ [5.3]

50. $f(x) = \frac{3}{4}x^3 - x$ [4.1]

51. $f(x) = 3$ [1.3]

52. $f(x) = 2 - x - x^2$ [3.2]

Synthesis

In Exercises 53 and 54, three solutions of the equation $y = ax^2 + bx + c$ are given. Use a system of three equations in three variables and Gaussian elimination or Gauss–Jordan elimination to find the constants a, b, and c and write the equation.

53. $(-3, 12)$, $(-1, -7)$, and $(1, -2)$

54. $(-1, 0)$, $(1, -3)$, and $(3, -22)$

55. Find two different row-echelon forms of
$$\begin{bmatrix} 1 & 5 \\ 3 & 2 \end{bmatrix}.$$

56. Consider the system of equations
$x - y + 3z = -8,$
$2x + 3y - z = 5,$
$3x + 2y + 2kz = -3k.$

For what value(s) of k, if any, will the system have:
a) no solution?
b) exactly one solution?
c) infinitely many solutions?

Solve using matrices.

57. $y = x + z,$
$3y + 5z = 4,$
$x + 4 = y + 3z$

58. $x + y = 2z,$
$2x - 5z = 4,$
$x - z = y + 8$

59. $x - 4y + 2z = 7,$
$3x + y + 3z = -5$

60. $x - y - 3z = 3,$
$-x + 3y + z = -7$

61. $4x + 5y = 3,$
$-2x + y = 9,$
$3x - 2y = -15$

62. $2x - 3y = -1,$
$-x + 2y = -2,$
$3x - 5y = 1$

6.4 Matrix Operations

- Add, subtract, and multiply matrices when possible.
- Write a matrix equation equivalent to a system of equations.

In addition to solving systems of equations, matrices are useful in many other types of applications. In this section, we study matrices and some of their properties.

A capital letter is generally used to name a matrix, and lower-case letters with double subscripts generally denote its entries. For example, a_{47}, read "*a* sub four seven," indicates the entry in the fourth row and the seventh column. A general term is represented by a_{ij}. The notation a_{ij} indicates the entry in row *i* and column *j*. In general, we can write a matrix as

$$\mathbf{A} = [a_{ij}] = \begin{bmatrix} a_{11} & a_{12} & a_{13} & \cdots & a_{1n} \\ a_{21} & a_{22} & a_{23} & \cdots & a_{2n} \\ a_{31} & a_{32} & a_{33} & \cdots & a_{3n} \\ \vdots & \vdots & \vdots & & \vdots \\ a_{m1} & a_{m2} & a_{m3} & \cdots & a_{mn} \end{bmatrix}.$$

The matrix above has m rows and n columns; that is, its order is $m \times n$.

Two matrices are **equal** if they have the same order and corresponding entries are equal.

Matrix Addition and Subtraction

To add or subtract matrices, we add or subtract their corresponding entries. The matrices must have the same order for this to be possible.

ADDITION AND SUBTRACTION OF MATRICES

Given two $m \times n$ matrices $\mathbf{A} = [a_{ij}]$ and $\mathbf{B} = [b_{ij}]$, their sum is

$$\mathbf{A} + \mathbf{B} = [a_{ij} + b_{ij}]$$

and their difference is

$$\mathbf{A} - \mathbf{B} = [a_{ij} - b_{ij}].$$

Addition of matrices is both commutative and associative.

EXAMPLE 1 Find $\mathbf{A} + \mathbf{B}$ for each of the following.

a) $\mathbf{A} = \begin{bmatrix} -5 & 0 \\ 4 & \frac{1}{2} \end{bmatrix}, \quad \mathbf{B} = \begin{bmatrix} 6 & -3 \\ 2 & 3 \end{bmatrix}$

b) $\mathbf{A} = \begin{bmatrix} 1 & 3 \\ -1 & 5 \\ 6 & 0 \end{bmatrix}, \quad \mathbf{B} = \begin{bmatrix} -1 & -2 \\ 1 & -2 \\ -3 & 1 \end{bmatrix}$

Technology Connection

We can use a graphing calculator to add matrices. In Example 1(a), we enter **A** and **B** and then find **A** + **B**.

```
[A]+[B]
                [[1  -3]
                 [6 3.5]]
```

We can do the same for the matrices in Example 1(b).

```
[A]+[B]
                [[0 1]
                 [0 3]
                 [3 1]]
```

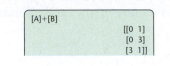

Solution We have a pair of 2 × 2 matrices in part (a) and a pair of 3 × 2 matrices in part (b). Since each pair of matrices has the same order, we can add the corresponding entries.

a) $$\mathbf{A}+\mathbf{B}=\begin{bmatrix}-5 & 0\\ 4 & \frac{1}{2}\end{bmatrix}+\begin{bmatrix}6 & -3\\ 2 & 3\end{bmatrix}=\begin{bmatrix}-5+6 & 0+(-3)\\ 4+2 & \frac{1}{2}+3\end{bmatrix}=\begin{bmatrix}1 & -3\\ 6 & 3\frac{1}{2}\end{bmatrix}$$

b) $$\mathbf{A}+\mathbf{B}=\begin{bmatrix}1 & 3\\ -1 & 5\\ 6 & 0\end{bmatrix}+\begin{bmatrix}-1 & -2\\ 1 & -2\\ -3 & 1\end{bmatrix}=\begin{bmatrix}1+(-1) & 3+(-2)\\ -1+1 & 5+(-2)\\ 6+(-3) & 0+1\end{bmatrix}=\begin{bmatrix}0 & 1\\ 0 & 3\\ 3 & 1\end{bmatrix}$$

Now Try Exercise 5.

EXAMPLE 2 Find **C** − **D** for each of the following.

a) $\mathbf{C}=\begin{bmatrix}1 & 2\\ -2 & 0\\ -3 & -1\end{bmatrix}$, $\mathbf{D}=\begin{bmatrix}1 & -1\\ 1 & 3\\ 2 & 3\end{bmatrix}$

b) $\mathbf{C}=\begin{bmatrix}5 & -6\\ -3 & 4\end{bmatrix}$, $\mathbf{D}=\begin{bmatrix}-4\\ 1\end{bmatrix}$

Technology Connection

We can use a graphing calculator to subtract matrices. In Example 2(a), we enter **C** and **D** and then find **C** − **D**.

```
[C]-[D]
                [[0    3]
                 [-3  -3]
                 [-5  -4]]
```

Solution

a) Since the order of each matrix is 3 × 2, we can subtract corresponding entries:

$$\mathbf{C}-\mathbf{D}=\begin{bmatrix}1 & 2\\ -2 & 0\\ -3 & -1\end{bmatrix}-\begin{bmatrix}1 & -1\\ 1 & 3\\ 2 & 3\end{bmatrix}=\begin{bmatrix}1-1 & 2-(-1)\\ -2-1 & 0-3\\ -3-2 & -1-3\end{bmatrix}=\begin{bmatrix}0 & 3\\ -3 & -3\\ -5 & -4\end{bmatrix}.$$

b) **C** is a 2 × 2 matrix and **D** is a 2 × 1 matrix. Since the matrices do not have the same order, we cannot subtract.

Now Try Exercise 13.

The **opposite**, or **additive inverse**, of a matrix is obtained by replacing each entry with its opposite.

EXAMPLE 3 Find −**A** and **A** + (−**A**) for

$$\mathbf{A}=\begin{bmatrix}1 & 0 & 2\\ 3 & -1 & 5\end{bmatrix}.$$

Solution To find $-\mathbf{A}$, we replace each entry of $\mathbf{A}$ with its opposite.

$$-\mathbf{A} = \begin{bmatrix} -1 & 0 & -2 \\ -3 & 1 & -5 \end{bmatrix},$$

$$\mathbf{A} + (-\mathbf{A}) = \begin{bmatrix} 1 & 0 & 2 \\ 3 & -1 & 5 \end{bmatrix} + \begin{bmatrix} -1 & 0 & -2 \\ -3 & 1 & -5 \end{bmatrix}$$
$$= \begin{bmatrix} 0 & 0 & 0 \\ 0 & 0 & 0 \end{bmatrix}$$

A matrix having 0's for all its entries is called a **zero matrix**. When a zero matrix is added to a second matrix of the same order, the second matrix is unchanged. Thus a zero matrix is an **additive identity**. For example,

$$\begin{bmatrix} 2 & 3 & -4 \\ 0 & 6 & 5 \end{bmatrix} + \begin{bmatrix} 0 & 0 & 0 \\ 0 & 0 & 0 \end{bmatrix} = \begin{bmatrix} 2 & 3 & -4 \\ 0 & 6 & 5 \end{bmatrix}.$$

The matrix

$$\begin{bmatrix} 0 & 0 & 0 \\ 0 & 0 & 0 \end{bmatrix}$$

is the additive identity for any 2×3 matrix.

Scalar Multiplication

When we find the product of a number and a matrix, we obtain a **scalar product**.

> **SCALAR PRODUCT**
>
> The **scalar product** of a number k and a matrix $\mathbf{A}$ is the matrix denoted $k\mathbf{A}$, obtained by multiplying each entry of $\mathbf{A}$ by the number k. The number k is called a **scalar**.

Technology Connection

Scalar products, like those in Example 4, can be found using a graphing calculator.

```
3[A]
                [[-9   0]
                 [ 12 15]]
(-1)[A]
                [[3   0 ]
                 [-4 -5]]
```

EXAMPLE 4 Find 3A and $(-1)\mathbf{A}$ for

$$\mathbf{A} = \begin{bmatrix} -3 & 0 \\ 4 & 5 \end{bmatrix}.$$

Solution We have

$$3\mathbf{A} = 3\begin{bmatrix} -3 & 0 \\ 4 & 5 \end{bmatrix} = \begin{bmatrix} 3(-3) & 3 \cdot 0 \\ 3 \cdot 4 & 3 \cdot 5 \end{bmatrix} = \begin{bmatrix} -9 & 0 \\ 12 & 15 \end{bmatrix},$$

$$(-1)\mathbf{A} = -1\begin{bmatrix} -3 & 0 \\ 4 & 5 \end{bmatrix} = \begin{bmatrix} -1(-3) & -1 \cdot 0 \\ -1 \cdot 4 & -1 \cdot 5 \end{bmatrix} = \begin{bmatrix} 3 & 0 \\ -4 & -5 \end{bmatrix}.$$

Now Try Exercise 9.

The properties of matrix addition and scalar multiplication are similar to the properties of addition and multiplication of real numbers.

PROPERTIES OF MATRIX ADDITION AND SCALAR MULTIPLICATION

For any $m \times n$ matrices **A**, **B**, and **C** and any scalars k and l:

$\mathbf{A} + \mathbf{B} = \mathbf{B} + \mathbf{A}$.	Commutative property of addition
$\mathbf{A} + (\mathbf{B} + \mathbf{C}) = (\mathbf{A} + \mathbf{B}) + \mathbf{C}$.	Associative property of addition
$(kl)\mathbf{A} = k(l\mathbf{A})$.	Associative property of scalar multiplication
$k(\mathbf{A} + \mathbf{B}) = k\mathbf{A} + k\mathbf{B}$.	Distributive property
$(k + l)\mathbf{A} = k\mathbf{A} + l\mathbf{A}$.	Distributive property

There exists a unique matrix **0** such that:

$\mathbf{A} + \mathbf{0} = \mathbf{0} + \mathbf{A} = \mathbf{A}$. Additive identity property

There exists a unique matrix $-\mathbf{A}$ such that:

$\mathbf{A} + (-\mathbf{A}) = -\mathbf{A} + \mathbf{A} = \mathbf{0}$. Additive inverse property

EXAMPLE 5 ***Production.*** Waterworks, Inc., manufactures three types of kayaks in its two plants. The following table lists the number of each style produced at each plant in April.

	Whitewater Kayak	Ocean Kayak	Crossover Kayak
Madison Plant	150	120	100
Greensburg Plant	180	90	130

a) Write a 2×3 matrix **A** that represents the information in the table.

b) The manufacturer increased production by 20% in May. Find a matrix **M** that represents the increased production figures.

c) Find the matrix $\mathbf{A} + \mathbf{M}$ and tell what it represents.

Solution

a) Write the entries in the table in a 2×3 matrix **A**.

$$\mathbf{A} = \begin{bmatrix} 150 & 120 & 100 \\ 180 & 90 & 130 \end{bmatrix}$$

b) The production in May will be represented by $\mathbf{A} + 20\%\mathbf{A}$, or $\mathbf{A} + 0.2\mathbf{A}$, or $1.2\mathbf{A}$. Thus,

$$\mathbf{M} = (1.2)\begin{bmatrix} 150 & 120 & 100 \\ 180 & 90 & 130 \end{bmatrix} = \begin{bmatrix} 180 & 144 & 120 \\ 216 & 108 & 156 \end{bmatrix}.$$

c) $$\mathbf{A} + \mathbf{M} = \begin{bmatrix} 150 & 120 & 100 \\ 180 & 90 & 130 \end{bmatrix} + \begin{bmatrix} 180 & 144 & 120 \\ 216 & 108 & 156 \end{bmatrix} = \begin{bmatrix} 330 & 264 & 220 \\ 396 & 198 & 286 \end{bmatrix}$$

The matrix $\mathbf{A} + \mathbf{M}$ represents the total production of each of the three types of kayaks at each plant in April and May.

Now Try Exercise 29.

▸ Products of Matrices

Matrix multiplication is defined in such a way that it can be used in solving systems of equations and in many applications.

MATRIX MULTIPLICATION

For an $m \times n$ matrix $\mathbf{A} = [a_{ij}]$ and an $n \times p$ matrix $\mathbf{B} = [b_{ij}]$, the **product** $\mathbf{AB} = [c_{ij}]$ is an $m \times p$ matrix, where

$$c_{ij} = a_{i1} \cdot b_{1j} + a_{i2} \cdot b_{2j} + a_{i3} \cdot b_{3j} + \cdots + a_{in} \cdot b_{nj}.$$

In other words, the entry c_{ij} in **AB** is obtained by multiplying the entries in row i of **A** by the corresponding entries in column j of **B** and adding the results.

Note that we can multiply two matrices only when the number of columns in the first matrix is equal to the number of rows in the second matrix.

EXAMPLE 6 For

$$\mathbf{A} = \begin{bmatrix} 3 & 1 & -1 \\ 2 & 0 & 3 \end{bmatrix}, \quad \mathbf{B} = \begin{bmatrix} 1 & 6 \\ 3 & -5 \\ -2 & 4 \end{bmatrix}, \quad \text{and} \quad \mathbf{C} = \begin{bmatrix} 4 & -6 \\ 1 & 2 \end{bmatrix},$$

find each of the following.

a) **AB** **b)** **BA**

c) **BC** **d)** **AC**

Solution

a) **A** is a 2×3 matrix and **B** is a 3×2 matrix, so **AB** will be a 2×2 matrix.

$$\mathbf{AB} = \begin{bmatrix} 3 & 1 & -1 \\ 2 & 0 & 3 \end{bmatrix} \begin{bmatrix} 1 & 6 \\ 3 & -5 \\ -2 & 4 \end{bmatrix}$$

$$= \begin{bmatrix} 3 \cdot 1 + 1 \cdot 3 + (-1)(-2) & 3 \cdot 6 + 1(-5) + (-1)(4) \\ 2 \cdot 1 + 0 \cdot 3 + 3(-2) & 2 \cdot 6 + 0(-5) + 3 \cdot 4 \end{bmatrix}$$

$$= \begin{bmatrix} 8 & 9 \\ -4 & 24 \end{bmatrix}$$

b) **B** is a 3×2 matrix and **A** is a 2×3 matrix, so **BA** will be a 3×3 matrix.

$$\mathbf{BA} = \begin{bmatrix} 1 & 6 \\ 3 & -5 \\ -2 & 4 \end{bmatrix} \begin{bmatrix} 3 & 1 & -1 \\ 2 & 0 & 3 \end{bmatrix}$$

$$= \begin{bmatrix} 1 \cdot 3 + 6 \cdot 2 & 1 \cdot 1 + 6 \cdot 0 & 1(-1) + 6 \cdot 3 \\ 3 \cdot 3 + (-5)(2) & 3 \cdot 1 + (-5)(0) & 3(-1) + (-5)(3) \\ -2 \cdot 3 + 4 \cdot 2 & -2 \cdot 1 + 4 \cdot 0 & -2(-1) + 4 \cdot 3 \end{bmatrix} = \begin{bmatrix} 15 & 1 & 17 \\ -1 & 3 & -18 \\ 2 & -2 & 14 \end{bmatrix}$$

Note in parts (a) and (b) that $\mathbf{AB} \neq \mathbf{BA}$. Multiplication of matrices is generally not commutative.

Technology Connection

Matrix multiplication can be performed on a graphing calculator. The products in Examples 6(a) and 6(b) are shown below.

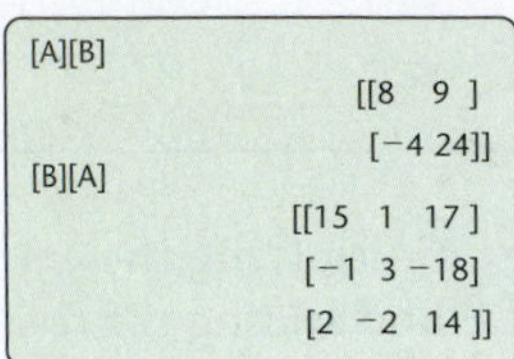

When the product **AC** in Example 6(d) is entered on a graphing calculator, an ERROR message is returned, indicating that the dimensions of the matrices are mismatched.

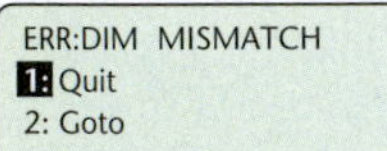

c) **B** is a 3×2 matrix and **C** is a 2×2 matrix, so **BC** will be a 3×2 matrix.

$$\mathbf{BC} = \begin{bmatrix} 1 & 6 \\ 3 & -5 \\ -2 & 4 \end{bmatrix} \begin{bmatrix} 4 & -6 \\ 1 & 2 \end{bmatrix}$$

$$= \begin{bmatrix} 1 \cdot 4 + 6 \cdot 1 & 1(-6) + 6 \cdot 2 \\ 3 \cdot 4 + (-5)(1) & 3(-6) + (-5)(2) \\ -2 \cdot 4 + 4 \cdot 1 & -2(-6) + 4 \cdot 2 \end{bmatrix} = \begin{bmatrix} 10 & 6 \\ 7 & -28 \\ -4 & 20 \end{bmatrix}$$

d) The product **AC** is not defined because the number of columns of **A**, 3, is not equal to the number of rows of **C**, 2.

Now Try Exercises 23 and 25.

EXAMPLE 7 *Bakery Profit.* Two of the items sold at Sweet Treats Bakery are gluten-free bagels and gluten-free doughnuts. The following table lists the number of dozens of each product that are sold at the bakery's three stores one week.

	Main Street Store	Avon Road Store	Dalton Avenue Store
Bagels (in dozens)	25	30	20
Doughnuts (in dozens)	40	35	15

The bakery's profit on one dozen bagels is \$5, and its profit on one dozen doughnuts is \$6. Use matrices to find the total profit on these items at each store for the given week.

Solution We can write the table showing the sales of the products as a 2×3 matrix:

$$\mathbf{S} = \begin{bmatrix} 25 & 30 & 20 \\ 40 & 35 & 15 \end{bmatrix}.$$

The profit per dozen for each product can also be written as a matrix:

$$\mathbf{P} = [5 \quad 6].$$

Then the total profit at each store is given by the matrix product **PS**:

$$\mathbf{PS} = [5 \quad 6] \begin{bmatrix} 25 & 30 & 20 \\ 40 & 35 & 15 \end{bmatrix}$$

$$= [5 \cdot 25 + 6 \cdot 40 \quad 5 \cdot 30 + 6 \cdot 35 \quad 5 \cdot 20 + 6 \cdot 15]$$

$$= [365 \quad 360 \quad 190].$$

The total profit on the sale of gluten-free bagels and gluten-free doughnuts for the given week was \$365 at the Main Street store, \$360 at the Avon Road store, and \$190 at the Dalton Avenue store.

Now Try Exercise 33.

A matrix that consists of a single row, like **P** in Example 7, is called a **row matrix**. Similarly, a matrix that consists of a single column, like

$$\begin{bmatrix} 8 \\ -3 \\ 5 \end{bmatrix},$$

is called a **column matrix**.

We have already seen that matrix multiplication is generally not commutative. Nevertheless, matrix multiplication does have some properties that are similar to those for multiplication of real numbers.

PROPERTIES OF MATRIX MULTIPLICATION

For matrices A, B, and C, assuming that the indicated operations are possible:

$\mathbf{A}(\mathbf{BC}) = (\mathbf{AB})\mathbf{C}.$ Associative property of multiplication

$\mathbf{A}(\mathbf{B} + \mathbf{C}) = \mathbf{AB} + \mathbf{AC}.$ Distributive property

$(\mathbf{B} + \mathbf{C})\mathbf{A} = \mathbf{BA} + \mathbf{CA}.$ Distributive property

Matrix Equations

We can write a matrix equation equivalent to a system of equations.

EXAMPLE 8 Write a matrix equation equivalent to the following system of equations:

$$\begin{aligned} 4x + 2y - z &= 3, \\ 9x \quad\quad + z &= 5, \\ 4x + 5y - 2z &= 1. \end{aligned}$$

Solution We write the coefficients on the left in a matrix. We then write the product of that matrix and the column matrix containing the variables and set the result equal to the column matrix containing the constants on the right:

$$\begin{bmatrix} 4 & 2 & -1 \\ 9 & 0 & 1 \\ 4 & 5 & -2 \end{bmatrix} \begin{bmatrix} x \\ y \\ z \end{bmatrix} = \begin{bmatrix} 3 \\ 5 \\ 1 \end{bmatrix}.$$

If we let

$$\mathbf{A} = \begin{bmatrix} 4 & 2 & -1 \\ 9 & 0 & 1 \\ 4 & 5 & -2 \end{bmatrix}, \quad \mathbf{X} = \begin{bmatrix} x \\ y \\ z \end{bmatrix}, \quad \text{and} \quad \mathbf{B} = \begin{bmatrix} 3 \\ 5 \\ 1 \end{bmatrix},$$

we can write this matrix equation as $\mathbf{AX} = \mathbf{B}$.

Now Try Exercise 39.

6.4 Exercise Set

Find x and y.

1. $[5 \quad x] = [y \quad -3]$

2. $\begin{bmatrix} 6x \\ 25 \end{bmatrix} = \begin{bmatrix} -9 \\ 5y \end{bmatrix}$

3. $\begin{bmatrix} 3 & 2x \\ y & -8 \end{bmatrix} = \begin{bmatrix} 3 & -2 \\ 1 & -8 \end{bmatrix}$

4. $\begin{bmatrix} x - 1 & 4 \\ y + 3 & -7 \end{bmatrix} = \begin{bmatrix} 0 & 4 \\ -2 & -7 \end{bmatrix}$

For Exercises 5–20, let

$$\mathbf{A} = \begin{bmatrix} 1 & 2 \\ 4 & 3 \end{bmatrix}, \quad \mathbf{B} = \begin{bmatrix} -3 & 5 \\ 2 & -1 \end{bmatrix},$$

$$\mathbf{C} = \begin{bmatrix} 1 & -1 \\ -1 & 1 \end{bmatrix}, \quad \mathbf{D} = \begin{bmatrix} 1 & 1 \\ 1 & 1 \end{bmatrix},$$

$$\mathbf{E} = \begin{bmatrix} 1 & 3 \\ 2 & 6 \end{bmatrix}, \quad \mathbf{F} = \begin{bmatrix} 3 & 3 \\ -1 & -1 \end{bmatrix},$$

$$\mathbf{0} = \begin{bmatrix} 0 & 0 \\ 0 & 0 \end{bmatrix}, \quad \mathbf{I} = \begin{bmatrix} 1 & 0 \\ 0 & 1 \end{bmatrix}.$$

Find each of the following.

5. $\mathbf{A} + \mathbf{B}$ **6.** $\mathbf{B} + \mathbf{A}$ **7.** $\mathbf{E} + \mathbf{0}$

8. $2\mathbf{A}$ **9.** $3\mathbf{F}$ **10.** $(-1)\mathbf{D}$

11. $3\mathbf{F} + 2\mathbf{A}$ **12.** $\mathbf{A} - \mathbf{B}$

13. $\mathbf{B} - \mathbf{A}$ **14.** $\mathbf{AB}$

15. $\mathbf{BA}$ **16.** $\mathbf{0F}$

17. $\mathbf{CD}$ **18.** $\mathbf{EF}$

19. $\mathbf{AI}$ **20.** $\mathbf{IA}$

Find the product, if possible.

21. $\begin{bmatrix} -1 & 0 & 7 \\ 3 & -5 & 2 \end{bmatrix} \begin{bmatrix} 6 \\ -4 \\ 1 \end{bmatrix}$

22. $\begin{bmatrix} 6 & -1 & 2 \end{bmatrix} \begin{bmatrix} 1 & 4 \\ -2 & 0 \\ 5 & -3 \end{bmatrix}$

23. $\begin{bmatrix} -2 & 4 \\ 5 & 1 \\ -1 & -3 \end{bmatrix} \begin{bmatrix} 3 & -6 \\ -1 & 4 \end{bmatrix}$

24. $\begin{bmatrix} 2 & -1 & 0 \\ 0 & 5 & 4 \end{bmatrix} \begin{bmatrix} -3 & 1 & 0 \\ 0 & 2 & -1 \\ 5 & 0 & 4 \end{bmatrix}$

25. $\begin{bmatrix} 1 \\ -5 \\ 3 \end{bmatrix} \begin{bmatrix} -6 & 5 & 8 \\ 0 & 4 & -1 \end{bmatrix}$

26. $\begin{bmatrix} 2 & 0 & 0 \\ 0 & -1 & 0 \\ 0 & 0 & 3 \end{bmatrix} \begin{bmatrix} 0 & -4 & 3 \\ 2 & 1 & 0 \\ -1 & 0 & 6 \end{bmatrix}$

27. $\begin{bmatrix} 1 & -4 & 3 \\ 0 & 8 & 0 \\ -2 & -1 & 5 \end{bmatrix} \begin{bmatrix} 3 & 0 & 0 \\ 0 & -4 & 0 \\ 0 & 0 & 1 \end{bmatrix}$

28. $\begin{bmatrix} 4 \\ -5 \end{bmatrix} \begin{bmatrix} 2 & 0 \\ 6 & -7 \\ 0 & -3 \end{bmatrix}$

29. *Produce.* The produce manager at Dugan's Market orders 40 lb of tomatoes, 20 lb of zucchini, and 30 lb of onions from a local farmer one week.

a) Write a 1×3 matrix **A** that represents the amount of each item ordered.

b) The following week, the produce manager increases his order by 10%. Find a matrix **B** that represents this order.

c) Find $\mathbf{A} + \mathbf{B}$ and tell what the entries represent.

30. *Budget.* For the month of June, Maggie budgets \$320 for food, \$140 for clothes, and \$80 for entertainment.

a) Write a 1×3 matrix **B** that represents the amount budgeted for each of these items.

b) After receiving a raise, Maggie increases the amount budgeted for each item in July by 5%. Find a matrix **R** that represents the new amounts.

c) Find $\mathbf{B} + \mathbf{R}$ and tell what the entries represent.

31. *Nutrition.* A 3-oz serving of roasted, skinless chicken breast contains 140 Cal, 27 g of protein, 3 g of fat, 13 mg of calcium, and 64 mg of sodium. One-half cup of potato salad contains 180 Cal, 4 g of protein, 11 g of fat, 24 mg of calcium, and 662 mg of sodium. One broccoli spear contains 50 Cal, 5 g of protein, 1 g of fat, 82 mg of calcium, and 20 mg of sodium. (*Source: Home and Garden Bulletin No. 72,* U.S. Government Printing Office, Washington, D.C. 20402)

a) Write 1×5 matrices **C**, **P**, and **B** that represent the nutritional values of each food.

b) Find $\mathbf{C} + 2\mathbf{P} + 3\mathbf{B}$ and tell what the entries represent.

32. *Nutrition.* One slice of cheese pizza contains 290 Cal, 15 g of protein, 9 g of fat, and 39 g of carbohydrates. One-half cup of gelatin dessert contains 70 Cal, 2 g of protein, 0 g of fat, and 17 g of carbohydrates. One cup of whole milk contains 150 Cal, 8 g of protein, 8 g of fat, and 11 g of carbohydrates. (*Source: Home and Garden Bulletin No. 72,* U.S. Government Printing Office, Washington, D.C. 20402)

a) Write 1×4 matrices **P**, **G**, and **M** that represent the nutritional values of each food.

b) Find $3\mathbf{P} + 2\mathbf{G} + 2\mathbf{M}$ and tell what the entries represent.

33. *Food Service Management.* The food service manager at a large hospital is concerned about maintaining reasonable food costs. The following table lists the cost per serving, in dollars, for items on four menus.

Menu	Meat	Potato	Vegetable	Salad	Dessert
1	1.50	0.30	0.36	0.45	0.64
2	1.55	0.28	0.48	0.57	0.75
3	1.62	0.52	0.65	0.38	0.53
4	1.70	0.43	0.40	0.42	0.68

On a particular day, a dietician orders 65 meals from menu 1, 48 from menu 2, 93 from menu 3, and 57 from menu 4.

a) Write the information in the table as a 4×5 matrix **M**.
b) Write a row matrix **N** that represents the number of each menu ordered.
c) Find the product **NM**.
d) State what the entries of **NM** represent.

34. *Food Service Management.* A college food service manager uses a table like the one below to list the number of units of ingredients, by weight, required for various menu items.

	White Cake	Bread	Coffee Cake	Sugar Cookies
Flour	1	2.5	0.75	0.5
Milk	0	0.5	0.25	0
Eggs	0.75	0.25	0.5	0.5
Butter	0.5	0	0.5	1

The cost per unit of each ingredient is 25 cents for flour, 34 cents for milk, 54 cents for eggs, and 83 cents for butter.

a) Write the information in the table as a 4×4 matrix **M**.
b) Write a row matrix **C** that represents the cost per unit of each ingredient.
c) Find the product **CM**.
d) State what the entries of **CM** represent.

35. *Production Cost.* Karin supplies two small campus coffee shops with homemade chocolate chip cookies, oatmeal cookies, and peanut butter cookies. The following table shows the number of each type of cookie, in dozens, that Karin sold in one week.

	Mugsy's Coffee Shop	The Coffee Club
Chocolate Chip	8	15
Oatmeal	6	10
Peanut Butter	4	3

Karin spends \$4 for the ingredients for one dozen chocolate chip cookies, \$2.50 for the ingredients for one dozen oatmeal cookies, and \$3 for the ingredients for one dozen peanut butter cookies.

a) Write the information in the table as a 3×2 matrix **S**.
b) Write a row matrix **C** that represents the cost, per dozen, of the ingredients for each type of cookie.
c) Find the product **CS**.
d) State what the entries of **CS** represent.

36. *Profit.* A manufacturer produces exterior plywood, interior plywood, and fiberboard, which are shipped to two distributors. The following table shows the number of units of each type of product that are shipped to each warehouse.

	Distributor 1	Distributor 2
Exterior Plywood	900	500
Interior Plywood	450	1000
Fiberboard	600	700

The profits from each unit of exterior plywood, interior plywood, and fiberboard are \$8, \$10, and \$7, respectively.

a) Write the information in the table as a 3×2 matrix **M**.
b) Write a row matrix **P** that represents the profit per unit of each type of product.
c) Find the product **PM**.
d) State what the entries of **PM** represent.

37. *Profit.* In Exercise 35, suppose that Karin's profits on one dozen chocolate chip, oatmeal, and peanut butter cookies are \$7.50, \$4.80, and \$6.25, respectively.

a) Write a row matrix **P** that represents this information.

b) Use the matrices **S** and **P** to find Karin's total profit from each coffee shop.

38. *Production Cost.* In Exercise 36, suppose that the manufacturer's production costs for each unit of exterior plywood, interior plywood, and fiberboard are \$20, \$25, and \$15, respectively.

a) Write a row matrix **C** that represents this information.

b) Use the matrices **M** and **C** to find the total production cost for the products shipped to each distributor.

Write a matrix equation equivalent to the system of equations.

39. $2x - 3y = 7,$
$x + 5y = -6$

40. $-x + y = 3,$
$5x - 4y = 16$

41. $x + y - 2z = 6,$
$3x - y + z = 7,$
$2x + 5y - 3z = 8$

42. $3x - y + z = 1,$
$x + 2y - z = 3,$
$4x + 3y - 2z = 11$

43. $3x - 2y + 4z = 17,$
$2x + y - 5z = 13$

44. $3x + 2y + 5z = 9,$
$4x - 3y + 2z = 10$

45. $-4w + x - y + 2z = 12,$
$w + 2x - y - z = 0,$
$-w + x + 4y - 3z = 1,$
$2w + 3x + 5y - 7z = 9$

46. $12w + 2x + 4y - 5z = 2,$
$-w + 4x - y + 12z = 5,$
$2w - x + 4y = 13,$
$2x + 10y + z = 5$

▶ Skill Maintenance

In Exercises 47–50:

a) *Find the vertex.*

b) *Find the axis of symmetry.*

c) *Determine whether there is a maximum or a minimum value and find that value.*

d) *Graph the function.* [3.3]

47. $f(x) = x^2 - x - 6$

48. $f(x) = 2x^2 - 5x - 3$

49. $f(x) = -x^2 - 3x + 2$

50. $f(x) = -3x^2 + 4x + 4$

▶ Synthesis

For Exercises 51–54, let

$$\mathbf{A} = \begin{bmatrix} -1 & 0 \\ 2 & 1 \end{bmatrix} \quad \textit{and} \quad \mathbf{B} = \begin{bmatrix} 1 & -1 \\ 0 & 2 \end{bmatrix}.$$

51. Show that
$$(\mathbf{A} + \mathbf{B})(\mathbf{A} - \mathbf{B}) \neq \mathbf{A}^2 - \mathbf{B}^2,$$
where
$$\mathbf{A}^2 = \mathbf{AA} \quad \text{and} \quad \mathbf{B}^2 = \mathbf{BB}.$$

52. Show that
$$(\mathbf{A} + \mathbf{B})(\mathbf{A} + \mathbf{B}) \neq \mathbf{A}^2 + 2\mathbf{AB} + \mathbf{B}^2.$$

53. Show that
$$(\mathbf{A} + \mathbf{B})(\mathbf{A} - \mathbf{B}) = \mathbf{A}^2 + \mathbf{BA} - \mathbf{AB} - \mathbf{B}^2.$$

54. Show that
$$(\mathbf{A} + \mathbf{B})(\mathbf{A} + \mathbf{B}) = \mathbf{A}^2 + \mathbf{BA} + \mathbf{AB} + \mathbf{B}^2.$$

Mid-Chapter Mixed Review

Determine whether the statement is true or false.

1. For a system of two linear equations in two variables, if the graphs of the equations are parallel lines, then the system of equations has infinitely many solutions. [6.1]

2. One of the properties of a matrix written in row-echelon form is that all the rows consisting entirely of 0's are at the bottom of the matrix. [6.3]

3. We can multiply two matrices only when the number of columns in the first matrix is equal to the number of rows in the second matrix. [6.4]

4. Addition of matrices is not commutative. [6.4]

Solve. [6.1], [6.2]

5. $2x + y = -4,$
$x = y - 5$

6. $x + y = 4,$
$y = 2 - x$

7. $2x - 3y = 8,$
$3x + 2y = -1$

8. $x - 3y = 1,$
$6y = 2x - 2$

9. $x + 2y + 3z = 4,$
$x - 2y + z = 2,$
$2x - 6y + 4z = 7$

10. *e-Commerce.* computerwarehouse.com charges \$8 to ship orders up to 10 lb, \$12 for orders from 10 lb up to 15 lb, and \$15 for orders of 15 lb or more. One day, shipping charges for 150 orders totaled \$1620. The number of orders under 10 lb was three times the number of orders weighing 15 lb or more. Find the number of packages shipped at each rate. [6.2]

Solve the system of equations using Gaussian elimination or Gauss–Jordan elimination. [6.3]

11. $2x + y = 5,$
$3x + 2y = 6$

12. $3x + 2y - 3z = -2,$
$2x + 3y + 2z = -2,$
$x + 4y + 4z = 1$

For Exercises 13–20, let

$$\mathbf{A} = \begin{bmatrix} 3 & -1 \\ 5 & 4 \end{bmatrix}, \quad \mathbf{B} = \begin{bmatrix} -2 & 6 \\ 1 & -3 \end{bmatrix}, \quad \mathbf{C} = \begin{bmatrix} -4 & 1 & -1 \\ 2 & 3 & -2 \end{bmatrix}, \quad \textit{and} \quad \mathbf{D} = \begin{bmatrix} -2 & 3 & 0 \\ 1 & -1 & 2 \\ -3 & 4 & 1 \end{bmatrix}.$$

Find each of the following. [6.4]

13. $\mathbf{A} + \mathbf{B}$

14. $\mathbf{B} - \mathbf{A}$

15. $4\mathbf{D}$

16. $2\mathbf{A} + 3\mathbf{B}$

17. $\mathbf{AB}$

18. $\mathbf{BA}$

19. $\mathbf{BC}$

20. $\mathbf{DC}$

21. Write a matrix equation equivalent to the following system of equations: [6.4]
$2x - y + 3z = 7,$
$x + 2y - z = 3,$
$3x - 4y + 2z = 5.$

Collaborative Discussion and Writing

22. Explain in your own words when the elimination method for solving a system of equations is preferable to the substitution method. [6.1]

23. Given two linear equations in three variables, $Ax + By + Cz = D$ and $Ex + Fy + Gz = H$, explain how you could find a third equation such that the system contains dependent equations. [6.2]

24. Explain in your own words why the following augmented matrix represents a system of dependent equations. [6.3]

$$\left[\begin{array}{rrr|r} 1 & -3 & 2 & -5 \\ 0 & 1 & -4 & 8 \\ 0 & 0 & 0 & 0 \end{array}\right]$$

25. Is it true that if $\mathbf{AB} = \mathbf{0}$, for matrices $\mathbf{A}$ and $\mathbf{B}$, then $\mathbf{A} = \mathbf{0}$ or $\mathbf{B} = \mathbf{0}$? Why or why not? [6.4]

6.5 Inverses of Matrices

- Find the inverse of a square matrix, if it exists.
- Use inverses of matrices to solve systems of equations.

In this section, we continue our study of matrix algebra, finding the **multiplicative inverse**, or simply **inverse**, of a square matrix, if it exists. Then we use such inverses to solve systems of equations.

The Identity Matrix

Recall that, for real numbers, $a \cdot 1 = 1 \cdot a = a$; 1 is the multiplicative identity. A multiplicative identity matrix is very similar to the number 1.

IDENTITY MATRIX

For any positive integer n, the $n \times n$ **identity matrix** is an $n \times n$ matrix with 1's on the main diagonal and 0's elsewhere and is denoted by

$$\mathbf{I} = \begin{bmatrix} 1 & 0 & 0 \cdots 0 \\ 0 & 1 & 0 \cdots 0 \\ 0 & 0 & 1 \cdots 0 \\ \vdots & \vdots & \vdots \quad \vdots \\ 0 & 0 & 0 \cdots 1 \end{bmatrix}$$

Then $\mathbf{AI} = \mathbf{IA} = \mathbf{A}$, for any $n \times n$ matrix $\mathbf{A}$.

EXAMPLE 1 For

$$\mathbf{A} = \begin{bmatrix} 4 & -7 \\ -3 & 2 \end{bmatrix} \quad \text{and} \quad \mathbf{I} = \begin{bmatrix} 1 & 0 \\ 0 & 1 \end{bmatrix},$$

find each of the following.

a) **AI** **b)** **IA**

Solution

a) $$\mathbf{AI} = \begin{bmatrix} 4 & -7 \\ -3 & 2 \end{bmatrix}\begin{bmatrix} 1 & 0 \\ 0 & 1 \end{bmatrix}$$

$$= \begin{bmatrix} 4\cdot 1 - 7\cdot 0 & 4\cdot 0 - 7\cdot 1 \\ -3\cdot 1 + 2\cdot 0 & -3\cdot 0 + 2\cdot 1 \end{bmatrix} = \begin{bmatrix} 4 & -7 \\ -3 & 2 \end{bmatrix} = \mathbf{A}$$

b) $$\mathbf{IA} = \begin{bmatrix} 1 & 0 \\ 0 & 1 \end{bmatrix}\begin{bmatrix} 4 & -7 \\ -3 & 2 \end{bmatrix}$$

$$= \begin{bmatrix} 1\cdot 4 + 0(-3) & 1(-7) + 0\cdot 2 \\ 0\cdot 4 + 1(-3) & 0(-7) + 1\cdot 2 \end{bmatrix}$$

$$= \begin{bmatrix} 4 & -7 \\ -3 & 2 \end{bmatrix} = \mathbf{A}$$

JUST IN TIME 2

The Inverse of a Matrix

Recall that for every nonzero real number a, there is a multiplicative inverse $1/a$, or a^{-1}, such that $a \cdot a^{-1} = a^{-1} \cdot a = 1$. The multiplicative inverse of a matrix behaves in a similar manner.

> **INVERSE OF A MATRIX**
>
> For an $n \times n$ matrix **A**, if there is a matrix $\mathbf{A}^{-1}$ for which $\mathbf{A}^{-1} \cdot \mathbf{A} = \mathbf{I} = \mathbf{A} \cdot \mathbf{A}^{-1}$, then $\mathbf{A}^{-1}$ is the **inverse** of **A**.

We read $\mathbf{A}^{-1}$ as "**A** inverse." Note that not every matrix has an inverse.

EXAMPLE 2 Verify that

$$\mathbf{B} = \begin{bmatrix} 4 & -3 \\ 3 & -2 \end{bmatrix} \quad \text{is the inverse of} \quad \mathbf{A} = \begin{bmatrix} -2 & 3 \\ -3 & 4 \end{bmatrix}.$$

Solution We show that **BA** = **I** = **AB**.

$$\mathbf{BA} = \begin{bmatrix} 4 & -3 \\ 3 & -2 \end{bmatrix}\begin{bmatrix} -2 & 3 \\ -3 & 4 \end{bmatrix} = \begin{bmatrix} 1 & 0 \\ 0 & 1 \end{bmatrix} = \mathbf{I}.$$

$$\mathbf{AB} = \begin{bmatrix} -2 & 3 \\ -3 & 4 \end{bmatrix}\begin{bmatrix} 4 & -3 \\ 3 & -2 \end{bmatrix} = \begin{bmatrix} 1 & 0 \\ 0 & 1 \end{bmatrix} = \mathbf{I}.$$

Now Try Exercise 1.

We can find the inverse of a square matrix, if it exists, by using row-equivalent operations as in the Gauss–Jordan elimination method. For example, consider the matrix

$$\mathbf{A} = \begin{bmatrix} -2 & 3 \\ -3 & 4 \end{bmatrix}.$$

To find its inverse, we first form an **augmented matrix** consisting of **A** on the left side and the 2×2 identity matrix on the right side:

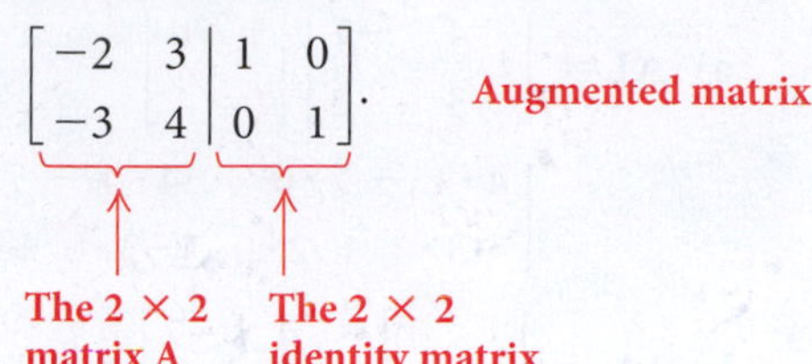

Then we attempt to transform the augmented matrix to one of the form

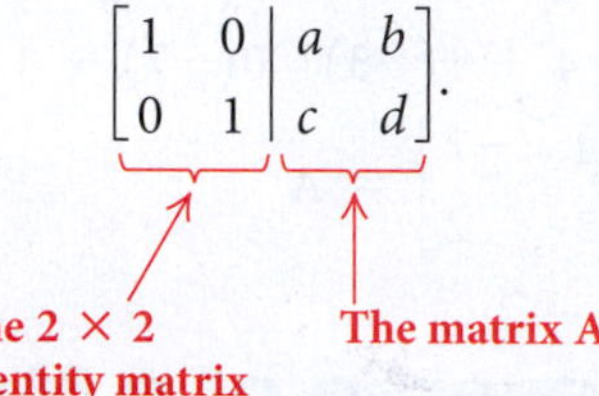

If we can do this, the matrix on the right, $\begin{bmatrix} a & b \\ c & d \end{bmatrix}$, is $\mathbf{A}^{-1}$.

EXAMPLE 3 Find $\mathbf{A}^{-1}$, where

$$\mathbf{A} = \begin{bmatrix} -2 & 3 \\ -3 & 4 \end{bmatrix}.$$

Solution First, we write the augmented matrix. Then we transform it to the desired form.

$$\left[\begin{array}{cc|cc} -2 & 3 & 1 & 0 \\ -3 & 4 & 0 & 1 \end{array}\right]$$

$$\left[\begin{array}{cc|cc} 1 & -\frac{3}{2} & -\frac{1}{2} & 0 \\ -3 & 4 & 0 & 1 \end{array}\right] \quad \text{New row 1} = -\tfrac{1}{2}(\text{row 1})$$

$$\left[\begin{array}{cc|cc} 1 & -\frac{3}{2} & -\frac{1}{2} & 0 \\ 0 & -\frac{1}{2} & -\frac{3}{2} & 1 \end{array}\right] \quad \text{New row 2} = 3(\text{row 1}) + \text{row 2}$$

$$\left[\begin{array}{cc|cc} 1 & -\frac{3}{2} & -\frac{1}{2} & 0 \\ 0 & 1 & 3 & -2 \end{array}\right] \quad \text{New row 2} = -2(\text{row 2})$$

$$\left[\begin{array}{cc|cc} 1 & 0 & 4 & -3 \\ 0 & 1 & 3 & -2 \end{array}\right] \quad \text{New row 1} = \tfrac{3}{2}(\text{row 2}) + \text{row 1}$$

Thus,

$$\mathbf{A}^{-1} = \begin{bmatrix} 4 & -3 \\ 3 & -2 \end{bmatrix},$$

which we verified in Example 2.

Now Try Exercise 5.

Technology Connection

The x^{-1} key on a graphing calculator can also be used to find the inverse of a matrix like the one in Example 3.

```
[A]⁻¹
                [[4  -3]
                 [3  -2]]
```

EXAMPLE 4 Find $\mathbf{A}^{-1}$, where

$$\mathbf{A} = \begin{bmatrix} 1 & 2 & -1 \\ 3 & 5 & 3 \\ 2 & 4 & 3 \end{bmatrix}.$$

Solution First, we write the augmented matrix. Then we transform it to the desired form.

$$\left[\begin{array}{ccc|ccc} 1 & 2 & -1 & 1 & 0 & 0 \\ 3 & 5 & 3 & 0 & 1 & 0 \\ 2 & 4 & 3 & 0 & 0 & 1 \end{array}\right]$$

$$\left[\begin{array}{ccc|ccc} 1 & 2 & -1 & 1 & 0 & 0 \\ 0 & -1 & 6 & -3 & 1 & 0 \\ 0 & 0 & 5 & -2 & 0 & 1 \end{array}\right] \quad \begin{array}{l} \text{New row 2} = -3(\text{row 1}) + \text{row 2} \\ \text{New row 3} = -2(\text{row 1}) + \text{row 3} \end{array}$$

$$\left[\begin{array}{ccc|ccc} 1 & 2 & -1 & 1 & 0 & 0 \\ 0 & -1 & 6 & -3 & 1 & 0 \\ 0 & 0 & 1 & -\frac{2}{5} & 0 & \frac{1}{5} \end{array}\right] \quad \text{New row 3} = \tfrac{1}{5}(\text{row 3})$$

$$\left[\begin{array}{ccc|ccc} 1 & 2 & 0 & \frac{3}{5} & 0 & \frac{1}{5} \\ 0 & -1 & 0 & -\frac{3}{5} & 1 & -\frac{6}{5} \\ 0 & 0 & 1 & -\frac{2}{5} & 0 & \frac{1}{5} \end{array}\right] \quad \begin{array}{l} \text{New row 1} = \text{row 3} + \text{row 1} \\ \text{New row 2} = -6(\text{row 3}) + \text{row 2} \end{array}$$

$$\left[\begin{array}{ccc|ccc} 1 & 0 & 0 & -\frac{3}{5} & 2 & -\frac{11}{5} \\ 0 & -1 & 0 & -\frac{3}{5} & 1 & -\frac{6}{5} \\ 0 & 0 & 1 & -\frac{2}{5} & 0 & \frac{1}{5} \end{array}\right] \quad \text{New row 1} = 2(\text{row 2}) + \text{row 1}$$

$$\left[\begin{array}{ccc|ccc} 1 & 0 & 0 & -\frac{3}{5} & 2 & -\frac{11}{5} \\ 0 & 1 & 0 & \frac{3}{5} & -1 & \frac{6}{5} \\ 0 & 0 & 1 & -\frac{2}{5} & 0 & \frac{1}{5} \end{array}\right] \quad \text{New row 2} = -1(\text{row 2})$$

Thus,

$$\mathbf{A}^{-1} = \begin{bmatrix} -\frac{3}{5} & 2 & -\frac{11}{5} \\ \frac{3}{5} & -1 & \frac{6}{5} \\ -\frac{2}{5} & 0 & \frac{1}{5} \end{bmatrix}.$$

Now Try Exercise 11.

Technology Connection

When we try to find the inverse of a singular matrix using a graphing calculator, the calculator returns an error message similar to ERR: SINGULAR MATRIX.

If a matrix has an inverse, we say that it is **invertible**, or **nonsingular**. When we cannot obtain the identity matrix on the left using the Gauss–Jordan method, then no inverse exists. This occurs when we obtain a row consisting entirely of 0's in either of the two matrices in the augmented matrix. In this case, we say that **A** is a **singular matrix**.

Solving Systems of Equations

MATRIX EQUATIONS
REVIEW SECTION 6.4

We can write a system of n linear equations in n variables as a matrix equation $\mathbf{AX} = \mathbf{B}$. If **A** has an inverse, then the system of equations has a unique solution that can be found by solving for **X**, as follows:

$$\begin{aligned} \mathbf{AX} &= \mathbf{B} \\ \mathbf{A}^{-1}(\mathbf{AX}) &= \mathbf{A}^{-1}\mathbf{B} && \text{Multiplying by } \mathbf{A}^{-1} \text{ on the left on both sides} \\ (\mathbf{A}^{-1}\mathbf{A})\mathbf{X} &= \mathbf{A}^{-1}\mathbf{B} && \text{Using the associative property of matrix multiplication} \\ \mathbf{IX} &= \mathbf{A}^{-1}\mathbf{B} && \mathbf{A}^{-1}\mathbf{A} = \mathbf{I} \\ \mathbf{X} &= \mathbf{A}^{-1}\mathbf{B}. && \mathbf{IX} = \mathbf{X} \end{aligned}$$

MATRIX SOLUTIONS OF SYSTEMS OF EQUATIONS

For a system of n linear equations in n variables, $\mathbf{AX} = \mathbf{B}$, if $\mathbf{A}$ is an invertible matrix, then the unique solution of the system is given by

$$\mathbf{X} = \mathbf{A}^{-1}\mathbf{B}.$$

Since matrix multiplication is not commutative in general, care must be taken to multiply *on the left* by $\mathbf{A}^{-1}$.

EXAMPLE 5 Use an inverse matrix to solve the following system of equations:

$$-2x + 3y = 4,$$
$$-3x + 4y = 5.$$

Solution We write an equivalent matrix equation, $\mathbf{AX} = \mathbf{B}$:

$$\underset{\mathbf{A}}{\begin{bmatrix} -2 & 3 \\ -3 & 4 \end{bmatrix}} \cdot \underset{\mathbf{X}}{\begin{bmatrix} x \\ y \end{bmatrix}} = \underset{\mathbf{B}}{\begin{bmatrix} 4 \\ 5 \end{bmatrix}}$$

In Example 3, we found that

$$\mathbf{A}^{-1} = \begin{bmatrix} 4 & -3 \\ 3 & -2 \end{bmatrix}.$$

We also verified this in Example 2. Now we have

$$\mathbf{X} = \mathbf{A}^{-1}\mathbf{B}$$

$$\begin{bmatrix} x \\ y \end{bmatrix} = \begin{bmatrix} 4 & -3 \\ 3 & -2 \end{bmatrix}\begin{bmatrix} 4 \\ 5 \end{bmatrix} = \begin{bmatrix} 1 \\ 2 \end{bmatrix}.$$

The solution of the system of equations is $(1, 2)$. Now Try Exercise 25.

Technology Connection

To use a graphing calculator to solve the system of equations in Example 5, we enter $\mathbf{A}$ and $\mathbf{B}$ and then enter the notation $\mathbf{A}^{-1}\mathbf{B}$ on the home screen.

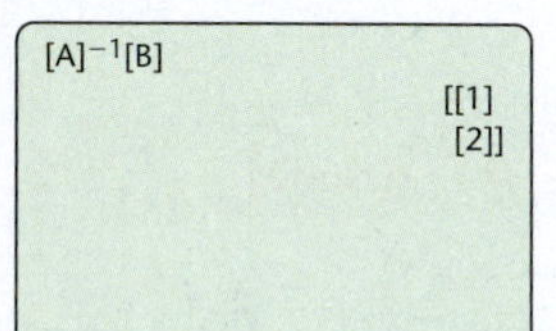

6.5 Exercise Set

Determine whether $\mathbf{B}$ *is the inverse of* $\mathbf{A}$.

1. $\mathbf{A} = \begin{bmatrix} 1 & -3 \\ -2 & 7 \end{bmatrix}$, $\mathbf{B} = \begin{bmatrix} 7 & 3 \\ 2 & 1 \end{bmatrix}$

2. $\mathbf{A} = \begin{bmatrix} 3 & 2 \\ 4 & 3 \end{bmatrix}$, $\mathbf{B} = \begin{bmatrix} 3 & -2 \\ -4 & 3 \end{bmatrix}$

3. $\mathbf{A} = \begin{bmatrix} -1 & -1 & 6 \\ 1 & 0 & -2 \\ 1 & 0 & -3 \end{bmatrix}$, $\mathbf{B} = \begin{bmatrix} 2 & 3 & 2 \\ 3 & 3 & 4 \\ 1 & 1 & 1 \end{bmatrix}$

4. $\mathbf{A} = \begin{bmatrix} -2 & 0 & -3 \\ 5 & 1 & 7 \\ -3 & 0 & 4 \end{bmatrix}$, $\mathbf{B} = \begin{bmatrix} 4 & 0 & -3 \\ 1 & 1 & 1 \\ -3 & 0 & 2 \end{bmatrix}$

Use the Gauss–Jordan method to find $\mathbf{A}^{-1}$, *if it exists. Check your answers by finding* $\mathbf{A}^{-1}\mathbf{A}$ *and* $\mathbf{A}\mathbf{A}^{-1}$.

5. $\mathbf{A} = \begin{bmatrix} 3 & 2 \\ 5 & 3 \end{bmatrix}$

6. $\mathbf{A} = \begin{bmatrix} 3 & 5 \\ 1 & 2 \end{bmatrix}$

7. $\mathbf{A} = \begin{bmatrix} 6 & 9 \\ 4 & 6 \end{bmatrix}$

8. $\mathbf{A} = \begin{bmatrix} -4 & -6 \\ 2 & 3 \end{bmatrix}$

9. $\mathbf{A} = \begin{bmatrix} 4 & -3 \\ 1 & -2 \end{bmatrix}$

10. $\mathbf{A} = \begin{bmatrix} 0 & -1 \\ 1 & 0 \end{bmatrix}$

11. $\mathbf{A} = \begin{bmatrix} 3 & 1 & 0 \\ 1 & 1 & 1 \\ 1 & -1 & 2 \end{bmatrix}$

12. $\mathbf{A} = \begin{bmatrix} 1 & 0 & 1 \\ 2 & 1 & 0 \\ 1 & -1 & 1 \end{bmatrix}$

13. $\mathbf{A} = \begin{bmatrix} 1 & -4 & 8 \\ 1 & -3 & 2 \\ 2 & -7 & 10 \end{bmatrix}$

14. $\mathbf{A} = \begin{bmatrix} -2 & 5 & 3 \\ 4 & -1 & 3 \\ 7 & -2 & 5 \end{bmatrix}$

15. $\mathbf{A} = \begin{bmatrix} 2 & 3 & 2 \\ 3 & 3 & 4 \\ -1 & -1 & -1 \end{bmatrix}$

16. $\mathbf{A} = \begin{bmatrix} 1 & 2 & 3 \\ 2 & -1 & -2 \\ -1 & 3 & 3 \end{bmatrix}$

17. $\mathbf{A} = \begin{bmatrix} 1 & 2 & -1 \\ -2 & 0 & 1 \\ 1 & -1 & 0 \end{bmatrix}$

18. $\mathbf{A} = \begin{bmatrix} 7 & -1 & -9 \\ 2 & 0 & -4 \\ -4 & 0 & 6 \end{bmatrix}$

19. $\mathbf{A} = \begin{bmatrix} 1 & 3 & -1 \\ 0 & 2 & -1 \\ 1 & 1 & 0 \end{bmatrix}$

20. $\mathbf{A} = \begin{bmatrix} -1 & 0 & -1 \\ -1 & 1 & 0 \\ 0 & 1 & 1 \end{bmatrix}$

21. $\mathbf{A} = \begin{bmatrix} 1 & 2 & 3 & 4 \\ 0 & 1 & 3 & -5 \\ 0 & 0 & 1 & -2 \\ 0 & 0 & 0 & -1 \end{bmatrix}$

22. $\mathbf{A} = \begin{bmatrix} -2 & -3 & 4 & 1 \\ 0 & 1 & 1 & 0 \\ 0 & 4 & -6 & 1 \\ -2 & -2 & 5 & 1 \end{bmatrix}$

23. $\mathbf{A} = \begin{bmatrix} 1 & -14 & 7 & 38 \\ -1 & 2 & 1 & -2 \\ 1 & 2 & -1 & -6 \\ 1 & -2 & 3 & 6 \end{bmatrix}$

24. $\mathbf{A} = \begin{bmatrix} 10 & 20 & -30 & 15 \\ 3 & -7 & 14 & -8 \\ -7 & -2 & -1 & 2 \\ 4 & 4 & -3 & 1 \end{bmatrix}$

In Exercises 25–28, a system of equations is given, together with the inverse of the coefficient matrix. Use the inverse of the coefficient matrix to solve the system of equations.

25. $11x + 3y = -4,$
$7x + 2y = 5;$ $\quad \mathbf{A}^{-1} = \begin{bmatrix} 2 & -3 \\ -7 & 11 \end{bmatrix}$

26. $8x + 5y = -6,$
$5x + 3y = 2;$ $\quad \mathbf{A}^{-1} = \begin{bmatrix} -3 & 5 \\ 5 & -8 \end{bmatrix}$

27. $3x + y = 2,$
$2x - y + 2z = -5,$
$x + y + z = 5;$ $\quad \mathbf{A}^{-1} = \frac{1}{9}\begin{bmatrix} 3 & 1 & -2 \\ 0 & -3 & 6 \\ -3 & 2 & 5 \end{bmatrix}$

28. $y - z = -4,$
$4x + y = -3,$
$3x - y + 3z = 1;$ $\quad \mathbf{A}^{-1} = \frac{1}{5}\begin{bmatrix} -3 & 2 & -1 \\ 12 & -3 & 4 \\ 7 & -3 & 4 \end{bmatrix}$

Solve the system of equations using the inverse of the coefficient matrix of the equivalent matrix equation.

29. $4x + 3y = 2,$
$x - 2y = 6$

30. $2x - 3y = 7,$
$4x + y = -7$

31. $5x + y = 2,$
$3x - 2y = -4$

32. $x - 6y = 5,$
$-x + 4y = -5$

33. $x + z = 1,$
$2x + y = 3,$
$x - y + z = 4$

34. $x + 2y + 3z = -1,$
$2x - 3y + 4z = 2,$
$-3x + 5y - 6z = 4$

35. $2x + 3y + 4z = 2,$
$x - 4y + 3z = 2,$
$5x + y + z = -4$

36. $x + y = 2,$
$3x + 2z = 5,$
$2x + 3y - 3z = 9$

37. $2w - 3x + 4y - 5z = 0,$
$3w - 2x + 7y - 3z = 2,$
$w + x - y + z = 1,$
$-w - 3x - 6y + 4z = 6$

38. $5w - 4x + 3y - 2z = -6,$
$w + 4x - 2y + 3z = -5,$
$2w - 3x + 6y - 9z = 14,$
$3w - 5x + 2y - 4z = -3$

39. *Cranberry Production.* In 2012, a total of 660 million lb of cranberries were produced in Wisconsin and Massachusetts. Wisconsin grew 240 million lb more than Massachusetts grew. (*Source:* U.S. Department of Agriculture) How many pounds of cranberries were grown in each state?

40. *Lunch Cost.* Coworkers Jan and Richard purchase lunch from a food truck. Jan buys 1 beef taco and 2 fruit cups for \$5.25. Richard buys 3 beef tacos and 1 fruit cup for \$8.25. Find the price of each item.

41. *Landscaping Cost.* Green-Up Landscaping bought 4 tons of topsoil, 3 tons of mulch, and 6 tons of pea gravel for \$2825. The next week, the firm bought 5 tons of topsoil, 2 tons of mulch, and 5 tons of pea gravel for \$2663. Pea gravel costs \$17 less per ton than topsoil. Find the price per ton for each item.

42. *Investment.* Trevor receives \$230 per year in simple interest from three investments totaling \$8500. Part is invested at 2.2%, part at 2.65%, and the rest at 3.05%. There is \$1500 more invested at 3.05% than at 2.2%. Find the amount invested at each rate.

Skill Maintenance

Use synthetic division to find the function values.

43. $f(x) = x^3 - 6x^2 + 4x - 8$; find $f(-2)$ [4.3]

44. $f(x) = 2x^4 - x^3 + 5x^2 + 6x - 4$; find $f(3)$ [4.3]

Solve.

45. $2x^2 + x = 7$ [3.2]

46. $\frac{1}{x+1} - \frac{6}{x-1} = 1$ [3.4]

47. $\sqrt{2x+1} - 1 = \sqrt{2x-4}$ [3.4]

48. $x - \sqrt{x} - 6 = 0$ [3.4]

Factor the polynomial $f(x)$. [4.3]

49. $f(x) = x^3 - 3x^2 - 6x + 8$

50. $f(x) = x^4 + 2x^3 - 16x^2 - 2x + 15$

Synthesis

State the conditions under which $\mathbf{A}^{-1}$ exists. Then find a formula for $\mathbf{A}^{-1}$.

51. $\mathbf{A} = [x]$

52. $\mathbf{A} = \begin{bmatrix} x & 0 \\ 0 & y \end{bmatrix}$

53. $\mathbf{A} = \begin{bmatrix} 0 & 0 & x \\ 0 & y & 0 \\ z & 0 & 0 \end{bmatrix}$

54. $\mathbf{A} = \begin{bmatrix} x & 1 & 1 & 1 \\ 0 & y & 0 & 0 \\ 0 & 0 & z & 0 \\ 0 & 0 & 0 & w \end{bmatrix}$

6.6 Determinants and Cramer's Rule

- Evaluate determinants of square matrices.
- Use Cramer's rule to solve systems of equations.

Determinants of Square Matrices

With every square matrix, we associate a number called its *determinant*.

DETERMINANT OF A 2 × 2 MATRIX

The **determinant** of the matrix $\begin{bmatrix} a & c \\ b & d \end{bmatrix}$ is denoted $\begin{vmatrix} a & c \\ b & d \end{vmatrix}$ and is defined as

$$\begin{vmatrix} a & c \\ b & d \end{vmatrix} = ad - bc.$$

JUST IN TIME 22

EXAMPLE 1 Evaluate: $\begin{vmatrix} \sqrt{2} & -3 \\ -4 & -\sqrt{2} \end{vmatrix}$.

Solution

$$\begin{vmatrix} \sqrt{2} & -3 \\ -4 & -\sqrt{2} \end{vmatrix}$$

The arrows indicate the products involved.

$$= \sqrt{2}(-\sqrt{2}) - (-4)(-3)$$
$$= -2 - 12$$
$$= -14$$

Now Try Exercise 1.

We now consider a way to evaluate determinants of square matrices of order 3×3 or higher.

Evaluating Determinants Using Cofactors

Often we first find minors and cofactors of matrices in order to evaluate determinants.

> **MINOR**
>
> For a square matrix $\mathbf{A} = [a_{ij}]$, the **minor** M_{ij} of an entry a_{ij} is the determinant of the matrix formed by deleting the ith row and the jth column of $\mathbf{A}$.

EXAMPLE 2 For the matrix

$$\mathbf{A} = [a_{ij}] = \begin{bmatrix} -8 & 0 & 6 \\ 4 & -6 & 7 \\ -1 & -3 & 5 \end{bmatrix},$$

find each of the following.

a) M_{11} b) M_{23}

Solution

a) For M_{11}, we delete the first row and the first column and find the determinant of the 2×2 matrix formed by the remaining entries.

$$\begin{bmatrix} -8 & 0 & 6 \\ 4 & -6 & 7 \\ -1 & -3 & 5 \end{bmatrix} \qquad M_{11} = \begin{vmatrix} -6 & 7 \\ -3 & 5 \end{vmatrix}$$
$$= (-6) \cdot 5 - (-3) \cdot 7$$
$$= -30 - (-21)$$
$$= -30 + 21$$
$$= -9$$

b) For M_{23}, we delete the second row and the third column and find the determinant of the 2×2 matrix formed by the remaining entries.

$$\begin{bmatrix} -8 & 0 & 6 \\ 4 & -6 & 7 \\ -1 & -3 & 5 \end{bmatrix} \qquad M_{23} = \begin{vmatrix} -8 & 0 \\ -1 & -3 \end{vmatrix}$$
$$= -8(-3) - (-1)0$$
$$= 24$$

Now Try Exercise 9.

COFACTOR

For a square matrix $\mathbf{A} = [a_{ij}]$, the **cofactor** A_{ij} of an entry a_{ij} is given by

$$A_{ij} = (-1)^{i+j} M_{ij},$$

where M_{ij} is the minor of a_{ij}.

EXAMPLE 3 For the matrix given in Example 2, find each of the following.

a) A_{11} **b)** A_{23}

Solution

a) In Example 2, we found that $M_{11} = -9$. Then

$$A_{11} = (-1)^{1+1}(-9) = (1)(-9) = -9.$$

b) In Example 2, we found that $M_{23} = 24$. Then

$$A_{23} = (-1)^{2+3}(24) = (-1)(24) = -24.$$

Now Try Exercise 11.

Note that minors and cofactors are *numbers*. They are *not matrices*.

Consider the matrix $\mathbf{A}$ given by

$$\mathbf{A} = \begin{bmatrix} a_{11} & a_{12} & a_{13} \\ a_{21} & a_{22} & a_{23} \\ a_{31} & a_{32} & a_{33} \end{bmatrix}.$$

The determinant of the matrix, denoted $|\mathbf{A}|$, can be found by multiplying each element of the first column by its cofactor and adding:

$$|\mathbf{A}| = a_{11}A_{11} + a_{21}A_{21} + a_{31}A_{31}.$$

Because

$$A_{11} = (-1)^{1+1}M_{11} = M_{11},$$
$$A_{21} = (-1)^{2+1}M_{21} = -M_{21},$$

and $A_{31} = (-1)^{3+1}M_{31} = M_{31},$

we can write

$$|\mathbf{A}| = a_{11} \cdot \begin{vmatrix} a_{22} & a_{23} \\ a_{32} & a_{33} \end{vmatrix} - a_{21} \cdot \begin{vmatrix} a_{12} & a_{13} \\ a_{32} & a_{33} \end{vmatrix} + a_{31} \cdot \begin{vmatrix} a_{12} & a_{13} \\ a_{22} & a_{23} \end{vmatrix}.$$

It can be shown that we can determine $|\mathbf{A}|$ by choosing *any* row or column, multiplying each element in that row or column by its cofactor, and adding. This is called *expanding* across a row or down a column. We just expanded down the first column. We now define the determinant of a square matrix of any order.

DETERMINANT OF ANY SQUARE MATRIX

For any square matrix **A** of order $n \times n$ $(n > 1)$, we define the **determinant** of **A**, denoted $|\mathbf{A}|$, as follows. Choose any row or column. Multiply each element in that row or column by its cofactor and add the results. The determinant of a 1×1 matrix is simply the element of the matrix. The value of a determinant will be the same no matter which row or column is chosen.

EXAMPLE 4 Evaluate $|\mathbf{A}|$ by expanding across the third row.

$$\mathbf{A} = \begin{bmatrix} -8 & 0 & 6 \\ 4 & -6 & 7 \\ -1 & -3 & 5 \end{bmatrix}$$

Solution We have

$$\begin{aligned} |\mathbf{A}| &= (-1)A_{31} + (-3)A_{32} + 5A_{33} \\ &= (-1)(-1)^{3+1} \cdot \begin{vmatrix} 0 & 6 \\ -6 & 7 \end{vmatrix} + (-3)(-1)^{3+2} \cdot \begin{vmatrix} -8 & 6 \\ 4 & 7 \end{vmatrix} \\ &\quad + 5(-1)^{3+3} \cdot \begin{vmatrix} -8 & 0 \\ 4 & -6 \end{vmatrix} \\ &= (-1) \cdot 1 \cdot [0 \cdot 7 - (-6)6] + (-3)(-1)[-8 \cdot 7 - 4 \cdot 6] \\ &\quad + 5 \cdot 1 \cdot [-8(-6) - 4 \cdot 0] \\ &= -[36] + 3[-80] + 5[48] \\ &= -36 - 240 + 240 = -36. \end{aligned}$$

The value of this determinant is -36 no matter which row or column we expand on.

Now Try Exercise 13.

Technology Connection

Determinants can be evaluated on a graphing calculator. After entering a matrix, we select the determinant operation from the MATRIX MATH menu and enter the name of the matrix. The calculator will return the value of the determinant of the matrix. For example, for

$$\mathbf{A} = \begin{bmatrix} 1 & 6 & -1 \\ -3 & -5 & 3 \\ 0 & 4 & 2 \end{bmatrix},$$

we have

```
det ([A])
                26
```

Cramer's Rule

Determinants can be used to solve systems of linear equations. Consider a system of two linear equations:

$$a_1x + b_1y = c_1,$$
$$a_2x + b_2y = c_2.$$

Solving this system using the elimination method, we obtain

$$x = \frac{c_1b_2 - c_2b_1}{a_1b_2 - a_2b_1} \quad \text{and} \quad y = \frac{a_1c_2 - a_2c_1}{a_1b_2 - a_2b_1}.$$

The numerators and the denominators of these expressions can be written as determinants:

$$x = \frac{\begin{vmatrix} c_1 & b_1 \\ c_2 & b_2 \end{vmatrix}}{\begin{vmatrix} a_1 & b_1 \\ a_2 & b_2 \end{vmatrix}} \quad \text{and} \quad y = \frac{\begin{vmatrix} a_1 & c_1 \\ a_2 & c_2 \end{vmatrix}}{\begin{vmatrix} a_1 & b_1 \\ a_2 & b_2 \end{vmatrix}}.$$

If we let

$$D = \begin{vmatrix} a_1 & b_1 \\ a_2 & b_2 \end{vmatrix}, \quad D_x = \begin{vmatrix} c_1 & b_1 \\ c_2 & b_2 \end{vmatrix}, \quad \text{and} \quad D_y = \begin{vmatrix} a_1 & c_1 \\ a_2 & c_2 \end{vmatrix},$$

we have

$$x = \frac{D_x}{D} \quad \text{and} \quad y = \frac{D_y}{D}.$$

This procedure for solving systems of equations is known as *Cramer's rule.*

> **CRAMER'S RULE FOR 2 × 2 SYSTEMS**
>
> The solution of the system of equations
>
> $$a_1x + b_1y = c_1,$$
> $$a_2x + b_2y = c_2$$
>
> is given by
>
> $$x = \frac{D_x}{D}, \qquad y = \frac{D_y}{D},$$
>
> where
>
> $$D = \begin{vmatrix} a_1 & b_1 \\ a_2 & b_2 \end{vmatrix}, \quad D_x = \begin{vmatrix} c_1 & b_1 \\ c_2 & b_2 \end{vmatrix}, \quad D_y = \begin{vmatrix} a_1 & c_1 \\ a_2 & c_2 \end{vmatrix}, \quad \text{and} \quad D \neq 0.$$

Technology Connection

To use Cramer's rule to solve the system of equations in Example 5 on a graphing calculator, we first enter the matrices corresponding to D, D_x, and D_y. We enter

$$\mathbf{A} = \begin{bmatrix} 2 & 5 \\ 5 & -2 \end{bmatrix},$$

$$\mathbf{B} = \begin{bmatrix} 7 & 5 \\ -3 & -2 \end{bmatrix},$$

and

$$\mathbf{C} = \begin{bmatrix} 2 & 7 \\ 5 & -3 \end{bmatrix}.$$

Then

$$x = \frac{\det(\mathbf{B})}{\det(\mathbf{A})}$$

and

$$y = \frac{\det(\mathbf{C})}{\det(\mathbf{A})}.$$

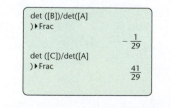

```
det ([B])/det([A]
)▸Frac
                    -1/29
det ([C])/det([A]
)▸Frac
                    41/29
```

Note that the denominator D contains the coefficients of x and y, in the same position as in the original equations. For x, the numerator is obtained by replacing the x-coefficients in D (the a's) with the c's. For y, the numerator is obtained by replacing the y-coefficients in D (the b's) with the c's.

EXAMPLE 5 Solve using Cramer's rule:

$$2x + 5y = 7,$$
$$5x - 2y = -3.$$

Solution We have

$$x = \frac{\begin{vmatrix} 7 & 5 \\ -3 & -2 \end{vmatrix}}{\begin{vmatrix} 2 & 5 \\ 5 & -2 \end{vmatrix}} = \frac{7(-2) - (-3)5}{2(-2) - 5 \cdot 5} = \frac{1}{-29} = -\frac{1}{29},$$

$$y = \frac{\begin{vmatrix} 2 & 7 \\ 5 & -3 \end{vmatrix}}{\begin{vmatrix} 2 & 5 \\ 5 & -2 \end{vmatrix}} = \frac{2(-3) - 5 \cdot 7}{-29} = \frac{-41}{-29} = \frac{41}{29}.$$

The solution is $\left(-\frac{1}{29}, \frac{41}{29}\right)$.

Now Try Exercise 29.

Cramer's rule works only when a system of equations has a unique solution. This occurs when $D \neq 0$. If $D = 0$ and D_x and D_y are also 0, then the equations are dependent. If $D = 0$ and D_x and/or D_y is not 0, then the system is inconsistent.

Cramer's rule can be extended to a system of n linear equations in n variables. We consider a 3×3 system.

CRAMER'S RULE FOR 3 × 3 SYSTEMS

The solution of the system of equations

$$a_1x + b_1y + c_1z = d_1,$$
$$a_2x + b_2y + c_2z = d_2,$$
$$a_3x + b_3y + c_3z = d_3$$

is given by

$$x = \frac{D_x}{D}, \quad y = \frac{D_y}{D}, \quad z = \frac{D_z}{D},$$

where

$$D = \begin{vmatrix} a_1 & b_1 & c_1 \\ a_2 & b_2 & c_2 \\ a_3 & b_3 & c_3 \end{vmatrix}, \quad D_x = \begin{vmatrix} d_1 & b_1 & c_1 \\ d_2 & b_2 & c_2 \\ d_3 & b_3 & c_3 \end{vmatrix},$$

$$D_y = \begin{vmatrix} a_1 & d_1 & c_1 \\ a_2 & d_2 & c_2 \\ a_3 & d_3 & c_3 \end{vmatrix}, \quad D_z = \begin{vmatrix} a_1 & b_1 & d_1 \\ a_2 & b_2 & d_2 \\ a_3 & b_3 & d_3 \end{vmatrix}, \text{ and } D \neq 0.$$

Note that the determinant D_x is obtained from D by replacing the x-coefficients with d_1, d_2, and d_3. D_y and D_z are obtained in a similar manner. As with a system of two equations, Cramer's rule cannot be used if $D = 0$. If $D = 0$ and D_x, D_y, and D_z are 0, then the equations are dependent. If $D = 0$ and one of D_x, D_y, or D_z is not 0, then the system is inconsistent.

EXAMPLE 6 Solve using Cramer's rule:

$$x - 3y + 7z = 13,$$
$$x + y + z = 1,$$
$$x - 2y + 3z = 4.$$

Solution We have

$$D = \begin{vmatrix} 1 & -3 & 7 \\ 1 & 1 & 1 \\ 1 & -2 & 3 \end{vmatrix} = -10, \quad D_x = \begin{vmatrix} 13 & -3 & 7 \\ 1 & 1 & 1 \\ 4 & -2 & 3 \end{vmatrix} = 20,$$

$$D_y = \begin{vmatrix} 1 & 13 & 7 \\ 1 & 1 & 1 \\ 1 & 4 & 3 \end{vmatrix} = -6, \quad D_z = \begin{vmatrix} 1 & -3 & 13 \\ 1 & 1 & 1 \\ 1 & -2 & 4 \end{vmatrix} = -24.$$

Then

$$x = \frac{D_x}{D} = \frac{20}{-10} = -2, \quad y = \frac{D_y}{D} = \frac{-6}{-10} = \frac{3}{5}, \quad z = \frac{D_z}{D} = \frac{-24}{-10} = \frac{12}{5}.$$

The solution is $\left(-2, \frac{3}{5}, \frac{12}{5}\right)$.

In practice, it is not necessary to evaluate D_z. When we have found values for x and y, we can substitute them into one of the equations to find z.

Now Try Exercise 37.

6.6 Exercise Set

Evaluate the determinant.

1. $\begin{vmatrix} 5 & 3 \\ -2 & -4 \end{vmatrix}$

2. $\begin{vmatrix} -8 & 6 \\ -1 & 2 \end{vmatrix}$

3. $\begin{vmatrix} 4 & -7 \\ -2 & 3 \end{vmatrix}$

4. $\begin{vmatrix} -9 & -6 \\ 5 & 4 \end{vmatrix}$

5. $\begin{vmatrix} -2 & -\sqrt{5} \\ -\sqrt{5} & 3 \end{vmatrix}$

6. $\begin{vmatrix} \sqrt{5} & -3 \\ 4 & 2 \end{vmatrix}$

7. $\begin{vmatrix} x & 4 \\ x & x^2 \end{vmatrix}$

8. $\begin{vmatrix} y^2 & -2 \\ y & 3 \end{vmatrix}$

Use the following matrix for Exercises 9–16:

$$\mathbf{A} = \begin{bmatrix} 7 & -4 & -6 \\ 2 & 0 & -3 \\ 1 & 2 & -5 \end{bmatrix}.$$

9. Find M_{11}, M_{32}, and M_{22}.

10. Find M_{13}, M_{31}, and M_{23}.

11. Find A_{11}, A_{32}, and A_{22}.

12. Find A_{13}, A_{31}, and A_{23}.

13. Evaluate $|\mathbf{A}|$ by expanding across the second row.

14. Evaluate $|\mathbf{A}|$ by expanding down the second column.

15. Evaluate $|\mathbf{A}|$ by expanding down the third column.

16. Evaluate $|\mathbf{A}|$ by expanding across the first row.

Use the following matrix for Exercises 17–22:

$$\mathbf{A} = \begin{bmatrix} 1 & 0 & 0 & -2 \\ 4 & 1 & 0 & 0 \\ 5 & 6 & 7 & 8 \\ -2 & -3 & -1 & 0 \end{bmatrix}.$$

17. Find M_{12} and M_{44}.

18. Find M_{41} and M_{33}.

19. Find A_{22} and A_{34}.

20. Find A_{24} and A_{43}.

21. Evaluate $|\mathbf{A}|$ by expanding across the first row.

22. Evaluate $|\mathbf{A}|$ by expanding down the third column.

Evaluate the determinant.

23. $\begin{vmatrix} 3 & 1 & 2 \\ -2 & 3 & 1 \\ 3 & 4 & -6 \end{vmatrix}$

24. $\begin{vmatrix} 3 & -2 & 1 \\ 2 & 4 & 3 \\ -1 & 5 & 1 \end{vmatrix}$

25. $\begin{vmatrix} x & 0 & -1 \\ 2 & x & x^2 \\ -3 & x & 1 \end{vmatrix}$

26. $\begin{vmatrix} x & 1 & -1 \\ x^2 & x & x \\ 0 & x & 1 \end{vmatrix}$

Solve using Cramer's rule.

27. $-2x + 4y = 3,$
$3x - 7y = 1$

28. $5x - 4y = -3,$
$7x + 2y = 6$

29. $2x - y = 5,$
$x - 2y = 1$

30. $3x + 4y = -2,$
$5x - 7y = 1$

31. $2x + 9y = -2,$
$4x - 3y = 3$

32. $2x + 3y = -1,$
$3x + 6y = -0.5$

33. $2x + 5y = 7,$
$3x - 2y = 1$

34. $3x + 2y = 7,$
$2x + 3y = -2$

35. $3x + 2y - z = 4,$
$3x - 2y + z = 5,$
$4x - 5y - z = -1$

36. $3x - y + 2z = 1,$
$x - y + 2z = 3,$
$-2x + 3y + z = 1$

37. $3x + 5y - z = -2,$
$x - 4y + 2z = 13,$
$2x + 4y + 3z = 1$

38. $3x + 2y + 2z = 1,$
$5x - y - 6z = 3,$
$2x + 3y + 3z = 4$

39. $x - 3y - 7z = 6,$
$2x + 3y + z = 9,$
$4x + y = 7$

40. $x - 2y - 3z = 4,$
$3x - 2z = 8,$
$2x + y + 4z = 13$

41. $6y + 6z = -1,$
$8x + 6z = -1,$
$4x + 9y = 8$

42. $3x + 5y = 2,$
$2x - 3z = 7,$
$4y + 2z = -1$

▶ Skill Maintenance

Determine whether the function is one-to-one, and if it is, find a formula for $f^{-1}(x)$. [5.1]

43. $f(x) = 3x + 2$

44. $f(x) = x^2 - 4$

45. $f(x) = |x| + 3$

46. $f(x) = \sqrt[3]{x} + 1$

Simplify. Write answers in the form $a + bi$, where a and b are real numbers. [3.1]

47. $(3 - 4i) - (-2 - i)$

48. $(5 + 2i) + (1 - 4i)$

49. $(1 - 2i)(6 + 2i)$

50. $\dfrac{3 + i}{4 - 3i}$

▶ Synthesis

Solve.

51. $\begin{vmatrix} y & 2 \\ 3 & y \end{vmatrix} = y$

52. $\begin{vmatrix} x & -3 \\ -1 & x \end{vmatrix} \geq 0$

53. $\begin{vmatrix} 2 & x & 1 \\ 1 & 2 & -1 \\ 3 & 4 & -2 \end{vmatrix} = -6$

54. $\begin{vmatrix} m + 2 & -3 \\ m + 5 & -4 \end{vmatrix} = 3m - 5$

Rewrite the expression using a determinant. Answers may vary.

55. $a^2 + b^2$

56. $\frac{1}{2}h(a + b)$

57. $2\pi r^2 + 2\pi rh$

58. $x^2y^2 - Q^2$

6.7 Systems of Inequalities and Linear Programming

- ▶ Graph linear inequalities.
- ▶ Graph systems of linear inequalities.
- ▶ Solve linear programming problems.

A graph of an inequality is a drawing that represents its solutions. We have already seen that an inequality in one variable can be graphed on the number line. An inequality in two variables can be graphed on a coordinate plane.

SOLVE LINEAR INEQUALITIES
REVIEW SECTION 1.6

▶ Graphs of Linear Inequalities

A statement like $5x - 4y < 20$ is a linear inequality in two variables.

LINEAR INEQUALITY IN TWO VARIABLES

A **linear inequality in two variables** is an inequality that can be written in the form

$$Ax + By < C,$$

where A, B, and C are real numbers and A and B are not both zero. The symbol $<$ may be replaced with $\leq$, $>$, or $\geq$.

A solution of a linear inequality in two variables is an ordered pair (x, y) for which the inequality is true. For example, $(1, 3)$ is a solution of $5x - 4y < 20$ because $5 \cdot 1 - 4 \cdot 3 < 20$, or $-7 < 20$, is true. On the other hand, $(2, -6)$ is not a solution of $5x - 4y < 20$ because $5 \cdot 2 - 4 \cdot (-6) \not< 20$, or $34 \not< 20$.

The **solution set** of an inequality is the set of all ordered pairs that make it true. The **graph of an inequality** represents its solution set.

EXAMPLE 1 Graph: $y < x + 3$.

Solution We begin by graphing the **related equation** $y = x + 3$. We use a dashed line because the inequality symbol is $<$. This indicates that the line itself is not in the solution set of the inequality.

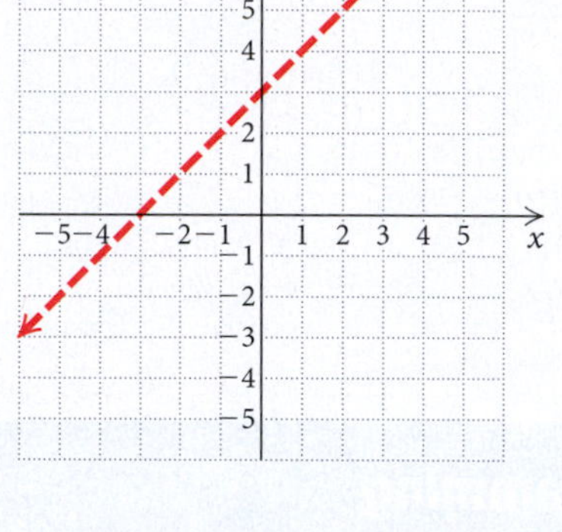

Note that the line divides the coordinate plane into two regions called **half-planes**. One of these half-planes satisfies the inequality. Either *all* points in a half-plane are in the solution set of the inequality or *none* is.

To determine which half-plane satisfies the inequality, we try a test point in either region. The point $(0, 0)$ is usually a convenient choice so long as it does not lie on the line.

$$\begin{array}{c|c} y & x + 3 \\ \hline 0 \; ? & 0 + 3 \\ 0 & 3 \end{array} \qquad \text{TRUE} \qquad 0 < 3 \text{ is true.}$$

Since $(0, 0)$ satisfies the inequality, so do all points in the half-plane that contains $(0, 0)$. We shade this region to show the solution set of the inequality.

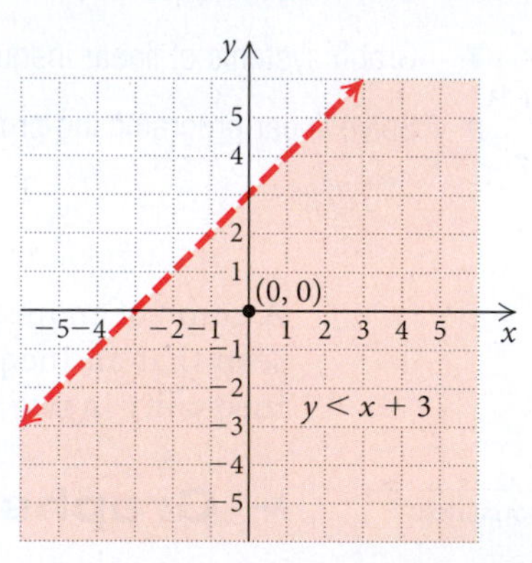

Now Try Exercise 13.

Technology Connection

One way to graph the inequality in Example 1 on a graphing calculator is to first enter the related equation, $y = x + 3$. Then select the "shade below" graph style. Note that we must keep in mind that the line $y = x + 3$ is not included in the solution set.

Some calculators have an application Inequalz on the APPS menu that can be used to graph an inequality. When this application is used, the inequality $y < x + 3$ is entered directly and the graph of the related equation appears as a dashed line.

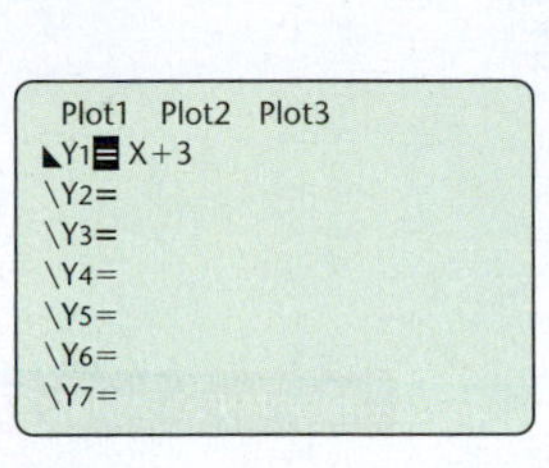

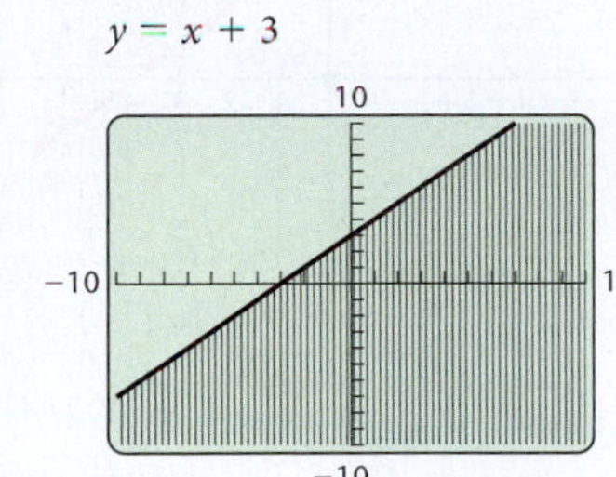

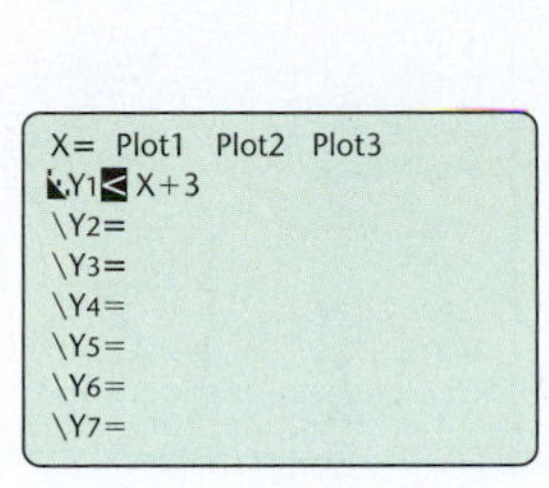

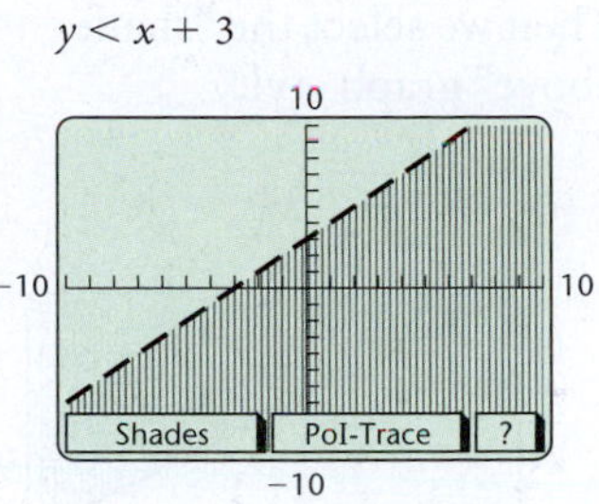

In general, we use the following procedure to graph linear inequalities in two variables by hand.

> To graph a linear inequality in two variables:
>
> 1. Replace the inequality symbol with an equals sign and graph this related equation. If the inequality symbol is $<$ or $>$, draw the line dashed. If the inequality symbol is $\leq$ or $\geq$, draw the line solid.
> 2. The graph consists of a half-plane on one side of the line and, if the line is solid, the line as well. To determine which half-plane to shade, test a point not on the line in the original inequality. If that point is a solution, shade the half-plane containing that point. If not, shade the opposite half-plane.

EXAMPLE 2 Graph: $3x + 4y \geq 12$.

Solution

1. First, we graph the related equation $3x + 4y = 12$. We use a solid line because the inequality symbol is $\geq$. This indicates that the line is included in the solution set.
2. To determine which half-plane to shade, we test a point in either region. We choose $(0, 0)$.

$$\begin{array}{r|l} 3x + 4y & \geq 12 \\ \hline 3\cdot 0 + 4\cdot 0 & ?\ 12 \\ 0 & 12 \end{array} \quad \text{FALSE} \qquad 0 \geq 12 \text{ is false.}$$

Technology Connection

To graph the inequality in Example 2 on a graphing calculator, we first solve the related equation for y and enter it in the form $y = \dfrac{-3x + 12}{4}$. Then we select the "shade above" graph style.

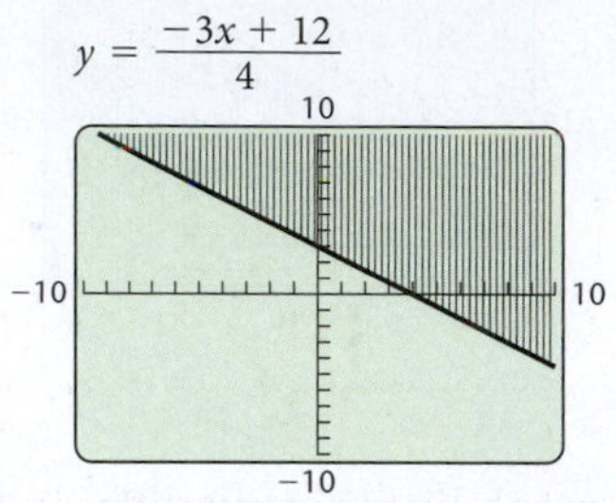

To use the Inequalz application on the APP menu, we solve the inequality for y and enter

$$y \geq \frac{-3x + 12}{4}.$$

Because $(0, 0)$ is *not* a solution, all the points in the half-plane that does *not* contain $(0, 0)$ are solutions. We shade that region, as shown in the following figure.

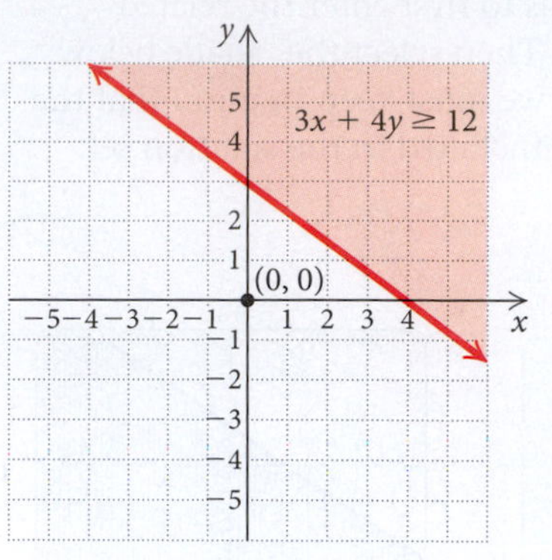

Now Try Exercise 17.

EXAMPLE 3 Graph $x > -3$ on a plane.

Solution

1. First, we graph the related equation $x = -3$. We use a dashed line because the inequality symbol is $>$. This indicates that the line is not included in the solution set.
2. The inequality tells us that all points (x, y) for which $x > -3$ are solutions. These are the points to the right of the line. We can also use a test point to determine the solutions. We choose $(5, 1)$.

$$\begin{array}{c|c} x > -3 & \\ \hline 5 \ ? \ -3 & \text{TRUE} \end{array} \qquad 5 > -3 \text{ is true.}$$

Because $(5, 1)$ is a solution, we shade the region containing that point—that is, the region to the right of the dashed line.

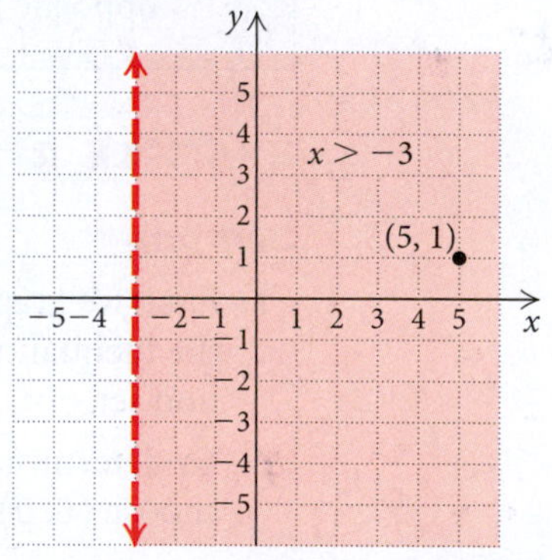

Now Try Exercise 23.

Technology Connection

We can graph inequalities, such as the one in Example 3, on a calculator that has the Inequalz application on the APPS menu.

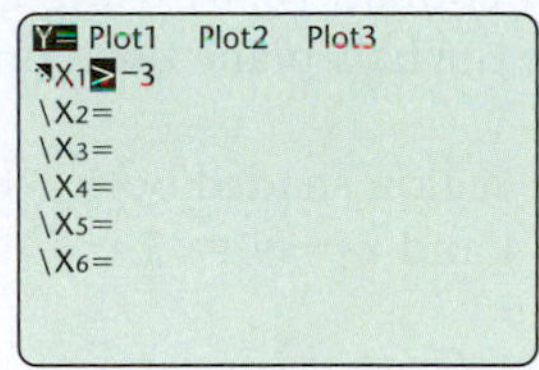

$x > -3$

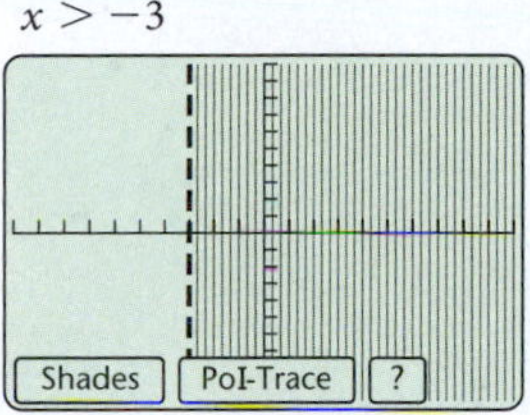

We can graph the inequality $y \leq 4$ in Example 4 using Inequalz or by first graphing $y = 4$ and then using the "shade below" graph style.

$y = 4$

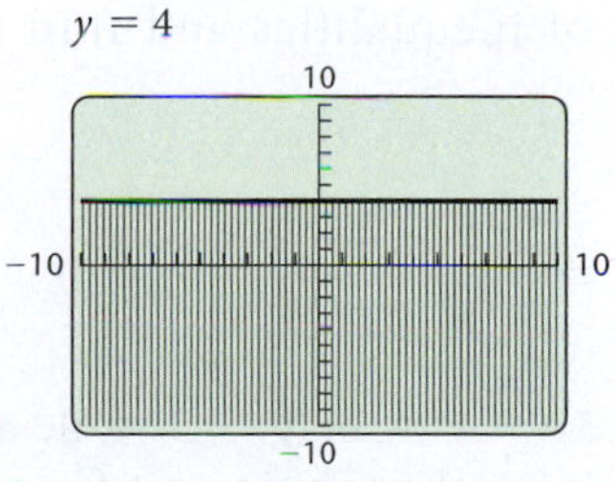

EXAMPLE 4 Graph $y \leq 4$ on a plane.

Solution

1. First, we graph the related equation $y = 4$. We use a solid line because the inequality symbol is $\leq$.
2. The inequality tells us that all points (x, y) for which $y \leq 4$ are solutions of the inequality. These are the points on or below the line. We can also use a test point to determine the solutions. We choose $(-2, 5)$.

$$\begin{array}{c|c} y & \leq 4 \\ \hline 5 & ? \; 4 \end{array} \quad \text{FALSE} \qquad 5 \leq 4 \text{ is false.}$$

Because $(-2, 5)$ is not a solution, we shade the half-plane that does not contain that point.

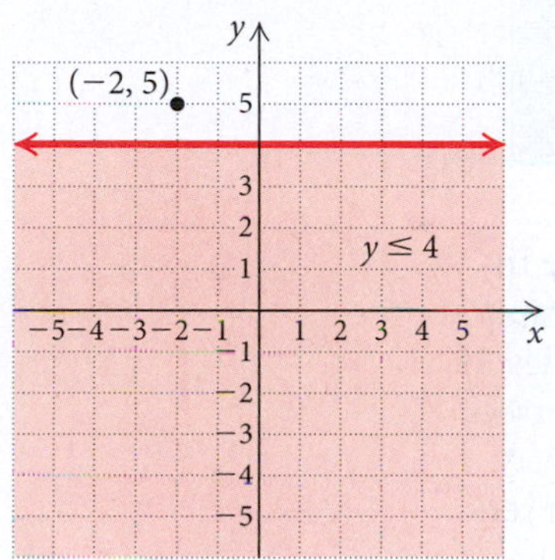

Now Try Exercise 25.

Systems of Linear Inequalities

A system of inequalities consists of two or more inequalities considered simultaneously. For example,

$$x + y \leq 4,$$
$$x - y \geq 2$$

is a system of *two linear inequalities in two variables.*

A solution of a system of inequalities is an ordered pair that is a solution of each inequality in the system. To graph a system of linear inequalities, we graph each inequality and determine the region that is common to *all* the solution sets.

Technology Connection

We can use different shading patterns on a graphing calculator to graph the system of inequalities in Example 5. The solution set is the region shaded using both patterns.

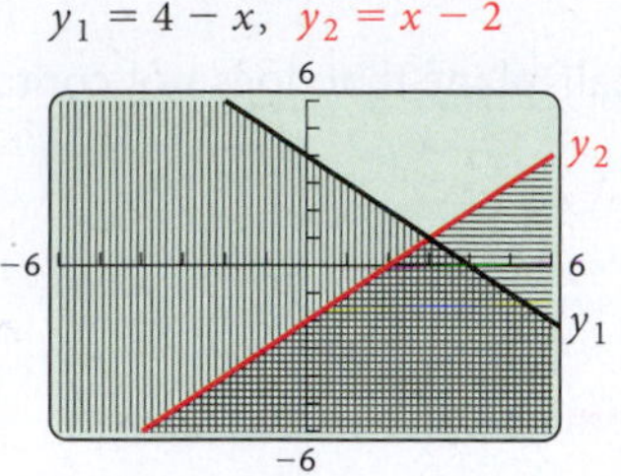

We can also use the Inequalz application to graph this system of inequalities. If we choose the Ineq Intersection option from the Shades menu, only the solution set is shaded, as shown in the following figure.

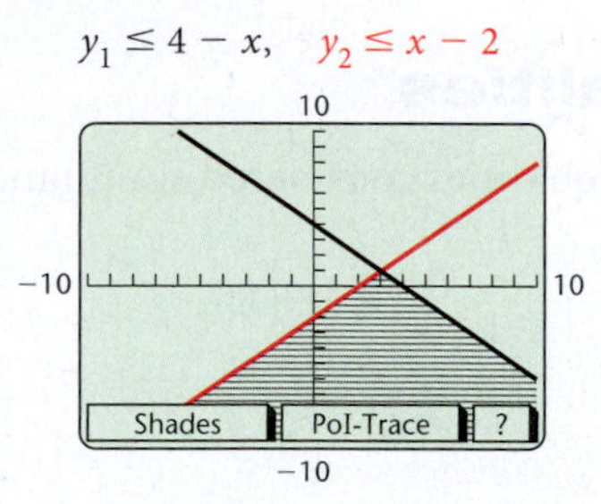

EXAMPLE 5 Graph the solution set of the system

$$x + y \le 4,$$
$$x - y \ge 2.$$

Solution We graph $x + y \le 4$ by first graphing the equation $x + y = 4$ using a solid line. Next, we choose $(0, 0)$ as a test point and find that it is a solution of $x + y \le 4$, so we shade the half-plane containing $(0, 0)$ using red. Next, we graph $x - y = 2$ using a solid line. We find that $(0, 0)$ is not a solution of $x - y \ge 2$, so we shade the half-plane that does not contain $(0, 0)$ using green. The arrows near the ends of each line help to indicate the half-plane that contains each solution set.

The solution set of the system of equations is the region shaded both red and green, or brown, including parts of the lines $x + y = 4$ and $x - y = 2$.

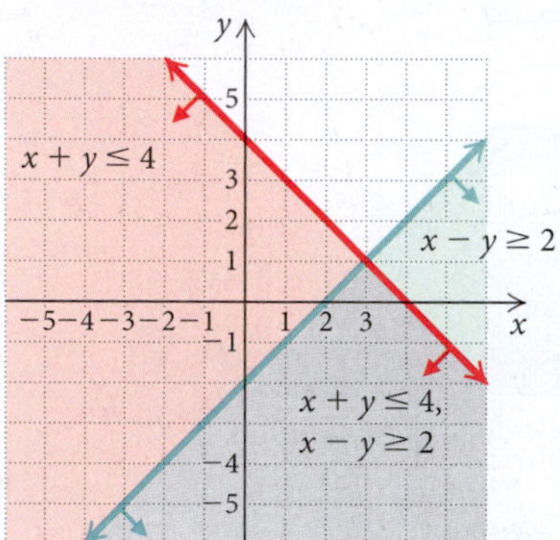

A system of inequalities may have a graph that consists of a polygon and its interior. As we will see later in this section, in many applications we will need to know the vertices of such a polygon.

EXAMPLE 6 Graph the following system of inequalities and find the coordinates of any vertices formed:

$$3x - y \le 6, \quad (1)$$
$$y - 3 \le 0, \quad (2)$$
$$x + y \ge 0. \quad (3)$$

Solution We graph the related equations $3x - y = 6$, $y - 3 = 0$, and $x + y = 0$ using solid lines. The half-plane containing the solution set for each inequality is indicated by the arrows near the ends of each line. We shade the region common to all three solution sets.

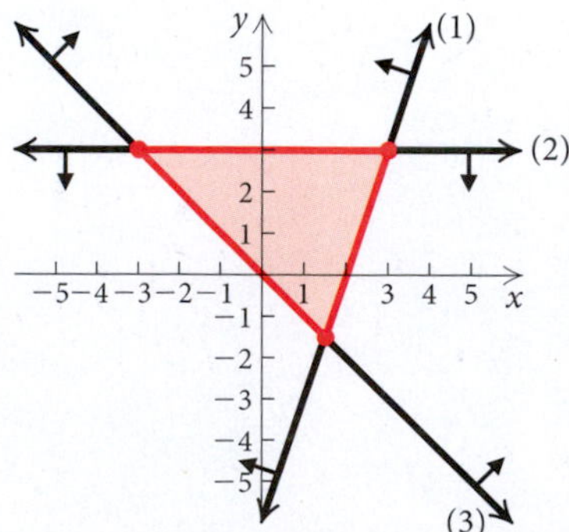

To find the vertices, we solve three systems of equations. The system of equations from inequalities (1) and (2) is

$$3x - y = 6,$$
$$y - 3 = 0.$$

Solving, we obtain the vertex $(3, 3)$.

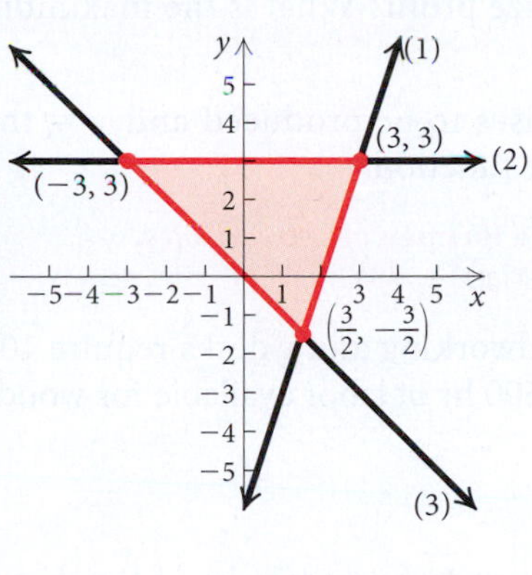

The system of equations from inequalities (1) and (3) is

$$3x - y = 6,$$
$$x + y = 0.$$

Solving, we obtain the vertex $\left(\frac{3}{2}, -\frac{3}{2}\right)$.

The system of equations from inequalities (2) and (3) is

$$y - 3 = 0,$$
$$x + y = 0.$$

Solving, we obtain the vertex $(-3, 3)$.

Applications: Linear Programming

In many applications, we want to find a maximum value or a minimum value. In business, for example, we might want to maximize profit and minimize cost. **Linear programming** can tell us how to do this.

In our study of linear programming, we will consider linear functions of two variables that are to be maximized or minimized subject to several conditions, or **constraints**. These constraints are expressed as inequalities. The solution set of the system of inequalities made up of the constraints contains all the **feasible solutions** of a linear programming problem. The function that we want to maximize or minimize is called the **objective function**.

It can be shown that the maximum and minimum values of the objective function occur at a vertex of the region of feasible solutions. Thus we have the following procedure.

Linear Programming Procedure

To find the maximum or minimum value of a linear objective function subject to a set of constraints:

1. Graph the region of feasible solutions.
2. Determine the coordinates of the vertices of the region.
3. Evaluate the objective function at each vertex. The largest and smallest of those values are the maximum and minimum values of the function, respectively.

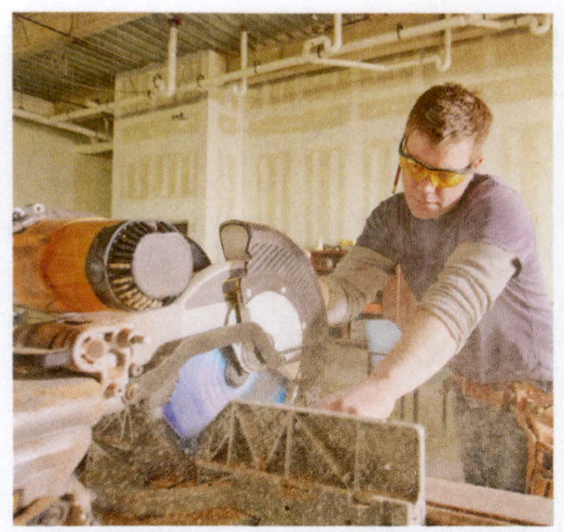

EXAMPLE 7 ***Maximizing Profit.*** Dovetail Carpentry Shop makes bookcases and desks. Each bookcase requires 5 hr of woodworking and 4 hr of finishing. Each desk requires 10 hr of woodworking and 3 hr of finishing. Each month the shop has 600 hr of labor available for woodworking and 240 hr for finishing. The profit on each bookcase is \$40 and on each desk is \$75. How many of each product should be made each month in order to maximize profit? What is the maximum profit?

Solution We let $x =$ the number of bookcases to be produced and $y =$ the number of desks. Then the profit P is given by the function

$$P = 40x + 75y.$$

To emphasize that P is a function of two variables, we sometimes write $P(x, y) = 40x + 75y$.

We know that x bookcases require $5x$ hr of woodworking and y desks require $10y$ hr of woodworking. Since there is no more than 600 hr of labor available for woodworking, we have one constraint:

$$5x + 10y \leq 600.$$

Similarly, the bookcases and the desks require $4x$ hr and $3y$ hr of finishing, respectively. There is no more than 240 hr of labor available for finishing, so we have a second constraint:

$$4x + 3y \leq 240.$$

We also know that $x \geq 0$ and $y \geq 0$ because the carpentry shop cannot make a negative number of either product.

Thus we want to maximize the objective function.

$$P = 40x + 75y$$

subject to the constraints

$$5x + 10y \leq 600,$$
$$4x + 3y \leq 240,$$
$$x \geq 0,$$
$$y \geq 0.$$

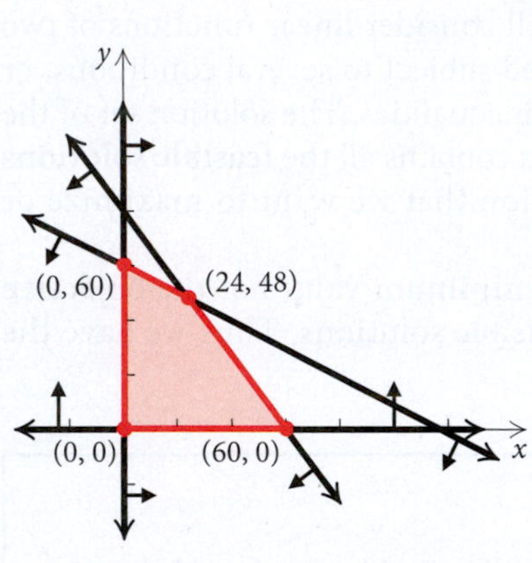

We graph the system of inequalities and determine the vertices, as shown in the figure at left.

Next, we evaluate the objective function P at each vertex.

Vertices (x, y)	Profit $P = 40x + 75y$	
$(0, 0)$	$P = 40 \cdot 0 + 75 \cdot 0 = 0$	
$(60, 0)$	$P = 40 \cdot 60 + 75 \cdot 0 = 2400$	
$(24, 48)$	$P = 40 \cdot 24 + 75 \cdot 48 = 4560$	← Maximum
$(0, 60)$	$P = 40 \cdot 0 + 75 \cdot 60 = 4500$	

The carpentry shop will make a maximum profit of \$4560 when 24 bookcases and 48 desks are produced and sold.

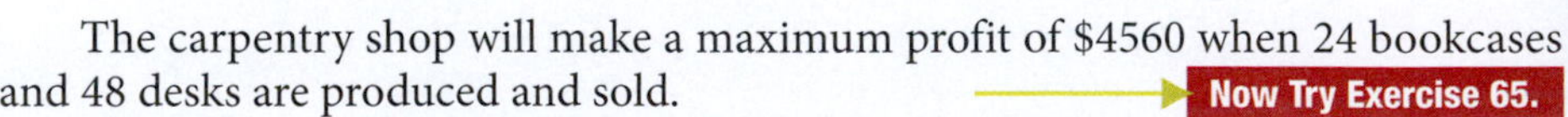

Technology Connection

We can create a table in which an objective function is evaluated at each vertex of a system of inequalities if the system has been graphed using the Inequalz application in the APPS menu. Refer to the online *Graphing Calculator Manual* that accompanies the text for details.

6.7 Exercise Set

In Exercises 1–8, match the inequality with one of the graphs (a)–(h) that follow.

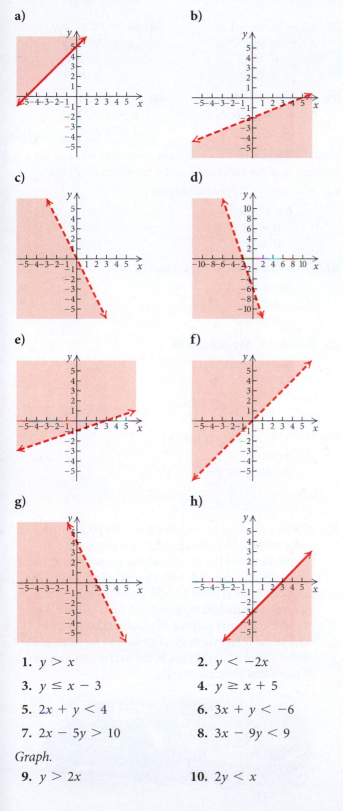

a)

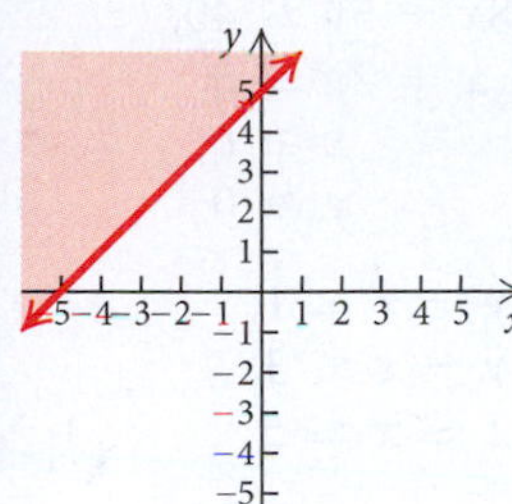

b)

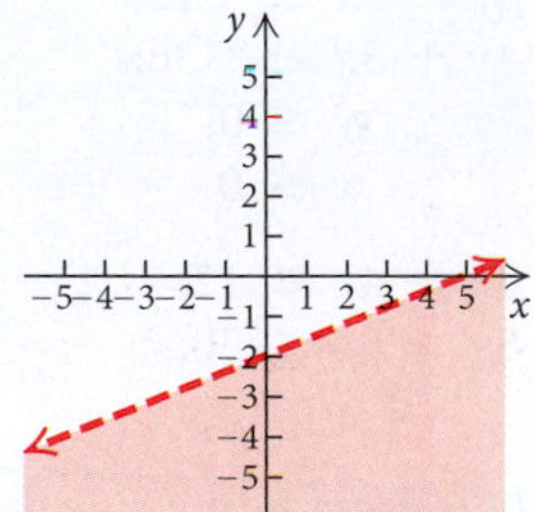

c)

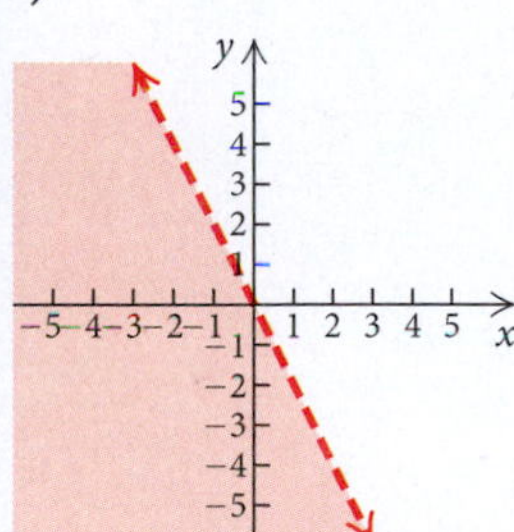

d)

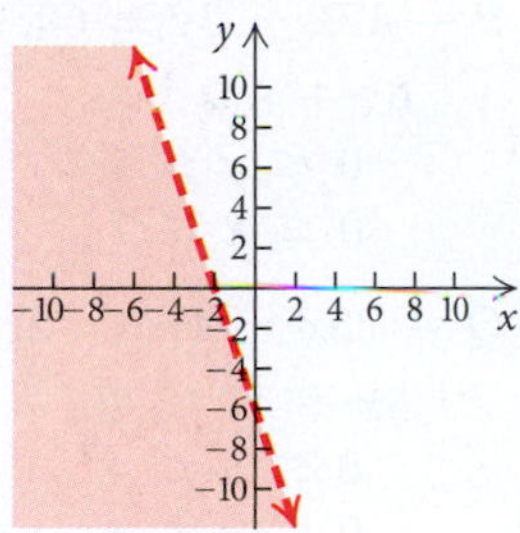

e)

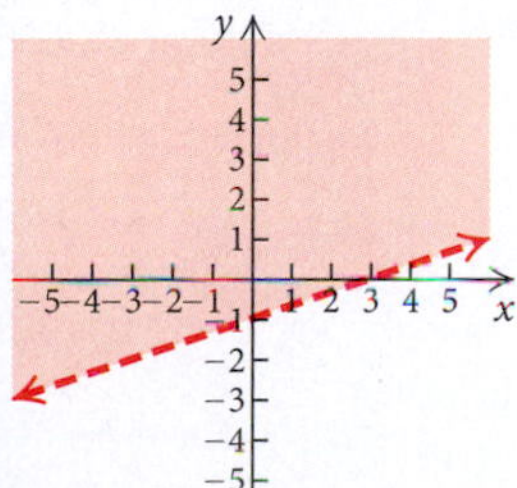

f)

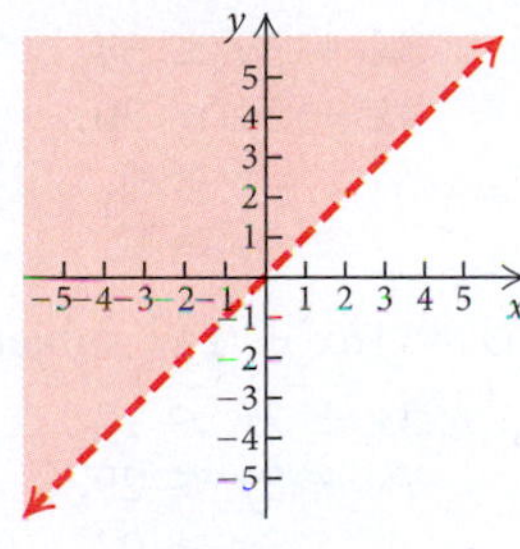

g)

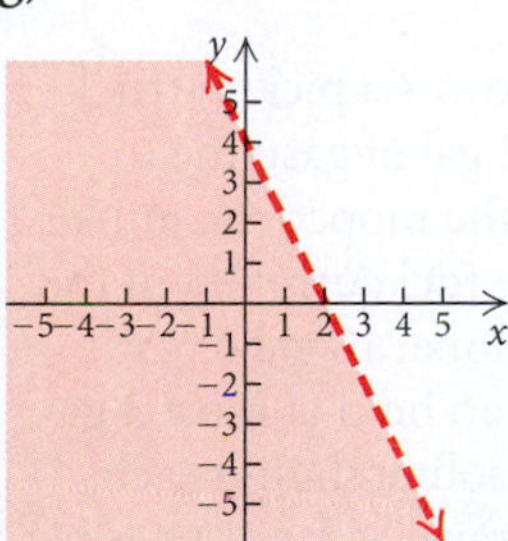

h)

1. $y > x$

2. $y < -2x$

3. $y \leq x - 3$

4. $y \geq x + 5$

5. $2x + y < 4$

6. $3x + y < -6$

7. $2x - 5y > 10$

8. $3x - 9y < 9$

Graph.

9. $y > 2x$

10. $2y < x$

11. $y + x \geq 0$

12. $y - x < 0$

13. $y > x - 3$

14. $y \leq x + 4$

15. $x + y < 4$

16. $x - y \geq 5$

17. $3x - 2y \leq 6$

18. $2x - 5y < 10$

19. $3y + 2x \geq 6$

20. $2y + x \leq 4$

21. $3x - 2 \leq 5x + y$

22. $2x - 6y \geq 8 + 2y$

23. $x < -4$

24. $y > -3$

25. $y \geq 5$

26. $x \leq 5$

27. $-4 < y < -1$
(*Hint*: Think of this as $-4 < y$ *and* $y < -1$.)

28. $-3 \leq x \leq 3$

29. $y \geq |x|$

30. $y \leq |x + 2|$

In Exercises 31–36, match the system of inequalities with one of the graphs (a)–(f) that follow.

a)

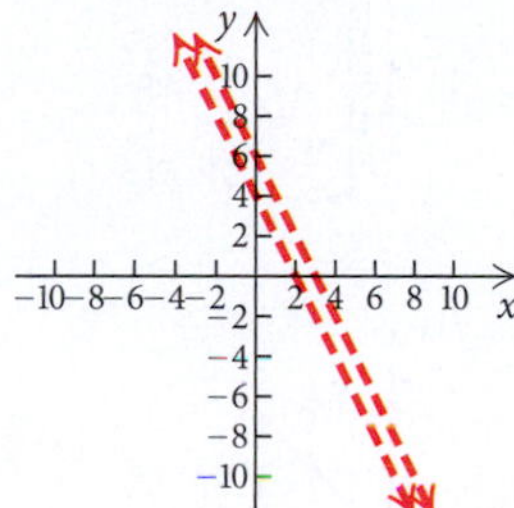

b)

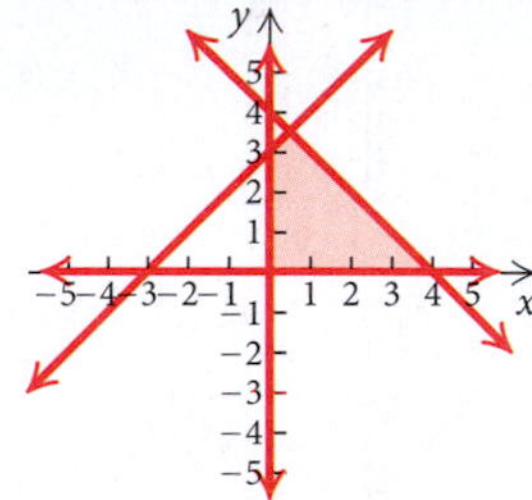

c)

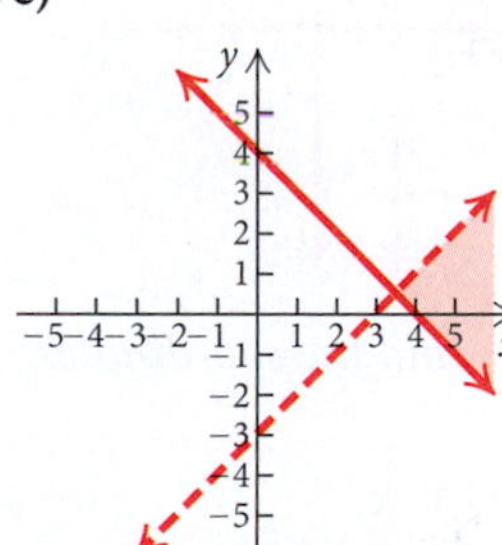

d)

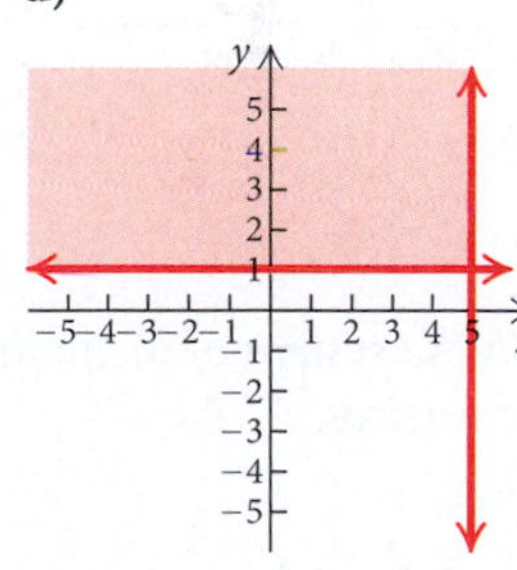

e)

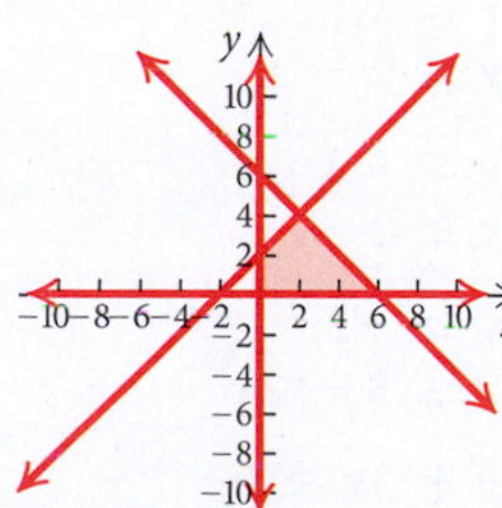

f)

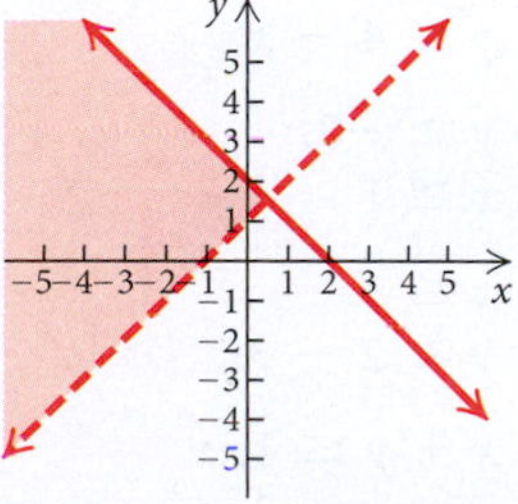

31. $y > x + 1,$
$y \leq 2 - x$

32. $y < x - 3,$
$y \geq 4 - x$

33. $2x + y < 4,$
$4x + 2y > 12$

34. $x \le 5,$
$y \ge 1$

35. $x + y \le 4,$
$x - y \ge -3,$
$x \ge 0,$
$y \ge 0$

36. $x - y \ge -2,$
$x + y \le 6,$
$x \ge 0,$
$y \ge 0$

Find a system of inequalities with the given graph. Answers may vary.

37.

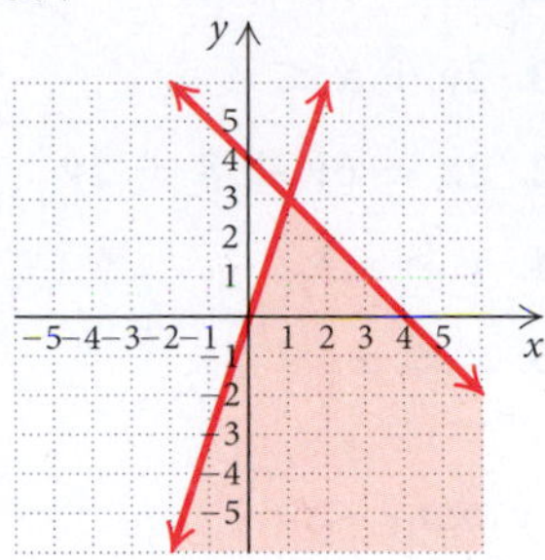

38.

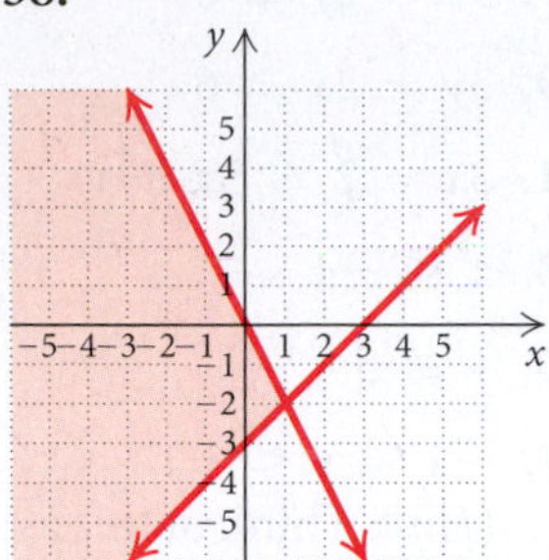

39.

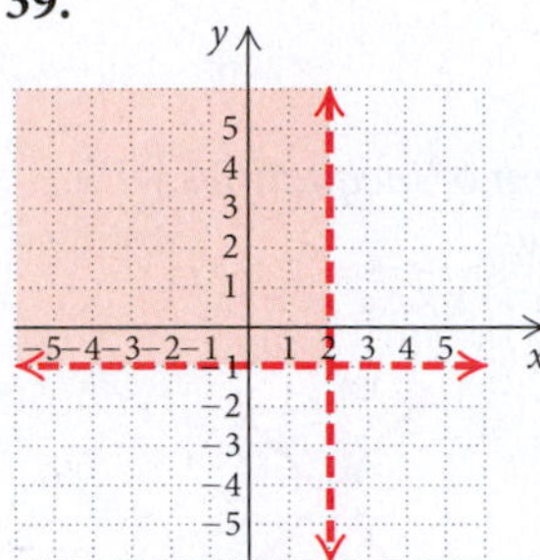

40.

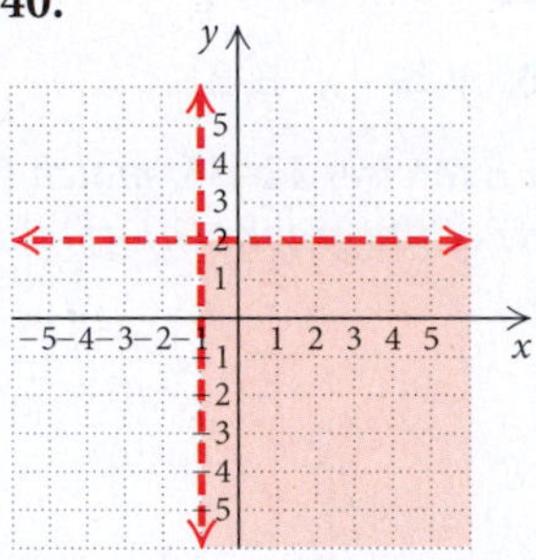

41.

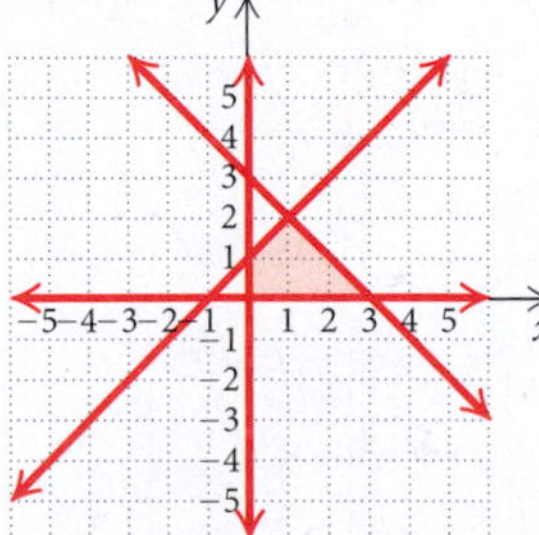

42.

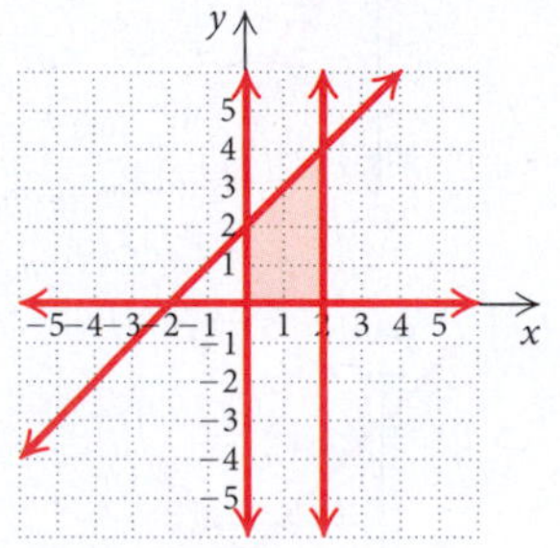

Graph the system of inequalities. Then find the coordinates of the vertices.

43. $y \le x,$
$y \ge 3 - x$

44. $y \le x,$
$y \ge 5 - x$

45. $y \ge x,$
$y \le 4 - x$

46. $y \ge x,$
$y \le 2 - x$

47. $y \ge -3,$
$x \ge 1$

48. $y \le -2,$
$x \ge 2$

49. $x \le 3,$
$y \ge 2 - 3x$

50. $x \ge -2,$
$y \le 3 - 2x$

51. $x + y \le 1,$
$x - y \le 2$

52. $y + 3x \ge 0,$
$y + 3x \le 2$

53. $2y - x \le 2,$
$y + 3x \ge -1$

54. $y \le 2x + 1,$
$y \ge -2x + 1,$
$x - 2 \le 0$

55. $x - y \le 2,$
$x + 2y \ge 8,$
$y - 4 \le 0$

56. $x + 2y \le 12,$
$2x + y \le 12,$
$x \ge 0,$
$y \ge 0$

57. $4y - 3x \ge -12,$
$4y + 3x \ge -36,$
$y \le 0,$
$x \le 0$

58. $8x + 5y \le 40,$
$x + 2y \le 8,$
$x \ge 0,$
$y \ge 0$

59. $3x + 4y \ge 12,$
$5x + 6y \le 30,$
$1 \le x \le 3$

60. $y - x \ge 1,$
$y - x \le 3,$
$2 \le x \le 5$

Find the maximum value and the minimum value of the function and the values of x and y for which they occur.

61. $P = 17x - 3y + 60$, subject to
$6x + 8y \le 48,$
$0 \le y \le 4,$
$0 \le x \le 7.$

62. $Q = 28x - 4y + 72$, subject to
$5x + 4y \ge 20,$
$0 \le y \le 4,$
$0 \le x \le 3.$

63. $F = 5x + 36y$, subject to
$5x + 3y \le 34,$
$3x + 5y \le 30,$
$x \ge 0,$
$y \ge 0.$

64. $G = 16x + 14y$, subject to
$3x + 2y \le 12,$
$7x + 5y \le 29,$
$x \ge 0,$
$y \ge 0.$

65. *Maximizing Mileage.* Jazmin owns a pickup truck and a moped. She can afford 12 gal of gasoline to be split between the truck and the moped. Jazmin's truck gets 20 mpg and, with the fuel currently in the tank, can hold at most an additional 10 gal of gas. Her moped gets 100 mpg and can hold at most 3 gal of gas. How many gallons of gasoline should each vehicle use if Jazmin wants to travel as far as possible on the 12 gal of gas? What is the maximum number of miles that she can travel?

66. *Maximizing Income.* Golden Harvest Foods makes jumbo biscuits and regular biscuits. The oven can cook at most 200 biscuits per day. Each jumbo biscuit requires 2 oz of flour, each regular biscuit requires 1 oz of flour, and there are 300 oz of flour available. The income from each jumbo biscuit is \$0.45 and from each regular biscuit is \$0.30. How many of each size biscuit should be made in order to maximize income? What is the maximum income?

67. *Maximizing Profit.* Waterbrook Farm includes 240 acres of cropland. The farm owner wishes to plant this acreage in both corn and soybeans. The profit per acre in corn production is \$325 and in soybeans is \$180. A total of 320 hr of labor is available. Each acre of corn requires 2 hr of labor, whereas each acre of soybean requires 1 hr of labor. How should the land be divided between corn and soybeans in order to yield the maximum profit? What is the maximum profit?

68. *Maximizing Profit.* Norris Mill can convert logs into lumber and plywood. In a given week, the mill can turn out 400 units of production, of which at least 100 units of lumber and at least 150 units of plywood are required by regular customers. The profit is \$25 per unit of lumber and \$38 per unit of plywood. How many units of each should the mill produce in order to maximize the profit? What is the maximum profit?

69. *Minimizing Cost.* An animal feed to be mixed from soybean meal and oats must contain at least 120 lb of protein, 24 lb of fat, and 10 lb of mineral ash. Each 100-lb sack of soybean meal costs \$20 and contains 50 lb of protein, 8 lb of fat, and 5 lb of mineral ash. Each 100-lb sack of oats costs \$8 and contains 15 lb of protein, 5 lb of fat, and 1 lb of mineral ash. How many sacks of each should be used in order to satisfy the minimum requirements at minimum cost? What is the minimum cost?

70. *Minimizing Cost.* Suppose that in the preceding exercise the oats were replaced by alfalfa, which costs \$10 per 100-lb sack and contains 20 lb of protein, 6 lb of fat, and 8 lb of mineral ash. How much of each would now be required in order to minimize the cost? What is the minimum cost?

71. *Maximizing Income.* Francisco is planning to invest up to \$40,000 in corporate and municipal bonds. The least he is allowed to invest in corporate bonds is \$6000, and he does not want to invest more than \$22,000 in corporate bonds. He also does not want to invest more than \$30,000 in municipal bonds. The interest is 3% on corporate bonds and $4\frac{1}{4}\%$ on municipal bonds. This is simple interest for one year. How much should he invest in each type of bond in order to maximize his income? What is the maximum income?

72. *Maximizing Income.* Mila is planning to invest up to \$22,000 in certificates of deposit at City Bank and People's Bank. She wants to invest at least \$2000 but no more than \$14,000 at City Bank. People's Bank does not insure more than a \$15,000 investment, so she will invest no more than that in People's Bank. The interest is $2\frac{1}{2}\%$ at City Bank and $1\frac{3}{4}\%$ at People's Bank. This is simple interest for one year. How much should she invest in each bank in order to maximize her income? What is the maximum income?

73. *Minimizing Transportation Cost.* An airline with two types of airplanes, P_1 and P_2, has contracted with a tour group to provide transportation for a minimum of 2000 first-class, 1500 tourist-class, and 2400 economy-class passengers. For a certain trip, airplane P_1 costs \$12 thousand to operate and can accommodate 40 first-class, 40 tourist-class, and 120 economy-class passengers, whereas airplane P_2 costs \$10 thousand to operate and can accommodate 80 first-class, 30 tourist-class, and 40 economy-class passengers. How many of each type of airplane should be used in order to minimize the operating cost? What is the minimum operating cost?

74. *Minimizing Transportation Cost.* Suppose that in the preceding exercise a new airplane P_3 becomes available, having an operating cost for the same trip of \$15 thousand and accommodating 40 first-class, 40 tourist-class, and 80 economy-class passengers. If airplane P_1 were replaced by airplane P_3, how many of P_2 and P_3 should be used in order to minimize the operating cost? What is the minimum operating cost?

75. *Maximizing Profit.* It takes Fena Tailoring 3 hr of cutting and 6 hr of sewing to make a tiered silk organza bridal dress. It takes 6 hr of cutting and 3 hr of sewing to make a lace sheath bridal dress. The shop has at most 27 hr per week available for cutting and at most 36 hr per week for sewing. The profit is $320 on an organza dress and $305 on a lace dress. How many of each kind of bridal dress should be made each week in order to maximize profit? What is the maximum profit?

76. *Maximizing Profit.* Cambridge Metal Works manufactures two sizes of gears. The smaller gear requires 4 hr of machining and 1 hr of polishing and yields a profit of $45. The larger gear requires 1 hr of machining and 1 hr of polishing and yields a profit of $30. The firm has available at most 24 hr per day for machining and 9 hr per day for polishing. How many of each type of gear should be produced each day in order to maximize profit? What is the maximum profit?

77. *Minimizing Nutrition Cost.* Suppose that it takes 12 units of carbohydrates and 6 units of protein to satisfy Jacob's minimum weekly requirements. A particular type of meat contains 2 units of carbohydrates and 2 units of protein per pound. A particular cheese contains 3 units of carbohydrates and 1 unit of protein per pound. The meat costs $3.50 per pound and the cheese costs $4.60 per pound. How many pounds of each are needed in order to minimize the cost and still meet the minimum requirements? What is the minimum cost?

78. *Minimizing Salary Cost.* The Spring Hill school board is analyzing education costs for Hill Top School. It wants to hire teachers and teacher's aides to make up a faculty that satisfies its needs at minimum cost. The average annual salary for a teacher is $53,000 and for a teacher's aide is $23,600. The school building can accommodate a faculty of no more than 50 but needs at least 20 faculty members to function properly. The school must have at least 12 aides, but the number of teachers must be at least twice the number of aides in order to accommodate the expectations of the community. How many teachers and teacher's aides should be hired in order to minimize salary costs? What is the minimum salary cost?

79. *Maximizing Animal Support in a Forest.* A certain area of forest is populated by two species of animal, which scientists refer to as A and B for simplicity. The forest supplies two kinds of food, referred to as F_1 and F_2. For one year, each member of species A requires 1 unit of F_1 and 0.5 unit of F_2. Each member of species B requires 0.2 unit of F_1 and 1 unit of F_2. The forest can normally supply at most 600 units of F_1 and 525 units of F_2 per year. What is the maximum total number of these animals that the forest can support?

80. *Maximizing Animal Support in a Forest.* Refer to Exercise 79. If there is a wet spring, then supplies of food increase to 1080 units of F_1 and 810 units of F_2. In this case, what is the maximum total number of these animals that the forest can support?

▶ Skill Maintenance

Solve.

81. $-5 \le x + 2 < 4$ [1.6]

82. $|x - 3| \ge 2$ [3.5]

83. $x^2 - 2x \le 3$ [4.6]

84. $\dfrac{x - 1}{x + 2} > 4$ [4.6]

▶ Synthesis

Graph the system of inequalities.

85. $y \ge x^2 - 2,$
$y \le 2 - x^2$

86. $y < x + 1,$
$y \ge x^2$

Graph the inequality.

87. $|x + y| \le 1$

88. $|x| + |y| \le 1$

89. $|x| > |y|$

90. $|x - y| > 0$

91. *Allocation of Resources.* Comfort-by-Design Furniture produces chairs and sofas. Each chair requires 20 ft of wood, 1 lb of foam rubber, and 2 yd^2 of fabric. Each sofa requires 100 ft of wood, 50 lb of foam rubber, and 20 yd^2 of fabric. The manufacturer has in stock 1900 ft of wood, 500 lb of foam rubber, and 240 yd^2 of fabric. The chairs can be sold for $200 each and the sofas for $750 each. How many of each should be produced in order to maximize income? What is the maximum income?

6.8 Partial Fractions

▶ Decompose rational expressions into partial fractions.

There are situations in calculus in which it is useful to write a rational expression as a sum of two or more simpler rational expressions. In the equation

$$\frac{4x - 13}{2x^2 + x - 6} = \frac{3}{x + 2} + \frac{-2}{2x - 3},$$

for example, each fraction on the right side is called a **partial fraction**. The expression on the right side is the **partial fraction decomposition** of the rational expression on the left side. In this section, we learn how such decompositions are created.

▶ Partial Fraction Decompositions

The procedure for finding the partial fraction decomposition of a rational expression involves factoring its denominator into linear factors and quadratic factors.

Procedure for Decomposing a Rational Expression into Partial Fractions

Consider any rational expression $P(x)/Q(x)$ such that $P(x)$ and $Q(x)$ have no common factor other than 1 or -1.

1. If the degree of $P(x)$ is greater than or equal to the degree of $Q(x)$, divide to express $P(x)/Q(x)$ as a quotient + remainder$/Q(x)$ and follow steps (2)–(5) to decompose the resulting rational expression.
2. If the degree of $P(x)$ is less than the degree of $Q(x)$, factor $Q(x)$ into linear factors of the form $(px + q)^n$ and/or quadratic factors of the form $(ax^2 + bx + c)^m$. Any quadratic factor $ax^2 + bx + c$ must be *irreducible*, meaning that it cannot be factored into linear factors with rational coefficients.
3. Assign to each linear factor $(px + q)^n$ the sum of n partial fractions:
$$\frac{A_1}{px + q} + \frac{A_2}{(px + q)^2} + \cdots + \frac{A_n}{(px + q)^n}.$$
4. Assign to each quadratic factor $(ax^2 + bx + c)^m$ the sum of m partial fractions:
$$\frac{B_1x + C_1}{ax^2 + bx + c} + \frac{B_2x + C_2}{(ax^2 + bx + c)^2} + \cdots + \frac{B_mx + C_m}{(ax^2 + bx + c)^m}.$$
5. Apply algebraic methods, as illustrated in the following examples, to find the constants in the numerators of the partial fractions.

JUST IN TIME 20

EXAMPLE 1 Decompose into partial fractions:

$$\frac{4x - 13}{2x^2 + x - 6}.$$

Solution The degree of the numerator is less than the degree of the denominator. We begin by factoring the denominator: $(x + 2)(2x - 3)$. We find constants A and B such that

$$\frac{4x - 13}{(x + 2)(2x - 3)} = \frac{A}{x + 2} + \frac{B}{2x - 3}.$$

To determine A and B, we add the expressions on the right:

$$\frac{4x - 13}{(x + 2)(2x - 3)} = \frac{A(2x - 3) + B(x + 2)}{(x + 2)(2x - 3)}.$$

Next, we equate the numerators:

$$4x - 13 = A(2x - 3) + B(x + 2).$$

Since the last equation containing A and B is true for all x, we can substitute any value of x and still have a true equation. If we choose $x = \frac{3}{2}$, then $2x - 3 = 0$ and A will be eliminated when we make the substitution. This gives us

$$4\left(\tfrac{3}{2}\right) - 13 = A\left(2 \cdot \tfrac{3}{2} - 3\right) + B\left(\tfrac{3}{2} + 2\right)$$
$$-7 = 0 + \tfrac{7}{2}B.$$

Solving, we obtain $B = -2$.

If we choose $x = -2$, then $x + 2 = 0$ and B will be eliminated when we make the substitution. This gives us

$$4(-2) - 13 = A[2(-2) - 3] + B(-2 + 2)$$
$$-21 = -7A + 0.$$

Solving, we obtain $A = 3$.

The decomposition is as follows:

$$\frac{4x - 13}{2x^2 + x - 6} = \frac{3}{x + 2} + \frac{-2}{2x - 3}, \quad \text{or} \quad \frac{3}{x + 2} - \frac{2}{2x - 3}.$$

To check, we can add to see if we get the expression on the left.

Now Try Exercise 3.

Technology Connection

We can use the TABLE feature on a graphing calculator to check a partial fraction decomposition. To check the decomposition in Example 1, we compare values of

$$y_1 = \frac{4x - 13}{2x^2 + x - 6}$$

and

$$y_2 = \frac{3}{x + 2} - \frac{2}{2x - 3}$$

for the same values of x. Since $y_1 = y_2$ for the given values of x as we scroll through the table, the decomposition appears to be correct.

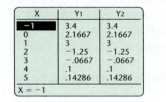

X	Y1	Y2
−1	3.4	3.4
0	2.1667	2.1667
1	3	3
2	−1.25	−1.25
3	−.0667	−.0667
4	.1	.1
5	.14286	.14286

X = −1

EXAMPLE 2 Decompose into partial fractions:

$$\frac{7x^2 - 29x + 24}{(2x - 1)(x - 2)^2}.$$

Solution The degree of the numerator is 2 and the degree of the denominator is 3, so the degree of the numerator is less than the degree of the denominator. The denominator is given in factored form. The decomposition has the following form:

$$\frac{7x^2 - 29x + 24}{(2x - 1)(x - 2)^2} = \frac{A}{2x - 1} + \frac{B}{x - 2} + \frac{C}{(x - 2)^2}.$$

As in Example 1, we add the expressions on the right:

$$\frac{7x^2 - 29x + 24}{(2x - 1)(x - 2)^2} = \frac{A(x - 2)^2 + B(2x - 1)(x - 2) + C(2x - 1)}{(2x - 1)(x - 2)^2}.$$

Then we equate the numerators. This gives us

$$7x^2 - 29x + 24 = A(x - 2)^2 + B(2x - 1)(x - 2) + C(2x - 1).$$

Since the equation containing A, B, and C is true for all x, we can substitute any value of x and still have a true equation. In order to have $2x - 1 = 0$, we let $x = \frac{1}{2}$. This gives us

$$7\left(\tfrac{1}{2}\right)^2 - 29\cdot\tfrac{1}{2} + 24 = A\left(\tfrac{1}{2} - 2\right)^2 + 0 + 0$$
$$\tfrac{45}{4} = \tfrac{9}{4}A.$$

Solving, we obtain $A = 5$.

In order to have $x - 2 = 0$, we let $x = 2$. Substituting gives us

$$7(2)^2 - 29(2) + 24 = 0 + 0 + C(2\cdot 2 - 1)$$
$$-6 = 3C.$$

Solving, we obtain $C = -2$.

To find B, we choose any value for x except $\frac{1}{2}$ or 2 and replace A with 5 and C with -2. We let $x = 1$:

$$7\cdot 1^2 - 29\cdot 1 + 24 = 5(1-2)^2 + B(2\cdot 1 - 1)(1-2) + (-2)(2\cdot 1 - 1)$$
$$2 = 5 - B - 2$$
$$B = 1.$$

The decomposition is as follows:

$$\frac{7x^2 - 29x + 24}{(2x-1)(x-2)^2} = \frac{5}{2x-1} + \frac{1}{x-2} - \frac{2}{(x-2)^2}.$$

Now Try Exercise 7.

POLYNOMIAL DIVISION

REVIEW SECTION 4.3

EXAMPLE 3 Decompose into partial fractions:

$$\frac{6x^3 + 5x^2 - 7}{3x^2 - 2x - 1}.$$

Solution The degree of the numerator is greater than that of the denominator. Therefore, we divide and find an equivalent expression:

$$\begin{array}{r} 2x + 3 \\ 3x^2 - 2x - 1\overline{)6x^3 + 5x^2 - 7} \\ \underline{6x^3 - 4x^2 - 2x} \\ 9x^2 + 2x - 7 \\ \underline{9x^2 - 6x - 3} \\ 8x - 4 \end{array}$$

The original expression is thus equivalent to

$$2x + 3 + \frac{8x - 4}{3x^2 - 2x - 1}.$$

We decompose the fraction to get

$$\frac{8x - 4}{(3x+1)(x-1)} = \frac{5}{3x+1} + \frac{1}{x-1}.$$

The final result is

$$2x + 3 + \frac{5}{3x+1} + \frac{1}{x-1}.$$

Now Try Exercise 17.

Systems of equations can also be used to decompose rational expressions. Let's reconsider Example 2.

EXAMPLE 4 Decompose into partial fractions:

$$\frac{7x^2 - 29x + 24}{(2x - 1)(x - 2)^2}.$$

Solution The decomposition has the following form:

$$\frac{A}{2x - 1} + \frac{B}{x - 2} + \frac{C}{(x - 2)^2}.$$

We first add as in Example 2:

$$\frac{7x^2 - 29x + 24}{(2x - 1)(x - 2)^2} = \frac{A}{2x - 1} + \frac{B}{x - 2} + \frac{C}{(x - 2)^2}$$
$$= \frac{A(x - 2)^2 + B(2x - 1)(x - 2) + C(2x - 1)}{(2x - 1)(x - 2)^2}.$$

Then we equate numerators:

$$\begin{aligned} 7x^2 - 29x + 24 &\\ &= A(x - 2)^2 + B(2x - 1)(x - 2) + C(2x - 1) \\ &= A(x^2 - 4x + 4) + B(2x^2 - 5x + 2) + C(2x - 1) \\ &= Ax^2 - 4Ax + 4A + 2Bx^2 - 5Bx + 2B + 2Cx - C, \end{aligned}$$

or, combining like terms,

$$\begin{aligned} 7x^2 - 29x + 24 &\\ &= (A + 2B)x^2 + (-4A - 5B + 2C)x + (4A + 2B - C). \end{aligned}$$

Next, we equate corresponding coefficients:

$7 = A + 2B,$ The coefficients of the x^2-terms must be the same.

$-29 = -4A - 5B + 2C,$ The coefficients of the x-terms must be the same.

$24 = 4A + 2B - C.$ The constant terms must be the same.

SYSTEMS OF EQUATIONS IN THREE VARIABLES

REVIEW SECTION 6.2, 6.5, OR 6.6

We now have a system of three equations. You should confirm that the solution of the system is

$$A = 5, \quad B = 1, \quad \text{and} \quad C = -2.$$

The decomposition is as follows:

$$\frac{7x^2 - 29x + 24}{(2x - 1)(x - 2)^2} = \frac{5}{2x - 1} + \frac{1}{x - 2} - \frac{2}{(x - 2)^2}.$$

Now Try Exercise 15.

EXAMPLE 5 Decompose into partial fractions:

$$\frac{11x^2 - 8x - 7}{(2x^2 - 1)(x - 3)}.$$

Solution The decomposition has the following form:

$$\frac{11x^2 - 8x - 7}{(2x^2 - 1)(x - 3)} = \frac{Ax + B}{2x^2 - 1} + \frac{C}{x - 3}.$$

Adding and equating numerators, we get

$$\begin{aligned} 11x^2 - 8x - 7 &= (Ax + B)(x - 3) + C(2x^2 - 1) \\ &= Ax^2 - 3Ax + Bx - 3B + 2Cx^2 - C, \end{aligned}$$

or $\quad 11x^2 - 8x - 7 = (A + 2C)x^2 + (-3A + B)x + (-3B - C).$

We then equate corresponding coefficients:

$$11 = A + 2C, \quad \text{The coefficients of the } x^2\text{-terms}$$
$$-8 = -3A + B, \quad \text{The coefficients of the } x\text{-terms}$$
$$-7 = -3B - C. \quad \text{The constant terms}$$

We solve this system of three equations and obtain

$$A = 3, \quad B = 1, \quad \text{and} \quad C = 4.$$

The decomposition is as follows:

$$\frac{11x^2 - 8x - 7}{(2x^2 - 1)(x - 3)} = \frac{3x + 1}{2x^2 - 1} + \frac{4}{x - 3}.$$

Now Try Exercise 13.

6.8 Exercise Set

Decompose into partial fractions.

1. $\dfrac{x + 7}{(x - 3)(x + 2)}$
2. $\dfrac{2x}{(x + 1)(x - 1)}$
3. $\dfrac{7x - 1}{6x^2 - 5x + 1}$
4. $\dfrac{13x + 46}{12x^2 - 11x - 15}$
5. $\dfrac{3x^2 - 11x - 26}{(x^2 - 4)(x + 1)}$
6. $\dfrac{5x^2 + 9x - 56}{(x - 4)(x - 2)(x + 1)}$
7. $\dfrac{9}{(x + 2)^2(x - 1)}$
8. $\dfrac{x^2 - x - 4}{(x - 2)^3}$
9. $\dfrac{2x^2 + 3x + 1}{(x^2 - 1)(2x - 1)}$
10. $\dfrac{x^2 - 10x + 13}{(x^2 - 5x + 6)(x - 1)}$
11. $\dfrac{x^4 - 3x^3 - 3x^2 + 10}{(x + 1)^2(x - 3)}$
12. $\dfrac{10x^3 - 15x^2 - 35x}{x^2 - x - 6}$
13. $\dfrac{-x^2 + 2x - 13}{(x^2 + 2)(x - 1)}$
14. $\dfrac{26x^2 + 208x}{(x^2 + 1)(x + 5)}$
15. $\dfrac{6 + 26x - x^2}{(2x - 1)(x + 2)^2}$
16. $\dfrac{5x^3 + 6x^2 + 5x}{(x^2 - 1)(x + 1)^3}$
17. $\dfrac{6x^3 + 5x^2 + 6x - 2}{2x^2 + x - 1}$
18. $\dfrac{2x^3 + 3x^2 - 11x - 10}{x^2 + 2x - 3}$
19. $\dfrac{2x^2 - 11x + 5}{(x - 3)(x^2 + 2x - 5)}$
20. $\dfrac{3x^2 - 3x - 8}{(x - 5)(x^2 + x - 4)}$
21. $\dfrac{-4x^2 - 2x + 10}{(3x + 5)(x + 1)^2}$
22. $\dfrac{26x^2 - 36x + 22}{(x - 4)(2x - 1)^2}$
23. $\dfrac{36x + 1}{12x^2 - 7x - 10}$
24. $\dfrac{-17x + 61}{6x^2 + 39x - 21}$
25. $\dfrac{-4x^2 - 9x + 8}{(3x^2 + 1)(x - 2)}$
26. $\dfrac{11x^2 - 39x + 16}{(x^2 + 4)(x - 8)}$

Skill Maintenance

Find the zeros of the polynomial function.

27. $f(x) = x^3 + x^2 + 9x + 9$ [4.1], [4.3], [4.4]

28. $f(x) = x^3 - 3x^2 + x - 3$ [4.1], [4.3], [4.4]

29. $f(x) = x^3 + x^2 - 3x - 2$ [4.4]

30. $f(x) = x^4 - x^3 - 5x^2 - x - 6$ [4.3]

31. $f(x) = x^3 + 5x^2 + 5x - 3$ [4.3]

Synthesis

Decompose into partial fractions.

32. $\dfrac{9x^3 - 24x^2 + 48x}{(x - 2)^4(x + 1)}$

[*Hint*: Let the expression equal

$$\frac{A}{x + 1} + \frac{P(x)}{(x - 2)^4}$$

and find $P(x)$].

33. $\dfrac{x}{x^4 - a^4}$

34. $\dfrac{1}{e^{-x} + 3 + 2e^x}$

35. $\dfrac{1 + \ln x^2}{(\ln x + 2)(\ln x - 3)^2}$

Chapter 6 Summary and Review

STUDY GUIDE

KEY TERMS AND CONCEPTS	EXAMPLES
SECTION 6.1: SYSTEMS OF EQUATIONS IN TWO VARIABLES	
A **system of two linear equations in two variables** is composed of two linear equations that are considered simultaneously. The **solutions** of the system of equations are all ordered pairs that make *both* equations true. A system of equations is **consistent** if it has at least one solution. A system of equations that has no solution is **inconsistent**. The equations are **dependent** if one equation can be obtained by multiplying on both sides of the other equation by a constant. Otherwise, the equations are **independent**. Systems of two equations in two variables can be solved graphically.	Solve: $x + y = 2,$ $y = x - 4.$ 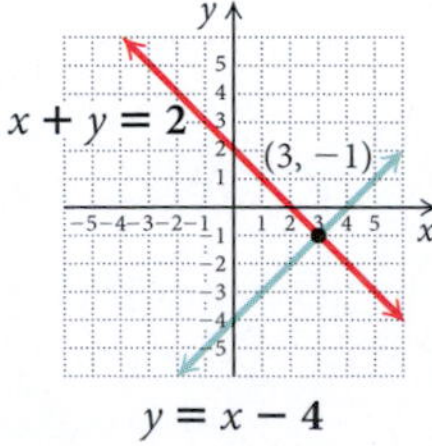 The solution is the point of intersection, $(3, -1)$. The system is consistent. The equations are independent.

	Solve: $x + y = 2,$ $x + y = -2.$ 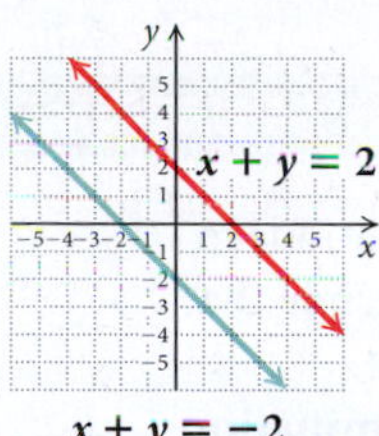 The graphs do not intersect, so there is no solution. The system is inconsistent. The equations are independent. Solve: $x + y = 2,$ $3x + 3y = 6.$ 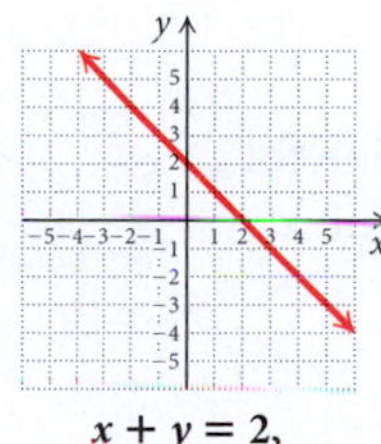 The graphs are the same. There are infinitely many common points, so there are infinitely many solutions. The solutions are of the form $(x, 2 - x)$ or $(2 - y, y)$. The system is consistent. The equations are dependent.
Systems of two equations in two variables can be solved using substitution.	Solve: $x = y - 5,$ $2x + 3y = 5.$ *Substitute and solve for y:* $2(y - 5) + 3y = 5$ $2y - 10 + 3y = 5$ $5y - 10 = 5$ $5y = 15$ $y = 3.$ *Back-substitute and solve for x:* $x = y - 5$ $x = 3 - 5$ $x = -2.$ The solution is $(-2, 3)$.
Systems of two equations in two variables can be solved using elimination.	Solve: $3x + y = -1,$ $x - 3y = 8.$ *Eliminate y and solve for x:* $9x + 3y = -3$ $x - 3y = 8$ $10x = 5$ $x = \frac{1}{2}.$ *Back-substitute and solve for y:* $3x + y = -1$ $3 \cdot \frac{1}{2} + y = -1$ $\frac{3}{2} + y = -1$ $y = -\frac{5}{2}.$ The solution is $\left(\frac{1}{2}, -\frac{5}{2}\right)$.

Some applied problems can be solved by translating to a system of two equations in two variables.	See Examples 6–8 on pp. 399–403.

SECTION 6.2: SYSTEMS OF EQUATIONS IN THREE VARIABLES

A **solution** of a system of equations in three variables is an ordered triple that makes *all* three equations true.

We can use **Gaussian elimination** to solve a system of three equations in three variables by using the operations listed on p. 409 to transform the original system to one of the form

$$\begin{aligned} Ax + By + Cz &= D, \\ Ey + Fz &= G, \\ Hz &= K. \end{aligned}$$

Then we solve the third equation for z and back-substitute to find y and x.

As we see in Example 1 on p. 409, Gaussian elimination can be used to transform the system of equations

$$\begin{aligned} x - 2y + 3z &= 11, \\ 4x + 2y - 3z &= 4, \\ 3x + 3y - z &= 4 \end{aligned}$$

to the equivalent form

$$\begin{aligned} x - 2y + 3z &= 11, \\ 10y - 15z &= -40, \\ 35z &= 70. \end{aligned}$$

Solve for z: $35z = 70$

$$z = 2.$$

Back-substitute to find y and x:

$$\begin{aligned} 10y - 15 \cdot 2 &= -40 \qquad & x - 2(-1) + 3 \cdot 2 &= 11 \\ 10y - 30 &= -40 & x + 2 + 6 &= 11 \\ 10y &= -10 & x + 8 &= 11 \\ y &= -1. & x &= 3. \end{aligned}$$

The solution is $(3, -1, 2)$.

Some applied problems can be solved by translating to a system of three equations in three variables.	See Example 3 on p. 412.

We can use a system of three equations to model a situation with a quadratic function.

Find a quadratic function that fits the data points $(0, -5)$, $(1, -4)$, and $(2, 1)$.

We substitute in the function $f(x) = ax^2 + bx + c$:

For $(0, -5)$: $\quad -5 = a \cdot 0^2 + b \cdot 0 + c,$

For $(1, -4)$: $\quad -4 = a \cdot 1^2 + b \cdot 1 + c,$

For $(2, 1)$: $\quad 1 = a \cdot 2^2 + b \cdot 2 + c.$

We now have a system of equations:

$$\begin{aligned} c &= -5, \\ a + b + c &= -4, \\ 4a + 2b + c &= 1. \end{aligned}$$

Solving this system of equations gives $(2, -1, -5)$. Thus,

$$f(x) = 2x^2 - x - 5.$$

SECTION 6.3: MATRICES AND SYSTEMS OF EQUATIONS

A **matrix** (pl., **matrices**) is a rectangular array of numbers called **entries**, or **elements**, of the matrix.	Row 1 → Row 2 → $\begin{bmatrix} 3 & -2 & 5 \\ -1 & 4 & -3 \end{bmatrix}$ ↑ Column 1 ↑ Column 2 ↑ Column 3 This matrix has 2 rows and 3 columns. Its **order** is 2×3.
We can apply the **row-equivalent operations** on p. 419 to use **Gaussian elimination** with matrices to solve systems of equations.	Solve: $x - 2y = 8,$ $2x + y = 1.$ We write the augmented matrix and transform it to **row-echelon form** or **reduced row-echelon form**: $\left[\begin{array}{cc\|c} 1 & -2 & 8 \\ 2 & 1 & 1 \end{array}\right] \rightarrow \left[\begin{array}{cc\|c} 1 & -2 & 8 \\ 0 & 1 & -3 \end{array}\right] \rightarrow \left[\begin{array}{cc\|c} 1 & 0 & 2 \\ 0 & 1 & -3 \end{array}\right]$ ↑ Row-echelon form ↑ Reduced row-echelon form Thus we have $x = 2, y = -3$. The solution is $(2, -3)$.

SECTION 6.4: MATRIX OPERATIONS

Matrices of the same order can be added or subtracted by adding or subtracting their corresponding entries.	Find each of the following. $\begin{bmatrix} 3 & -4 \\ -1 & 2 \end{bmatrix} + \begin{bmatrix} -5 & -1 \\ 3 & 0 \end{bmatrix} = \begin{bmatrix} 3 + (-5) & -4 + (-1) \\ -1 + 3 & 2 + 0 \end{bmatrix} = \begin{bmatrix} -2 & -5 \\ 2 & 2 \end{bmatrix}$ $\begin{bmatrix} 3 & -4 \\ -1 & 2 \end{bmatrix} - \begin{bmatrix} -5 & -1 \\ 3 & 0 \end{bmatrix} = \begin{bmatrix} 3 - (-5) & -4 - (-1) \\ -1 - 3 & 2 - 0 \end{bmatrix} = \begin{bmatrix} 8 & -3 \\ -4 & 2 \end{bmatrix}$
The **scalar product** of a number k and a matrix **A** is the matrix k**A** obtained by multiplying each entry of **A** by k. The number k is called a **scalar**. The properties of matrix addition and scalar multiplication are given on p. 428.	For $\mathbf{A} = \begin{bmatrix} 2 & 3 & -1 \\ -4 & -2 & 5 \end{bmatrix}$, find 2**A**. $2\mathbf{A} = 2\begin{bmatrix} 2 & 3 & -1 \\ -4 & -2 & 5 \end{bmatrix} = \begin{bmatrix} 2 \cdot 2 & 2 \cdot 3 & 2 \cdot (-1) \\ 2 \cdot (-4) & 2 \cdot (-2) & 2 \cdot 5 \end{bmatrix} = \begin{bmatrix} 4 & 6 & -2 \\ -8 & -4 & 10 \end{bmatrix}$
For an $m \times n$ matrix $\mathbf{A} = [a_{ij}]$ and an $n \times p$ matrix $\mathbf{B} = [b_{ij}]$, the **product** $\mathbf{AB} = [c_{ij}]$ is an $m \times p$ matrix, where $c_{ij} = a_{i1} \cdot b_{1j} + a_{i2} \cdot b_{2j} + a_{i3} \cdot b_{3j} + \cdots + a_{in} \cdot b_{nj}.$ The properties of matrix multiplication are given on p. 431.	For $\mathbf{A} = \begin{bmatrix} 4 & -1 & 3 \\ 0 & -2 & 1 \end{bmatrix}$ and $\mathbf{B} = \begin{bmatrix} -3 & 1 \\ 3 & 4 \\ 2 & -1 \end{bmatrix}$, find **AB**. $\mathbf{AB} = \begin{bmatrix} 4 & -1 & 3 \\ 0 & -2 & 1 \end{bmatrix} \begin{bmatrix} -3 & 1 \\ 3 & 4 \\ 2 & -1 \end{bmatrix}$ $= \begin{bmatrix} 4 \cdot (-3) + (-1) \cdot 3 + 3 \cdot 2 & 4 \cdot 1 + (-1) \cdot 4 + 3 \cdot (-1) \\ 0 \cdot (-3) + (-2) \cdot 3 + 1 \cdot 2 & 0 \cdot 1 + (-2) \cdot 4 + 1 \cdot (-1) \end{bmatrix}$ $= \begin{bmatrix} -9 & -3 \\ -4 & -9 \end{bmatrix}$

We can write a matrix equation equivalent to a system of equations.

Write a matrix equation equivalent to the system of equations:

$$2x - 3y = 6,$$
$$x - 4y = 1.$$

This system of equations can be written as

$$\begin{bmatrix} 2 & -3 \\ 1 & -4 \end{bmatrix} \begin{bmatrix} x \\ y \end{bmatrix} = \begin{bmatrix} 6 \\ 1 \end{bmatrix}.$$

SECTION 6.5: INVERSES OF MATRICES

The $n \times n$ **identity matrix I** is an $n \times n$ matrix with 1's on the main diagonal and 0's elsewhere.

For any $n \times n$ matrix **A**,

$$\mathbf{AI} = \mathbf{IA} = \mathbf{A}.$$

For an $n \times n$ matrix **A**, if there is a matrix $\mathbf{A}^{-1}$ for which $\mathbf{A}^{-1} \cdot \mathbf{A} = \mathbf{I} = \mathbf{A} \cdot \mathbf{A}^{-1}$, then $\mathbf{A}^{-1}$ is the **inverse** of **A**.

The inverse of an $n \times n$ matrix **A** can be found by first writing an augmented matrix consisting of **A** on the left side and the $n \times n$ identity matrix on the right side. Then row-equivalent operations are used to transform the augmented matrix to a matrix with the $n \times n$ identity matrix on the left side and the inverse on the right side.

See Examples 3 and 4 on pp. 438 and 439.

For a system of n linear equations in n variables, $\mathbf{AX} = \mathbf{B}$, if **A** has an inverse, then the solution of the system of equations is given by

$$\mathbf{X} = \mathbf{A}^{-1}\mathbf{B}.$$

Since matrix multiplication is not commutative, in general, **B** *must* be multiplied *on the left* by $\mathbf{A}^{-1}$.

Use an inverse matrix to solve the following system of equations:

$$x - y = 1,$$
$$x - 2y = -1.$$

First, we write an equivalent matrix equation:

$$\begin{bmatrix} 1 & -1 \\ 1 & -2 \end{bmatrix} \cdot \begin{bmatrix} x \\ y \end{bmatrix} = \begin{bmatrix} 1 \\ -1 \end{bmatrix}.$$

$$\mathbf{A} \cdot \mathbf{X} = \mathbf{B}$$

Then we find $\mathbf{A}^{-1}$ and multiply *on the left* by $\mathbf{A}^{-1}$:

$$\mathbf{X} = \mathbf{A}^{-1} \cdot \mathbf{B}$$

$$\begin{bmatrix} x \\ y \end{bmatrix} = \begin{bmatrix} 2 & -1 \\ 1 & -1 \end{bmatrix} \begin{bmatrix} 1 \\ -1 \end{bmatrix} = \begin{bmatrix} 3 \\ 2 \end{bmatrix}.$$

The solution is $(3, 2)$.

SECTION 6.6: DETERMINANTS AND CRAMER'S RULE

Determinant of a 2 × 2 Matrix

The determinant of the matrix $\begin{bmatrix} a & c \\ b & d \end{bmatrix}$ is denoted by $\begin{vmatrix} a & c \\ b & d \end{vmatrix}$ and is defined as

$$\begin{vmatrix} a & c \\ b & d \end{vmatrix} = ad - bc.$$

Evaluate: $\begin{vmatrix} 3 & -4 \\ 2 & 1 \end{vmatrix}$.

$$\begin{vmatrix} 3 & -4 \\ 2 & 1 \end{vmatrix} = 3 \cdot 1 - 2(-4) = 3 + 8 = 11$$

The **determinant** of any **square matrix** can be found by *expanding across a row* or *down a column*. See p. 445.

See Example 4 on p. 445.

We can use determinants to solve systems of linear equations.

Cramer's rule for a 2×2 system is given on p. 446. Cramer's rule for a 3×3 system is given on p. 447.

Solve: $2x - 3y = 2,$
$6x + 6y = 1.$

$$x = \frac{\begin{vmatrix} 2 & -3 \\ 1 & 6 \end{vmatrix}}{\begin{vmatrix} 2 & -3 \\ 6 & 6 \end{vmatrix}}, \qquad y = \frac{\begin{vmatrix} 2 & 2 \\ 6 & 1 \end{vmatrix}}{\begin{vmatrix} 2 & -3 \\ 6 & 6 \end{vmatrix}}$$

$$x = \frac{15}{30} = \frac{1}{2}, \qquad y = \frac{-10}{30} = -\frac{1}{3}.$$

The solution is $\left(\frac{1}{2}, -\frac{1}{3}\right)$.

SECTION 6.7: SYSTEMS OF INEQUALITIES AND LINEAR PROGRAMMING

To graph a linear inequality in two variables:

1. Graph the related equation. Draw a dashed line if the inequality symbol is $<$ or $>$. Draw a solid line if the inequality symbol is $\leq$ or $\geq$.
2. Use a test point to determine which half-plane to shade.

Graph: $x + y > 2$.

1. Graph $x + y = 2$ using a dashed line.
2. Test a point not on the line. We use $(0, 0)$.

$$\begin{array}{c|c} x + y & 2 \\ \hline 0 + 0 & ?\ 2 \\ 0 & \end{array} \quad \text{FALSE}$$

Since $0 > 2$ is false, we shade the half-plane that does not contain $(0, 0)$.

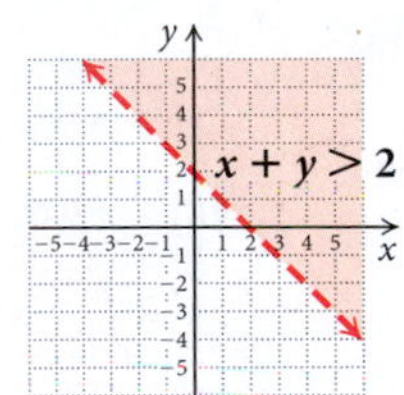

To graph a system of inequalities, graph each inequality and determine the region that is common to all the solution sets.

Graph the solution set of the system

$$x - y \leq 3,$$
$$2x + y \geq 4.$$

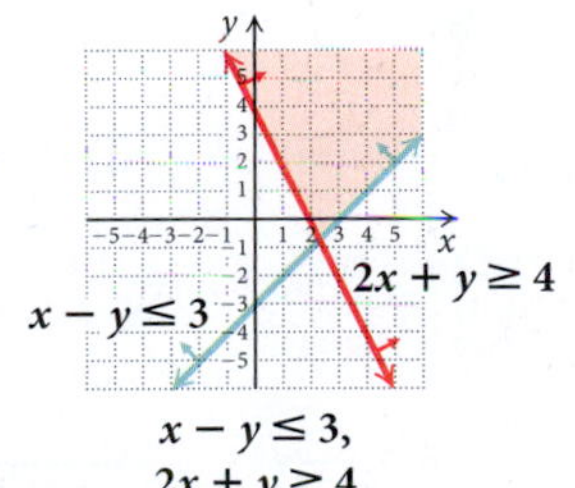

The maximum or minimum value of an **objective function** over a region of **feasible solutions** is the maximum or minimum value of the function at a vertex of that region.

Maximize $G = 8x - 5y$ subject to

$$x + y \leq 3,$$
$$x \geq 0,$$
$$y \geq 1.$$

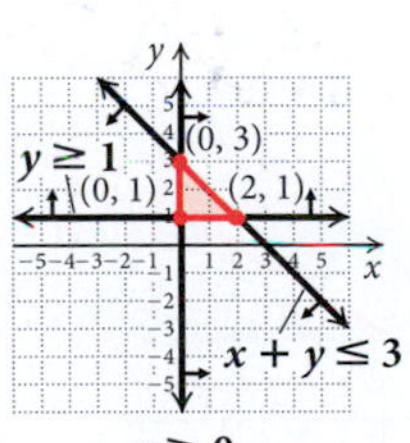

Vertex	$G = 8x - 5y$	
$(0, 1)$	$G = 8 \cdot 0 - 5 \cdot 1 = -5$	
$(0, 3)$	$G = 8 \cdot 0 - 5 \cdot 3 = -15$	
$(2, 1)$	$G = 8 \cdot 2 - 5 \cdot 1 = 11$	← Maximum

SECTION 6.8: PARTIAL FRACTIONS

The procedure for **decomposing a rational expression into partial fractions** is given on p. 461.	See Examples 1–5 on pp. 462–465.

REVIEW EXERCISES

Determine whether the statement is true or false.

1. A system of equations with exactly one solution is consistent and has independent equations. [6.1]

2. A system of two linear equations in two variables can have exactly two solutions. [6.1]

3. For any $m \times n$ matrices **A** and **B**, $\mathbf{A} + \mathbf{B} = \mathbf{B} + \mathbf{A}$. [6.4]

4. In general, matrix multiplication is commutative. [6.4]

In Exercises 5–12, match the equations or inequalities with one of the graphs (a)–(h) that follow.

a)

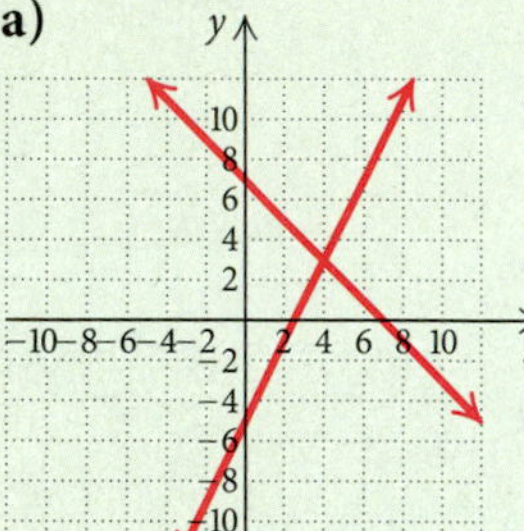

b)

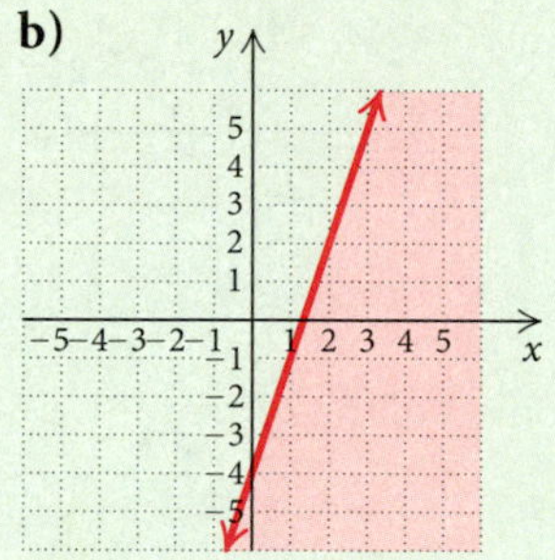

c)

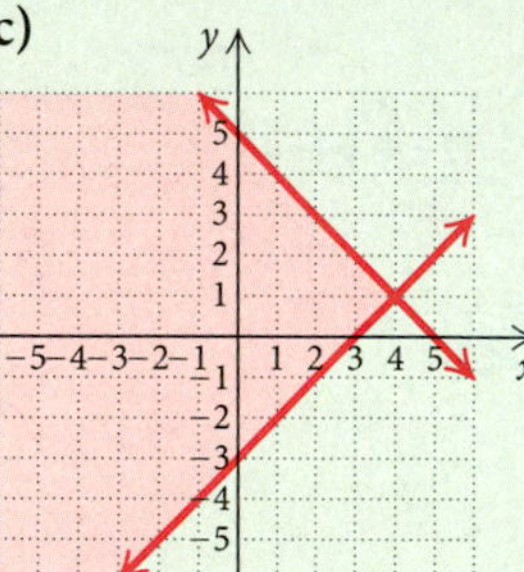

d)

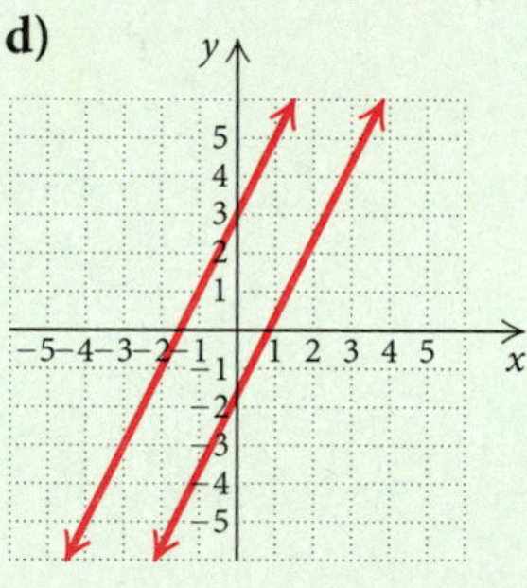

e)

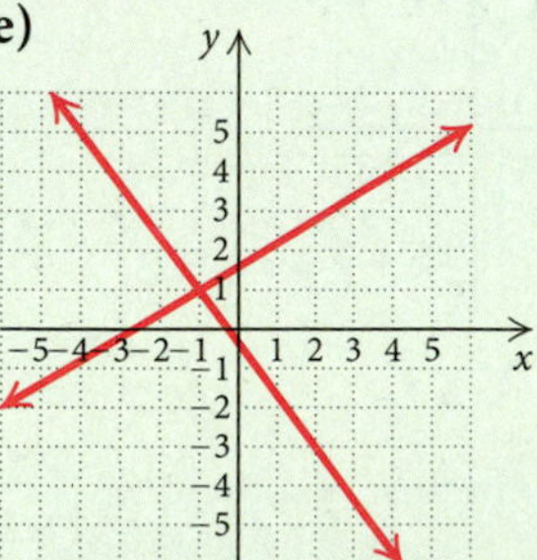

f)

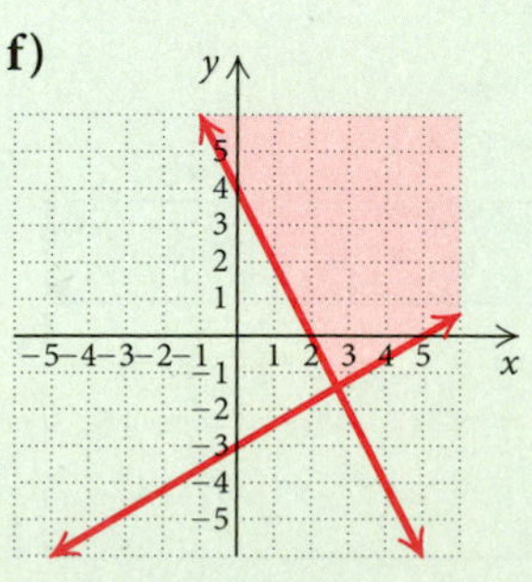

g)

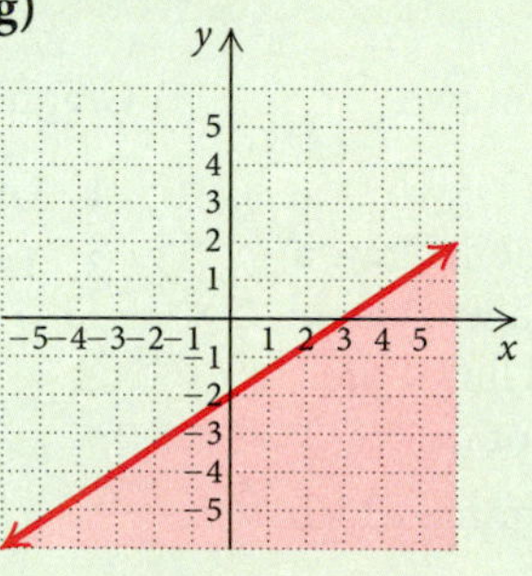

h)

5. $x + y = 7,$ $2x - y = 5$ [6.1]

6. $3x - 5y = -8,$ $4x + 3y = -1$ [6.1]

7. $y = 2x - 1,$ $4x - 2y = 2$ [6.1]

8. $6x - 3y = 5,$ $y = 2x + 3$ [6.1]

9. $y \leq 3x - 4$ [6.7]

10. $2x - 3y \geq 6$ [6.7]

11. $x - y \leq 3,$ $x + y \leq 5$ [6.7]

12. $2x + y \geq 4,$ $3x - 5y \leq 15$ [6.7]

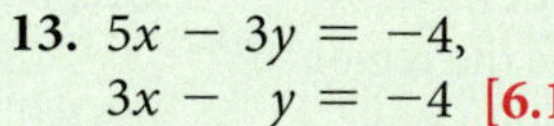

Solve.

13. $5x - 3y = -4,$ $3x - y = -4$ [6.1]

14. $2x + 3y = 2,$ $5x - y = -29$ [6.1]

15. $x + 5y = 12,$ $5x + 25y = 12$ [6.1]

16. $x + y = -2,$ $-3x - 3y = 6$ [6.1]

17. $x + 5y - 3z = 4,$ $3x - 2y + 4z = 3,$ $2x + 3y - z = 5$ [6.2]

18. $2x - 4y + 3z = -3,$ $-5x + 2y - z = 7,$ $3x + 2y - 2z = 4$ [6.2]

19. $x - y = 5,$ $y - z = 6,$ $-w + z = 7,$ $w + x = 8$ [6.2]

20. Classify each of the systems in Exercises 13–19 as either consistent or inconsistent. [6.1], [6.2]

21. Classify each of the systems in Exercises 13–19 as having either dependent equations or independent equations. [6.1], [6.2]

Solve the system of equations using Gaussian elimination or Gauss–Jordan elimination. [6.3]

22. $x + 2y = 5,$
$2x - 5y = -8$

23. $3x + 4y + 2z = 3,$
$5x - 2y - 13z = 3,$
$4x + 3y - 3z = 6$

24. $3x + 5y + z = 0,$
$2x - 4y - 3z = 0,$
$x + 3y + z = 0$

25. $w + x + y + z = -2,$
$-3w - 2x + 3y + 2z = 10,$
$2w + 3x + 2y - z = -12,$
$2w + 4x - y + z = 1$

26. *Coins.* The value of 75 coins, consisting of nickels and dimes, is \$5.95. How many of each kind of coin are there? [6.1]

27. *Investment.* The Davidson family invested \$5000, part at 3% and the remainder at 3.5%. The annual income from both investments is \$167. What is the amount invested at each rate? [6.1]

28. *Nutrition.* A dietician must plan a breakfast menu that provides 460 Cal, 9 g of fat, and 55 mg of calcium. One plain bagel contains 200 Cal, 2 g of fat, and 29 mg of calcium. A one-tablespoon serving of cream cheese contains 100 Cal, 10 g of fat, and 24 mg of calcium. One banana contains 105 Cal, 1 g of fat, and 7 g of calcium. (*Source: Home and Garden Bulletin No. 72*, U.S. Government Printing Office, Washington D.C. 20402) How many servings of each are required to provide the desired nutritional values? [6.2]

29. *Test Scores.* A student has a total of 226 on three tests. The sum of the scores on the first and second tests exceeds the score on the third test by 62. The first score exceeds the second by 6. Find the three scores. [6.2]

30. *Employed Civilians.* The following table lists the number of persons, ages 16 and older, employed in the United States, represented in terms of the number of years after 2008.

Year, x	Persons Ages 16 and Older Employed (in millions)
2008, 0	145
2010, 2	139
2012, 4	142

Source: Bureau of Labor Statistics, U.S. Department of Labor

a) Use a system of equations to fit a quadratic function $f(x) = ax^2 + bx + c$ to the data.

b) Use the function to estimate the number of persons employed in 2014.

For Exercises 31–38, let

$$\mathbf{A} = \begin{bmatrix} 1 & -1 & 0 \\ 2 & 3 & -2 \\ -2 & 0 & 1 \end{bmatrix}, \quad \mathbf{B} = \begin{bmatrix} -1 & 0 & 6 \\ 1 & -2 & 0 \\ 0 & 1 & -3 \end{bmatrix},$$

and

$$\mathbf{C} = \begin{bmatrix} -2 & 0 \\ 1 & 3 \end{bmatrix}.$$

Find each of the following, if possible. [6.4]

31. $\mathbf{A} + \mathbf{B}$

32. $-3\mathbf{A}$

33. $-\mathbf{A}$

34. $\mathbf{AB}$

35. $\mathbf{B} + \mathbf{C}$

36. $\mathbf{A} - \mathbf{B}$

37. $\mathbf{BA}$

38. $\mathbf{A} + 3\mathbf{B}$

39. *Food Service Management.* The following table lists the cost per serving, in dollars, for items on four menus that are served at an NFL football training camp.

Menu	Meat	Potato	Vegetable	Salad	Dessert
1	2.25	0.38	0.55	0.33	0.85
2	3.09	0.42	0.46	0.48	0.51
3	2.40	0.31	0.59	0.36	0.64
4	1.80	0.29	0.34	0.55	0.52

On a particular day, a dietician orders 41 meals from menu 1, 18 from menu 2, 39 from menu 3, and 36 from menu 4.

a) Write the information in the table as a 4×5 matrix **M**. [6.4]

b) Write a row matrix **N** that represents the number of each menu ordered. [6.4]

c) Find the product **NM**. [6.4]

d) State what the entries of **NM** represent. [6.4]

Find $\mathbf{A}^{-1}$*, if it exists.* [6.5]

40. $\mathbf{A} = \begin{bmatrix} -2 & 0 \\ 1 & 3 \end{bmatrix}$

41. $\mathbf{A} = \begin{bmatrix} 0 & 0 & 3 \\ 0 & -2 & 0 \\ 4 & 0 & 0 \end{bmatrix}$

42. $\mathbf{A} = \begin{bmatrix} 1 & 0 & 0 & 0 \\ 0 & 4 & -5 & 0 \\ 0 & 2 & 2 & 0 \\ 0 & 0 & 0 & 1 \end{bmatrix}$

43. Write a matrix equation equivalent to this system of equations:

$$3x - 2y + 4z = 13,$$
$$x + 5y - 3z = 7,$$
$$2x - 3y + 7z = -8. \quad [6.4]$$

Solve the system of equations using the inverse of the coefficient matrix of the equivalent matrix equation. [6.5]

44. $2x + 3y = 5,$
$3x + 5y = 11$

45. $5x - y + 2z = 17,$
$3x + 2y - 3z = -16,$
$4x - 3y - z = 5$

46. $w - x - y + z = -1,$
$2w + 3x - 2y - z = 2,$
$-w + 5x + 4y - 2z = 3,$
$3w - 2x + 5y + 3z = 4$

Evaluate the determinant. [6.6]

47. $\begin{vmatrix} 1 & -2 \\ 3 & 4 \end{vmatrix}$

48. $\begin{vmatrix} \sqrt{3} & -5 \\ -3 & -\sqrt{3} \end{vmatrix}$

49. $\begin{vmatrix} 2 & -1 & 1 \\ 1 & 2 & -1 \\ 3 & 4 & -3 \end{vmatrix}$

50. $\begin{vmatrix} 1 & -1 & 2 \\ -1 & 2 & 0 \\ -1 & 3 & 1 \end{vmatrix}$

Solve using Cramer's rule. [6.6]

51. $5x - 2y = 19,$
$7x + 3y = 15$

52. $x + y = 4,$
$4x + 3y = 11$

53. $3x - 2y + z = 5,$
$4x - 5y - z = -1,$
$3x + 2y - z = 4$

54. $2x - y - z = 2,$
$3x + 2y + 2z = 10,$
$x - 5y - 3z = -2$

Graph. [6.7]

55. $y \le 3x + 6$

56. $4x - 3y \ge 12$

57. Graph this system of inequalities and find the coordinates of any vertices formed. [6.7]

$$2x + y \ge 9,$$
$$4x + 3y \ge 23,$$
$$x + 3y \ge 8,$$
$$x \ge 0,$$
$$y \ge 0$$

58. Find the maximum value and the minimum value of $T = 6x + 10y$ subject to

$$x + y \le 10,$$
$$5x + 10y \ge 50,$$
$$x \ge 2,$$
$$y \ge 0. \quad [6.7]$$

59. *Maximizing a Test Score.* Jackson is taking a test that contains questions in group A worth 7 points each and questions in group B worth 12 points each. The total number of questions answered must be at least 8. If Jackson knows that group A questions take 8 min each and group B questions take 10 min each and the maximum time for the test is 80 min, how many questions from each group must he answer correctly in order to maximize his score? What is the maximum score? [6.7]

Decompose into partial fractions. [6.8]

60. $\dfrac{5}{(x + 2)^2(x + 1)}$

61. $\dfrac{-8x + 23}{2x^2 + 5x - 12}$

62. Solve: $2x + y = 7,$
$x - 2y = 6.$ [6.1]

A. x and y are both positive numbers.
B. x and y are both negative numbers.
C. x is positive and y is negative.
D. x is negative and y is positive.

63. Which of the following is *not* a row-equivalent operation on a matrix? [6.3]

A. Interchange any two columns.
B. Interchange any two rows.
C. Add two rows.
D. Multiply each entry in a row by -3.

64. The graph of the given system of inequalities is which of the following? [6.7]

$$x + y \le 3,$$
$$x - y \le 4$$

A.

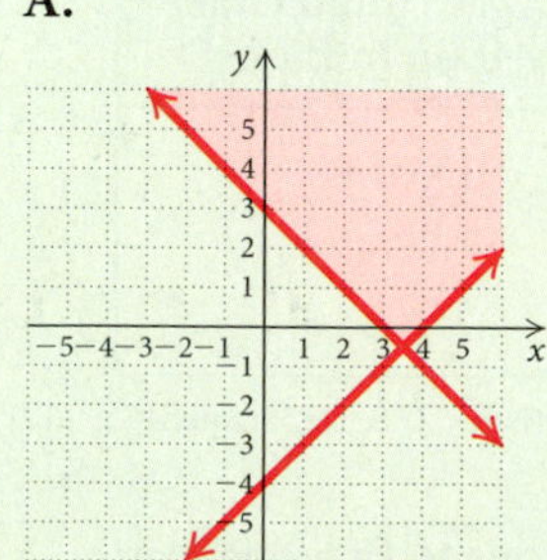

B.

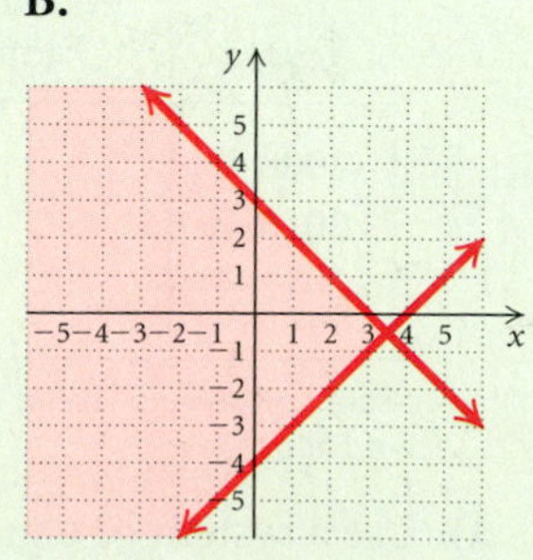

C.

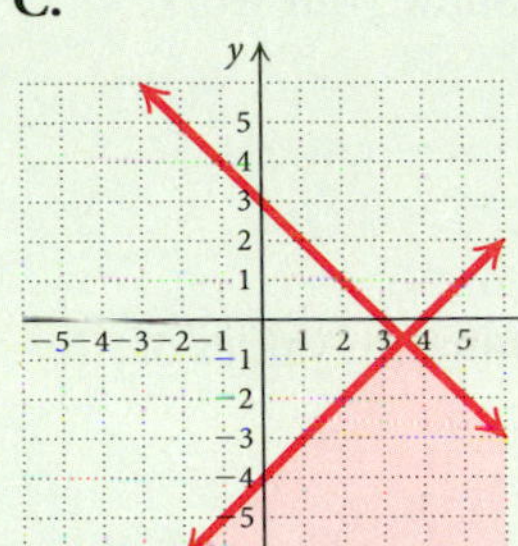

D.

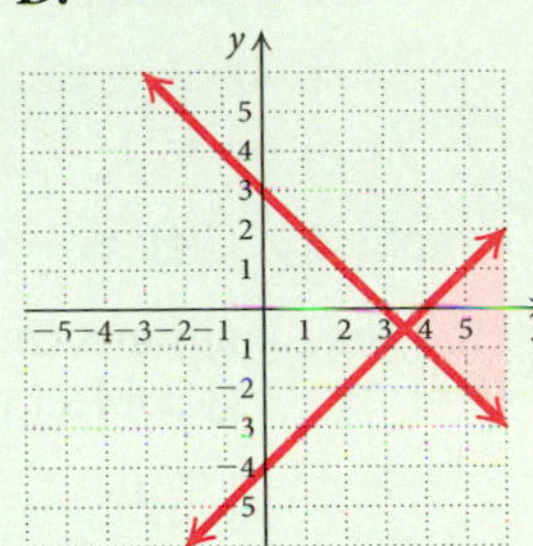

▶ Synthesis

65. One year, Lucia invested a total of $40,000, part at 4%, part at 5%, and the rest at $5\frac{1}{2}$%. The total amount of interest received on the investments was $1990. The interest received on the $5\frac{1}{2}$% investment was $590 more than the interest received on the 4% investment. How much was invested at each rate? [6.2]

Solve.

66. $\frac{2}{3x} + \frac{4}{5y} = 8,$
$\frac{5}{4x} - \frac{3}{2y} = -6$ [6.1]

67. $\frac{3}{x} - \frac{4}{y} + \frac{1}{z} = -2,$
$\frac{5}{x} + \frac{1}{y} - \frac{2}{z} = 1,$
$\frac{7}{x} + \frac{3}{y} + \frac{2}{z} = 19$ [6.2]

Graph. [6.7]

68. $|x| - |y| \le 1$

69. $|xy| > 1$

▶ Collaborative Discussion and Writing

70. Dylon solves the equation $2x + 5 = 3x - 7$ by finding the point of intersection of the graphs of $y_1 = 2x + 5$ and $y_2 = 3x - 7$. He finds the same point when he solves the system of equations

$y = 2x + 5,$
$y = 3x - 7.$

Explain the difference between the solution of the equation and the solution of the system of equations. [6.1]

71. For square matrices **A** and **B**, is it true, in general, that $(\mathbf{AB})^2 = \mathbf{A}^2\mathbf{B}^2$? Explain. [6.4]

72. Given the system of equations

$a_1x + b_1y = c_1,$
$a_2x + b_2y = c_2,$

explain why the equations are dependent or the system is inconsistent when

$$\begin{vmatrix} a_1 & b_1 \\ a_2 & b_2 \end{vmatrix} = 0.$$ [6.6]

73. If the lines $a_1x + b_1y = c_1$ and $a_2x + b_2y = c_2$ are parallel, what can you say about the values of

$\begin{vmatrix} a_1 & b_1 \\ a_2 & b_2 \end{vmatrix}$, $\begin{vmatrix} c_1 & b_1 \\ c_2 & b_2 \end{vmatrix}$, and $\begin{vmatrix} a_1 & c_1 \\ a_2 & c_2 \end{vmatrix}$? [6.6]

74. Describe how the graph of a linear inequality differs from the graph of a linear equation. [6.7]

75. What would you say to a classmate who tells you that the partial fraction decomposition of

$$\frac{3x^2 - 8x + 9}{(x + 3)(x^2 - 5x + 6)}$$

is

$$\frac{2}{x + 3} + \frac{x - 1}{x^2 - 5x + 6}?$$

Explain. [6.8]

6 Chapter Test

Solve. Use any method. Also determine whether the system is consistent or inconsistent and whether the equations are dependent or independent.

1. $3x + 2y = 1,$
$2x - y = -11$

2. $2x - y = 3,$
$2y = 4x - 6$

3. $x - y = 4,$
$3y = 3x - 8$

4. $2x - 3y = 8,$
$5x - 2y = 9$

Solve.

5. $4x + 2y + z = 4,$
$3x - y + 5z = 4,$
$5x + 3y - 3z = -2$

6. *Ticket Sales.* One evening, 620 tickets were sold for Clearview Community College's talent show. Tickets cost $8 each for students and $12 each for nonstudents. Total receipts were \$5592. How many of each type of ticket were sold?

7. Hui, Ashlyn, and Sheriann can process 352 telephone orders per day. Hui and Ashlyn together can process 224 orders per day while Hui and Sheriann together can process 248 orders per day. How many orders can each of them process alone?

For Exercises 8–13, let

$$\mathbf{A} = \begin{bmatrix} 1 & -1 & 3 \\ -2 & 5 & 2 \end{bmatrix}, \quad \mathbf{B} = \begin{bmatrix} -5 & 1 \\ -2 & 4 \end{bmatrix},$$

and

$$\mathbf{C} = \begin{bmatrix} 3 & -4 \\ -1 & 0 \end{bmatrix}.$$

Find each of the following, if possible.

8. $\mathbf{B} + \mathbf{C}$ **9.** $\mathbf{A} - \mathbf{C}$ **10.** $\mathbf{CB}$

11. $\mathbf{AB}$ **12.** $2\mathbf{A}$ **13.** $\mathbf{C}^{-1}$

14. *Food Service Management.* The following table lists the cost per serving, in dollars, for items on three lunch menus served at a senior citizens' center.

Menu	Main Dish	Side Dish	Dessert
1	1.55	1.00	0.99
2	1.70	0.95	1.01
3	1.65	0.99	0.96

On a particular day, 26 Menu 1 meals, 18 Menu 2 meals, and 23 Menu 3 meals are served.

a) Write the information in the table as a 3×3 matrix $\mathbf{M}$.

b) Write a row matrix $\mathbf{N}$ that represents the number of each menu served.

c) Find the product $\mathbf{NM}$.

d) State what the entries of $\mathbf{NM}$ represent.

15. Write a matrix equation equivalent to the system of equations

$$3x - 4y + 2z = -8,$$
$$2x + 3y + z = 7,$$
$$x - 5y - 3z = 3.$$

16. Solve the system of equations using the inverse of the coefficient matrix of the equivalent matrix equation.

$$3x + 2y + 6z = 2,$$
$$x + y + 2z = 1,$$
$$2x + 2y + 5z = 3$$

Evaluate the determinant.

17. $\begin{vmatrix} 3 & -5 \\ 8 & 7 \end{vmatrix}$ **18.** $\begin{vmatrix} 2 & -1 & 4 \\ -3 & 1 & -2 \\ 5 & 3 & -1 \end{vmatrix}$

19. Solve using Cramer's rule. Show your work.

$$5x + 2y = -1,$$
$$7x + 6y = 1$$

20. Graph: $3x + 4y \leq -12$.

21. Find the maximum and minimum values of $Q = 2x + 3y$ subject to

$$x + y \leq 6,$$
$$2x - 3y \geq -3,$$
$$x \geq 1,$$
$$y \geq 0.$$

22. *Maximizing Profit.* Jane's Cakes prepares pound cakes and carrot cakes. In a given week, at most 100 cakes can be prepared, of which 25 pound cakes and 15 carrot cakes are required by regular customers. The profit from each pound cake is \$6, and the profit from each carrot cake is \$8. How many of each type of cake should be prepared in order to maximize the profit? What is the maximum profit?

23. Decompose into partial fractions:

$$\frac{3x - 11}{x^2 + 2x - 3}.$$

24. The graph of the given system of inequalities is which of the following?

$$x + 2y \geq 4,$$
$$x - y \leq 2$$

A.

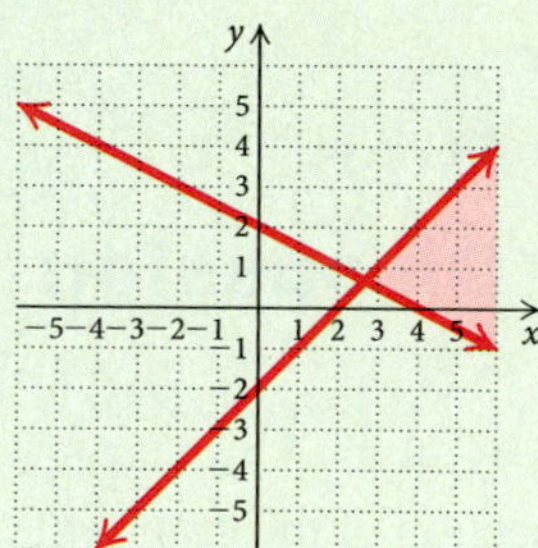

B.

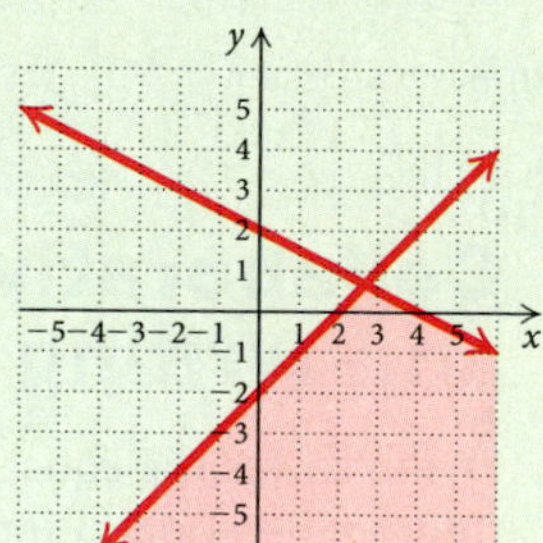

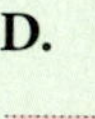

C.

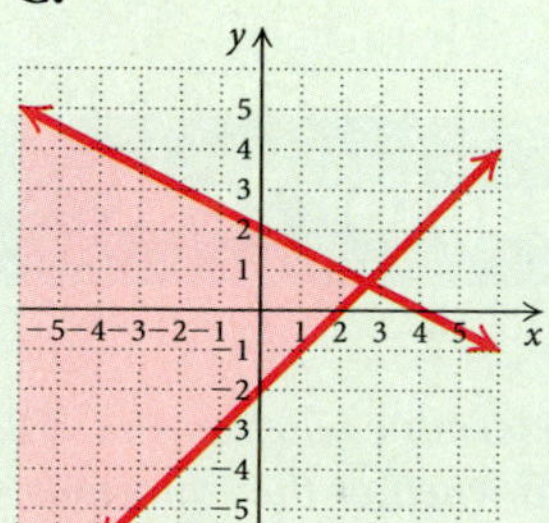

D.

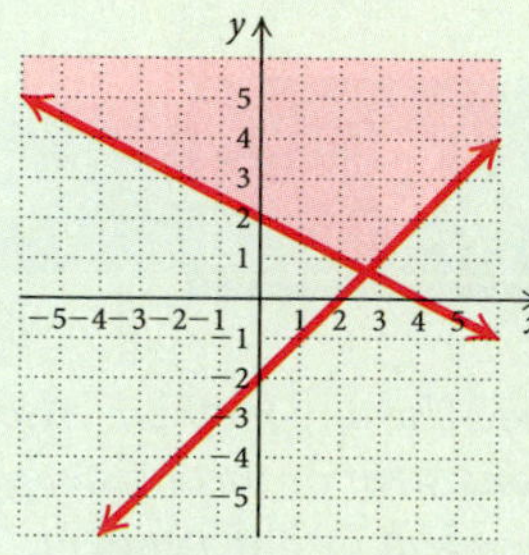

Synthesis

25. Three solutions of the equation $Ax - By = Cz - 8$ are $(2, -2, 2)$, $(-3, -1, 1)$, and $(4, 2, 9)$. Find A, B, and C.

CHAPTER

7

Conic Sections

PPLICATION This problem appears as Exercise 54 in Section 7.2.

The lighting of the National Christmas Tree located on the Ellipse, a large grassy area south of the White House, marks the beginning of the holiday season in Washington, D.C. This area of the lawn is actually an ellipse with major axis of length 1048 ft and minor axis of length 898 ft. Assuming that a coordinate system is superimposed on the area in such a way that the center is at the origin and the major and minor axes are on the x- and y-axes of the coordinate system, respectively, find an equation of the ellipse.

7.1 The Parabola

- Given an equation of a parabola, complete the square, if necessary, and then find the vertex, the focus, and the directrix and graph the parabola.

A **conic section** is formed when a right circular cone with two parts, called *nappes*, is intersected by a plane. One of four types of curves can be formed: a parabola, a circle, an ellipse, or a hyperbola.

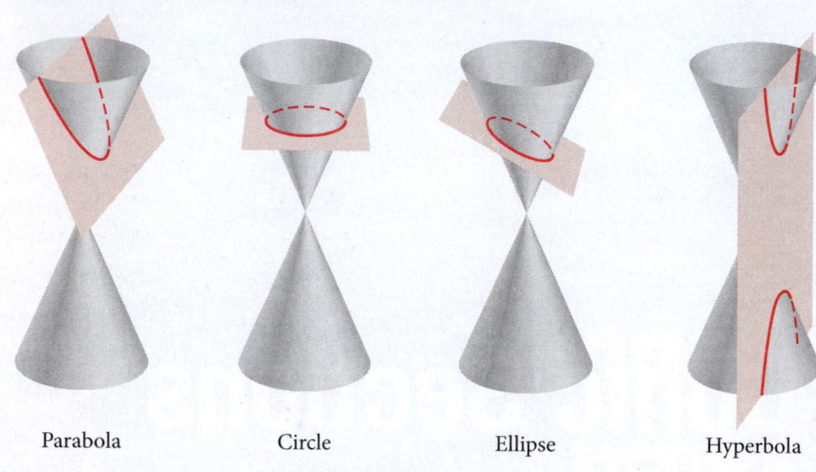

Conic Sections

Conic sections can be defined algebraically using second-degree equations of the form $Ax^2 + Bxy + Cy^2 + Dx + Ey + F = 0$. In addition, they can be defined geometrically as a set of points that satisfy certain conditions.

Parabolas

The graph of the quadratic function $f(x) = ax^2 + bx + c, a \neq 0$, is a parabola. A parabola can be defined geometrically.

> **PARABOLA**
>
> A **parabola** is the set of all points in a plane equidistant from a fixed line (the **directrix**) and a fixed point not on the line (the **focus**).

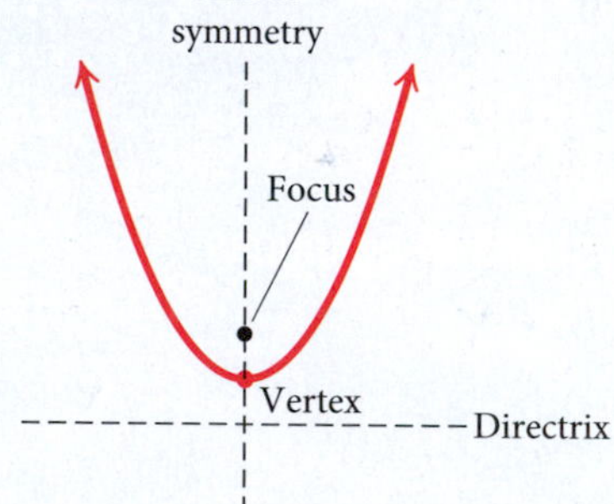

The line that is perpendicular to the directrix and contains the focus is the **axis of symmetry**. The **vertex** is the midpoint of the segment between the focus and the directrix. (See the figure at left.)

Let's derive the standard equation of a parabola with vertex $(0, 0)$ and directrix $y = -p$, where $p > 0$. We place the coordinate axes as shown in Fig. 1. The y-axis is the axis of symmetry and contains the focus F. The distance from the focus to the vertex is the same as the distance from the vertex to the directrix. Thus the coordinates of F are $(0, p)$.

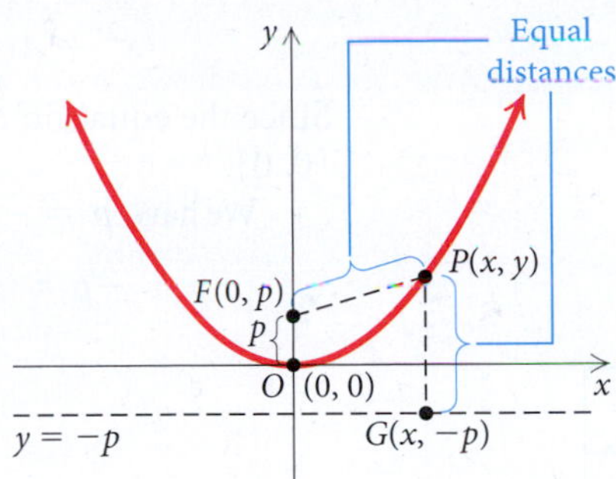

Figure 1.

Let $P(x, y)$ be any point on the parabola and consider $\overline{PG}$ perpendicular to the line $y = -p$. The coordinates of G are $(x, -p)$. By the definition of a parabola,

$PF = PG$. **The distance from P to the focus is the same as the distance from P to the directrix.**

Then using the distance formula, we have

$$\sqrt{(x-0)^2 + (y-p)^2} = \sqrt{(x-x)^2 + [y-(-p)]^2}$$

$$x^2 + y^2 - 2py + p^2 = y^2 + 2py + p^2$$ **Squaring both sides and squaring the binomials**

$$x^2 = 4py.$$

We have shown that if $P(x, y)$ is on the parabola shown in Fig. 1, then its coordinates satisfy this equation. The converse is also true, but we will not prove it here.

Note that if $p > 0$, as above, the graph opens up. If $p < 0$, the graph opens down.

The equation of a parabola with vertex $(0, 0)$ and directrix $x = -p$ is derived similarly. Such a parabola opens either to the right $(p > 0)$, as shown in Fig. 2, or to the left $(p < 0)$.

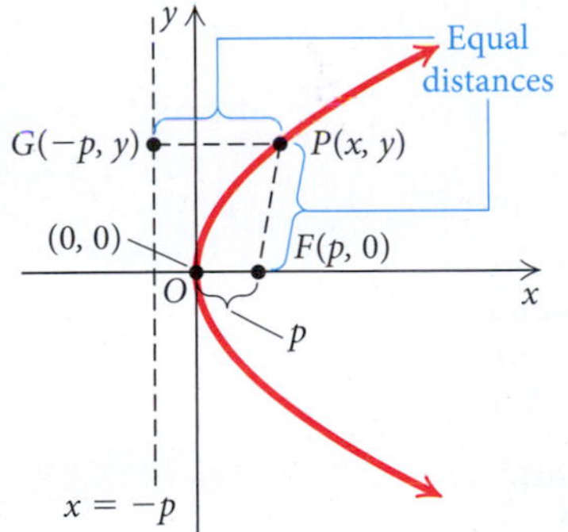

Figure 2.

STANDARD EQUATION OF A PARABOLA WITH VERTEX AT THE ORIGIN

The standard equation of a parabola with vertex $(0, 0)$ and directrix $y = -p$ is

$$x^2 = 4py.$$

The focus is $(0, p)$, and the y-axis is the axis of symmetry.

The standard equation of a parabola with vertex $(0, 0)$ and directrix $x = -p$ is

$$y^2 = 4px.$$

The focus is $(p, 0)$, and the x-axis is the axis of symmetry.

EXAMPLE 1 Find the vertex, the focus, and the directrix of the parabola $y = -\frac{1}{12}x^2$. Then graph the parabola.

Solution We write $y = -\frac{1}{12}x^2$ in the form $x^2 = 4py$:

$$-\frac{1}{12}x^2 = y \qquad \text{Given equation}$$

$$x^2 = -12y \qquad \text{Multiplying by } -12 \text{ on both sides}$$

$$x^2 = 4(-3)y. \qquad \text{Standard form}$$

Since the equation can be written in the form $x^2 = 4py$, we know that the vertex is $(0, 0)$.

We have $p = -3$, so the focus is $(0, p)$, or $(0, -3)$. The directrix is

$$y = -p = -(-3) = 3.$$

x	y
0	0
± 1	$-\frac{1}{12}$
± 2	$-\frac{1}{3}$
± 3	$-\frac{3}{4}$
± 4	$-\frac{4}{3}$

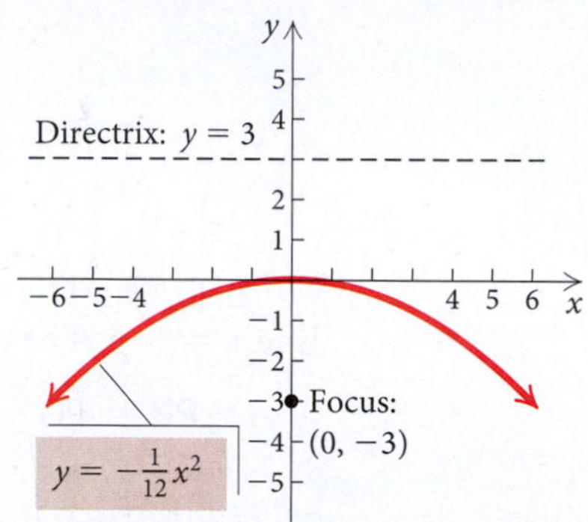

Now Try Exercise 7.

EXAMPLE 2 Find an equation of the parabola with vertex $(0, 0)$ and focus $(5, 0)$. Then graph the parabola.

Solution The focus is on the x-axis so the line of symmetry is the x-axis. Thus the equation is of the type

$$y^2 = 4px.$$

Since the focus $(5, 0)$ is 5 units to the right of the vertex, $p = 5$ and the equation is

$$y^2 = 4(5)x, \quad \text{or} \quad y^2 = 20x.$$

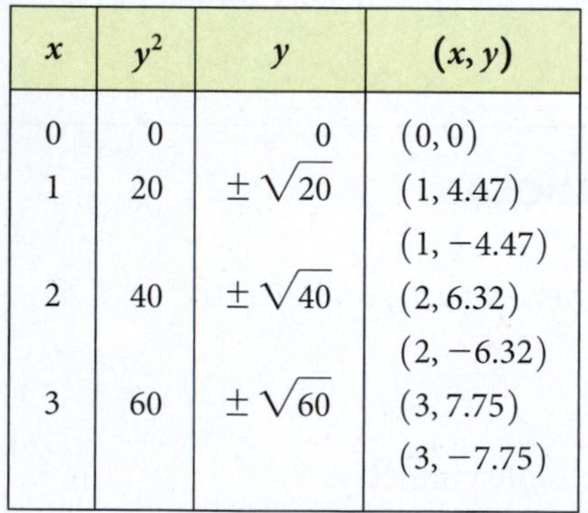

x	y^2	y	(x, y)
0	0	0	$(0, 0)$
1	20	$\pm\sqrt{20}$	$(1, 4.47)$
			$(1, -4.47)$
2	40	$\pm\sqrt{40}$	$(2, 6.32)$
			$(2, -6.32)$
3	60	$\pm\sqrt{60}$	$(3, 7.75)$
			$(3, -7.75)$

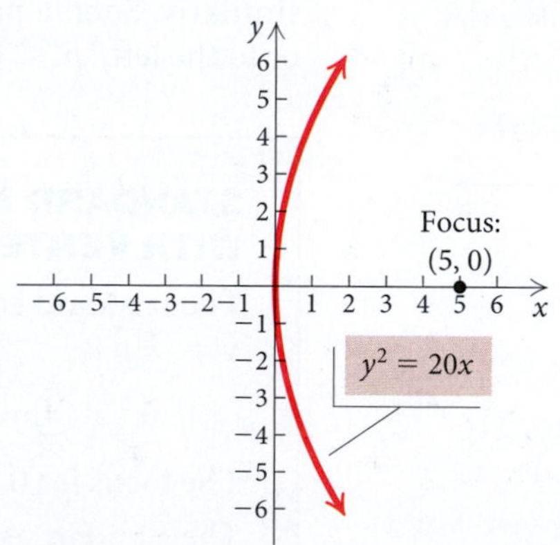

Now Try Exercise 17.

Technology Connection

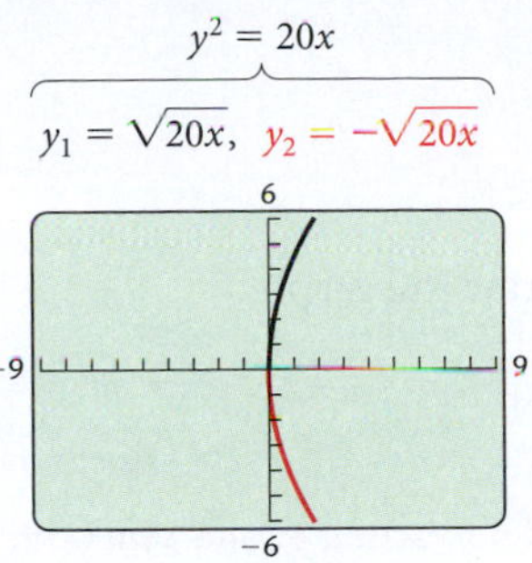

We can use a graphing calculator to graph parabolas. Consider the parabola in Example 2. It might be necessary to solve the equation for y before entering it in the calculator:

$$y^2 = 20x$$
$$y = \pm\sqrt{20x}.$$

We now graph $y_1 = \sqrt{20x}$ and $y_2 = -\sqrt{20x}$ or $y_1 = \sqrt{20x}$ and $y_2 = -y_1$ in a squared viewing window.

On some graphing calculators, the Conics application from the APPS menu can be used to graph parabolas. This method will be discussed following Example 4.

Finding Standard Form by Completing the Square

If a parabola with vertex at the origin is translated horizontally $|h|$ units and vertically $|k|$ units, it has an equation as follows.

STANDARD EQUATION OF A PARABOLA WITH VERTEX (h, k) AND VERTICAL AXIS OF SYMMETRY

The standard equation of a parabola with vertex (h, k) and vertical axis of symmetry is

$$(x - h)^2 = 4p(y - k),$$

where the vertex is (h, k), the focus is $(h, k + p)$, and the directrix is $y = k - p$.

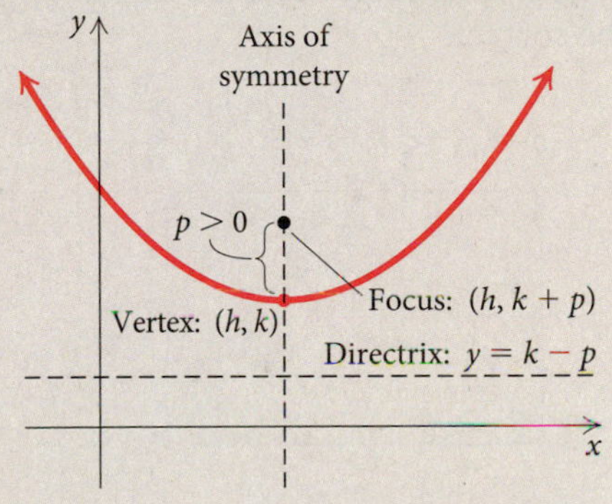

(When $p < 0$, the parabola opens down.)

STANDARD EQUATION OF A PARABOLA WITH VERTEX (h, k) AND HORIZONTAL AXIS OF SYMMETRY

The standard equation of a parabola with vertex (h, k) and horizontal axis of symmetry is

$$(y - k)^2 = 4p(x - h),$$

where the vertex is (h, k), the focus is $(h + p, k)$, and the directrix is $x = h - p$.

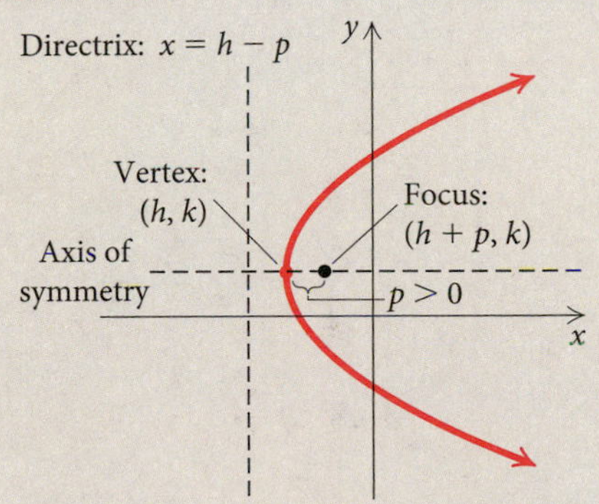

(When $p < 0$, the parabola opens to the left.)

COMPLETING THE SQUARE

REVIEW SECTION 3.2

We can complete the square on equations of the form

$$y = ax^2 + bx + c \quad \text{or} \quad x = ay^2 + by + c$$

in order to write them in standard form.

JUST IN TIME 13

EXAMPLE 3 For the parabola

$$x^2 + 6x + 4y + 5 = 0,$$

find the vertex, the focus, and the directrix. Then draw the graph.

Solution We first complete the square:

$$x^2 + 6x + 4y + 5 = 0$$

$$x^2 + 6x = -4y - 5 \quad \text{Subtracting } 4y \text{ and 5 on both sides}$$

$$x^2 + 6x + 9 = -4y - 5 + 9 \quad \text{Adding 9 on both sides to complete the square on the left side}$$

$$x^2 + 6x + 9 = -4y + 4$$

$$(x + 3)^2 = -4(y - 1) \quad \text{Factoring}$$

$$[x - (-3)]^2 = 4(-1)(y - 1). \quad \text{Writing standard form: } (x - h)^2 = 4p(y - k)$$

We see that $h = -3$, $k = 1$, and $p = -1$, so we have the following:

Vertex (h, k): $(-3, 1)$;

Focus $(h, k + p)$: $(-3, 1 + (-1))$, or $(-3, 0)$;

Directrix $y = k - p$: $y = 1 - (-1)$, or $y = 2$.

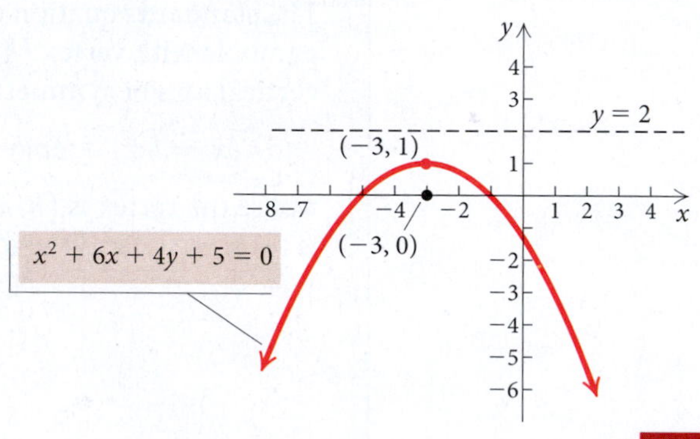

Technology Connection

We can check the graph in Example 3 on a graphing calculator using a squared viewing window. It might be necessary to solve for y first:

$$x^2 + 6x + 4y + 5 = 0$$

$$4y = -x^2 - 6x - 5$$

$$y = \tfrac{1}{4}(-x^2 - 6x - 5).$$

The graph at right appears to be correct.

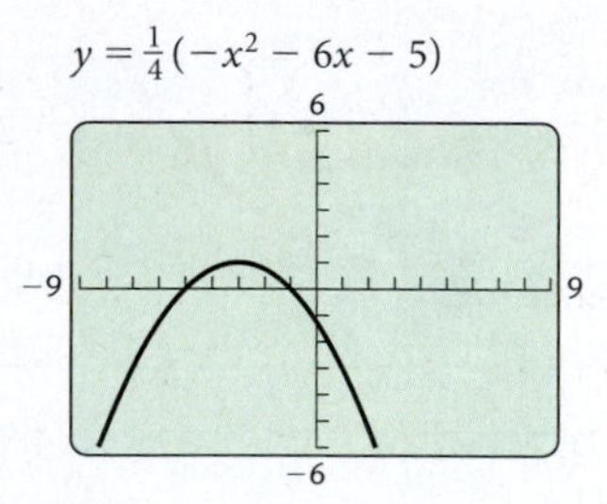

Now Try Exercise 25.

EXAMPLE 4 For the parabola

$$y^2 - 2y - 8x - 31 = 0,$$

find the vertex, the focus, and the directrix. Then draw the graph.

Solution We first complete the square:

$$y^2 - 2y - 8x - 31 = 0$$

$$y^2 - 2y = 8x + 31 \quad \text{Adding } 8x \text{ and 31 on both sides}$$

$$y^2 - 2y + 1 = 8x + 31 + 1 \quad \text{Adding 1 on both sides to complete the square on the left side}$$

$$y^2 - 2y + 1 = 8x + 32$$

$$(y - 1)^2 = 8(x + 4) \quad \text{Factoring}$$

$$(y - 1)^2 = 4(2)[x - (-4)]. \quad \text{Writing standard form: } (y - k)^2 = 4p(x - h)$$

Technology Connection

When the equation of a parabola is written in standard form, we can use the Conics PARABOLA APP to graph it. The standard form of the equation of the parabola in Example 4 is $(y - 1)^2 = 4(2)[x - (-4)]$. We enter -4 for H, 1 for K, and 2 for P.

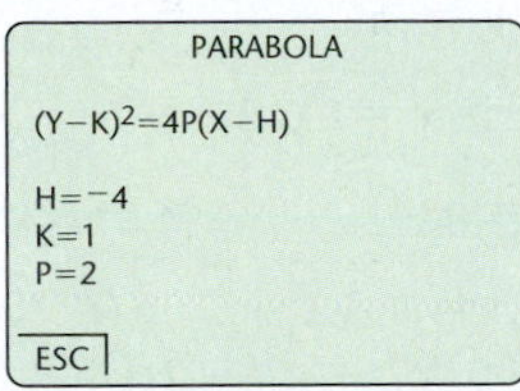

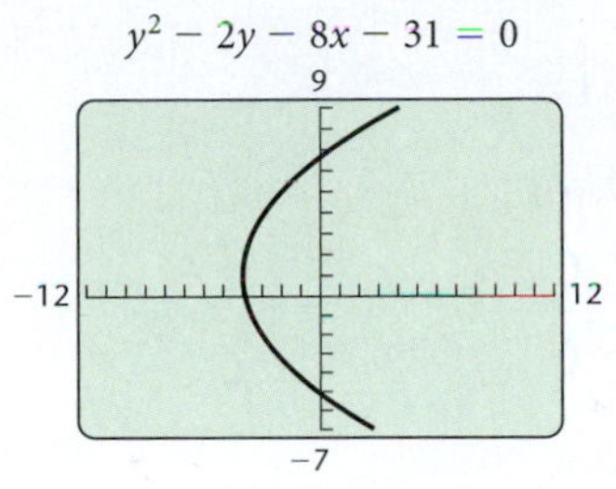

We see that $h = -4$, $k = 1$, and $p = 2$, so we have the following:

Vertex (h, k): $(-4, 1)$;

Focus $(h + p, k)$: $(-4 + 2, 1)$, or $(-2, 1)$;

Directrix $x = h - p$: $x = -4 - 2$, or $x = -6$.

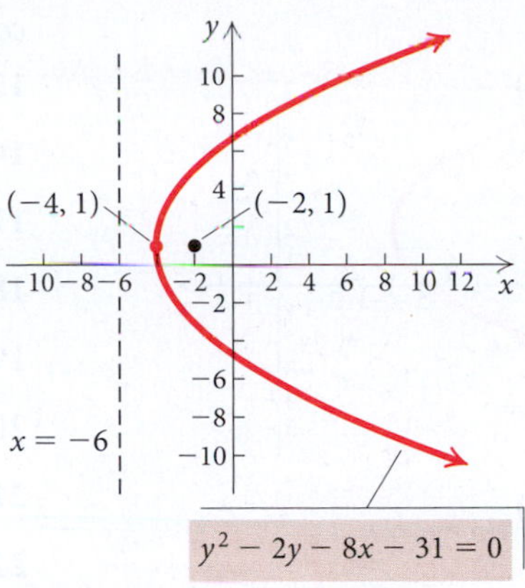

Now Try Exercise 31.

Applications

Parabolas have many applications. For example, cross sections of car headlights, flashlights, and searchlights are parabolas. The bulb is located at the focus and light from that point is reflected outward parallel to the axis of symmetry. Satellite dishes and field microphones used at sporting events often have parabolic cross sections. Incoming radio waves or sound waves parallel to the axis are reflected into the focus.

Similarly, in solar cooking, a parabolic mirror is mounted on a rack with a cooking pot hung in the focal area. Incoming sun rays parallel to the axis are reflected into the focus, producing a temperature high enough for cooking.

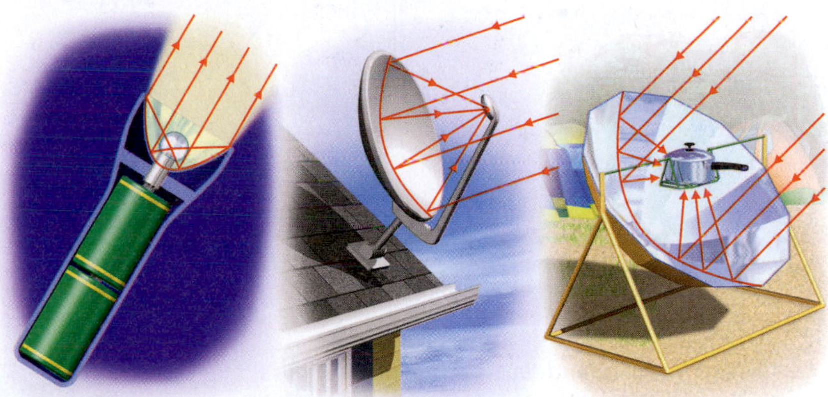

7.1 Exercise Set

In Exercises 1–6, match the equation with one of the graphs (a)–(f) that follow.

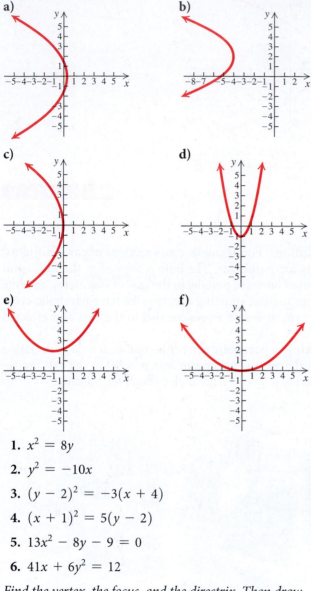

a)

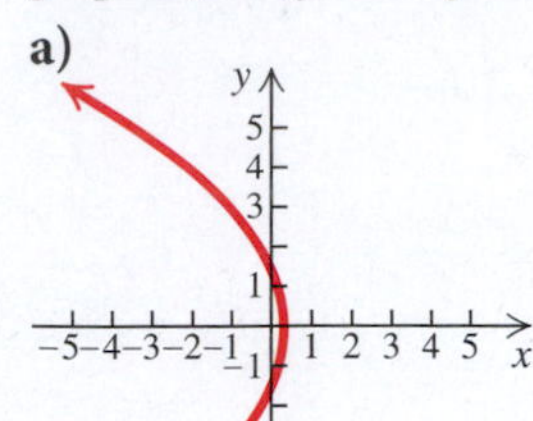

b)

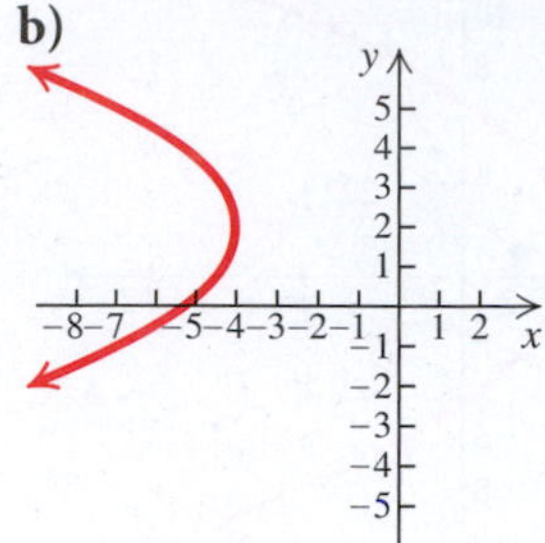

c)

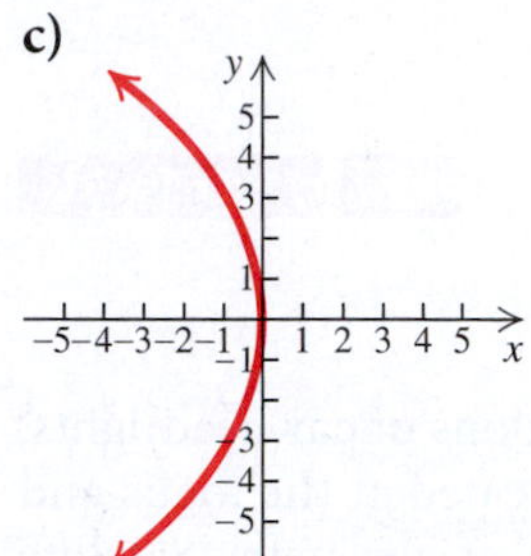

d)

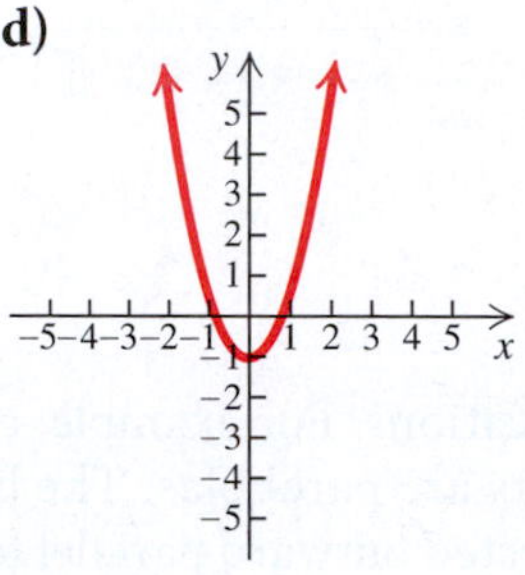

e)

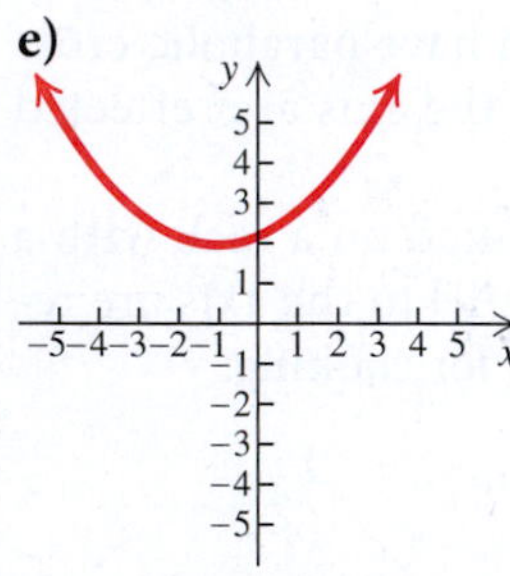

f)

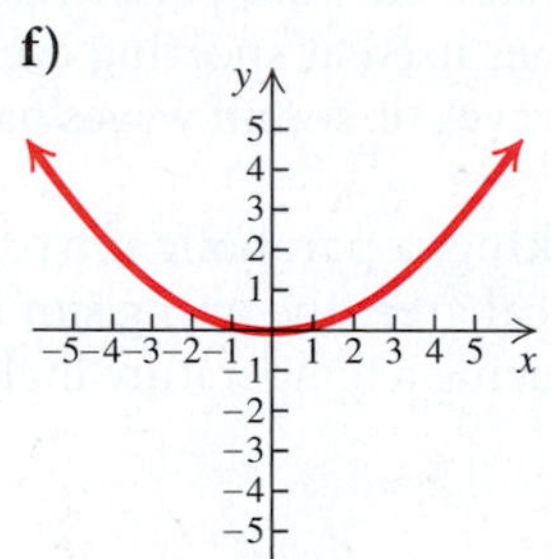

1. $x^2 = 8y$

2. $y^2 = -10x$

3. $(y - 2)^2 = -3(x + 4)$

4. $(x + 1)^2 = 5(y - 2)$

5. $13x^2 - 8y - 9 = 0$

6. $41x + 6y^2 = 12$

Find the vertex, the focus, and the directrix. Then draw the graph.

7. $x^2 = 20y$

8. $x^2 = 16y$

9. $y^2 = -6x$

10. $y^2 = -2x$

11. $x^2 - 4y = 0$

12. $y^2 + 4x = 0$

13. $x = 2y^2$

14. $y = \frac{1}{2}x^2$

Find an equation of a parabola satisfying the given conditions.

15. Vertex $(0, 0)$, focus $(-3, 0)$

16. Vertex $(0, 0)$, focus $(0, 10)$

17. Focus $(7, 0)$, directrix $x = -7$

18. Focus $\left(0, \frac{1}{4}\right)$, directrix $y = -\frac{1}{4}$

19. Focus $(0, -\pi)$, directrix $y = \pi$

20. Focus $\left(-\sqrt{2}, 0\right)$, directrix $x = \sqrt{2}$

21. Focus $(3, 2)$, directrix $x = -4$

22. Focus $(-2, 3)$, directrix $y = -3$

Find the vertex, the focus, and the directrix. Then draw the graph.

23. $(x + 2)^2 = -6(y - 1)$

24. $(y - 3)^2 = -20(x + 2)$

25. $x^2 + 2x + 2y + 7 = 0$

26. $y^2 + 6y - x + 16 = 0$

27. $x^2 - y - 2 = 0$

28. $x^2 - 4x - 2y = 0$

29. $y = x^2 + 4x + 3$

30. $y = x^2 + 6x + 10$

31. $y^2 - y - x + 6 = 0$

32. $y^2 + y - x - 4 = 0$

33. *Satellite Dish.* An engineer designs a satellite dish with a parabolic cross section. The dish is 15 ft wide at the opening, and the focus is placed 4 ft from the vertex.

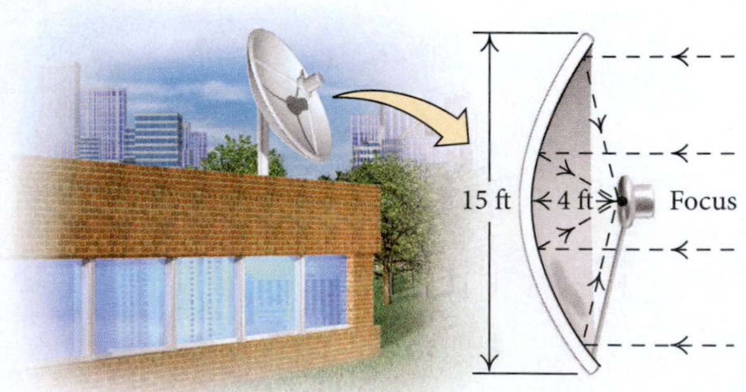

a) Position a coordinate system with the origin at the vertex and the x-axis on the parabola's axis of symmetry and find an equation of the parabola.
b) Find the depth of the satellite dish at the vertex.

34. *Flashlight Mirror.* A heavy-duty flashlight mirror has a parabolic cross section with diameter 6 in. and depth 1 in.

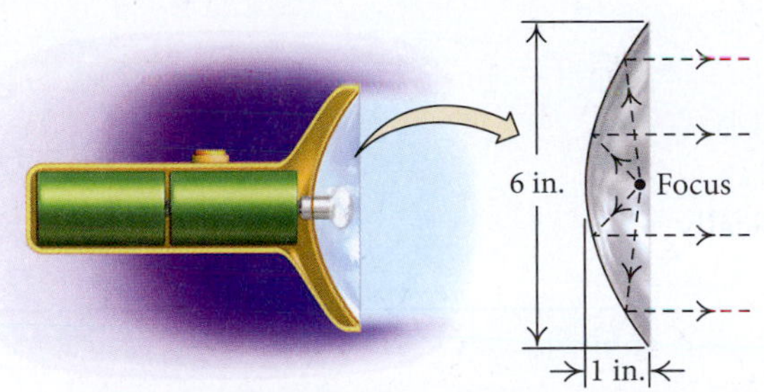

a) Position a coordinate system with the origin at the vertex and the x-axis on the parabola's axis of symmetry and find an equation of the parabola.
b) How far from the vertex should the bulb be positioned if it is to be placed at the focus?

35. *Spotlight.* A spotlight has a parabolic cross section that is 4 ft wide at the opening and 1.5 ft deep at the vertex. How far from the vertex is the focus?

36. *Ultrasound Receiver.* Information Unlimited designed and sells the Ultrasonic Receiver, which detects sounds unable to be heard by the human ear. The HT90P can detect mechanical and electrical sounds such as leaking gases, air, corona, and motor friction noises. It can also be used to hear bats, insects, and even beading water. The receiver has a parabolic cross section and is 2.625 in. deep. The focus is 3.287 in. from the vertex. (*Source:* Information Unlimited, Amherst, NH, Robert Iannini, President) Find the diameter of the outside edge of the receiver.

Skill Maintenance

Consider the following linear equations. Without graphing them, answer the questions below.

a) $y = 2x$
b) $y = \frac{1}{3}x + 5$
c) $y = -3x - 2$
d) $y = -0.9x + 7$
e) $y = -5x + 3$
f) $y = x + 4$
g) $8x - 4y = 7$
h) $3x + 6y = 2$

37. Which has/have x-intercept $\left(\frac{2}{3}, 0\right)$? [1.1]

38. Which has/have y-intercept $(0, 7)$? [1.1], [1.4]

39. Which slant up from left to right? [1.3]

40. Which has the least steep slant? [1.3]

41. Which has/have slope $\frac{1}{3}$? [1.3]

42. Which, if any, contain the point $(3, 7)$? [1.1]

43. Which, if any, are parallel? [1.4]

44. Which, if any, are perpendicular? [1.4]

Synthesis

45. Find an equation of the parabola with a vertical axis of symmetry and vertex $(-1, 2)$ and containing the point $(-3, 1)$.

46. Find an equation of a parabola with a horizontal axis of symmetry and vertex $(-2, 1)$ and containing the point $(-3, 5)$.

47. *Suspension Bridge.* The parabolic cables of a 200-ft portion of the roadbed of a suspension bridge are positioned as shown below. Vertical cables are to be spaced every 20 ft along this portion of the roadbed. Calculate the lengths of these vertical cables.

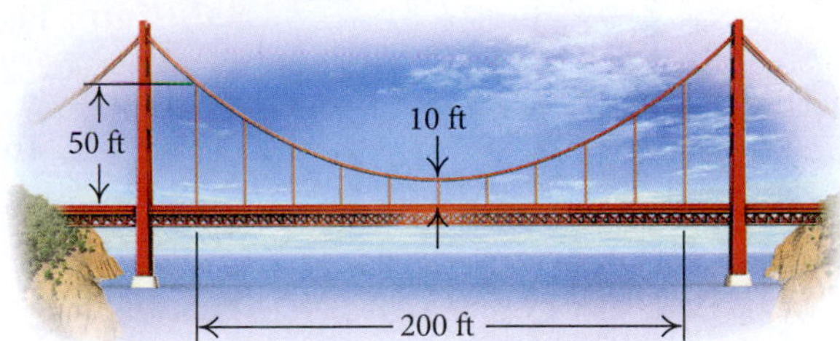

7.2 The Circle and the Ellipse

- Given an equation of a circle, complete the square, if necessary, and then find the center and the radius and graph the circle.
- Given an equation of an ellipse, complete the square, if necessary, and then find the center, the vertices, and the foci and graph the ellipse.

Circles

We can define a circle geometrically.

CIRCLE

A **circle** is the set of all points in a plane that are at a fixed distance from a fixed point (the **center**) in the plane.

CIRCLES

REVIEW SECTION 1.1

Recall the standard equation of a circle with center (h, k) and radius r.

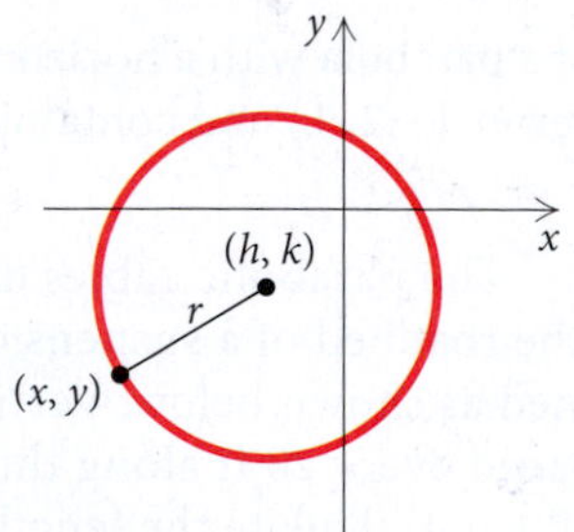

STANDARD EQUATION OF A CIRCLE

The standard equation of a circle with center (h, k) and radius r is

$$(x - h)^2 + (y - k)^2 = r^2.$$

EXAMPLE 1 For the circle

$$x^2 + y^2 - 16x + 14y + 32 = 0,$$

find the center and the radius. Then graph the circle.

Solution First, we complete the square twice:

$$x^2 + y^2 - 16x + 14y + 32 = 0$$

$$x^2 - 16x \quad + y^2 + 14y \quad = -32$$

$$x^2 - 16x + 64 + y^2 + 14y + 49 = -32 + 64 + 49$$

$\left[\frac{1}{2}(-16)\right]^2 = (-8)^2 = 64$ and $\left(\frac{1}{2} \cdot 14\right)^2 = 7^2 = 49$; adding 64 and 49 on both sides to complete the square twice on the left side

$$(x - 8)^2 + (y + 7)^2 = 81$$

$$(x - 8)^2 + [y - (-7)]^2 = 9^2. \qquad \text{Writing standard form}$$

The center is $(8, -7)$ and the radius is 9. We graph the circle as shown below.

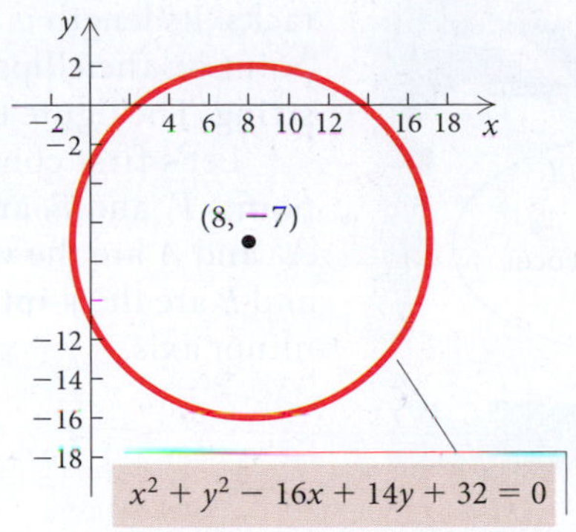

Now Try Exercise 7.

Technology Connection

When we use the Conics CIRCLE APP to graph a circle, it is not necessary to write the equation in standard form or to solve it for y first. We enter the coefficients of x^2, y^2, x, and y and also the constant term when the equation is written in the form $ax^2 + ay^2 + bx + cy + d = 0$. For the circle in Example 1, we enter 1 for A, -16 for B, 14 for C, and 32 for D.

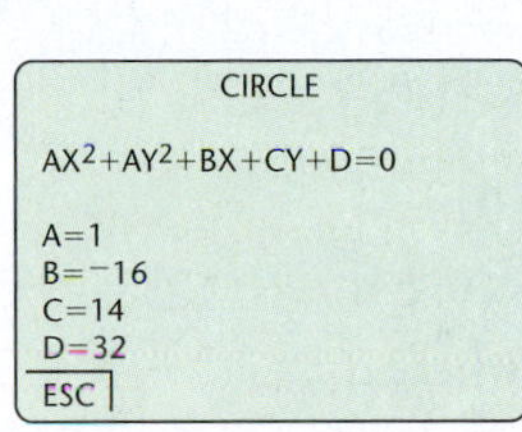

$x^2 + y^2 - 16x + 14y + 32 = 0$

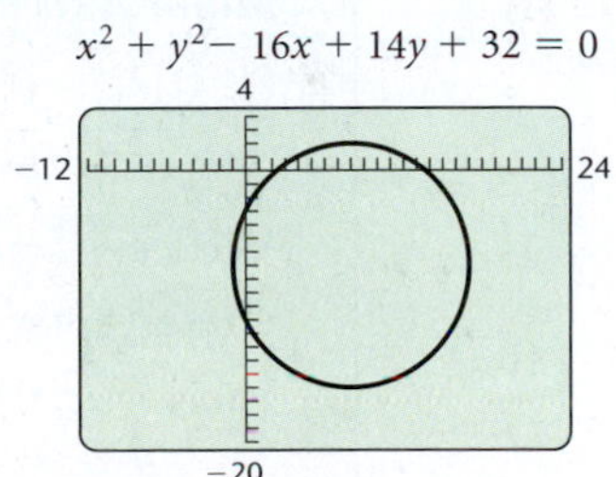

Some graphing calculators have a DRAW feature that provides a quick way to graph a circle when the center and the radius are known. This feature is described on p. 12.

Ellipses

We have studied two conic sections, the parabola and the circle. Now we turn our attention to a third, the *ellipse*.

ELLIPSE

An **ellipse** is the set of all points in a plane, the sum of whose distances from two fixed points (the **foci**) is constant. The **center** of an ellipse is the midpoint of the segment between the foci.

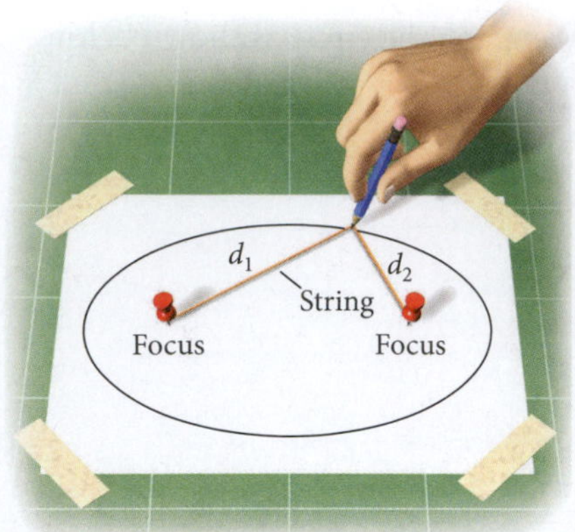

We can draw an ellipse by first placing two thumbtacks in a piece of cardboard. These are the foci (singular, *focus*). We then attach a piece of string to the tacks. Its length is the constant sum of the distances $d_1 + d_2$ from the foci to any point on the ellipse. Next, we trace a curve with a pencil held tight against the string. The figure traced is an ellipse.

Let's first consider the ellipse shown below with center at the origin. The points F_1 and F_2 are the foci. The segment $\overline{A'A}$ is the **major axis**, and the points A' and A are the **vertices**. The segment $\overline{B'B}$ is the **minor axis**, and the points B' and B are the ***y*-intercepts**. Note that the major axis of an ellipse is longer than the minor axis.

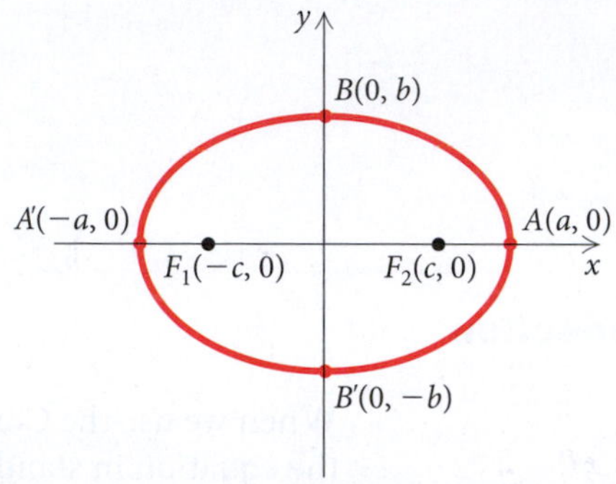

STANDARD EQUATION OF AN ELLIPSE WITH CENTER AT THE ORIGIN

Major Axis Horizontal

$$\frac{x^2}{a^2} + \frac{y^2}{b^2} = 1, \quad a > b > 0$$

Vertices: $(-a, 0), (a, 0)$

y-intercepts: $(0, -b), (0, b)$

Foci: $(-c, 0), (c, 0)$, where $c^2 = a^2 - b^2$

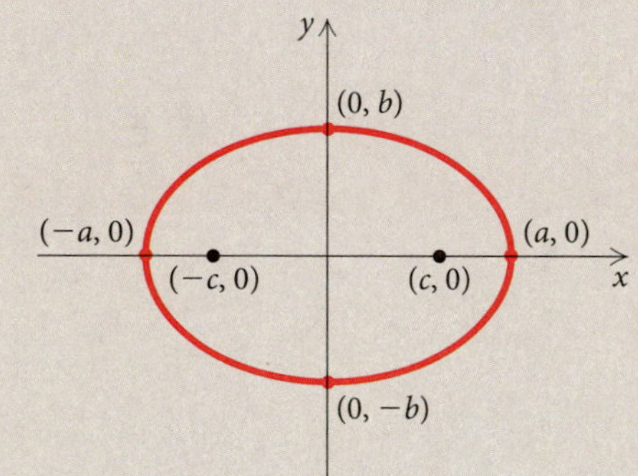

Major Axis Vertical

$$\frac{x^2}{b^2} + \frac{y^2}{a^2} = 1, \quad a > b > 0$$

Vertices: $(0, -a), (0, a)$

x-intercepts: $(-b, 0), (b, 0)$

Foci: $(0, -c), (0, c)$, where $c^2 = a^2 - b^2$

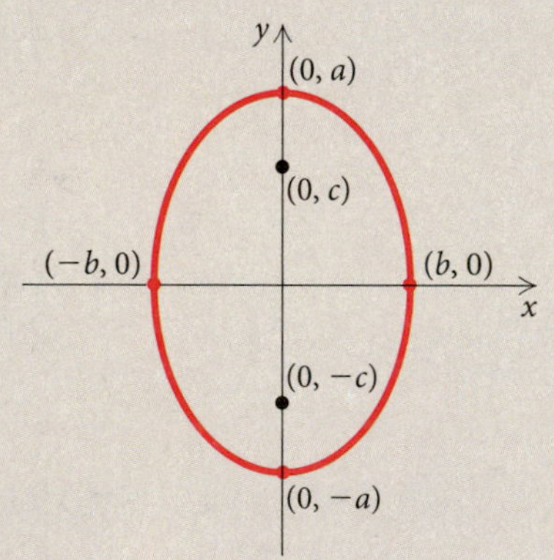

EXAMPLE 2 Find the standard equation of the ellipse with vertices $(-5, 0)$ and $(5, 0)$ and foci $(-3, 0)$ and $(3, 0)$. Then graph the ellipse.

Solution Since the foci are on the x-axis and the origin is the midpoint of the segment between them, the major axis is horizontal and $(0, 0)$ is the center of the ellipse. Thus the equation is of the form

$$\frac{x^2}{a^2} + \frac{y^2}{b^2} = 1.$$

Since the vertices are $(-5, 0)$ and $(5, 0)$ and the foci are $(-3, 0)$ and $(3, 0)$, we know that $a = 5$ and $c = 3$. These values can be used to find b^2:

$$c^2 = a^2 - b^2$$
$$3^2 = 5^2 - b^2$$
$$9 = 25 - b^2$$
$$b^2 = 16.$$

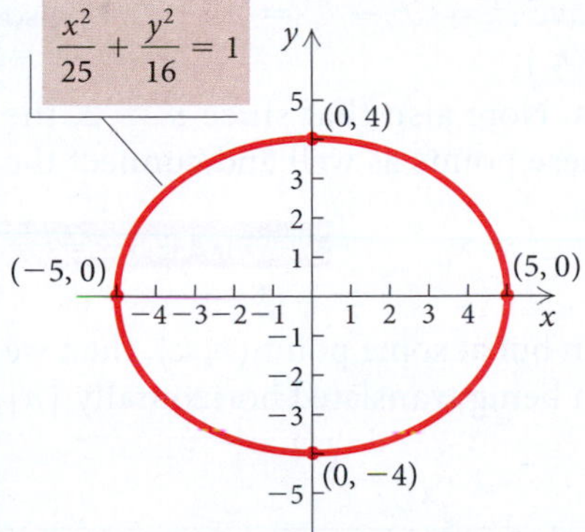

Thus the equation of the ellipse is

$$\frac{x^2}{5^2} + \frac{y^2}{4^2} = 1, \quad \text{or} \quad \frac{x^2}{25} + \frac{y^2}{16} = 1.$$

To graph the ellipse, we plot the vertices $(-5, 0)$ and $(5, 0)$. Since $b^2 = 16$, we know that $b = 4$ and the y-intercepts are $(0, -4)$ and $(0, 4)$. We plot these points as well and connect the four points we have plotted with a smooth curve.

Now Try Exercise 31.

Technology Connection

When the equation of an ellipse is written in standard form, we can use the Conics ELLIPSE APP on a graphing calculator to graph it. The standard equation of the ellipse in Example 2 is

$$\frac{x^2}{5^2} + \frac{y^2}{4^2} = 1.$$

Note that the center is $(0, 0)$. We enter 5 for A, 4 for B, 0 for H, and 0 for K.

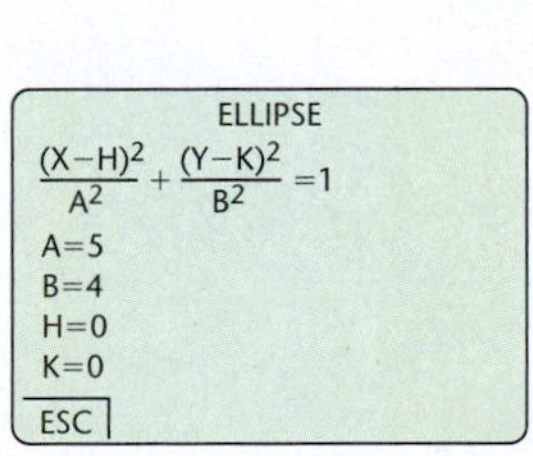

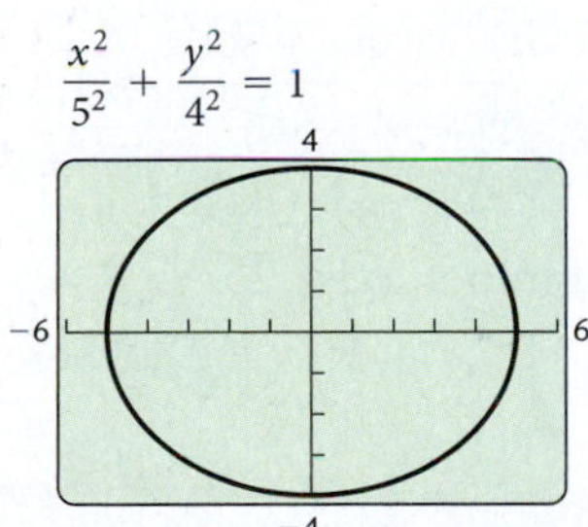

EXAMPLE 3 For the ellipse

$$9x^2 + 4y^2 = 36,$$

find the vertices and the foci. Then draw the graph.

Solution We first find standard form:

$$9x^2 + 4y^2 = 36$$

$$\frac{9x^2}{36} + \frac{4y^2}{36} = \frac{36}{36} \quad \text{Dividing by 36 on both sides to get 1 on the right side}$$

$$\frac{x^2}{4} + \frac{y^2}{9} = 1$$

$$\frac{x^2}{2^2} + \frac{y^2}{3^2} = 1. \quad \text{Writing standard form}$$

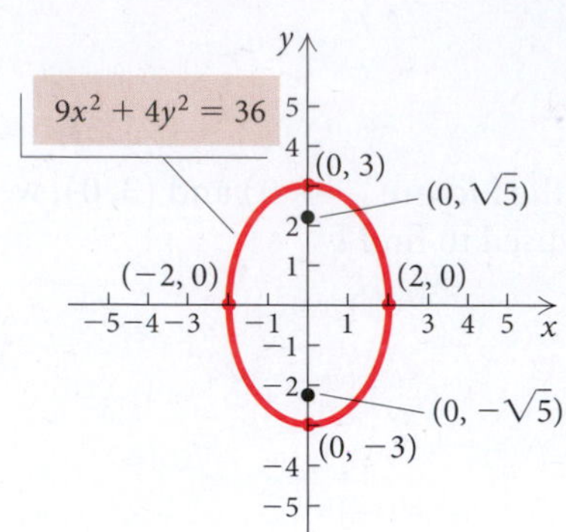

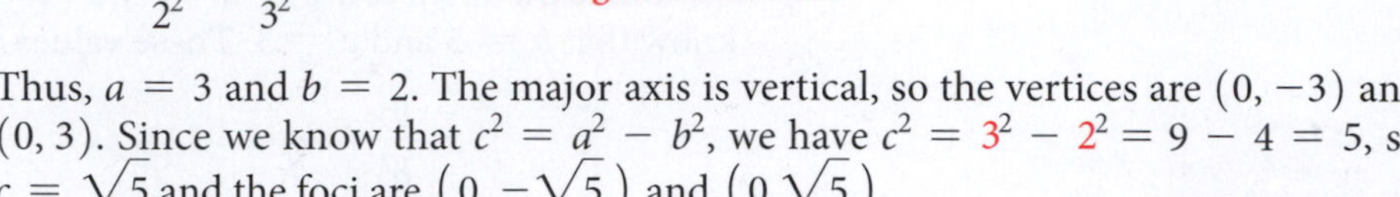

Thus, $a = 3$ and $b = 2$. The major axis is vertical, so the vertices are $(0, -3)$ and $(0, 3)$. Since we know that $c^2 = a^2 - b^2$, we have $c^2 = 3^2 - 2^2 = 9 - 4 = 5$, so $c = \sqrt{5}$ and the foci are $(0, -\sqrt{5})$ and $(0, \sqrt{5})$.

To graph the ellipse, we plot the vertices. Note also that since $b = 2$, the x-intercepts are $(-2, 0)$ and $(2, 0)$. We plot these points as well and connect the four points we have plotted with a smooth curve.

Now Try Exercise 25.

If the center of an ellipse is not at the origin but at some point (h, k), then we can think of an ellipse with center at the origin being translated horizontally $|h|$ units and vertically $|k|$ units.

STANDARD EQUATION OF AN ELLIPSE WITH CENTER AT (h, k)

Major Axis Horizontal

$$\frac{(x - h)^2}{a^2} + \frac{(y - k)^2}{b^2} = 1, \quad a > b > 0$$

Vertices: $(h - a, k)$, $(h + a, k)$

Length of minor axis: $2b$

Foci: $(h - c, k)$, $(h + c, k)$, where $c^2 = a^2 - b^2$

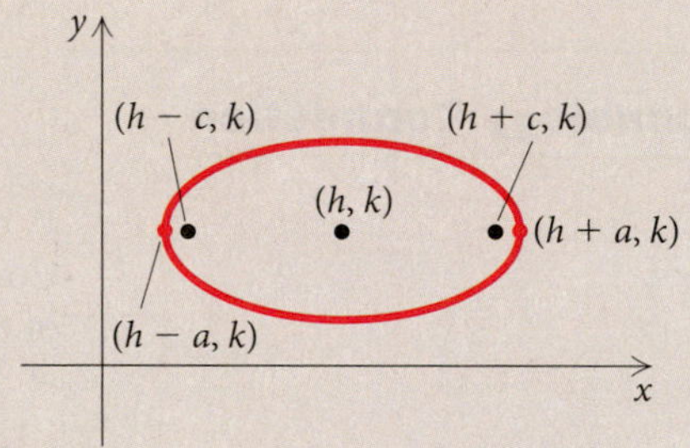

Major Axis Vertical

$$\frac{(x - h)^2}{b^2} + \frac{(y - k)^2}{a^2} = 1, \quad a > b > 0$$

Vertices: $(h, k - a)$, $(h, k + a)$

Length of minor axis: $2b$

Foci: $(h, k - c)$, $(h, k + c)$, where $c^2 = a^2 - b^2$

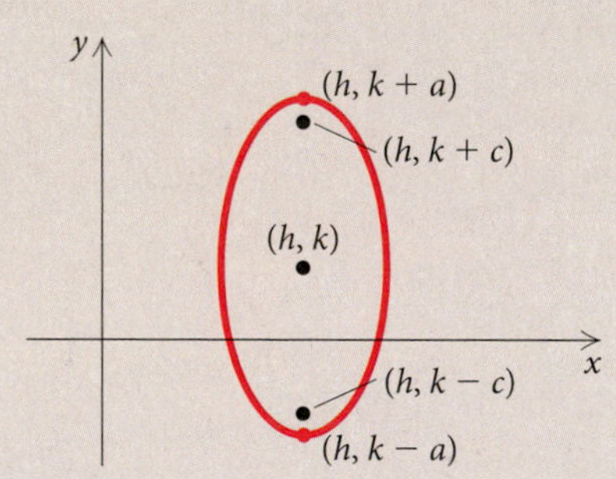

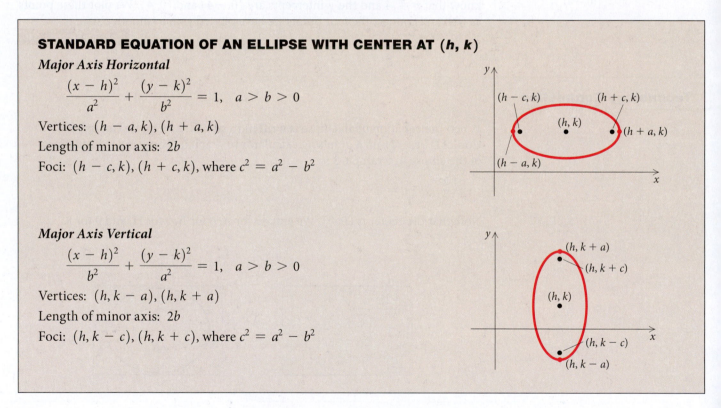

EXAMPLE 4 For the ellipse

$$4x^2 + y^2 + 24x - 2y + 21 = 0,$$

find the center, the vertices, and the foci. Then draw the graph.

Solution First, we complete the square twice to get standard form:

$$\begin{aligned}
4x^2 + y^2 + 24x - 2y + 21 &= 0 \\
4(x^2 + 6x \qquad) + (y^2 - 2y \qquad) &= -21 \\
4(x^2 + 6x + 9 - 9) + (y^2 - 2y + 1 - 1) &= -21 && \text{Completing the square twice} \\
4(x^2 + 6x + 9) + 4(-9) + (y^2 - 2y + 1) + (-1) &= -21 \\
4(x + 3)^2 - 36 + (y - 1)^2 - 1 &= -21 \\
4(x + 3)^2 + (y - 1)^2 &= 16 && \text{Adding 37 on both sides} \\
\tfrac{1}{16}[4(x + 3)^2 + (y - 1)^2] &= \tfrac{1}{16}\cdot 16 \\
\frac{(x + 3)^2}{4} + \frac{(y - 1)^2}{16} &= 1 \\
\frac{[x - (-3)]^2}{2^2} + \frac{(y - 1)^2}{4^2} &= 1. && \text{Writing standard form}
\end{aligned}$$

The center is $(-3, 1)$. Note that $a = 4$ and $b = 2$. The major axis is vertical, so the vertices are 4 units above and below the center:

$$(-3, 1 + 4) \text{ and } (-3, 1 - 4), \quad \text{or} \quad (-3, 5) \text{ and } (-3, -3).$$

We know that $c^2 = a^2 - b^2$, so $c^2 = 4^2 - 2^2 = 16 - 4 = 12$ and $c = \sqrt{12}$, or $2\sqrt{3}$. Then the foci are $2\sqrt{3}$ units above and below the center:

$$\left(-3, 1 + 2\sqrt{3}\right) \quad \text{and} \quad \left(-3, 1 - 2\sqrt{3}\right).$$

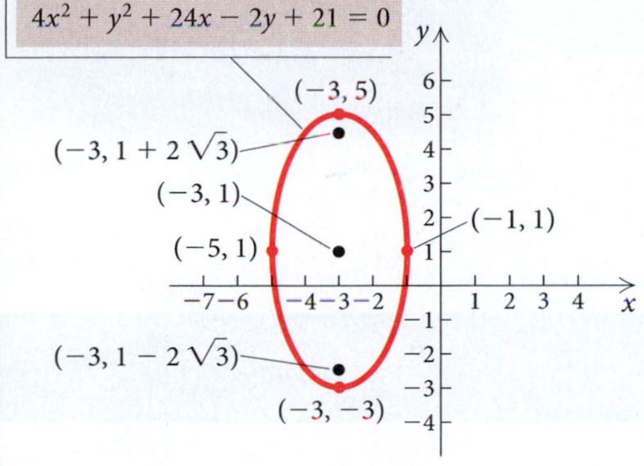

To graph the ellipse, we plot the vertices. Note also that since $b = 2$, two other points on the graph are the endpoints of the minor axis, 2 units right and left of the center:

$$(-3 + 2, 1) \quad \text{and} \quad (-3 - 2, 1),$$

or

$$(-1, 1) \quad \text{and} \quad (-5, 1).$$

We plot these points as well and connect the four points with a smooth curve, as shown, at left.

Now Try Exercise 43.

Technology Connection

When the equation of an ellipse is written in standard form, we can use the Conics ELLIPSE APP to graph it. The standard equation of the ellipse in Example 4 is

$$\frac{[x - (-3)]^2}{2^2} + \frac{(y - 1)^2}{4^2} = 1.$$

Note that the center is $(-3, 1)$. We enter 4 for A, 2 for B, -3 for H, and 1 for K.

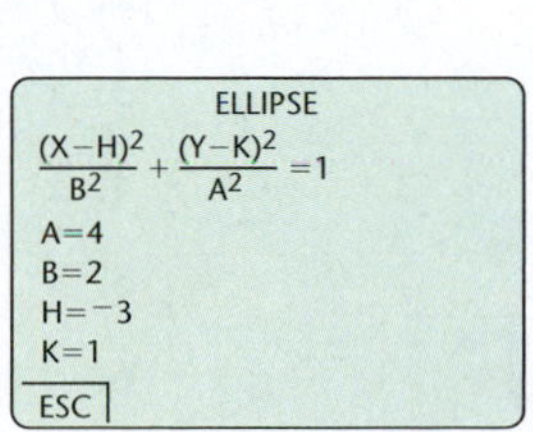

$4x^2 + y^2 + 24x - 2y + 21 = 0$

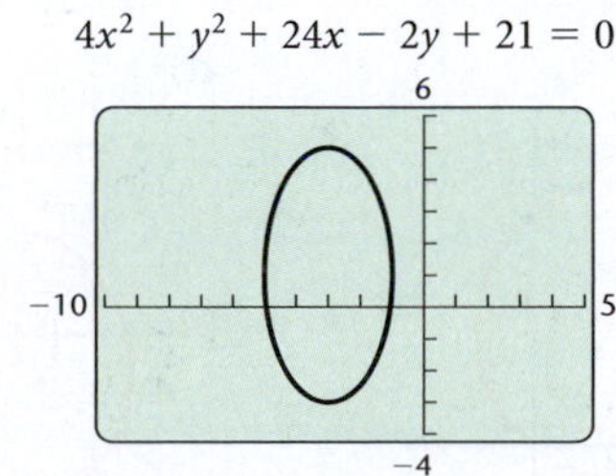

▶ Applications

An exciting medical application of an ellipse is a device called a *lithotripter.* One type of this device uses electromagnetic technology to generate a shock wave to pulverize kidney stones. The wave originates at one focus of an ellipse and is reflected to the kidney stone, which is positioned at the other focus. Recovery time following the use of this technique is much shorter than with conventional surgery.

Ellipses have many other applications. Planets travel around the sun in elliptical orbits with the sun at one focus, for example, and satellites travel around the earth in elliptical orbits as well.

A room with an ellipsoidal ceiling is known as a *whispering gallery.* In such a room, a word whispered at one focus can be clearly heard at the other. Whispering galleries are found in the rotunda of the Capitol Building in Washington, D.C., and in St. Paul's Cathedral in London.

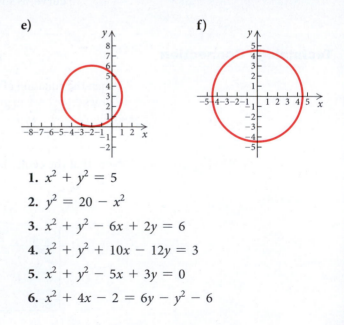

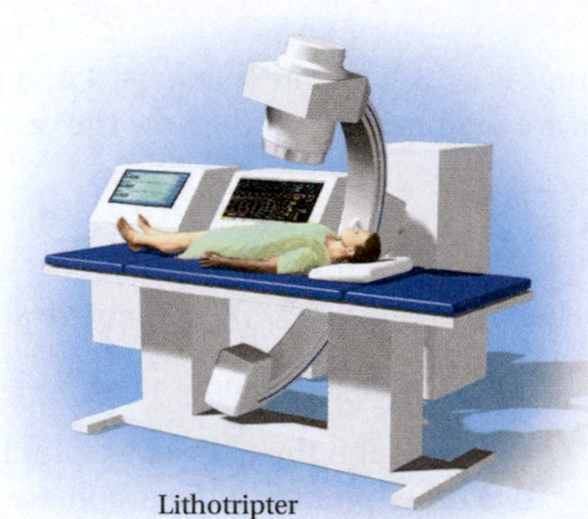
Lithotripter

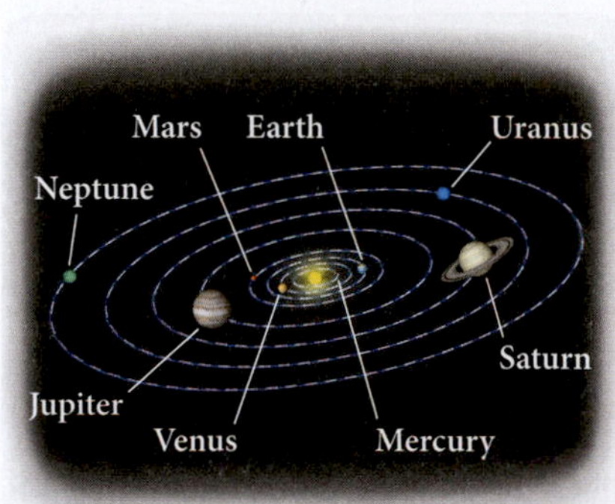

7.2 Exercise Set

In Exercises 1–6, match the equation with one of the graphs (a)–(f) that follow.

a)

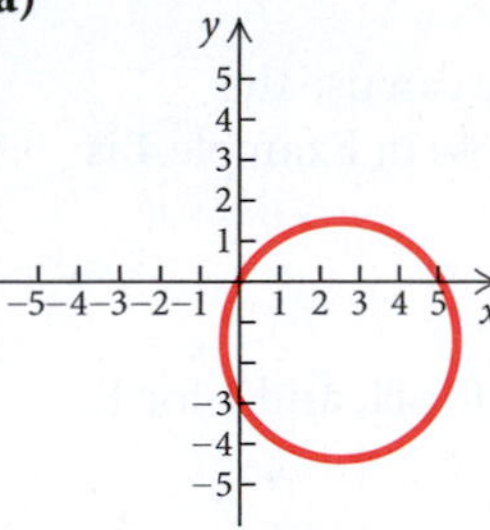

b)

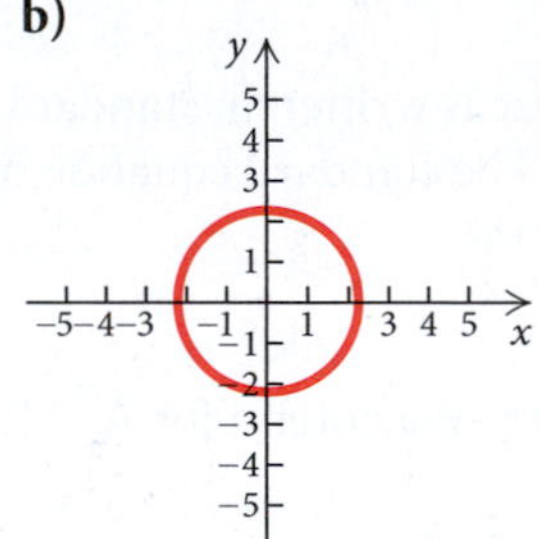

c)

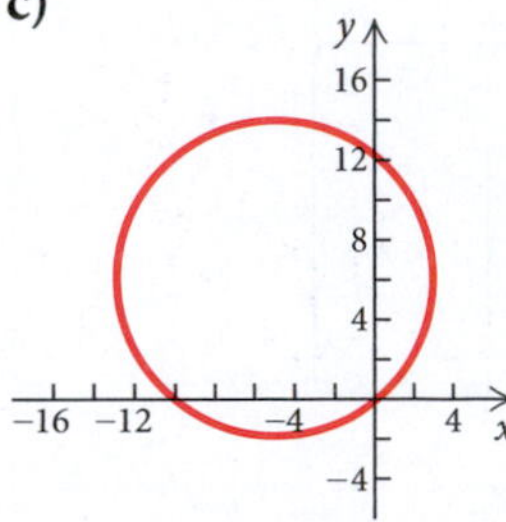

d)

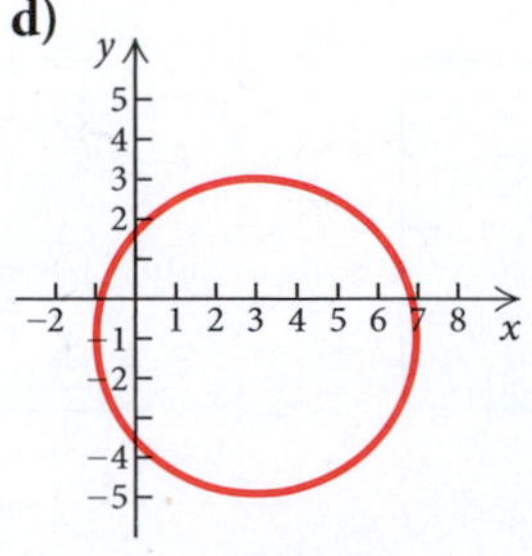

e)

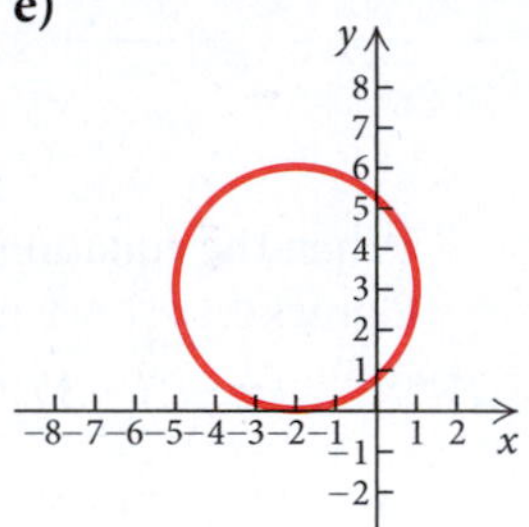

f)

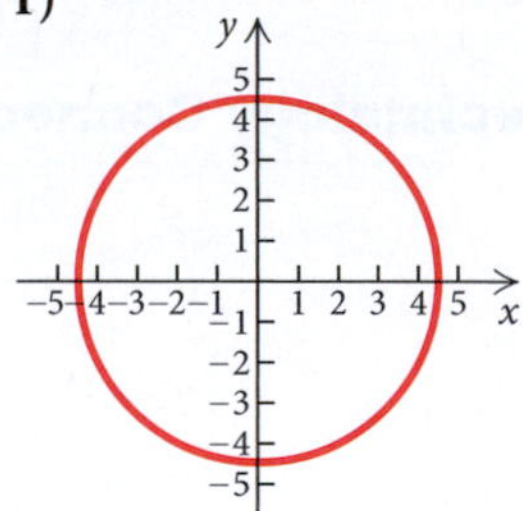

1. $x^2 + y^2 = 5$

2. $y^2 = 20 - x^2$

3. $x^2 + y^2 - 6x + 2y = 6$

4. $x^2 + y^2 + 10x - 12y = 3$

5. $x^2 + y^2 - 5x + 3y = 0$

6. $x^2 + 4x - 2 = 6y - y^2 - 6$

Find the center and the radius of the circle with the given equation. Then draw the graph.

7. $x^2 + y^2 - 14x + 4y = 11$

8. $x^2 + y^2 + 2x - 6y = -6$

9. $x^2 + y^2 + 6x - 2y = 6$

10. $x^2 + y^2 - 4x + 2y = 4$

11. $x^2 + y^2 + 4x - 6y - 12 = 0$

12. $x^2 + y^2 - 8x - 2y - 19 = 0$

13. $x^2 + y^2 - 6x - 8y + 16 = 0$

14. $x^2 + y^2 - 2x + 6y + 1 = 0$

15. $x^2 + y^2 + 6x - 10y = 0$

16. $x^2 + y^2 - 7x - 2y = 0$

17. $x^2 + y^2 - 9x = 7 - 4y$

18. $y^2 - 6y - 1 = 8x - x^2 + 3$

In Exercises 19–22, match the equation with one of the graphs (a)–(d) that follow.

a)

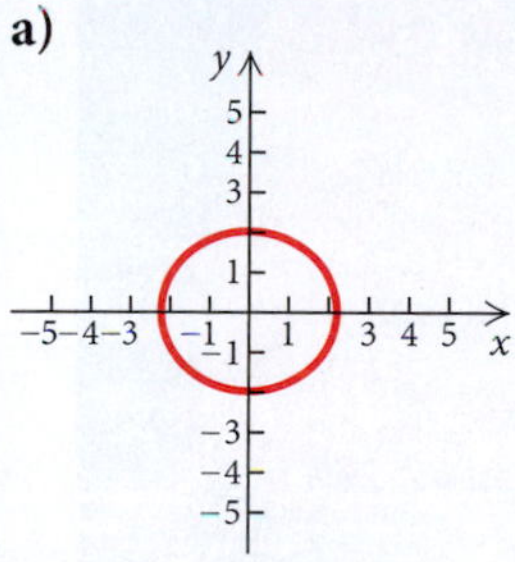

b)

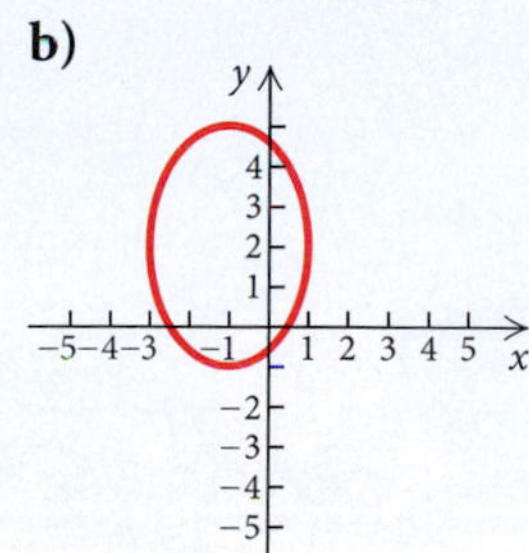

c)

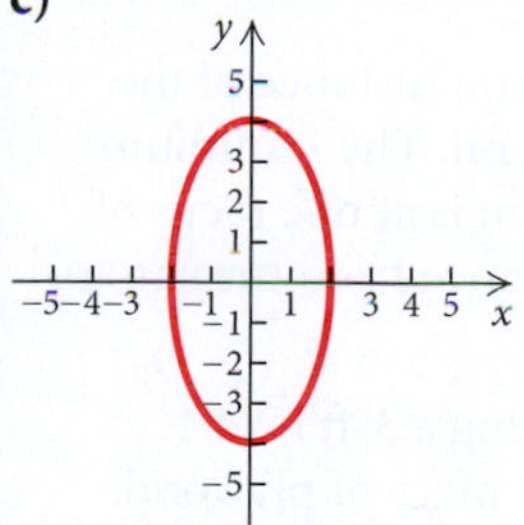

d)

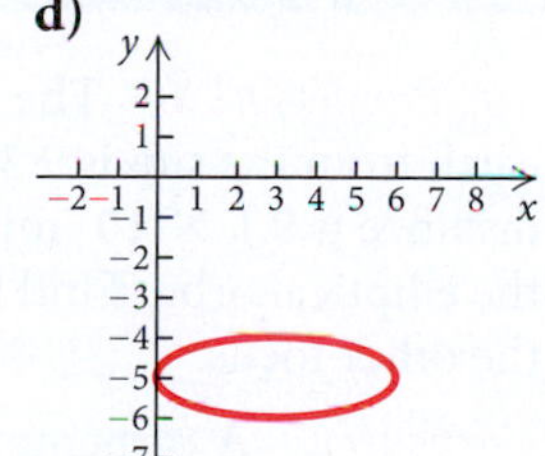

19. $16x^2 + 4y^2 = 64$

20. $4x^2 + 5y^2 = 20$

21. $x^2 + 9y^2 - 6x + 90y = -225$

22. $9x^2 + 4y^2 + 18x - 16y = 11$

Find the vertices and the foci of the ellipse with the given equation. Then draw the graph.

23. $\dfrac{x^2}{4} + \dfrac{y^2}{1} = 1$

24. $\dfrac{x^2}{25} + \dfrac{y^2}{36} = 1$

25. $16x^2 + 9y^2 = 144$

26. $9x^2 + 4y^2 = 36$

27. $2x^2 + 3y^2 = 6$

28. $5x^2 + 7y^2 = 35$

29. $4x^2 + 9y^2 = 1$

30. $25x^2 + 16y^2 = 1$

Find an equation of an ellipse satisfying the given conditions.

31. Vertices: $(-7, 0)$ and $(7, 0)$;
foci: $(-3, 0)$ and $(3, 0)$

32. Vertices: $(0, -6)$ and $(0, 6)$;
foci: $(0, -4)$ and $(0, 4)$

33. Vertices: $(0, -8)$ and $(0, 8)$;
length of minor axis: 10

34. Vertices: $(-5, 0)$ and $(5, 0)$;
length of minor axis: 6

35. Foci: $(-2, 0)$ and $(2, 0)$;
length of major axis: 6

36. Foci: $(0, -3)$ and $(0, 3)$;
length of major axis: 10

Find the center, the vertices, and the foci of the ellipse. Then draw the graph.

37. $\dfrac{(x - 1)^2}{9} + \dfrac{(y - 2)^2}{4} = 1$

38. $\dfrac{(x - 1)^2}{1} + \dfrac{(y - 2)^2}{4} = 1$

39. $\dfrac{(x + 3)^2}{25} + \dfrac{(y - 5)^2}{36} = 1$

40. $\dfrac{(x - 2)^2}{16} + \dfrac{(y + 3)^2}{25} = 1$

41. $3(x + 2)^2 + 4(y - 1)^2 = 192$

42. $4(x - 5)^2 + 3(y - 4)^2 = 48$

43. $4x^2 + 9y^2 - 16x + 18y - 11 = 0$

44. $x^2 + 2y^2 - 10x + 8y + 29 = 0$

45. $4x^2 + y^2 - 8x - 2y + 1 = 0$

46. $9x^2 + 4y^2 + 54x - 8y + 49 = 0$

*The **eccentricity** of an ellipse is defined as $e = c/a$. For an ellipse, $0 < c < a$, so $0 < e < 1$. When e is close to 0, an ellipse appears to be nearly circular. When e is close to 1, an ellipse is very flat.*

47. Note the shapes of the ellipses in Examples 2 and 4. Which ellipse has the smaller eccentricity? Confirm your answer by computing the eccentricity of each ellipse.

48. Which ellipse has the smaller eccentricity? (Assume that the coordinate systems have the same scale.)

a) **b)**

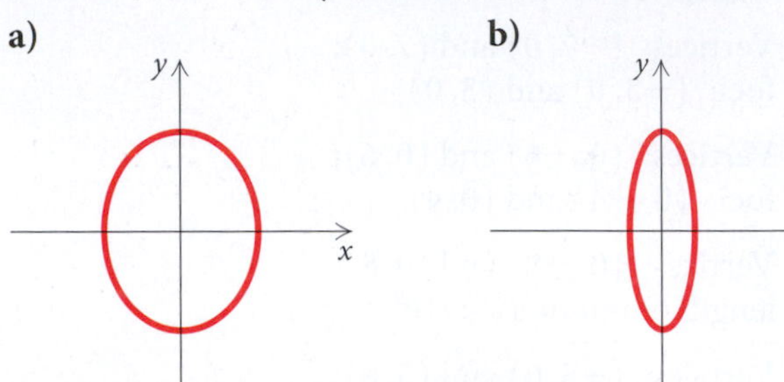

49. Find an equation of an ellipse with vertices $(0, -4)$ and $(0, 4)$ and $e = \frac{1}{4}$.

50. Find an equation of an ellipse with vertices $(-3, 0)$ and $(3, 0)$ and $e = \frac{7}{10}$.

51. *Bridge Supports.* The bridge support shown in the following figure is the top half of an ellipse. Assuming that a coordinate system is superimposed on the drawing in such a way that point Q, the center of the ellipse, is at the origin, find an equation of the ellipse.

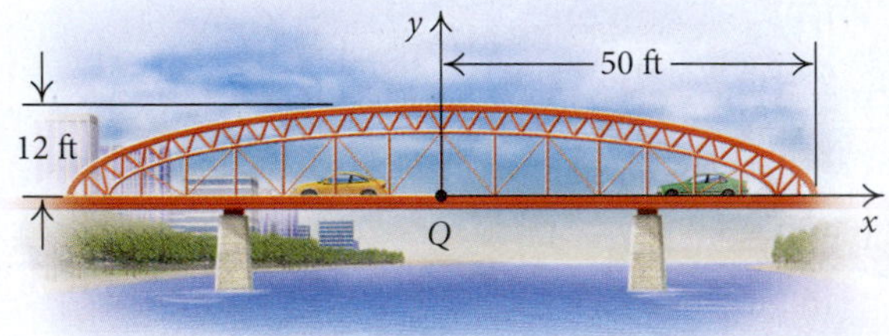

52. *Whispering Gallery.* An art museum is adding a new exhibit room in the shape of an ellipse. The director wants to mark the foci so that a tour guide can stand at one focus and without speaking loudly can be clearly heard by a group touring the museum. If the room is 64 ft long and each focus is 5 ft from the outside wall along the major axis, how high is the ceiling? Round to the nearest tenth of a foot.

53. *Whispering Gallery.* A whispering gallery, often elliptical in shape, has acoustic properties such that a whisper made at one point can be heard at other distant points. A science museum is designing a new exhibit hall that will illustrate a whispering gallery. The hall will be 90 ft in length with the ceiling 30 ft high at the center. How far are the foci from the center of the ellipse? Round to the nearest tenth of a foot.

54. *The Ellipse.* The lighting of the National Christmas Tree located on the Ellipse, a large grassy area south of the White House, marks the beginning of the holiday season in Washington, D.C. This area of the lawn is actually an ellipse with major axis of length 1048 ft and minor axis of length 898 ft. Assuming that a coordinate system is superimposed on the area in such a way that the center is at the origin and the major and minor axes are on the x- and y-axes of the coordinate system, respectively, find an equation of the ellipse.

55. *The Earth's Orbit.* The maximum distance of the earth from the sun is 9.3×10^7 mi. The minimum distance is 9.1×10^7 mi. The sun is at one focus of the elliptical orbit. Find the distance from the sun to the other focus.

56. *Carpentry.* A carpenter is cutting a 3-ft by 4-ft elliptical sign from a 3-ft by 4-ft piece of plywood. The ellipse will be drawn using a string attached to the board at the foci of the ellipse.

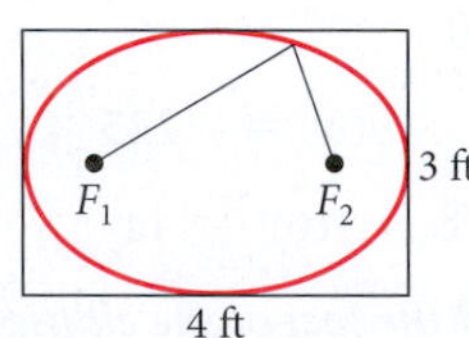

a) How far from the ends of the board should the string be attached?

b) How long should the string be?

Skill Maintenance

Vocabulary Reinforcement

In each of Exercises 57–64, fill in the blank with the correct term. Some of the given choices will not be used.

piecewise function
linear equation
factor
remainder
solution
zero
x-intercept
y-intercept
parabola
circle
ellipse
midpoint
distance
one real-number solution
two different real-number solutions
two different imaginary-number solutions

57. The ______________ between two points (x_1, y_1) and (x_2, y_2) is given by $\left(\frac{x_1 + x_2}{2}, \frac{y_1 + y_2}{2}\right)$. [1.1]

58. An input c of a function f is a(n) ______________ of the function if $f(c) = 0$. [1.5]

59. A(n) ______________ of the graph of an equation is a point $(0, b)$. [1.1]

60. For a quadratic equation $ax^2 + bx + c = 0$, if $b^2 - 4ac > 0$, the equation has ______________. [3.2]

61. Given a polynomial $f(x)$, then $f(c)$ is the ______________ that would be obtained by dividing $f(x)$ by $x - c$. [4.3]

62. A(n) ______________ is the set of all points in a plane the sum of whose distances from two fixed points is constant. [7.2]

63. A(n) ______________ is the set of all points in a plane equidistant from a fixed line and a fixed point not on the line. [7.1]

64. A(n) ______________ is the set of all points in a plane that are at a fixed distance from a fixed point in the plane. [7.2]

Synthesis

Find an equation of an ellipse satisfying the given conditions.

65. Vertices: $(3, -4), (3, 6)$;
endpoints of minor axis: $(1, 1), (5, 1)$

66. Vertices: $(-1, -1), (-1, 5)$;
endpoints of minor axis: $(-3, 2), (1, 2)$

67. Vertices: $(-3, 0)$ and $(3, 0)$;
passing through $\left(2, \frac{22}{3}\right)$

68. Center: $(-2, 3)$; major axis vertical;
length of major axis: 4;
length of minor axis: 1

69. *Bridge Arch.* A bridge with a semielliptical arch spans a river as shown here. What is the clearance 6 ft from the riverbank?

Mid-Chapter Mixed Review

Determine whether the statement is true or false.

1. The graph of $(x + 3)^2 = 8(y - 2)$ is a parabola with vertex $(-3, 2)$. [7.1]

2. A parabola must open up or down. [7.1]

3. The graph of $(x - 4)^2 + (y + 1)^2 = 9$ is a circle with radius 9. [7.2]

4. The major axis of the ellipse $\frac{x^2}{4} + \frac{y^2}{16} = 1$ is vertical. [7.2]

In Exercises 5–12, match the equation with one of the graphs (a)–(h) that follow. [7.1], [7.2]

a) 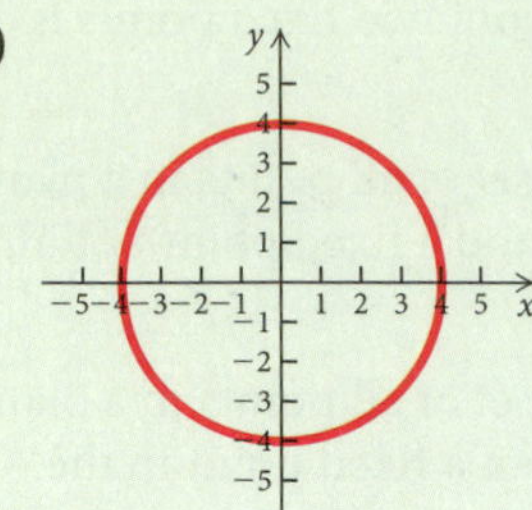

b) 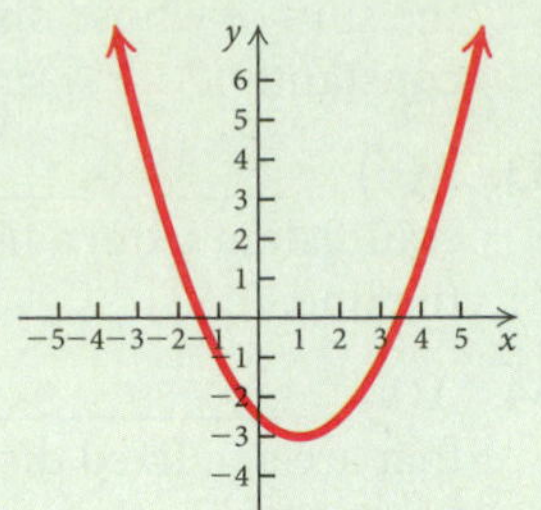

c) 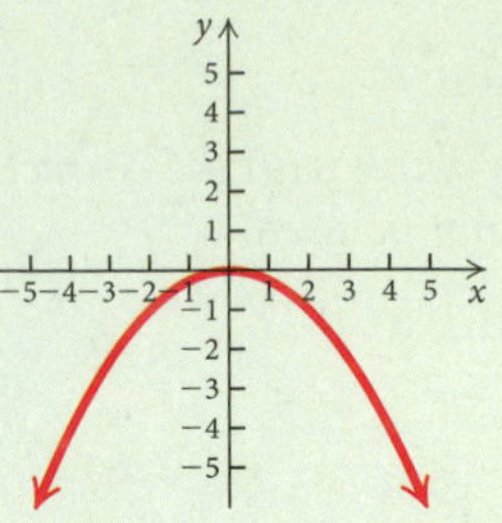

d) 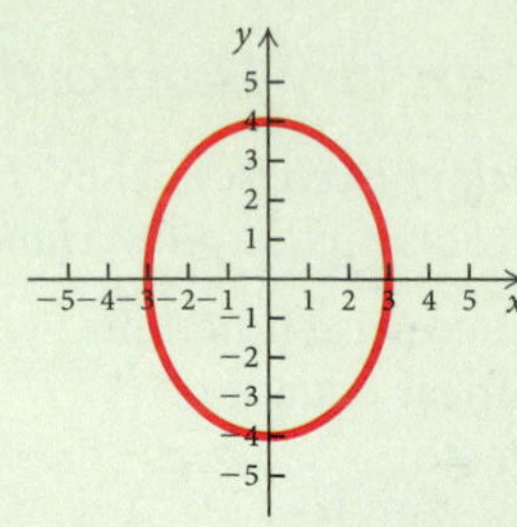

e) 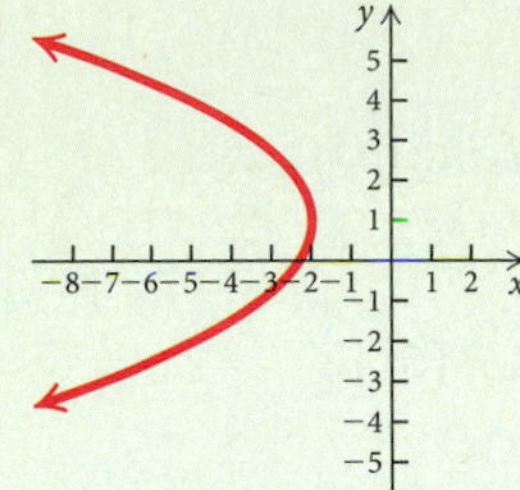

f) 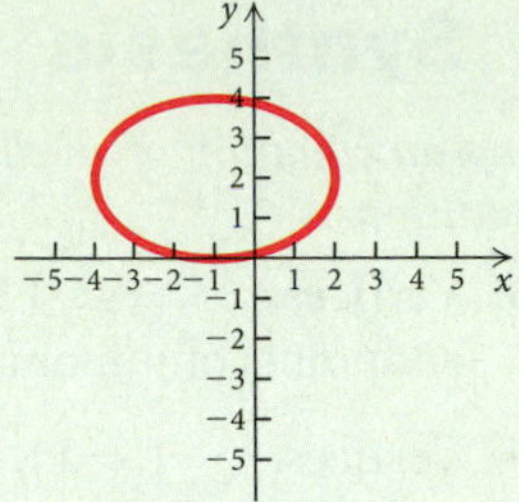

g) 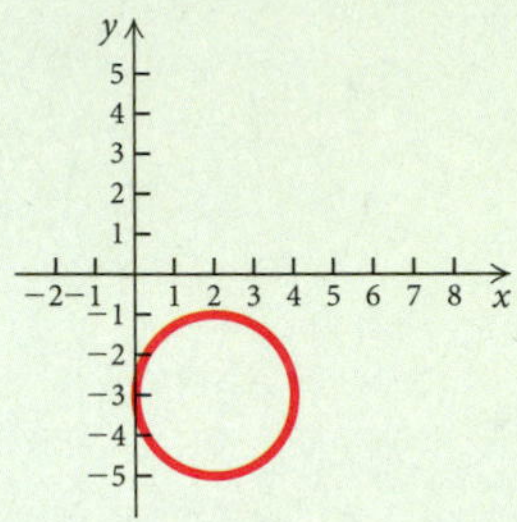

h) 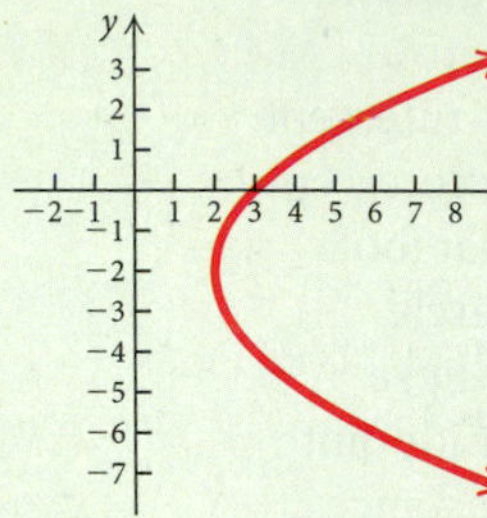

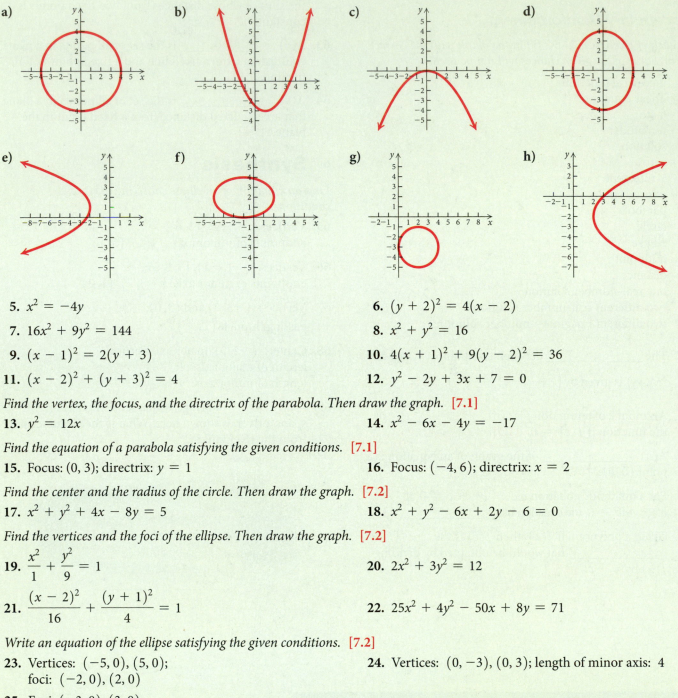

5. $x^2 = -4y$

6. $(y + 2)^2 = 4(x - 2)$

7. $16x^2 + 9y^2 = 144$

8. $x^2 + y^2 = 16$

9. $(x - 1)^2 = 2(y + 3)$

10. $4(x + 1)^2 + 9(y - 2)^2 = 36$

11. $(x - 2)^2 + (y + 3)^2 = 4$

12. $y^2 - 2y + 3x + 7 = 0$

Find the vertex, the focus, and the directrix of the parabola. Then draw the graph. [7.1]

13. $y^2 = 12x$

14. $x^2 - 6x - 4y = -17$

Find the equation of a parabola satisfying the given conditions. [7.1]

15. Focus: $(0, 3)$; directrix: $y = 1$

16. Focus: $(-4, 6)$; directrix: $x = 2$

Find the center and the radius of the circle. Then draw the graph. [7.2]

17. $x^2 + y^2 + 4x - 8y = 5$

18. $x^2 + y^2 - 6x + 2y - 6 = 0$

Find the vertices and the foci of the ellipse. Then draw the graph. [7.2]

19. $\dfrac{x^2}{1} + \dfrac{y^2}{9} = 1$

20. $2x^2 + 3y^2 = 12$

21. $\dfrac{(x - 2)^2}{16} + \dfrac{(y + 1)^2}{4} = 1$

22. $25x^2 + 4y^2 - 50x + 8y = 71$

Write an equation of the ellipse satisfying the given conditions. [7.2]

23. Vertices: $(-5, 0)$, $(5, 0)$;
foci: $(-2, 0)$, $(2, 0)$

24. Vertices: $(0, -3)$, $(0, 3)$; length of minor axis: 4

25. Foci: $(-3, 0)$, $(3, 0)$;
length of major axis: 8

Collaborative Discussion and Writing

26. Is a parabola always the graph of a function? Why or why not? [7.1]

27. Explain how the distance formula is used to find the standard equation of a parabola. [7.1]

28. Explain why function notation is not used in Section 7.2. [7.2]

29. Is the center of an ellipse part of the graph of the ellipse? Why or why not? [7.2]

7.3 The Hyperbola

▶ Given an equation of a hyperbola, complete the square, if necessary, and then find the center, the vertices, and the foci and graph the hyperbola.

The last type of conic section that we will study is the *hyperbola*.

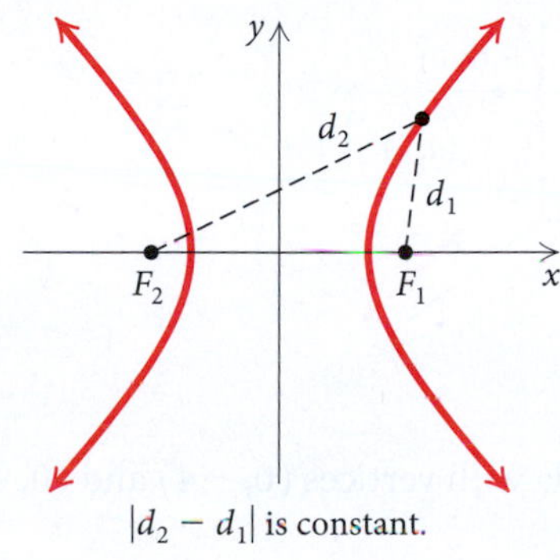

$|d_2 - d_1|$ is constant.

HYPERBOLA

A **hyperbola** is the set of all points in a plane for which the absolute value of the difference of the distances from two fixed points (the **foci**) is constant. The midpoint of the segment between the foci is the **center** of the hyperbola.

▶ Standard Equations of Hyperbolas

We first consider the equation of a hyperbola with center at the origin. In the figure at left, F_1 and F_2 are the foci. The segment $\overline{V_2V_1}$ is the **transverse axis** and the points V_2 and V_1 are the **vertices**.

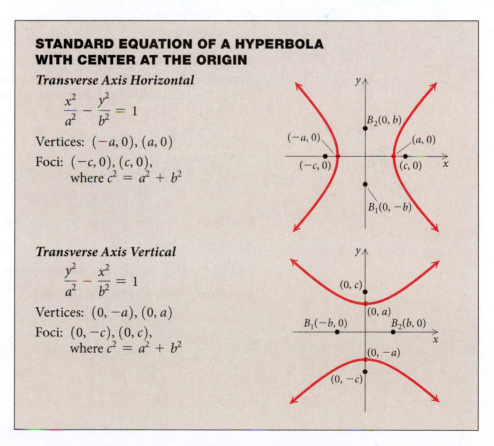

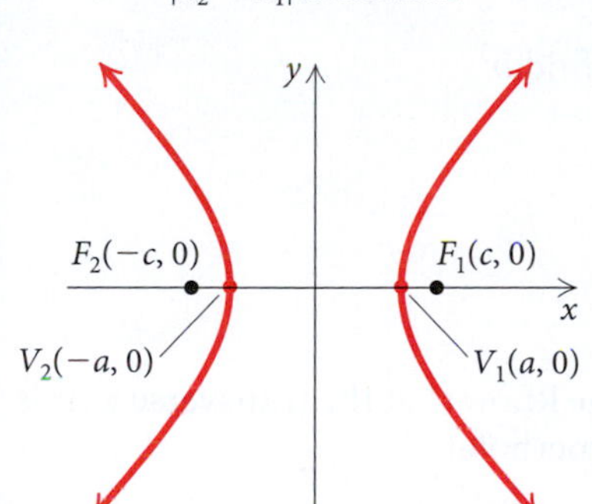

STANDARD EQUATION OF A HYPERBOLA WITH CENTER AT THE ORIGIN

Transverse Axis Horizontal

$$\frac{x^2}{a^2} - \frac{y^2}{b^2} = 1$$

Vertices: $(-a, 0), (a, 0)$

Foci: $(-c, 0), (c, 0)$,
where $c^2 = a^2 + b^2$

Transverse Axis Vertical

$$\frac{y^2}{a^2} - \frac{x^2}{b^2} = 1$$

Vertices: $(0, -a), (0, a)$

Foci: $(0, -c), (0, c)$,
where $c^2 = a^2 + b^2$

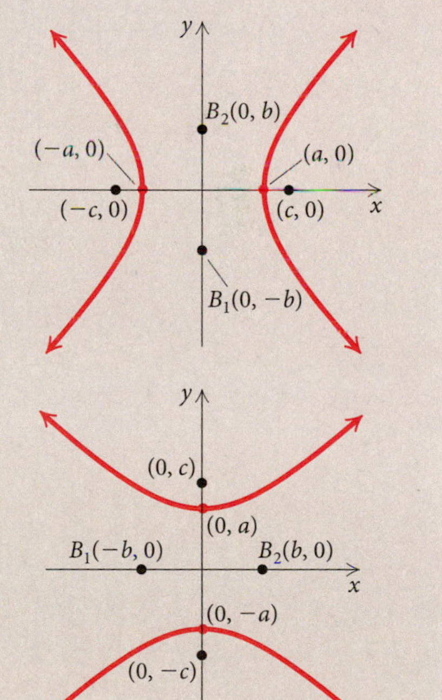

The segment $\overline{B_1B_2}$ is the **conjugate axis** of the hyperbola.

To graph a hyperbola with a horizontal transverse axis, it is helpful to begin by graphing the lines $y = -(b/a)x$ and $y = (b/a)x$. These are the **asymptotes** of the hyperbola. For a hyperbola with a vertical transverse axis, the asymptotes are $y = -(a/b)x$ and $y = (a/b)x$. As $|x|$ gets larger and larger, the graph of the hyperbola gets closer and closer to the asymptotes.

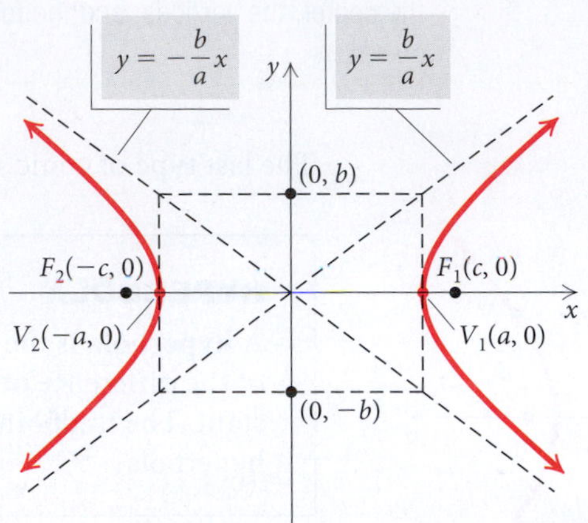

EXAMPLE 1 Find an equation of the hyperbola with vertices $(0, -4)$ and $(0, 4)$ and foci $(0, -6)$ and $(0, 6)$.

Solution We know that $a = 4$ and $c = 6$. We find b^2:

$$c^2 = a^2 + b^2$$
$$6^2 = 4^2 + b^2$$
$$36 = 16 + b^2$$
$$20 = b^2.$$

Since the vertices and the foci are on the y-axis, we know that the transverse axis is vertical. We can now write the equation of the hyperbola:

$$\frac{y^2}{a^2} - \frac{x^2}{b^2} = 1$$
$$\frac{y^2}{16} - \frac{x^2}{20} = 1.$$

Now Try Exercise 7.

EXAMPLE 2 For the hyperbola given by

$$9x^2 - 16y^2 = 144,$$

find the vertices, the foci, and the asymptotes. Then graph the hyperbola.

Solution First, we find standard form:

$$9x^2 - 16y^2 = 144$$
$$\frac{1}{144}(9x^2 - 16y^2) = \frac{1}{144} \cdot 144 \quad \text{Multiplying by } \tfrac{1}{144} \text{ to get 1 on the right side}$$
$$\frac{x^2}{16} - \frac{y^2}{9} = 1$$
$$\frac{x^2}{4^2} - \frac{y^2}{3^2} = 1. \quad \text{Writing standard form}$$

The hyperbola has a horizontal transverse axis, so the vertices are $(-a, 0)$ and $(a, 0)$, or $(-4, 0)$ and $(4, 0)$. From the standard form of the equation, we know that $a^2 = 4^2$, or 16, and $b^2 = 3^2$, or 9. We find the foci:

$$c^2 = a^2 + b^2$$
$$c^2 = 16 + 9$$
$$c^2 = 25$$
$$c = 5.$$

Thus the foci are $(-5, 0)$ and $(5, 0)$.

Next, we find the asymptotes:

$$y = -\frac{b}{a}x = -\frac{3}{4}x \quad \text{and} \quad y = \frac{b}{a}x = \frac{3}{4}x.$$

To draw the graph, we sketch the asymptotes first. This is easily done by drawing the rectangle with horizontal sides passing through $(0, 3)$ and $(0, -3)$ and vertical sides through $(4, 0)$ and $(-4, 0)$. Then we draw and extend the diagonals of this rectangle. The two extended diagonals are the asymptotes of the hyperbola. Next, we plot the vertices and draw the branches of the hyperbola outward from the vertices toward the asymptotes.

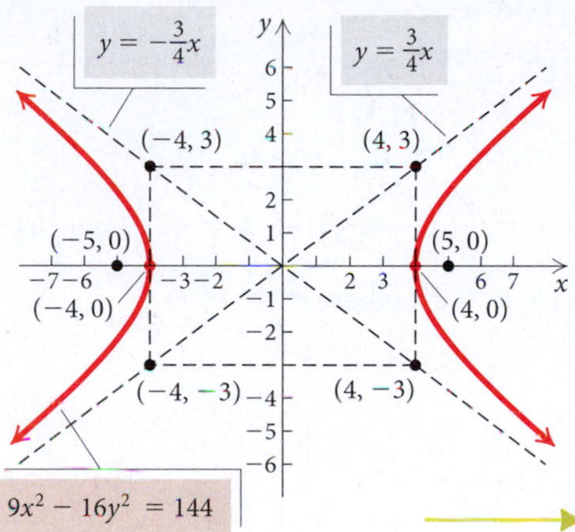

Now Try Exercise 17.

Technology Connection

When the equation of a hyperbola is written in standard form, we can use the Conics HYPERBOLA APP to graph it. The standard equation of the hyperbola in Example 2 is

$$\frac{x^2}{4^2} - \frac{y^2}{3^2} = 1.$$

Note that the center is $(0, 0)$. We enter 4 for A, 3 for B, 0 for H, and 0 for K.

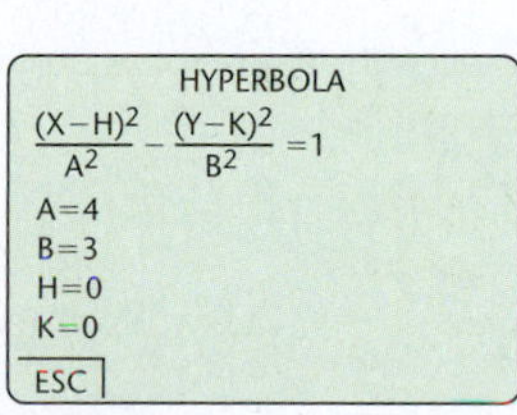

$9x^2 - 16y^2 = 144$

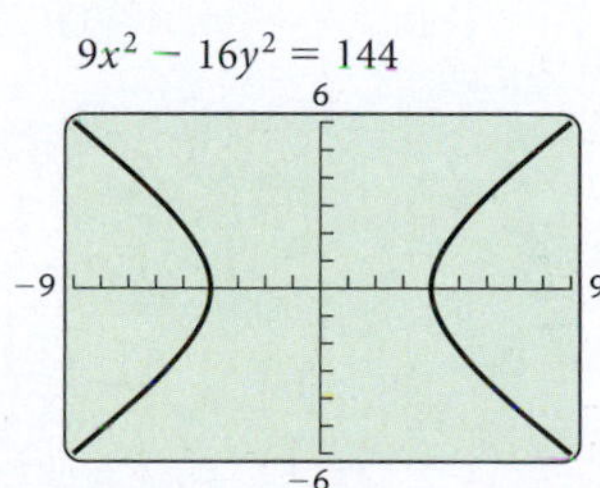

If a hyperbola with center at the origin is translated horizontally $|h|$ units and vertically $|k|$ units, the center is at the point (h, k).

STANDARD EQUATION OF A HYPERBOLA WITH CENTER AT (h, k)

Transverse Axis Horizontal

$$\frac{(x-h)^2}{a^2} - \frac{(y-k)^2}{b^2} = 1$$

Vertices: $(h-a, k), (h+a, k)$

Asymptotes: $y - k = \frac{b}{a}(x-h)$, $y - k = -\frac{b}{a}(x-h)$

Foci: $(h-c, k), (h+c, k)$, where $c^2 = a^2 + b^2$

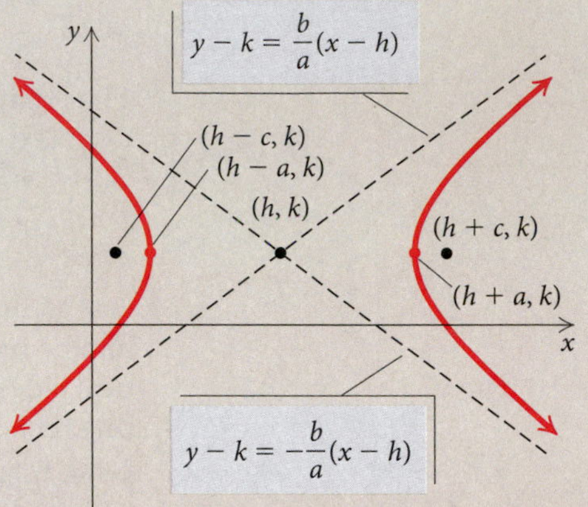

Transverse Axis Vertical

$$\frac{(y-k)^2}{a^2} - \frac{(x-h)^2}{b^2} = 1$$

Vertices: $(h, k-a), (h, k+a)$

Asymptotes: $y - k = \frac{a}{b}(x-h)$, $y - k = -\frac{a}{b}(x-h)$

Foci: $(h, k-c), (h, k+c)$, where $c^2 = a^2 + b^2$

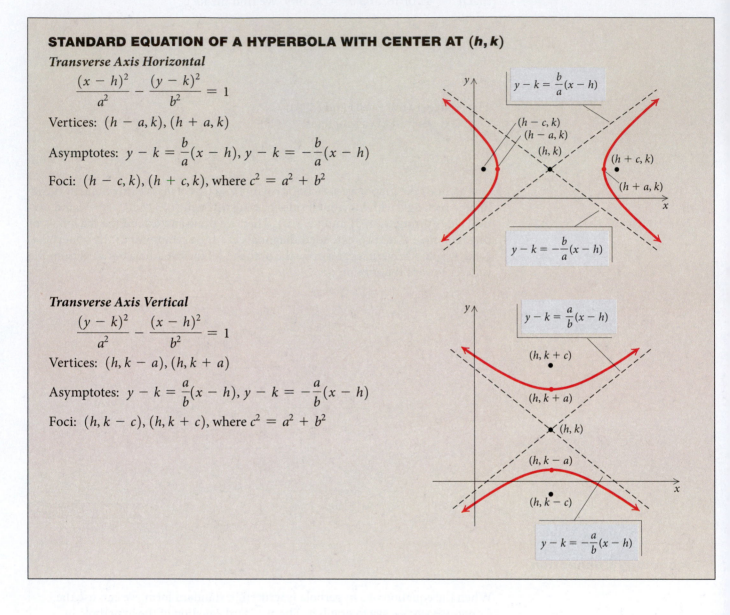

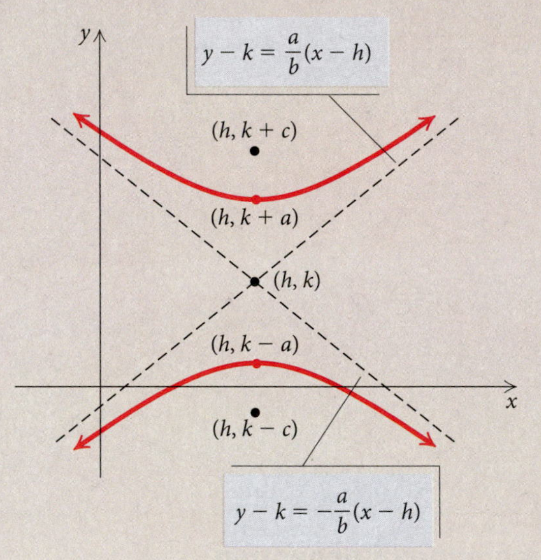

EXAMPLE 3 For the hyperbola given by

$$4y^2 - x^2 + 24y + 4x + 28 = 0,$$

find the center, the vertices, the foci, and the asymptotes. Then draw the graph.

Solution First, we complete the square to get standard form:

$$\begin{aligned} 4y^2 - x^2 + 24y + 4x + 28 &= 0 \\ 4(y^2 + 6y \qquad\quad) - (x^2 - 4x \qquad\quad) &= -28 \\ 4(y^2 + 6y + 9 - 9) - (x^2 - 4x + 4 - 4) &= -28 \\ 4(y^2 + 6y + 9) + 4(-9) - (x^2 - 4x + 4) - (-4) &= -28 \\ 4(y^2 + 6y + 9) - 36 - (x^2 - 4x + 4) + 4 &= -28 \\ 4(y^2 + 6y + 9) - (x^2 - 4x + 4) &= -28 + 36 - 4 \\ 4(y + 3)^2 - (x - 2)^2 &= 4 \\ \frac{(y + 3)^2}{1} - \frac{(x - 2)^2}{4} &= 1 \quad \text{Dividing by 4} \\ \frac{[y - (-3)]^2}{1^2} - \frac{(x - 2)^2}{2^2} &= 1. \quad \text{Standard form} \end{aligned}$$

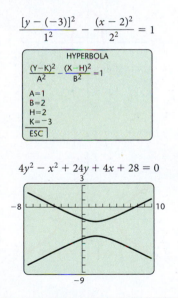

The center is $(2, -3)$. Note that $a = 1$ and $b = 2$. The transverse axis is vertical, so the vertices are 1 unit below and above the center:

$$(2, -3 - 1) \text{ and } (2, -3 + 1), \quad \text{or} \quad (2, -4) \text{ and } (2, -2).$$

We know that $c^2 = a^2 + b^2$, so $c^2 = 1^2 + 2^2 = 1 + 4 = 5$ and $c = \sqrt{5}$. Thus the foci are $\sqrt{5}$ units below and above the center:

$$\left(2, -3 - \sqrt{5}\right) \text{ and } \left(2, -3 + \sqrt{5}\right).$$

The asymptotes are

$$y - (-3) = \frac{1}{2}(x - 2) \text{ and } y - (-3) = -\frac{1}{2}(x - 2),$$

or

$$y + 3 = \frac{1}{2}(x - 2) \text{ and } y + 3 = -\frac{1}{2}(x - 2).$$

We sketch the asymptotes, plot the vertices, and draw the graph.

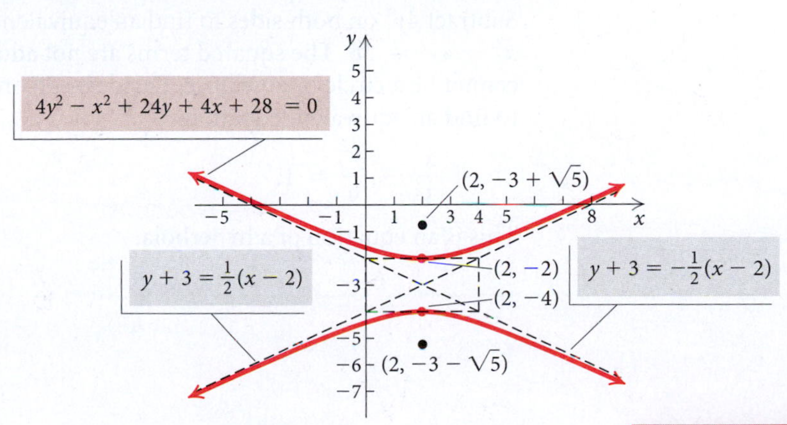

Now Try Exercise 29.

Technology Connection

When the equation of a hyperbola is written in standard form, we can use the Conics HYPERBOLA APP to graph it. The hyperbola in Example 3 is shown below.

$$\frac{[y - (-3)]^2}{1^2} - \frac{(x - 2)^2}{2^2} = 1$$

HYPERBOLA
$\frac{(Y-K)^2}{A^2} - \frac{(X-H)^2}{B^2} = 1$
A=1
B=2
H=2
K=−3
ESC

$4y^2 - x^2 + 24y + 4x + 28 = 0$

Window: −8 to 10, −9 to 3

CONNECTING THE CONCEPTS

Classifying Equations of Conic Sections

EQUATION	TYPE OF CONIC SECTION	GRAPH
$x - 4 + 4y = y^2$	Only one variable is squared, so this cannot be a circle, an ellipse, or a hyperbola. Find an equivalent equation: $x = (y - 2)^2$. This is an equation of a parabola.	 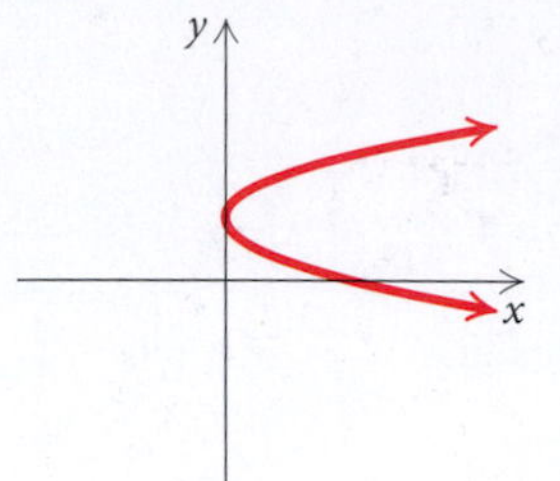
$3x^2 + 3y^2 = 75$	Both variables are squared, so this cannot be a parabola. The squared terms are added, so this cannot be a hyperbola. Divide by 3 on both sides to find an equivalent equation: $x^2 + y^2 = 25$. This is an equation of a circle.	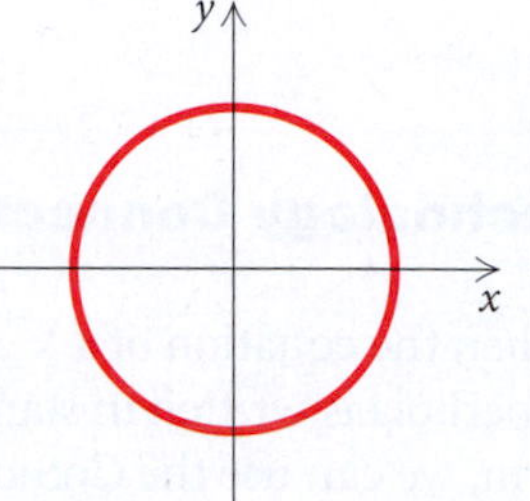
$y^2 = 16 - 4x^2$	Both variables are squared, so this cannot be a parabola. Add $4x^2$ on both sides to find an equivalent equation: $4x^2 + y^2 = 16$. The squared terms are added, so this cannot be a hyperbola. The coefficients of x^2 and y^2 are not the same, so this is not a circle. Divide by 16 on both sides to find an equivalent equation: $\frac{x^2}{4} + \frac{y^2}{16} = 1$. This is an equation of an ellipse.	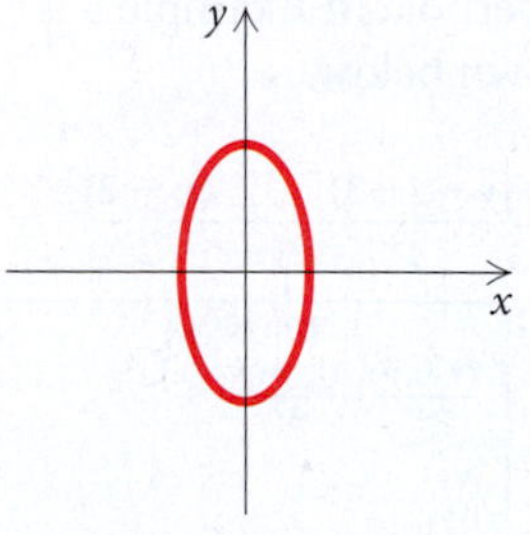
$x^2 = 4y^2 + 36$	Both variables are squared, so this cannot be a parabola. Subtract $4y^2$ on both sides to find an equivalent equation: $x^2 - 4y^2 = 36$. The squared terms are not added, so this cannot be a circle or an ellipse. Divide by 36 on both sides to find an equivalent equation: $\frac{x^2}{36} - \frac{y^2}{9} = 1$. This is an equation of a hyperbola.	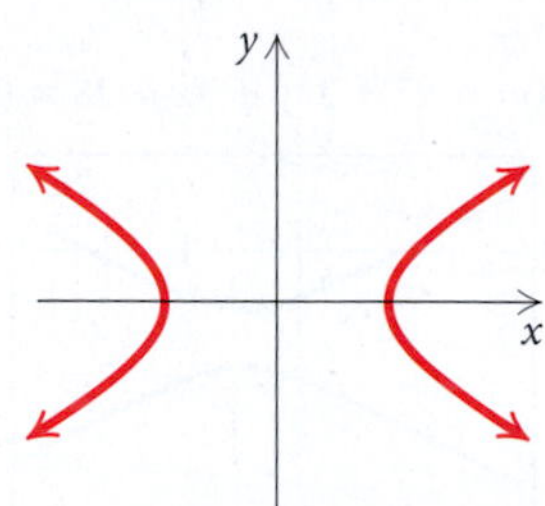

▶ Applications

Some comets travel in hyperbolic paths with the sun at one focus. Such comets pass by the sun only one time, unlike those with elliptical orbits, which reappear at intervals. We also see hyperbolas in architecture, such as in a cross section of a planetarium, an amphitheater, or a cooling tower for a steam or nuclear power plant.

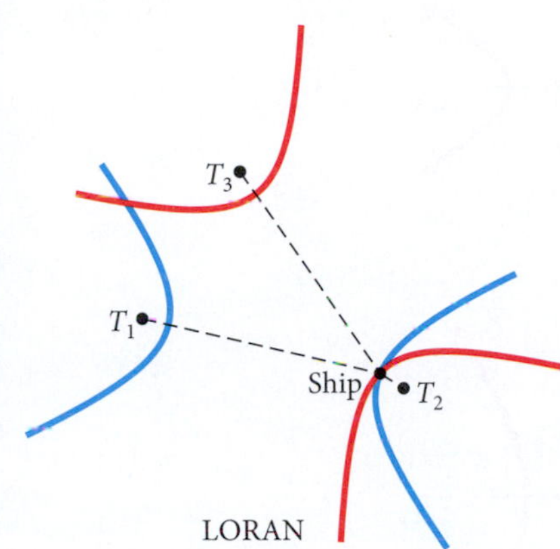

Another application of hyperbolas is in the long-range navigation system LORAN. This system uses transmitting stations in three locations to send out simultaneous signals to a ship or an aircraft. The difference in the arrival times of the signals from one pair of transmitters is recorded on the ship or aircraft. This difference is also recorded for signals from another pair of transmitters. For each pair, a computation is performed to determine the difference in the distances from each member of the pair to the ship or aircraft. If each pair of differences is kept constant, two hyperbolas can be drawn. Each has one of the pairs of transmitters as foci, and the ship or aircraft lies on the intersection of two of their branches.

7.3 Exercise Set

In Exercises 1–6, match the equation with one of the graphs (a)–(f) that follow.

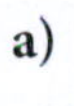
a)

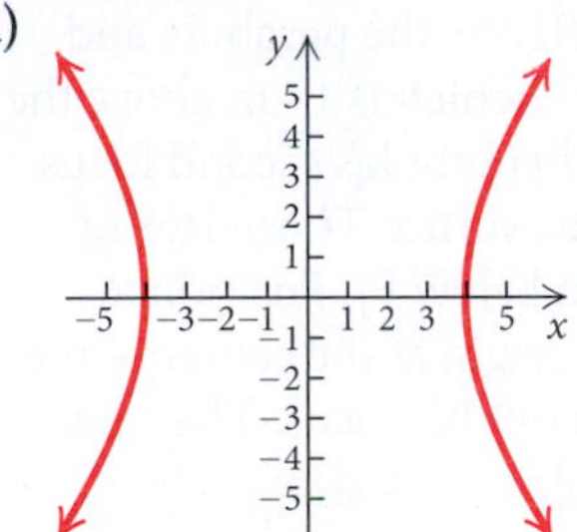

b)

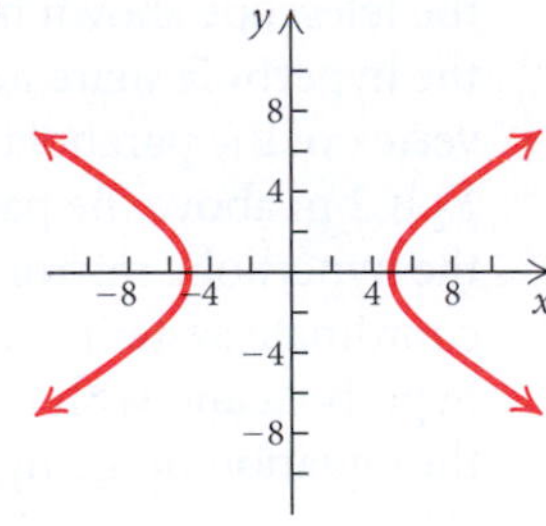

c)

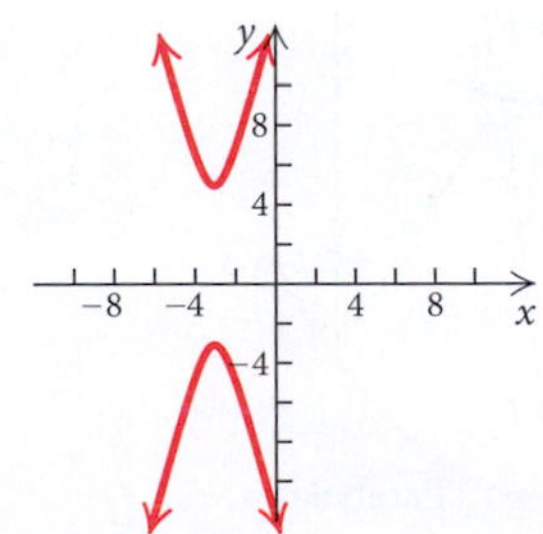

d)

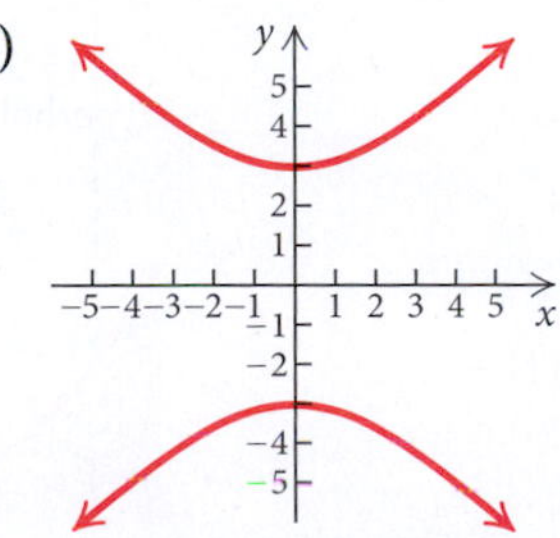

e)

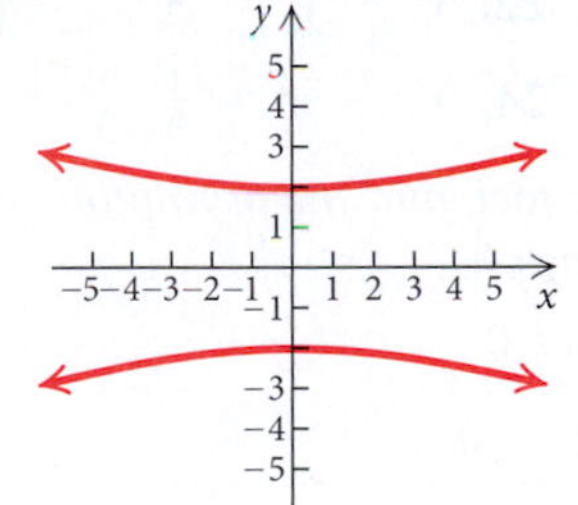

f)

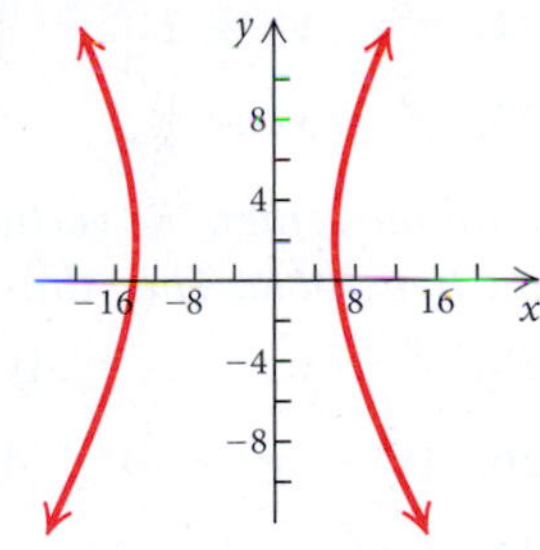

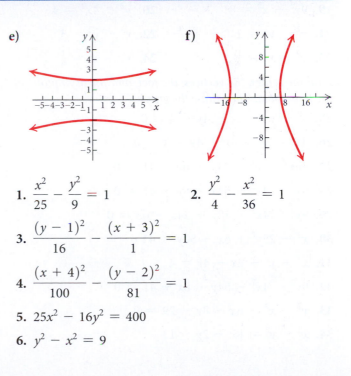

1. $\dfrac{x^2}{25} - \dfrac{y^2}{9} = 1$

2. $\dfrac{y^2}{4} - \dfrac{x^2}{36} = 1$

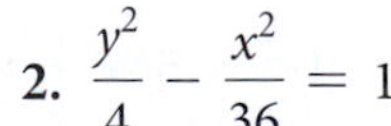

3. $\dfrac{(y-1)^2}{16} - \dfrac{(x+3)^2}{1} = 1$

4. $\dfrac{(x+4)^2}{100} - \dfrac{(y-2)^2}{81} = 1$

5. $25x^2 - 16y^2 = 400$

6. $y^2 - x^2 = 9$

Find an equation of a hyperbola satisfying the given conditions.

7. Vertices at $(0, 3)$ and $(0, -3)$; foci at $(0, 5)$ and $(0, -5)$

8. Vertices at $(1, 0)$ and $(-1, 0)$; foci at $(2, 0)$ and $(-2, 0)$

9. Asymptotes $y = \frac{3}{2}x$, $y = -\frac{3}{2}x$; one vertex $(2, 0)$

10. Asymptotes $y = \frac{5}{4}x$, $y = -\frac{5}{4}x$; one vertex $(0, 3)$

Find the center, the vertices, the foci, and the asymptotes. Then draw the graph.

11. $\dfrac{x^2}{4} - \dfrac{y^2}{4} = 1$

12. $\dfrac{x^2}{1} - \dfrac{y^2}{9} = 1$

13. $\dfrac{(x-2)^2}{9} - \dfrac{(y+5)^2}{1} = 1$

14. $\dfrac{(x-5)^2}{16} - \dfrac{(y+2)^2}{9} = 1$

15. $\dfrac{(y+3)^2}{4} - \dfrac{(x+1)^2}{16} = 1$

16. $\dfrac{(y+4)^2}{25} - \dfrac{(x+2)^2}{16} = 1$

17. $x^2 - 4y^2 = 4$

18. $4x^2 - y^2 = 16$

19. $9y^2 - x^2 = 81$

20. $y^2 - 4x^2 = 4$

21. $x^2 - y^2 = 2$

22. $x^2 - y^2 = 3$

23. $y^2 - x^2 = \frac{1}{4}$

24. $y^2 - x^2 = \frac{1}{9}$

Find the center, the vertices, the foci, and the asymptotes of the hyperbola. Then draw the graph.

25. $x^2 - y^2 - 2x - 4y - 4 = 0$

26. $4x^2 - y^2 + 8x - 4y - 4 = 0$

27. $36x^2 - y^2 - 24x + 6y - 41 = 0$

28. $9x^2 - 4y^2 + 54x + 8y + 41 = 0$

29. $9y^2 - 4x^2 - 18y + 24x - 63 = 0$

30. $x^2 - 25y^2 + 6x - 50y = 41$

31. $x^2 - y^2 - 2x - 4y = 4$

32. $9y^2 - 4x^2 - 54y - 8x + 41 = 0$

33. $y^2 - x^2 - 6x - 8y - 29 = 0$

34. $x^2 - y^2 = 8x - 2y - 13$

The ***eccentricity*** *of a hyperbola is defined as $e = c/a$. For a hyperbola, $c > a > 0$, so $e > 1$. When e is close to 1, a hyperbola appears to be very narrow. As the eccentricity increases, the hyperbola becomes "wider."*

35. Note the shapes of the hyperbolas in Examples 2 and 3. Which hyperbola has the larger eccentricity? Confirm your answer by computing the eccentricity of each hyperbola.

36. Which hyperbola has the larger eccentricity? (Assume that the coordinate systems have the same scale.)

a)

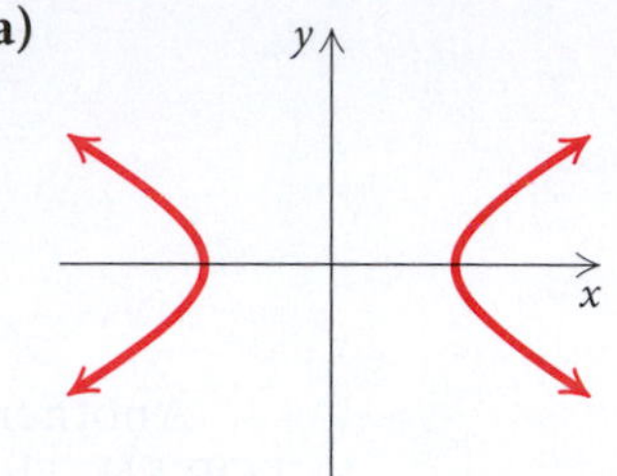

b)

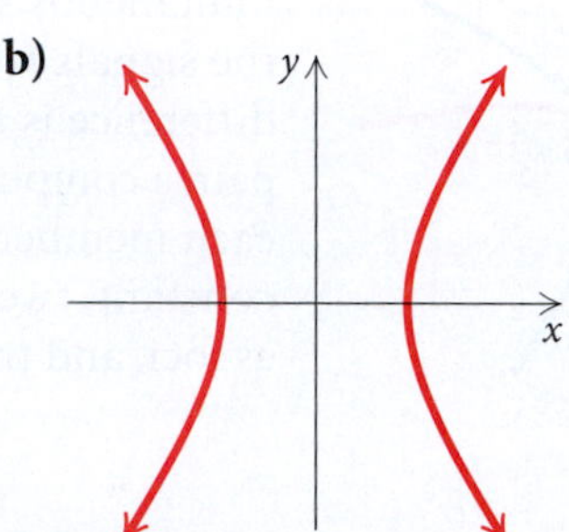

37. Find an equation of a hyperbola with vertices $(3, 7)$ and $(-3, 7)$ and $e = \frac{5}{3}$.

38. Find an equation of a hyperbola with vertices $(-1, 3)$ and $(-1, 7)$ and $e = 4$.

39. *Hyperbolic Mirror.* Certain telescopes contain both a parabolic mirror and a hyperbolic mirror. In the telescope shown in the figure, the parabola and the hyperbola share focus F_1, which is 14 m above the vertex of the parabola. The hyperbola's second focus F_2 is 2 m above the parabola's vertex. The vertex of the hyperbolic mirror is 1 m below F_1. Position a coordinate system with the origin at the center of the hyperbola and with the foci on the y-axis. Then find the equation of the hyperbola.

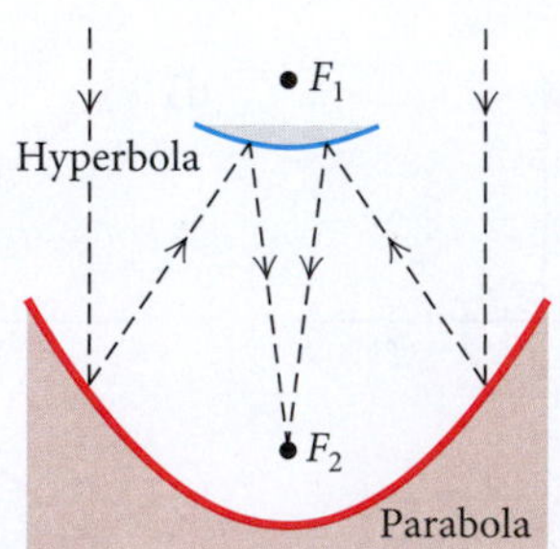

40. *Nuclear Cooling Tower.* A cross section of a nuclear cooling tower is a hyperbola with equation

$$\frac{x^2}{90^2} - \frac{y^2}{130^2} = 1.$$

The tower is 450 ft tall and the distance from the top of the tower to the center of the hyperbola is half the distance from the base of the tower to the center of the hyperbola. Find the diameter of the top and the base of the tower.

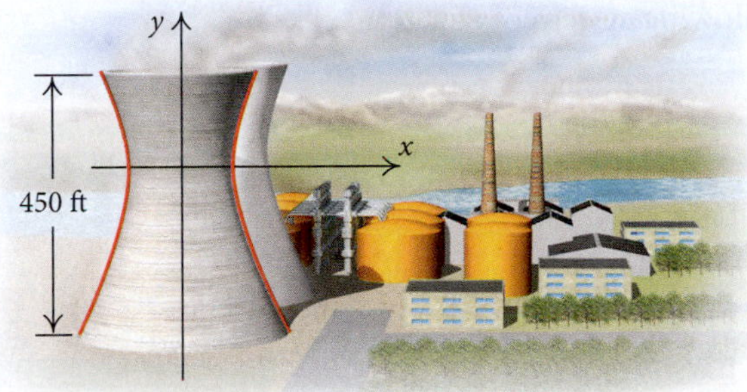

▸ Skill Maintenance

In Exercises 41–44, given the function:

a) *Determine whether it is one-to-one.* [5.1]

b) *If it is one-to-one, find a formula for the inverse.* [5.1]

41. $f(x) = 2x - 3$

42. $f(x) = x^3 + 2$

43. $f(x) = \dfrac{5}{x - 1}$

44. $f(x) = \sqrt{x + 4}$

Solve. [6.1], [6.3], [6.5], [6.6]

45. $x + y = 5,$
$x - y = 7$

46. $3x - 2y = 5,$
$5x + 2y = 3$

47. $2x - 3y = 7,$
$3x + 5y = 1$

48. $3x + 2y = -1,$
$2x + 3y = 6$

▸ Synthesis

Find an equation of a hyperbola satisfying the given conditions.

49. Vertices at $(3, -8)$ and $(3, -2)$; asymptotes $y = 3x - 14$, $y = -3x + 4$

50. Vertices at $(-9, 4)$ and $(-5, 4)$; asymptotes $y = 3x + 25$, $y = -3x - 17$

51. *Navigation.* Two radio transmitters positioned 300 mi apart along the shore send simultaneous signals to a ship that is 200 mi offshore and sailing parallel to the shoreline. The signal from transmitter S reaches the ship 200 microseconds later than the signal from transmitter T. The signals travel at a speed of 186,000 miles per second, or 0.186 mile per microsecond. Find the equation of the hyperbola with foci S and T on which the ship is located. (*Hint:* For any point on the hyperbola, the absolute value of the difference of its distances from the foci is $2a$.)

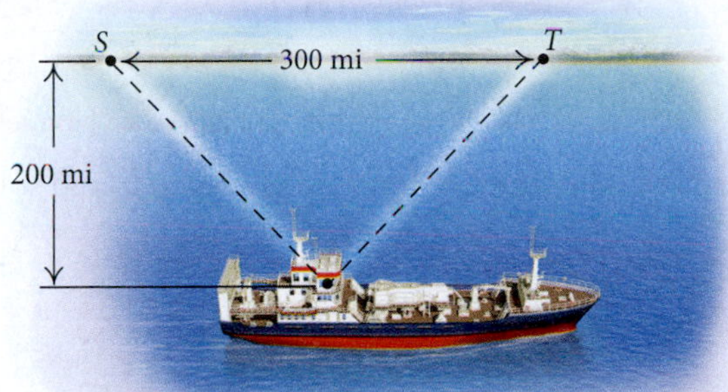

7.4 Nonlinear Systems of Equations and Inequalities

- ▸ Solve a nonlinear system of equations.
- ▸ Use nonlinear systems of equations to solve applied problems.
- ▸ Graph nonlinear systems of inequalities.

The systems of equations that we have studied so far have been composed of linear equations. Now we consider systems of two equations in two variables in which at least one equation is not linear.

Nonlinear Systems of Equations

The graphs of the equations in a nonlinear system of equations can have no point of intersection or one or more points of intersection. The coordinates of each point of intersection represent a solution of the system of equations. When no point of intersection exists, the system of equations has no real-number solution.

Solutions of nonlinear systems of equations can be found using the substitution method or the elimination method. The substitution method is preferable for a system consisting of one linear equation and one nonlinear equation. The elimination method is preferable in most, but not all, cases when both equations are nonlinear.

EXAMPLE 1 Solve the following system of equations:

$x^2 + y^2 = 25,$ (1) The graph is a circle.

$3x - 4y = 0.$ (2) The graph is a line.

Algebraic Solution

We use the substitution method. First, we solve equation (2) for x:

$3x - 4y = 0$ (2)

$3x = 4y$

$x = \frac{4}{3}y.$ (3) We could have solved for y instead.

Next, we substitute $\frac{4}{3}y$ for x in equation (1) and solve for y:

$\left(\frac{4}{3}y\right)^2 + y^2 = 25$

$\frac{16}{9}y^2 + y^2 = 25$

$\frac{25}{9}y^2 = 25$

$y^2 = 9$ Multiplying by $\frac{9}{25}$

$y = \pm 3.$

Now we substitute these numbers for y in equation (3) and solve for x:

$x = \frac{4}{3}(3) = 4,$ The pair (4, 3) appears to be a solution.

$x = \frac{4}{3}(-3) = -4.$ The pair (−4, −3) appears to be a solution.

Check: For (4, 3):

$x^2 + y^2 = 25$			$3x - 4y = 0$		
$4^2 + 3^2$	? 25		$3(4) - 4(3)$	? 0	
$16 + 9$			$12 - 12$		
25	25	TRUE	0	0	TRUE

For (−4, −3):

$x^2 + y^2 = 25$			$3x - 4y = 0$		
$(-4)^2 + (-3)^2$	? 25		$3(-4) - 4(-3)$	? 0	
$16 + 9$			$-12 + 12$		
25	25	TRUE	0	0	TRUE

The pairs (4, 3) and (−4, −3) check, so they are the solutions.

Visualizing the Solution

The ordered pairs corresponding to the points of intersection of the graphs of the equations are the solutions of the system of equations.

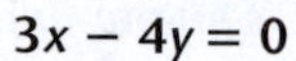

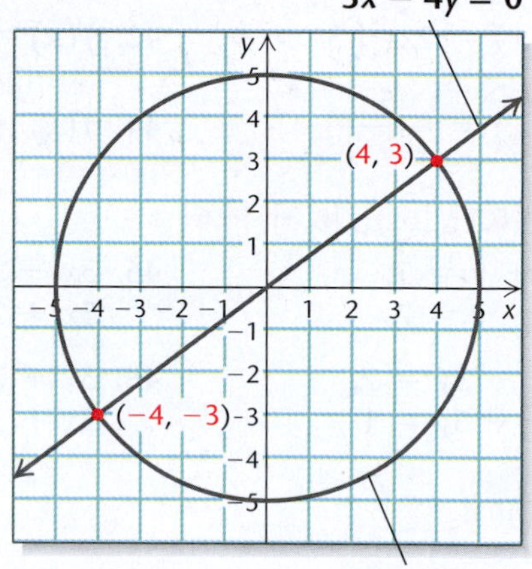

We see that the solutions are (4, 3) and (−4, −3).

Now Try Exercise 7.

Technology Connection

To solve the system of equations in Example 1 on a graphing calculator, we graph both equations in the same viewing window. Note that there are two points of intersection. We can find their coordinates using the INTERSECT feature.

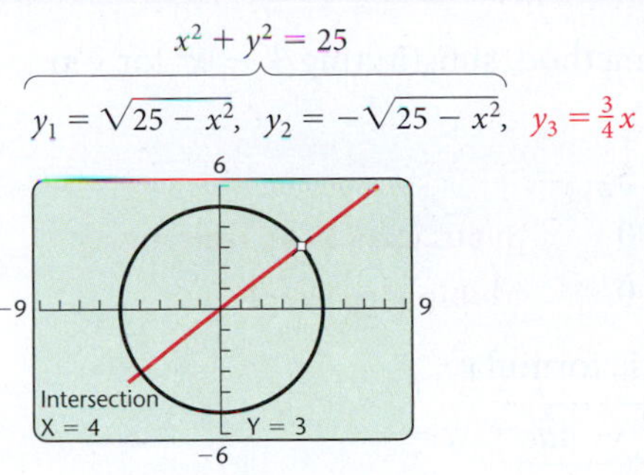

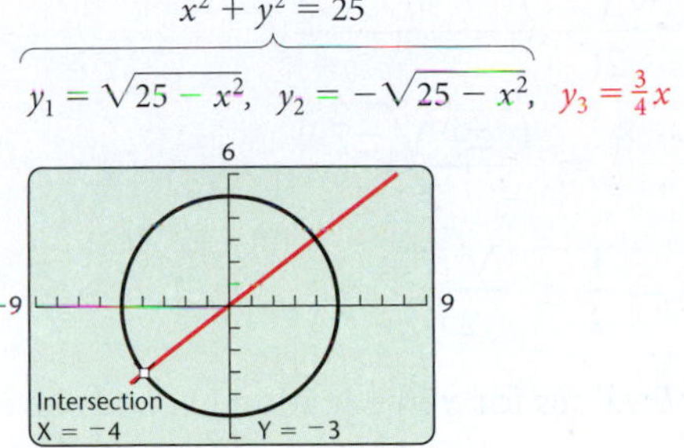

The solutions are $(4, 3)$ and $(-4, -3)$.

In the solution in Example 1, suppose that to find x we had substituted 3 and -3 in equation (1) rather than equation (3). If $y = 3, y^2 = 9$, and if $y = -3, y^2 = 9$, so both substitutions can be performed at the same time:

$$
\begin{aligned}
x^2 + y^2 &= 25 \qquad \textbf{(1)} \\
x^2 + (\pm 3)^2 &= 25 \\
x^2 + 9 &= 25 \\
x^2 &= 16 \\
x &= \pm 4.
\end{aligned}
$$

Each y-value produces two values for x. Thus, if $y = 3, x = 4$ or $x = -4$, and if $y = -3, x = 4$ or $x = -4$. The possible solutions are $(4, 3)$, $(-4, 3)$, $(4, -3)$, and $(-4, -3)$. A check reveals that $(4, -3)$ and $(-4, 3)$ are not solutions of equation (2). Since a circle and a line can intersect in at most two points, it is clear that there can be at most two real-number solutions.

EXAMPLE 2 Solve the following system of equations:

$$x + y = 5, \quad (1) \qquad \text{The graph is a line.}$$
$$y = 3 - x^2. \quad (2) \qquad \text{The graph is a parabola.}$$

Algebraic Solution

We use the substitution method, substituting $3 - x^2$ for y in equation (1):

$$x + 3 - x^2 = 5$$
$$-x^2 + x - 2 = 0 \qquad \text{Subtracting 5 and rearranging}$$
$$x^2 - x + 2 = 0. \qquad \text{Multiplying by } -1$$

Next, we use the quadratic formula:

$$x = \frac{-b \pm \sqrt{b^2 - 4ac}}{2a}$$
$$= \frac{-(-1) \pm \sqrt{(-1)^2 - 4(1)(2)}}{2(1)}$$
$$= \frac{1 \pm \sqrt{1 - 8}}{2} = \frac{1 \pm \sqrt{-7}}{2}$$
$$= \frac{1 \pm i\sqrt{7}}{2} = \frac{1}{2} \pm \frac{\sqrt{7}}{2}i.$$

Now, we substitute these values for x in equation (1) and solve for y:

$$\frac{1}{2} + \frac{\sqrt{7}}{2}i + y = 5$$
$$y = 5 - \frac{1}{2} - \frac{\sqrt{7}}{2}i$$
$$= \frac{9}{2} - \frac{\sqrt{7}}{2}i$$

and

$$\frac{1}{2} - \frac{\sqrt{7}}{2}i + y = 5$$
$$y = 5 - \frac{1}{2} + \frac{\sqrt{7}}{2}i$$
$$= \frac{9}{2} + \frac{\sqrt{7}}{2}i.$$

The solutions are

$$\left(\frac{1}{2} + \frac{\sqrt{7}}{2}i, \frac{9}{2} - \frac{\sqrt{7}}{2}i\right) \text{ and } \left(\frac{1}{2} - \frac{\sqrt{7}}{2}i, \frac{9}{2} + \frac{\sqrt{7}}{2}i\right).$$

There are no real-number solutions.

Visualizing the Solution

We graph the equations, as shown below.

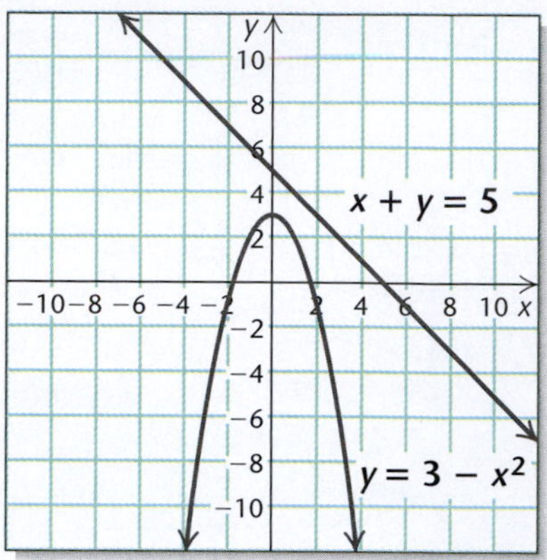

Note that there are no points of intersection. This indicates that there are no real-number solutions.

Now Try Exercise 17.

EXAMPLE 3 Solve the following system of equations:

$$2x^2 + 5y^2 = 39, \quad (1)$$ The graph is an ellipse.

$$3x^2 - y^2 = -1. \quad (2)$$ The graph is a hyperbola.

Algebraic Solution

We use the elimination method. First, we multiply equation (2) by 5 and add to eliminate the y^2-term:

$$2x^2 + 5y^2 = 39 \quad (1)$$
$$15x^2 - 5y^2 = -5 \quad \text{Multiplying (2) by 5}$$
$$17x^2 = 34 \quad \text{Adding}$$
$$x^2 = 2$$
$$x = \pm\sqrt{2}.$$

If $x = \sqrt{2}$, $x^2 = 2$, and if $x = -\sqrt{2}$, $x^2 = 2$. Thus substituting $\sqrt{2}$ or $-\sqrt{2}$ for x in equation (2) gives us

$$3(\pm\sqrt{2})^2 - y^2 = -1$$
$$3 \cdot 2 - y^2 = -1$$
$$6 - y^2 = -1$$
$$-y^2 = -7$$
$$y^2 = 7$$
$$y = \pm\sqrt{7}.$$

Thus, for $x = \sqrt{2}$, we have $y = \sqrt{7}$ or $y = -\sqrt{7}$, and for $x = -\sqrt{2}$, we have $y = \sqrt{7}$ or $y = -\sqrt{7}$. The possible solutions are $(\sqrt{2}, \sqrt{7})$, $(\sqrt{2}, -\sqrt{7})$, $(-\sqrt{2}, \sqrt{7})$, $(-\sqrt{2}, -\sqrt{7})$. All four pairs check, so they are the solutions.

Visualizing the Solution

We graph the equations and note that there are four points of intersection.

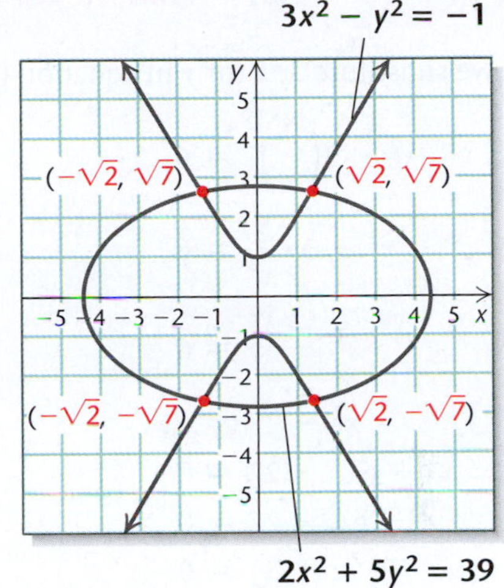

The coordinates of the points of intersection are $(\sqrt{2}, \sqrt{7})$, $(\sqrt{2}, -\sqrt{7})$, $(-\sqrt{2}, \sqrt{7})$, and $(-\sqrt{2}, -\sqrt{7})$. These are the solutions of the system of equations.

Now Try Exercise 27.

Technology Connection

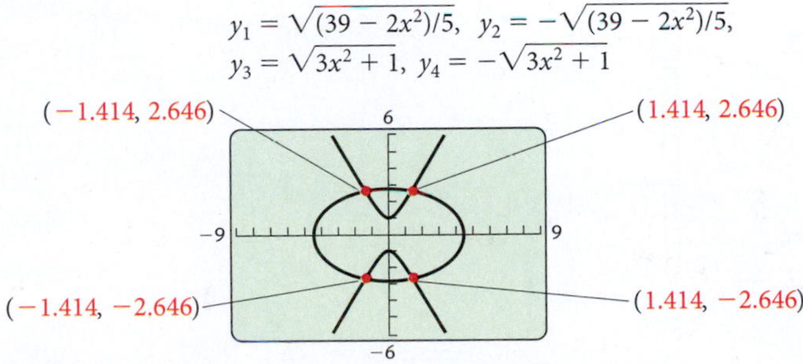

To solve the system of equations in Example 3 on a graphing calculator, we graph the equations in the same viewing window. We can use the INTERSECT feature to find the coordinates of the four points of intersection.

Note that the algebraic solution yields exact solutions, whereas the graphical solution yields decimal approximations of the solutions on most graphing calculators.

The solutions are approximately (1.414, 2.646), (1.414, −2.646), (−1.414, 2.646), and (−1.414, −2.646).

EXAMPLE 4 Solve the following system of equations:

$$x^2 - 3y^2 = 6, \qquad (1)$$
$$xy = 3. \qquad (2)$$

Algebraic Solution

We use the substitution method. First, we solve equation (2) for y:

$$xy = 3 \qquad (2)$$
$$y = \frac{3}{x}. \qquad (3) \qquad \text{Dividing by } x$$

Next, we substitute $3/x$ for y in equation (1) and solve for x:

$$x^2 - 3\left(\frac{3}{x}\right)^2 = 6$$
$$x^2 - 3\cdot\frac{9}{x^2} = 6$$
$$x^2 - \frac{27}{x^2} = 6$$
$$x^4 - 27 = 6x^2 \qquad \text{Multiplying by } x^2$$
$$x^4 - 6x^2 - 27 = 0$$
$$u^2 - 6u - 27 = 0 \qquad \text{Letting } u = x^2$$
$$(u - 9)(u + 3) = 0 \qquad \text{Factoring}$$
$$u - 9 = 0 \quad or \quad u + 3 = 0 \qquad \text{Principle of zero products}$$
$$u = 9 \quad or \quad u = -3$$
$$x^2 = 9 \quad or \quad x^2 = -3 \qquad \text{Substituting } x^2 \text{ for } u$$
$$x = \pm 3 \quad or \quad x = \pm i\sqrt{3}.$$

Since $y = 3/x$,

when $x = 3$, $\quad y = \frac{3}{3} = 1;$

when $x = -3$, $\quad y = \frac{3}{-3} = -1;$

when $x = i\sqrt{3}$, $\quad y = \frac{3}{i\sqrt{3}} = \frac{3}{i\sqrt{3}}\cdot\frac{-i\sqrt{3}}{-i\sqrt{3}} = -i\sqrt{3};$

when $x = -i\sqrt{3}$, $\quad y = \frac{3}{-i\sqrt{3}} = \frac{3}{-i\sqrt{3}}\cdot\frac{i\sqrt{3}}{i\sqrt{3}} = i\sqrt{3}.$

The pairs $(3, 1)$, $(-3, -1)$, $(i\sqrt{3}, -i\sqrt{3})$, and $(-i\sqrt{3}, i\sqrt{3})$ check, so they are the solutions.

Visualizing the Solution

The coordinates of the points of intersection of the graphs of the equations give us the real-number solutions of the system of equations. These graphs do not show us the imaginary-number solutions.

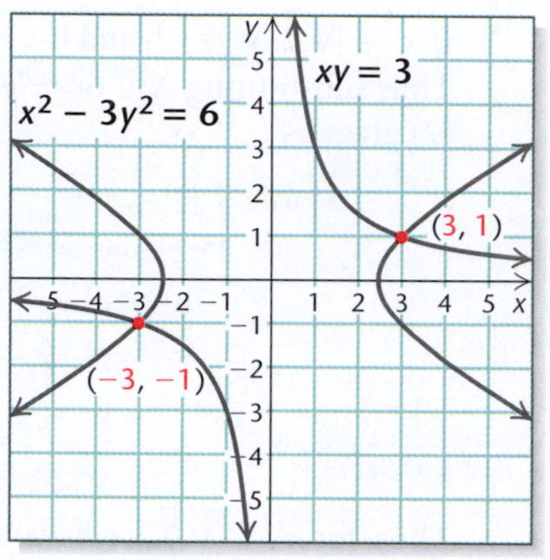

Now Try Exercise 19.

Modeling and Problem Solving

EXAMPLE 5 ***Dimensions of a Piece of Land.*** For a student recreation building at Southport Community College, an architect wants to lay out a rectangular piece of land that has a perimeter of 204 m and an area of 2565 m^2. Find the dimensions of the piece of land.

Solution

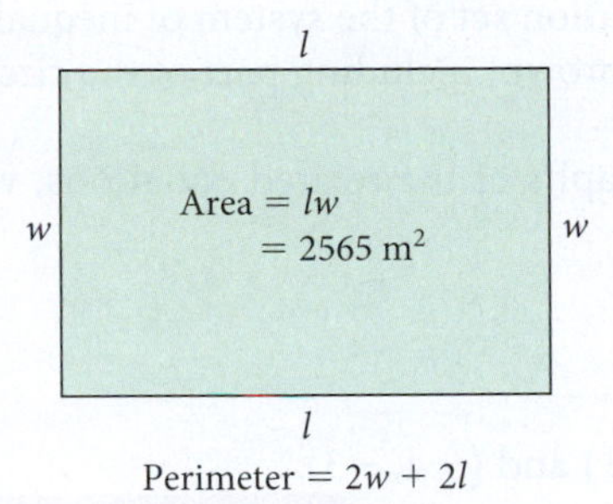

1. **Familiarize.** We make a drawing and label it, letting $l =$ the length of the piece of land, in meters, and $w =$ the width, in meters.

2. **Translate.** We now have the following:

$$\text{Perimeter: } 2w + 2l = 204, \quad (1)$$
$$\text{Area: } lw = 2565. \quad (2)$$

3. **Carry out.** We solve the system of equations

$$2w + 2l = 204,$$
$$lw = 2565.$$

Solving the second equation for l gives us $l = 2565/w$. We then substitute $2565/w$ for l in equation (1) and solve for w:

$$2w + 2\left(\frac{2565}{w}\right) = 204$$
$$2w^2 + 2(2565) = 204w \quad \text{Multiplying by } w$$
$$2w^2 - 204w + 2(2565) = 0$$
$$w^2 - 102w + 2565 = 0 \quad \text{Multiplying by } \tfrac{1}{2}$$
$$(w - 57)(w - 45) = 0$$
$$w - 57 = 0 \quad or \quad w - 45 = 0 \quad \text{Principle of zero products}$$
$$w = 57 \quad or \quad w = 45.$$

If $w = 57$, then $l = 2565/w = 2565/57 = 45$. If $w = 45$, then $l = 2565/w = 2565/45 = 57$. Since length is generally considered to be longer than width, we have the solution $l = 57$ and $w = 45$, or $(57, 45)$.

4. **Check.** If $l = 57$ and $w = 45$, the perimeter is $2 \cdot 45 + 2 \cdot 57$, or 204. The area is $57 \cdot 45$, or 2565. The numbers check.

5. **State.** The length of the piece of land is 57 m and the width is 45 m.

Now Try Exercise 61.

Nonlinear Systems of Inequalities

SYSTEMS OF INEQUALITIES
REVIEW SECTION 6.7

Recall that a solution of a system of inequalities is an ordered pair that is a solution of each inequality in the system. We graphed systems of linear inequalities in Section 6.7. Now we graph a system of nonlinear inequalities.

EXAMPLE 6 Graph the solution set of the system

$$x^2 + y^2 \le 25,$$
$$3x - 4y > 0.$$

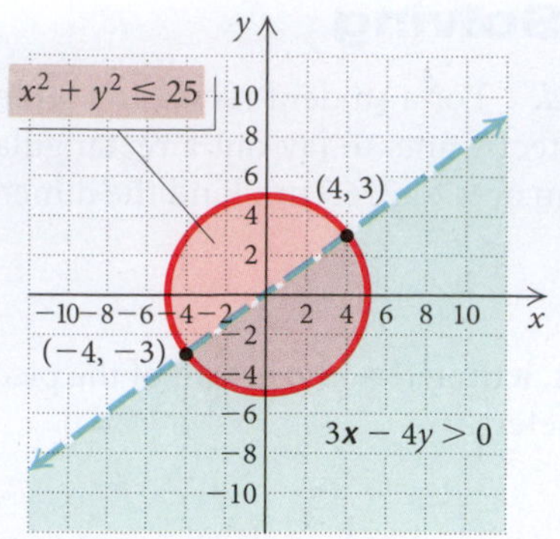

Solution We graph $x^2 + y^2 \leq 25$ by first graphing the related equation of the circle $x^2 + y^2 = 25$. We use a solid line since the inequality symbol is $\leq$. Next, we choose $(0, 0)$ as a test point and find that it is a solution of $x^2 + y^2 \leq 25$, so we shade the region that contains $(0, 0)$ using red. This is the region inside the circle. Now we graph the line $3x - 4y = 0$ using a dashed line since the inequality symbol is $>$. The point $(0, 0)$ is on the line, so we choose another test point, say, $(0, 2)$. We find that this point is not a solution of $3x - 4y > 0$, so we shade the half-plane that does not contain $(0, 2)$ using green. The solution set of the system of inequalities is the region shaded both red and green, or brown, including part of the circle $x^2 + y^2 = 25$.

To find the points of intersection of the graphs of the related equations, we solve the system composed of those equations:

$$x^2 + y^2 = 25,$$
$$3x - 4y = 0.$$

In Example 1, we found that these points are $(4, 3)$ and $(-4, -3)$.

Now Try Exercise 75.

Technology Connection

To use a graphing calculator to graph the system of inequalities in Example 7, we first graph $y_1 = 4 - x^2$ and $y_2 = 2 - x$. Using the test point $(0, 0)$ for each inequality, we find that we should shade below y_1 and above y_2. We can find the points of intersection of the graphs of the related equations, $(-1, 3)$ and $(2, 0)$, using the INTERSECT feature.

$y_1 = 4 - x^2$, $y_2 = 2 - x$

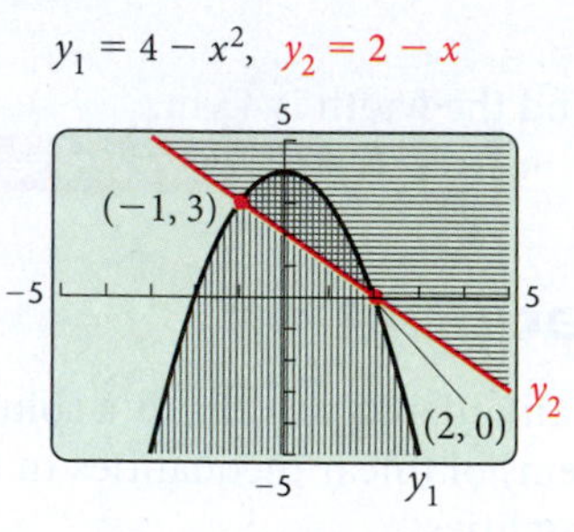

EXAMPLE 7 Graph the solution set of the system

$$y \leq 4 - x^2,$$
$$x + y \geq 2.$$

Solution We graph $y \leq 4 - x^2$ by first graphing the equation of the parabola $y = 4 - x^2$. We use a solid line since the inequality symbol is $\leq$. Next, we choose $(0, 0)$ as a test point and find that it is a solution of $y \leq 4 - x^2$, so we shade the region that contains $(0, 0)$ using red. Now we graph the line $x + y = 2$, again using a solid line since the inequality symbol is $\geq$. We test the point $(0, 0)$ and find that it is not a solution of $x + y \geq 2$, so we shade the half-plane that does not contain $(0, 0)$ using green. The solution set of the system of inequalities is the region shaded both red and green, or brown, including part of the parabola $y = 4 - x^2$ and part of the line $x + y = 2$.

Solving the system of equations

$$y = 4 - x^2,$$
$$x + y = 2,$$

we find that the points of intersection of the graphs of the related equations are $(-1, 3)$ and $(2, 0)$.

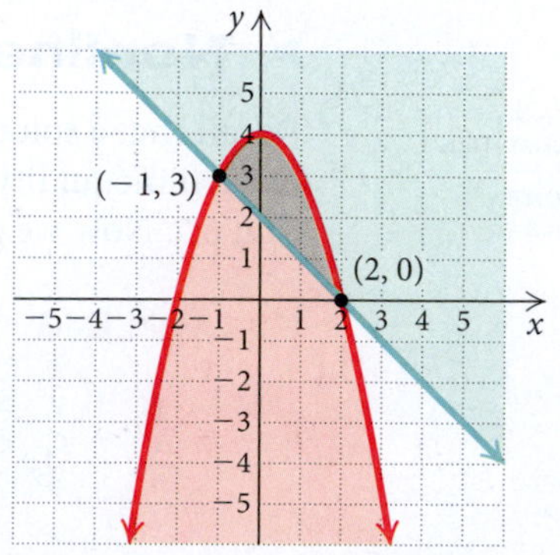

Now Try Exercise 81.

Visualizing the Graph

Match the equation or system of equations with its graph.

1. $y = x^3 - 3x$

2. $y = x^2 + 2x - 3$

3. $y = \dfrac{x - 1}{x^2 - x - 2}$

4. $y = -3x + 2$

5. $x + y = 3,$
   $2x + 5y = 3$

6. $9x^2 - 4y^2 = 36,$
   $x^2 + y^2 = 9$

7. $5x^2 + 5y^2 = 20$

8. $4x^2 + 16y^2 = 64$

9. $y = \log_2 x$

10. $y = 2^x$

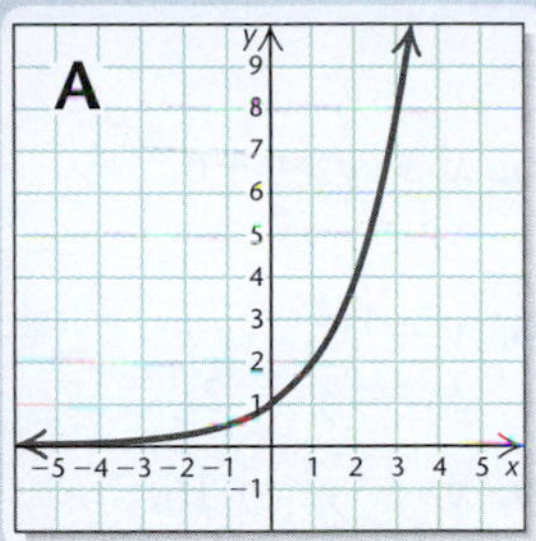

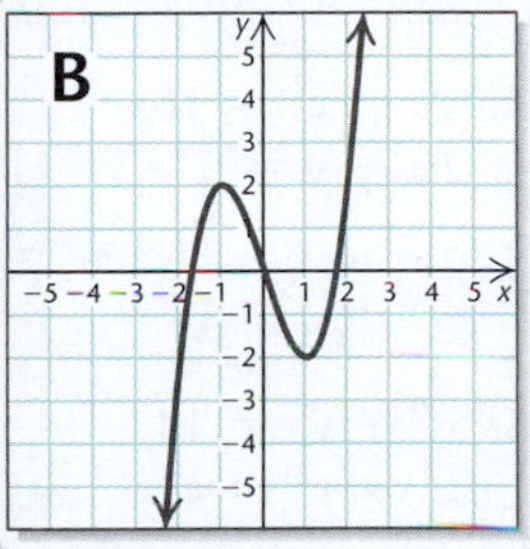

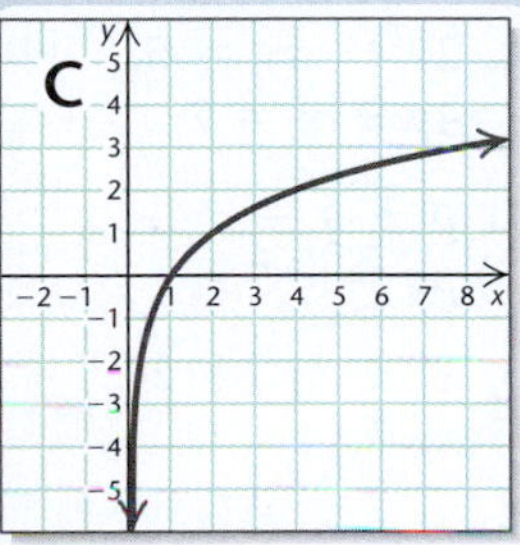

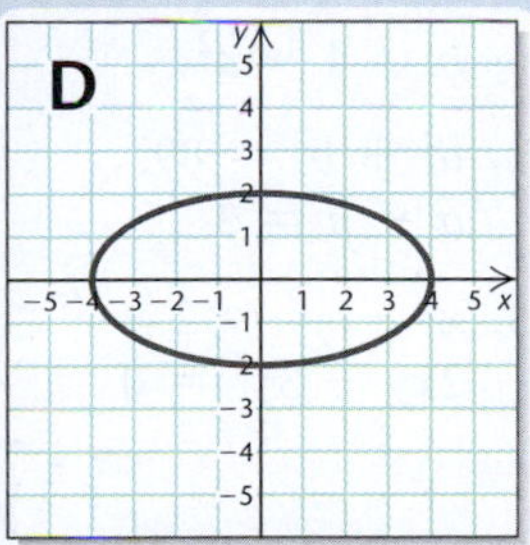

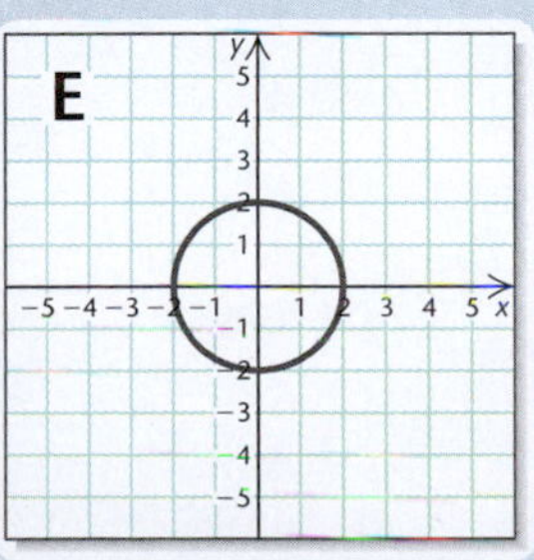

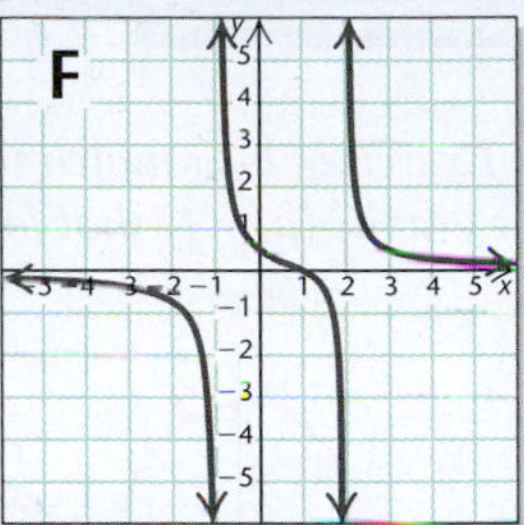

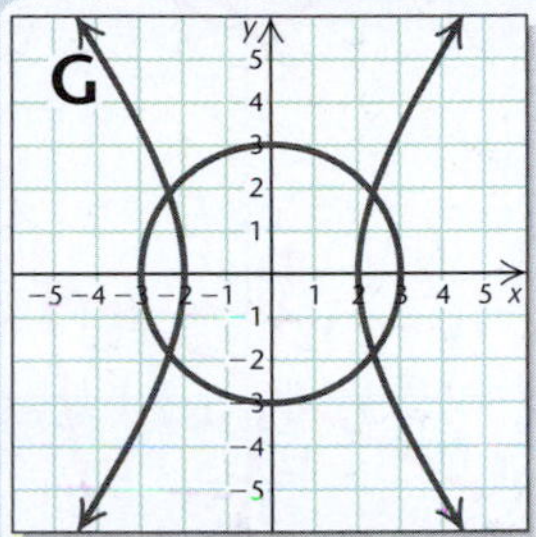

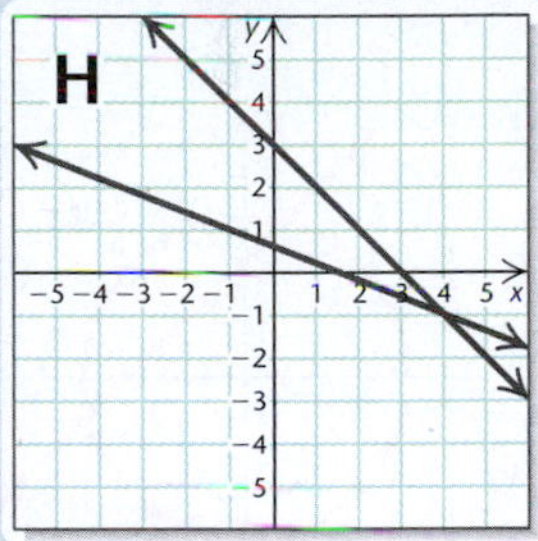

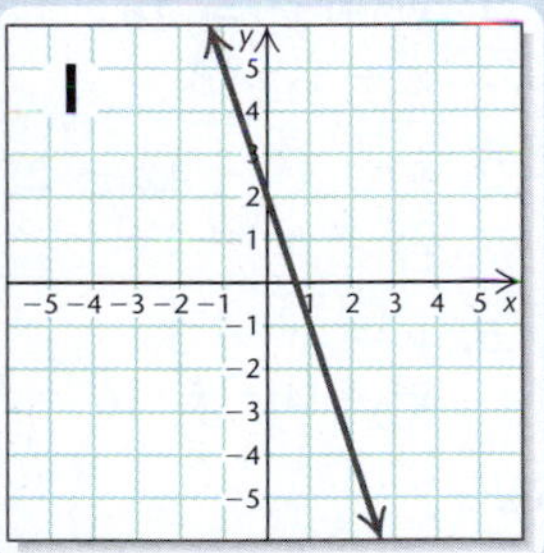

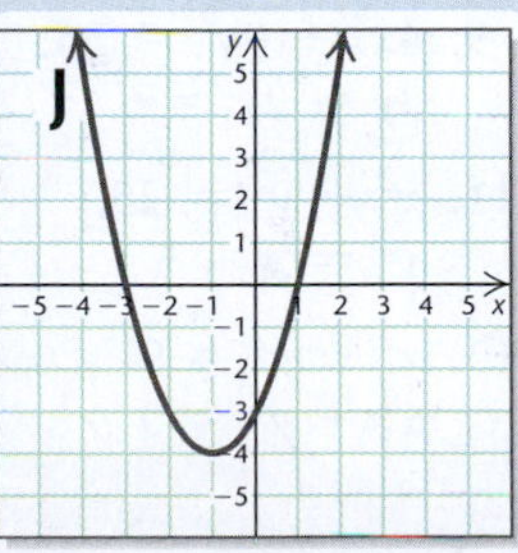

Answers on page A-40

7.4 Exercise Set

In Exercises 1–6, match the system of equations with one of the graphs (a)–(f) that follow.

a)

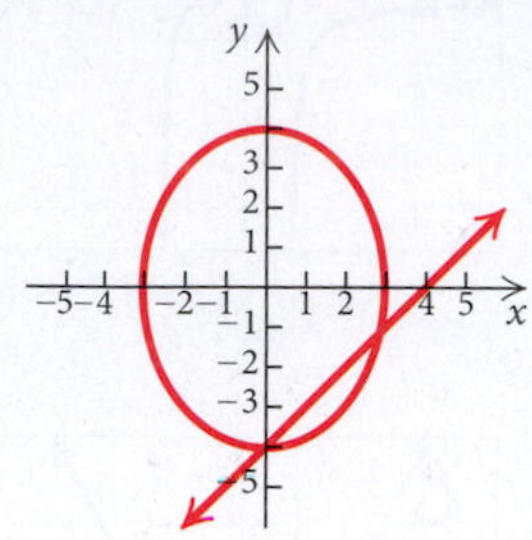

b)

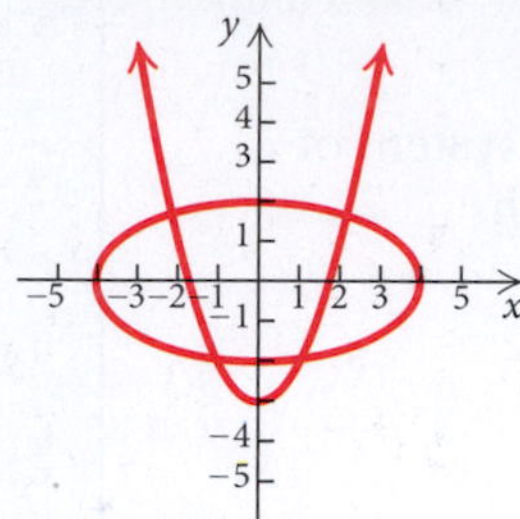

c)

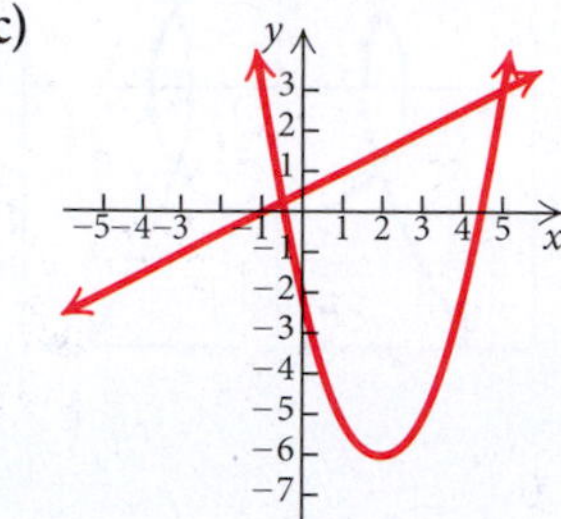

d)

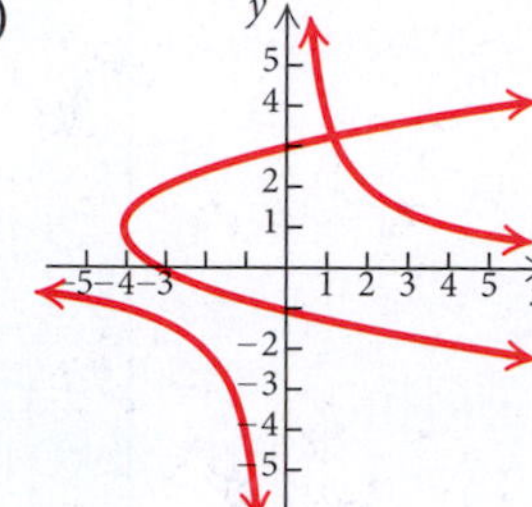

e)

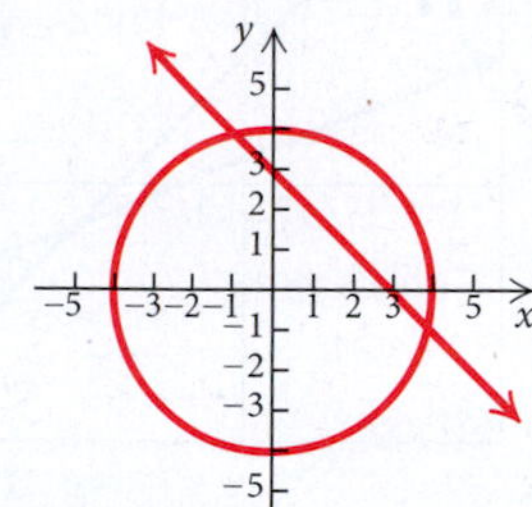

f)

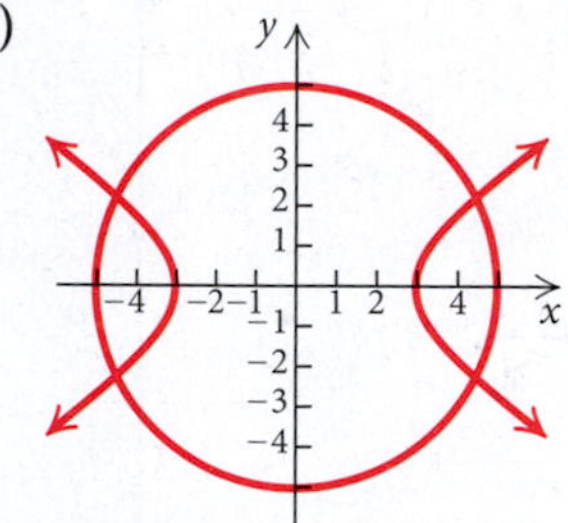

1. $x^2 + y^2 = 16$, $x + y = 3$

2. $16x^2 + 9y^2 = 144$, $x - y = 4$

3. $y = x^2 - 4x - 2$, $2y - x = 1$

4. $4x^2 - 9y^2 = 36$, $x^2 + y^2 = 25$

5. $y = x^2 - 3$, $x^2 + 4y^2 = 16$

6. $y^2 - 2y = x + 3$, $xy = 4$

Solve.

7. $x^2 + y^2 = 25$, $y - x = 1$

8. $x^2 + y^2 = 100$, $y - x = 2$

9. $4x^2 + 9y^2 = 36$, $3y + 2x = 6$

10. $9x^2 + 4y^2 = 36$, $3x + 2y = 6$

11. $x^2 + y^2 = 25$, $y^2 = x + 5$

12. $y = x^2$, $x = y^2$

13. $x^2 + y^2 = 9$ $x^2 - y^2 = 9$

14. $y^2 - 4x^2 = 4$, $4x^2 + y^2 = 4$

15. $y^2 - x^2 = 9$, $2x - 3 = y$

16. $x + y = -6$, $xy = -7$

17. $y^2 = x + 3$, $2y = x + 4$

18. $y = x^2$, $3x = y + 2$

19. $x^2 + y^2 = 25$, $xy = 12$

20. $x^2 - y^2 = 16$, $x + y^2 = 4$

21. $x^2 + y^2 = 4$, $16x^2 + 9y^2 = 144$

22. $x^2 + y^2 = 25$, $25x^2 + 16y^2 = 400$

23. $x^2 + 4y^2 = 25$, $x + 2y = 7$

24. $y^2 - x^2 = 16$, $2x - y = 1$

25. $x^2 - xy + 3y^2 = 27$, $x - y = 2$

26. $2y^2 + xy + x^2 = 7$, $x - 2y = 5$

27. $x^2 + y^2 = 16$, $y^2 - 2x^2 = 10$

28. $x^2 + y^2 = 14$, $x^2 - y^2 = 4$

29. $x^2 + y^2 = 5$, $xy = 2$

30. $x^2 + y^2 = 20$, $xy = 8$

31. $3x + y = 7$, $4x^2 + 5y = 56$

32. $2y^2 + xy = 5$, $4y + x = 7$

33. $a + b = 7$, $ab = 4$

34. $p + q = -4$, $pq = -5$

35. $x^2 + y^2 = 13$, $xy = 6$

36. $x^2 + 4y^2 = 20$, $xy = 4$

37. $x^2 + y^2 + 6y + 5 = 0$, $x^2 + y^2 - 2x - 8 = 0$

38. $2xy + 3y^2 = 7$, $3xy - 2y^2 = 4$

39. $2a + b = 1$, $b = 4 - a^2$

40. $4x^2 + 9y^2 = 36$, $x + 3y = 3$

41. $a^2 + b^2 = 89$, $a - b = 3$

42. $xy = 4$, $x + y = 5$

43. $xy - y^2 = 2$, $2xy - 3y^2 = 0$

44. $4a^2 - 25b^2 = 0$, $2a^2 - 10b^2 = 3b + 4$

45. $m^2 - 3mn + n^2 + 1 = 0$, $3m^2 - mn + 3n^2 = 13$

46. $ab - b^2 = -4$, $ab - 2b^2 = -6$

47. $x^2 + y^2 = 5$, $x - y = 8$

48. $4x^2 + 9y^2 = 36$, $y - x = 8$

49. $a^2 + b^2 = 14,$
$ab = 3\sqrt{5}$

50. $x^2 + xy = 5,$
$2x^2 + xy = 2$

51. $x^2 + y^2 = 25,$
$9x^2 + 4y^2 = 36$

52. $x^2 + y^2 = 1,$
$9x^2 - 16y^2 = 144$

53. $5y^2 - x^2 = 1,$
$xy = 2$

54. $x^2 - 7y^2 = 6,$
$xy = 1$

In Exercises 55–58, determine whether the statement is true or false.

55. A nonlinear system of equations can have both real-number solutions and imaginary-number solutions.

56. If the graph of a nonlinear system of equations consists of a line and a parabola, then the system has two real-number solutions.

57. If the graph of a nonlinear system of equations consists of a line and a circle, then the system has at most two real-number solutions.

58. If the graph of a nonlinear system of equations consists of a line and an ellipse, then it is possible for the system to have exactly one real-number solution.

59. *Photo Dimensions.* Hailey's Frame Shop has been commissioned to frame 5 black-and-white photos for an island resort. Each photo has a perimeter of 68 in. and a diagonal of 26 in. Find the dimensions of the photos.

60. *Sign Dimensions.* Alison's Advertising is building a rectangular sign with an area of 2 yd^2 and a perimeter of 6 yd. Find the dimensions of the sign.

61. *Graphic Design.* Marcia Graham, owner of Graham's Graphics, is designing an advertising brochure for the Art League's spring show. Each page of the brochure is rectangular with an area of 20 in^2 and a perimeter of 18 in. Find the dimensions of the brochure.

62. *Landscaping.* Green Leaf Landscaping is planting a rectangular wildflower garden with a perimeter of 6 m and a diagonal of $\sqrt{5}$ m. Find the dimensions of the garden.

63. *Fencing.* Clark's Country Pet Resort is fencing a new play area for dogs. The manager has purchased 210 yd of fence to enclose a rectangular pen. The area of the pen must be 2250 yd^2. What are the dimensions of the pen?

64. *Carpentry.* Ted Hansen of Hansen Woodworking Designs has been commissioned to make a rectangular tabletop with an area of $\sqrt{2}$ m^2 and a diagonal of $\sqrt{3}$ m for the Decorators' Show House. Find the dimensions of the tabletop.

65. *Banner Design.* A rectangular banner with an area of $\sqrt{3}$ m^2 is being designed to advertise an exhibit at the Davis Gallery. The length of a diagonal is 2 m. Find the dimensions of the banner.

66. *Investment.* Luke made an investment for 1 year that earned \$72 simple interest. If the principal had been \$240 more and the interest rate 1% less, the interest would have been the same. Find the principal and the interest rate.

67. *Seed Test Plots.* The Burton Seed Company has two square test plots. The sum of their areas is 832 ft^2 and the difference of their areas is 320 ft^2. Find the length of a side of each plot.

68. *Office Dimensions.* The diagonal of the floor of a rectangular office cubicle is 1 ft longer than the length of the cubicle and 3 ft longer than twice the width. Find the dimensions of the cubicle.

In Exercises 69–74, match the system of inequalities with one of the graphs (a)–(f) that follow.

a)
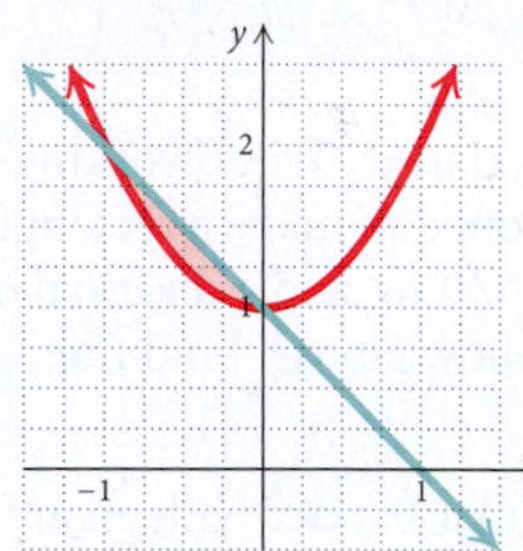

b)
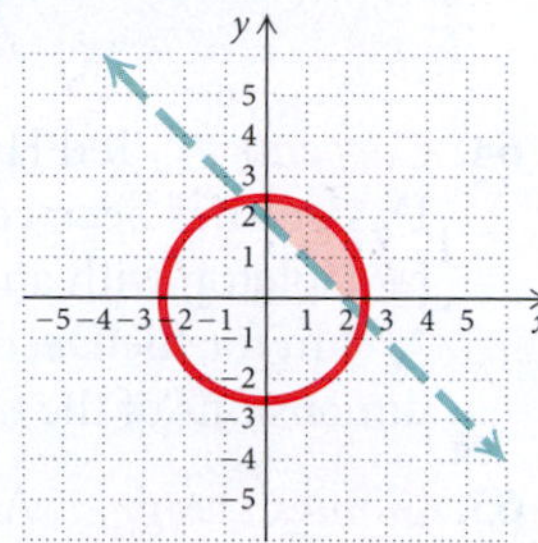

c)
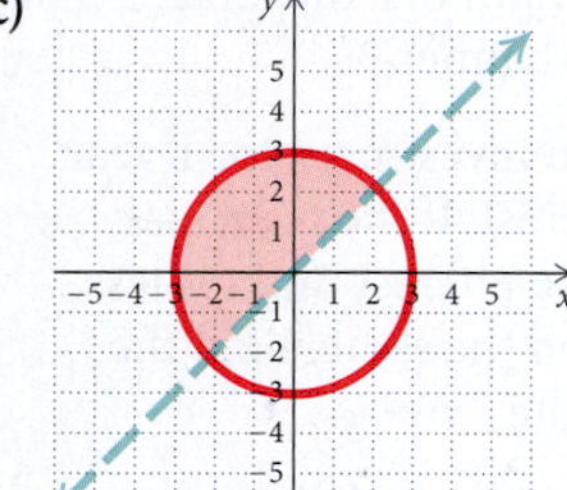

d)
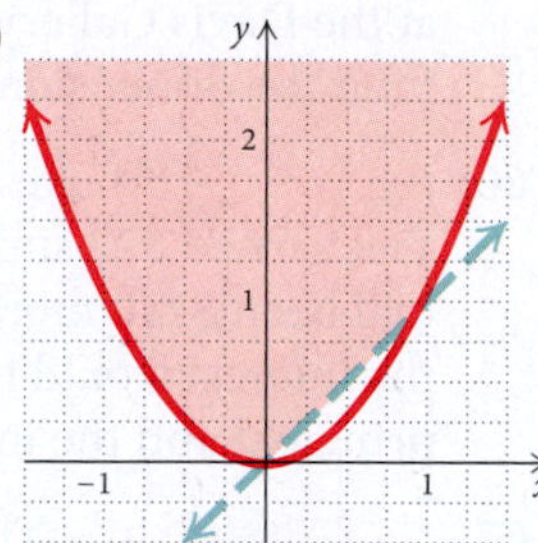

e)
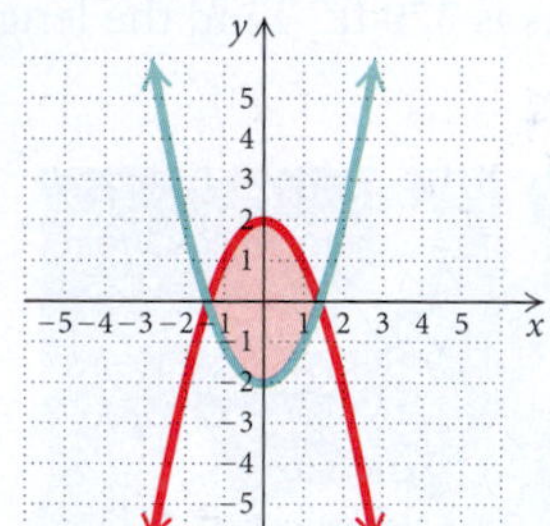

f)
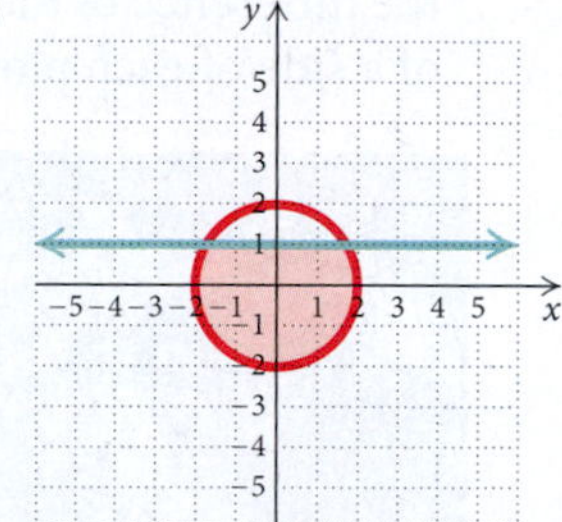

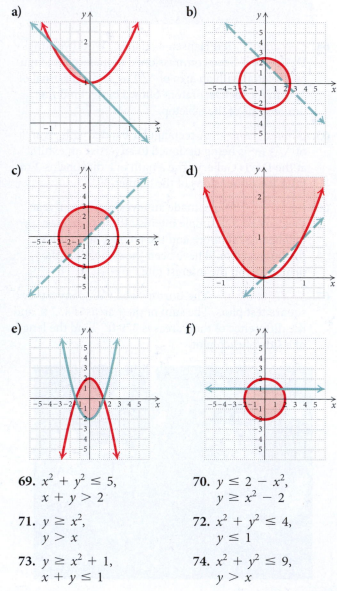

69. $x^2 + y^2 \le 5$,
$x + y > 2$

70. $y \le 2 - x^2$,
$y \ge x^2 - 2$

71. $y \ge x^2$,
$y > x$

72. $x^2 + y^2 \le 4$,
$y \le 1$

73. $y \ge x^2 + 1$,
$x + y \le 1$

74. $x^2 + y^2 \le 9$,
$y > x$

Graph the system of inequalities. Then find the coordinates of the points of intersection of the graphs of the related equations.

75. $x^2 + y^2 \le 16$,
$y < x$

76. $x^2 + y^2 \le 10$,
$y > x$

77. $x^2 \le y$,
$x + y \ge 2$

78. $x \ge y^2$,
$x - y \le 2$

79. $x^2 + y^2 \le 25$,
$x - y > 5$

80. $x^2 + y^2 \ge 9$,
$x - y > 3$

81. $y \ge x^2 - 3$,
$y \le 2x$

82. $y \le 3 - x^2$,
$y \ge x + 1$

83. $y \ge x^2$,
$y < x + 2$

84. $y \le 1 - x^2$,
$y > x - 1$

Skill Maintenance

Solve. [5.5]

85. $2^{3x} = 64$

86. $5^x = 27$

87. $\log_3 x = 4$

88. $\log(x - 3) + \log x = 1$

Synthesis

89. Find an equation of the circle that passes through the points $(2, 4)$ and $(3, 3)$ and whose center is on the line $3x - y = 3$.

90. Find an equation of the circle that passes through the points $(2, 3)$, $(4, 5)$, and $(0, -3)$.

91. Find an equation of an ellipse centered at the origin that passes through the points $\left(1, \sqrt{3}/2\right)$ and $\left(\sqrt{3}, 1/2\right)$.

92. Find an equation of a hyperbola of the type
$$\frac{x^2}{b^2} - \frac{y^2}{a^2} = 1$$
that passes through the points $\left(-3, -3\sqrt{5}/2\right)$ and $(-3/2, 0)$.

93. Show that a hyperbola does not intersect its asymptotes. That is, solve the system of equations
$$\frac{x^2}{a^2} - \frac{y^2}{b^2} = 1,$$
$$y = \frac{b}{a}x \quad \left(\text{or } y = -\frac{b}{a}x\right).$$

94. *Numerical Relationship.* Find two numbers whose product is 2 and the sum of whose reciprocals is $\frac{33}{8}$.

95. *Numerical Relationship.* The sum of two numbers is 1, and their product is 1. Find the sum of their cubes. There is a method to solve this problem that is easier than solving a nonlinear system of equations. Can you discover it?

96. *Box Dimensions.* Four squares with sides 5 in. long are cut from the corners of a rectangular metal sheet that has an area of 340 in^2. The edges are bent up to form an open box with a volume of 350 in^3. Find the dimensions of the box.

Solve.

97. $x^3 + y^3 = 72,$
$x + y = 6$

98. $a + b = \frac{5}{6},$
$\frac{a}{b} + \frac{b}{a} = \frac{13}{6}$

99. $p^2 + q^2 = 13,$
$\frac{1}{pq} = -\frac{1}{6}$

100. $e^x - e^{x+y} = 0,$
$e^y - e^{x-y} = 0$

Chapter 7 Summary and Review

STUDY GUIDE

KEY TERMS AND CONCEPTS	EXAMPLES

SECTION 7.1: THE PARABOLA

Standard Equation of a Parabola with Vertex $(0, 0)$ and Vertical Axis of Symmetry

The standard equation of a parabola with vertex $(0, 0)$ and directrix $y = -p$ is

$$x^2 = 4py.$$

The focus is $(0, p)$ and the y-axis is the axis of symmetry.

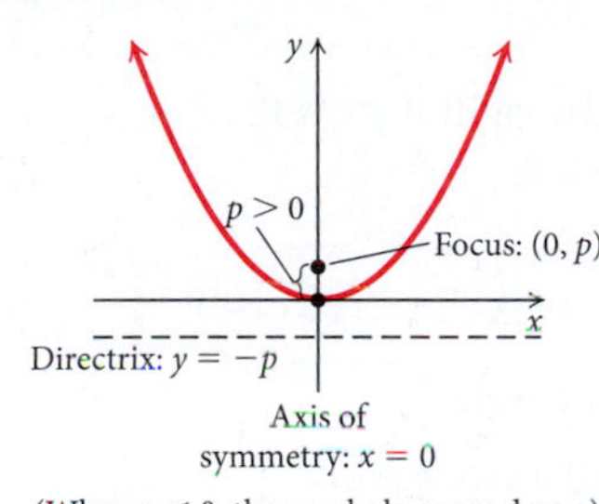

(When $p < 0$, the parabola opens down.)

See also Example 1 on p. 480.

Standard Equation of a Parabola with Vertex $(0, 0)$ and Horizontal Axis of Symmetry

The standard equation of a parabola with vertex $(0, 0)$ and directrix $x = -p$ is

$$y^2 = 4px.$$

The focus is $(p, 0)$ and the x-axis is the axis of symmetry.

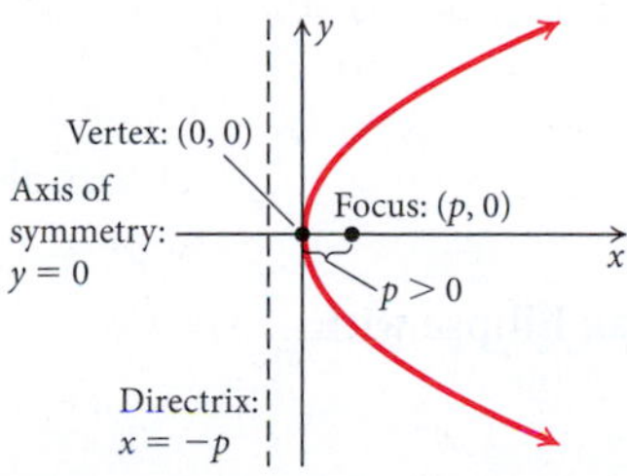

(When $p < 0$, the parabola opens to the left.)

See also Example 2 on p. 480.

Standard Equation of a Parabola with Vertex (h, k) and Vertical Axis of Symmetry

The standard equation of a parabola with vertex (h, k) and vertical axis of symmetry is

$$(x - h)^2 = 4p(y - k),$$

where the vertex is (h, k), the focus is $(h, k + p)$, and the directrix is $y = k - p$.

The parabola opens up if $p > 0$. It opens down if $p < 0$.

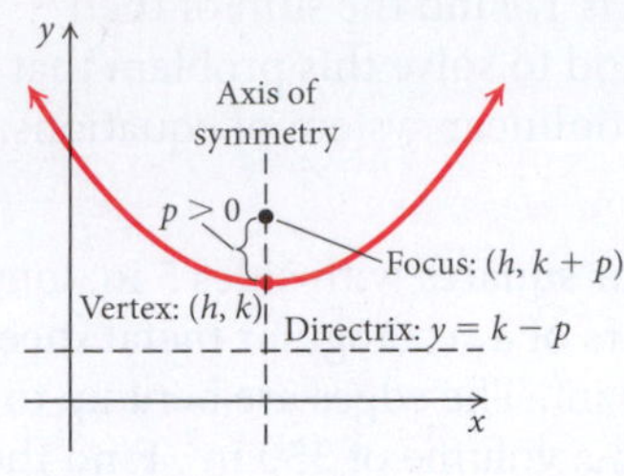

(When $p < 0$, the parabola opens down.)

See also Example 3 on p. 482.

Standard Equation of a Parabola with Vertex (h, k) and Horizontal Axis of Symmetry

The standard equation of a parabola with vertex (h, k) and horizontal axis of symmetry is

$$(y - k)^2 = 4p(x - h),$$

where the vertex is (h, k), the focus is $(h + p, k)$, and the directrix is $x = h - p$.

The parabola opens to the right if $p > 0$. It opens to the left if $p < 0$.

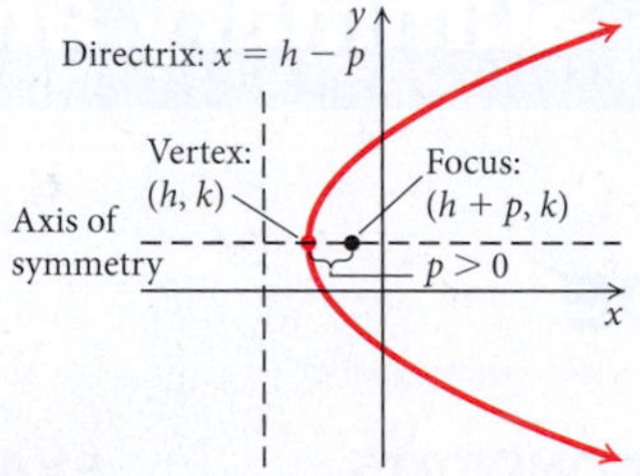

(When $p < 0$, the parabola opens to the left.)

See also Example 4 on p. 482.

SECTION 7.2: THE CIRCLE AND THE ELLIPSE

Standard Equation of a Circle

The standard equation of a circle with center (h, k) and radius r is

$$(x - h)^2 + (y - k)^2 = r^2.$$

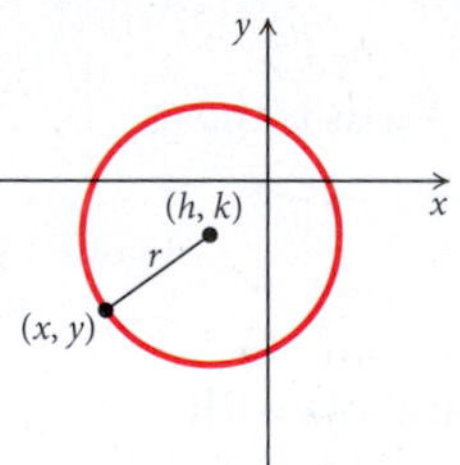

See also Example 1 on p. 486.

Standard Equation of an Ellipse with Center at the Origin

Major Axis Horizontal

$$\frac{x^2}{a^2} + \frac{y^2}{b^2} = 1, \quad a > b > 0$$

Vertices: $(-a, 0), (a, 0)$

y-intercepts: $(0, -b), (0, b)$

Foci: $(-c, 0), (c, 0)$, where $c^2 = a^2 - b^2$

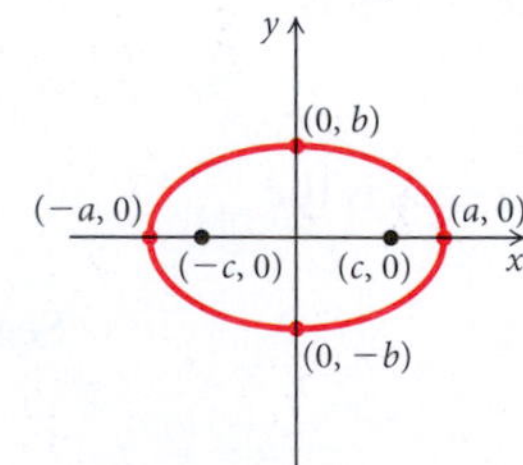

Major Axis Vertical

$$\frac{x^2}{b^2} + \frac{y^2}{a^2} = 1, \quad a > b > 0$$

Vertices: $(0, -a), (0, a)$

x-intercepts: $(-b, 0), (b, 0)$

Foci: $(0, -c), (0, c)$, where $c^2 = a^2 - b^2$

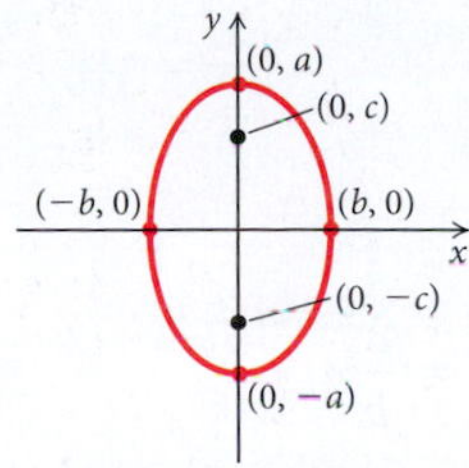

See also Examples 2 and 3 on pp. 489 and 490.

Standard Equation of an Ellipse with Center at (h, k)

Major Axis Horizontal

$$\frac{(x-h)^2}{a^2} + \frac{(y-k)^2}{b^2} = 1, \quad a > b > 0$$

Vertices: $(h - a, k), (h + a, k)$

Length of minor axis: $2b$

Foci: $(h - c, k), (h + c, k)$, where $c^2 = a^2 - b^2$

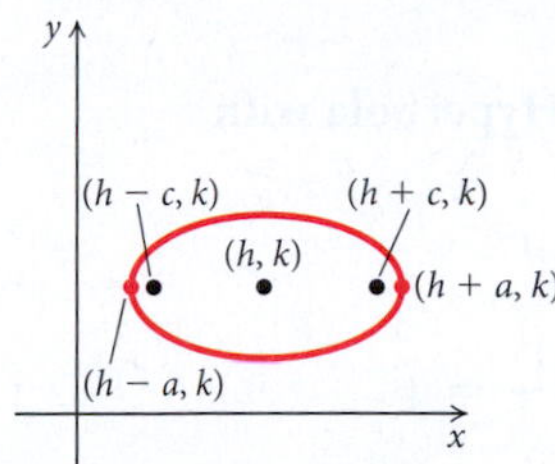

Major Axis Vertical

$$\frac{(x-h)^2}{b^2} + \frac{(y-k)^2}{a^2} = 1, \quad a > b > 0$$

Vertices: $(h, k - a), (h, k + a)$

Length of minor axis: $2b$

Foci: $(h, k - c), (h, k + c)$, where $c^2 = a^2 - b^2$

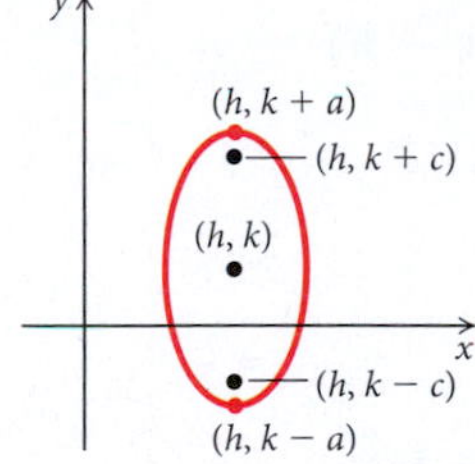

See also Example 4 on p. 490.

SECTION 7.3: THE HYPERBOLA

Standard Equation of a Hyperbola with Center at the Origin

Transverse Axis Horizontal

$$\frac{x^2}{a^2} - \frac{y^2}{b^2} = 1$$

Vertices: $(-a, 0), (a, 0)$

Asymptotes: $y = -\frac{b}{a}x, y = \frac{b}{a}x$

Foci: $(-c, 0), (c, 0)$, where $c^2 = a^2 + b^2$

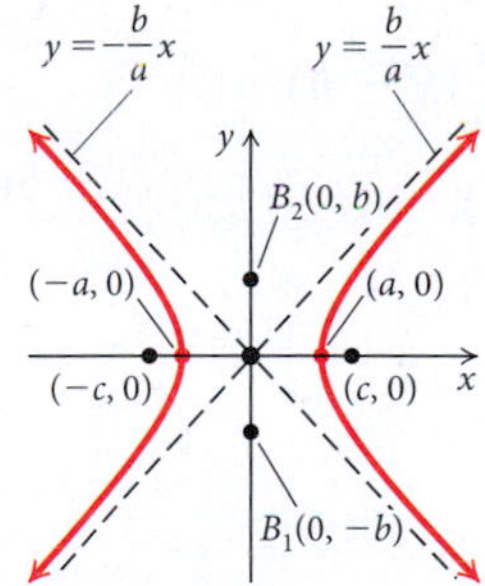

Transverse Axis Vertical

$$\frac{y^2}{a^2} - \frac{x^2}{b^2} = 1$$

Vertices: $(0, -a)$, $(0, a)$

Asymptotes: $y = -\frac{a}{b}x$, $y = \frac{a}{b}x$

Foci: $(0, -c)$, $(0, c)$, where $c^2 = a^2 + b^2$

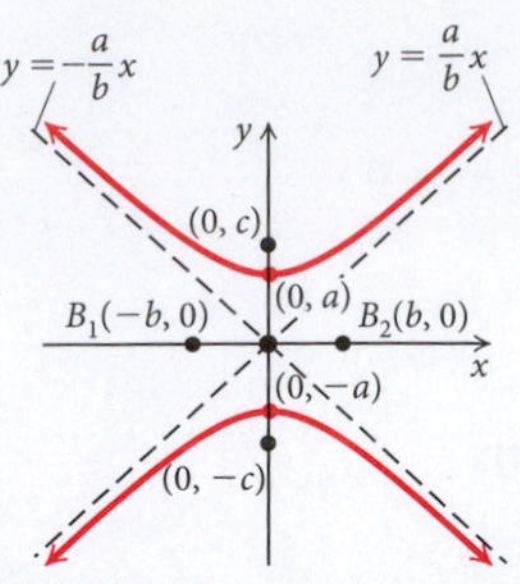

See also Examples 1 and 2 on p. 498.

Standard Equation of a Hyperbola with Center at (h, k)

Transverse Axis Horizontal

$$\frac{(x - h)^2}{a^2} - \frac{(y - k)^2}{b^2} = 1$$

Vertices: $(h - a, k)$, $(h + a, k)$

Asymptotes: $y - k = \frac{b}{a}(x - h)$,

$$y - k = -\frac{b}{a}(x - h)$$

Foci: $(h - c, k)$, $(h + c, k)$, where $c^2 = a^2 + b^2$

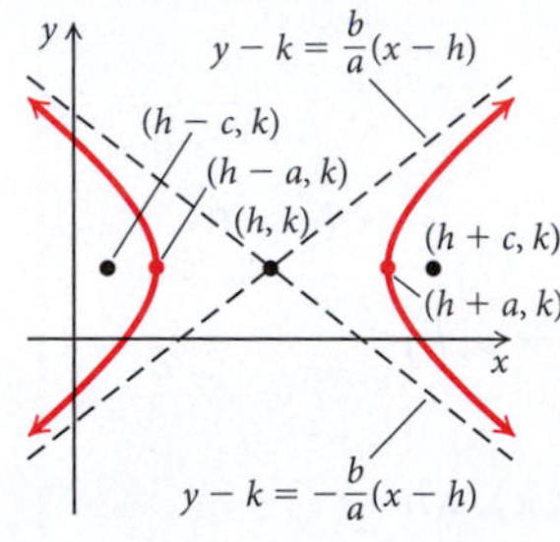

Transverse Axis Vertical

$$\frac{(y - k)^2}{a^2} - \frac{(x - h)^2}{b^2} = 1$$

Vertices: $(h, k - a)$, $(h, k + a)$

Asymptotes: $y - k = \frac{a}{b}(x - h)$,

$$y - k = -\frac{a}{b}(x - h)$$

Foci: $(h, k - c)$, $(h, k + c)$, where $c^2 = a^2 + b^2$

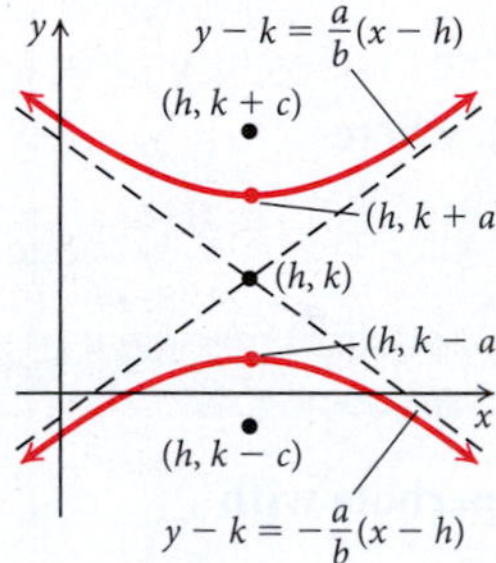

See also Example 3 on p. 501.

SECTION 7.4: NONLINEAR SYSTEMS OF EQUATIONS AND INEQUALITIES

Substitution or elimination can be used to solve **systems of equations containing at least one nonlinear equation.**	Solve: $x^2 - y = 2,$ (1) **The graph is a parabola.** $x - y = -4.$ (2) **The graph is a line.** $x = y - 4$ **Solving equation (2) for x** $(y - 4)^2 - y = 2$ **Substituting for x in equation (1)** $y^2 - 8y + 16 - y = 2$ $y^2 - 9y + 14 = 0$ $(y - 2)(y - 7) = 0$ $y - 2 = 0$ *or* $y - 7 = 0$ $y = 2$ *or* $y = 7$ If $y = 2$, then $x = 2 - 4 = -2$. If $y = 7$, then $x = 7 - 4 = 3$. The pairs $(-2, 2)$ and $(3, 7)$ check, so they are the solutions.
Some applied problems translate to a nonlinear system of equations.	See Example 5 on p. 511.
To graph a **nonlinear system of inequalities**, graph each inequality in the system and then shade the region where their solution sets overlap. To find the point(s) of intersection of the graphs of the related equations, solve the system of equations composed of those equations.	Graph: $x^2 - y \leq 2,$ $x - y > -4.$ 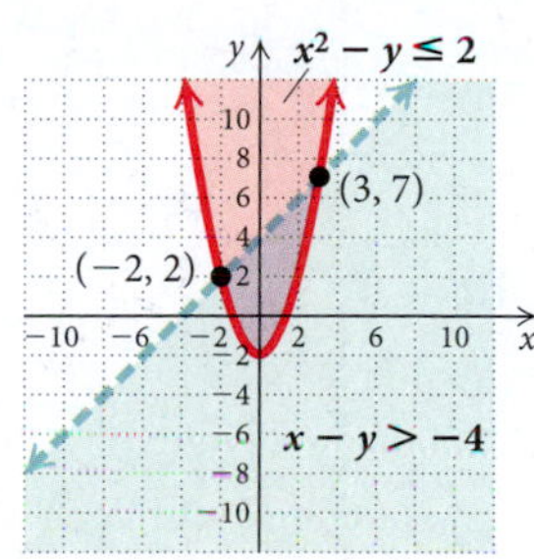 To find the points of intersection of the graphs of the related equations, solve the system of equations $x^2 - y = 2,$ $x - y = -4.$ We saw in the example above that these points are $(-2, 2)$ and $(3, 7)$.

REVIEW EXERCISES

Determine whether the statement is true or false.

1. The graph of $x + y^2 = 1$ is a parabola that opens to the left. [7.1]
2. The graph of $\frac{(x-2)^2}{4} + \frac{(y+3)^2}{9} = 1$ is an ellipse with center $(-2, 3)$. [7.2]
3. The hyperbola $\frac{x^2}{5} - \frac{y^2}{10} = 1$ has a horizontal transverse axis. [7.3]
4. Every nonlinear system of equations has at least one real-number solution. [7.4]
5. The graph of a nonlinear system of equations shows all the solutions of the system of equations. [7.4]

In Exercises 6–13, match the equation with one of the graphs (a)–(h) that follow.

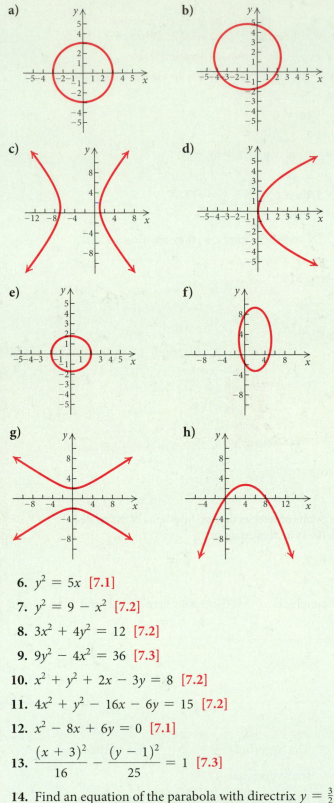

a)
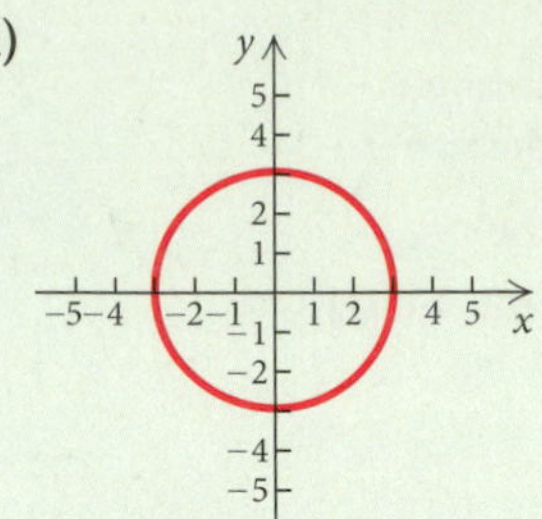

b)
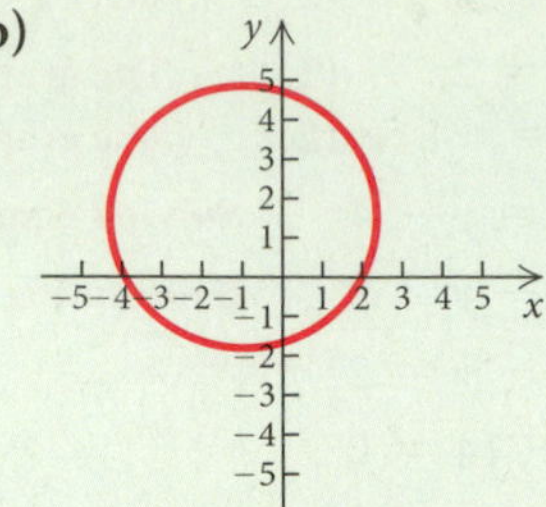

c)
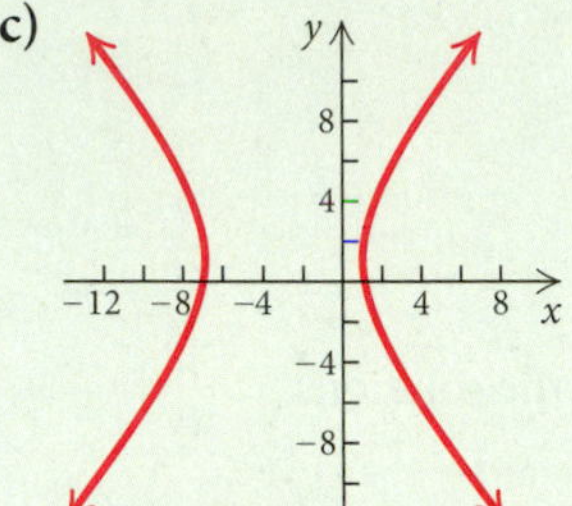

d)
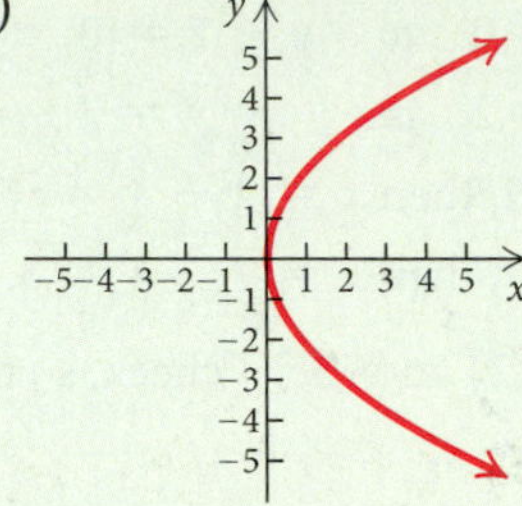

e)
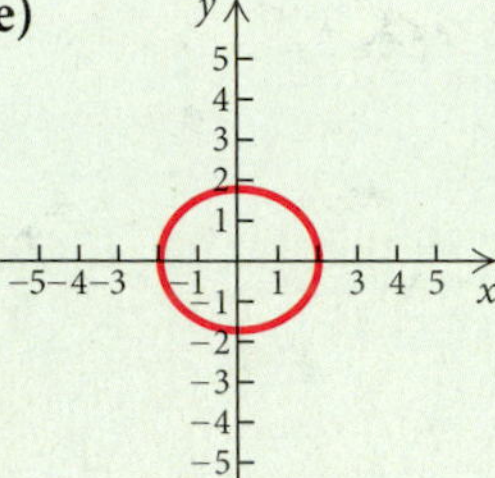

f)
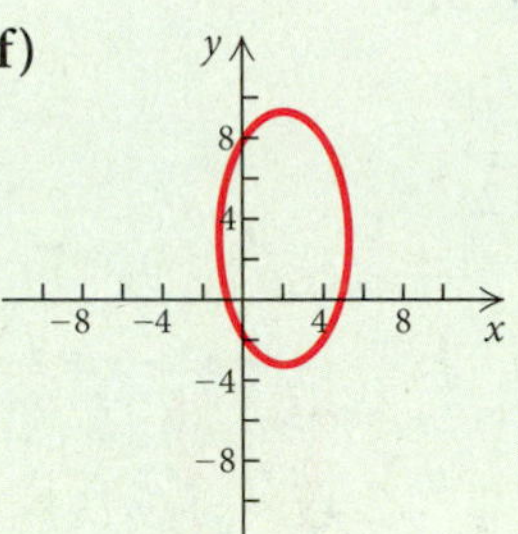

g)
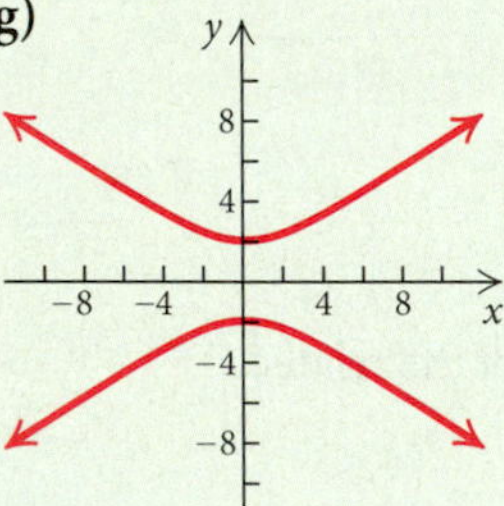

h)
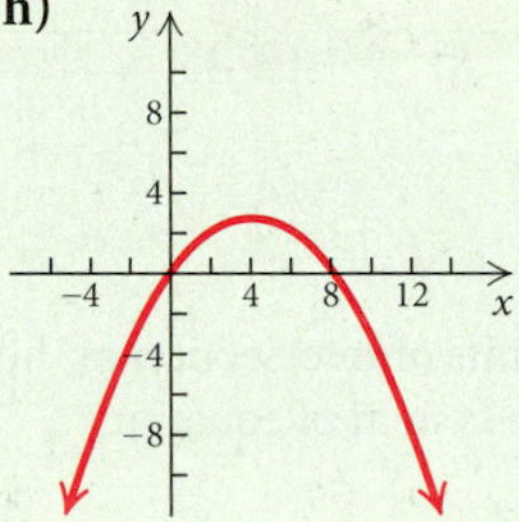

6. $y^2 = 5x$ [7.1]

7. $y^2 = 9 - x^2$ [7.2]

8. $3x^2 + 4y^2 = 12$ [7.2]

9. $9y^2 - 4x^2 = 36$ [7.3]

10. $x^2 + y^2 + 2x - 3y = 8$ [7.2]

11. $4x^2 + y^2 - 16x - 6y = 15$ [7.2]

12. $x^2 - 8x + 6y = 0$ [7.1]

13. $\frac{(x+3)^2}{16} - \frac{(y-1)^2}{25} = 1$ [7.3]

14. Find an equation of the parabola with directrix $y = \frac{3}{2}$ and focus $\left(0, -\frac{3}{2}\right)$. [7.1]

15. Find the focus, the vertex, and the directrix of the parabola given by
$$y^2 = -12x. \text{ [7.1]}$$

16. Find the vertex, the focus, and the directrix of the parabola given by
$$x^2 + 10x + 2y + 9 = 0. \text{ [7.1]}$$

17. Find the center, the vertices, and the foci of the ellipse given by
$$16x^2 + 25y^2 - 64x + 50y - 311 = 0.$$
Then draw the graph. [7.2]

18. Find an equation of the ellipse having vertices $(0, -4)$ and $(0, 4)$ with minor axis of length 6. [7.2]

19. Find the center, the vertices, the foci, and the asymptotes of the hyperbola given by
$$x^2 - 2y^2 + 4x + y - \tfrac{1}{8} = 0. \text{ [7.3]}$$

20. *Spotlight.* A spotlight has a parabolic cross section that is 2 ft wide at the opening and 1.5 ft deep at the vertex. How far from the vertex is the focus? [7.1]

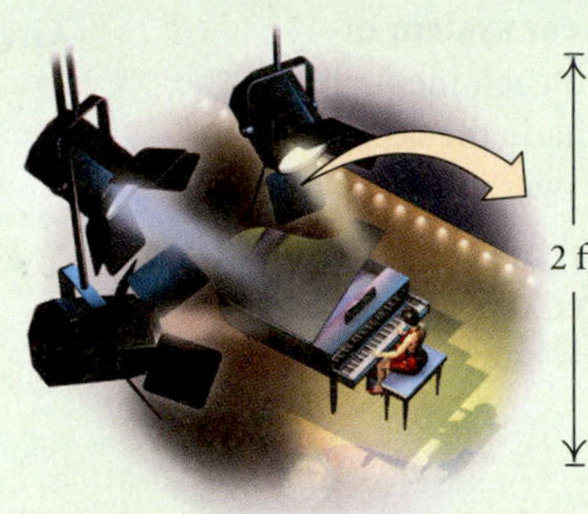

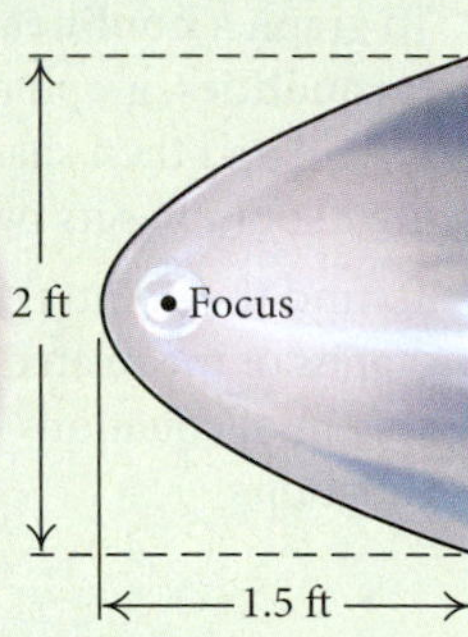

Solve. [7.4]

21. $x^2 - 16y = 0,$
$x^2 - y^2 = 64$

22. $4x^2 + 4y^2 = 65,$
$6x^2 - 4y^2 = 25$

23. $x^2 - y^2 = 33,$
$x + y = 11$

24. $x^2 - 2x + 2y^2 = 8,$
$2x + y = 6$

25. $x^2 - y = 3,$
$2x - y = 3$

26. $x^2 + y^2 = 25,$
$x^2 - y^2 = 7$

27. $x^2 - y^2 = 3,$
$y = x^2 - 3$

28. $x^2 + y^2 = 18,$
$2x + y = 3$

29. $x^2 + y^2 = 100,$
$2x^2 - 3y^2 = -120$

30. $x^2 + 2y^2 = 12,$
$xy = 4$

31. *Numerical Relationship.* The sum of two numbers is 11, and the sum of their squares is 65. Find the numbers. [7.4]

32. *Dimensions of a Rectangle.* A rectangle has a perimeter of 38 m and an area of 84 m^2. What are the dimensions of the rectangle? [7.4]

33. *Numerical Relationship.* Find two positive integers whose sum is 12 and the sum of whose reciprocals is $\frac{3}{8}$. [7.4]

34. *Perimeter.* The perimeter of a square is 12 cm more than the perimeter of another square. The area of the first square exceeds the area of the other by 39 cm^2. Find the perimeter of each square. [7.4]

35. *Radius of a Circle.* The sum of the areas of two circles is $130\pi \text{ ft}^2$. The difference of the areas is $112\pi \text{ ft}^2$. Find the radius of each circle. [7.4]

Graph the system of inequalities. Then find the coordinates of the points of intersection of the graphs of the related equations. [7.4]

36. $y \leq 4 - x^2$, $x - y \leq 2$

37. $x^2 + y^2 \leq 16$, $x + y < 4$

38. $y \geq x^2 - 1$, $y < 1$

39. $x^2 + y^2 \leq 9$, $x \leq -1$

40. The vertex of the parabola $y^2 - 4y - 12x - 8 = 0$ is which of the following? [7.1]

A. $(1, -2)$ B. $(-1, 2)$
C. $(2, -1)$ D. $(-2, 1)$

41. Which of the following cannot be a number of solutions possible for a system of equations representing an ellipse and a straight line? [7.4]

A. 0 B. 1
C. 2 D. 4

42. The graph of $x^2 + 4y^2 = 4$ is which of the following? [7.2], [7.3]

A.

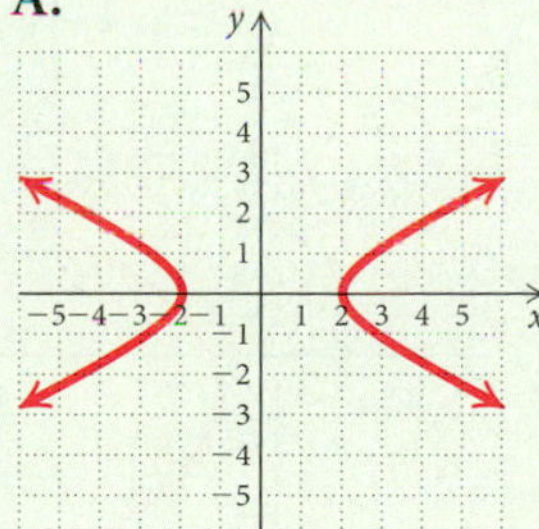

B.

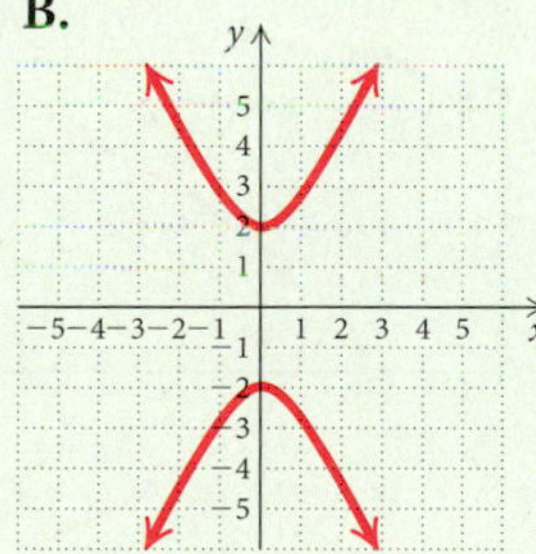

C.

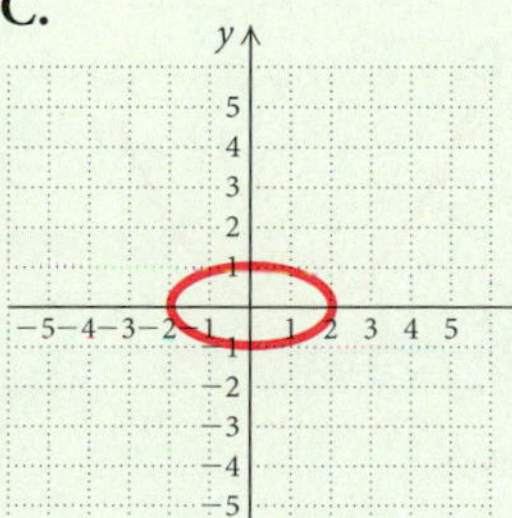

D.

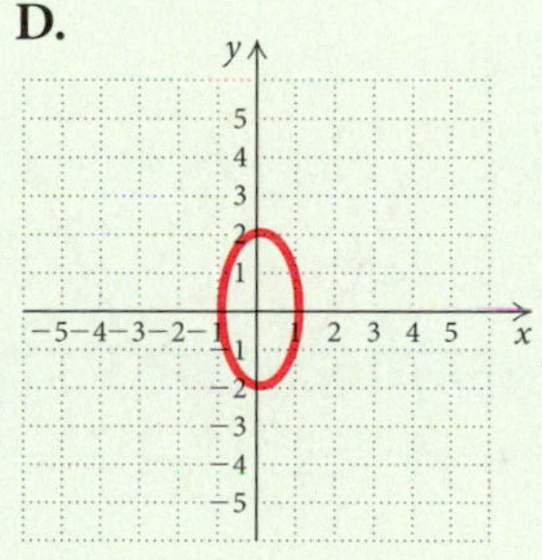

▶ Synthesis

43. Find two numbers whose product is 4 and the sum of whose reciprocals is $\frac{65}{56}$. [7.4]

44. Find an equation of the circle that passes through the points $(10, 7)$, $(-6, 7)$, and $(-8, 1)$. [7.2], [7.4]

45. Find an equation of the ellipse containing the point $\left(-1/2, 3\sqrt{3}/2\right)$ and with vertices $(0, -3)$ and $(0, 3)$. [7.2]

46. *Navigation.* Two radio transmitters positioned 400 mi apart along the shore send simultaneous signals to a ship that is 250 mi offshore and sailing parallel to the shoreline. The signal from transmitter *A* reaches the ship 300 microseconds before the signal from transmitter *B*. The signals travel at a speed of 186,000 miles per second, or 0.186 mile per microsecond. Find the equation of the hyperbola with foci *A* and *B* on which the ship is located. (*Hint*: For any point on the hyperbola, the absolute value of the difference of its distances from the foci is $2a$.) [7.3]

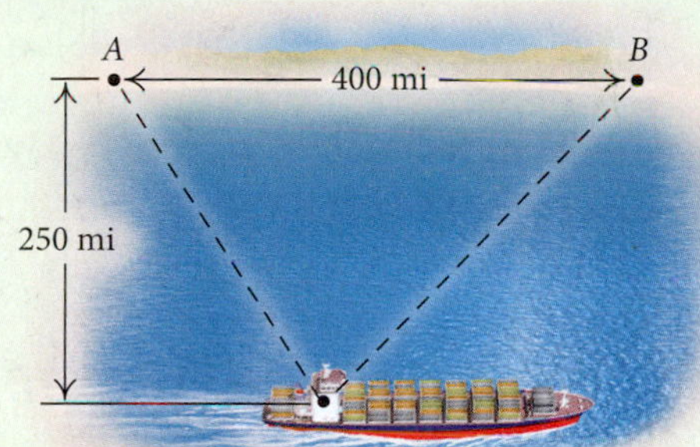

▶ Collaborative Discussion and Writing

47. Is a circle a special type of ellipse? Why or why not? [7.2]

48. Are the asymptotes of a hyperbola part of the graph of the hyperbola? Why or why not? [7.3]

49. What would you say to a classmate who tells you that it is always possible to visualize all of the solutions of a nonlinear system of equations? [7.4]

7 Chapter Test

In Exercises 1–4, match the equation with one of the graphs (a)–(d) that follow.

a) 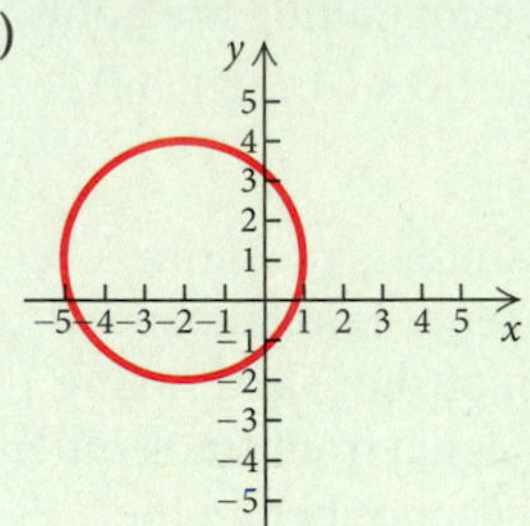

b) 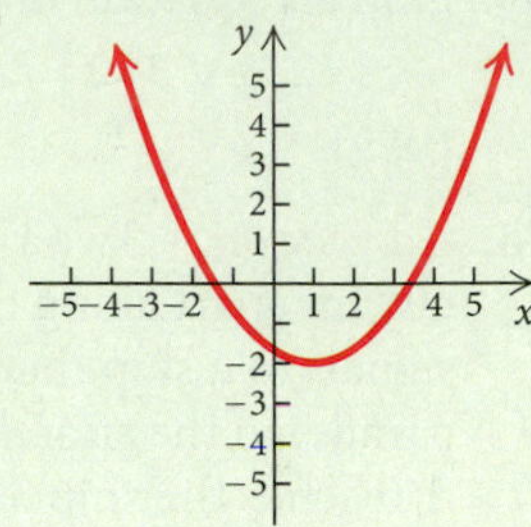

c) 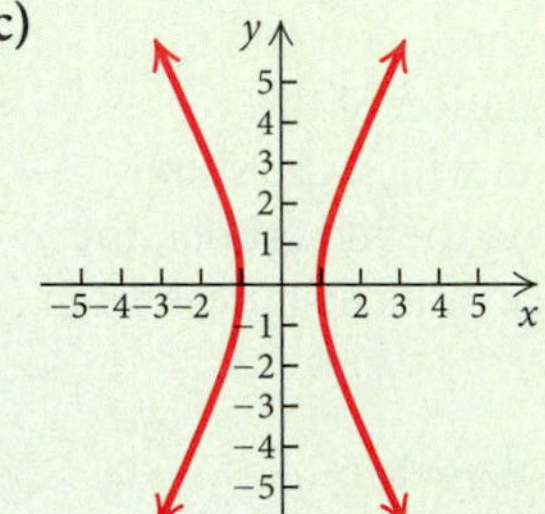

d) 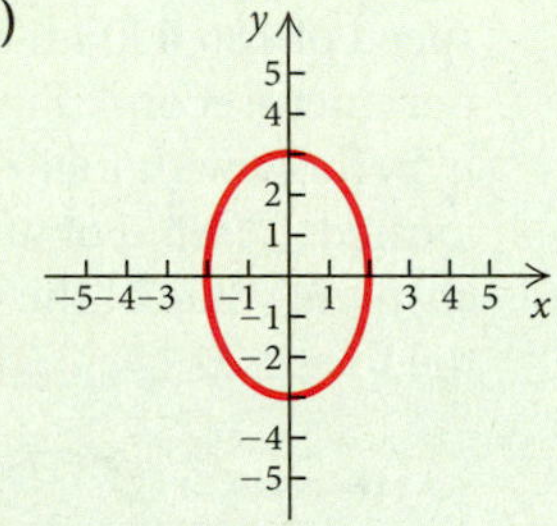

1. $4x^2 - y^2 = 4$
2. $x^2 - 2x - 3y = 5$
3. $x^2 + 4x + y^2 - 2y - 4 = 0$
4. $9x^2 + 4y^2 = 36$

Find the vertex, the focus, and the directrix of the parabola. Then draw the graph.

5. $x^2 = 12y$
6. $y^2 + 2y - 8x - 7 = 0$
7. Find an equation of the parabola with focus $(0, 2)$ and directrix $y = -2$.
8. Find the center and the radius of the circle given by $x^2 + y^2 + 2x - 6y - 15 = 0$. Then draw the graph.

Find the center, the vertices, and the foci of the ellipse. Then draw the graph.

9. $9x^2 + 16y^2 = 144$
10. $\dfrac{(x + 1)^2}{4} + \dfrac{(y - 2)^2}{9} = 1$
11. Find an equation of the ellipse having vertices $(0, -5)$ and $(0, 5)$ and with minor axis of length 4.

Find the center, the vertices, the foci, and the asymptotes of the hyperbola. Then draw the graph.

12. $4x^2 - y^2 = 4$
13. $\dfrac{(y - 2)^2}{4} - \dfrac{(x + 1)^2}{9} = 1$
14. Find the asymptotes of the hyperbola given by $2y^2 - x^2 = 18$.
15. *Satellite Dish.* A satellite dish has a parabolic cross section that is 18 in. wide at the opening and 6 in. deep at the vertex. How far from the vertex is the focus?

Solve.

16. $2x^2 - 3y^2 = -10,$
    $x^2 + 2y^2 = 9$
17. $x^2 + y^2 = 13,$
    $x + y = 1$
18. $x + y = 5,$
    $xy = 6$
19. *Landscaping.* Leisurescape is planting a rectangular flower garden with a perimeter of 18 ft and a diagonal of $\sqrt{41}$ ft. Find the dimensions of the garden.
20. *Fencing.* It will take 210 ft of fencing to enclose a rectangular playground with an area of 2700 ft^2. Find the dimensions of the playground.
21. Graph the system of inequalities. Then find the coordinates of the points of intersection of the graphs of the related equations.
    $y \geq x^2 - 4,$
    $y < 2x - 1.$
22. The graph of $(y - 1)^2 = 4(x + 1)$ is which of the following?

A. 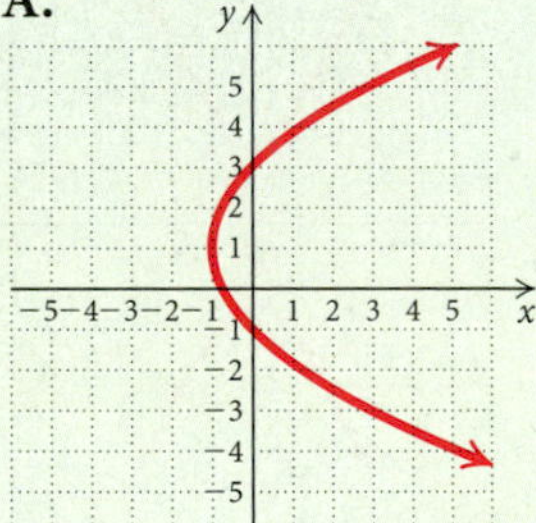

B. 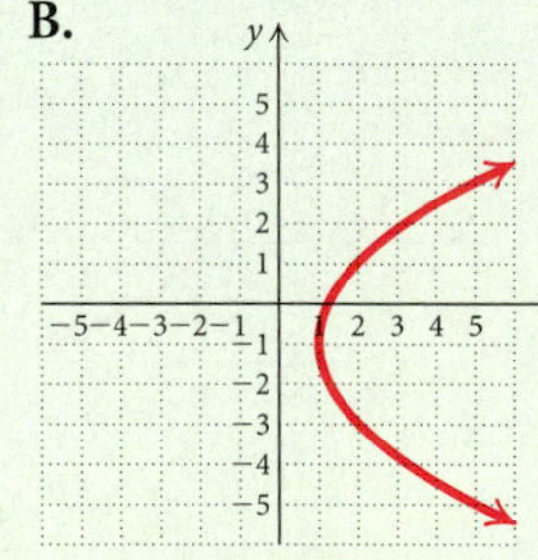

C. 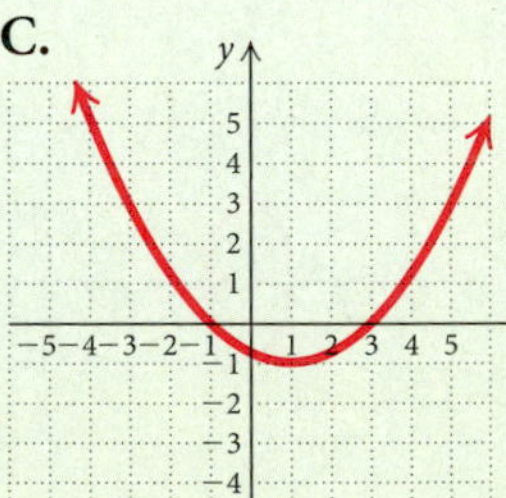

D. 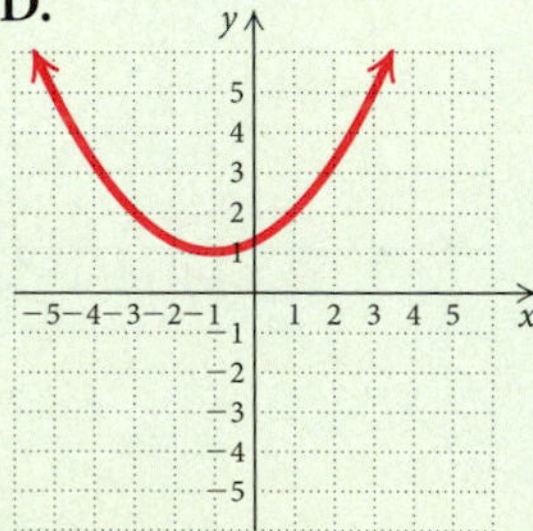

Synthesis

23. Find an equation of the circle for which the endpoints of a diameter are $(1, 1)$ and $(5, -3)$.

CHAPTER

8

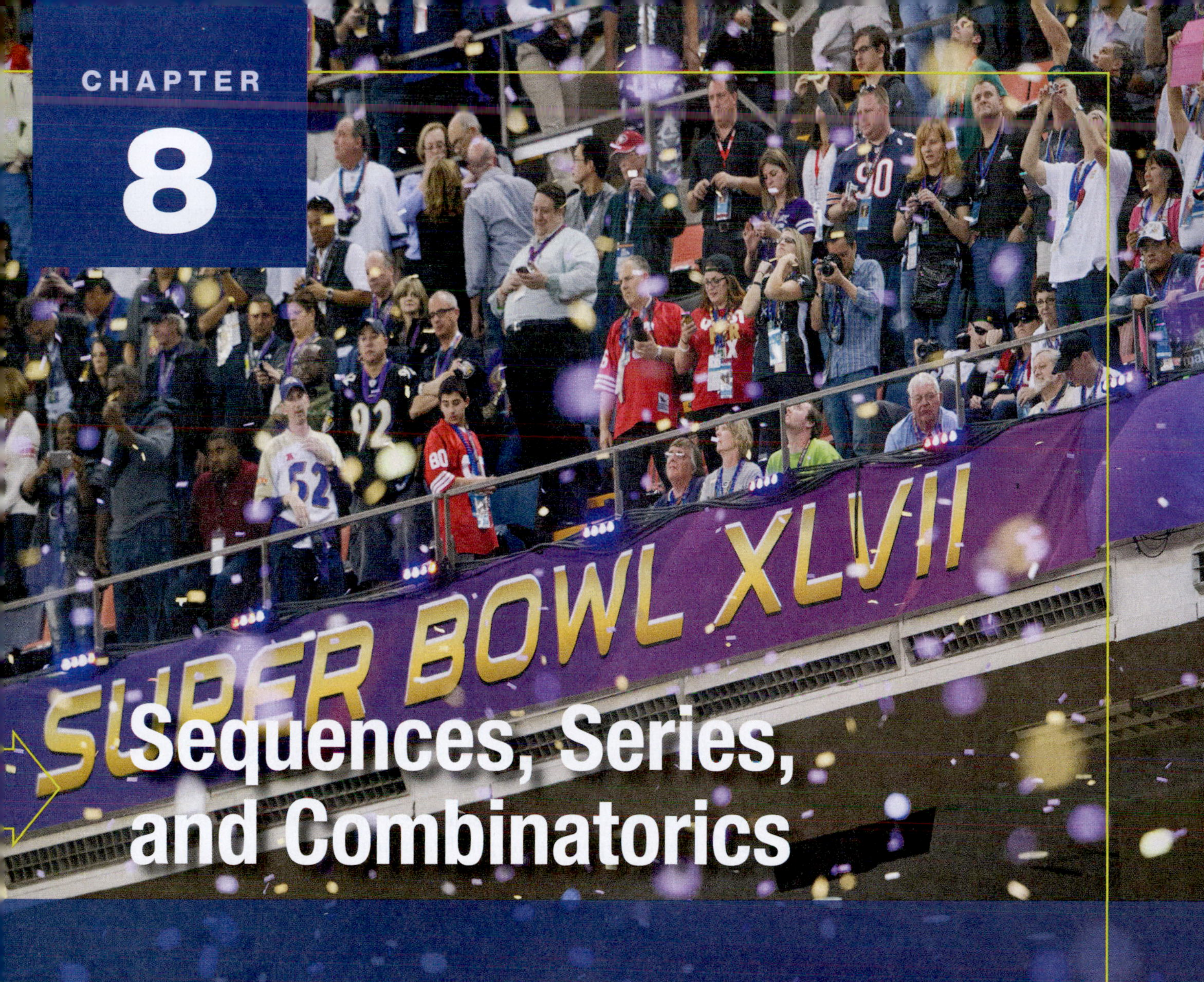

Sequences, Series, and Combinatorics

APPLICATION This problem appears as Example 10 in Section 8.3.

Large sporting events have a significant impact on the economy of the host city. Super Bowl XLVII, hosted by New Orleans, generated a \$480-million net impact for the region (*Source:* NewOrleansSaints.com, posted April 18, 2013, Marius M. Mihai, Research Analyst of the Division of Business and Economic Research at the University of New Orleans (DBER)). Assume that 60% of that amount is spent again in the area, and then 60% of that amount is spent again, and so on. This is known as the *economic multiplier effect*. Find the total effect on the economy.

8.1 Sequences and Series

- Find terms of sequences given the nth term.
- Look for a pattern in a sequence and try to determine a general term.
- Convert between sigma notation and other notation for a series.
- Construct the terms of a recursively defined sequence.

In this section, we discuss sets or lists of numbers, considered in order, and their sums.

Sequences

Suppose that \$1000 is invested at 4%, compounded annually. The amounts to which the account will grow after 1 year, 2 years, 3 years, 4 years, and so on, form the following sequence of numbers:

(1)	(2)	(3)	(4)
↓	↓	↓	↓
\$1040.00	\$1081.60	\$1124.86	\$1169.86.

We can think of this as a function that pairs 1 with \$1040.00, 2 with \$1081.60, 3 with \$1124.86, and so on. A **sequence** is thus a *function*, where the domain is a set of consecutive positive integers beginning with 1.

If we continue to compute the amounts of money in the account forever, we obtain an **infinite sequence** with function values

\$1040.00, \$1081.60, \$1124.86, \$1169.86, \$1216.65, \$1265.32, ...

The dots ". . . " at the end indicate that the sequence goes on without stopping. If we stop after a certain number of years, we obtain a **finite sequence**:

\$1040.00, \$1081.60, \$1124.86, \$1169.86.

SEQUENCES

An **infinite sequence** is a function having for its domain the set of positive integers, $\{1, 2, 3, 4, 5, \dots\}$.

A **finite sequence** is a function having for its domain a set of positive integers, $\{1, 2, 3, 4, 5, \dots, n\}$, for some positive integer n.

Consider the sequence given by the formula

$$a(n) = 2^n, \quad \text{or} \quad a_n = 2^n.$$

Some of the function values, also known as the **terms** of the sequence, follow:

$$a_1 = 2^1 = 2,$$
$$a_2 = 2^2 = 4,$$
$$a_3 = 2^3 = 8,$$
$$a_4 = 2^4 = 16,$$
$$a_5 = 2^5 = 32.$$

The first term of the sequence is denoted as a_1, the fifth term as a_5, and the nth term, or **general term**, as a_n. This sequence can also be denoted as

$$2, 4, 8, \ldots, \quad \text{or as} \quad 2, 4, 8, \ldots, 2^n, \ldots.$$

EXAMPLE 1 Find the first 4 terms and the 23rd term of the sequence whose general term is given by $a_n = (-1)^n n^2$.

Solution We have $a_n = (-1)^n n^2$, so

$$\begin{aligned} a_1 &= (-1)^1 \cdot 1^2 = -1, \\ a_2 &= (-1)^2 \cdot 2^2 = 4, \\ a_3 &= (-1)^3 \cdot 3^2 = -9, \\ a_4 &= (-1)^4 \cdot 4^2 = 16, \\ a_{23} &= (-1)^{23} \cdot 23^2 = -529. \end{aligned}$$

Now Try Exercise 1.

Note in Example 1 that the power $(-1)^n$ causes the signs of the terms to alternate between positive and negative, depending on whether n is even or odd. This kind of sequence is called an **alternating sequence**.

Technology Connection

X	Y1	
1	−1	
2	4	
3	−9	
4	16	
23	−529	

X =

seq(X/(X+1),X,1,5)▸Frac
{1/2 2/3 3/4 4/5 5/6}

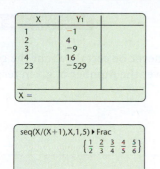

We can use a graphing calculator to find the desired terms of the sequence in Example 1. We enter $y_1 = (-1)^x x^2$. We then set up a table in ASK mode and enter 1, 2, 3, 4, and 23 as values for x.

We can also use the SEQ feature to find the terms of a sequence. Suppose, for example, that we want to find the first 5 terms of the sequence whose general term is given by $a_n = n/(n + 1)$. We select SEQ from the LIST OPS menu and enter the general term, the variable, and the numbers of the first and last terms desired. The calculator will write the terms horizontally as a list. The list can also be written in fraction notation. The first 5 terms of the sequence are $1/2, 2/3, 3/4, 4/5$, and $5/6$.

We can graph a sequence just as we graph other functions. Consider the function given by $f(x) = x + 1$ and the sequence whose general term is given by $a_n = n + 1$. The graph of $f(x) = x + 1$ is shown on the left below. Since the domain of a sequence is a set of positive integers, the graph of a sequence is a set of points that are not connected. Thus if we use only positive integers for inputs of $f(x) = x + 1$, we have the graph of the sequence $a_n = n + 1$, as shown on the right below.

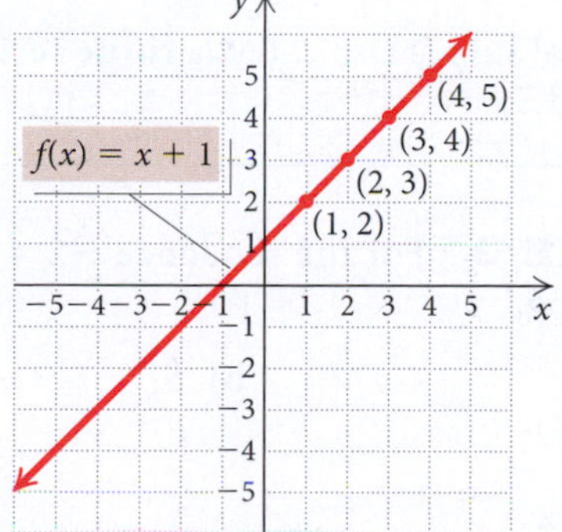

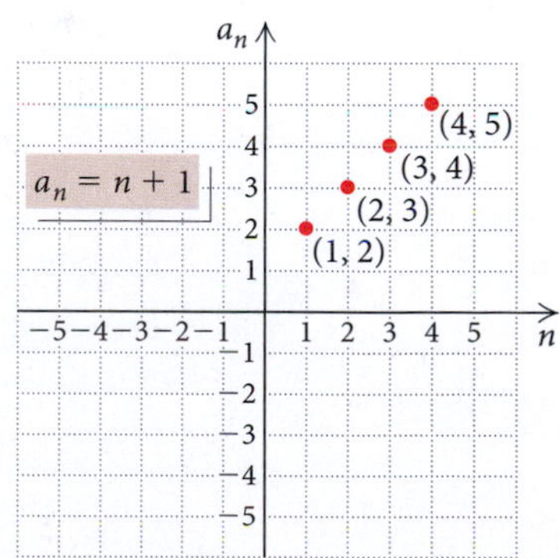

Finding the General Term

When only the first few terms of a sequence are known, we do not know for sure what the general term is, but we might be able to make a prediction by looking for a pattern.

EXAMPLE 2 For each of the following sequences, predict the general term.

a) $1, \sqrt{2}, \sqrt{3}, 2, \ldots$ b) $-1, 3, -9, 27, -81, \ldots$

c) $2, 4, 8, \ldots$

Solution

a) These are square roots of consecutive integers, so the general term might be $\sqrt{n}$.

b) These are powers of 3 with alternating signs, so the general term might be $(-1)^n 3^{n-1}$.

c) If we see the pattern of powers of 2, we will see 16 as the next term and guess 2^n for the general term. Then the sequence could be written with more terms as

$$2, 4, 8, 16, 32, 64, 128, \ldots.$$

If we see that we can get the second term by adding 2, the third term by adding 4, and the next term by adding 6, and so on, we will see 14 as the next term. A general term for the sequence is $n^2 - n + 2$, and the sequence can be written with more terms as

$$2, 4, 8, 14, 22, 32, 44, 58, \ldots.$$

Now Try Exercise 19.

Example 2(c) illustrates that, in fact, you can never be certain about the general term when only a few terms are given. The fewer the number of given terms, the greater the uncertainty.

Sums and Series

SERIES

Given the infinite sequence

$$a_1, a_2, a_3, a_4, \ldots, a_n, \ldots,$$

the sum of the terms

$$a_1 + a_2 + a_3 + \cdots + a_n + \cdots$$

is called an **infinite series**. A **partial sum** is the sum of the first n terms:

$$a_1 + a_2 + a_3 + \cdots + a_n.$$

A partial sum is also called a **finite series**, or ***n*th partial sum**, and is denoted S_n.

EXAMPLE 3 For the sequence $-2, 4, -6, 8, -10, 12, -14, \ldots$, find each of the following.

a) S_1 b) S_4 c) S_5

Solution

a) $S_1 = -2$

b) $S_4 = -2 + 4 + (-6) + 8 = 4$

c) $S_5 = -2 + 4 + (-6) + 8 + (-10) = -6$

Now Try Exercise 29.

Technology Connection

```
cumSum(seq(n² − 3,n,1,4))
            {−2 −1 5 18}
```

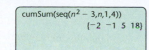

We can use a graphing calculator to find partial sums of a sequence when a formula for the general term is known. Suppose, for example, that we want to find S_1, S_2, S_3, and S_4 for the sequence whose general term is given by $a_n = n^2 - 3$. We can use the CUMSUM feature from the LIST OPS menu. The calculator will write the partial sums as a list. (Note that the calculator can be set in either FUNCTION mode or SEQUENCE mode. Here we show SEQUENCE mode.)

We have $S_1 = -2$, $S_2 = -1$, $S_3 = 5$, and $S_4 = 18$.

Sigma Notation

The Greek letter Σ (sigma) can be used to denote a sum when the general term of a sequence is a formula. For example, the sum of the first four terms of the sequence 3, 5, 7, 9, . . . , $2k + 1$, . . . can be named as follows, using what is called **sigma notation**, or **summation notation**:

$$\sum_{k=1}^{4}(2k + 1).$$

This is read "the sum as k goes from 1 to 4 of $2k + 1$." The letter k is called the **index of summation**. The index of summation might start at a number other than 1, and letters other than k can be used.

EXAMPLE 4 Find and evaluate each of the following sums.

a) $\sum_{k=1}^{5} k^3$ b) $\sum_{k=0}^{4} (-1)^k 5^k$ c) $\sum_{i=8}^{11} \left(2 + \frac{1}{i}\right)$

Solution

a) We replace k with 1, 2, 3, 4, and 5. Then we add the results.

$$\sum_{k=1}^{5} k^3 = 1^3 + 2^3 + 3^3 + 4^3 + 5^3$$
$$= 1 + 8 + 27 + 64 + 125$$
$$= 225$$

b) $$\sum_{k=0}^{4} (-1)^k 5^k = (-1)^0 5^0 + (-1)^1 5^1 + (-1)^2 5^2 + (-1)^3 5^3 + (-1)^4 5^4$$
$$= 1 - 5 + 25 - 125 + 625 = 521$$

c) $$\sum_{i=8}^{11} \left(2 + \frac{1}{i}\right) = \left(2 + \frac{1}{8}\right) + \left(2 + \frac{1}{9}\right) + \left(2 + \frac{1}{10}\right) + \left(2 + \frac{1}{11}\right)$$
$$= 8\frac{1691}{3960}$$

Now Try Exercise 33.

Technology Connection

We can combine the SUM and SEQ features on a graphing calculator to add the terms of a sequence. We find the sum in Example 4(a) as shown below.

```
sum(seq(n³,n,1,5))
                225
```

EXAMPLE 5 Write sigma notation for each sum.

a) $1 + 2 + 4 + 8 + 16 + 32 + 64$

b) $-2 + 4 - 6 + 8 - 10$

c) $x + \frac{x^2}{2} + \frac{x^3}{3} + \frac{x^4}{4} + \cdots$

Solution

a) $1 + 2 + 4 + 8 + 16 + 32 + 64$

This is the sum of powers of 2, beginning with 2^0, or 1, and ending with 2^6, or 64. Sigma notation is $\sum_{k=0}^{6} 2^k$.

b) $-2 + 4 - 6 + 8 - 10$

Disregarding the alternating signs, we see that this is the sum of the first 5 even integers. Note that $2k$ is a formula for the kth positive even integer, and $(-1)^k = -1$ when k is odd and $(-1)^k = 1$ when k is even. Thus the general term is $(-1)^k(2k)$. The sum begins with $k = 1$ and ends with $k = 5$, so sigma notation is $\sum_{k=1}^{5}(-1)^k(2k)$.

c) $x + \dfrac{x^2}{2} + \dfrac{x^3}{3} + \dfrac{x^4}{4} + \cdots$

The general term is x^k/k, beginning with $k = 1$. This is also an infinite series. We use the symbol ∞ for infinity and write the series using sigma notation: $\sum_{k=1}^{\infty}(x^k/k)$.

Now Try Exercise 51.

Recursive Definitions

A sequence can be defined **recursively** or by using a **recursion formula**. Such a definition lists the first term, or the first few terms, and then describes how to determine the remaining terms from the given terms.

EXAMPLE 6 Find the first 5 terms of the sequence defined by

$$a_1 = 5, \quad a_{n+1} = 2a_n - 3, \quad \text{for } n \geq 1.$$

Solution We have

$$a_1 = 5,$$
$$a_2 = 2a_1 - 3 = 2 \cdot 5 - 3 = 7,$$
$$a_3 = 2a_2 - 3 = 2 \cdot 7 - 3 = 11,$$
$$a_4 = 2a_3 - 3 = 2 \cdot 11 - 3 = 19,$$
$$a_5 = 2a_4 - 3 = 2 \cdot 19 - 3 = 35.$$

Now Try Exercise 61.

Technology Connection

Many graphing calculators have the capability to work with recursively defined sequences when they are set in SEQUENCE mode. In Example 6, for instance, the function could be entered as $u(n) = 2 * u(n - 1) - 3$ with $u(n\text{Min}) = 5$. We can read the terms of the sequence from a table.

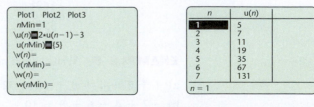

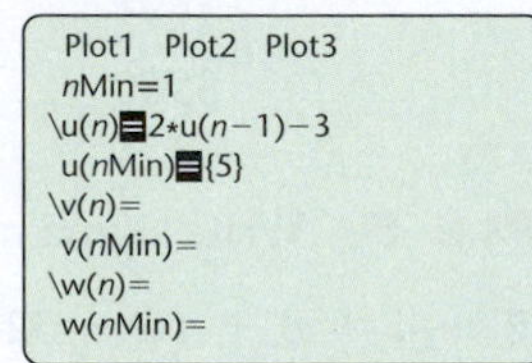

n	$u(n)$
1	5
2	7
3	11
4	19
5	35
6	67
7	131

$n = 1$

8.1 Exercise Set

In each of the following, the nth term of a sequence is given. Find the first 4 terms, a_{10}, and a_{15}.

1. $a_n = 4n - 1$

2. $a_n = (n - 1)(n - 2)(n - 3)$

3. $a_n = \frac{n}{n - 1}, n \geq 2$

4. $a_n = n^2 - 1, n \geq 3$

5. $a_n = \frac{n^2 - 1}{n^2 + 1}$

6. $a_n = \left(-\frac{1}{2}\right)^{n-1}$

7. $a_n = (-1)^n n^2$

8. $a_n = (-1)^{n-1}(3n - 5)$

9. $a_n = 5 + \frac{(-2)^{n+1}}{2^n}$

10. $a_n = \frac{2n - 1}{n^2 + 2n}$

Find the indicated term of the given sequence.

11. $a_n = 5n - 6$; a_8

12. $a_n = (3n - 4)(2n + 5)$; a_7

13. $a_n = (2n + 3)^2$; a_6

14. $a_n = (-1)^{n-1}(4.6n - 18.3)$; a_{12}

15. $a_n = 5n^2(4n - 100)$; a_{11}

16. $a_n = \left(1 + \frac{1}{n}\right)^2$; a_{80}

17. $a_n = \ln e^n$; a_{67}

18. $a_n = 2 - \frac{1000}{n}$; a_{100}

Predict the general term, or nth term, a_n, of the sequence. Answers may vary.

19. 2, 4, 6, 8, 10, . . .

20. 3, 9, 27, 81, 243, . . .

21. −2, 6, −18, 54, . . .

22. −2, 3, 8, 13, 18, . . .

23. $\frac{2}{3}, \frac{3}{4}, \frac{4}{5}, \frac{5}{6}, \frac{6}{7}, \ldots$

24. $\sqrt{2}, 2, \sqrt{6}, 2\sqrt{2}, \sqrt{10}, \ldots$

25. $1 \cdot 2, 2 \cdot 3, 3 \cdot 4, 4 \cdot 5, \ldots$

26. −1, −4, −7, −10, −13, . . .

27. 0, log 10, log 100, log 1000, . . .

28. $\ln e^2, \ln e^3, \ln e^4, \ln e^5, \ldots$

Find the indicated partial sums for the sequence.

29. 1, 2, 3, 4, 5, 6, 7, . . . ; S_3 and S_7

30. 1, −3, 5, −7, 9, −11, . . . ; S_2 and S_5

31. 2, 4, 6, 8, . . . ; S_4 and S_5

32. $1, \frac{1}{4}, \frac{1}{9}, \frac{1}{16}, \frac{1}{25}, \ldots$; S_1 and S_5

Find and evaluate the sum.

33. $\sum_{k=1}^{5} \frac{1}{2k}$

34. $\sum_{i=1}^{6} \frac{1}{2i + 1}$

35. $\sum_{i=0}^{6} 2^i$

36. $\sum_{k=4}^{7} \sqrt{2k - 1}$

37. $\sum_{k=7}^{10} \ln k$

38. $\sum_{k=1}^{4} \pi k$

39. $\sum_{k=1}^{8} \frac{k}{k + 1}$

40. $\sum_{i=1}^{5} \frac{i - 1}{i + 3}$

41. $\sum_{i=1}^{5} (-1)^i$

42. $\sum_{k=0}^{5} (-1)^{k+1}$

43. $\sum_{k=1}^{8} (-1)^{k+1} 3k$

44. $\sum_{k=0}^{7} (-1)^k 4^{k+1}$

45. $\sum_{k=0}^{6} \frac{2}{k^2 + 1}$

46. $\sum_{i=1}^{10} i(i + 1)$

47. $\sum_{k=0}^{5} (k^2 - 2k + 3)$

48. $\sum_{k=1}^{10} \frac{1}{k(k + 1)}$

49. $\sum_{i=0}^{10} \frac{2^i}{2^i + 1}$

50. $\sum_{k=0}^{3} (-2)^{2k}$

Write sigma notation. Answers may vary.

51. $5 + 10 + 15 + 20 + 25 + \cdots$

52. $7 + 14 + 21 + 28 + 35 + \cdots$

53. $2 - 4 + 8 - 16 + 32 - 64$

54. $3 + 6 + 9 + 12 + 15$

55. $-\frac{1}{2} + \frac{2}{3} - \frac{3}{4} + \frac{4}{5} - \frac{5}{6} + \frac{6}{7}$

56. $\frac{1}{1^2} + \frac{1}{2^2} + \frac{1}{3^2} + \frac{1}{4^2} + \frac{1}{5^2}$

57. $4 - 9 + 16 - 25 + \cdots + (-1)^n n^2$

58. $9 - 16 + 25 + \cdots + (-1)^{n+1} n^2$

59. $\frac{1}{1 \cdot 2} + \frac{1}{2 \cdot 3} + \frac{1}{3 \cdot 4} + \frac{1}{4 \cdot 5} + \cdots$

60. $\frac{1}{1 \cdot 2^2} + \frac{1}{2 \cdot 3^2} + \frac{1}{3 \cdot 4^2} + \frac{1}{4 \cdot 5^2} + \cdots$

Find the first 4 terms of the recursively defined sequence.

61. $a_1 = 4, \ a_{n+1} = 1 + \frac{1}{a_n}$

62. $a_1 = 256, \ a_{n+1} = \sqrt{a_n}$

63. $a_1 = 6561, \ a_{n+1} = (-1)^n \sqrt{a_n}$

64. $a_1 = e^Q, \ a_{n+1} = \ln a_n$

65. $a_1 = 2, a_2 = 3, \ a_{n+1} = a_n + a_{n-1}$

66. $a_1 = -10, a_2 = 8, \ a_{n+1} = a_n - a_{n-1}$

67. *Compound Interest.* Suppose that \$1000 is invested at 6.2%, compounded annually. The value of the investment after n years is given by the sequence model

$$a_n = \$1000(1.062)^n, \quad n = 1, 2, 3, \ldots.$$

a) Find the first 10 terms of the sequence.
b) Find the value of the investment after 20 years.

68. *Salvage Value.* The value of a post-hole digger is \$5200. Its salvage value each year is 75% of its value the year before. Give a sequence that lists the salvage value of the machine for each year of a 10-year period.

69. *Wage Sequence.* Adahy is paid \$9.80 per hour for working at Red Freight Limited. Each year he receives a \$1.10 hourly raise. Give a sequence that lists Adahy's hourly wage over a 10-year period.

70. *Bacteria Growth.* Suppose that a single cell of bacteria divides into two every 15 min. Suppose that the same rate of division is maintained for 4 hr. Give a sequence that lists the number of cells after successive 15-min periods.

71. *Fibonacci Sequence: Rabbit Population Growth.* One of the most famous recursively defined sequences is the **Fibonacci sequence**. In 1202, the Italian mathematician Leonardo da Pisa, also called Fibonacci, proposed the following model for rabbit population growth. Suppose that every month each mature pair of rabbits in the population produces a new pair that begins reproducing after two months, and also suppose that no rabbits die. Beginning with one pair of newborn rabbits, the population can be modeled by the following recursively defined sequence:

$$a_1 = 1, \quad a_2 = 1, \quad a_n = a_{n-1} + a_{n-2}, \text{ for } n \geq 3,$$

where a_n is the total number of pairs of rabbits in month n. Find the first 7 terms of the Fibonacci sequence.

Skill Maintenance

Solve. [6.1], [6.3], [6.5], [6.6]

72. $3x - 2y = 3,$
$2x + 3y = -11$

73. *Harvesting Pumpkins.* A total of 23,400 acres of pumpkins were harvested in Illinois and Ohio in 2012. The number of acres of pumpkins harvested in Ohio was 9000 fewer than the number of acres of pumpkins harvested in Illinois. (*Source:* U. S. Department of Agriculture) Find the number of acres of pumpkins harvested in Illinois and in Ohio in 2012.

Find the center and the radius of the circle with the given equation. [7.2]

74. $x^2 + y^2 - 6x + 4y = 3$

75. $x^2 + y^2 + 5x - 8y = 2$

Synthesis

Find the first 5 terms of the sequence, and then find S_5.

76. $a_n = \frac{1}{2^n} \log 1000^n$

77. $a_n = i^n, i = \sqrt{-1}$

78. $a_n = \ln(1 \cdot 2 \cdot 3 \cdots n)$

For each sequence, find a formula for S_n.

79. $a_n = \ln n$

80. $a_n = \frac{1}{n} - \frac{1}{n+1}$

8.2 Arithmetic Sequences and Series

- For any arithmetic sequence, find the nth term when n is given and n when the nth term is given; and given two terms, find the common difference and construct the sequence.
- Find the sum of the first n terms of an arithmetic sequence.

A sequence in which each term after the first is found by adding the same number to the preceding term is an **arithmetic sequence**.

Arithmetic Sequences

The sequence 2, 5, 8, 11, 14, 17, . . . is arithmetic because adding 3 to any term produces the next term. In other words, the difference between any term and the preceding one is 3. Arithmetic sequences are also called *arithmetic progressions*.

ARITHMETIC SEQUENCE

A sequence is **arithmetic** if there exists a number d, called the **common difference**, such that $a_{n+1} = a_n + d$ for any integer $n \geq 1$.

EXAMPLE 1 For each of the following arithmetic sequences, identify the first term, a_1, and the common difference, d.

a) 4, 9, 14, 19, 24, . . .

b) 34, 27, 20, 13, 6, −1, −8, . . .

c) $2, 2\frac{1}{2}, 3, 3\frac{1}{2}, 4, 4\frac{1}{2}, \ldots$

Solution The first term, a_1, is the first term listed. To find the common difference, d, we choose any term beyond the first and subtract the preceding term from it.

SEQUENCE	FIRST TERM, a_1	COMMON DIFFERENCE, d
a) 4, 9, 14, 19, 24, . . .	4	5 $(9 - 4 = 5)$
b) 34, 27, 20, 13, 6, −1, −8, . . .	34	−7 $(27 - 34 = -7)$
c) $2, 2\frac{1}{2}, 3, 3\frac{1}{2}, 4, 4\frac{1}{2}, \ldots$	2	$\frac{1}{2}$ $\left(2\frac{1}{2} - 2 = \frac{1}{2}\right)$

Note that we obtained the common difference by subtracting a_1 from a_2. Had we subtracted a_2 from a_3 or a_3 from a_4, we would have obtained the same values for d. Thus we can check by adding d to each term in a sequence to see if we progress correctly to the next term.

Check:

a) $4 + 5 = 9, \quad 9 + 5 = 14, \quad 14 + 5 = 19, \quad 19 + 5 = 24$

b) $34 + (-7) = 27, \quad 27 + (-7) = 20, \quad 20 + (-7) = 13,$
$13 + (-7) = 6, \quad 6 + (-7) = -1, \quad -1 + (-7) = -8$

c) $2 + \frac{1}{2} = 2\frac{1}{2}, \quad 2\frac{1}{2} + \frac{1}{2} = 3, \quad 3 + \frac{1}{2} = 3\frac{1}{2}, \quad 3\frac{1}{2} + \frac{1}{2} = 4,$
$4 + \frac{1}{2} = 4\frac{1}{2}$

Now Try Exercise 1.

To find a formula for the general, or nth, term of any arithmetic sequence, we denote the common difference by d, write out the first few terms, and look for a pattern:

$$\begin{aligned} &a_1, \\ &a_2 = a_1 + d, \\ &a_3 = a_2 + d = (a_1 + d) + d = a_1 + 2d, && \text{Substituting for } a_2 \\ &a_4 = a_3 + d = (a_1 + 2d) + d = a_1 + 3d. && \text{Substituting for } a_3 \end{aligned}$$

Note that the coefficient of d in each case is 1 less than the subscript.

Generalizing, we obtain the following formula.

nTH TERM OF AN ARITHMETIC SEQUENCE

The **nth term** of an arithmetic sequence is given by

$$a_n = a_1 + (n - 1)d, \quad \text{for any integer } n \geq 1.$$

EXAMPLE 2 Find the 14th term of the arithmetic sequence 4, 7, 10, 13, ...

Solution We first note that $a_1 = 4$, $d = 7 - 4$, or 3, and $n = 14$. Then using the formula for the nth term, we obtain

$$\begin{aligned} a_n &= a_1 + (n - 1)d \\ a_{14} &= 4 + (14 - 1) \cdot 3 && \text{Substituting} \\ &= 4 + 13 \cdot 3 = 4 + 39 \\ &= 43. \end{aligned}$$

The 14th term is 43. **Now Try Exercise 9.**

EXAMPLE 3 In the sequence of Example 2, which term is 301? That is, find n if $a_n = 301$.

Solution We substitute 301 for a_n, 4 for a_1, and 3 for d in the formula for the nth term and solve for n:

$$\begin{aligned} a_n &= a_1 + (n - 1)d \\ 301 &= 4 + (n - 1) \cdot 3 && \text{Substituting} \\ 301 &= 4 + 3n - 3 \\ 301 &= 3n + 1 && \text{Solving for } n \\ 300 &= 3n \\ 100 &= n. \end{aligned}$$

The term 301 is the 100th term of the sequence. **Now Try Exercise 15.**

Given two terms and their places in an arithmetic sequence, we can construct the sequence.

EXAMPLE 4 The 3rd term of an arithmetic sequence is 8, and the 16th term is 47. Find a_1 and d and construct the sequence.

Solution We know that $a_3 = 8$ and $a_{16} = 47$. Thus we would have to add d 13 times to get from 8 to 47. That is,

$$8 + 13d = 47.$$ **a_3 and a_{16} are 16 − 3, or 13, terms apart.**

Solving $8 + 13d = 47$, we obtain

$$13d = 39$$
$$d = 3.$$

Since $a_3 = 8$, we subtract d twice to get a_1. Thus,

$$a_1 = 8 - 2 \cdot 3 = 2.$$ **a_1 and a_3 are 3 − 1, or 2, terms apart.**

The sequence is 2, 5, 8, 11, ... Note that we could also subtract d 15 times from a_{16} in order to find a_1.

Now Try Exercise 23.

In general, d should be subtracted $n - 1$ times from a_n in order to find a_1.

Sum of the First *n* Terms of an Arithmetic Sequence

Consider the arithmetic sequence

$$3, 5, 7, 9, \ldots .$$

When we add the first 4 terms of the sequence, we get S_4, which is

$$3 + 5 + 7 + 9, \quad \text{or} \quad 24.$$

This sum is called an **arithmetic series**. To find a formula for the sum of the first n terms, S_n, of an arithmetic sequence, we first denote an arithmetic sequence, as follows:

This term is two terms back from the last. If you add d to this term, the result is the next-to-last term, $a_n - d$.

$$a_1, \quad (a_1 + d), \quad (a_1 + 2d), \quad \ldots, \quad (a_n - 2d), \quad (a_n - d), \quad a_n.$$

This is the next-to-last term. If you add d to this term, the result is a_n.

Then S_n is given by

$$S_n = a_1 + (a_1 + d) + (a_1 + 2d) + \cdots + (a_n - 2d) + (a_n - d) + a_n. \quad (1)$$

Reversing the order of the addition gives us

$$S_n = a_n + (a_n - d) + (a_n - 2d) + \cdots + (a_1 + 2d) + (a_1 + d) + a_1. \quad (2)$$

If we add corresponding terms of each side of equations (1) and (2), we get

$$2S_n = [a_1 + a_n] + [(a_1 + d) + (a_n - d)] + [(a_1 + 2d) + (a_n - 2d)] + \cdots + [(a_n - 2d) + (a_1 + 2d)] + [(a_n - d) + (a_1 + d)] + [a_n + a_1].$$

In the expression for $2S_n$, there are n expressions in square brackets. Each of these expressions is equivalent to $a_1 + a_n$. Thus the expression for $2S_n$ can be written in simplified form as

$$2S_n = [a_1 + a_n] + [a_1 + a_n] + [a_1 + a_n] + \cdots + [a_n + a_1] + [a_n + a_1] + [a_n + a_1].$$

Since $a_1 + a_n$ is being added n times, it follows that

$$2S_n = n(a_1 + a_n),$$

from which we get the following formula.

SUM OF THE FIRST n TERMS

The sum of the first n terms of an arithmetic sequence is given by

$$S_n = \frac{n}{2}(a_1 + a_n).$$

EXAMPLE 5 Find the sum of the first 100 natural numbers.

Solution The sum is

$$1 + 2 + 3 + \cdots + 99 + 100.$$

This is the sum of the first 100 terms of the arithmetic sequence for which

$$a_1 = 1, \quad a_n = 100, \quad \text{and} \quad n = 100.$$

Thus substituting into the formula

$$S_n = \frac{n}{2}(a_1 + a_n),$$

we get

$$S_{100} = \frac{100}{2}(1 + 100) = 50(101) = 5050.$$

The sum of the first 100 natural numbers is 5050.

Now Try Exercise 27.

EXAMPLE 6 Find the sum of the first 15 terms of the arithmetic sequence 4, 7, 10, 13, ...

Solution Note that $a_1 = 4$, $d = 3$, and $n = 15$. Before using the formula

$$S_n = \frac{n}{2}(a_1 + a_n),$$

we find the last term, a_{15}:

$$a_{15} = 4 + (15 - 1)3 \qquad \text{Substituting into the formula } a_n = a_1 + (n-1)d$$
$$= 4 + 14 \cdot 3 = 46.$$

Thus,

$$S_{15} = \frac{15}{2}(4 + 46) = \frac{15}{2}(50) = 375.$$

The sum of the first 15 terms is 375.

Now Try Exercise 25.

EXAMPLE 7 Find the sum: $\sum_{k=1}^{130}(4k + 5)$.

Solution It is helpful to first write out a few terms:

$$9 + 13 + 17 + \cdots.$$

It appears that this is an arithmetic series coming from an arithmetic sequence with $a_1 = 9$, $d = 4$, and $n = 130$. Before using the formula

$$S_n = \frac{n}{2}(a_1 + a_n),$$

we find the last term, a_{130}:

$$\begin{aligned} a_{130} &= 4 \cdot 130 + 5 \qquad \text{The } k\text{th term is } 4k + 5. \\ &= 520 + 5 \\ &= 525. \end{aligned}$$

Thus,

$$\begin{aligned} S_{130} &= \frac{130}{2}(9 + 525) \qquad \text{Substituting into } S_n = \frac{n}{2}(a_1 + a_n) \\ &= 34{,}710. \end{aligned}$$

Now Try Exercise 33.

Applications

The translation of some applications and problem-solving situations may involve arithmetic sequences or series. We consider some examples.

EXAMPLE 8 ***Hourly Wages.*** Kendall accepts a job, starting with an hourly wage of \$14.25, and is promised a raise of 15¢ per hour every 2 months for 5 years. At the end of 5 years, what will Kendall's hourly wage be?

Solution It helps to first write down the hourly wage for several 2-month time periods:

Beginning: \$14.25,
After 2 months: \$14.40,
After 4 months: \$14.55,
and so on.

What appears is a sequence of numbers: 14.25, 14.40, 14.55, ... This sequence is arithmetic, because adding 0.15 each time gives us the next term.

We want to find the last term of an arithmetic sequence, so we use the formula $a_n = a_1 + (n - 1)d$. We know that $a_1 = 14.25$ and $d = 0.15$, but what is n? That is, how many terms are in the sequence? Each year there are 12/2, or 6 raises, since Kendall gets a raise every 2 months. There are 5 years, so the total number of raises will be $5 \cdot 6$, or 30. Thus there will be 31 terms: the original wage and 30 increased rates.

Substituting in the formula $a_n = a_1 + (n - 1)d$ gives us

$$\begin{aligned} a_{31} &= 14.25 + (31 - 1) \cdot 0.15 \\ &= 18.75. \end{aligned}$$

Thus, at the end of 5 years, Kendall's hourly wage will be \$18.75.

Now Try Exercise 43.

The calculations in Example 8 could be done in a number of ways. There is often a variety of ways in which a problem can be solved. In this chapter, we concentrate on the use of sequences and series and their related formulas.

EXAMPLE 9 ***Total in a Stack.*** A stack of electric poles has 30 poles in the bottom row. There are 29 poles in the second row, 28 in the next row, and so on. How many poles are in the stack if there are 5 poles in the top row?

Solution A drawing will help in this case. The following figure shows the ends of the poles and the way in which they stack.

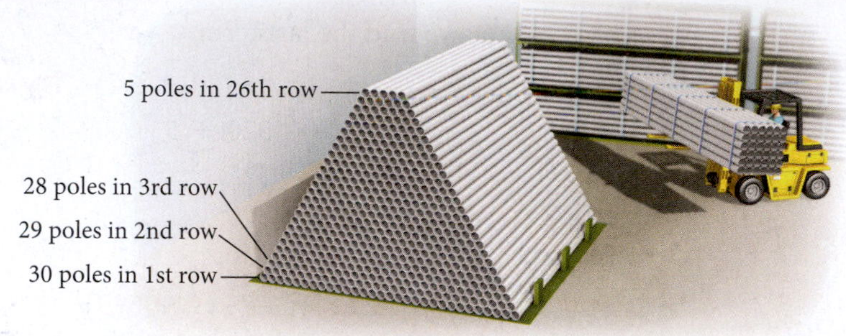

Since the number of poles decreases from 30 in a row up to 5 in the top row, there must be 26 rows. We want the sum

$$30 + 29 + 28 + \cdots + 5.$$

Thus we have an arithmetic series. We use the formula

$$S_n = \frac{n}{2}(a_1 + a_n),$$

with $n = 26$, $a_1 = 30$, and $a_{26} = 5$.

Substituting, we get

$$S_{26} = \frac{26}{2}(30 + 5) = 455.$$

There are 455 poles in the stack. Now Try Exercise 39.

8.2 Exercise Set

Find the first term and the common difference.

1. 3, 8, 13, 18, . . .
2. \$1.08, \$1.16, \$1.24, \$1.32, . . .
3. 9, 5, 1, −3, . . .
4. −8, −5, −2, 1, 4, . . .
5. $\frac{3}{2}, \frac{9}{4}, 3, \frac{15}{4}, \ldots$
6. $\frac{3}{5}, \frac{1}{10}, -\frac{2}{5}, \ldots$
7. \$316, \$313, \$310, \$307, . . .
8. Find the 11th term of the arithmetic sequence 0.07, 0.12, 0.17, ...
9. Find the 12th term of the arithmetic sequence 2, 6, 10, ...
10. Find the 17th term of the arithmetic sequence 7, 4, 1, ...
11. Find the 14th term of the arithmetic sequence $3, \frac{7}{3}, \frac{5}{3}, \ldots$.
12. Find the 13th term of the arithmetic sequence \$1200, \$964.32, \$728.64, ...

13. Find the 10th term of the arithmetic sequence \$2345.78, \$2967.54, \$3589.30, ...

14. In the sequence of Exercise 8, what term is the number 1.67?

15. In the sequence of Exercise 9, what term is the number 106?

16. In the sequence of Exercise 10, what term is -296?

17. In the sequence of Exercise 11, what term is -27?

18. Find a_{20} when $a_1 = 14$ and $d = -3$.

19. Find a_1 when $d = 4$ and $a_8 = 33$.

20. Find d when $a_1 = 8$ and $a_{11} = 26$.

21. Find n when $a_1 = 25$, $d = -14$, and $a_n = -507$.

22. In an arithmetic sequence, $a_{17} = -40$ and $a_{28} = -73$. Find a_1 and d. Write the first 5 terms of the sequence.

23. In an arithmetic sequence, $a_{17} = \frac{25}{3}$ and $a_{32} = \frac{95}{6}$. Find a_1 and d. Write the first 5 terms of the sequence.

24. Find the sum of the first 14 terms of the series $11 + 7 + 3 + \cdots$.

25. Find the sum of the first 20 terms of the series $5 + 8 + 11 + 14 + \cdots$.

26. Find the sum of the first 300 natural numbers.

27. Find the sum of the first 400 even natural numbers.

28. Find the sum of the odd numbers 1 to 199, inclusive.

29. Find the sum of the multiples of 7 from 7 to 98, inclusive.

30. Find the sum of all multiples of 4 that are between 14 and 523.

31. If an arithmetic series has $a_1 = 2$, $d = 5$, and $n = 20$, what is S_n?

32. If an arithmetic series has $a_1 = 7$, $d = -3$, and $n = 32$, what is S_n?

Find the sum.

33. $\sum_{k=1}^{40} (2k + 3)$

34. $\sum_{k=5}^{20} 8k$

35. $\sum_{k=0}^{19} \frac{k - 3}{4}$

36. $\sum_{k=2}^{50} (2000 - 3k)$

37. $\sum_{k=12}^{57} \frac{7 - 4k}{13}$

38. $\sum_{k=101}^{200} (1.14k - 2.8) - \sum_{k=1}^{5} \left(\frac{k + 4}{10}\right)$

39. *Total Savings.* If 10¢ is saved on October 1, 20¢ is saved on October 2, 30¢ on October 3, and so on, how much is saved altogether during the 31 days of October?

40. *Stacking Poles.* How many poles will be in a stack of telephone poles if there are 50 in the first layer, 49 in the second, and so on, with 6 in the top layer?

41. *Auditorium Seating.* Auditoriums are often built with more seats per row as the rows move toward the back. Suppose that the first balcony of a theater has 28 seats in the first row, 32 in the second, 36 in the third, and so on, for 20 rows. How many seats are in the first balcony altogether?

42. *Investment Return.* Brett sets up an investment situation for a client that will return \$5000 the first year, \$6125 the second year, \$7250 the third year, and so on, for 25 years. How much is received from the investment altogether?

43. *Parachutist Free Fall.* When a parachutist jumps from an airplane, the distances, in feet, that the parachutist falls in each successive second before pulling the ripcord to release the parachute are as follows:

16, 48, 80, 112, 144, ...

Is this sequence arithmetic? What is the common difference? What is the total distance fallen in 10 sec?

44. *Lightning Distance.* The following table lists the distance, in miles, from lightning d_n when thunder is heard n seconds after lightning is seen. Is this sequence arithmetic? What is the common difference?

n (in seconds)	d_n (in miles)
5	1
6	1.2
7	1.4
8	1.6
9	1.8
10	2

45. *Garden Plantings.* A gardener is making a planting in the shape of a trapezoid. It will have 35 plants in the front row, 31 in the second row, 27 in the third row, and so on. If the pattern is consistent, how many plants will there be in the last row? How many plants are there altogether?

46. *Band Formation.* A formation of a marching band has 10 marchers in the front row, 12 in the second row, 14 in the third row, and so on, for 8 rows. How many marchers are in the last row? How many marchers are there altogether?

47. *Raw Material Production.* In a manufacturing process, it took 3 units of raw materials to produce 1 unit of a product. The raw material needs thus formed the sequence

$$3, 6, 9, \ldots, 3n, \ldots.$$

Is this sequence arithmetic? What is the common difference?

Skill Maintenance

Solve. [6.1], [6.3], [6.5], [6.6]

48. $7x - 2y = 4,$
$x + 3y = 17$

49. $2x + y + 3z = 12,$
$x - 3y + 2z = 11,$
$5x + 2y - 4z = -4$

50. Find the vertices and the foci of the ellipse with the equation $9x^2 + 16y^2 = 144$. [7.2]

51. Find an equation of the ellipse with vertices $(0, -5)$ and $(0, 5)$ and minor axis of length 4. [7.2]

Synthesis

52. *Straight-Line Depreciation.* A company buys an office machine for \$5200 on January 1 of a given year. The machine is expected to last for 8 years, at the end of which time its **trade-in value**, or **salvage value**, will be \$1100. If the company's accountant figures the decline in value to be the same each year, then its **book values**, or **salvage values**, after t years, $0 \le t \le 8$, form an arithmetic sequence given by

$$a_t = C - t\left(\frac{C - S}{N}\right),$$

where C is the original cost of the item (\$5200), N is the number of years of expected life (8), and S is the salvage value (\$1100).

a) Find the formula for a_t for the straight-line depreciation of the office machine.

b) Find the salvage value after 0 year, 1 year, 2 years, 3 years, 4 years, 7 years, and 8 years.

53. Find a formula for the sum of the first n odd natural numbers:

$$1 + 3 + 5 + \cdots + (2n - 1).$$

54. Find three numbers in an arithmetic sequence such that the sum of the first and the third is 10 and the product of the first and the second is 15.

55. Find the first term and the common difference for the arithmetic sequence for which

$$a_2 = 40 - 3q \quad \text{and} \quad a_4 = 10p + q.$$

*If p, m, and q form an arithmetic sequence, it can be shown that $m = (p + q)/2$. The number m is the **arithmetic mean**, or **average**, of p and q. Given two numbers p and q, if we find k other numbers $m_1, m_2, \ldots, m_k$ such that*

$$p, m_1, m_2, \ldots, m_k, q$$

forms an arithmetic sequence, we say that we have "inserted k arithmetic means between p and q."

56. Insert three arithmetic means between -3 and 5.

57. Insert four arithmetic means between 4 and 13.

8.3 Geometric Sequences and Series

- Identify the common ratio of a geometric sequence, and find a given term and the sum of the first n terms.
- Find the sum of an infinite geometric series, if it exists.

A sequence in which each term after the first is found by multiplying the preceding term by the same number is a **geometric sequence**.

Geometric Sequences

Consider the sequence:

$$2,\ 6,\ 18,\ 54,\ 162,\ \ldots.$$

Note that multiplying each term by 3 produces the next term. We call the number 3 the **common ratio** because it can be found by dividing any term by the preceding term. A geometric sequence is also called a *geometric progression.*

GEOMETRIC SEQUENCE

A sequence is **geometric** if there is a number r, called the **common ratio**, such that

$$\frac{a_{n+1}}{a_n} = r, \quad \text{or} \quad a_{n+1} = a_n r, \quad \text{for any integer } n \geq 1.$$

EXAMPLE 1 For each of the following geometric sequences, identify the common ratio.

a) $3,\ 6,\ 12,\ 24,\ 48,\ \ldots$

b) $1,\ -\frac{1}{2},\ \frac{1}{4},\ -\frac{1}{8},\ \ldots$

c) \$5200, \$3900, \$2925, \$2193.75, . . .

d) \$1000, \$1060, \$1123.60, . . .

Solution

SEQUENCE	COMMON RATIO
a) $3,\ 6,\ 12,\ 24,\ 48,\ \ldots$	$2 \quad \left(\frac{6}{3} = 2,\ \frac{12}{6} = 2, \text{ and so on}\right)$
b) $1,\ -\frac{1}{2},\ \frac{1}{4},\ -\frac{1}{8},\ \ldots$	$-\frac{1}{2} \quad \left(\frac{-\frac{1}{2}}{1} = -\frac{1}{2},\ \frac{\frac{1}{4}}{-\frac{1}{2}} = -\frac{1}{2}, \text{ and so on}\right)$
c) \$5200, \$3900, \$2925, \$2193.75, . . .	$0.75 \quad \left(\frac{\$3900}{\$5200} = 0.75,\ \frac{\$2925}{\$3900} = 0.75, \text{ and so on}\right)$
d) \$1000, \$1060, \$1123.60, . . .	$1.06 \quad \left(\frac{\$1060}{\$1000} = 1.06,\ \frac{\$1123.60}{\$1060} = 1.06, \text{ and so on}\right)$

Now Try Exercise 1.

We now find a formula for the general, or nth, term of a geometric sequence. Let a_1 be the first term and r the common ratio. The first few terms are as follows:

$$\begin{aligned} &a_1, \\ &a_2 = a_1 r, \\ &a_3 = a_2 r = (a_1 r)r = a_1 r^2, && \text{Substituting } a_1 r \text{ for } a_2 \\ &a_4 = a_3 r = (a_1 r^2)r = a_1 r^3. && \text{Substituting } a_1 r^2 \text{ for } a_3 \end{aligned}$$

Note that the exponent is 1 less than the subscript.

Generalizing, we obtain the following.

nTH TERM OF A GEOMETRIC SEQUENCE

The **nth term** of a geometric sequence is given by

$$a_n = a_1 r^{n-1}, \quad \text{for any integer } n \geq 1.$$

EXAMPLE 2 Find the 7th term of the geometric sequence 4, 20, 100, ...

Solution We first note that

$$a_1 = 4 \quad \text{and} \quad n = 7.$$

To find the common ratio, we can divide any term (other than the first) by the preceding term. Since the second term is 20 and the first is 4, we get

$$r = \frac{20}{4}, \quad \text{or} \quad 5.$$

Then using the formula $a_n = a_1 r^{n-1}$, we have

$$a_7 = 4 \cdot 5^{7-1} = 4 \cdot 5^6 = 4 \cdot 15{,}625 = 62{,}500.$$

Thus the 7th term is 62,500.

Now Try Exercise 11.

EXAMPLE 3 Find the 10th term of the geometric sequence 64, −32, 16, −8, ...

Solution We first note that

$$a_1 = 64, \quad n = 10, \quad \text{and} \quad r = \frac{-32}{64}, \quad \text{or} \quad -\frac{1}{2}.$$

Then using the formula $a_n = a_1 r^{n-1}$, we have

$$a_{10} = 64 \cdot \left(-\frac{1}{2}\right)^{10-1} = 64 \cdot \left(-\frac{1}{2}\right)^9 = 2^6 \cdot \left(-\frac{1}{2^9}\right) = -\frac{1}{2^3} = -\frac{1}{8}.$$

Thus the 10th term is $-\frac{1}{8}$.

Now Try Exercise 15.

Sum of the First n Terms of a Geometric Sequence

Next, we develop a formula for the sum S_n of the first n terms of a geometric sequence:

$$a_1, a_1 r, a_1 r^2, a_1 r^3, \ldots, a_1 r^{n-1}, \ldots.$$

The associated **geometric series** is given by

$$S_n = a_1 + a_1r + a_1r^2 + a_1r^3 + \cdots + a_1r^{n-1}. \tag{1}$$

We want to find a formula for this sum. If we multiply by r on both sides of equation (1), we have

$$rS_n = a_1r + a_1r^2 + a_1r^3 + a_1r^4 + \cdots + a_1r^n. \tag{2}$$

Subtracting equation (2) from equation (1), we see that the differences of the terms shown in red are 0, leaving

$$S_n - rS_n = a_1 - a_1r^n,$$

or

$$S_n(1 - r) = a_1(1 - r^n). \quad \text{Factoring}$$

Dividing by $1 - r$ on both sides gives us the following formula.

SUM OF THE FIRST n TERMS

The sum of the first n terms of a geometric sequence is given by

$$S_n = \frac{a_1(1 - r^n)}{1 - r}, \quad \text{for any } r \neq 1.$$

EXAMPLE 4 Find the sum of the first 7 terms of the geometric sequence 3, 15, 75, 375, ...

Solution We first note that

$$a_1 = 3, \quad n = 7, \quad \text{and} \quad r = \frac{15}{3}, \text{ or } 5.$$

Then using the formula

$$S_n = \frac{a_1(1 - r^n)}{1 - r},$$

we have

$$S_7 = \frac{3(1 - 5^7)}{1 - 5}$$
$$= \frac{3(1 - 78{,}125)}{-4}$$
$$= 58{,}593.$$

Thus the sum of the first 7 terms is 58,593.

Now Try Exercise 23.

EXAMPLE 5 Find the sum: $\sum_{k=1}^{11} (0.3)^k$.

Solution This is a geometric series with $a_1 = 0.3$, $r = 0.3$, and $n = 11$. Thus,

$$S_{11} = \frac{0.3(1 - 0.3^{11})}{1 - 0.3}$$
$$\approx 0.42857.$$

Now Try Exercise 41.

Infinite Geometric Series

The sum of the terms of an infinite geometric sequence is an **infinite geometric series**. For some geometric sequences, S_n gets close to a specific number as n gets large. For example, consider the infinite series

$$\frac{1}{2} + \frac{1}{4} + \frac{1}{8} + \frac{1}{16} + \cdots + \frac{1}{2^n} + \cdots.$$

We can visualize S_n by considering the area of a square. For S_1, we shade half the square. For S_2, we shade half the square plus half the remaining half, or $\frac{1}{4}$. For S_3, we shade the parts shaded in S_2 plus half the remaining part. We see that the values of S_n will get close to 1 (shading the complete square).

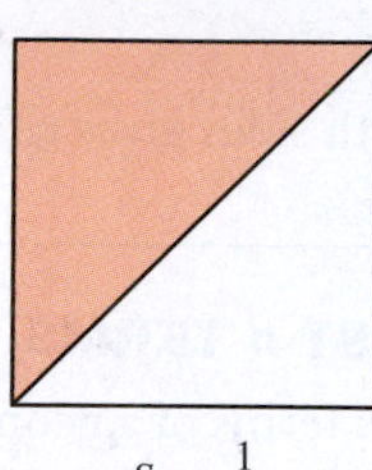
$S_1 = \frac{1}{2}$

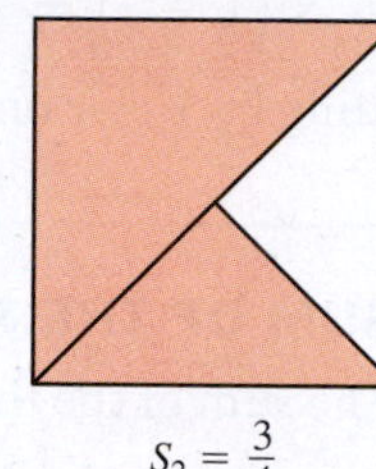
$S_2 = \frac{3}{4}$

$S_3 = \frac{7}{8}$

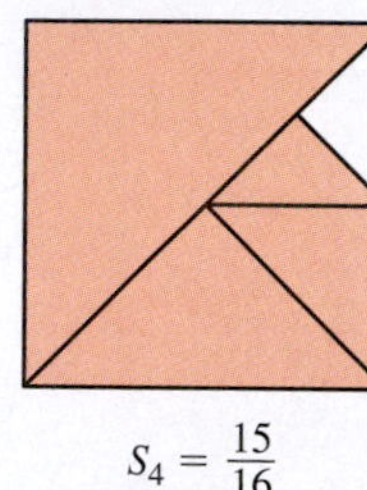
$S_4 = \frac{15}{16}$

TABLE 1

n	S_n
1	0.5
5	0.96875
10	0.9990234375
20	0.9999990463
30	0.9999999991

We examine some partial sums. Note that each of the partial sums in Table 1 is less than 1, but S_n gets very close to 1 as n gets large. We say that 1 is the **limit** of S_n and also that 1 is the **sum of the infinite geometric sequence**. The sum of an infinite geometric sequence is denoted S_∞. In this case, $S_\infty = 1$.

Some infinite sequences do not have sums. Consider the infinite geometric series

$$2 + 4 + 8 + 16 + \cdots + 2^n + \cdots.$$

We again examine some partial sums. Note in Table 2 that as n gets large, S_n gets large without bound. This sequence does not have a sum.

TABLE 2

n	S_n
1	2
5	62
10	2,046
20	2,097,150
30	2,147,483,646

It can be shown (but we will not do so here) that the sum of an infinite geometric series exists if and only if $|r| < 1$ (that is, the absolute value of the common ratio is less than 1).

To find a formula for the sum of an infinite geometric series, we first consider the sum of the first n terms:

$$S_n = \frac{a_1(1 - r^n)}{1 - r} = \frac{a_1 - a_1 r^n}{1 - r}.$$ Using the distributive law

For $|r| < 1$, values of r^n get close to 0 as n gets large. As r^n gets close to 0, so does $a_1 r^n$. Thus, S_n gets close to $a_1/(1 - r)$.

LIMIT OR SUM OF AN INFINITE GEOMETRIC SERIES

When $|r| < 1$, the limit or sum of an infinite geometric series is given by

$$S_\infty = \frac{a_1}{1 - r}.$$

EXAMPLE 6 Determine whether each of the following infinite geometric series has a limit. If a limit exists, find it.

a) $1 + 3 + 9 + 27 + \cdots$

b) $-2 + 1 - \frac{1}{2} + \frac{1}{4} - \frac{1}{8} + \cdots$

Solution

a) Here $r = 3$, so $|r| = |3| = 3$. Since $|r| > 1$, the series *does not* have a limit.

b) Here $r = -\frac{1}{2}$, so $|r| = \left|-\frac{1}{2}\right| = \frac{1}{2}$. Since $|r| < 1$, the series *does* have a limit. We find the limit:

$$S_\infty = \frac{a_1}{1-r} = \frac{-2}{1-\left(-\frac{1}{2}\right)} = \frac{-2}{\frac{3}{2}} = -\frac{4}{3}.$$

Now Try Exercises 33 and 37.

EXAMPLE 7 Find fraction notation for 0.78787878 . . . , or $0.\overline{78}$.

Solution We can express this as

$$0.78 + 0.0078 + 0.000078 + \cdots.$$

Then we see that this is an infinite geometric series, where $a_1 = 0.78$ and $r = 0.01$. Since $|r| < 1$, this series has a limit:

$$S_\infty = \frac{a_1}{1-r} = \frac{0.78}{1-0.01} = \frac{0.78}{0.99} = \frac{78}{99}, \quad \text{or} \quad \frac{26}{33}.$$

Thus fraction notation for 0.78787878. . . is $\frac{26}{33}$. You can check this on your calculator.

Now Try Exercise 51.

Applications

The translation of some applications and problem-solving situations may involve geometric sequences or series. Examples 9 and 10, in particular, show applications in business and economics.

EXAMPLE 8 ***A Daily Doubling Salary.*** Suppose someone offered you a job for the month of September (30 days) under the following conditions. You will be paid \$0.01 for the first day, \$0.02 for the second, \$0.04 for the third, and so on, doubling your previous day's salary each day. How much would you earn altogether for the month? (Would you take the job? Make a conjecture before reading further.)

Solution You earn \$0.01 the first day, \$0.01(2) the second day, \$0.01(2)(2) the third day, and so on. The amount earned is the geometric series

$$\$0.01 + \$0.01(2) + \$0.01(2^2) + \$0.01(2^3) + \cdots + \$0.01(2^{29}),$$

where $a_1 = 0.01$, $r = 2$, and $n = 30$. Using the formula

$$S_n = \frac{a_1(1-r^n)}{1-r},$$

we have

$$S_{30} = \frac{\$0.01(1-2^{30})}{1-2} = \$10{,}737{,}418.23.$$

The pay exceeds \$10.7 million for the month.

Now Try Exercise 57.

EXAMPLE 9 ***The Amount of an Annuity.*** An **annuity** is a sequence of equal payments, made at equal time intervals, that earn interest. Fixed deposits in a savings account are an example of an annuity. Suppose that to save money to buy a car, Jacob deposits \$2000 at the *end* of each of 5 years in an account that pays 3% interest, compounded annually. The total amount in the account at the end of 5 years is called the **amount of the annuity**. Find that amount.

Solution The following time diagram can help visualize the problem. Note that no deposit is made until the end of the first year.

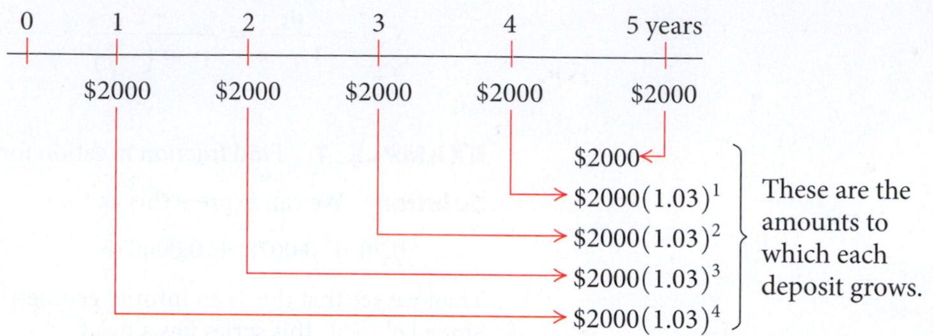

The amount of the annuity is the geometric series

$$\$2000 + \$2000(1.03)^1 + \$2000(1.03)^2 + \$2000(1.03)^3 + \$2000(1.03)^4,$$

where $a_1 = \$2000$, $n = 5$, and $r = 1.03$. Using the formula

$$S_n = \frac{a_1(1 - r^n)}{1 - r},$$

we have

$$S_5 = \frac{\$2000(1 - 1.03^5)}{1 - 1.03} \approx \$10{,}618.27.$$

The amount of the annuity is \$10,618.27.

Now Try Exercise 61.

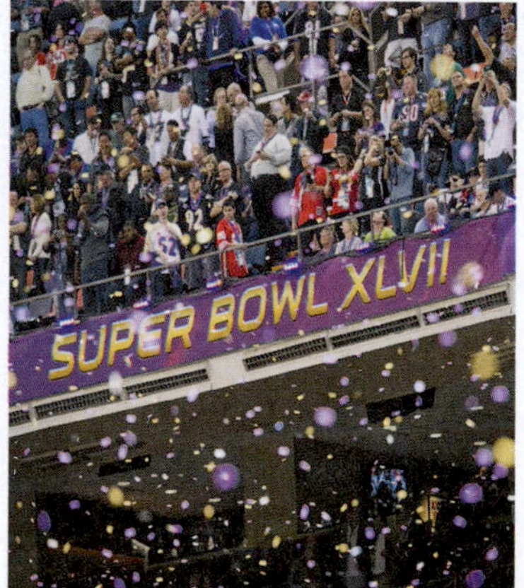

EXAMPLE 10 ***The Economic Multiplier.*** Large sporting events have a significant impact on the economy of the host city. Super Bowl XLVII, hosted by New Orleans, generated a \$480-million net impact for the region (*Source:* NewOrleansSaints.com, posted April 18, 2013, Marius M. Mihai, Research Analyst of the Division of Business and Economic Research at the University of New Orleans (DBER)). Assume that 60% of that amount is spent again in the area, and then 60% of that amount is spent again, and so on. This is known as the *economic multiplier effect*. Find the total effect on the economy.

Solution The total economic effect is given by the infinite series

$$\$480{,}000{,}000 + \$480{,}000{,}000(0.6) + \$480{,}000{,}000(0.6)^2 + \cdots.$$

Since $|r| = |0.6| = 0.6 < 1$, the series has a sum. Using the formula for the sum of an infinite geometric series, we have

$$S_\infty = \frac{a_1}{1 - r} = \frac{\$480{,}000{,}000}{1 - 0.6} = \$1{,}200{,}000{,}000.$$

The total effect of the spending on the economy is \$1,200,000,000.

Now Try Exercise 67.

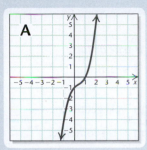

Visualizing the Graph

A

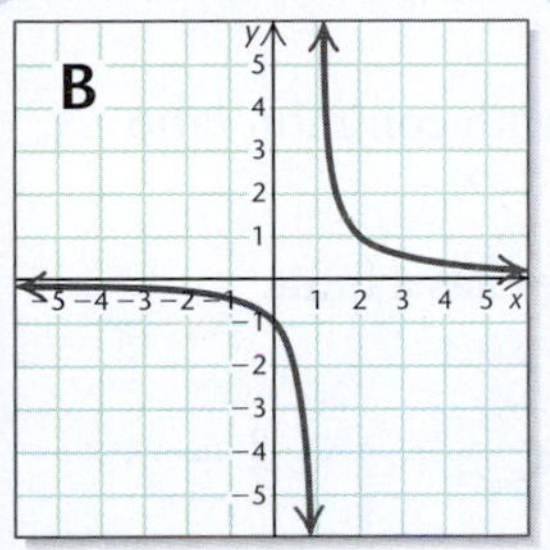

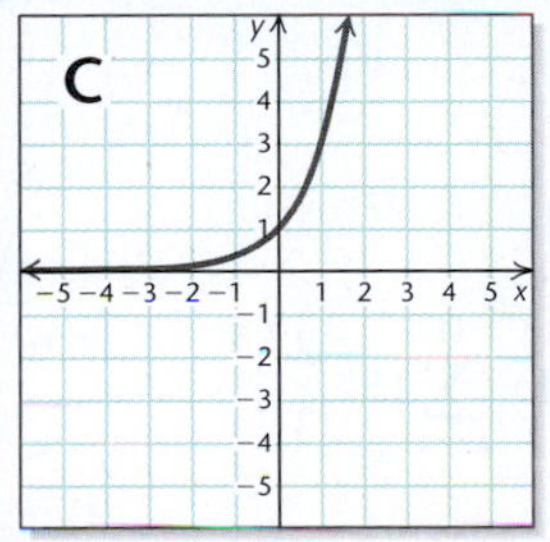

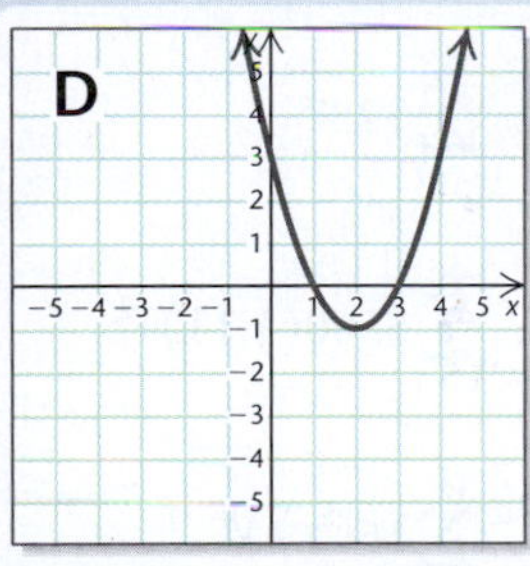

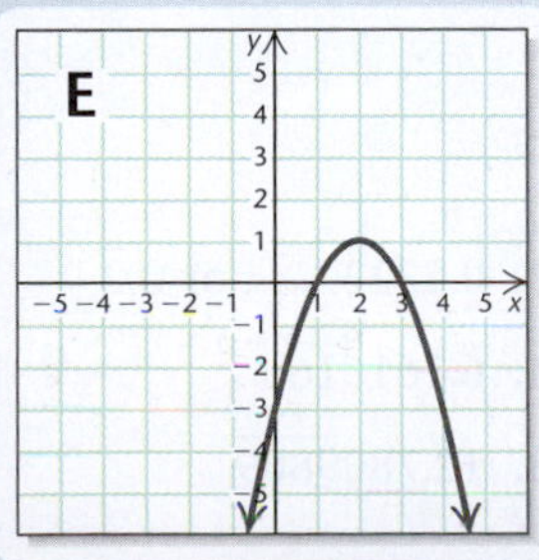

Match the equation with its graph.

1. $(x - 1)^2 + (y + 2)^2 = 9$

2. $y = x^3 - x^2 + x - 1$

3. $f(x) = 3^x$

4. $f(x) = x$

5. $a_n = n$

6. $y = \log (x + 3)$

7. $f(x) = -(x - 2)^2 + 1$

8. $f(x) = (x - 2)^2 - 1$

9. $y = \dfrac{1}{x - 1}$

10. $y = -3x + 4$

Answers on page A-42

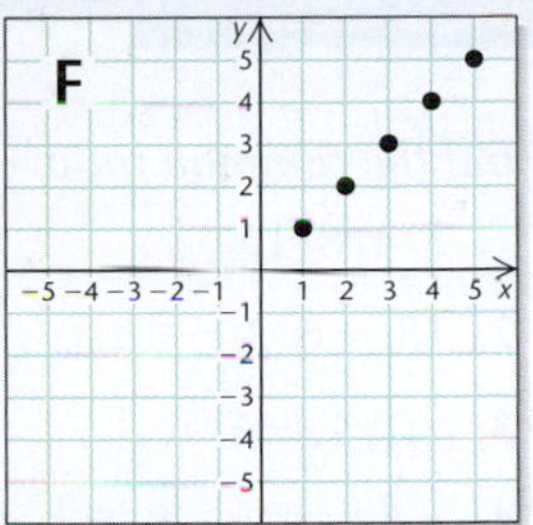

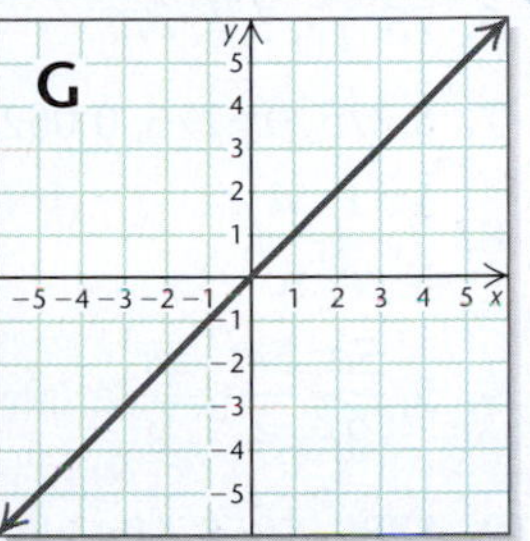

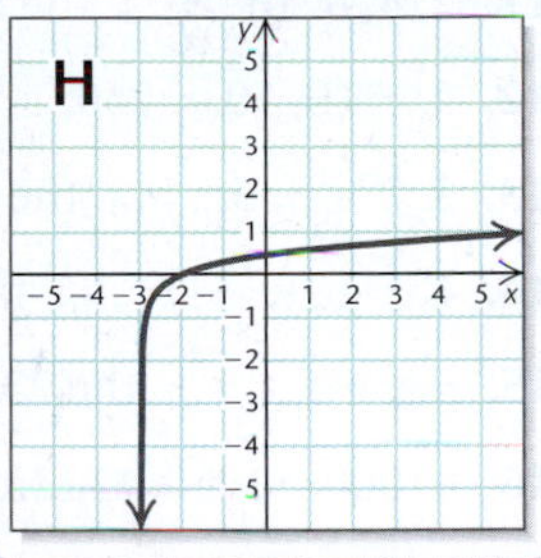

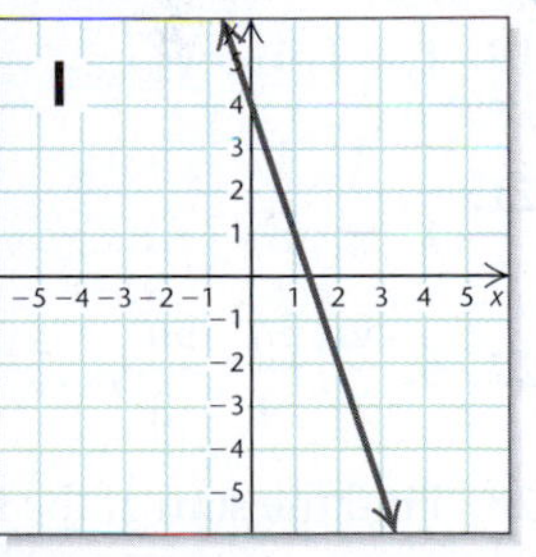

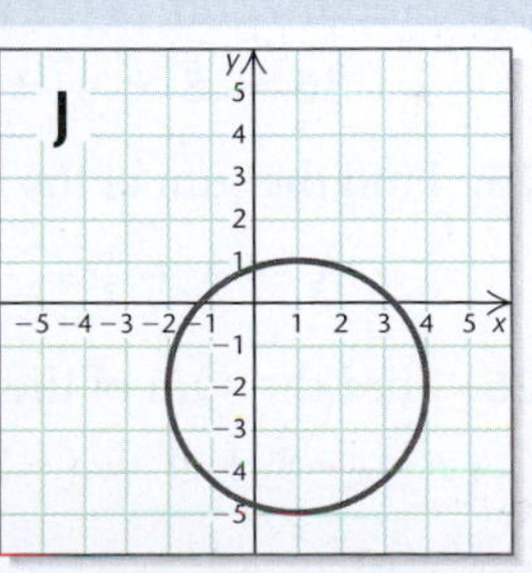

8.3 Exercise Set

Find the common ratio.

1. $2, 4, 8, 16, \ldots$
2. $18, -6, 2, -\frac{2}{3}, \ldots$
3. $-1, 1, -1, 1, \ldots$
4. $-8, -0.8, -0.08, -0.008, \ldots$
5. $\frac{2}{3}, -\frac{4}{3}, \frac{8}{3}, -\frac{16}{3}, \ldots$
6. $75, 15, 3, \frac{3}{5}, \ldots$
7. $6.275, 0.6275, 0.06275, \ldots$
8. $\frac{1}{x}, \frac{1}{x^2}, \frac{1}{x^3}, \ldots$
9. $5, \frac{5a}{2}, \frac{5a^2}{4}, \frac{5a^3}{8}, \ldots$
10. \$780, \$858, \$943.80, \$1038.18, . . .

Find the indicated term.

11. $2, 4, 8, 16, \ldots$; the 7th term
12. $2, -10, 50, -250, \ldots$; the 9th term
13. $2, 2\sqrt{3}, 6, \ldots$; the 9th term
14. $1, -1, 1, -1, \ldots$; the 57th term
15. $\frac{7}{625}, -\frac{7}{25}, \ldots$; the 23rd term
16. \$1000, \$1060, \$1123.60, . . . ; the 5th term

Find the nth, or general, term.

17. $1, 3, 9, \ldots$
18. $25, 5, 1, \ldots$
19. $1, -1, 1, -1, \ldots$
20. $-2, 4, -8, \ldots$
21. $\frac{1}{x}, \frac{1}{x^2}, \frac{1}{x^3}, \ldots$
22. $5, \frac{5a}{2}, \frac{5a^2}{4}, \frac{5a^3}{8}, \ldots$
23. Find the sum of the first 7 terms of the geometric series $6 + 12 + 24 + \cdots$.
24. Find the sum of the first 10 terms of the geometric series $16 - 8 + 4 - \cdots$.
25. Find the sum of the first 9 terms of the geometric series $\frac{1}{18} - \frac{1}{6} + \frac{1}{2} - \cdots$.
26. Find the sum of the geometric series $-8 + 4 + (-2) + \cdots + \left(-\frac{1}{32}\right)$.

Determine whether the statement is true or false.

27. The sequence $2, -2\sqrt{2}, 4, -4\sqrt{2}, 8, \ldots$ is geometric.
28. The sequence with general term $3n$ is geometric.
29. The sequence with general term 2^n is geometric.
30. Multiplying a term of a geometric sequence by the common ratio produces the next term of the sequence.
31. An infinite geometric series with common ratio -0.75 has a sum.
32. Every infinite geometric series has a limit.

Find the sum, if it exists.

33. $4 + 2 + 1 + \cdots$
34. $7 + 3 + \frac{9}{7} + \cdots$
35. $25 + 20 + 16 + \cdots$
36. $100 - 10 + 1 - \frac{1}{10} + \cdots$
37. $8 + 40 + 200 + \cdots$
38. $-6 + 3 - \frac{3}{2} + \frac{3}{4} - \cdots$
39. $0.6 + 0.06 + 0.006 + \cdots$
40. $\sum_{k=0}^{10} 3^k$
41. $\sum_{k=1}^{11} 15\left(\frac{2}{3}\right)^k$
42. $\sum_{k=0}^{50} 200(1.08)^k$
43. $\sum_{k=1}^{\infty} \left(\frac{1}{2}\right)^{k-1}$
44. $\sum_{k=1}^{\infty} 2^k$
45. $\sum_{k=1}^{\infty} 12.5^k$
46. $\sum_{k=1}^{\infty} 400(1.0625)^k$
47. $\sum_{k=1}^{\infty} \$500(1.11)^{-k}$
48. $\sum_{k=1}^{\infty} \$1000(1.06)^{-k}$
49. $\sum_{k=1}^{\infty} 16(0.1)^{k-1}$
50. $\sum_{k=1}^{\infty} \frac{8}{3}\left(\frac{1}{2}\right)^{k-1}$

Find fraction notation.

51. $0.131313\ldots$, or $0.\overline{13}$
52. $0.2222\ldots$, or $0.\overline{2}$
53. $8.999\overline{9}$
54. $6.1616\overline{16}$
55. $3.4125\overline{125}$
56. $12.7809\overline{809}$

57. *Daily Doubling Salary.* Suppose that someone offered you a job for the month of February (28 days) under the following conditions. You will be paid \$0.01 the 1st day, \$0.02 the 2nd, \$0.04 the 3rd, and so on, doubling your previous day's salary each day. How much would you earn altogether for the month?

58. *Bouncing Ping-Pong Ball.* A ping-pong ball is dropped from a height of 16 ft and always rebounds $\frac{1}{4}$ of the distance fallen.

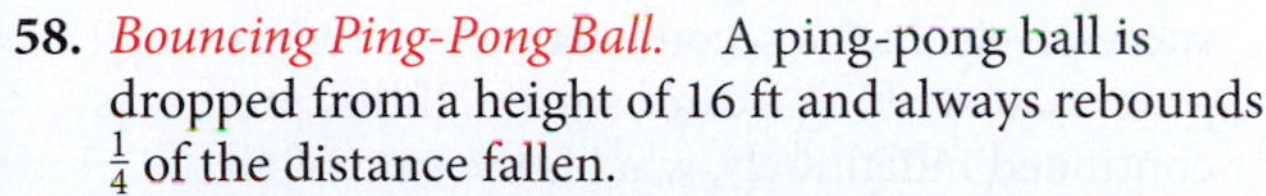

a) How high does it rebound the 6th time?

b) Find the total sum of the rebound heights of the ball.

59. *Bungee Jumping.* A bungee jumper always rebounds 60% of the distance fallen. A bungee jump is made using a cord that stretches to 200 ft.

a) After jumping and then rebounding 9 times, how far has a bungee jumper traveled upward (the total rebound distance)?

b) About how far will a jumper have traveled upward (bounced) before coming to rest?

60. *Population Growth.* A coastal town has a present population of 32,100, and the population is increasing by 3% each year.

a) What will the population be in 15 years?

b) How long will it take for the population to double?

61. *Amount of an Annuity.* To save for the down payment on a house, the Clines make a sequence of 10 yearly deposits of \$3200 each in a savings account on which interest is compounded annually at 4.6%. Find the amount of the annuity.

62. *Amount of an Annuity.* To create a college fund, a parent makes a sequence of 18 yearly deposits of \$1000 each in a savings account on which interest is compounded annually at 3.2%. Find the amount of the annuity.

63. *Doubling the Thickness of Paper.* A piece of paper is 0.01 in. thick. It is cut and stacked repeatedly in such a way that its thickness is doubled each time for 20 times. How thick is the result?

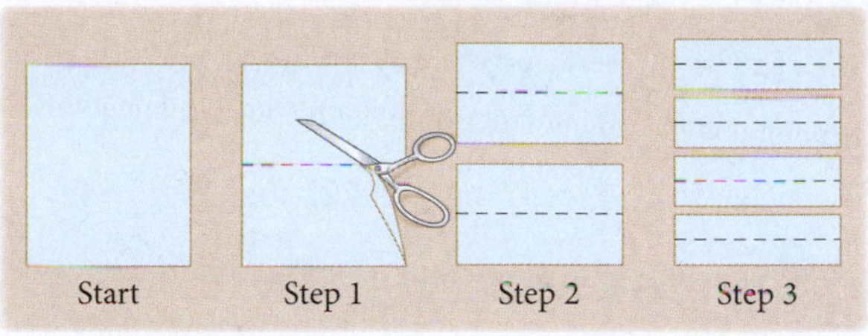

64. *Amount of an Annuity.* A sequence of yearly payments of P dollars is invested at the end of each of N years at interest rate i, compounded annually. The total amount in the account, or the amount of the annuity, is V.

a) Show that

$$V = \frac{P[(1 + i)^N - 1]}{i}.$$

b) Suppose that interest is compounded n times per year and deposits are made every compounding period. Show that the formula for V is then given by

$$V = \frac{P\left[\left(1 + \frac{i}{n}\right)^{nN} - 1\right]}{i/n}.$$

65. *Amount of an Annuity.* A sequence of payments of \$300 is invested over 12 years at the end of each quarter at 5.1%, compounded quarterly. Find the amount of the annuity. Use the formula in Exercise 64(b).

66. *Amount of an Annuity.* A sequence of yearly payments of \$750 is invested at the end of each of 10 years at 4.75%, compounded annually. Find the amount of the annuity. Use the formula in Exercise 64(a).

67. *The Economic Multiplier.* Suppose that the government is making a \$13,000,000,000 expenditure to stimulate the economy. If 85% of this is spent again, and so on, what is the total effect on the economy?

68. *Advertising Effect.* Gigi's Cupcake Truck is about to open for business in a city of 3,000,000 people, traveling to several curbside locations in the city each day to sell cupcakes. The owners plan an advertising campaign that they think will induce 30% of the people to buy their cupcakes. They estimate that if those people like the product, they will induce 30% · 30% · 3,000,000 more to buy the product, and those will induce 30% · 30% · 30% · 3,000,000 and so on. In all, how many people will buy Gigi's cupcakes as a result of the advertising campaign? What percentage of the population is this?

Skill Maintenance

For each pair of functions, find $(f \circ g)(x)$ and $(g \circ f)(x)$. [2.3]

69. $f(x) = x^2$, $g(x) = 4x + 5$

70. $f(x) = x - 1$, $g(x) = x^2 + x + 3$

Solve. [5.5]

71. $5^x = 35$

72. $\log_2 x = -4$

Synthesis

73. Prove that $\sqrt{3} - \sqrt{2}$, $4 - \sqrt{6}$, and $6\sqrt{3} - 2\sqrt{2}$ form a geometric sequence.

74. Assume that $a_1, a_2, a_3, \ldots$ is a geometric sequence. Prove that $\ln a_1, \ln a_2, \ln a_3, \ldots$ is an arithmetic sequence.

75. Consider the sequence $x + 3, x + 7, 4x - 2, \ldots$.

a) If the sequence is arithmetic, find x and then determine each of the 3 terms and the 4th term.

b) If the sequence is geometric, find x and then determine each of the 3 terms and the 4th term.

76. Find the sum of the first n terms of

$$1 + x + x^2 + \cdots.$$

77. Find the sum of the first n terms of

$$x^2 - x^3 + x^4 - x^5 + \cdots.$$

78. The sides of a square are 16 cm long. A second square is inscribed by joining the midpoints of the sides, successively. In the second square, we repeat the process, inscribing a third square. If this process is continued indefinitely, what is the sum of all the areas of all the squares? (*Hint*: Use an infinite geometric series.)

8.4 Mathematical Induction

- Prove infinite sequences of statements using mathematical induction.

In this section, we learn to prove a sequence of mathematical statements using a procedure called *mathematical induction.*

Proving Infinite Sequences of Statements

Infinite sequences of statements occur often in mathematics. In an infinite sequence of statements, there is a statement for each natural number. For example, consider the sequence of statements represented by the following:

"The sum of the first n positive odd integers is n^2," or
$1 + 3 + 5 + \cdots + (2n - 1) = n^2$.

Let's think of this as $S(n)$, or S_n. Substituting natural numbers for n gives a sequence of statements. We list the first four:

S_1: $1 = 1^2$;

S_2: $1 + 3 = 4 = 2^2$;

S_3: $1 + 3 + 5 = 9 = 3^2$;

S_4: $1 + 3 + 5 + 7 = 16 = 4^2$.

The fact that the statement is true for $n = 1, 2, 3$, and 4 might tempt us to conclude that the statement is true for any natural number n, but we cannot be sure that this is the case. We can, however, use the principle of mathematical induction to prove that the statement is true for all natural numbers.

THE PRINCIPLE OF MATHEMATICAL INDUCTION

We can prove an infinite sequence of statements S_n by showing the following.

(1) *Basis step.* S_1 is true.
(2) *Induction step.* For all natural numbers k, $S_k \rightarrow S_{k+1}$.

Mathematical induction is analogous to lining up a sequence of dominoes. The induction step tells us that if any one domino is knocked over, then the one next to it will be hit and knocked over. The basis step tells us that the first domino can indeed be knocked over. Note that in order for all dominoes to fall, *both* conditions must be satisfied.

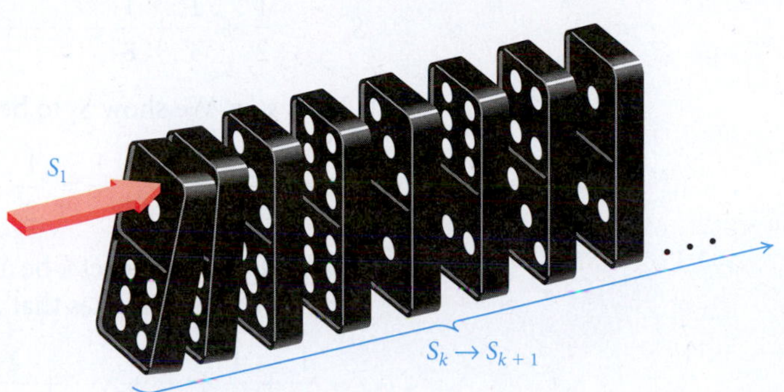

When you are learning to do proofs by mathematical induction, it is helpful to first write out S_n, S_1, S_k, and S_{k+1}. This helps to identify what is to be assumed and what is to be deduced.

EXAMPLE 1 Prove: For every natural number n,

$$1 + 3 + 5 + \cdots + (2n - 1) = n^2.$$

Proof. We first write out S_n, S_1, S_k, and S_{k+1}.

S_n: $1 + 3 + 5 + \cdots + (2n - 1) = n^2$

S_1: $1 = 1^2$

S_k: $1 + 3 + 5 + \cdots + (2k - 1) = k^2$

S_{k+1}: $1 + 3 + 5 + \cdots + (2k - 1) + [2(k + 1) - 1] = (k + 1)^2$

(1) *Basis step.* S_1, as listed, is true since $1 = 1^2$, or $1 = 1$.

(2) *Induction step.* We let k be any natural number. We assume S_k to be true and try to show that it implies that S_{k+1} is true. Now S_k is

$$1 + 3 + 5 + \cdots + (2k - 1) = k^2.$$

Starting with the left side of S_{k+1} and substituting k^2 for $1 + 3 + 5 + \cdots + (2k - 1)$, we have

$$\underbrace{1 + 3 + \cdots + (2k - 1)} + [2(k + 1) - 1]$$

$$= k^2 + [2(k + 1) - 1] \quad \text{We assume } S_k \text{ is true.}$$
$$= k^2 + 2k + 2 - 1$$
$$= k^2 + 2k + 1$$
$$= (k + 1)^2.$$

We have shown that for all natural numbers k, $S_k \rightarrow S_{k+1}$. This completes the induction step. It and the basis step tell us that the proof is complete.

Now Try Exercise 5.

EXAMPLE 2 Prove: For every natural number n,

$$\frac{1}{2} + \frac{1}{4} + \frac{1}{8} + \cdots + \frac{1}{2^n} = \frac{2^n - 1}{2^n}.$$

Proof. We first list S_n, S_1, S_k, and S_{k+1}.

$$S_n: \quad \frac{1}{2} + \frac{1}{4} + \frac{1}{8} + \cdots + \frac{1}{2^n} = \frac{2^n - 1}{2^n}$$

$$S_1: \quad \frac{1}{2^1} = \frac{2^1 - 1}{2^1}$$

$$S_k: \quad \frac{1}{2} + \frac{1}{4} + \frac{1}{8} + \cdots + \frac{1}{2^k} = \frac{2^k - 1}{2^k}$$

$$S_{k+1}: \quad \frac{1}{2} + \frac{1}{4} + \frac{1}{8} + \cdots + \frac{1}{2^k} + \frac{1}{2^{k+1}} = \frac{2^{k+1} - 1}{2^{k+1}}$$

(1) *Basis step.* We show S_1 to be true as follows:

$$\frac{2^1 - 1}{2^1} = \frac{2 - 1}{2} = \frac{1}{2}.$$

(2) *Induction step.* We let k be any natural number. We assume S_k to be true and try to show that it implies that S_{k+1} is true. Now S_k is

$$\frac{1}{2} + \frac{1}{4} + \frac{1}{8} + \cdots + \frac{1}{2^k} = \frac{2^k - 1}{2^k}.$$

We start with the left side of S_{k+1}. Since we assume that S_k is true, we can substitute

$$\frac{2^k - 1}{2^k} \quad \text{for} \quad \frac{1}{2} + \frac{1}{4} + \cdots + \frac{1}{2^k}.$$

We have

$$\underbrace{\frac{1}{2} + \frac{1}{4} + \frac{1}{8} + \cdots + \frac{1}{2^k}} + \frac{1}{2^{k+1}}$$

$$= \frac{2^k - 1}{2^k} + \frac{1}{2^{k+1}} = \frac{2^k - 1}{2^k} \cdot \frac{2}{2} + \frac{1}{2^{k+1}}$$

$$= \frac{(2^k - 1) \cdot 2 + 1}{2^{k+1}}$$

$$= \frac{2^{k+1} - 2 + 1}{2^{k+1}}$$

$$= \frac{2^{k+1} - 1}{2^{k+1}}.$$

We have shown that for all natural numbers k, $S_k \rightarrow S_{k+1}$. This completes the induction step. It and the basis step tell us that the proof is complete.

Now Try Exercise 15.

EXAMPLE 3 Prove: For every natural number n, $n < 2^n$.

Proof. We first list S_n, S_1, S_k, and S_{k+1}.

S_n: $n < 2^n$
S_1: $1 < 2^1$
S_k: $k < 2^k$
S_{k+1}: $k + 1 < 2^{k+1}$

(1) *Basis step.* S_1, as listed, is true since $2^1 = 2$ and $1 < 2$.

(2) *Induction step.* We let k be any natural number. We assume S_k to be true and try to show that it implies that S_{k+1} is true. Now

$k < 2^k$ **This is S_k.**
$2k < 2 \cdot 2^k$ **Multiplying by 2 on both sides**
$2k < 2^{k+1}$ **Adding exponents on the right**
$k + k < 2^{k+1}$. **Rewriting $2k$ as $k + k$**

Since k is any natural number, we know that $1 \leq k$. Thus,

$k + 1 \leq k + k$. **Adding k on both sides of $1 \leq k$**

Putting the results $k + 1 \leq k + k$ and $k + k < 2^{k+1}$ together gives us

$k + 1 < 2^{k+1}$. **This is S_{k+1}.**

We have shown that for all natural numbers k, $S_k \rightarrow S_{k+1}$. This completes the induction step. It and the basis step tell us that the proof is complete.

Now Try Exercise 11.

8.4 Exercise Set

List the first five statements in the sequence that can be obtained from each of the following. Determine whether each of the five statements is true or false.

1. $n^2 < n^3$
2. $n^2 - n + 41$ is prime. Find a value for n for which the statement is false.
3. A polygon of n sides has $[n(n - 3)]/2$ diagonals.
4. The sum of the angles of a polygon of n sides is $(n - 2) \cdot 180°$.

Use mathematical induction to prove each of the following.

5. $2 + 4 + 6 + \cdots + 2n = n(n + 1)$
6. $4 + 8 + 12 + \cdots + 4n = 2n(n + 1)$
7. $1 + 5 + 9 + \cdots + (4n - 3) = n(2n - 1)$
8. $3 + 6 + 9 + \cdots + 3n = \dfrac{3n(n + 1)}{2}$
9. $2 + 4 + 8 + \cdots + 2^n = 2(2^n - 1)$
10. $2 \leq 2^n$
11. $n < n + 1$
12. $3^n < 3^{n+1}$
13. $2n \leq 2^n$
14. $\dfrac{1}{1 \cdot 2} + \dfrac{1}{2 \cdot 3} + \cdots + \dfrac{1}{n(n + 1)} = \dfrac{n}{n + 1}$
15. $\dfrac{1}{1 \cdot 2 \cdot 3} + \dfrac{1}{2 \cdot 3 \cdot 4} + \dfrac{1}{3 \cdot 4 \cdot 5} + \cdots + \dfrac{1}{n(n + 1)(n + 2)} = \dfrac{n(n + 3)}{4(n + 1)(n + 2)}$
16. If x is any real number greater than 1, then for any natural number n, $x \leq x^n$.

The following formulas can be used to find sums of powers of natural numbers. Use mathematical induction to prove each formula.

17. $1 + 2 + 3 + \cdots + n = \dfrac{n(n + 1)}{2}$
18. $1^2 + 2^2 + 3^2 + \cdots + n^2 = \dfrac{n(n + 1)(2n + 1)}{6}$
19. $1^3 + 2^3 + 3^3 + \cdots + n^3 = \dfrac{n^2(n + 1)^2}{4}$

20. $1^4 + 2^4 + 3^4 + \cdots + n^4$
$$= \frac{n(n + 1)(2n + 1)(3n^2 + 3n - 1)}{30}$$

Use mathematical induction to prove each of the following.

21. $\sum_{i=1}^{n} i(i + 1) = \frac{n(n + 1)(n + 2)}{3}$

22. $\left(1 + \frac{1}{1}\right)\left(1 + \frac{1}{2}\right)\left(1 + \frac{1}{3}\right)\cdots\left(1 + \frac{1}{n}\right)$
$$= n + 1$$

23. The sum of n terms of an arithmetic sequence:

$$a_1 + (a_1 + d) + (a_1 + 2d) + \cdots + [a_1 + (n - 1)d]$$
$$= \frac{n}{2}[2a_1 + (n - 1)d]$$

▶ Skill Maintenance

Solve.

24. $2x - 3y = 1$,
$3x - 4y = 3$ [6.1], [6.3], [6.5], [6.6]

25. *Investment.* Clarise received \$104 in simple interest one year from three investments. Part is invested at 1.5%, part at 2%, and part at 3%. The amount invested at 2% is twice the amount invested at 1.5%. There is \$400 more invested at 3% than at 2%. Find the amount invested at each rate. [6.2], [6.3], [6.5], [6.6]

▶ Synthesis

Use mathematical induction to prove each of the following.

26. The sum of n terms of a geometric sequence:

$$a_1 + a_1r + a_1r^2 + \cdots + a_1r^{n-1} = \frac{a_1 - a_1r^n}{1 - r}$$

27. $x + y$ is a factor of $x^{2n} - y^{2n}$.

Prove each of the following using mathematical induction. Do the basis step for $n = 2$.

28. For every natural number $n \geq 2$,
$$2n + 1 < 3^n.$$

29. For every natural number $n \geq 2$,
$$\log_a (b_1b_2\cdots b_n)$$
$$= \log_a b_1 + \log_a b_2 + \cdots + \log_a b_n.$$

Prove each of the following for any complex numbers z_1, $z_2, \ldots, z_n$, where $i^2 = -1$ and $\bar{z}$ is the conjugate of z.

30. $\overline{z^n} = \bar{z}^n$

31. $\overline{z_1 + z_2 + \cdots + z_n} = \overline{z_1} + \overline{z_2} + \cdots + \overline{z_n}$

32. *The Tower of Hanoi Problem.* There are three pegs on a board. On one peg are n disks, each smaller than the one on which it rests. The problem is to move this pile of disks to another peg. The final order must be the same, but you can move only one disk at a time and can never place a larger disk on a smaller one.

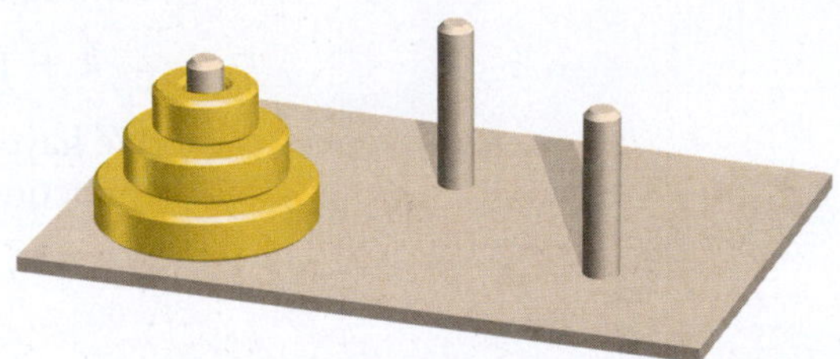

a) What is the *fewest* number of moves needed to move 3 disks? 4 disks? 2 disks? 1 disk?

b) Conjecture a formula for the *fewest* number of moves needed to move n disks. Prove it by mathematical induction.

Mid-Chapter Mixed Review

Determine whether the statement is true or false.

1. The general term of the sequence $1, -2, 3, -4, \ldots$ can be expressed as $a_n = n$. [8.1]

2. To find the common difference of an arithmetic sequence, choose any term except the first and then subtract the preceding term from it. [8.2]

3. The sequence $7, 3, -1, -5, \ldots$ is geometric. [8.2], [8.3]

4. If we can show that $S_k \rightarrow S_{k+1}$ for some natural number k, then we know that S_n is true for all natural numbers n. [8.4]

In each of the following, the nth term of a sequence is given. Find the first 4 terms, a_9, and a_{14}. [8.1]

5. $a_n = 3n + 5$

6. $a_n = (-1)^{n+1}(n - 1)$

Predict the general term, or nth term, a_n, of the sequence. Answers may vary. [8.1]

7. 3, 6, 9, 12, 15, . . .

8. $-1, 4, -9, 16, -25, \ldots$

9. Find the partial sum S_4 for the sequence $1, \frac{1}{2}, \frac{1}{4}, \frac{1}{8}, \frac{1}{16}, \ldots$. [8.1]

10. Find and evaluate the sum $\sum_{k=1}^{5} k(k + 1)$. [8.1]

11. Write sigma notation for the sum $-4 + 8 - 12 + 16 - 20 + \cdots$. [8.1]

12. Find the first 4 terms of the sequence defined by $a_1 = 2, a_{n+1} = 4a_n - 2$. [8.1]

13. Find the common difference of the arithmetic sequence 12, 7, 2, −3, ... [8.2]

14. Find the 10th term of the arithmetic sequence 4, 6, 8, 10, ... [8.2]

15. In the sequence in Exercise 14, what term is the number 44? [8.2]

16. Find the sum of the first 16 terms of the arithmetic series $6 + 11 + 16 + 21 + \cdots$. [8.2]

17. Find the common ratio of the geometric sequence 16, −8, 4, −2, 1, ... [8.3]

18. Find **(a)** the 8th term and **(b)** the sum of the first 10 terms of the geometric sequence $\frac{1}{16}, \frac{1}{8}, \frac{1}{4}, \frac{1}{2}, 1, \ldots$. [8.3]

Find the sum, if it exists. [8.3]

19. $-8 + 4 - 2 + 1 - \cdots$

20. $\sum_{k=0}^{\infty} 5^k$

21. *Landscaping.* A landscaper is planting a triangular flower bed with 36 plants in the first row, 30 plants in the second row, 24 in the third row, and so on, for a total of 6 rows. How many plants will be planted in all? [8.2]

22. *Amount of an Annuity.* To save money for adding a bedroom to their home, at the end of each of 4 years the Davidsons deposit \$1500 in an account that pays 4% interest, compounded annually. Find the total amount of the annuity. [8.3]

23. Prove: For every natural number n,
$1 + 4 + 7 + \cdots + (3n - 2) = \frac{1}{2}n(3n - 1)$. [8.4]

Collaborative Discussion and Writing

24. The sum of the first n terms of an arithmetic sequence can be given by

$$S_n = \frac{n}{2}[2a_1 + (n - 1)d].$$

Compare this formula to

$$S_n = \frac{n}{2}(a_1 + a_n).$$

Discuss the reasons for the use of one formula over the other. [8.2]

25. It is said that as a young child, the mathematician Karl F. Gauss (1777–1855) was able to compute the sum $1 + 2 + 3 + \cdots + 100$ very quickly in his head to the amazement of a teacher. Explain how Gauss might have done this had he possessed some knowledge of arithmetic sequences and series. Then give a formula for the sum of the first n natural numbers. [8.2]

26. Write a problem for a classmate to solve. Devise the problem so that a geometric series is involved and the solution is "The total amount in the bank is $900(1.08)^{40}$, or about \$19,552." [8.3]

27. Write an explanation of the idea behind mathematical induction for a fellow student. [8.4]

8.5 Combinatorics: Permutations

- Evaluate factorial notation and permutation notation and solve related applied problems.

In order to study probability, it is first necessary that we learn about **combinatorics**, the theory of counting.

Permutations

In this section, we will consider the part of combinatorics called *permutations*.

> The study of permutations involves *order* and *arrangements*.

EXAMPLE 1 How many 3-letter code symbols can be formed with the letters A, B, C *without* repetition (that is, using each letter only once)?

Solution Consider placing the letters in these boxes.

We can select any of the 3 letters for the first letter in the symbol. Once this letter has been selected, the second must be selected from the 2 remaining letters. After this, the third letter is already determined, since only 1 possibility is left. That is, we can place any of the 3 letters in the first box, either of the remaining 2 letters in the second box, and the only remaining letter in the third box. The possibilities can be determined using a **tree diagram**, as shown below.

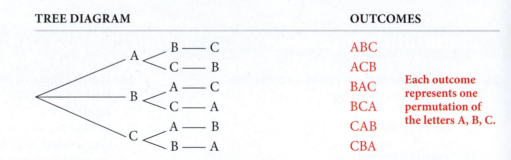

TREE DIAGRAM			OUTCOMES
A	B	C	ABC
	C	B	ACB
B	A	C	BAC
	C	A	BCA
C	A	B	CAB
	B	A	CBA

Each outcome represents one permutation of the letters A, B, C.

We see that there are 6 possibilities. The set of all the possibilities is

$$\{\text{ABC, ACB, BAC, BCA, CAB, CBA}\}.$$

This is the set of all *permutations* of the letters A, B, C. ■

Suppose that we perform an experiment such as selecting letters (as in the preceding example), flipping a coin, or drawing a card. The results are called **outcomes**. An **event** is a set of outcomes. The following principle enables us to count actions that are combined to form an event.

THE FUNDAMENTAL COUNTING PRINCIPLE

Given a combined action, or *event*, in which the first action can be performed in n_1 ways, the second action can be performed in n_2 ways, and so on, the total number of ways in which the combined action can be performed is the product

$$n_1 \cdot n_2 \cdot n_3 \cdot \cdots \cdot n_k.$$

Thus, in Example 1, there are 3 choices for the first letter, 2 for the second letter, and 1 for the third letter, making a total of $3 \cdot 2 \cdot 1$, or 6 possibilities.

EXAMPLE 2 How many 3-letter code symbols can be formed with the letters A, B, C, D, E *with* repetition (that is, allowing letters to be repeated)?

Solution Since repetition is allowed, there are 5 choices for the first letter, 5 choices for the second, and 5 for the third. Thus, by the fundamental counting principle, there are $5 \cdot 5 \cdot 5$, or 125 code symbols.

PERMUTATION

A **permutation** of a set of n objects is an ordered arrangement of all n objects.

We can use the fundamental counting principle to count the number of permutations of the objects in a set. Consider, for example, a set of 4 objects

$$\{A, B, C, D\}.$$

To find the number of ordered arrangements of the set, we select a first letter: There are 4 choices. Then we select a second letter: There are 3 choices. Then we select a third letter: There are 2 choices. Finally, there is 1 choice for the last selection. Thus, by the fundamental counting principle, there are $4 \cdot 3 \cdot 2 \cdot 1$, or 24, permutations of a set of 4 objects.

We can find a formula for the total number of permutations of all objects in a set of n objects. We have n choices for the first selection, $n - 1$ choices for the second, $n - 2$ for the third, and so on. For the nth selection, there is only 1 choice.

THE TOTAL NUMBER OF PERMUTATIONS OF *n* OBJECTS

The total number of permutations of n objects, denoted ${}_nP_n$, is given by

$${}_nP_n = n(n-1)(n-2)\cdots 3 \cdot 2 \cdot 1.$$

Technology Connection

We can find the total number of permutations of n objects, as in Example 3, using the ${}_nP_r$ operation from the MATH PRB (probability) menu on a graphing calculator.

```
4 nPr 4
                24
7 nPr 7
              5040
```

EXAMPLE 3 Find each of the following.

a) ${}_4P_4$

b) ${}_7P_7$

Solution

Start with 4.

a) ${}_4P_4 = 4 \cdot 3 \cdot 2 \cdot 1 = 24$

4 factors

b) ${}_7P_7 = 7 \cdot 6 \cdot 5 \cdot 4 \cdot 3 \cdot 2 \cdot 1 = 5040$

Now Try Exercise 1.

EXAMPLE 4 In how many ways can 9 packages be placed in 9 mailboxes, one package in a box?

Solution We have

$${}_9P_9 = 9 \cdot 8 \cdot 7 \cdot 6 \cdot 5 \cdot 4 \cdot 3 \cdot 2 \cdot 1 = 362{,}880.$$

Now Try Exercise 23.

Factorial Notation

We will use products such as $7 \cdot 6 \cdot 5 \cdot 4 \cdot 3 \cdot 2 \cdot 1$ so often that it is convenient to adopt a notation for them. For the product

$$7 \cdot 6 \cdot 5 \cdot 4 \cdot 3 \cdot 2 \cdot 1,$$

we write 7!, read "7 factorial."

We now define factorial notation for natural numbers and for 0.

FACTORIAL NOTATION

For any natural number n,

$$n! = n(n-1)(n-2) \cdots 3 \cdot 2 \cdot 1.$$

For the number 0,

$$0! = 1.$$

We define 0! as 1 so that certain formulas can be stated concisely and with a consistent pattern.

Here are some examples of factorial notation.

Technology Connection

We can evaluate factorial notation using the ! operation from the MATH PRB (probability) menu.

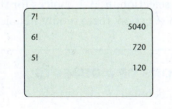

7!	
	5040
6!	
	720
5!	
	120

$$\begin{aligned}
7! &= 7 \cdot 6 \cdot 5 \cdot 4 \cdot 3 \cdot 2 \cdot 1 = 5040\\
6! &= 6 \cdot 5 \cdot 4 \cdot 3 \cdot 2 \cdot 1 = 720\\
5! &= 5 \cdot 4 \cdot 3 \cdot 2 \cdot 1 = 120\\
4! &= 4 \cdot 3 \cdot 2 \cdot 1 = 24\\
3! &= 3 \cdot 2 \cdot 1 = 6\\
2! &= 2 \cdot 1 = 2\\
1! &= 1 = 1\\
0! &= 1 = 1
\end{aligned}$$

We now see that the following statement is true.

$${}_nP_n = n!$$

We will often need to manipulate factorial notation. For example, note that

$$\begin{aligned}
8! &= 8 \cdot 7 \cdot 6 \cdot 5 \cdot 4 \cdot 3 \cdot 2 \cdot 1\\
&= 8 \cdot (7 \cdot 6 \cdot 5 \cdot 4 \cdot 3 \cdot 2 \cdot 1) = 8 \cdot 7!.
\end{aligned}$$

Generalizing, we get the following.

For any natural number n, $n! = n(n-1)!$.

By using this result repeatedly, we can further manipulate factorial notation.

EXAMPLE 5 Rewrite 7! with a factor of 5!.

Solution We have $7! = 7 \cdot 6! = 7 \cdot 6 \cdot 5!$.

In general, we have the following.

For any natural numbers k and n, with $k < n$,

$$n! = \underbrace{n(n-1)(n-2)\cdots[n-(k-1)]}_{k \text{ factors}} \cdot \underbrace{(n-k)}_{n-k \text{ factors}}!.$$

Permutations of *n* Objects Taken *k* at a Time

Consider a set of 5 objects

$$\{A, B, C, D, E\}.$$

How many ordered arrangements can be formed using 3 objects from the set without repetition? Examples of such an arrangement are EBA, CAB, and BCD. There are 5 choices for the first object, 4 choices for the second, and 3 choices for the third. By the fundamental counting principle, there are

$5 \cdot 4 \cdot 3$, or 60 *permutations* of a set of 5 objects taken 3 at a time.

Note that

$$5 \cdot 4 \cdot 3 = \frac{5 \cdot 4 \cdot 3 \cdot 2 \cdot 1}{2 \cdot 1}, \quad \text{or} \quad \frac{5!}{2!}.$$

PERMUTATION OF *n* OBJECTS TAKEN *k* AT A TIME

A **permutation** of a set of n objects taken k at a time is an ordered arrangement of k objects taken from the set.

Consider a set of n objects and the selection of an ordered arrangement of k of them. There would be n choices for the first object. Then there would remain $n - 1$ choices for the second, $n - 2$ choices for the third, and so on. We make k choices in all, so there are k factors in the product. By the fundamental counting principle, the total number of permutations is

$$\underbrace{n(n-1)(n-2)\cdots[n-(k-1)]}_{k \text{ factors}}.$$

We can express this in another way by multiplying by 1, as follows:

$$n(n-1)(n-2)\cdots[n-(k-1)] \cdot \frac{(n-k)!}{(n-k)!}$$
$$= \frac{n(n-1)(n-2)\cdots[n-(k-1)](n-k)!}{(n-k)!}$$
$$= \frac{n!}{(n-k)!}.$$

This gives us the following.

THE NUMBER OF PERMUTATIONS OF n OBJECTS TAKEN k AT A TIME

The number of permutations of a set of n objects taken k at a time, denoted ${}_nP_k$, is given by

$$ {}_nP_k = \underbrace{n(n-1)(n-2)\cdots[n-(k-1)]}_{k \text{ factors}} \qquad (1)$$

$$ = \frac{n!}{(n-k)!}. \qquad (2)$$

Technology Connection

We can evaluate computations like the one in Example 6 using the ${}_nP_r$ operation from the MATH PRB menu on a graphing calculator.

EXAMPLE 6 Compute ${}_8P_4$ using both forms of the formula.

Solution Using form (1), we have

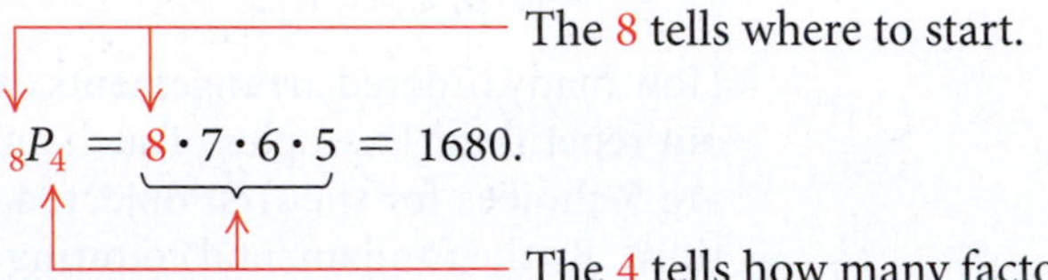

$$ {}_8P_4 = \underbrace{8\cdot7\cdot6\cdot5}_{} = 1680.$$

Using form (2), we have

$$ {}_8P_4 = \frac{8!}{(8-4)!} = \frac{8!}{4!}$$
$$ = \frac{8\cdot7\cdot6\cdot5\cdot4!}{4!} = \frac{8\cdot7\cdot6\cdot5\cdot\cancel{4!}}{\cancel{4!}}$$
$$ = 8\cdot7\cdot6\cdot5 = 1680.$$

Now Try Exercise 3.

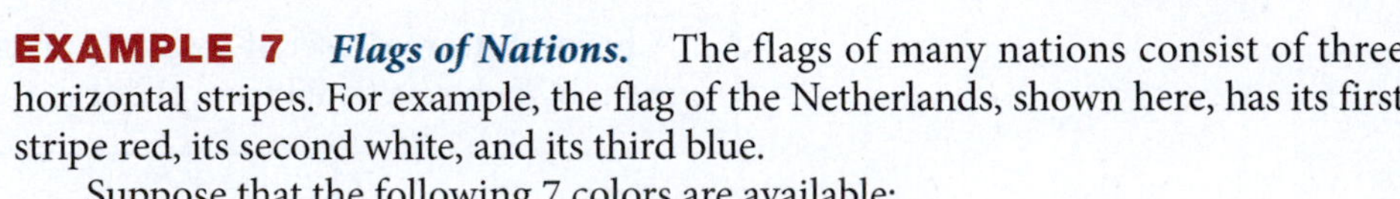

EXAMPLE 7 ***Flags of Nations.*** The flags of many nations consist of three horizontal stripes. For example, the flag of the Netherlands, shown here, has its first stripe red, its second white, and its third blue.

Suppose that the following 7 colors are available:

$$\{\text{black, yellow, red, white, blue, orange, green}\}.$$

How many different flags of three horizontal stripes can be made without repetition of colors in a flag? (This assumes that the order in which the stripes appear is considered.)

Solution We are determining the number of permutations of 7 objects taken 3 at a time. There is no repetition of colors. Using form (1), we get

$$ {}_7P_3 = 7\cdot6\cdot5 = 210.$$

Now Try Exercise 37(a).

EXAMPLE 8 ***Batting Orders.*** A baseball manager arranges the batting order as follows: The 4 infielders will bat first. Then the 3 outfielders, the catcher, and the pitcher will follow, not necessarily in that order. How many different batting orders are possible?

Solution The infielders can bat in ${}_4P_4$ different ways, the rest in ${}_5P_5$ different ways. Then by the fundamental counting principle, we have

$$ {}_4P_4\cdot{}_5P_5 = 4!\cdot5!, \quad \text{or} \quad 2880 \text{ possible batting orders.}$$

Now Try Exercise 31.

If we allow repetition, a situation like the following can occur.

EXAMPLE 9 How many 5-letter code symbols can be formed with the letters A, B, C, and D if we allow a letter to occur more than once?

Solution We can select each of the 5 letters in 4 ways. That is, we can select the first letter in 4 ways, the second in 4 ways, and so on. Thus there are 4^5, or 1024 arrangements.

Now Try Exercise 37(b).

The number of distinct arrangements of n objects taken k at a time, allowing repetition, is n^k.

Permutations of Sets with Nondistinguishable Objects

Consider a set of 7 marbles, 4 of which are blue and 3 of which are red. When they are lined up, one red marble will look just like any other red marble. In this sense, we say that the red marbles are nondistinguishable and, similarly, the blue marbles are nondistinguishable.

We know that there are 7! permutations of this set. Many of them will look alike, however. We develop a formula for finding the number of distinguishable permutations.

Consider a set of n objects in which n_1 are of one kind, n_2 are of a second kind, . . . , and n_k are of a kth kind. The total number of permutations of the set is $n!$, but this includes many that are nondistinguishable. Let N be the total number of distinguishable permutations. For each of these N permutations, there are $n_1!$ actual, nondistinguishable permutations, obtained by permuting the objects of the first kind. For each of these $N \cdot n_1!$ permutations, there are $n_2!$ nondistinguishable permutations, obtained by permuting the objects of the second kind, and so on. By the fundamental counting principle, the total number of permutations, including those that are nondistinguishable, is

$$N \cdot n_1! \cdot n_2! \cdot \cdots \cdot n_k!.$$

Then we have $N \cdot n_1! \cdot n_2! \cdot \cdots \cdot n_k! = n!$. Solving for N, we obtain

$$N = \frac{n!}{n_1! \cdot n_2! \cdot \cdots \cdot n_k!}.$$

Now, to finish our problem with the marbles, we have

$$N = \frac{7!}{4!3!}$$

$$= \frac{7 \cdot 6 \cdot 5 \cdot 4!}{4! \cdot 3 \cdot 2 \cdot 1} = \frac{7 \cdot \cancel{3} \cdot \cancel{2} \cdot 5 \cdot \cancel{4!}}{\cancel{4!} \cdot \cancel{3} \cdot \cancel{2} \cdot 1} = \frac{7 \cdot 5}{1}, \quad \text{or} \quad 35$$

distinguishable permutations of the marbles.

In general, we have the following.

> For a set of n objects in which n_1 are of one kind, n_2 are of another kind, . . . , and n_k are of a kth kind, the number of distinguishable permutations is
>
> $$\frac{n!}{n_1! \cdot n_2! \cdot \cdots \cdot n_k!}.$$

EXAMPLE 10 In how many distinguishable ways can the letters of the word CINCINNATI be arranged?

Solution There are 2 C's, 3 I's, 3 N's, 1 A, and 1 T for a total of 10 letters. Thus,

$$N = \frac{10!}{2! \cdot 3! \cdot 3! \cdot 1! \cdot 1!}, \quad \text{or} \quad 50{,}400.$$

The letters of the word CINCINNATI can be arranged in 50,400 distinguishable ways.

Now Try Exercise 35.

8.5 Exercise Set

Evaluate.

1. ${}_6P_6$
2. ${}_4P_3$
3. ${}_{10}P_7$
4. ${}_{10}P_3$
5. $5!$
6. $7!$
7. $0!$
8. $1!$
9. $\frac{9!}{5!}$
10. $\frac{9!}{4!}$
11. $(8 - 3)!$
12. $(8 - 5)!$
13. $\frac{10!}{7!\,3!}$
14. $\frac{7!}{(7 - 2)!}$
15. ${}_8P_0$
16. ${}_{13}P_1$
17. ${}_{52}P_4$
18. ${}_{52}P_5$
19. ${}_nP_3$
20. ${}_nP_2$
21. ${}_nP_1$
22. ${}_nP_0$

In each of Exercises 23–41, give your answer using permutation notation, factorial notation, or other operations. Then evaluate.

How many permutations are there of the letters in each of the following words, if all the letters are used without repetition?

23. CREDIT
24. FRUIT
25. EDUCATION
26. TOURISM

27. How many permutations are there of the letters of the word EDUCATION if the letters are taken 4 at a time?

28. How many permutations are there of the letters of the word TOURISM if the letters are taken 5 at a time?

29. How many 5-digit numbers can be formed using the digits 2, 4, 6, 8, and 9 without repetition? with repetition?

30. In how many ways can 7 athletes be arranged in a straight line?

31. *Program Planning.* A program is planned to have 5 musical numbers and 4 speeches. In how many ways can this be done if a musical number and a speech are to alternate and a musical number is to come first?

32. A professor is going to grade her 24 students on a curve. She will give 3 A's, 5 B's, 9 C's, 4 D's, and 3 F's. In how many ways can she do this?

33. *Phone Numbers.* How many 7-digit phone numbers can be formed with the digits 0, 1, 2, 3, 4, 5, 6, 7, 8, and 9, assuming that the first number cannot be 0 or 1? Accordingly, how many telephone numbers can there be within a given area code, before the area needs to be split with a new area code?

34. How many distinguishable code symbols can be formed from the letters of the word BUSINESS? BIOLOGY? MATHEMATICS?

35. Suppose the expression $a^2b^3c^4$ is rewritten without exponents. In how many distinguishable ways can this be done?

36. *Coin Arrangements.* A penny, a nickel, a dime, and a quarter are arranged in a straight line.

a) Considering just the coins, in how many ways can they be lined up?
b) Considering the coins and heads and tails, in how many ways can they be lined up?

37. How many code symbols can be formed using 5 out of 6 letters of A, B, C, D, E, F if the letters:
a) are not repeated?
b) can be repeated?
c) are not repeated but must begin with D?
d) are not repeated but must begin with DE?

38. *License Plates.* A state forms its license plates by first listing a number that corresponds to the county in which the owner of the car resides. (The names of the counties are alphabetized and the number is its location in that order.) Then the plate lists a letter of the alphabet, and this is followed by a number from 1 to 9999. How many such plates are possible if there are 80 counties?

39. *Zip Codes.* A U.S. postal zip code is a five-digit number.
a) How many zip codes are possible if any of the digits 0 to 9 can be used?
b) If each post office has its own zip code, how many possible post offices can there be?

40. *Zip-Plus-4 Codes.* A zip-plus-4 postal code uses a 9-digit number like 75247-5456. How many 9-digit zip-plus-4 postal codes are possible?

41. *Social Security Numbers.* A social security number is a 9-digit number like 243-47-0825.
a) How many different social security numbers can there be?
b) There are about 311 million people in the United States. Can each person have a unique social security number?

Skill Maintenance

Find the zero(s) of the function.

42. $f(x) = 4x - 9$ [1.5]

43. $f(x) = x^2 + x - 6$ [3.2]

44. $f(x) = 2x^2 - 3x - 1$ [3.2]

45. $f(x) = x^3 - 4x^2 - 7x + 10$ [4.4]

Synthesis

Solve for n.

46. ${}_nP_5 = 7 \cdot {}_nP_4$

47. ${}_nP_4 = 8 \cdot {}_{n-1}P_3$

48. ${}_nP_5 = 9 \cdot {}_{n-1}P_4$

49. ${}_nP_4 = 8 \cdot {}_nP_3$

50. Show that $n! = n(n-1)(n-2)(n-3)!$.

51. *Single-Elimination Tournaments.* In a single-elimination sports tournament consisting of n teams, a team is eliminated when it loses one game. How many games are required to complete the tournament?

52. *Double-Elimination Tournaments.* In a double-elimination softball tournament consisting of n teams, a team is eliminated when it loses two games. At most, how many games are required to complete the tournament?

8.6 Combinatorics: Combinations

▶ Evaluate combination notation and solve related applied problems.

We now consider counting techniques in which order is not considered.

▶ Combinations

We sometimes make a selection from a set *without regard to order.* Such a selection is called a *combination.* If you play cards, for example, you know that in most situations the *order* in which you hold cards is not important. That is,

The hand
is "equivalent"
to these hands.

Each hand contains the same combination of three cards.

EXAMPLE 1 Find all the combinations of 3 letters taken from the set of 5 letters $\{A, B, C, D, E\}$.

Solution The combinations are

$$\{A, B, C\}, \quad \{A, B, D\},$$
$$\{A, B, E\}, \quad \{A, C, D\},$$
$$\{A, C, E\}, \quad \{A, D, E\},$$
$$\{B, C, D\}, \quad \{B, C, E\},$$
$$\{B, D, E\}, \quad \{C, D, E\}.$$

There are 10 combinations of the 5 letters taken 3 at a time. ■

When we find all the combinations from a set of 5 objects taken 3 at a time, we are finding all the 3-element subsets. When a set is named, the order of the elements is *not* considered. Thus,

$$\{A, C, B\} \quad \text{names the same set as} \quad \{A, B, C\}.$$

COMBINATION; COMBINATION NOTATION

A **combination** containing k objects chosen from a set of n objects, $k \leq n$, is denoted using **combination notation** ${}_nC_k$.

We want to derive a general formula for ${}_nC_k$ for any $k \leq n$. First, it is true that ${}_nC_n = 1$, because a set with n objects has only 1 subset with n objects, the set itself. Second, ${}_nC_1 = n$, because a set with n objects has n subsets with 1 element each. Finally, ${}_nC_0 = 1$, because a set with n objects has only one subset with 0 elements, namely, the empty set $\varnothing$. To consider other possibilities, let's return to Example 1 and compare the number of combinations with the number of permutations.

	COMBINATIONS		PERMUTATIONS						
${}_5C_3$ of these	{A, B, C}	→	ABC	BCA	CAB	CBA	BAC	ACB	$3! \cdot {}_5C_3$ of these
	{A, B, D}	→	ABD	BDA	DAB	DBA	BAD	ADB	
	{A, B, E}	→	ABE	BEA	EAB	EBA	BAE	AEB	
	{A, C, D}	→	ACD	CDA	DAC	DCA	CAD	ADC	
	{A, C, E}	→	ACE	CEA	EAC	ECA	CAE	AEC	
	{A, D, E}	→	ADE	DEA	EAD	EDA	DAE	AED	
	{B, C, D}	→	BCD	CDB	DBC	DCB	CBD	BDC	
	{B, C, E}	→	BCE	CEB	EBC	ECB	CBE	BEC	
	{B, D, E}	→	BDE	DEB	EBD	EDB	DBE	BED	
	{C, D, E}	→	CDE	DEC	ECD	EDC	DCE	CED	

Note that each combination of 3 objects yields 6, or 3!, permutations:

$$3! \cdot {}_5C_3 = 60 = {}_5P_3 = 5 \cdot 4 \cdot 3,$$

so

$${}_5C_3 = \frac{{}_5P_3}{3!} = \frac{5 \cdot 4 \cdot 3}{3 \cdot 2 \cdot 1} = 10.$$

In general, the number of combinations of n objects taken k at a time, ${}_nC_k$, times the number of permutations of these objects, $k!$, must equal the number of permutations of n objects taken k at a time:

$$\begin{aligned} k! \cdot {}_nC_k &= {}_nP_k \\ {}_nC_k &= \frac{{}_nP_k}{k!} \\ &= \frac{1}{k!} \cdot {}_nP_k \\ &= \frac{1}{k!} \cdot \frac{n!}{(n-k)!} = \frac{n!}{k!(n-k)!}. \end{aligned}$$

COMBINATIONS OF *n* OBJECTS TAKEN *k* AT A TIME

The total number of combinations of n objects taken k at a time, denoted ${}_nC_k$, is given by

$${}_nC_k = \frac{n!}{k!(n-k)!}, \quad (1)$$

or

$${}_nC_k = \frac{{}_nP_k}{k!} = \frac{n(n-1)(n-2)\cdots[n-(k-1)]}{k!}. \quad (2)$$

Another kind of notation for ${}_nC_k$ is **binomial coefficient notation**. The reason for such terminology will be seen later.

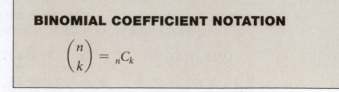

BINOMIAL COEFFICIENT NOTATION

$$\binom{n}{k} = {}_nC_k$$

You should be able to use either notation and either form of the formula.

EXAMPLE 2 Evaluate $\binom{7}{5}$, using forms (1) and (2).

Solution

a) By form (1),

$$\binom{7}{5} = \frac{7!}{5!\,(7-5)!} = \frac{7!}{5!2!}$$
$$= \frac{7 \cdot 6 \cdot 5!}{5! \cdot 2!} = \frac{7 \cdot 6 \cdot \cancel{5!}}{\cancel{5!} \cdot 2!} = \frac{7 \cdot 6}{2 \cdot 1} = 21.$$

b) By form (2),

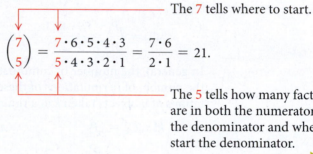

The 7 tells where to start.

$$\binom{7}{5} = \frac{7 \cdot 6 \cdot 5 \cdot 4 \cdot 3}{5 \cdot 4 \cdot 3 \cdot 2 \cdot 1} = \frac{7 \cdot 6}{2 \cdot 1} = 21.$$

The 5 tells how many factors there are in both the numerator and the denominator and where to start the denominator.

Now Try Exercise 11.

Technology Connection

We can do computations like the one in Example 2 using the ${}_nC_r$ operation from the MATH PRB (probability) menu on a graphing calculator.

7 nCr 5
21

Be sure to keep in mind that $\binom{n}{k}$ does not mean $n \div k$, or n/k.

EXAMPLE 3 Evaluate $\binom{n}{0}$ and $\binom{n}{2}$.

Solution We use form (1) for the first expression and form (2) for the second. Then

$$\binom{n}{0} = \frac{n!}{0!(n-0)!} = \frac{n!}{1 \cdot n!} = 1,$$

using form (1), and

$$\binom{n}{2} = \frac{n(n-1)}{2!} = \frac{n(n-1)}{2}, \quad \text{or} \quad \frac{n^2 - n}{2},$$

using form (2).

Now Try Exercise 19.

Note that

$$\binom{7}{2} = \frac{7 \cdot 6}{2 \cdot 1} = 21.$$

Using the result of Example 2 gives us

$$\binom{7}{5} = \binom{7}{2}.$$

This says that the number of 5-element subsets of a set of 7 objects is the same as the number of 2-element subsets of a set of 7 objects. When 5 elements are chosen from a set, one also chooses *not* to include 2 elements. To see this, consider the set {A, B, C, D, E, F, G}:

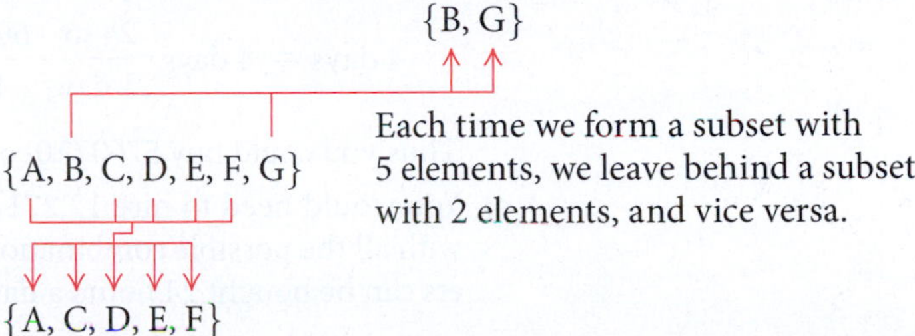
{B, G}
{A, B, C, D, E, F, G}
Each time we form a subset with 5 elements, we leave behind a subset with 2 elements, and vice versa.
{A, C, D, E, F}

In general, we have the following. This result provides an alternative way to compute combinations.

SUBSETS OF SIZE *k* AND OF SIZE *n* − *k*

$$\binom{n}{k} = \binom{n}{n-k} \quad \text{and} \quad {}_nC_k = {}_nC_{n-k}$$

The number of subsets of size k of a set with n objects is the same as the number of subsets of size $n - k$. The number of combinations of n objects taken k at a time is the same as the number of combinations of n objects taken $n - k$ at a time.

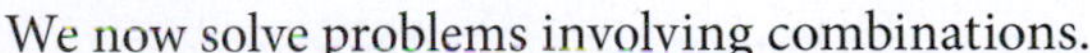

We now solve problems involving combinations.

$3,950,000

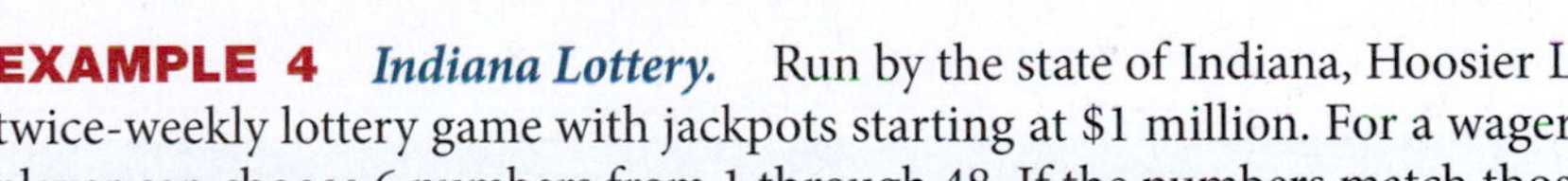

EXAMPLE 4 ***Indiana Lottery.*** Run by the state of Indiana, Hoosier Lotto is a twice-weekly lottery game with jackpots starting at $1 million. For a wager of $1, a player can choose 6 numbers from 1 through 48. If the numbers match those drawn by the state, the player wins the jackpot. (*Source:* www.hoosierlottery.com)

a) How many 6-number combinations are there?

b) Suppose it takes you 10 min to pick your numbers and buy a game ticket. How many tickets can you buy in 4 days?

c) How many people would you have to hire for 4 days to buy tickets with all the possible combinations and ensure that you win?

Solution

a) No order is implied here. You pick any 6 different numbers from 1 through 48. Thus the number of combinations is

$$ {}_{48}C_6 = \binom{48}{6} = \frac{48!}{6!(48-6)!} = \frac{48!}{6!\,42!} $$

$$ = \frac{48 \cdot 47 \cdot 46 \cdot 45 \cdot 44 \cdot 43 \cdot \cancel{42!}}{6 \cdot 5 \cdot 4 \cdot 3 \cdot 2 \cdot 1 \cdot \cancel{42!}} $$

$$ = \frac{48 \cdot 47 \cdot 46 \cdot 45 \cdot 44 \cdot 43}{6 \cdot 5 \cdot 4 \cdot 3 \cdot 2 \cdot 1} = 12{,}271{,}512. $$

b) First, we find the number of minutes in 4 days:

$$ 4 \text{ days} = 4\ \cancel{\text{days}} \cdot \frac{24\ \cancel{\text{hr}}}{1\ \cancel{\text{day}}} \cdot \frac{60 \text{ min}}{1\ \cancel{\text{hr}}} = 5760 \text{ min}. $$

Thus you could buy 5760/10, or 576 tickets in 4 days.

c) You would need to hire 12,271,512/576, or about 21,305 people, to buy tickets with all the possible combinations and ensure a win. (This presumes lottery tickets can be bought 24 hours a day.)

Now Try Exercise 23.

EXAMPLE 5 How many committees can be formed from a group of 5 governors and 7 senators if each committee consists of 3 governors and 4 senators?

Solution The 3 governors can be selected in ${}_5C_3$ ways and the 4 senators can be selected in ${}_7C_4$ ways. If we use the fundamental counting principle, it follows that the number of possible committees is

$$ {}_5C_3 \cdot {}_7C_4 = \frac{5!}{3!\,2!} \cdot \frac{7!}{4!\,3!} $$

$$ = \frac{5 \cdot 4 \cdot 3!}{3! \cdot 2 \cdot 1} \cdot \frac{7 \cdot 6 \cdot 5 \cdot 4!}{4! \cdot 3 \cdot 2 \cdot 1} $$

$$ = \frac{5 \cdot \cancel{2} \cdot 2 \cdot \cancel{3!}}{\cancel{3!} \cdot \cancel{2} \cdot 1} \cdot \frac{7 \cdot \cancel{3} \cdot \cancel{2} \cdot 5 \cdot \cancel{4!}}{\cancel{4!} \cdot \cancel{3} \cdot \cancel{2} \cdot 1} $$

$$ = 10 \cdot 35 $$

$$ = 350. $$

Now Try Exercise 27.

CONNECTING THE CONCEPTS

Permutations and Combinations

PERMUTATIONS	COMBINATIONS
Permutations involve order and arrangements of objects.	Combinations do not involve order or arrangements of objects.
Given 5 books, we can arrange 3 of them on a shelf in ${}_5P_3$, or 60 ways.	Given 5 books, we can select 3 of them in ${}_5C_3$, or 10 ways.
Placing the books in different orders produces different arrangements.	The order in which the books are chosen does not matter.

8.6 Exercise Set

Evaluate.

1. ${}_{13}C_2$

2. ${}_{9}C_6$

3. $\binom{13}{11}$

4. $\binom{9}{3}$

5. $\binom{7}{1}$

6. $\binom{8}{8}$

7. $\frac{{}_5P_3}{3!}$

8. $\frac{{}_{10}P_5}{5!}$

9. $\binom{6}{0}$

10. $\binom{6}{1}$

11. $\binom{6}{2}$

12. $\binom{6}{3}$

13. $\binom{7}{0} + \binom{7}{1} + \binom{7}{2} + \binom{7}{3} + \binom{7}{4} + \binom{7}{5} + \binom{7}{6} + \binom{7}{7}$

14. $\binom{6}{0} + \binom{6}{1} + \binom{6}{2} + \binom{6}{3} + \binom{6}{4} + \binom{6}{5} + \binom{6}{6}$

15. ${}_{52}C_4$

16. ${}_{52}C_5$

17. $\binom{27}{11}$

18. $\binom{37}{8}$

19. $\binom{n}{1}$

20. $\binom{n}{3}$

21. $\binom{m}{m}$

22. $\binom{t}{4}$

In each of the following exercises, give an expression for the answer using permutation notation, combination notation, factorial notation, or other operations. Then evaluate.

23. *Key Club Officers.* There are 36 students in a high school Key Club, a service organization for teens. How many sets of 4 officers can be selected?

24. *League Games.* How many games can be played in a 9-team sports league if each team plays all other teams once? twice?

25. *Test Options.* On a test, a student is to select 10 out of 13 questions. In how many ways can this be done?

26. *Senate Committees.* Suppose that the Senate of the United States consists of 58 Democrats and 42 Republicans. How many committees can be formed consisting of 6 Democrats and 4 Republicans?

27. *Test Options.* Of the first 10 questions on a test, a student must answer 7. Of the second 5 questions, the student must answer 3. In how many ways can this be done?

28. *Lines and Triangles from Points.* How many lines are determined by 8 points, no 3 of which are collinear? How many triangles are determined by the same points?

29. *Poker Hands.* How many 5-card poker hands are possible with a 52-card deck?

30. *Bridge Hands.* How many 13-card bridge hands are possible with a 52-card deck?

31. *Baskin-Robbins Ice Cream.* Burt Baskin and Irv Robbins began making ice cream in 1945. Initially they developed 31 flavors—one for each day of the month. (*Source:* Baskin-Robbins)
    - **a)** How many 2-dip cones are possible using the 31 original flavors if order of flavors is to be considered and no flavor is repeated?
    - **b)** How many 2-dip cones are possible if order is to be considered and a flavor can be repeated?
    - **c)** How many 2-dip cones are possible if order is not considered and no flavor is repeated?

32. *Powerball®.* Powerball® is a biweekly lottery game in which 5 white balls are drawn from a drum of 59 balls numbered 1–59 and 1 red ball is drawn from a drum of 35 balls numbered 1–35. To win the jackpot, a player must select numbers to match in any order the 5 white balls and the 1 red ball. (*Source:* www.powerball.com) How many 6-number combinations are there?

Skill Maintenance

Solve.

33. $3x - 7 = 5x + 10$ [1.5]

34. $2x^2 - x = 3$ [3.2]

35. $x^2 + 5x + 1 = 0$ [3.2]

36. $x^3 + 3x^2 - 10x = 24$ [4.4]

Synthesis

37. *Flush.* A flush in poker consists of a 5-card hand with all cards of the same suit. How many 5-card hands (flushes) are there that consist of all diamonds?

38. *Full House.* A full house in poker consists of three of a kind and a pair (two of a kind). How many full houses are there that consist of 3 aces and 2 queens? (See Section 8.8 for a description of a 52-card deck.)

39. *League Games.* How many games are played in a league with n teams if each team plays each other team once? twice?

40. There are n points on a circle. How many quadrilaterals can be inscribed with these points as vertices?

Solve for n.

41. $\binom{n}{n-2} = 6$

42. $\binom{n+1}{3} = 2 \cdot \binom{n}{2}$

43. $\binom{n+2}{4} = 6 \cdot \binom{n}{2}$

44. $\binom{n}{3} = 2 \cdot \binom{n-1}{2}$

45. How many line segments are determined by the n vertices of an n-gon? Of these, how many are diagonals? Use mathematical induction to prove the result for the diagonals.

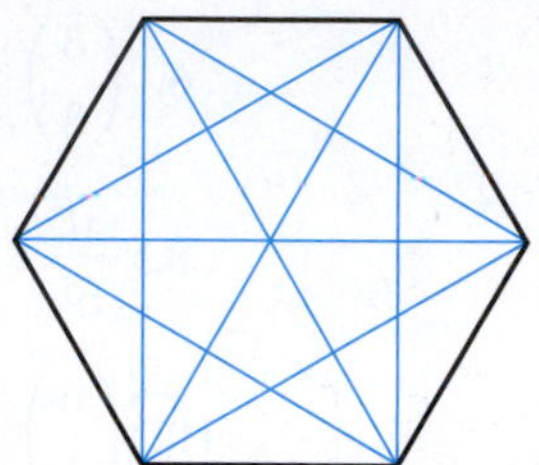

46. Prove that

$$\binom{n}{k-1} + \binom{n}{k} = \binom{n+1}{k}$$

for any natural numbers n and k, $k \leq n$.

8.7 The Binomial Theorem

- Expand a power of a binomial using Pascal's triangle or factorial notation.
- Find a specific term of a binomial expansion.
- Find the total number of subsets of a set of n objects.

In this section, we consider ways of expanding a binomial $(a + b)^n$.

Binomial Expansion Using Pascal's Triangle

Consider the following expanded powers of $(a + b)^n$, where $a + b$ is any binomial and n is a whole number. Look for patterns.

$$\begin{aligned}
(a + b)^0 &= 1 \\
(a + b)^1 &= a + b \\
(a + b)^2 &= a^2 + 2ab + b^2 \\
(a + b)^3 &= a^3 + 3a^2b + 3ab^2 + b^3 \\
(a + b)^4 &= a^4 + 4a^3b + 6a^2b^2 + 4ab^3 + b^4 \\
(a + b)^5 &= a^5 + 5a^4b + 10a^3b^2 + 10a^2b^3 + 5ab^4 + b^5
\end{aligned}$$

Each expansion is a polynomial. There are some patterns to be noted.

1. There is one more term than the power of the exponent, n. That is, there are $n + 1$ terms in the expansion of $(a + b)^n$.
2. In each term, the sum of the exponents is n, the power to which the binomial is raised.
3. The exponents of a start with n, the power of the binomial, and decrease to 0. The last term has no factor of a. The first term has no factor of b, so powers of b start with 0 and increase to n.
4. The coefficients start at 1 and increase through certain values about "half"-way and then decrease through these same values back to 1.

Let's explore the coefficients further. Suppose that we want to find an expansion of $(a + b)^6$. The patterns we just noted indicate that there are 7 terms in the expansion:

$$a^6 + c_1a^5b + c_2a^4b^2 + c_3a^3b^3 + c_4a^2b^4 + c_5ab^5 + b^6.$$

How can we determine the value of each coefficient, c_i? We can do so in two ways. The first method involves writing the coefficients in a triangular array, as follows. This is known as **Pascal's triangle**:

$$\begin{array}{lccccccccccc}
(a+b)^0: & & & & & & 1 & & & & & \\
(a+b)^1: & & & & & 1 & & 1 & & & & \\
(a+b)^2: & & & & 1 & & 2 & & 1 & & & \\
(a+b)^3: & & & 1 & & 3 & & 3 & & 1 & & \\
(a+b)^4: & & 1 & & 4 & & 6 & & 4 & & 1 & \\
(a+b)^5: & 1 & & 5 & & 10 & & 10 & & 5 & & 1
\end{array}$$

There are many patterns in the triangle. Find as many as you can.

Perhaps you discovered a way to write the next row of numbers, given the numbers in the row above it. There are always 1's on the outside. Each remaining number is the sum of the two numbers above it. Let's try to find an expansion for $(a + b)^6$ by adding another row using the patterns we have discovered:

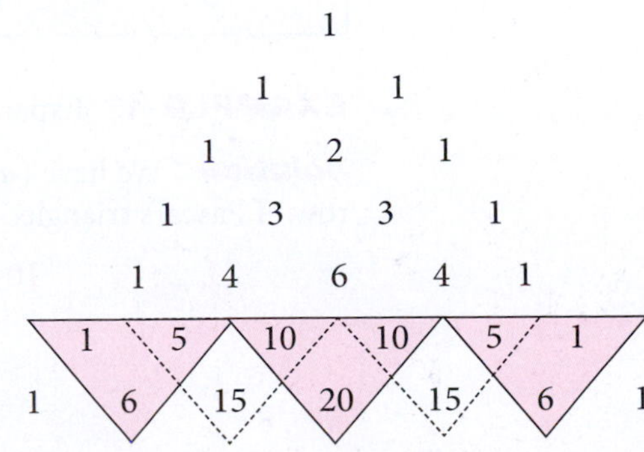

We see that in the last row

the 1st and last numbers are **1**;
the 2nd number is 1 + 5, or **6**;
the 3rd number is 5 + 10, or **15**;
the 4th number is 10 + 10, or **20**;
the 5th number is 10 + 5, or **15**; and
the 6th number is 5 + 1, or **6**.

Thus the expansion for $(a + b)^6$ is

$$(a + b)^6 = 1a^6 + 6a^5b + 15a^4b^2 + 20a^3b^3 + 15a^2b^4 + 6ab^5 + 1b^6.$$

To find an expansion for $(a + b)^8$, we complete two more rows of Pascal's triangle:

```
                1
              1   1
            1   2   1
          1   3   3   1
        1   4   6   4   1
      1   5  10  10   5   1
    1   6  15  20  15   6   1
  1   7  21  35  35  21   7   1
1   8  28  56  70  56  28   8   1
```

Thus the expansion of $(a + b)^8$ is

$$(a + b)^8 = a^8 + 8a^7b + 28a^6b^2 + 56a^5b^3 + 70a^4b^4 + 56a^3b^5 + 28a^2b^6 + 8ab^7 + b^8.$$

We can generalize our results as follows.

THE BINOMIAL THEOREM USING PASCAL'S TRIANGLE

For any binomial $a + b$ and any natural number n,

$$(a + b)^n = c_0a^nb^0 + c_1a^{n-1}b^1 + c_2a^{n-2}b^2 + \cdots + c_{n-1}a^1b^{n-1} + c_na^0b^n,$$

where the numbers $c_0, c_1, c_2, \ldots, c_{n-1}, c_n$ are from the $(n + 1)$st row of Pascal's triangle.

EXAMPLE 1 Expand: $(u - v)^5$.

Solution We have $(a + b)^n$, where $a = u$, $b = -v$, and $n = 5$. We use the 6th row of Pascal's triangle:

1 5 10 10 5 1

Then we have

$$\begin{aligned}(u-v)^5 &= [u+(-v)]^5\\ &= 1(u)^5 + 5(u)^4(-v)^1 + 10(u)^3(-v)^2 + 10(u)^2(-v)^3\\ &\quad + 5(u)(-v)^4 + 1(-v)^5\\ &= u^5 - 5u^4v + 10u^3v^2 - 10u^2v^3 + 5uv^4 - v^5.\end{aligned}$$

Note that the signs of the terms alternate between + and −. When the power of $-v$ is odd, the sign is −.

Now Try Exercise 5.

EXAMPLE 2 Expand: $\left(2t + \frac{3}{t}\right)^4$.

Solution We have $(a+b)^n$, where $a = 2t$, $b = 3/t$, and $n = 4$. We use the 5th row of Pascal's triangle:

1 4 6 4 1

Then we have

$$\begin{aligned}\left(2t+\frac{3}{t}\right)^4 &= 1(2t)^4 + 4(2t)^3\left(\frac{3}{t}\right)^1 + 6(2t)^2\left(\frac{3}{t}\right)^2 + 4(2t)^1\left(\frac{3}{t}\right)^3 + 1\left(\frac{3}{t}\right)^4\\ &= 1(16t^4) + 4(8t^3)\left(\frac{3}{t}\right) + 6(4t^2)\left(\frac{9}{t^2}\right) + 4(2t)\left(\frac{27}{t^3}\right) + 1\left(\frac{81}{t^4}\right)\\ &= 16t^4 + 96t^2 + 216 + 216t^{-2} + 81t^{-4}.\end{aligned}$$

Now Try Exercise 9.

Binomial Expansion Using Factorial Notation

Suppose that we want to find the expansion of $(a+b)^{11}$. The disadvantage in using Pascal's triangle is that we must compute all the preceding rows of the triangle to obtain the row needed for the expansion. The following method avoids this. It also enables us to find a specific term—say, the 8th term—without computing all the other terms of the expansion. This method is useful in such courses as finite mathematics, calculus, and statistics, and it uses the *binomial coefficient notation* $\binom{n}{k}$ developed in Section 8.6.

We can restate the binomial theorem as follows.

THE BINOMIAL THEOREM USING FACTORIAL NOTATION

For any binomial $a + b$ and any natural number n,

$$\begin{aligned}(a+b)^n &= \binom{n}{0}a^nb^0 + \binom{n}{1}a^{n-1}b^1 + \binom{n}{2}a^{n-2}b^2 + \cdots\\ &\quad + \binom{n}{n-1}a^1b^{n-1} + \binom{n}{n}a^0b^n\\ &= \sum_{k=0}^{n}\binom{n}{k}a^{n-k}b^k.\end{aligned}$$

The binomial theorem can be proved by mathematical induction. (See Exercise 57.) This form shows why $\binom{n}{k}$ is called a *binomial coefficient*.

EXAMPLE 3 Expand: $(x^2 - 2y)^5$.

Solution We have $(a + b)^n$, where $a = x^2$, $b = -2y$, and $n = 5$. Then using the binomial theorem, we have

$$
\begin{aligned}
(x^2 - 2y)^5 &= \binom{5}{0}(x^2)^5 + \binom{5}{1}(x^2)^4(-2y) + \binom{5}{2}(x^2)^3(-2y)^2 \\
&\quad + \binom{5}{3}(x^2)^2(-2y)^3 + \binom{5}{4}x^2(-2y)^4 + \binom{5}{5}(-2y)^5 \\
&= \frac{5!}{0!\,5!}x^{10} + \frac{5!}{1!\,4!}x^8(-2y) + \frac{5!}{2!\,3!}x^6(4y^2) + \frac{5!}{3!\,2!}x^4(-8y^3) \\
&\quad + \frac{5!}{4!\,1!}x^2(16y^4) + \frac{5!}{5!\,0!}(-32y^5) \\
&= 1 \cdot x^{10} + 5x^8(-2y) + 10x^6(4y^2) + 10x^4(-8y^3) \\
&\quad + 5x^2(16y^4) + 1 \cdot (-32y^5) \\
&= x^{10} - 10x^8y + 40x^6y^2 - 80x^4y^3 + 80x^2y^4 - 32y^5.
\end{aligned}
$$

Now Try Exercise 11.

EXAMPLE 4 Expand: $\left(\frac{2}{x} + 3\sqrt{x}\right)^4$.

Solution We have $(a + b)^n$, where $a = 2/x$, $b = 3\sqrt{x}$, and $n = 4$. Then using the binomial theorem, we have

$$
\begin{aligned}
\left(\frac{2}{x} + 3\sqrt{x}\right)^4 &= \binom{4}{0}\left(\frac{2}{x}\right)^4 + \binom{4}{1}\left(\frac{2}{x}\right)^3(3\sqrt{x}) + \binom{4}{2}\left(\frac{2}{x}\right)^2(3\sqrt{x})^2 \\
&\quad + \binom{4}{3}\left(\frac{2}{x}\right)(3\sqrt{x})^3 + \binom{4}{4}(3\sqrt{x})^4 \\
&= \frac{4!}{0!\,4!}\left(\frac{16}{x^4}\right) + \frac{4!}{1!\,3!}\left(\frac{8}{x^3}\right)(3x^{1/2}) \\
&\quad + \frac{4!}{2!\,2!}\left(\frac{4}{x^2}\right)(9x) + \frac{4!}{3!\,1!}\left(\frac{2}{x}\right)(27x^{3/2}) \\
&\quad + \frac{4!}{4!\,0!}(81x^2) \\
&= \frac{16}{x^4} + \frac{96}{x^{5/2}} + \frac{216}{x} + 216x^{1/2} + 81x^2.
\end{aligned}
$$

Now Try Exercise 13.

Finding a Specific Term

Suppose that we want to determine only a particular term of an expansion. The method we have developed will allow us to find such a term without computing all the rows of Pascal's triangle or all the preceding coefficients.

Note that in the binomial theorem, $\binom{n}{0}a^nb^0$ gives us the 1st term, $\binom{n}{1}a^{n-1}b^1$ gives us the 2nd term, $\binom{n}{2}a^{n-2}b^2$ gives us the 3rd term, and so on. This can be generalized as follows.

FINDING THE $(k + 1)$ST TERM

The $(k + 1)$st term of $(a + b)^n$ is $\binom{n}{k}a^{n-k}b^k$.

EXAMPLE 5 Find the 5th term in the expansion of $(2x - 5y)^6$.

Solution First, we note that $5 = 4 + 1$. Thus, $k = 4$, $a = 2x$, $b = -5y$, and $n = 6$. Then the 5th term of the expansion is

$$\binom{6}{4}(2x)^{6-4}(-5y)^4, \quad \text{or} \quad \frac{6!}{4!\,2!}(2x)^2(-5y)^4, \quad \text{or} \quad 37{,}500x^2y^4.$$

Now Try Exercise 21.

EXAMPLE 6 Find the 8th term in the expansion of $(3x - 2)^{10}$.

Solution First, we note that $8 = 7 + 1$. Thus, $k = 7$, $a = 3x$, $b = -2$, and $n = 10$. Then the 8th term of the expansion is

$$\binom{10}{7}(3x)^{10-7}(-2)^7, \quad \text{or} \quad \frac{10!}{7!\,3!}(3x)^3(-2)^7, \quad \text{or} \quad -414{,}720x^3.$$

Now Try Exercise 25.

Total Number of Subsets

Suppose that a set has n objects. The number of subsets containing k elements is $\binom{n}{k}$ by a result of Section 8.6. The total number of subsets of a set is the number of subsets with 0 elements, plus the number of subsets with 1 element, plus the number of subsets with 2 elements, and so on. The total number of subsets of a set with n elements is

$$\binom{n}{0} + \binom{n}{1} + \binom{n}{2} + \cdots + \binom{n}{n}.$$

Now consider the expansion of $(1 + 1)^n$:

$$\begin{aligned}(1 + 1)^n &= \binom{n}{0}\cdot 1^n + \binom{n}{1}\cdot 1^{n-1}\cdot 1^1 + \binom{n}{2}\cdot 1^{n-2}\cdot 1^2 \\ &\quad + \cdots + \binom{n}{n}\cdot 1^n \\ &= \binom{n}{0} + \binom{n}{1} + \binom{n}{2} + \cdots + \binom{n}{n}.\end{aligned}$$

Thus the total number of subsets is $(1 + 1)^n$, or 2^n. We have proved the following.

TOTAL NUMBER OF SUBSETS

The total number of subsets of a set with n elements is 2^n.

EXAMPLE 7 The set $\{A, B, C, D, E\}$ has how many subsets?

Solution The set has 5 elements, so the number of subsets is 2^5, or 32.

Now Try Exercise 31.

EXAMPLE 8 Wendy's, a national restaurant chain, offers the following toppings for its hamburgers:

$$\{\text{catsup, mustard, mayonnaise, tomato, lettuce, onions, pickle}\}.$$

In how many different ways can Wendy's serve hamburgers, excluding size of hamburger or number of patties?

Solution The toppings on each hamburger are the elements of a subset of the set of all possible toppings, the empty set being a plain hamburger. The total number of possible hamburgers is

$$\binom{7}{0} + \binom{7}{1} + \binom{7}{2} + \cdots + \binom{7}{7} = 2^7 = 128.$$

Thus Wendy's serves hamburgers in 128 different ways.

Now Try Exercise 33.

8.7 Exercise Set

Expand.

1. $(x + 5)^4$
2. $(x - 1)^4$
3. $(x - 3)^5$
4. $(x + 2)^9$
5. $(x - y)^5$
6. $(x + y)^8$
7. $(5x + 4y)^6$
8. $(2x - 3y)^5$
9. $\left(2t + \dfrac{1}{t}\right)^7$
10. $\left(3y - \dfrac{1}{y}\right)^4$
11. $(x^2 - 1)^5$
12. $(1 + 2q^3)^8$
13. $(\sqrt{5} + t)^6$
14. $(x - \sqrt{2})^6$
15. $\left(a - \dfrac{2}{a}\right)^9$
16. $(1 + 3)^n$
17. $(\sqrt{2} + 1)^6 - (\sqrt{2} - 1)^6$
18. $(1 - \sqrt{2})^4 + (1 + \sqrt{2})^4$
19. $(x^{-2} + x^2)^4$
20. $\left(\dfrac{1}{\sqrt{x}} - \sqrt{x}\right)^6$

Find the indicated term of the binomial expansion.

21. 3rd; $(a + b)^7$
22. 6th; $(x + y)^8$
23. 6th; $(x - y)^{10}$
24. 5th; $(p - 2q)^9$
25. 12th; $(a - 2)^{14}$
26. 11th; $(x - 3)^{12}$
27. 5th; $(2x^3 - \sqrt{y})^8$
28. 4th; $\left(\dfrac{1}{b^2} + \dfrac{b}{3}\right)^7$
29. Middle; $(2u - 3v^2)^{10}$
30. Middle two; $(\sqrt{x} + \sqrt{3})^5$

Determine the number of subsets of each of the following.

31. A set of 7 elements
32. A set of 6 members
33. The set of letters of the Greek alphabet, which contains 24 letters
34. The set of letters of the English alphabet, which contains 26 letters
35. What is the degree of $(x^5 + 3)^4$?
36. What is the degree of $(2 - 5x^3)^7$?

Expand each of the following, where $i^2 = -1$.

37. $(3 + i)^5$
38. $(1 + i)^6$
39. $(\sqrt{2} - i)^4$
40. $\left(\dfrac{\sqrt{3}}{2} - \dfrac{1}{2}i\right)^{11}$
41. Find a formula for $(a - b)^n$. Use sigma notation.

42. Expand and simplify:
$$\frac{(x+h)^{13} - x^{13}}{h}.$$

43. Expand and simplify:
$$\frac{(x+h)^{n} - x^{n}}{h}.$$
Use sigma notation.

▶ Skill Maintenance

Given that $f(x) = x^2 + 1$ *and* $g(x) = 2x - 3$, *find each of the following.*

44. $(f + g)(x)$ [2.2]

45. $(fg)(x)$ [2.2]

46. $(f \circ g)(x)$ [2.3]

47. $(g \circ f)(x)$ [2.3]

▶ Synthesis

Solve for x.

48. $\sum_{k=0}^{8} \binom{8}{k} x^{8-k} 3^k = 0$

49. $\sum_{k=0}^{4} \binom{4}{k} (-1)^k x^{4-k} 6^k = 81$

50. Find the ratio of the 4th term of
$$\left(p^2 - \frac{1}{2}p\sqrt[3]{q}\right)^5$$
to the 3rd term.

51. Find the term of
$$\left(\sqrt[3]{x} - \frac{1}{\sqrt{x}}\right)^7$$
containing $1/x^{1/6}$.

52. *Money Combinations.* A money clip contains one each of the following bills: \$1, \$2, \$5, \$10, \$20, \$50, and \$100. How many different sums of money can be formed using the bills?

Find the sum.

53. ${}_{100}C_0 + {}_{100}C_1 + \cdots + {}_{100}C_{100}$

54. ${}_{n}C_0 + {}_{n}C_1 + \cdots + {}_{n}C_n$

Simplify.

55. $\sum_{k=0}^{23} \binom{23}{k} (\log_a x)^{23-k} (\log_a t)^k$

56. $\sum_{k=0}^{15} \binom{15}{k} i^{30-2k}$

57. Use mathematical induction and the property
$$\binom{n}{r-1} + \binom{n}{r} = \binom{n+1}{r}$$
to prove the binomial theorem.

8.8 Probability

▶ Compute the probability of a simple event.

When a coin is tossed, we can reason that the chance, or the likelihood, that it will fall heads is 1 out of 2—that is, the **probability** that it will fall heads is $\frac{1}{2}$. Of course, this does not mean that if a coin is tossed 10 times it will necessarily fall heads 5 times. If the coin is a "fair coin" and it is tossed a great many times, however, it will fall heads very nearly half of the time. Here we give an introduction to two kinds of probability, **experimental** and **theoretical**.

Experimental Probability and Theoretical Probability

If we toss a coin a great number of times—say, 1000—and count the number of times it falls heads, we can determine the probability that it will fall heads. If it falls heads 503 times, we would calculate the probability of its falling heads to be

$$\frac{503}{1000}, \quad \text{or} \quad 0.503.$$

This is an **experimental** determination of probability. Such a determination of probability is discovered by the observation and study of data and is quite common and very useful. Here, for example, are some probabilities that have been determined *experimentally*:

1. 60% of all college freshmen entering four-year colleges graduate in 6 years (*Source:* www.satprepct.com, College Planning Partnership's Blog, February 24, 2011, Sam Rosensohn).
2. The probability that a woman will be diagnosed with breast cancer in her lifetime is $\frac{1}{8}$ (*Source:* National Cancer Institute).
3. Anyone who reaches the age of 65 has a 0.4 probability of entering a nursing home during the remaining years of life (*Source:* "Facing the Future," Russ Banham, *Wall Street Journal*).

If we consider a coin and reason that it is just as likely to fall heads as to fall tails, we would calculate the probability that it will fall heads to be $\frac{1}{2}$. This is a **theoretical** determination of probability. Here are some other probabilities that have been determined *theoretically*, using mathematics:

1. If there are 30 people in a room, the probability that two of them have the same birthday (excluding year) is 0.706.
2. While on a trip, you meet someone and, after a period of conversation, discover that you have a common acquaintance. The typical reaction, "It's a small world!", is actually not appropriate, because the probability of such an occurrence is quite high—just over 22%.

In summary, experimental probabilities are determined by making observations and gathering data. Theoretical probabilities are determined by reasoning mathematically. Examples of experimental and theoretical probability like those above, especially those we do not expect, lead us to see the value of a study of probability. You might ask, "What is the *true* probability?" In fact, there is none. Experimentally, we can determine probabilities within certain limits. These may or may not agree with the probabilities that we obtain theoretically. There are situations in which it is much easier to determine one of these types of probabilities than the other. For example, it would be quite difficult to arrive at the probability of catching a cold using theoretical probability.

Computing Experimental Probabilities

We first consider experimental determination of probability. The basic principle we use in computing such probabilities is as follows.

PRINCIPLE *P* (EXPERIMENTAL)

Given an experiment in which n observations are made, if a situation, or event, E occurs m times out of n observations, then we say that the **experimental probability** of the event, $P(E)$, is given by

$$P(E) = \frac{m}{n}.$$

EXAMPLE 1 ***Television Ratings.*** There are an estimated 114,200,000 households in the United States that have at least one television. Each week, viewing information is collected and reported. One week, 28,510,000 households tuned in to the 2013 Grammy Awards ceremony on CBS, and 14,204,000 households tuned in to the action series "NCIS" on CBS (*Source:* Nielsen Media Research). What is the probability that a television household tuned in to the Grammy Awards ceremony during the given week? to "NCIS"?

Solution The probability that a television household was tuned in to the Grammy Awards ceremony is P, where

$$P = \frac{28{,}510{,}000}{114{,}200{,}000} \approx 0.2496 \approx 24.96\%.$$

The probability that a television household was tuned in to "NCIS" is P, where

$$P = \frac{14{,}204{,}000}{114{,}200{,}000} \approx 0.1244 \approx 12.44\%.$$

Now Try Exercise 1.

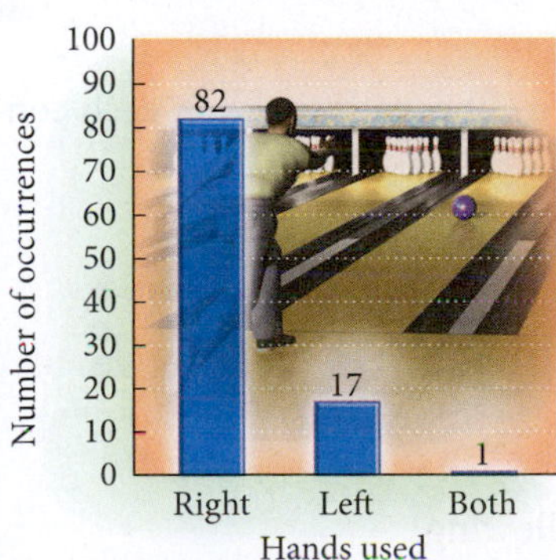

EXAMPLE 2 ***Sociological Survey.*** The authors of this text conducted an experimental survey to determine the number of people who are left-handed, right-handed, or both. The results are shown in the graph at left.

a) Determine the probability that a person is right-handed.

b) Determine the probability that a person is left-handed.

c) Determine the probability that a person is ambidextrous (uses both hands with equal ability).

d) There are 120 bowlers in most tournaments held by the Professional Bowlers Association. On the basis of the data in this experiment, how many of the bowlers would you expect to be left-handed?

Solution

a) The number of people who are right-handed is 82, the number who are left-handed is 17, and the number who are ambidextrous is 1. The total number of observations is $82 + 17 + 1$, or 100. Thus the probability that a person is right-handed is P, where

$$P = \frac{82}{100}, \quad \text{or} \quad 0.82, \quad \text{or} \quad 82\%.$$

b) The probability that a person is left-handed is P, where

$$P = \frac{17}{100}, \quad \text{or} \quad 0.17, \quad \text{or} \quad 17\%.$$

c) The probability that a person is ambidextrous is P, where

$$P = \frac{1}{100}, \quad \text{or} \quad 0.01, \quad \text{or} \quad 1\%.$$

d) There are 120 bowlers, and from part (b) we can expect 17% to be left-handed. Since

$$17\% \text{ of } 120 = 0.17 \cdot 120 = 20.4,$$

we can expect that about 20 of the bowlers will be left-handed.

Now Try Exercise 3.

Theoretical Probability

Suppose that we perform an experiment such as flipping a coin, throwing a dart, drawing a card from a deck, or checking an item off an assembly line for quality. Each possible result of such an experiment is called an **outcome**. The set of all possible outcomes is called the **sample space**. An **event** is a set of outcomes, that is, a subset of the sample space.

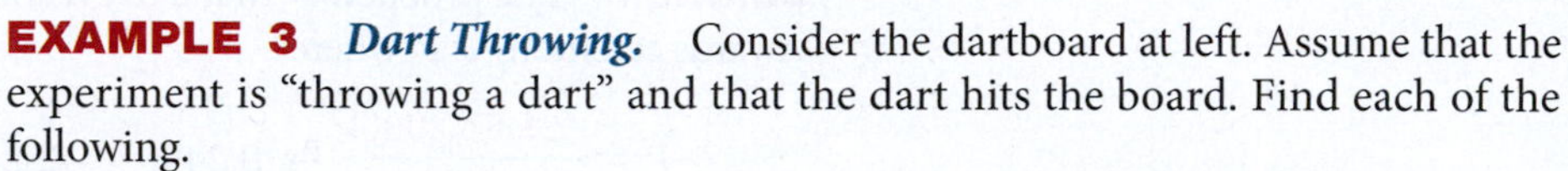

EXAMPLE 3 ***Dart Throwing.*** Consider the dartboard at left. Assume that the experiment is "throwing a dart" and that the dart hits the board. Find each of the following.

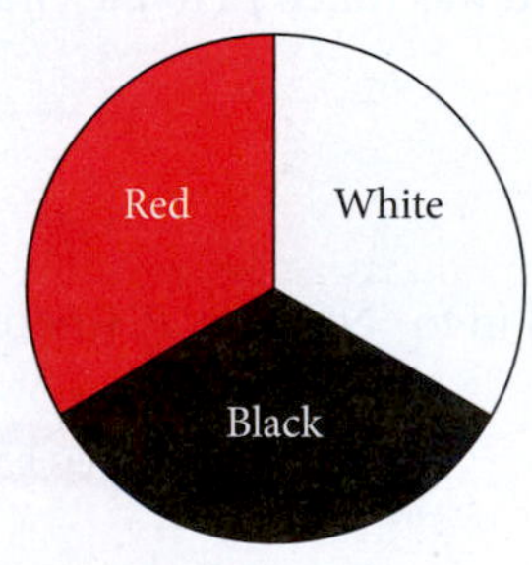

a) The outcomes

b) The sample space

Solution

a) The outcomes are *hitting black* (B), *hitting red* (R), and *hitting white* (W).

b) The sample space is {*hitting black*, *hitting red*, *hitting white*}, which can be stated simply as {B, R, W}.

EXAMPLE 4 ***Die Rolling.*** A die (pl., dice) is a cube, with six faces, each containing a number of dots from 1 to 6 on each side.

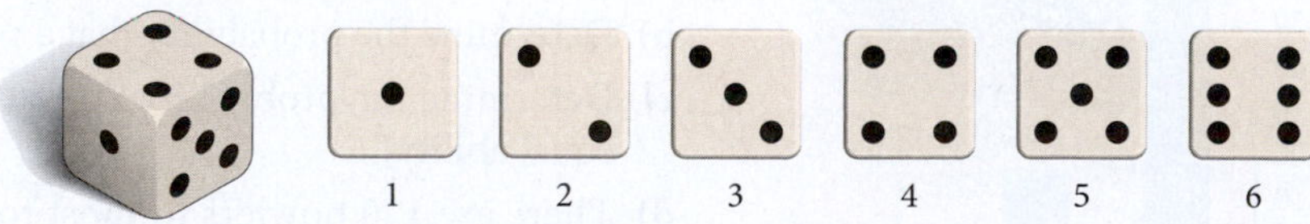

Suppose that a die is rolled. Find each of the following.

a) The outcomes

b) The sample space

Solution

a) The outcomes are 1, 2, 3, 4, 5, 6.

b) The sample space is $\{1, 2, 3, 4, 5, 6\}$.

We denote the probability that an event E occurs as $P(E)$. For example, "a coin falling heads" may be denoted H. Then $P(H)$ represents the probability of the coin falling heads. When all the outcomes of an experiment have the same probability of occurring, we say that they are *equally likely.* To see the distinction between events that are equally likely and those that are not, consider the dartboards shown below.

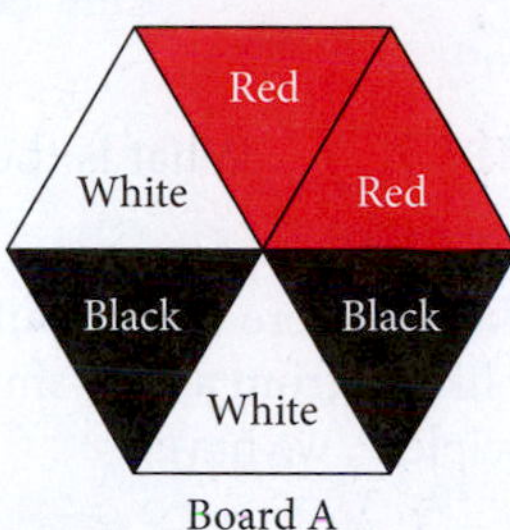

Board A

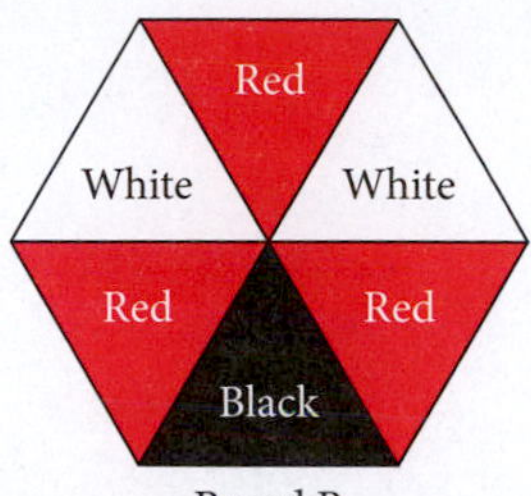

Board B

For board A, the events *hitting black*, *hitting red*, and *hitting white* are equally likely, because the black, red, and white areas are the same. For board B, however, the areas are not the same so these events are not equally likely.

> **PRINCIPLE *P* (THEORETICAL)**
>
> If an event E can occur m ways out of n possible equally likely outcomes of a sample space S, then the **theoretical probability** of the event, $P(E)$, is given by
>
> $$P(E) = \frac{m}{n}.$$

EXAMPLE 5 Suppose that we select, without looking, one marble from a bag containing 3 red marbles and 4 green marbles. What is the probability of selecting a red marble?

Solution There are 7 equally likely ways of selecting any marble, and since the number of ways of getting a red marble is 3, we have

$$P(\text{selecting a red marble}) = \frac{3}{7}.$$

EXAMPLE 6 What is the probability of rolling an even number on a die?

Solution The event is rolling an *even* number. It can occur 3 ways (rolling 2, 4, or 6). The number of equally likely outcomes is 6. By Principle P,

$$P(\text{even}) = \frac{3}{6}, \quad \text{or} \quad \frac{1}{2}.$$

Now Try Exercise 7.

We will use a number of examples related to a standard bridge deck of 52 cards. Such a deck is made up as shown in the following figure.

EXAMPLE 7 What is the probability of drawing an ace from a well-shuffled deck of cards?

Solution There are 52 outcomes (the number of cards in the deck), they are equally likely (from a well-shuffled deck), and there are 4 ways to obtain an ace, so by Principle P, we have

$$P(\text{drawing an ace}) = \frac{4}{52}, \quad \text{or} \quad \frac{1}{13}.$$

Now Try Exercise 9(a).

The following are some results that follow from Principle P.

PROBABILITY PROPERTIES

a) If an event E cannot occur, then $P(E) = 0$.
b) If an event E is certain to occur, then $P(E) = 1$.
c) The probability that an event E will occur is a number from 0 to 1: $0 \leq P(E) \leq 1$.

For example, in coin tossing, the event that a coin will land on its edge has probability 0. The event that a coin falls either heads or tails has probability 1.

In the following examples, we use the combinatorics that we studied in Sections 8.5 and 8.6 to calculate theoretical probabilities.

EXAMPLE 8 Suppose that 2 cards are drawn from a well-shuffled deck of 52 cards. What is the probability that both of them are spades?

Solution The number of ways n of drawing 2 cards from a well-shuffled deck of 52 cards is ${}_{52}C_2$. Since 13 of the 52 cards are spades, the number of ways m of drawing 2 spades is ${}_{13}C_2$. Thus,

$$P(\text{drawing 2 spades}) = \frac{m}{n} = \frac{{}_{13}C_2}{{}_{52}C_2} = \frac{78}{1326} = \frac{1}{17}.$$

Now Try Exercise 11.

EXAMPLE 9 Suppose that 3 people are selected at random from a group that consists of 6 men and 4 women. What is the probability that 1 man and 2 women are selected?

Technology Connection

We can use the ${}_nC_r$ operation from the MATH PRB menu and the ▶ Frac operation from the MATH MATH menu to compute the probabilities in Examples 8 and 9 on a graphing calculator.

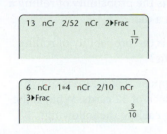

```
13 nCr 2/52 nCr 2▶Frac
                  1/17

6 nCr 1*4 nCr 2/10 nCr
3▶Frac
                  3/10
```

Solution The number of ways of selecting 3 people from a group of 10 is ${}_{10}C_3$. One man can be selected in ${}_6C_1$ ways, and 2 women can be selected in ${}_4C_2$ ways. By the fundamental counting principle, the number of ways of selecting 1 man and 2 women is ${}_6C_1 \cdot {}_4C_2$. Thus the probability that 1 man and 2 women are selected is

$$P = \frac{{}_6C_1 \cdot {}_4C_2}{{}_{10}C_3} = \frac{3}{10}.$$

Now Try Exercise 13.

EXAMPLE 10 ***Rolling Two Dice.*** What is the probability of getting a total of 8 on a roll of a pair of dice?

Solution On each die, there are 6 possible outcomes. The outcomes are paired so there are $6 \cdot 6$, or 36, possible ways in which the two can fall. (Assuming that the dice are different colors—say, one red and one blue—can help in visualizing this.)

6	(1, 6)	(2, 6)	(3, 6)	(4, 6)	(5, 6)	(6, 6)
5	(1, 5)	(2, 5)	(3, 5)	(4, 5)	(5, 5)	(6, 5)
4	(1, 4)	(2, 4)	(3, 4)	(4, 4)	(5, 4)	(6, 4)
3	(1, 3)	(2, 3)	(3, 3)	(4, 3)	(5, 3)	(6, 3)
2	(1, 2)	(2, 2)	(3, 2)	(4, 2)	(5, 2)	(6, 2)
1	(1, 1)	(2, 1)	(3, 1)	(4, 1)	(5, 1)	(6, 1)
	1	2	3	4	5	6

The pairs that total 8 are as shown in the figure above. There are 5 possible ways of getting a total of 8, so the probability is $\frac{5}{36}$.

Now Try Exercise 19.

8.8 Exercise Set

1. *Select a Number.* In a survey conducted by the authors, 100 people were polled and asked to select a number from 1 to 5. The results are shown in the following table.

Number Chosen	1	2	3	4	5
Number Who Chose That Number	18	24	23	23	12

a) What is the probability that the number chosen is 1? 2? 3? 4? 5?

b) What general conclusion might be made from the results of the experiment?

2. *Mason Dots®.* Made by the Tootsie Industries of Chicago, Illinois, Mason Dots® is a gumdrop candy. A box was opened by the authors and was found to contain the following number of gumdrops:

Orange	9
Lemon	8
Strawberry	7
Grape	6
Lime	5
Cherry	4

If we take one gumdrop out of the box, what is the probability of getting lemon? lime? orange? grape? strawberry? licorice?

3. *Marketing via E-mail.* In the second quarter of 2013, the probability that a marketing e-mail would be opened was 28.5% (*Source:* Q2 2013 Email Trends and Benchmarks, Epsilon). A business sent a marketing e-mail to 18,200 subscribers. How many of these e-mails can the business expect will be opened?

4. *Linguistics.* An experiment was conducted by the authors to determine the relative occurrence of various letters of the English alphabet. The front page of a newspaper was considered. In all, there were 9136 letters. The number of occurrences of each letter of the alphabet is listed in the following table.

Letter	Number of Occurrences	Probability
A	853	$853/9136 \approx 9.3\%$
B	136	
C	273	
D	286	
E	1229	
F	173	
G	190	
H	399	
I	539	
J	21	
K	57	
L	417	
M	231	
N	597	
O	705	
P	238	
Q	4	
R	609	
S	745	
T	789	
U	240	
V	113	
W	127	
X	20	
Y	124	
Z	21	$21/9136 \approx 0.2\%$

a) Complete the table of probabilities with the percentage, to the nearest tenth of a percent, of the occurrence of each letter.
b) What is the probability of a vowel occurring?
c) What is the probability of a consonant occurring?

5. *Marbles.* Suppose that we select, without looking, one marble from a bag containing 4 red marbles and 10 green marbles. What is the probability of selecting each of the following?

a) A red marble
b) A green marble
c) A purple marble
d) A red marble or a green marble

6. *Selecting Coins.* Suppose that we select, without looking, one coin from a bag containing 5 pennies, 3 dimes, and 7 quarters. What is the probability of selecting each of the following?

a) A dime
b) A quarter
c) A nickel
d) A penny, a dime, or a quarter

7. *Rolling a Die.* What is the probability of rolling a number less than 4 on a die?

8. *Rolling a Die.* What is the probability of rolling either a 1 or a 6 on a die?

9. *Drawing a Card.* Suppose that a card is drawn from a well-shuffled deck of 52 cards. What is the probability of drawing each of the following?

a) A queen
b) An ace or a 10
c) A heart
d) A black 6

10. *Drawing a Card.* Suppose that a card is drawn from a well-shuffled deck of 52 cards. What is the probability of drawing each of the following?

a) A 7
b) A jack or a king
c) A black ace
d) A red card

11. *Drawing Cards.* Suppose that 3 cards are drawn from a well-shuffled deck of 52 cards. What is the probability that they are all aces?

12. *Drawing Cards.* Suppose that 4 cards are drawn from a well-shuffled deck of 52 cards. What is the probability that they are all red?

13. *Production Unit.* The sales force of a business consists of 10 men and 10 women. A production unit of 4 people is set up at random. What is the probability that 2 men and 2 women are chosen?

14. *Coin Drawing.* A sack contains 7 dimes, 5 nickels, and 10 quarters, and 8 coins are drawn at random. What is the probability of getting 4 dimes, 3 nickels, and 1 quarter?

Five-Card Poker Hands. Suppose that 5 cards are drawn from a deck of 52 cards. What is the probability of drawing each of the following?

15. 3 sevens and 2 kings

16. 5 aces

17. 5 spades

18. 4 aces and 1 five

19. *Tossing Three Coins.* Three coins are flipped. An outcome might be HTH.
 a) Find the sample space.
 What is the probability of getting each of the following?
 b) Exactly one head
 c) At most two tails
 d) At least one head
 e) Exactly two tails

Roulette. An American roulette wheel contains 38 slots numbered 00, 0, 1, 2, 3, . . . , 35, 36. Eighteen of the slots numbered 1–36 are colored red and 18 are colored black. The 00 and 0 slots are considered to be uncolored. The wheel is spun, and a ball is rolled around the rim until it falls into a slot. What is the probability that the ball falls in each of the following?

20. A red slot
21. A black slot
22. The 00 slot
23. The 0 slot
24. Either the 00 or the 0 slot
25. A red slot or a black slot
26. The number 24
27. An odd-numbered slot
28. *Dartboard.* The following figure shows a dartboard. A dart is thrown and hits the board. Find the following probabilities.
 a) P (red)
 b) P (green)
 c) P (blue)
 d) P (yellow)

Skill Maintenance

In each of Exercises 29–36, fill in the blank with the correct term. Some of the given choices will be used more than once. Others will not be used.

range
domain
function
inverse function
composite function
direct variation
inverse variation
factor
solution
zero
y-intercept
one-to-one
rational
permutation
combination
arithmetic sequence
geometric sequence

29. A(n) ______________ of a function is an input for which the output is 0. [1.5]

30. A function is ______________ if different inputs have different outputs. [5.1]

31. A(n) ______________ is a correspondence between a first set, called the ______________, and a second set, called the ______________, such that each member of the ______________ corresponds to exactly one member of the ______________. [1.2]

32. The first coordinate of an x-intercept of a function is a(n) ______________ of the function. [1.5]

33. A selection made from a set without regard to order is a(n) ______________. [8.6]

34. If we have a function $f(x) = k/x$, where k is a positive constant, we have ______________. [2.5]

35. For a polynomial function $f(x)$, if $f(c) = 0$, then $x - c$ is a(n) ______________ of the polynomial. [4.3]

36. We have $\dfrac{a_{n+1}}{a_n} = r$, for any integer $n \geq 1$, in a(n) ______________. [8.3]

Synthesis

Five-Card Poker Hands. Suppose that 5 cards are drawn from a deck of 52 cards. For the following exercises, give both a reasoned expression and an answer.

37. *Two Pairs.* A hand with *two pairs* is a hand like Q-Q-3-3-A.
    a) How many are there?
    b) What is the probability of getting two pairs?

38. *Full House.* A *full house* consists of 3 of a kind and a pair such as Q-Q-Q-4-4.
    a) How many full houses are there?
    b) What is the probability of getting a full house?

39. *Three of a Kind.* A *three-of-a-kind* is a 5-card hand in which exactly 3 of the cards are of the same denomination and the other 2 are not a pair, such as Q-Q-Q-10-7.
    a) How many three-of-a-kind hands are there?
    b) What is the probability of getting three of a kind?

40. *Four of a Kind.* A *four-of-a-kind* is a 5-card hand in which 4 of the cards are of the same denomination, such as J-J-J-J-6, 7-7-7-7-A, or 2-2-2-2-5.
    a) How many four-of-a-kind hands are there?
    b) What is the probability of getting four of a kind?

Chapter 8 Summary and Review

STUDY GUIDE

KEY TERMS AND CONCEPTS	EXAMPLES
SECTION 8.1: SEQUENCES AND SERIES	
An **infinite sequence** is a function having for its domain the set of positive integers $\{1, 2, 3, 4, 5, \ldots\}$. A **finite sequence** is a function having for its domain a set of positive integers $\{1, 2, 3, 4, 5, \ldots, n\}$ for some positive integer n.	The first four terms of the sequence whose general term is given by $a_n = 3n + 2$ are $a_1 = 3 \cdot 1 + 2 = 5,$ $a_2 = 3 \cdot 2 + 2 = 8,$ $a_3 = 3 \cdot 3 + 2 = 11,$ and $a_4 = 3 \cdot 4 + 2 = 14.$
The sum of the terms of an infinite sequence is an **infinite series**. A **partial sum** is the sum of the first n terms. It is also called a **finite series** or the ***n*th partial sum** and is denoted S_n.	For the sequence above, $S_4 = 5 + 8 + 11 + 14$, or 38. We can denote this sum using **sigma notation** as $\sum_{k=1}^{4} (3k + 2).$
A sequence can be defined **recursively** by listing the first term, or the first few terms, and then using a **recursion formula** to determine the remaining terms from the given term.	The first four terms of the recursively defined sequence $a_1 = 3, \; a_{n+1} = (a_n - 1)^2$ are $a_1 = 3,$ $a_2 = (a_1 - 1)^2 = (3 - 1)^2 = 4,$ $a_3 = (a_2 - 1)^2 = (4 - 1)^2 = 9,$ and $a_4 = (a_3 - 1)^2 = (9 - 1)^2 = 64.$

SECTION 8.2: ARITHMETIC SEQUENCES AND SERIES

For an arithmetic sequence:

$a_{n+1} = a_n + d;$ **d is the common difference.**

$a_n = a_1 + (n - 1)d;$ **The nth term**

$S_n = \frac{n}{2}(a_1 + a_n).$ **The sum of the first n terms**

For the arithmetic sequence 5, 8, 11, 14, . . . :

$a_1 = 5;$

$d = 3 \quad (8 - 5 = 3, 11 - 8 = 3, \text{and so on});$

$a_6 = 5 + (6 - 1)3 = 5 + 15 = 20;$

$S_6 = \frac{6}{2}(5 + 20) = 3(25) = 75.$

SECTION 8.3: GEOMETRIC SEQUENCES AND SERIES

For a geometric sequence:

$a_{n+1} = a_n r$ **r is the common ratio.**

$a_n = a_1 r^{n-1};$ **The nth term**

$S_n = \frac{a_1(1 - r^n)}{1 - r};$ **The sum of the first n terms**

$S_\infty = \frac{a_1}{1 - r}, \quad |r| < 1.$ **The limit, or sum, of an infinite geometric series**

For the geometric sequence $12, -6, 3, -\frac{3}{2}, \ldots$:

$a_1 = 12;$

$r = -\frac{1}{2} \quad \left(\frac{-6}{12} = -\frac{1}{2}, \frac{3}{-6} = -\frac{1}{2}, \text{ and so on}\right);$

$a_6 = 12\left(-\frac{1}{2}\right)^{6-1} = 12\left(-\frac{1}{2^5}\right) = -\frac{3}{8};$

$S_6 = \frac{12\left[1 - \left(-\frac{1}{2}\right)^6\right]}{1 - \left(-\frac{1}{2}\right)} = \frac{12\left(1 - \frac{1}{64}\right)}{\frac{3}{2}} = \frac{63}{8};$

$|r| = \left|-\frac{1}{2}\right| = \frac{1}{2} < 1$, so we have

$S_\infty = \frac{12}{1 - \left(-\frac{1}{2}\right)} = \frac{12}{\frac{3}{2}} = 8.$

SECTION 8.4: MATHEMATICAL INDUCTION

The Principle of Mathematical Induction

We can prove an infinite sequence of statements S_n by showing the following.

(1) *Basis step.* S_1 is true.

(2) *Induction step.* For all natural numbers k, $S_k \rightarrow S_{k+1}$.

See Examples 1–3 on pp. 551–553.

SECTION 8.5: COMBINATORICS: PERMUTATIONS

The Fundamental Counting Principle

Given a combined action, or *event*, in which the first action can be performed in n_1 ways, the second action can be performed in n_2 ways, and so on, the total number of ways in which the combined action can be performed is the product

$$n_1 \cdot n_2 \cdot n_3 \cdot \cdots \cdot n_k.$$

The product $n(n - 1)(n - 2) \cdots 3 \cdot 2 \cdot 1$, for any natural number n, can also be written in **factorial notation** as $n!$. For the number 0, $0! = 1$.

(*continued*)

The total number of permutations, or ordered arrangements, of n objects, denoted ${}_nP_n$, is given by $${}_nP_n = n(n-1)(n-2)\cdots 3\cdot 2\cdot 1, \quad \text{or} \quad n!.$$	In how many ways can 7 books be arranged in a straight line? We have $${}_7P_7 = 7! = 7\cdot 6\cdot 5\cdot 4\cdot 3\cdot 2\cdot 1 = 5040.$$
The Number of Permutations of n Objects Taken k at a Time $${}_nP_k = n(n-1)(n-2)\cdots[n-(k-1)] \quad \textbf{(1)}$$ $$= \frac{n!}{(n-k)!} \quad \textbf{(2)}$$	Compute ${}_7P_4$. Using form (1), we have $${}_7P_4 = 7\cdot 6\cdot 5\cdot 4 = 840.$$ Using form (2), we have $${}_7P_4 = \frac{7!}{(7-4)!} = \frac{7\cdot 6\cdot 5\cdot 4\cdot 3!}{3!} = \frac{7\cdot 6\cdot 5\cdot 4\cdot \cancel{3!}}{\cancel{3!}} = 840.$$
The number of distinct arrangements of n objects taken k at a time, allowing repetition, is n^k.	The number of 4-number code symbols that can be formed with the numbers 5, 6, 7, 8, and 9, if we allow a number to occur more than once, is 5^4, or 625.
For a set of n objects in which n_1 are of one kind, n_2 are of another kind, . . . , and n_k are of a kth kind, the number of distinguishable permutations is $$\frac{n!}{n_1!\cdot n_2!\cdot\cdots\cdot n_k!}.$$	Find the number of distinguishable code symbols that can be formed using the letters in the word MISSISSIPPI. There are 1 M, 4 I's, 4 S's, and 2 P's, for a total of 11 letters, so we have $$\frac{11!}{1!\,4!\,4!\,2!}, \quad \text{or} \quad 34{,}650.$$

SECTION 8.6: COMBINATORICS: COMBINATIONS

The Number of Combinations of n Objects Taken k at a Time $${}_nC_k = \frac{n!}{k!(n-k)!} \quad \textbf{(1)}$$ $$= \frac{{}_nP_k}{k!}$$ $$= \frac{n(n-1)(n-2)\cdots[n-(k-1)]}{k!}. \quad \textbf{(2)}$$ We can also use **binomial coefficient notation**: $$\binom{n}{k} = {}_nC_k.$$	Compute: ${}_6C_4$, or $\binom{6}{4}$. Using form (1), we have $$\binom{6}{4} = \frac{6!}{4!(6-4)!} = \frac{6!}{4!\,2!} = \frac{6\cdot 5\cdot 4!}{4!\,2!} = \frac{6\cdot 5\cdot \cancel{4!}}{\cancel{4!}\cdot 2\cdot 1} = 15.$$ Using form (2), we have $$\binom{6}{4} = \frac{{}_6P_4}{4!} = \frac{6\cdot 5\cdot 4\cdot 3}{4\cdot 3\cdot 2\cdot 1} = 15.$$

SECTION 8.7: THE BINOMIAL THEOREM

The Binomial Theorem Using Pascal's Triangle For any binomial $a + b$ and any natural number n, $(a+b)^n = c_0a^nb^0 + c_1a^{n-1}b^1 + c_2a^{n-2}b^2 + \cdots + c_{n-1}a^1b^{n-1} + c_na^0b^n$, where the numbers $c_0, c_1, c_2, \ldots, c_{n-1}, c_n$ are from the $(n+1)$st row of Pascal's triangle. (See Pascal's triangle on p. 572.)	Expand: $(x-2)^4$. We have $a = x$, $b = -2$, and $n = 4$. We use the fourth row of Pascal's triangle. $(x-2)^4 = 1\cdot x^4 + 4\cdot x^3(-2)^1 + 6\cdot x^2(-2)^2 + 4\cdot x^1(-2)^3 + 1(-2)^4$ $= x^4 + 4x^3(-2) + 6x^2\cdot 4 + 4x(-8) + 16$ $= x^4 - 8x^3 + 24x^2 - 32x + 16$
The Binomial Theorem Using Factorial Notation For any binomial $a + b$ and any natural number n, $(a+b)^n = \binom{n}{0}a^nb^0 + \binom{n}{1}a^{n-1}b^1 + \binom{n}{2}a^{n-2}b^2 + \cdots + \binom{n}{n-1}a^1b^{n-1} + \binom{n}{n}a^0b^n$ $= \sum_{k=0}^{n}\binom{n}{k}a^{n-k}b^k$.	Expand: $(x^2+3)^3$. We have $a = x^2$, $b = 3$, and $n = 3$. $(x^2+3)^3 = \binom{3}{0}(x^2)^3 + \binom{3}{1}(x^2)^2(3) + \binom{3}{2}(x^2)3^2 + \binom{3}{3}3^3$ $= \frac{3!}{0!\,3!}x^6 + \frac{3!}{1!\,2!}(x^4)(3) + \frac{3!}{2!\,1!}(x^2)(9) + \frac{3!}{3!\,0!}(27)$ $= 1\cdot x^6 + 3\cdot 3x^4 + 3\cdot 9x^2 + 1\cdot 27$ $= x^6 + 9x^4 + 27x^2 + 27$
The $(k+1)$st term of $(a+b)^n$ is $\binom{n}{k}a^{n-k}b^k$.	The third term of $(x^2+3)^3$ is $\binom{3}{2}(x^2)^{3-2}\cdot 3^2 = 3\cdot x^2\cdot 9 = 27x^2$. $(k = 2)$
The **total number of subsets** of a set with n elements is 2^n.	How many subsets does the set $\{W, X, Y, Z\}$ have? The set has 4 elements, so we have 2^4, or 16.

SECTION 8.8: PROBABILITY

Principle *P* (Experimental) Given an experiment in which n observations are made, if a situation, or event, E occurs m times out of n observations, then we say that the **experimental probability** of the event, $P(E)$, is given by $P(E) = \frac{m}{n}$.	From a batch of 1000 gears, 35 were found to be defective. The probability that a defective gear is produced is $\frac{35}{1000} = 0.035$, or 3.5%.

Principle P (Theoretical)
If an event E can occur m ways out of n possible equally likely outcomes of a sample space S, then the **theoretical probability** of the event, $P(E)$, is given by

$$P(E) = \frac{m}{n}.$$

What is the probability of drawing 2 red marbles and 1 green marble from a bag containing 5 red marbles, 6 green marbles, and 4 white marbles?

Number of ways of drawing 3 marbles from a bag of 15: ${}_{15}C_3$

Number of ways of drawing 2 red marbles from 5 red marbles: ${}_5C_2$

Number of ways of drawing 1 green marble from 6 green marbles: ${}_6C_1$

Probability that 2 red marbles and 1 green marble are drawn:

$$\frac{{}_5C_2 \cdot {}_6C_1}{{}_{15}C_3} = \frac{10 \cdot 6}{455} = \frac{12}{91}$$

REVIEW EXERCISES

Determine whether the statement is true or false.

1. A sequence is a function. [8.1]
2. An infinite geometric series with $r = -1$ has a limit. [8.3]
3. Permutations involve order and arrangements of objects. [8.5]
4. The total number of subsets of a set with n elements is n^2. [8.7]
5. Find the first 4 terms, a_{11}, and a_{23}:
$$a_n = (-1)^n\left(\frac{n^2}{n^4 + 1}\right).$$ [8.1]
6. Predict the general, or nth, term. Answers may vary.
$2, -5, 10, -17, 26, \ldots$ [8.1]
7. Find and evaluate:
$$\sum_{k=1}^{4} \frac{(-1)^{k+1}3^k}{3^k - 1}.$$ [8.1]
8. Write sigma notation. Answers may vary.
$0 + 3 + 8 + 15 + 24 + 35 + 48$ [8.1]
9. Find the 10th term of the arithmetic sequence
$\frac{3}{4}, \frac{13}{12}, \frac{17}{12}, \ldots.$ [8.2]
10. Find the 6th term of the arithmetic sequence
$a - b, a, a + b, \ldots.$ [8.2]
11. Find the sum of the first 18 terms of the arithmetic sequence
$4, 7, 10, \ldots.$ [8.2]
12. Find the sum of the first 200 natural numbers. [8.2]
13. The 1st term in an arithmetic sequence is 5, and the 17th term is 53. Find the 3rd term. [8.2]
14. The common difference in an arithmetic sequence is 3. The 10th term is 23. Find the first term. [8.2]
15. For a geometric sequence, $a_1 = -2$, $r = 2$, and $a_n = -64$. Find n and S_n. [8.3]
16. For a geometric sequence, $r = \frac{1}{2}$ and $S_5 = \frac{31}{2}$. Find a_1 and a_5. [8.3]

Find the sum, if it exists, of each infinite geometric series. [8.3]

17. $25 + 27.5 + 30.25 + 33.275 + \cdots$
18. $0.27 + 0.0027 + 0.000027 + \cdots$
19. $\frac{1}{2} - \frac{1}{6} + \frac{1}{18} - \cdots$
20. Find fraction notation for $2.\overline{43}$. [8.3]
21. Insert four arithmetic means between 5 and 9. [8.2]
22. *Bouncing Golfball.* A golfball is dropped to the pavement from a height of 30 ft. It always rebounds three-fourths of the distance that it drops. How far (up and down) will the ball have traveled when it hits the pavement for the 6th time? [8.3]
23. *The Amount of an Annuity.* To create a college fund, a parent makes a sequence of 18 yearly deposits of $2000 each in a savings account on which interest is compounded annually at 2.8%. Find the amount of the annuity. [8.3]

24. *Total Gift.* Suppose you receive 10¢ on the first day of the year, 12¢ on the 2nd day, 14¢ on the 3rd day, and so on.

    a) How much will you receive on the 365th day? [8.2]

    b) What is the sum of these 365 gifts? [8.2]

25. *The Economic Multiplier.* Suppose that the government is making a $24,000,000,000 expenditure for travel to Mars. If 73% of this amount is spent again, and so on, what is the total effect on the economy? [8.3]

Use mathematical induction to prove each of the following. [8.4]

26. For every natural number n,

$$1 + 4 + 7 + \cdots + (3n - 2) = \frac{n(3n - 1)}{2}.$$

27. For every natural number n,

$$1 + 3 + 3^2 + \cdots + 3^{n-1} = \frac{3^n - 1}{2}.$$

28. For every natural number $n \geq 2$,

$$\left(1 - \frac{1}{2}\right)\left(1 - \frac{1}{3}\right)\cdots\left(1 - \frac{1}{n}\right) = \frac{1}{n}.$$

29. *Book Arrangements.* In how many ways can 6 books be arranged on a shelf? [8.5]

30. *Flag Displays.* If 9 different signal flags are available, how many different displays are possible using 4 flags in a row? [8.5]

31. *Prize Choices.* The winner of a contest can choose any 8 of 15 prizes. How many different sets of prizes can be chosen? [8.6]

32. *Fraternity–Sorority Names.* The Greek alphabet contains 24 letters. How many fraternity or sorority names can be formed using 3 different letters? [8.5]

33. *Letter Arrangements.* In how many distinguishable ways can the letters of the word TENNESSEE be arranged? [8.5]

34. *Floor Plans.* A manufacturer of houses has 1 floor plan but achieves variety by having 3 different roofs, 4 different ways of attaching the garage, and 3 different types of entrances. Find the number of different houses that can be produced. [8.5]

35. *Code Symbols.* How many code symbols can be formed using 5 out of 6 of the letters of G, H, I, J, K, L if the letters:

    a) cannot be repeated? [8.5]

    b) can be repeated? [8.5]

    c) cannot be repeated but must begin with K? [8.5]

    d) cannot be repeated but must end with IGH? [8.5]

36. Determine the number of subsets of a set containing 8 members. [8.7]

Expand. [8.7]

37. $(m + n)^7$

38. $(x - \sqrt{2})^5$

39. $(x^2 - 3y)^4$

40. $\left(a + \frac{1}{a}\right)^8$

41. $(1 + 5i)^6$, where $i^2 = -1$

42. Find the 4th term of $(a + x)^{12}$. [8.7]

43. Find the 12th term of $(2a - b)^{18}$. Do not multiply out the factorials. [8.7]

44. *Rolling Dice.* What is the probability of getting a 10 on a roll of a pair of dice? on a roll of 1 die? [8.8]

45. *Drawing a Card.* From a deck of 52 cards, 1 card is drawn at random. What is the probability that it is a club? [8.8]

46. *Drawing Three Cards.* From a deck of 52 cards, 3 are drawn at random without replacement. What is the probability that 2 are aces and 1 is a king? [8.8]

47. *Election Poll.* Three people were running for mayor in an election campaign. A poll was conducted to see which candidate was favored. During the polling, 86 favored candidate A, 97 favored B, and 23 favored C. Assuming that the poll is a valid indicator of the election results, what is the probability that the election will be won by A? B? C? [8.8]

48. Which of the following is the 25th term of the arithmetic sequence 12, 10, 8, 6, . . . ? [8.2]

A. -38 B. -36
C. 32 D. 60

49. What is the probability of getting a total of 4 on a roll of a pair of dice? [8.8]

A. $\frac{1}{12}$ B. $\frac{1}{9}$
C. $\frac{1}{6}$ D. $\frac{5}{36}$

50. The graph of the sequence whose general term is $a_n = n - 1$ is which of the following? [8.1]

A.

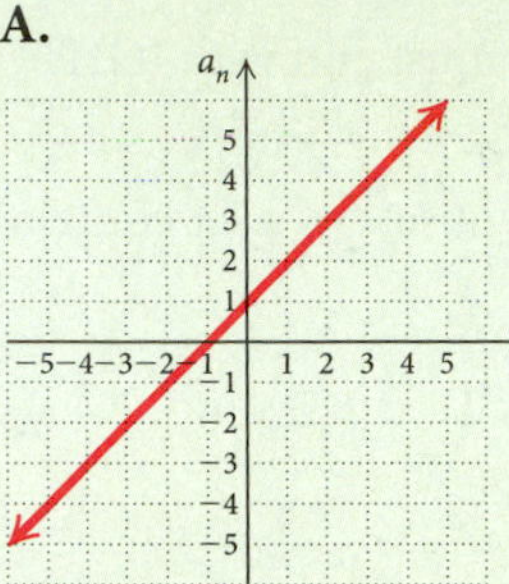

B.

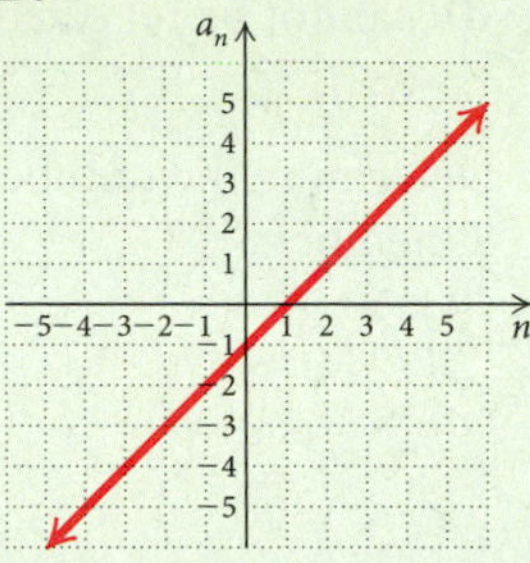

C.

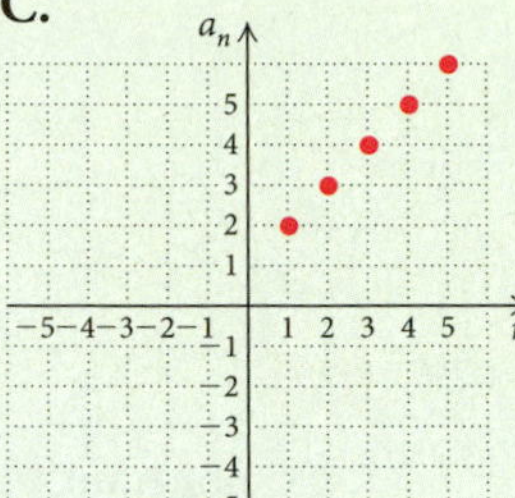

D.

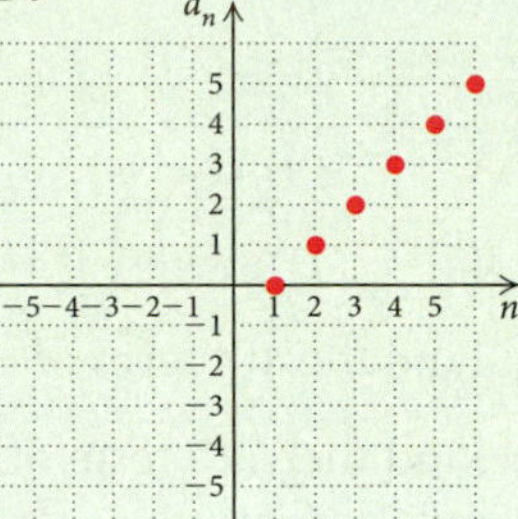

Synthesis

51. Suppose that $a_1, a_2, \ldots, a_n$ is an arithmetic sequence. Is $b_1, b_2, \ldots, b_n$ an arithmetic sequence if:

a) $b_n = |a_n|$? [8.2]
b) $b_n = a_n + 8$? [8.2]
c) $b_n = 7a_n$? [8.2]
d) $b_n = \dfrac{1}{a_n}$? [8.2]
e) $b_n = \log a_n$? [8.2]
f) $b_n = a_n^3$? [8.2]

52. Suppose that $a_1, a_2, \ldots, a_n$ and $b_1, b_2, \ldots, b_n$ are geometric sequences. Prove that $c_1, c_2, \ldots, c_n$ is a geometric sequence, where $c_n = a_n b_n$. [8.3]

53. Write the first 3 terms of the infinite geometric series with $r = -\frac{1}{3}$ and $S_\infty = \frac{3}{8}$. [8.3]

54. The zeros of this polynomial function form an arithmetic sequence. Find them. [8.2]

$$f(x) = x^4 - 4x^3 - 4x^2 + 16x$$

55. Simplify:

$$\sum_{k=0}^{10} (-1)^k \binom{10}{k} (\log x)^{10-k} (\log y)^k. \quad [8.6]$$

Solve for n. [8.6]

56. $\binom{n}{6} = 3 \cdot \binom{n-1}{5}$

57. $\binom{n}{n-1} = 36$

58. Solve for a:

$$\sum_{k=0}^{5} \binom{5}{k} 9^{5-k} a^k = 0. \quad [8.7]$$

Collaborative Discussion and Writing

59. *Circular Arrangements.* In how many ways can the numbers on a clock face be arranged? See if you can derive a formula for the number of distinct circular arrangements of n objects. Explain your reasoning. [8.5]

60. How "long" is 15? Suppose you own 15 books and decide to make up all the possible arrangements of the books on a shelf. About how long, in years, would it take you if you were to make one arrangement per second? Write out the reasoning you used for this problem in the form of a paragraph. [8.5]

61. Explain why a "combination" lock should really be called a "permutation" lock. [8.6]

62. Give the reasoning that you might use with a fellow student to explain that

$$\binom{n}{k} = \binom{n}{n-k}. \quad [8.6]$$

8 Chapter Test

1. For the sequence whose nth term is $a_n = (-1)^n(2n + 1)$, find a_{21}.

2. Find the first 5 terms of the sequence with general term
$$a_n = \frac{n+1}{n+2}.$$

3. Find and evaluate:
$$\sum_{k=1}^{4}(k^2 + 1).$$

Write sigma notation. Answers may vary.

4. $4 + 8 + 12 + 16 + 20 + 24$

5. $2 + 4 + 8 + 16 + 32 + \cdots$

6. Find the first 4 terms of the recursively defined sequence
$$a_1 = 3, \quad a_{n+1} = 2 + \frac{1}{a_n}.$$

7. Find the 15th term of the arithmetic sequence $2, 5, 8, \ldots.$

8. The 1st term of an arithmetic sequence is 8 and the 21st term is 108. Find the 7th term.

9. Find the sum of the first 20 terms of the series $17 + 13 + 9 + \cdots.$

10. Find the sum: $\sum_{k=1}^{25}(2k + 1)$.

11. Find the 11th term of the geometric sequence $10, -5, \frac{5}{2}, -\frac{5}{4}, \ldots.$

12. For a geometric sequence, $r = 0.2$ and $S_4 = 1248$. Find a_1.

Find the sum, if it exists.

13. $\sum_{k=1}^{8} 2^k$

14. $18 + 6 + 2 + \cdots$

15. Find fraction notation for $0.\overline{56}$.

16. *Salvage Value.* The value of an office machine is \$10,000. Its salvage value each year is 80% of its value the year before. Give a sequence that lists the salvage value of the machine for each year of a 6-year period.

17. *Hourly Wage.* William accepts a job, starting with an hourly wage of \$12.25, and is promised a raise of 30¢ per hour every three months for 4 years. What will William's hourly wage be at the end of the 4-year period?

18. *Amount of an Annuity.* To create a college fund, a parent makes a sequence of 18 equal yearly deposits of \$2500 in a savings account on which interest is compounded annually at 5.6%. Find the amount of the annuity.

19. Use mathematical induction to prove that, for every natural number n,
$$2 + 5 + 8 + \cdots + (3n - 1) = \frac{n(3n + 1)}{2}.$$

Evaluate.

20. ${}_{15}P_6$

21. ${}_{21}C_{10}$

22. $\binom{n}{4}$

23. How many 4-digit numbers can be formed using the digits 1, 3, 5, 6, 7, and 9 without repetition

24. How many code symbols can be formed using 4 of the 6 letters A, B, C, X, Y, Z if the letters:
    a) can be repeated
    b) are not repeated and must begin with Z?

25. *Scuba Club Officers.* The Bay Woods Scuba Club has 28 members. How many sets of 4 officers can be selected from this group?

26. *Test Options.* On a test with 20 questions, a student must answer 8 of the first 12 questions and 4 of the last 8. In how many ways can this be done?

27. Expand: $(x + 1)^5$.

28. Find the 5th term of the binomial expansion $(x - y)^7$.

29. Determine the number of subsets of a set containing 9 members.

30. *Marbles.* Suppose that we select, without looking, one marble from a bag containing 6 red marbles and 8 blue marbles. What is the probability of selecting a blue marble?

31. *Drawing Coins.* Ethan has 6 pennies, 5 dimes, and 4 quarters in his pocket. Six coins are drawn at random. What is the probability of getting 1 penny, 2 dimes, and 3 quarters?

32. The graph of the sequence whose general term is $a_n = 2n - 2$ is which of the following?

A.

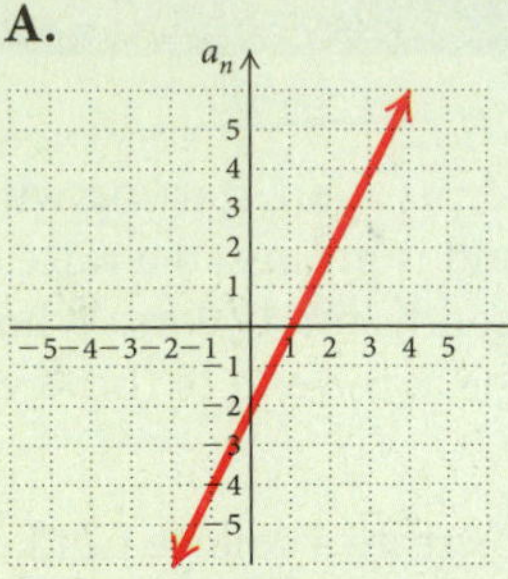

B.

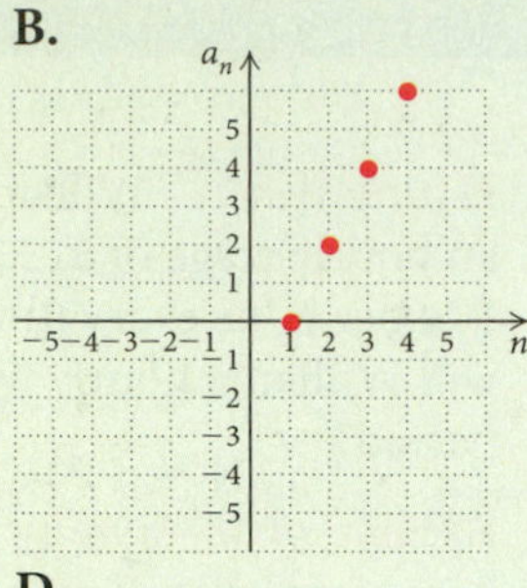

C.

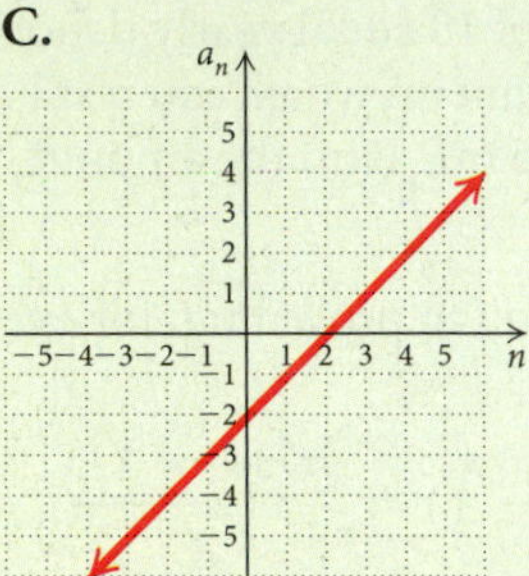

D.

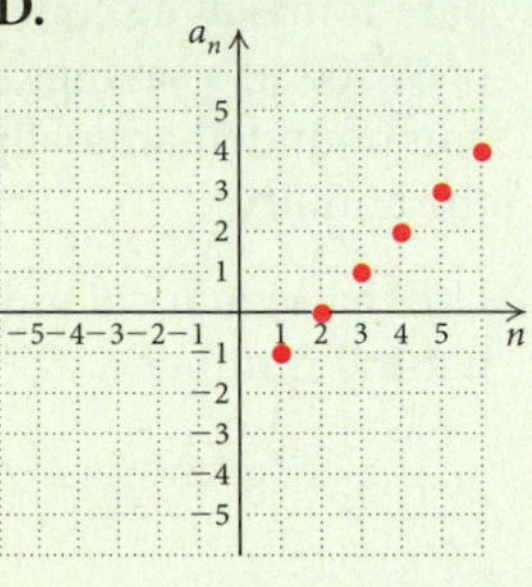

Synthesis

33. Solve for n: ${}_nP_7 = 9 \cdot {}_nP_6$.

JUST-IN-TIME Review

Throughout this text, there are Just-In-Time icons, numbered 1–25, that refer to the following 25 intermediate-algebra topics. Each mini-review lesson is accompanied by several exercises. All answers are provided in the answer section at the back of the text.

1. **Real Numbers**
2. **Properties of Real Numbers**
3. **Absolute Value**
4. **Operations with Real Numbers**
5. **Order on the Number Line**
6. **Interval Notation**
7. **Integers as Exponents**
8. **Scientific Notation**
9. **Order of Operations**
10. **Introduction to Polynomials**
11. **Add and Subtract Polynomials**
12. **Multiply Polynomials**
13. **Factor Polynomials**
14. **Equation-Solving Principles**
15. **Inequality-Solving Principles**
16. **Principle of Zero Products**
17. **Principle of Square Roots**
18. **Simplify Rational Expressions**
19. **Multiply and Divide Rational Expressions**
20. **Add and Subtract Rational Expressions**
21. **Simplify Complex Rational Expressions**
22. **Simplify Radical Expressions**
23. **Rationalizing Denominators**
24. **Rational Exponents**
25. **Pythagorean Theorem**

JUST IN TIME 1

REAL NUMBERS

Some frequently used sets of real numbers and the relationships among them are shown in the following diagram.

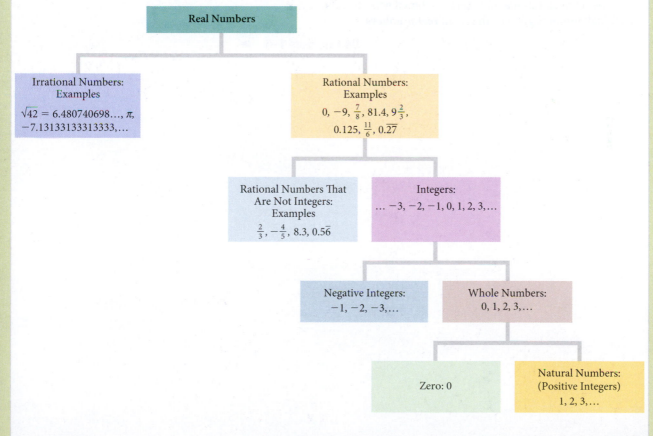

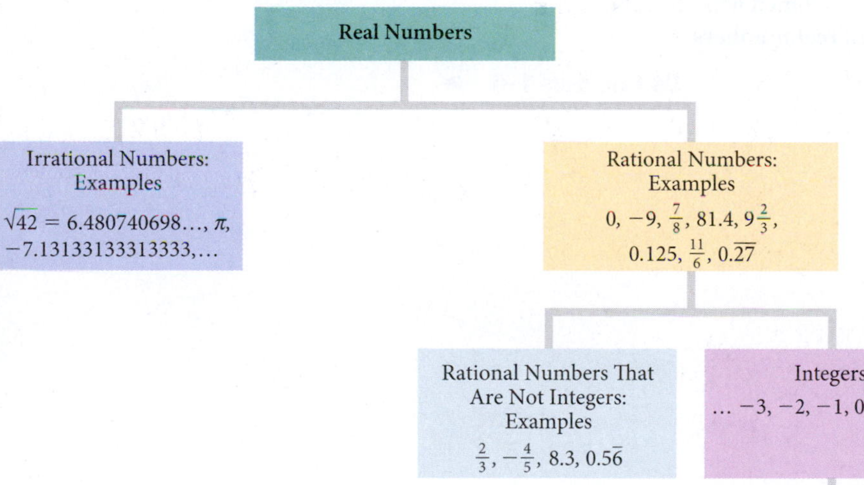

(*continued*)

JUST-IN-TIME Review

Numbers that can be expressed in the form p/q, where p and q are integers and $q \neq 0$, are **rational numbers**. Decimal notation for rational numbers either *terminates* (ends) or *repeats*. Each of the following is a rational number:

$$0, \quad -17, \quad \frac{13}{4}, \quad \sqrt{25} = 5;$$

$$\frac{1}{4} = 0.25 \qquad \text{(terminating decimal);}$$

$$-\frac{5}{11} = -0.454545\ldots = -0.\overline{45} \qquad \text{(repeating decimal);}$$

$$\frac{5}{6} = 0.8333\ldots = 0.8\overline{3} \qquad \text{(repeating decimal).}$$

The real numbers that are not rational are **irrational numbers**. Decimal notation for irrational numbers neither terminates nor repeats. Each of the following is an irrational number. Note in each that there is no repeating block of digits.

$$\sqrt{2} = 1.414213562\ldots,$$

$$-6.12122122212222\ldots,$$

$$\pi = 3.1415926535\ldots$$

$\left(\frac{22}{7}\right.$ and 3.14 are rational *approximations* of the irrational number $\left.\pi.\right)$

The set of all rational numbers combined with the set of all irrational numbers gives us the set of **real numbers.**

Do Exercises 1–6. ►

In Exercises 1–6, consider the numbers
$\frac{2}{3}$, 6, $\sqrt{3}$, -2.45, $\sqrt[6]{26}$, $18.\overline{4}$, -11, $\sqrt[3]{27}$, $5\frac{1}{6}$, $7.151551555\ldots$, $-\sqrt{35}$, $\sqrt[5]{3}$, $-\frac{8}{7}$, 0, $\sqrt{16}$.

1. Which are rational numbers?
2. Which are rational numbers but not integers?
3. Which are irrational numbers?
4. Which are integers?
5. Which are whole numbers?
6. Which are real numbers?

JUST IN TIME 2

PROPERTIES OF REAL NUMBERS

Properties of Real Numbers

For any real numbers a, b, and c:

$a + b = b + a$ and $ab = ba$	Commutative properties of addition and multiplication
$a + (b + c) = (a + b) + c$ and $a(bc) = (ab)c$	Associative properties of addition and multiplication
$a + 0 = 0 + a = a$	Additive identity property
$-a + a = a + (-a) = 0$	Additive inverse property
$a \cdot 1 = 1 \cdot a = a$	Multiplicative identity property
$a \cdot \frac{1}{a} = \frac{1}{a} \cdot a = 1 \ (a \neq 0)$	Multiplicative inverse property
$a(b + c) = ab + ac$ and $a(b - c) = ab - ac$	Distributive properties

EXAMPLES Name the property illustrated in each sentence.

1. $8 \cdot 5 = 5 \cdot 8$	Commutative property of multiplication
2. $14 + (-14) = 0$	Additive inverse property
3. $2(a - b) = 2a - 2b$	Distributive property
4. $5 + (m + n) = (5 + m) + n$	Associative property of addition
5. $6 \cdot 1 = 1 \cdot 6 = 6$	Multiplicative identity property

Do Exercises 1–10. ▶

Name the property illustrated by the sentence.

1. $-24 + 24 = 0$

2. $7(xy) = (7x)y$

3. $9(r - s) = 9r - 9s$

4. $11 + z = z + 11$

5. $-20 \cdot 1 = -20$

6. $5(x + y) = (x + y)5$

7. $q + 0 = q$

8. $75 \cdot \frac{1}{75} = 1$

9. $(x + y) + w = x + (y + w)$

10. $8(a + b) = 8a + 8b$

JUST IN TIME 3

ABSOLUTE VALUE

The **absolute value** of a number a, denoted $|a|$, is its distance from 0 on the number line. For example, $|-5| = 5$, because the distance of -5 from 0 is 5. For any real number a,

$$|a| = a \quad \text{if } a \geq 0$$

and

$$|a| = -a \quad \text{if } a < 0.$$

EXAMPLES Simplify.

1. $|-10| = 10$
2. $|0| = 0$
3. $\left|\frac{4}{9}\right| = \frac{4}{9}$

Absolute value can be used to find the distance between two points on the number line. For any real numbers a and b, the distance between a and b is $|a - b|$ or, equivalently, $|b - a|$.

EXAMPLE 4 Find the distance between -2 and 3.

$|-2 - 3| = |-5| = 5,$

or equivalently,

$|3 - (-2)| = |3 + 2| = |5| = 5$

Do Exercises 1–8. ►

Simplify.

1. $|-98|$
2. $|0|$
3. $|4.7|$
4. $\left|-\frac{2}{3}\right|$

Find the distance between the given pair of points on the number line.

5. $-7,\ 13$
6. $2,\ 14.6$
7. $-39,\ -28$
8. $-\frac{3}{4},\ \frac{15}{8}$

JUST IN TIME 4

OPERATIONS WITH REAL NUMBERS

Rules for Operations with Real Numbers

Addition

- *Positive Numbers*: Add the same way that we add arithmetic numbers. The answer is positive.
- *Negative Numbers*: Add absolute values. The answer is negative.
- *A Positive Number and a Negative Number*: If the numbers have different absolute values, subtract the smaller absolute value from the larger. If the positive number has the greater absolute value, the answer is positive. If the negative number has the greater absolute value, the answer is negative.

EXAMPLES Add.

1. $9 + (-29) = -20$
2. $-9 + (-29) = -38$
3. $-9 + 29 = 20$

Subtraction

- To subtract, add the opposite, or additive inverse, of the number being subtracted.

EXAMPLES Subtract.

4. $15 - 6 = 15 + (-6) = 9$
5. $15 - (-6) = 15 + 6 = 21$
6. $-15 - 6 = -15 + (-6) = -21$
7. $-15 - (-6) = -15 + 6 = -9$

Multiplication and Division, where the divisor is nonzero

- Multiply or divide the absolute values. If the signs are the same, the answer is positive. If the signs are different, the answer is negative.

EXAMPLES Multiply or divide.

8. $-5 \cdot 20 = -100$
9. $32 \div (-4) = -8$
10. $-32 \div 4 = -8$
11. $-32 \div (-4) = 8$
12. $-5 \cdot (-20) = 100$
13. $5 \cdot 20 = 100$

Do Exercises 1–15. ▶

Compute and simplify.

1. $8 - (-11)$
2. $-\frac{3}{10} \cdot \left(-\frac{1}{3}\right)$
3. $15 \div (-3)$
4. $-4 - (-1)$
5. $7 \cdot (-50)$
6. $-0.5 - 5$
7. $-3 + 27$
8. $-400 \div (-40)$
9. $4.2 \cdot (-3)$
10. $-13 - (-33)$
11. $-60 + 45$
12. $\frac{1}{2} - \frac{2}{3}$
13. $-24 \div 3$
14. $-6 + (-16)$
15. $-\frac{1}{2} \div \left(-\frac{5}{8}\right)$

JUST IN TIME 5

ORDER ON THE NUMBER LINE

The real numbers are modeled using a **number line,** as shown below. Each point on the line represents a real number, and every real number is represented by a point on the line.

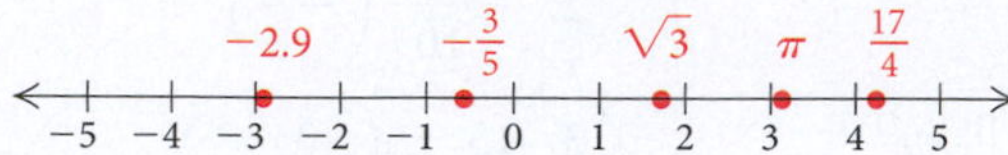

The order of the real numbers can be determined from the number line. If a number a is to the left of a number b, then a **is less than** b $(a < b)$. Similarly, a **is greater than** b $(a > b)$ if a is to the right of b on the number line. For example, we see from the number line above that $-2.9 < -\frac{3}{5}$, because -2.9 is to the left of $-\frac{3}{5}$. Also, $\frac{17}{4} > \sqrt{3}$, because $\frac{17}{4}$ is to the right of $\sqrt{3}$.

The statement $a \leq b$, read "a is less than or equal to b," is true if either $a < b$ is true or $a = b$ is true. A similar statement holds for $a \geq b$.

Do Exercises 1–6. ►

Classify the inequality as true or false.

1. $9 < -9$
2. $-10 \leq -1$
3. $-\sqrt{26} < -5$
4. $\sqrt{6} \geq \sqrt{6}$
5. $-30 > -25$
6. $-\frac{4}{5} > -\frac{5}{4}$

JUST IN TIME 6

INTERVAL NOTATION

Sets of real numbers can be expressed using **interval notation**. For example, for real numbers a and b such that $a < b$, the **open interval** (a, b) is the set of real numbers between, but not including, a and b.

Some intervals extend without bound in one or both directions. The interval $[a, \infty)$, for example, begins at a and extends to the right without bound. The bracket indicates that a is included in the interval.

SET NOTATION	INTERVAL NOTATION	GRAPH
$\{x \mid a < x < b\}$	(a, b)	a, b
$\{x \mid a \le x \le b\}$	$[a, b]$	a, b
$\{x \mid a \le x < b\}$	$[a, b)$	a, b
$\{x \mid a < x \le b\}$	$(a, b]$	a, b
$\{x \mid x > a\}$	(a, ∞)	a
$\{x \mid x \ge a\}$	$[a, \infty)$	a
$\{x \mid x < b\}$	$(-\infty, b)$	b
$\{x \mid x \le b\}$	$(-\infty, b]$	b
$\{x \mid x \text{ is a real number}\}$	$(-\infty, \infty)$	

Do Exercises 1–10.

Write interval notation.

1. $\{x \mid -5 \le x \le 5\}$

2. $\{x \mid -3 < x \le -1\}$

3. $\{x \mid x \le -2\}$

4. $\{x \mid x > 3.8\}$

5. $\{x \mid 7 < x\}$

6. $\{x \mid -2 < x < 2\}$

Write interval notation for the graph.

7. −5 −4 −3 −2 −1 0 1 2 3 4 5

8. 1.7
−5 −4 −3 −2 −1 0 1 2 3 4 5

9. −5 −4 −3 −2 −1 0 1 2 3 4 5

10. $\sqrt{5}$
−5 −4 −3 −2 −1 0 1 2 3 4 5

JUST IN TIME 7

INTEGERS AS EXPONENTS

When a positive integer is used as an *exponent*, it indicates the number of times that a factor appears in a product. For example, 7^3 means $7 \cdot 7 \cdot 7$, where 7 is the **base** and 3 is the **exponent**.

For any nonzero numbers a and b and any integers m and n,

$$a^0 = 1, \quad a^{-m} = \frac{1}{a^m}, \quad \text{and} \quad \frac{a^{-m}}{b^{-n}} = \frac{b^n}{a^m}.$$

Properties of Exponents

For any real numbers a and b and any integers m and n, assuming 0 is not raised to a nonpositive power:

$a^m \cdot a^n = a^{m+n}$ — Product rule

$\dfrac{a^m}{a^n} = a^{m-n} \; (a \neq 0)$ — Quotient rule

$(a^m)^n = a^{mn}$ — Power rule

$(ab)^m = a^m b^m$ — Raising a product to a power

$\left(\dfrac{a}{b}\right)^m = \dfrac{a^m}{b^m} \; (b \neq 0)$ — Raising a quotient to a power

EXAMPLES Simplify each of the following.

1. $4^2 \cdot 4^{-5} = 4^{2+(-5)} = 4^{-3}$, or $\dfrac{1}{4^3}$
2. $\left(\dfrac{7}{9}\right)^0 = 1$
3. $(8^2)^{-5} = 8^{2(-5)} = 8^{-10}$, or $\dfrac{1}{8^{10}}$
4. $\dfrac{x^{11}}{x^4} = x^{11-4} = x^7$
5. $\left(\dfrac{a}{b}\right)^3 = \dfrac{a^3}{b^3}$
6. $(cd)^{-2} = c^{-2}d^{-2}$, or $\dfrac{1}{c^2d^2}$

Do Exercises 1–10. ►

Simplify.

1. 3^{-6}
2. $\dfrac{1}{(0.2)^{-5}}$
3. $\dfrac{w^{-4}}{z^{-9}}$
4. $\left(\dfrac{z}{y}\right)^2$
5. 100^0
6. $\dfrac{a^5}{a^{-3}}$
7. $(2xy^3)(-3x^{-5}y)$
8. $x^{-4} \cdot x^{-7}$
9. $(mn)^{-6}$
10. $(t^{-5})^4$

JUST IN TIME 8

SCIENTIFIC NOTATION

We can use scientific notation to name both very large and very small positive numbers and to perform computations. **Scientific notation** for a number is an expression of the type $N \times 10^m$, where $1 \le N < 10$, N is in decimal notation, and m is an integer.

EXAMPLES Convert to scientific notation.

1. $9{,}460{,}000{,}000{,}000 = 9.46 \times 10^{12}$ — Large number; exponent is positive.
2. $0.0648 = 6.48 \times 10^{-2}$ — Small number; exponent is negative.

Convert to decimal notation.

3. $5.4 \times 10^7 = 54{,}000{,}000$
4. $3.819 \times 10^{-3} = 0.003819$

Do Exercises 1–8. ►

Convert to scientific notation.

1. 18,500,000
2. 0.000786
3. 0.0000000023
4. 8,927,000,000

Convert to decimal notation.

5. 4.3×10^{-8}
6. 5.17×10^6
7. 6.203×10^{11}
8. 2.94×10^{-5}

JUST IN TIME 9

ORDER OF OPERATIONS

Recall that to simplify the expression $3 + 4 \cdot 5$, first we multiply 4 and 5 to get 20 and then we add 3 to get 23. Mathematicians have agreed on the following procedure, or rules for order of operations.

Rules for Order of Operations

1. Do all calculations within grouping symbols before operations outside. When nested grouping symbols are present, work from the inside out.
2. Evaluate all exponential expressions.
3. Do all multiplications and divisions in order from left to right.
4. Do all additions and subtractions in order from left to right.

EXAMPLES

1. $8(5-3)^3 - 20$
$= 8 \cdot 2^3 - 20$
$= 8 \cdot 8 - 20$
$= 64 - 20$
$= 44$

2. $\dfrac{10 \div (8-6) + 9 \cdot 4}{2^5 + 3^2}$
$= \dfrac{10 \div 2 + 9 \cdot 4}{32 + 9}$
$= \dfrac{5 + 36}{41} = \dfrac{41}{41} = 1$

Do Exercises 1–6. ►

Calculate.

1. $3 + 18 \div 6 - 3$
2. $5 \cdot 3 + 8 \cdot 3^2 + 4(6-2)$
3. $5(3 - 8 \cdot 3^2 + 4 \cdot 6 - 2)$
4. $16 \div 4 \cdot 4 \div 2 \cdot 256$
5. $2^6 \cdot 2^{-3} \div 2^{10} \div 2^{-8}$
6. $\dfrac{4(8-6)^2 - 4 \cdot 3 + 2 \cdot 8}{3^1 + 19^0}$

JUST IN TIME 10 INTRODUCTION TO POLYNOMIALS

Polynomials are a type of algebraic expression that you will often encounter in your study of algebra. Some examples of polynomials are

$3x - 4y$, $5y^3 - \frac{7}{3}y^2 + 3y - 2$, $-2.3a^4$, 16, and $z^6 - \sqrt{5}$.

Algebraic expressions like $8x - 13$, $x^2 + 3x - 4$, and $3a^5 - 11 + a$ are **polynomials in one variable.** Algebraic expressions like $3ab^3 - 8$ and $5x^4y^2 - 3x^3y^8 + 7xy^2 + 6$ are **polynomials in several variables.** The **degree of a term** is the sum of the exponents of the variables in that term. The **degree of a polynomial** is the degree of the term of highest degree.

A polynomial with just one term, like $-9y^6$, is a **monomial.** If a polynomial has two terms, like $x^2 + 4$, it is a **binomial.** A polynomial with three terms, like $4x^2 - 4xy + 1$, is a **trinomial.**

EXAMPLES Determine the degree of the polynomial.

1. $2x^3 - 1$ Degree: 3
2. -5 $(-5 = -5x^0)$ Degree: 0
3. $w^2 - 3.5 + 4w^5 = 4w^5 + w^2 - 3.5$ Degree: 5
4. $7xy^3 - 16x^2y^4$ Degree: $2 + 4$, or 6

Do Exercises 1–8.

Determine the degree of the polynomial.

1. $5 - x^6$
2. $x^2y^5 - x^7y + 4$
3. $2a^4 - 3 + a^2$
4. -41
5. $4x - x^3 + 0.1x^8 - 2x^5$

Classify the polynomial as a monomial, a binomial, or a trinomial.

6. $x - 3$
7. $14y^5$
8. $2y - \frac{1}{4}y^2 + 8$

JUST IN TIME 11 ADD AND SUBTRACT POLYNOMIALS

If two terms of an expression have the same variables raised to the same powers, they are called **like terms**, or **similar terms**. We can **combine**, or **collect, like terms** using the distributive property. For example, $3y^2$ and $5y^2$ are like terms and $3y^2 + 5y^2 = (3 + 5)y^2$. We add or subtract polynomials by combining like terms.

EXAMPLES Add or subtract each of the following.

1. $(-5x^3 + 3x^2 - x) + (12x^3 - 7x^2 + 3)$
   $= (-5x^3 + 12x^3) + (3x^2 - 7x^2) - x + 3$
   $= (-5 + 12)x^3 + (3 - 7)x^2 - x + 3 = 7x^3 - 4x^2 - x + 3$
2. $(6x^2y^3 - 9xy) - (5x^2y^3 - 4xy)$
   $= (6x^2y^3 - 9xy) + (-5x^2y^3 + 4xy)$
   $= 6x^2y^3 - 9xy - 5x^2y^3 + 4xy = x^2y^3 - 5xy$

Do Exercises 1–5.

Add or subtract.

1. $(8y - 1) - (3 - y)$
2. $(3x^2 - 2x - x^3 + 2) - (5x^2 - 8x - x^3 + 4)$
3. $(2x + 3y + z - 7) + (4x - 2y - z + 8) + (-3x + y - 2z - 4)$
4. $(3ab^2 + 4a^2b - 2ab + 6) + (-ab^2 - 5a^2b + 8ab + 4)$
5. $(5x^2 + 4xy - 3y^2 + 2) - (9x^2 - 4xy + 2y^2 - 1)$

JUST IN TIME 12

MULTIPLY POLYNOMIALS

To multiply monomials, we first multiply their coefficients, and then we multiply their variables.

EXAMPLES

1. $(-2x^3)(5x^4) = (-2 \cdot 5)(x^3 \cdot x^4) = -10x^7$
2. $(3yz^2)(8y^3z^5) = (3 \cdot 8)(y \cdot y^3)(z^2 \cdot z^5) = 24y^4z^7$

We can find the product of two binomials by multiplying the **F**irst terms, then the **O**uter terms, then the **I**nner terms, then the **L**ast terms. Then we combine like terms, if possible. This procedure is sometimes called **FOIL**.

EXAMPLE 3 Multiply: $(2x - 7)(3x + 4)$.

$$(2x - 7)(3x + 4) = \overset{F}{6x^2} + \overset{O}{8x} - \overset{I}{21x} - \overset{L}{28}$$
$$= 6x^2 - 13x - 28.$$

Special Products of Binomials

$(A + B)^2 = A^2 + 2AB + B^2$	Square of a sum
$(A - B)^2 = A^2 - 2AB + B^2$	Square of a difference
$(A + B)(A - B) = A^2 - B^2$	Product of a sum and a difference

EXAMPLES

4. $(4x + 1)^2 = (4x)^2 + 2 \cdot 4x \cdot 1 + 1^2$
   $= 16x^2 + 8x + 1$
5. $(3y^2 - 2)^2 = (3y^2)^2 - 2 \cdot 3y^2 \cdot 2 + 2^2$
   $= 9y^4 - 12y^2 + 4$
6. $(x^2 + 3y)(x^2 - 3y) = (x^2)^2 - (3y)^2$
   $= x^4 - 9y^2$

Do Exercises 1–10. ▶

Multiply.

1. $(3a^2)(-7a^4)$
2. $(y - 3)(y + 5)$
3. $(x + 6)(x + 3)$
4. $(2a + 3)(a + 5)$
5. $(2x + 3y)(2x + y)$
6. $(x + 3)^2$
7. $(5x - 3)^2$
8. $(2x + 3y)^2$
9. $(n + 6)(n - 6)$
10. $(3y + 4)(3y - 4)$

JUST IN TIME 13

FACTOR POLYNOMIALS

When a polynomial is to be factored, we should always look first to factor out a factor that is common to all the terms using the distributive property. We generally look for the constant common factor with the largest absolute value and for variables with the largest exponent common to all the terms.

EXAMPLE 1 Factor: $15 + 10x - 5x^2$.

$$15 + 10x - 5x^2 = 5 \cdot 3 + 5 \cdot 2x - 5 \cdot x^2 = 5(3 + 2x - x^2)$$

In some polynomials, pairs of terms have a common binomial factor that can be removed in a process called **factoring by grouping**.

EXAMPLE 2 Factor: $x^3 + 3x^2 - 5x - 15$.

$$\begin{aligned} x^3 + 3x^2 - 5x - 15 &= (x^3 + 3x^2) + (-5x - 15) \\ &= x^2(x + 3) - 5(x + 3) \\ &= (x + 3)(x^2 - 5) \end{aligned}$$

Some trinomials can be factored into the product of two binomials. To factor a trinomial of the form $x^2 + bx + c$, we look for binomial factors of the form $(x + p)(x + q)$, where $p \cdot q = c$ and $p + q = b$. That is, we look for two numbers p and q whose sum is the coefficient of the middle term of the polynomial, b, and whose product is the constant term, c.

EXAMPLES Factor.

3. $x^2 + 5x + 6 = (x + 2)(x + 3)$
4. $x^4 - 6x^3 + 8x^2 = x^2(x^2 - 6x + 8) = x^2(x - 2)(x - 4)$

To factor trinomials of the type $ax^2 + bx + c, a \neq 1$, using the **FOIL method:**

1. Factor out the largest common factor.
2. Find two First terms whose product is ax^2:

$$(\square x + \quad)(\square x + \quad) = ax^2 + bx + c.$$

FOIL

3. Find two Last terms whose product is c:

$$(\quad x + \square)(\quad x + \square) = ax^2 + bx + c$$

FOIL

4. Repeat steps (2) and (3) until a combination is found for which the sum of the Outer product and the Inner product is bx:

$$(\square x + \square)(\square x + \square) = ax^2 + bx + c.$$

I O FOIL

(continued)

EXAMPLES Factor.

5. $3x^2 - 10x - 8 = (3x + 2)(x - 4)$

6. $12y^2 + 44y - 45 = (2y + 9)(6y - 5)$

Special Factorizations

EXAMPLES

7. $A^2 - B^2 = (A + B)(A - B)$:

$$x^2 - 16 = x^2 - 4^2 = (x + 4)(x - 4)$$

8. $A^2 + 2AB + B^2 = (A + B)^2$:

$$x^2 + 8x + 16 = x^2 + 2 \cdot x \cdot 4 + 4^2 = (x + 4)^2$$

9. $A^2 - 2AB + B^2 = (A - B)^2$

$$\begin{aligned} 25y^2 - 30y + 9 &= (5y)^2 - 2 \cdot 5y \cdot 3 + 3^2 \\ &= (5y - 3)^2 \end{aligned}$$

10. $A^3 + B^3 = (A + B)(A^2 - AB + B^2)$:

$$x^3 + 27 = x^3 + 3^3 = (x + 3)(x^2 - 3x + 9)$$

11. $A^3 - B^3 = (A - B)(A^2 + AB + B^2)$:

$$\begin{aligned} 16y^3 - 250 = 2(8y^3 - 125) &= 2[(2y)^3 - 5^3] \\ &= 2(2y - 5)(4y^2 + 10y + 25) \end{aligned}$$

Do Exercises 1–20. ▶

Factor out the largest common factor.

1. $3x + 18$
2. $2z^3 - 8z^2$

Factor by grouping.

3. $3x^3 - x^2 + 18x - 6$
4. $t^3 + 6t^2 - 2t - 12$

Factor the trinomial.

5. $w^2 - 7w + 10$
6. $t^2 + 8t + 15$
7. $2n^2 - 20n - 48$
8. $y^4 - 9y^3 + 14y^2$
9. $2n^2 + 9n - 56$
10. $2y^2 + y - 6$

Factor the difference of squares.

11. $z^2 - 81$
12. $16x^2 - 9$
13. $7pq^4 - 7py^4$

Factor the square of a binomial.

14. $x^2 + 12x + 36$
15. $9z^2 - 12z + 4$
16. $a^3 + 24a^2 + 144a$

Factor the sum or the difference of cubes.

17. $x^3 + 64$
18. $m^3 - 216$
19. $3a^5 - 24a^2$
20. $t^6 + 1$

JUST IN TIME 14 EQUATION-SOLVING PRINCIPLES

For any real numbers a, b, and c,

The Addition Principle

If $a = b$ is true, then $a + c = b + c$ is true.

The Multiplication Principle

If $a = b$ is true, then $ac = bc$ is true.

EXAMPLES Solve.

1.
$$\begin{aligned} y - 11 &= 12 \\ y - 11 + 11 &= 12 + 11 \\ y &= 23 \end{aligned}$$

2.
$$\begin{aligned} 15c &= 90 \\ \frac{1}{15} \cdot 15c &= \frac{1}{15} \cdot 90 \\ c &= 6 \end{aligned}$$

3.
$$\begin{aligned} \frac{1}{4}x + 5 &= 8 \\ \frac{1}{4}x + 5 - 5 &= 8 - 5 \\ \frac{1}{4}x &= 3 \\ 4 \cdot \frac{1}{4}x &= 4 \cdot 3 \\ x &= 12 \end{aligned}$$

4.
$$\begin{aligned} 2x + 3 &= 1 - 6(x - 1) \\ 2x + 3 &= 1 - 6x + 6 \\ 2x + 3 &= 7 - 6x \\ 8x + 3 &= 7 \\ 8x &= 4 \\ x &= \frac{1}{2} \end{aligned}$$

Do Exercises 1–8. ►

Solve.

1. $7t = 70$
2. $x - 5 = 7$
3. $3x + 4 = -8$
4. $6x - 15 = 45$
5. $7y - 1 = 23 - 5y$
6. $3m - 7 = -13 + m$
7. $2(x + 7) = 5x + 14$
8. $5y - 4(2y - 10) = 25$

JUST IN TIME 15

INEQUALITY-SOLVING PRINCIPLES

For any real numbers a, b, and c:

The Addition Principle for Inequalities

If $a < b$ is true, then $a + c < b + c$ is true.

The Multiplication Principle for Inequalities

a) If $a < b$ and $c > 0$ are true, then $ac < bc$ is true.

b) If $a < b$ and $c < 0$ are true, then $ac > bc$ is true.

(When both sides of an inequality are multiplied by a negative number, the inequality sign must be reversed.)

Similar statements hold for $a \le b$.

EXAMPLES Solve.

1.
$$a + 9 \le -50$$
$$a + 9 - 9 \le -50 - 9$$
$$a \le -59$$

2.
$$-5x > 4$$
$$-\tfrac{1}{5}(-5x) < -\tfrac{1}{5}(4)$$
$$x < -\tfrac{4}{5}$$

3.
$$2y - 1 < 5$$
$$2y - 1 + 1 < 5 + 1$$
$$2y < 6$$
$$\tfrac{1}{2} \cdot 2y < \tfrac{1}{2} \cdot 6$$
$$y < 3$$

4.
$$4 - 3x \ge 13$$
$$-4 + 4 - 3x \ge -4 + 13$$
$$-3x \ge 9$$
$$\frac{-3x}{-3} \le \frac{9}{-3}$$
$$x \le -3$$

Do Exercises 1–6. ►

Solve.

1. $p + 25 \ge -100$

2. $-\frac{2}{3}x > 6$

3. $9x - 1 < 17$

4. $-x - 16 \ge 40$

5. $\frac{1}{3}y - 6 < 3$

6. $8 - 2w \le -14$

JUST IN TIME 16

PRINCIPLE OF ZERO PRODUCTS

The product of two numbers is 0 if one or both of the numbers is 0. Furthermore, *if any product is* 0, *then a factor must be* 0. For example:

If $7x = 0$, then we know that $x = 0$.

If $x(2x - 9) = 0$, then we know that $x = 0$ *or* $2x - 9 = 0$.

If $(x + 3)(x - 2) = 0$, then we know that $x + 3 = 0$ *or* $x - 2 = 0$.

The Principle of Zero Products

If $ab = 0$ is true, then $a = 0$ or $b = 0$, and if $a = 0$ or $b = 0$, then $ab = 0$.

Some quadratic equations can be solved using the principle of zero products.

EXAMPLES Solve.

1. $x^2 - 3x - 4 = 0$

$(x + 1)(x - 4) = 0$

$x + 1 = 0$ *or* $x - 4 = 0$

$x = -1$ *or* $x = 4$

The solutions are -1 and 4.

2. $2x^2 + 5x - 3 = 0$

$(x + 3)(2x - 1) = 0$

$x + 3 = 0$ *or* $2x - 1 = 0$

$x = -3$ *or* $2x = 1$

$x = -3$ *or* $x = \frac{1}{2}$

The solutions are -3 and $\frac{1}{2}$.

Do Exercises 1–7. ►

Solve.

1. $(a + 7)(a - 1) = 0$
2. $(5y + 3)(y - 4) = 0$
3. $6x^2 + 7x - 5 = 0$
4. $t(t - 8) = 0$
5. $x^2 - 8x - 33 = 0$
6. $x^2 + 13x = 30$
7. $12x^2 - 7x - 12 = 0$

JUST IN TIME 17

PRINCIPLE OF SQUARE ROOTS

The principle of square roots can be used to solve some quadratic equations.

The Principle of Square Roots

If $x^2 = k$, then $x = \sqrt{k}$ or $x = -\sqrt{k}$.

EXAMPLES Solve.

1. $s^2 - 144 = 0$

$s^2 = 144$

$s = -\sqrt{144}$ *or* $s = \sqrt{144}$

$s = -12$ *or* $s = 12$

The solutions are -12 and 12, or ± 12.

2. $3x^2 - 21 = 0$

$3x^2 = 21$

$x^2 = 7$

$x = -\sqrt{7}$ *or* $x = \sqrt{7}$

The solutions are $-\sqrt{7}$ and $\sqrt{7}$, or $\pm\sqrt{7}$.

Do Exercises 1–6. ►

Solve.

1. $x^2 - 36 = 0$

2. $2y^2 - 20 = 0$

3. $6z^2 = 18$

4. $3t^2 - 15 = 0$

5. $z^2 - 1 = 24$

6. $5x^2 - 75 = 0$

JUST IN TIME 18 SIMPLIFY RATIONAL EXPRESSIONS

A **rational expression** is the quotient of two polynomials. The **domain** of an algebraic expression is the set of all real numbers for which the expression is defined. Since division by 0 is not defined, any number that makes the denominator 0 is not in the domain of a rational expression.

EXAMPLE 1 Find the domain of

$$\frac{x^2 - 4}{x^2 - 4x - 5}.$$

We solve the equation $x^2 - 4x - 5 = 0$, or $(x + 1)(x - 5) = 0$, to find the numbers that are not in the domain. The solutions are -1 and 5. Since the denominator is 0 when $x = -1$ *or* $x = 5$, the domain is the set of all real numbers except -1 and 5.

EXAMPLES Simplify.

1.
$$\begin{aligned}\frac{9x^2 + 6x - 3}{12x^2 - 12} &= \frac{3(3x^2 + 2x - 1)}{12(x^2 - 1)}\\ &= \frac{3(x + 1)(3x - 1)}{3 \cdot 4(x + 1)(x - 1)}\\ &= \frac{3(x + 1)}{3(x + 1)} \cdot \frac{3x - 1}{4(x - 1)}\\ &= 1 \cdot \frac{3x - 1}{4(x - 1)}\\ &= \frac{3x - 1}{4(x - 1)}\end{aligned}$$

Canceling is a shortcut that is often used to remove a factor of 1.

2.
$$\begin{aligned}\frac{2 - x}{x^2 + x - 6} &= \frac{2 - x}{(x + 3)(x - 2)}\\ &= \frac{-1(x - 2)}{(x + 3)(x - 2)}\\ &= \frac{-1\cancel{(x - 2)}}{(x + 3)\cancel{(x - 2)}}\\ &= \frac{-1}{x + 3}, \text{ or } -\frac{1}{x + 3}\end{aligned}$$

Do Exercises 1–6. ►

Find the domain of the rational expression.

1. $\dfrac{3x - 3}{x(x - 1)}$

2. $\dfrac{y + 6}{y^2 + 4y - 21}$

Simplify.

3. $\dfrac{x^2 - 4}{x^2 - 4x + 4}$

4. $\dfrac{x^2 + 2x - 3}{x^2 - 9}$

5. $\dfrac{x^3 - 6x^2 + 9x}{x^3 - 3x^2}$

6. $\dfrac{6y^2 + 12y - 48}{3y^2 - 9y + 6}$

JUST IN TIME 19

MULTIPLY AND DIVIDE RATIONAL EXPRESSIONS

To multiply rational expressions, we multiply numerators and multiply denominators and, if possible, simplify the result. To divide rational expressions, we multiply the dividend by the reciprocal of the divisor and, if possible, simplify the result; that is,

$$\frac{a}{b}\cdot\frac{c}{d}=\frac{ac}{bd} \quad \text{and} \quad \frac{a}{b}\div\frac{c}{d}=\frac{a}{b}\cdot\frac{d}{c}=\frac{ad}{bc}.$$

EXAMPLES Multiply or divide.

1. $$\frac{a^2-4}{16a}\cdot\frac{20a^2}{a+2}=\frac{(a^2-4)(20a^2)}{16a(a+2)}$$
$$=\frac{(a+2)(a-2)\cdot 4\cdot 5\cdot a\cdot a}{4\cdot 4\cdot a\cdot(a+2)}$$
$$=\frac{5a(a-2)}{4}$$

2. $$\frac{x-2}{12}\div\frac{x^2-4x+4}{3x^3+15x^2}=\frac{x-2}{12}\cdot\frac{3x^3+15x^2}{x^2-4x+4}$$
$$=\frac{(x-2)(3x^3+15x^2)}{12(x^2-4x+4)}$$
$$=\frac{(x-2)(3)(x^2)(x+5)}{3\cdot 4(x-2)(x-2)}$$
$$=\frac{x^2(x+5)}{4(x-2)}$$

Do Exercises 1–6. ▶

Multiply or divide and, if possible, simplify.

1. $\dfrac{r-s}{r+s}\cdot\dfrac{r^2-s^2}{(r-s)^2}$

2. $\dfrac{m^2-n^2}{r+s}\div\dfrac{m-n}{r+s}$

3. $\dfrac{4x^2+9x+2}{x^2+x-2}\cdot\dfrac{x^2-1}{3x^2+x-2}$

4. $\dfrac{a^2-a-2}{a^2-a-6}\div\dfrac{a^2-2a}{2a+a^2}$

5. $\dfrac{3x+12}{2x-8}\div\dfrac{(x+4)^2}{(x-4)^2}$

6. $\dfrac{x^2-y^2}{x^3-y^3}\cdot\dfrac{x^2+xy+y^2}{x^2+2xy+y^2}$

JUST IN TIME 20

ADD AND SUBTRACT RATIONAL EXPRESSIONS

When rational expressions have the same denominator, we can add or subtract by adding or subtracting the numerators and retaining the common denominator. If the denominators differ, we must find equivalent rational expressions that have a common denominator. In general, it is most efficient to find the **least common denominator (LCD)** of the expressions.

To find the least common denominator of rational expressions, factor each denominator and form the product that uses each factor the greatest number of times it occurs in any factorization.

EXAMPLE 1 Add.

$$\frac{x^2 - 4x + 4}{2x^2 - 3x + 1} + \frac{x + 4}{2x - 2}$$

$$= \frac{x^2 - 4x + 4}{(2x - 1)(x - 1)} + \frac{x + 4}{2(x - 1)}$$

The LCD is $(2x - 1)(x - 1)(2)$, or $2(2x - 1)(x - 1)$.

$$= \frac{x^2 - 4x + 4}{(2x - 1)(x - 1)} \cdot \frac{2}{2} + \frac{x + 4}{2(x - 1)} \cdot \frac{2x - 1}{2x - 1}$$

$$= \frac{2x^2 - 8x + 8}{(2x - 1)(x - 1)(2)} + \frac{2x^2 + 7x - 4}{2(x - 1)(2x - 1)}$$

$$= \frac{4x^2 - x + 4}{2(2x - 1)(x - 1)}$$

EXAMPLE 2 Subtract.

$$\frac{x}{x^2 + 11x + 30} - \frac{5}{x^2 + 9x + 20}$$

$$= \frac{x}{(x + 5)(x + 6)} - \frac{5}{(x + 5)(x + 4)}$$

The LCD is $(x + 5)(x + 6)(x + 4)$.

$$= \frac{x}{(x + 5)(x + 6)} \cdot \frac{x + 4}{x + 4} - \frac{5}{(x + 5)(x + 4)} \cdot \frac{x + 6}{x + 6}$$

$$= \frac{x^2 + 4x}{(x + 5)(x + 6)(x + 4)} - \frac{5x + 30}{(x + 5)(x + 4)(x + 6)}$$

$$= \frac{x^2 + 4x - (5x + 30)}{(x + 5)(x + 6)(x + 4)} = \frac{x^2 + 4x - 5x - 30}{(x + 5)(x + 6)(x + 4)}$$

$$= \frac{x^2 - x - 30}{(x + 5)(x + 6)(x + 4)} = \frac{\cancel{(x + 5)}(x - 6)}{\cancel{(x + 5)}(x + 6)(x + 4)}$$

$$= \frac{(x - 6)}{(x + 6)(x + 4)}$$

Do Exercises 1–6. ►

Add or subtract and, if possible, simplify.

1. $\dfrac{a - 3b}{a + b} + \dfrac{a + 5b}{a + b}$

2. $\dfrac{x^2 - 5}{3x^2 - 5x - 2} + \dfrac{x + 1}{3x - 6}$

3. $\dfrac{a^2 + 1}{a^2 - 1} - \dfrac{a - 1}{a + 1}$

4. $\dfrac{9x + 2}{3x^2 - 2x - 8} + \dfrac{7}{3x^2 + x - 4}$

5. $\dfrac{y}{y^2 - y - 20} - \dfrac{2}{y + 4}$

6. $\dfrac{3y}{y^2 - 7y + 10} - \dfrac{2y}{y^2 - 8y + 15}$

JUST IN TIME 21

SIMPLIFY COMPLEX RATIONAL EXPRESSIONS

A **complex rational expression** has rational expressions in its numerator or its denominator or both.

To simplify a complex rational expression: First, add or subtract, if necessary, to get a single rational expression in the numerator and in the denominator. Then divide by multiplying by the reciprocal of the denominator.

EXAMPLE

$$\frac{\frac{1}{a}+\frac{1}{b}}{\frac{1}{a^3}+\frac{1}{b^3}} = \frac{\frac{1}{a}\cdot\frac{b}{b}+\frac{1}{b}\cdot\frac{a}{a}}{\frac{1}{a^3}\cdot\frac{b^3}{b^3}+\frac{1}{b^3}\cdot\frac{a^3}{a^3}}$$

← The LCD is ab.

← The LCD is a^3b^3.

$$= \frac{\frac{b}{ab}+\frac{a}{ab}}{\frac{b^3}{a^3b^3}+\frac{a^3}{a^3b^3}}$$

$$= \frac{\frac{b+a}{ab}}{\frac{b^3+a^3}{a^3b^3}}$$

$$= \frac{b+a}{ab}\cdot\frac{a^3b^3}{b^3+a^3}$$

$$= \frac{(b+a)(a)(b)(a^2b^2)}{(a)(b)(b+a)(b^2-ba+a^2)}$$

$$= \frac{a^2b^2}{b^2-ba+a^2}$$

Do Exercises 1–5. ►

Simplify.

1. $\dfrac{\frac{x}{y}-\frac{y}{x}}{\frac{1}{y}+\frac{1}{x}}$

2. $\dfrac{\frac{a-b}{b}}{\frac{a^2-b^2}{ab}}$

3. $\dfrac{w+\frac{8}{w^2}}{1+\frac{2}{w}}$

4. $\dfrac{\frac{x^2-y^2}{xy}}{\frac{x-y}{y}}$

5. $\dfrac{\frac{a}{b}-\frac{b}{a}}{\frac{1}{a}-\frac{1}{b}}$

Note: $b-a=-1(a-b)$.

JUST IN TIME 22

SIMPLIFY RADICAL EXPRESSIONS

The symbol $\sqrt{a}$ denotes the nonnegative square root of a, and the symbol $\sqrt[3]{a}$ denotes the real-number cube root of a. The symbol $\sqrt[n]{a}$ denotes the nth root of a; that is, a number whose nth power is a. The symbol $\sqrt[n]{\ }$ is called a **radical**, and the expression under the radical is called the **radicand**. The number n (which is omitted when it is 2) is called the **index**.

Any real number has only one real-number odd root. Any positive number has two square roots, one positive and one negative. Similarly, for any even index, a positive number has two real-number roots. The positive root is called the **principal root**.

EXAMPLES Simplify.

1. $\sqrt{36} = 6$ because $6 \cdot 6 = 36$.
2. $-\sqrt{36} = -6$
3. $\sqrt[3]{-8} = -2$
4. $\sqrt[5]{\dfrac{32}{243}} = \dfrac{2}{3}$
5. $\sqrt[4]{-16}$ is not a real number.

Properties of Radicals

Let a and b be any real numbers or expressions for which the given roots exist. For any natural numbers m and n $(n \neq 1)$:

1. If n is even, $\sqrt[n]{a^n} = |a|$.
2. If n is odd, $\sqrt[n]{a^n} = a$.
3. $\sqrt[n]{a} \cdot \sqrt[n]{b} = \sqrt[n]{ab}$.
4. $\sqrt[n]{\dfrac{a}{b}} = \dfrac{\sqrt[n]{a}}{\sqrt[n]{b}}$ $(b \neq 0)$.
5. $\sqrt[n]{a^m} = (\sqrt[n]{a})^m$.

Here, we assume that no radicands are formed by raising negative quantities to even powers and, consequently, we will not use absolute-value notation when we simplify radical expressions involving variables.

EXAMPLES Simplify.

6. $\sqrt{(-5)^2} = |-5| = 5$
7. $\sqrt[3]{(-5)^3} = -5$
8. $\sqrt[4]{4} \cdot \sqrt[4]{5} = \sqrt[4]{4 \cdot 5} = \sqrt[4]{20}$
9. $\sqrt[3]{8^5} = (\sqrt[3]{8})^5 = 2^5 = 32$
10. $\sqrt{50} = \sqrt{25 \cdot 2} = \sqrt{25} \cdot \sqrt{2} = 5\sqrt{2}$
11. $\dfrac{\sqrt{72}}{\sqrt{6}} = \sqrt{\dfrac{72}{6}} = \sqrt{12} = \sqrt{4 \cdot 3} = \sqrt{4} \cdot \sqrt{3} = 2\sqrt{3}$
12. $\sqrt{216x^5y^3} = \sqrt{36 \cdot 6 \cdot x^4 \cdot x \cdot y^2 \cdot y} = \sqrt{36x^4y^2}\sqrt{6xy}$
    $= 6x^2y\sqrt{6xy}$
13. $8\sqrt{50} - 3\sqrt{8} = 8\sqrt{25 \cdot 2} - 3\sqrt{4 \cdot 2} = 8 \cdot 5\sqrt{2} - 3 \cdot 2\sqrt{2}$
    $= 40\sqrt{2} - 6\sqrt{2} = 34\sqrt{2}$
14. $(5 - \sqrt{2})(4 + 3\sqrt{2}) = 20 + 15\sqrt{2} - 4\sqrt{2} - 3(\sqrt{2})^2$
    $= 20 + 11\sqrt{2} - 6 = 14 + 11\sqrt{2}$

Do Exercises 1–20. ►

Simplify. Assume that no radicands were formed by raising negative quantities to even powers.

1. $\sqrt{(-21)^2}$
2. $\sqrt{9y^2}$
3. $\sqrt{(a-2)^2}$
4. $\sqrt[3]{-27x^3}$
5. $\sqrt[4]{81x^8}$
6. $\sqrt[5]{32}$
7. $\sqrt[4]{48x^6y^4}$
8. $\sqrt{15}\sqrt{35}$
9. $\dfrac{\sqrt{40xy}}{\sqrt{8x}}$
10. $\dfrac{\sqrt[3]{3x^2}}{\sqrt[3]{24x^5}}$
11. $\sqrt{x^2 - 4x + 4}$
12. $\sqrt{2x^3y}\sqrt{12xy}$
13. $\sqrt[3]{3x^2y}\sqrt[3]{36x}$
14. $5\sqrt{2} + 3\sqrt{32}$
15. $7\sqrt{12} - 2\sqrt{3}$
16. $2\sqrt{32} + 3\sqrt{8} - 4\sqrt{18}$
17. $6\sqrt{20} - 4\sqrt{45} + \sqrt{80}$
18. $(2 + \sqrt{3})(5 + 2\sqrt{3})$
19. $(\sqrt{8} + 2\sqrt{5})(\sqrt{8} - 2\sqrt{5})$
20. $(1 + \sqrt{3})^2$

JUST IN TIME 23

RATIONALIZING DENOMINATORS

There are times when we need to remove the radicals in a denominator. This is called **rationalizing the denominator**. It is done by multiplying by 1 in such a way as to obtain a perfect *n*th power.

EXAMPLES Rationalize the denominator.

1. $\sqrt{\frac{3}{2}} = \sqrt{\frac{3}{2}\cdot\frac{2}{2}} = \sqrt{\frac{6}{4}} = \frac{\sqrt{6}}{\sqrt{4}} = \frac{\sqrt{6}}{2}$

2. $\frac{2}{\sqrt{3}} = \frac{2}{\sqrt{3}}\cdot\frac{\sqrt{3}}{\sqrt{3}} = \frac{2\sqrt{3}}{3}$

3. $\frac{\sqrt[3]{7}}{\sqrt[3]{9}} = \frac{\sqrt[3]{7}}{\sqrt[3]{9}}\cdot\frac{\sqrt[3]{3}}{\sqrt[3]{3}} = \frac{\sqrt[3]{21}}{\sqrt[3]{27}} = \frac{\sqrt[3]{21}}{3}$

Pairs of expressions of the form $a\sqrt{b} + c\sqrt{d}$ and $a\sqrt{b} - c\sqrt{d}$ are called **conjugates**. The product of such a pair contains no radicals and can be used to rationalize a denominator or a numerator.

EXAMPLE 4 Rationalize the denominator: $\frac{7}{3 + \sqrt{5}}$.

$$\begin{aligned}\frac{7}{3 + \sqrt{5}} &= \frac{7}{3 + \sqrt{5}}\cdot\frac{3 - \sqrt{5}}{3 - \sqrt{5}}\\ &= \frac{21 - 7\sqrt{5}}{3^2 - 3\sqrt{5} + 3\sqrt{5} - (\sqrt{5})^2}\\ &= \frac{21 - 7\sqrt{5}}{9 - 5}\\ &= \frac{21 - 7\sqrt{5}}{4}\end{aligned}$$

Do Exercises 1–7. ►

Rationalize the denominator.

1. $\frac{4}{\sqrt{11}}$

2. $\sqrt{\frac{3}{7}}$

3. $\frac{\sqrt[3]{7}}{\sqrt[3]{2}}$

4. $\sqrt[3]{\frac{16}{9}}$

5. $\frac{3}{\sqrt{30} - 4}$

6. $\frac{1 - \sqrt{2}}{\sqrt{3} - \sqrt{6}}$

7. $\frac{6}{\sqrt{m} - \sqrt{n}}$

JUST IN TIME 24

RATIONAL EXPONENTS

For any real number a and any natural numbers m and n, $n \neq 1$, for which $\sqrt[n]{a}$ exists:

$$a^{1/n} = \sqrt[n]{a}, \quad a^{m/n} = \left(\sqrt[n]{a}\right)^m = \sqrt[n]{a^m}, \quad \text{and} \quad a^{-m/n} = \frac{1}{a^{m/n}}.$$

EXAMPLES Convert to radical notation and, if possible, simplify.

1. $m^{1/6} = \sqrt[6]{m}$
2. $7^{3/4} = \sqrt[4]{7^3}$, or $\left(\sqrt[4]{7}\right)^3$
3. $8^{-5/3} = \dfrac{1}{8^{5/3}} = \dfrac{1}{\left(\sqrt[3]{8}\right)^5} = \dfrac{1}{2^5} = \dfrac{1}{32}$

EXAMPLES Convert to exponential notation.

4. $\left(\sqrt[4]{7xy}\right)^5 = (7xy)^{5/4}$
5. $\sqrt[6]{x^3} = x^{3/6} = x^{1/2}$

EXAMPLES Simplify and then, if appropriate, write radical notation.

6. $x^{5/6} \cdot x^{2/3} = x^{5/6+2/3} = x^{9/6} = x^{3/2} = \sqrt{x^3}$
$= \sqrt{x^2}\sqrt{x} = x\sqrt{x}$
7. $(x+3)^{5/2}(x+3)^{-1/2} = (x+3)^{5/2-1/2} = (x+3)^2$

Do Exercises 1–11. ►

Convert to radical notation and, if possible, simplify.

1. $y^{5/6}$
2. $x^{2/3}$
3. $16^{3/4}$
4. $4^{7/2}$
5. $125^{-1/3}$
6. $32^{-4/5}$

Convert to exponential notation.

7. $\sqrt[12]{y^4}$
8. $\sqrt{x^5}$

Simplify and then, if appropriate, write radical notation.

9. $x^{1/2} \cdot x^{2/3}$
10. $(a-2)^{9/4}(a-2)^{-1/4}$
11. $(m^{1/2}n^{5/2})^{2/3}$

JUST IN TIME 25

PYTHAGOREAN THEOREM

A **right triangle** is a triangle with a 90° angle, as shown in the following figure. The small square in the corner indicates the 90° angle.

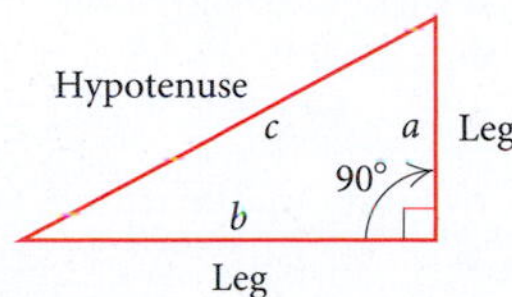

In a right triangle, the longest side is called the **hypotenuse**. It is also the side opposite the right angle. The other two sides are called **legs**. We generally use the letters a and b for the lengths of the legs and c for the length of the hypotenuse. They are related as follows.

The Pythagorean Theorem

In any right triangle, if a and b are the lengths of the legs and c is the length of the hypotenuse, then

$$a^2 + b^2 = c^2.$$

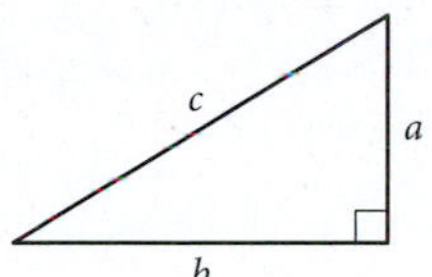

The equation $a^2 + b^2 = c^2$ is called the **Pythagorean theorem**.

EXAMPLE 1 Find the length of the hypotenuse of this right triangle. Give an exact answer and an approximation to three decimal places.

$$4^2 + 5^2 = c^2$$
$$16 + 25 = c^2$$
$$41 = c^2$$
$$c = \sqrt{41}$$
$$c \approx 6.403$$

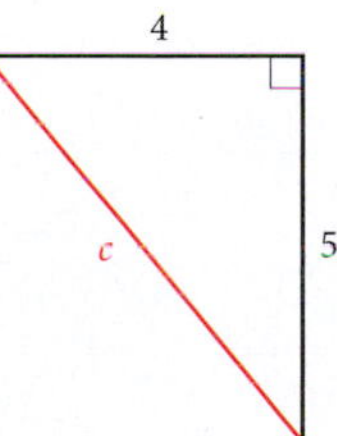

EXAMPLE 2 Find the length of leg b of this right triangle. Give an exact answer and an approximation to three decimal places.

$$10^2 + b^2 = 12^2$$
$$100 + b^2 = 144$$
$$b^2 = 144 - 100$$
$$b^2 = 44$$
$$b = \sqrt{44}$$
$$b \approx 6.633$$

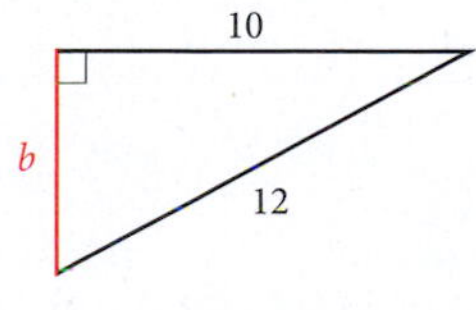

Do Exercises 1–5. ▶

Find the length of the third side of each right triangle. Where appropriate, give both an exact answer and an approximation to three decimal places.

1.
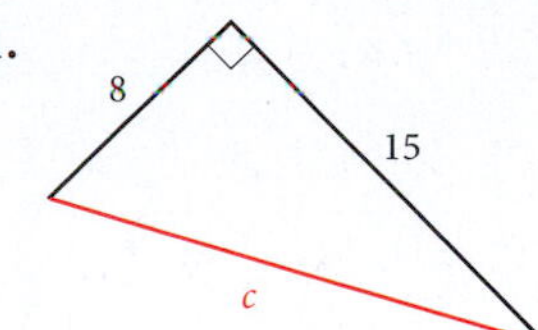

2.
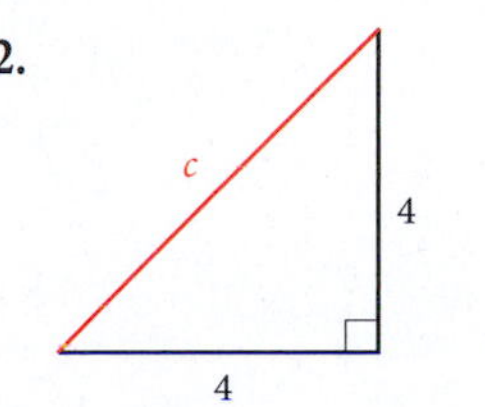

3.
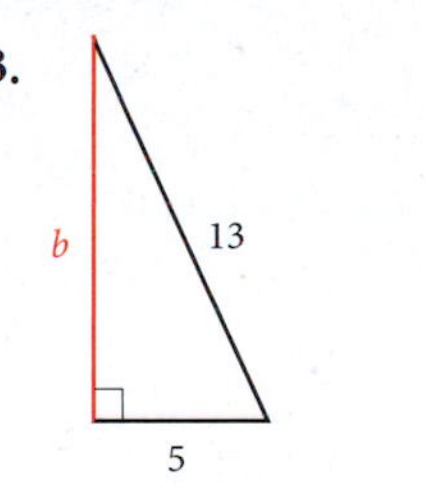

4.
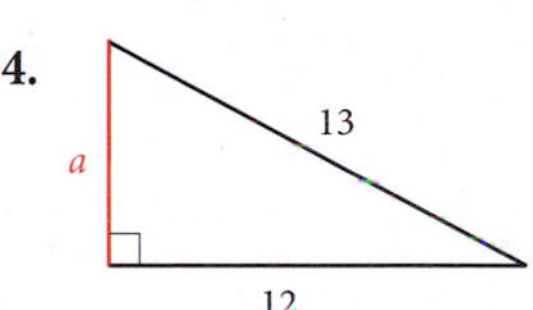

5.
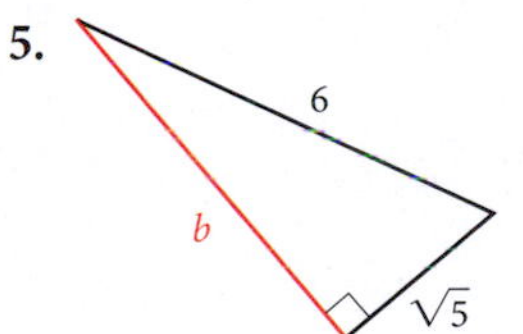

Answers

Chapter 1

Visualizing the Graph

1. H **2.** B **3.** D **4.** A **5.** G **6.** I **7.** C
8. J **9.** F **10.** E

Exercise Set 1.1

1. *A*: $(-5, 4)$; *B*: $(2, -2)$; *C*: $(0, -5)$; *D*: $(3, 5)$; *E*: $(-5, -4)$; *F*: $(3, 0)$

3.

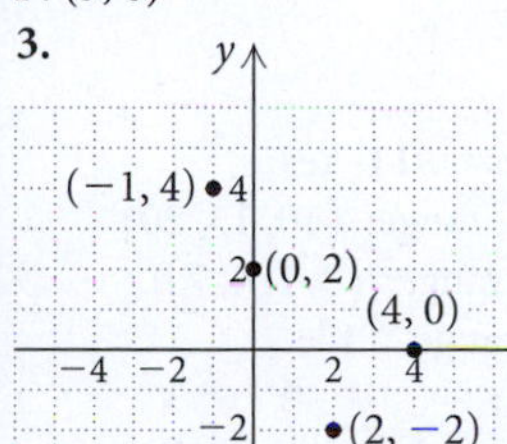

5.

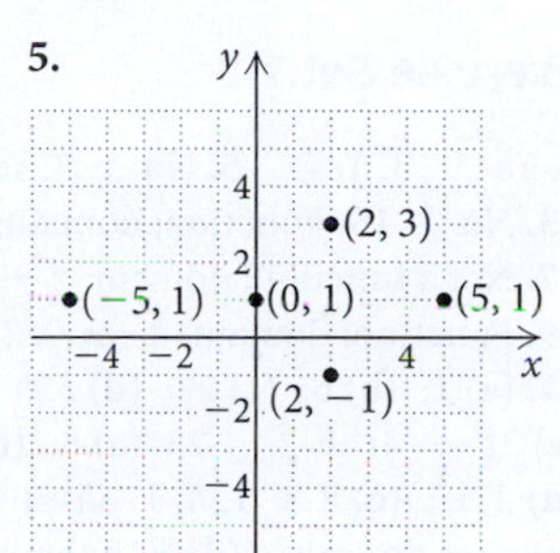

7. (2008, 10), (2009, 15), (2010, 9), (2011, 9), (2012, 12), (2013, 5) **9.** Yes; no **11.** Yes; no
13. No; yes **15.** No; yes

17. *x*-intercept: $(-3, 0)$; *y*-intercept: $(0, 5)$;

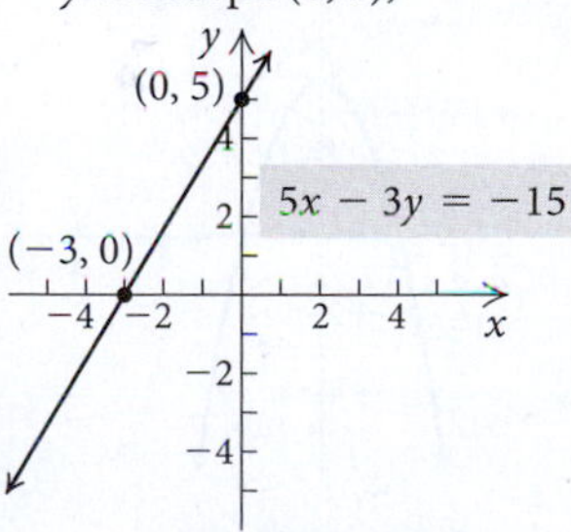

19. *x*-intercept: $(2, 0)$; *y*-intercept: $(0, 4)$;

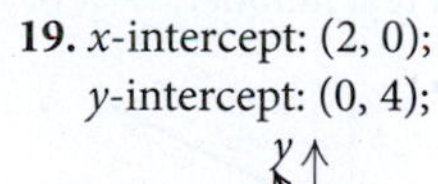

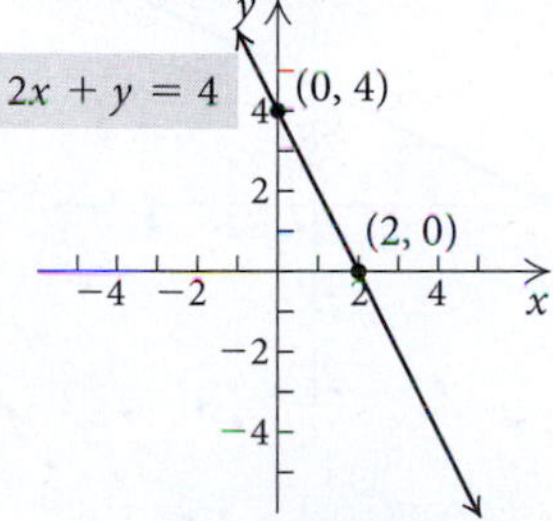

21. *x*-intercept: $(-4, 0)$; *y*-intercept: $(0, 3)$;

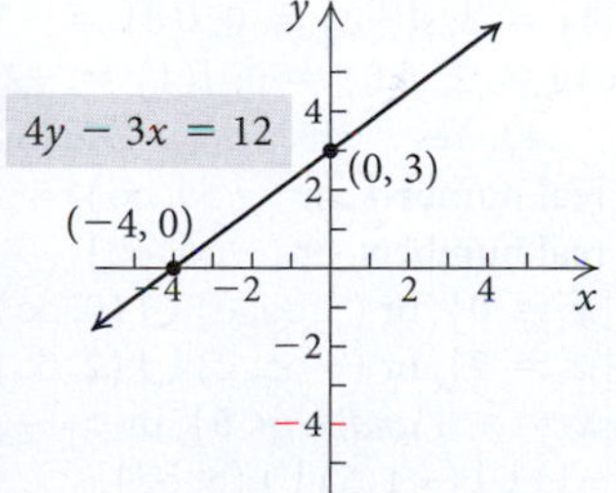

23.

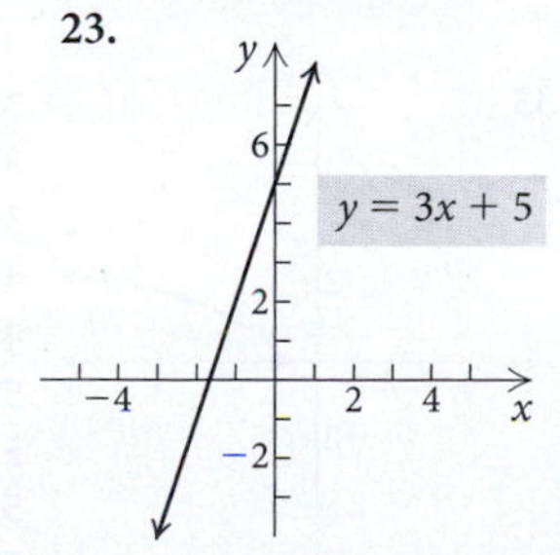

25.

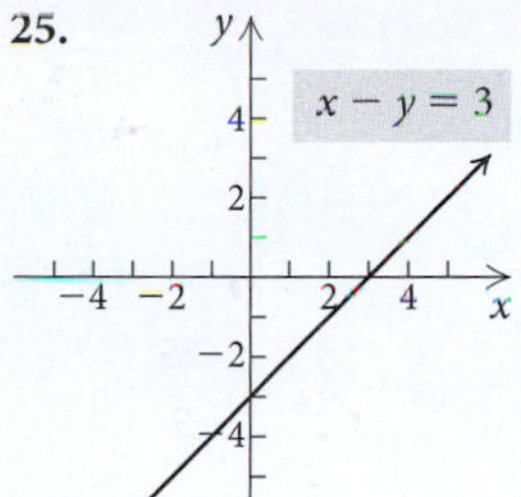

27.

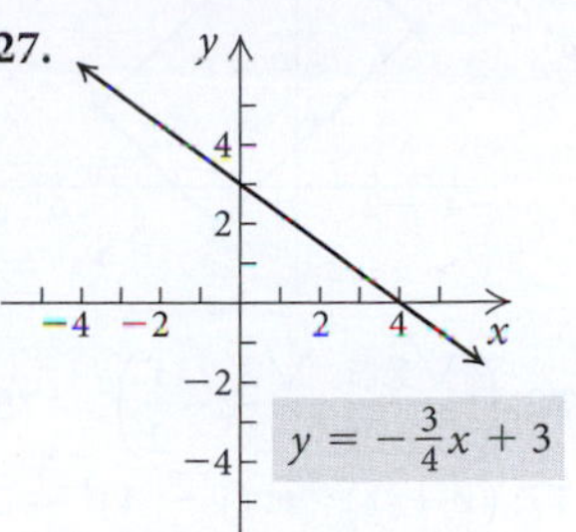

29.

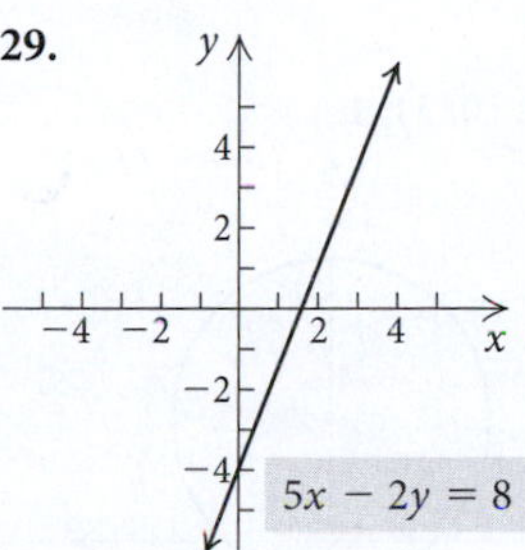

31.

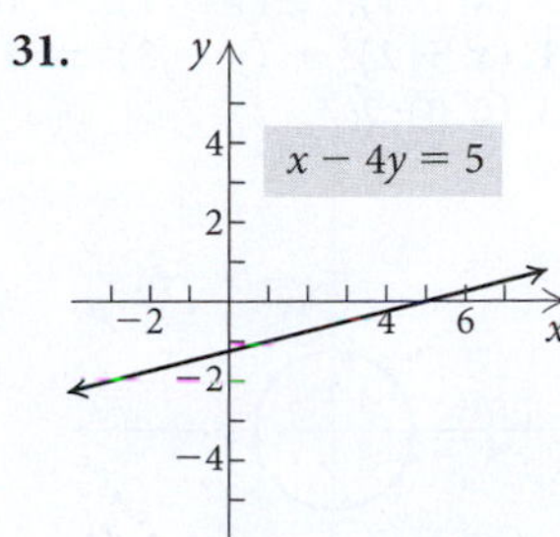

33.

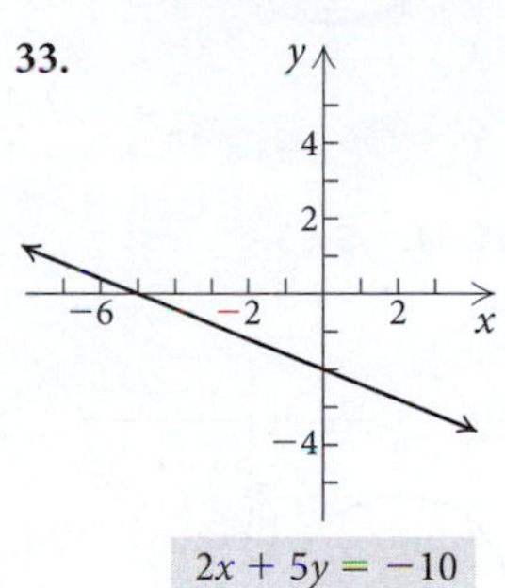

35.

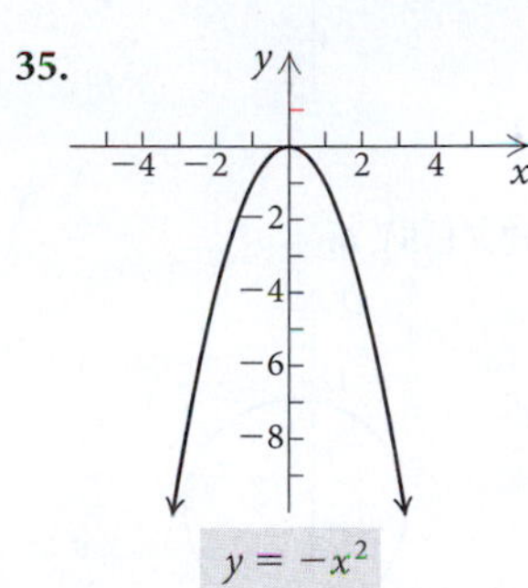

37.

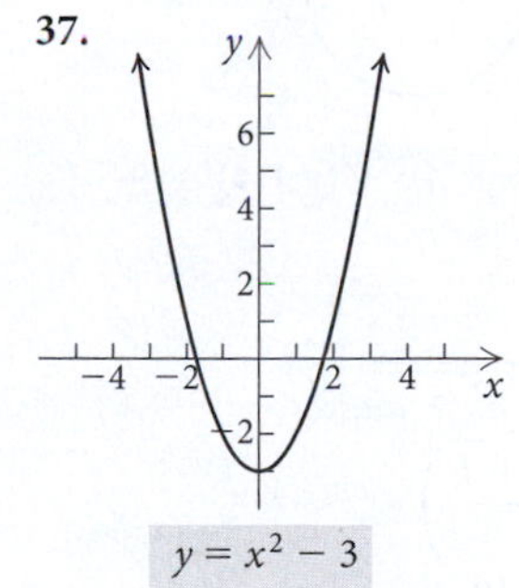

39.

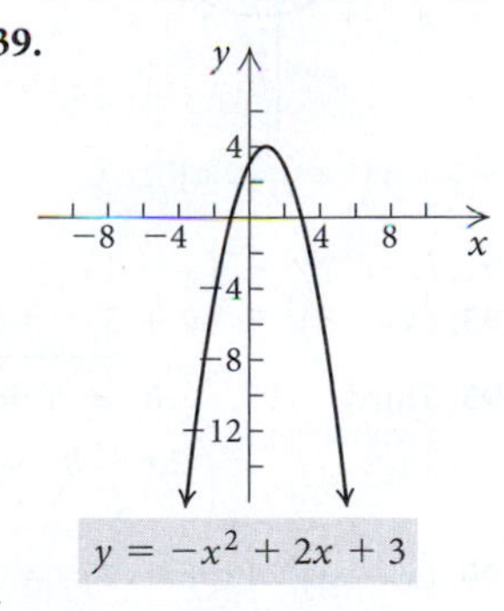

41. $\sqrt{10}$, 3.162 **43.** 13 **45.** $\sqrt{45}$, 6.708 **47.** 16
49. $\frac{14}{3}$ **51.** $\sqrt{128.05}$, 11.316 **53.** $\sqrt{a^2 + b^2}$ **55.** 6.5

57. Yes **59.** No **61.** $(-4, -6)$ **63.** $\left(-\frac{1}{5}, \frac{1}{4}\right)$
65. $(4.95, -4.95)$ **67.** $\left(-6, \frac{13}{2}\right)$ **69.** $\left(-\frac{5}{12}, \frac{13}{40}\right)$
71. $\left(-\frac{1}{2}, \frac{3}{2}\right), \left(\frac{7}{2}, \frac{1}{2}\right), \left(\frac{5}{2}, \frac{9}{2}\right), \left(-\frac{3}{2}, \frac{11}{2}\right)$; no

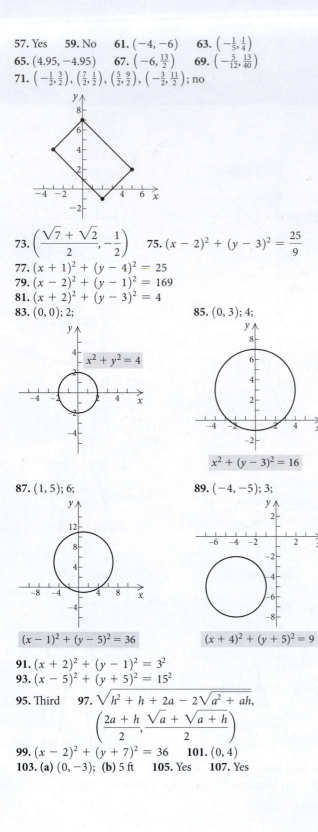

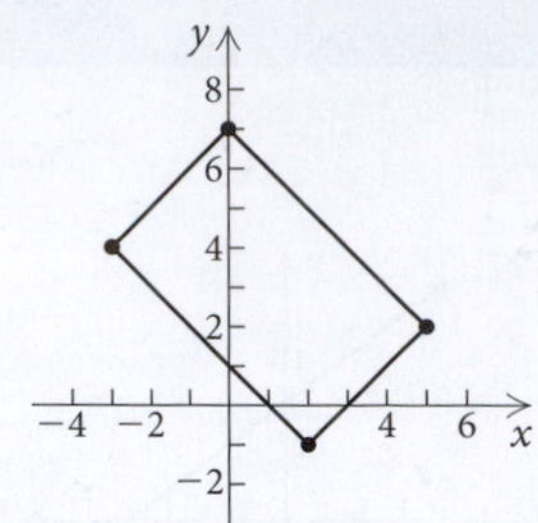

73. $\left(\dfrac{\sqrt{7} + \sqrt{2}}{2}, -\dfrac{1}{2}\right)$ **75.** $(x - 2)^2 + (y - 3)^2 = \dfrac{25}{9}$
77. $(x + 1)^2 + (y - 4)^2 = 25$
79. $(x - 2)^2 + (y - 1)^2 = 169$
81. $(x + 2)^2 + (y - 3)^2 = 4$
83. $(0, 0)$; 2;

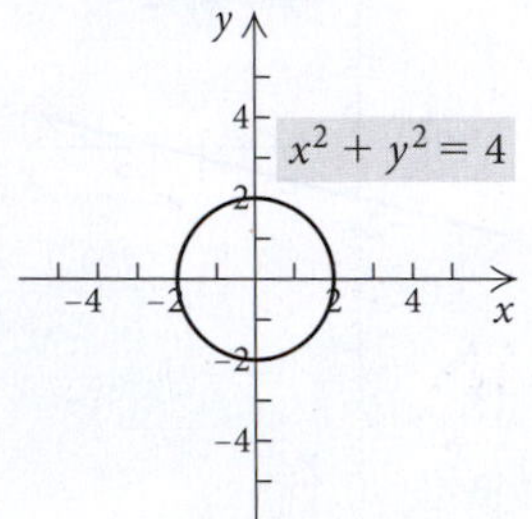

85. $(0, 3)$; 4;

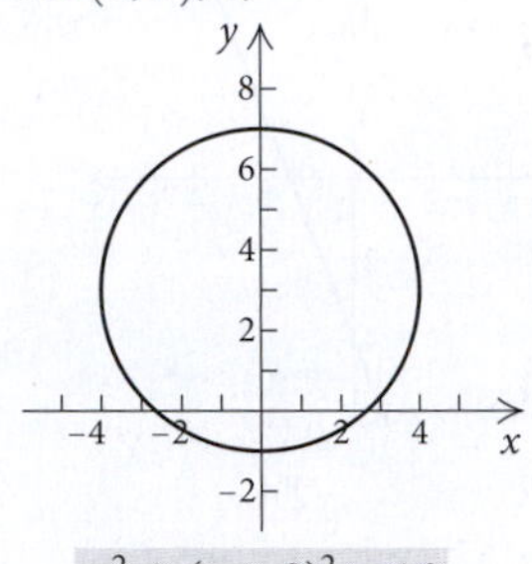

$x^2 + (y - 3)^2 = 16$

87. $(1, 5)$; 6;

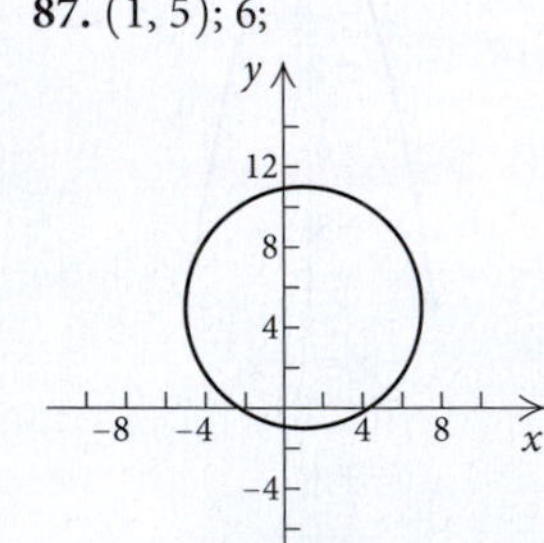

$(x - 1)^2 + (y - 5)^2 = 36$

89. $(-4, -5)$; 3;

$(x + 4)^2 + (y + 5)^2 = 9$

91. $(x + 2)^2 + (y - 1)^2 = 3^2$
93. $(x - 5)^2 + (y + 5)^2 = 15^2$
95. Third **97.** $\sqrt{h^2 + h + 2a - 2\sqrt{a^2 + ah}}$,
$\left(\dfrac{2a + h}{2}, \dfrac{\sqrt{a} + \sqrt{a + h}}{2}\right)$
99. $(x - 2)^2 + (y + 7)^2 = 36$ **101.** $(0, 4)$
103. **(a)** $(0, -3)$; **(b)** 5 ft **105.** Yes **107.** Yes

109. Let $P_1 = (x_1, y_1)$, $P_2 = (x_2, y_2)$, and $M = \left(\dfrac{x_1 + x_2}{2}, \dfrac{y_1 + y_2}{2}\right)$. Let $d(AB)$ denote the distance from point A to point B.

$$d(P_1M) = \sqrt{\left(\frac{x_1 + x_2}{2} - x_1\right)^2 + \left(\frac{y_1 + y_2}{2} - y_1\right)^2}$$
$$= \frac{1}{2}\sqrt{(x_2 - x_1)^2 + (y_2 - y_1)^2};$$
$$d(P_2M) = \sqrt{\left(\frac{x_1 + x_2}{2} - x_2\right)^2 + \left(\frac{y_1 + y_2}{2} - y_2\right)^2}$$
$$= \frac{1}{2}\sqrt{(x_1 - x_2)^2 + (y_1 - y_2)^2}$$
$$= \frac{1}{2}\sqrt{(x_2 - x_1)^2 + (y_2 - y_1)^2} = d(P_1M)$$

Exercise Set 1.2

1. Yes **3.** Yes **5.** No **7.** Yes **9.** Yes **11.** Yes
13. No **15.** Function; domain: $\{2, 3, 4\}$; range: $\{10, 15, 20\}$
17. Not a function; domain: $\{-7, -2, 0\}$; range: $\{3, 1, 4, 7\}$
19. Function; domain: $\{-2, 0, 2, 4, -3\}$; range: $\{1\}$
21. **(a)** 1; **(b)** 6; **(c)** 22; **(d)** $3x^2 + 2x + 1$;
(e) $3t^2 - 4t + 2$ **23.** **(a)** 8; **(b)** -8; **(c)** $-x^3$;
(d) $27y^3$; **(e)** $8 + 12h + 6h^2 + h^3$ **25.** **(a)** $\frac{1}{8}$; **(b)** 0;
(c) does not exist; **(d)** $\frac{81}{53}$, or approximately 1.5283;
(e) $\dfrac{x + h - 4}{x + h + 3}$ **27.** 0; does not exist; does not exist as a real number; $\dfrac{1}{\sqrt{3}}$, or $\dfrac{\sqrt{3}}{3}$

29.

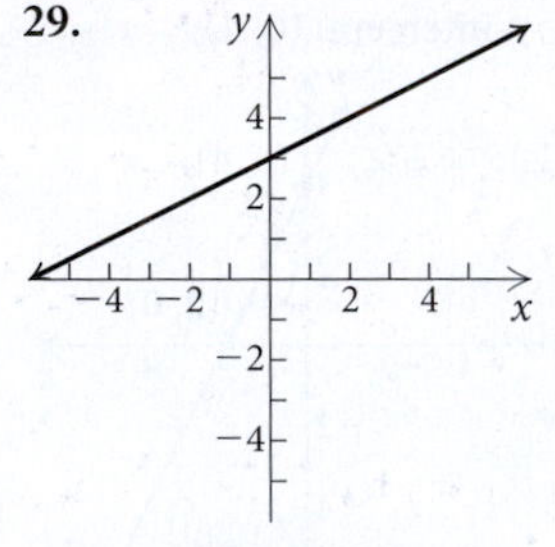

$f(x) = \frac{1}{2}x + 3$

31.

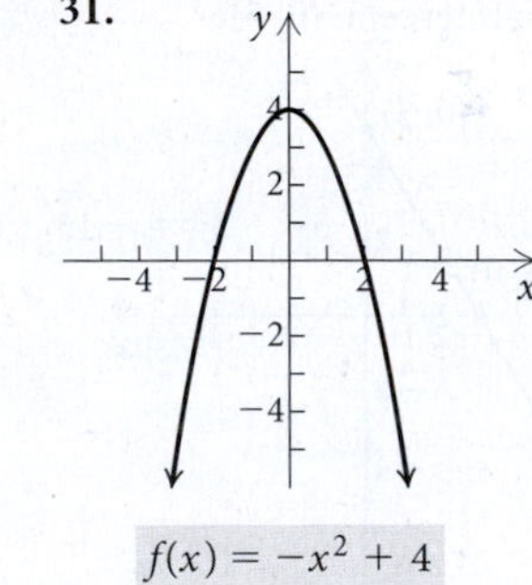

$f(x) = -x^2 + 4$

33.

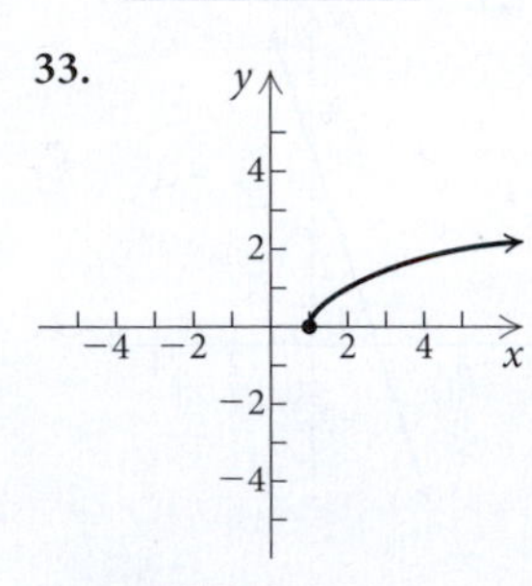

$f(x) = \sqrt{x - 1}$

35. $h(1) = -2$; $h(3) = 2$; $h(4) = 1$
37. $s(-4) = 3$; $s(-2) = 0$; $s(0) = -3$
39. $f(-1) = 2$; $f(0) = 0$; $f(1) = -2$
41. No **43.** Yes **45.** Yes **47.** No
49. All real numbers, or $(-\infty, \infty)$
51. All real numbers, or $(-\infty, \infty)$
53. $\{x | x \neq 0\}$, or $(-\infty, 0) \cup (0, \infty)$
55. $\{x | x \neq 2\}$, or $(-\infty, 2) \cup (2, \infty)$
57. $\{x | x \neq -1 \text{ and } x \neq 5\}$, or $(-\infty, -1) \cup (-1, 5) \cup (5, \infty)$

59. All real numbers, or $(-\infty, \infty)$
61. $\{x \mid x \neq 0 \text{ and } x \neq 7\}$, or $(-\infty, 0) \cup (0, 7) \cup (7, \infty)$
63. All real numbers, or $(-\infty, \infty)$
65. Domain: $[0, 5]$; range: $[0, 3]$
67. Domain: $[-2\pi, 2\pi]$; range: $[-1, 1]$
69. Domain: $(-\infty, \infty)$; range: $\{-3\}$
71. Domain: $[-5, 3]$; range: $[-2, 2]$
73. Domain: all real numbers, or $(-\infty, \infty)$; range: $[0, \infty)$
75. Domain: all real numbers, or $(-\infty, \infty)$; range: all real numbers, or $(-\infty, \infty)$ **77.** Domain: $(-\infty, 3) \cup (3, \infty)$; range: $(-\infty, 0) \cup (0, \infty)$ **79.** Domain: all real numbers, or $(-\infty, \infty)$; range: all real numbers, or $(-\infty, \infty)$ **81.** Domain: $(-\infty, 7]$; range: $[0, \infty)$ **83.** Domain: all real numbers, or $(-\infty, \infty)$; range: $(-\infty, 3]$ **85. (a)** 2018: \$25.21; 2025: \$28.23; **(b)** about 49 years after 1985, or in 2034
87. 645 m; 0 m **89.** $(-3, -2)$, yes; $(2, -3)$, no
90. $(0, -7)$, no; $(8, 11)$, yes **91.** $\left(\frac{4}{5}, -2\right)$, yes; $\left(\frac{11}{5}, \frac{1}{10}\right)$, yes

92.

$y = (x-1)^2$

93.

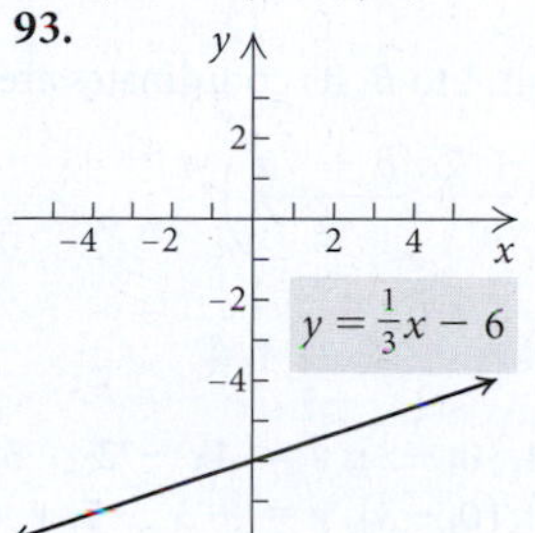

94.

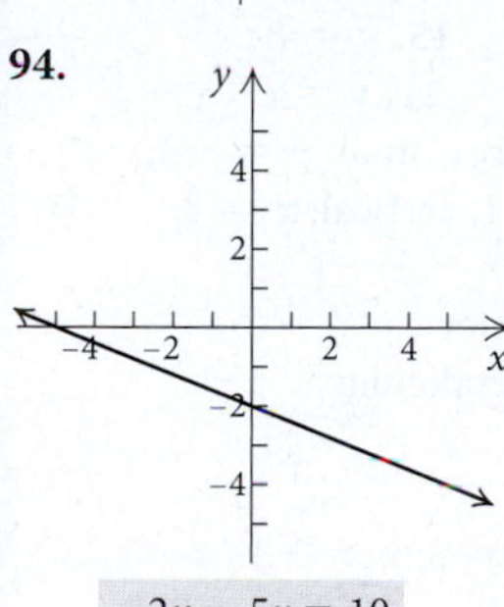

95.

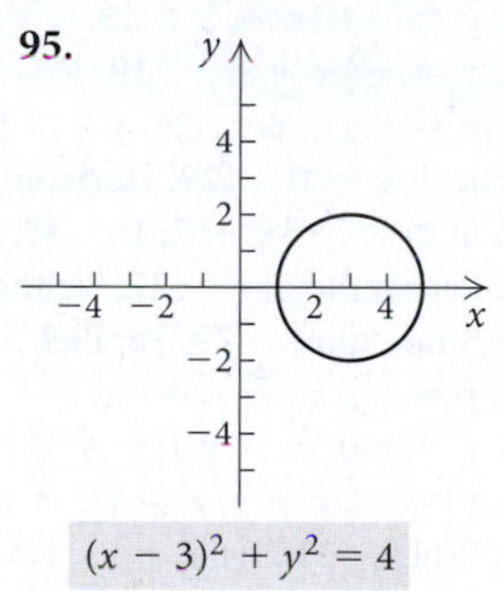

97. $\{x \mid x \geq -1 \text{ and } x \neq 0\}$, or $[-1, 0) \cup (0, \infty)$
99. $\{x \mid 0 \leq x \leq 4\}$, or $[0, 4]$

101.

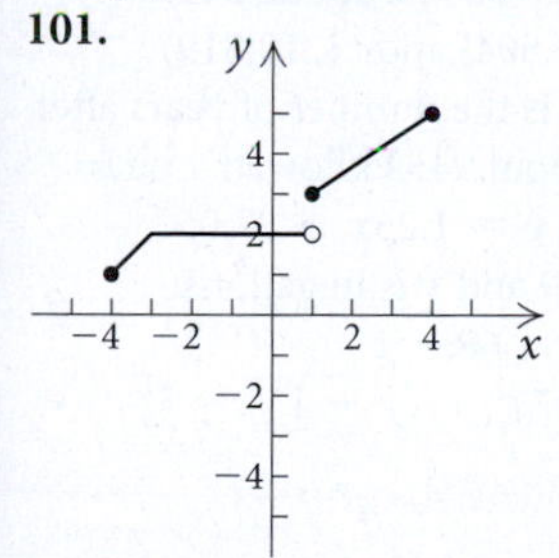

103. (a) $f(x) = -7$; **(b)** $f(x) = 2x - 1$; **(c)** $f(x) = 7$

Visualizing the Graph

1. E **2.** D **3.** A **4.** J **5.** C **6.** F **7.** H
8. G **9.** B **10.** I

Exercise Set 1.3

1. (a) Yes; **(b)** yes; **(c)** yes **3. (a)** Yes; **(b)** no; **(c)** no **5.** $\frac{6}{5}$ **7.** $-\frac{3}{5}$ **9.** 0 **11.** $\frac{1}{5}$ **13.** Not defined **15.** 0.3 **17.** 0 **19.** $-\frac{6}{5}$ **21.** $-\frac{1}{3}$ **23.** Not defined **25.** -2 **27.** 5 **29.** 0 **31.** 1.3 **33.** Not defined **35.** $-\frac{1}{2}$ **37.** -1 **39.** 0 **41.** The average rate of change in sales of electric bicycles from 2013 to 2020 is expected to be \$0.343 billion, or \$343 million. **43.** The average rate of change in the population of Cleveland, Ohio, over the 12-year period was about -7290 people per year. **45.** The average rate of change in the number of acres used for growing almonds in California from 2003 to 2012 was about 28,889 acres per year.
47. The average rate of change in per-capita consumption of whole milk from 1970 to 2011 was about -0.5 gal per year.
49. $\frac{3}{5}$; $(0, -7)$ **51.** Slope is not defined; there is no y-intercept.
53. $-\frac{1}{2}$; $(0, 5)$ **55.** $-\frac{3}{2}$; $(0, 5)$ **57.** 0; $(0, -6)$
59. $\frac{4}{5}$; $\left(0, \frac{8}{5}\right)$ **61.** $\frac{1}{4}$; $\left(0, -\frac{1}{2}\right)$

63.

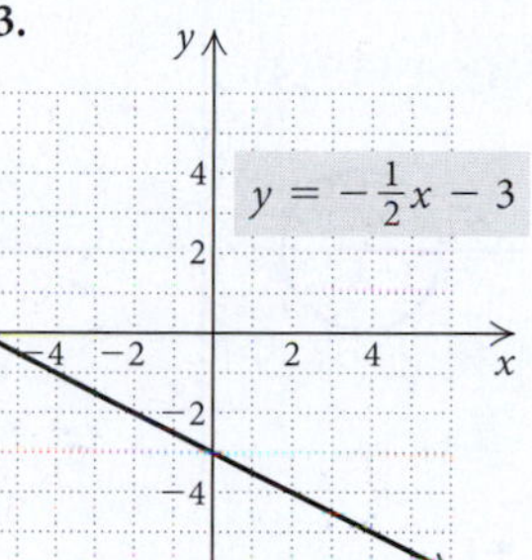

65.

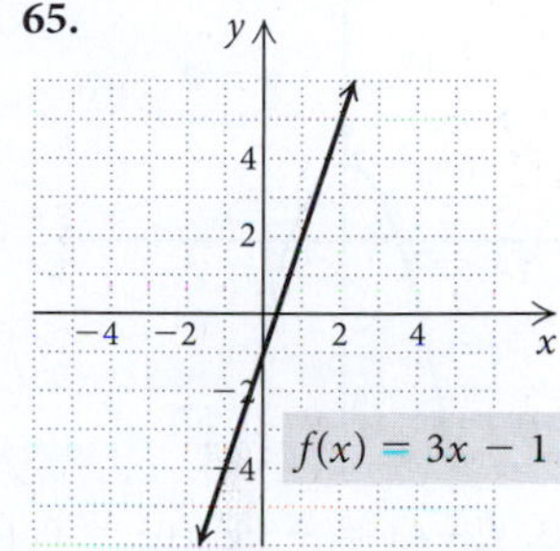

67.

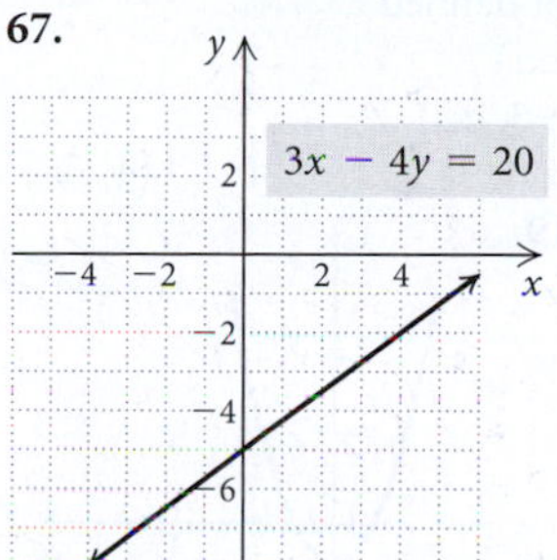

69.

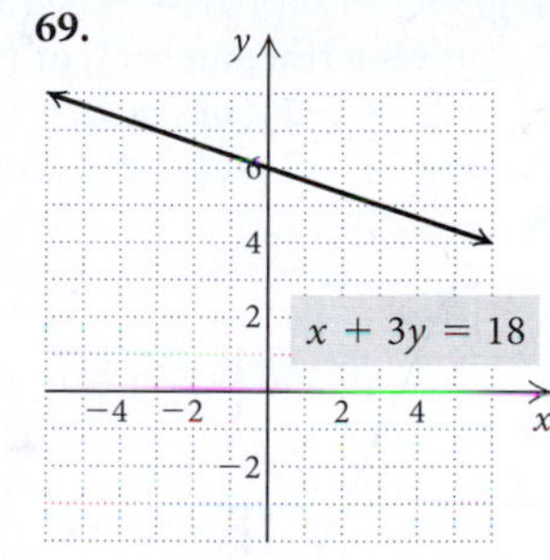

71. 1 atm, 2 atm, $31\frac{10}{33}$ atm, $152\frac{17}{33}$ atm, $213\frac{4}{33}$ atm
73. (a) $\frac{11}{10}$. For each mile per hour faster that the car travels, it takes $\frac{11}{10}$ ft longer to stop; **(b)** 6 ft, 11.5 ft, 22.5 ft, 55.5 ft, 72 ft; **(c)** $\{r \mid r > 0\}$, or $(0, \infty)$. If r is allowed to be 0, the function says that a stopped car has a reaction distance of $\frac{1}{2}$ ft.
75. $C(t) = 2250 + 3380t$; $C(20) = \$69{,}850$
77. $C(t) = 750 + 15x$; $C(32) = \$1230$
79. $-\frac{5}{4}$ **80.** 10 **81.** 40 **82.** $a^2 + 3a$
83. $a^2 + 2ah + h^2 - 3a - 3h$ **85.** $2a + h$
87. False **89.** $f(x) = x + b$

Mid-Chapter Mixed Review: Chapter 1

1. False **2.** True **3.** False **4.** x-intercept: $(5, 0)$; y-intercept: $(0, -8)$ **5.** $\sqrt{605} = 11\sqrt{5} \approx 24.6$; $\left(-\frac{5}{2}, -4\right)$
6. $\sqrt{2} \approx 1.4$; $\left(-\frac{1}{4}, -\frac{3}{10}\right)$ **7.** $(x + 5)^2 + (y - 2)^2 = 169$
8. Center: $(3, -1)$; radius: 2
9.

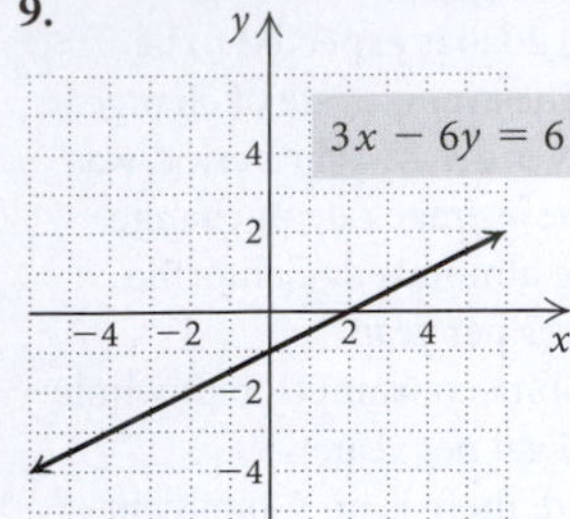

10.

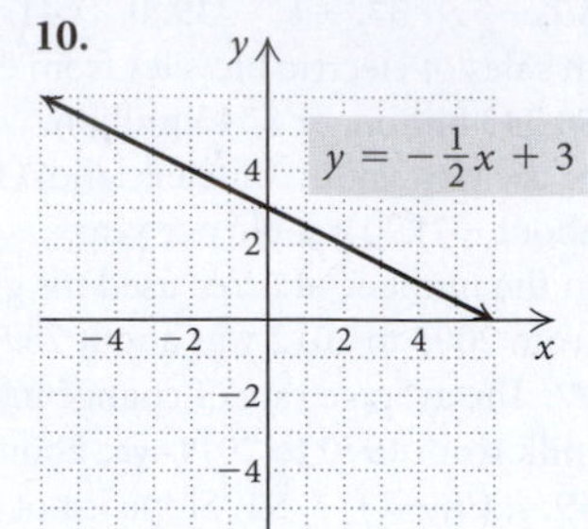

11.

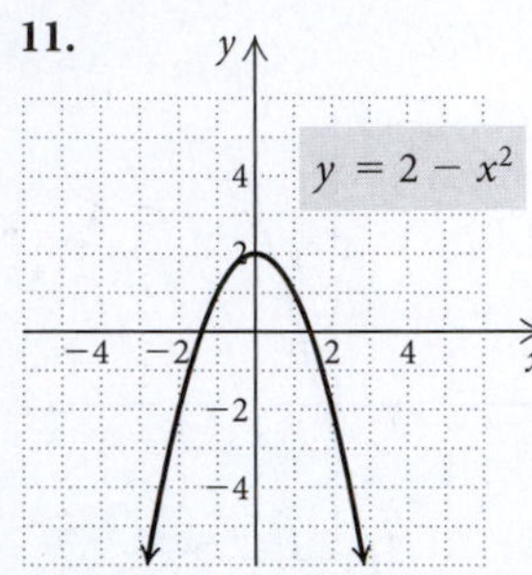

12.

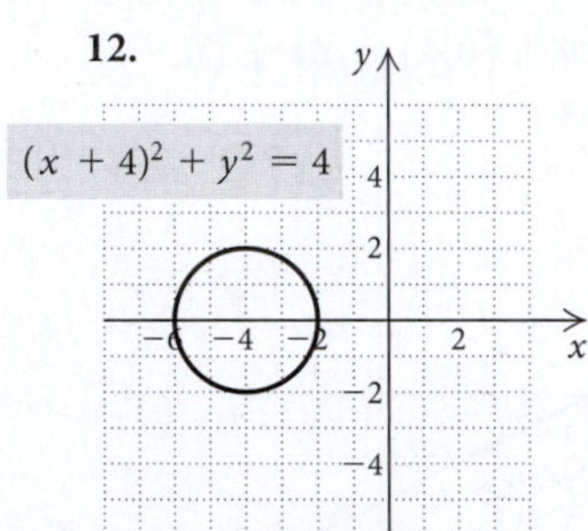

13. $f(-4) = -36$; $f(0) = 0$; $f(1) = -1$
14. $g(-6) = 0$; $g(0) = -2$; $g(3)$ is not defined
15. $\{x \mid x \text{ is a real number}\}$, or $(-\infty, \infty)$
16. $\{x \mid x \neq -5\}$, or $(-\infty, -5) \cup (-5, \infty)$
17. $\{x \mid x \neq -3 \text{ and } x \neq 1\}$, or $(-\infty, -3) \cup (-3, 1) \cup (1, \infty)$
18.

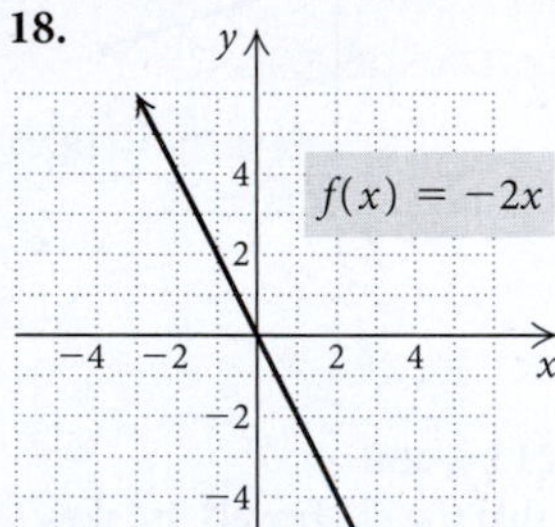

19.

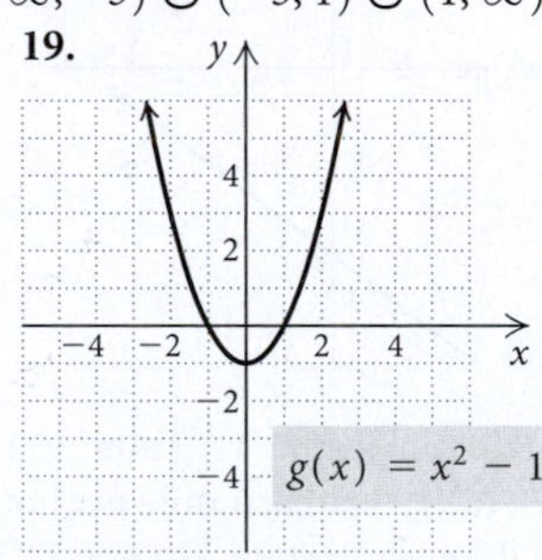

20. Domain: $[-4, 3)$; range: $[-4, 5)$ **21.** Not defined
22. $-\frac{1}{4}$ **23.** 0 **24.** Slope: $-\frac{1}{9}$; y-intercept: $(0, 12)$
25. Slope: 0; y-intercept: $(0, -6)$
26. Slope is not defined; there is no y-intercept
27. Slope: $\frac{3}{16}$; y-intercept: $\left(0, \frac{1}{16}\right)$ **28.** The sign of the slope indicates the slant of a line. A line that slants up from left to right has positive slope because corresponding changes in x and y have the same sign. A line that slants down from left to right has negative slope, because corresponding changes in x and y have opposite signs. A horizontal line has zero slope, because there is no change in y for a given change in x. A vertical line has undefined slope, because there is no change in x for a given change in y and division by 0 is not defined. The larger the absolute value of slope, the steeper the line. This is because a larger absolute value corresponds to a greater change in y, compared to the change in x, than a smaller absolute value. **29.** A vertical line $(x = a)$ crosses the graph more than once; thus, $x = a$ fails the vertical line test.
30. The domain of a function is the set of all inputs of the function. The range is the set of all outputs. The range depends on the domain. **31.** Let $A = (a, b)$ and $B = (c, d)$. The coordinates of a point C one-half of the way from A to B are $\left(\frac{a + c}{2}, \frac{b + d}{2}\right)$. A point D that is one-half of the way from C to B is $\frac{1}{2} + \frac{1}{2} \cdot \frac{1}{2}$, or $\frac{3}{4}$ of the way from A to B. Its coordinates are $\left(\frac{\frac{a + c}{2} + c}{2}, \frac{\frac{b + d}{2} + d}{2}\right)$, or $\left(\frac{a + 3c}{4}, \frac{b + 3d}{4}\right)$. Then a point E that is one-half of the way from D to B is $\frac{3}{4} + \frac{1}{2} \cdot \frac{1}{4}$, or $\frac{7}{8}$ of the way from A to B. Its coordinates are $\left(\frac{\frac{a + 3c}{4} + c}{2}, \frac{\frac{b + 3d}{4} + d}{2}\right)$, or $\left(\frac{a + 7c}{8}, \frac{b + 7d}{8}\right)$.

Exercise Set 1.4

1. 4, $(0, -2)$; $y = 4x - 2$ **3.** -1, $(0, 0)$; $y = -x$
5. 0, $(0, -3)$; $y = -3$ **7.** $y = \frac{2}{9}x + 4$ **9.** $y = -4x - 7$
11. $y = -4.2x + \frac{3}{4}$ **13.** $y = \frac{2}{9}x + \frac{19}{3}$ **15.** $y = 8$
17. $y = -\frac{3}{5}x - \frac{17}{5}$ **19.** $y = -3x + 2$ **21.** $y = -\frac{1}{2}x + \frac{7}{2}$
23. $y = \frac{2}{3}x - 6$ **25.** $y = 7.3$ **27.** Horizontal: $y = -3$; vertical: $x = 0$ **29.** Horizontal: $y = -1$; vertical: $x = \frac{2}{11}$
31. $h(x) = -3x + 7$; 1 **33.** $f(x) = \frac{2}{5}x - 1$; -1
35. Perpendicular **37.** Neither parallel nor perpendicular **39.** Parallel **41.** Perpendicular
43. $y = \frac{2}{7}x + \frac{29}{7}$; $y = -\frac{7}{2}x + \frac{31}{2}$
45. $y = -0.3x - 2.1$; $y = \frac{10}{3}x + \frac{70}{3}$
47. $y = -\frac{3}{4}x + \frac{1}{4}$; $y = \frac{4}{3}x - 6$ **49.** $x = 3$; $y = -3$
51. True **53.** True **55.** False **57.** No **59.** Yes
61. (a) Using $(1, 1.319)$ and $(7, 2.749)$ gives us $y = 0.238x + 1.081$, where x is the number of years after 2006 and y is in billions; **(b)** 2017: about 3.699 billion Internet users; 2020: about 4.413 billion Internet users **63.** Using $(0, 11{,}504)$ and $(3, 10{,}819)$ gives us $y = -228x + 11{,}504$, where x is the number of years after 2010 and y is in kilowatt-hours; 2019: about 9452 kilowatt-hours
65. Using $(1, 28.3)$ and $(3, 30.8)$ gives us $y = 1.25x + 27.05$, where x is the number of years after 2009 and y is in gallons; 2017: about 37.1 gal **67.** Not defined **68.** -1
69. $x^2 + (y - 3)^2 = 6.25$ **70.** $(x + 7)^2 + (y + 1)^2 = \frac{81}{25}$
71. -7.75 **73.** 6.7% grade; $y = 0.067x$

Exercise Set 1.5

1. 4 **3.** All real numbers, or $(-\infty, \infty)$ **5.** $-\frac{3}{4}$
7. -9 **9.** 6 **11.** No solution **13.** $\frac{11}{5}$ **15.** $\frac{35}{6}$
17. 8 **19.** -4 **21.** 6 **23.** -1 **25.** $\frac{4}{5}$

27. $-\frac{3}{2}$ **29.** $-\frac{2}{3}$ **31.** $\frac{1}{2}$ **33.** About 51,075 words
35. \$1300 **37.** 26°, 130°, 24° **39.** \$3.455 billion **41.** 3 hr
43. 25.2% **45.** \$5000 **47.** About 287,000 students
49. Length: 100 yd; width: 65 yd **51.** Length: 93 m; width: 68 m
53. 2.5 hr **55.** \$2400 at 3%; \$2600 at 4%
57. IBM: 6809 patents; Samsung: 4676 patents
59. 10,040 ft **61.** 74.25 lb **63.** 4.5 hr **65.** \$20.50
67. Italy: 30%; Spain: 20%; United States: 8% **69.** -5
71. $\frac{11}{2}$ **73.** 16 **75.** -12 **77.** 6 **79.** 20 **81.** 25
83. 15 **85.** (a) $(4, 0)$; (b) 4 **87.** (a) $(-2, 0)$; (b) -2
89. (a) $(-4, 0)$; (b) -4 **91.** $y = -\frac{3}{4}x + \frac{13}{4}$
92. $y = -\frac{3}{4}x + \frac{1}{4}$ **93.** 13 **94.** $\left(-1, \frac{1}{2}\right)$
95. $f(-3) = \frac{1}{2}; f(0) = 0; f(3)$ does not exist
96. $m = 7$; y-intercept: $\left(0, -\frac{1}{2}\right)$ **97.** Yes **99.** No
101. $-\frac{2}{3}$ **103.** No; the 6-oz cup costs about 6.4% more per ounce. **105.** 11.25 mi

Exercise Set 1.6

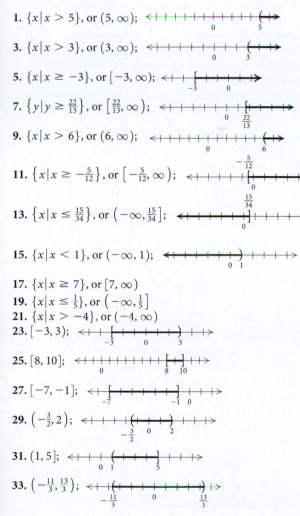

1. $\{x \mid x > 5\}$, or $(5, \infty)$;

3. $\{x \mid x > 3\}$, or $(3, \infty)$;

5. $\{x \mid x \geq -3\}$, or $[-3, \infty)$;

7. $\left\{y \mid y \geq \frac{22}{13}\right\}$, or $\left[\frac{22}{13}, \infty\right)$;

9. $\{x \mid x > 6\}$, or $(6, \infty)$;

11. $\left\{x \mid x \geq -\frac{5}{12}\right\}$, or $\left[-\frac{5}{12}, \infty\right)$;

13. $\left\{x \mid x \leq \frac{15}{34}\right\}$, or $\left(-\infty, \frac{15}{34}\right]$;

15. $\{x \mid x < 1\}$, or $(-\infty, 1)$;

17. $\{x \mid x \geq 7\}$, or $[7, \infty)$
19. $\left\{x \mid x \leq \frac{1}{5}\right\}$, or $\left(-\infty, \frac{1}{5}\right]$
21. $\{x \mid x > -4\}$, or $(-4, \infty)$
23. $[-3, 3)$;

25. $[8, 10]$;

27. $[-7, -1]$;

29. $\left(-\frac{3}{2}, 2\right)$;

31. $(1, 5]$;

33. $\left(-\frac{11}{3}, \frac{13}{3}\right)$;

35. $(-\infty, -2] \cup (1, \infty)$;

37. $\left(-\infty, -\frac{7}{2}\right] \cup \left[\frac{1}{2}, \infty\right)$;

39. $(-\infty, 9.6) \cup (10.4, \infty)$;

41. $\left(-\infty, -\frac{57}{4}\right] \cup \left[-\frac{55}{4}, \infty\right)$;

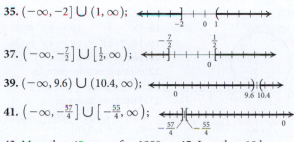

43. More than 45 years after 1980 **45.** Less than 10 hr
47. \$5000 **49.** \$300,000 **51.** Sales greater than \$18,000
53. Function; domain; range; domain; exactly one; range
54. Midpoint formula **55.** x-intercept
56. Constant; identity **57.** $\left(-\frac{1}{4}, \frac{5}{9}\right]$ **59.** $\left(-\frac{1}{8}, \frac{1}{2}\right)$

Review Exercises: Chapter 1

1. True **2.** True **3.** False **4.** False **5.** True
6. False **7.** Yes; no **8.** Yes; no
9. x-intercept: $(3, 0)$; y-intercept: $(0, -2)$;
10. x-intercept: $(2, 0)$; y-intercept: $(0, 5)$;

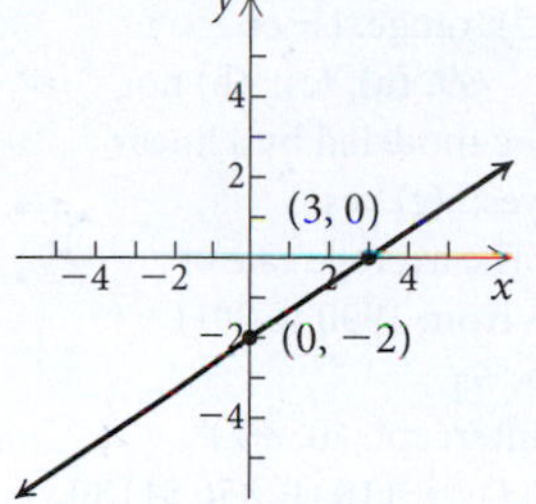

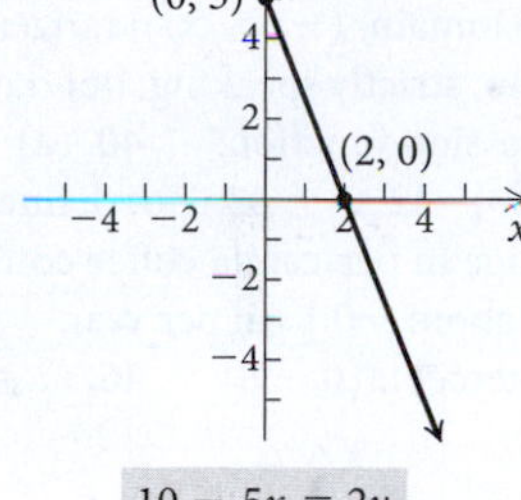

11.

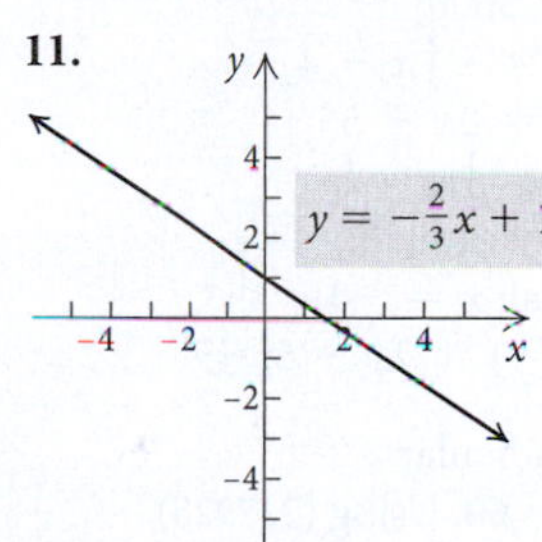

12.

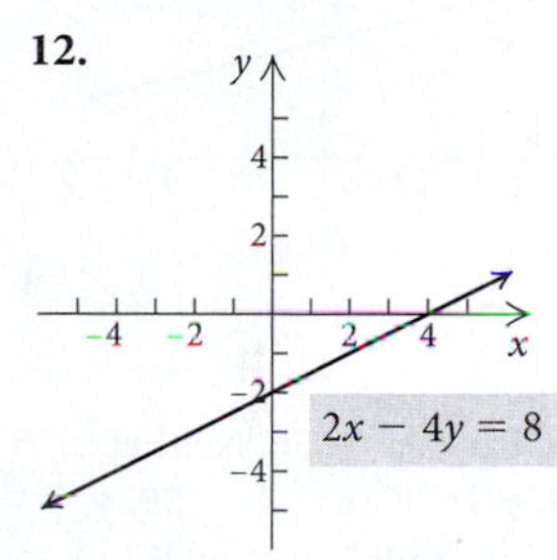

13.

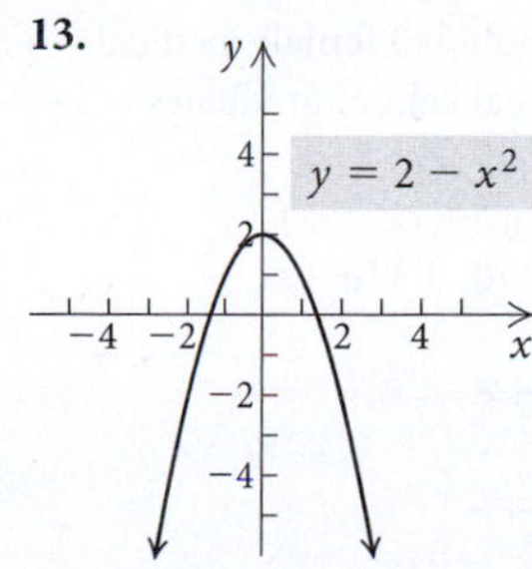

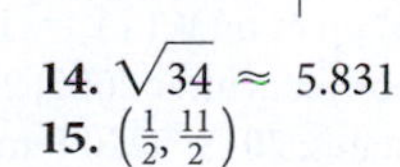

14. $\sqrt{34} \approx 5.831$
15. $\left(\frac{1}{2}, \frac{11}{2}\right)$

16. Center: $(-1, 3)$; radius: 3;

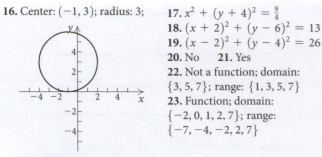

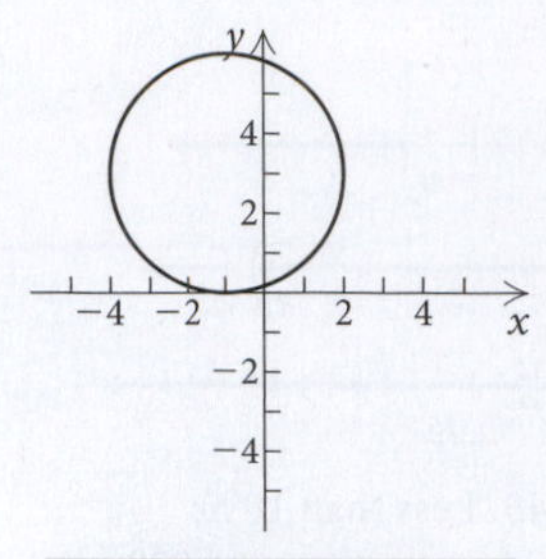

$(x + 1)^2 + (y - 3)^2 = 9$

17. $x^2 + (y + 4)^2 = \frac{9}{4}$
18. $(x + 2)^2 + (y - 6)^2 = 13$
19. $(x - 2)^2 + (y - 4)^2 = 26$
20. No **21.** Yes
22. Not a function; domain: $\{3, 5, 7\}$; range: $\{1, 3, 5, 7\}$
23. Function; domain: $\{-2, 0, 1, 2, 7\}$; range: $\{-7, -4, -2, 2, 7\}$

24. (a) -3; (b) 9; (c) $a^2 - 3a - 1$; (d) $x^2 + x - 3$
25. (a) 0; (b) $\dfrac{x - 6}{x + 6}$; (c) does not exist; (d) $-\frac{5}{3}$
26. $f(2) = -1$; $f(-4) = -3$; $f(0) = -1$ **27.** No
28. Yes **29.** No **30.** Yes **31.** All real numbers, or $(-\infty, \infty)$ **32.** $\{x | x \neq 0\}$, or $(-\infty, 0) \cup (0, \infty)$
33. $\{x | x \neq 5 \text{ and } x \neq 1\}$, or $(-\infty, 1) \cup (1, 5) \cup (5, \infty)$
34. $\{x | x \neq -4 \text{ and } x \neq 4\}$, or $(-\infty, -4) \cup (-4, 4) \cup (4, \infty)$
35. Domain: $[-4, 4]$; range: $[0, 4]$ **36.** Domain: $(-\infty, \infty)$; range: $[0, \infty)$ **37.** Domain: $(-\infty, \infty)$; range: $(-\infty, \infty)$
38. Domain: $(-\infty, \infty)$; range: $[0, \infty)$ **39.** (a) Yes; (b) no; (c) no, strictly speaking, but data might be modeled by a linear regression function. **40.** (a) Yes; (b) yes; (c) yes
41. $\frac{5}{3}$ **42.** 0 **43.** Not defined **44.** The average rate of change in per-capita coffee consumption from 1990 to 2011 was about -0.1 gal per year. **45.** $m = -\frac{7}{11}$; y-intercept: $(0, -6)$ **46.** $m = -2$; y-intercept: $(0, -7)$
47.

$y = -\frac{1}{4}x + 3$

48. $C(t) = 110 + 85t$; \$1130
49. (a) 70°C, 220°C, 10,020°C; (b) $[0, 5600]$
50. $y = -\frac{2}{3}x - 4$
51. $y = 3x + 5$
52. $y = \frac{1}{3}x - \frac{1}{3}$
53. Horizontal: $y = \frac{2}{5}$; vertical: $x = -4$
54. $h(x) = 2x - 5$; -5

55. Parallel **56.** Neither **57.** Perpendicular
58. $y = -\frac{2}{3}x - \frac{1}{3}$ **59.** $y = \frac{3}{2}x - \frac{5}{2}$ **60.** Using (2, 7925) and (6, 8396) gives us $W(x) = 117.75x + 7689.5$, where x is the number of years after 2005; 2008: about 8043 female medical school graduates; 2018: 9220 female medical school graduates
61. $\frac{3}{2}$ **62.** -6 **63.** -1 **64.** -21 **65.** $\frac{95}{24}$
66. No solution **67.** All real numbers, or $(-\infty, \infty)$
68. 568 million quarters **69.** \$2300 **70.** 3.4 hr
71. 3 **72.** 4 **73.** 0.2, or $\frac{1}{5}$ **74.** 4
75. $(-\infty, 12)$;
76. $(-\infty, -4]$;
77. $\left[-\frac{4}{3}, \frac{4}{3}\right]$;
78. $\left(\frac{2}{5}, 2\right]$;
79. $\left(-\infty, -\frac{1}{2}\right) \cup (3, \infty)$;

80. $\left(-\infty, -\frac{5}{3}\right] \cup [1, \infty)$;

81. Years after 2019 **82.** Fahrenheit temperatures less than 113° **83.** B **84.** B **85.** C **86.** $\left(\frac{5}{2}, 0\right)$ **87.** $\{x | x < 0\}$, or $(-\infty, 0)$ **88.** $\{x | x \neq -3 \text{ and } x \neq 0 \text{ and } x \neq 3\}$, or $(-\infty, -3) \cup (-3, 0) \cup (0, 3) \cup (3, \infty)$ **89.** Think of the slopes as $\dfrac{-3/5}{1}$ and $\dfrac{1/2}{1}$. The graph of $f(x)$ changes $\frac{3}{5}$ unit vertically for each unit of horizontal change, whereas the graph of $g(x)$ changes $\frac{1}{2}$ unit vertically for each unit of horizontal change. Since $\frac{3}{5} > \frac{1}{2}$, the graph of $f(x) = -\frac{3}{5}x + 4$ is steeper than the graph of $g(x) = \frac{1}{2}x - 6$. **90.** If an equation contains no fractions, using the addition principle before using the multiplication principle eliminates the need to add or subtract fractions. **91.** The solution set of a disjunction is a union of sets, so it is only possible for a disjunction to have no solution when the solution set of each inequality is the empty set.
92. The graph of $f(x) = mx + b$, $m \neq 0$, is a straight line that is not horizontal. The graph of such a line intersects the x-axis exactly once. Thus the function has exactly one zero.
93. By definition, the notation $3 < x < 4$ indicates that $3 < x$ *and* $x < 4$. The disjunction $x < 3$ *or* $x > 4$ cannot be written $3 > x > 4$, or $4 < x < 3$, because it is not possible for x to be greater than 4 *and* less than 3. **94.** A function is a correspondence between two sets in which each member of the first set corresponds to exactly one member of the second set.

Test: Chapter 1

1. [1.1] Yes **2.** [1.1] x-intercept: $(-2, 0)$; y-intercept: $(0, 5)$;

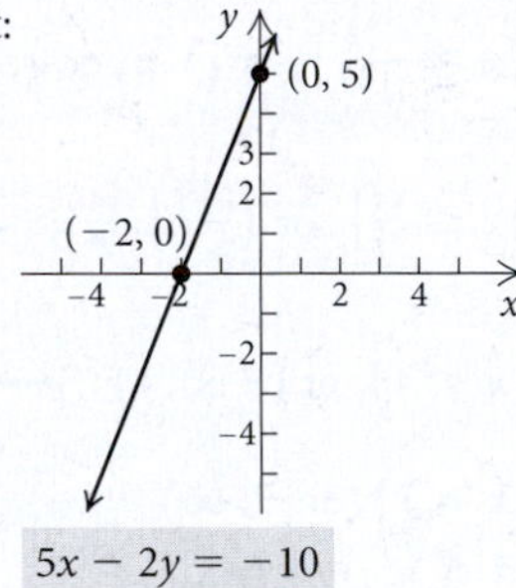

$5x - 2y = -10$

3. [1.1] $\sqrt{45} \approx 6.708$ **4.** [1.1] $\left(-3, \frac{9}{2}\right)$ **5.** [1.1] Center: $(-4, 5)$; radius: 6 **6.** [1.1] $(x + 1)^2 + (y - 2)^2 = 5$
7. [1.2] (a) Yes; (b) $\{-4, 3, 1, 0\}$; (c) $\{7, 0, 5\}$
8. [1.2] (a) 8; (b) $2a^2 + 7a + 11$ **9.** [1.2] (a) Does not exist; (b) 0 **10.** [1.2] 0 **11.** [1.2] (a) No; (b) yes
12. [1.2] $\{x | x \neq 4\}$, or $(-\infty, 4) \cup (4, \infty)$ **13.** [1.2] All real numbers, or $(-\infty, \infty)$ **14.** [1.2] $\{x | -5 \leq x \leq 5\}$, or $[-5, 5]$
15. [1.2] (a)

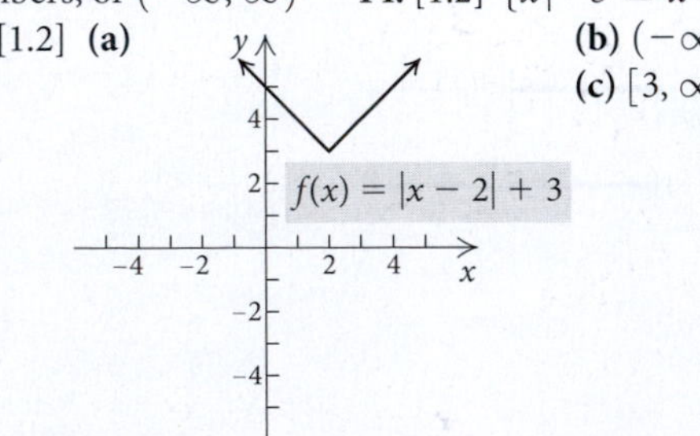

(b) $(-\infty, \infty)$; (c) $[3, \infty)$

16. [1.3] Not defined **17.** [1.3] $-\frac{11}{6}$ **18.** [1.3] 0 **19.** [1.3] The average rate of change in the percent of 12th graders who smoke daily from 1995 to 2012 was about -0.7% per year.

20. [1.3] Slope: $\frac{3}{2}$; y-intercept: $\left(0, \frac{5}{2}\right)$
21. [1.3] $C(t) = 65 + 48t$; \$173 **22.** [1.4] $y = -\frac{5}{8}x - 5$
23. [1.4] $y - 4 = -\frac{3}{4}(x - (-5))$, or $y - (-2) = -\frac{3}{4}(x - 3)$, or $y = -\frac{3}{4}x + \frac{1}{4}$ **24.** [1.4] $x = -\frac{3}{8}$ **25.** [1.4] Perpendicular
26. [1.4] $y - 3 = -\frac{1}{2}(x + 1)$, or $y = -\frac{1}{2}x + \frac{5}{2}$ **27.** [1.4] $y - 3 = 2(x + 1)$, or $y = 2x + 5$ **28.** [1.4] Using (2, 507.03) and (12, 666.99) gives us $y = 15.996x + 475.038$, where x is the number of years after 2000; 2016: \$730.97; 2020: \$794.96
29. [1.5] -1 **30.** [1.5] All real numbers, or $(-\infty, \infty)$
31. [1.5] -60 **32.** [1.5] $\frac{21}{11}$ **33.** [1.5] Length: 60 m; width: 45 m
34. [1.5] \$1.80 **35.** [1.5] -3
36. [1.6] $(-\infty, -3]$;
37. [1.6] $(-5, 3)$;
38. [1.6] $(-\infty, 2] \cup [4, \infty)$;
39. [1.6] More than 5.7 hr **40.** [1.3] B **41.** [1.2] -2

Chapter 2

Exercise Set 2.1

1. **(a)** $(-5, 1)$; **(b)** $(3, 5)$; **(c)** $(1, 3)$ **3.** **(a)** $(-3, -1), (3, 5)$; **(b)** $(1, 3)$; **(c)** $(-5, -3)$ **5.** **(a)** $(-\infty, -8), (-3, -2)$; **(b)** $(-8, -6)$; **(c)** $(-6, -3), (-2, \infty)$ **7.** Domain: $[-5, 5]$; range: $[-3, 3]$ **9.** Domain: $[-5, -1] \cup [1, 5]$; range: $[-4, 6]$
11. Domain: $(-\infty, \infty)$; range: $(-\infty, 3]$ **13.** Relative maximum: 3.25 at $x = 2.5$; increasing: $(-\infty, 2.5)$; decreasing: $(2.5, \infty)$
15. Relative maximum: 2.370 at $x = -0.667$; relative minimum: 0 at $x = 2$; increasing: $(-\infty, -0.667), (2, \infty)$; decreasing: $(-0.667, 2)$
17. Increasing: $(0, \infty)$; decreasing: $(-\infty, 0)$; relative minimum: 0 at $x = 0$ **19.** Increasing: $(-\infty, 0)$; decreasing: $(0, \infty)$; relative maximum: 5 at $x = 0$ **21.** Increasing: $(3, \infty)$; decreasing: $(-\infty, 3)$; relative minimum: 1 at $x = 3$ **23.** $A(x) = x(240 - x)$, or $240x - x^2$ **25.** $h(d) = \sqrt{d^2 - 3500^2}$ **27.** $A(w) = 10w - \dfrac{w^2}{2}$
29. $d(s) = \dfrac{14}{s}$ **31.** **(a)** $A(x) = x(240 - 4x)$, or $240x - 4x^2$; **(b)** $\{x \mid 0 < x < 60\}$; **(c)** 120 ft by 30 ft
33. **(a)** $V(x) = x(12 - 2x)(12 - 2x)$, or $4x(6 - x)^2$; **(b)** $\{x \mid 0 < x < 6\}$; **(c)** 8 cm by 8 cm by 2 cm **35.** $g(-4) = 0$; $g(0) = 4$; $g(1) = 5$; $g(3) = 5$ **37.** $h(-5) = 1$; $h(0) = 1$; $h(1) = 3$; $h(4) = 6$
39.

41.

43.

45.

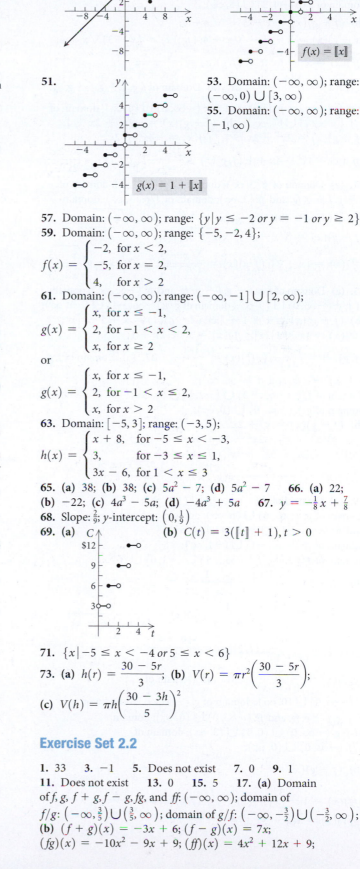

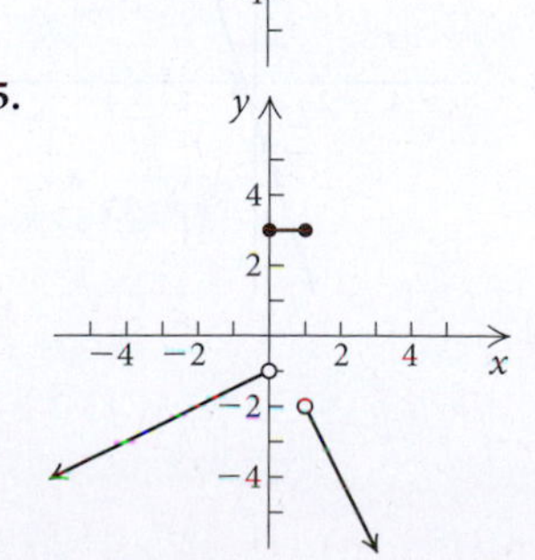

47.

49.

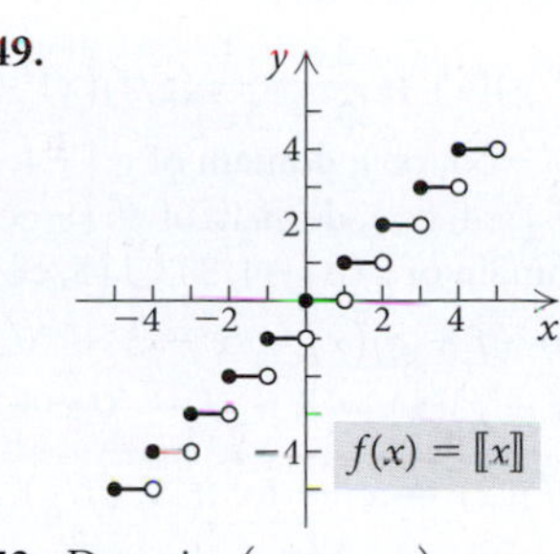

51.

53. Domain: $(-\infty, \infty)$; range: $(-\infty, 0) \cup [3, \infty)$
55. Domain: $(-\infty, \infty)$; range: $[-1, \infty)$

57. Domain: $(-\infty, \infty)$; range: $\{y \mid y \leq -2 \text{ or } y = -1 \text{ or } y \geq 2\}$
59. Domain: $(-\infty, \infty)$; range: $\{-5, -2, 4\}$;

$$f(x) = \begin{cases} -2, & \text{for } x < 2, \\ -5, & \text{for } x = 2, \\ 4, & \text{for } x > 2 \end{cases}$$

61. Domain: $(-\infty, \infty)$; range: $(-\infty, -1] \cup [2, \infty)$;

$$g(x) = \begin{cases} x, & \text{for } x \leq -1, \\ 2, & \text{for } -1 < x < 2, \\ x, & \text{for } x \geq 2 \end{cases}$$

or

$$g(x) = \begin{cases} x, & \text{for } x \leq -1, \\ 2, & \text{for } -1 < x \leq 2, \\ x, & \text{for } x > 2 \end{cases}$$

63. Domain: $[-5, 3]$; range: $(-3, 5)$;

$$h(x) = \begin{cases} x + 8, & \text{for } -5 \leq x < -3, \\ 3, & \text{for } -3 \leq x \leq 1, \\ 3x - 6, & \text{for } 1 < x \leq 3 \end{cases}$$

65. **(a)** 38; **(b)** 38; **(c)** $5a^2 - 7$; **(d)** $5a^2 - 7$ **66.** **(a)** 22; **(b)** -22; **(c)** $4a^3 - 5a$; **(d)** $-4a^3 + 5a$ **67.** $y = -\frac{1}{8}x + \frac{7}{8}$
68. Slope: $\frac{2}{9}$; y-intercept: $\left(0, \frac{1}{9}\right)$
69. **(a)** **(b)** $C(t) = 3(⟦t⟧ + 1), t > 0$

71. $\{x \mid -5 \leq x < -4 \text{ or } 5 \leq x < 6\}$
73. **(a)** $h(r) = \dfrac{30 - 5r}{3}$; **(b)** $V(r) = \pi r^2\left(\dfrac{30 - 5r}{3}\right)$; **(c)** $V(h) = \pi h\left(\dfrac{30 - 3h}{5}\right)^2$

Exercise Set 2.2

1. 33 **3.** -1 **5.** Does not exist **7.** 0 **9.** 1
11. Does not exist **13.** 0 **15.** 5 **17.** **(a)** Domain of f, g, $f + g$, $f - g$, fg, and ff: $(-\infty, \infty)$; domain of f/g: $\left(-\infty, \frac{3}{5}\right) \cup \left(\frac{3}{5}, \infty\right)$; domain of g/f: $\left(-\infty, -\frac{3}{2}\right) \cup \left(-\frac{3}{2}, \infty\right)$; **(b)** $(f + g)(x) = -3x + 6$; $(f - g)(x) = 7x$; $(fg)(x) = -10x^2 - 9x + 9$; $(ff)(x) = 4x^2 + 12x + 9$;

$(f/g)(x) = \dfrac{2x+3}{3-5x}$; $(g/f)(x) = \dfrac{3-5x}{2x+3}$ **19. (a)** Domain of f: $(-\infty, \infty)$; domain of g: $[-4, \infty)$; domain of $f+g$, $f-g$, and fg: $[-4, \infty)$; domain of ff: $(-\infty, \infty)$; domain of f/g: $(-4, \infty)$; domain of g/f: $[-4, 3) \cup (3, \infty)$;
(b) $(f+g)(x) = x - 3 + \sqrt{x+4}$;
$(f-g)(x) = x - 3 - \sqrt{x+4}$; $(fg)(x) = (x-3)\sqrt{x+4}$;
$(ff)(x) = x^2 - 6x + 9$; $(f/g)(x) = \dfrac{x-3}{\sqrt{x+4}}$;
$(g/f)(x) = \dfrac{\sqrt{x+4}}{x-3}$ **21. (a)** Domain of f, g, $f+g$, $f-g$, fg, and ff: $(-\infty, \infty)$; domain of f/g: $(-\infty, 0) \cup (0, \infty)$; domain of g/f: $\left(-\infty, \frac{1}{2}\right) \cup \left(\frac{1}{2}, \infty\right)$; **(b)** $(f+g)(x) = -2x^2 + 2x - 1$;
$(f-g)(x) = 2x^2 + 2x - 1$; $(fg)(x) = -4x^3 + 2x^2$;
$(ff)(x) = 4x^2 - 4x + 1$; $(f/g)(x) = \dfrac{2x-1}{-2x^2}$; $(g/f)(x) = \dfrac{-2x^2}{2x-1}$
23. (a) Domain of f: $[3, \infty)$; domain of g: $[-3, \infty)$; domain of $f+g$, $f-g$, fg, and ff: $[3, \infty)$; domain of f/g: $[3, \infty)$; domain of g/f: $(3, \infty)$; **(b)** $(f+g)(x) = \sqrt{x-3} + \sqrt{x+3}$;
$(f-g)(x) = \sqrt{x-3} - \sqrt{x+3}$; $(fg)(x) = \sqrt{x^2-9}$;
$(ff)(x) = |x-3|$; $(f/g)(x) = \dfrac{\sqrt{x-3}}{\sqrt{x+3}}$; $(g/f)(x) = \dfrac{\sqrt{x+3}}{\sqrt{x-3}}$
25. (a) Domain of f, g, $f+g$, $f-g$, fg, and ff: $(-\infty, \infty)$; domain of f/g: $(-\infty, 0) \cup (0, \infty)$; domain of g/f: $(-\infty, -1) \cup (-1, \infty)$;
(b) $(f+g)(x) = x + 1 + |x|$; $(f-g)(x) = x + 1 - |x|$;
$(fg)(x) = (x+1)|x|$; $(ff)(x) = x^2 + 2x + 1$;
$(f/g)(x) = \dfrac{x+1}{|x|}$; $(g/f)(x) = \dfrac{|x|}{x+1}$ **27. (a)** Domain of f, g, $f+g$, $f-g$, fg, and ff: $(-\infty, \infty)$;
domain of f/g: $(-\infty, -3) \cup \left(-3, \frac{1}{2}\right) \cup \left(\frac{1}{2}, \infty\right)$;
domain of g/f: $(-\infty, 0) \cup (0, \infty)$;
(b) $(f+g)(x) = x^3 + 2x^2 + 5x - 3$;
$(f-g)(x) = x^3 - 2x^2 - 5x + 3$;
$(fg)(x) = 2x^5 + 5x^4 - 3x^3$; $(ff)(x) = x^6$;
$(f/g)(x) = \dfrac{x^3}{2x^2+5x-3}$; $(g/f)(x) = \dfrac{2x^2+5x-3}{x^3}$
29. (a) Domain of f: $(-\infty, -1) \cup (-1, \infty)$;
domain of g: $(-\infty, 6) \cup (6, \infty)$; domain of $f+g$, $f-g$, and fg: $(-\infty, -1) \cup (-1, 6) \cup (6, \infty)$;
domain of ff: $(-\infty, -1) \cup (-1, \infty)$;
domain of f/g and g/f: $(-\infty, -1) \cup (-1, 6) \cup (6, \infty)$;
(b) $(f+g)(x) = \dfrac{4}{x+1} + \dfrac{1}{6-x}$;
$(f-g)(x) = \dfrac{4}{x+1} - \dfrac{1}{6-x}$; $(fg)(x) = \dfrac{4}{(x+1)(6-x)}$;
$(ff)(x) = \dfrac{16}{(x+1)^2}$; $(f/g)(x) = \dfrac{4(6-x)}{x+1}$;
$(g/f)(x) = \dfrac{x+1}{4(6-x)}$ **31. (a)** Domain of f: $(-\infty, 0) \cup (0, \infty)$; domain of g: $(-\infty, \infty)$; domain of $f+g$, $f-g$, fg, and ff: $(-\infty, 0) \cup (0, \infty)$; domain of f/g: $(-\infty, 0) \cup (0, 3) \cup (3, \infty)$; domain of g/f: $(-\infty, 0) \cup (0, \infty)$;
(b) $(f+g)(x) = \dfrac{1}{x} + x - 3$; $(f-g)(x) = \dfrac{1}{x} - x + 3$;
$(fg)(x) = 1 - \dfrac{3}{x}$; $(ff)(x) = \dfrac{1}{x^2}$; $(f/g)(x) = \dfrac{1}{x(x-3)}$;
$(g/f)(x) = x(x-3)$

33. (a) Domain of f: $(-\infty, 2) \cup (2, \infty)$; domain of g: $[1, \infty)$; domain of $f+g$, $f-g$, and fg: $[1, 2) \cup (2, \infty)$; domain of ff: $(-\infty, 2) \cup (2, \infty)$; domain of f/g: $(1, 2) \cup (2, \infty)$; domain of g/f: $[1, 2) \cup (2, \infty)$;
(b) $(f+g)(x) = \dfrac{3}{x-2} + \sqrt{x-1}$;
$(f-g)(x) = \dfrac{3}{x-2} - \sqrt{x-1}$;
$(fg)(x) = \dfrac{3\sqrt{x-1}}{x-2}$; $(ff)(x) = \dfrac{9}{(x-2)^2}$;
$(f/g)(x) = \dfrac{3}{(x-2)\sqrt{x-1}}$; $(g/f)(x) = \dfrac{(x-2)\sqrt{x-1}}{3}$
35. Domain of F: $[2, 11]$; domain of G: $[1, 9]$; domain of $F+G$: $[2, 9]$ **37.** $[2, 3) \cup (3, 9]$
39.

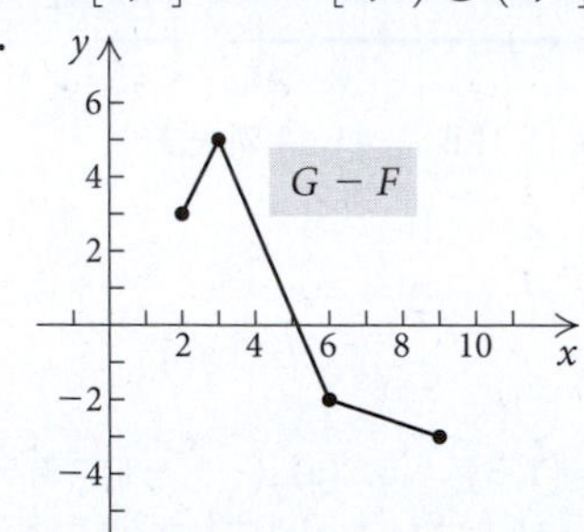

41. Domain of F: $[0, 9]$; domain of G: $[3, 10]$; domain of $F+G$: $[3, 9]$ **43.** $[3, 6) \cup (6, 8) \cup (8, 9]$
45.

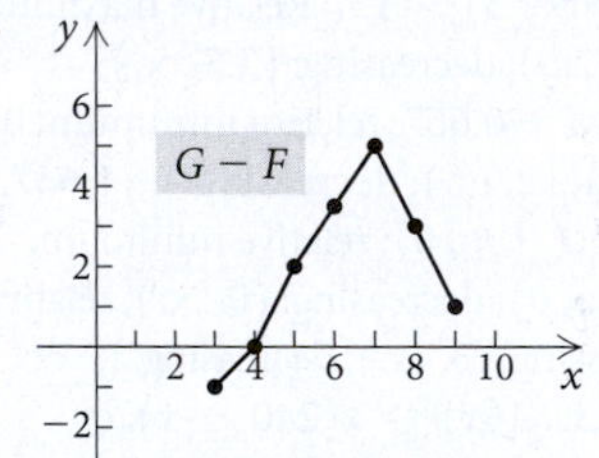

47. (a) $P(x) = -0.4x^2 + 57x - 13$; **(b)** $R(100) = 2000$; $C(100) = 313$; $P(100) = 1687$
49. 3 **51.** 6 **53.** $\frac{1}{3}$ **55.** $\dfrac{-1}{3x(x+h)}$, or $-\dfrac{1}{3x(x+h)}$
57. $\dfrac{1}{4x(x+h)}$ **59.** $2x + h$ **61.** $-2x - h$
63. $6x + 3h - 2$ **65.** $\dfrac{5|x+h| - 5|x|}{h}$
67. $3x^2 + 3xh + h^2$ **69.** $\dfrac{7}{(x+h+3)(x+3)}$

71. **72.**

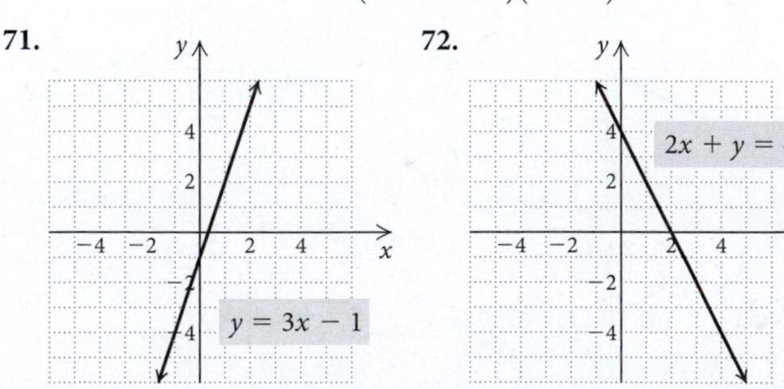

73. 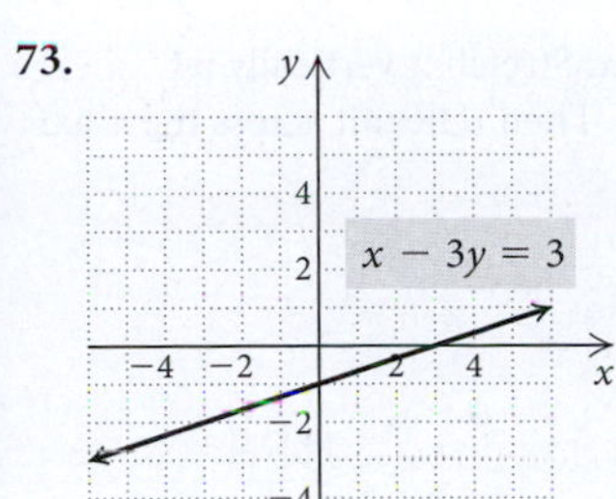

74. 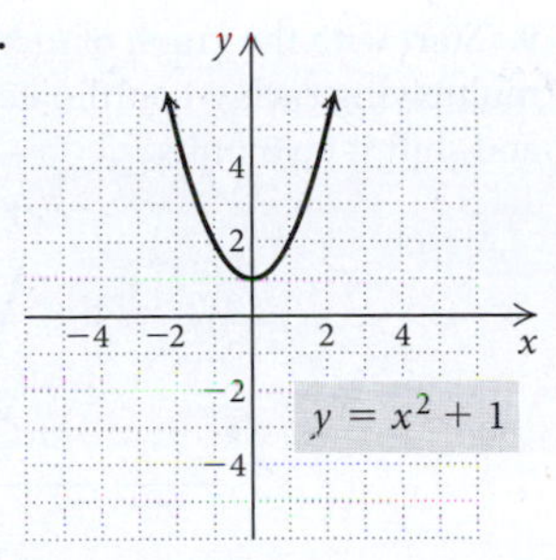

75. $f(x) = \frac{1}{x+7}, g(x) = \frac{1}{x-3}$; answers may vary

77. $(-\infty, -1) \cup (-1, 1) \cup (1, \frac{7}{3}) \cup (\frac{7}{3}, 3) \cup (3, \infty)$

Exercise Set 2.3

1. -8 **3.** 64 **5.** 218 **7.** -80 **9.** -6 **11.** 512 **13.** -32 **15.** x^9 **17.** $(f \circ g)(x) = (g \circ f)(x) = x$; domain of $f \circ g$ and $g \circ f$: $(-\infty, \infty)$ **19.** $(f \circ g)(x) = 3x^2 - 2x$; $(g \circ f)(x) = 3x^2 + 4x$; domain of $f \circ g$ and $g \circ f$: $(-\infty, \infty)$ **21.** $(f \circ g)(x) = 16x^2 - 24x + 6$; $(g \circ f)(x) = 4x^2 - 15$; domain of $f \circ g$ and $g \circ f$: $(-\infty, \infty)$ **23.** $(f \circ g)(x) = \frac{4x}{x-5}$; $(g \circ f)(x) = \frac{1-5x}{4}$; domain of $f \circ g$: $(-\infty, 0) \cup (0, 5) \cup (5, \infty)$; domain of $g \circ f$: $(-\infty, \frac{1}{5}) \cup (\frac{1}{5}, \infty)$ **25.** $(f \circ g)(x) = (g \circ f)(x) = x$; domain of $f \circ g$ and $g \circ f$: $(-\infty, \infty)$ **27.** $(f \circ g)(x) = 2\sqrt{x+1}$; $(g \circ f)(x) = \sqrt{2x+1}$; domain of $f \circ g$: $[0, \infty)$; domain of $g \circ f$: $[-\frac{1}{2}, \infty)$ **29.** $(f \circ g)(x) = 20$; $(g \circ f)(x) = 0.05$; domain of $f \circ g$ and $g \circ f$: $(-\infty, \infty)$ **31.** $(f \circ g)(x) = |x|$; $(g \circ f)(x) = x$; domain of $f \circ g$: $(-\infty, \infty)$; domain of $g \circ f$: $[-5, \infty)$ **33.** $(f \circ g)(x) = 5 - x$; $(g \circ f)(x) = \sqrt{1 - x^2}$; domain of $f \circ g$: $(-\infty, 3]$; domain of $g \circ f$: $[-1, 1]$ **35.** $(f \circ g)(x) = (g \circ f)(x) = x$; domain of $f \circ g$: $(-\infty, -1) \cup (-1, \infty)$; domain of $g \circ f$: $(-\infty, 0) \cup (0, \infty)$ **37.** $(f \circ g)(x) = x^3 - 2x^2 - 4x + 6$; $(g \circ f)(x) = x^3 - 5x^2 + 3x + 8$; domain of $f \circ g$ and $g \circ f$: $(-\infty, \infty)$ **39.** $f(x) = x^5$; $g(x) = 4 + 3x$ **41.** $f(x) = \frac{1}{x}$; $g(x) = (x-2)^4$ **43.** $f(x) = \frac{x-1}{x+1}$; $g(x) = x^3$ **45.** $f(x) = x^6$; $g(x) = \frac{2+x^3}{2-x^3}$ **47.** $f(x) = \sqrt{x}$; $g(x) = \frac{x-5}{x+2}$ **49.** $f(x) = x^3 - 5x^2 + 3x - 1$; $g(x) = x + 2$ **51.** **(a)** $r(t) = 3t$; **(b)** $A(r) = \pi r^2$; **(c)** $(A \circ r)(t) = 9\pi t^2$; the function gives the area of the ripple in terms of time t. **53.** $f(x) = x + 1$ **55.** (c) **56.** None **57.** (b), (d), (f), and (h) **58.** (b) **59.** (a) **60.** (c) and (g) **61.** (c) and (g) **62.** (a) and (f) **63.** Only $(c \circ p)(a)$ makes sense. It represents the cost of the grass seed required to seed a lawn with area a.

Mid-Chapter Mixed Review: Chapter 2

1. True **2.** False **3.** True **4.** **(a)** $(2, 4)$; **(b)** $(-5, -3), (4, 5)$; **(c)** $(-3, -1)$ **5.** Relative maximum: 6.30 at $x = -1.29$; relative minimum: -2.30 at $x = 1.29$; increasing: $(-\infty, -1.29), (1.29, \infty)$; decreasing: $(-1.29, 1.29)$ **6.** Domain: $[-5, -1] \cup [2, 5]$; range: $[-3, 5]$ **7.** $A(h) = \frac{h^2}{2} + 2h$ **8.** -10; -8; 1; 3

9. 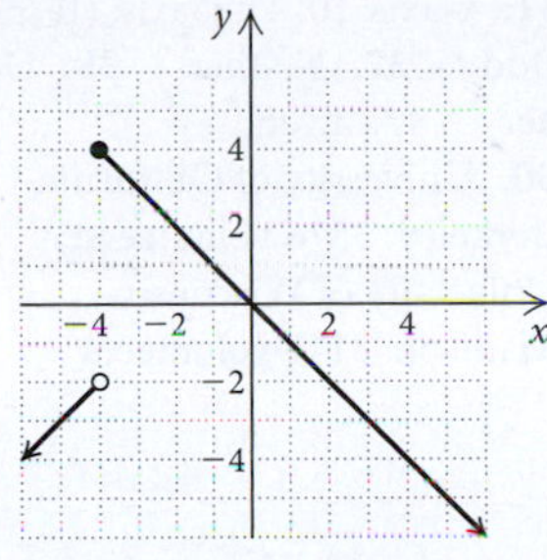

10. 1 **11.** -4 **12.** 5 **13.** Does not exist

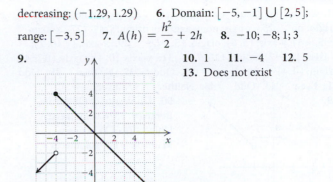

14. **(a)** Domain of $f, g, f + g, f - g, fg$, and ff: $(-\infty, \infty)$; domain of f/g: $(-\infty, -4) \cup (-4, \infty)$; domain of g/f: $\left(-\infty, -\frac{5}{2}\right) \cup \left(-\frac{5}{2}, \infty\right)$; **(b)** $(f + g)(x) = x + 1$; $(f - g)(x) = 3x + 9$; $(fg)(x) = -2x^2 - 13x - 20$; $(ff)(x) = 4x^2 + 20x + 25$; $(f/g)(x) = \frac{2x+5}{-x-4}$; $(g/f)(x) = \frac{-x-4}{2x+5}$ **15.** **(a)** Domain of f: $(-\infty, \infty)$; domain of $g, f + g, f - g$, and fg: $[-2, \infty)$; domain of ff: $(-\infty, \infty)$; domain of f/g: $(-2, \infty)$; domain of g/f: $[-2, 1) \cup (1, \infty)$; **(b)** $(f + g)(x) = x - 1 + \sqrt{x+2}$; $(f - g)(x) = x - 1 - \sqrt{x+2}$; $(fg)(x) = (x-1)\sqrt{x+2}$; $(ff)(x) = x^2 - 2x + 1$; $(f/g)(x) = \frac{x-1}{\sqrt{x+2}}$; $(g/f)(x) = \frac{\sqrt{x+2}}{x-1}$ **16.** 4 **17.** $-2x - h$ **18.** 6 **19.** 28 **20.** -24 **21.** 102 **22.** $(f \circ g)(x) = 3x + 2$; $(g \circ f)(x) = 3x + 4$; domain of $f \circ g$ and $g \circ f$: $(-\infty, \infty)$ **23.** $(f \circ g)(x) = 3\sqrt{x} + 2$; $(g \circ f)(x) = \sqrt{3x+2}$; domain of $f \circ g$: $[0, \infty)$; domain of $g \circ f$: $[-\frac{2}{3}, \infty)$ **24.** The graph of $y = (h - g)(x)$ will be the same as the graph of $y = h(x)$ moved down b units. **25.** Under the given conditions, $(f + g)(x)$ and $(f/g)(x)$ have different domains if $g(x) = 0$ for one or more real numbers x. **26.** If f and g are linear functions, then any real number can be an input for each function. Thus the domain of $f \circ g$ = the domain of $g \circ f = (-\infty, \infty)$. **27.** This approach is not valid. Consider Exercise 23 in Section 2.3, for example. Since $(f \circ g)(x) = \frac{4x}{x-5}$, an examination of only this composed function would lead to the incorrect conclusion that the domain of $f \circ g$ is $(-\infty, 5) \cup (5, \infty)$. However, we must also exclude from the domain of $f \circ g$ those values of x that are not in the domain of g. Thus the domain of $f \circ g$ is $(-\infty, 0) \cup (0, 5) \cup (5, \infty)$.

Exercise Set 2.4

1. x-axis: no; y-axis: yes; origin: no **3.** x-axis: yes; y-axis: no; origin: no **5.** x-axis: no; y-axis: no; origin: yes **7.** x-axis: no; y-axis: yes; origin: no **9.** x-axis: no; y-axis: no; origin: no **11.** x-axis: no; y-axis: yes; origin: no **13.** x-axis: no; y-axis: no; origin: yes **15.** x-axis: no; y-axis: no; origin: yes **17.** x-axis: yes; y-axis: yes; origin: yes **19.** x-axis: no; y-axis: yes; origin: no

21. x-axis: yes; y-axis: yes; origin: yes **23.** x-axis: no; y-axis: no; origin: no **25.** x-axis: no; y-axis: no; origin: yes **27.** x-axis: $(-5, -6)$; y-axis: $(5, 6)$; origin: $(5, -6)$ **29.** x-axis: $(-10, 7)$; y-axis: $(10, -7)$; origin: $(10, 7)$ **31.** x-axis: $(0, 4)$; y-axis: $(0, -4)$; origin: $(0, 4)$ **33.** Even **35.** Odd **37.** Neither **39.** Odd **41.** Even **43.** Odd **45.** Neither **47.** Even

49.

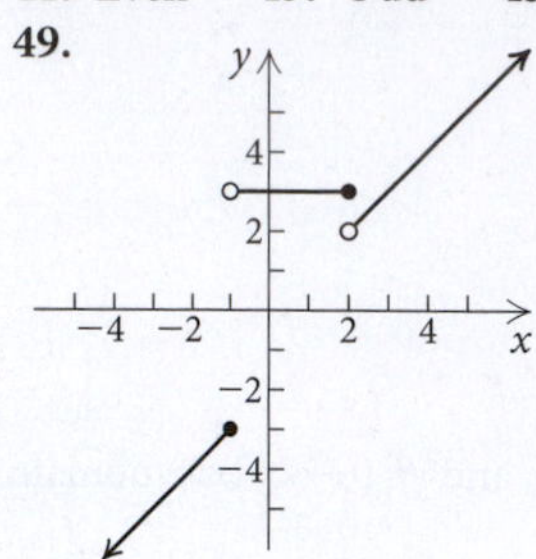

50. University of California–Berkeley: 3576 volunteers; University of Wisconsin–Madison: 3112 volunteers

51. Odd **53.** x-axis: yes; y-axis: no; origin: no

55. $E(-x) = \dfrac{f(-x) + f(-(-x))}{2} = \dfrac{f(-x) + f(x)}{2} = E(x)$

57. (a) $E(x) + O(x) = \dfrac{f(x) + f(-x)}{2} + \dfrac{f(x) - f(-x)}{2} = \dfrac{2f(x)}{2} = f(x)$; **(b)** $f(x) = \dfrac{-22x^2 + \sqrt{x} + \sqrt{-x} - 20}{2} + \dfrac{8x^3 + \sqrt{x} - \sqrt{-x}}{2}$ **59.** True

Visualizing the Graph

1. C **2.** B **3.** A **4.** E **5.** G **6.** D **7.** H **8.** I **9.** F

Exercise Set 2.5

1. Start with the graph of $f(x) = x^2$. Shift it right 3 units.

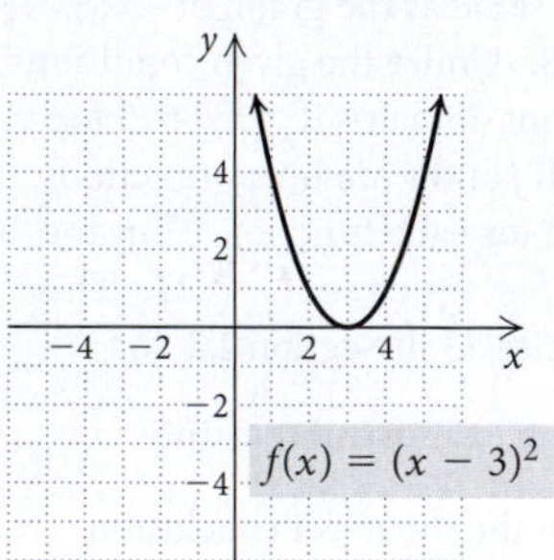

3. Start with the graph of $g(x) = x$. Shift it down 3 units.

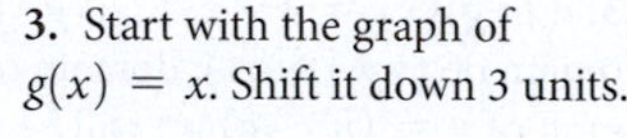

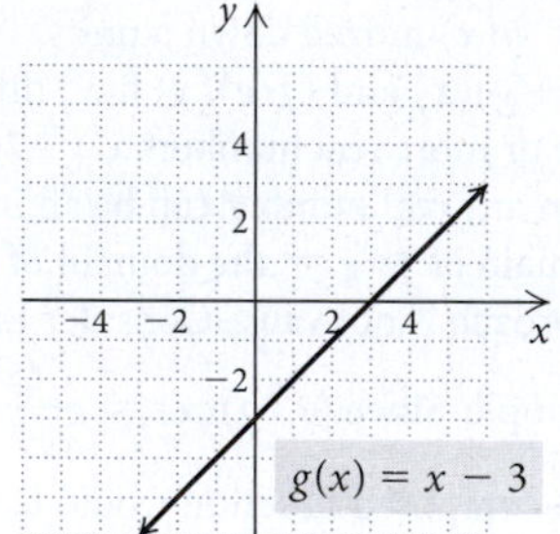

5. Start with the graph of $h(x) = \sqrt{x}$. Reflect it across the x-axis.

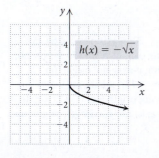

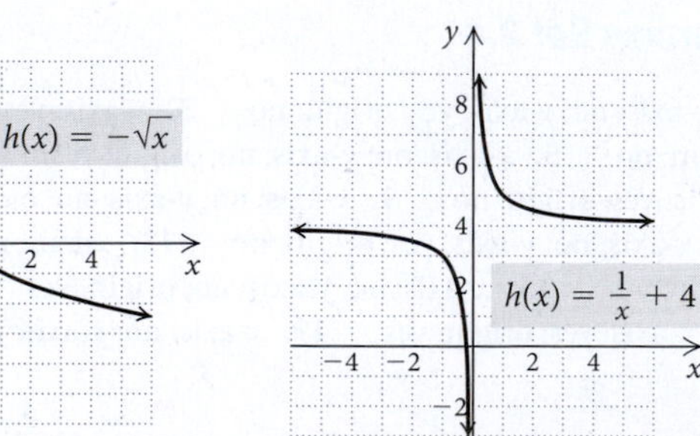

7. Start with the graph of $h(x) = \dfrac{1}{x}$. Shift it up 4 units.

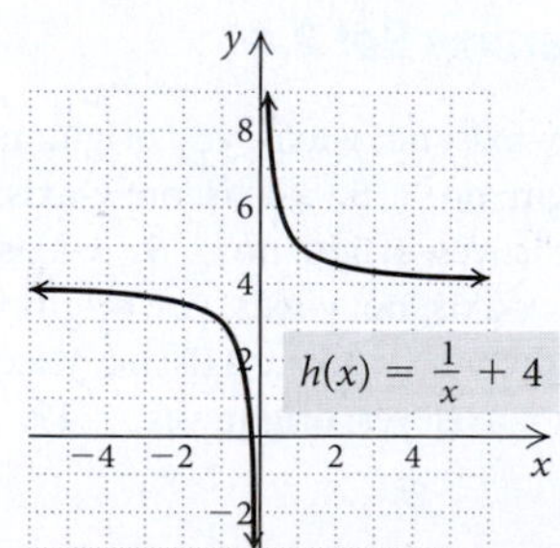

9. Start with the graph of $h(x) = x$. Stretch it vertically by multiplying each y-coordinate by 3. Then reflect it across the x-axis and shift it up 3 units.

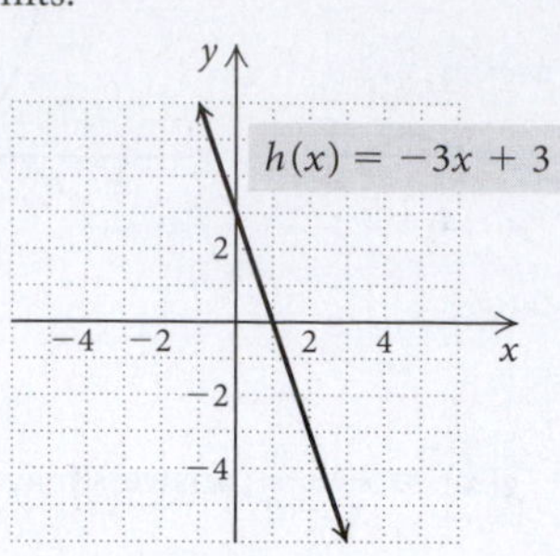

11. Start with the graph of $h(x) = |x|$. Shrink it vertically by multiplying each y-coordinate by $\frac{1}{2}$. Then shift it down 2 units.

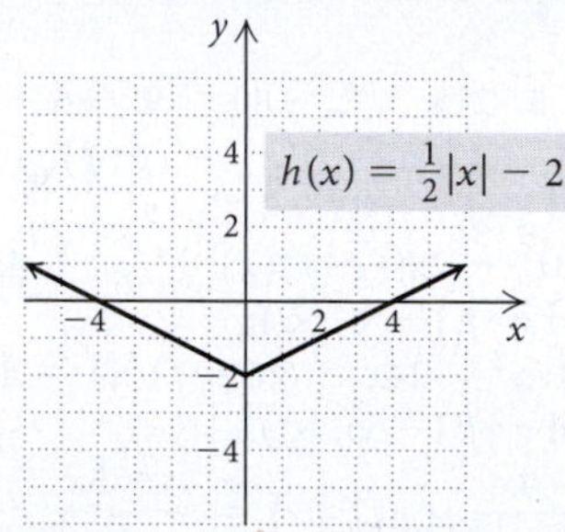

13. Start with the graph of $g(x) = x^3$. Shift it right 2 units. Then reflect it across the x-axis.

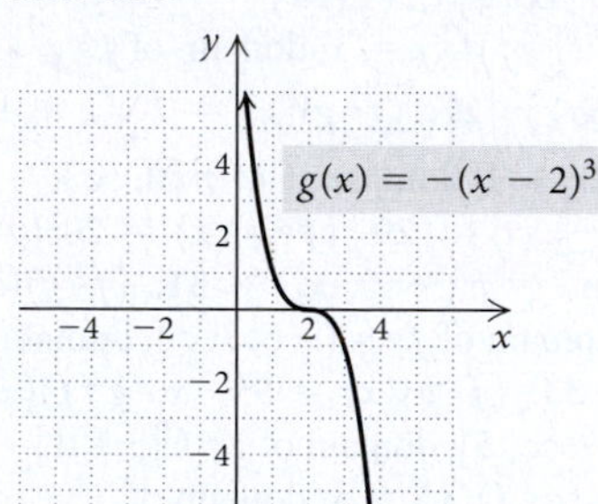

15. Start with the graph of $g(x) = x^2$. Shift it left 1 unit. Then shift it down 1 unit.

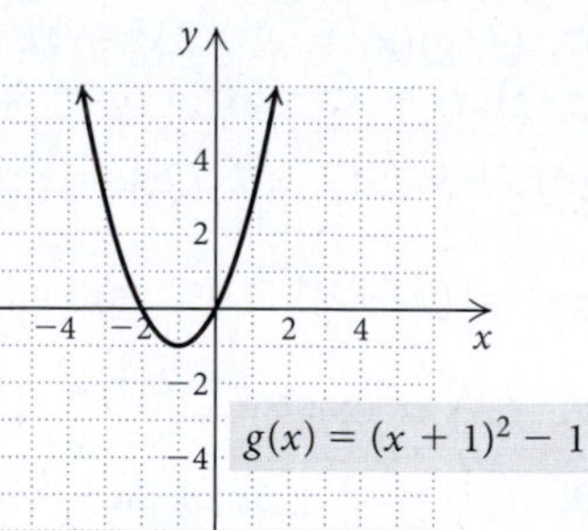

17. Start with the graph of $g(x) = x^3$. Shrink it vertically by multiplying each y-coordinate by $\frac{1}{3}$. Then shift it up 2 units.

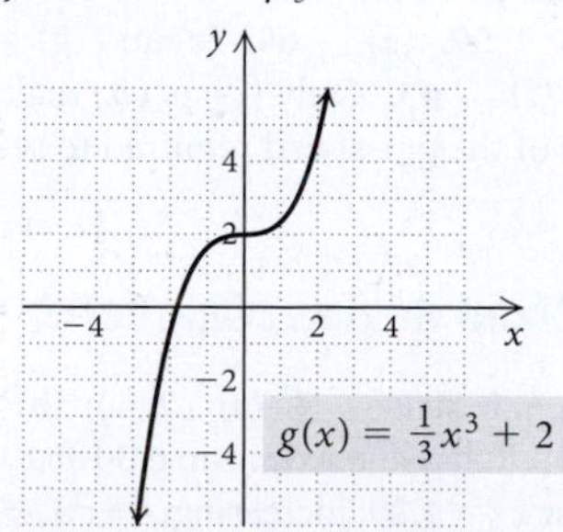

19. Start with the graph of $f(x) = \sqrt{x}$. Shift it left 2 units.

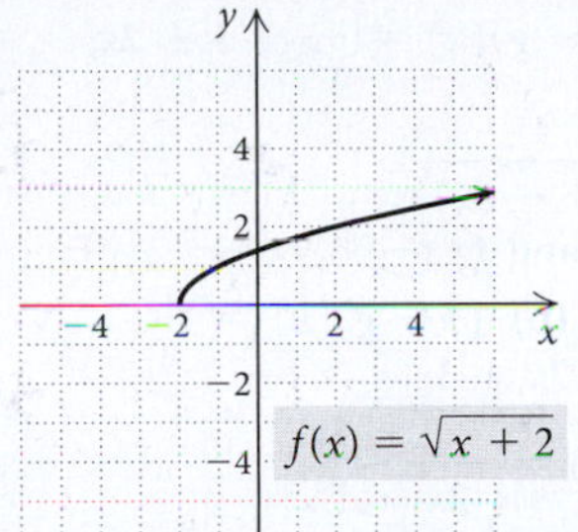

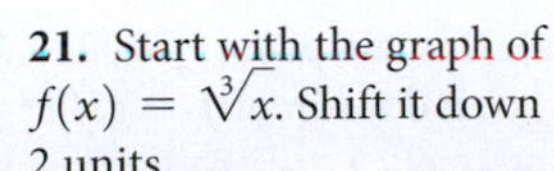

21. Start with the graph of $f(x) = \sqrt[3]{x}$. Shift it down 2 units.

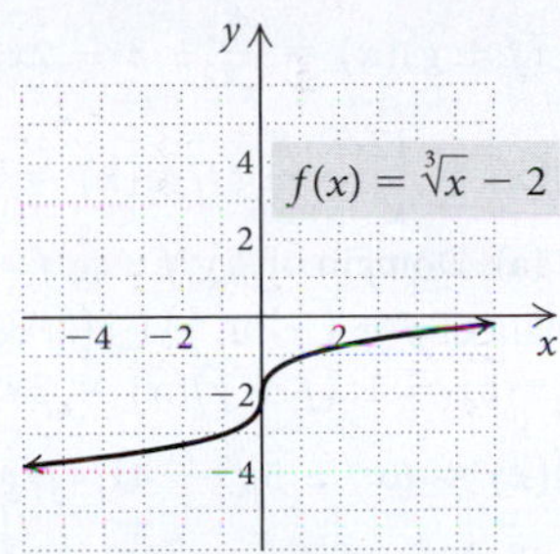

23. Start with the graph of $g(x) = |x|$. Shrink it horizontally by multiplying each x-coordinate by $\frac{1}{3}$ (or dividing each x-coordinate by 3). **25.** Start with the graph of $h(x) = \frac{1}{x}$. Stretch it vertically by multiplying each y-coordinate by 2. **27.** Start with the graph of $f(x) = \sqrt{x}$. Stretch it vertically by multiplying each y-coordinate by 3. Then shift it down 5 units. **29.** Start with the graph of $g(x) = |x|$. Stretch it horizontally by multiplying each x-coordinate by 3. Then shift it down 4 units. **31.** Start with the graph of $f(x) = x^2$. Shift it right 5 units, shrink it vertically by multiplying each y-coordinate by $\frac{1}{4}$, and then reflect it across the x-axis. **33.** Start with the graph of $f(x) = \frac{1}{x}$. Shift it left 3 units, then up 2 units. **35.** Start with the graph of $h(x) = x^2$. Shift it right 3 units. Then reflect it across the x-axis and shift it up 5 units. **37.** $(-12, 2)$ **39.** $(12, 4)$ **41.** $(-12, 2)$ **43.** $(-12, 16)$ **45.** B **47.** A **49.** $f(x) = -(x - 8)^2$ **51.** $f(x) = |x + 7| + 2$ **53.** $f(x) = \frac{1}{2x} - 3$ **55.** $f(x) = -(x - 3)^2 + 4$ **57.** $f(x) = \sqrt{-(x + 2)} - 1$

59.

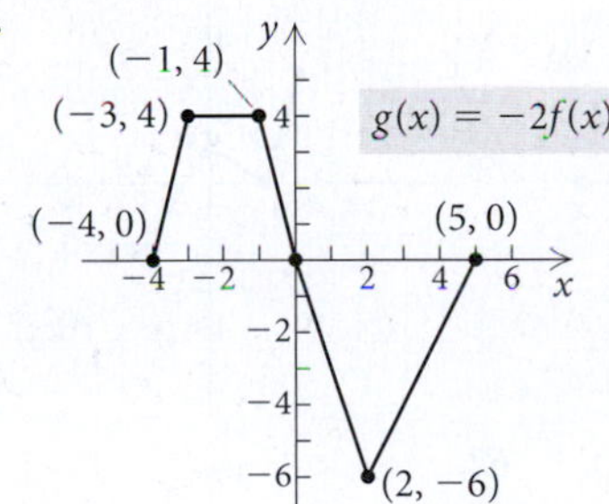

61.

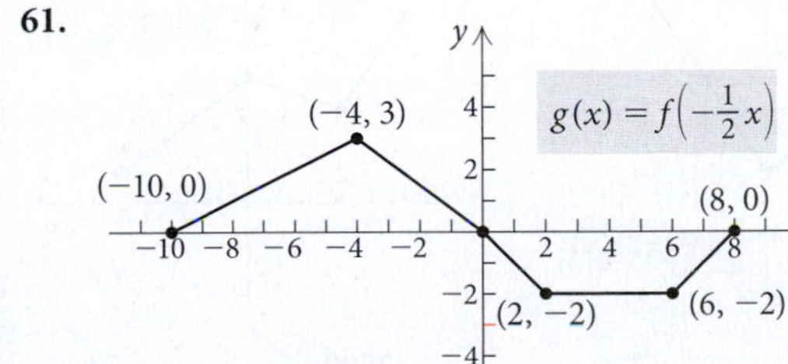

63.

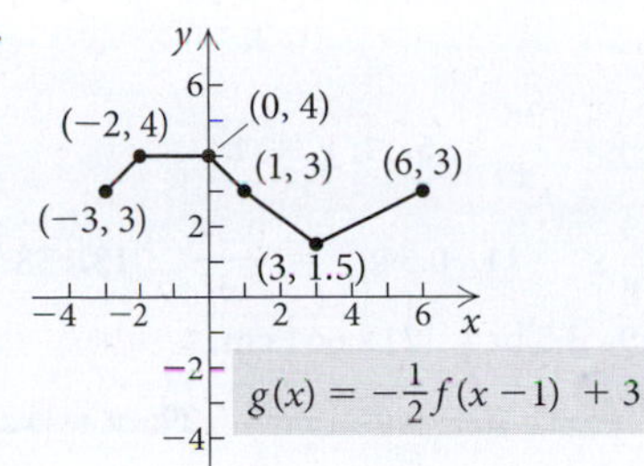

65.

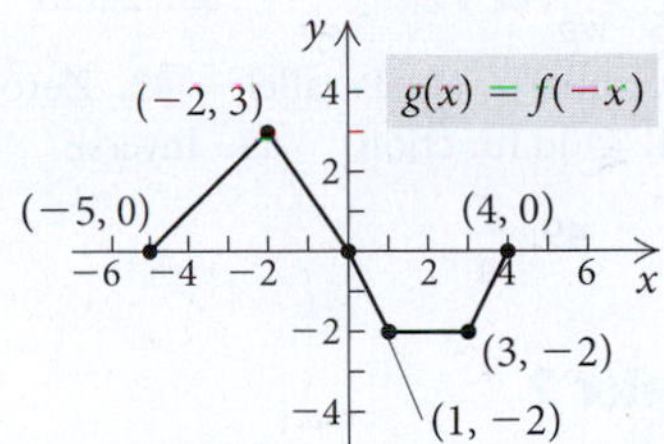

67.

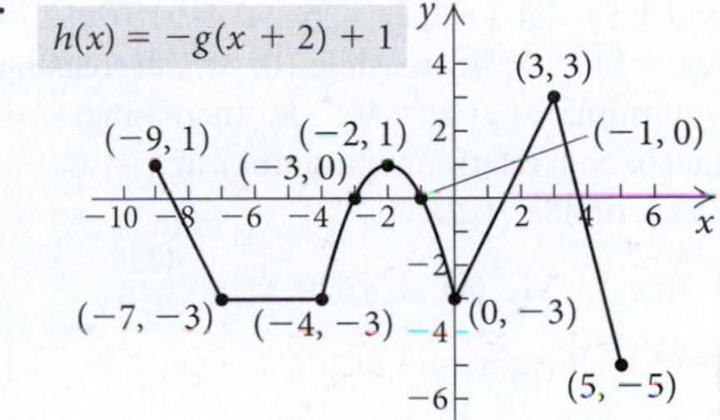

69.

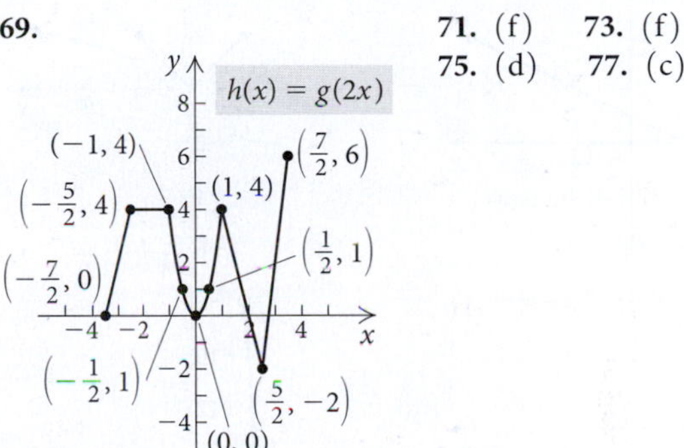

71. (f) **73.** (f) **75.** (d) **77.** (c)

79. $f(-x) = 2(-x)^4 - 35(-x)^3 + 3(-x) - 5 = 2x^4 + 35x^3 - 3x - 5 = g(x)$ **81.** $g(x) = x^3 - 3x^2 + 2$ **83.** $k(x) = (x + 1)^3 - 3(x + 1)^2$ **85.** x-axis, no; y-axis, yes; origin, no **86.** x-axis, yes; y-axis, no; origin, no **87.** x-axis, no; y-axis, no; origin, yes **88.** 40,504 pages **89.** 1123 guns **90.** About 29,700 acres

91.

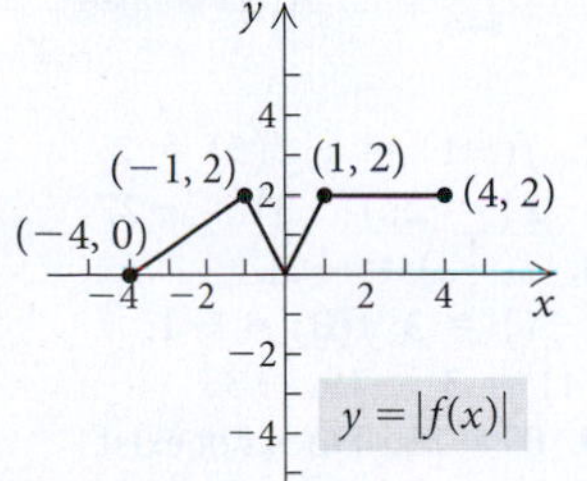

93.

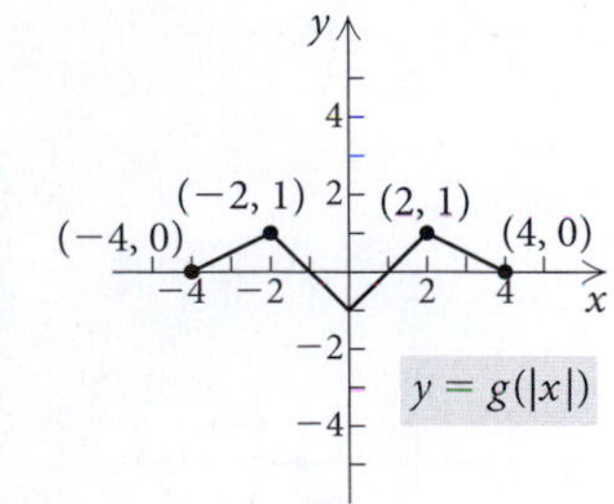

95. 5

Exercise Set 2.6

1. 4.5; $y = 4.5x$ **3.** 36; $y = \frac{36}{x}$ **5.** 4; $y = 4x$ **7.** 4; $y = \frac{4}{x}$ **9.** $\frac{3}{8}$; $y = \frac{3}{8}x$ **11.** 0.54; $y = \frac{0.54}{x}$ **13.** \$8.25 **15.** $5\frac{5}{7}$ hr **17.** 90 g **19.** 3.5 hr **21.** $66\frac{2}{3}$ cm **23.** 1.92 ft **25.** $y = \frac{0.0015}{x^2}$ **27.** $y = 15x^2$ **29.** $y = xz$ **31.** $y = \frac{3}{10}xz^2$ **33.** $y = \frac{1}{5} \cdot \frac{xz}{wp}$, or $y = \frac{xz}{5wp}$ **35.** 2.5 m **37.** 36 mph **39.** 98 earned runs **41.** Parallel **42.** Zero **43.** Relative minimum **44.** Odd function **45.** Inverse variation **47.** \$3.56; \$3.53 **49.** $\frac{\pi}{4}$

Review Exercises: Chapter 2

1. True **2.** False **3.** True **4.** True **5. (a)** $(-4, -2)$; **(b)** $(2, 5)$; **(c)** $(-2, 2)$ **6. (a)** $(-1, 0)$, $(2, \infty)$; **(b)** $(0, 2)$; **(c)** $(-\infty, -1)$ **7.** Increasing: $(0, \infty)$; decreasing: $(-\infty, 0)$; relative minimum: -1 at $x = 0$ **8.** Increasing: $(-\infty, 0)$; decreasing: $(0, \infty)$; relative maximum: 2 at $x = 0$ **9.** $A(x) = x(48 - 2x)$, or $48x - 2x^2$ **10.** $A(x) = 2x\sqrt{4 - x^2}$ **11. (a)** $A(x) = x^2 + \frac{432}{x}$; **(b)** $(0, \infty)$; **(c)** $x = 6$ in., height $= 3$ in.

12.

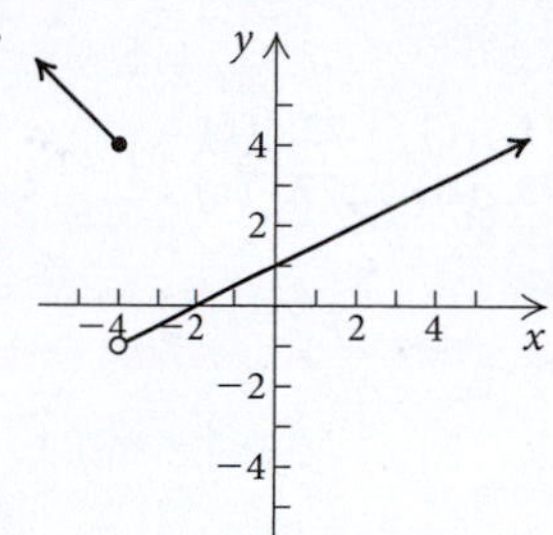

13.

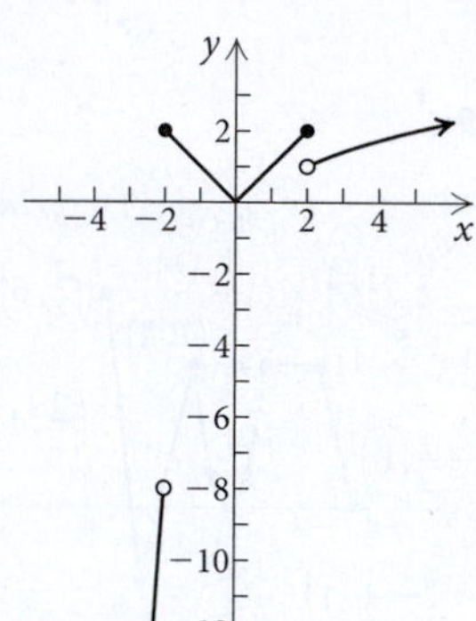

14.

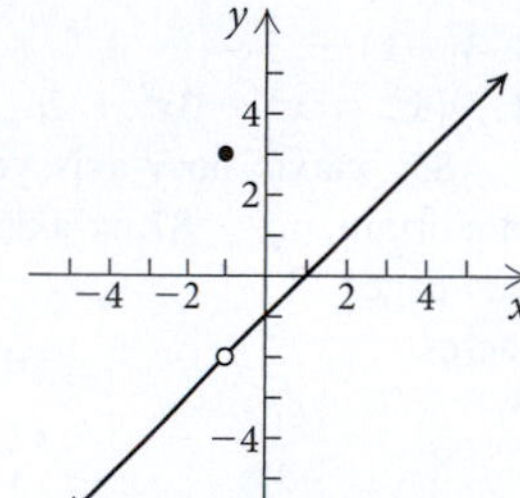

15.

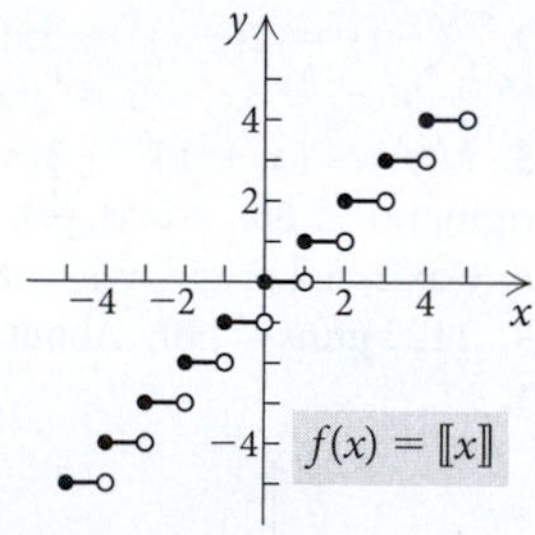

16.

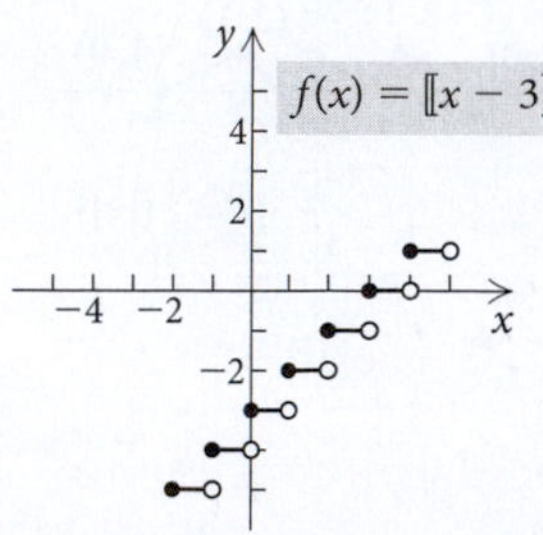

17. $f(-1) = 1$; $f(5) = 2$; $f(-2) = 2$; $f(-3) = -27$ **18.** $f(-2) = -3$; $f(-1) = 3$; $f(0) = -1$; $f(4) = 3$ **19.** -33 **20.** 0 **21.** Does not exist

22. (a) Domain of f: $(-\infty, 0) \cup (0, \infty)$; domain of g: $(-\infty, \infty)$; domain of $f + g$, $f - g$, and fg: $(-\infty, 0) \cup (0, \infty)$; domain of f/g: $(-\infty, 0) \cup \left(0, \frac{3}{2}\right) \cup \left(\frac{3}{2}, \infty\right)$; **(b)** $(f + g)(x) = \frac{4}{x^2} + 3 - 2x$; $(f - g)(x) = \frac{4}{x^2} - 3 + 2x$; $(fg)(x) = \frac{12}{x^2} - \frac{8}{x}$; $(f/g)(x) = \frac{4}{x^2(3 - 2x)}$ **23. (a)** Domain of f, g, $f + g$, $f - g$, and fg: $(-\infty, \infty)$; domain of f/g: $\left(-\infty, \frac{1}{2}\right) \cup \left(\frac{1}{2}, \infty\right)$; **(b)** $(f + g)(x) = 3x^2 + 6x - 1$; $(f - g)(x) = 3x^2 + 2x + 1$; $(fg)(x) = 6x^3 + 5x^2 - 4x$; $(f/g)(x) = \frac{3x^2 + 4x}{2x - 1}$ **24.** $P(x) = -0.5x^2 + 105x - 6$ **25.** 2 **26.** $-2x - h$ **27.** $\frac{-4}{x(x + h)}$, or $-\frac{4}{x(x + h)}$ **28.** 9 **29.** 5 **30.** 128 **31.** 580 **32.** 7 **33.** -509 **34.** $4x - 3$ **35.** $-24 + 27x^3 - 9x^6 + x^9$ **36. (a)** $(f \circ g)(x) = \frac{4}{(3 - 2x)^2}$; $(g \circ f)(x) = 3 - \frac{8}{x^2}$; **(b)** domain of $f \circ g$: $\left(-\infty, \frac{3}{2}\right) \cup \left(\frac{3}{2}, \infty\right)$; domain of $g \circ f$: $(-\infty, 0) \cup (0, \infty)$ **37. (a)** $(f \circ g)(x) = 12x^2 - 4x - 1$; $(g \circ f)(x) = 6x^2 + 8x - 1$; **(b)** domain of $f \circ g$ and $g \circ f$: $(-\infty, \infty)$ **38.** $f(x) = \sqrt{x}$, $g(x) = 5x + 2$; answers may vary. **39.** $f(x) = 4x^2 + 9$, $g(x) = 5x - 1$; answers may vary. **40.** x-axis: yes; y-axis: yes; origin: yes **41.** x-axis: yes; y-axis: yes; origin: yes **42.** x-axis: no; y-axis: no; origin: no **43.** x-axis: no; y-axis: yes; origin: no **44.** x-axis: no; y-axis: no; origin: yes **45.** x-axis: no; y-axis: yes; origin: no **46.** Even **47.** Even **48.** Odd **49.** Even **50.** Even **51.** Neither **52.** Odd **53.** Even **54.** Even **55.** Odd **56.** $f(x) = (x + 3)^2$ **57.** $f(x) = -\sqrt{x - 3} + 4$ **58.** $f(x) = 2|x - 3|$

59.

$y = f(x - 1)$

60.

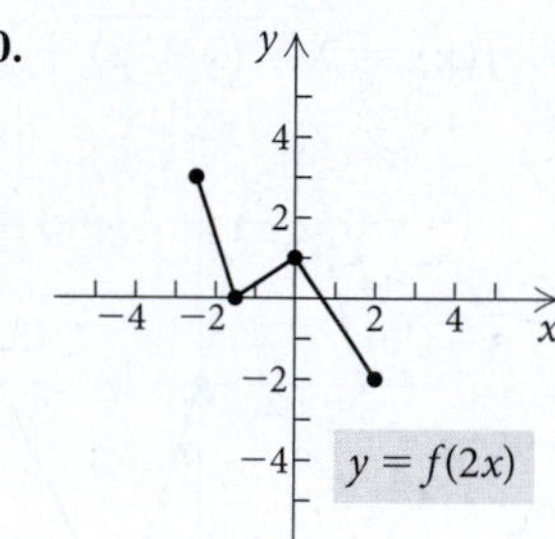

61.

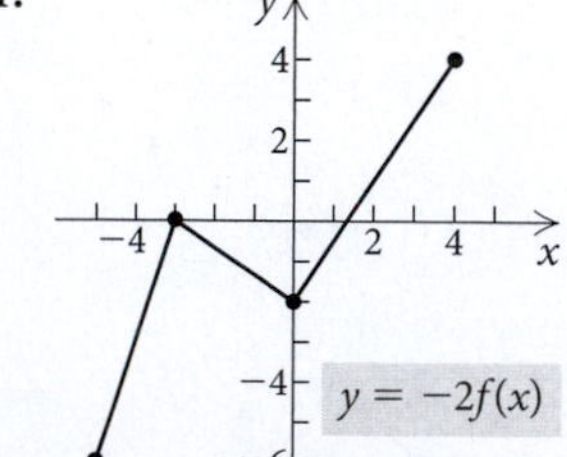

62.

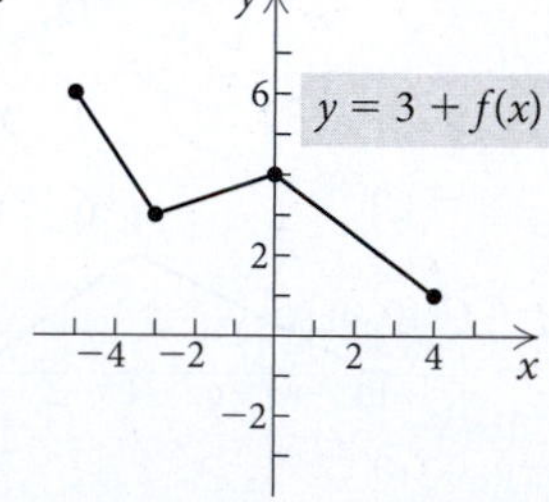

63. $y = 4x$ **64.** $y = \frac{2}{3}x$ **65.** $y = \frac{2500}{x}$ **66.** $y = \frac{54}{x}$ **67.** $y = \frac{48}{x^2}$ **68.** $y = \frac{1}{10} \cdot \frac{xz^2}{w}$ **69.** 20 min **70.** 75 **71.** 500 watts **72.** A **73.** C **74.** B **75.** Let $f(x)$ and $g(x)$ be odd functions. Then by definition, $f(-x) = -f(x)$, or

$f(x) = -f(-x)$, and $g(-x) = -g(x)$, or $g(x) = -g(-x)$. Thus, $(f + g)(x) = f(x) + g(x) = -f(-x) + [-g(-x)] = -[f(-x) + g(-x)] = -(f + g)(-x)$ and $f + g$ is odd.
76. Reflect the graph of $y = f(x)$ across the x-axis and then across the y-axis. **77.** **(a)** $4x^3 - 2x + 9$; **(b)** $4x^3 + 24x^2 + 46x + 35$; **(c)** $4x^3 - 2x + 42$. **(a)** adds 2 to each function value; **(b)** adds 2 to each input before finding a function value; **(c)** adds the output for 2 to the output for x **78.** In the graph of $y = f(cx)$, the constant c stretches or shrinks the graph of $y = f(x)$ horizontally. The constant c in $y = cf(x)$ stretches or shrinks the graph of $y = f(x)$ vertically. For $y = f(cx)$, the x-coordinates of $y = f(x)$ are divided by c; for $y = cf(x)$, the y-coordinates of $y = f(x)$ are multiplied by c.
79. The graph of $f(x) = 0$ is symmetric with respect to the x-axis, the y-axis, and the origin. This function is both even and odd. **80.** If all the exponents are even numbers, then $f(x)$ is an even function. If $a_0 = 0$ and all the exponents are odd numbers, then $f(x)$ is an odd function. **81.** Let $y(x) = kx^2$. Then $y(2x) = k(2x)^2 = k \cdot 4x^2 = 4 \cdot kx^2 = 4 \cdot y(x)$. Thus doubling x causes y to be quadrupled. **82.** Let $y = k_1 x$ and $x = \frac{k_2}{z}$. Then $y = k_1 \cdot \frac{k_2}{z}$, or $y = \frac{k_1 k_2}{z}$, so y varies inversely as z.

Test: Chapter 2

1. [2.1] **(a)** $(-5, -2)$; **(b)** $(2, 5)$; **(c)** $(-2, 2)$
2. [2.1] Increasing: $(-\infty, 0)$; decreasing: $(0, \infty)$; relative maximum: 2 at $x = 0$ **3.** [2.1] $A(b) = \frac{1}{2}b(4b - 6)$, or $2b^2 - 3b$
4. [2.1]

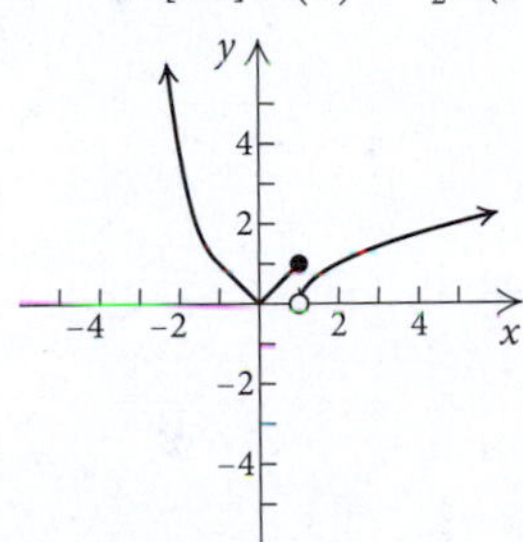

5. [2.1] $f\left(-\frac{7}{8}\right) = \frac{7}{8}; f(5) = 2; f(-4) = 16$ **6.** [2.2] 66
7. [2.2] 6 **8.** [2.2] -1 **9.** [2.2] 0 **10.** [2.1] $(-\infty, \infty)$
11. [2.1] $[3, \infty)$ **12.** [2.2] $[3, \infty)$ **13.** [2.2] $[3, \infty)$
14. [2.2] $[3, \infty)$ **15.** [2.2] $(3, \infty)$
16. [2.2] $(f + g)(x) = x^2 + \sqrt{x - 3}$
17. [2.2] $(f - g)(x) = x^2 - \sqrt{x - 3}$
18. [2.2] $(fg)(x) = x^2\sqrt{x - 3}$
19. [2.2] $(f/g)(x) = \dfrac{x^2}{\sqrt{x - 3}}$ **20.** [2.2] $\frac{1}{2}$
21. [2.2] $4x + 2h - 1$ **22.** [2.3] 83 **23.** [2.3] 0
24. [2.3] 4 **25.** [2.3] $16x + 15$
26. [2.3] $(f \circ g)(x) = \sqrt{x^2 - 4}$; $(g \circ f)(x) = x - 4$ **27.** [2.3] Domain of $(f \circ g)(x)$: $(-\infty, -2] \cup [2, \infty)$; domain of $(g \circ f)(x)$: $[5, \infty)$ **28.** [2.3] $f(x) = x^4$; $g(x) = 2x - 7$; answers may vary **29.** [2.4] x-axis: no; y-axis: yes; origin: no
30. [2.4] Odd **31.** [2.5] $f(x) = (x - 2)^2 - 1$
32. [2.5] $f(x) = (x + 2)^2 - 3$

33. [2.5]

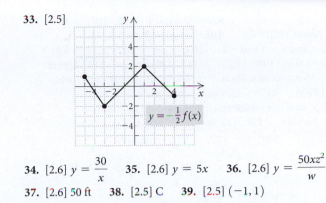

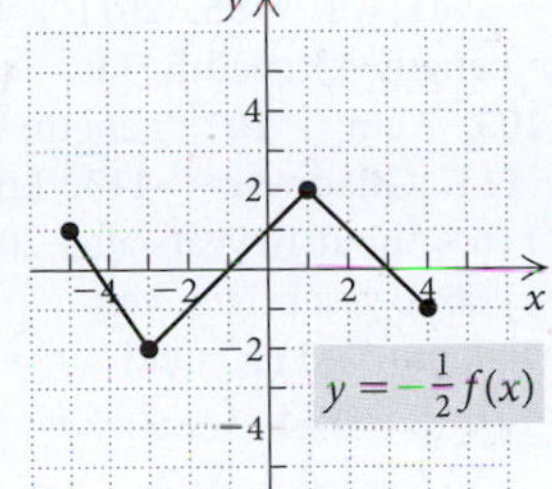

34. [2.6] $y = \dfrac{30}{x}$ **35.** [2.6] $y = 5x$ **36.** [2.6] $y = \dfrac{50xz^2}{w}$
37. [2.6] 50 ft **38.** [2.5] C **39.** [2.5] $(-1, 1)$

Chapter 3

Exercise Set 3.1

1. $\sqrt{3}i$ **3.** $5i$ **5.** $-\sqrt{33}i$ **7.** $-9i$ **9.** $7\sqrt{2}i$
11. $2 + 11i$ **13.** $5 - 12i$ **15.** $4 + 8i$ **17.** $-4 - 2i$
19. $5 + 9i$ **21.** $5 + 4i$ **23.** $5 + 7i$ **25.** $11 - 5i$
27. $-1 + 5i$ **29.** $2 - 12i$ **31.** -12 **33.** -45
35. $35 + 14i$ **37.** $6 + 16i$ **39.** $13 - i$ **41.** $-11 + 16i$
43. $-10 + 11i$ **45.** $-31 - 34i$ **47.** $-14 + 23i$ **49.** 41
51. 13 **53.** 74 **55.** $12 + 16i$ **57.** $-45 - 28i$
59. $-8 - 6i$ **61.** $2i$ **63.** $-7 + 24i$ **65.** $\frac{15}{146} + \frac{33}{146}i$
67. $\frac{10}{13} - \frac{15}{13}i$ **69.** $-\frac{14}{13} + \frac{5}{13}i$ **71.** $\frac{11}{25} - \frac{27}{25}i$
73. $\dfrac{-4\sqrt{3} + 10}{41} + \dfrac{5\sqrt{3} + 8}{41}i$ **75.** $-\frac{1}{2} + \frac{1}{2}i$
77. $-\frac{1}{2} - \frac{13}{2}i$ **79.** $-i$ **81.** $-i$ **83.** 1 **85.** i
87. 625 **89.** $y = -2x + 1$ **90.** All real numbers, or $(-\infty, \infty)$ **91.** $\left(-\infty, -\frac{5}{3}\right) \cup \left(-\frac{5}{3}, \infty\right)$
92. $x^2 - 3x - 1$ **93.** $\frac{8}{11}$ **94.** $2x + h - 3$ **95.** True
97. True **99.** $a^2 + b^2$ **101.** $x^2 - 6x + 25$

Exercise Set 3.2

1. $\frac{2}{3}, \frac{3}{2}$ **3.** $-2, 10$ **5.** $-1, \frac{2}{3}$ **7.** $-\sqrt{3}, \sqrt{3}$ **9.** $-\sqrt{7}, \sqrt{7}$
11. $-\sqrt{2}i, \sqrt{2}i$ **13.** $-4i, 4i$ **15.** $0, 3$ **17.** $-\frac{1}{3}, 0, 2$
19. $-1, -\frac{1}{7}, 1$ **21.** **(a)** $(-4, 0), (2, 0)$; **(b)** $-4, 2$
23. **(a)** $(-1, 0), (3, 0)$; **(b)** $-1, 3$ **25.** **(a)** $(-2, 0), (2, 0)$; **(b)** $-2, 2$ **27.** **(a)** $(1, 0)$; **(b)** 1 **29.** $-7, 1$
31. $4 \pm \sqrt{7}$ **33.** $-4 \pm 3i$ **35.** $-2, \frac{1}{3}$ **37.** $-3, 5$
39. $-1, \frac{2}{5}$ **41.** $\dfrac{5 \pm \sqrt{7}}{3}$ **43.** $-\dfrac{1}{2} \pm \dfrac{\sqrt{7}}{2}i$ **45.** $\dfrac{4 \pm \sqrt{31}}{5}$
47. $\dfrac{5}{6} \pm \dfrac{\sqrt{23}}{6}i$ **49.** $4 \pm \sqrt{11}$ **51.** $\dfrac{-1 \pm \sqrt{61}}{6}$
53. $\dfrac{5 \pm \sqrt{17}}{4}$ **55.** $-\dfrac{1}{5} \pm \dfrac{3}{5}i$ **57.** 144; two real
59. -7; two imaginary **61.** 49; two real **63.** $-5, -1$
65. $\dfrac{3 \pm \sqrt{21}}{2}$; $-0.791, 3.791$ **67.** $\dfrac{5 \pm \sqrt{21}}{2}$; $0.209, 4.791$
69. $-1 \pm \sqrt{6}$; $-3.449, 1.449$ **71.** $\dfrac{1}{4} \pm \dfrac{\sqrt{31}}{4}i$
73. $\dfrac{1 \pm \sqrt{13}}{6}$; $-0.434, 0.768$ **75.** $\dfrac{1 \pm \sqrt{6}}{5}$; $-0.290, 0.690$
77. $\dfrac{-3 \pm \sqrt{57}}{8}$; $-1.319, 0.569$ **79.** $\pm 1, \pm\sqrt{2}$
81. $\pm\sqrt{2}, \pm\sqrt{5}i$ **83.** $\pm 1, \pm\sqrt{5}i$ **85.** 16 **87.** $-8, 64$

89. 1, 16 **91.** $\frac{5}{2}$, 3 **93.** $-\frac{3}{2}, -1, \frac{1}{2}, 1$ **95.** 2011 **97.** 1995 **99.** About 10.216 sec **101.** Length: 4 ft; width: 3 ft **103.** 4 and 9; −9 and −4 **105.** 2 cm **107.** Length: 8 ft; width: 6 ft **109.** Linear **111.** Quadratic **113.** Linear **115.** About \$3.95 million **116.** About 16 years after 2004, or in 2020 **117.** x-axis: yes; y-axis: yes; origin: yes **118.** x-axis: no; y-axis: yes; origin: no **119.** Odd **120.** Neither **121.** (a) 2; (b) $\frac{11}{2}$ **123.** (a) 2; (b) $1 - i$ **125.** 1 **127.** $-\sqrt{7}, -\frac{3}{2}, 0, \frac{1}{3}, \sqrt{7}$ **129.** $\dfrac{-1 \pm \sqrt{1 + 4\sqrt{2}}}{2}$ **131.** $3 \pm \sqrt{5}$ **133.** 19 **135.** $-2 \pm \sqrt{2}, \dfrac{1}{2} \pm \dfrac{\sqrt{7}}{2}i$

Visualizing the Graph

1. C **2.** B **3.** A **4.** J **5.** F **6.** D **7.** I **8.** G **9.** H **10.** E

Exercise Set 3.3

1. (a) $\left(-\frac{1}{2}, -\frac{9}{4}\right)$; (b) $x = -\frac{1}{2}$; (c) minimum: $-\frac{9}{4}$
3. (a) $(4, -4)$; (b) $x = 4$; (c) minimum: -4;
(d)

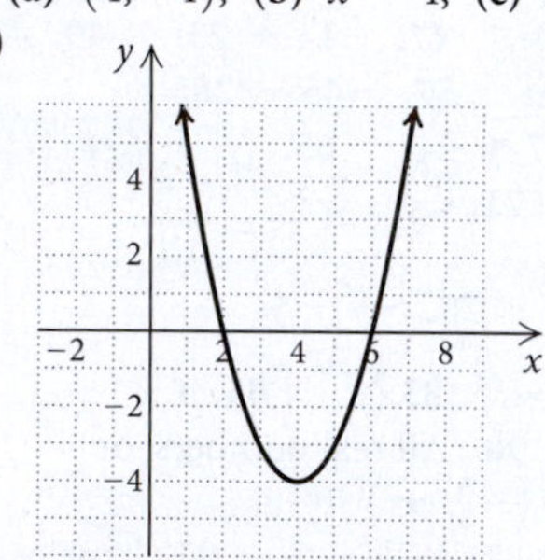

$f(x) = x^2 - 8x + 12$

5. (a) $\left(\frac{7}{2}, -\frac{1}{4}\right)$; (b) $x = \frac{7}{2}$; (c) minimum: $-\frac{1}{4}$;
(d)

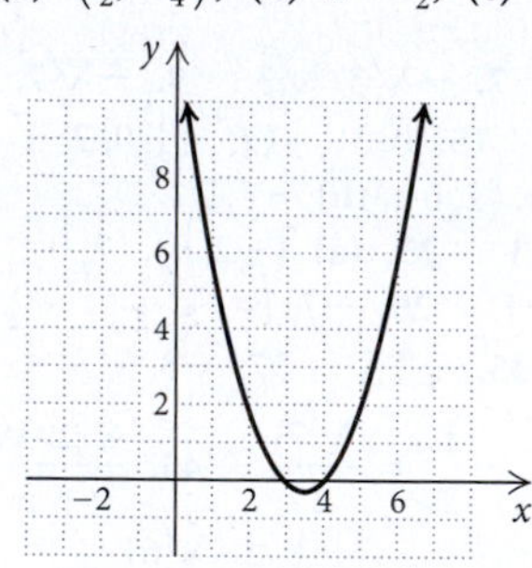

$f(x) = x^2 - 7x + 12$

7. (a) $(-2, 1)$; (b) $x = -2$; (c) minimum: 1;
(d)

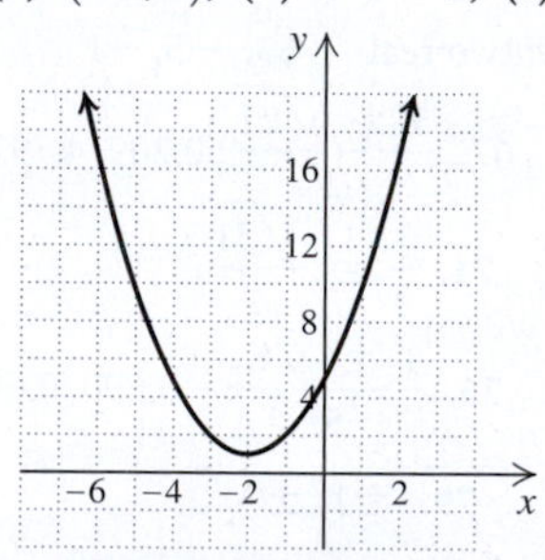

$f(x) = x^2 + 4x + 5$

9. (a) $(-4, -2)$; (b) $x = -4$; (c) minimum: -2;
(d)

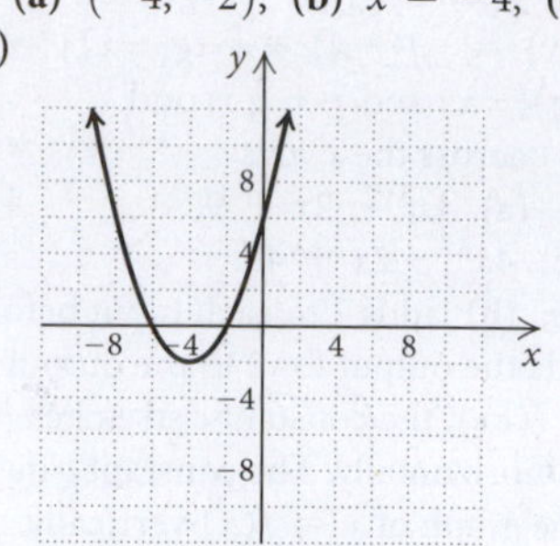

$g(x) = \dfrac{x^2}{2} + 4x + 6$

11. (a) $\left(-\frac{3}{2}, \frac{7}{2}\right)$; (b) $x = -\frac{3}{2}$; (c) minimum: $\frac{7}{2}$;
(d)

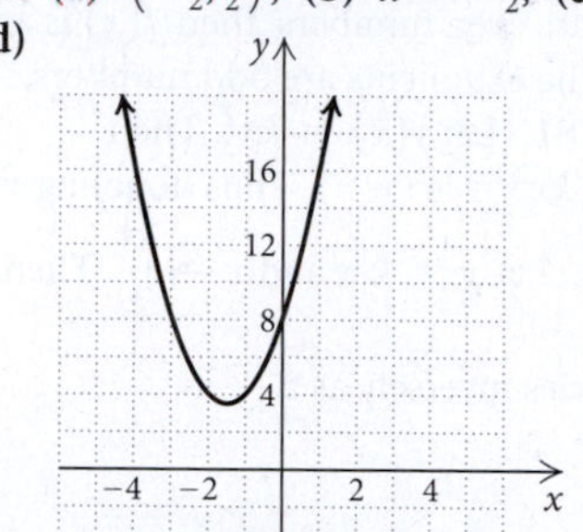

$g(x) = 2x^2 + 6x + 8$

13. (a) $(-3, 12)$; (b) $x = -3$; (c) maximum: 12;
(d)

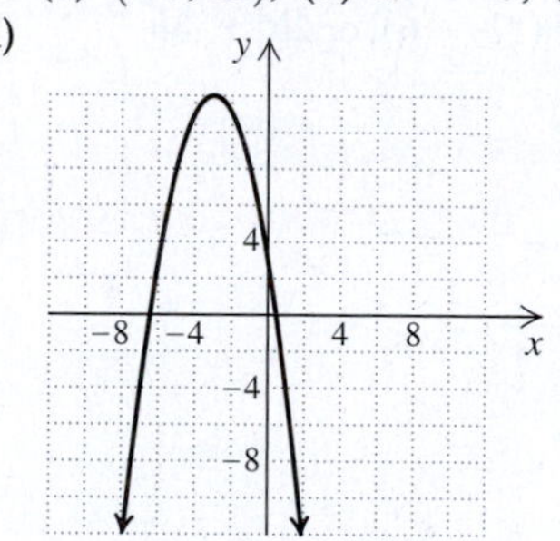

$f(x) = -x^2 - 6x + 3$

15. (a) $\left(\frac{1}{2}, \frac{3}{2}\right)$; (b) $x = \frac{1}{2}$; (c) maximum: $\frac{3}{2}$;
(d)

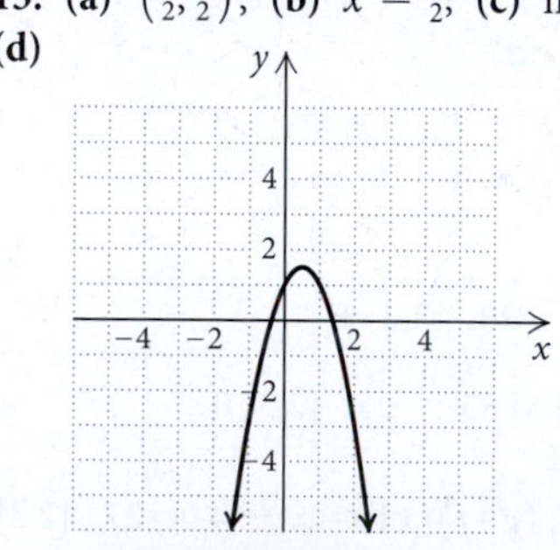

$g(x) = -2x^2 + 2x + 1$

17. (f) **19.** (b) **21.** (h) **23.** (c) **25.** True **27.** False **29.** True **31.** (a) $(3, -4)$; (b) minimum: -4; (c) $[-4, \infty)$; (d) increasing: $(3, \infty)$; decreasing: $(-\infty, 3)$
33. (a) $(-1, -18)$; (b) minimum: -18; (c) $[-18, \infty)$; (d) increasing: $(-1, \infty)$; decreasing: $(-\infty, -1)$

35. (a) $\left(5, \frac{9}{2}\right)$; (b) maximum: $\frac{9}{2}$; (c) $\left(-\infty, \frac{9}{2}\right]$; (d) increasing: $(-\infty, 5)$; decreasing: $(5, \infty)$ **37.** (a) $(-1, 2)$; (b) minimum: 2; (c) $[2, \infty)$; (d) increasing: $(-1, \infty)$; decreasing: $(-\infty, -1)$ **39.** (a) $\left(-\frac{3}{2}, 18\right)$; (b) maximum: 18; (c) $(-\infty, 18]$; (d) increasing: $\left(-\infty, -\frac{3}{2}\right)$; decreasing: $\left(-\frac{3}{2}, \infty\right)$
41. 0.625 sec; 12.25 ft **43.** 3.75 sec; 305 ft **45.** 4.5 in.
47. Base: 10 cm; height: 10 cm **49.** 350 doghouses
51. \$797; 40 units **53.** 4800 yd^2 **55.** About 60.5 ft
57. 3 **58.** $4x + 2h - 1$
59.

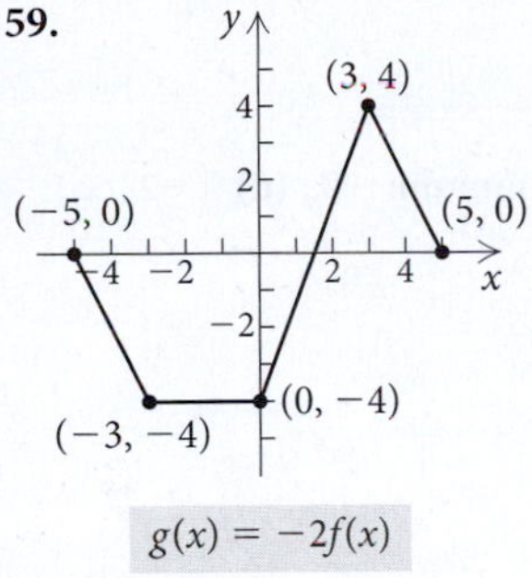

60.

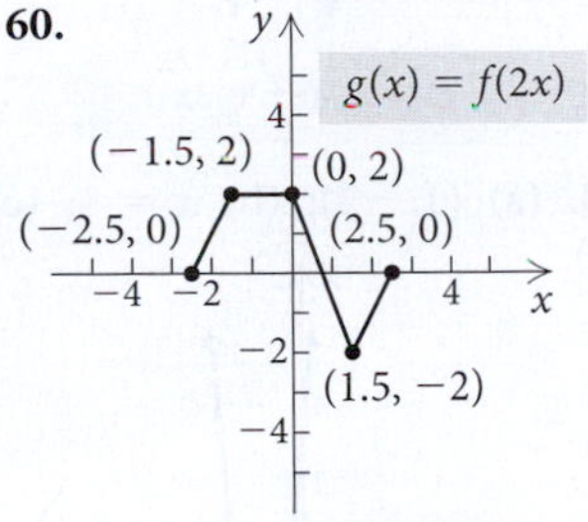

61. -236.25
63.

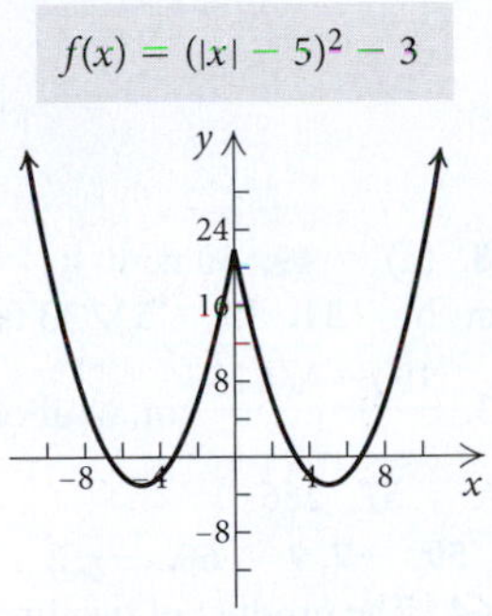

65. Pieces should be $\frac{24\pi}{4 + \pi}$ in. and $\frac{96}{4 + \pi}$ in.

Mid-Chapter Mixed Review: Chapter 3

1. True **2.** False **3.** True **4.** False **5.** $6i$ **6.** $\sqrt{5}i$
7. $-4i$ **8.** $4\sqrt{2}i$ **9.** $-1 + i$ **10.** $-7 + 5i$
11. $23 + 2i$ **12.** $-\frac{1}{29} - \frac{17}{29}i$ **13.** i **14.** 1 **15.** $-i$
16. -64 **17.** $-4, 1$ **18.** $-2, -\frac{3}{2}$ **19.** $\pm\sqrt{6}$
20. $\pm 10i$ **21.** $4x^2 - 8x - 3 = 0$; $4x^2 - 8x = 3$; $x^2 - 2x = \frac{3}{4}$; $x^2 - 2x + 1 = \frac{3}{4} + 1$; $(x - 1)^2 = \frac{7}{4}$; $x - 1 = \pm\frac{\sqrt{7}}{2}$; $x = 1 \pm \frac{\sqrt{7}}{2} = \frac{2 \pm \sqrt{7}}{2}$
22. (a) 29; two real; (b) $\frac{3 \pm \sqrt{29}}{2}$; $-1.193, 4.193$
23. (a) 0; one real; (b) $\frac{3}{2}$ **24.** (a) -8; two nonreal; (b) $-\frac{1}{3} \pm \frac{\sqrt{2}}{3}i$ **25.** $\pm 1, \pm\sqrt{6}i$ **26.** $\frac{1}{4}, 4$ **27.** 5 and 7; -7 and -5 **28.** (a) $(3, -2)$; (b) $x = 3$; (c) minimum: -2; (d) $[-2, \infty)$; (e) increasing: $(3, \infty)$; decreasing: $(-\infty, 3)$;
(f)

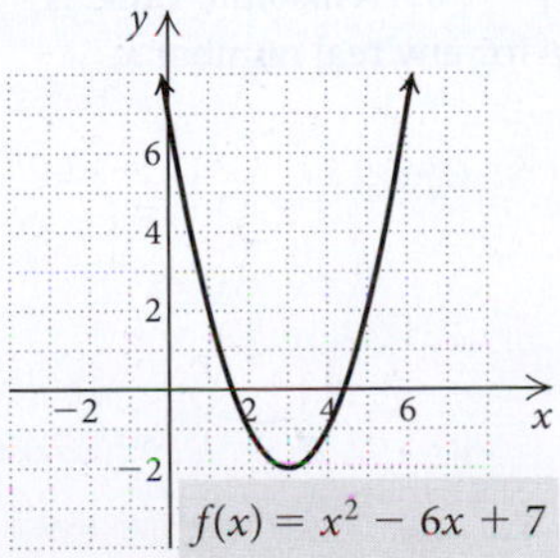

29. (a) $(-1, -3)$; (b) $x = -1$; (c) maximum: -3; (d) $(-\infty, -3]$; (e) increasing: $(-\infty, -1)$; decreasing: $(-1, \infty)$;
(f)

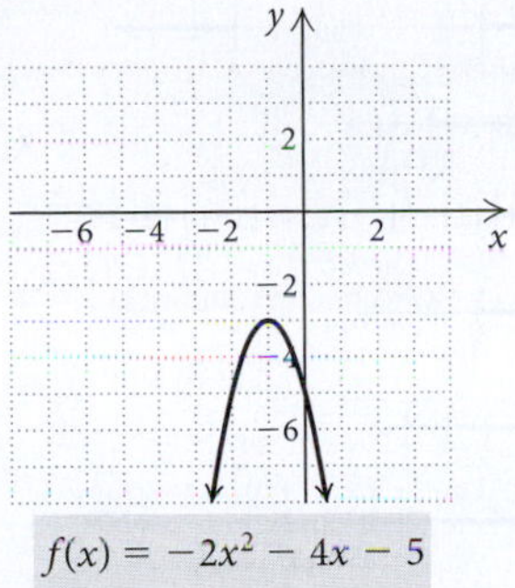

30. Base: 8 in., height: 8 in. **31.** The sum of two imaginary numbers is not always an imaginary number. For example, $(2 + i) + (3 - i) = 5$, a real number. **32.** Use the discriminant. If $b^2 - 4ac < 0$, there are no x-intercepts. If $b^2 - 4ac = 0$, there is one x-intercept. If $b^2 - 4ac > 0$, there are two x-intercepts. **33.** Completing the square was used in Section 3.2 to solve quadratic equations. It was used again in Section 3.3 to write quadratic functions in the form $f(x) = a(x - h)^2 + k$. **34.** The x-intercepts of $g(x)$ are also $(x_1, 0)$ and $(x_2, 0)$. This is true because $f(x)$ and $g(x)$ have the same zeros. Consider $g(x) = 0$, or $-ax^2 - bx - c = 0$. Multiplying by -1 on both sides, we get an equivalent equation $ax^2 + bx + c = 0$, or $f(x) = 0$.

Exercise Set 3.4

1. $\frac{20}{9}$ **3.** 286 **5.** 6 **7.** 6 **9.** 4 **11.** 2, 3 **13.** $-1, 6$
15. $\frac{1}{2}, 5$ **17.** 7 **19.** No solution **21.** $-\frac{69}{14}$ **23.** $-\frac{37}{18}$
25. 2 **27.** No solution **29.** $\{x \mid x$ is a real number *and* $x \neq 0$ *and* $x \neq 6\}$ **31.** $\frac{5}{3}$ **33.** $\frac{9}{2}$ **35.** 3 **37.** -4
39. -5 **41.** $\pm\sqrt{2}$ **43.** No solution **45.** 6
47. -1 **49.** $\frac{35}{2}$ **51.** -98 **53.** -6 **55.** 5 **57.** 7
59. 2 **61.** $-1, 2$ **63.** 7 **65.** 7 **67.** No solution
69. 1 **71.** 3, 7 **73.** 5 **75.** -1 **77.** -8 **79.** 81
81. $T_1 = \frac{P_1V_1T_2}{P_2V_2}$ **83.** $C = \frac{1}{LW^2}$ **85.** $R_2 = \frac{RR_1}{R_1 - R}$
87. $P = \frac{A}{I^2 + 2I + 1}$, or $\frac{A}{(I + 1)^2}$ **89.** $p = \frac{Fm}{m - F}$
91. 7.5 **92.** 3 **93.** China: 53,800,000 metric tons; United States: 10,508,000 metric tons **94.** 119,771 baseball players
95. $3 \pm 2\sqrt{2}$ **97.** -1 **99.** 0, 1

Exercise Set 3.5

1. $-7, 7$ **3.** 0 **5.** $-\frac{5}{6}, \frac{5}{6}$ **7.** No solution **9.** $-\frac{1}{3}, \frac{1}{3}$
11. $-3, 3$ **13.** $-3, 5$ **15.** $-8, 4$ **17.** $-1, -\frac{1}{3}$
19. $-24, 44$ **21.** $-2, 4$ **23.** $-13, 7$ **25.** $-\frac{4}{3}, \frac{2}{3}$
27. $-\frac{3}{4}, \frac{9}{4}$ **29.** $-13, 1$ **31.** 0, 1
33. $(-7, 7)$;

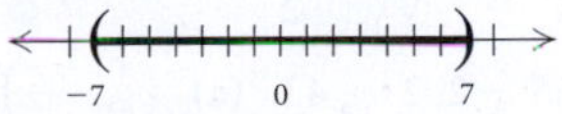

35. $[-2, 2]$;

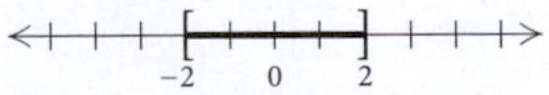

37. $(-\infty, -4.5] \cup [4.5, \infty)$;

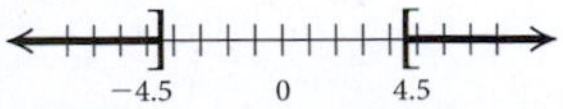

39. $(-\infty, -3) \cup (3, \infty)$;

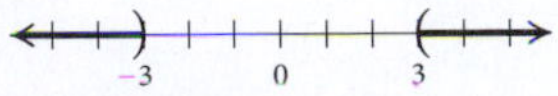

41. $\left(-\frac{1}{3}, \frac{1}{3}\right)$;

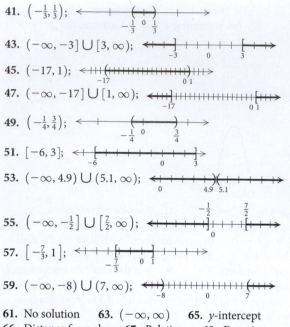

43. $(-\infty, -3] \cup [3, \infty)$;

45. $(-17, 1)$;

47. $(-\infty, -17] \cup [1, \infty)$;

49. $\left(-\frac{1}{4}, \frac{3}{4}\right)$;

51. $[-6, 3]$;

53. $(-\infty, 4.9) \cup (5.1, \infty)$;

55. $\left(-\infty, -\frac{1}{2}\right] \cup \left[\frac{7}{2}, \infty\right)$;

57. $\left[-\frac{7}{3}, 1\right]$;

59. $(-\infty, -8) \cup (7, \infty)$;

61. No solution **63.** $(-\infty, \infty)$ **65.** y-intercept **66.** Distance formula **67.** Relation **68.** Function **69.** Horizontal lines **70.** Parallel **71.** Decreasing **72.** Symmetric with respect to the y-axis **73.** $\left(-\infty, \frac{1}{2}\right)$ **75.** No solution **77.** $\left(-\infty, -\frac{8}{3}\right) \cup (-2, \infty)$

Review Exercises: Chapter 3

1. True **2.** True **3.** False **4.** False **5.** $-\frac{5}{2}, \frac{1}{3}$ **6.** $-5, 1$ **7.** $-2, \frac{4}{3}$ **8.** $-\sqrt{3}, \sqrt{3}$ **9.** $-\sqrt{10}i, \sqrt{10}i$ **10.** 1 **11.** $-5, 3$ **12.** $\dfrac{1 \pm \sqrt{41}}{4}$ **13.** $-\dfrac{1}{3} \pm \dfrac{2\sqrt{2}}{3}i$ **14.** $\frac{27}{7}$ **15.** $-\frac{1}{2}, \frac{9}{4}$ **16.** $0, 3$ **17.** 5 **18.** $1, 7$ **19.** $-8, 1$

20. $(-\infty, -3] \cup [3, \infty)$;

21. $\left(-\frac{14}{3}, 2\right)$;

22. $\left(-\frac{2}{3}, 1\right)$;

23. $(-\infty, -6] \cup [-2, \infty)$;

24. $P = \dfrac{MN}{M + N}$ **25.** $-2\sqrt{10}i$ **26.** $-4\sqrt{15}$ **27.** $-\frac{7}{8}$ **28.** $2 - i$ **29.** $1 - 4i$ **30.** $-18 - 26i$ **31.** $\frac{11}{10} + \frac{3}{10}i$ **32.** $-i$ **33.** $x^2 - 3x + \frac{9}{4} = 18 + \frac{9}{4}$; $\left(x - \frac{3}{2}\right)^2 = \frac{81}{4}$; $x = \frac{3}{2} \pm \frac{9}{2}$; $-3, 6$ **34.** $x^2 - 4x = 2$; $x^2 - 4x + 4 = 2 + 4$; $(x - 2)^2 = 6$; $x = 2 \pm \sqrt{6}$; $2 - \sqrt{6}, 2 + \sqrt{6}$ **35.** $-4, \frac{2}{3}$ **36.** $1 - 3i, 1 + 3i$ **37.** $-2, 5$ **38.** 1 **39.** $\pm\sqrt{\dfrac{3 \pm \sqrt{5}}{2}}$ **40.** $-\sqrt{3}, 0, \sqrt{3}$ **41.** $-2, -\frac{2}{3}, 3$ **42.** $-5, -2, 2$ **43.** **(a)** $\left(\frac{3}{8}, -\frac{7}{16}\right)$; **(b)** $x = \frac{3}{8}$; **(c)** maximum: $-\frac{7}{16}$; **(d)** $\left(-\infty, -\frac{7}{16}\right]$;

(e)

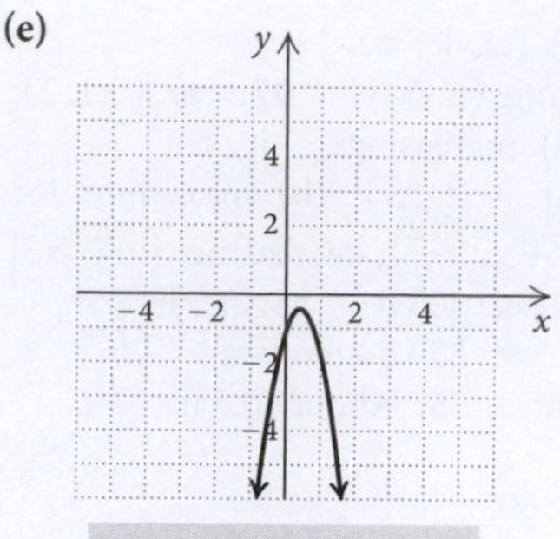

$f(x) = -4x^2 + 3x - 1$

44. **(a)** $(1, -2)$; **(b)** $x = 1$; **(c)** minimum: -2; **(d)** $[-2, \infty)$;

(e)

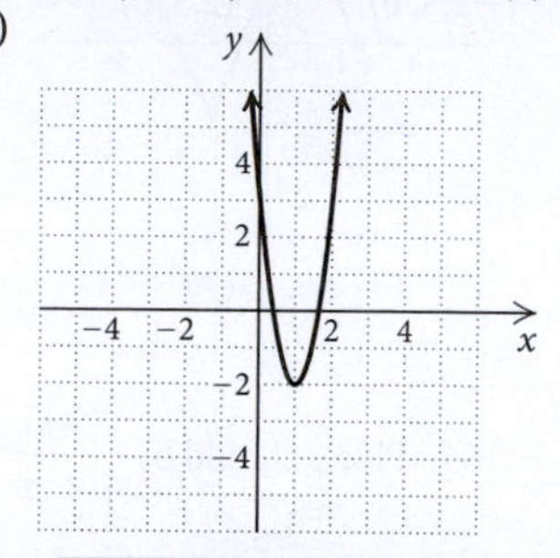

$f(x) = 5x^2 - 10x + 3$

45. (d) **46.** (c) **47.** (b) **48.** (a) **49.** 30 ft, 40 ft **50.** Rebecca: 15 km/h; Harry: 8 km/h **51.** $35 - 5\sqrt{33}$ ft, or about 6.3 ft **52.** 6 ft by 6 ft **53.** $\dfrac{15 - \sqrt{115}}{2}$ cm, or about 2.1 cm **54.** B **55.** B **56.** A **57.** 256 **58.** $4 \pm \sqrt[4]{243}$, or 0.052, 7.948 **59.** $-7, 9$ **60.** $-\frac{1}{4}, 2$ **61.** -1 **62.** 9% **63.** ± 6 **64.** The product of two imaginary numbers is not always an imaginary number. For example, $i \cdot i = i^2 = -1$, a real number. **65.** No; consider the quadratic formula $x = \dfrac{-b \pm \sqrt{b^2 - 4ac}}{2a}$. If $b^2 - 4ac = 0$, then $x = \dfrac{-b}{2a}$, so there is one real zero. If $b^2 - 4ac > 0$, then $\sqrt{b^2 - 4ac}$ is a real number and there are two real zeros. If $b^2 - 4ac < 0$, then $\sqrt{b^2 - 4ac}$ is an imaginary number and there are two imaginary zeros. Thus a quadratic function cannot have one real zero and one imaginary zero. **66.** You can conclude that $|a_1| = |a_2|$ since these constants determine how wide the parabolas are. Nothing can be concluded about the h's and the k's. **67.** When both sides of an equation are multiplied by the LCD, the resulting equation might not be equivalent to the original equation. One or more of the possible solutions of the resulting equation might make a denominator of the original equation 0. **68.** When both sides of an equation are raised to an even power, the resulting equation might not be equivalent to the original equation. For example, the solution set of $x = -2$ is $\{-2\}$, but the solution set of $x^2 = (-2)^2$, or $x^2 = 4$, is $\{-2, 2\}$. **69.** Absolute value is nonnegative. **70.** $|x| \geq 0 > p$ for any real number x.

Test: Chapter 3

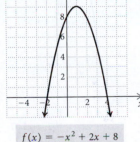

1. [3.2] $\frac{1}{2}, -5$ **2.** [3.2] $-\sqrt{6}, \sqrt{6}$ **3.** [3.2] $-2i, 2i$
4. [3.2] $-1, 3$ **5.** [3.2] $\frac{5 \pm \sqrt{13}}{2}$ **6.** [3.2] $\frac{3}{4} \pm \frac{\sqrt{23}}{4}i$
7. [3.2] 16 **8.** [3.4] $-1, \frac{13}{6}$ **9.** [3.4] 5 **10.** [3.4] 5
11. [3.5] $-11, 3$ **12.** [3.5] $-\frac{1}{2}, 2$
13. [3.5] $[-7, 1]$;
14. [3.5] $(-2, 3)$;
15. [3.5] $(-\infty, -7) \cup (-3, \infty)$;
16. [3.5] $(-\infty, -2] \cup [5, \infty)$;
17. [3.4] $B = \frac{AC}{A - C}$ **18.** [3.4] $n = \frac{R^2}{3p}$
19. [3.2] $x^2 + 4x = 1; x^2 + 4x + 4 = 1 + 4; (x + 2)^2 = 5;$ $x = -2 \pm \sqrt{5}; -2 - \sqrt{5}, -2 + \sqrt{5}$
20. [3.2] About 11.4 sec **21.** [3.1] $\sqrt{43}\,i$ **22.** [3.1] $-5i$
23. [3.1] $3 - 5i$ **24.** [3.1] $10 + 5i$ **25.** [3.1] $\frac{1}{10} - \frac{1}{5}i$
26. [3.1] i **27.** [3.2] $-\frac{1}{4}, 3$ **28.** [3.2] $\frac{1 \pm \sqrt{57}}{4}$
29. [3.3] **(a)** $(1, 9)$; **(b)** $x = 1$; **(c)** maximum: 9; **(d)** $(-\infty, 9]$;
(e)

$f(x) = -x^2 + 2x + 8$

30. [3.3] 20 ft by 40 ft **31.** [3.3] C **32.** [3.3], [3.4] $-\frac{4}{9}$

Chapter 4

Exercise Set 4.1

1. $\frac{1}{2}x^3; \frac{1}{2}; 3$; cubic **3.** $0.9x; 0.9; 1$; linear **5.** $305x^4; 305; 4$; quartic **7.** $x^4; 1; 4$; quartic **9.** $4x^3; 4; 3$; cubic
11. (d) **13.** (b) **15.** (c) **17.** (a) **19.** (c)
21. (d) **23.** Yes; no; no **25.** No; yes; yes
27. -3, multiplicity 2; 1, multiplicity 1 **29.** 4, multiplicity 3; -6, multiplicity 1 **31.** ± 3, each has multiplicity 3
33. 0, multiplicity 3; 1, multiplicity 2; -4, multiplicity 1
35. 3, multiplicity 2; -4, multiplicity 3; 0, multiplicity 4
37. $\pm\sqrt{3}, \pm 1$, each has multiplicity 1 **39.** $-3, -1, 1$, each has multiplicity 1 **41.** $\pm 2, \frac{1}{2}$, each has multiplicity 1 **43.** False
45. True **47.** 2008: 1.7 million albums; 2012: 4.9 million albums; 2016: 7.9 million albums **49.** 26, 64, and 80
51. 5 sec **53.** 2003: 684,025 admissions; 2006: 739,119 admissions; 2011: 665,806 admissions **55.** 6.3% **57.** 5
58. $6\sqrt{2}$ **59.** Center: $(3, -5)$; radius: 7
60. Center: $(-4, 3)$; radius: $2\sqrt{2}$

61. $\{y \mid y \geq 3\}$, or $[3, \infty)$ **62.** $\{x \mid x > \frac{5}{3}\}$, or $(\frac{5}{3}, \infty)$
63. $\{x \mid x \leq -13$ *or* $x \geq 1\}$, or $(-\infty, -13] \cup [1, \infty)$
64. $\{x \mid -\frac{11}{12} \leq x \leq \frac{5}{12}\}$, or $[-\frac{11}{12}, \frac{5}{12}]$ **65.** $16; x^{16}$

Visualizing the Graph

1. H **2.** D **3.** J **4.** B **5.** A **6.** C **7.** I
8. E **9.** G **10.** F

Exercise Set 4.2

1. (a) 5; **(b)** 5; **(c)** 4 **3. (a)** 10; **(b)** 10; **(c)** 9
5. (a) 3; **(b)** 3; **(c)** 2 **7.** (d) **9.** (f) **11.** (b)

13.

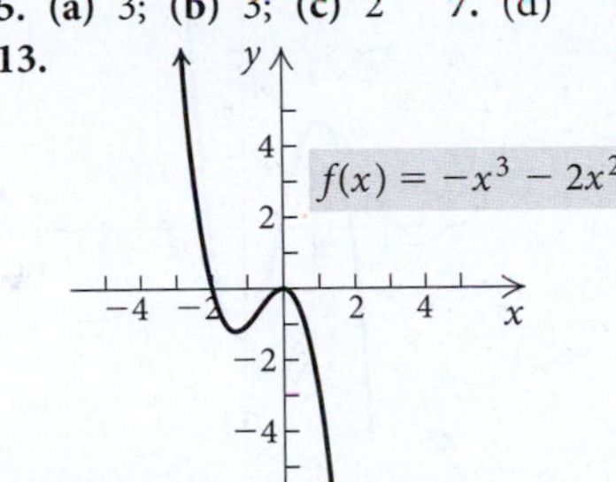

15.

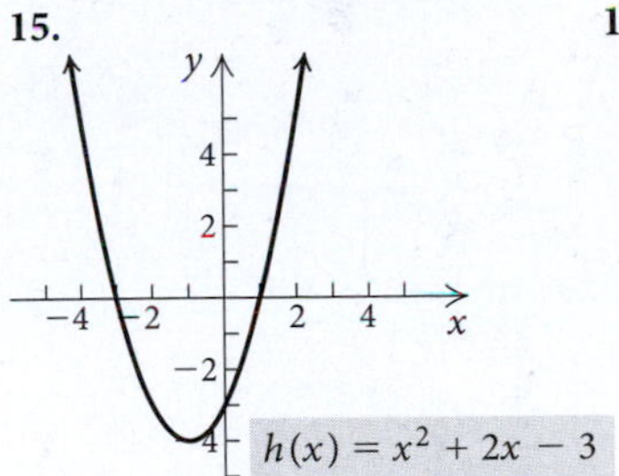

17.

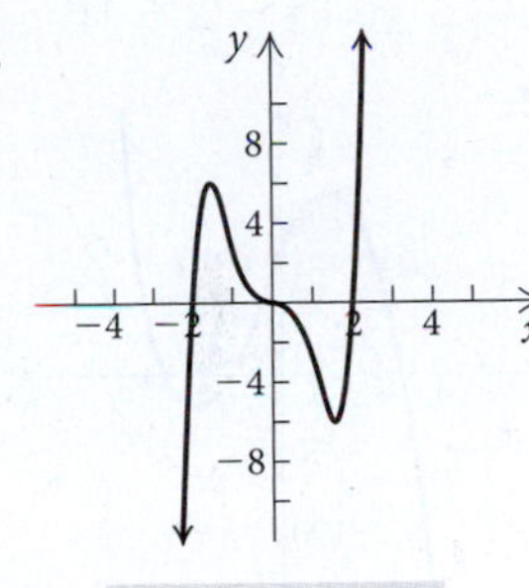

$h(x) = x^5 - 4x^3$

19. **21.**

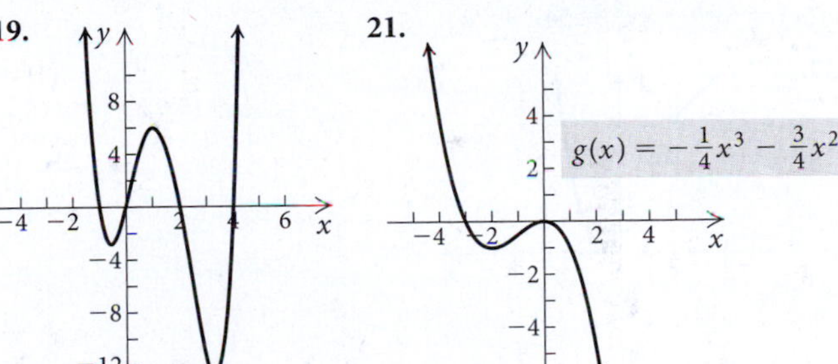

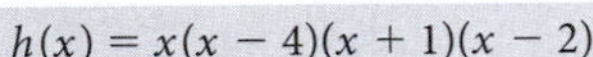

$h(x) = x(x - 4)(x + 1)(x - 2)$

23.

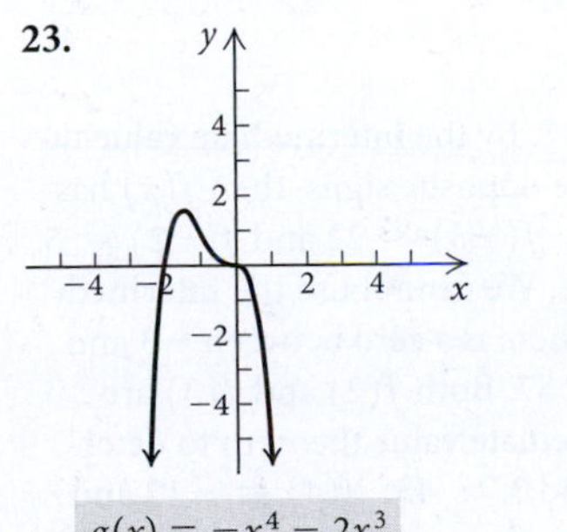

$g(x) = -x^4 - 2x^3$

25.

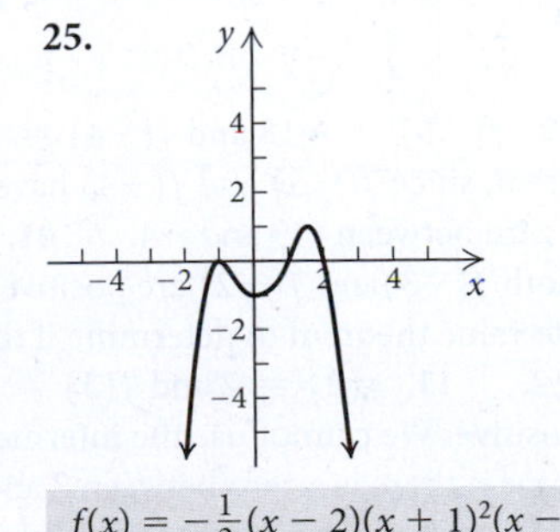

$f(x) = -\frac{1}{2}(x - 2)(x + 1)^2(x - 1)$

27.

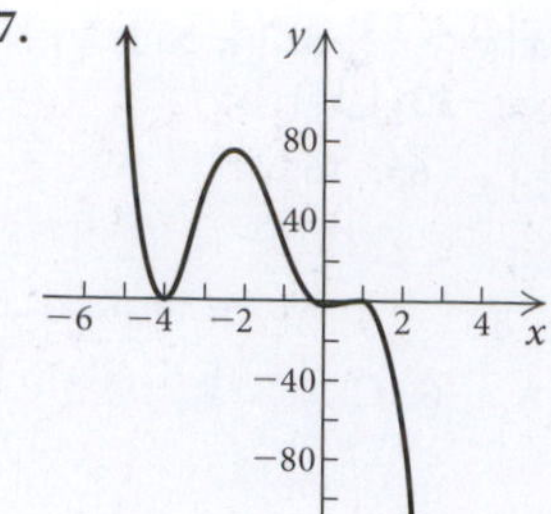

$g(x) = -x(x-1)^2(x+4)^2$

29.

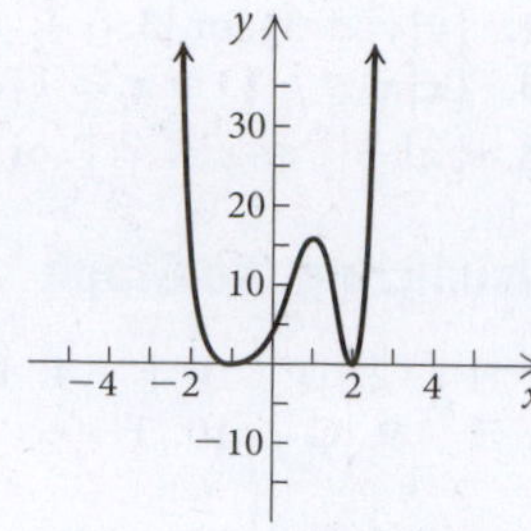

$f(x) = (x-2)^2(x+1)^4$

31.

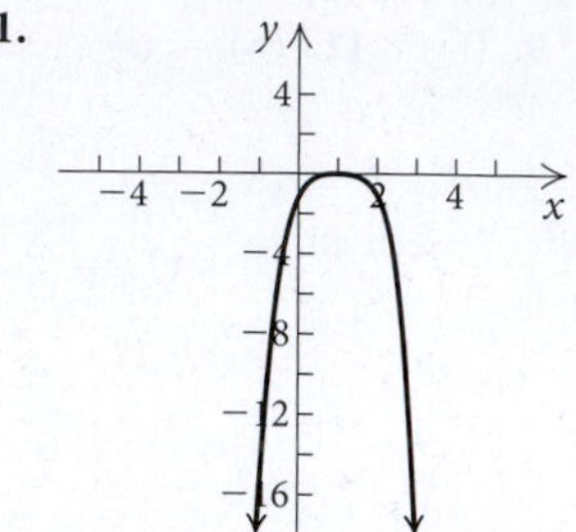

$g(x) = -(x-1)^4$

33.

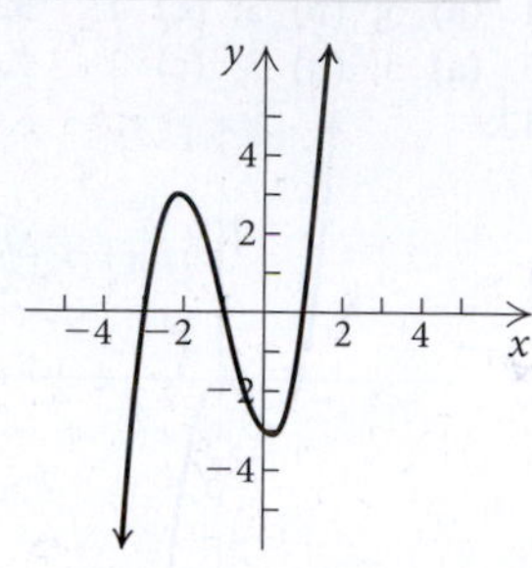

$h(x) = x^3 + 3x^2 - x - 3$

35.

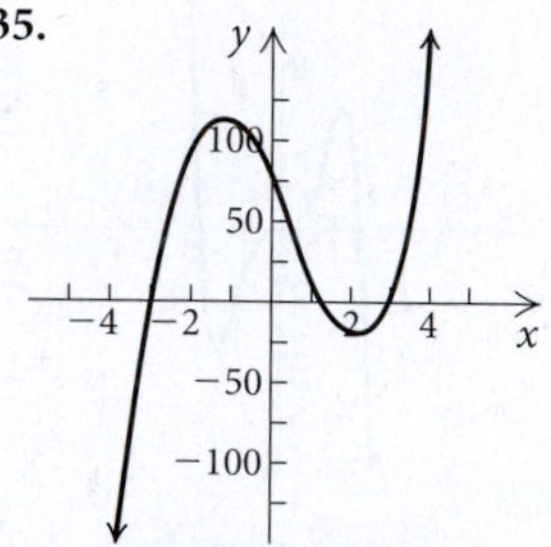

$f(x) = 6x^3 - 8x^2 - 54x + 72$

37.

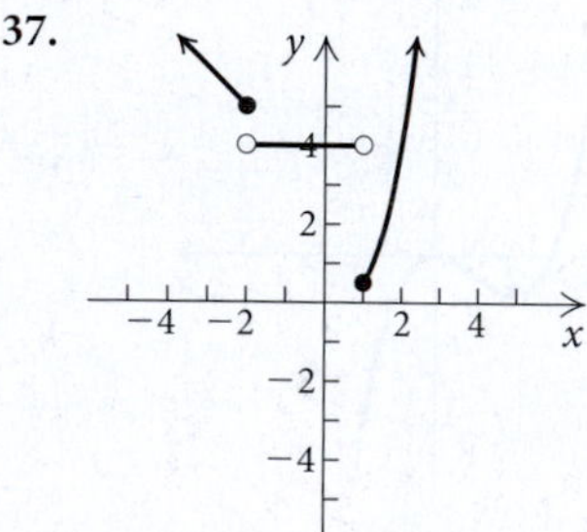

$$g(x) = \begin{cases} -x+3, & \text{for } x \le -2, \\ 4, & \text{for } -2 < x < 1, \\ \frac{1}{2}x^3, & \text{for } x \ge 1 \end{cases}$$

39. $f(-5) = -18$ and $f(-4) = 7$. By the intermediate value theorem, since $f(-5)$ and $f(-4)$ have opposite signs, then $f(x)$ has a zero between -5 and -4. **41.** $f(-3) = 22$ and $f(-2) = 5$. Both $f(-3)$and $f(-2)$ are positive. We cannot use the intermediate value theorem to determine if there is a zero between -3 and -2. **43.** $f(2) = 2$ and $f(3) = 57$. Both $f(2)$ and $f(3)$ are positive. We cannot use the intermediate value theorem to determine if there is a zero between 2 and 3. **45.** $f(4) = -12$ and $f(5) = 4$. By the intermediate value theorem, since $f(4)$ and $f(5)$ have opposite signs, then $f(x)$ has a zero between 4 and 5.
47. (d) **48.** (f) **49.** (e) **50.** (a) **51.** (b) **52.** (c)
53. $\frac{9}{10}$ **54.** $-3, 0, 4$ **55.** $-\frac{5}{3}, \frac{11}{2}$ **56.** $\frac{196}{25}$

Exercise Set 4.3

1. (a) No; (b) yes; (c) no **3.** (a) Yes; (b) no; (c) yes
5. $P(x) = (x+2)(x^2 - 2x + 4) - 16$
7. $P(x) = (x+9)(x^2 - 3x + 2) + 0$
9. $P(x) = (x+2)(x^3 - 2x^2 + 2x - 4) + 11$
11. $Q(x) = 2x^3 + x^2 - 3x + 10, R(x) = -42$
13. $Q(x) = x^2 - 4x + 8, R(x) = -24$
15. $Q(x) = 3x^2 - 4x + 8, R(x) = -18$
17. $Q(x) = x^4 + 3x^3 + 10x^2 + 30x + 89, R(x) = 267$
19. $Q(x) = x^3 + x^2 + x + 1, R(x) = 0$
21. $Q(x) = 2x^3 + x^2 + \frac{7}{2}x + \frac{7}{4}, R(x) = -\frac{1}{8}$
23. 0; -60; 0 **25.** 10; 80; 998 **27.** 5,935,988; -772
29. 0; 0; 65; $1 - 12\sqrt{2}$ **31.** Yes; no **33.** Yes; yes
35. No; yes **37.** No; no
39. $f(x) = (x-1)(x+2)(x+3)$; $1, -2, -3$
41. $f(x) = (x-2)(x-5)(x+1)$; $2, 5, -1$
43. $f(x) = (x-2)(x-3)(x+4)$; $2, 3, -4$
45. $f(x) = (x-3)^3(x+2)$; $3, -2$
47. $f(x) = (x-1)(x-2)(x-3)(x+5)$; $1, 2, 3, -5$

49.

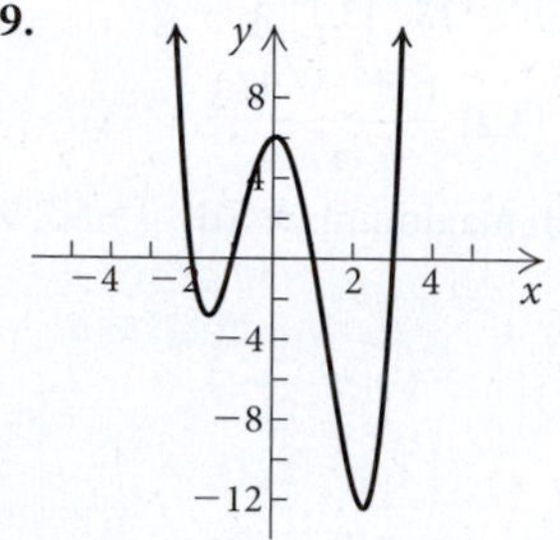

$f(x) = x^4 - x^3 - 7x^2 + x + 6$

51.

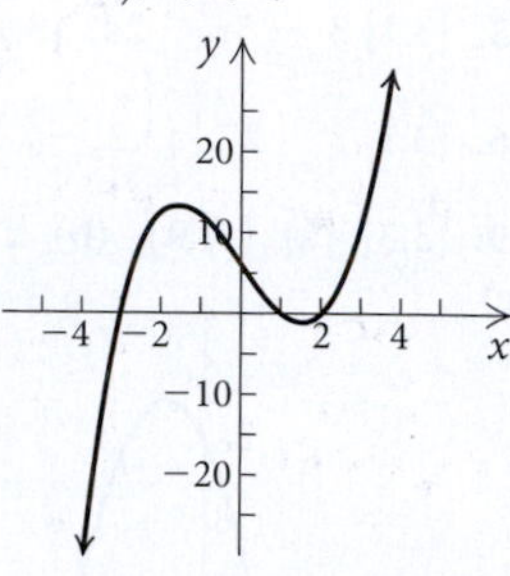

$f(x) = x^3 - 7x + 6$

53.

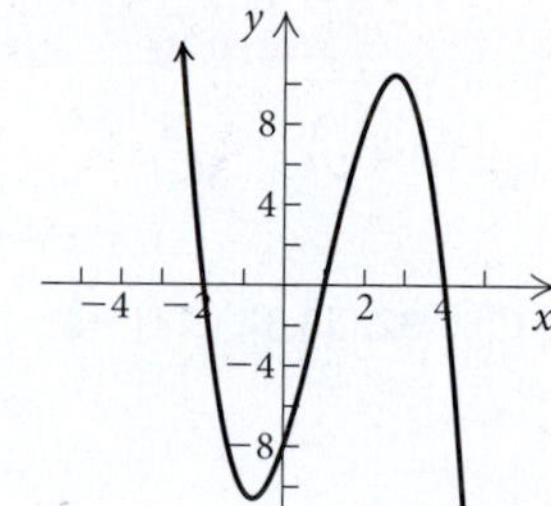

$f(x) = -x^3 + 3x^2 + 6x - 8$

55. $\frac{5}{4} \pm \frac{\sqrt{71}}{4}i$
56. $-1, \frac{3}{7}$ **57.** $-5, 0$
58. 10 **59.** $-3, -2$
60. $f(x) = 0.172x + 2.69$; 1995: \$5.27; 2018: \$9.23
61. $b = 15$ in., $h = 15$ in.

63. (a) $x + 4, x + 3, x - 2, x - 5$; (b) $P(x) = (x+4)(x+3)(x-2)(x-5)$; (c) yes; two examples are $f(x) = c \cdot P(x)$ for any nonzero constant c; and $g(x) = (x-a)P(x)$; (d) no **65.** $\frac{14}{3}$
67. $0, -6$ **69.** Answers may vary. One possibility is $P(x) = x^{15} - x^{14}$. **71.** $x^2 + 2ix + (2 - 4i)$, R $-6 - 2i$
73. $x - 3 + i$, R $6 - 3i$

Mid-Chapter Mixed Review: Chapter 4

1. False **2.** True **3.** True **4.** False
5. 5; multiplicity 6 **6.** $-5, -\frac{1}{2}, 5$; each has multiplicity 1
7. $\pm 1, \pm\sqrt{2}$; each has multiplicity 1 **8.** 3, multiplicity 2; -4, multiplicity 1 **9.** (d) **10.** (a) **11.** (b) **12.** (c)
13. $f(-2) = -13$ and $f(0) = 3$. By the intermediate value theorem, since $f(-2)$ and $f(0)$ have opposite signs, then $f(x)$ has at least one zero between -2 and 0. **14.** $f\left(-\frac{1}{2}\right) = 2\frac{3}{8}$ and $f(1) = 2$. Both $f\left(-\frac{1}{2}\right)$ and $f(1)$ are positive. We cannot use the intermediate value theorem to determine if there is a zero between

$-\frac{1}{2}$ and 1. **15.** $P(x) = (x - 1)(x^3 - 5x^2 - 5x - 4) - 6$
16. $Q(x) = 3x^3 + 5x^2 + 12x + 18, R(x) = 42$
17. $Q(x) = x^4 - x^3 + x^2 - x + 1, R(x) = -6$
18. $g(-5) = -380$ **19.** $f\left(\frac{1}{2}\right) = -15$
20. $f(-\sqrt{2}) = 20 - \sqrt{2}$ **21.** Yes; no
22. Yes; yes **23.** $h(x) = (x - 1)(x - 8)(x + 7); -7, 1, 8$
24. $g(x) = (x + 1)(x - 2)(x - 4)(x + 3); -3, -1, 2, 4$
25. The range of a polynomial function with an odd degree is $(-\infty, \infty)$. The range of a polynomial function with an even degree is $[s, \infty)$ for some real number s if $a_n > 0$ and is $(-\infty, s]$ for some real number s if $a_n < 0$. **26.** Since we can find $f(0)$ for any polynomial function $f(x)$, it is not possible for the graph of a polynomial function to have no y-intercept. It is possible for a polynomial function to have no x-intercepts. For instance, a function of the form $f(x) = x^2 + a, a > 0$, has no x-intercepts. There are other examples as well. **27.** The zeros of a polynomial function are the first coordinates of the points at which the graph of the function crosses or is tangent to the x-axis. **28.** For a polynomial $P(x)$ of degree n, when we have $P(x) = d(x) \cdot Q(x) + R(x)$, where the degree of $d(x)$ is 1, then the degree of $Q(x)$ must be $n - 1$.

Exercise Set 4.4

1. $f(x) = x^3 - 6x^2 - x + 30$ **3.** $f(x) = x^3 + 3x^2 + 4x + 12$
5. $f(x) = x^3 - 3x^2 - 2x + 6$ **7.** $f(x) = x^3 - 6x - 4$
9. $f(x) = x^3 + 2x^2 + 29x + 148$ **11.** $f(x) = x^3 - \frac{5}{3}x^2 - \frac{2}{3}x$
13. $f(x) = x^5 + 2x^4 - 2x^2 - x$
15. $f(x) = x^4 + 3x^3 + 3x^2 + x$ **17.** $-\sqrt{3}$ **19.** $i, 2 + \sqrt{5}$
21. $-3i$ **23.** $-4 + 3i, 2 + \sqrt{3}$ **25.** $-\sqrt{5}, 4i$ **27.** $2 + i$
29. $-3 - 4i, 4 + \sqrt{5}$ **31.** $4 + i$
33. $f(x) = x^3 - 4x^2 + 6x - 4$ **35.** $f(x) = x^2 + 16$
37. $f(x) = x^3 - 5x^2 + 16x - 80$
39. $f(x) = x^4 - 2x^3 - 3x^2 + 10x - 10$
41. $f(x) = x^4 + 4x^2 - 45$ **43.** $-\sqrt{2}, \sqrt{2}$ **45.** $i, 2, 3$
47. $1 + 2i, 1 - 2i$ **49.** ± 1 **51.** $\pm 1, \pm\frac{1}{2}, \pm 2, \pm 4, \pm 8$
53. $\pm 1, \pm 2, \pm\frac{1}{3}, \pm\frac{1}{5}, \pm\frac{2}{3}, \pm\frac{2}{5}, \pm\frac{1}{15}, \pm\frac{2}{15}$
55. **(a)** Rational: -3; other: $\pm\sqrt{2}$;
(b) $f(x) = (x + 3)(x + \sqrt{2})(x - \sqrt{2})$
57. **(a)** Rational: $\frac{1}{3}$; other: $\pm\sqrt{5}$;
(b) $f(x) = 3\left(x - \frac{1}{3}\right)(x + \sqrt{5})(x - \sqrt{5})$
59. **(a)** Rational: $-2, 1$; other: none;
(b) $f(x) = (x + 2)(x - 1)^2$
61. **(a)** Rational: $-\frac{3}{2}$; other: $\pm 3i$;
(b) $f(x) = 2\left(x + \frac{3}{2}\right)(x + 3i)(x - 3i)$
63. **(a)** Rational: $-\frac{1}{5}, 1$; other: $\pm 2i$;
(b) $f(x) = 5\left(x + \frac{1}{5}\right)(x - 1)(x + 2i)(x - 2i)$, or $(5x + 1)(x - 1)(x + 2i)(x - 2i)$
65. **(a)** Rational: $-2, -1$; other: $3 \pm \sqrt{13}$;
(b) $f(x) = (x + 2)(x + 1)(x - 3 - \sqrt{13})(x - 3 + \sqrt{13})$
67. **(a)** Rational: 2; other: $1 \pm \sqrt{3}$;
(b) $f(x) = (x - 2)(x - 1 - \sqrt{3})(x - 1 + \sqrt{3})$
69. **(a)** Rational: -2; other: $1 \pm \sqrt{3}i$;
(b) $f(x) = (x + 2)(x - 1 - \sqrt{3}i)(x - 1 + \sqrt{3}i)$
71. **(a)** Rational: $\frac{1}{2}$; other: $\frac{1 \pm \sqrt{5}}{2}$;
(b) $f(x) = \frac{1}{3}\left(x - \frac{1}{2}\right)\left(x - \frac{1 + \sqrt{5}}{2}\right)\left(x - \frac{1 - \sqrt{5}}{2}\right)$

73. $1, -3$ **75.** No rational zeros **77.** No rational zeros
79. $-2, 1, 2$ **81.** 3 or 1; 0 **83.** 0; 3 or 1 **85.** 2 or 0; 2 or 0
87. 1; 1 **89.** 1; 0 **91.** 2 or 0; 2 or 0 **93.** 3 or 1;1 **95.** 1; 1
97.

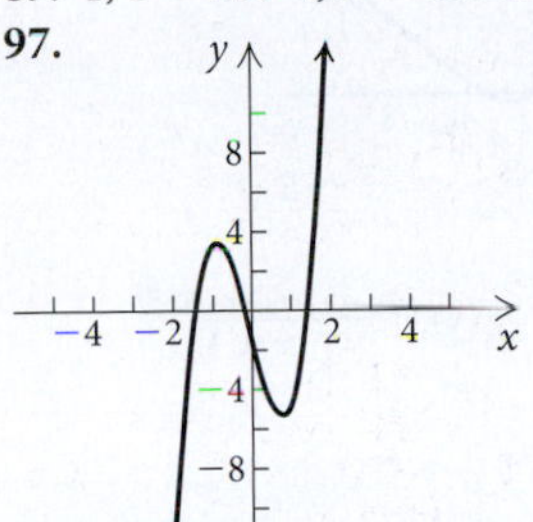

99.

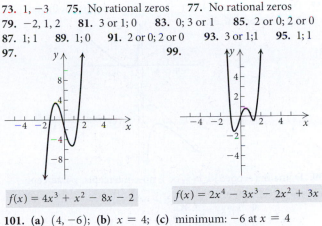

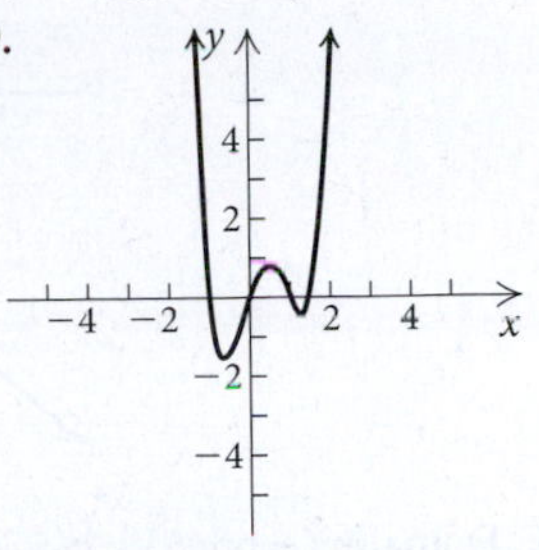

101. **(a)** $(4, -6)$; **(b)** $x = 4$; **(c)** minimum: -6 at $x = 4$
102. **(a)** $(1, -4)$; **(b)** $x = 1$; **(c)** minimum: -4 at $x = 1$
103. 10 **104.** $-3, 11$ **105.** Cubic; $-x^3$; -1; 3; as $x \to \infty$, $g(x) \to -\infty$, and as $x \to -\infty$, $g(x) \to \infty$ **106.** Quadratic; $-x^2$; -1; 2; as $x \to \infty$, $f(x) \to -\infty$, and as $x \to -\infty$, $f(x) \to -\infty$
107. Constant; $-\frac{4}{9}$; $-\frac{4}{9}$; zero degree; for all x, $f(x) = -\frac{4}{9}$
108. Linear; x; 1; 1; as $x \to \infty$, $h(x) \to \infty$, and as $x \to -\infty$, $h(x) \to -\infty$ **109.** Quartic; x^4; 1; 4; as $x \to \infty$, $g(x) \to \infty$, and as $x \to -\infty$, $g(x) \to \infty$
110. Cubic; x^3; 1; 3; as $x \to \infty$, $h(x) \to \infty$, and as $x \to -\infty$, $h(x) \to -\infty$ **111.** **(a)** $-1, \frac{1}{2}, 3$; **(b)** $0, \frac{3}{2}, 4$;
(c) $-3, -\frac{3}{2}, 1$; **(d)** $-\frac{1}{2}, \frac{1}{4}, \frac{3}{2}$ **113.** $-8, -\frac{3}{2}, 4, 7, 15$

Visualizing the Graph

1. A **2.** C **3.** D **4.** H **5.** G **6.** F **7.** B
8. I **9.** J **10.** E

Exercise Set 4.5

1. $\{x | x \neq 2\}$, or $(-\infty, 2) \cup (2, \infty)$
3. $\{x | x \neq 1 \text{ and } x \neq 5\}$, or $(-\infty, 1) \cup (1, 5) \cup (5, \infty)$
5. $\{x | x \neq -5\}$, or $(-\infty, -5) \cup (-5, \infty)$
7. (d); $x = 2, x = -2, y = 0$ **9.** (e); $x = 2, x = -2, y = 0$
11. (c); $x = 2, x = -2, y = 8x$ **13.** $x = 0$ **15.** $x = 2$
17. $x = 4, x = -6$ **19.** $x = \frac{3}{2}, x = -1$ **21.** $y = \frac{3}{4}$
23. $y = 0$ **25.** No horizontal asymptote **27.** $y = x + 1$
29. $y = x$ **31.** $y = x - 3$ **33.** Domain: $(-\infty, 0) \cup (0, \infty)$; no x-intercepts, no y-intercept;

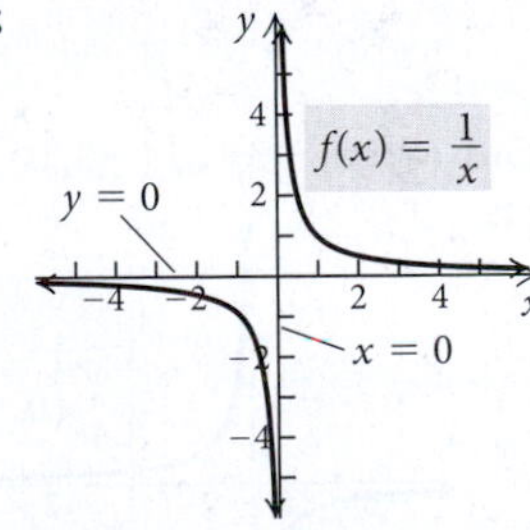

35. Domain: $(-\infty, 0) \cup (0, \infty)$; no x-intercepts, no y-intercept;

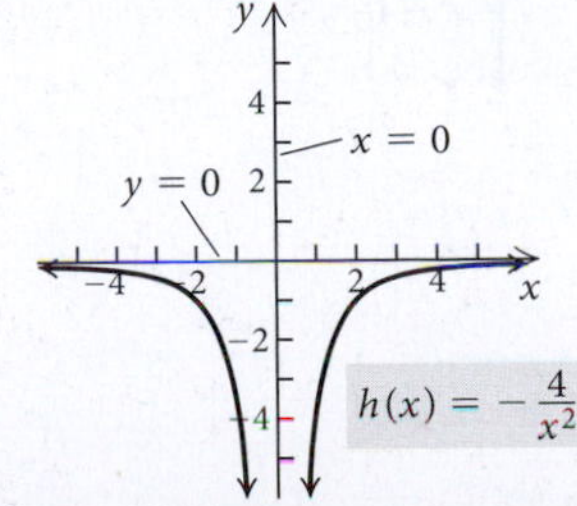

37. Domain: $(-\infty, -1) \cup (-1, \infty)$; x-intercepts: $(1, 0)$ and $(3, 0)$; y-intercept: $(0, 3)$;

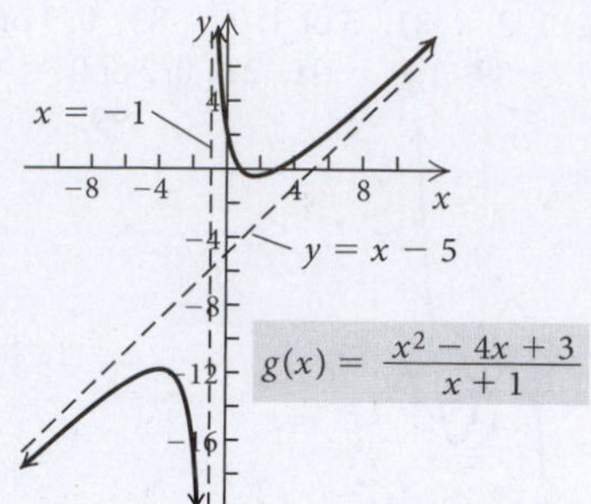

39. Domain: $(-\infty, 5) \cup (5, \infty)$; no x-intercepts, y-intercept: $\left(0, \frac{2}{5}\right)$;

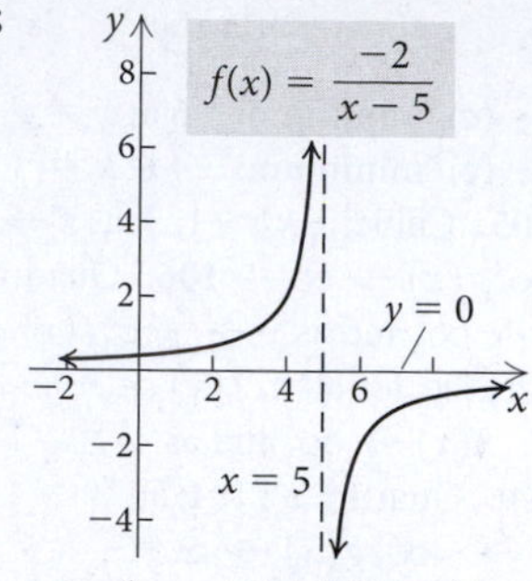

41. Domain: $(-\infty, 0) \cup (0, \infty)$; x-intercept: $\left(-\frac{1}{2}, 0\right)$, no y-intercept;

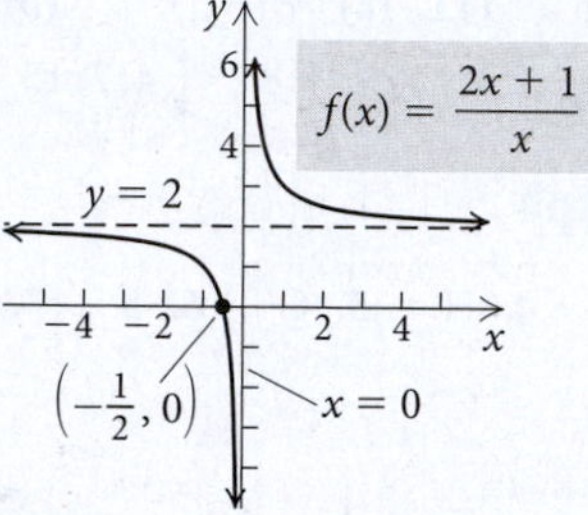

43. Domain: $(-\infty, -3) \cup (-3, 3) \cup (3, \infty)$; no x-intercepts, y-intercept: $\left(0, -\frac{1}{3}\right)$;

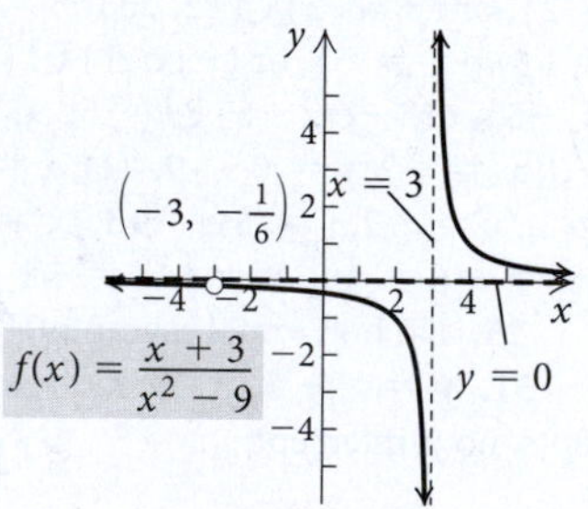

45. Domain: $(-\infty, -3) \cup (-3, 0) \cup (0, \infty)$; no x-intercepts, no y-intercept;

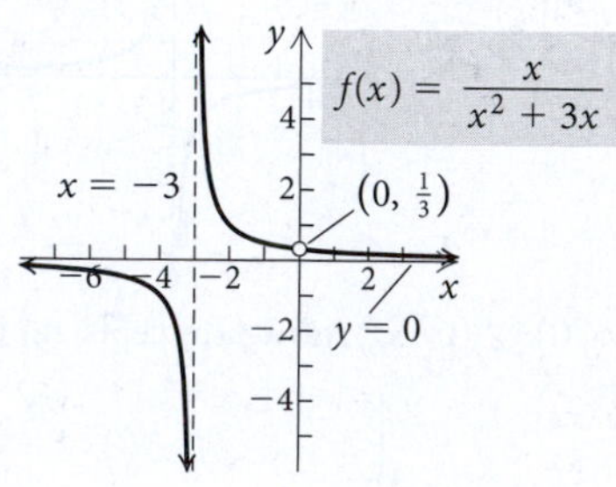

47. Domain: $(-\infty, 2) \cup (2, \infty)$; no x-intercepts, y-intercept: $\left(0, \frac{1}{4}\right)$;

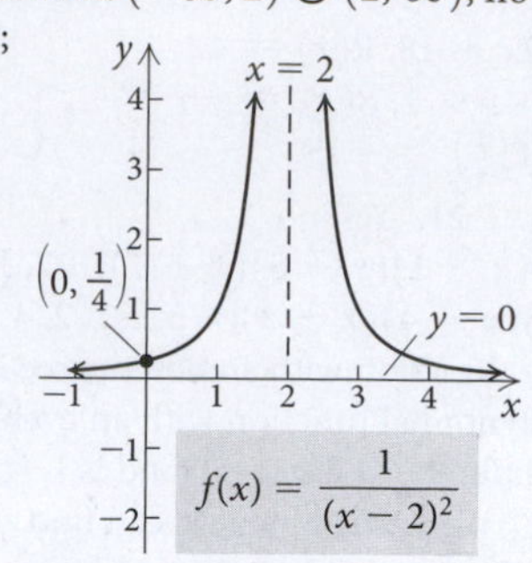

49. Domain: $(-\infty, -3) \cup (-3, -1) \cup (-1, \infty)$; x-intercept: $(1, 0)$, y-intercept: $(0, -1)$;

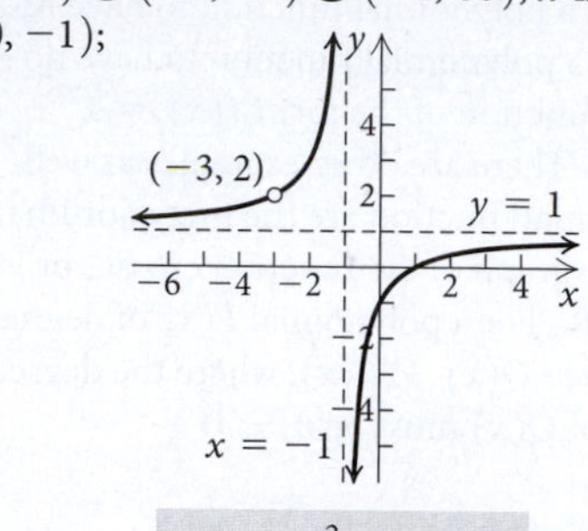

51. Domain: $(-\infty, \infty)$; no x-intercepts, y-intercept: $\left(0, \frac{1}{3}\right)$;

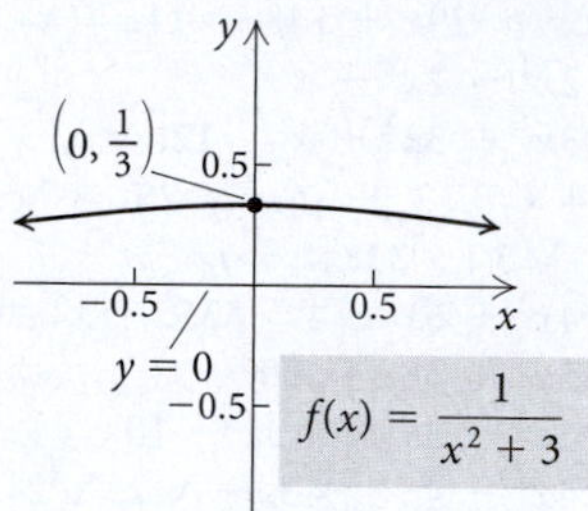

53. Domain: $(-\infty, 2) \cup (2, \infty)$; x-intercept: $(-2, 0)$, y-intercept: $(0, 2)$;

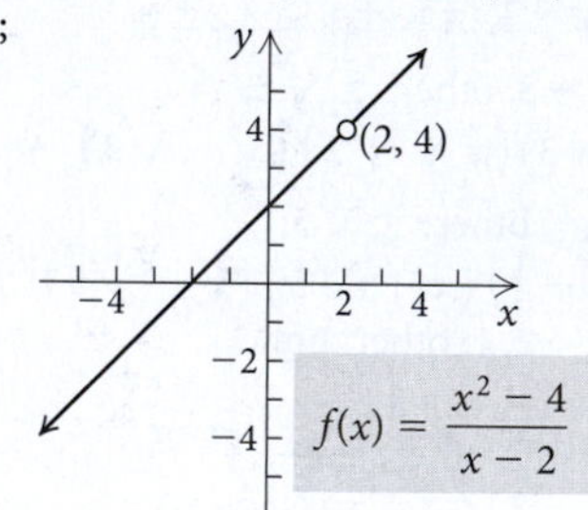

55. Domain: $(-\infty, -2) \cup (-2, \infty)$; x-intercept: $(1, 0)$, y-intercept: $\left(0, -\frac{1}{2}\right)$;

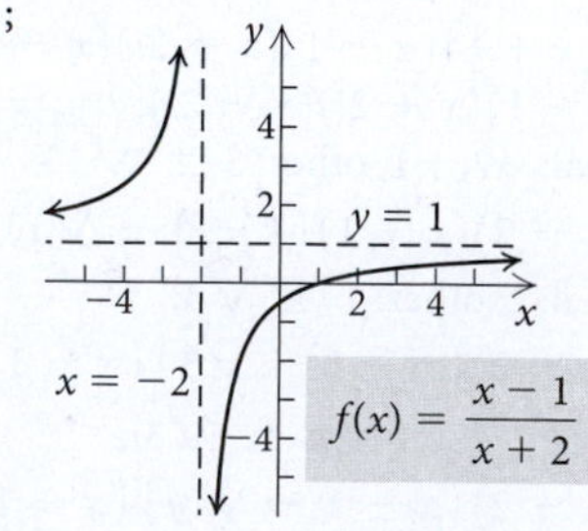

57. Domain: $(-\infty, -\frac{1}{2}) \cup (-\frac{1}{2}, 0) \cup (0, 3) \cup (3, \infty)$; x-intercept: $(-3, 0)$, no y-intercept;

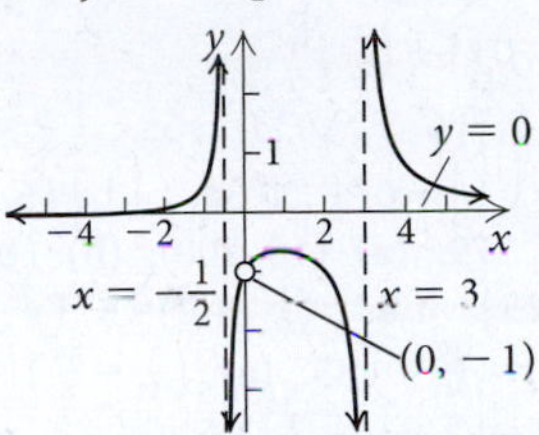

$f(x) = \dfrac{x^2 + 3x}{2x^3 - 5x^2 - 3x}$

59. Domain: $(-\infty, -1) \cup (-1, \infty)$; x-intercepts: $(-3, 0)$ and $(3, 0)$, y-intercept: $(0, -9)$;

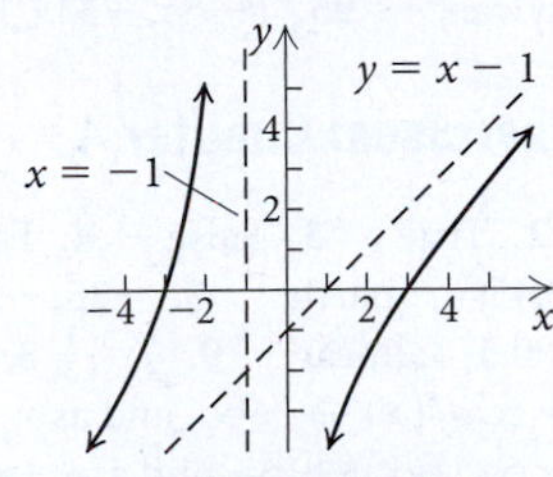

$f(x) = \dfrac{x^2 - 9}{x + 1}$

61. Domain: $(-\infty, \infty)$; x-intercepts: $(-2, 0)$ and $(1, 0)$, y-intercept: $(0, -2)$;

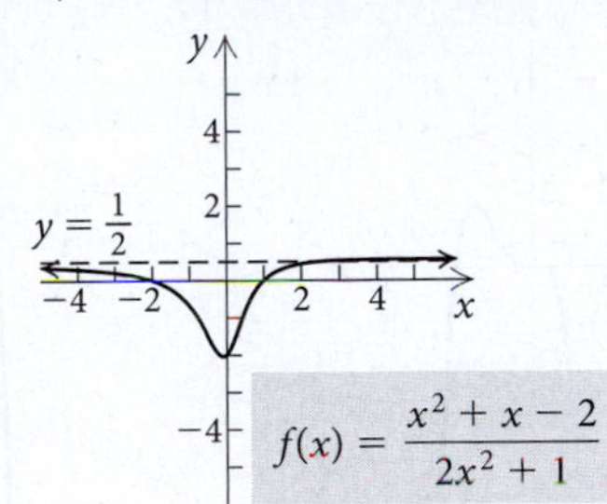

$f(x) = \dfrac{x^2 + x - 2}{2x^2 + 1}$

63. Domain: $(-\infty, 1) \cup (1, \infty)$; x-intercept: $(-\frac{2}{3}, 0)$, y-intercept: $(0, 2)$;

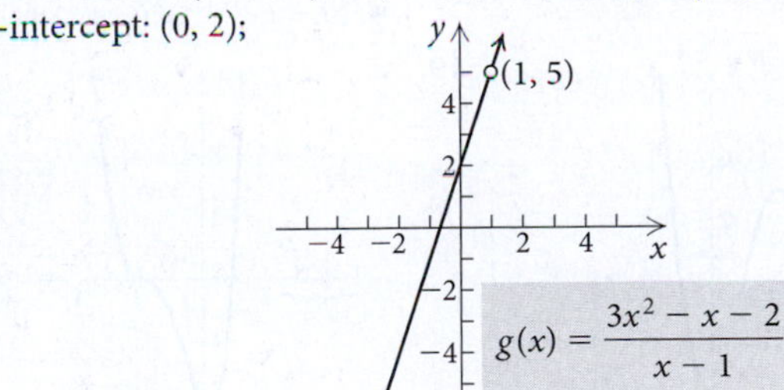

$g(x) = \dfrac{3x^2 - x - 2}{x - 1}$

65. Domain: $(-\infty, -1) \cup (-1, 3) \cup (3, \infty)$; x-intercept: $(1, 0)$, y-intercept: $(0, \frac{1}{3})$;

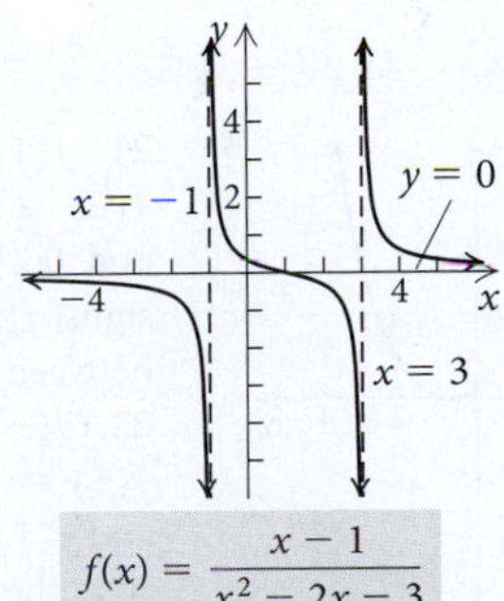

$f(x) = \dfrac{x - 1}{x^2 - 2x - 3}$

67. Domain: $(-\infty, -4) \cup (-4, 2) \cup (2, \infty)$; x-intercept: $(\frac{1}{3}, 0)$, y-intercept: $(0, \frac{1}{2})$;

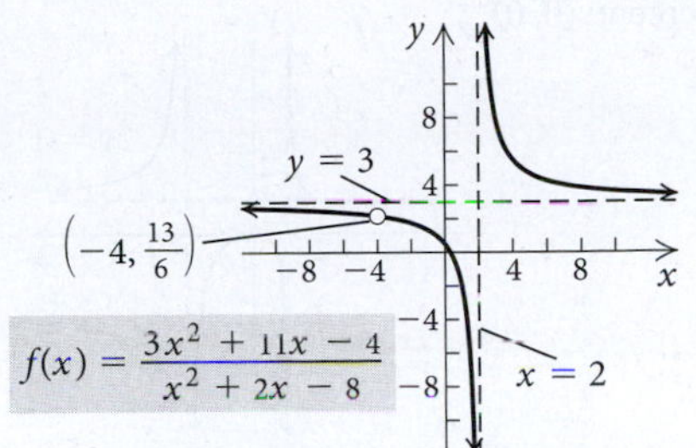

$f(x) = \dfrac{3x^2 + 11x - 4}{x^2 + 2x - 8}$

69. Domain: $(-\infty, -1) \cup (-1, \infty)$; x-intercept: $(3, 0)$, y-intercept: $(0, -3)$;

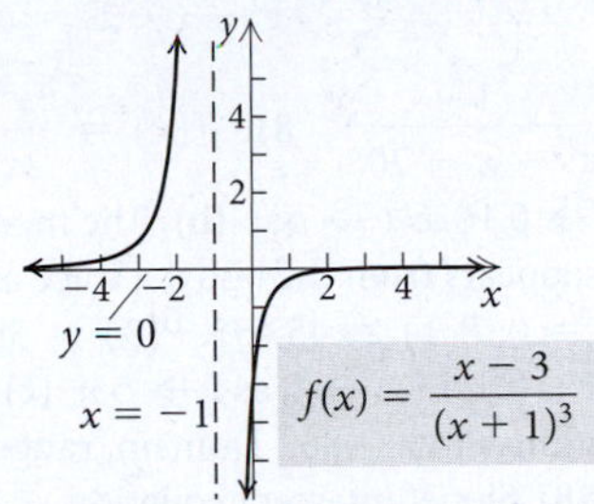

$f(x) = \dfrac{x - 3}{(x + 1)^3}$

71. Domain: $(-\infty, 0) \cup (0, \infty)$; x-intercept: $(-1, 0)$, no y-intercept;

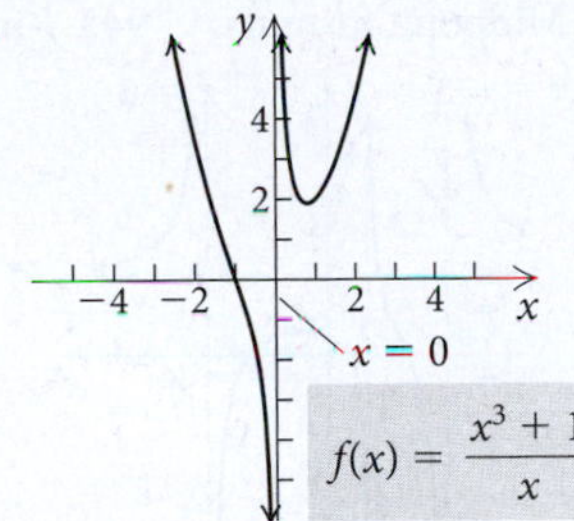

$f(x) = \dfrac{x^3 + 1}{x}$

73. Domain: $(-\infty, -2) \cup (-2, 7) \cup (7, \infty)$; x-intercepts: $(-5, 0)$, $(0, 0)$, and $(3, 0)$, y-intercept: $(0, 0)$;

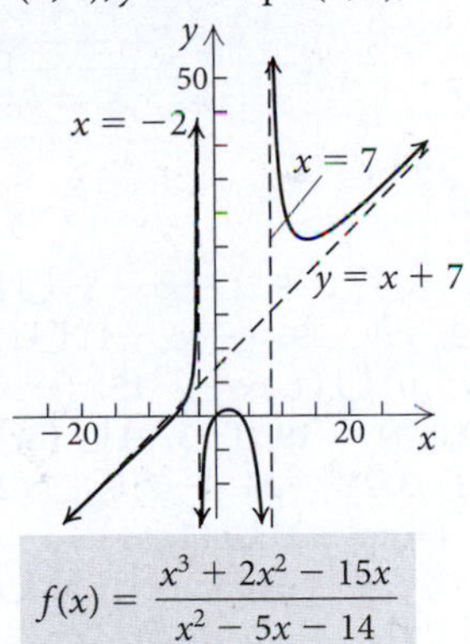

$f(x) = \dfrac{x^3 + 2x^2 - 15x}{x^2 - 5x - 14}$

75. Domain: $(-\infty, \infty)$; x-intercept: $(0, 0)$, y-intercept: $(0, 0)$;

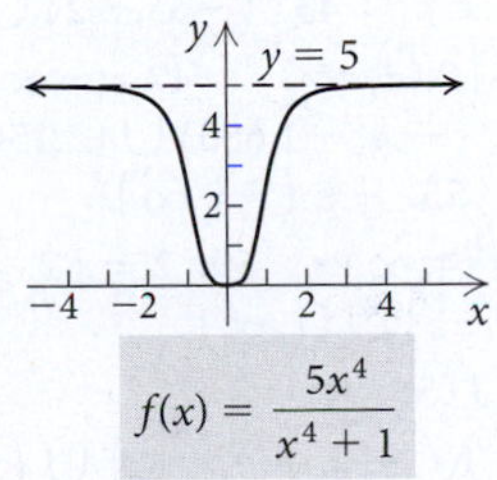

$f(x) = \dfrac{5x^4}{x^4 + 1}$

77. Domain: $(-\infty, -1) \cup (-1, 2) \cup (2, \infty)$; x-intercept: $(0, 0)$, y-intercept: $(0, 0)$;

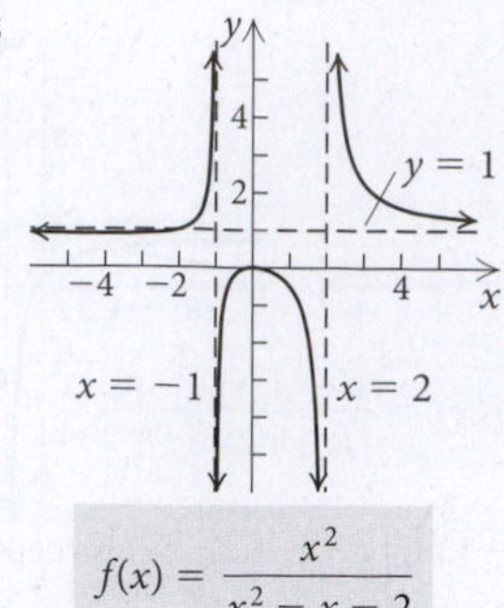

$f(x) = \dfrac{x^2}{x^2 - x - 2}$

79. $f(x) = \dfrac{1}{x^2 - x - 20}$ **81.** $f(x) = \dfrac{3x^2 + 12x + 12}{2x^2 - 2x - 40}$
83. (a) $N(t) \to 0.16$ as $t \to \infty$; **(b)** The medication never completely disappears from the body; a trace amount remains.
85. (a) $P(0) = 0$; $P(1) = 45{,}455$; $P(3) = 55{,}556$; $P(8) = 29{,}197$; **(b)** $P(t) \to 0$ as $t \to \infty$; **(c)** In time, no one lives in this community. **86.** Domain, range, domain, range
87. Slope **88.** Slope–intercept equation **89.** Point–slope equation **90.** x-intercept **91.** $f(-x) = -f(x)$
92. Vertical lines **93.** Midpoint formula **94.** y-intercept
95. $y = x^3 + 4$ **97.**

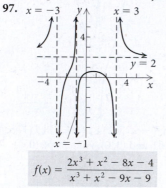

$f(x) = \dfrac{2x^3 + x^2 - 8x - 4}{x^3 + x^2 - 9x - 9}$

Exercise Set 4.6

1. $\{-5, 3\}$ **3.** $[-5, 3]$ **5.** $(-\infty, -5] \cup [3, \infty)$
7. $(-\infty, -4) \cup (2, \infty)$ **9.** $(-\infty, -4) \cup [2, \infty)$
11. $\{0\}$ **13.** $(-5, 0] \cup (1, \infty)$ **15.** $(-\infty, -5) \cup (0, 1)$
17. $(-\infty, -3) \cup (0, 3)$ **19.** $(-3, 0) \cup (3, \infty)$
21. $(-\infty, -5) \cup (-3, 2)$ **23.** $(-2, 0] \cup (2, \infty)$
25. $(-4, 1)$ **27.** $(-\infty, -2) \cup (1, \infty)$
29. $(-\infty, -1] \cup [3, \infty)$ **31.** $(-\infty, -5) \cup (5, \infty)$
33. $(-\infty, -2] \cup [2, \infty)$ **35.** $(-\infty, 3) \cup (3, \infty)$
37. $\varnothing$ **39.** $\left(-\infty, -\frac{5}{4}\right] \cup [0, 3]$
41. $[-3, -1] \cup [1, \infty)$ **43.** $(-\infty, -2) \cup (1, 3)$
45. $\left[-\sqrt{2}, -1\right] \cup \left[\sqrt{2}, \infty\right)$ **47.** $(-\infty, -1] \cup \left[\frac{3}{2}, 2\right]$
49. $(-\infty, 5]$ **51.** $(-\infty, -1.680) \cup (2.154, 5.526)$
53. -4; $(-4, \infty)$ **55.** $-\frac{5}{2}$; $\left(-\frac{5}{2}, \infty\right)$
57. $0, 4$; $(-\infty, 0] \cup (4, \infty)$ **59.** $2, \frac{7}{2}$; $\left(2, \frac{7}{2}\right]$
61. $-3, -\frac{1}{5}, 1$; $\left(-3, -\frac{1}{5}\right] \cup (1, \infty)$
63. $2, \frac{46}{11}, 5$; $\left(2, \frac{46}{11}\right) \cup (5, \infty)$
65. $1 - \sqrt{2}, 0, 1 + \sqrt{2}$; $\left(1 - \sqrt{2}, 0\right) \cup \left(1 + \sqrt{2}, \infty\right)$
67. $-3, 1, 3, \frac{11}{3}$; $(-\infty, -3) \cup (1, 3) \cup \left[\frac{11}{3}, \infty\right)$
69. 0; $(-\infty, \infty)$ **71.** $-3, \dfrac{1 - \sqrt{61}}{6}, -\dfrac{1}{2}, 0, \dfrac{1 + \sqrt{61}}{6}$;
$\left(-3, \dfrac{1 - \sqrt{61}}{6}\right) \cup \left(-\dfrac{1}{2}, 0\right) \cup \left(\dfrac{1 + \sqrt{61}}{6}, \infty\right)$
73. $-1, 0, \frac{2}{7}, \frac{7}{2}$; $(-1, 0) \cup \left(\frac{2}{7}, \frac{7}{2}\right)$
75. $-6 - \sqrt{33}, -5, -6 + \sqrt{33}, 1, 5$; $\left[-6 - \sqrt{33}, -5\right) \cup \left[-6 + \sqrt{33}, 1\right) \cup (5, \infty)$
77. $(0.408, 2.449)$ **79. (a)** $(10, 200)$; **(b)** $(0, 10) \cup (200, \infty)$
81. $\{n \mid 9 \le n \le 23\}$ **83.** $(x + 2)^2 + (y - 4)^2 = 9$
84. $x^2 + (y + 3)^2 = \frac{49}{16}$ **85. (a)** $\left(\frac{3}{4}, -\frac{55}{8}\right)$; **(b)** maximum: $-\frac{55}{8}$ when $x = \frac{3}{4}$; **(c)** $\left(-\infty, -\frac{55}{8}\right]$
86. (a) $(5, -23)$; **(b)** minimum: -23 when $x = 5$; **(c)** $[-23, \infty)$ **87.** $\left[-\sqrt{5}, \sqrt{5}\right]$ **89.** $\left[-\frac{3}{2}, \frac{3}{2}\right]$
91. $\left(-\infty, -\frac{1}{4}\right) \cup \left(\frac{1}{2}, \infty\right)$ **93.** $x^2 + x - 12 < 0$; answers may vary **95.** $(-\infty, -3) \cup (7, \infty)$

Review Exercises: Chapter 4

1. True **2.** True **3.** False **4.** False **5.** False
6. $0.45x^4$, 0.45, 4, quartic **7.** -25, -25, 0, constant
8. $-0.5x$, -0.5, 1, linear **9.** $\frac{1}{3}x^3$, $\frac{1}{3}$, 3, cubic
10. As $x \to \infty$, $f(x) \to -\infty$, and as $x \to -\infty$, $f(x) \to -\infty$.
11. As $x \to \infty$, $f(x) \to \infty$, and as $x \to -\infty$, $f(x) \to -\infty$.
12. $\frac{2}{3}$, multiplicity 1; -2, multiplicity 3; 5, multiplicity 2
13. ± 1, ± 5, each has multiplicity 1 **14.** ± 3, -4, each has multiplicity 1 **15. (a)** 4%; **(b)** 5%

16.

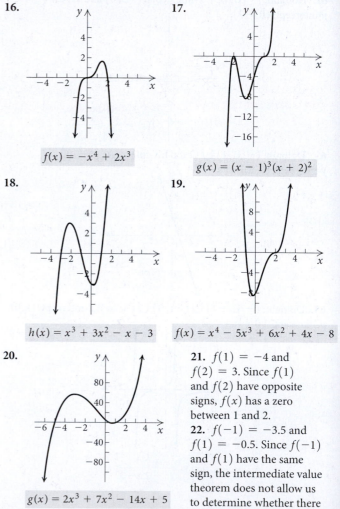

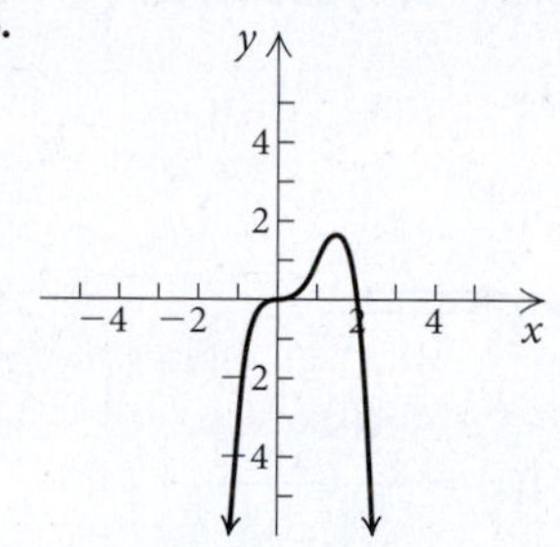

$f(x) = -x^4 + 2x^3$

17.

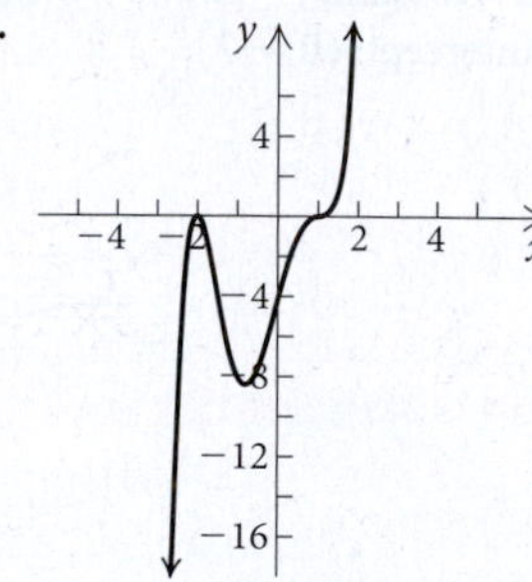

$g(x) = (x - 1)^3(x + 2)^2$

18.

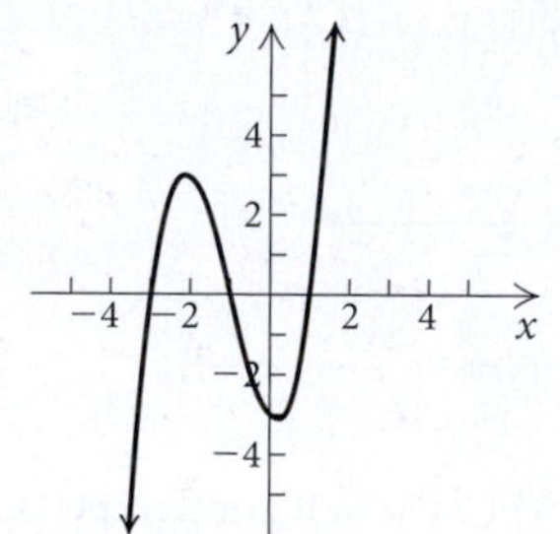

$h(x) = x^3 + 3x^2 - x - 3$

19.

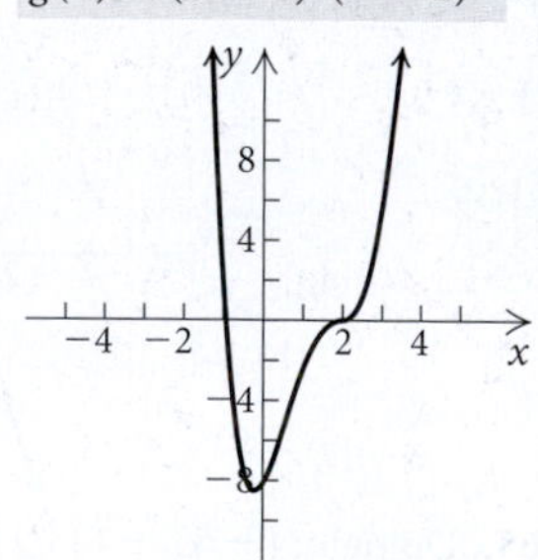

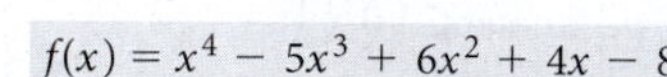

$f(x) = x^4 - 5x^3 + 6x^2 + 4x - 8$

20.

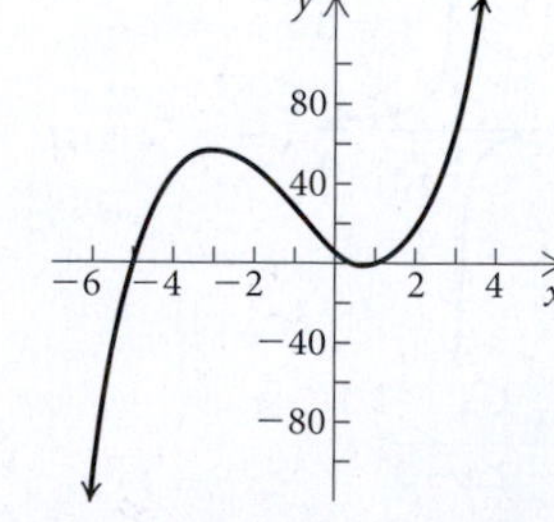

$g(x) = 2x^3 + 7x^2 - 14x + 5$

21. $f(1) = -4$ and $f(2) = 3$. Since $f(1)$ and $f(2)$ have opposite signs, $f(x)$ has a zero between 1 and 2.
22. $f(-1) = -3.5$ and $f(1) = -0.5$. Since $f(-1)$ and $f(1)$ have the same sign, the intermediate value theorem does not allow us to determine whether there is a zero between -1 and 1.

23. $Q(x) = 6x^2 + 16x + 52, R(x) = 155;$
$P(x) = (x - 3)(6x^2 + 16x + 52) + 155$
24. $Q(x) = x^3 - 3x^2 + 3x - 2, R(x) = 7;$
$P(x) = (x + 1)(x^3 - 3x^2 + 3x - 2) + 7$
25. $x^2 + 7x + 22$, R 120 **26.** $x^3 + x^2 + x + 1$, R 0
27. $x^4 - x^3 + x^2 - x - 1$, R 1 **28.** 36 **29.** 0
30. $-141{,}220$ **31.** Yes, no **32.** No, yes **33.** Yes, no
34. No, yes **35.** $f(x) = (x - 1)^2(x + 4); -4, 1$
36. $f(x) = (x - 2)(x + 3)^2; -3, 2$
37. $f(x) = (x - 2)^2(x - 5)(x + 5); -5, 2, 5$
38. $f(x) = (x - 1)(x + 1)(x - \sqrt{2})(x + \sqrt{2});$
$-\sqrt{2}, -1, 1, \sqrt{2}$ **39.** $f(x) = x^3 + 3x^2 - 6x - 8$
40. $f(x) = x^3 + x^2 - 4x + 6$
41. $f(x) = x^3 - \frac{5}{2}x^2 + \frac{1}{2}$, or $2x^3 - 5x^2 + 1$
42. $f(x) = x^4 + \frac{29}{2}x^3 + \frac{135}{2}x^2 + \frac{175}{2}x - \frac{125}{2}$, or
$2x^4 + 29x^3 + 135x^2 + 175x - 125$
43. $f(x) = x^5 + 4x^4 - 3x^3 - 18x^2$
44. $-\sqrt{5}, -i$ **45.** $1 - \sqrt{3}, \sqrt{3}$ **46.** $\sqrt{2}$
47. $f(x) = x^2 - 11$ **48.** $f(x) = x^3 - 6x^2 + x - 6$
49. $f(x) = x^4 - 5x^3 + 4x^2 + 2x - 8$
50. $f(x) = x^4 - x^2 - 20$ **51.** $f(x) = x^3 + \frac{8}{3}x^2 - x$
52. $\pm\frac{1}{4}, \pm\frac{1}{2}, \pm\frac{3}{4}, \pm 1, \pm\frac{3}{2}, \pm 2, \pm 3, \pm 4, \pm 6, \pm 12$
53. $\pm\frac{1}{3}, \pm 1$ **54.** $\pm 1, \pm 2, \pm 3, \pm 4, \pm 6, \pm 8, \pm 12, \pm 24$
55. **(a)** Rational: $0, -2, \frac{1}{3}, 3$; other: none;
(b) $f(x) = 3x\left(x - \frac{1}{3}\right)(x + 2)^2(x - 3)$
56. **(a)** Rational: 2; other: $\pm\sqrt{3}$;
(b) $f(x) = (x - 2)(x + \sqrt{3})(x - \sqrt{3})$
57. **(a)** Rational: $-1, 1$; other: $3 \pm i$;
(b) $f(x) = (x + 1)(x - 1)(x - 3 - i)(x - 3 + i)$
58. **(a)** Rational: -5; other: $1 \pm \sqrt{2}$;
(b) $f(x) = (x + 5)(x - 1 - \sqrt{2})(x - 1 + \sqrt{2})$
59. **(a)** Rational: $\frac{2}{3}, 1$; other: none;
(b) $f(x) = 3\left(x - \frac{2}{3}\right)(x - 1)^2$
60. **(a)** Rational: 2; other: $1 \pm \sqrt{5}$;
(b) $f(x) = (x - 2)^3(x - 1 + \sqrt{5})(x - 1 - \sqrt{5})$
61. **(a)** Rational: $-4, 0, 3, 4$; other: none;
(b) $f(x) = x^2(x + 4)^2(x - 3)(x - 4)$
62. **(a)** Rational: $\frac{5}{2}, 1$; other: none;
(b) $f(x) = 2\left(x - \frac{5}{2}\right)(x - 1)^4$, or $(2x - 5)(x - 1)^4$
63. 3 or 1; 0 **64.** 4 or 2 or 0; 2 or 0 **65.** 3 or 1; 0
66. Domain: $(-\infty, -2) \cup (-2, \infty)$; x-intercepts: $(-\sqrt{5}, 0)$ and $(\sqrt{5}, 0)$, y-intercept: $\left(0, -\frac{5}{2}\right)$

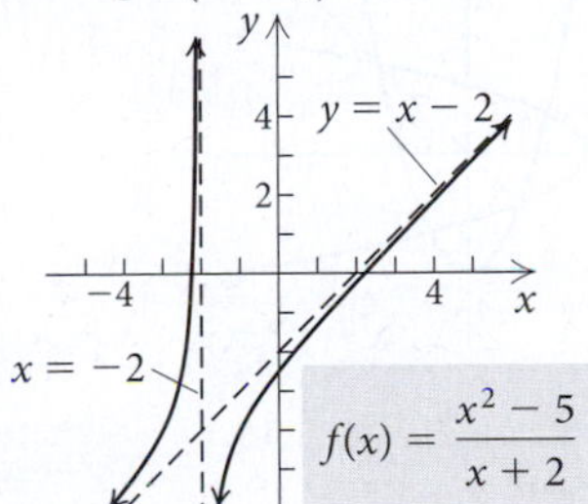

67. Domain: $(-\infty, 2) \cup (2, \infty)$;
x-intercepts: none,
y-intercept: $\left(0, \frac{5}{4}\right)$

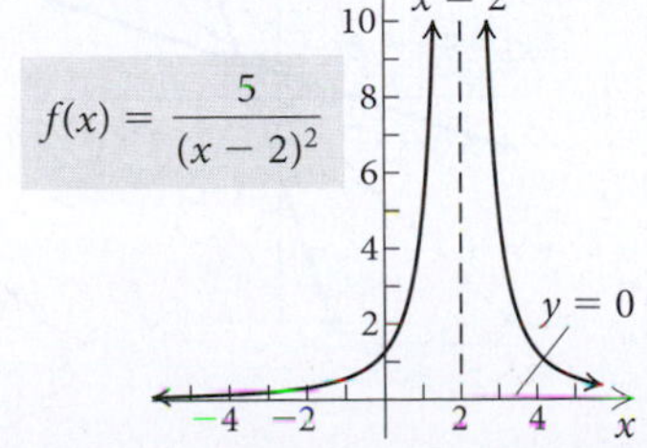

68. Domain: $(-\infty, -4) \cup (-4, 5) \cup (5, \infty)$; x-intercepts: $(-3, 0)$ and $(2, 0)$, y-intercept: $\left(0, \frac{3}{10}\right)$

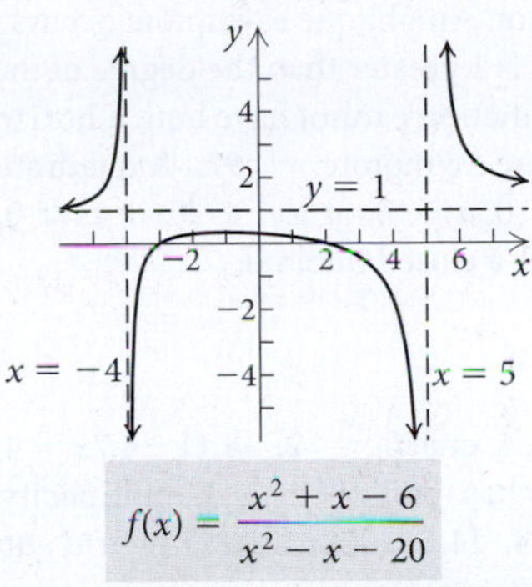

69. Domain: $(-\infty, -3) \cup (-3, 5) \cup (5, \infty)$; x-intercept: $(2, 0)$, y-intercept: $\left(0, \frac{2}{15}\right)$

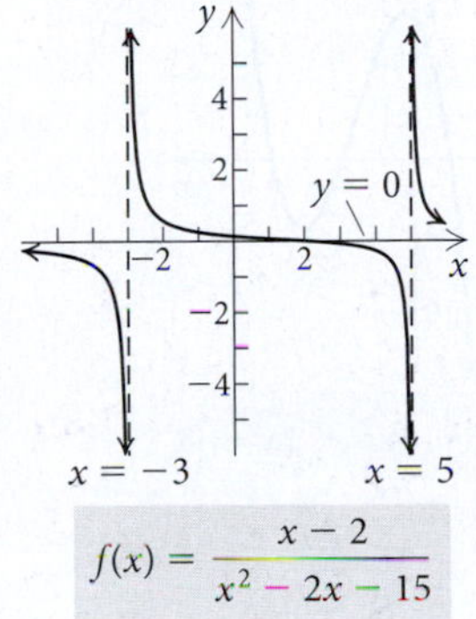

70. $f(x) = \dfrac{1}{x^2 - x - 6}$ **71.** $f(x) = \dfrac{4x^2 + 12x}{x^2 - x - 6}$
72. **(a)** $N(t) \to 0.0875$ as $t \to \infty$; **(b)** The medication never completely disappears from the body; a trace amount remains.
73. $(-3, 3)$ **74.** $\left(-\infty, -\frac{1}{2}\right) \cup (2, \infty)$ **75.** $[-4, 1] \cup [2, \infty)$
76. $\left(-\infty, -\frac{14}{3}\right) \cup (-3, \infty)$ **77.** **(a)** $t = 7$; **(b)** $(2, 3)$
78. $\left[\dfrac{5 - \sqrt{15}}{2}, \dfrac{5 + \sqrt{15}}{2}\right]$ **79.** A **80.** C **81.** B
82. $(-\infty, -1 - \sqrt{6}] \cup [-1 + \sqrt{6}, \infty)$
83. $\left(-\infty, -\frac{1}{2}\right) \cup \left(\frac{1}{2}, \infty\right)$ **84.** $\{1 + i, 1 - i, i, -i\}$
85. $(-\infty, 2)$ **86.** $(x - 1)\left(x + \dfrac{1}{2} - \dfrac{\sqrt{3}}{2}i\right)\left(x + \dfrac{1}{2} + \dfrac{\sqrt{3}}{2}i\right)$
87. 7 **88.** -4 **89.** $(-\infty, -5] \cup [2, \infty)$
90. $(-\infty, 1.1] \cup [2, \infty)$ **91.** $\left(-1, \frac{3}{7}\right)$
92. A polynomial function is a function that can be defined by a polynomial expression. A rational function is a function that can be defined as a quotient of two polynomials. **93.** No; since imaginary zeros of polynomials with rational coefficients occur in conjugate pairs, a third-degree polynomial with rational coefficients can have at most two imaginary zeros. Thus there must be at least one real zero. **94.** Vertical asymptotes occur at any x-values that make the denominator zero. The graph of a rational function does not cross any vertical asymptotes. Horizontal asymptotes occur when the degree of the numerator is less than or equal to the degree of the denominator. Oblique asymptotes occur when the degree of the numerator is 1 greater than the degree of the denominator. Graphs of rational functions may cross horizontal or oblique asymptotes. **95.** If $P(x)$ is an even function, then $P(-x) = P(x)$ and thus $P(-x)$ has the same number of sign changes as $P(x)$. Hence, $P(x)$ has one negative real zero also.

96. A horizontal asymptote occurs when the degree of the numerator of a rational function is less than or equal to the degree of the denominator. An oblique asymptote occurs when the degree of the numerator is 1 greater than the degree of the denominator. Thus a rational function cannot have both a horizontal asymptote and an oblique asymptote. **97.** A quadratic inequality $ax^2 + bx + c \le 0, a > 0$, or $ax^2 + bx + c \ge 0, a < 0$, has a solution set that is a closed interval.

Test: Chapter 4

1. [4.1] $-x^4, -1, 4$; quartic **2.** [4.1] $-4.7x, -4.7, 1$; linear **3.** [4.1] $0, \frac{5}{3}$, each has multiplicity 1; 3, multiplicity 2; -1, multiplicity 3 **4.** [4.1] 2008: 329,277 hybrid automobiles; 2011: 275,779 hybrid automobiles
5. [4.2]

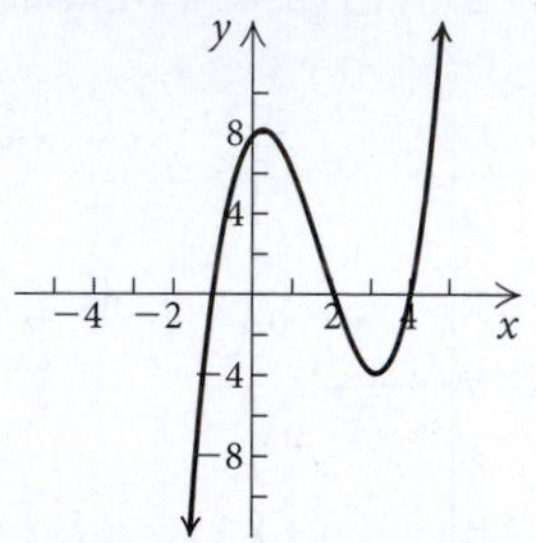

$f(x) = x^3 - 5x^2 + 2x + 8$

6. [4.2]

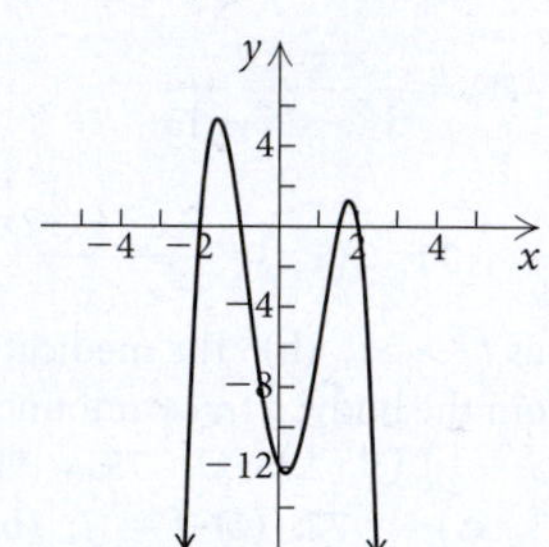

$f(x) = -2x^4 + x^3 + 11x^2 - 4x - 12$

7. [4.2] $f(0) = 3$ and $f(2) = -17$. Since $f(0)$ and $f(2)$ have opposite signs, $f(x)$ has a zero between 0 and 2.
8. [4.2] $g(-2) = 5$ and $g(-1) = 1$. Both $g(-2)$ and $g(-1)$ are positive. We cannot use the intermediate value theorem to determine if there is a zero between -2 and -1.
9. [4.3] $Q(x) = x^3 + 4x^2 + 4x + 6, R(x) = 1$; $P(x) = (x - 1)(x^3 + 4x^2 + 4x + 6) + 1$
10. [4.3] $3x^2 + 15x + 63$, R 322 **11.** [4.3] -115
12. [4.3] Yes **13.** [4.4] $f(x) = x^4 - 27x^2 - 54x$
14. [4.4] $-\sqrt{3}, 2 + i$
15. [4.4] $f(x) = x^3 + 10x^2 + 9x + 90$
16. [4.4] $f(x) = x^5 - 2x^4 - x^3 + 6x^2 - 6x$
17. [4.4] $\pm 1, \pm 2, \pm 3, \pm 4, \pm 6, \pm 12, \pm\frac{1}{2}, \pm\frac{3}{2}$
18. [4.4] $\pm\frac{1}{10}, \pm\frac{1}{5}, \pm\frac{1}{2}, \pm 1, \pm\frac{5}{2}, \pm 5$
19. [4.4] **(a)** Rational: -1; other: $\pm\sqrt{5}$;
(b) $f(x) = (x + 1)(x - \sqrt{5})(x + \sqrt{5})$
20. [4.4] **(a)** Rational: $-\frac{1}{2}$, 1, 2, 3; other: none;
(b) $f(x) = 2\left(x + \frac{1}{2}\right)(x - 1)(x - 2)(x - 3)$
21. [4.4] **(a)** Rational: -4; other: $\pm 2i$;
(b) $f(x) = (x - 2i)(x + 2i)(x + 4)$
22. [4.4] **(a)** Rational: $\frac{2}{3}$, 1; other: none;
(b) $f(x) = 3\left(x - \frac{2}{3}\right)(x - 1)^3$

23. [4.4] 2 or 0; 2 or 0 **24.** [4.5] Domain: $(-\infty, 3) \cup (3, \infty)$; x-intercepts: none, y-intercept: $\left(0, \frac{2}{9}\right)$;

$f(x) = \dfrac{2}{(x - 3)^2}$

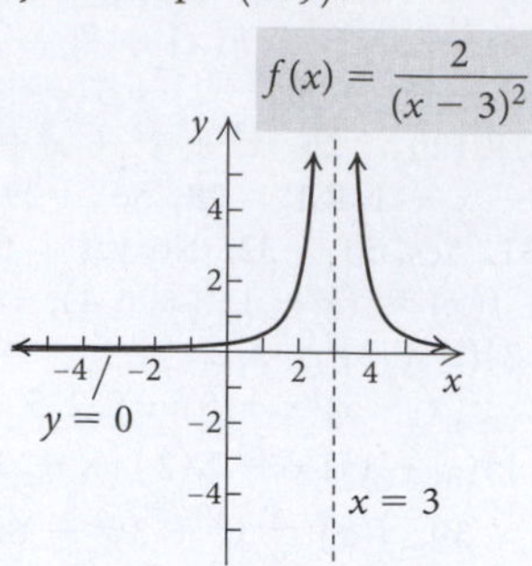

25. [4.5] Domain: $(-\infty, -1) \cup (-1, 4) \cup (4, \infty)$; x-intercept: $(-3, 0)$, y-intercept: $\left(0, -\frac{3}{4}\right)$;

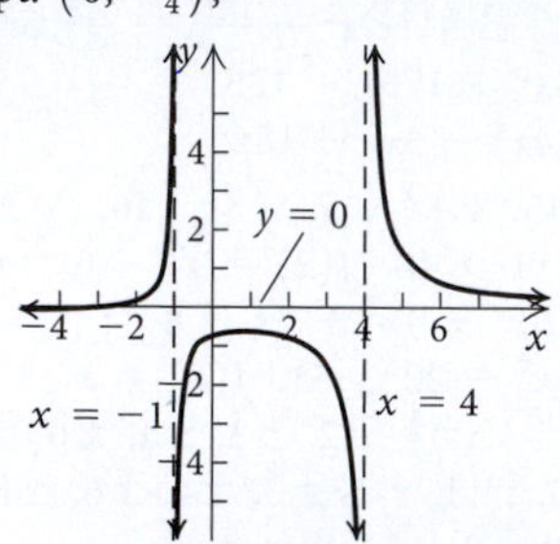

$f(x) = \dfrac{x + 3}{x^2 - 3x - 4}$

26. [4.5] Answers may vary; $f(x) = \dfrac{x + 4}{x^2 - x - 2}$
27. [4.6] $\left(-\infty, -\frac{1}{2}\right) \cup (3, \infty)$ **28.** [4.6] $(-\infty, 4) \cup \left[\frac{13}{2}, \infty\right)$
29. **(a)** [4.1] 6 sec; **(b)** [4.1], [4.6] $(1, 3)$ **30.** [4.2] D
31. [4.1], [4.6] $(-\infty, -4] \cup [3, \infty)$

Chapter 5

Exercise Set 5.1

1. $\{(8, 7), (8, -2), (-4, 3), (-8, 8)\}$
3. $\{(-1, -1), (4, -3)\}$ **5.** $x = 4y - 5$
7. $y^3x = -5$ **9.** $y = x^2 - 2x$
11.

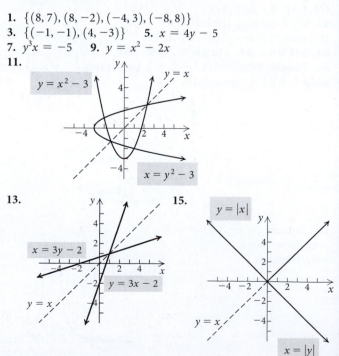

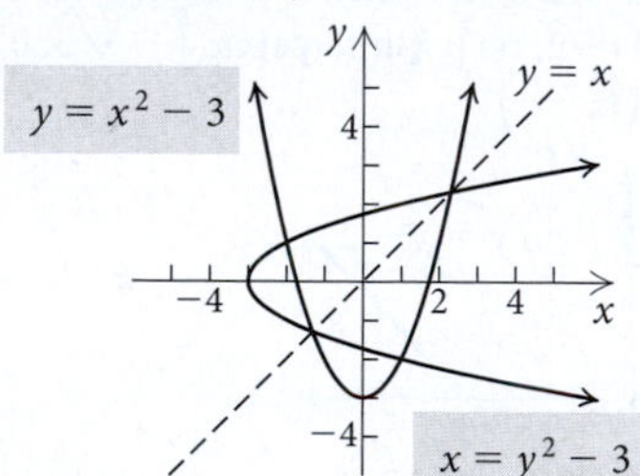

13.

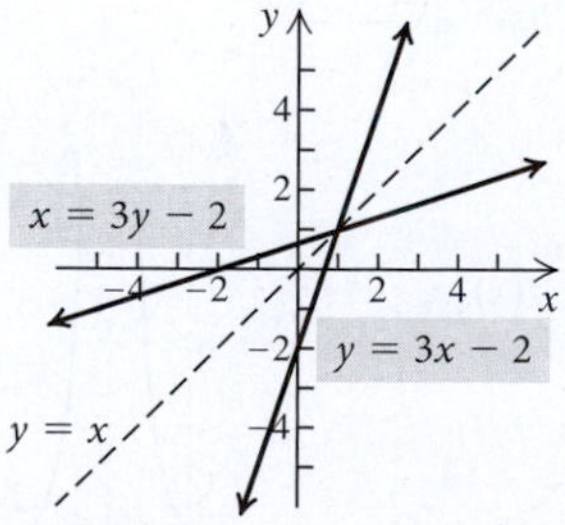

y = x

x = 3y − 2

y = 3x − 2

15.

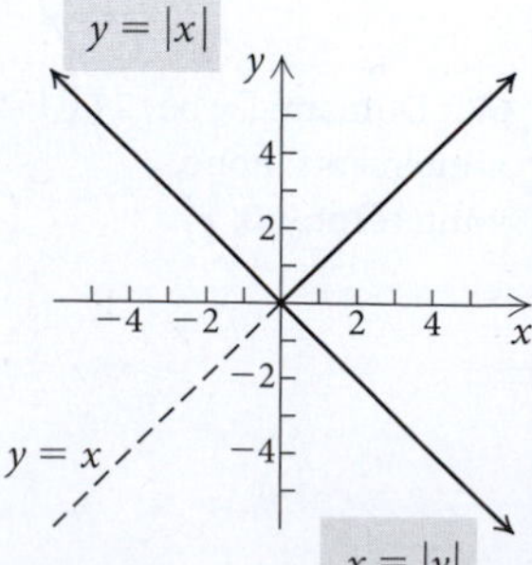

17. Assume $f(a) = f(b)$ for any numbers a and b in the domain of f. Since $f(a) = \frac{1}{3}a - 6$ and $f(b) = \frac{1}{3}b - 6$, we have

$\frac{1}{3}a - 6 = \frac{1}{3}b - 6$

$\frac{1}{3}a = \frac{1}{3}b$ **Adding 6**

$a = b.$ **Multiplying by 3**

Thus, if $f(a) = f(b)$, then $a = b$ and f is one-to-one.

19. Assume $f(a) = f(b)$ for any numbers a and b in the domain of f. Since $f(a) = a^3 + \frac{1}{2}$ and $f(b) = b^3 + \frac{1}{2}$, we have

$a^3 + \frac{1}{2} = b^3 + \frac{1}{2}$

$a^3 = b^3$ **Subtracting $\frac{1}{2}$**

$a = b.$ **Taking the cube root**

Thus, if $f(a) = f(b)$, then $a = b$ and f is one-to-one.

21. Find two numbers a and b for which $a \neq b$ and $g(a) = g(b)$. Two such numbers are -2 and 2, because $g(-2) = g(2) = -3$. Thus, g is not one-to-one. **23.** Find two numbers a and b for which $a \neq b$ and $g(a) = g(b)$. Two such numbers are -1 and 1, because $g(-1) = g(1) = 0$. Thus, g is not one-to-one.

25. Yes **27.** No **29.** No **31.** Yes **33.** Yes **35.** No **37.** No **39.** Yes **41.** No **43.** No

45. **(a)** One-to-one; **(b)** $f^{-1}(x) = x - 4$

47. **(a)** One-to-one; **(b)** $f^{-1}(x) = \dfrac{x + 1}{2}$

49. **(a)** One-to-one; **(b)** $f^{-1}(x) = \dfrac{4}{x} - 7$

51. **(a)** One-to-one; **(b)** $f^{-1}(x) = \dfrac{3x + 4}{x - 1}$

53. **(a)** One-to-one; **(b)** $f^{-1}(x) = \sqrt[3]{x + 1}$

55. **(a)** Not one-to-one; **(b)** does not have an inverse that is a function **57.** **(a)** One-to-one; **(b)** $f^{-1}(x) = \sqrt{\dfrac{x + 2}{5}}$

59. **(a)** One-to-one; **(b)** $f^{-1}(x) = x^2 - 1, x \geq 0$ **61.** $\frac{1}{3}x$

63. $-x$ **65.** $x^3 + 5$

67.

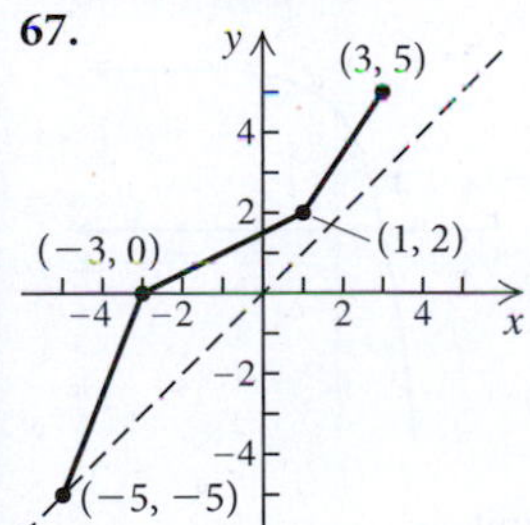

69.

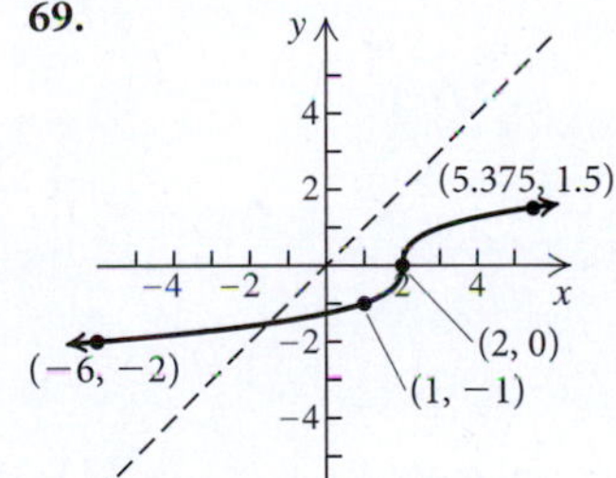

71.

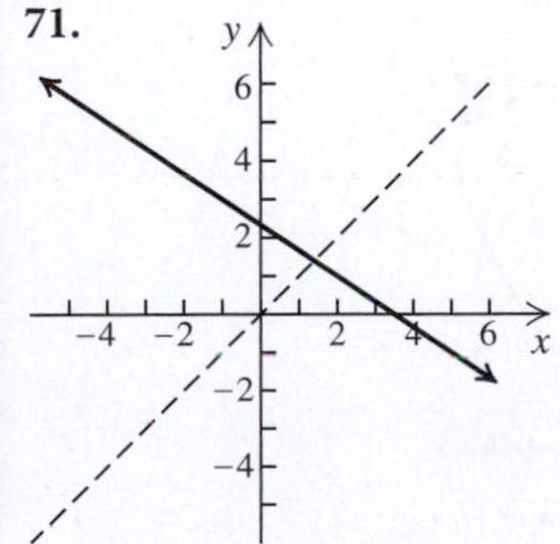

73. $f^{-1}(f(x)) = f^{-1}\left(\frac{7}{8}x\right) = \frac{8}{7} \cdot \frac{7}{8}x = x$;
$f(f^{-1}(x)) = f\left(\frac{8}{7}x\right) = \frac{7}{8} \cdot \frac{8}{7}x = x$

75. $f^{-1}(f(x)) = f^{-1}\left(\dfrac{1 - x}{x}\right) = \dfrac{1}{\dfrac{1 - x}{x} + 1} = \dfrac{1}{\dfrac{1 - x + x}{x}} = \dfrac{1}{\dfrac{1}{x}} = 1 \cdot \dfrac{x}{1} = x$; $f(f^{-1}(x)) = f\left(\dfrac{1}{x + 1}\right) = \dfrac{1 - \dfrac{1}{x + 1}}{\dfrac{1}{x + 1}} = \dfrac{\dfrac{x + 1 - 1}{x + 1}}{\dfrac{1}{x + 1}} = \dfrac{x}{x + 1} \cdot \dfrac{x + 1}{1} = x$

77. $f^{-1}(f(x)) = f^{-1}\left(\dfrac{2}{5}x + 1\right) = \dfrac{5\left(\dfrac{2}{5}x + 1\right) - 5}{2} = \dfrac{2x + 5 - 5}{2} = \dfrac{2x}{2} = x$; $f(f^{-1}(x)) = f\left(\dfrac{5x - 5}{2}\right) = \dfrac{2}{5}\left(\dfrac{5x - 5}{2}\right) + 1 = x - 1 + 1 = x$

79. $f^{-1}(x) = \frac{1}{5}x + \frac{3}{5}$; domain of f and f^{-1}: $(-\infty, \infty)$; range of f and f^{-1}: $(-\infty, \infty)$;

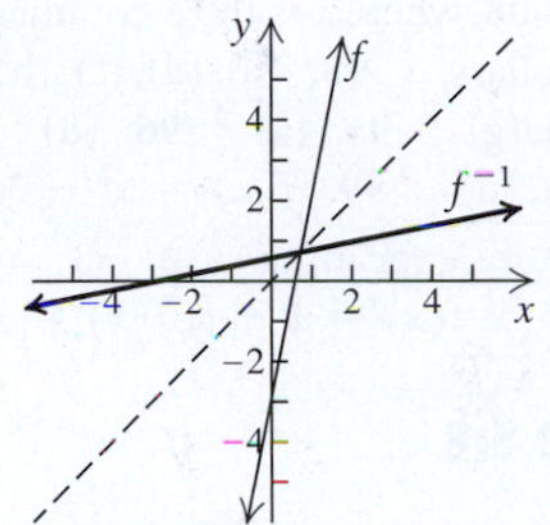

81. $f^{-1}(x) = \dfrac{2}{x}$; domain of f and f^{-1}: $(-\infty, 0) \cup (0, \infty)$; range of f and f^{-1}: $(-\infty, 0) \cup (0, \infty)$;

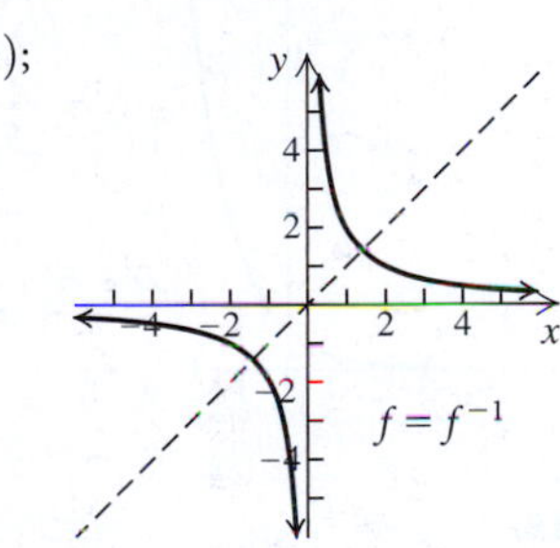

83. $f^{-1}(x) = \sqrt[3]{3x + 6}$; domain of f and f^{-1}: $(-\infty, \infty)$; range of f and f^{-1}: $(-\infty, \infty)$;

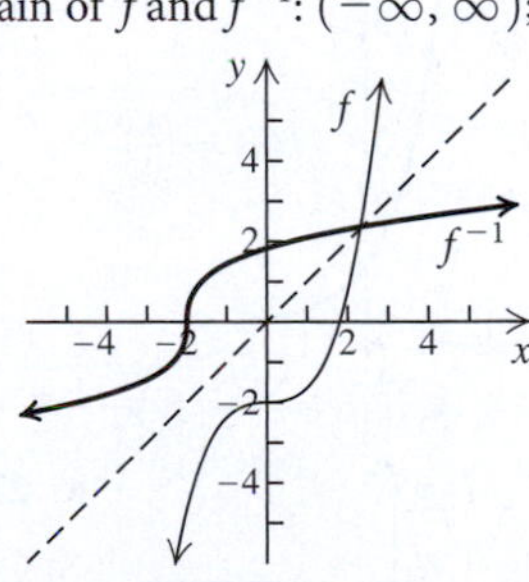

85. $f^{-1}(x) = \dfrac{3x + 1}{x - 1}$; domain of f: $(-\infty, 3) \cup (3, \infty)$; range of f: $(-\infty, 1) \cup (1, \infty)$; domain of f^{-1}: $(-\infty, 1) \cup (1, \infty)$; range of f^{-1}: $(-\infty, 3) \cup (3, \infty)$;

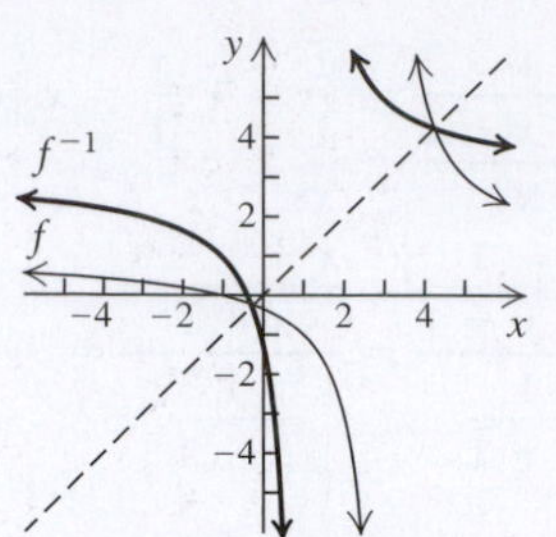

87. 5; a **89.** **(a)** \$38, \$16.40, \$11; **(b)** $C^{-1}(x) = \dfrac{72}{x - 2}$; $C^{-1}(x)$ represents the number of players in the group lesson, where x is the cost per player, in dollars; **(c)** 1 player, 4 players, 8 players **91.** **(a)** 2010: \$40.86 billion; 2013: \$60.6 billion; **(b)** $H^{-1}(x) = \dfrac{x - 27.7}{6.58}$; $H^{-1}(x)$ represents the number of years after 2008, where x is the e-commerce holiday season sales, in billions of dollars. **93.** (b), (d), (f), (h) **94.** (a), (c), (e), (g) **95.** (a) **96.** (d) **97.** (f) **98.** (a), (b), (c), (d) **99.** $f(x) = x^2 - 3$, for inputs $x \geq 0$; $f^{-1}(x) = \sqrt{x + 3}$, for inputs $x \geq -3$ **101.** Answers may vary; $f(x) = 3/x, f(x) = 1 - x, f(x) = x$

Exercise Set 5.2

1. 54.5982 **3.** 0.0856 **5.** (f) **7.** (e) **9.** (a)

11. $f(x) = 3^x$

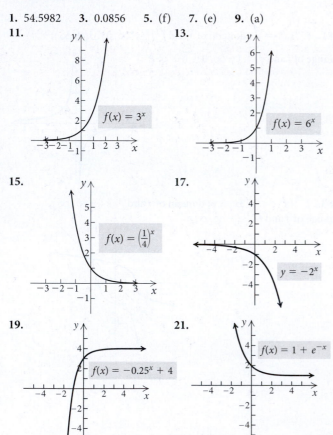

13.

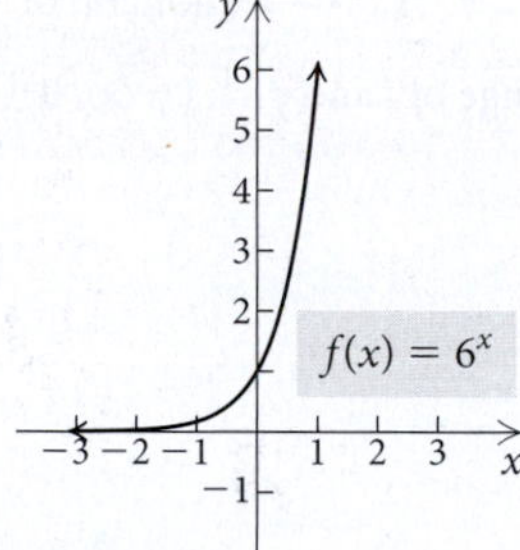

15. $f(x) = \left(\frac{1}{4}\right)^x$

17.

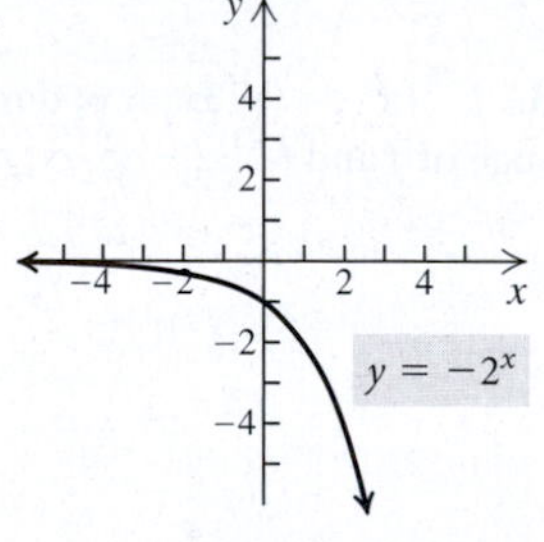

19.

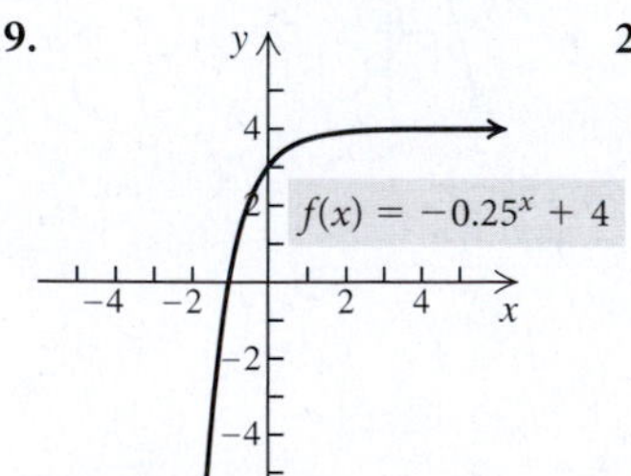

21.

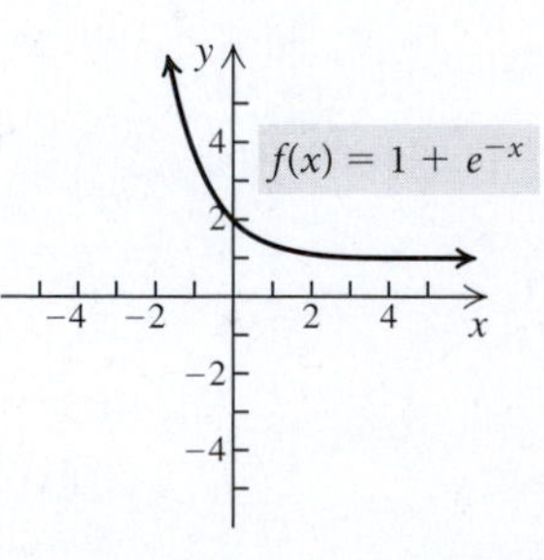

23.

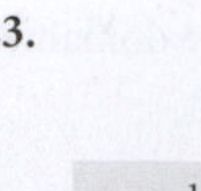

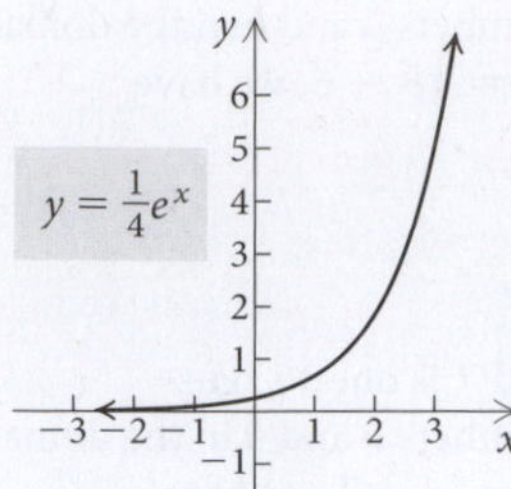

25.

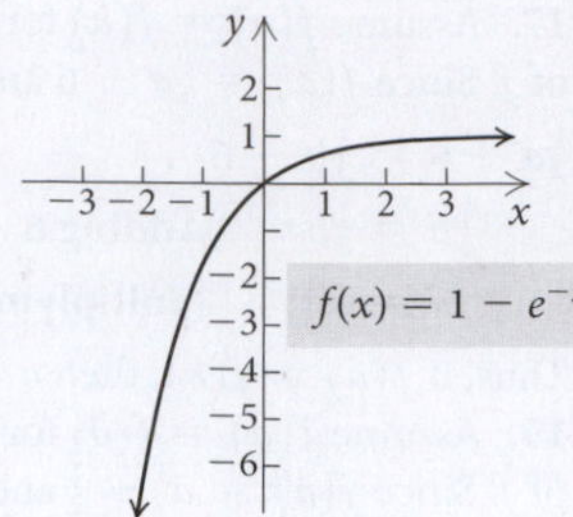

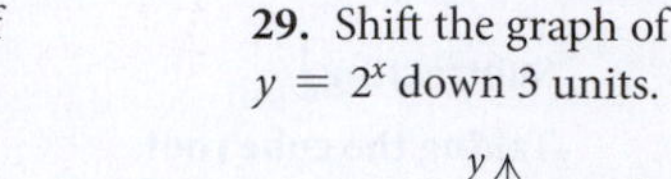

27. Shift the graph of $y = 2^x$ left 1 unit.

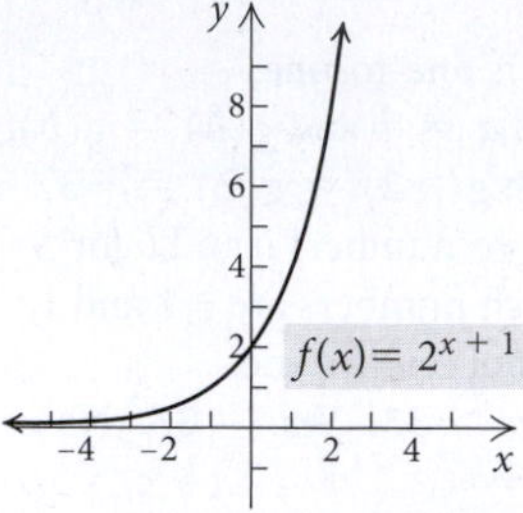

29. Shift the graph of $y = 2^x$ down 3 units.

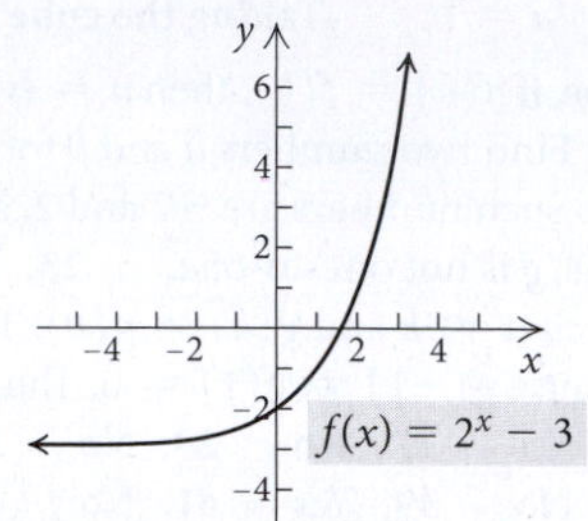

31. Shift the graph of $y = 2^x$ left 1 unit, reflect it across the y-axis, and shift it up 2 units.

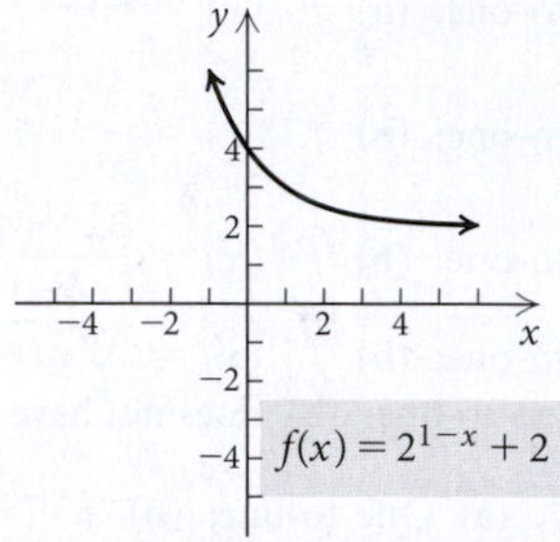

33. Reflect the graph of $y = 3^x$ across the y-axis and then across the x-axis and then shift it up 4 units.

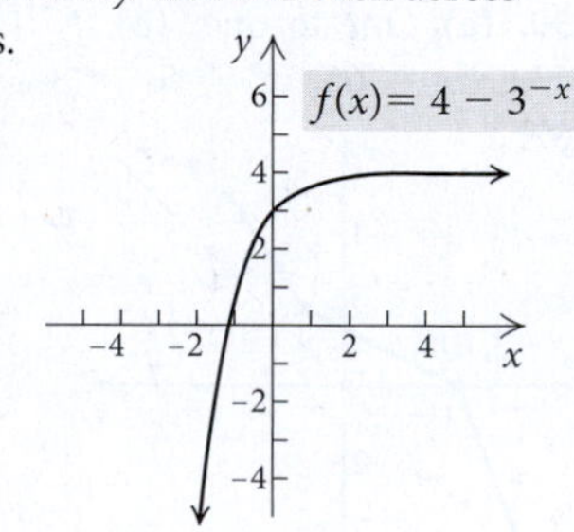

35. Shift the graph of $y = \left(\frac{3}{2}\right)^x$ right 1 unit.

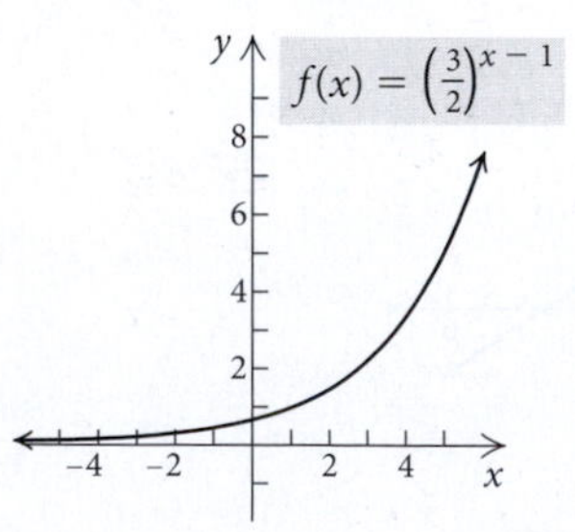

37. Shift the graph of $y = 2^x$ left 3 units and then down 5 units.

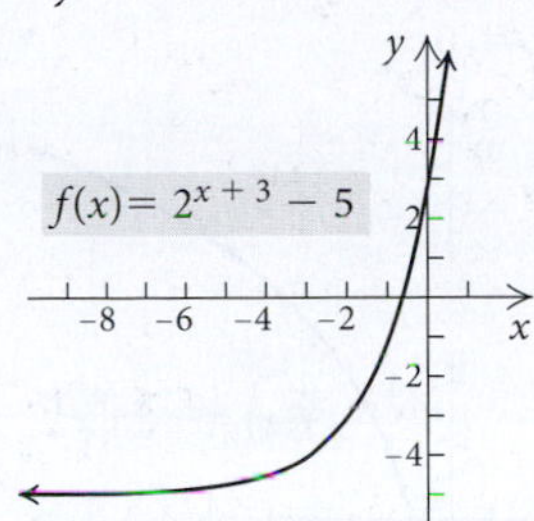

39. Shift the graph of $y = 2^x$ right 1 unit, stretch it vertically, and shift it up 1 unit.

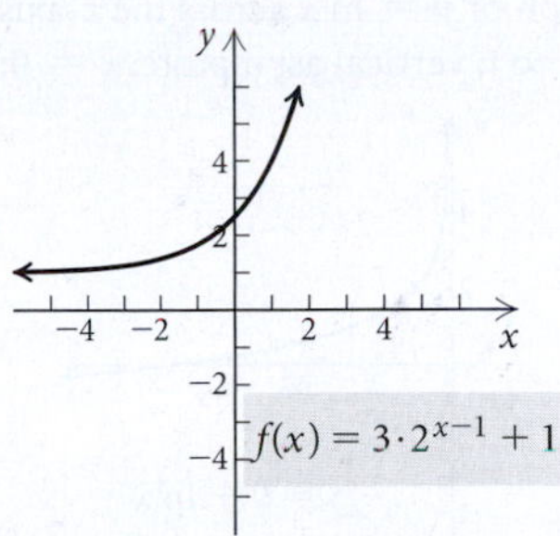

41. Shrink the graph of $y = e^x$ horizontally.

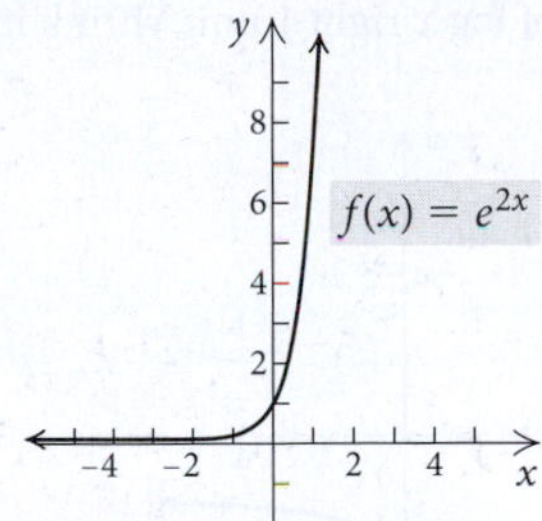

43. Reflect the graph of $y = e^x$ across the x-axis, shift it up 1 unit, and shrink it vertically.

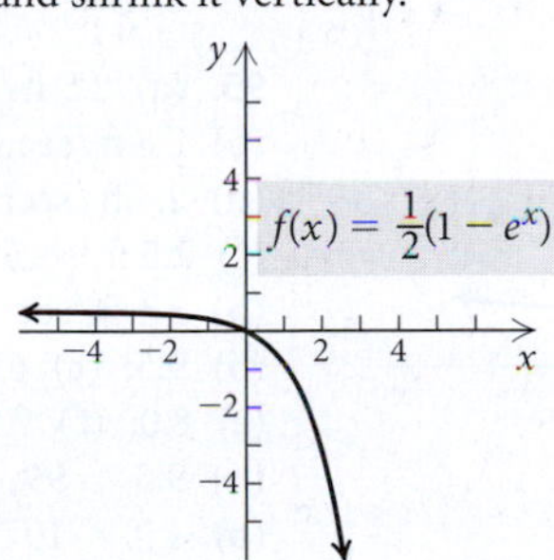

45. Shift the graph of $y = e^x$ left 1 unit and then reflect it across the y-axis.

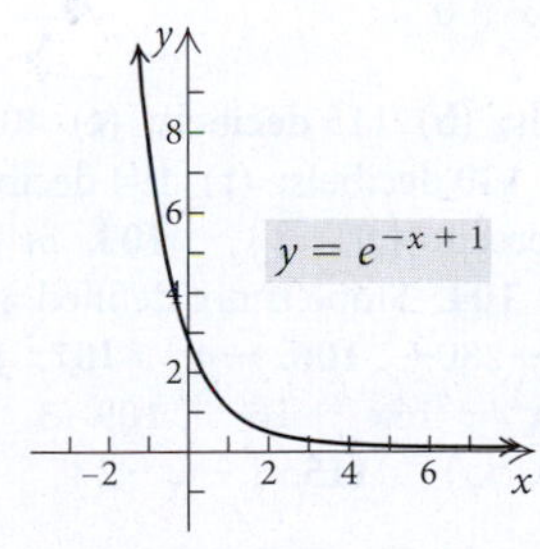

47. Reflect the graph of $y = e^x$ across the y-axis, then across the x-axis, then shift it up 1 unit, and then stretch it vertically.

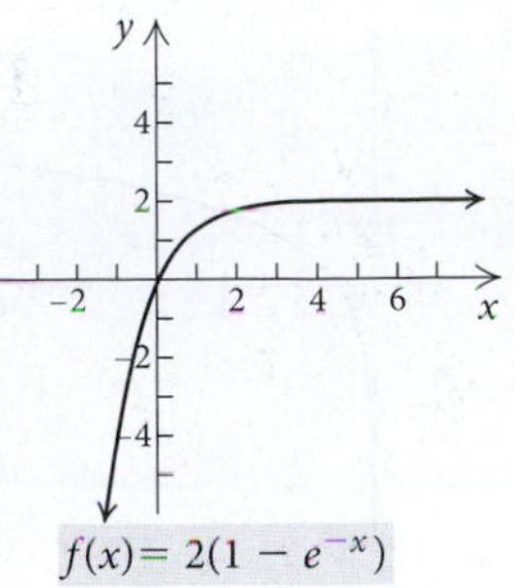

49.

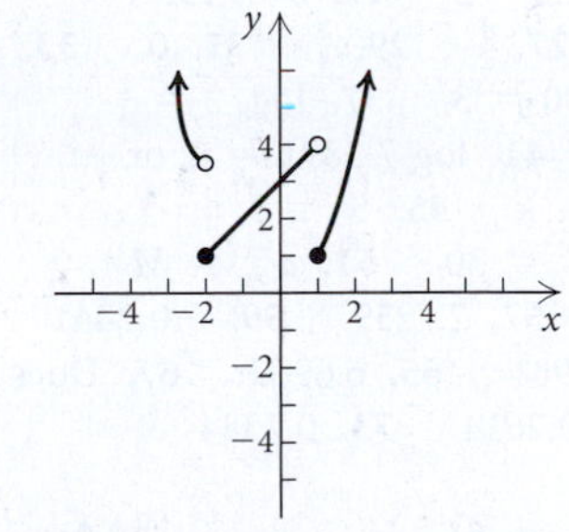

$$f(x) = \begin{cases} e^{-x} - 4, & \text{for } x < -2, \\ x + 3, & \text{for } -2 \le x < 1, \\ x^2, & \text{for } x \ge 1 \end{cases}$$

51. **(a)** $A(t) = 82{,}000(1.01125)^{4t}$; **(b)** \$82,000; \$89,677.22; \$102,561.54; \$128,278.90 **53.** \$4930.86 **55.** \$3247.30 **57.** \$153,610.15 **59.** \$76,305.59 **61.** \$26,086.69 **63.** 1998: 322,420 vehicles; 2010: 938,297 vehicles; 2018: 1,912,580 vehicles **65.** 2011: \$234 million; 2015: \$5844 million, or \$5.844 billion **67.** 2005: 3 million users; 2009: 17 million users; 2012: 54 million users **69.** 2020: 101,234 centenarians; 2050: 414,387 centenarians **71.** 1982: \$48 billion; 1995: \$109 billion; 2010: \$284 billion **73.** \$6982; \$5935; \$5044; \$3098; \$1903 **75.** About 63% **77.** $31 - 22i$ **78.** $\frac{1}{2} - \frac{1}{2}i$ **79.** $\left(-\frac{1}{2}, 0\right), (7, 0); -\frac{1}{2}, 7$ **80.** $(1, 0); 1$ **81.** $(-1, 0), (0, 0), (1, 0); -1, 0, 1$ **82.** $(-4, 0), (0, 0), (3, 0); -4, 0, 3$ **83.** $-8, 0, 2$ **84.** $\dfrac{5 \pm \sqrt{97}}{6}$ **85.** $\pi^7; 70^{80}$

Visualizing the Graph

1. J **2.** F **3.** H **4.** B **5.** E **6.** A **7.** C **8.** I **9.** D **10.** G

Exercise Set 5.3

1.

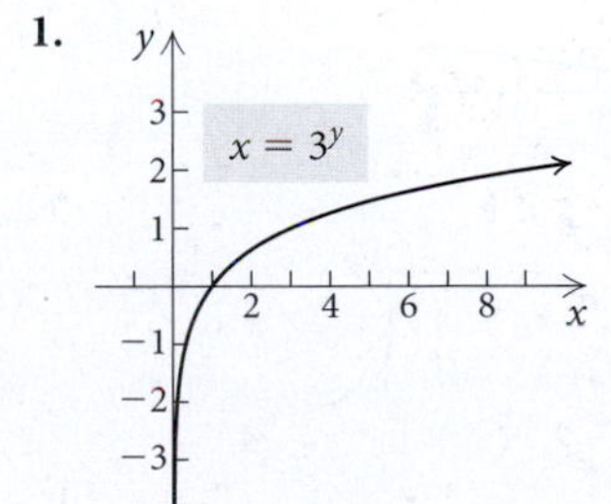

3.

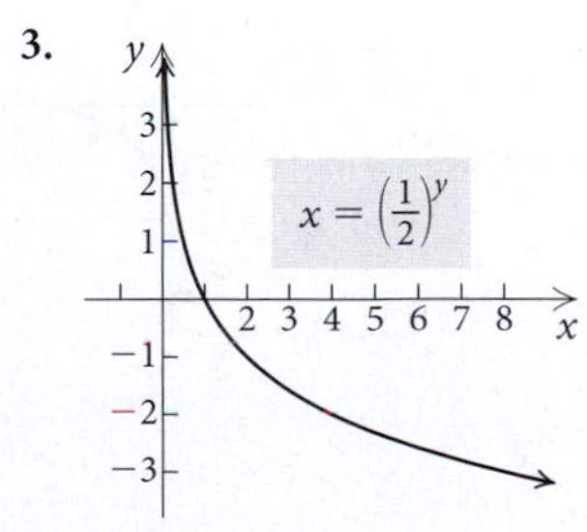

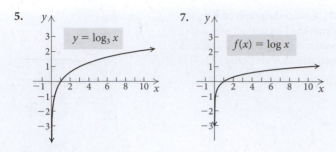

5. $y = \log_3 x$

7. $f(x) = \log x$

9. 4 **11.** 3 **13.** -3 **15.** -2 **17.** 0 **19.** 1
21. 4 **23.** $\frac{1}{4}$ **25.** -7 **27.** $\frac{1}{2}$ **29.** $\frac{3}{4}$ **31.** 0 **33.** $\frac{1}{2}$
35. $\log_{10} 1000 = 3$, or $\log 1000 = 3$ **37.** $\log_8 2 = \frac{1}{3}$
39. $\log_e t = 3$, or $\ln t = 3$ **41.** $\log_e 7.3891 = 2$, or $\ln 7.3891 = 2$ **43.** $\log_p 3 = k$ **45.** $5^1 = 5$
47. $10^{-2} = 0.01$ **49.** $e^{3.4012} = 30$ **51.** $a^{-x} = M$
53. $a^x = T^3$ **55.** 0.4771 **57.** 2.7259 **59.** -0.2441
61. Does not exist **63.** 0.6931 **65.** 6.6962 **67.** Does not exist **69.** 3.3219 **71.** -0.2614 **73.** 0.7384
75. 2.2619 **77.** 0.5880

79.

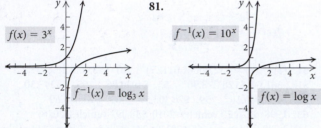

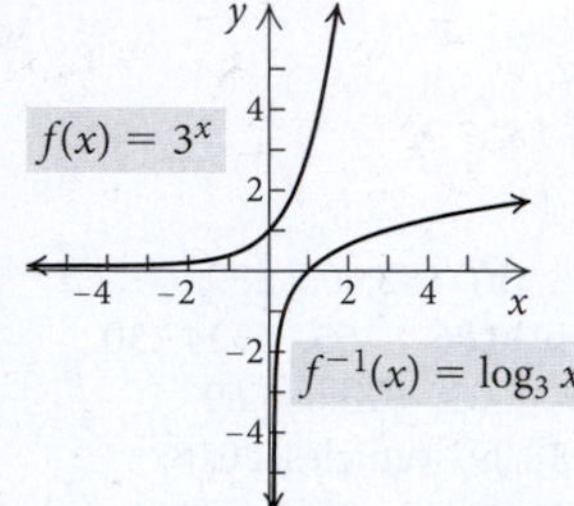

81. $f^{-1}(x) = 10^x$; $f(x) = \log x$

83. Shift the graph of $y = \log_2 x$ left 3 units. Domain: $(-3, \infty)$; vertical asymptote: $x = -3$;

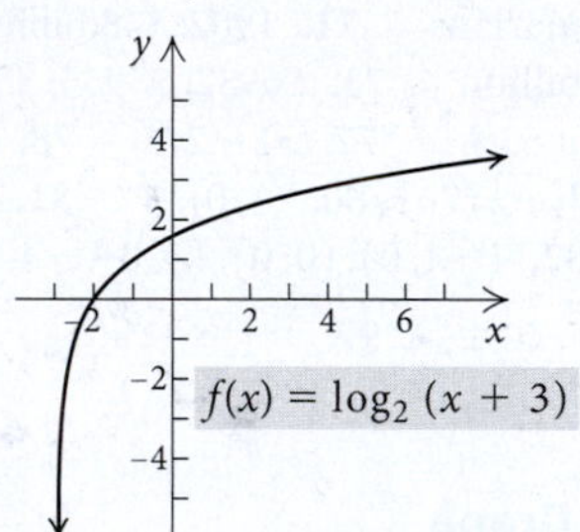

85. Shift the graph of $y = \log_3 x$ down 1 unit. Domain: $(0, \infty)$; vertical asymptote: $x = 0$;

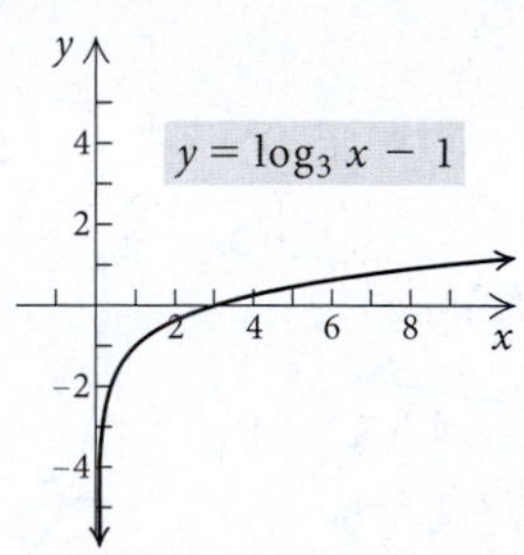

87. Stretch the graph of $y = \ln x$ vertically. Domain: $(0, \infty)$; vertical asymptote: $x = 0$;

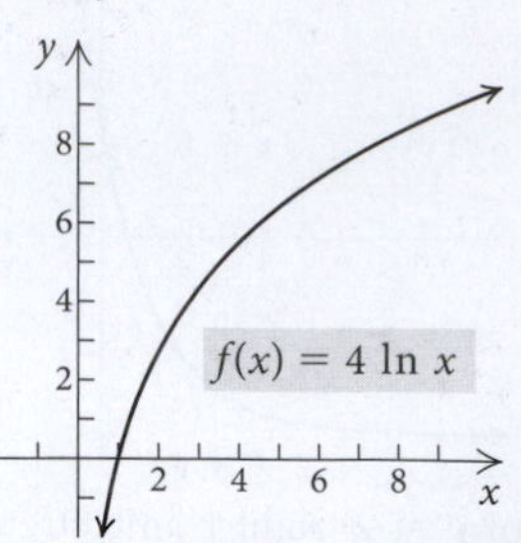

89. Reflect the graph of $y = \ln x$ across the x-axis and shift it up 2 units. Domain: $(0, \infty)$; vertical asymptote: $x = 0$;

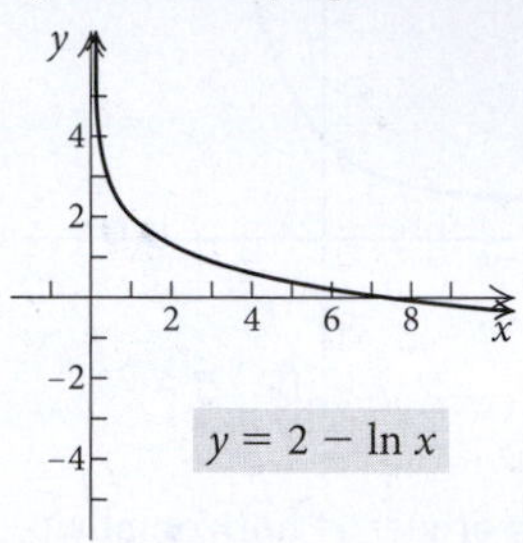

91. Shift the graph of $\log x$ right 1 unit, shrink it vertically, and shift it down 2 units.

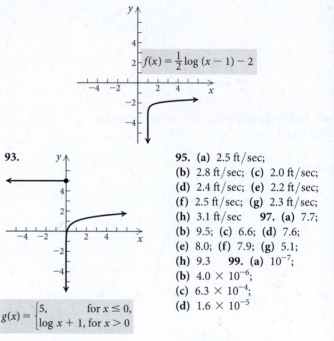

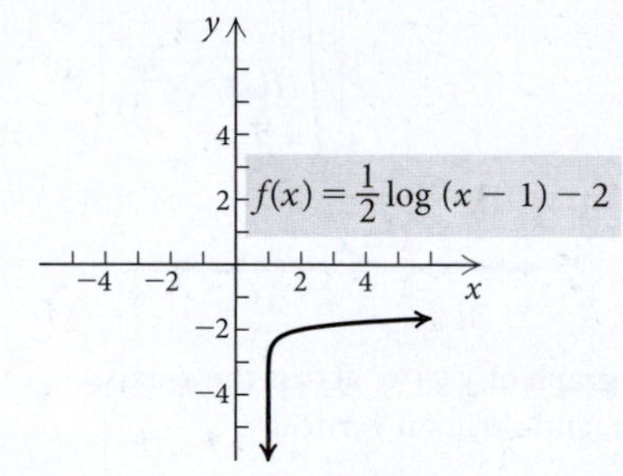

93.

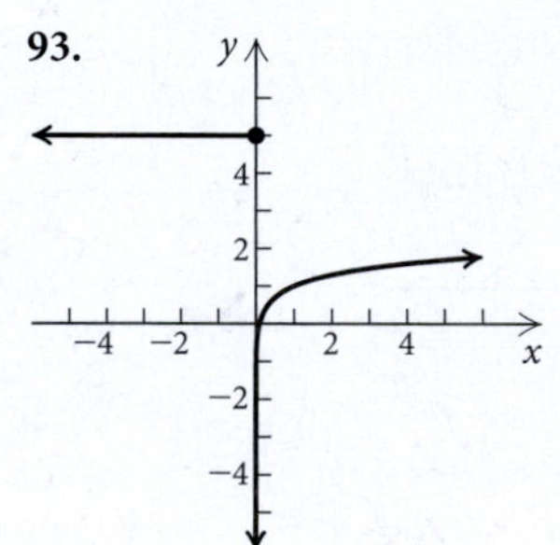

95. **(a)** 2.5 ft/sec; **(b)** 2.8 ft/sec; **(c)** 2.0 ft/sec; **(d)** 2.4 ft/sec; **(e)** 2.2 ft/sec; **(f)** 2.5 ft/sec; **(g)** 2.3 ft/sec; **(h)** 3.1 ft/sec **97.** **(a)** 7.7; **(b)** 9.5; **(c)** 6.6; **(d)** 7.6; **(e)** 8.0; **(f)** 7.9; **(g)** 5.1; **(h)** 9.3 **99.** **(a)** 10^{-7}; **(b)** 4.0×10^{-6}; **(c)** 6.3×10^{-4}; **(d)** 1.6×10^{-5}

101. **(a)** 140 decibels; **(b)** 115 decibels; **(c)** 40 decibels; **(d)** 65 decibels; **(e)** 120 decibels; **(f)** 194 decibels
102. $m = \frac{3}{10}$; y-intercept: $\left(0, -\frac{7}{5}\right)$ **103.** $m = 0$; y-intercept: $(0, 6)$ **104.** Slope is not defined; no y-intercept **105.** -280 **106.** -4 **107.** $f(x) = x^3 - 7x$
108. $f(x) = x^3 - x^2 + 16x - 16$ **109.** 3 **111.** $(0, \infty)$
113. $(-\infty, 0) \cup (0, \infty)$ **115.** $\left(-\frac{5}{2}, -2\right)$ **117.** (d)
119. (b)

Mid-Chapter Mixed Review: Chapter 5

1. False **2.** True **3.** False **4.** Yes; $f^{-1}(x) = -\dfrac{2}{x}$
5. No **6.** Yes; $f^{-1}(x) = \dfrac{5}{x} + 2$
7. $(f^{-1} \circ f)(x) = f^{-1}(\sqrt{x-5}) = (\sqrt{x-5})^2 + 5 = x - 5 + 5 = x$; $(f \circ f^{-1})(x) = f(x^2 + 5) = \sqrt{(x^2+5) - 5} = \sqrt{x^2} = x$ **8.** $f^{-1}(x) = \sqrt[3]{x} - 2$; domain of f: $(-\infty, \infty)$, range of f^1: $(-\infty, \infty)$; domain of f^{-1}: $(-\infty, \infty)$, range of f^{-1}: $(-\infty, \infty)$ **9.** (d) **10.** (h)
11. (c) **12.** (g) **13.** (b) **14.** (f) **15.** (e) **16.** (a)
17. \$4185.57 **18.** 0 **19.** $-\frac{4}{5}$ **20.** -2 **21.** 2 **22.** 0
23. -4 **24.** 0 **25.** 3 **26.** $\frac{1}{4}$ **27.** 1 **28.** $\ln 0.0025 = -6$
29. $10^r = T$ **30.** 2.7268 **31.** 2.0115 **32.** For an even function f, $f(x) = f(-x)$, so we have $f(x) = f(-x)$ but $x \neq -x$ (for $x \neq 0$). Thus, f is not one-to-one and hence it does not have an inverse. **33.** The most interest will be earned in the eighth year, because the principal is greatest during that year.
34. In $f(x) = x^3$, the variable x is the base. The range of f is $(-\infty, \infty)$. In $g(x) = 3^x$, the variable x is the exponent. The range of g is $(0, \infty)$. The graph of f does not have an asymptote. The graph of g has an asymptote $y = 0$.
35. If $\log b < 0$, then $0 < b < 1$.

Exercise Set 5.4

1. $\log_3 81 + \log_3 27 = 4 + 3 = 7$
3. $\log_5 5 + \log_5 125 = 1 + 3 = 4$ **5.** $\log_t 8 + \log_t Y$
7. $\ln x + \ln y$ **9.** $3 \log_b t$ **11.** $8 \log y$ **13.** $-6 \log_c K$
15. $\frac{1}{3} \ln 4$ **17.** $\log_t M - \log_t 8$ **19.** $\log x - \log y$
21. $\ln r - \ln s$ **23.** $\log_a 6 + \log_a x + 5 \log_a y + 4 \log_a z$
25. $2 \log_b p + 5 \log_b q - 4 \log_b m - 9$
27. $\ln 2 - \ln 3 - 3 \ln x - \ln y$ **29.** $\frac{3}{2} \log r + \frac{1}{2} \log t$
31. $3 \log_a x - \frac{5}{2} \log_a p - 4 \log_a q$
33. $2 \log_a m + 3 \log_a n - \frac{3}{4} - \frac{5}{4} \log_a b$
35. $\log_a 150$ **37.** $\log 100 = 2$ **39.** $\log m^3 \sqrt{n}$
41. $\log_a x^{-5/2} y^4$, or $\log_a \dfrac{y^4}{x^{5/2}}$ **43.** $\ln x$ **45.** $\ln (x - 2)$
47. $\log \dfrac{x-7}{x-2}$ **49.** $\ln \dfrac{x}{(x^2 - 25)^3}$ **51.** $\ln \dfrac{2^{11/5} x^9}{y^8}$
53. -0.74 **55.** 1.991 **57.** 0.356 **59.** 4.827
61. -1.792 **63.** 0.099 **65.** 3 **67.** $|x - 4|$ **69.** $4x$
71. w **73.** $8t$ **75.** $\frac{1}{2}$ **77.** Quartic **78.** Exponential
79. Linear (constant) **80.** Exponential **81.** Rational
82. Logarithmic **83.** Cubic **84.** Rational **85.** Linear
86. Quadratic **87.** 4 **89.** $\log_a (x^3 - y^3)$
91. $\frac{1}{2} \log_a (x - y) - \frac{1}{2} \log_a (x + y)$ **93.** 7 **95.** True
97. True **99.** True **101.** -2 **103.** 3
105. $e^{-xy} = \dfrac{a}{b}$ **107.** $\log_a \left(\dfrac{x + \sqrt{x^2 - 5}}{5} \cdot \dfrac{x - \sqrt{x^2 - 5}}{x - \sqrt{x^2 - 5}} \right) = \log_a \dfrac{5}{5(x - \sqrt{x^2 - 5})} = -\log_a (x - \sqrt{x^2 - 5})$

Exercise Set 5.5

1. 4 **3.** $\frac{3}{2}$ **5.** 5.044 **7.** $\frac{5}{2}$ **9.** $-3, \frac{1}{2}$ **11.** 0.959
13. 0 **15.** 0 **17.** 6.908 **19.** 84.191 **21.** -1.710
23. 2.844 **25.** $-1.567, 1.567$ **27.** 1.869
29. $-1.518, 0.825$ **31.** 625 **33.** 0.0001 **35.** e
37. $-\frac{1}{3}$ **39.** $\frac{22}{3}$ **41.** 10 **43.** 4 **45.** $\frac{1}{63}$ **47.** 2
49. $\frac{2}{5}$ **51.** 5 **53.** $\frac{21}{8}$ **55.** $\frac{8}{7}$ **57.** 6 **59.** 6.192
61. 0 **63.** (a) $(0, -6)$; (b) $x = 0$; (c) minimum: -6 when $x = 0$ **64.** (a) $(3, 1)$; (b) $x = 3$; (c) maximum: 1 when $x = 3$ **65.** (a) $(-1, -5)$; (b) $x = -1$; (c) maximum: -5 when $x = -1$ **66.** (a) $(2, 4)$; (b) $x = 2$; (c) minimum: 4 when $x = 2$ **67.** $\dfrac{\ln 2}{2}$, or 0.347
69. $1, e^4$ or $1, 54.598$ **71.** $\frac{1}{3}, 27$ **73.** $1, e^2$ or $1, 7.389$
75. $0, \dfrac{\ln 2}{\ln 5}$ or $0, 0.431$ **77.** e^{-2}, e^2 or $0.135, 7.389$
79. $\frac{7}{4}$ **81.** $a = \frac{2}{3} b$

Exercise Set 5.6

1. (a) $P(t) = 6.18e^{0.0214t}$, where t is the number of years after 2012 and P is in millions; (b) 7.0 million; (c) about 12.1 years after 2012; (d) about 32.4 years **3.** (a) 0.90%; (b) 1.63%; (c) 20.9 years; (d) 62.4 years; (e) 0.18%; (f) 29.9 years; (g) 54.2 years; (h) 0.46%; (i) 2.64%; (j) 177.7 years
5. About 819 years after 2013 **7.** (a) $P(t) = 10{,}000e^{0.054t}$; (b) \$10,554.85; \$11,140.48; \$13,099.64; \$17,160.07; (c) about 12.8 years **9.** About 12,320 years
11. (a) 22.4% per minute; (b) 3.1% per year; (c) 60.3 days; (d) 10.7 years; (e) 2.4% per year; (f) 1.0% per year; (g) 0.0029% per year **13.** (a) $k \approx 0.0069$, $M(t) = 72.2e^{-0.0069t}$; (b) 2015: 49.4%; 2018: 48.4%; (c) in 2046
15. (a) $k \approx 0.0536$, $C(t) = 1.85e^{0.0536t}$; (b) 7.07 million barrels of oil per day; (c) 12.9 years; (d) 36.4 years after 1980
17. (a) 167; (b) 500; 1758; 3007; 3449; 3495; (c) as $t \to \infty$, $N(t) \to 3500$; the number approaches 3500 but never actually reaches it. **19.** 46.7°F **21.** 59.6°F
23. Multiplication principle for inequalities **24.** Product rule
25. Principle of zero products **26.** Principle of square roots
27. Power rule **28.** Multiplication principle for equations
29. \$166.16 **31.** \$19,609.67
33. $t = -\dfrac{L}{R}\left[\ln \left(1 - \dfrac{iR}{V} \right) \right]$ **35.** Linear

Review Exercises: Chapter 5

1. True **2.** False **3.** False **4.** True **5.** False **6.** True
7. $\{(-2.7, 1.3), (-3, 8), (3, -5), (-3, 6), (-5, 7)\}$
8. (a) $x = -2y + 3$; (b) $x = 3y^2 + 2y - 1$; (c) $0.8y^3 - 5.4x^2 = 3y$ **9.** No **10.** No **11.** Yes **12.** Yes
13. (a) Yes; (b) $f^{-1}(x) = \dfrac{-x + 2}{3}$ **14.** (a) Yes; (b) $f^{-1}(x) = \dfrac{x + 2}{x - 1}$ **15.** (a) Yes; (b) $f^{-1}(x) = x^2 + 6$, $x \geq 0$ **16.** (a) Yes; (b) $f^{-1}(x) = \sqrt[3]{x + 8}$
17. (a) No **18.** (a) Yes; (b) $f^{-1}(x) = \ln x$
19. $f^{-1}(f(x)) = f^{-1}(6x - 5) = \dfrac{6x - 5 + 5}{6} = \dfrac{6x}{6} = x$; $f(f^{-1}(x)) = f\left(\dfrac{x + 5}{6}\right) = 6\left(\dfrac{x + 5}{6}\right) - 5 = x + 5 - 5 = x$

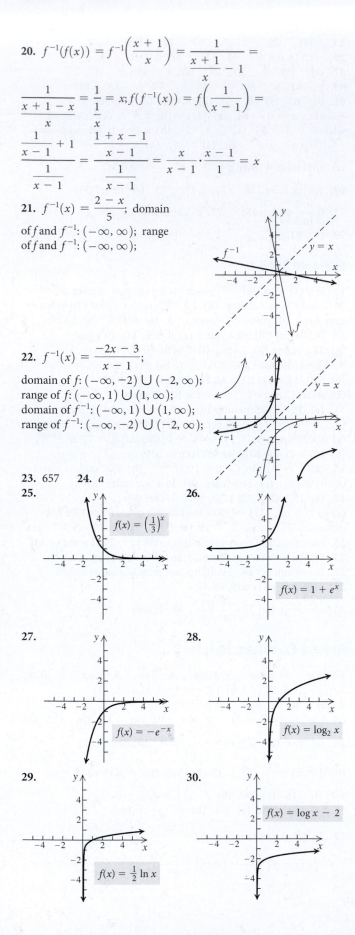

20. $f^{-1}(f(x)) = f^{-1}\left(\dfrac{x+1}{x}\right) = \dfrac{1}{\dfrac{x+1}{x} - 1} =$

$\dfrac{1}{\dfrac{x+1-x}{x}} = \dfrac{1}{\dfrac{1}{x}} = x$; $f(f^{-1}(x)) = f\left(\dfrac{1}{x-1}\right) =$

$\dfrac{\dfrac{1}{x-1} + 1}{\dfrac{1}{x-1}} = \dfrac{\dfrac{1+x-1}{x-1}}{\dfrac{1}{x-1}} = \dfrac{x}{x-1} \cdot \dfrac{x-1}{1} = x$

21. $f^{-1}(x) = \dfrac{2-x}{5}$; domain of f and f^{-1}: $(-\infty, \infty)$; range of f and f^{-1}: $(-\infty, \infty)$;

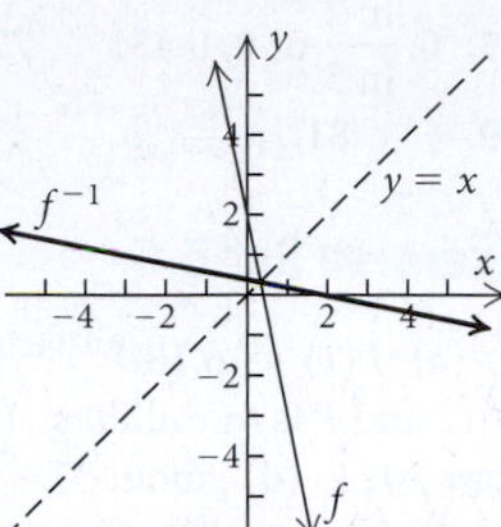

22. $f^{-1}(x) = \dfrac{-2x-3}{x-1}$;
domain of f: $(-\infty, -2) \cup (-2, \infty)$;
range of f: $(-\infty, 1) \cup (1, \infty)$;
domain of f^{-1}: $(-\infty, 1) \cup (1, \infty)$;
range of f^{-1}: $(-\infty, -2) \cup (-2, \infty)$;

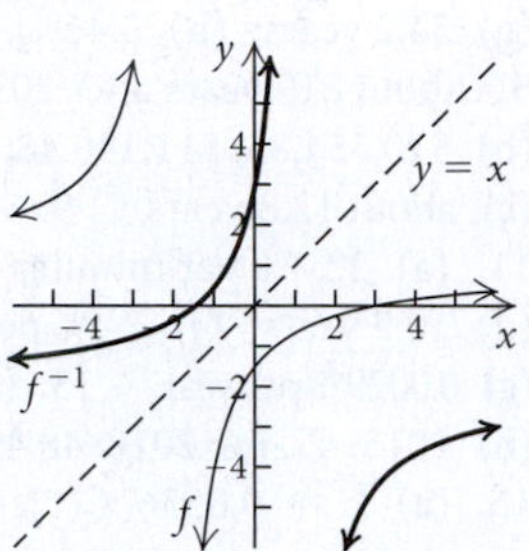

23. 657 **24.** a

25.

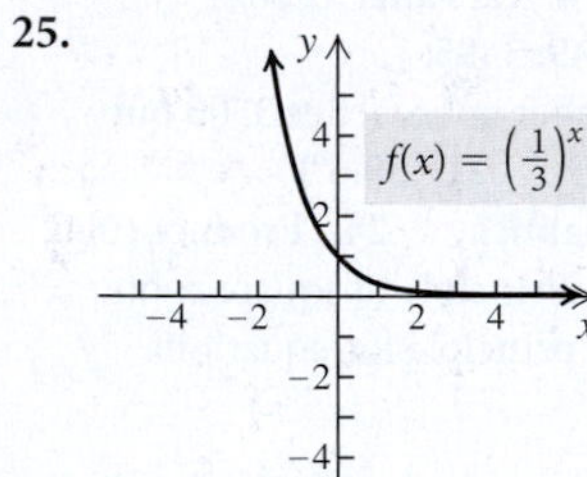

26. $f(x) = 1 + e^x$

27. $f(x) = -e^{-x}$

28. $f(x) = \log_2 x$

29. $f(x) = \frac{1}{2}\ln x$

30. $f(x) = \log x - 2$

31. (c) **32.** (a) **33.** (b) **34.** (f) **35.** (e) **36.** (d) **37.** 3 **38.** 5 **39.** 1 **40.** 0 **41.** $\frac{1}{4}$ **42.** $\frac{1}{2}$ **43.** 0 **44.** 1 **45.** $\frac{1}{3}$ **46.** -2 **47.** $4^2 = x$ **48.** $a^k = Q$ **49.** $\log_4 \frac{1}{64} = -3$ **50.** $\ln 80 = x$, or $\log_e 80 = x$ **51.** 1.0414 **52.** -0.6308 **53.** 1.0986 **54.** -3.6119 **55.** Does not exist **56.** Does not exist **57.** 1.9746 **58.** 0.5283 **59.** $\log_b \dfrac{x^3\sqrt{z}}{y^4}$ **60.** $\ln(x^2 - 4)$ **61.** $\frac{1}{4}\ln w + \frac{1}{2}\ln r$ **62.** $\frac{2}{3}\log M - \frac{1}{3}\log N$ **63.** 0.477 **64.** 1.699 **65.** -0.699 **66.** 0.233 **67.** $-5k$ **68.** $-6t$ **69.** 16 **70.** $\frac{1}{5}$ **71.** 4.382 **72.** 2 **73.** $\frac{1}{2}$ **74.** 5 **75.** 4 **76.** 9 **77.** 1 **78.** 3.912 **79.** **(a)** $A(t) = 30{,}000(1.0105)^{4t}$; **(b)** \$30,000; \$38,547.20; \$49,529.56, \$63,640.87 **80.** 2005: 59.8 GW; 2010: 189.9 GW; 2016: 760.1 Gw **81.** 15.4 years **82.** 2.7% **83.** About 2623 years **84.** 5.6 **85.** 6.3 **86.** 30 decibels **87.** **(a)** 2.2 ft/sec; **(b)** 8,553,143 **88.** **(a)** $k \approx 0.1392$; **(b)** $S(t) = 0.035e^{0.1392t}$, where t is the number of years after 1940 and S is in billions of dollars; **(c)** 1970: \$2.279 billion; 2000: \$148.353 billion; 2015: \$1197.023 billion, or about \$1.197 trillion; **(d)** in 2019 **89.** **(a)** $P(t) = 15.2e^{0.0167t}$, where t is the number of years after 2013 and P is in millions; **(b)** 2017: 16.3 million; 2020: 17.1 million; **(c)** about 10 years after 2013; **(d)** 41.5 years **90.** D **91.** A **92.** D **93.** B **94.** $\frac{1}{64}$, 64 **95.** 1 **96.** 16 **97.** Measure the atmospheric pressure P at the top of the building. Substitute that value in the equation $P = 14.7e^{-0.00005a}$, and solve for the height, or altitude, a of the top of the building. Also measure the atmospheric pressure at the base of the building and solve for the altitude of the base. Then subtract to find the height of the building. **98.** Reflect the graph of $f(x) = \ln x$ across the line $y = x$ to obtain the graph of $h(x) = e^x$. Then shift this graph right 2 units to obtain the graph of $g(x) = e^{x-2}$. **99.** The inverse of a function $f(x)$ is written $f^{-1}(x)$, whereas $[f(x)]^{-1}$ means $\dfrac{1}{f(x)}$. **100.** $\log_a ab^3 \neq (\log_a a)(\log_a b^3)$. If the first step had been correct, then the second step would be as well. The correct procedure follows: $\log_a ab^3 = \log_a a + \log_a b^3 = 1 + 3\log_a b$.

Test: Chapter 5

1. [5.1] $\{(5, -2), (3, 4), (-1, 0), (-3, -6)\}$ **2.** [5.1] No **3.** [5.1] Yes **4.** [5.1] **(a)** Yes; **(b)** $f^{-1}(x) = \sqrt[3]{x-1}$ **5.** [5.1] **(a)** Yes; **(b)** $f^{-1}(x) = 1 - x$ **6.** [5.1] **(a)** Yes; **(b)** $f^{-1}(x) = \dfrac{2x}{1+x}$ **7.** [5.1] **(a)** No **8.** [5.1] $f^{-1}(f(x)) = f^{-1}(-4x+3) = \dfrac{3-(-4x+3)}{4} = \dfrac{4x}{4} = x$; $f(f^{-1}(x)) = f\left(\dfrac{3-x}{4}\right) = -4\left(\dfrac{3-x}{4}\right) + 3 = -3 + x + 3 = x$ **9.** [5.1] $f^{-1}(x) = \dfrac{4x+1}{x}$;
domain of f: $(-\infty, 4) \cup (4, \infty)$;
range of f: $(-\infty, 0) \cup (0, \infty)$;
domain of f^{-1}: $(-\infty, 0) \cup (0, \infty)$;
range of f^{-1}: $(-\infty, 4) \cup (4, \infty)$;

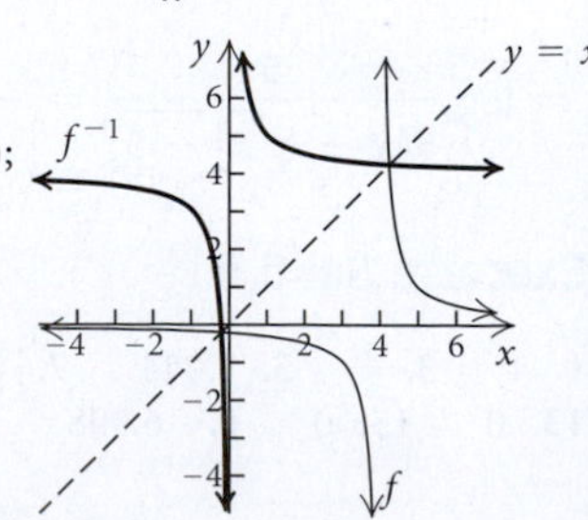

10. [5.2]

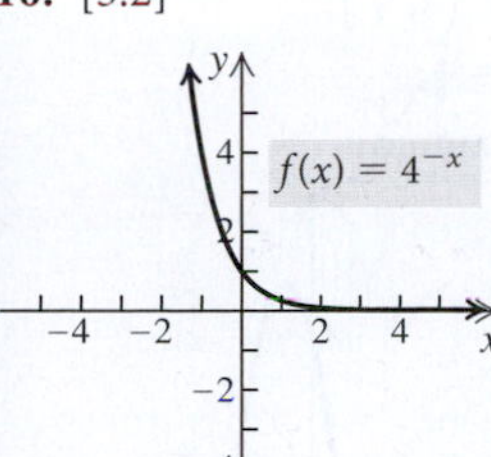

11. [5.3]

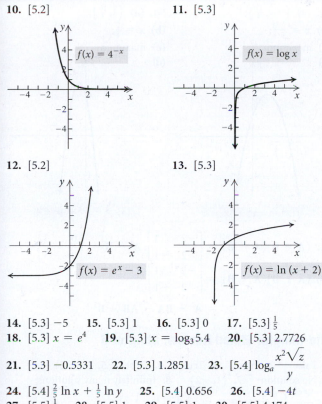

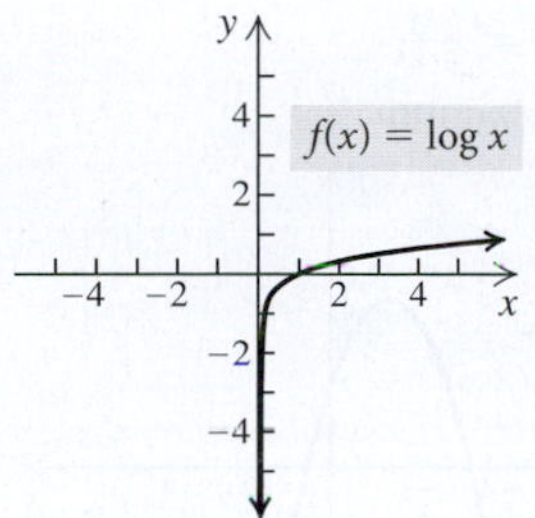

12. [5.2]

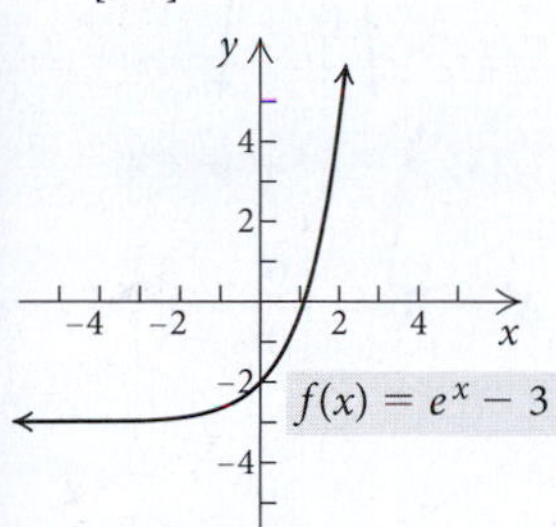

13. [5.3]

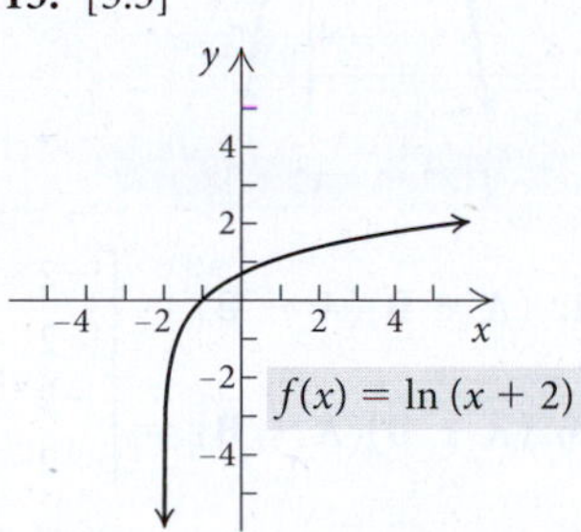

14. [5.3] -5 **15.** [5.3] 1 **16.** [5.3] 0 **17.** [5.3] $\frac{1}{5}$ **18.** [5.3] $x = e^4$ **19.** [5.3] $x = \log_3 5.4$ **20.** [5.3] 2.7726 **21.** [5.3] -0.5331 **22.** [5.3] 1.2851 **23.** [5.4] $\log_a \dfrac{x^2\sqrt{z}}{y}$ **24.** [5.4] $\frac{2}{5}\ln x + \frac{1}{5}\ln y$ **25.** [5.4] 0.656 **26.** [5.4] $-4t$ **27.** [5.5] $\frac{1}{2}$ **28.** [5.5] 1 **29.** [5.5] 1 **30.** [5.5] 4.174 **31.** [5.3] 6.6 **32.** [5.6] 0.0154 **33.** [5.6] **(a)** 4.5%; **(b)** $P(t) = 1000e^{0.045t}$; **(c)** \$1433.33; **(d)** 15.4 years **34.** [5.2] C **35.** [5.5] $\frac{27}{8}$

Chapter 6

Visualizing the Graph

1. C **2.** G **3.** D **4.** J **5.** A **6.** F **7.** I **8.** B **9.** H **10.** E

Exercise Set 6.1

1. (c) **3.** (f) **5.** (b) **7.** $(-1, 3)$ **9.** $(-1, 1)$ **11.** No solution **13.** $(-2, 4)$ **15.** Infinitely many solutions; $\left(x, \dfrac{x-1}{2}\right)$ or $(2y + 1, y)$ **17.** $(5, 4)$ **19.** $(1, -3)$ **21.** $(2, -2)$ **23.** No solution **25.** $\left(\frac{39}{11}, -\frac{1}{11}\right)$ **27.** $\left(\frac{1}{2}, \frac{3}{4}\right)$ **29.** Infinitely many solutions; $(x, 3x - 5)$ or $\left(\frac{1}{3}y + \frac{5}{3}, y\right)$ **31.** $(1, 3)$; consistent, independent **33.** $(-4, -2)$; consistent, independent **35.** Infinitely many solutions; $(4y + 2, y)$ or $\left(x, \frac{1}{4}x - \frac{1}{2}\right)$; consistent, dependent **37.** No solution; inconsistent, independent **39.** $(1, 1)$; consistent, independent **41.** $(-3, 0)$; consistent, independent **43.** $(10, 8)$; consistent, independent **45.** True **47.** False **49.** True **51.** Liposuction: 363,912 surgeries; breast augmentation: 313,327 surgeries **53.** Boston: \$1881; San Francisco: \$3023 **55.** Wisconsin: 14,957 Amish; Ohio: 59,103 Amish **57.** Standard: 76 packages; express: 44 packages **59.** 4%: \$6000; 5%: \$9000 **61.** Embroidered floral scarves: 16; sheer chevron scarves: 23 **63.** 1.5 servings of spaghetti, 2 servings of lettuce **65.** Boat: 20 km/h; stream: 3 km/h **67.** 2 hr **69.** (15, \$200) **71.** 140 **73.** 6000 **75.** 115,017 registered snowmobiles **76.** About 22,960 adoptions **77.** $-2, 6$ **78.** $-1, 5$ **79.** 15 **80.** 1, 3 **81.** City: 66 mi; highway: 248 mi **83.** First train: 36 km/h; second train: 54 km/h **85.** $A = \frac{1}{10}, B = -\frac{7}{10}$

Exercise Set 6.2

1. $(3, -2, 1)$ **3.** $(-3, 2, 1)$ **5.** $\left(2, \frac{1}{2}, -2\right)$ **7.** No solution **9.** Infinitely many solutions; $\left(\dfrac{11y + 19}{5}, y, \dfrac{9y + 11}{5}\right)$ **11.** $\left(\frac{1}{2}, \frac{2}{3}, -\frac{5}{6}\right)$ **13.** $(-1, 4, 3)$ **15.** $(1, -2, 4, -1)$ **17.** Russian Federation: 80 medals; Ukraine: 25 medals; United States: 18 medals **19.** In a restaurant: 78 meals; in a car: 34 meals; at home: 58 meals **21.** Bacon: \$142.4 million; Warhol: \$105.4 million; Koons: \$58.4 million **23.** Dogs: 69.9 million; cats: 74.1 million; birds: 8.3 million **25.** United States: \$31.5 billion; Great Britain: \$17.9 billion; Germany: \$14.1 billion **27.** $1\frac{1}{4}$ servings of beef, 1 baked potato, $\frac{3}{4}$ serving of strawberries **29.** 2%: \$300; 3%: \$800; 4%: \$2400 **31.** Orange juice: \$2.00; bagel: \$2.50; coffee: \$1.80 **33.** **(a)** $f(x) = \frac{3}{40}x^2 - \frac{7}{4}x + 43$; **(b)** 2007: 33.625% 2014: 40.8% **35.** **(a)** $f(x) = -7x^2 + 55x + 291$; **(b)** 333,000 deportations **37.** Perpendicular **38.** The leading-term test **39.** Vertical line **40.** One-to-one function **41.** Rational function **42.** Inverse variation **43.** Vertical asymptote **44.** Horizontal asymptote **45.** $\left(-1, \frac{1}{5}, -\frac{1}{2}\right)$ **47.** 180° **49.** $3x + 4y + 2z = 12$ **51.** $y = -4x^3 + 5x^2 - 3x + 1$ **53.** Adults: 5; students: 1; children: 94

Exercise Set 6.3

1. 3×2 **3.** 1×4 **5.** 3×3 **7.** $\left[\begin{array}{cc|c} 2 & -1 & 7 \\ 1 & 4 & -5 \end{array}\right]$

9. $\left[\begin{array}{ccc|c} 1 & -2 & 3 & 12 \\ 2 & 0 & -4 & 8 \\ 0 & 3 & 1 & 7 \end{array}\right]$ **11.** $\begin{aligned} 3x - 5y &= 1, \\ x + 4y &= -2 \end{aligned}$

13. $\begin{aligned} 2x + y - 4z &= 12, \\ 3x + 5z &= -1, \\ x - y + z &= 2 \end{aligned}$

15. $\left(\frac{3}{2}, \frac{5}{2}\right)$ **17.** $\left(-\frac{63}{29}, -\frac{114}{29}\right)$ **19.** $\left(-1, \frac{5}{2}\right)$ **21.** $(0, 3)$ **23.** No solution **25.** Infinitely many solutions; $(3y - 2, y)$ **27.** $(-1, 2, -2)$ **29.** $\left(\frac{3}{2}, -4, 3\right)$ **31.** $(-1, 6, 3)$ **33.** Infinitely many solutions; $\left(\frac{1}{2}z + \frac{1}{2}, -\frac{1}{2}z - \frac{1}{2}, z\right)$ **35.** Infinitely many solutions; $(r - 2, -2r + 3, r)$ **37.** No solution **39.** $(1, -3, -2, -1)$ **41.** 8%: \$8000; 10%: \$12,000; 12%: \$10,000 **43.** 49¢: 160 stamps; 21¢: 40 stamps **45.** Exponential **46.** Linear **47.** Rational **48.** Quartic **49.** Logarithmic **50.** Cubic **51.** Linear **52.** Quadratic **53.** $y = 3x^2 + \frac{5}{2}x - \frac{15}{2}$ **55.** $\begin{bmatrix} 1 & 5 \\ 0 & 1 \end{bmatrix}, \begin{bmatrix} 1 & 0 \\ 0 & 1 \end{bmatrix}$ **57.** $\left(-\frac{4}{3}, -\frac{1}{3}, 1\right)$ **59.** Infinitely many solutions; $\left(-\frac{14}{13}z - 1, \frac{3}{13}z - 2, z\right)$ **61.** $(-3, 3)$

Exercise Set 6.4

1. $x = -3, y = 5$ **3.** $x = -1, y = 1$ **5.** $\begin{bmatrix} -2 & 7 \\ 6 & 2 \end{bmatrix}$

7. $\begin{bmatrix} 1 & 3 \\ 2 & 6 \end{bmatrix}$ **9.** $\begin{bmatrix} 9 & 9 \\ -3 & -3 \end{bmatrix}$ **11.** $\begin{bmatrix} 11 & 13 \\ 5 & 3 \end{bmatrix}$

13. $\begin{bmatrix} -4 & 3 \\ -2 & -4 \end{bmatrix}$ **15.** $\begin{bmatrix} 17 & 9 \\ -2 & 1 \end{bmatrix}$ **17.** $\begin{bmatrix} 0 & 0 \\ 0 & 0 \end{bmatrix}$

19. $\begin{bmatrix} 1 & 2 \\ 4 & 3 \end{bmatrix}$ **21.** $\begin{bmatrix} 1 \\ 40 \end{bmatrix}$ **23.** $\begin{bmatrix} -10 & 28 \\ 14 & -26 \\ 0 & -6 \end{bmatrix}$

25. Not defined **27.** $\begin{bmatrix} 3 & 16 & 3 \\ 0 & -32 & 0 \\ -6 & 4 & 5 \end{bmatrix}$

29. **(a)** $[40 \quad 20 \quad 30]$; **(b)** $[44 \quad 22 \quad 33]$; **(c)** $[84 \quad 42 \quad 63]$; the total amount of each type of produce ordered for both weeks

31. **(a)** $\mathbf{C} = [140 \quad 27 \quad 3 \quad 13 \quad 64]$, $\mathbf{P} = [180 \quad 4 \quad 11 \quad 24 \quad 662]$, $\mathbf{B} = [50 \quad 5 \quad 1 \quad 82 \quad 20]$; **(b)** $[650 \quad 50 \quad 28 \quad 307 \quad 1448]$; the total nutritional value of a meal of 1 serving of chicken, 1 cup of potato salad, and 3 broccoli spears

33. **(a)** $\begin{bmatrix} 1.50 & 0.30 & 0.36 & 0.45 & 0.64 \\ 1.55 & 0.28 & 0.48 & 0.57 & 0.75 \\ 1.62 & 0.52 & 0.65 & 0.38 & 0.53 \\ 1.70 & 0.43 & 0.40 & 0.42 & 0.68 \end{bmatrix}$;
(b) $[65 \quad 48 \quad 93 \quad 57]$;
(c) $[419.46 \quad 105.81 \quad 129.69 \quad 115.89 \quad 165.65]$;
(d) the total cost, in dollars, for each item for the day's meals

35. **(a)** $\begin{bmatrix} 8 & 15 \\ 6 & 10 \\ 4 & 3 \end{bmatrix}$; **(b)** $[4 \quad 2.50 \quad 3]$; **(c)** $[59 \quad 94]$;
(d) the total cost, in dollars, of ingredients for each coffee shop

37. **(a)** $[7.50 \quad 4.80 \quad 6.25]$; **(b)** $\mathbf{PS} = [113.80 \quad 179.25]$

39. $\begin{bmatrix} 2 & -3 \\ 1 & 5 \end{bmatrix} \begin{bmatrix} x \\ y \end{bmatrix} = \begin{bmatrix} 7 \\ -6 \end{bmatrix}$

41. $\begin{bmatrix} 1 & 1 & -2 \\ 3 & -1 & 1 \\ 2 & 5 & -3 \end{bmatrix} \begin{bmatrix} x \\ y \\ z \end{bmatrix} = \begin{bmatrix} 6 \\ 7 \\ 8 \end{bmatrix}$

43. $\begin{bmatrix} 3 & -2 & 4 \\ 2 & 1 & -5 \end{bmatrix} \begin{bmatrix} x \\ y \\ z \end{bmatrix} = \begin{bmatrix} 17 \\ 13 \end{bmatrix}$

45. $\begin{bmatrix} -4 & 1 & -1 & 2 \\ 1 & 2 & -1 & -1 \\ -1 & 1 & 4 & -3 \\ 2 & 3 & 5 & -7 \end{bmatrix} \begin{bmatrix} w \\ x \\ y \\ z \end{bmatrix} = \begin{bmatrix} 12 \\ 0 \\ 1 \\ 9 \end{bmatrix}$

47. **(a)** $\left(\frac{1}{2}, -\frac{25}{4}\right)$;
(b) $x = \frac{1}{2}$;
(c) minimum: $-\frac{25}{4}$;
(d)

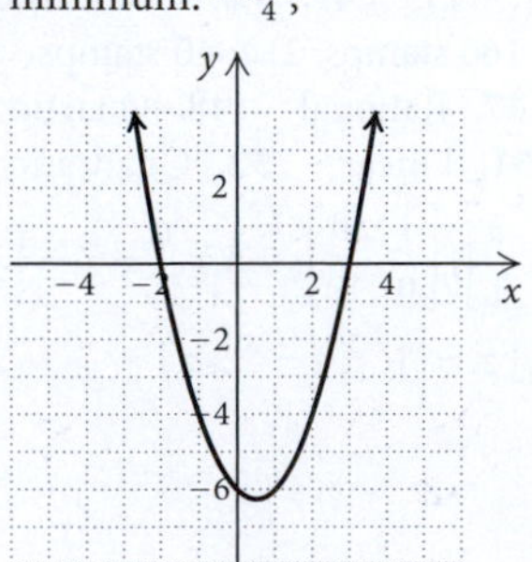

$f(x) = x^2 - x - 6$

48. **(a)** $\left(\frac{5}{4}, -\frac{49}{8}\right)$;
(b) $x = \frac{5}{4}$;
(c) minimum: $-\frac{49}{8}$;
(d)

$f(x) = 2x^2 - 5x - 3$

49. **(a)** $\left(-\frac{3}{2}, \frac{17}{4}\right)$;
(b) $x = -\frac{3}{2}$;
(c) maximum: $\frac{17}{4}$;
(d)

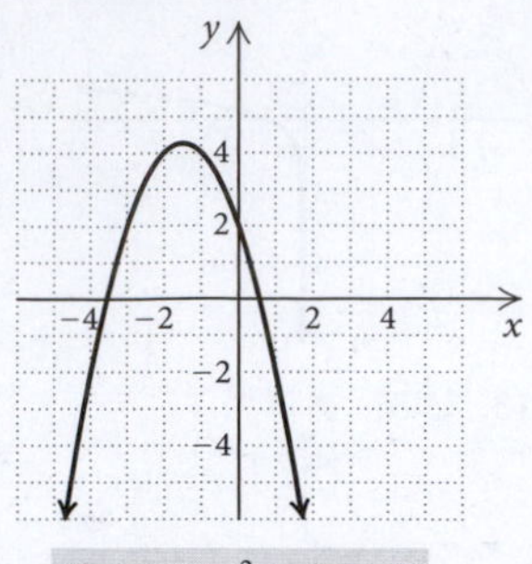

$f(x) = -x^2 - 3x + 2$

50. **(a)** $\left(\frac{2}{3}, \frac{16}{3}\right)$;
(b) $x = \frac{2}{3}$;
(c) maximum: $\frac{16}{3}$;
(d)

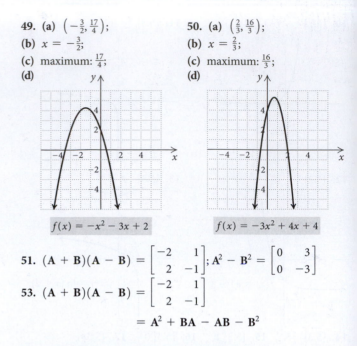

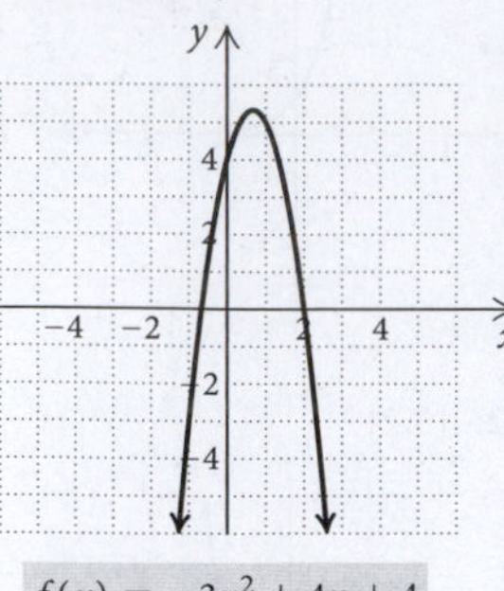

$f(x) = -3x^2 + 4x + 4$

51. $(\mathbf{A} + \mathbf{B})(\mathbf{A} - \mathbf{B}) = \begin{bmatrix} -2 & 1 \\ 2 & -1 \end{bmatrix}$; $\mathbf{A}^2 - \mathbf{B}^2 = \begin{bmatrix} 0 & 3 \\ 0 & -3 \end{bmatrix}$

53. $(\mathbf{A} + \mathbf{B})(\mathbf{A} - \mathbf{B}) = \begin{bmatrix} -2 & 1 \\ 2 & -1 \end{bmatrix}$
$= \mathbf{A}^2 + \mathbf{BA} - \mathbf{AB} - \mathbf{B}^2$

Mid-Chapter Mixed Review: Chapter 6

1. False **2.** True **3.** True **4.** False **5.** $(-3, 2)$
6. No solution **7.** $(1, -2)$ **8.** Infinitely many solutions; $\left(x, \dfrac{x-1}{3}\right)$ or $(3y + 1, y)$ **9.** $\left(\frac{1}{3}, -\frac{1}{6}, \frac{4}{3}\right)$

10. Under 10 lb: 60 packages; 10 lb up to 15 lb: 70 packages; 15 lb or more: 20 packages **11.** $(4, -3)$ **12.** $(-3, 2, -1)$

13. $\begin{bmatrix} 1 & 5 \\ 6 & 1 \end{bmatrix}$ **14.** $\begin{bmatrix} -5 & 7 \\ -4 & -7 \end{bmatrix}$ **15.** $\begin{bmatrix} -8 & 12 & 0 \\ 4 & -4 & 8 \\ -12 & 16 & 4 \end{bmatrix}$

16. $\begin{bmatrix} 0 & 16 \\ 13 & -1 \end{bmatrix}$ **17.** $\begin{bmatrix} -7 & 21 \\ -6 & 18 \end{bmatrix}$ **18.** $\begin{bmatrix} 24 & 26 \\ -12 & -13 \end{bmatrix}$

19. $\begin{bmatrix} 20 & 16 & -10 \\ -10 & -8 & 5 \end{bmatrix}$ **20.** Not defined

21. $\begin{bmatrix} 2 & -1 & 3 \\ 1 & 2 & -1 \\ 3 & -4 & 2 \end{bmatrix} \begin{bmatrix} x \\ y \\ z \end{bmatrix} = \begin{bmatrix} 7 \\ 3 \\ 5 \end{bmatrix}$

22. When a variable is not alone on one side of an equation or when solving for a variable is difficult or produces an expression containing fractions, the elimination method is preferable to the substitution method. **23.** Add a nonzero multiple of one equation to a nonzero multiple of the other equation, where the multiples are not opposites. **24.** The last row of the matrix corresponds to the equation $0 = 0$, which is true for all values of x, y, and z. Therefore, the equations are dependent.

25. No; for example, let $A = \begin{bmatrix} 1 & -1 \\ -1 & 1 \end{bmatrix}$ and $B = \begin{bmatrix} 1 & 1 \\ 1 & 1 \end{bmatrix}$; then $AB = \begin{bmatrix} 0 & 0 \\ 0 & 0 \end{bmatrix}$ and neither A nor B is $\begin{bmatrix} 0 & 0 \\ 0 & 0 \end{bmatrix}$.

Exercise Set 6.5

1. Yes **3.** No **5.** $\begin{bmatrix} -3 & 2 \\ 5 & -3 \end{bmatrix}$ **7.** Does not exist

9. $\begin{bmatrix} \frac{2}{5} & -\frac{3}{5} \\ \frac{1}{5} & -\frac{4}{5} \end{bmatrix}$ **11.** $\begin{bmatrix} \frac{3}{8} & -\frac{1}{4} & \frac{1}{8} \\ -\frac{1}{8} & \frac{3}{4} & -\frac{3}{8} \\ -\frac{1}{4} & \frac{1}{2} & \frac{1}{4} \end{bmatrix}$ **13.** Does not exist

15. $\begin{bmatrix} -1 & -1 & -6 \\ 1 & 0 & 2 \\ 0 & 1 & 3 \end{bmatrix}$ **17.** $\begin{bmatrix} 1 & 1 & 2 \\ 1 & 1 & 1 \\ 2 & 3 & 4 \end{bmatrix}$

19. Does not exist **21.** $\begin{bmatrix} 1 & -2 & 3 & 8 \\ 0 & 1 & -3 & 1 \\ 0 & 0 & 1 & -2 \\ 0 & 0 & 0 & -1 \end{bmatrix}$

23. $\begin{bmatrix} 0.25 & 0.25 & 1.25 & -0.25 \\ 0.5 & 1.25 & 1.75 & -1 \\ -0.25 & -0.25 & -0.75 & 0.75 \\ 0.25 & 0.5 & 0.75 & -0.5 \end{bmatrix}$ **25.** $(-23, 83)$

27. $(-1, 5, 1)$ **29.** $(2, -2)$ **31.** $(0, 2)$ **33.** $(3, -3, -2)$ **35.** $(-1, 0, 1)$ **37.** $(1, -1, 0, 1)$ **39.** Wisconsin: 450 lb; Massachusetts: 210 lb **41.** Topsoil: \$239; mulch: \$179; pea gravel: \$222 **43.** -48 **44.** 194 **45.** $\frac{-1 \pm \sqrt{57}}{4}$ **46.** $-3, -2$ **47.** 4 **48.** 9 **49.** $(x + 2)(x - 1)(x - 4)$ **50.** $(x + 5)(x + 1)(x - 1)(x - 3)$

51. $\mathbf{A}^{-1}$ exists if and only if $x \neq 0$. $\mathbf{A}^{-1} = \begin{bmatrix} \frac{1}{x} \end{bmatrix}$

53. $\mathbf{A}^{-1}$ exists if and only if $xyz \neq 0$. $\mathbf{A}^{-1} = \begin{bmatrix} 0 & 0 & \frac{1}{z} \\ 0 & \frac{1}{y} & 0 \\ \frac{1}{x} & 0 & 0 \end{bmatrix}$

Exercise Set 6.6

1. -14 **3.** -2 **5.** -11 **7.** $x^3 - 4x$ **9.** $M_{11} = 6, M_{32} = -9, M_{22} = -29$ **11.** $A_{11} = 6, A_{32} = 9, A_{22} = -29$ **13.** -10 **15.** -10 **17.** $M_{12} = 32, M_{44} = 7$ **19.** $A_{22} = -10, A_{34} = 1$ **21.** 110 **23.** -109 **25.** $-x^4 + x^2 - 5x$ **27.** $\left(-\frac{25}{2}, -\frac{11}{2}\right)$ **29.** $(3, 1)$ **31.** $\left(\frac{1}{2}, -\frac{1}{3}\right)$ **33.** $(1, 1)$ **35.** $\left(\frac{3}{2}, \frac{13}{14}, \frac{33}{14}\right)$ **37.** $(3, -2, 1)$ **39.** $(1, 3, -2)$ **41.** $\left(\frac{1}{2}, \frac{2}{3}, -\frac{5}{6}\right)$ **43.** $f^{-1}(x) = \frac{x - 2}{3}$ **44.** Not one-to-one **45.** Not one-to-one **46.** $f^{-1}(x) = (x - 1)^3$ **47.** $5 - 3i$ **48.** $6 - 2i$ **49.** $10 - 10i$ **50.** $\frac{9}{25} + \frac{13}{25}i$ **51.** $3, -2$ **53.** 4

55. Answers may vary.
$\begin{vmatrix} a & b \\ -b & a \end{vmatrix}$

657. Answers may vary.
$\begin{vmatrix} 2\pi r & 2\pi r \\ -h & r \end{vmatrix}$

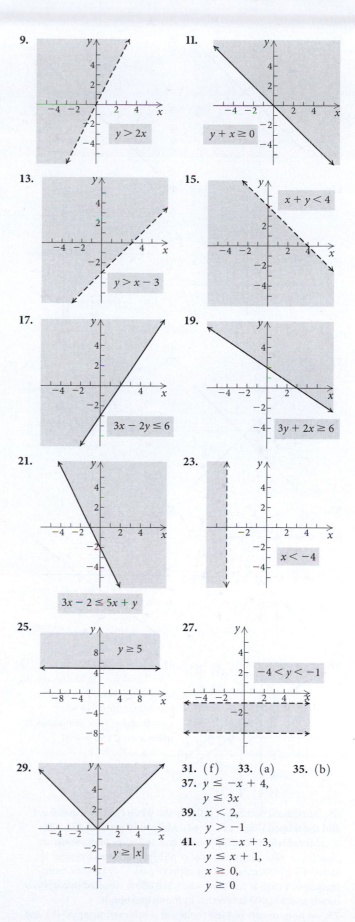

Exercise Set 6.7

1. (f) **3.** (h) **5.** (g) **7.** (b)

9. $y > 2x$

11. $y + x \geq 0$

13. $y > x - 3$

15.

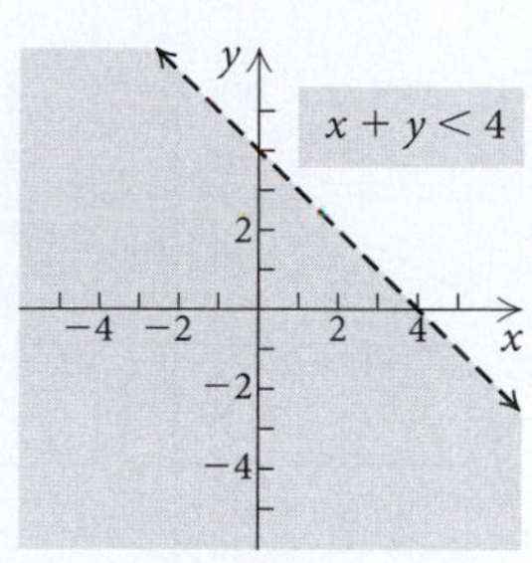

17.

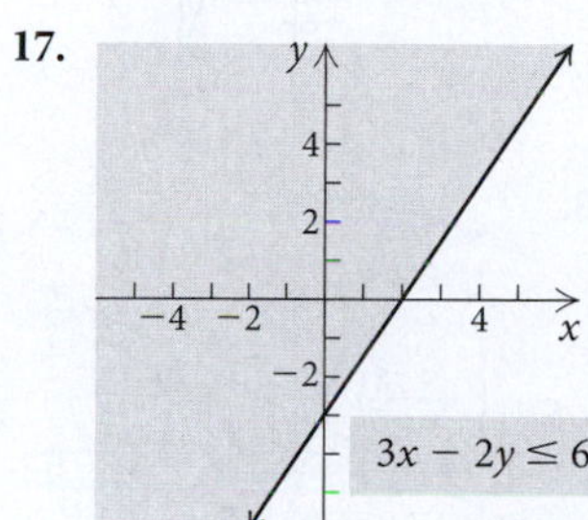

19.

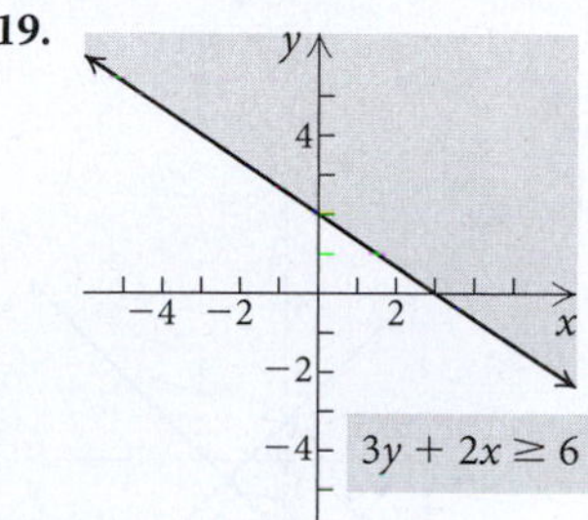

21.

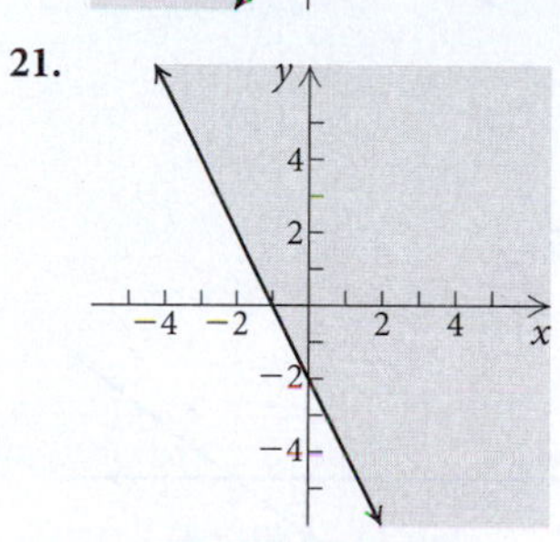

23.

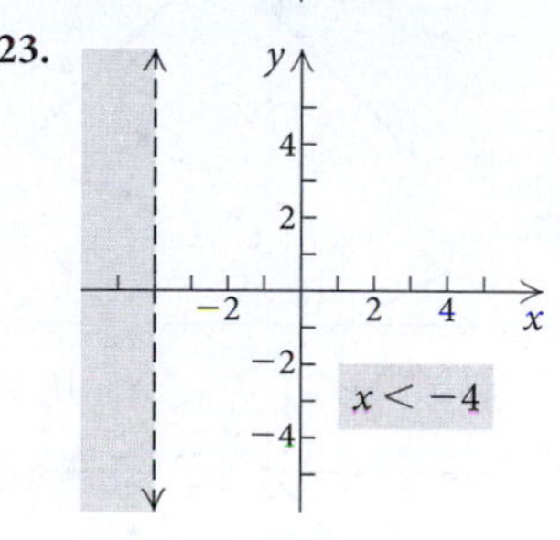

25.

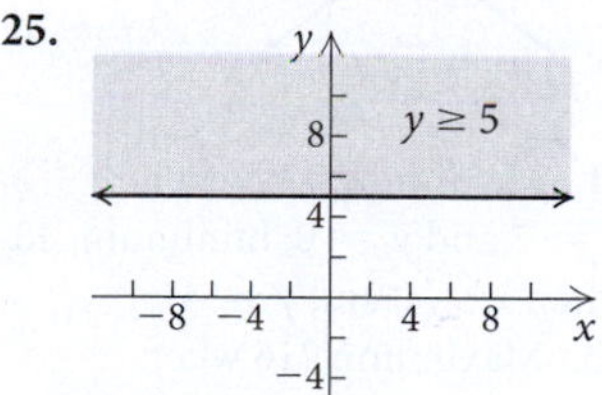

27.

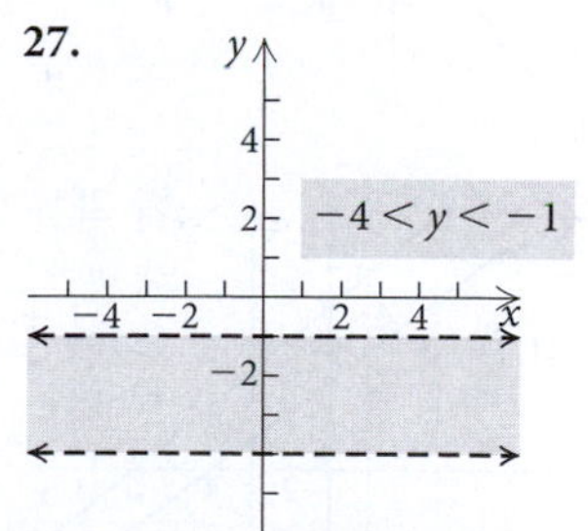

29.

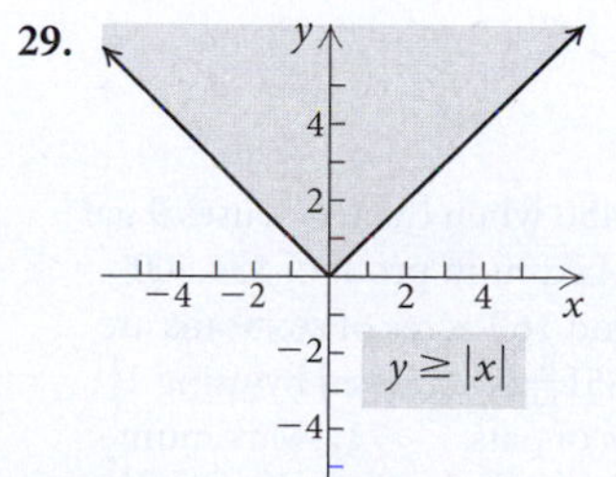

31. (f) **33.** (a) **35.** (b)

37. $y \leq -x + 4,$
$y \leq 3x$

39. $x < 2,$
$y > -1$

41. $y \leq -x + 3,$
$y \leq x + 1,$
$x \geq 0,$
$y \geq 0$

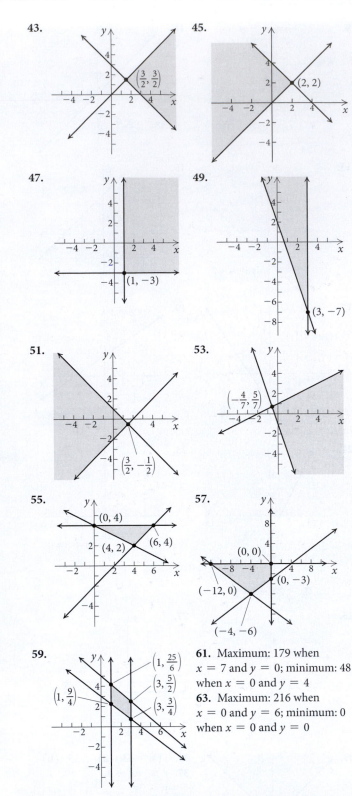

43.

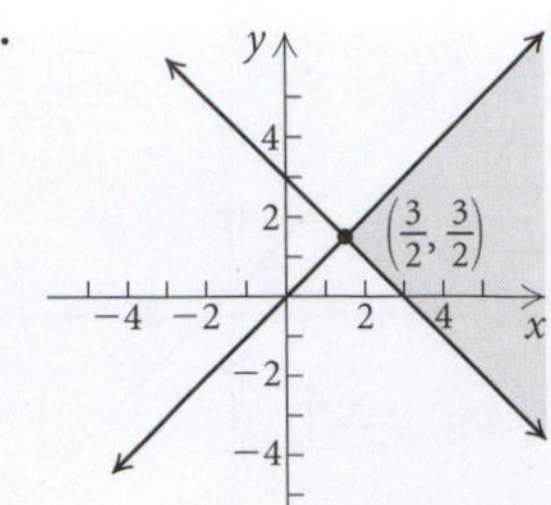

45.

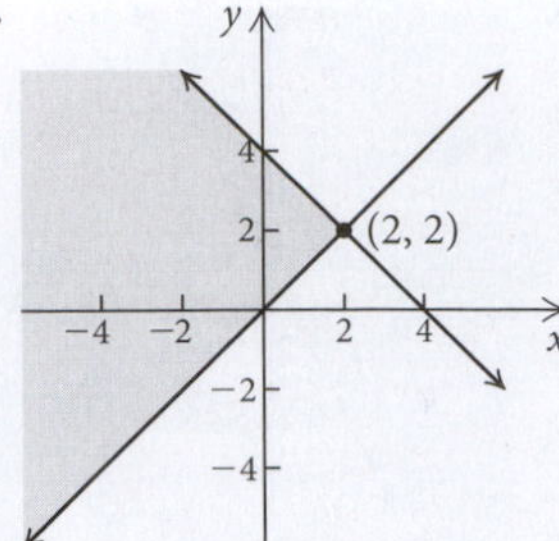

47.

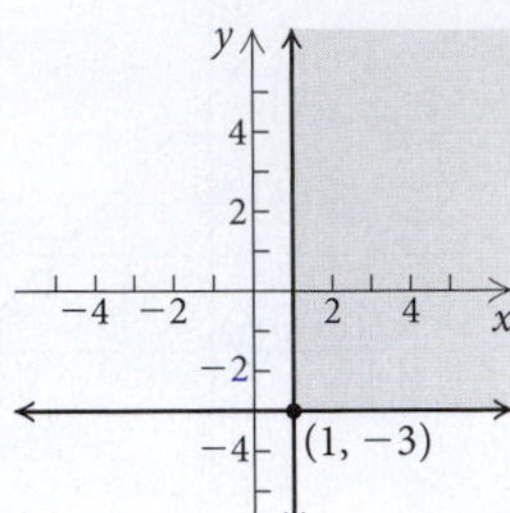

49.

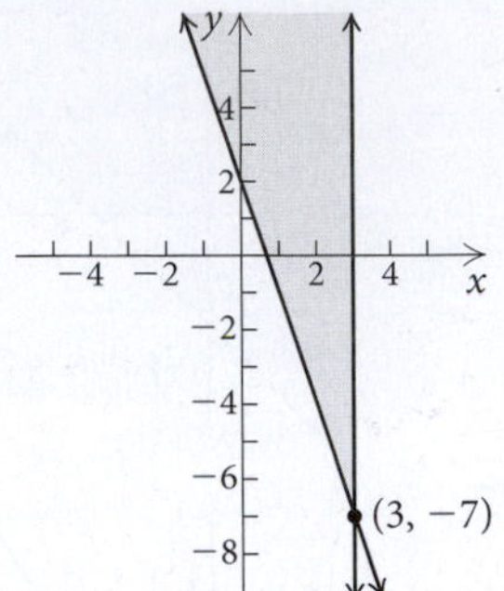

51.

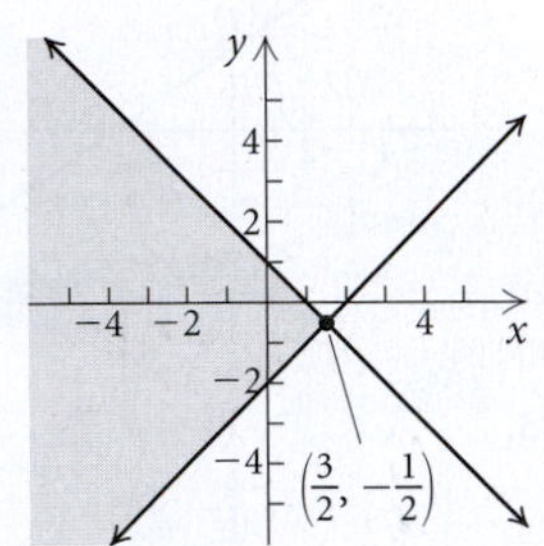

53.

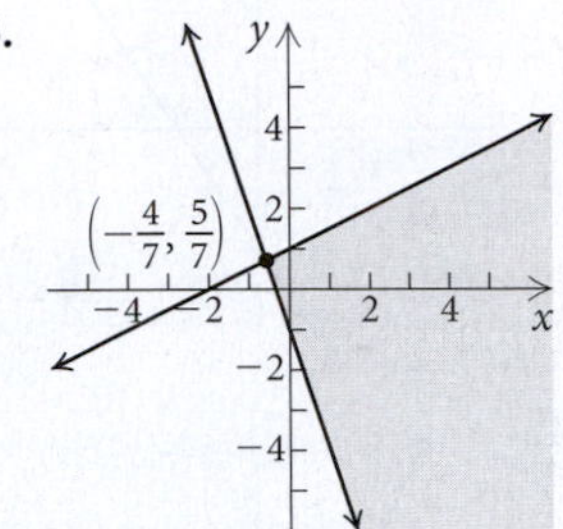

55.

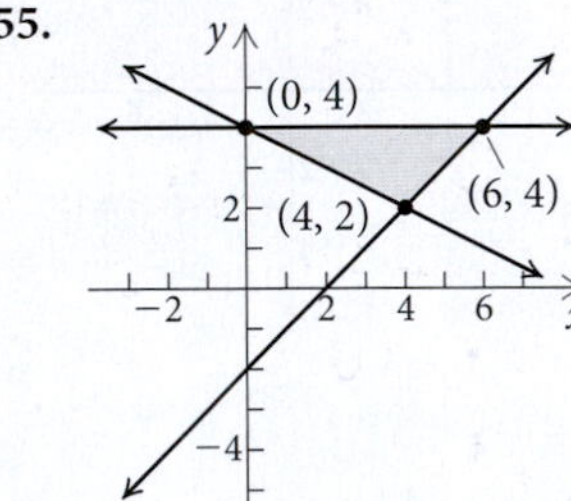

57.

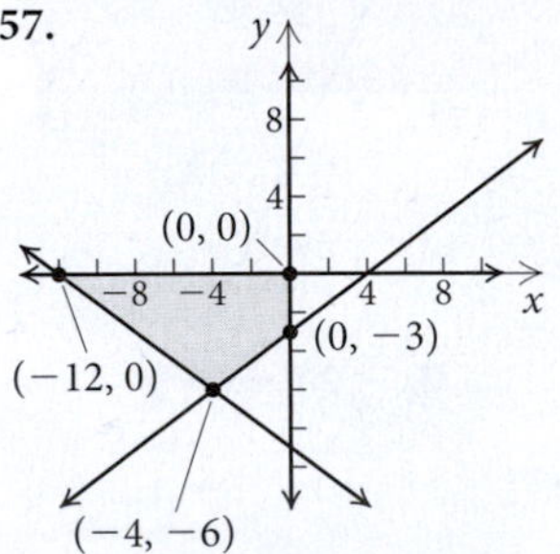

59.

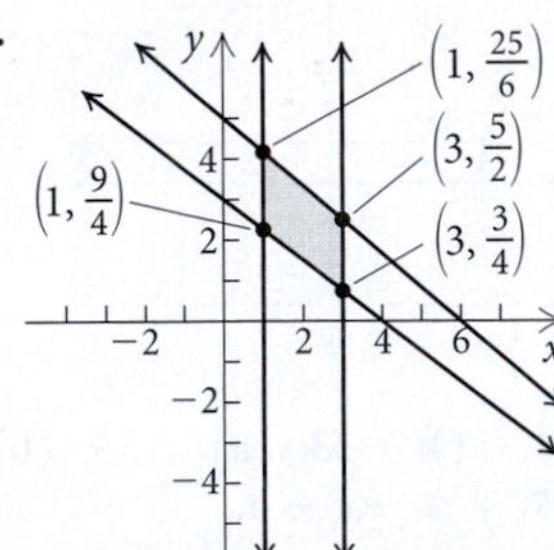

61. Maximum: 179 when $x = 7$ and $y = 0$; minimum: 48 when $x = 0$ and $y = 4$ **63.** Maximum: 216 when $x = 0$ and $y = 6$; minimum: 0 when $x = 0$ and $y = 0$

65. Maximum number of miles is 480 when the truck uses 9 gal and the moped uses 3 gal. **67.** Maximum profit of \$54,800 is achieved when 80 acres of corn and 160 acres of soybeans are planted. **69.** Minimum cost of \$51$\frac{9}{13}$ is achieved by using 1$\frac{11}{13}$ sacks of soybean meal and 1$\frac{11}{13}$ sacks of oats. **71.** Maximum income of \$1575 is achieved when \$10,000 is invested in corporate bonds and \$30,000 is invested in municipal bonds. **73.** Minimum cost of \$460 thousand is achieved using 30 P_1's and 10 P_2's. **75.** Maximum profit per week of \$2210 is achieved when 5 silk organza dresses and 2 lace dresses are made. **77.** Minimum weekly cost of \$19.05 is achieved when 1.5 lb of meat and 3 lb of cheese are used. **79.** Maximum total number of 800 is achieved when there are 550 of A and 250 of B.

81. $\{x \mid -7 \le x < 2\}$, or $[-7, 2)$
82. $\{x \mid x \le 1 \text{ or } x \ge 5\}$, or $(-\infty, 1] \cup [5, \infty)$
83. $\{x \mid -1 \le x \le 3\}$, or $[-1, 3]$
84. $\{x \mid -3 < x < -2\}$, or $(-3, -2)$

85.

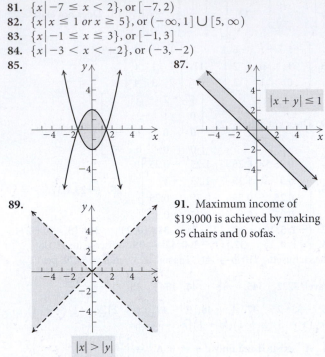

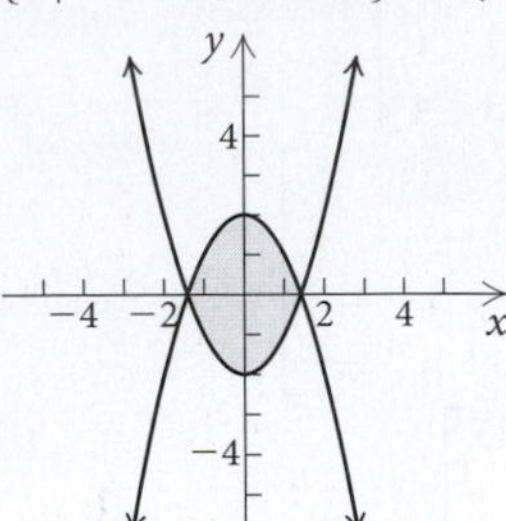

87.

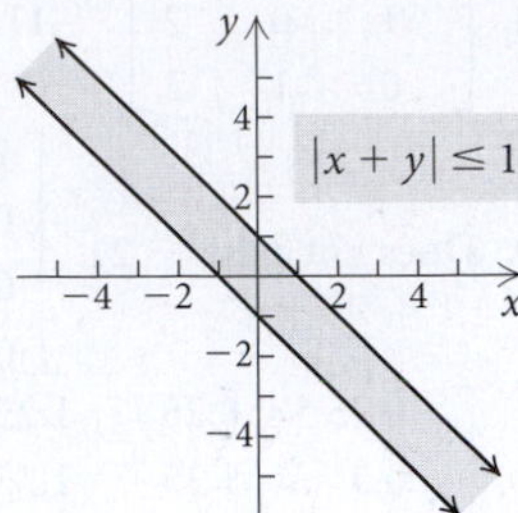

89.

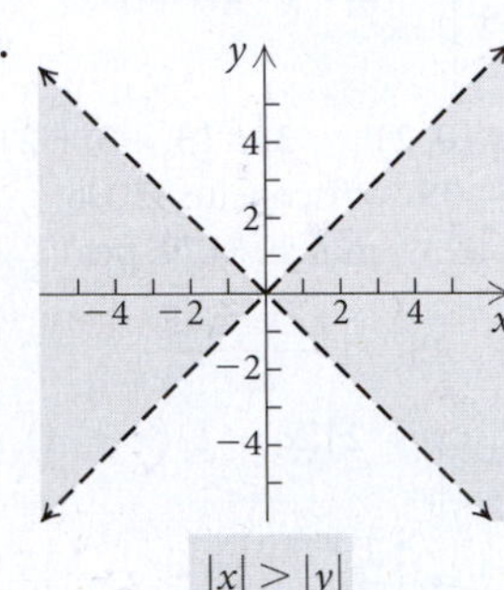

91. Maximum income of \$19,000 is achieved by making 95 chairs and 0 sofas.

Exercise Set 6.8

1. $\frac{2}{x-3} - \frac{1}{x+2}$ **3.** $\frac{5}{2x-1} - \frac{4}{3x-1}$

5. $\frac{2}{x+2} - \frac{3}{x-2} + \frac{4}{x+1}$

7. $-\frac{3}{(x+2)^2} - \frac{1}{x+2} + \frac{1}{x-1}$ **9.** $\frac{3}{x-1} - \frac{4}{2x-1}$

11. $x - 2 + \frac{\frac{17}{16}}{x+1} - \frac{\frac{11}{4}}{(x+1)^2} - \frac{\frac{17}{16}}{x-3}$

13. $\frac{3x+5}{x^2+2} - \frac{4}{x-1}$ **15.** $\frac{3}{2x-1} - \frac{2}{x+2} + \frac{10}{(x+2)^2}$

17. $3x + 1 + \frac{2}{2x-1} + \frac{3}{x+1}$

19. $-\frac{1}{x-3} + \frac{3x}{x^2+2x-5}$

21. $\frac{5}{3x+5} - \frac{3}{x+1} + \frac{4}{(x+1)^2}$ **23.** $\frac{8}{4x-5} + \frac{3}{3x+2}$

25. $\frac{2x-5}{3x^2+1} - \frac{2}{x-2}$ **27.** $-1, \pm 3i$ **28.** $3, \pm i$

29. $-2, \frac{1 \pm \sqrt{5}}{2}$ **30.** $-2, 3, \pm i$ **31.** $-3, -1 \pm \sqrt{2}$

33. $-\frac{\frac{1}{2a^2}x}{x^2+a^2} + \frac{\frac{1}{4a^2}}{x-a} + \frac{\frac{1}{4a^2}}{x+a}$

35. $-\frac{3}{25(\ln x + 2)} + \frac{3}{25(\ln x - 3)} + \frac{7}{5(\ln x - 3)^2}$

Review Exercises: Chapter 6

1. True **2.** False **3.** True **4.** False **5.** (a) **6.** (e) **7.** (h) **8.** (d) **9.** (b) **10.** (g) **11.** (c) **12.** (f) **13.** $(-2, -2)$ **14.** $(-5, 4)$ **15.** No solution **16.** Infinitely many solutions; $(-y - 2, y)$ or $(x, -x - 2)$ **17.** $(3, -1, -2)$ **18.** No solution **19.** $(-5, 13, 8, 2)$ **20.** Consistent: 13, 14, 16, 17, 19; the others are inconsistent. **21.** Dependent: 16; the others are independent. **22.** $(1, 2)$ **23.** $(-3, 4, -2)$ **24.** Infinitely many solutions; $\left(\frac{z}{2}, -\frac{z}{2}, z\right)$ **25.** $(-4, 1, -2, 3)$ **26.** Nickels: 31; dimes: 44 **27.** 3%: \$1600; 3.5%: \$3400 **28.** 1 bagel, $\frac{1}{2}$ serving of cream cheese, 2 bananas **29.** 75, 69, 82 **30. (a)** $f(x) = 1.125x^2 - 5.25x + 145$; **(b)** 154 million persons employed

31. $\begin{bmatrix} 0 & -1 & 6 \\ 3 & 1 & -2 \\ -2 & 1 & -2 \end{bmatrix}$ **32.** $\begin{bmatrix} -3 & 3 & 0 \\ -6 & -9 & 6 \\ 6 & 0 & -3 \end{bmatrix}$

33. $\begin{bmatrix} -1 & 1 & 0 \\ -2 & -3 & 2 \\ 2 & 0 & -1 \end{bmatrix}$ **34.** $\begin{bmatrix} -2 & 2 & 6 \\ 1 & -8 & 18 \\ 2 & 1 & -15 \end{bmatrix}$

35. Not defined **36.** $\begin{bmatrix} 2 & -1 & -6 \\ 1 & 5 & -2 \\ -2 & -1 & 4 \end{bmatrix}$

37. $\begin{bmatrix} -13 & 1 & 6 \\ -3 & -7 & 4 \\ 8 & 3 & -5 \end{bmatrix}$ **38.** $\begin{bmatrix} -2 & -1 & 18 \\ 5 & -3 & -2 \\ -2 & 3 & -8 \end{bmatrix}$

39. (a) $\begin{bmatrix} 2.25 & 0.38 & 0.55 & 0.33 & 0.85 \\ 3.09 & 0.42 & 0.46 & 0.48 & 0.51 \\ 2.40 & 0.31 & 0.59 & 0.36 & 0.64 \\ 1.80 & 0.29 & 0.34 & 0.55 & 0.52 \end{bmatrix}$;
(b) $[41 \quad 18 \quad 39 \quad 36]$;
(c) $[306.27 \quad 45.67 \quad 66.08 \quad 56.01 \quad 87.71]$;
(d) the total cost, in dollars, for each item for the day's meals

40. $\begin{bmatrix} -\frac{1}{2} & 0 \\ \frac{1}{6} & \frac{1}{3} \end{bmatrix}$ **41.** $\begin{bmatrix} 0 & 0 & \frac{1}{4} \\ 0 & -\frac{1}{2} & 0 \\ \frac{1}{3} & 0 & 0 \end{bmatrix}$

42. $\begin{bmatrix} 1 & 0 & 0 & 0 \\ 0 & \frac{1}{9} & \frac{5}{18} & 0 \\ 0 & -\frac{1}{9} & \frac{2}{9} & 0 \\ 0 & 0 & 0 & 1 \end{bmatrix}$ **43.** $\begin{bmatrix} 3 & -2 & 4 \\ 1 & 5 & -3 \\ 2 & -3 & 7 \end{bmatrix}\begin{bmatrix} x \\ y \\ z \end{bmatrix} = \begin{bmatrix} 13 \\ 7 \\ -8 \end{bmatrix}$

44. $(-8, 7)$ **45.** $(1, -2, 5)$ **46.** $(2, -1, 1, -3)$ **47.** 10 **48.** -18 **49.** -6 **50.** -1 **51.** $(3, -2)$ **52.** $(-1, 5)$ **53.** $\left(\frac{3}{2}, \frac{13}{14}, \frac{33}{14}\right)$ **54.** $(2, -1, 3)$

55.

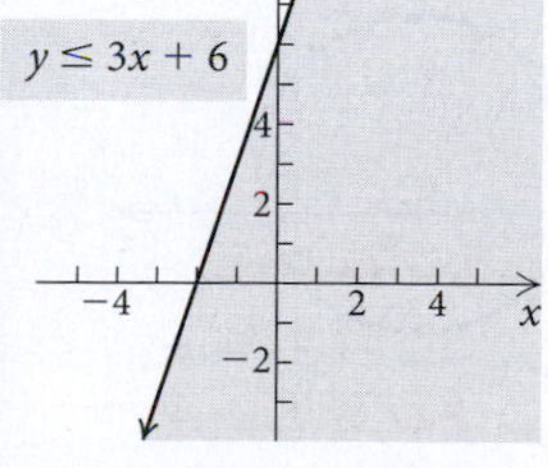

$y \le 3x + 6$

56.

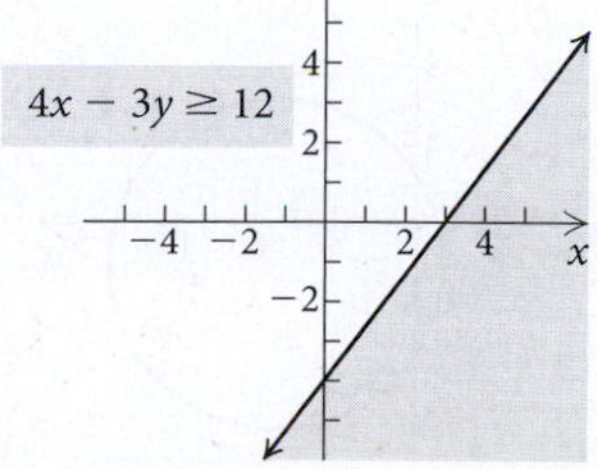

$4x - 3y \ge 12$

57.

(0, 9), (2, 5), (5, 1), (8, 0)

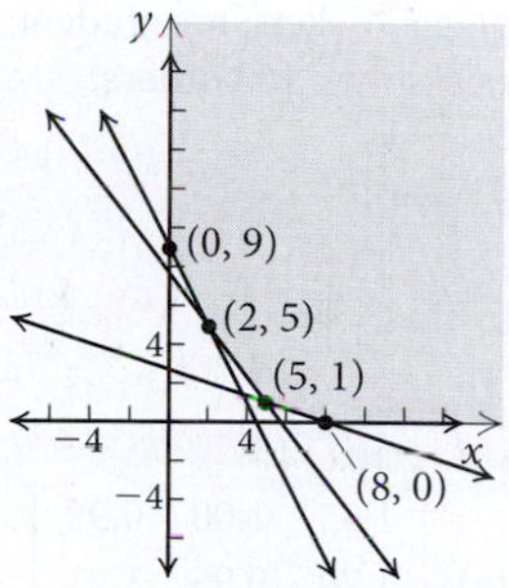

58. Minimum: 52 when $x = 2$ and $y = 4$; maximum: 92 when $x = 2$ and $y = 8$ **59.** Maximum score of 96 is achieved when 0 group A questions and 8 group B questions are answered.

60. $\frac{5}{x + 1} - \frac{5}{x + 2} - \frac{5}{(x + 2)^2}$ **61.** $\frac{2}{2x - 3} - \frac{5}{x + 4}$

62. C **63.** A **64.** B **65.** 4%: \$10,000; 5%: \$12,000; $5\frac{1}{2}$%: \$18,000 **66.** $\left(\frac{5}{18}, \frac{1}{7}\right)$ **67.** $\left(1, \frac{1}{2}, \frac{1}{3}\right)$

68.

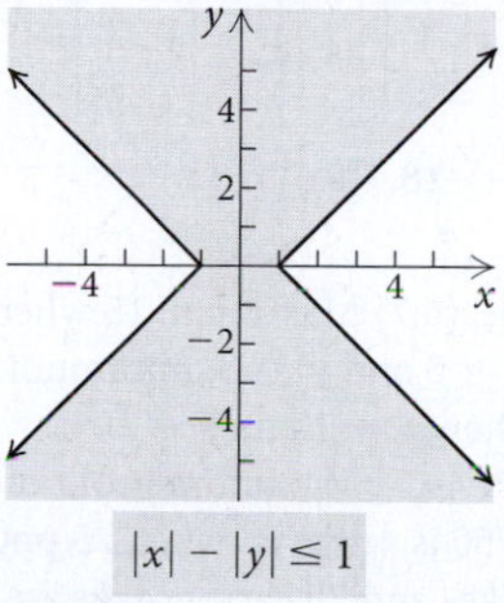

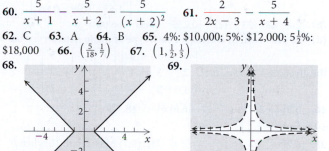

$|x| - |y| \le 1$

69.

$|xy| > 1$

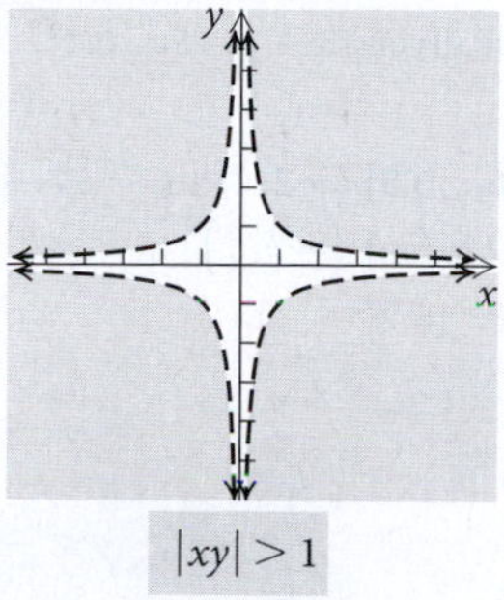

70. The solution of the equation $2x + 5 = 3x - 7$ is the first coordinate of the point of intersection of the graphs of $y_1 = 2x + 5$ and $y_2 = 3x - 7$. The solution of the system of equations $y = 2x + 5$, $y = 3x - 7$ is the ordered pair that is the point of intersection of y_1 and y_2. **71.** In general, $(\mathbf{AB})^2 \ne \mathbf{A}^2\mathbf{B}^2$. $(\mathbf{AB})^2 = \mathbf{ABAB}$ and $\mathbf{A}^2\mathbf{B}^2 = \mathbf{AABB}$. Since matrix multiplication is not commutative, $\mathbf{BA} \ne \mathbf{AB}$, so $(\mathbf{AB})^2 \ne \mathbf{A}^2\mathbf{B}^2$.

72. If $\begin{vmatrix} a_1 & b_1 \\ a_2 & b_2 \end{vmatrix} = 0$, then $a_1 = ka_2$ and $b_1 = kb_2$ for some number k. This means that the equations $a_1x + b_1y = c_1$ and $a_2x + b_2y = c_2$ are dependent if $c_1 = kc_2$, or the system is inconsistent if $c_1 \ne kc_2$. **73.** If $a_1x + b_1y = c_1$ and $a_2x + b_2y = c_2$ are parallel lines, then $a_1 = ka_2$, $b_1 = kb_2$, and $c_1 \ne kc_2$, for some number k. Then $\begin{vmatrix} a_1 & b_1 \\ a_2 & b_2 \end{vmatrix} = 0$, $\begin{vmatrix} c_1 & b_1 \\ c_2 & b_2 \end{vmatrix} \ne 0$, and $\begin{vmatrix} a_1 & c_1 \\ a_2 & c_2 \end{vmatrix} \ne 0$.

74. The graph of a linear equation consists of a set of points on a line. The graph of a linear inequality consists of the set of points in a half-plane and might also include the points on the line that is the boundary of the half-plane. **75.** The denominator of the second fraction, $x^2 - 5x + 6$, can be factored into linear factors with real coefficients: $(x - 3)(x - 2)$. Thus the given expression is not a partial fraction decomposition.

Test: Chapter 6

1. [6.1] $(-3, 5)$; consistent, independent

2. [6.1] Infinitely many solutions; $(x, 2x - 3)$ or $\left(\frac{y + 3}{2}, y\right)$; consistent, dependent **3.** [6.1] No solution; inconsistent, independent **4.** [6.1] $(1, -2)$; consistent, independent

5. [6.2] $(-1, 3, 2)$ **6.** [6.1] Student: 462 tickets; nonstudent: 158 tickets **7.** [6.2] Hui: 120 orders; Ashlyn: 104 orders; Sheriann: 128 orders **8.** [6.4] $\begin{bmatrix} -2 & -3 \\ -3 & 4 \end{bmatrix}$

9. [6.4] Not defined **10.** [6.4] $\begin{bmatrix} -7 & -13 \\ 5 & -1 \end{bmatrix}$

11. [6.4] Not defined **12.** [6.4] $\begin{bmatrix} 2 & -2 & 6 \\ -4 & 10 & 4 \end{bmatrix}$

13. [6.5] $\begin{bmatrix} 0 & -1 \\ -\frac{1}{4} & -\frac{3}{4} \end{bmatrix}$ **14.** [6.4] **(a)** $\begin{bmatrix} 1.55 & 1.00 & 0.99 \\ 1.70 & 0.95 & 1.01 \\ 1.65 & 0.99 & 0.96 \end{bmatrix}$; **(b)** $[26 \quad 18 \quad 23]$; **(c)** $[108.85 \quad 65.87 \quad 66.00]$; **(d)** the total cost, in dollars, for each type of menu item served on the given day **15.** [6.4] $\begin{bmatrix} 3 & -4 & 2 \\ 2 & 3 & 1 \\ 1 & -5 & -3 \end{bmatrix} \begin{bmatrix} x \\ y \\ z \end{bmatrix} = \begin{bmatrix} -8 \\ 7 \\ 3 \end{bmatrix}$

16. [6.5] $(-2, 1, 1)$ **17.** [6.6] 61 **18.** [6.6] -33

19. [6.6] $\left(-\frac{1}{2}, \frac{3}{4}\right)$

20. [6.7]

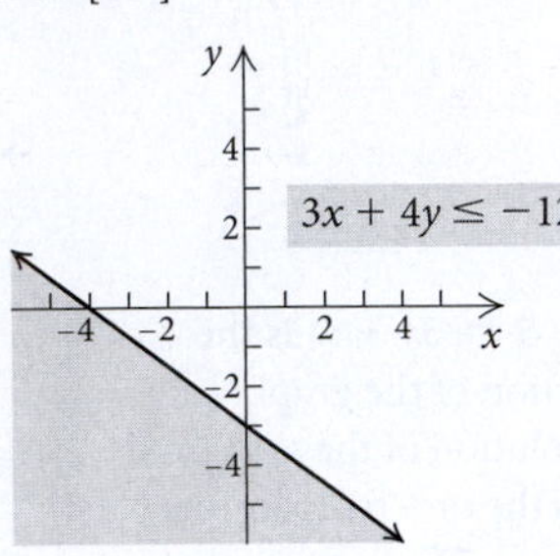

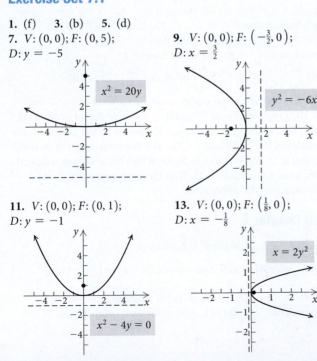

21. [6.7] Maximum: 15 when $x = 3$ and $y = 3$; minimum: 2 when $x = 1$ and $y = 0$

22. [6.7] Maximum profit of \$750 is achieved when 25 pound cakes and 75 carrot cakes are prepared.

23. [6.8] $-\dfrac{2}{x-1} + \dfrac{5}{x+3}$

24. [6.7] D

25. [6.2] $A = 1, B = -3, C = 2$

Chapter 7

Exercise Set 7.1

1. (f) **3.** (b) **5.** (d)

7. $V: (0, 0)$; $F: (0, 5)$; $D: y = -5$

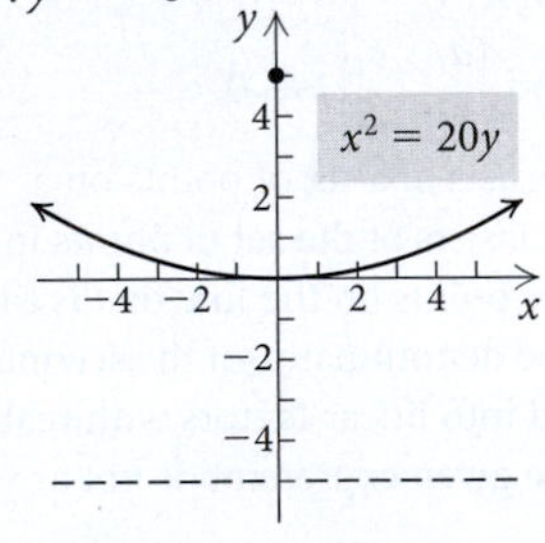

9. $V: (0, 0)$; $F: \left(-\frac{3}{2}, 0\right)$; $D: x = \frac{3}{2}$

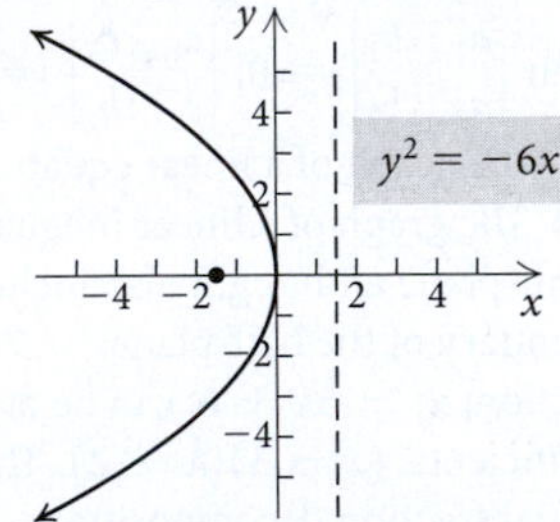

11. $V: (0, 0)$; $F: (0, 1)$; $D: y = -1$

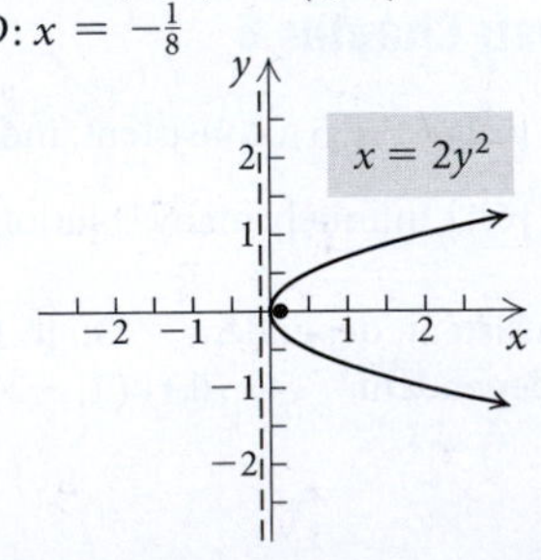

13. $V: (0, 0)$; $F: \left(\frac{1}{8}, 0\right)$; $D: x = -\frac{1}{8}$

15. $y^2 = -12x$ **17.** $y^2 = 28x$ **19.** $x^2 = -4\pi y$

21. $(y-2)^2 = 14\left(x + \frac{1}{2}\right)$

23. $V: (-2, 1)$; $F: \left(-2, -\frac{1}{2}\right)$; $D: y = \frac{5}{2}$

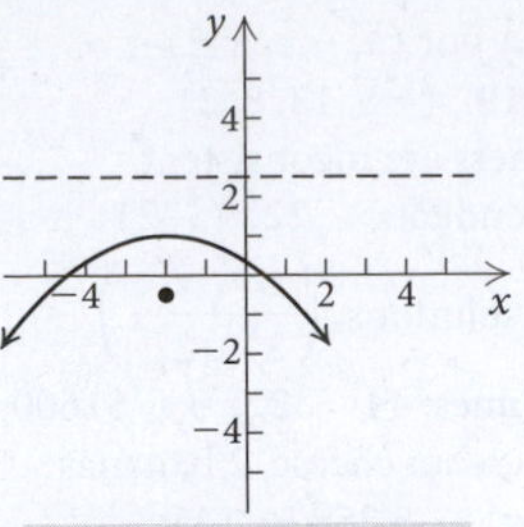

25. $V: (-1, -3)$; $F: \left(-1, -\frac{7}{2}\right)$; $D: y = -\frac{5}{2}$

27. $V: (0, -2)$; $F: \left(0, -1\frac{3}{4}\right)$; $D: y = -2\frac{1}{4}$

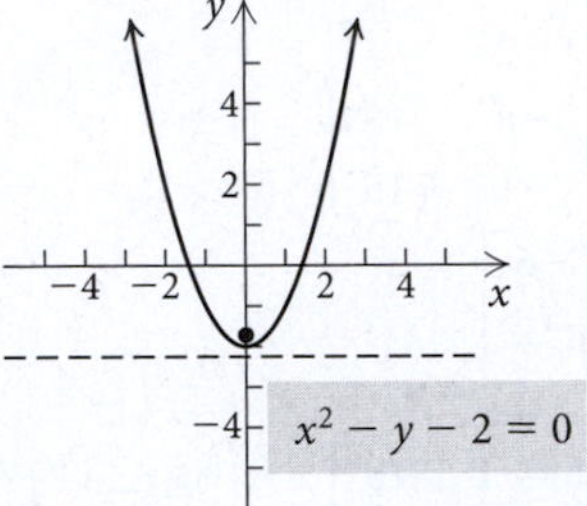

29. $V: (-2, -1)$; $F: \left(-2, -\frac{3}{4}\right)$; $D: y = -1\frac{1}{4}$

31. $V: \left(5\frac{3}{4}, \frac{1}{2}\right)$; $F: \left(6, \frac{1}{2}\right)$; $D: x = 5\frac{1}{2}$

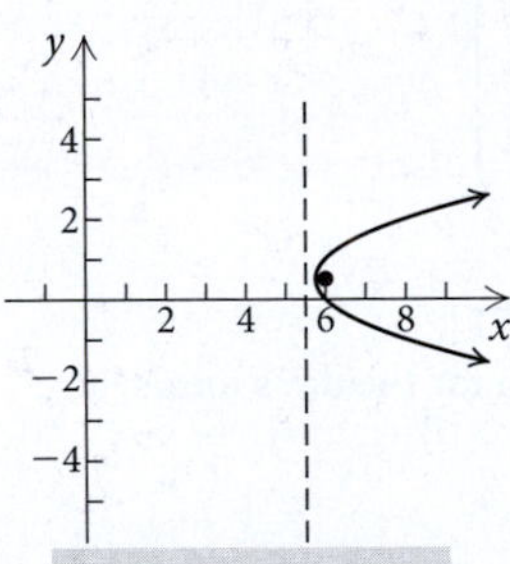

33. (a) $y^2 = 16x$; **(b)** $3\frac{33}{64}$ ft

35. $\frac{2}{3}$ ft, or 8 in. **37.** (h)

38. (d) **39.** (a), (b), (f), (g)

40. (b) **41.** (b) **42.** (f)

43. (a) and (g)

44. (a) and (h); (g) and (h); (b) and (c)

45. $(x+1)^2 = -4(y-2)$

47. 10 ft, 11.6 ft, 16.4 ft, 24.4 ft, 35.6 ft, 50 ft

Exercise Set 7.2

1. (b) **3.** (d) **5.** (a)

7. $(7, -2)$; 8

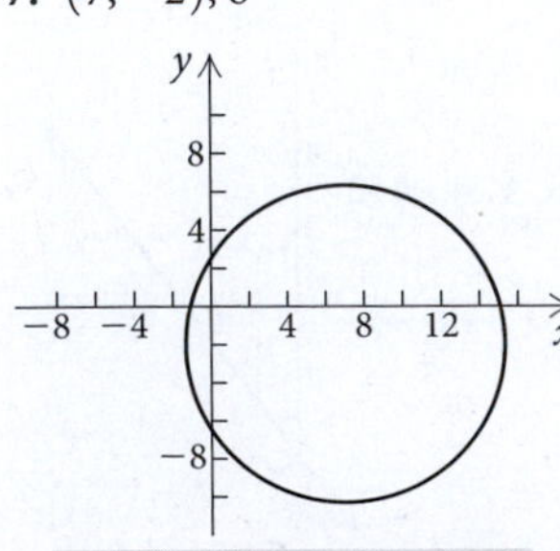

9. $(-3, 1)$; 4

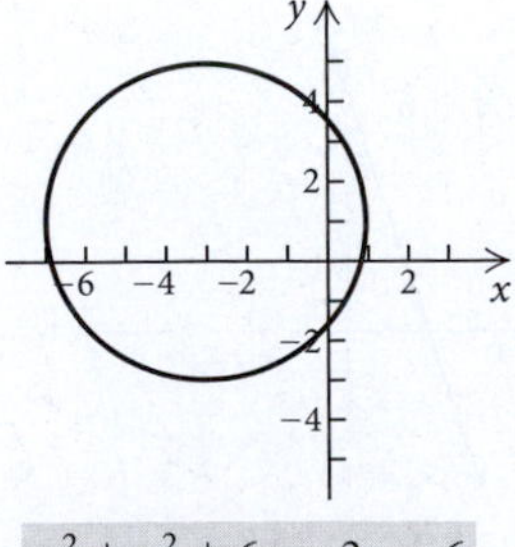

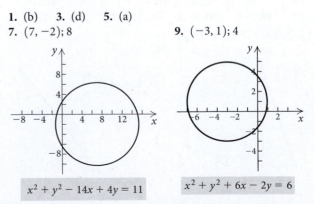

11. $(-2, 3)$; 5

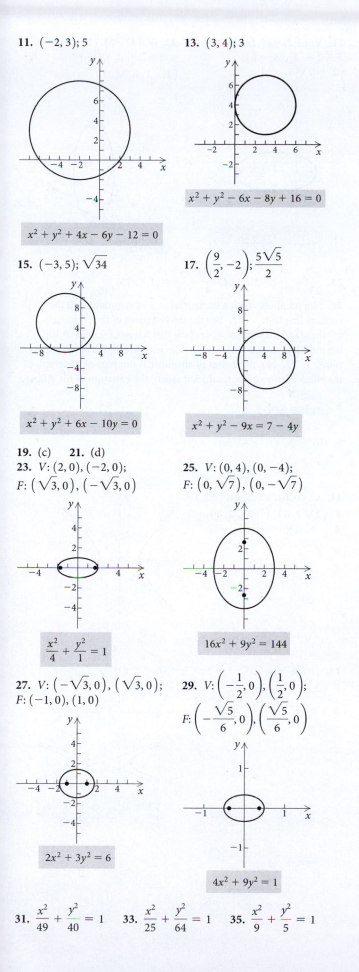

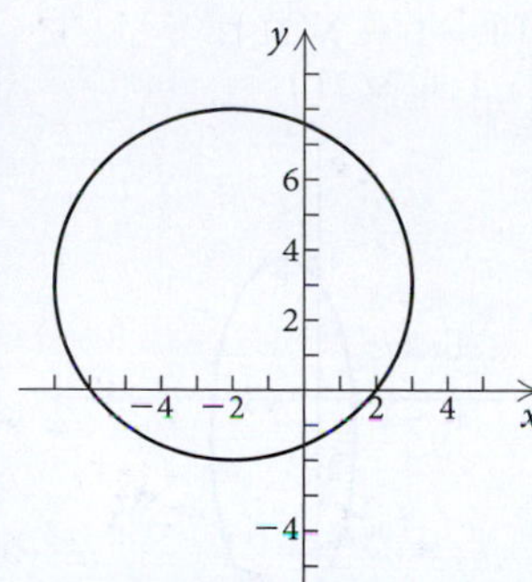

$x^2 + y^2 + 4x - 6y - 12 = 0$

13. $(3, 4)$; 3

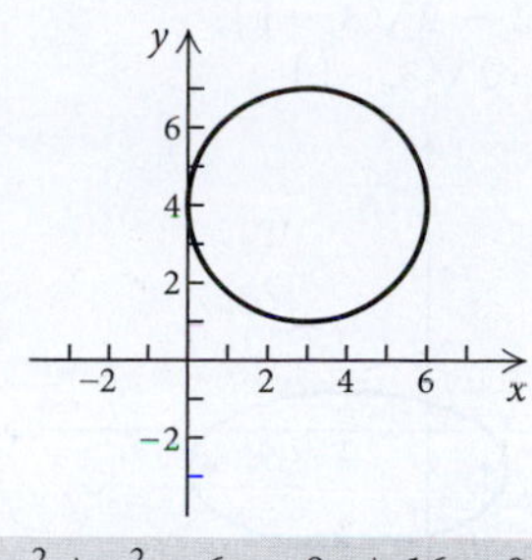

$x^2 + y^2 - 6x - 8y + 16 = 0$

15. $(-3, 5)$; $\sqrt{34}$

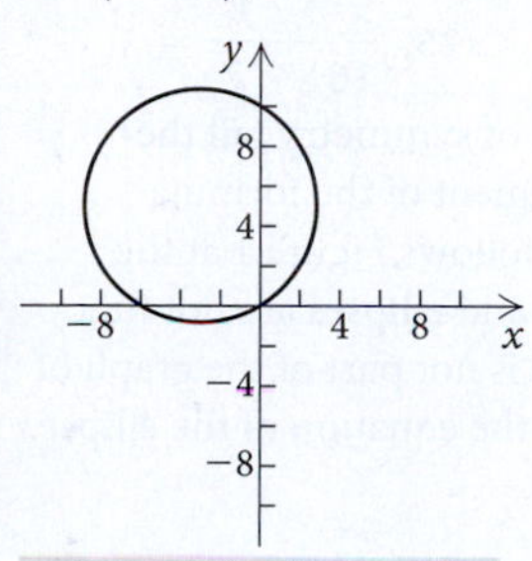

$x^2 + y^2 + 6x - 10y = 0$

17. $\left(\frac{9}{2}, -2\right)$; $\frac{5\sqrt{5}}{2}$

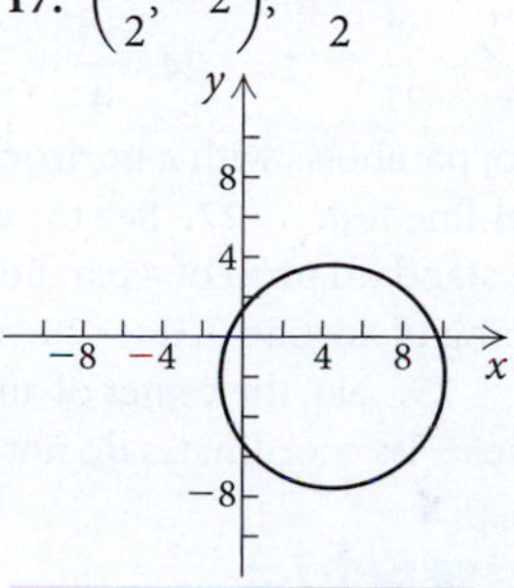

$x^2 + y^2 - 9x = 7 - 4y$

19. (c) **21.** (d)

23. V: $(2, 0)$, $(-2, 0)$; F: $(\sqrt{3}, 0)$, $(-\sqrt{3}, 0)$

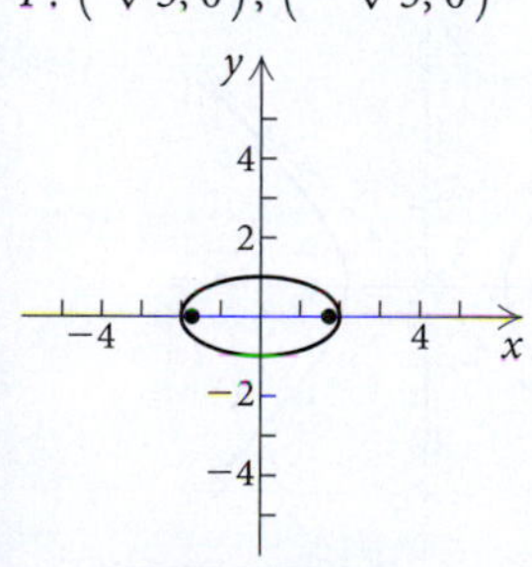

$\frac{x^2}{4} + \frac{y^2}{1} = 1$

25. V: $(0, 4)$, $(0, -4)$; F: $(0, \sqrt{7})$, $(0, -\sqrt{7})$

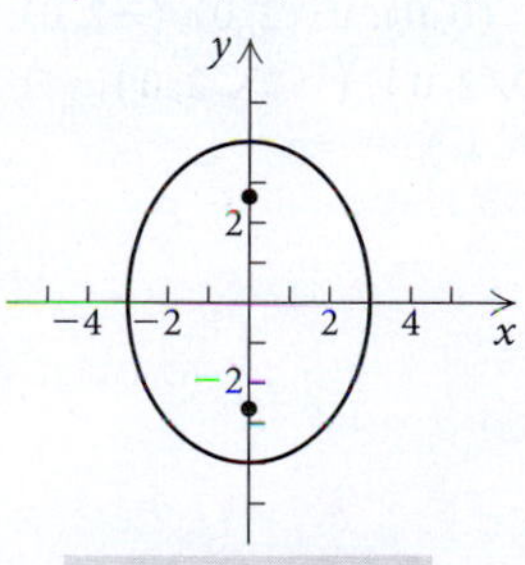

$16x^2 + 9y^2 = 144$

27. V: $(-\sqrt{3}, 0)$, $(\sqrt{3}, 0)$; F: $(-1, 0)$, $(1, 0)$

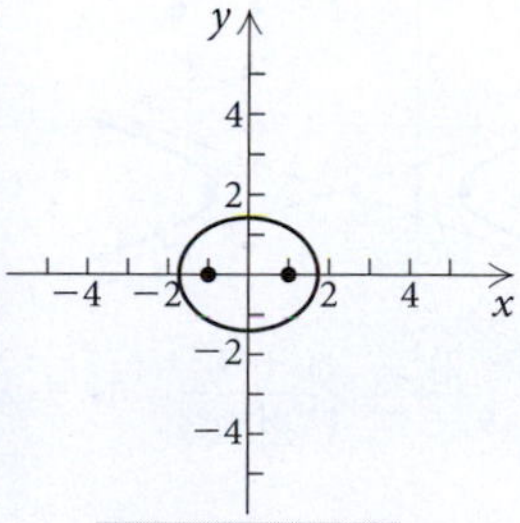

$2x^2 + 3y^2 = 6$

29. V: $\left(-\frac{1}{2}, 0\right)$, $\left(\frac{1}{2}, 0\right)$; F: $\left(-\frac{\sqrt{5}}{6}, 0\right)$, $\left(\frac{\sqrt{5}}{6}, 0\right)$

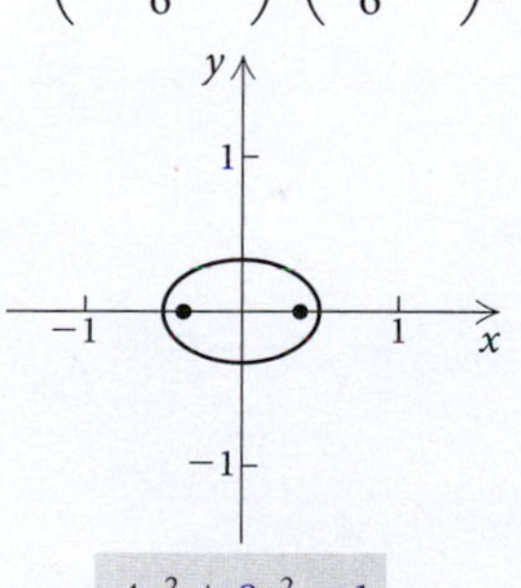

$4x^2 + 9y^2 = 1$

31. $\frac{x^2}{49} + \frac{y^2}{40} = 1$ **33.** $\frac{x^2}{25} + \frac{y^2}{64} = 1$ **35.** $\frac{x^2}{9} + \frac{y^2}{5} = 1$

37. C: $(1, 2)$; V: $(4, 2)$, $(-2, 2)$; F: $(1 + \sqrt{5}, 2)$, $(1 - \sqrt{5}, 2)$

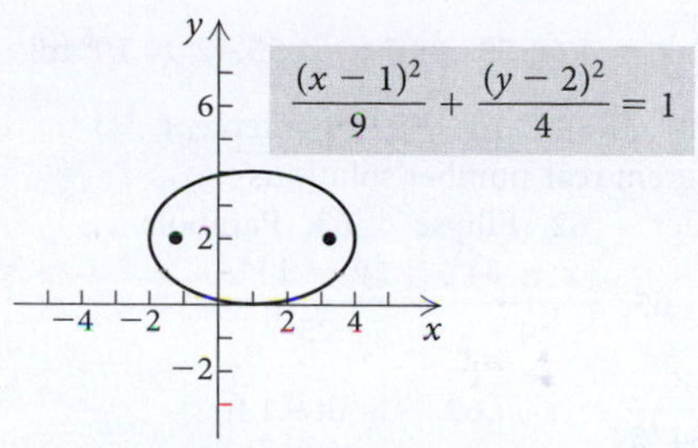

$\frac{(x-1)^2}{9} + \frac{(y-2)^2}{4} = 1$

39. C: $(-3, 5)$; V: $(-3, 11)$, $(-3, -1)$; F: $(-3, 5 + \sqrt{11})$, $(-3, 5 - \sqrt{11})$

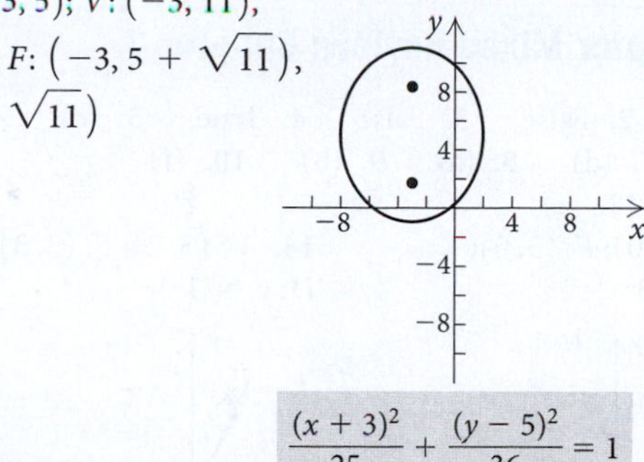

$\frac{(x+3)^2}{25} + \frac{(y-5)^2}{36} = 1$

41. C: $(-2, 1)$; V: $(-10, 1)$, $(6, 1)$; F: $(-6, 1)$, $(2, 1)$

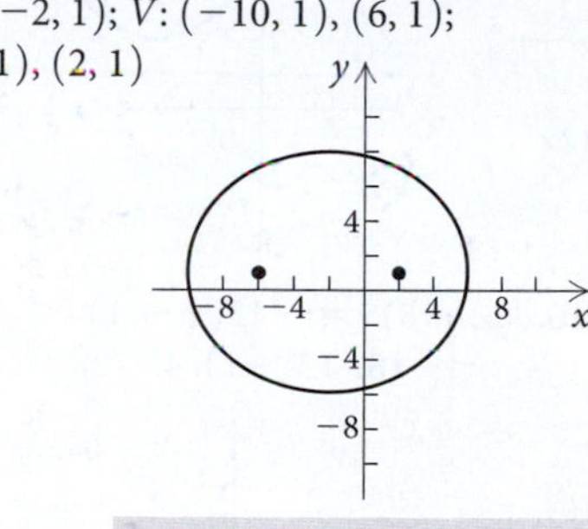

$3(x + 2)^2 + 4(y - 1)^2 = 192$

43. C: $(2, -1)$; V: $(-1, -1)$, $(5, -1)$; F: $(2 + \sqrt{5}, -1)$, $(2 - \sqrt{5}, -1)$

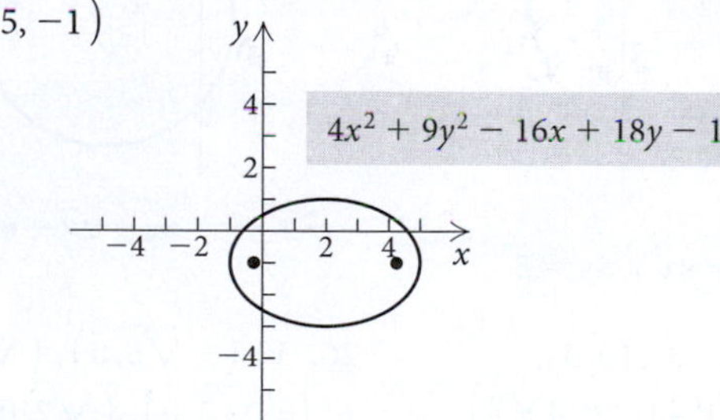

$4x^2 + 9y^2 - 16x + 18y - 11 = 0$

45. C: $(1, 1)$; V: $(1, 3)$, $(1, -1)$; F: $(1, 1 + \sqrt{3})$, $(1, 1 - \sqrt{3})$

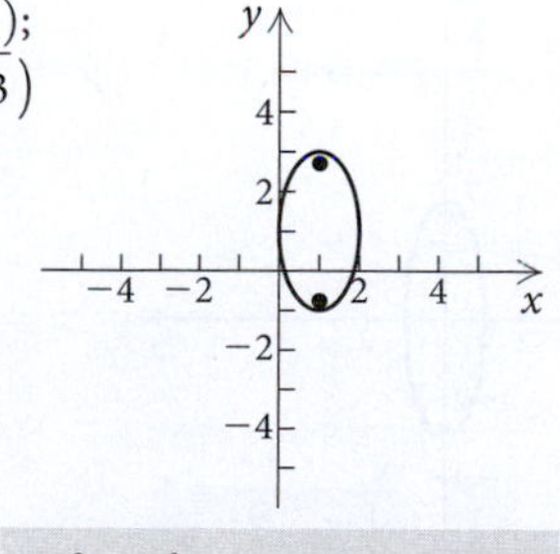

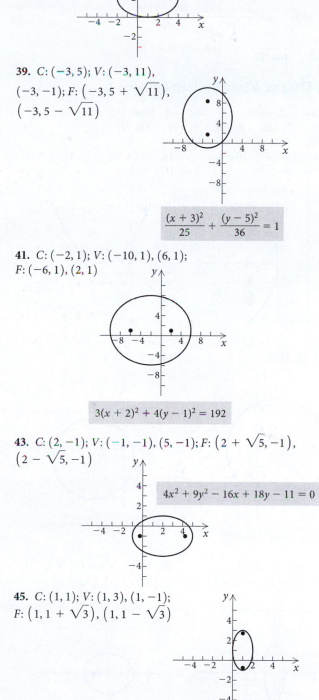

$4x^2 + y^2 - 8x - 2y + 1 = 0$

47. Example 2; $\frac{3}{5} < \frac{\sqrt{12}}{4}$ **49.** $\frac{x^2}{15} + \frac{y^2}{16} = 1$

51. $\frac{x^2}{50^2} + \frac{y^2}{12^2} = 1$ **53.** 33.5 ft **55.** 2×10^6 mi

57. Midpoint **58.** Zero **59.** *y*-intercept

60. Two different real-number solutions

61. Remainder **62.** Ellipse **63.** Parabola

64. Circle **65.** $\frac{(x-3)^2}{4} + \frac{(y-1)^2}{25} = 1$

67. $\frac{x^2}{9} + \frac{y^2}{484/5} = 1$ **69.** About 9.1 ft

Mid-Chapter Mixed Review: Chapter 7

1. True **2.** False **3.** False **4.** True **5.** (c)
6. (h) **7.** (d) **8.** (a) **9.** (b) **10.** (f)
11. (g) **12.** (e)

13. V: $(0, 0)$; F: $(3, 0)$; D: $x = -3$

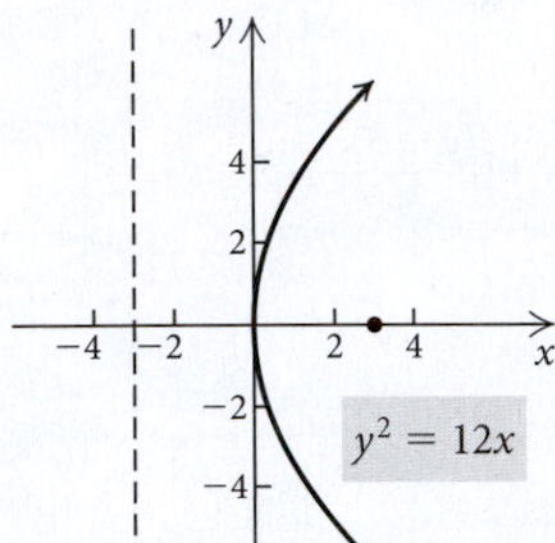

14. V: $(3, 2)$; F: $(3, 3)$; D: $y = 1$

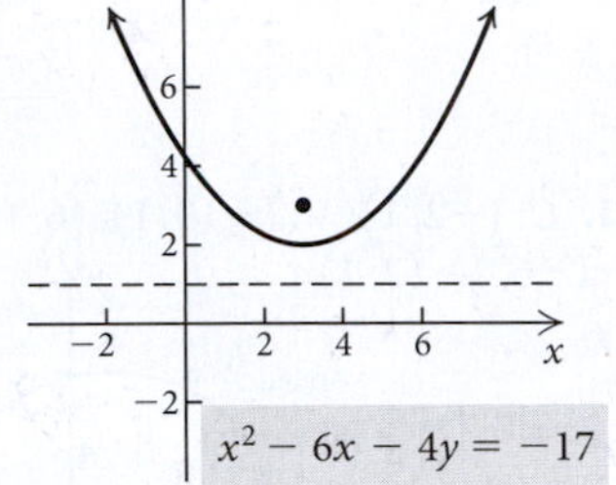

15. $x^2 = 4(y - 2)$ **16.** $(y - 6)^2 = -12(x + 1)$

17. $(-2, 4)$; 5

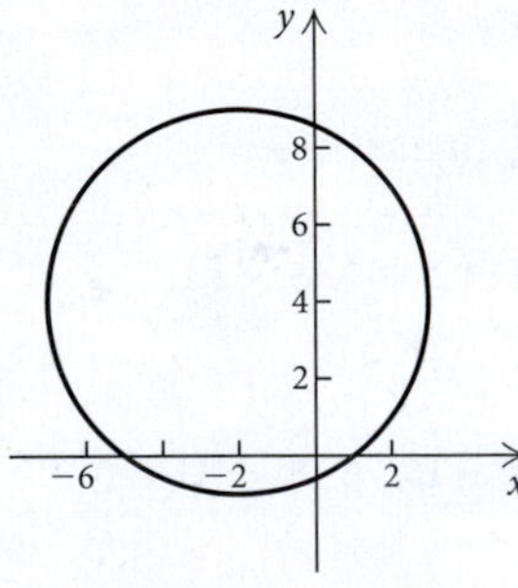

18. $(3, -1)$; 4

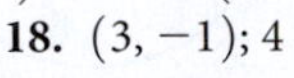

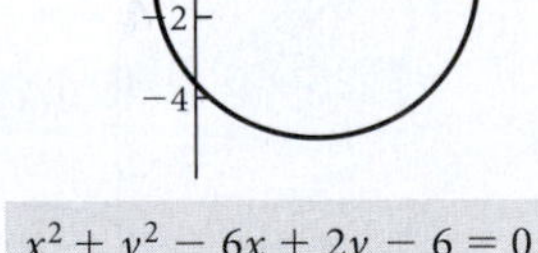

19. V: $(0, -3)$, $(0, 3)$; F: $(0, -2\sqrt{2})$, $(0, 2\sqrt{2})$

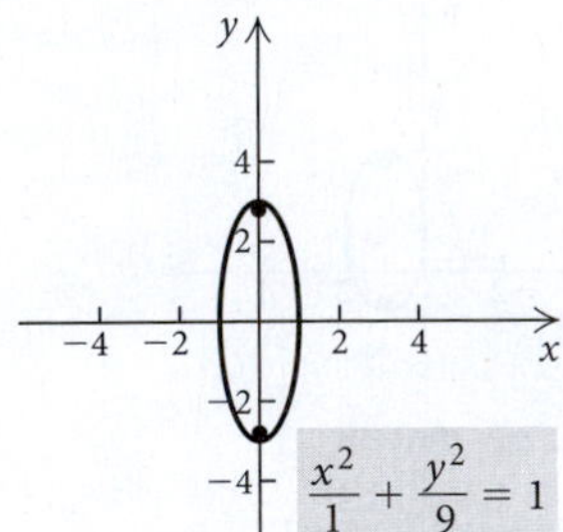

20. V: $(-\sqrt{6}, 0)$, $(\sqrt{6}, 0)$; F: $(-\sqrt{2}, 0)$, $(\sqrt{2}, 0)$

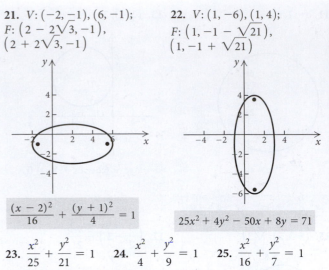

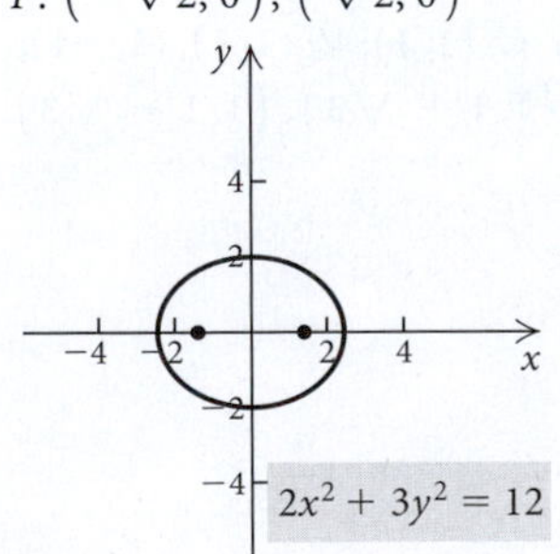

21. V: $(-2, -1)$, $(6, -1)$; F: $(2 - 2\sqrt{3}, -1)$, $(2 + 2\sqrt{3}, -1)$

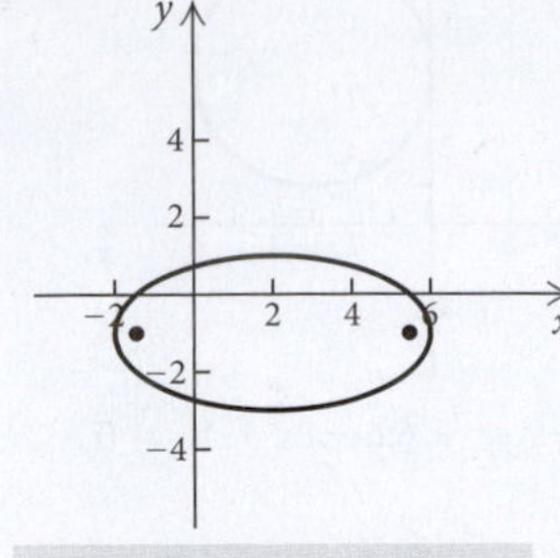

22. V: $(1, -6)$, $(1, 4)$; F: $(1, -1 - \sqrt{21})$, $(1, -1 + \sqrt{21})$

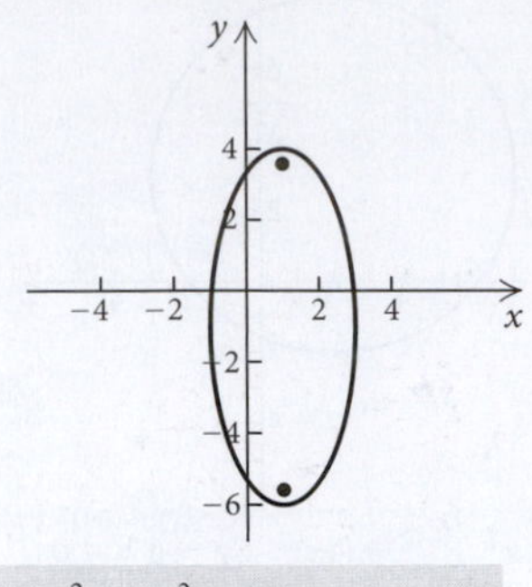

23. $\frac{x^2}{25} + \frac{y^2}{21} = 1$ **24.** $\frac{x^2}{4} + \frac{y^2}{9} = 1$ **25.** $\frac{x^2}{16} + \frac{y^2}{7} = 1$

26. No; parabolas with a horizontal axis of symmetry fail the vertical-line test. **27.** See the development of the formula for the standard form of a parabola that follows Figure 1 at the beginning of Section 7.1. **28.** Circles and ellipses are not functions. **29.** No; the center of an ellipse is not part of the graph of the ellipse. Its coordinates do not satisfy the equation of the ellipse.

Exercise Set 7.3

1. (b) **3.** (c) **5.** (a) **7.** $\frac{y^2}{9} - \frac{x^2}{16} = 1$

9. $\frac{x^2}{4} - \frac{y^2}{9} = 1$

11. C: $(0, 0)$; V: $(2, 0)$, $(-2, 0)$; F: $(2\sqrt{2}, 0)$, $(-2\sqrt{2}, 0)$; A: $y = x$, $y = -x$

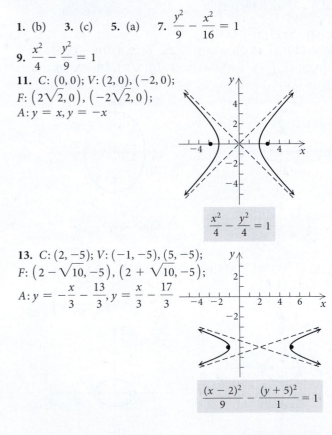

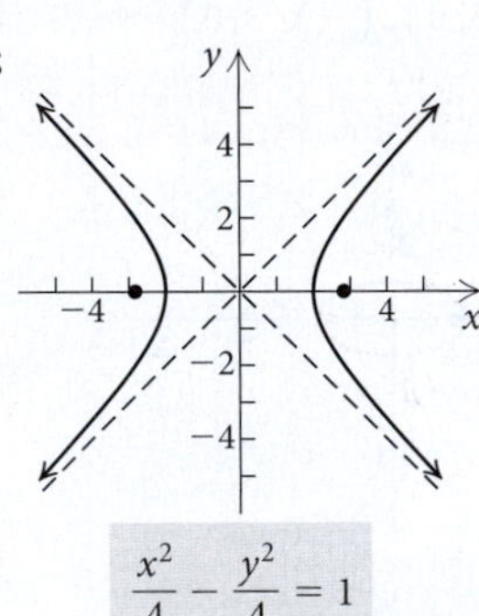

13. C: $(2, -5)$; V: $(-1, -5)$, $(5, -5)$; F: $(2 - \sqrt{10}, -5)$, $(2 + \sqrt{10}, -5)$; A: $y = -\frac{x}{3} - \frac{13}{3}$, $y = \frac{x}{3} - \frac{17}{3}$

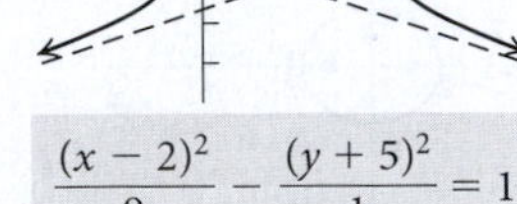

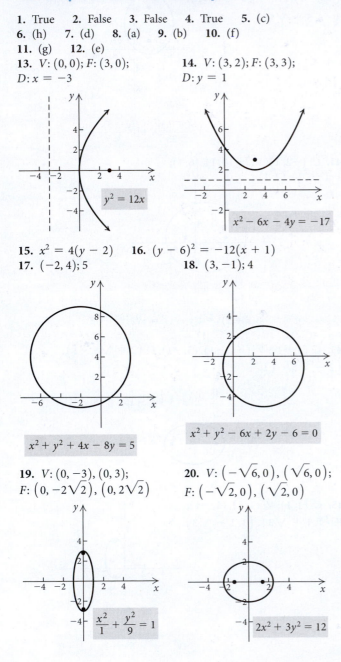

15. $C: (-1, -3)$; $V: (-1, -1), (-1, -5)$; $F: \left(-1, -3 + 2\sqrt{5}\right), \left(-1, -3 - 2\sqrt{5}\right)$; $A: y = \frac{1}{2}x - \frac{5}{2}, y = -\frac{1}{2}x - \frac{7}{2}$

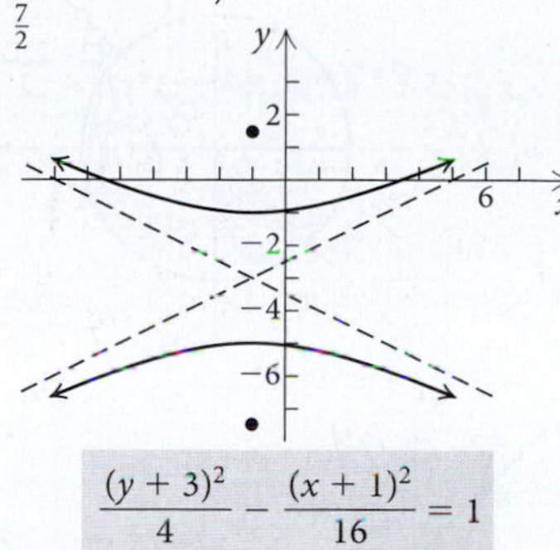

17. $C: (0, 0)$; $V: (-2, 0), (2, 0)$; $F: \left(-\sqrt{5}, 0\right), \left(\sqrt{5}, 0\right)$; $A: y = -\frac{1}{2}x, y = \frac{1}{2}x$

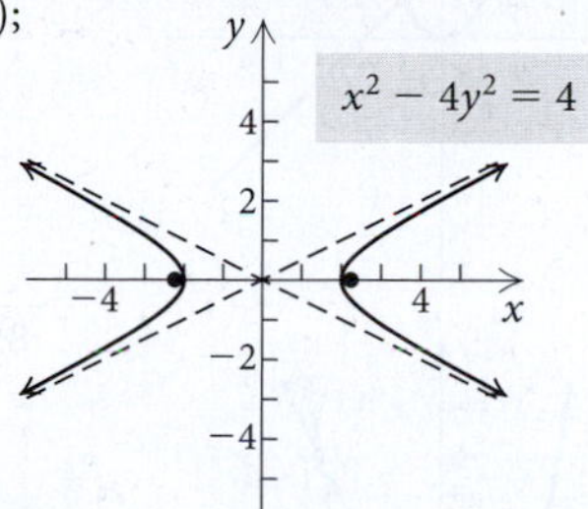

19. $C: (0, 0)$; $V: (0, -3), (0, 3)$; $F: \left(0, -3\sqrt{10}\right), \left(0, 3\sqrt{10}\right)$; $A: y = \frac{1}{3}x, y = -\frac{1}{3}x$

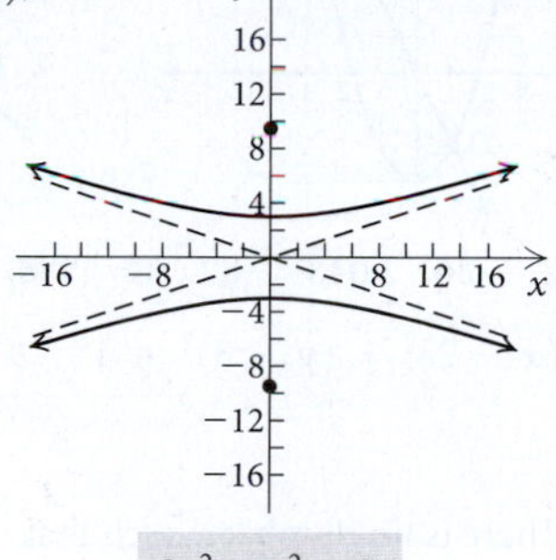

21. $C: (0, 0)$; $V: \left(-\sqrt{2}, 0\right), \left(\sqrt{2}, 0\right)$; $F: (-2, 0), (2, 0)$; $A: y = x, y = -x$

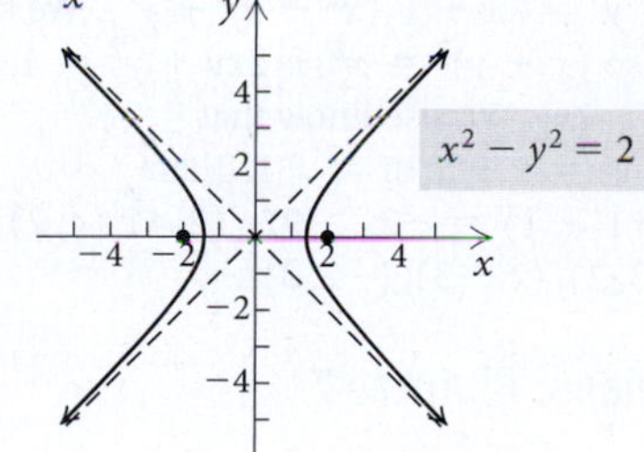

23. $C: (0, 0)$; $V: \left(0, -\frac{1}{2}\right), \left(0, \frac{1}{2}\right)$; $F: \left(0, -\frac{\sqrt{2}}{2}\right), \left(0, \frac{\sqrt{2}}{2}\right)$; $A: y = x, y = -x$

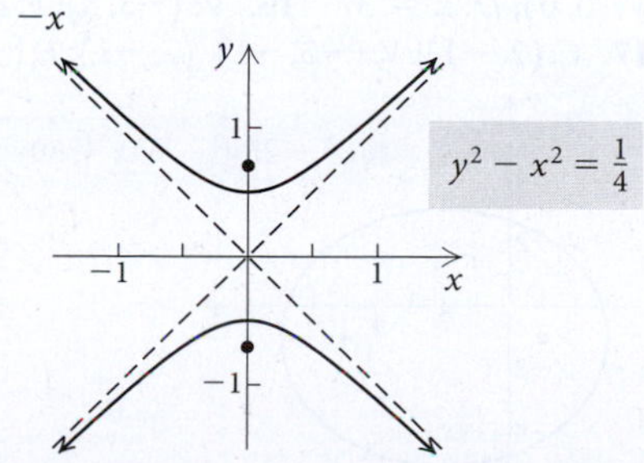

25. $C: (1, -2)$; $V: (0, -2), (2, -2)$; $F: \left(1 - \sqrt{2}, -2\right), \left(1 + \sqrt{2}, -2\right)$; $A: y = -x - 1, y = x - 3$

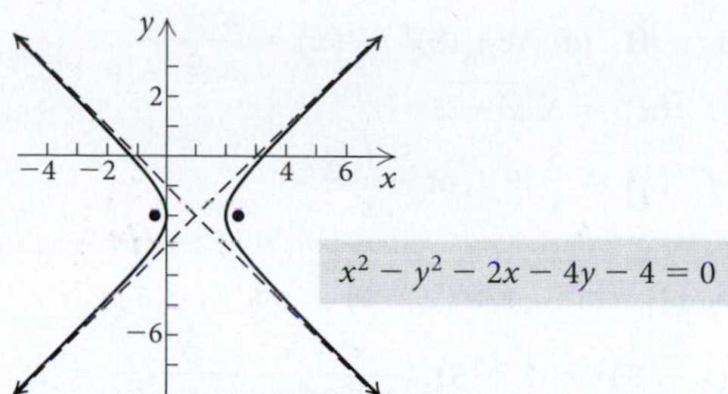

27. $C: \left(\frac{1}{3}, 3\right)$; $V: \left(-\frac{2}{3}, 3\right), \left(\frac{4}{3}, 3\right)$; $F: \left(\frac{1}{3} - \sqrt{37}, 3\right), \left(\frac{1}{3} + \sqrt{37}, 3\right)$; $A: y = 6x + 1, y = -6x + 5$

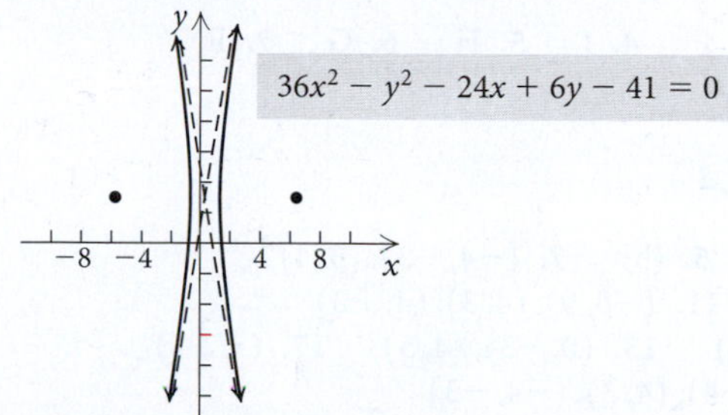

29. $C: (3, 1)$; $V: (3, 3), (3, -1)$; $F: \left(3, 1 + \sqrt{13}\right), \left(3, 1 - \sqrt{13}\right)$; $A: y = \frac{2}{3}x - 1, y = -\frac{2}{3}x + 3$

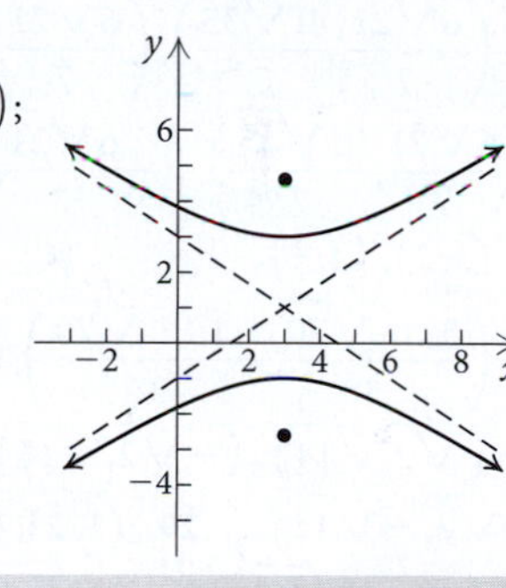

31. $C: (1, -2)$; $V: (2, -2), (0, -2)$; $F: (1 + \sqrt{2}, -2), (1 - \sqrt{2}, -2)$; $A: y = x - 3, y = -x - 1$

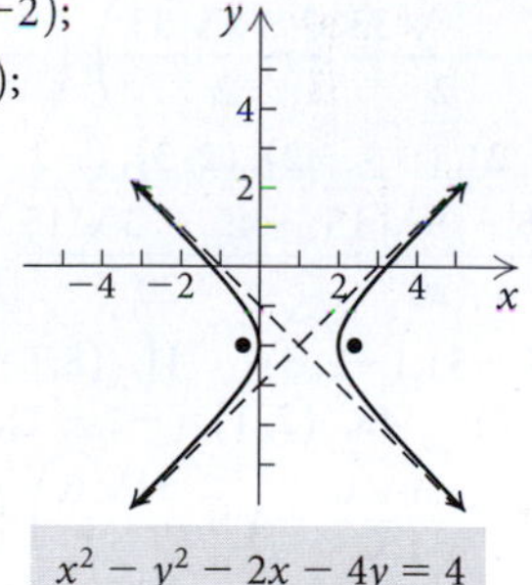

33. $C: (-3, 4)$; $V: (-3, 10), (-3, -2)$; $F: \left(-3, 4 + 6\sqrt{2}\right), \left(-3, 4 - 6\sqrt{2}\right)$; $A: y = x + 7, y = -x + 1$

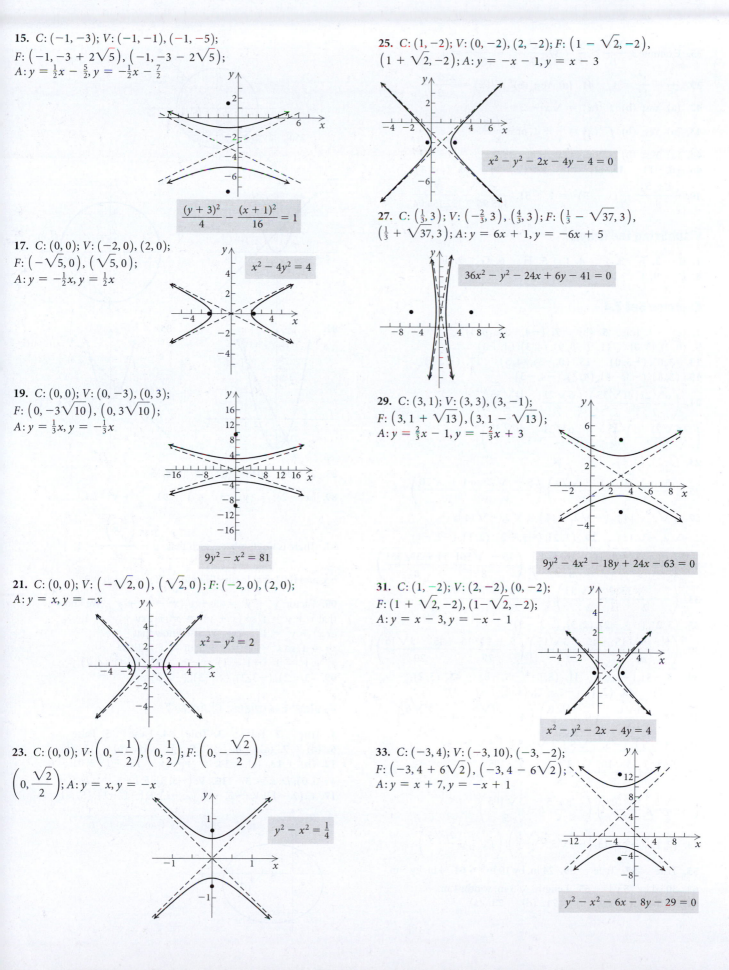

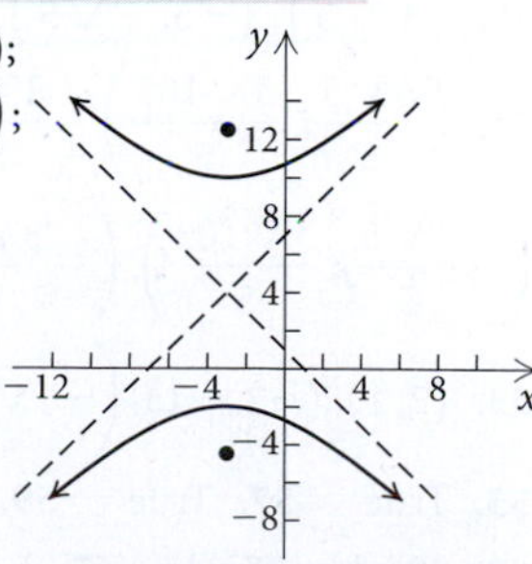

35. Example 3; $\frac{\sqrt{5}}{1} > \frac{5}{4}$ **37.** $\frac{x^2}{9} - \frac{(y-7)^2}{16} = 1$

39. $\frac{y^2}{25} - \frac{x^2}{11} = 1$ **41.** (a) Yes; (b) $f^{-1}(x) = \frac{x+3}{2}$

42. (a) Yes; (b) $f^{-1}(x) = \sqrt[3]{x-2}$

43. (a) Yes; (b) $f^{-1}(x) = \frac{5}{x} + 1$, or $\frac{5+x}{x}$

44. (a) Yes; (b) $f^{-1}(x) = x^2 - 4, x \geq 0$

45. $(6, -1)$ **46.** $(1, -1)$ **47.** $(2, -1)$ **48.** $(-3, 4)$

49. $\frac{(y+5)^2}{9} - (x-3)^2 = 1$ **51.** $\frac{x^2}{345.96} - \frac{y^2}{22{,}154.04} = 1$

Visualizing the Graph

1. B **2.** J **3.** F **4.** I **5.** H **6.** G **7.** E **8.** D **9.** C **10.** A

Exercise Set 7.4

1. (e) **3.** (c) **5.** (b) **7.** $(-4, -3), (3, 4)$
9. $(0, 2), (3, 0)$ **11.** $(-5, 0), (4, 3), (4, -3)$
13. $(3, 0), (-3, 0)$ **15.** $(0, -3), (4, 5)$ **17.** $(-2, 1)$
19. $(3, 4), (-3, -4), (4, 3), (-4, -3)$

21. $\left(\frac{6\sqrt{21}}{7}, \frac{4i\sqrt{35}}{7}\right), \left(\frac{6\sqrt{21}}{7}, -\frac{4i\sqrt{35}}{7}\right), \left(-\frac{6\sqrt{21}}{7}, \frac{4i\sqrt{35}}{7}\right), \left(-\frac{6\sqrt{21}}{7}, -\frac{4i\sqrt{35}}{7}\right)$

23. $(3, 2), \left(4, \frac{3}{2}\right)$

25. $\left(\frac{5+\sqrt{70}}{3}, \frac{-1+\sqrt{70}}{3}\right), \left(\frac{5-\sqrt{70}}{3}, \frac{-1-\sqrt{70}}{3}\right)$

27. $(\sqrt{2}, \sqrt{14}), (-\sqrt{2}, \sqrt{14}), (\sqrt{2}, -\sqrt{14}), (-\sqrt{2}, -\sqrt{14})$ **29.** $(1, 2), (-1, -2), (2, 1), (-2, -1)$

31. $\left(\frac{15+\sqrt{561}}{8}, \frac{11-3\sqrt{561}}{8}\right), \left(\frac{15-\sqrt{561}}{8}, \frac{11+3\sqrt{561}}{8}\right)$

33. $\left(\frac{7-\sqrt{33}}{2}, \frac{7+\sqrt{33}}{2}\right), \left(\frac{7+\sqrt{33}}{2}, \frac{7-\sqrt{33}}{2}\right)$

35. $(3, 2), (-3, -2), (2, 3), (-2, -3)$

37. $\left(\frac{5-9\sqrt{15}}{20}, \frac{-45+3\sqrt{15}}{20}\right), \left(\frac{5+9\sqrt{15}}{20}, \frac{-45-3\sqrt{15}}{20}\right)$

39. $(3, -5), (-1, 3)$ **41.** $(8, 5), (-5, -8)$ **43.** $(3, 2), (-3, -2)$ **45.** $(2, 1), (-2, -1), (1, 2), (-1, -2)$

47. $\left(4 + \frac{3i\sqrt{6}}{2}, -4 + \frac{3i\sqrt{6}}{2}\right), \left(4 - \frac{3i\sqrt{6}}{2}, -4 - \frac{3i\sqrt{6}}{2}\right)$

49. $(3, \sqrt{5}), (-3, -\sqrt{5}), (\sqrt{5}, 3), (-\sqrt{5}, -3)$

51. $\left(\frac{8\sqrt{5}}{5}i, \frac{3\sqrt{105}}{5}\right), \left(\frac{8\sqrt{5}}{5}i, -\frac{3\sqrt{105}}{5}\right), \left(-\frac{8\sqrt{5}}{5}i, \frac{3\sqrt{105}}{5}\right), \left(-\frac{8\sqrt{5}}{5}i, -\frac{3\sqrt{105}}{5}\right)$

53. $(2, 1), (-2, -1), \left(-i\sqrt{5}, \frac{2i\sqrt{5}}{5}\right), \left(i\sqrt{5}, -\frac{2i\sqrt{5}}{5}\right)$

55. True **57.** True **59.** 24 in. by 10 in. **61.** 4 in. by 5 in.
63. 30 yd by 75 yd **65.** Length: $\sqrt{3}$ m; width: 1 m
67. 16 ft, 24 ft **69.** (b) **71.** (d) **73.** (a)

75.

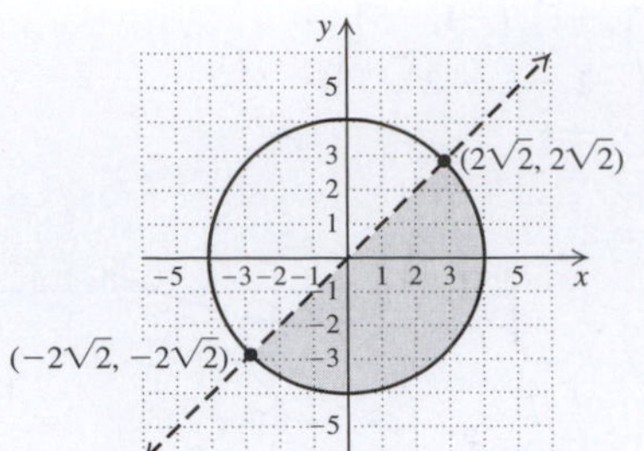

77.

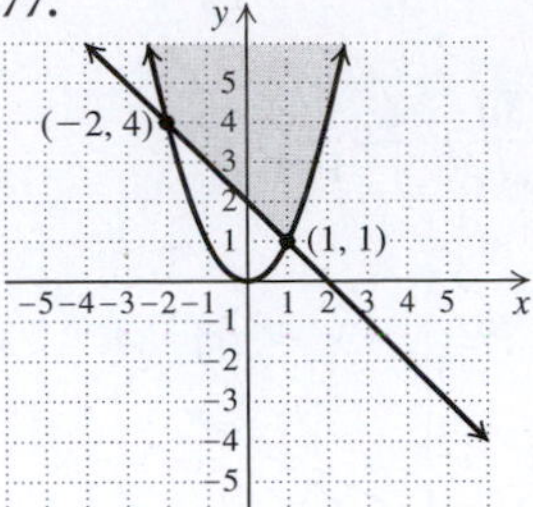

79.

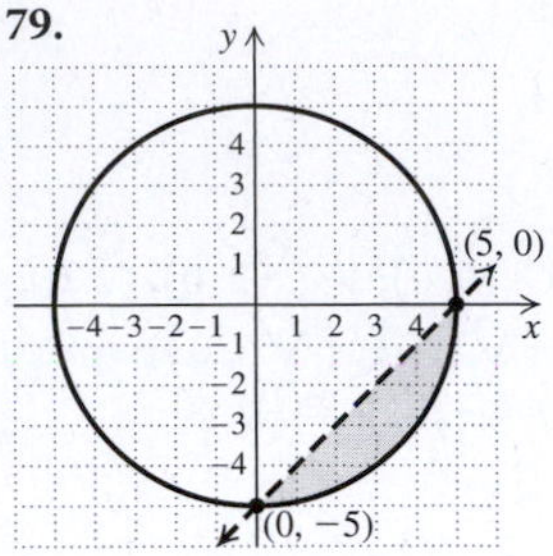

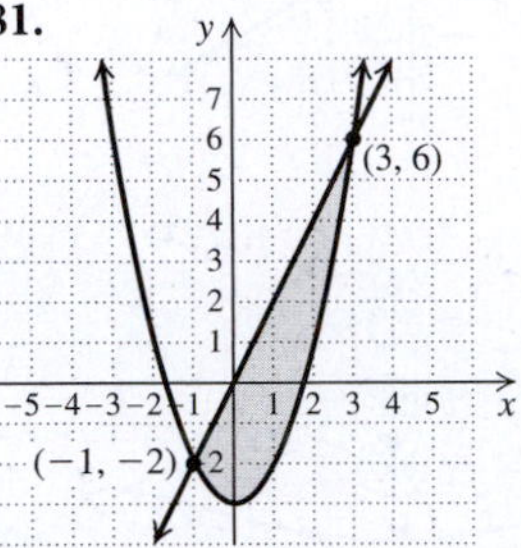

81.

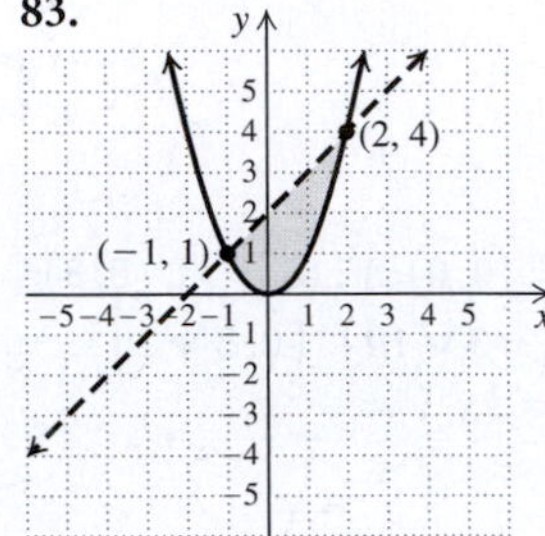

83.

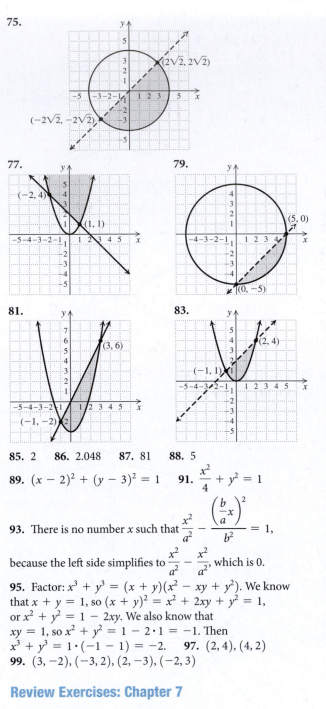

85. 2 **86.** 2.048 **87.** 81 **88.** 5

89. $(x-2)^2 + (y-3)^2 = 1$ **91.** $\frac{x^2}{4} + y^2 = 1$

93. There is no number x such that $\frac{x^2}{a^2} - \frac{\left(\frac{b}{a}x\right)^2}{b^2} = 1$, because the left side simplifies to $\frac{x^2}{a^2} - \frac{x^2}{a^2}$, which is 0.

95. Factor: $x^3 + y^3 = (x+y)(x^2 - xy + y^2)$. We know that $x + y = 1$, so $(x+y)^2 = x^2 + 2xy + y^2 = 1$, or $x^2 + y^2 = 1 - 2xy$. We also know that $xy = 1$, so $x^2 + y^2 = 1 - 2 \cdot 1 = -1$. Then $x^3 + y^3 = 1 \cdot (-1 - 1) = -2$. **97.** $(2, 4), (4, 2)$
99. $(3, -2), (-3, 2), (2, -3), (-2, 3)$

Review Exercises: Chapter 7

1. True **2.** False **3.** True **4.** False **5.** False
6. (d) **7.** (a) **8.** (e) **9.** (g) **10.** (b) **11.** (f)
12. (h) **13.** (c) **14.** $x^2 = -6y$ **15.** $F: (-3, 0)$; $V: (0, 0)$; $D: x = 3$ **16.** $V: (-5, 8)$; $F: \left(-5, \frac{15}{2}\right)$; $D: y = \frac{17}{2}$
17. $C: (2, -1)$; $V: (-3, -1), (7, -1)$; $F: (-1, -1), (5, -1)$

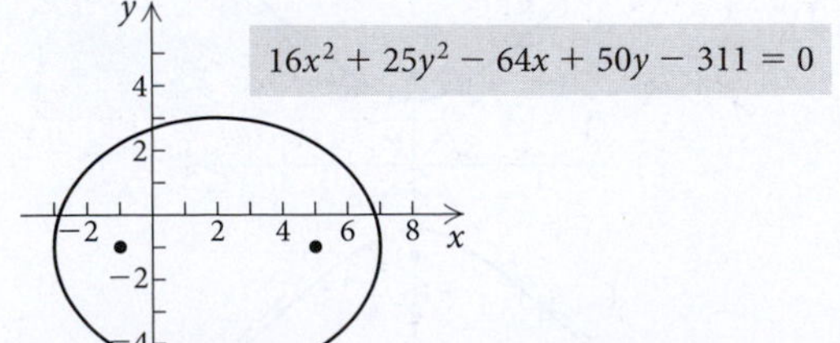

18. $\frac{x^2}{9} + \frac{y^2}{16} = 1$

19. $C: \left(-2, \frac{1}{4}\right)$; $V: \left(0, \frac{1}{4}\right), \left(-4, \frac{1}{4}\right)$; $F: \left(-2 + \sqrt{6}, \frac{1}{4}\right), \left(-2 - \sqrt{6}, \frac{1}{4}\right)$; $A: y - \frac{1}{4} = \frac{\sqrt{2}}{2}(x + 2), y - \frac{1}{4} = -\frac{\sqrt{2}}{2}(x + 2)$

20. 0.167 ft **21.** $(-8\sqrt{2}, 8), (8\sqrt{2}, 8)$

22. $\left(3, \frac{\sqrt{29}}{2}\right), \left(-3, \frac{\sqrt{29}}{2}\right), \left(3, -\frac{\sqrt{29}}{2}\right), \left(-3, -\frac{\sqrt{29}}{2}\right)$ **23.** $(7, 4)$ **24.** $(2, 2), \left(\frac{32}{9}, -\frac{10}{9}\right)$

25. $(0, -3), (2, 1)$ **26.** $(4, 3), (4, -3), (-4, 3), (-4, -3)$

27. $(-\sqrt{3}, 0), (\sqrt{3}, 0), (-2, 1), (2, 1)$ **28.** $\left(-\frac{3}{5}, \frac{21}{5}\right)$, $(3, -3)$ **29.** $(6, 8), (6, -8), (-6, 8), (-6, -8)$

30. $(2, 2), (-2, -2), (2\sqrt{2}, \sqrt{2}), (-2\sqrt{2}, -\sqrt{2})$

31. 7, 4 **32.** 7 m by 12 m **33.** 4, 8 **34.** 32 cm, 20 cm

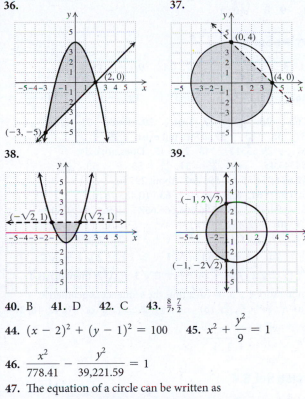

35. 11 ft, 3 ft

36.

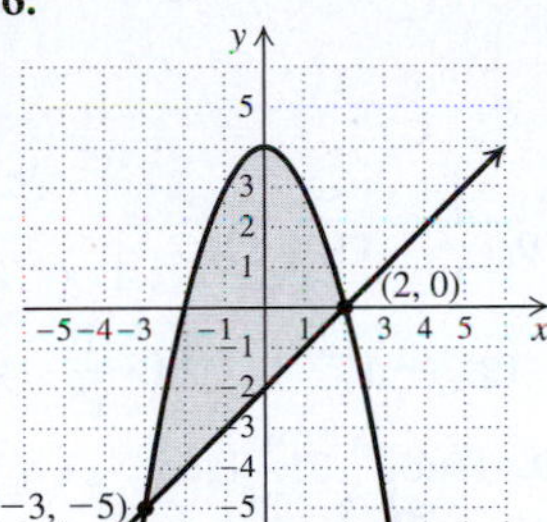

37.

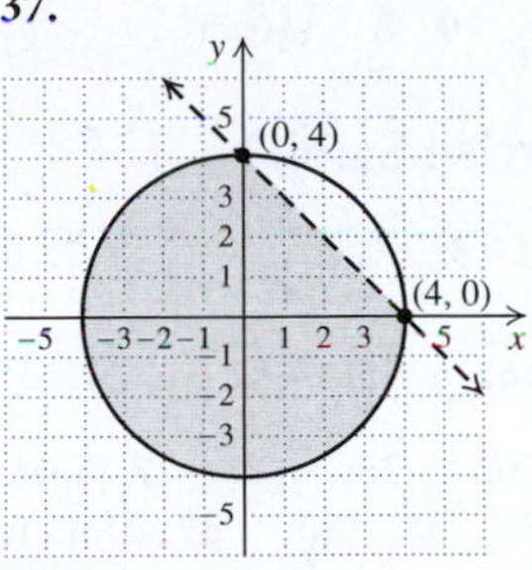

38.

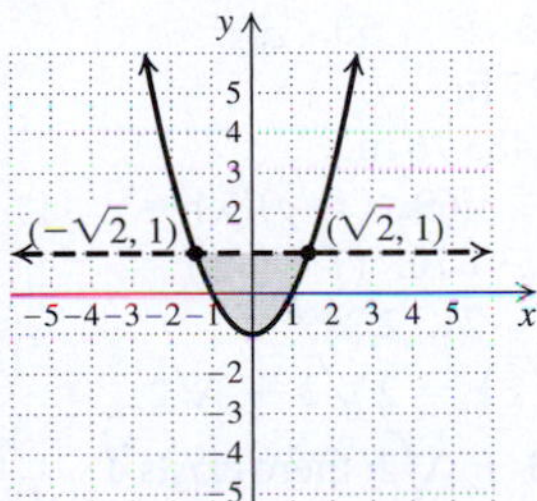

39.

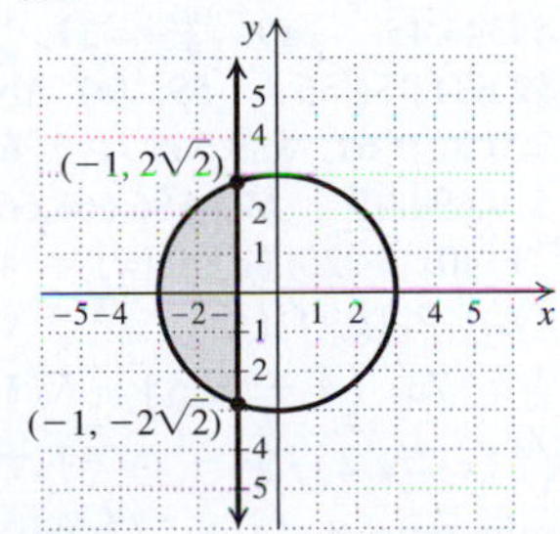

40. B **41.** D **42.** C **43.** $\frac{8}{7}, \frac{7}{2}$

44. $(x - 2)^2 + (y - 1)^2 = 100$ **45.** $x^2 + \frac{y^2}{9} = 1$

46. $\frac{x^2}{778.41} - \frac{y^2}{39{,}221.59} = 1$

47. The equation of a circle can be written as

$$\frac{(x - h)^2}{a^2} + \frac{(y - k)^2}{b^2} = 1,$$

where $a = b = r$, the radius of the circle. In an ellipse, $a > b$, so a circle is not a special type of ellipse. **48.** No; the asymptotes of a hyperbola are not part of the graph of the hyperbola. The coordinates of points on the asymptotes do not satisfy the equation of the hyperbola. **49.** Although we can always visualize the real-number solutions, we cannot visualize the imaginary-number solutions.

Test: Chapter 7

1. [7.3] (c) **2.** [7.1] (b) **3.** [7.2] (a) **4.** [7.2] (d)

5. [7.1] $V: (0, 0)$; $F: (0, 3)$; $D: y = -3$

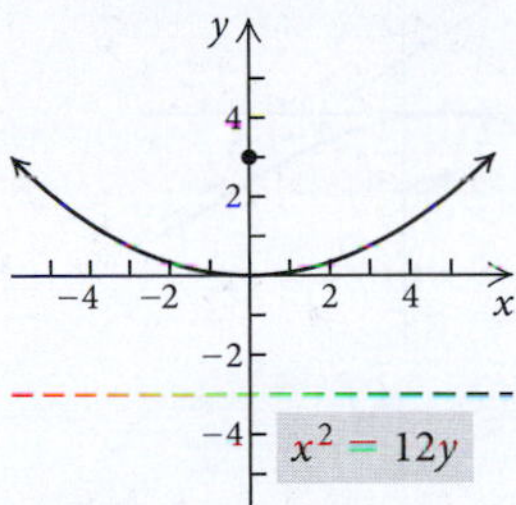

6. [7.1] $V: (-1, -1)$; $F: (1, -1)$; $D: x = -3$

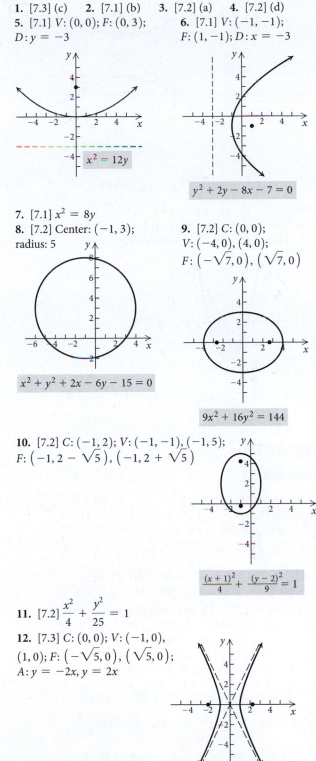

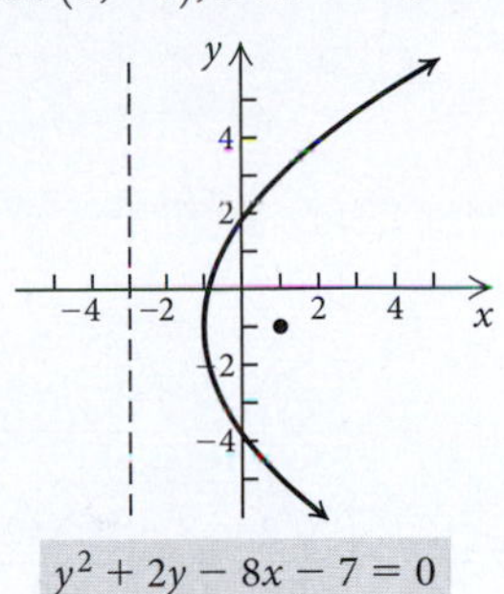

7. [7.1] $x^2 = 8y$

8. [7.2] Center: $(-1, 3)$; radius: 5

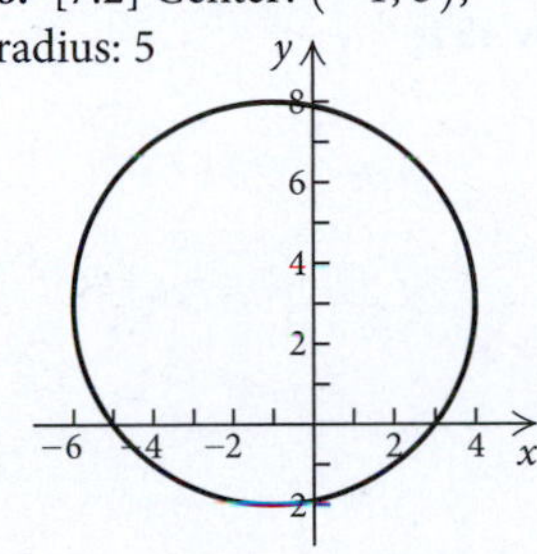

9. [7.2] $C: (0, 0)$; $V: (-4, 0), (4, 0)$; $F: (-\sqrt{7}, 0), (\sqrt{7}, 0)$

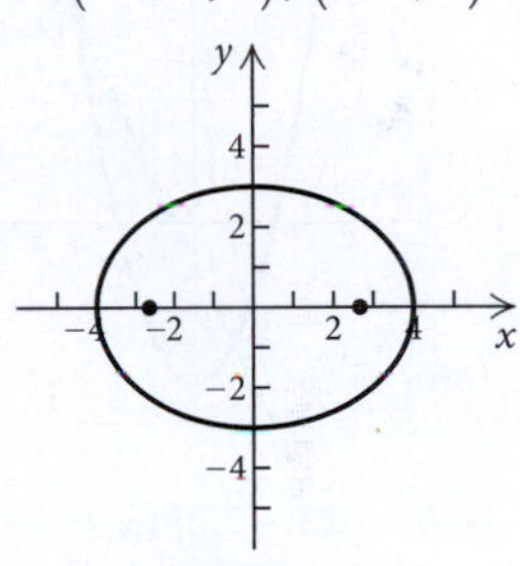

10. [7.2] $C: (-1, 2)$; $V: (-1, -1), (-1, 5)$; $F: (-1, 2 - \sqrt{5}), (-1, 2 + \sqrt{5})$

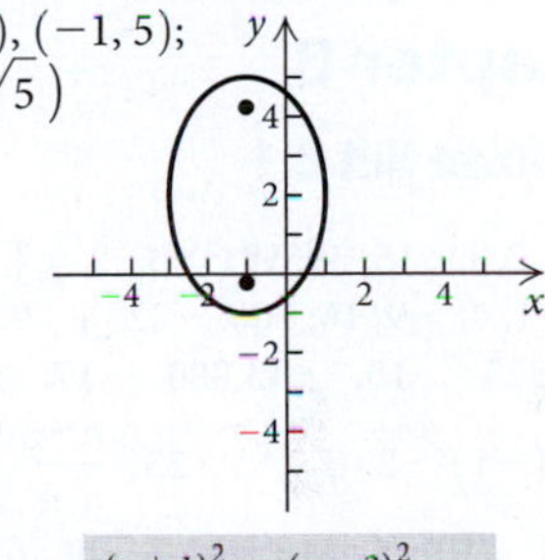

11. [7.2] $\frac{x^2}{4} + \frac{y^2}{25} = 1$

12. [7.3] $C: (0, 0)$; $V: (-1, 0), (1, 0)$; $F: (-\sqrt{5}, 0), (\sqrt{5}, 0)$; $A: y = -2x, y = 2x$

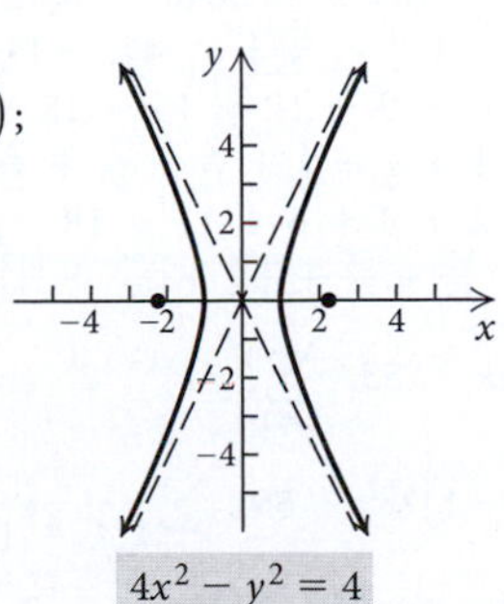

13. [7.3] $C: (-1, 2)$; $V: (-1, 0), (-1, 4)$;
$F: \left(-1, 2 - \sqrt{13}\right), \left(-1, 2 + \sqrt{13}\right)$;
$A: y = -\frac{2}{3}x + \frac{4}{3}, y = \frac{2}{3}x + \frac{8}{3}$

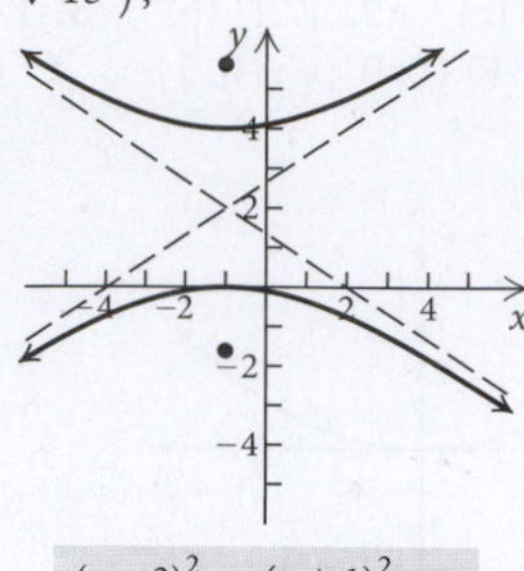

$$\frac{(y-2)^2}{4} - \frac{(x+1)^2}{9} = 1$$

14. [7.3] $y = \dfrac{\sqrt{2}}{2}x, y = -\dfrac{\sqrt{2}}{2}x$ **15.** [7.1] $\frac{27}{8}$ in.
16. [7.4] $(1, 2), (1, -2), (-1, 2), (-1, -2)$
17. [7.4] $(3, -2), (-2, 3)$ **18.** [7.4] $(2, 3), (3, 2)$
19. [7.4] 5 ft by 4 ft **20.** [7.4] 60 ft by 45 ft
21. [7.4]

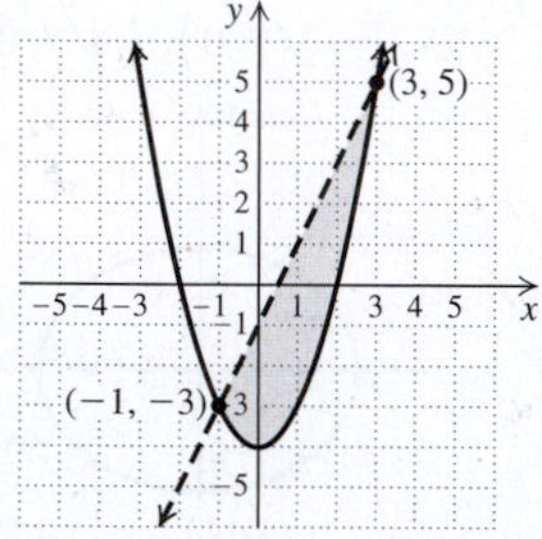

22. [7.1] A **23.** [7.2] $(x - 3)^2 + (y + 1)^2 = 8$

Chapter 8

Exercise Set 8.1

1. 3, 7, 11, 15; 39; 59 **3.** $2, \frac{3}{2}, \frac{4}{3}, \frac{5}{4}; \frac{10}{9}; \frac{15}{14}$ **5.** $0, \frac{3}{5}, \frac{4}{5}, \frac{15}{17}; \frac{99}{101}; \frac{112}{113}$
7. $-1, 4, -9, 16; 100; -225$ **9.** 7, 3, 7, 3; 3; 7 **11.** 34
13. 225 **15.** $-33{,}880$ **17.** 67 **19.** $2n$
21. $(-1)^n \cdot 2 \cdot 3^{n-1}$ **23.** $\dfrac{n+1}{n+2}$ **25.** $n(n + 1)$
27. $\log 10^{n-1}$, or $n - 1$ **29.** 6; 28 **31.** 20; 30
33. $\frac{1}{2} + \frac{1}{4} + \frac{1}{6} + \frac{1}{8} + \frac{1}{10} = \frac{137}{120}$ **35.** $1 + 2 + 4 + 8 + 16 + 32 + 64 = 127$ **37.** $\ln 7 + \ln 8 + \ln 9 + \ln 10 = \ln(7 \cdot 8 \cdot 9 \cdot 10) = \ln 5040 \approx 8.5252$ **39.** $\frac{1}{2} + \frac{2}{3} + \frac{3}{4} + \frac{4}{5} + \frac{5}{6} + \frac{6}{7} + \frac{7}{8} + \frac{8}{9} = \frac{15{,}551}{2520}$ **41.** $-1 + 1 - 1 + 1 - 1 + 1 - 1 = -1$
43. $3 - 6 + 9 - 12 + 15 - 18 + 21 - 24 = -12$
45. $2 + 1 + \frac{2}{5} + \frac{1}{5} + \frac{2}{17} + \frac{1}{13} + \frac{2}{37} = \frac{157{,}351}{40{,}885}$
47. $3 + 2 + 3 + 6 + 11 + 18 = 43$ **49.** $\frac{1}{2} + \frac{2}{3} + \frac{4}{5} + \frac{8}{9} + \frac{16}{17} + \frac{32}{33} + \frac{64}{65} + \frac{128}{129} + \frac{256}{257} + \frac{512}{513} + \frac{1024}{1025} \approx 9.736$
51. $\displaystyle\sum_{k=1}^{\infty} 5k$ **53.** $\displaystyle\sum_{k=1}^{6} (-1)^{k+1} 2^k$ **55.** $\displaystyle\sum_{k=1}^{6} (-1)^k \frac{k}{k+1}$
57. $\displaystyle\sum_{k=2}^{n} (-1)^k k^2$ **59.** $\displaystyle\sum_{k=1}^{\infty} \frac{1}{k(k+1)}$ **61.** $4, 1\frac{1}{4}, 1\frac{4}{5}, 1\frac{5}{9}$
63. $6561, -81, 9i, -3\sqrt{i}$ **65.** 2, 3, 5, 8 **67. (a)** 1062, 1127.84, 1197.77, 1272.03, 1350.90, 1434.65, 1523.60, 1618.07, 1718.39, 1824.93; **(b)** \$3330.35 **69.** \$9.80, \$10.90, \$12.00, \$13.10, \$14.20, \$15.30, \$16.40, \$17.50, \$18.60, \$19.70
71. 1, 1, 2, 3, 5, 8, 13 **72.** $(-1, -3)$
73. Illinois: 16,200 acres; Ohio: 7200 acres **74.** $(3, -2)$; 4
75. $\left(-\frac{5}{2}, 4\right); \dfrac{\sqrt{97}}{2}$ **77.** $i, -1, -i, 1, i; i$
79. $\ln(1 \cdot 2 \cdot 3 \cdot \cdots \cdot n)$

Exercise Set 8.2

1. $a_1 = 3, d = 5$ **3.** $a_1 = 9, d = -4$ **5.** $a_1 = \frac{3}{2}, d = \frac{3}{4}$
7. $a_1 = \$316, d = -\$3$ **9.** $a_{12} = 46$ **11.** $a_{14} = -\frac{17}{3}$
13. $a_{10} = \$7941.62$ **15.** 27th term **17.** 46th term
19. $a_1 = 5$ **21.** $n = 39$ **23.** $a_1 = \frac{1}{3}; d = \frac{1}{2}; \frac{1}{3}, \frac{5}{6}, \frac{4}{3}, \frac{11}{6}, \frac{7}{3}$
25. 670 **27.** 160,400 **29.** 735 **31.** 990 **33.** 1760
35. $\frac{65}{2}$ **37.** $-\frac{6026}{13}$ **39.** 4960¢, or \$49.60 **41.** 1320 seats
43. Yes; 32; 1600 ft **45.** 3 plants; 171 plants **47.** Yes; 3
48. $(2, 5)$ **49.** $(2, -1, 3)$ **50.** $(-4, 0), (4, 0); \left(-\sqrt{7}, 0\right), \left(\sqrt{7}, 0\right)$ **51.** $\dfrac{x^2}{4} + \dfrac{y^2}{25} = 1$ **53.** n^2
55. $a_1 = 60 - 5p - 5q; d = 5p + 2q - 20$ **57.** $5\frac{4}{5}, 7\frac{3}{5}, 9\frac{2}{5}, 11\frac{1}{5}$

Visualizing the Graph

1. J **2.** A **3.** C **4.** G **5.** F **6.** H **7.** E
8. D **9.** B **10.** I

Exercise Set 8.3

1. 2 **3.** -1 **5.** -2 **7.** 0.1 **9.** $\dfrac{a}{2}$ **11.** 128
13. 162 **15.** $7(5)^{40}$ **17.** 3^{n-1} **19.** $(-1)^{n-1}$ **21.** $\dfrac{1}{x^n}$
23. 762 **25.** $\frac{4921}{18}$ **27.** True **29.** True **31.** True
33. 8 **35.** 125 **37.** Does not exist **39.** $\frac{2}{3}$
41. $S_{11} \approx 29.65317$ **43.** 2 **45.** Does not exist
47. \$4545.45 **49.** $\frac{160}{9}$ **51.** $\frac{13}{99}$ **53.** 9 **55.** $\frac{34{,}091}{9990}$
57. \$2,684,354.55 **59. (a)** About 297 ft; **(b)** 300 ft **61.** \$39,505.71 **63.** 10,485.76 in.
65. \$19,694.01 **67.** \$86,666,666,667 **69.** $(f \circ g)(x) = 16x^2 + 40x + 25$; $(g \circ f)(x) = 4x^2 + 5$ **70.** $(f \circ g)(x) = x^2 + x + 2$; $(g \circ f)(x) = x^2 - x + 3$ **71.** 2.209
72. $\frac{1}{16}$ **73.** $\left(4 - \sqrt{6}\right)/\left(\sqrt{3} - \sqrt{2}\right) = 2\sqrt{3} + \sqrt{2}$, $\left(6\sqrt{3} - 2\sqrt{2}\right)/\left(4 - \sqrt{6}\right) = 2\sqrt{3} + \sqrt{2}$; there exists a common ratio, $2\sqrt{3} + \sqrt{2}$; thus the sequence is geometric.
75. (a) $\frac{13}{3}; \frac{22}{3}, \frac{34}{3}, \frac{46}{3}, \frac{58}{3}$; **(b)** $-\frac{11}{3}; -\frac{2}{3}, \frac{10}{3}, -\frac{50}{3}, \frac{250}{3}$ or 5; 8, 12, 18, 27
77. $S_n = \dfrac{x^2(1 - (-x)^n)}{x + 1}$

Exercise Set 8.4

1. $1^2 < 1^3$, false; $2^2 < 2^3$, true; $3^2 < 3^3$, true; $4^2 < 4^3$, true; $5^2 < 5^3$, true **3.** A polygon of 3 sides has $\dfrac{3(3-3)}{2}$ diagonals. True; A polygon of 4 sides has $\dfrac{4(4-3)}{2}$ diagonals. True; A polygon of 5 sides has $\dfrac{5(5-3)}{2}$ diagonals. True; A polygon of 6 sides has $\dfrac{6(6-3)}{2}$ diagonals. True; A polygon of 7 sides has $\dfrac{7(7-3)}{2}$ diagonals. True.

5. S_n: $2 + 4 + 6 + \cdots + 2n = n(n + 1)$
S_1: $2 = 1(1 + 1)$
S_k: $2 + 4 + 6 + \cdots + 2k = k(k + 1)$
S_{k+1}: $2 + 4 + 6 + \cdots + 2k + 2(k + 1)$
$= (k + 1)(k + 2)$
(1) *Basis step:* S_1 true by substitution.
(2) *Induction step:* Assume S_k. Deduce S_{k+1}.
Starting with the left side of S_{k+1}, we have
$2 + 4 + 6 + \cdots + 2k + 2(k + 1)$
$= k(k + 1) + 2(k + 1)$ **By S_k**
$= (k + 1)(k + 2).$ **Distributive law**

7. S_n: $1 + 5 + 9 + \cdots + (4n - 3) = n(2n - 1)$
S_1: $1 = 1(2 \cdot 1 - 1)$
S_k: $1 + 5 + 9 + \cdots + (4k - 3) = k(2k - 1)$
S_{k+1}: $1 + 5 + 9 + \cdots + (4k - 3) + [4(k + 1) - 3]$
$= (k + 1)[2(k + 1) - 1]$
$= (k + 1)(2k + 1)$
(1) *Basis step:* S_1 true by substitution.
(2) *Induction step:* Assume S_k. Deduce S_{k+1}.
Starting with the left side of S_{k+1}, we have
$1 + 5 + 9 + \cdots + (4k - 3) + [4(k + 1) - 3]$
$= k(2k - 1) + [4(k + 1) - 3]$ **By S_k**
$= 2k^2 - k + 4k + 4 - 3$
$= 2k^2 + 3k + 1$
$= (k + 1)(2k + 1).$

9. S_n: $2 + 4 + 8 + \cdots + 2^n = 2(2^n - 1)$
S_1: $2 = 2(2 - 1)$
S_k: $2 + 4 + 8 + \cdots + 2^k = 2(2^k - 1)$
S_{k+1}: $2 + 4 + 8 + \cdots + 2^k + 2^{k+1} = 2(2^{k+1} - 1)$
(1) *Basis step:* S_1 true by substitution.
(2) *Induction step:* Assume S_k. Deduce S_{k+1}.
Starting with the left side of S_{k+1}, we have
$\underbrace{2 + 4 + 8 + \cdots + 2^k} + 2^{k+1}$
$= \quad 2(2^k - 1) \quad + 2^{k+1}$ **By S_k**
$= 2^{k+1} - 2 + 2^{k+1}$
$= 2 \cdot 2^{k+1} - 2$
$= 2(2^{k+1} - 1).$

11. S_n: $n < n + 1$
S_1: $1 < 1 + 1$
S_k: $k < k + 1$
S_{k+1}: $k + 1 < (k + 1) + 1$
(1) *Basis step:* Since $1 < 1 + 1$, S_1 is true.
(2) *Induction step:* Assume S_k. Deduce S_{k+1}. Now
$k < k + 1$ **By S_k**
$k + 1 < k + 1 + 1$ **Adding 1**
$k + 1 < k + 2.$ **Simplifying**

13. S_n: $2n \le 2^n$
S_1: $2 \cdot 1 \le 2^1$
S_k: $2k \le 2^k$
S_{k+1}: $2(k + 1) \le 2^{k+1}$
(1) *Basis step:* Since $2 = 2$, S_1 is true.
(2) *Induction step:* Let k be any natural number.
Assume S_k. Deduce S_{k+1}.
$2k \le 2^k$ **By S_k**
$2 \cdot 2k \le 2 \cdot 2^k$ **Multiplying by 2**
$4k \le 2^{k+1}$
Since $1 \le k$, $k + 1 \le k + k$, or $k + 1 \le 2k$.
Then $2(k + 1) \le 4k$. **Multiplying by 2**
Thus, $2(k + 1) \le 4k \le 2^{k+1}$, so $2(k + 1) \le 2^{k+1}$.

15. S_n: $\dfrac{1}{1 \cdot 2 \cdot 3} + \dfrac{1}{2 \cdot 3 \cdot 4} + \dfrac{1}{3 \cdot 4 \cdot 5} + \cdots$
$+ \dfrac{1}{n(n + 1)(n + 2)} = \dfrac{n(n + 3)}{4(n + 1)(n + 2)}$
S_1: $\dfrac{1}{1 \cdot 2 \cdot 3} = \dfrac{1(1 + 3)}{4(1 + 1)(1 + 2)}$
S_k: $\dfrac{1}{1 \cdot 2 \cdot 3} + \dfrac{1}{2 \cdot 3 \cdot 4} + \cdots + \dfrac{1}{k(k + 1)(k + 2)}$
$= \dfrac{k(k + 3)}{4(k + 1)(k + 2)}$
S_{k+1}: $\dfrac{1}{1 \cdot 2 \cdot 3} + \dfrac{1}{2 \cdot 3 \cdot 4} + \cdots + \dfrac{1}{k(k + 1)(k + 2)}$
$+ \dfrac{1}{(k + 1)(k + 2)(k + 3)}$
$= \dfrac{(k + 1)(k + 1 + 3)}{4(k + 1 + 1)(k + 1 + 2)} = \dfrac{(k + 1)(k + 4)}{4(k + 2)(k + 3)}$
(1) *Basis step:* Since $\dfrac{1}{1 \cdot 2 \cdot 3} = \dfrac{1}{6}$ and
$\dfrac{1(1 + 3)}{4(1 + 1)(1 + 2)} = \dfrac{1 \cdot 4}{4 \cdot 2 \cdot 3} = \dfrac{1}{6}$, S_1 is true.
(2) *Induction step:* Assume S_k. Deduce S_{k+1}.
Add $\dfrac{1}{(k + 1)(k + 2)(k + 3)}$ on both sides of S_k and simplify the right side. Only the right side is shown here.
$\dfrac{k(k + 3)}{4(k + 1)(k + 2)} + \dfrac{1}{(k + 1)(k + 2)(k + 3)}$
$= \dfrac{k(k + 3)(k + 3) + 4}{4(k + 1)(k + 2)(k + 3)}$
$= \dfrac{k^3 + 6k^2 + 9k + 4}{4(k + 1)(k + 2)(k + 3)}$
$= \dfrac{(k + 1)^2(k + 4)}{4(k + 1)(k + 2)(k + 3)}$
$= \dfrac{(k + 1)(k + 4)}{4(k + 2)(k + 3)}$

17. S_n: $1 + 2 + 3 + \cdots + n = \dfrac{n(n + 1)}{2}$
S_1: $1 = \dfrac{1(1 + 1)}{2}$
S_k: $1 + 2 + 3 + \cdots + k = \dfrac{k(k + 1)}{2}$
S_{k+1}: $1 + 2 + 3 + \cdots + k + (k + 1) = \dfrac{(k + 1)(k + 2)}{2}$
(1) *Basis step:* S_1 true by substitution.
(2) *Induction step:* Assume S_k. Deduce S_{k+1}.
Starting with the left side of S_{k+1}, we have
$\underbrace{1 + 2 + 3 + \cdots + k} + (k + 1)$
$= \dfrac{k(k + 1)}{2} + (k + 1)$ **By S_k**
$= \dfrac{k(k + 1) + 2(k + 1)}{2}$ **Adding**
$= \dfrac{(k + 1)(k + 2)}{2}.$ **Distributive law**

19. S_n: $1^3 + 2^3 + 3^3 + \cdots + n^3 = \dfrac{n^2(n+1)^2}{4}$

S_1: $1^3 = \dfrac{1^2(1+1)^2}{4}$

S_k: $1^3 + 2^3 + 3^3 + \cdots + k^3 = \dfrac{k^2(k+1)^2}{4}$

S_{k+1}: $1^3 + 2^3 + 3^3 + \cdots + k^3 + (k+1)^3$
$= \dfrac{(k+1)^2[(k+1)+1]^2}{4}$

(1) *Basis step:* S_1: $1^3 = \dfrac{1^2(1+1)^2}{4} = 1$. True.

(2) *Induction step:* Assume S_k. Deduce S_{k+1}.

$$1^3 + 2^3 + \cdots + k^3 = \frac{k^2(k+1)^2}{4} \qquad S_k$$

$$1^3 + 2^3 + \cdots + k^3 + (k+1)^3 = \frac{k^2(k+1)^2}{4} + (k+1)^3$$

Adding $(k+1)^3$

$$= \frac{k^2(k+1)^2 + 4(k+1)^3}{4}$$
$$= \frac{(k+1)^2}{4}[k^2 + 4(k+1)]$$
$$= \frac{(k+1)^2}{4}(k^2 + 4k + 4)$$
$$= \frac{(k+1)^2(k+2)^2}{4}$$

21. S_n: $2 + 6 + 12 + \cdots + n(n+1) = \dfrac{n(n+1)(n+2)}{3}$

S_1: $1(1+1) = \dfrac{1(1+1)(1+2)}{3}$

S_k: $2 + 6 + 12 + \cdots + k(k+1) = \dfrac{k(k+1)(k+2)}{3}$

S_{k+1}:
$2 + 6 + 12 + \cdots + k(k+1) + (k+1)[(k+1)+1]$
$= \dfrac{(k+1)[(k+1)+1][(k+1)+2]}{3}$

(1) *Basis step:* S_1: $1(1+1) = \dfrac{1(1+1)(1+2)}{3}$. True.

(2) *Induction step:* Assume S_k:

$$2 + 6 + 12 + \cdots + k(k+1) = \frac{k(k+1)(k+2)}{3}.$$

Then $2 + 6 + 12 + \cdots + k(k+1) + (k+1)(k+1+1)$

$$= \frac{k(k+1)(k+2)}{3} + (k+1)(k+2)$$
$$= \frac{k(k+1)(k+2) + 3(k+1)(k+2)}{3}$$
$$= \frac{(k+1)(k+2)(k+3)}{3}$$
$$= \frac{(k+1)(k+1+1)(k+1+2)}{3}.$$

23. S_n: $a_1 + (a_1 + d) + (a_1 + 2d) + \cdots$
$+ [a_1 + (n-1)d] = \dfrac{n}{2}[2a_1 + (n-1)d]$

S_1: $a_1 = \dfrac{1}{2}[2a_1 + (1-1)d]$

S_k: $a_1 + (a_1 + d) + (a_1 + 2d) + \cdots$
$+ [a_1 + (k-1)d] = \dfrac{k}{2}[2a_1 + (k-1)d]$

S_{k+1}: $a_1 + (a_1 + d) + (a_1 + 2d) + \cdots$
$+ [a_1 + (k-1)d] + [a_1 + ((k+1)-1)d]$
$= \dfrac{k+1}{2}[2a_1 + ((k+1)-1)d]$

(1) *Basis step:* Since $\frac{1}{2}[2a_1 + (1-1)d] = \frac{1}{2}\cdot 2a_1 = a_1$, S_1 is true.

(2) *Induction step:* Assume S_k. Deduce S_{k+1}. Starting with the left side of S_{k+1}, we have

$$\underbrace{a_1 + (a_1 + d) + \cdots + [a_1 + (k-1)d]} + [a_1 + kd]$$
$$= \frac{k}{2}[2a_1 + (k-1)d] + [a_1 + kd]$$

By S_k

$$= \frac{k[2a_1 + (k-1)d]}{2} + \frac{2[a_1 + kd]}{2}$$
$$= \frac{2ka_1 + k(k-1)d + 2a_1 + 2kd}{2}$$
$$= \frac{2a_1(k+1) + k(k-1)d + 2kd}{2}$$
$$= \frac{2a_1(k+1) + (k-1+2)kd}{2}$$
$$= \frac{2a_1(k+1) + (k+1)kd}{2}$$
$$= \frac{k+1}{2}[2a_1 + kd].$$

24. $(5, 3)$ **25.** 1.5%: \$800; 2%: \$1600; 3%: \$2000

27. S_n: $x + y$ is a factor of $x^{2n} - y^{2n}$.
S_1: $x + y$ is a factor of $x^2 - y^2$.
S_k: $x + y$ is a factor of $x^{2k} - y^{2k}$.
S_{k+1}: $x + y$ is a factor of $x^{2(k+1)} - y^{2(k+1)}$.

(1) *Basis step:* S_1: $x + y$ is a factor of $x^2 - y^2$. True.
S_2: $x + y$ is a factor of $x^4 - y^4$. True.

(2) *Induction step:* Assume S_{k-1}: $x + y$ is a factor of $x^{2(k-1)} - y^{2(k-1)}$. Then $x^{2(k-1)} - y^{2(k-1)} = (x+y)Q(x)$ for some polynomial $Q(x)$.

Assume S_k: $x + y$ is a factor of $x^{2k} - y^{2k}$. Then $x^{2k} - y^{2k} = (x+y)P(x)$ for some polynomial $P(x)$.

$x^{2(k+1)} - y^{2(k+1)}$
$= (x^{2k} - y^{2k})(x^2 + y^2) - (x^{2(k-1)} - y^{2(k-1)})(x^2y^2)$
$= (x+y)P(x)(x^2 + y^2) - (x+y)Q(x)(x^2y^2)$
$= (x+y)[P(x)(x^2 + y^2) - Q(x)(x^2y^2)]$

so $x + y$ is a factor of $x^{2(k+1)} - y^{2(k+1)}$.

29. S_2: $\log_a(b_1b_2) = \log_a b_1 + \log_a b_2$
S_k: $\log_a(b_1b_2\cdots b_k) = \log_a b_1 + \log_a b_2 + \cdots + \log_a b_k$
S_{k+1}: $\log_a(b_1b_2\cdots b_{k+1}) = \log_a b_1 + \log_a b_2 + \cdots + \log_a b_{k+1}$

(1) *Basis step*: S_2 is true by the properties of logarithms.
(2) *Induction step*: Let k be a natural number $k \geq 2$. Assume S_k. Deduce S_{k+1}.

$\log_a(b_1 b_2 \cdots b_{k+1})$ **Left side of S_{k+1}**
$= \log_a(b_1 b_2 \cdots b_k) + \log_a b_{k+1}$ **By S_2**
$= \log_a b_1 + \log_a b_2 + \cdots + \log_a b_k + \log_a b_{k+1}$

31. S_2: $\overline{z_1 + z_2} = \overline{z_1} + \overline{z_2}$:
$\overline{(a + bi) + (c + di)} = \overline{(a + c) + (b + d)i}$
$= (a + c) - (b + d)i$
$\overline{(a + bi)} + \overline{(c + di)} = a - bi + c - di$
$= (a + c) - (b + d)i.$

S_k: $\overline{z_1 + z_2 + \cdots + z_k} = \overline{z_1} + \overline{z_2} + \cdots + \overline{z_k}$.
$\overline{(z_1 + z_2 + \cdots + z_k) + z_{k+1}}$
$= \overline{(z_1 + z_2 + \cdots + z_k)} + \overline{z_{k+1}}$ **By S_2**
$= \overline{z_1} + \overline{z_2} + \cdots + \overline{z_k} + \overline{z_{k+1}}$ **By S_k**

Mid-Chapter Mixed Review: Chapter 8

1. False **2.** True **3.** False **4.** False **5.** 8, 11, 14, 17; 32; 47 **6.** 0, −1, 2, −3; 8; −13 **7.** $a_n = 3n$ **8.** $a_n = (-1)^n n^2$ **9.** $1\frac{7}{8}$, or $\frac{15}{8}$ **10.** $2 + 6 + 12 + 20 + 30 = 70$ **11.** $\sum_{k=1}^{\infty} (-1)^k 4k$ **12.** 2, 6, 22, 86 **13.** −5 **14.** 22 **15.** 21 **16.** 696 **17.** $-\frac{1}{2}$ **18.** (a) 8; (b) $\frac{1023}{16}$, or 63.9375 **19.** $-\frac{16}{3}$ **20.** Does not exist **21.** 126 plants **22.** \$6369.70

23. S_n: $1 + 4 + 7 + \cdots + (3n - 2) = \frac{1}{2}n(3n - 1)$
S_1: $3 \cdot 1 - 2 = \frac{1}{2} \cdot 1(3 \cdot 1 - 1)$
S_k: $1 + 4 + 7 + \cdots + (3k - 2) = \frac{1}{2}k(3k - 1)$
S_{k+1}: $1 + 4 + 7 + \cdots + (3k - 2) + [3(k + 1) - 2]$
$= \frac{1}{2}(k + 1)[3(k + 1) - 1]$
$= \frac{1}{2}(k + 1)(3k + 2)$

(1) *Basis step*: S_1: $3 \cdot 1 - 2 = \frac{1}{2} \cdot 1(3 \cdot 1 - 1)$. True
(2) *Induction step*: Assume S_k:
$1 + 4 + 7 + \cdots + (3k - 2) = \frac{1}{2}k(3k - 1)$.

Then $1 + 4 + 7 + \cdots + (3k - 2) + [3(k + 1) - 2]$
$= \frac{1}{2}k(3k - 1) + [3(k + 1) - 2]$
$= \frac{3}{2}k^2 - \frac{1}{2}k + 3k + 1$
$= \frac{3}{2}k^2 + \frac{5}{2}k + 1$
$= \frac{1}{2}(3k^2 + 5k + 2)$
$= \frac{1}{2}(k + 1)(3k + 2)$.

24. The first formula can be derived from the second by substituting $a_1 + (n - 1)d$ for a_n. When the first and last terms of the sum are known, the second formula is the better one to use. If the last term is not known, the first formula allows us to compute the sum in one step without first finding a_n.

25. $1 + 2 + 3 + \cdots + 100$
$= (1 + 100) + (2 + 99) + (3 + 98) + \cdots + (50 + 51)$
$= \underbrace{101 + 101 + 101 + \cdots + 101}_{\text{50 addends of 101}}$
$= 50 \cdot 101$
$= 5050$

A formula for the first n natural numbers is $\frac{n}{2}(1 + n)$.

26. Answers may vary. One possibility is given. Casey invests \$900 at 8% interest, compounded annually. How much will be in the account at the end of 40 years? **27.** We can prove an infinite sequence of statements S_n by showing that a basis statement S_1 is true and then that for all natural numbers k, if S_k is true, then S_{k+1} is true.

Exercise Set 8.5

1. 720 **3.** 604,800 **5.** 120 **7.** 1 **9.** 3024 **11.** 120 **13.** 120 **15.** 1 **17.** 6,497,400 **19.** $n(n - 1)(n - 2)$ **21.** n **23.** $6! = 720$ **25.** $9! = 362,880$ **27.** ${}_9P_4 = 3024$ **29.** ${}_5P_5 = 120$; $5^5 = 3125$ **31.** ${}_5P_5 \cdot {}_4P_4 = 2880$ **33.** $8 \cdot 10^6 = 8,000,000$; 8 million **35.** $\frac{9!}{2!\,3!\,4!} = 1260$ **37.** (a) ${}_6P_5 = 720$; (b) $6^5 = 7776$; (c) $1 \cdot {}_5P_4 = 120$; (d) $1 \cdot 1 \cdot {}_4P_3 = 24$ **39.** (a) 10^5, or 100,000; (b) 100,000 **41.** (a) $10^9 = 1,000,000,000$; (b) yes **42.** $\frac{9}{4}$, or 2.25 **43.** −3, 2 **44.** $\frac{3 \pm \sqrt{17}}{4}$ **45.** −2, 1, 5 **47.** 8 **49.** 11 **51.** $n - 1$

Exercise Set 8.6

1. 78 **3.** 78 **5.** 7 **7.** 10 **9.** 1 **11.** 15 **13.** 128 **15.** 270,725 **17.** 13,037,895 **19.** n **21.** 1 **23.** ${}_{36}C_4 = 58,905$ **25.** ${}_{13}C_{10} = 286$ **27.** $\binom{10}{7} \cdot \binom{5}{3} = 1200$ **29.** $\binom{52}{5} = 2,598,960$ **31.** (a) ${}_{31}P_2 = 930$; (b) $31^2 = 961$; (c) ${}_{31}C_2 = 465$ **33.** $-\frac{17}{2}$ **34.** $-1, \frac{3}{2}$ **35.** $\frac{-5 \pm \sqrt{21}}{2}$ **36.** −4, −2, 3 **37.** $\binom{13}{5} = 1287$ **39.** $\binom{n}{2}$; $2\binom{n}{2}$ **41.** 4 **43.** 7

45. Line segments:
$${}_nC_2 = \frac{n!}{2!(n - 2)!} = \frac{n(n - 1)(n - 2)!}{2 \cdot 1 \cdot (n - 2)!} = \frac{n(n - 1)}{2}$$

Diagonals: The n line segments that form the sides of the n-gon are not diagonals. Thus the number of diagonals is
$${}_nC_2 - n = \frac{n(n - 1)}{2} - n$$
$$= \frac{n^2 - n - 2n}{2} = \frac{n^2 - 3n}{2}$$
$$= \frac{n(n - 3)}{2}, n \geq 4.$$

Let D_n be the number of diagonals of an n-gon. Prove the result above for diagonals using mathematical induction.

S_n: $D_n = \frac{n(n - 3)}{2}$, for $n = 4, 5, 6, \ldots$
S_4: $D_4 = \frac{4 \cdot 1}{2}$
S_k: $D_k = \frac{k(k - 3)}{2}$
S_{k+1}: $D_{k+1} = \frac{(k + 1)(k - 2)}{2}$

(1) *Basis step*: S_4 is true (a quadrilateral has 2 diagonals).
(2) *Induction step*: Assume S_k. Note that when an additional vertex V_{k+1} is added to the k-gon, we gain k segments, 2 of which are sides of the $(k + 1)$-gon, and a former side $\overline{V_1 V_k}$ becomes a diagonal. Thus the additional number of diagonals is $k - 2 + 1$, or $k - 1$. Then the new total of diagonals is $D_k + (k - 1)$, or

$D_{k+1} = D_k + (k - 1)$
$= \frac{k(k - 3)}{2} + (k - 1)$ **By S_k**
$= \frac{(k + 1)(k - 2)}{2}$

Exercise Set 8.7

1. $x^4 + 20x^3 + 150x^2 + 500x + 625$
3. $x^5 - 15x^4 + 90x^3 - 270x^2 + 405x - 243$
5. $x^5 - 5x^4y + 10x^3y^2 - 10x^2y^3 + 5xy^4 - y^5$
7. $15{,}625x^6 + 75{,}000x^5y + 150{,}000x^4y^2 + 160{,}000x^3y^3 + 96{,}000x^2y^4 + 30{,}720xy^5 + 4096y^6$
9. $128t^7 + 448t^5 + 672t^3 + 560t + 280t^{-1} + 84t^{-3} + 14t^{-5} + t^{-7}$ **11.** $x^{10} - 5x^8 + 10x^6 - 10x^4 + 5x^2 - 1$
13. $125 + 150\sqrt{5}\,t + 375t^2 + 100\sqrt{5}\,t^3 + 75t^4 + 6\sqrt{5}\,t^5 + t^6$ **15.** $a^9 - 18a^7 + 144a^5 - 672a^3 + 2016a - 4032a^{-1} + 5376a^{-3} - 4608a^{-5} + 2304a^{-7} - 512a^{-9}$
17. $140\sqrt{2}$ **19.** $x^{-8} + 4x^{-4} + 6 + 4x^4 + x^8$ **21.** $21a^5b^2$
23. $-252x^5y^5$ **25.** $-745{,}472a^3$ **27.** $1120x^{12}y^2$
29. $-1{,}959{,}552u^5v^{10}$ **31.** 2^7, or 128 **33.** 2^{24}, or 16,777,216
35. 20 **37.** $-12 + 316i$ **39.** $-7 - 4\sqrt{2}i$
41. $\sum_{k=0}^{n}\binom{n}{k}(-1)^k a^{n-k}b^k$ **43.** $\sum_{k=1}^{n}\binom{n}{k}x^{n-k}h^{k-1}$
44. $x^2 + 2x - 2$ **45.** $2x^3 - 3x^2 + 2x - 3$
46. $4x^2 - 12x + 10$ **47.** $2x^2 - 1$ **49.** $3, 9, 6 \pm 3i$
51. $-\dfrac{35}{x^{1/6}}$ **53.** 2^{100} **55.** $[\log_a (xt)]^{23}$

57. **(1)** *Basis step*: Since $a + b = (a + b)^1$, S_1 is true.
(2) *Induction step*: Let S_k be the statement of the binomial theorem with n replaced by k. Multiply both sides of S_k by $(a + b)$ to obtain

$$(a+b)^{k+1} = \left[a^k + \cdots + \binom{k}{r-1}a^{k-(r-1)}b^{r-1} + \binom{k}{r}a^{k-r}b^r + \cdots + b^k\right](a+b)$$
$$= a^{k+1} + \cdots + \left[\binom{k}{r-1} + \binom{k}{r}\right]a^{(k+1)-r}b^r + \cdots + b^{k+1}$$
$$= a^{k+1} + \cdots + \binom{k+1}{r}a^{(k+1)-r}b^r + \cdots + b^{k+1}.$$

This proves S_{k+1}, assuming S_k. Hence S_n is true for $n = 1, 2, 3, \ldots$.

Exercise Set 8.8

1. **(a)** 0.18, 0.24, 0.23, 0.23, 0.12; **(b)** Opinions may vary, but it seems that people tend not to pick the first or last numbers.
3. 5187 e-mails **5.** **(a)** $\frac{2}{7}$; **(b)** $\frac{5}{7}$; **(c)** 0; **(d)** 1 **7.** $\frac{1}{2}$
9. **(a)** $\frac{1}{13}$; **(b)** $\frac{2}{13}$; **(c)** $\frac{1}{4}$; **(d)** $\frac{1}{26}$ **11.** $\frac{1}{5525}$ **13.** $\frac{135}{323}$
15. $\frac{1}{108{,}290}$ **17.** $\frac{33}{66{,}640}$ **19.** **(a)** HHH, HHT, HTH, HTT, THH, THT, TTH, TTT; **(b)** $\frac{3}{8}$; **(c)** $\frac{7}{8}$; **(d)** $\frac{7}{8}$; **(e)** $\frac{3}{8}$ **21.** $\frac{9}{19}$
23. $\frac{1}{38}$ **25.** $\frac{18}{19}$ **27.** $\frac{9}{19}$ **29.** Zero **30.** One-to-one
31. Function; domain; range; domain; range **32.** Zero
33. Combination **34.** Inverse variation **35.** Factor
36. Geometric sequence
37. **(a)** $\binom{13}{2}\cdot\binom{4}{2}\cdot\binom{4}{2}\cdot\binom{44}{1} = 123{,}552$
(b) 0.0475 **39.** **(a)** $13\cdot\binom{4}{3}\cdot\binom{48}{2} - 3744$, or 54,912;
(b) $\dfrac{54{,}912}{\binom{52}{5}} \approx 0.0211$

Review Exercises: Chapter 8

1. True **2.** False **3.** True **4.** False
5. $-\frac{1}{2}, \frac{4}{17}, -\frac{9}{82}, \frac{16}{257}; -\frac{121}{14{,}642}, -\frac{529}{279{,}842}$ **6.** $(-1)^{n+1}(n^2 + 1)$
7. $\frac{3}{2} - \frac{9}{8} + \frac{27}{26} - \frac{81}{80} = \frac{417}{1040}$ **8.** $\sum_{k=1}^{7}(k^2 - 1)$ **9.** $\frac{15}{4}$
10. $a + 4b$ **11.** 531 **12.** 20,100 **13.** 11 **14.** -4
15. $n = 6$, $S_n = -126$ **16.** $a_1 = 8$, $a_5 = \frac{1}{2}$ **17.** Does not exist **18.** $\frac{3}{11}$ **19.** $\frac{3}{8}$ **20.** $\frac{241}{99}$ **21.** $5\frac{4}{5}, 6\frac{3}{5}, 7\frac{2}{5}, 8\frac{1}{5}$
22. 167.3 ft **23.** \$45,993.04 **24.** **(a)** \$7.38; **(b)** \$1365.10
25. \$88,888,888,889

26.
S_n: $1 + 4 + 7 + \cdots + (3n - 2) = \dfrac{n(3n-1)}{2}$
S_1: $1 = \dfrac{1(3-1)}{2}$
S_k: $1 + 4 + 7 + \cdots + (3k - 2) = \dfrac{k(3k-1)}{2}$
S_{k+1}: $1 + 4 + 7 + \cdots + (3k - 2) + [3(k+1) - 2]$
$= 1 + 4 + 7 + \cdots + (3k - 2) + (3k + 1)$
$= \dfrac{(k+1)(3k+2)}{2}$

(1) *Basis step*: $\dfrac{1(3-1)}{2} = \dfrac{2}{2} = 1$ is true.
(2) *Induction step*: Assume S_k. Add $(3k + 1)$ on both sides.
$1 + 4 + 7 + \cdots + (3k - 2) + (3k + 1)$
$= \dfrac{k(3k-1)}{2} + (3k+1) = \dfrac{k(3k-1)}{2} + \dfrac{2(3k+1)}{2}$
$= \dfrac{3k^2 - k + 6k + 2}{2} = \dfrac{3k^2 + 5k + 2}{2}$
$= \dfrac{(k+1)(3k+2)}{2}$

27.
S_n: $1 + 3 + 3^2 + \cdots + 3^{n-1} = \dfrac{3^n - 1}{2}$
S_1: $1 = \dfrac{3^1 - 1}{2}$
S_k: $1 + 3 + 3^2 + \cdots + 3^{k-1} = \dfrac{3^k - 1}{2}$
S_{k+1}: $1 + 3 + 3^2 + \cdots + 3^{(k+1)-1} = \dfrac{3^{k+1} - 1}{2}$

(1) *Basis step*: $\dfrac{3^1 - 1}{2} = \dfrac{2}{2} = 1$ is true.
(2) *Induction step*: Assume S_k. Add 3^k on both sides.
$1 + 3 + \cdots + 3^{k-1} + 3^k$
$= \dfrac{3^k - 1}{2} + 3^k = \dfrac{3^k - 1}{2} + 3^k\cdot\dfrac{2}{2}$
$= \dfrac{3\cdot 3^k - 1}{2} = \dfrac{3^{k+1} - 1}{2}$

28.
S_n: $\left(1 - \frac{1}{2}\right)\left(1 - \frac{1}{3}\right)\cdots\left(1 - \frac{1}{n}\right) = \frac{1}{n}$
S_2: $\left(1 - \frac{1}{2}\right) = \frac{1}{2}$
S_k: $\left(1 - \frac{1}{2}\right)\left(1 - \frac{1}{3}\right)\cdots\left(1 - \frac{1}{k}\right) = \frac{1}{k}$
S_{k+1}: $\left(1 - \frac{1}{2}\right)\left(1 - \frac{1}{3}\right)\cdots\left(1 - \frac{1}{k}\right)\left(1 - \frac{1}{k+1}\right) = \frac{1}{k+1}$.

(1) *Basis step*: S_2 is true by substitution.
(2) *Induction step*: Assume S_k. Deduce S_{k+1}. Starting with the left side of S_{k+1}, we have

$$\underbrace{\left(1-\frac{1}{2}\right)\left(1-\frac{1}{3}\right)\cdots\left(1-\frac{1}{k}\right)}\left(1-\frac{1}{k+1}\right)$$

$$=\frac{1}{k}\cdot\left(1-\frac{1}{k+1}\right) \quad \textbf{By } S_k$$
$$=\frac{1}{k}\cdot\left(\frac{k+1-1}{k+1}\right)$$
$$=\frac{1}{k}\cdot\frac{k}{k+1}$$
$$=\frac{1}{k+1}. \quad \textbf{Simplifying}$$

29. $6! = 720$ **30.** $9\cdot8\cdot7\cdot6 = 3024$ **31.** $\binom{15}{8} = 6435$
32. $24\cdot23\cdot22 = 12{,}144$ **33.** $\frac{9!}{1!\,4!\,2!\,2!} = 3780$
34. $3\cdot4\cdot3 = 36$ **35. (a)** ${}_6P_5 = 720$; **(b)** $6^5 = 7776$; **(c)** ${}_5P_4 = 120$; **(d)** ${}_3P_2 = 6$ **36.** 2^8, or 256
37. $m^7 + 7m^6n + 21m^5n^2 + 35m^4n^3 + 35m^3n^4 + 21m^2n^5 + 7mn^6 + n^7$ **38.** $x^5 - 5\sqrt{2}x^4 + 20x^3 - 20\sqrt{2}x^2 + 20x - 4\sqrt{2}$ **39.** $x^8 - 12x^6y + 54x^4y^2 - 108x^2y^3 + 81y^4$ **40.** $a^8 + 8a^6 + 28a^4 + 56a^2 + 70 + 56a^{-2} + 28a^{-4} + 8a^{-6} + a^{-8}$ **41.** $-6624 + 16{,}280i$
42. $220a^9x^3$ **43.** $-\binom{18}{11}128a^7b^{11}$ **44.** $\frac{1}{12}$; 0 **45.** $\frac{1}{4}$
46. $\frac{6}{5525}$ **47.** $\frac{86}{206} \approx 0.42, \frac{97}{206} \approx 0.47, \frac{23}{206} \approx 0.11$ **48.** B
49. A **50.** D **51. (a)** No (unless a_n is all positive or all negative); **(b)** yes; **(c)** yes; **(d)** no (unless a_n is constant); **(e)** no (unless a_n is constant); **(f)** no (unless a_n is constant)
52. $\frac{a_{k+1}}{a_k} = r_1, \frac{b_{k+1}}{b_k} = r_2$, so $\frac{a_{k+1}b_{k+1}}{a_kb_k} = r_1r_2$, a constant.
53. $\frac{1}{2}, -\frac{1}{6}, \frac{1}{18}$ **54.** $-2, 0, 2, 4$ **55.** $\left(\log\frac{x}{y}\right)^{10}$ **56.** 18
57. 36 **58.** -9 **59.** For each circular arrangement of the numbers on a clock face, there are 12 distinguishable ordered arrangements on a line. The number of arrangements of 12 objects on a line is ${}_{12}P_{12}$, or 12!. Thus the number of circular permutations is $\frac{{}_{12}P_{12}}{12} = \frac{12!}{12} = 11! = 39{,}916{,}800$. In general, for each circular arrangement of n objects, there are n distinguishable ordered arrangements on a line. The total number of arrangements of n objects on a line is ${}_nP_n$, or $n!$. Thus the number of circular permutations is $\frac{n!}{n} = \frac{n(n-1)!}{n} = (n-1)!$. **60.** Put the following in the form of a paragraph. First find the number of seconds in a year (365 days): $365 \text{ days}\cdot\frac{24 \text{ hr}}{1 \text{ day}}\cdot\frac{60 \text{ min}}{1 \text{ hr}}\cdot\frac{60 \text{ sec}}{1 \text{ min}} = 31{,}536{,}000$ sec. The number of arrangements possible is 15!. The time is $\frac{15!}{31{,}536{,}000} \approx 41{,}466$ years. **61.** Order is considered in a combination lock. **62.** Choosing k objects from a set of n objects is equivalent to not choosing the other $n-k$ objects.

Test: Chapter 8

1. [8.1] -43 **2.** [8.1] $\frac{2}{3}, \frac{3}{4}, \frac{4}{5}, \frac{5}{6}, \frac{6}{7}$
3. [8.1] $2 + 5 + 10 + 17 = 34$ **4.** [8.1] $\sum_{k=1}^{6} 4k$
5. [8.1] $\sum_{k=1}^{\infty} 2^k$ **6.** [8.1] $3, 2\frac{1}{3}, 2\frac{3}{7}, 2\frac{7}{17}$ **7.** [8.2] 44
8. [8.2] 38 **9.** [8.2] -420 **10.** [8.2] 675 **11.** [8.3] $\frac{5}{512}$
12. [8.3] 1000 **13.** [8.3] 510 **14.** [8.3] 27 **15.** [8.3] $\frac{56}{99}$
16. [8.1] \$10,000, \$8000, \$6400, \$5120, \$4096, \$3276.80
17. [8.2] \$17.05 **18.** [8.3] \$74,399.77
19. [8.4]

$$S_n:\quad 2 + 5 + 8 + \cdots + (3n-1) = \frac{n(3n+1)}{2}$$
$$S_1:\quad 2 = \frac{1(3\cdot1+1)}{2}$$
$$S_k:\quad 2 + 5 + 8 + \cdots + (3k-1) = \frac{k(3k+1)}{2}$$
$$S_{k+1}:\quad 2 + 5 + 8 + \cdots + (3k-1) + [3(k+1)-1] = \frac{(k+1)[3(k+1)+1]}{2}$$

(1) *Basis step*: $\frac{1(3\cdot1+1)}{2} = \frac{1\cdot4}{2} = 2$, so S_1 is true.
(2) *Induction step*:

$$2 + 5 + 8 + \cdots + (3k-1) + [3(k+1)-1]$$
$$= \frac{k(3k+1)}{2} + [3k+3-1] \quad \textbf{By } S_k$$
$$= \frac{3k^2}{2} + \frac{k}{2} + 3k + 2$$
$$= \frac{3k^2}{2} + \frac{7k}{2} + 2 = \frac{3k^2+7k+4}{2}$$
$$= \frac{(k+1)(3k+4)}{2} = \frac{(k+1)[3(k+1)+1]}{2}$$

20. [8.5] 3,603,600 **21.** [8.6] 352,716
22. [8.6] $\frac{n(n-1)(n-2)(n-3)}{24}$ **23.** [8.5] ${}_6P_4 = 360$
24. [8.5] **(a)** $6^4 = 1296$; **(b)** ${}_5P_3 = 60$
25. [8.6] ${}_{28}C_4 = 20{,}475$ **26.** [8.6] ${}_{12}C_8\cdot{}_8C_4 = 34{,}650$
27. [8.7] $x^5 + 5x^4 + 10x^3 + 10x^2 + 5x + 1$
28. [8.7] $35x^3y^4$ **29.** [8.7] $2^9 = 512$ **30.** [8.8] $\frac{4}{7}$
31. [8.8] $\frac{48}{1001}$ **32.** [8.1] B **33.** [8.5] 15

Just-In-Time

1. Real Numbers

1. $\frac{2}{3}, 6, -2.45, 18.\overline{4}, -11, \sqrt[3]{27}, 5\frac{1}{6}, -\frac{8}{7}, 0, \sqrt{16}$
2. $\frac{2}{3}, -2.45, 18.\overline{4}, 5\frac{1}{6}, -\frac{8}{7}$
3. $\sqrt{3}, \sqrt[6]{26}, 7.151551555\ldots, -\sqrt{35}, \sqrt[5]{3}$
4. $6, -11, \sqrt[3]{27}, 0, \sqrt{16}$ **5.** $6, \sqrt[3]{27}, 0, \sqrt{16}$ **6.** All of them

2. Properties of Real Numbers

1. Additive inverse property **2.** Associative property of multiplication **3.** Distributive property **4.** Commutative property of addition **5.** Multiplicative identity property **6.** Commutative property of multiplication **7.** Additive identity property **8.** Multiplicative inverse property **9.** Associative property of addition **10.** Distributive property

3. Absolute Value

1. 98 **2.** 0 **3.** 4.7 **4.** $\frac{2}{3}$ **5.** 20 **6.** 12.6 **7.** 11 **8.** $\frac{21}{8}$

4. Operations with Real Numbers

1. 19 **2.** $\frac{1}{10}$ **3.** -5 **4.** -3 **5.** -350 **6.** -5.5 **7.** 24 **8.** 10 **9.** -12.6 **10.** 20 **11.** -15 **12.** $-\frac{1}{6}$ **13.** -8 **14.** -22 **15.** $\frac{4}{5}$

5. Order on the Number Line

1. False **2.** True **3.** True **4.** True **5.** False **6.** True

6. Interval Notation

1. $[-5, 5]$ **2.** $(-3, -1]$ **3.** $(-\infty, -2]$ **4.** $(3.8, \infty)$ **5.** $(7, \infty)$ **6.** $(-2, 2)$ **7.** $(-4, 5)$ **8.** $[1.7, \infty)$ **9.** $(-5, -2]$ **10.** $(-\infty, \sqrt{5})$

7. Integers as Exponents

1. $\frac{1}{3^6}$ **2.** $(0.2)^5$ **3.** $\frac{z^9}{w^4}$ **4.** $\frac{z^2}{y^2}$ **5.** 1 **6.** a^8 **7.** $-6x^{-4}y^4$, or $-\frac{6y^4}{x^4}$ **8.** x^{-11}, or $\frac{1}{x^{11}}$ **9.** $m^{-6}n^{-6}$, or $\frac{1}{m^6n^6}$ **10.** t^{-20}, or $\frac{1}{t^{20}}$

8. Scientific Notation

1. 1.85×10^7 **2.** 7.86×10^{-4} **3.** 2.3×10^{-9} **4.** 8.927×10^9 **5.** 0.000000043 **6.** 5,170,000 **7.** 620,300,000,000 **8.** 0.0000294

9. Order of Operations

1. 3 **2.** 103 **3.** -235 **4.** 2048 **5.** 2 **6.** 5

10. Introduction to Polynomials

1. 6 **2.** 8 **3.** 4 **4.** 0 **5.** 8 **6.** Binomial **7.** Monomial **8.** Trinomial

11. Add and Subtract Polynomials

1. $9y - 4$ **2.** $-2x^2 + 6x - 2$ **3.** $3x + 2y - 2z - 3$ **4.** $2ab^2 - a^2b + 6ab + 10$ **5.** $-4x^2 + 8xy - 5y^2 + 3$

12. Multiply Polynomials

1. $-21a^6$ **2.** $y^2 + 2y - 15$ **3.** $x^2 + 9x + 18$ **4.** $2a^2 + 13a + 15$ **5.** $4x^2 + 8xy + 3y^2$ **6.** $x^2 + 6x + 9$ **7.** $25x^2 - 30x + 9$ **8.** $4x^2 + 12xy + 9y^2$ **9.** $n^2 - 36$ **10.** $9y^2 - 16$

13. Factor Polynomials

1. $3(x + 6)$ **2.** $2z^2(z - 4)$ **3.** $(3x - 1)(x^2 + 6)$ **4.** $(t + 6)(t^2 - 2)$ **5.** $(w - 5)(w - 2)$ **6.** $(t + 3)(t + 5)$ **7.** $2(n - 12)(n + 2)$ **8.** $y^2(y - 2)(y - 7)$ **9.** $(2n - 7)(n + 8)$ **10.** $(2y - 3)(y + 2)$ **11.** $(z + 9)(z - 9)$ **12.** $(4x + 3)(4x - 3)$ **13.** $7p(q^2 + y^2)(q + y)(q - y)$ **14.** $(x + 6)^2$ **15.** $(3z - 2)^2$ **16.** $a(a + 12)^2$ **17.** $(x + 4)(x^2 - 4x + 16)$ **18.** $(m - 6)(m^2 + 6m + 36)$ **19.** $3a^2(a - 2)(a^2 + 2a + 4)$ **20.** $(t^2 + 1)(t^4 - t^2 + 1)$

14. Equation-Solving Principles

1. 10 **2.** 12 **3.** -4 **4.** 10 **5.** 2 **6.** -3 **7.** 0 **8.** 5

15. Inequality-Solving Principles

1. $p \geq -125$ **2.** $x < -9$ **3.** $x < 2$ **4.** $x \leq -56$ **5.** $y < 27$ **6.** $w \geq 11$

16. Principle of Zero Products

1. $-7, 1$ **2.** $-\frac{3}{5}, 4$ **3.** $-\frac{5}{3}, \frac{1}{2}$ **4.** 0, 8 **5.** $-3, 11$ **6.** $-15, 2$ **7.** $-\frac{3}{4}, \frac{4}{3}$

17. Principle of Square Roots

1. -6 and 6, or ± 6 **2.** $-\sqrt{10}$ and $\sqrt{10}$, or $\pm\sqrt{10}$ **3.** $-\sqrt{3}$ and $\sqrt{3}$, or $\pm\sqrt{3}$ **4.** $-\sqrt{5}$ and $\sqrt{5}$, or $\pm\sqrt{5}$ **5.** -5 and 5, or ± 5 **6.** $-\sqrt{15}$ and $\sqrt{15}$, or $\pm\sqrt{15}$

18. Simplify Rational Expressions

1. The set of all real numbers except 0 and 1. **2.** The set of all real numbers except -7 and 3. **3.** $\frac{x + 2}{x - 2}$ **4.** $\frac{x - 1}{x - 3}$ **5.** $\frac{x - 3}{x}$ **6.** $\frac{2(y + 4)}{y - 1}$

19. Multiply and Divide Rational Expressions

1. 1 **2.** $m + n$ **3.** $\frac{4x + 1}{3x - 2}$ **4.** $\frac{a + 1}{a - 3}$ **5.** $\frac{3(x - 4)}{2(x + 4)}$ **6.** $\frac{1}{x + y}$

20. Add and Subtract Rational Expressions

1. 2 **2.** $\frac{2(3x^2 + 2x - 7)}{3(3x + 1)(x - 2)}$ **3.** $\frac{2a}{(a + 1)(a - 1)}$
4. $\frac{3x - 4}{(x - 2)(x - 1)}$ **5.** $\frac{-y + 10}{(y + 4)(y - 5)}$ **6.** $\frac{y}{(y - 2)(y - 3)}$

21. Simplify Complex Rational Expressions

1. $x - y$ **2.** $\frac{a}{a + b}$ **3.** $\frac{w^2 - 2w + 4}{w}$ **4.** $\frac{x + y}{x}$
5. $-a - b$

22. Simplify Radical Expressions

1. 21 **2.** $3y$ **3.** $a - 2$ **4.** $-3x$ **5.** $3x^2$ **6.** 2
7. $2xy\sqrt[4]{3x^2}$ **8.** $5\sqrt{21}$ **9.** $\sqrt{5y}$ **10.** $\frac{1}{2x}$ **11.** $x - 2$
12. $2x^2y\sqrt{6}$ **13.** $3x\sqrt[3]{4y}$ **14.** $17\sqrt{2}$ **15.** $12\sqrt{3}$
16. $2\sqrt{2}$ **17.** $4\sqrt{5}$ **18.** $16 + 9\sqrt{3}$ **19.** -12
20. $4 + 2\sqrt{3}$

23. Rationalizing Denominators and Numerators

1. $\frac{4\sqrt{11}}{11}$ **2.** $\frac{\sqrt{21}}{7}$ **3.** $\frac{\sqrt[3]{28}}{2}$ **4.** $\frac{2\sqrt[3]{6}}{3}$
5. $\frac{3\sqrt{30} + 12}{14}$ **6.** $\frac{\sqrt{3}}{3}$ **7.** $\frac{6\sqrt{m} + 6\sqrt{n}}{m - n}$

24. Rational Exponents

1. $\sqrt[6]{y^5}$ **2.** $\sqrt[3]{x^2}$ **3.** 8 **4.** 128 **5.** $\frac{1}{5}$ **6.** $\frac{1}{16}$ **7.** $y^{1/3}$
8. $x^{5/2}$ **9.** $x\sqrt[6]{x}$ **10.** $(a - 2)^2$ **11.** $n\sqrt[3]{mn^2}$

25. Pythagorean Theorem

1. 17 **2.** $\sqrt{32} \approx 5.657$ **3.** 12 **4.** 5 **5.** $\sqrt{31} \approx 5.568$

Photo Credits

CHAPTER 1: 1, Kzenon/Shutterstock **8,** Fresh_water/Fotolia **17,** Rivision/Fotolia **18,** Pete Marovich/ZUMA Press/Alamy **19,** Leo Mason/Alamy Live News **26,** JTB Media Creation, Inc./Alamy **31,** Edelweiss/Fotolia **38 (top),** Ruben Sprich/Reuters **38 (middle),** Olesya Shelomova/Fotolia **42,** Rob/Fotolia **45,** epa european pressphoto agency b.v./Alamy **46 (left),** Deyan Georgiev/Fotolia **46 (right),** Catmando/Fotolia **47,** Glow Images **66,** Jeff Greenberg 3 of 6/Alamy **67,** Ian G. Dagnall/Alamy **70,** Braden Gunem/Alamy **71,** Stuart Monk/Fotolia **75,** Donvictorio/Shutterstock **77 (bottom left),** Bepsphoto/Fotolia **77 (top right),** Moreno Soppelsa/Shutterstock **77 (bottom right),** Kzenon/Shutterstock **82,** Jhk2303/Fotolia **83,** Spaceport9/Fotolia **93,** Andres Rodriguez/Fotolia **CHAPTER 2: 97,** Blend Images/Glow Images **107,** Taina Sohlman/Fotolia **108,** Gladskikh Tatiana/Shutterstock **118,** Pkstock/Fotolia **124,** Africa Studio/Fotolia **125,** Blend Images/Glow Images **127 (left),** WavebreakMediaMicro/Fotolia **127 (top middle),** Sor Jong Ang/Fotolia **127 (bottom middle),** Velirina/Fotolia **127 (right),** Charlie Harding/Robert Harding **133,** Chor Sokunthea/Reuters **137 (top left),** Pob_anythink/Fotolia **137 (bottom left),** Bonzodog/Fotolia **137 (middle),** Donvanstaden/Fotolia **137 (top right),** Gabriele Maltinti/Fotolia **137 (bottom right),** CnOra/Fotolia **148,** Ekina/Fotolia **150,** Eléonore H/Fotolia **152,** Kalichka/Fotolia **153 (top left),** Mark Lennihan/AP Photo **153 (bottom left),** Kristoffer Tripplaar/Alamy **153 (right),** Peter Mooji/Shutterstock **CHAPTER 3: 167,** Wilfredo Lee/Associated Press **183 (top),** Shanghai Daily/Imaginechina/Associated Press **183 (bottom),** Christian Delbert/Fotolia **187,** Wilfredo Lee/Associated Press **188 (top),** SeanPavonePhoto/Fotolia **188 (bottom),** Riccardo Valsecchi/Demotix/Corbis **197,** Qushe/Shutterstock **210,** Dario Sabljak/Fotolia **CHAPTER 4: 225, 236,** Warren Goldswain/Shutterstock **237 (left),** Selezenj/Fotolia **237 (right),** Sakhorn/Shutterstock **279,** Monika Wisniewska/Fotolia **294,** TATSIANAMA/Shutterstock **CHAPTER 5: 311,** Igor Mojzes/Fotolia **321,** David Madison/Digital Vision/Getty Images **322,** Yanik Chauvin/Fotolia **331 (left),** Viappy/Fotolia **331 (right),** Kolett/Fotolia **332 (top left),** Gladskikh Tatiana/Shutterstock **332 (bottom left),** Igor Mojzes/Fotolia **332 (right),** NetPhotos/Alamy **333,** Blend Images/Glow Images **343,** US Air Force Photo/Alamy **346,** HUAJI/Shutterstock **347,** Stephen Rees/Shutterstock **372,** Wiktor Bubniak/Fotolia **375,** BibleLandPictures/Alamy **376,** Andrey_Popov/Shutterstock **377 (left),** Steve Healey **377 (right),** Nyiragongo/Fotolia **378 (left),** Petoo/Fotolia **378 (top right),** Jonathan Little/Alamy **378 (bottom right),** Cozyta/Fotolia **379 (left),** Associated Press **379 (right),** Rjack/Fotolia **CHAPTER 6: 393,** epa European pressphoto agency b.v./Alamy **406 (left),** Monkey Business Images/Shutterstock **406 (right),** SOMATUSCANI/Fotolia **407,** Alarsonphoto/Fotolia **408,** Adrian Sherratt/Alamy **415 (top left),** epa european pressphoto agency b.v./Alamy **415 (bottom left),** Stockyimages/Fotolia **415 (right),** Don Emmert/AFP/Getty Images/Newscom **424,** Dennis MacDonald/Alamy **428,** Richard Cummins/SuperStock **432,** GWImages/Shutterstock **433,** Aleph Studio/Shutterstock **441,** Lijuan Guo/Shutterstock **456,** Aastock/Shutterstock **459,** Sima/Fotolia **460,** Furmananna/Fotolia **CHAPTER 7: 477,** Wiskerke/Alamy **492,** Jason Tench/Shutterstock **494,** Wiskerke/Alamy **503,** Frank Oppermann/Shutterstock **515 (left),** Pavel Losevsky/Shutterstock **515 (top right),** Stefan Wackerhagen/Glow Images **515 (bottom right),** DenisNata/Fotolia **CHAPTER 8: 525,** Jeff Lewis/Icon SMI 910/Jeff Lewis/Icon SMI/Newscom **532,** Carola Schubbel/Fotolia **539,** Epitavi/Shutterstock **540,** Greg Ashman/Icon SMI CCJ/Greg Ashman/Icon SMI/Newscom **546,** Jeff Lewis/Icon SMI 910/Jeff Lewis/Icon SMI/Newscom **567,** ZUMA Press, Inc./Alamy **579,** Abaca Press/KRT/Newscom **585,** Vojtech Vlk/Fotolia **591,** David Coleman/Alamy

Index of Applications

Education

Engineering

Environment

Finance

General Interest

Geometry

Government

Health/Medicine

Labor

Physics

Social Sciences

Sports/Entertainment

Statistics/Demographics

Technology

Transportation

Index

H

I

T

Geometry

Plane Geometry

Rectangle
Area: $A = lw$
Perimeter: $P = 2l + 2w$

Square
Area: $A = s^2$
Perimeter: $P = 4s$

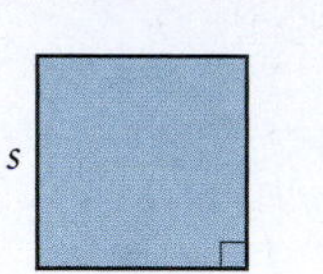

Triangle
Area: $A = \frac{1}{2}bh$

Sum of Angle Measures
$A + B + C = 180°$

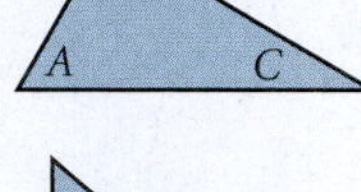

Right Triangle
Pythagorean theorem (equation):
$$a^2 + b^2 = c^2$$

Parallelogram
Area: $A = bh$

Trapezoid
Area: $A = \frac{1}{2}h(a + b)$

Circle
Area: $A = \pi r^2$
Circumference:
$$C = \pi d = 2\pi r$$

Solid Geometry

Rectangular Solid
Volume: $V = lwh$

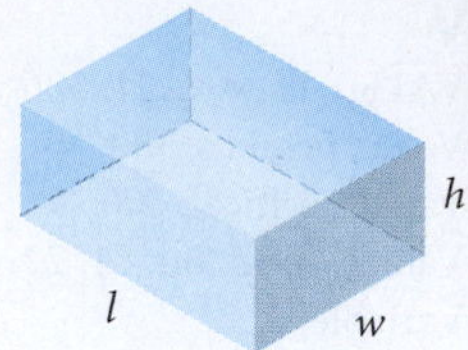

Cube
Volume: $V = s^3$

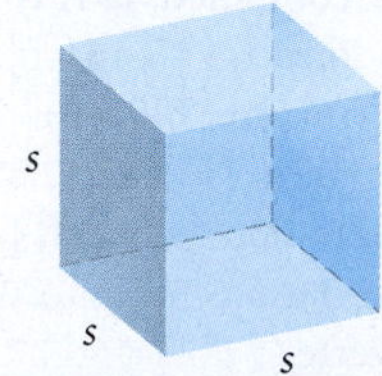

Right Circular Cylinder
Volume: $V = \pi r^2 h$
Lateral surface area:
$$L = 2\pi rh$$
Total surface area:
$$S = 2\pi rh + 2\pi r^2$$

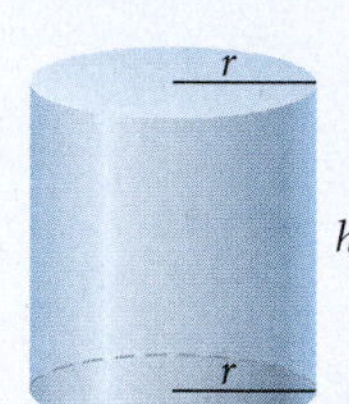

Right Circular Cone
Volume: $V = \frac{1}{3}\pi r^2 h$
Lateral surface area:
$$L = \pi rs$$
Total surface area:
$$S = \pi r^2 + \pi rs$$
Slant height:

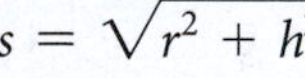
$$s = \sqrt{r^2 + h^2}$$

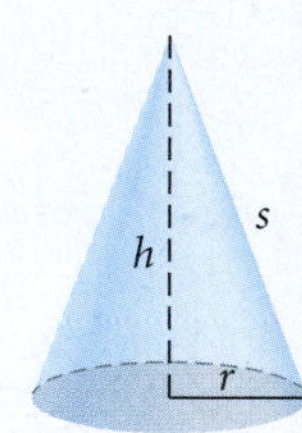

Sphere
Volume: $V = \frac{4}{3}\pi r^3$
Surface area: $S = 4\pi r^2$

Algebra

Properties of Real Numbers

Commutative: $a + b = b + a$; $ab = ba$

Associative: $a + (b + c) = (a + b) + c$; $a(bc) = (ab)c$

Additive Identity: $a + 0 = 0 + a = a$

Additive Inverse: $-a + a = a + (-a) = 0$

Multiplicative Identity: $a \cdot 1 = 1 \cdot a = a$

Multiplicative Inverse: $a \cdot \frac{1}{a} = \frac{1}{a} \cdot a = 1, a \neq 0$

Distributive: $a(b + c) = ab + ac$

Exponents and Radicals

$a^m \cdot a^n = a^{m+n}$ $\qquad \frac{a^m}{a^n} = a^{m-n}$

$(a^m)^n = a^{mn}$ $\qquad (ab)^m = a^m b^m$

$\left(\frac{a}{b}\right)^m = \frac{a^m}{b^m}$ $\qquad a^{-n} = \frac{1}{a^n}$

If n is even, $\sqrt[n]{a^n} = |a|$.

If n is odd, $\sqrt[n]{a^n} = a$.

$\sqrt[n]{a} \cdot \sqrt[n]{b} = \sqrt[n]{ab}, \quad a, b \geq 0$

$\sqrt[n]{\frac{a}{b}} = \frac{\sqrt[n]{a}}{\sqrt[n]{b}}$

$\sqrt[n]{a^m} = (\sqrt[n]{a})^m = a^{m/n}$

Special-Product Formulas

$(a + b)(a - b) = a^2 - b^2$

$(a + b)^2 = a^2 + 2ab + b^2$

$(a - b)^2 = a^2 - 2ab + b^2$

$(a + b)^3 = a^3 + 3a^2b + 3ab^2 + b^3$

$(a - b)^3 = a^3 - 3a^2b + 3ab^2 - b^3$

$(a + b)^n = \sum_{k=0}^{n} \binom{n}{k} a^{n-k} b^k$, where

$$\binom{n}{k} = \frac{n!}{k!\,(n-k)!} = \frac{n(n-1)(n-2)\cdots[n-(k-1)]}{k!}$$

Factoring Formulas

$a^2 - b^2 = (a + b)(a - b)$

$a^2 + 2ab + b^2 = (a + b)^2$

$a^2 - 2ab + b^2 = (a - b)^2$

$a^3 + b^3 = (a + b)(a^2 - ab + b^2)$

$a^3 - b^3 = (a - b)(a^2 + ab + b^2)$

Interval Notation

$(a, b) = \{x \mid a < x < b\}$

$[a, b] = \{x \mid a \leq x \leq b\}$

$(a, b] = \{x \mid a < x \leq b\}$

$[a, b) = \{x \mid a \leq x < b\}$

$(-\infty, a) = \{x \mid x < a\}$

$(a, \infty) = \{x \mid x > a\}$

$(-\infty, a] = \{x \mid x \leq a\}$

$[a, \infty) = \{x \mid x \geq a\}$

Absolute Value

$|a| \geq 0$

For $a > 0$,

$|X| = a \rightarrow X = -a$ *or* $X = a$,

$|X| < a \rightarrow -a < X < a$,

$|X| > a \rightarrow X < -a$ *or* $X > a$.

Equation-Solving Principles

$a = b \rightarrow a + c = b + c$

$a = b \rightarrow ac = bc$

$a = b \rightarrow a^n = b^n$

$ab = 0 \leftrightarrow a = 0$ *or* $b = 0$

$x^2 = k \rightarrow x = \sqrt{k}$ *or* $x = -\sqrt{k}$

Inequality-Solving Principles

$a < b \rightarrow a + c < b + c$

$a < b$ and $c > 0 \rightarrow ac < bc$

$a < b$ and $c < 0 \rightarrow ac > bc$

(*Algebra continued*)

Algebra *(continued)*

The Distance Formula

The distance from (x_1, y_1) to (x_2, y_2) is given by

$$d = \sqrt{(x_2 - x_1)^2 + (y_2 - y_1)^2}.$$

The Midpoint Formula

The midpoint of the line segment from (x_1, y_1) to (x_2, y_2) is given by

$$\left(\frac{x_1 + x_2}{2}, \frac{y_1 + y_2}{2}\right).$$

Formulas Involving Lines

The slope of the line containing points (x_1, y_1) and (x_2, y_2) is given by

$$m = \frac{y_2 - y_1}{x_2 - x_1}.$$

Slope–intercept equation:	$y = f(x) = mx + b$
Horizontal line:	$y = b$ or $f(x) = b$
Vertical line:	$x = a$
Point–slope equation:	$y - y_1 = m(x - x_1)$

The Quadratic Formula

The solutions of $ax^2 + bx + c = 0, a \neq 0$, are given by

$$x = \frac{-b \pm \sqrt{b^2 - 4ac}}{2a}.$$

Compound Interest Formulas

Compounded n times per year: $A = P\left(1 + \frac{i}{n}\right)^{nt}$

Compounded continuously: $P(t) = P_0e^{kt}$

Properties of Exponential and Logarithmic Functions

$\log_a x = y \leftrightarrow x = a^y$ $\qquad a^x = a^y \leftrightarrow x = y$

$\log_a MN = \log_a M + \log_a N$ $\qquad \log_a M^p = p \log_a M$

$$\log_a \frac{M}{N} = \log_a M - \log_a N$$

$$\log_b M = \frac{\log_a M}{\log_a b}$$

$\log_a a = 1$ $\qquad \log_a 1 = 0$

$\log_a a^x = x$ $\qquad a^{\log_a x} = x$

Conic Sections

Circle: $(x - h)^2 + (y - k)^2 = r^2$

Ellipse: $\frac{(x - h)^2}{a^2} + \frac{(y - k)^2}{b^2} = 1$, $\quad \frac{(x - h)^2}{b^2} + \frac{(y - k)^2}{a^2} = 1$

Parabola: $(x - h)^2 = 4p(y - k)$, $\quad (y - k)^2 = 4p(x - h)$

Hyperbola: $\frac{(x - h)^2}{a^2} - \frac{(y - k)^2}{b^2} = 1$, $\quad \frac{(y - k)^2}{a^2} - \frac{(x - h)^2}{b^2} = 1$

Arithmetic Sequences and Series

$a_1, \quad a_1 + d, \quad a_1 + 2d, \quad a_1 + 3d, \ldots$

$a_{n+1} = a_n + d$ $\qquad a_n = a_1 + (n - 1)d$

$$S_n = \frac{n}{2}(a_1 + a_n)$$

Geometric Sequences and Series

$a_1, \quad a_1r, \quad a_1r^2, \quad a_1r^3, \ldots$

$a_{n+1} = a_nr$ $\qquad a_n = a_1r^{n-1}$

$S_n = \frac{a_1(1 - r^n)}{1 - r}$ $\qquad S_\infty = \frac{a_1}{1 - r}, |r| < 1$

A Library of Functions

Linear function

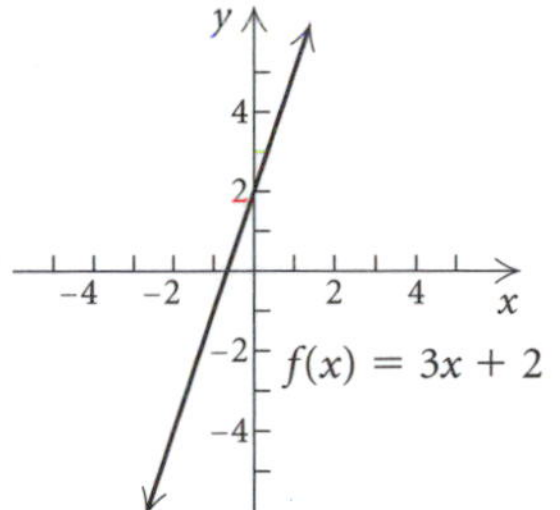

Linear function

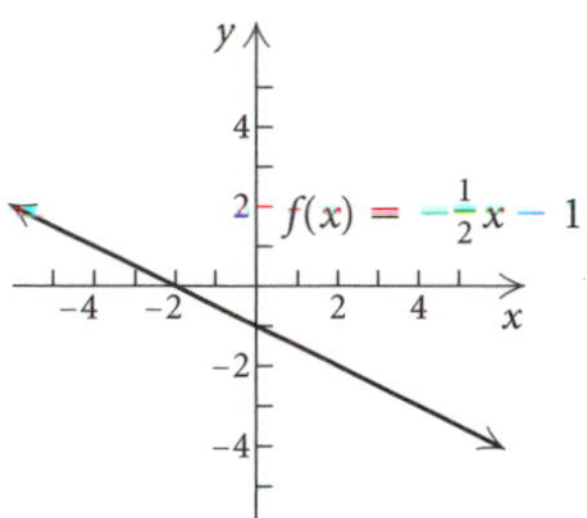

Constant function

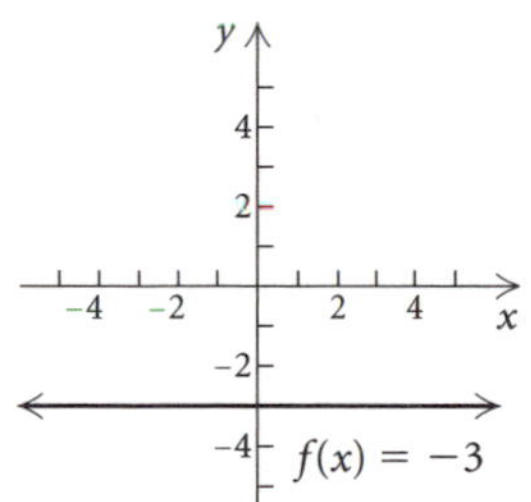

Absolute-value function

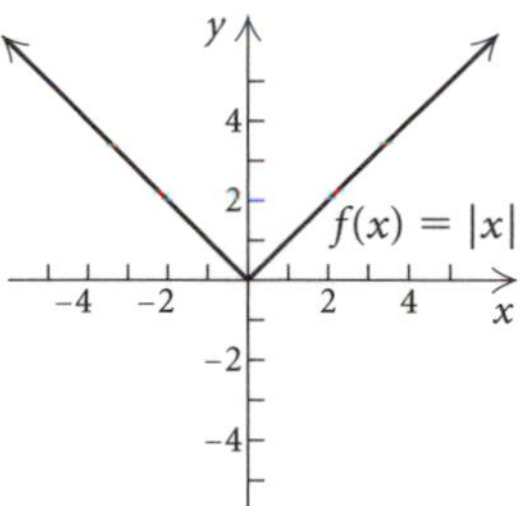

Squaring function

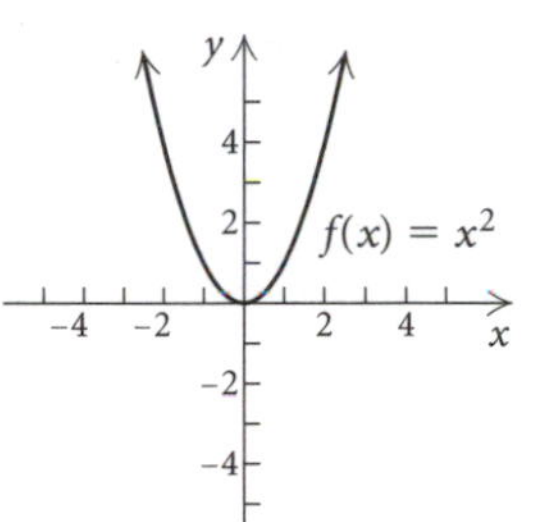

Quadratic function

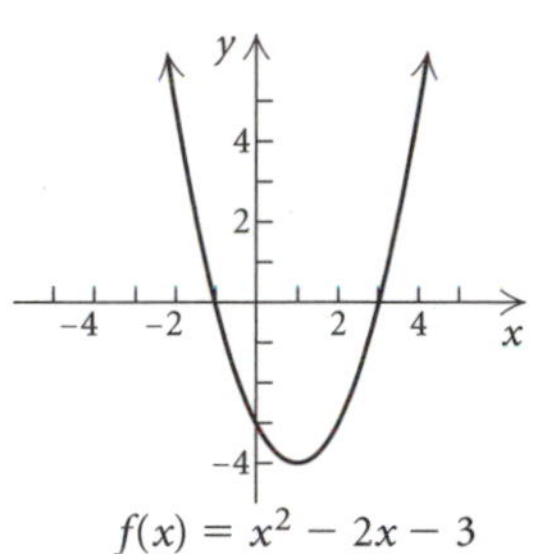

Quadratic function

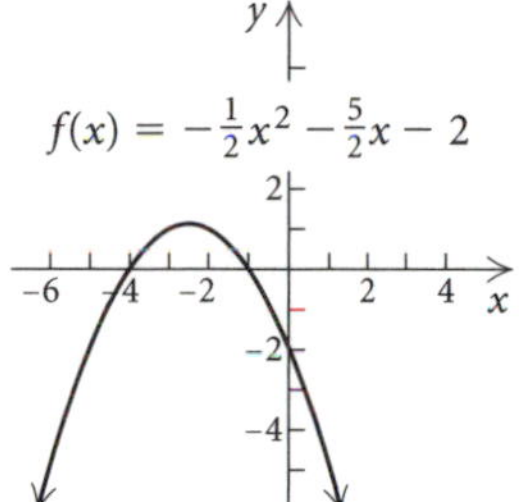

Square-root function

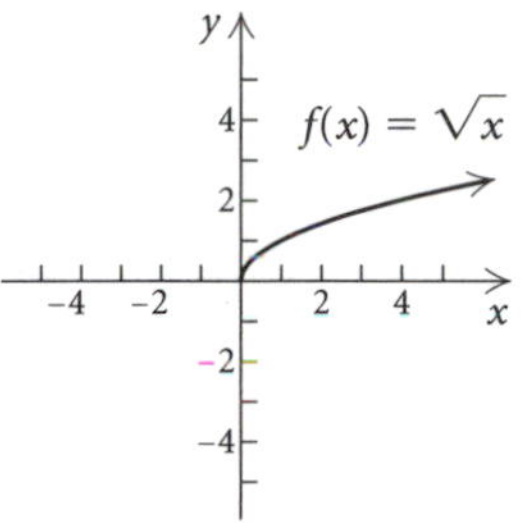

Cubing function

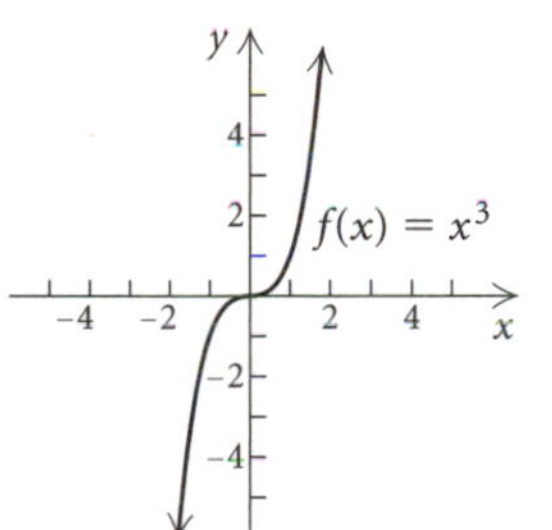

Cube root function

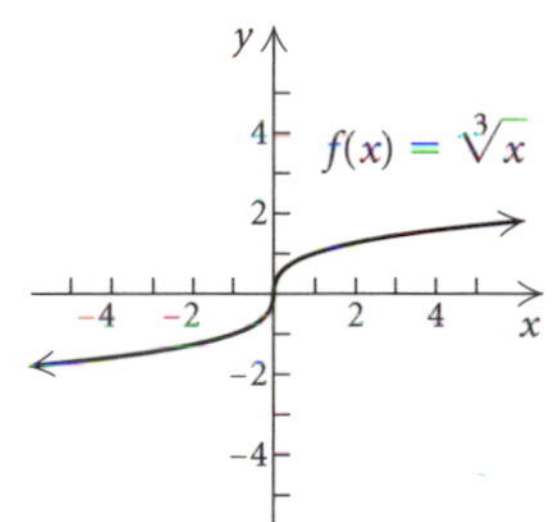

Greatest integer function

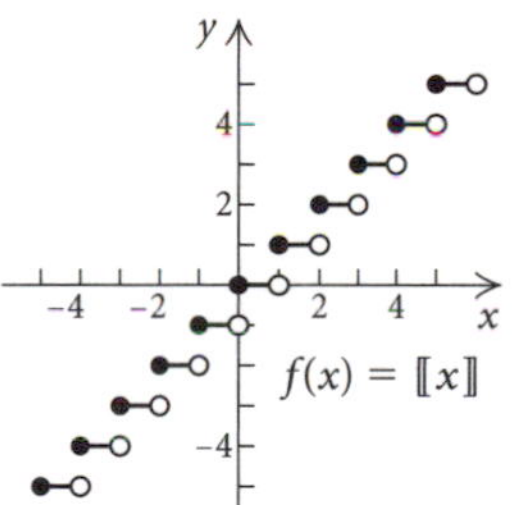

Rational function

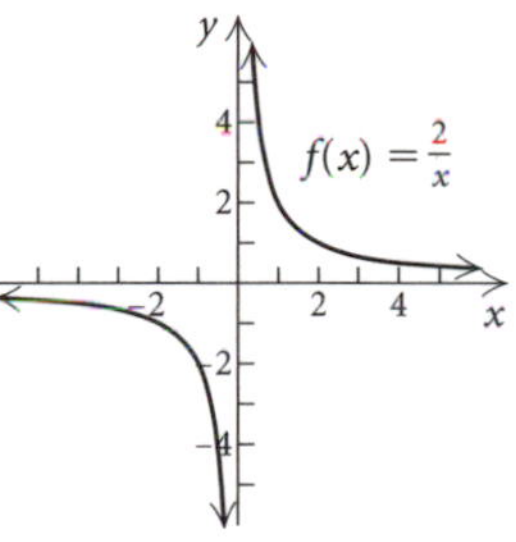

Exponential function

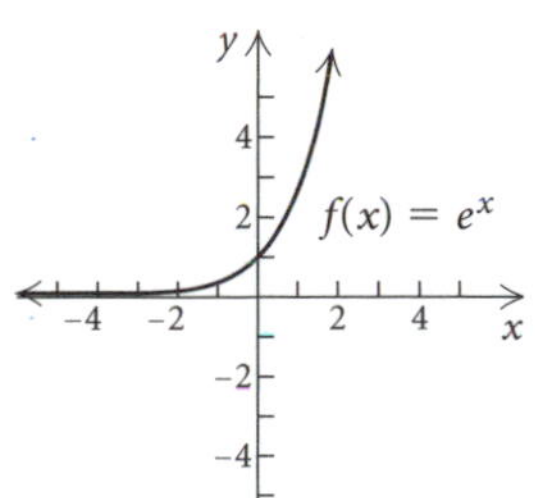

Exponential function

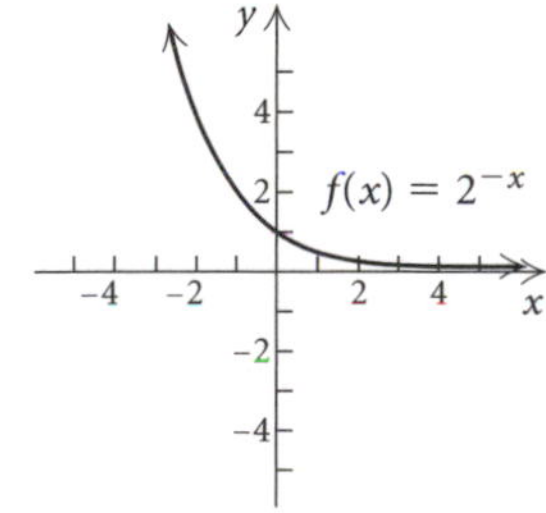

Logarithmic function

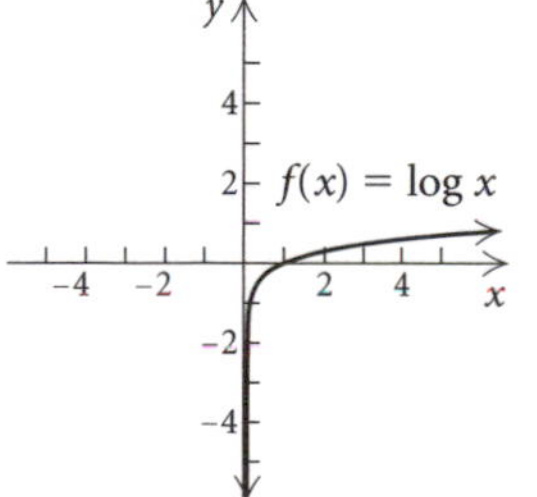

Logistic function

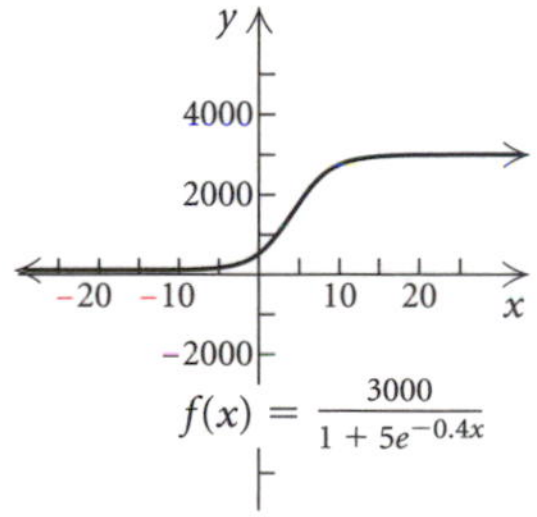